D1784693

Microscopy of Semiconducting Materials 1997

Other titles in the Series

The Institute of Physics Conference Series regularly features papers presented at important conferences and symposia highlighting new developments in physics and related fields.

Microscopy of Semiconducting Materials 1997

Proceedings of the Royal Microscopical Society Conference held at Oxford University, 7–10 April 1997

Edited by A G Cullis and J L Hutchison

S
M M
X

Institute of Physics Conference Series Number 157
Institute of Physics Publishing, Bristol and Philadelphia

Copyright ©1997 by IOP Publishing Ltd and individual contributors. All rights reserved. No part of this publication may be reproduced, stored in a retrieval system or transmitted in any form or by any means, electronic, mechanical, photocopying, recording or otherwise, without the written permission of the publisher, except as stated below. Single photocopies of single articles may be made for private study or research. Illustrations and short extracts from the text of individual contributions may be copied provided that the source is acknowledged, the permission of the authors is obtained and IOP Publishing Ltd is notified. Multiple copying is permitted in accordance with the terms of licences issued by the Copyright Licensing Agency under the terms of its agreement with the Committee of Vice-Chancellors and Principals. Authorization to photocopy items for internal or personal use, or the internal or personal use of specific clients in the USA, is granted by IOP Publishing Ltd to libraries and other users registered with the Copyright Clearance Center (CCC) Transactional Reporting Service, provided that the base fee of $19.50 per copy is paid directly to CCC, 27 Congress Street, Salem, MA 01970, USA.
0305-2346/97 $19.50+.00

CODEN IPHSAC 157 1-709 (1997)

British Library Cataloguing in Publication Data

A catalogue record for this book is available from the British Library.

ISBN 0 7503 0464 2

Library of Congress Cataloging-in-Publication Data are available

Conference Chairmen
    A G Cullis and J L Hutchison

Honorary Editors
    A G Cullis and J L Hutchison

Scientific Sponsors
    The Royal Microscopical Society
    The Institute of Physics
    The Materials Research Society

Published by Institute of Physics Publishing, wholly owned by The Institute of Physics, London
Institute of Physics Publishing, Dirac House, Temple Back, Bristol BS1 6BE, UK
US Editorial Office: Institute of Physics Publishing, The Public Ledger Building, Suite 1035, 150 South Independence Mall West, Philadelphia, PA 19106, USA

Printed in the UK by Galliard (Printers) Ltd, Great Yarmouth, Norfolk

Contents

vi

Dislocations and boundaries

Epitaxial layers: defect behaviour and strain relief

Epitaxial layers: wide-bandgap nitrides

Epitaxial layers: general growth phenomena

Processed silicon, diamond and specimen preparation methods

Scanning probe microscopy

Advanced scanning electron and optical microscopy

Preface

This volume contains the invited and contributed papers presented at the conference on 'Microscopy of Semiconducting Materials' held on 7–10 April 1997 at the University of Oxford. The event was organized under the scientific sponsorship of the Royal Microscopical Society, the Electron Microscopy and Analysis Group of the Institute of Physics and the Materials Research Society. This conference was the tenth in the series which focuses on the state-of-the-art in semiconductor studies carried out by all forms of microscopy. An initial Symposium celebrated the centenary of the discovery of the electron and, overall, delegates from more than 20 countries gave a comprehensive account of the most recent developments in experimental and theoretical work which have taken place world-wide.

Advances in semiconductor materials understanding underpin future electronic device technology. This strong linkage, together with the key role of electron-optical instrumentation, was emphasized in the 'Electron Centenary' symposium. The work presented in the main conference scientific sessions demonstrated the dynamic activity taking place in all areas of semiconductor studies from the latest developments in fundamental research through to applied investigations directly supporting device production. As the scale of device structures continues to shrink, ever more reliance will need to be placed upon electron and related forms of microscopy to overcome materials problems encountered in fabrication. This situation is evident in the work presented in this volume.

Each camera-ready manuscript submitted for publication in these Proceedings has been reviewed by at least two referees and modified accordingly. The editors are very grateful to the following scientific referees for their rapid and meticulous work:

M A Al-Khafaji, U Bangert, H Bender, G R Booker, G A D Briggs, P D Brown, D Cherns, A Cornet, K Durose, C Frigeri, S Galloway, D Gerthsen, F Glas, P J Goodhew, E Grünbaum, A Gustafsson, D B Holt, C J Humphreys, W Jäger, A Jakubowicz, D E Jesson, A R Lang, R E Mallard, C D Marsh, R Murray, A G Norman, C E Norman, B Pécz, D D Perovic, A J Pidduck, P Pongratz, M Prutton, J-L Rouvière, G Salviati, T-Y Seong, J Spence, A E Staton-Bevan, H P Strunk, R T Tung, J Vanhellemont, G Wakefield, T Walther, P R Wilshaw and A C Wright.

The conference organizers are particularly pleased to acknowledge the financial support provided by the Royal Society, JEOL (UK) Ltd and Oxford Instruments Ltd. Special thanks are due to K Hale (RMS) for supporting the conference organization over an extended period. The organizers would also like to thank C Sealy and A M Blood (University of Oxford) for assistance in correcting proof copies of many manuscripts.

A G Cullis
J L Hutchison
September 1997

Inst. Phys. Conf. Ser. No 157
Paper presented at Microsc. Semicond. Mater. Conf., Oxford, 7–10 April 1997
© *1997 IOP Publishing Ltd*

1

The materials basis behind the telecommunications revolution

W F Brinkman

Lucent Technologies, Bell Laboratories, 700 Mountain Avenue, Murray Hill, NJ 07974, USA

ABSTRACT: A selected review of some recent materials advances which will drive the telecommunications revolution is presented. Included are lithographic materials, projection electron lithography, understanding transient enhanced diffusion, ultra thin oxides, sol-gel preparation for optical fibers, silicon optical bench technology, fiber Bragg gratings and electro absorption modulator lasers.

1. INTRODUCTION

The changes in telecommunications since the invention of the transistor have, without doubt, profoundly impacted the world in which we live. These changes are faster and perhaps more far reaching than those of the original industrial revolution. Society is just beginning to use the results of this revolution as it is by no means over - neither in the technical progress driving it nor in services it will provide. It is natural to ask how can we understand its origins, impetus and future. Neglecting a lengthy historical discussion, we could submit that the telecommunications revolution owes its origins and its future to the remarkable past progress and current dynamics of two fields: microelectronics and photonics, and that both of these fields can be characterized by logarithmic growth of the sort originally shown by Gordon Moore for integrated circuits. For electronics, for instance for DRAMS, the number of bits doubles every 18 months. This is partly due to increasing chip size, but mainly due to a decrease in linear dimensions which can be fabricated. For photonics the product of the bit rate and repeater spacing of an optical fibre has doubled every year for the past two decades.

Both of these "Moore relationships" have been driven by and paced by electronic materials progress and exciting recent results suggest that we can expect continued Moore behaviour and a continued "Telecommunications Revolution" well into the next century. A very large part of the new developments hinges on the control of materials at a remarkably small scale - often characterized by and understood through electron microscopy. We should like now to examine a selected group of recent accomplishments - accomplishments which hinge on materials and which are at the leading edge, with promise of driving technology into the next century.

2. MICROELECTRONICS

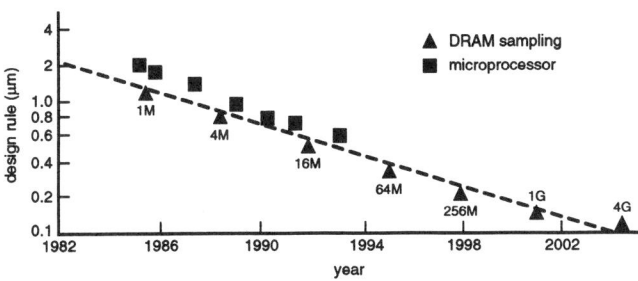

Fig. 1: Technology trends for DRAMS and microprocessors.

Figure 1 (Nisi 1995) shows the steady progress in Si chip component packing density and the commensurate shrinkage in feature sizes fabricated. It is important to point out that to achieve the packing density required by the early 21st century, not only must lithography shrink the features delineated through the use of improved resists, but also exposure radiation must move to shorter wavelengths and gate thickness must shrink proportionately. In addition doping must be controlled with unprecedented finesse. The challenges involved are becoming more difficult and expensive but we see no fundamental reason why the technology cannot be taken to minimum feature sizes of the order 0.05μm. Here we review a few advances that will enable us to continue the evolution implied by the Moore plot.

3. LITHOGRAPHIC MATERIALS

Significant advances are continually being made in microelectronic device fabrication, and especially in lithography, the technique used to generate the high resolution circuit elements characteristic of today's integrated circuits. Fifteen years ago, the state-of-the-art device contained up to 8000 transistor elements and had 5 to 6 μm minimum features. Today, devices with several million transistor cells are commercially available and are fabricated with minimum features of less than 0.35 μm. These accomplishments have been achieved using "conventional lithography" as the technology of choice. The steady decrease in integrated circuit linewidths has largely been fueled by improvements in the resolution of optical lithography. This, in turn, has been driven by the use of shorter and shorter ultraviolet wavelengths for the exposure tools. Deep-UV lithography at the 248 nm wavelength is just coming into production in the 0.25 μm generation. This is expected to be followed by a transition to 193 nm by a change in the excimer laser light source from KrF to ArF and ultimately possibly by flood e-beam exposure with tools like SCALPEL (see below).

Unfortunately, conventional photoresists are not appropriate for use with the non-conventional lithographic technologies that will be necessary for sub-0.35 μm lithography. The most notable deficiencies of the conventional novolac-quinonediazide resists are the sensitivity and absorption properties of the materials. For most resists, the quantum yield is significantly less than 1.0, and since the new lithographic tools in general have lower brightness sources, high sensitivity resists are required. Additionally, the absorption of conventional photoresists is too high to allow uniform imaging through practical resist film thicknesses (~ 1 μm). For 248 nm lithography, these challenges were accommodated by application of Chemically AMPlified, CAMP, resist technology. Photogeneration of a strong acid is followed by an acid catalyzed reaction that renders the exposed areas of the resist film soluble in aqueous base (Ito 1984) (Reichmanis 1991).

CAMP was the first commercially available deep-UV photoresist suitable for the manufacture of integrated circuit devices (Nalamasu 1993). It consists of a t-butoxycarbonyl-protected hydroxystyrene-sulfone matrix resin, plus a photosensitive nitrobenzyl ester photoacid generator (PAG). Upon irradiation, the PAG is photolyzed to produce a strong acid, which reacts in a subsequent heating step to develop an aqueous-base soluble hydroxystyrene-sulfone polymer. The initial, protected polymer is hydrophobic and insoluble in aqueous base, while the hydroxystyrene product is very soluble in such media. This material is classified as a "chemically amplified" resist simply because the photogenerated acid species catalyzes a cascade of chemical events such as the removal of the t-butoxycarbonyl protective groups. The inherent sensitivity associated with this and other positive-acting chemically amplified resists is based on the regeneration of acid, which then becomes available for additional de-blocking reactions during the post-exposure baking step. Experience with this material, quickly paved the way for more advanced resists with more robust process characteristics. Notable among these is the ARCH (Advanced Resist CHemically amplified) series of deep-UV resists (Mertesdorf 1995). ARCH is a state-of-the-art resist developed through research efforts at Bell Labs and Olin Microelectronic Materials. It is based on substituted polyvinylphenol matrix resin chemistry used in conjunction with a nitrobenzyl ester photoacid generator. This material demonstrates 0.22 μm linear resolution and is being used in the production of 0.25 μm devices. Figure 2 depicts 0.24 μm gate structures obtained in the ARCH II chemically-amplified deep-UV resist.

As device minimum features continue to decrease, new lithographic technologies will be required, necessitating development of new materials technologies. Research is already under way on new resists for the 21st century.

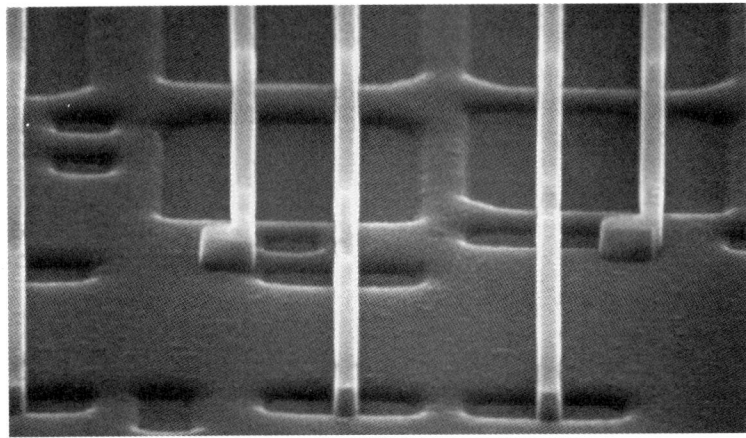

Fig. 2: 0.24 μm gates printed with ARCH2.

4. PROJECTION ELECTRON LITHOGRAPHY: SCALPEL

As design rules grow smaller, the wavelengths of the radiation used to delineate the features in integrated circuits will limit the size of the features. This is because diffraction limits the minimum features that can be printed to about that of the wavelength of the exposure radiation. The generally expected limit is features of about 0.13μm using deep UV sources. At shorter wavelengths conventional optics are no longer transparent.

The most promising means to circumvent these difficulties is to use projection electron beam lithography (Harriott 1996). It has been known for many years that electron beam direct-write lithography offers high resolution, large depth of focus, and wide process latitude. In fact, today most photomasks are patterned with e-beam tools. The problem has always been throughput which directly effects the economics of integrated circuit manufacturing. Projection lithography overcomes the throughput limitation by projecting the image of all or a large part of a circuit onto the wafer from a mask. This process is analogous to optical lithography except that the light beam is replaced by an electron beam and the mask is replaced by a special electron beam mask. If conventional "stencil" masks with opaque regions are used, significant heating can occur leading to unacceptable distortions. The Scattering with Angular Limitation Projection Electron-beam Lithography (SCALPEL) (Harriott 1996 & Liddie 1996) approach combines the high resolution and wide process latitude inherent in electron beam lithography and the throughput of projection systems, without the limitations of the stencil mask. In the SCALPEL system, a mask consisting of a low atomic number membrane and a high atomic number absorber pattern layer is uniformly illuminated by high energy (100 keV) electrons. The mask structure is such that a negligible amount of the beam's energy is absorbed in the mask. The portions of the beam which

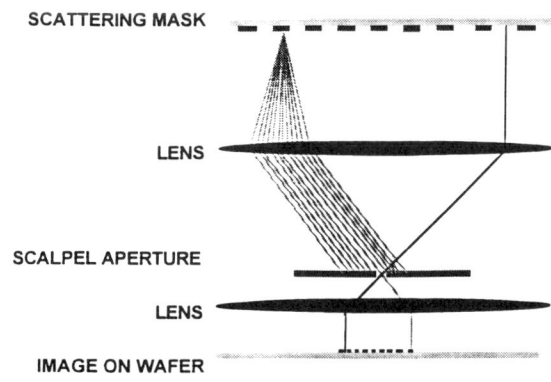

Fig. 3: SCALPEL Geometry.

pass through only the membrane suffer very little disturbance by scattering while the portions of the beam which pass through the high atomic number pattern layer are scattered through angles of a few milliradians. The SCALPEL principle is illustrated in Figure 3. The image contrast is determined by the energy of the electrons, the materials of the mask and their thicknesses, and the diameter of the aperture. Harriott, et al. (Liddie 1996). have addressed a step-and-scan strategy. In this approach, a relatively small (1 mm) electron beam illuminates part of the mask, and the reduced image of that portion of the mask is projected onto the wafer. To print the entire mask pattern, the mask and wafer are moved through the beam. Features as small as 0.08µm have been printed using SCALPEL. It has the potential of both high resolution and economically viable throughput.

5. TRANSIENT ENHANCED DIFFUSION

To reduce the size of a Si circuit, many features must undergo scaled reduction. As design rules become less than 0.5µm, an important scaling limitation is control over diffusion of dopants. Thus understanding and controlling material diffusion is key. The complication is that ion implantation of dopant which is required to achieve the doping levels needed causes damage in a material. Annealing is required to remove damage. The usual procedure is a "flash" anneal. During this anneal diffusion is enhanced over what it would be in an undamaged material - "transient enhanced diffusion" takes place. Minimizing transient enhanced diffusion (TED) is critical for fabricating tight design rules for Si integrated circuits. For example, at 0.18µm gate lengths (in ≈2001) the junction depth must be less than 100nm. Equilibrium diffusion lengths are ≈10nm for typical thermal budgets in manufacturing, TED typically leads to >100nm of diffusion even for very short low-temperature anneals resulting in "spill over" of dopant into adjacent regions. Predicting TED is critical in selection of implant energy and species. A critical step in designing models is understanding how defects aggregate. During ion implantation, silicon self-interstitials agglomerate to form "(311) defects", so-called because they lie on a (311) crystallographic plane. Their structure was first determined by Seiji Takeda and others at Osaka University in Japan, and they have been studied extensively by   Eaglesham and colleagues (1995) at Lucent Bell Labs using high-resolution transmission electron microscopy.

It has been found that (311) defects evaporate on heating. This occurs through the emission of the interstitials that comprise the defect into the bulk of the silicon. Evaporation occurs over a time scale comparable to that of TED, implying that interstitial emission from the evaporating defect causes TED. In addition, since these interstitial agglomerates are detected directly in electron microscope images, the number and size of (311) defects can be measured as a function of time and temperature. This allows us to calculate the rate at which interstitials are emitted to cause TED. Consequently, we now have the first direct measurement of the number of interstitials associated with a specific diffusion process, which removes a critical unknown parameter from the diffusion models. We can now use diffusion simulations that incorporate the exact measured rate of interstitial emission, and fit the remaining diffusion constants to reproduce the experimentally observed enhancement in diffusion, thus increasing our confidence in the predictive power of simulations.

This sort of modeling is beginning to put TED on a firm footing and gives us confidence that junction reduction can be continued into the next century.

6. THIN OXIDES

One of the most critical components behind the progress in scaling integrated circuit technology is the thin oxide insulator used to couple the gate of the MOS transistor to the conducting channel. Particularly important are minimum requirements in leakage, manufacturability and reliability.   The effects of direct tunneling leakage currents limit the ultimate thin oxides. Very thin oxides require very careful control of the interface of Si/SiO_2. Roughness of the silicon substrate contributes to increased leakage in the ultrathin (<2.5nm) regime, requiring careful control of interfaces during the processing. Optimal cleaning procedures which preserve smooth interfaces are required. Figure 4 shows a TEM of an oxide that is 1.5nm thick with an atomically flat interface. It is an example of the remarkable control recently achieved in thin oxides. Until recently it was assumed that the thinnest gate oxides that could be used would be about 3-4nm due to the device limitations on tunneling currents. New device studies (Momosa 1996) have shown that devices can be scaled such that oxides as thin as 1.5nm can be effectively used in MOS applications. The new devices were shown to have considerable direct tunneling currents through the gate oxide but could be effectively used in circuit applications. The control of the oxide thickness will have to be extremely accurate for these very thin oxides across wafer diameters of 200-300 mm.

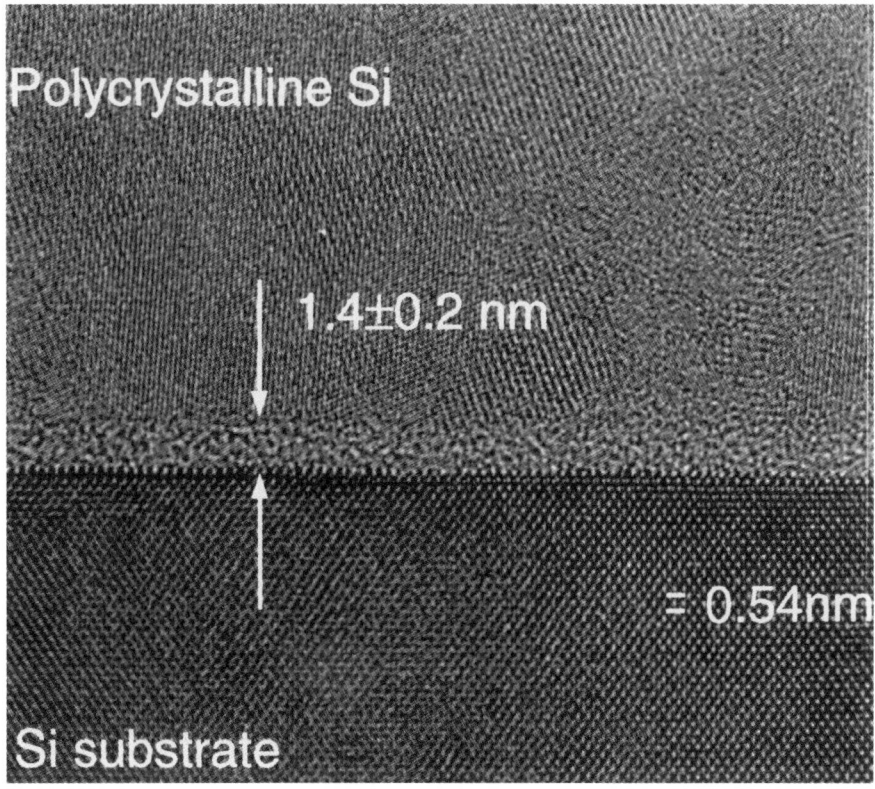

Fig. 4: Thin Gate Oxide.

Characterization of thin gate oxides and the electrical effects of the silicon/ SiO_2/Polysilicon structure used in the modern MOS devices is challenging. The effects due to quantum mechanical charge distributions change the way we think about the relationship of the gate oxide to the transistor characteristics. These quantum mechanical effects must be included to calculate the transistor characteristics, (Krisch 1996) as well as considered in the extrapolation of the lifetimes of the oxides.

Finally, dielectrics other than pure silicon dioxide are being considered for MOS gate stacks e.g. materials with higher dielectric permittivity so that in theory, thicker films (with lower tunneling leakage) could be used to obtain equivalent capacitive coupling to the channel. The search for such materials has been going on for a number of years motivated by the desire to reduce DRAM (Ohji 1995) capacitor area. However, due to the host of challenging requirements, no suitable alternative for MOS gates has yet been found. Some success has been achieved recently with oxides which incorporate moderate amounts of nitrogen, e.g. through incorporation in the growth or through implantation in the silicon before growth (Liu 1996). Because these films typically have only a monolayer or so of nitrogen, their permittivity looks much the same as pure silicon dioxide, but even small amounts of nitrogen have demonstrated improved reliability and have provided a barrier to dopant diffusion through the oxide.

6

7. PHOTONICS

A seminal materials achievement in photonics was the 10^{-4} reduction in fiber loss achieved in the first two decades of optical fiber preparation (Figure 5). This was the result of unprecedented purity in the optical preforms, made (sub ppb in critical impurities) by, for instance, chemical modified chemical vapor deposition (MacChesney 1989). Loss close to the intrinsic scattering loss is now standard in commercial fibers.

As stated earlier the "Moore's Law" for fiber optic communications has been doubling each year for over twenty years. Since the ultimate bit rate is set by the frequency of the optical carrier (200 THz) and rates in the latest experiments are ≈ 2.5Tb/s, we are a long way from the ultimate limit. Limits on bit rates are set by modulator considerations and fiber properties. Additional band width can be obtained by using several carrier frequencies on the same fiber, wave length division multiplexing (WDM). In fiber optics cost reduction is paramount. Glass fibers have changed from exotic laboratory objects to commodities in less than two decades.

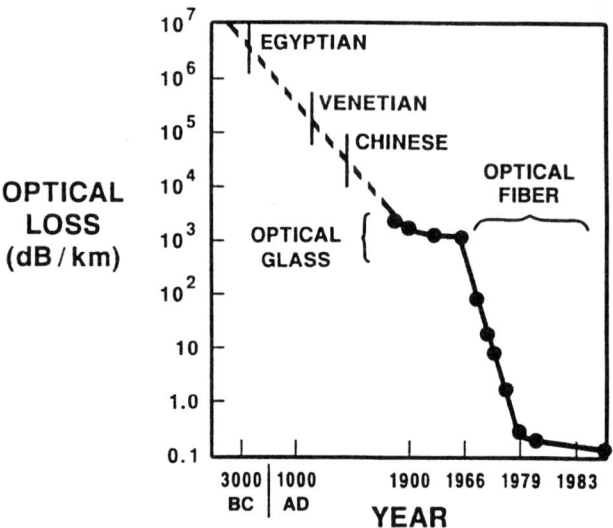

Fig. 5: Chronology of the improvement of glass transparency.

Optics is following the path of silicon by moving towards integration using silicon chips as the platform for both passive and active optical elements - literally constructing what would require a lab sized optical bench on a Si chip, the silicon optical bench. Low chirp optical transmitters which monolithically integrate distributed feed back lasers and electro absorption modulators made using novel selected area growth techniques and multiple quantum well structures are enabling efficient WDM systems. In the following we would like to describe some current leading edge research subjects, which will drive the "fiber optic Moore's Law."

8. SOL-GEL

Optical fiber cost has plummeted over the past decade, further decreases will accelerate novel applications like fiber to the home. One of the most time consuming and costly steps in fiber preparation is the relatively slow vapor phase deposition of SiO_2 and doped SiO_2 to prepare preforms. Cost reduction can be large if the volume of SiO_2 deposited is reduced by, for instance, using SiO_2 glass overcladding tubes on vapor deposited fiber optic preforms. In principal very cheap overcladding tubes could be made by using an inexpensive colloidal SiO_2 sol, changing pH in a controlled fashion to gel it, purifying, and drying, and firing to a glass. Indeed exploiting sol-gel technology to make large glass objects has been a holy grail for many years. Several barriers have thwarted efforts. The large shrinkage encountered when drying alkoxide and colloidal gels leads to cracking in large bodies; it is necessary to achieve exacting dimensional specifications; and it has been difficult to exclude or remove refractory particles which cause low strength breaks in optical fibers.

Currently, Lucent Technologies, Bell Labs has successfully developed a laboratory sol casting (MacChesney 1992), drying, purification and sintering process which produces large overcladding tubes for fabricating optical fiber preforms. It uses a fumed silica dispersed in a basic aqueous medium. This sol at a high pH is stable against gelation and can be centrifuged to remove large impurity particles. The sol is cast into tubular molds after a hydrolyzable ester is added to reduce the pH slowly and effect gelation. The gelled body is removed from the mold and dried over several days. Heat treatment involves three steps: 1) pyrolysis of organics; 2) treatment in reactive halogen atmospheres to purify the body; and 3) sintering in He at temperatures approaching 1500°C. Figure 6 shows a gel body partly sintered to a large tube.

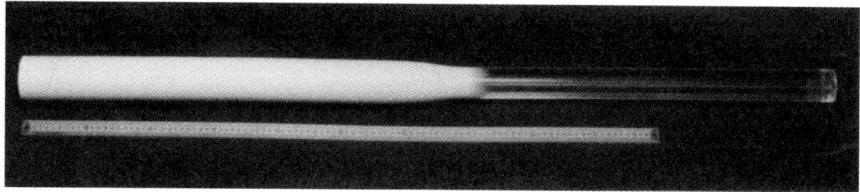

Fig. 6: Partly sintered (on left) SOL-GEL and resultant over cladding tube (on right)
(with meter stick).

Single mode fiber made by overcladding MCVD core-rods with gel-derived overcladding tubes gives performance equivalent to that of accepted commercial fibers: 1.3μm loss of < 0.35 dB/km, 1.55μm loss of < 0.20 dB/km, fiber break frequency caused by internal flaws of < one/megameter and fiber geometry which meets current specifications. The wider implication of this development is that net shape casting of colloidal gels can produce large tubes equivalent to those made by drawing from synthetic silica boules as well as other shaped bodies now made by cutting and welding; all with significant process and cost advantages.

9. SILICON OPTICAL BENCH TECHNOLOGY

At Bell Labs, the technology chosen for integrated optics is called "Silicon Optical Bench (SiOB)" (Presby 1996, Li 1997). We use a silicon wafer as a platform to build our passive and active optical circuits. In this way, we can take advantage of the excellent electrical, thermal and mechanical properties of silicon. We can also leverage off the silicon integrated circuit technology for a ready supply of high quality substrates, processing developments, and commercial processing equipment. Basically, the devices are made using processes that are fairly standard to make IC's: CVD deposition of dielectrics, photolithography and reactive ion etching (RIE), metallization and dicing. However, the issues are different. In the IC industry, we push the limit of lithography tools and resists. Gate oxides are getting thinner and thinner, approaching 50Å. In SiOB technology, the minimum feature size is rarely less than 1μm and the oxide layers are typically 10-20μms thick. The fabrication challenges involve depositing and annealing such thick layers; when the thermal mismatch with the substrate is extreme — both cracking of the oxide and bowing of the wafer can occur. In addition, transferring the core pattern to the thick oxide without undercut or roughness is important.

The first commercial use of SiOB technology was in producing the submounts and platforms used in packaging lasers and PIN photodetectors. Silicon is an excellent heat sink and the active components can be actively aligned and soldered right to the substrate, or passively aligned using a self-centered solder technique where the surface tension of the molten solder brings the component into alignment. Another advantage of the silicon substrate is the availability of well-controlled highly anisotropic etches, which make it possible to etch deep V-grooves into the substrate with great accuracy for passive alignment of a fiber to an active component. The presence of V-grooves in close proximity to metal pads necessitated the development of a non-planar lithography process using electrophoretic resist. Today, all of Lucent's laser products are packaged using silicon optical bench technology.

The typical *passive* waveguide structure starts with a lower cladding of pure SiO_2 that is 10-15μms thick and can be grown by high pressure steam oxidation (HiPOX) or by CVD. A core layer of a doped oxide (Ge, P, and/or B) is followed by CVD, flame hydrolysis, or electron beam deposition. The doping is chosen to give the desired Δ, normalized core/cladding index difference, for light guiding. After patterning and RIE to define the guides, an upper cladding layer of doped oxide, index-matched to the lower cladding, is deposited. Each of the layers may require some annealing to fully densify and form the glass. In addition, there may be flow layers of a "soft" glass that are deposited following the core definition to insure that there is adequate fill between closely spaced waveguides. Therefore, the softening temperature hierarchy of these glasses: lower cladding, core and upper cladding, is critical. Propagation loss in these guides can be as low as 0.025 dB/cm in P-glass. Once the wafer is diced and polished, connection to the waveguides is generally made by butt-coupling a fiber in a silicon piece part, which can be epoxied or soldered into place.

One of the most successful passive components made with silicon optical bench is the Waveguide Grating Router, a demultiplexer for dense wavelength division multiplexing (DWDM) applications. Multiple wavelengths enter the device on a single fiber and are routed onto separate fibers at the output via an elegant use of constructive and destructive interference. Although the device operates completely passively, the package must be temperature controlled in order to insure that the channel wavelengths do not vary by more than +/- 1Å. Typical router insertion losses are about 7 dB for an 8 channel device. Reducing that loss will take a combination of design refinements and processing improvements.

Active and passive silicon optical bench platforms come together in the development of an integrated transceiver (Scotti 1996). The driver here is cost: fiber-to-the-home (FTTH) will not be practical unless the cost of a subscriber line becomes comparable to copper. An integrated transceiver is an important step in reducing the cost of the optics layer of the optical network units (ONUs) that will be on each home. The transceiver design includes a 1.31μm/1.55μm WDM, a 1.31μm PIN detector, a 1.31μm Fabry-Perot laser for the upstream signal, and a reversing element so that the 1.55μm video signal can pass through the chip and exit on a separate fiber. Lasers, detectors, and fibers are passively aligned to the waveguides and the entire planar lightguide circuit (PLC) is packaged with the electronics in an electronics-style, 24-pin dual-inline ceramic package. The footprint of this device, shown in Fig. 7, is no larger than that of the bulk Astrotec laser package used in earlier versions. Passive attachment enables automated assembly and additional cost reductions should be realized by going to larger wafers, using wafer-scale assembly, and improving yields.

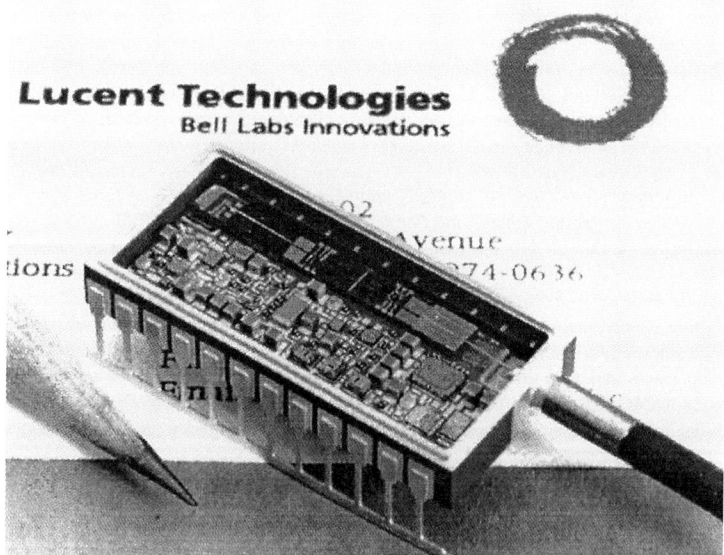

Fig. 7: Silicon Optical Bench Transceiver.

10. FIBER BRAGG GRATINGS

An exciting new technology that has tremendous potential for impacting fiber optic communications is UV-induced fiber Bragg gratings. The gratings are created directly in the germania-doped core of optical fibers by interfering two UV beams from the side after the coating has been removed. The UV light, usually from a KrF-excimer-pumped dye laser operating at 240nm, is absorbed by the germania defects in the core and the resultant periodic index of refraction variation is a Bragg reflector. The Bragg wavelength, λ_B, is given by:

$$\lambda_B = 2\, n_{eff}\, \Lambda = \lambda_{UV}\, /\, 2\, \sin(\alpha)$$

where Λ is the period of the index of refraction variation and α is the angle between the interfering UV beams. Since the fiber geometry is extremely well-controlled, the Bragg wavelength is precisely determined (to within a fraction of an angstrom) by α and λ_{UV}. The index change remains after the UV light is removed. The transmission spectrum of a grating shows effectively 100% transmission of the light except at the Bragg wavelength, where the light is reflected back down the fiber. These integrated reflectors can replace bulk optic components in transmitters, receivers, filters and amplifiers.

The first fiber gratings were written using the standing wave pattern formed by counter-propagating light from an Ar ion laser, however, the photosensitivity of ordinary transmission fiber was too weak to write the strong gratings of interest for applications. The invention of a sensitization process called "hydrogen loading" in 1993 (Lemaire) made it possible to write useful gratings in standard fiber, enabling a host of practical applications. In this process, the fiber is exposed to high pressure (300-11,000 psig) hydrogen or deuterium at low temperatures (20-100°C) for up to a week. Hydrogen loading makes any germania-doped fiber predictably photosensitive. Without H_2 loading, the index changes that are observed are on the order of 10^{-4}. With H_2 loading, index changes as large as 10^{-2} have been achieved.

Some of the applications envisioned for fiber gratings include fiber lasers, demultiplexers, and gain equalizers. The reliability requirements for components used in communications systems are notoriously demanding. Not surprisingly, then, the most frequently asked question about fiber Bragg gratings is, "how permanent are they?". Long term stability of the gratings is a major concern. A complete study of the degree of permanence of the gratings was reported by Erdogan (1994). The bottom line is that it is possible to design a set of conditions under which the aging of the UV-induced index change can be accelerated. This has important practical implications for establishing grating reliability. By eliminating higher activation energy traps with a short anneal at high temperatures, the decay of the UV-induced index change at lower temperatures over periods of years can be rendered insignificant.

11. ELECTRO ABSORPTION MODULATOR LASERS (EML's) AND SELECTIVE AREA GROWTH (SAG)

One of the keys to viable WDM systems is the availability of inexpensive low-chirp optical transmitters. By taking advantage of photonic integrated circuit technology, it is possible to produce monolithically integrated distributed feedback (DFB) laser/electroabsorption (EA) modulators (Aoki 1991, Tanbun-Ek 1994, Johnson 1997) (EML's) with low chirp, low drive voltage and high extinction ratio, in a single compact package as shown schematically in Fig. 8. Crucial to the fabrication of such monolithically integrated DFB laser/EA modulator is a selective-area growth (SAG) process (Aoki 1991, Tanbun-Ek 1994) used during the metallorganic chemical vapor deposition (MOCVD) of the multi-quantum well (MQW) active structure. In this process, the substrate surface is patterned by a pair of oxide masks as shown in the inset. Because of the enhanced growth rate of the wells and barriers in the vicinity of the oxide pattern, as well as a change in the composition (and hence strain) of the layers caused by the difference in diffusion and sticking coefficients of the group III precursors, the effective bandgap of the MQW active structure is different between the laser and modulator as shown by the transmission electron micrographs in Fig. 8. This thickness difference between the strip-window and broad regions depends on the widths of the oxide masks for a given strip opening. By properly designing the mask pattern, reproducible bandgap difference optimal for laser/modulator integration can be obtained. For EML fabrication, the laser

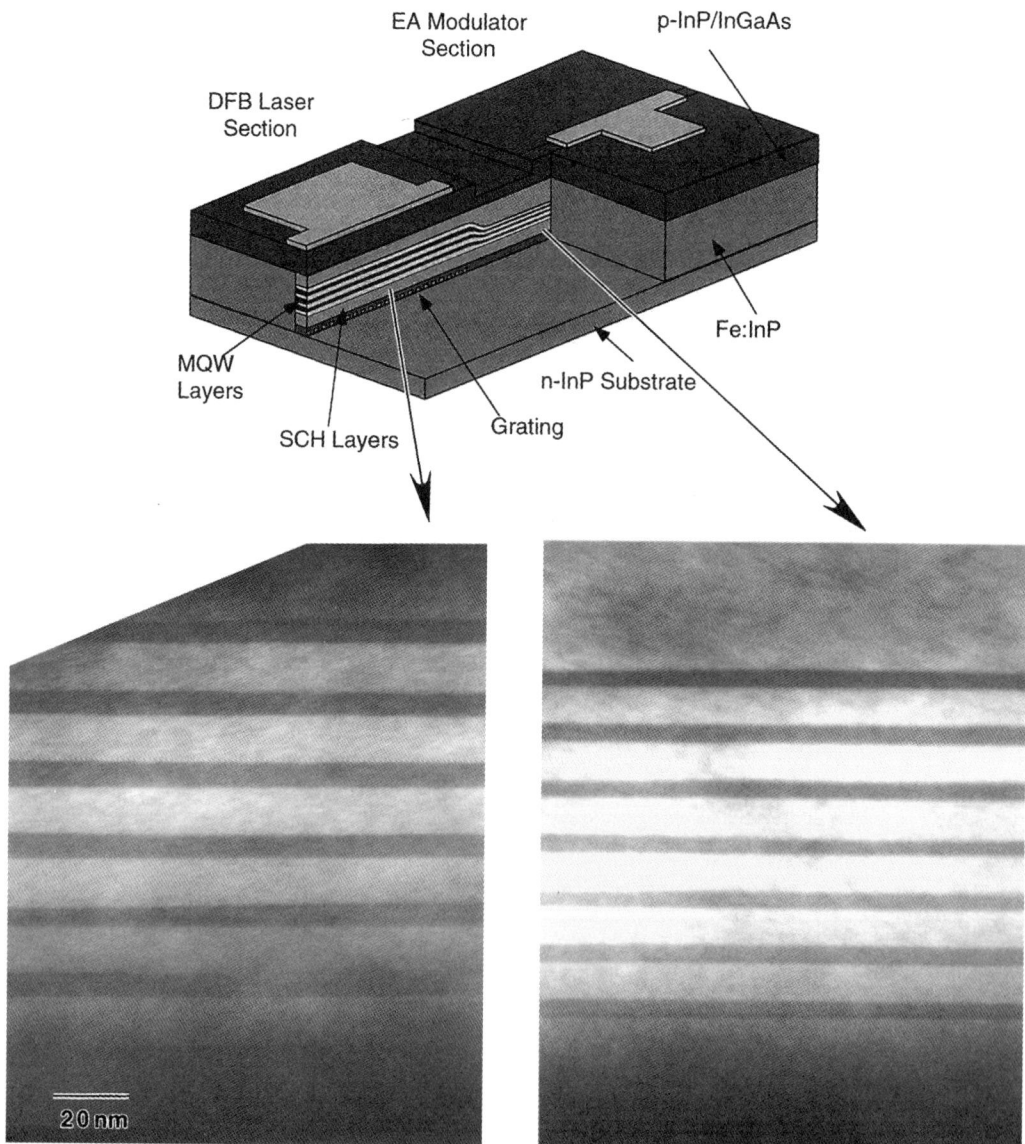

Fig. 8: Electro Absorbtion Modulator Laser.

section is formed in the thicker region while the modulator is formed in the thinner region so that the modulator is non-absorbing with zero external voltage biasing. Upon applying a reverse biasing voltage, the absorption edge of the MQW in the modulator shiftsto longer wavelength, thus absorbing at the lasing wavelength of the DFB laser. Since the DFB laser is operating under constant current (power) operation, there are no carrier fluctuations. Any carrier injection leads to fluctuation in the refractive index and hence wavelength change for the laser. As a result, EML is ideal as a low-chirp source critical for long distance WDM systems (Johnson 1997).

12. CONCLUSION

Telecommunications capabilities and usage have grown at an exponential rate for the past half century, driven by electronics and photonics. To a large degree electronic materials progress has enabled this revolution. Processing power on a chip to equal a main frame of a few decades ago and band width on a fiber to equal thousands of copper wire pairs is by no means the end. Feature sizes approaching the quantum limit and band widths going much closer to the optical limit are conceivable and,with the impetus of recent materials progress, are likely in the early part of the next century.

13. ACKNOWLEDGEMENTS

We would like to thank our colleagues for assisting us in describing their work and that of their colleagues and for comments on the manuscript, especially Mark Cardillo, Dave Eaglesham, Lloyd Harriott, Steve Hillenius, Eric Isaacs, Eric Monberg, Mark Pinto, Elsa Reichmanis, Won Tsang, and Alice White. Linda Garvey and Carol Ridings were an unusual help in preparing the manuscript.

REFERENCES

Aoki M, Sano M, Suzuki M, Takahaski M, Uomi K and Takai A, 1991 Electron. Lett. **27** p 2138.

Eaglesham D 1995 Physics World **8** (11), p 41 for an introduction and further references.

Erdogan T, Mizrahi V, Lemaire P J, Monroe D 1994, Decay of Ultraviolet-induced Fiber Bragg Gratings. J. Appl. Phys. Vol. **76**, p 73.

Harriott L R, et al 1996 J. Vac. Sci. Tech.

Ito H, Willson C G 1984 In Polymers in Electronics Symposium Series **242** American Chemical Society Washington, DC, pp 11-23.

Johnson J E, Morton P A, Park Y K, Ketelsen L J P, Grenko J A, Miller T J, Sputz S K, Tanbun-Ek T, Vandenberg J, Yadvish R D, Fullowan T R, Sciortino P F, Jr., Sergent A M and Tsang W T 1997 in High Speed Semiconductor Lasers for Communications, Kwon N S, Nagarajan R, Editors, Proc. SPIE **3038**, paper 05.

Krisch K, Bude J and Manchanda L 1996 Gate Capacitance Attenuation in MOS Devices with Thin Gate Dielectrics, IEEE EDL, EDL-**17**, p 8.

Lemaire P J, Atkins R M, Mizrahi V, Reed W A 1993, High Pressure H_2 Loading as a Technique for Achieving Ultrahigh UV Photosensitivity and Thermal Sensitivity in GeO_2 Doped Optical Fibers. Electronics Letters. Vol. **29**.

Li Yuan P, Henry C H, 1997, Silicon Optical Bench Waveguide Technology, Chapter 8 in Vol. **IIIB** of "optical Fiber Telecommunications", Kaminow I P and Koch T L, Editors, Academic Press.

Liddie J A, et al 1996 Japan J. Appl. Phys. **4**, p 6663.

Liu C, Ma Y, Cheung K, Fritzinger L, Becerro J, Luftman H, Vaidya H, Colonell J, Kamgar A, Minor J, Murray R, Lai W, Pai C, and Hillenius S, 1996 25A Gate Oxide without Boron Penetration for 0.25 μm and 0.3 μm PMOSFETS in VLSI Sym. Dig. Of Tech. Papers, p 18.

MacChesney J B 1989, A historical review is presented in J. Materials Education **11** (4), p 325.

MacChesney J B 1992, Optical Fiber Conference, Houston, Texas.

Mertesdorf C, et al 1995 Proceedings SPIE **2438**, p 84.

Momosa H, Ono M, Yoshitomi T, Ohguro T, Nakamura S, Saito M, Iwai H 1996 1.5 nm Direct Tunneling Gate Oxide Si MOSFETS, IEEE Trans. On Electron Devices, ED-43, **8** p 1233.

Nalamasu O, et al 1993 Proceedings SPIE **1925**, p 155.

Nisi Yoshio November 1995 Physics World **8** (11), p 31.

Ohji Y, Matsui Y, Itoga T, Hirayama M, Sugawara Y, Torii K, Miki H, Nakata M, Asano I, Iijima S, and Kawamoto Y 1995 TaO5 capacitors, dielectric material for Giga-bit DRAMS in IEDM Technical Dig., p 111.

Presby M, Editor, 1996, Selected Papers on Silica Integrated Optical Circuits, SPIE Optical Engineering Press, Washington DC, MS **125**.

Reichmanis E, et al 1991 Chem. Materials **3**, p 394.

Scotti R E, Anthony P J, Gates J V, Goodwin J Siebert D, 1996, An Integrated Opto-Electronic Transceiver Module, Integrated Photonics Research Conference Technical Digest Series **6**, pp 612-615.

Tanbun-Ek T, Chen Y K, Grenko J A, Byrne E K, Johnson J E, Logan R A, Tate A, Sergent A M, Wecht K W, Sciortino P F, Jr. and Chu S N G 1994, J. Crystal Growth **145** 902.

Inst. Phys. Conf. Ser. No 157
Paper presented at Microsc. Semicond. Mater. Conf., Oxford, 7–10 April 1997

The evolution of electron beam lithography and metrology for semiconductor technologies

T Matsuo

JEOL Ltd, 1-2 Musashino 3-chome, Akishima, Tokyo 196, Japan

ABSTRACT: Systematic development of electron beam lithography started as early as in 1965. The metrology of LSI patterns on wafers with electron beams was introduced into the semiconductor production lines around 1990. These two technologies are now recognised as essential technologies for advanced LSI production. The historical evolution and present state-of-the-art will be reviewed, and future development trends will be surveyed.

1. INTRODUCTION

It is 100 years since J J Thomson discovered, in 1897, that a beam in a cathode ray tube is substantially composed of electrons, which are fundamental particles that constitute all substances. In the period since that time, the technologies needed to control an electron beam in a vacuum by using electron lenses and deflectors have evolved greatly.

One of the most remarkable products of these technologies is the electron microscope. The transmission electron microscope has now been developed to an extent that a resolution of better than 1 Ångstrom can be attained by a commercial model. The scanning electron microscope is now also in wide use as a scientific instrument having the widest application scope. On the other hand, electron beam lithography and electron beam wafer inspection are, though not known so widely as the electron microscope, also very important for applications in the semiconductor field. These electron beam (EB) technologies are already indispensable for the production of integrated circuits. They have been dramatically evolved in the past 30 years, and their further development is now being looked forward to as we enter the 21st century.

2. APPLICATION FIELDS OF ELECTRON BEAMS

When an electron beam emitted in a vacuum is focused onto a workpiece at a position determined by a deflector, various phenomena take place, as shown in Fig.1. The application fields that utilize those phenomena include;
1) Scientific analysis that detects and particles in a material, as with the electron microscope.
2) Processing that utilizes the polymerization of polymer materials. This application includes electron beam lithography.
3) Various applications that utilize local temperature rise, such as melting furnaces, welders, evaporators, refiners and cutters.
4) Most familiar applications such as CRT displays, vacuum tubes, and microwave oscillators.

2.1 EB technologies for semiconductor fabrication

Fig. 2 shows two important applications of EB for semiconductor fabrication. One is the EB lithography technology for writing semi-conductor device patterns. The other is the technology for observing and inspecting devices on silicon wafers by means of EB.

(Electron Beams In Vacuum)

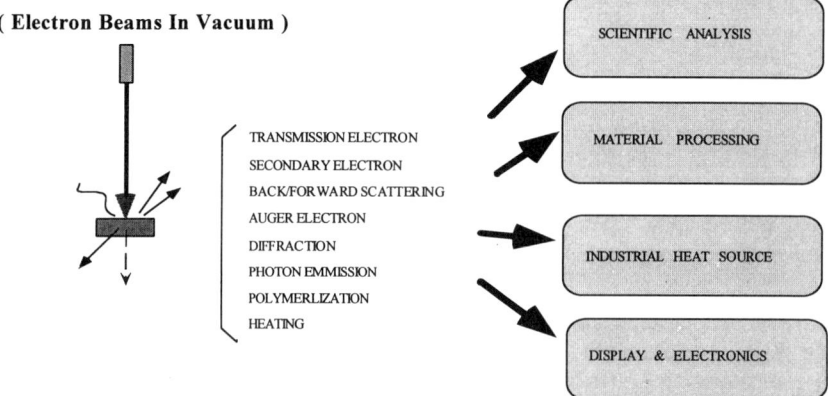

Fig. 1 Application fields of electron beams

The lithography technology is divided into two systems. One is writing of a stencil mask of a stepper, for mass production of semi-conductor devices. The other is direct writing on each silicon wafer without using a mask. Both are the fundamental technologies that support the present and future semiconductor industries. The observation and inspection of device patterns by means of EB, however, has not been widely used, but as the optical method came to its resolution limit with the development of the semiconductor technology, the EB metrology is now becoming the key technology that is indispensable for fabrication of next-generation devices.

2.2 Technologies from electron microscopy

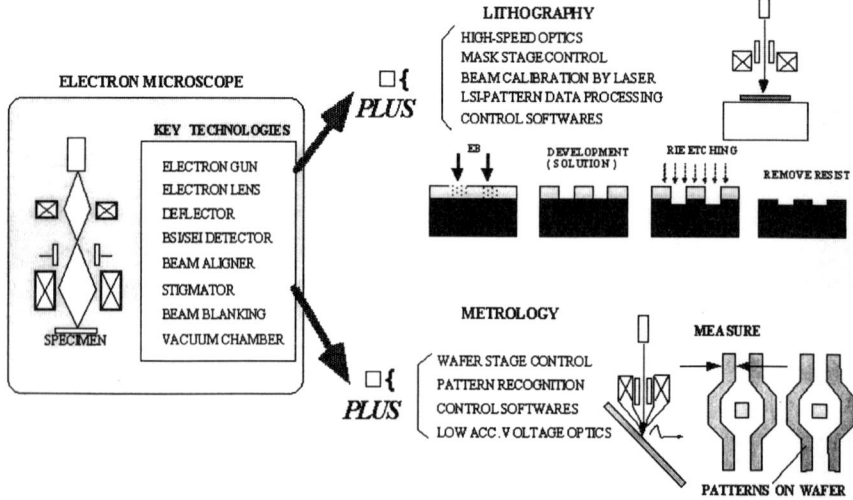

Fig. 2 Technologies from electron microscopy

These two EB technologies for semi-conductor production have been developed on the basis of the electron microscope technologies. As shown in Fig. 2, electron microscope technologies including the electron gun, electron lenses, deflectors, and vacuum technology are introduced into the EB technologies.
And various new technologies have been developed for lithography, such as an improved electron

optical system for higher-speed writing, a large-sized workpiece stage, beam calibration technology and control software. The present basic procedure of lithography, as shown in this figure, is to illuminate a focused EB on an EB resist composed of a thin polymer film on a silicon wafer, dissolve and remove the polymerized portions, and finally etch the silicon oxide film under the polymer film by using the resist pattern obtained, as from a mask.

EB metrology is also based on the electron microscope technologies. And new technologies such as low accelerating voltage high-resolution optics, a large-size wafer stage, signal processing, pattern recognition and control software have been added for the development of EB metrology. Signals detected from the material used reflect the topography and material distribution on the LSI pattern of the material surface. Therefore, size measurement and element analysis of foreign particles in the material are carried out by combining signal detection and beam scanning.

3. EVOLUTION OF ELECTRON BEAM LITHOGRAPHY

3.1 Technology roadmap of EB lithography

Shown in Fig. 3 is the "Technology roadmap" of the semiconductor industry's requirements for lithography. It was prepared at SEMATEC (US) and announced by SIA. This map shows the resolution and writing accuracy of lithography that have been achieved up to now, and also the indexes of lithography performance that will be needed by 2010. The reason why such strict lithography milestones are shown for many years ahead is that the development of lithography technology generally requires enormous manpower, cost and time. The development of EB lithography, also, requires the longest time and the largest cost and thus, governmental and international cooperation are required for healthy growth.

	1995	1998	2001	2004	2007	2010
	0.35µm	0.25µm	0.18µm	0.13µm	0.10µm	0.07µm
Function DRAM (bits)	64M	256M	1G	4G	16G	64G
microprocessor (logic transistors/cm^2)	4M	7M	13M	25M	50M	90M
ASIC(transistors/cm^2 auto layout)*	2M	4M	7M	12M	25M	40M
Resolution(µm)	0.35	0.25	0.18	0.13	0.10	0.07
Gate CD control at post etch (nm)	35	25	18	13	10	7
Overlay (nm)	100	75	50	40	30	20
Chip size DRAM (mm^2)	190	280	420	640	960	1400
microprocessor (mm^2)	250	300	360	430	520	620
DRAM (NM X mm)	10 x 20	12 x 24	15 x 30	18 x 36	22 x 44	28 x 50
microprocessor (mm X mm)	16 x 16	18 x 18	19 x 19	21 x 21	23 x 23	25 x 25
Minimum Field Size #DRAM/Field	2	2	1	1	1	1
(mm X mm)	22 x 22	26 x 26	26 x 30	26 x 36	26 x 44	28 x 50
(mm^2)	484	676	780	936	1144	1400
Depth of Focus (usable) (µm) (full field/±10% exposure)	1.0	0.8**	0.7**	TBD	TBD	TBD
Minimum mask count	18	20	20	22	22	24
Defect density lithography only (per layer/m^2 @ defect size µm)	690 @0.12	320 @0.08	135 @0.06	TBD	TBD	TBD
Mask size (inches) (Quartz)	6X6	6X6	6X6 Next size	Next size	Next size	Next size
* ASIC will use maximum available field size.						
** Assumes advanced techniques to maximize the usable depth of focus. Further analysis is needed.						

Fig. 3 Technical roadmap for semiconductor lithography

3.2 The first and more recent EB lithography

Fig. 4 shows line and space line patterns made with EB lithography in 1966 (Kanaya et al 1966) and in 1997. The first EB lithography system (Fig. 5) was constructed in 1967 by JEOL

16

and The Electrotechnical Laboratory and a FET transistor fabricated by this system was reported in 1968 (Tarui et al 1968). Next, I shall review the kind of technologies that have been developed during the last thirty years.

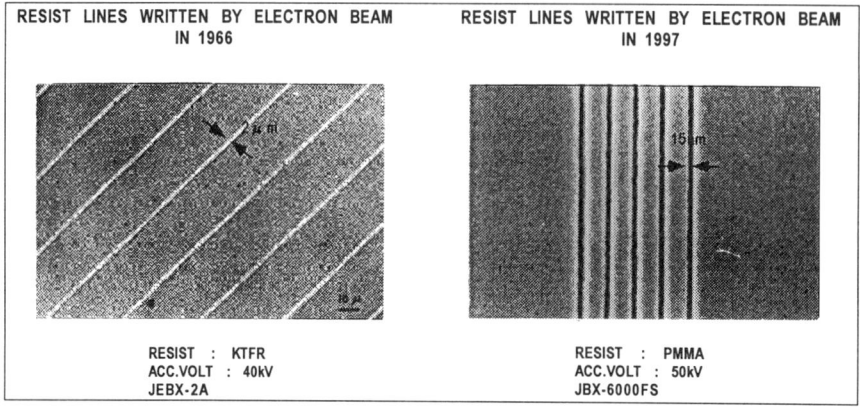

Fig. 4 Line and space patterns made by EB lithography

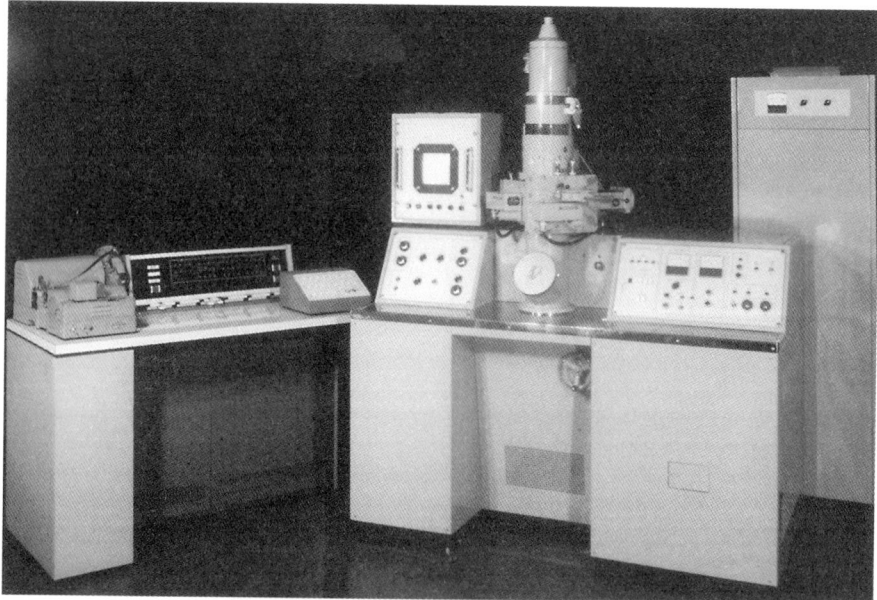

Fig. 5 The first EB lithography system in 1967 (JEOL)

3.3 Technologies developed for EB lithography

The history of EB lithography is the history of improving writing accuracy and writing speed. Technologies developed for EB lithography are classified roughly into three categories in Fig. 6.

(Pattern resolution)

One of the basic requirements for EB lithography is an increased beam current to increase writing speed. However, we have now already reached to a stage where the beam current is so large that space charge blurring due to the Coulomb effect restricts the resolution of written patterns. Consequently, efforts are now being made to reduce the distance between the beam source and the image plane. In the past, the column's optical path was several tens of centimeters long but it has recently been reduced to

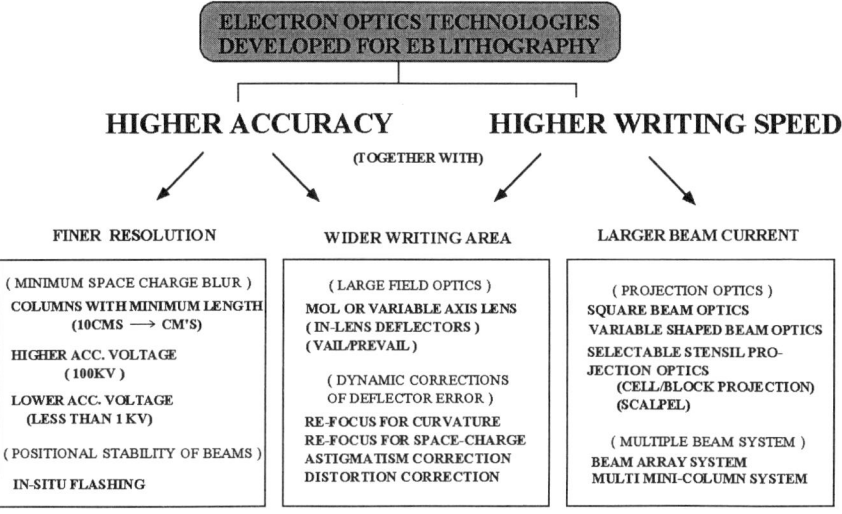

Fig. 6 Technologies developed for EB lithography

about twenty centimeters, and will soon be reduced to several centimeters. Beam scattering in the resist is also a factor that deteriorates pattern resolution. So, countermeasures are now taken by increasing the accelerating voltage up to 100kV or, inversely, decreasing it down to less than 1kV.

(Beam deflector)

The development of a deflector that enables a wider area to be scanned accurately with EB was another important evolution. Fig. 7 shows the evolution processes of the beam scanning technology for EB lithography. As for the scanner (deflector) position, the pre-lens system, where the scanner is placed in front of the objective lens as in the scanning electron microscope, was employed first, followed by the post-lens system, which allows accurate positioning, and finally the in-lens system, which incorporates the scanner in the objective lens. This in-lens system is indispensable for making the optical length the shortest. This system was made possible due largely to computer-aided off-axial electron beam trajectory simulation technology in an asymmetrical field. The deflector is usually composed of eight poles and the control voltage circuits for them are arranged near the electrodes, for high-speed control. Another improvement made possible by the in-lens concept is the expansion of the deflection range, MOL, which was firstly perceived by Ohiwa et al (1971), and was developed into a variable axial lens in by Pfeiffer (1975). This concept was further evolved by Pfeiffer et al (1981) into VAIL, which incorporates an electrostatic deflector and an immersion lens. And, further research is now under way to develop it to PRVAIL which combines VAIL with a project type imaging system. The advent of an in-lens MOL or VAL deflector system has increased the number of writing pixels available in one deflection range up to 50,000 x 50,000.

18

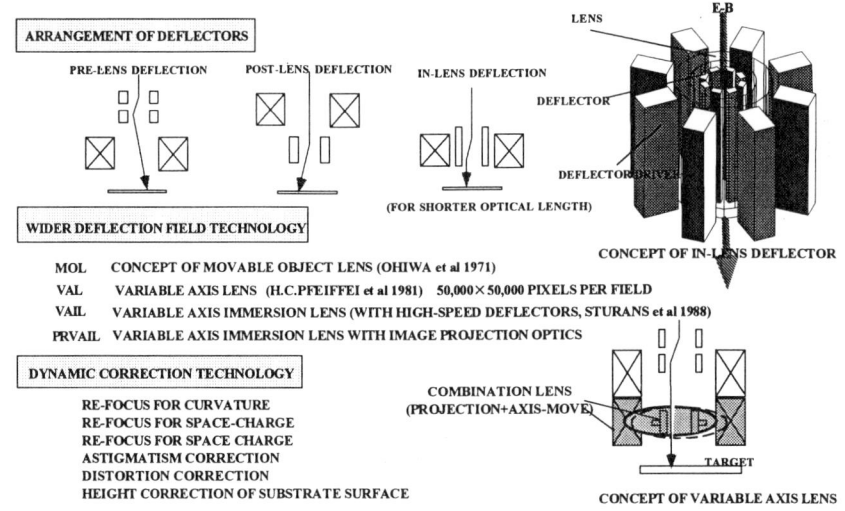

Fig. 7 Progress of beam deflector

(Beam shape)

A remarkable development in the EB lithography technology was in the beam shape (Fig. 8). The beam used in 1967 was a Gaussian spot beam from tungsten hair-pin cathode. In 1979, the emission source was changed to a thermally-assisted Shottkey emitter that achieved a current density as high as over 1000 A per cm^2. The square beam method, developed in 1975 (Pfeiffer), was evolved into the variable-shaped beam method in 1978 (Gotoh et al, Pfeiffer , Thomson et al), which enabled the lengths of the square sides of beam to be freely changed by electric control and the beam current was greatly increased.

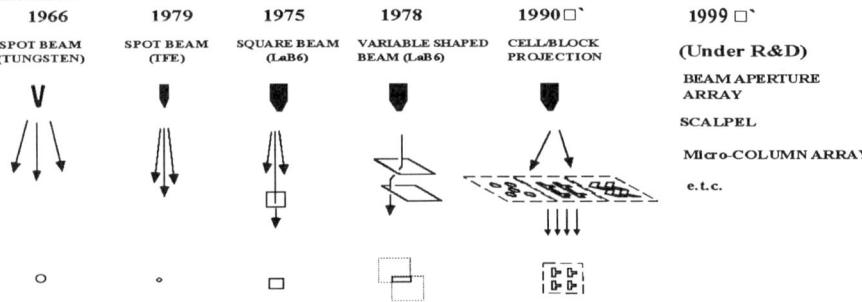

Fig. 8 Progress of beam shape

In 1990 a new method was developed which projects de-magnified images of cells or block stencils of small patterns (Nakayama et al 1990). This method is also capable of changing a variety of stencils electrically. EB systems of this method have recently started operation. They are used mainly for EB direct writing on wafers. New research has recently started on electron optics such as the beam aperture array, which generates many beams with a single column; SCALPEL (Harriott et al 1996), which allows a wider area to be projected; and the micro-column optics (Chang et al 1989), which writes many chips at same time by arranging many small columns. All of these methods are intended to give EB lithography a productivity equivalent to that of optical lithography by improving the writing speed.

(Accuracy)

The writing accuracy, together with the writing speed, are among the most important factors in EB lithography, but high writing accuracy cannot be obtained only by the electron optics technology. As shown in Fig. 9, the most significant factor that governs the accuracy was formerly electron optics.

ROADMAP FOR LITHOGRAPHY ACCURACIES

	1995	1998	2001	2004	2007	2010
RESOLUTION	0.35fÊm	0.25fÊm	0.18fÊm	0.13fÊm	0.10fÊm	0.07fÊm
PATTERN SIZE ACCURACY (Overall)	35nm	25nm	18nm	13nm	10nm	7nm
PATTERN SIZE ACCURACY [*1] (for E-B)	18nm	12nm	9nm	6nm	5nm	3nm
PATTERN POSITION ACCURACY (Overall)	100nm	75nm	50nm	40nm	30nm	20nm
PATTERN POSITION ACCURACY [*2] (for E-B)	50nm	35nm	25nm	20nm	15nm	10nm

[*1], [*2] : Assumes to be half of overall accuracy

TECHNOLOGIES DEVELOPED FOR HIGHER ACCURACY

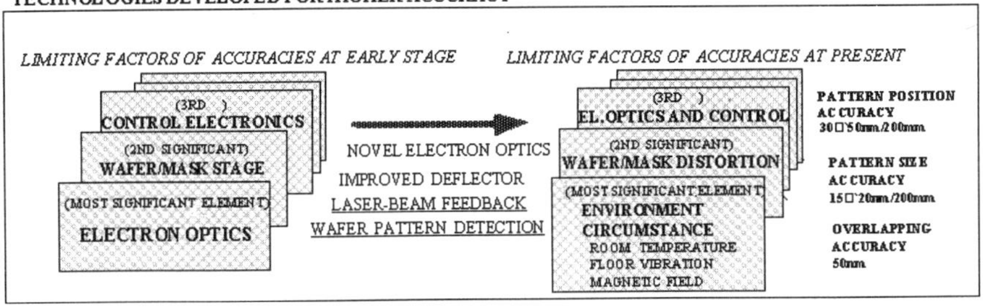

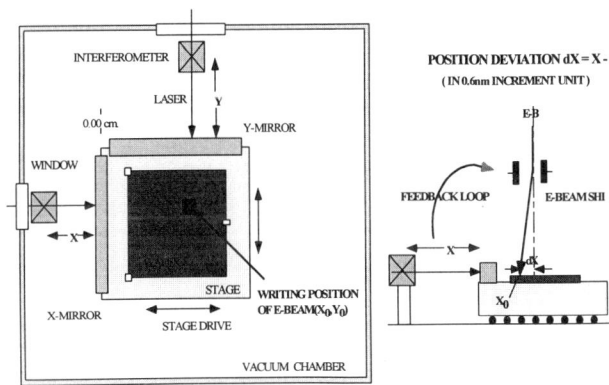

Fig. 9 Writing accuracy of EB lithography

But as several basic technologies besides the electron optics and deflector technologies have been developed, the most significant factors that govern the accuracy have shifted to the environment of EB lithography and to workpiece distortion. Two basic technologies which sustain the accuracy of EB lithography, besides electron optics, are laser-EB feedback loop technology and the correction technology for an overlay mark on the wafer.

Fig. 10 Laser- Beam feedback loop

All the existing EB lithography systems maintain the positional accuracy of writing, by using a configuration shown in Fig. 10.

The workpiece is firmly secured on its stage fitted with an optical mirror. The position of this stage is constantly monitored during writing by a laser measurement system, and if the position is deviated by stage vibration or drift, for example, the center position of the electron beam is corrected to cancel the positional error. If this loop is complete, the accuracy limit of laser measurement is determined by the stability of laser wavelength. The reliability thus attained is 10^{-8} to 10^{-9}. The measurement increment is 0.6 nm, and the sampling frequency is more than 10 MHz.

Since the required positional accuracy up to year 2010 is 10^{-7} according to the roadmap, this feedback loop also must be

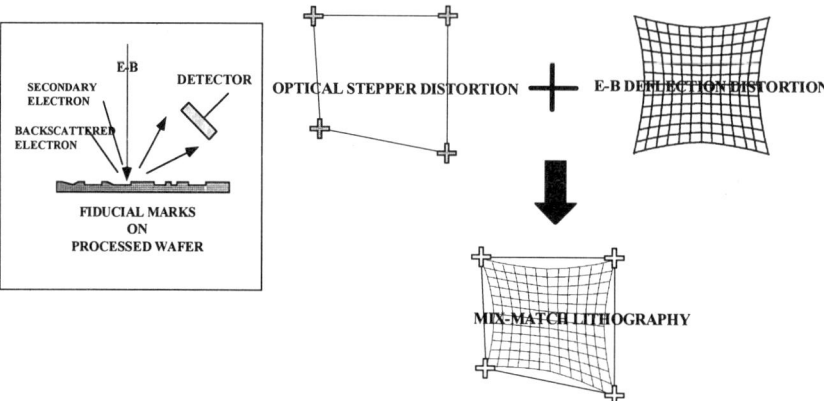

Fig. 11 Electron beam positioning to index marks on wafer

fully reliable in the future. It is necessary, however, to strictly maintain Abbe's condition that the measuring point and the writing point should agree with each other. The actual limit of positional accuracy is determined by errors that are not detected by laser measurement, such as the deviation between the mirror and workpiece, EB drift, deviation of the laser measurement standard (interferometer position), measurement errors caused by stage rotation and jitter. Another important technology that supports the writing accuracy is that for bring the overlay mark and beam position into agreement. Fig. 11 shows its principle. If a mark is made on the workpiece, a signal is generated from it when it is illuminated with a beam. Detecting the signal makes it possible to bring the beam and mark into strict positional agreement. Combining this system with the laser-EB loop mentioned before, allows the highest-level position measurement system to be constructed.

Use of this function offers an important merit obtainable only with EB lithography. This function is used to measure patterns written by an optical stepper. Also, it is used to measure the wiring distortion of the EB lithography system itself. The EB writing distortion is deformed so that the difference between these two become zero. As a result, pattern writing that strictly overlies on another lithography layer can be achieved. This is an accuracy defined as the overlay accuracy or registration accuracy, which is presently about 50 nm.

(Writing speed)

The writing speed is presently the most important subject in direct wafer writing. Mask writing, even if it takes several hours, sufficiently pays economically. Direct writing, however, does not seem to pay economically if the time required is more than five minutes per wafer, or 10 seconds per chip. Although the accuracy and resolution of EB lithography are already fully attractive, but the only reason why it has not yet been put to practical use is considered to be low writing speed.

The challenge for the writing speed up to now has been that for larger beam current. The reason of this is that the writing speed is almost proportional to the beam current. For example, no matter what the pattern or beam shape is like, if the beam current is 1 nA, the writing time can not be made less than 10,000 seconds per square centimeter or equivalent of pattern area , at a resist sensitivity of 10 μ C/cm^2.

Fig. 12 shows, from the top, the increase of the beam current. At first, the beam current was 0.5 nA at most with a spot beam, but it increased to 0.5µA with the development of the TFE emitter and variable shaped beam. It further increased to 5µA with cell projection.

Here, it became clear that even if the beam current is further increased, the coulomb effect results in lowering the resolution. The technology now under development to solve this problem is classified roughly into two approaches. One is to use an optical system that widens the beam so as to prevent it from being focused extremely, or to increase the accelerating voltage to more than 100 kV and thereby reduce the coulomb effect. SCALPEL and PREVAIL employ this approach.

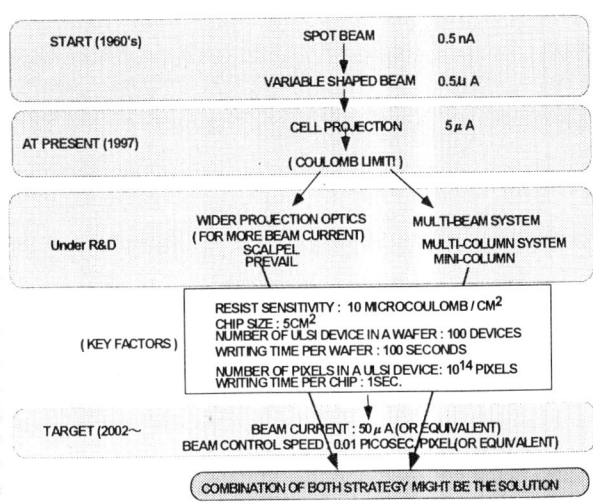

Fig. 12 Challenge; Hi-thruput EB lithography

The other approach is to increase the throughput by parallel writing with many beams onto a wafer. This approach includes the multi-column system and array-beam system.

Let us here assume a key factor which is given here. What if the writing time is made 1 second per chip, or 100 seconds per wafer? The answer is that the beam current must be 50µA, which is 10 times that in cell projection. How can this be made possible? I suppose that this may be difficult with only one of the two approaches I introduced above. For the present, it will be a general view that a combination of both approaches is needed.

4. EVOLUTION OF ELECTRON BEAM METROLOGY

4.1 Technical roadmap for metrology

SIA's technical roadmap for EB metrology till 2010 requires a reproducibility of several nanometers for critical dimension measurement and an overlay measurement reproducibility of less than 10 nm. The requirements of the semiconductor manufacturing process for metrology are shown in Fig. 13. As the gate dimension tolerance advances from the 35 nm generation to the 7 nm generation, the measurement resolution needs to be increased from 10 to 2 nm and the depth of field from 0.5 to 2.0 µm. Furthermore, the so-called "in-line and in-situ" inspection, which gives the least damage to wafers during wafer processing, becomes necessary.

4.2 Low voltage, high resolution electron optics

Because many objects for inspection are insulators and three-dimensional in structure, it is necessary to develop an optical system that allows high-resolution image observation to be carried out at voltages as low as 1kV and at a large angle. Fig. 14 shows the principle of a low accelerating voltage electron optical system that has been developed recently by Honda et al (1996). The beam emitted from the TFE emitter is once accelerated to 8kV by an aperture angle control lens (ACL) having a multi-stage acceleration mechanism. The beam is decelerated to 1kV just in front of the specimen, and focused into a 5-nanometer spot. This combined lens is a magnetic field lens that incorporates a multi-stage electrostatic retarding lens in it. The lens is capable of bringing the principal plane of the combined lens close to the

22

specimen, making it possible to greatly reduce the aberration coefficients. Fig. 15 shows a photograph, taken by this optics at 1kV. The resist side-wall and contact portion to the silicon substrate are clearly observed.

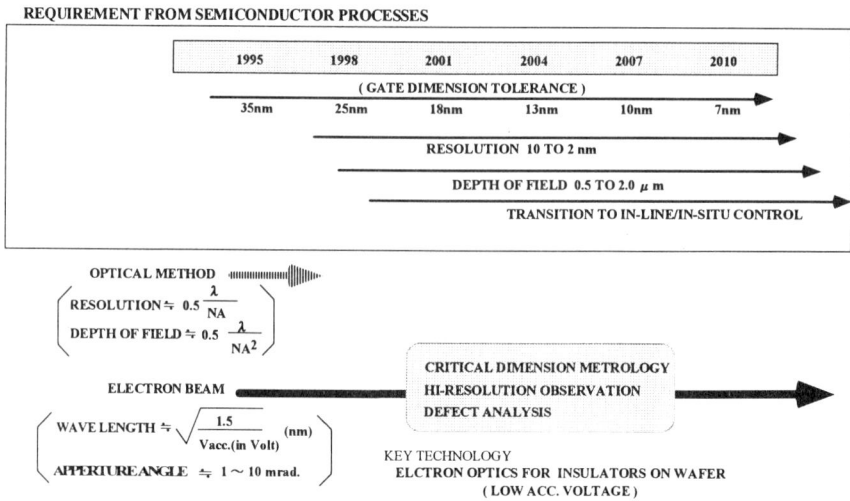

Fig. 13 Electron beam metrology

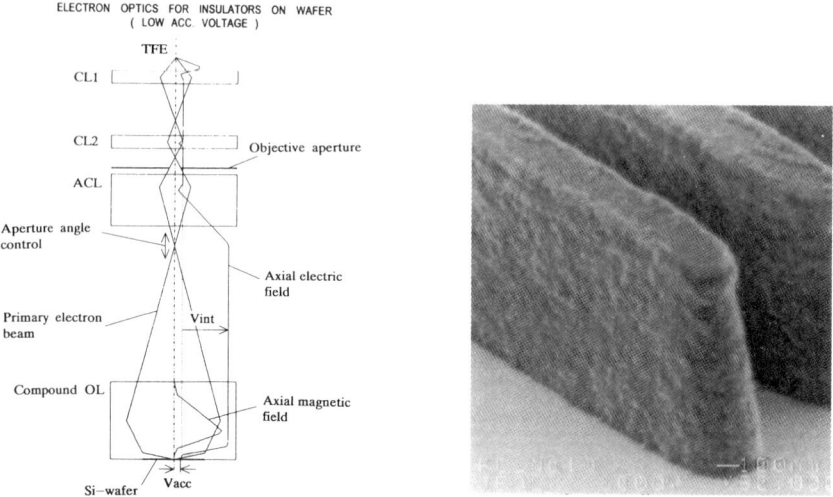

Fig. 14 Electron optics for wafer observation   Fig. 15 Observation of resist pattern

5. CHALLENGES FOR THE FUTURE

I have reviewed the evolution of electron beam technologies for semiconductor devices and I would like to summarize the future situation as follows;

(1) Trends from light optical techniques to EB techniques in semiconductor production will be accelerated.
(2) Challenges to the high speed EB lithography should be continued (the gap between the present and the target speeds is more than 10 times.)
(3) EB technologies for higher writing accuracy also to be continued.
(4) Pattern defect inspection system is the most important key technology for the next generation semiconductor device production.
(5) More international cooperation on EB technologies is required.

REFERENCES

Chang T H P 1989 J. Vac. Sci. Technol. **B6,** 1855

Gotoh E, Soma T and Idesawa M 1978 J. Vac. Sci. Technol. **15**, 883

Harriott L R and Beiger S D 1996 J. Vac. Sci. Technol. **B14**, 38258

Honda K and Takashima S et al. 1996 16th LSI testing Symp. 132 committee Japan, Nov.7-8

Kanaya K, Tanaka K and Yuasa T 1966 Sixth Intern. Congress for Electron Microscopy, Kyoto pp. 281-2

Nakayama Y and Saitou N et al 1990 J. Vac. Sci. Technol. **B8**, 1836

Ohiwa H, Gotoh E and Ono A 1971 Electron Comm. Japan Sec B. **54**, 44

Pfeiffer H C 1975 J. Vac. Sci. Technol. **12**, 1170

Pfeiffer H C 1978 J. Vac. Sci. Technol. **15**, 887

Pfeiffer H C and Langner G O 1981 J. Vac. Sci. Technol. **19**, 1058

Tarui Y, Denda S, Baba H, Miyauchi S and Tanaka K 1968 Trans. Inst. Electronics Comm. Eng, Japan **51**, C 74

Thomson M G R, Collier R J and Herriott D R, 1978 J. Vac. Sci. Technol. **15**, 891

Inst. Phys. Conf. Ser. No 157
Paper presented at Microsc. Semicond. Mater. Conf., Oxford, 7–10 April 1997
© *1997 IOP Publishing Ltd*

The structures of extended defects in Si and other materials studied by HRTEM

S Takeda

Department of Physics Graduate School of Science, Osaka University, 1-16 Machikane-yama-cho, Toyonaka, Osaka 560, Japan

ABSTRACT: The recent structural studies of point-defect-agglomerates in Si and related materials by high resolution transmission electron microscopy are complied in order to provide a more generalized view of the structures of extended defects. The structural data are presented on the planar defects created by electron irradiation and ion-implantation. An emphasis is placed on the extension of studies made possible only by structure determination by HRTEM; the mechanism of agglomeration is discussed, the unit structures of point defects are derived, and the electronic structures of the agglomerates are elucidated.

1. INTRODUCTION

Various kinds of extended defects are frequently observed when we examine Si based materials using the transmission electron microscope (TEM). The agglomerates of point defects are of particular interest, since they are created due to dynamic properties of the crystal lattice involving point defects and impurities. High resolution TEM (HRTEM) is a unique means of observing this co-operative behaviour of atoms occurring inside a crystal at the atomic level. Nevertheless, the atomic structures of various kinds of agglomerates have remained unclear and this has prevented us from understanding the dynamic phenomena at the atomic level.

We have shown that reasonably accurate structural data on planar agglomerates with longer life-times can be derived from HRTEM observations. With the data in hand, one can realize that the structures of agglomerates present the general structural nature of covalent bonding as in grain boundaries, reconstructed surfaces and dislocations. Also, the studies can be developed: for instance, the agglomeration mechanism of point defects can be elucidated at the atomic level; the possible unit structures of point defects can be deduced and the electronic structures of the agglomerates can be elucidated. This paper intends to compile our recent studies on the above mentioned subjects in Si and other materials by HRTEM and related techniques. After examining the agglomerates in the bulk, it is demonstrated that *in-situ* HRTEM can be utilized to create and observe novel defects near and at surfaces.

2. EXPERIMENTAL PROCEDURES

In order to obtain high-quality HRTEM data, which facilitates the structural analysis, sample preparation is very important. To analyse the atomic structure of a small defect embedded in crystal in a convincing manner, the specimen which includes a defect of interest is carefully thinned by Ar⁺ ion etching and chemical polishing. A defect suitable for HRTEM observation extends between the top and the bottom surfaces of the thinned specimen, which is less than ~10nm in thickness.

To observe the defect structures stably, we use a lower voltage HRTEM. For instance, Si specimens are examined using a microscope with 160keV electrons (a top-entry type JEOL JEM2000EX) at room temperature. More detailed description about experimental procedures can be seen elsewhere (Takeda 1991, Takeda Hirata, Muto, Hua, Hiraga and Kiritani 1991). The analysis of a defect structure does not proceed unless the HRTEM image of the non defective crystal nearby agrees with image simulations.

Analyses of atomic structures of defects using HRTEM and transmission electron diffraction

(TED) resemble   X-ray crystallography in some extent.   Some criteria commonly used in X-ray crystallography have naturally been introduced.   Of course, it should be stressed that, without any elaborate qualification , numerous achievements have been obtained so far by the imaging technique. The introduction of R-factor, or the reliable factor in HRTEM and TED is a simple example as used in the present study. Dynamical scattering of electrons brings an additional parameter, t the thickness of a specimen, in the R-factors, which may complicate an analysis but systematic coincidence of experiments with computations as a function of the additional parameter may confirm the reliability of the analysis. R-factors are used for both evaluating structure models and refining down the most appropriate one with displacing atom positions, etc.   In the studies of defects, the image contrast of a defective region is considered.   Therefore, the evaluation and refinement of a model structure can be performed using the R-factor defined in a local region, R_d.

3. STRUCTURAL STUDY OF POINT-DEFECT-AGGLOMERATES

3.1 The {113} planar defects in Si and Ge

The {113} defect is introduced in Si and Ge by various processes such as electron irradiation, ion-implantation and thermal annealing.   In its long history, the defect was identified as various structures.   It is now believed (Hutchison and Aseev, 1993, Eaglesham, Stolk, Cheng, Gossmann, Haynes and Poate, 1994) that the defect is an agglomerate of self-interstitials as suggested by the early researchers (Ferreira Lima and Howie, 1976 , Salisbury and Lerreto, 1979) and its atomic structure was analyzed by HRTEM observation.

The relative displacement vector of the planar defects, R was estimated in HRTEM images (Takeda, 1991, Takeda, Kohyama and Ibe, 1994) to be [116] a / 25. The accurate analysis of diffraction contrast by Ferreira Lima and Howie (1976) has yielded the same value.   Therefore, the actual contribution of HRTEM is regarded as the analysis of atomic configuration of the defects. According to the structure study by HRTEM (Takeda, 1991), the planar {113} defect is described as follows; the chains of additional Si atoms in the <110> direction are sandwiched by two reconstructed {113} surfaces, and constitute the 5-, 6- , 7- and 8-membered rings . The 6 membered-ring has the different atomic arrangement from the diamond-type structure, but is a part of the one-element Wurtzite (hexagonal) -type structure. The additional atoms (or agglomerated interstitials) inserted between the existing {113} net planes are indicated by the arrow heads in Fig. 1(b). The two kinds of structure units arrange on {113} with the short-ranged correlation. Clearly, the planar defect is not a stacking fault.   The {113} defect structure in electron-irradiated (Takeda et al. 1992) and He-implanted (Hutchison and Aseev, 1993) Ge is identical to that in Si.   After the structure data was obtained, it is noticed that the structure is a fascinating patch work joining together pieces of structure units found by Tan and Foll and Hu (1981), Bourett (1987) and Gibson, MacDonald and Unterwald (1985) .

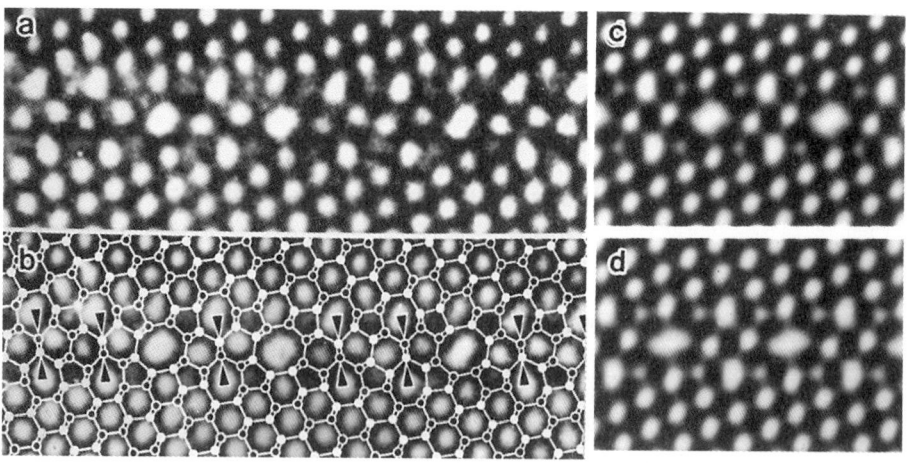

Fig. 1   The structure of the {113} defect in Si. (a) Observation and (b) the refined model pasted in(a).   (c) Simulations based on the refined model in (c) and on the raw model in (d).

It is often overlooked that structure studies of defects are useful from the technological point of view. However, the {113} defects play important roles in diffusion of doped atoms as recently illustrated (Eaglesham et al., 1994, 1996), and the planar density of self-interstitials, measured directly in HRTEM images (Table 1), has been utilized in the study as basic data.

Energy calculations (Kohyama and Takeda, 1992, 1995) yielded much information about the defects. It is shown that the defect structure is energetically favorable. The energy increase per self-interstitial against perfect crystal, E_{is} is estimated to be around 1eV (Table 1). The relaxed atomic positions which give the energy minimum in the framework of computation were obtained without the original bond topology changed. The bond length and angle distortions, Δr and $\Delta \theta$ in the refined structure are summarized in Table 1. The HRTEM image matching was improved by the refinement (Fig. 1(c) and (d), Takeda and Kohyama, 1993) and the R-factor estimated in Fig. 1(a) and (b) is reduced to 0.14. The R-factor in a plan-view electron diffraction of a {113} defect is estimated to be 0.30, and the systematic extinction of diffraction spots in the <100> pattern can be reproduced (Takeda, Kohyama and Ibe, 1994).

3. 2 The {100} planar defects in Ge

Ion implantation induces various extended defects in Si and Ge. In addition to the {113} defects, we found that another kind of self-interstitial agglomerate exists when germanium is implanted by D^+ (deuteron) at the certain temperature range. The defects are extended on {100} (Muto and Takeda, 1995). The structure analysis of the defect was carried out directly from HRTEM data with a similar experimental technique mentioned above . Clearly, an HRTEM image, taken with the [110] incidence, exhibits doubled periodicity along the [1T0] direction (Fig. 2 (a)). The relative displacement vector, R was measured to be $0.385[001]a$. The analysis of HRTEM images revealed that lattice reconstruction occurs on {100} involving only self-interstitials. Figure 2(b) illustrates the {100} defect structure viewed along <110>.

The {100} defect structure turned out to be identical to the Humble model of the {100} platelets in diamond (Humble, 1982). It is now believed that the platelet in diamond includes nitrogen in addition to self-interstitials (Humble, Lynch and Olsen, 1985). There are no other apparent presentations on the defect structure. In the paper by Ferreira Lima and Howie (1976), however, they described that "some attempts at model building initiated at an early stage indicated that a layer of interstitials could be accommodated on {100} with rather severely distorted bonds leading to a Burgers vector of 1/3<110> ". The structure is represented by sticks and balls (Ferreira Lima, 1975).

Energy increase , E_{is} in the model is close to that in the {113} defect in both Si (Table 1) and Ge, indicating the sufficient stability of the structure, even though the defect of this kind has not yet been observed in Si . The estimated distortions of bond length and angle, and the planar interstitial density are also tabulated in Table1. The relative displacement vector, R obtained by the energy minimization calculation agrees well with the measurement. The simulated image in Fig. 2(c) based on the refined structure shows an agreement with the observation in Fig. 2(a), and the R-factor on the area is measured to be 0.10. It is noted that the refined structure in Fig. 2(b) is assigned to the observed image upside down against the previous assignment.

It has not yet been reported that the defect of this kind is created by other processes in Si and Ge, therefore suggesting that an interesting interaction occurs between hydrogen and self-interstitials.

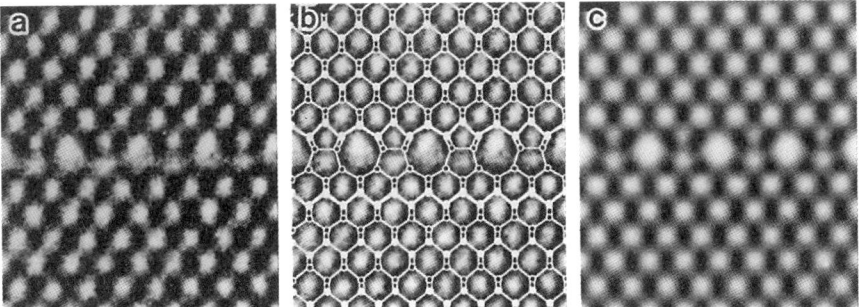

Fig. 2 The structure of the {100} defect in Ge. (a) Observation, (b) The refined model and (c) simulation.

28

3. 3 Hydrogen-induced platelets on {111} in Si

Implantation of hydrogen into semiconducting materials has extensively been studied in various aspects. TEM has revealed that implanted-hydrogen occasionally condenses on {111} in Si (Ponce, Johnson, , Tramontana, and Walker, 1987), Several structure models have been proposed: hydrogen saturating the broken bonds on {111}, the agglomerate of hydrogen on {111} involving vacancies or self-interstitials and the agglomerates of hydrogen molecules, H_2*. Hydrogen may not be detectable by TEM and related techniques, but an idea of the structural study is that the frame of silicon atoms may be revealed by HRTEM and consequently the positions of hydrogen atoms in the platelet are obtainable (Muto, Takeda and Hirata, 1995). The HRTEM images of the platelets viewed along <110> and <112> , taken near the Scherzer defocus were intensively analyzed. The relative displacement vector, R is normal to the plane of platelets, i. e. 0.81 <111>a/3. Thus, the models which include stacking faults are excluded, and the rest were considered; in the first model, the bonds break and the interior {111} surfaces are hydrogenated (the symmetrical model, Fig. 3(b)), and in the second one, the H_2 * molecules agglomerate on {111} (the asymmetric model). The image matching was best achieved in the refined symmetrical model (R_d = 0.34 in the <110> image). It is also shown that the variations of images of the platelets are systematically accounted for by the single model which includes atomic steps.

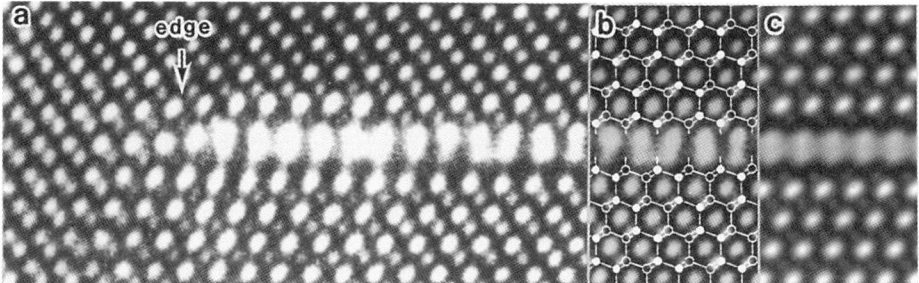

Fig. 3   The hydrogen-induced platelets in Si. (a) Observation, (b) the raw model and (c) simulation.

3. 4 General structural characteristic of planar extended defects

Having obtained the structural data about the planar extended defects, one may be aware of the close relationship between the structures of surfaces and planar defects. The periodic 8-membered rings along <110> in the {113} defect structure is well related to a {113} surface model, even though the accumulation of data on the surface seems in favor of the more complicated model (Dabrowski, Mussig and Wolff, 1994). Similarly, the {100} defect structure can be related to the {100} reconstructed surface ; the two 2 x 1 reconstructed surfaces, rotated along the surface normal by 90° each other, are connected without dangling bonds. As well known, the stacking faults on {111} are seen in both the Frank loop of self-interstitials and the Si 7 x 7 {111} surface structure. Furthermore, the hydrogen platelets are better described by the two hydrogenated {111} surfaces facing each other.

Table 1   Structural characteristic of planar interstitial defects in Si

Defect	Δ r(%) min.	Δ r(%) max.	Δ θ (degree) min.	Δ θ (degree) max.	Energy E_{is}(eV)	Planar density of self-interstitials		R	Atomic rings
{111}	0.	0.	0.	0.	------	$8/(\sqrt{3}a^2)$		<111>a/3	6_h, 6
{100}	-1.4	2.5	-14.0	17.4	0.74	$2/a^2$		0.37<100>a	5, 6, 7
{113} <IIO>	-1.7	2.2	-21.1	22.0	0.85	~ 0.62x	$8/(\sqrt{11}a^2)$	~ <116>a/25	5, 6_h, 6, 7, 8
{113} <IO>	-1.1	2.8	-19.0	22.0	1.12				5, 6_h, 6, 7, 8

The planar defect structures have much in common with the other internal defects in silicon and germanium. They consist of various kinds of atomic rings (Table 1). In particular, the {113} defect structure, consisting of the atomic rings extending along <110>, is similar to those in <110> tilt boundaries. The successive 5-membered rings in the {100} defect structure can be seen in the core model of a partial dislocation (Hirsch, 1979). The edge of the {113} defect is related to the reconstructed core model of the perfect dislocation (Hornstra, 1958). The hexagonal 6-membered rings in the {113} defect appear in the Si 7 x 7 {111} surface, the Frank loop, and the deformed structure introduced by indentation. (Pirouz, Chaim, Dahmen and Westmacott 1990). In these contexts, the planar defect structures present an atomic configuration dominated by the general nature of covalent bonding, minimising the number of dangling bonds.

4 POSSIBLE UNIT STRUCTURES OF INTERSTITIAL AGGLOMERATES

As shown above, HRTEM has effectively contributed to determining the bond topology of rather large agglomerates. Structure analyses of very small agglomerates consisting of several tens of point defects directly by conventional HRTEM have so far been no simple task. On the other hand, spectroscopy measurements such as electron paramagnetic resonance, infrared absorption spectroscopy and electronic measurements have yielded substantial knowledge about atomic structures, electronic structures and vibrational frequencies of several kinds of point-defect-clusters, which consist of only a few point defect and impurity atoms. However, even the determination of the size of a larger agglomerate has been virtually impossible by the spectroscopic techniques. Therefore, there has been a lack of knowledge about the agglomerates of point defects of medium size. The HRTEM studies of extended planar defects have shown that the lattice is reconstructed involving point defects. Furthermore, unlike the exterior surfaces, a planar defect terminates inside crystal surrounded by the closed interface, a part of which is, for instance, the reconstructed edge structure without dangling bonds as seen in the {113} and {100} defects. This knowledge leads to an underlying idea that a small cluster of point defects of medium size may be fully reconstructed. In addition, most of the impurities and point defects in Si such as oxygen and metal species are supersaturated in crystal, therefore possessing a strong tendency to agglomerate upon annealing. By solving the structure of a rather large agglomerate in detail, in particular the closed interface between the agglomerate and perfect crystal, it is possible to deduce the unit structure of point defects and impurities.

4.1 Line interstitial defect (LID) structure

The {113} defect structure possesses no dangling bonds in the {110} cross-section. The extended structure along <332> terminates with the 5 and 7-membered rings without dangling bonds. The narrowest structure in the <332> direction, in other words, the line structure along the <110> direction can be deduced, and it is called a line interstitial defect structure (LID). An HRTEM experiment has confirmed that the doubled line structure, LID2 exists as a part of an extremely extended {113} planar defect (Takeda, Kohyama and Ibe 1994). The estimated energy increases per self interstitials; E_{is} are 1.60eV and 1.16eV, respectively, for the single LID (Fig. 4(a) and LID2 structures, and they are smaller than that for an isolated self-interstitial (Table 2). There are several variations of line structures with relatively low energy. The LID and related structures leave unsaturated bonds at both ends, though.

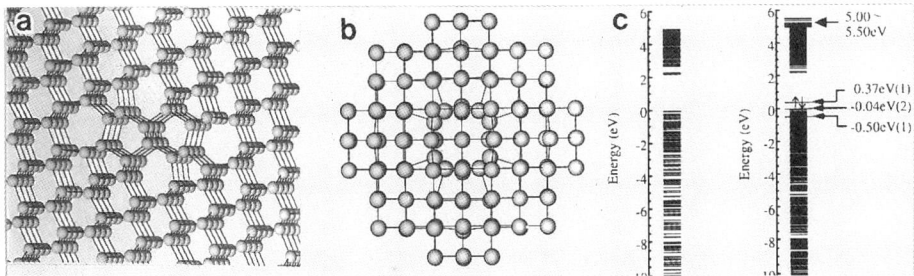

Fig. 4 (a) Single LID viewed along <110> and (b) I_4 viewed along <100>. (c) Electronic energy levels of I_4 (right) and the perfect crystal (left) for comparison.

4. 2 Four self-interstitial structure, I_4

Applying the idea above mentioned to the {100} defect, a cluster of four self-interstitials has been obtained (Fig. 4(b)). The interstitial cluster is fully reconstructed without dangling bonds at all and embedded in a crystal. Similarly, a series of clusters, I_n (n is a multiple of four) can easily be deduced. The energy calculation based on an empirical potential suggested that the cluster model is stable (Table 2). A more reliable electronic calculation of energy (Arai, Takeda, and Kohyama, 1996) has shown that the cluster is more stable than isolated four interstitials and possesses neither dangling bonds nor floating bonds. It exhibits no active electronic energy level but the defect localized states appears (0.37eV above the valence band edge) in the minimum band gap (Fig. 4 (c)). Extremely moveable self-interstitials in Si and Ge are always interesting, since they are definitely introduced by various processes such as thermal annealing, ion-implantation and electron irradiation. However, even the stable configuration of an isolated self-interstitial in a crystal lattice has not yet been clarified. The interstitial cluster, I_4 may exist in crystal since it is very stable once formed and is undetectable by any conventional spectroscopic techniques.

Table 2   Interstitial cluster models in Si.

Model	$\Delta r(\%)$		$\Delta \theta$ (degree)		Energy increase per interstitial
	min.	max.	min.	max.	E_{is}(eV)
LID	-2.9	2.1	-19.0	18.9	1.60
LID^2	-2.5	2.6	-21.2	22.6	1.16
I_4	-5.0	1.5	-22.2	22.8	2.10
a single interstitial					3.2 - 6.5

5 THE NUCLEATION AND GROWTH OF POINT-DEFECT-AGGLOMERATES

5. 1 Bulk defects

Observing silicon and germanium crystals by HRTEM, the {113} defect is frequently nucleated as shown in Fig. 5, which depicts filtered *in-situ* HRTEM images in order to show the defective part only (Takeda and Kamino, 1995). The observation of the phenomena by HRTEM itself has brought no scientific information, since the development of the defects under electron irradiation has already been fully described by *in situ* electron microscopy (for instance, Ferreira Lima and Howie, 1976, and a review by Aseev, Fedina, Hoehl and Bartsch, 1995) and the habit plane and the displacement vector were precisely determined. Based on the atomic structural data of the {113} defects, the growth mechanism in the <332> direction was discussed in a speculative manner (Takeda, Muto and Hirata, 1991) and in terms of energy (Kohyama and Takeda 1992). The LID structures are assumed to be nucleated, whether separated or interconnected, coherently on {113}. When the two single LIDs are nucleated successively, then the energy, E_{is} is reduced by 0.44 eV per self-interstitial (Table 2).

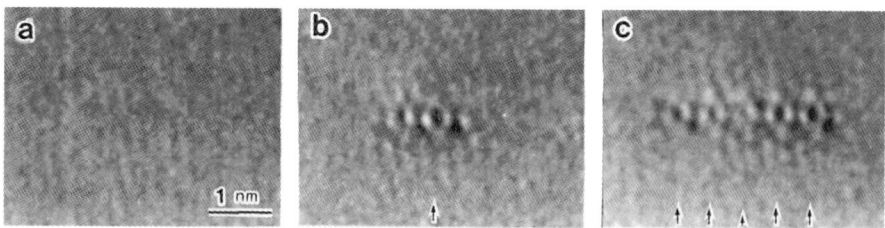

Fig. 5 *In-situ* HRTEM observation of the nucleation and growth of the {113} defects in Si.

With the data in hand, the HRTEM images in Fig. 5 can be interpreted at the atomic level. The twofold pattern, indicated by the arrows, in the early stage of growth (Fig. 5(b), is attributed to a LID structure formed inside the thin foil. The image pattern is accompanied with a fainter twofold pattern in its left side, suggesting that the LID structure is extended to the LID² structure in part in the thin foil. The defect further grew and exhibited its habit plane (Fig. 5(c)). The LIDs (marked by the longer arrows) are nucleated with the faint image pattern, denoted by the shorter arrow, in the middle of the sequence. The faint pattern may be due to the translational structure leading to the 8-membered ring, which is definitely needed for the extension of the {113} defects. At the stage, the defect includes at most 10^2 agglomerated self-interstitials since the thickness of the specimen is estimated to be about 10 nm.

Due to stress fields and impurities which are inevitably introduced in a specimen, the nucleation of the LID structures may step away from the {113} plane. Figure 6(a) summarizes the possible faulted development of a planar defect. The atomic step of the planar defect found by HRTEM is evidence of faulted nucleation (Fig. 6(b)). The steps in left and right are, respectively, ascending and descending steps in the [$\overline{33}2$] direction, and possess the ledge parallel to the [$1\overline{1}0$] direction. The height of both steps is represented by 2a [113]/11. The disturbed structures of the {113} defect have no rigorous habit plane. Once the faulted nucleation occurs, the successive LID formation process gives rise to further disturbed structures, thereby probably developing a hexagonal structure in part as shown in Fig. 6(a). These faulted nucleations have caused confusion in analysis of HRTEM images in the past.

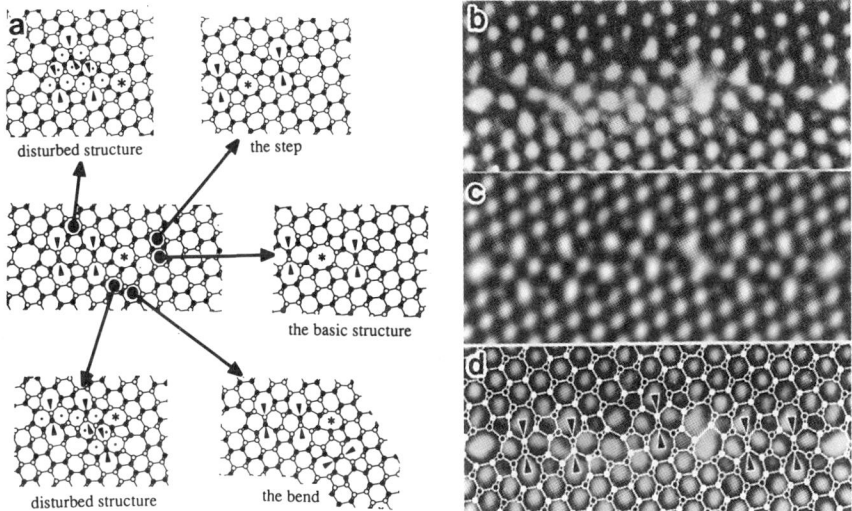

Fig. 6 Possible structural variations of the {113} defect in (a) , and (b) the atomic steps of the defect in Si found by HRTEM. Simulation in (c) is based on the model in (d) and seen in (a).

A question has still remained about the nucleation of the defects. Several authors have argued that impurities such as carbon, oxygen etc. affect the nucleation and growth rates of the defects . Nevertheless, no conclusive data have been presented on the effect of impurities. The most serious defect in previous studies arises from the uncontrolled surfaces of a thin crystal, and therefore, unintentional contamination may be involved in the phenomena. In a ultra-high vacuum (UHV)-TEM, the controlled clean surfaces of a thin Si foil can be provided, and the electron irradiation effect on the defect formation can be pursued in a controlled environment (Koto, Takeda, Ichihashi and Iijima, 1997). After the 7x7 surface reconstructed structure was formed on a thin silicon slab, electron irradiation commenced under the UHV condition. Intense electron irradiation gives rise to the numerous defects as in previous studies. The number of defects increases continuously with irradiation time. After the first irradiation was terminated, O_2 was injected and flowed in the specimen chamber, and the irradiation under the flow of O_2 proceeded. Just after O_2 was injected, the nucleation of the observable defects by TEM was extremely promoted.

32

It is well known that electron irradiation in Si introduces an oxygen-vacancy pair (A-center). The enhanced formation of the centers by the additional oxygen may disturb mutual annihilation of Frenkel pairs further, and interstitials are more accumulated in a thin foil and consequently their agglomeration is promoted. Alternatively, it is argued that the agglomeration of self-interstitials begins after the specific core structures, involving impurities such as oxygen, carbon and other impurities, are formed. These arguments were based on the data comparison between CZ and FZ silicon, between doped and undoped silicon and between furnace annealing in O_2 ambient and in vacuum. The intentional contamination of impurities was attempted not on clean surfaces but on uncontrolled oxides surfaces. The prominent change of the defect formation observed in UHV-TEM has shown that O_2 on surfaces affects definitely the nucleation and growth of the {113} defects.

5. 2 Defects near and at surfaces introduced by electron irradiation

5.2. 1 Electron diffraction channeling effects

Electron-diffraction channelling effect (Humphreys, Thomas, Lally and Fisher, 1971) on defect formation in the two beam condition is well known. Here, we show that the effect comes out during HRTEM observation (Takeda and Horiuch, 1994). Electrons represented by the s-type *bound* Bloch wave encounter atoms close to their nuclei and have a certain chance to create the displacement of atoms. The rate of the displacement damage is thus expected to be enhanced in the planar region distanced properly from the electron entrance surface where the s-type Bloch wave is dominant.

We observed single Si crystals in a high voltage (1MV) electron microscope with the <110> zone axis incidence at room temperature. Well-defined wedge-shaped crystals of Si were carefully selected for observation (Fig. 7(a)), since we expected the thickness dependence of defect formation. The characteristic spatial distribution of the {111} defects has been clearly observed (Fig. 7(b)). The defects on {111} were induced in the bands of nearly equal crystal thickness. After the nucleation, the defects gradually developed on {111} and exhibited clear image contrast under prolonged irradiation. The numbers of the {111} defects, counted after various irradiation periods, are plotted against the estimated crystal thickness in Fig. 7(c). The defects are introduced wherever the electron current density at the atom sites reaches a maximum near the electron exit surface (Fig. 7(d)).

Frenkel pairs are presumably generated heterogeneously inside a crystal in accordance with the heterogeneous current density distribution. However, due to the extremely fast migration of self-interstitials, the agglomeration of self-interstitlas takes place not always at the sites where the primary displacement occur. However, self-interstitals, created near surfaces, easily escape to the surfaces, and the considerable condensation of vacancies is expected. Since the vacancies in a Si crystal are known to be moveable at room temperature under electron irradiation, we conclude that vacancies are prominently agglomerated near the electron exit surface where the displacement events are enhanced by electron diffraction channeling effects.

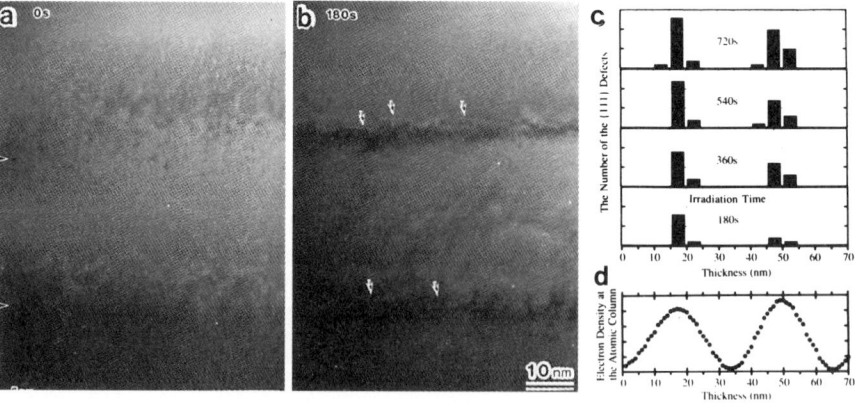

Fig. 7 Heterogeneous defect creation by diffraction-channeling effects in (a) and (b). The number density of the defects and the estimated electron current density at the electron exit surface as functions of thickness, respectively, in (c) and (d).

5. 2. 2 Surface hole formation under uniform intensity electron irradiation

Observing a parallel-sided silicon specimens in UHV condition, we have found that an array of holes of nanometer diameter is created on the electron exit surface (Takeda, Koto, Iijima and Ichihashi, 1997). Figure 8 shows an HRTEM image of holes created at the electron exit surface. The holes and grooves are generated regardless of the surfaces, i. e. {111} and {110} and of the content of oxygen, i. e. FZ and CZ. The flow of O_2 gas in the specimen chamber suppressed the hole formation, suggesting that the phenomenon can not be observable in a conventional TEM. Nucleation of the somewhat similar structures has been reported in ceramics and metals without proposing any plausible mechanism for the nucleation. Nevertheless, the observations indicate that, through migration of surface vacancies, the generation of holes and grooves occurs in oxides, metals and semiconductors regardless of the nature of cohesion.

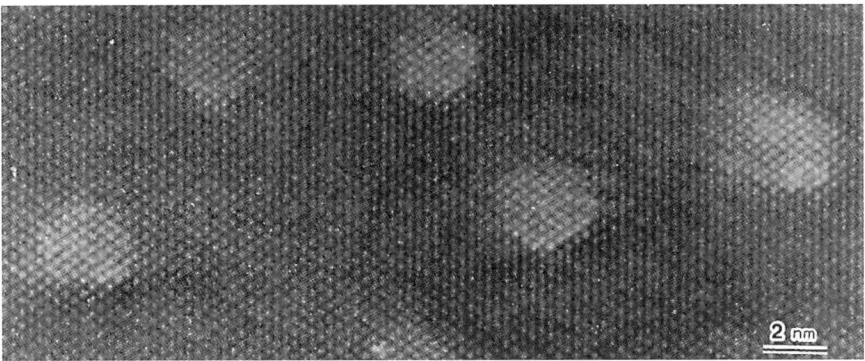

Fig. 8   Nanoholes on the electron exit surface in Si .

6. ASSOCIATED STUDIES OF EXTENDED DEFECTS

The structure studies by HRTEM lead to other information about the defect. The electronic structure of the defect is readily obtained by electronic computations. It is shown that the {113} defect has no active energy level in the band gap as expected but some defect-localized states appear near the valence and conduction band edges (Kohyama and Takeda, 1995). As suggested above, clusters of self-interstitials may be fully reconstructed in Si and other semiconductors. Therefore their identification is very difficult by electronic measurements. Exploring the band edges and inside of the valence bands by high-resolution electron energy loss spectroscopy (EELS), the reconstructed structures may be identified. The high resolution EELS spectra seem to correspond to the electronic structures of the {113} defects, even though the detailed studies are obviously needed.

A new approach to detect point defects under electron irradiation, which reinforcesHRTEM data, is *in-situ* optical spectroscopy in a transmission electron microscope (Ohno and Takeda, 1995) . Since several kinds of impurities in semiconductors act as a light emitting center , optical spectroscopy may be helpful in detecting impurities. The apparatus for *in-situ* optical spectroscopy in TEM, in which the laser probe of smaller than 10μm in diameter can be focussed on a specimen under TEM observation, has already been utilized to detect the migration of point defects involving impurity atoms in some semiconductors at low temperatures below 100K (Ohno and Takeda, 1996).

7. CONCLUSION

It has been demonstrated that HRTEM has contributed to analyzing the atomic structures of agglomerates of point defects inside of the crystal of Si and other semiconducting materials. According to the structural studies, a more generalized view of defects in Si and other materials is provided. An emphasis is placed on the development of studies made possible by the crucial structure determination. The electron irradiation effect on defect creation near and at surfaces has been observed by HRTEM. The studies have brought useful information towards understanding dynamic phenomena in semiconducting materials at the atomic level both in bulk and on surfaces.

34

ACKNOWLEDGMENTS

The present work was supported in part by a Grant-in-Aid for Scientific Research from the Ministry of Education, Science, Culture and Sports.

REFERENCES

Arai N, Takeda S and Kohyama M 1996 Submitted.
Aseev A L, Fedina L I, Hoehl D and Bartsch H 1995 Clusters of interstitial atoms in silicon and germanium (Berlin: Akademie Verlag GmbH)
Bourret A 1987 Proc. Microsc. of Semicond. Mater. Conf. (Oxford, 1987) eds A G Cullis and D B Holt (Bristol: Institute of Physics) Inst. Phys. Conf. Ser. **87** pp 39-48
Dabrowski J, Mussig H J and Wolff G 1994 Phys. Rev. Lett. **73** 1660
Eaglesham D J, Stolk P A, Gossmann H -J and Poate J M 1994 Appl. Phys. Lett. **65** 2305
Eaglesham D J, Stolk P A, Cheng J-Y, Gossmann H -J, Haynes T E and Poate J M 1995 Proc. Microsc. Semicond. Mater. Conf. (Oxford, 20-23 March, 1995) eds A G Cullis and A E Stanton-Bevan (Bristol: Institute of Physics) , Inst. Phys. Conf. Ser. **146** pp 451-6
Ferreira Lima C A 1975 PhD Thesis, University of Cambridge
Ferreira Lima C A and Howie A 1976 Philos Mag **34** 1057
Gibson J M, MacDonald M L and Unterwald F C 1985 Phys. Rev. Lett. **66** 1733
Hirsch P B 1979 J. de Phys. **40** C6-27
Hornstra J 1958 J. Phys. Chem. Solids. **5** 129
Humble P 1982 Proc R Soc Lond **A381** 65
Humble P Lynch D F and Olsen A 1985 Philos. Mag. **A52** 623
Humphreys C J, Thomas L E, Lally J S and Fisher R M 1971 Philos. Mag. **23** 87
Hutchison J L, Aseev A L and Fedina L I 1993 Proc. Microsc. Semicond. Mater. Conf. (Oxford, April 5-8 1993) eds A G Cullis, A E Staton-Bevan and J L Hutchison (Bristol: Institute of Physics) Inst. Phys. Conf. Ser. **134** pp 41-6
Kohyama M and Takeda S 1992 Phys. Rev. **B46** 12305-12315
Kohyama M and Takeda S 1995 Phys. Rev. **B51** 13111-13116
Koto K, Takeda S, Ichihashi T and Iijima S 1997 Submitted
Muto S and Takeda S 1995 Philos. Mag. Lett. **72** 99-104
Muto S, Takeda S and Hirata M 1995 Philos. Mag. **A72** 1057-1074
Ohno Y and Takeda S 1995 Rev. Sci. Inst. **66** 4866-4869
Ohno Y and Takeda S 1996 J. Electron Microscopy **45** 73-78
Pirouz P, Chaim R, Dahmen U and Westmacott K H 1990 Acta metall. mater. **38** 313
Ponce F A, Johnson N M, Tramontana J C and Walker J 1987, Proc. Microsc. of Semicond. Mater. Conf. (Oxford, 1987) eds A G Cullis and D B Holt (Bristol: Institute of Physics), Inst. Phys. Conf. Ser. **87** 49-54
Salisbury I G and Loretto M H 1979 Philos. Mag. **A39** 317-323
Takeda S 1991 Jpn. J. Appl. Phys. **30** L639-642
Takeda S, Hirata M, Muro S, Hua G-C, Hiraga K and Kiritani M 1991 Ultramicroscopy **39** 180-186
Takeda S, Muto S and Hirata M 1992 Proc. 16th Int. Conf. Defects in Semiconductors 16 (Lehigh, 22-26, July, 1991), eds G Davies, G G Deleo and M Stavola (Zurich:Trans Tec. Pub.) Materials Science Forum **83-87** pp 309-314
Takeda S and Kohyama M 1993 Proc. Microsc. Semicond. Mater. Conf. (Oxford, 5-8, April 1993) eds A G Cullis, A E Staton-Bevan and J L Hutchison (Bristol:Institute of Physics) Inst. Phys. Conf. Ser. **134** pp 33-6
Takeda S and Horiuchi S 1994 Ultramicroscopy **56** 144-162
Takeda S, Kohyama M and Ibe K 1994 Philos. Mag. **A70** 287-312
Takeda S and Kamino T 1995 Phys. Rev. **B51** 2148-2152
Takeda S, Koto K, Iijima S and Ichihashi T 1997 Submitted
Tan T Y, Foll H and Hu S M 1981 Philos. Mag. **A44** 127

Inst. Phys. Conf. Ser. No 157
Paper presented at Microsc. Semicond. Mater. Conf., Oxford, 7–10 April 1997
© *1997 IOP Publishing Ltd*

Defect structure of InSb grown within a synthetic opal matrix

V N Bogomolov, J L Hutchison*, S M Samoilovich, D A Kurdyukov, J Sloan*, L M Sorokin and G Wakefield*

Ioffe Physical-Technical Institute, Russian Academy of Sciences, St. Petersburg 194021, Russia.
*Department of Materials, University of Oxford, Parks Road, Oxford OX1 3PH, U.K.

ABSTRACT: Gem opal is a unique structure built up from regularly packed, uniform spheres of amorphous silica. The regular packing of spheres creates a regular array of interstitial voids analogous to those in close packed crystal structures. These may be used as 3-D templates for the growth of guest materials and this paper describes preliminary observations of InSb grown within such voids. Conventional TEM, electron diffraction and HREM have been used to investigate ordering within the InSb; EDX was also used to confirm its composition. Surprisingly, the InSb was single-crystal over large areas, as a result of the interconnecting channels linking the main voids being also filled, thus providing structural linking from one void to the next. The only lattice disorder appeared to be in the form of {111} twinning, with stepped boundaries.

1. INTRODUCTION

It has been known for a number of years that gem-quality opal is composed of amorphous silica. The spectacular play of colours for which the mineral is highly prized arises because the silica is in the form of very uniform amorphous spheres packed in regular arrays (Sanders 1974). The combination of "crystallographic" packing of the spheres and their size range, 100 - 300 nm, can give rise to optical diffraction, and internal "defects" such as stacking faults, grain boundaries, etc. cause the brilliant colour flashes. Regular packing of spheres implies an equivalent regular array of interstitial voids or cavities analogous to those encountered in close packed crystal structures. In simple cubic systems these are in octahedral and tetrahedral configurations, with dimensions determined by the atomic radius of the close packed atoms. These regular array of cavities are novel templates for 3-D growth of guest materials on a potentially interesting scale. We here describe the results of attempts to grow InSb within the voids of synthetic gem-opal, characterisation being carried out by TEM, EDX and HREM.

2. EXPERIMENTAL

Synthetic opal "crystals" were grown from aqueous sols of spherical silica particles by appropriate treatments involving sedimentation, autoclaving and annealing (Balchirev et al 1993, Bogomolov et al 1996). They were then impregnated by molten InSb under pressure. Specimens for TEM study were prepared by sawing 3 mm discs from bulk samples. These were glued to 3 mm copper washers, polished by dimpling and finally thinned to perforation by Ar$^+$ ion milling. Surprisingly, the rather delicate amorphous spheres survived the thinning procedures intact, as did the InSb filling. The specimens were then examined in a JEOL 4000EX(II) operating at 400 kV. The EDX spectra were obtained using a JEOL 2010 instrument equipped with a an ultra-thin window Si(Li) detector.

3. RESULTS

3.1. Initial observations of synthetic opal

Preliminary observations of unfilled synthetic opal confirmed the structure as being a regular array of spheres, about 230 nm in diameter, in a close packed structure. An example of this is shown in Figure 1. It can be seen that the spheres are in contact with one another, adhering by a thin (< 1 nm) layer, presumably amorphous silica. The mottled appearance of the spheres, rendered visible by out-of-focus phase contrast, is consistent with a porous structure and real-time observations at high resolution suggested that these pores may themselves be partially filled with a liquid. EDX indicated that the spheres were SiO_2, in this case hydrated as a result of the synthetic route employed.

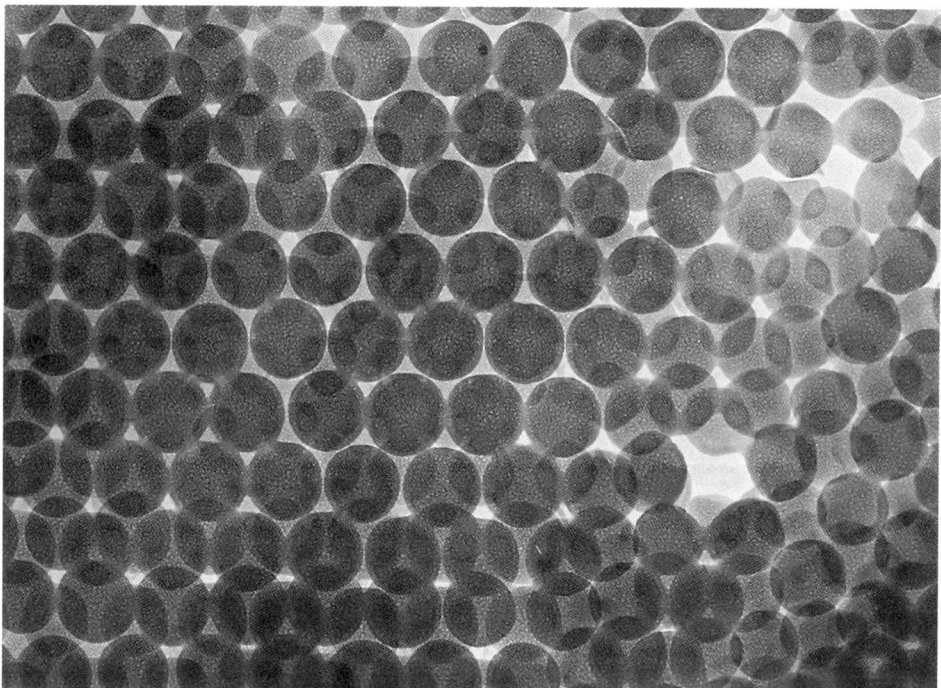

Figure 1. Gem-opal, showing regular array of amorphous, porous silica spheres.

The image also shows the remarkably regular packing of the spheres, with occasional "stacking faults" and other regions of disorder. The interstitial cavities are clearly visible (arrowed).

3.2. TEM, EDX and HREM of InSb-filled opal

Turning now to the attempts to fill the cavities with InSb, impregnated specimens were prepared for examination as before. Low-magnification images (primary magnification 5000x) confirmed immediately that large volumes of the lattice had indeed been filled. From images it was clear that the connecting channels were also filled, forming a 3-dimensional network of linked nodes. Selected area electron diffraction patterns obtained from such areas indicated, surprisingly, that the "filling" was largely single-crystal over such areas. The spot patterns were consistent with an fcc-type lattice. Identification was also carried out by EDX, which confirmed the guest material as stoichiometric InSb.

Electron diffraction having shown that the material was fcc, areas of InSb were selected which could be oriented with $<110>$ parallel to the electron beam. The zone-axis patterns thus obtained were consistent with a twinned lattice which could be seen by conventional TEM to extend over several adjacent cavities. At this stage HREM images were then obtained in order to assess the degree of twinning and any other disorder.

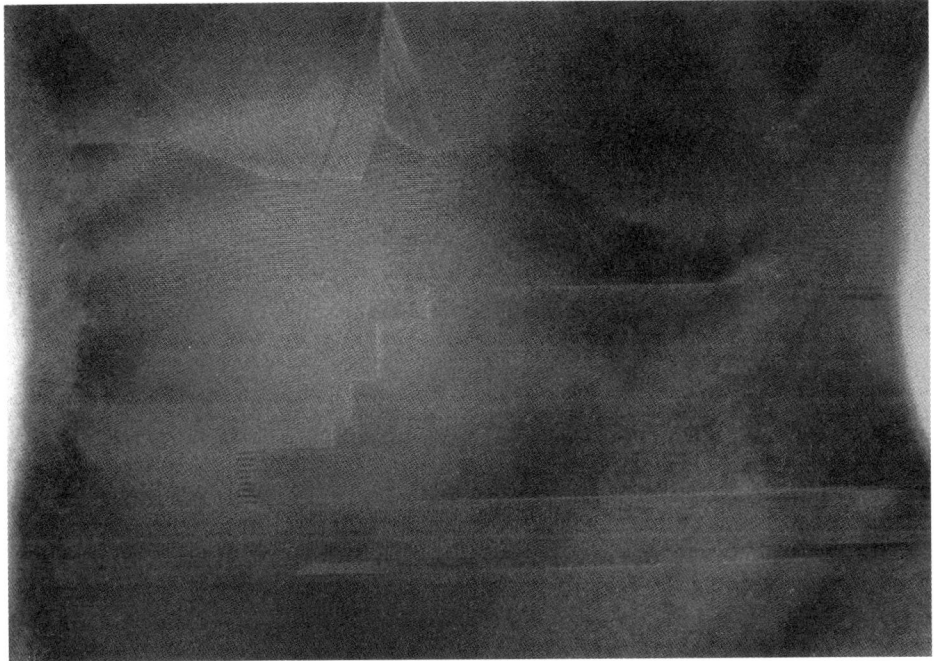

Figure 2 (a) Twinned InSb filling an octahedral cavity in synthetic opal.

Figure 2 (a) is an example of such a twinned lattice occurring within a filled cavity. A complex twin boundary is shown by arrows. It can be seen that the twinning is first-order, with an unusually stepped or serrated interface. A porion of this interface is seen enlarged in Figure 2 (b). From this image it is evident that the twin boundary is coherent along its length, although the atomic configuration of the steps has not yet been analysed in detail. The imaging conditions did not allow us to determine the polarity of the structures on either side of the twin boundary; this is currently under further investigation, the results of which will be reported elsewhere.

Figure 2 (b) Enlarged segment of one of the twin boundaries shown in Fig. 3(a)

4. CONCLUSIONS

Synthetic opal has been prepared and successfully impregnated by crystalline InSb, which penetrates the interstitial voids in the giant "lattice" of silica spheres. This guest material has been analysed by EDX and electron diffraction and shown to be single-crystalline InSb over considerable volumes of the opal matrix. TEM indicates that the connecting from one interstitial site to the next are also filled with crystalline InSb, in the same orientation. The only significant disorder appears to be {111} twinning, with saw-tooth twin boundaries.

ACKNOWLEDGEMENT

One of us (LMS) is grateful to the Royal Society for financial support.

REFERENCES

Balakirev V G, Bogomolov V N, Zhuralev V V, Kumerzov Y A, Petranovski V P, Romanov S G and Samoilovich L A 1993 Kristallographia **38** 111
Bogomolov V N, Kurdyukov D A, Prokofiev A V and Samoilovich S M 1996 JETP Letters **63** 520
Sanders J V 1974 Proc 8th Internationa Congress on ELectron Microsopy (Canberra) **1** 512

Inst. Phys. Conf. Ser. No 157
Paper presented at Microsc. Semicond. Mater. Conf., Oxford, 7–10 April 1997
© *1997 IOP Publishing Ltd*

Strain determination in mismatched semiconductor heterostructures by the digital analysis of lattice images

A Rosenauer, T Remmele, U Fischer, A Förster* and D Gerthsen

Laboratory for Electron Microscopy, University Karlsruhe, Kaiserstraße 12, 76128 Karlsruhe, FRG
*Institute for Thin Film and Ion Beam Technology, Research Center Jülich GmbH, Postfach 1913, 52425 Jülich, FRG

ABSTRACT: The evaluation program DALI (Digital Analysis of Lattice Images) is applied to determine strain and composition of $In_xGa_{1-x}As$ islands which were grown on (001)GaAs substrates and capped with 10 nm GaAs. Finite element calculations are used to find the reduction of the tetragonal distortion due to the small sample thickness below 40 nm which is determined by the combination of the real space method QUANTITEM and the consideration of local Fourier coefficients.

1. INTRODUCTION

The evaluation of lattice parameters from high-resolution transmission (HRTEM) micrographs yields structural information on an atomic scale. Local lattice parameters can be determined with DALI which has been introduced in a previous paper by Rosenauer et al. (1996). The procedure is similar to the evaluation procedure suggested by Bierwolf et al (1993). In pseudomorphic, mismatched semiconductur heterostructures, the tetragonal distortion of the unit cells can be used to obtain the composition of ternary semiconductor compounds. The precision is improved, if the elastic relaxation due to the small sample thickness below 40 nm is taken into account. In the next sections we will first show the procedure applied to evaluate local sample thicknesses and then present as an example $In_xGa_{1-x}As/GaAs(001)$ island structures with GaAs capping layer.

2. THICKNESS DETERMINATION

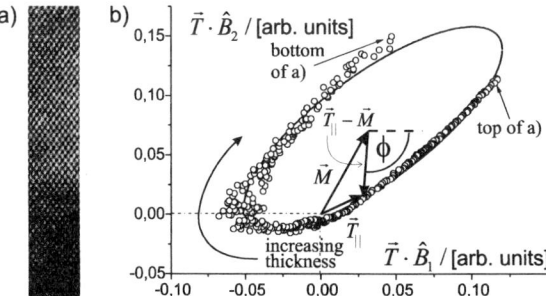

Fig. 1: *a) HRTEM image of a wedge shaped GaAs crystal in* $\langle 110 \rangle$ *projection. The open circles in b) indicate the tips of vectors* $\vec{T}_{||}$ *inside the plane (* $\hat{B}_1$, $\hat{B}_2$ *) evaluated from a). The solid line shows an ellipse fitted to the experimental data.*

The sample thickness in electron beam direction is determined by a method which is a combination of the QUANTITEM procedure proposed by Ourmazd et al (1993) and Kisielowski et al (1995) and the method suggested by Stenkamp and Strunk (1996). QUANTITEM detects the projected crystal potential which is a measure for the sample thickness. However, only relative thicknesses are obtained because the origin of the thickness scale is unknown. On the other hand, the consideration of local Fourier coefficients offers additional information.

Fig. 1a shows the HRTEM image of a wedge shaped GaAs crystal in $\langle 110 \rangle$-projection which is used to demonstrate the analysis procedure. In the first step, DALI is used to detect local brightness maxima and to decompose the image into non-primitve lattice cells. Each lattice cell Z is transformed into a quadratic cell Z' of $2^n \times 2^n$ (n is typically 5) pixels. Following the procedure described by Kisielowski et al (1995), the intensity values, $I'_{n,m}$, of the pixels, $Q'_{n,m}$, inside a cell, Z', are used to define the vector

$$\vec{R} = (\Gamma_{11}, \ldots \Gamma_{1N}, \Gamma_{21}, \ldots \Gamma_{2N}, \ldots, \Gamma_{NN}) \tag{1}$$

with $N = 2^n$. Three template image vectors, $\vec{R}^T_{1,2,3}$, are calculated by averaging the image vectors of the cells contained in three small regions of the experimental image displaying different image patterns. The vectors $\hat{B}_{1,2}$ given by

$$\hat{B}_1 := \frac{\vec{R}^T_1 - \vec{R}^T_2}{\left| \vec{R}^T_1 - \vec{R}^T_2 \right|} \quad \text{and} \quad \hat{B}_2 := \frac{\vec{V} - \hat{B}_1(\vec{V} \cdot \hat{B}_1)}{\left| \vec{V} - \hat{B}_1(\vec{V} \cdot \hat{B}_1) \right|} \quad \text{with} \quad \vec{V} = \frac{\vec{R}^T_3 - \vec{R}^T_2}{\left| \vec{R}^T_3 - \vec{R}^T_2 \right|} \tag{2}$$

span a two dimensional subspace, E, inside the N^2 dimensional image vector space. The tips of the evaluated image vectors may deviate slightly from the plane E. Therefore, we use $\vec{R}$ to define an in-plane vector, $\vec{T}_{||}$, and a vector, $\vec{T}_\perp$, perpendicular to E.

$$\vec{T}_{||} = \hat{B}_1(\vec{T} \cdot \hat{B}_1) + \hat{B}_2(\vec{T} \cdot \hat{B}_2) \quad \text{with} \quad \vec{T} = \frac{\vec{R}}{|\vec{R}|} - \frac{\vec{R}^T_2}{|\vec{R}^T_2|}, \quad \text{and} \quad \vec{T}_\perp = \vec{T} - \vec{T}_{||}. \tag{3}$$

The projection of the image vectors $\vec{R}$ into the subspace E yields a cloud of points (see Fig. 1b) which approximately form an ellipse (Kisielowski et al. (1995)). The angle, ϕ, between $\vec{T}_{||} - \vec{M}$ ($\vec{M}$ denotes the midpoint of the ellipse) and $\hat{B}_1$ is proportional to the sample thickness

$$d = \frac{\phi - \phi_0}{2\pi} \xi, \tag{4}$$

where ξ is the extinction distance of the undiffracted beam. However, the angle, ϕ_0, that represents the origin of the thickness scale is unknown.

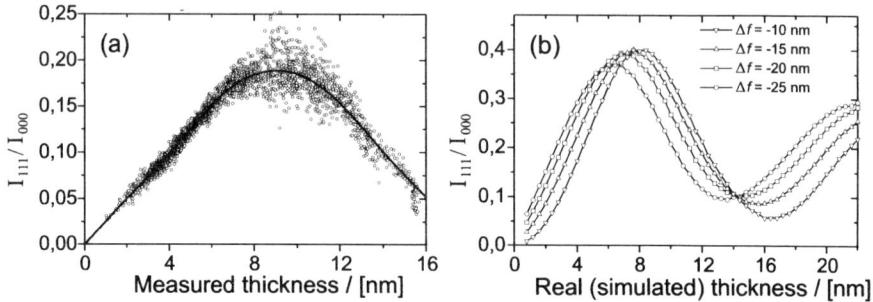

Fig.2: *a) The relation I_{111} / I_{000} evaluated for each image unit cell in Fig. 1a is plotted versus the measured thickness. b) I_{111} / I_{000} evaluated from HRTEM images simulated by EMS (Stadelmann (1987)) plotted versus the sample thickness (U=400 kV, C_s=1.0 mm, objective aperture diameter 10 nm^{-1}, spread of defocus 10 nm, beam semi convergence angle 1 mrad).*

For the calculation of ϕ_0, a fast Fourier transform algorithm is applied to each unit cell, Z', to obtain the intensities I_{111} and I_{000} of the Fourier coefficients for $\vec{g} = (111)$ and $\vec{g} = (000)$, respectively. The relation I_{111} / I_{000} is plotted versus the thickness, d, given by eq. (4) in Fig. 2a. The offset of the thickness scale, defined by ϕ_0, is obtained by a comparison with I_{111} / I_{000} calculated from simulated HRTEM images (see Fig. 2b). In the presented case, we used the characteristic behaviour $I_{111} / I_{000} \to 0$ for $d \to 0$.

3. EXPERIMENTAL PROCEDURES

The investigated samples consist of an $In_xGa_{1-x}As$ layer (nominal thickness 1.5 nm and nominal In-content $x=60\%$) and a 10 nm thick GaAs capping layer. The $In_xGa_{1-x}As$ layers were grown on GaAs (001) substrates using a Varian MOD Gen II system. To provide a clean surface, a 0.1 μm GaAs buffer layer was deposited at a substrate temperature of 610 °C, prior to the $In_xGa_{1-x}As$ growth which was carried out at 500°C.

The HRTEM was performed on cross-section specimens along the $\langle 110 \rangle$-projection using a PHILIPS CM 200 FEG/ST electron microscope with a Scherzer resolution of 0.24 nm. The HRTEM negatives were digitized with a CCD camera at a resolution of 1024 × 1024 picture elements (pixels). The digitized images are processed with DALI which allows the evaluation of local and averaged displacements and lattice parameters. After the noise reduction with a Wiener filter, local intensity maxima are used to generate a two-dimensional lattice. A reference region is chosen in the GaAs substrate. A regular reference lattice is generated to fit the intensity maxima positions in the reference region which is superimposed on the whole image. Local displacement vectors result from the calculation of distance vectors between intensity maxima positions and corresponding reference lattice positions.

The model for the finite element (FE) calculation is generated with MSC PATRAN according to the shape derived from the HRTEM image and from the evaluation of local sample thicknesses described above. The ABAQUS solver is used to calculate simulated displacements. In order to imitate the projection of atoms onto the image plane in the TEM, the calculated displacements are averaged along the viewing direction $\langle 110 \rangle$.

5. RESULTS AND DISCUSSION

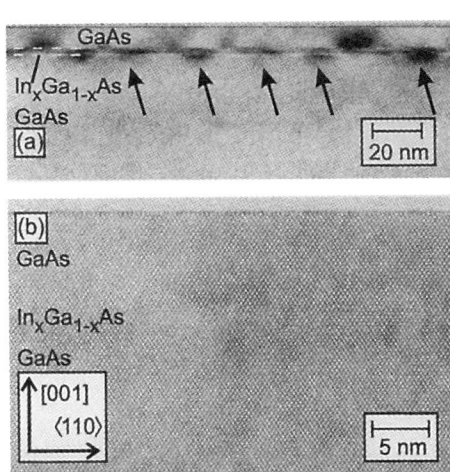

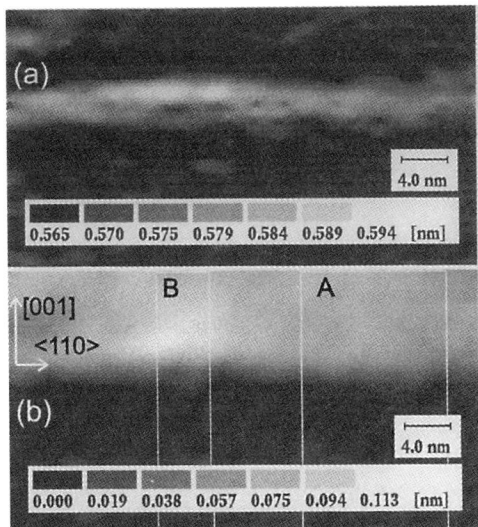

Fig. 3: *(a) TEM (004) dark field image of a sample with capping layer. The black arrows mark dark areas that are due to strain fields. (b) $\langle 110 \rangle$ HRTEM image of a region that contais substrate, layer and capping layer.*

Fig. 4: *(a) Grey scale map of local lattice parameters. The maximum lattice parameter of 0.594 nm corresponds to a strain of $\varepsilon_\perp = 0.051$. (b) Map of [001]-components of local displacements. The region A and B correspond to Fig. 5a and b, respectively*

42

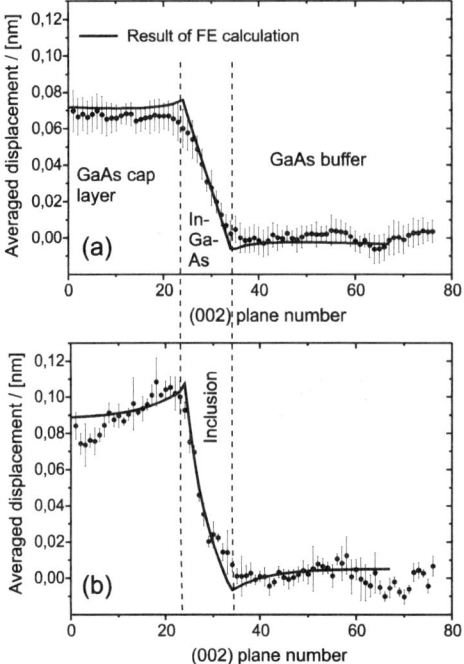

(a)

Result of FE calculation

GaAs cap layer

In-Ga-As

GaAs buffer

Averaged displacement / [nm]

(002) plane number

(b)

Inclusion

Averaged displacement / [nm]

(002) plane number

Fig. 5: *DALI evaluation of the [001]-components of the displacement averaged a) in the region of the pure InGaAs interlayer (A in Fig. 4b) and b) in the region of an inclusion with increased In content (B in Fig. 4b). The straight lines show the results obtained from the FE-calculation.*

Fig. 3a shows a [004] dark-field image of the $In_xGa_{1-x}As$ layer and the GaAs cap layer. The dark areas that are marked by black arrows are induced by strain fields. Their average distance is 15 nm. Islands that significantly surmount the $In_xGa_{1-x}As$ interlayer are not visible. Furthermore, the capping layer shows a very smooth surface which can also be seen in the HRTEM image in Fig. 3b. The image does not provide any chemical contrast. The sample thickness was evaluated from Fig. 3b with the method described in section 2 and reaches from 3 nm at the upper left corner to 11 nm at the bottom of the image. Fig. 4 displays the result of the DALI evaluation of Fig. 3. Fig. 4a shows a map of local lattice parameters which are a measure for the local In content according to Vegard's law due to the different lattice parameters $a_{InAs} = 0.60583$ nm of InAs and $a_{GaAs} = 0.565325$ nm of GaAs. Fig. 3a reveals that the composition is not homogeneous along the interlayer, (grey stripe in Fig. 3a), but contains an inclusion with increased In content.

Fig. 4b is a map of local displacements in [001]-direction. The displacements vanish inside the (black) region of the substrate. Due to the larger lattice parameter of InAs, increasing displacements are found in the $In_xGa_{1-x}As$ layer (black to middle grey). The bright spot emanates from the region of increased In-content that gives rise to a bending of the (001) lattice planes. Fig. 5a depicts the displacements averaged along ⟨110⟩ in region A without inclusion. The area with increasing displacements (in direction from buffer to cap layer) belongs to the $In_xGa_{1-x}As$ region. Its width of 10 monolayers, (ML) ≅ 2.8 nm, significantly exceeds the nominal $In_xGa_{1-x}As$ layer thickness of 1.5 nm. The averaged displacements for each ML are plotted in Fig. 5 as a function of the ML position for the areas A and B (see Fig. 4b). The solid line in Fig. 5a gives the result of the FE calculation which was performed with an In-content of 24 %. The best fit for the In-rich inclusion (Fig. 5b) was obtained with a FE-model with an In content that increases from the GaAs buffer to the GaAs cap from 24 % to 60 %.

In conclusion, interdiffusion and segregation of In during the overgrowth with GaAs is detected. In contrast to samples without capping layer, significant islanding is not found. Instead, we observed a broad $In_xGa_{1-x}As$ interlayer with fluctuating In content.

REFERENCES

Bierwolf R, Hohenstein M, Phiilipp F, Brandt O, Crook G E, Ploog K 1993 Ultramicroscopy **49,** 273

Kisielowski C, Schwander P, Baumann F H, Seibt M, Kim Y O, Ourmazd A 1995 Ultramicr. **58,** 131

Ourmazd A, Schwander P, Kisielowski C, Seibt M, Baumann F H, Kim Y O 1993 Inst. Phys. Conf. Ser. **134,** Section 1, 1

Rosenauer A, Kaiser S, Reisinger T, Zweck J, Gebhardt W 1996 OPTIK **102,** 63

Stadelmann P A 1987 Ultramicroscopy **51,** 131

Stenkamp D, Strunk H P 1996 Appl. Phys. A **62,** 369

Inst. Phys. Conf. Ser. No 157
Paper presented at Microsc. Semicond. Mater. Conf., Oxford, 7–10 April 1997
© 1997 IOP Publishing Ltd

New intermediate defect configuration in Si studied by in situ HREM irradiation

L Fedina [1,2] , **A Gutakovskii** [1], **A Aseev** [1], **J Van Landuyt** [2], **J Vanhellemont** [3,4]

1 Institute of Semiconductor Physics, pr.acad.Lavrentyeva 13, 630090 Novosibirsk, Russia
2 University of Antwerp, EMAT, Groenenborgerlaan 171, B 2020 Antwerpen, Belgium
3 IMEC, Kapeldreef 75,B-3001 Leuven,Belgium
4 Present address :Wacker Siltronic AG, P.O.Box 1140, D-84479,Burghausen,Germany

ABSTRACT : It was observed that <110> interstitial chains located in {111} planes at regular spacing are formed by agglomeration of self-interstitial atoms to the core of vacancy or interstitial Frank partial dislocation loops and by the insertion of interstitial chains between two perfect {111} planes. Based on experimental and calculated HREM images a structural model for the new intermediate defect configuration is proposed which includes a regular sequence of double five-membered and single eight-membered rings in which no dangling bonds are involved.

1. INTRODUCTION

The concept of intermediate defect configurations (IDC) has been introduced by T.Y.Tan in the early 80 s to interpret experimental observation on the structure of extended aggregates of point defects occurring in crystals with the diamond structure (Tan 1981). However, further studies of commonly formed interstitial clusters such as rodlike defects with {113}-habit planes by means of high resolution electron microscopy (HREM) resulted in more complicated structures of the defects compared to the ones predicted by Tan 1981and Takeda et al (1994). Apart from the predominant type of IDC which are the{113}-defects generally formed in silicon crystals at intermediate temperature independently of the process of point defect generation, there are few reports on the existence of point defect clusters with {111} habit planes (Beaufort et al 1994, Takeda and Horiuchi 1994, Muto et al 1995, Chou et al 1995). The displacement vector of {111} defect varies over a large range and the nature of the defect can be changed from interstitial to vacancy type.

The aim of the present study is the examination of the nucleation process and the detailed analysis of the atomic structure of {111}-defects at the initial formation stage by means of in situ electron irradiation in a HREM to reveal the nature of these defects.

2. EXPERIMENTAL DETAILS

Plan view samples were prepared from wafers of high purity n-type FZ-Si with (110) orientation. Both surfaces of the chemically thinned specimens were covered with 5 nm thick Si_3N_4 films in order to have a reproducible surface sink for the intrinsic point defects created by electron irradiation (Aseev et al 1994). In part of the Si samples Frank partial dislocation loops of extrinsic type have been introduced by means of B^+ ion implantation with an energy of 30 keV and a dose of 5×10^{14} cm^{-2} followed by subsequent annealing in dry oxygen ambient at T=1070°C.

In situ HREM experiments were carried out at room temperature in a JEM-4000EX microscope operated at 400 keV. The proposed atomic structure of point defect aggregates was obtained by comparison of experimental HREM images with simulated ones using the standard multislice methods.

3. RESULTS

Figure 1 shows the sequential stages of the formation of the point defect clusters on {111} during electron irradiation in a HREM at room temperature. Pronounced distortion of the {111} planes parallel to the defect plane clearly indicates the interstitial character of this defect. The displacement vector found by direct measurement of the atom image positions on the experimental HREM image in Fig.1b is about 0.18 ± 0.01 nm which is approximately equal to a/5<111>. From Fig.1c which is the HREM image superimposed with the atomic model of the defect,the regular sequence of double five- and single eight-membered atomic rings for interstitial chains inserted between two perfect {111}planes is revealed.This defect disappears during further electron irradiation in the HREM as illustrated in Fig. 1a,b.

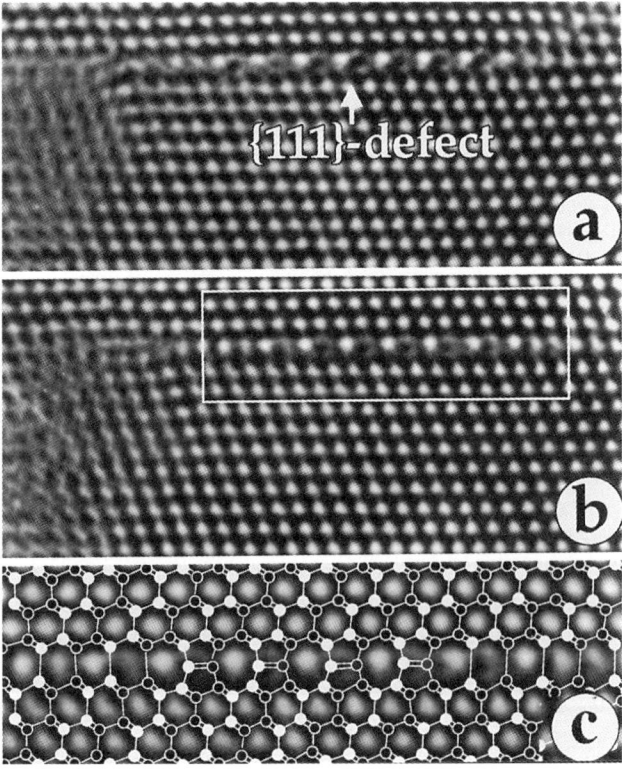

Fig.1. The successive stages of the disappearance of {111}-defect during electron irradiation in the HREM after: (a)-30, (b)-35, (c)- 45 minutes. d) The atomic model of the defect superimposed on the experimental HREM image.

Figure 2 reveals the sequential formation stages of a {111} defect after prolonged electron irradiation as a result of combined interaction of intrinsic point defects. The defect is characterized in its initial stage by a visible distortion of {111} planes parallel to the plane of the defect, which has the vacancy sign (see Fig.2a). The image shows also that several large bright dots appear in the defect plane and that the dark regions inside the defect plane shrink correspondingly. These HREM images were taken near the optimum "Scherzer" defocus condition, so that the atomic columns are imaged as dark dots. The image of this defect is therefore interpreted straightforwardly by removing Si atoms, in other words, by accumulating vacancies in the {111} plane. Further electron irradiation leads to a transformation of the atomic structure of the vacancy loop into a {111} defect yielding a HREM image without any measurable distortion of the {1̄1̄1̄} plane parallel to the defect plane (see Fig.2b). The structural model of this {111}-defect (Fig.3a) deduced from experimental HREM images reveals also in this stage a regular sequence of double five- and single eight-membered atomic rings in the defect plane viewed in (110) cross-section (2x1 reconstruction). These double interstitial atoms form the interstial chains in [110] direction perpendicular to the (110) crystal surface as we assumed. It is clearly seen that the double interstitial atoms inserted in the vacancy loop plane compensate for the displacement vector of the vacancy loop but the perfect crystal structure is not fully restored. For simulation of the HREM image of the {111}-defect it was assumed that uniform interstitial chains in <110> direction are inserted at regular spacing between two perfect {111} planes at a/3 <111> distance (the unit cell for the calculation is shown in Fig.3a). Figures 3b (part of experimental image) and 3c (calculated image) are in satisfactory agreement. Small differences between the calculated and experimental HREM images are probably due to the low relaxation rate at room temperature of silicon bonds in the disturbed area even at high intensity of electron irradiation as we assumed. This means that the positions of interstitial atoms may be slightly disoriented from the [110] direction perpendicular to the (110) crystal surface. One can thus conclude that for the present observation, the {111}-defect is initially a vacancy loop which is later filled partially with self-interstitials, accompanied by the formation of sequences of five- and eight-membered atomic rings. Comparison of the number of interstitials with that of vacancies in the plane of the {111}-defect results in a ratio of 50%. The insertion of self-interstitial atoms in the plane of the vacancy loop provides the stress relaxation of the crystal lattice around the defect of vacancy type due to the decrease of the displacement vector to about zero. The detailed analysis of the mechanism of {111} defect formation will be published elsewhere (L.Fedina et al 1997).

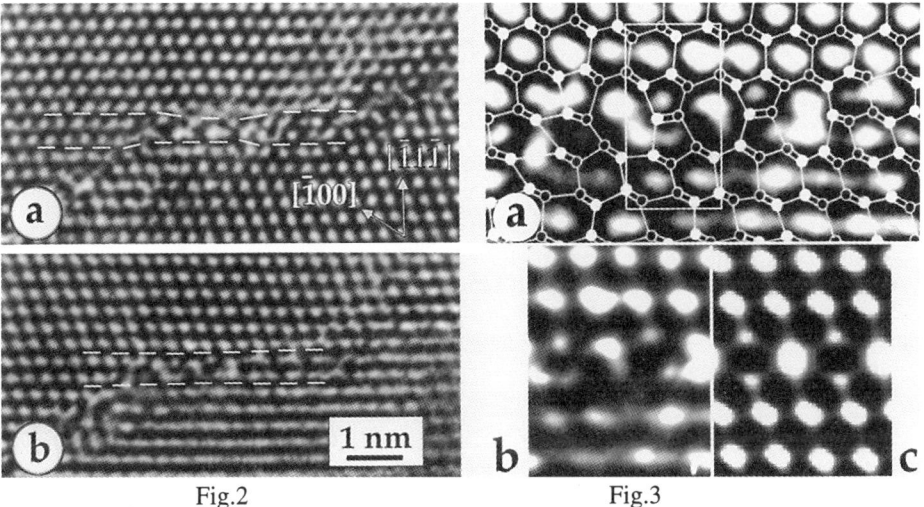

Fig.2.   Fig.3

Fig.2. HREM images of the same {111}-defect after 25 (a) and 35 (b) minutes of electron irradiation in the HREM.
Fig.3. (a) The atomic model of a {111}-defect in a further formation stage superimposed on the experimental image. The central part of experimental image (b) in comparison with the calculated one (c) made on the basis of this model.

To find other evidence of the possibility of formation of atomic configurations as found above due to the interaction of self-interstitial atoms with the core of Frank partial dislocations, Si specimens containing Frank dislocation loops of extrinsic type introduced by ion implantation were investigated. Figure 4 shows the images of the core of an interstitial Frank dislocation taken after several minutes of electron irradiation in a HREM. This defect reveals the same HREM image as the {111}-defects found by the atomic structure transformation of the vacancy loop and as isolated {111}-defect of interstial type (Fig.1,3). However, this defect can be characterized by a displacement vector similar to that of an interstitial Frank partial dislocation, (a/3<111>), i.e. 0.31 nm in the area close to the core of

46

the dislocation. At a large distance from the core of the Frank dislocation the displacement vector of the {111}-defect decreases gradually with distance to 0.18 nm as it was found for isolated {111}-defects of interstitial type (see Fig.4). The simulated HREM image of the {111}-defect obtained by using the above mentioned atomic model shows excellent agreement with the observed image which must be obtained in perfect (110) zone axis orientation of the defect area (Fig.5). No interaction of self-interstitial atoms with the core of extrinsic Frank dislocation loops was observed in the very thin (less than 10 nm) irradiated areas, because in these only defects of vacancy type are created . Prolonged irradiation leads to the disappearance of the {111}-defect and to the further transformation of the atomic structure in the core of the extrinsic Frank dislocation by means of Shockley dislocation gliding, as found recently (Fedina et al1994).

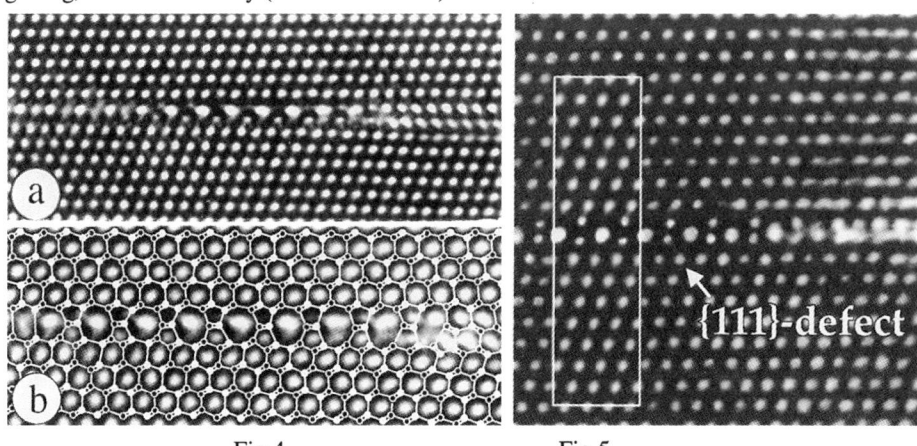

Fig.4   Fig.5

Fig.4. HREM images of an extrinsic Frank dislocation core after 1-2 minutes of irradiation in the high resolution electron microscope (a). The suggested atomic model of the dislocation core transformed by irradiation is superposed on the enlarged micrograph in (b)
Fig.5. Simulated image of a {111} defect superimposed on the experimental image of the {111}-defect created close to the core of an extrinsic Frank partial dislocation. The displacement vector of the defect used for simulation is a/3<111>.

ACKNOWLEDGMENTS

This work was supported by grant No.96-02-19432 from the Russian Foundation of Basic Research. This work was also partly sponsored by the contract G.0297.95 of the Belgian National Science Foundation (NFWO). L.F. is also grateful to the Belgian Science Policy Office (DWTC) for her fellowship at RUCA EMAT (Antwerp) in 1996.

REFERENCES

Aseev, A.L., Fedina, L., Hoehl, D. and Bartsch, H., 1994, Clusters of Interstitial Atoms in Silicon and Germanium, Academy Verlag, Berlin ,p152 .
Beaufort, M., Garem, H., and Lepinoux J., 1994, Phil. Mag., A, **69**, 881.
Chou, C.T., Cockayne, D.J.H.,Zou,J., Kringhoj, P., and Jagadish,C. 1995, Physical Review B, **52**, 17223-17230.
Fedina, L., Gutakovski A.,and Chuvilin, A., 1994, ICEM-13, Paris, 2A, 99.
Fedina, L., Gutakovskii, A., Aseev, A., Van Landuyt, J., Vanhellemont, 1997, J. Phil. Mag. A, to be published
Muto, S., Takeda, S., and Hirata, M., 1995, Phil. Mag., A, **72**, 1057.
Takeda, S., Kohyama, M., and Ibe, K., 1994, Phil.Mag.A, **4**, 287.
Takeda, S., and .Horiuchi S. 1994, Ultramicroscopy, **54**, 144.
Tan, T.Y., 1981, Phil. Mag., A, **44**, 101.

Inst. Phys. Conf. Ser. No 157
Paper presented at Microsc. Semicond. Mater. Conf., Oxford, 7–10 April 1997
© *1997 IOP Publishing Ltd*

A study of interdiffusion and germanium segregation in low-pressure chemical vapour deposition of SiGe/Si quantum wells

T Walther [1,2] **, C J Humphreys** [1]**, A G Cullis** [3] **and D J Robbins** [4]

[1] Dept Materials Science & Metallurgy, Cambridge University, Pembroke St, Cambridge CB2 3QZ, UK
[2] now at: CEA-Grenoble, DRFMC/SP2M, 17 rue des Martyrs, 38054 Grenoble, France
[3] Dept Electronic & Electrical Eng, Sheffield University, Mappin St, Sheffield S1 3JD, UK
[4] Defence Research Agency (DRA), St Andrews Rd, Malvern Worcs WR14 3PS, UK

ABSTRACT: The activation energy for interdiffusion of strained SiGe/Si quantum wells is determined using quantitative high-angle annular dark field imaging. The interface widths obtained from this and from electron energy-loss spectroscopy are larger and the interface asymmetries smaller than calculations using the standard segregation model suggest, indicating more than two monolayers are involved in this process. Increasing the growth temperature broadens mainly the SiGe-on-Si interface, growth interrupts locally decrease the Ge content and the Ge deposition rate is reduced when SiGe growth commences. These effects are explained by Ge-H bonds being weaker than Si-H bonds.

1. INTRODUCTION

Bulk diffusion, segregation and surface adsorption and desorption influence the compositional profiles of thin layers during epitaxy. These effects are temperature dependent. Bulk diffusivities can be determined from the layer broadening as a function of annealing. If diffusion is characterised by a diffusivity

$$D=D_0 \exp[-E/(k_B T)] \tag{1}$$

with Boltzmann constant k_B, temperature T, diffusion coefficient D_0 and activation energy E, then data from two different anneal temperatures are necessary to determine both, E and D_0.

Activation energies for Ge dopant diffusion in bulk Si at high temperatures lie between E=4.7 eV (McVay and DuCharme 1974) and 5.3 eV (Dorner et al 1984). However, the interaction between diffusing atoms and strain fields is neglected by the approximation of a dilute Si:Ge system, and the temperature range relevant to growth related problems is lower than the annealing temperatures used in most diffusion studies. The moderate annealing temperatures used here ensured that the layers were kept strained.

2. EXPERIMENTAL METHODS

2.1. Growth and specimen preparation for transmission electron microscopy

The SiGe/Si layers investigated were grown by low-pressure chemical vapour deposition (LPCVD), using silane and germane gas sources and hydrogen as carrier gas. One SiGe layer was grown at 700 °C, all other layers at 610 °C.

For the diffusion study two sections of a multi quantum well structure were enclosed in quartz tubes under argon atmosphere and annealed in furnaces at 710 and 810 °C, respectively, each for one hour. A ‹110› cross-sectional TEM specimen was prepared by glueing together back-thinned wafer pieces from the as-grown and both annealed samples. The hole produced by Ar+ ion milling was large enough to yield electron transparent areas from all three specimens so that the investigation could be performed under identical imaging conditions. Other specimens were prepared in a similar way.

2.2. Scanning transmission electron microscopy (STEM) with the VG HB501

STEM has been performed in a VG HB501 microscope operated at 100 kV with a sub-nm probe. The beam convergence semi-angle was 7.6 mrad. High-angle annular dark field (HAADF) images were recorded with a photomultiplier system whereby detector non-linearities near the upper end of the 8 bit range were carefully avoided. A purpose-built mask in front of the annular dark field detector limited the inner detection angle to about 200 mrad, thus ensuring the dominance of Rutherford scattering, with the intensity proportional to the thickness and the square of the mean atomic number. For energy-dispersive X-ray analysis (EDX) a windowless Si:Li detector and for electron energy-loss spectroscopy (EELS) a Gatan energy filter were used. In all cases the specimens were oriented edge-on in weakly diffracting conditions as strain effects dominate near zone axis orientations (Walther 1996).

Quantitative compositional data were obtained

a) in the HAADF mode: by linearly extrapolating in line scans the intensity of the Si into the SiGe region (thus compensating for the thickness influence on the image intensity) and converting for every point in the scans the intensity ratio of SiGe to Si into an average atomic number (Walther et al 1997);

b) from EDX: by scanning the electron beam along the growth direction while measuring the counts in the Si K, Ge K and Ge L lines and using experimentally determined k-factors for these lines (Walther 1996);

c) from EELS: by similarly obtaining loss spectra in the range of 1 to 2 keV and converting the ratio of the intensities of the Ge $L_{2,3}$ to Si K edges after power-law background subtraction (Egerton 1996) into compositions using the ionisation cross-sections calculated with the EL/P 3.0 software (Rez 1982).

3. EXPERIMENTAL RESULTS

3.1. Determination of the activation energy for interdiffusion

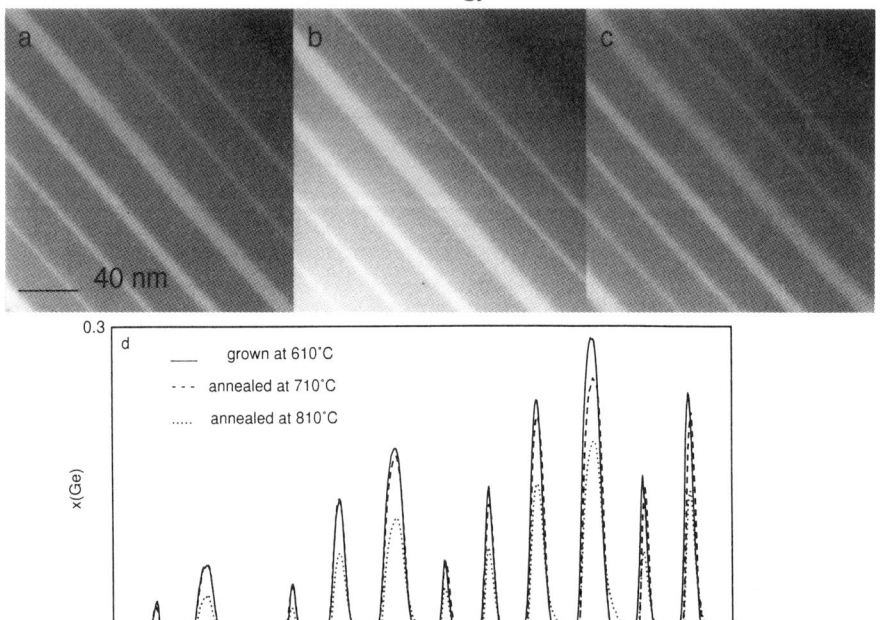

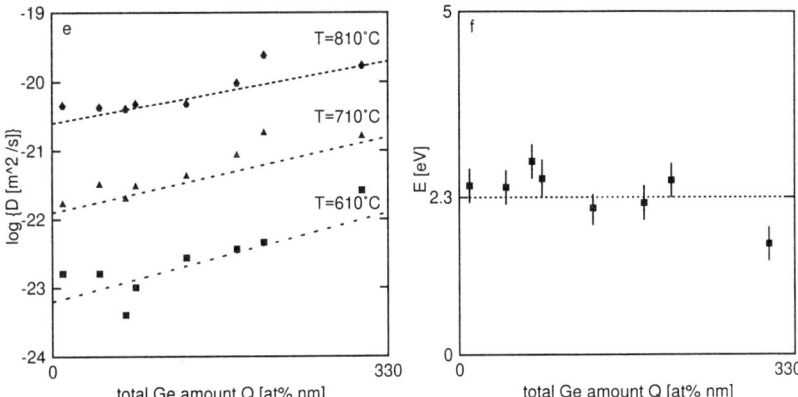

Fig. 1: HAADF images of $Si_{1-x}Ge_x$/Si quantum wells of nominal Ge compositions x=0.3 and 0.4 as grown at 610 °C (a) and annealed at 710 °C (b) and 810 °C (c) for one hour each. The [001] growth direction points to the top right where the wafer surface is visible. (d) Compositional line scan of all layers of the multi quantum well structure; (e) diffusivities and (f) activation energies calculated using Eqns.6, 7.

Figs. 1a-c show HAADF images of the top layers of a multi quantum well structure. The layer broadening with annealing is only obvious for the highest Ge content. Fig. 1d depicts the Ge compositions calculated from the HAADF intensity line scan of all layers, assuming Rutherford scattering. The decrease of the peak compositions of the SiGe quantum wells is more pronounced than the layer broadening. The asymmetric broadening of the layers near the free surface, yielding enhanced compositional tails of the trailing Si-on-SiGe interfaces, has not been completely understood so far. The native oxide layer on the wafer surface might have caused large surface stresses under the influence of which Si self-interstitials could be injected into the crystal (Antoniadis 1982, Hirth and Tiller 1984) where they are highly mobile and influence diffusion significantly (Fahey et al 1989).

For a layer of finite width w_0 at a temperature T_0, an anneal at a temperature T_i results in the convolution of the initial profile $x(z,T_0)$ along the z direction with a Gaussian envelope function whose standard deviation is given by the diffusion length

$$\lambda = 1/Q \int x(z) \, z^2 dz \ = \sqrt{2Dt} \tag{2}$$

where

$$Q = \int x(z) dz \tag{3}$$

is the total amount of Ge in the layer and constant during the anneal. In our case the initial as-grown profiles are themselves almost Gaussian in form, i.e. we may approximate

$$x_0(z,T_0) \approx \frac{Q}{\sqrt{2\pi}\sigma} \exp\left[-z^2/(2\sigma^2)\right] \tag{4}$$

with standard deviation (rms value) σ. The two annealed profiles then are given by

$$x_i(z,T_i) = \frac{Q}{\sqrt{2\pi(\sigma^2 + 2Dt)}} \exp\left[-z^2/(2\sigma^2 + 4Dt)\right] \quad \text{for i=1, 2.} \tag{5}$$

Using the relationship between height, width and integral under a Gaussian curve, E and D_0 can be calculated from the ratios of the peak concentrations $\hat{x}$ at the temperatures T_0=610 °C, T_1=710 °C and T_2=810 °C for a layer of initial width w_0 which has been annealed for a time t:

$$E = k_B \frac{T_1 T_2}{T_2 - T_1} \ln\left[\frac{(\hat{x}_0/\hat{x}_2)^2 - 1}{(\hat{x}_0/\hat{x}_1)^2 - 1}\right] \tag{6}$$

and

$$D_0 = \frac{w_0^2}{16t \ln2} \frac{\hat{x}_0^2}{\hat{x}_1^2 \hat{x}_2^2} \exp\left[\frac{E\ (T_1 + T_2)}{2 k_B T_1 T_2}\right] \tag{7}$$

The experimental data indicate that the ratios $\hat{x}_0/\hat{x}_i$ are, for a given anneal temperature T_i, identical for all quantum wells, with $\hat{x}_1/\hat{x}_0 = 0.93 \pm 0.02$ and $\hat{x}_2/\hat{x}_0 = 0.57 \pm 0.05$. This yields $E=2.3 \pm 0.3$ eV. Values of diffusivities and activation energies from eight quantum wells are plotted in Figs. 1e,f. For the thinnest layers with low Ge content, $\hat{x}_0 \approx \hat{x}_i$. Hence, the last term in Eqn. 6 diverges, and the values which scatter enormously have been omitted.

The activation energy determined above for strained SiGe/Si quantum wells is about half of the values reported for Ge dopant diffusion in Si or SiGe, but agrees with photo-luminescence (PL) studies of moderately annealed strained $Si_{1-x}Ge_x/Si$ layers grown by molecular beam epitaxy ($E=2.47 \pm 0.4$ eV for x=0.16; Sunamura et al 1993) or UHV-CVD ($E=2.42$ eV for x=0.29; Boucaud et al 1996) and an X-ray diffraction (XRD) study of annealed short-period Si/Ge superlattices ($E \leq 2.5$ eV ; Baribeau et al 1990). As both, PL and XRD, fit data to models and an anneal does not only change the peak composition but also the shape of the layers which in turn influences e.g. the confinement levels probed in PL, the coincidence with the value determined more directly from HAADF is not trivial. Also, we did not observe a significant dependence of the activation energy for interdiffusion on the amount of deposited Ge (Fig. 1f) and thus the layer width, in contrast to the above XRD study.

3.2. Ge surface segregation

The above diffusivities yield sub-nm diffusion lengths for typical growth temperatures and times. This means that the interdiffusion of our SiGe layers should be hardly discernible by TEM. A significant layer broadening may be due to Ge segregation, i.e. Ge atoms `floating' towards the free surface during epitaxy. In the case of layer-by-layer growth of SiGe/Si, this implies that Ge atoms from the sub-surface monolayer (ML) exchange sites with Si atoms in the top ML with a probability p which is much larger than the probability q=1-p of the reverse process. As the result a Ge enriched surface layer is being formed. This concept led to the development of a 2-state-exchange model for surface segregation (Harris et al 1984, Fukatsu et al 1991). With an activation frequency f, the exchange probabilities can be expressed as

$$p = f \, \exp\left[-E_{kin}/(k_B T)\right] \quad \text{and} \quad q = f \, \exp\left[-\left(E_{kin} + E_{seg}\right)/(k_B T)\right] \quad (8),$$

where $E_{kin} \approx 1.63$ eV is the kinetic barrier height between the two states a Ge atom can occupy and $E_{seg}=k_B T \ln(p/q)$ the Gibbs free energy of segregation which drives the segregation mechanism and depends on the chemical species terminating the surface (Copel and Tromp 1991). The model yields rate equations which can only be solved iteratively, but for the case of a Si cap on top of a dilute $Si_{1-x}Ge_x$ alloy, an approximate analytical solution exists in the form of an exponential decay of the Ge composition from the layer towards the surface:

$$x_i = x(1-p) \exp(-pt) \qquad \text{for ML \# i=0, 1, ..., m-1 from the SiGe layer} \quad (9)$$

Experimental segregation studies in the SiGe/Si system have been carried out on Si capped flat SiGe and Ge layers by surface-sensitive in-situ techniques such as Raman and ion scattering (Copel et al 1989), photoelectron spectroscopy (Fujita et al 1991), reflection high-energy electron diffraction (Ohtani et al 1993) and Auger electron spectroscopy (Butz and Kampers 1992, Li et al 1995). For a quantification the signals from MLs at different depths in the specimen need to be separated by modelling the depth penetration of the technique. Applying this to effectively deconvolute the measured profiles with the probe function then allows to study the decay of even rather weak Ge signals as a function of the thickness of the Si cap. The persistence of a strong Ge signal in this case has been interpreted in terms of Ge segregation. Values estimated for the floating probability range between p=0.94 (Copel and Tromp 1991) and 0.98 (Larsson and Hansson 1993). This implies that the probability for the reverse process which cannot be determined from these methods must be q<0.06. As the ratio p/q determines the segregation energy, an accurate knowledge of p *and* q is necessary. The simulations in Fig. 2 demonstrate that varying q influences the sharpness of the interfaces slightly. The layer profiles are furthermore insensitive to changes in p, if p≥0.8. TEM can therefore not provide reliable values for p, at least for x≈0.2, but can be used to determine q. It is thus complimentary to surface characterisation techniques.

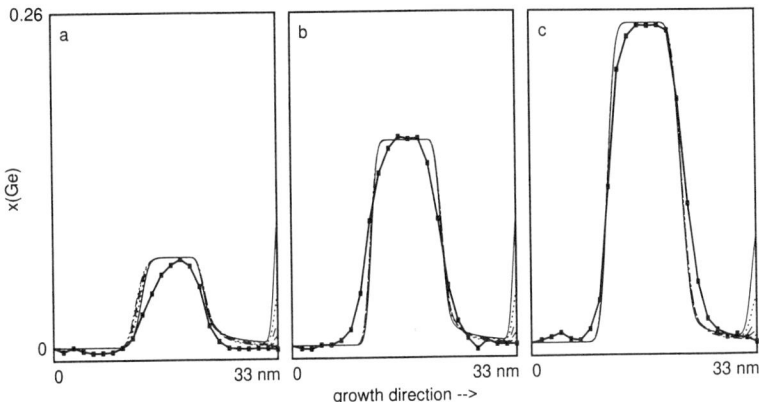

Fig. 2: Comparison of experimental compositional profiles of 10 nm wide SiGe/Si layers as determined by EELS (squares with thick lines) and simulations of Ge segregation for x=0.07 (a), 0.16 (b) and 0.25 (c), p=0.94, q=0.01 (thin solid lines), 0.02 (dotted), 0.03 (dashed), 0.04 (dot-dashed) and 0.05 (double-dot dashed). The simulated profiles have been calculated using the 2-state-exchange model and convoluting the profiles with Gaussians of rms widths of 0.9 nm (corresponding to an upper estimate of the diffusion length at 610 °C) and 0.3 nm (for a STEM probe of 0.7 nm FWHM). The spikes at the right sides are floating Ge surface layers as expected at free surfaces.

Detailed simulations predict that

1. larger q values decrease the interface widths for low Ge compositions but broaden the interfaces for higher Ge concentrations;
2. trailing Si-on-SiGe interfaces are always wider than leading SiGe-on-Si interfaces;
3. both interface widths decrease almost exponentially with increasing Ge composition.

The experimental interface widths are not only systematically larger than the simulations suggest (by about 3 nm whereby only 1 nm can be explained by diffusion and probe width influence), but for low Ge concentrations the leading is actually wider than the trailing interface (Fig. 2a) and the trailing interface width does not decrease significantly with increasing Ge composition, contrary to the model. It has already been suggested that more than only the topmost MLs need to be taken into account in properly modelling surface segregation (Lu et al 1994), but the evidence put forward so far has not been conclusive.

3.3. Rôle of the growth temperature and growth interrupts in LPCVD

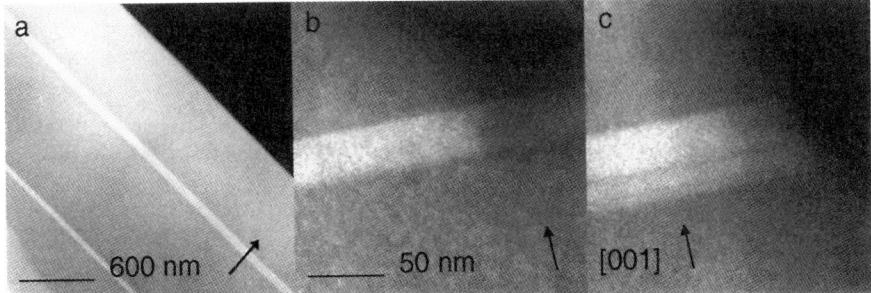

Fig. 3 HAADF images of a structure of two SiGe layers between Si spacers. Arrows denote the growth direction. The first SiGe layer is 37 nm wide and has been grown at 700 °C in 300 s (b), the second layer is 60 nm wide and has been grown at 610 °C in 480 s, with 10s and 30 s growth interrupts which are discernible as dark lines.

The leading SiGe-on-Si interface broadens considerably, by about a factor of 2, when the growth temperature is increased from 610 to 700 °C, whereas the trailing Si-on-SiGe interface is only weakly affected, as can be seen from a comparison of Figs. 2b,c. The diffusion study has shown that this amount of interface broadening cannot be explained by interdiffusion. Hence, segregation must be responsible for it. At the higher temperature more hydrogen desorbs from the surface, and as hydrogen is known to reduce Ge segregation by surface passivation (Copel and Tromp 1991), an increase of both interface widths is expected. The fact that the broadening is almost exclusively confined to the leading interface demonstrates that the temperature dependence of H desorption from SiGe surfaces is higher than from Si surfaces, i.e. the Ge–H bond is effectively weaker than the Si–H bond. This is consistent with an activation energy measured for hydrogen desorption from a Si(001) surface of E_{Si-H}=2.04±0.13 eV (Sinniah et al 1989), while the value for Ge(001) surfaces is only E_{Ge-H}=1.5±0.1 eV (Surnev and Tikhov 1984). Hence, the leading SiGe-on-Si interface width in CVD is governed by the desorption kinetics of the hydrogen.

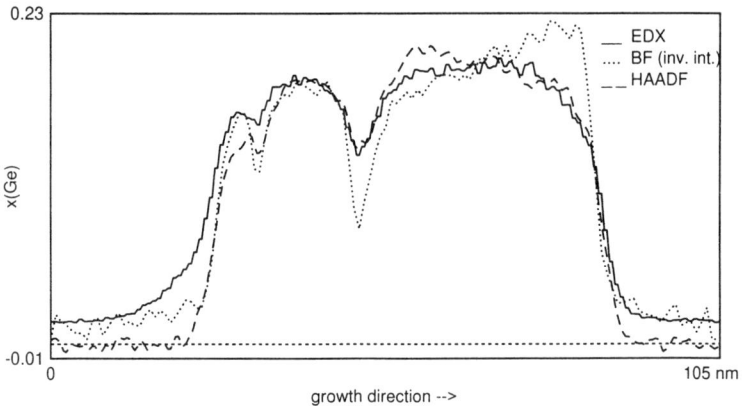

Fig. 4: Comparison of bright field (BF) intensity, HAADF and EDX compositional line scans of the SiGe layer with growth interrupts (see Fig. 3c). The BF intensity has been inverted and scaled to a minimum of 0 and a maximum of 0.22. The horizontal scale has been adjusted so that the positions of the minima of the curves coincide.

During the deposition of the second layer shown in Fig. 3c, the flow of silane and germane had been stopped twice, once for 10 s and another time for 30 s, while maintaining the hydrogen gas supply. Fig. 4 provides clear evidence for local Ge depletion during these growth interrupts. While the scan of the inverted BF intensity exhibits a strain related asymmetry and the EDX scan a broadening of at least 2 nm, particularly at the leading interface side shown on the left which is due to the effective probe size increasing with the specimen thickness towards the left, the agreement between EDX and HAADF is excellent (better than Δx=±0.01). Because of the finite probe size in STEM used in performing the EDX and HAADF experiments, it can be concluded that the Ge depletion is *at least* Δx=0.03±0.01 during the first and 0.07±0.02 during the second growth interrupt. The precise value might be larger, but the BF resolution is sufficiently good (about 0.5 nm) to give an upper estimate for the Ge depletion of $\Delta x \approx 0.12 \pm 0.03$. As in-situ ellipsometry indicated a discontinuous change of the dielectric function as a result of the growth interrupts, continued Si deposition due to e.g. rest gas in the chamber, leakage or Ge removal by surface etching can be ruled out as an explanation. It appears likely that hydrogen termination during the growth interrupt passivated the surface sufficiently to prevent Ge atoms from incorporation into the crystal lattice when growth was resumed, in agreement with the above observation that the Ge–H bond is less stable than the Si–H bond.

3.4. Compositional dependence of the growth rate in LPCVD

Lattice fringes from high-resolution electron micrographs have been used for a calibration of the magnification as the Si growth rate varied strongly across the wafer. The growth rate of thin SiGe layers is difficult to determine reliably as their widths cannot be measured to better than a few MLs by TEM which can already give an appreciable relative error. For SiGe layers narrower than about 8 nm it is more precise to calculate the growth rate from the amount of Ge deposited, Q, and to divide this value by the growth time, τ, and the peak composition, $\hat{x}$:

$$r_{SiGe}^{indirect} = Q/(\hat{x}\tau) \tag{10}$$

This approach effectively substitutes the measurement of the quantum well width by a measurement of the peak composition which is more accurate for thin SiGe layers of high Ge content. Fig. 5a depicts plots of the Ge deposition rate, Q/τ, as a function of the germane to silane flux ratio for various quantum wells. As the silane flow was identical for all SiGe layers, the amount of Ge deposited per time should depend only (and ideally linearly) on the germane flow rate chosen. While for a given quantum well thickness the Ge deposition rate increases indeed linearly with the flow ratio up to about $Q/\tau=0.02$ nm/s, it saturates for larger flow ratios. Furthermore, for a low germane to silane flow ratio the Ge deposition rate increases with the quantum well width, indicating that the probability for Ge incorporation into the layers during the early deposition stages is lower than the steady-state value reached after depositing more than 10 nm SiGe. In the plot of the SiGe growth rate versus the Ge peak composition as depicted in Fig. 5b, this leads to a maximum in the total growth rate of about r=0.10 nm/s for the widest SiGe layers which is reached for peak concentrations in the range of $\hat{x}=0.20\pm0.05$. The growth rate of the nominally 2 nm wide SiGe layers also reaches a maximum for $\hat{x} \approx 0.2$, but the maximum growth rate is higher than for the 10 nm wide quantum wells. The growth rate for the nominally 1 nm wide quantum wells is even higher, r $\approx$ 0.18±0.03 nm/s, and thus off the scale of Fig. 5b.

The observed maximum in the SiGe growth rate as a function of the Ge content agrees with experiments and simulations of the LPCVD process by Robbins et al (1991), although in this study the reported Ge composition which gave the maximum growth rate was significantly lower for a temperature of 610 °C. The model developed therein predicts maxima in the growth rates for x≤0.06 and is again based on the competition between different hydrogen desorption kinetics from Si and SiGe on the one and decreasing sticking probabilities for hydrides with increasing Ge composition on the other hand. Another study (Gu et al 1994) reported a peak in the growth rate of CVD $Si_{1-x}Ge_x$ for x≈0.2.

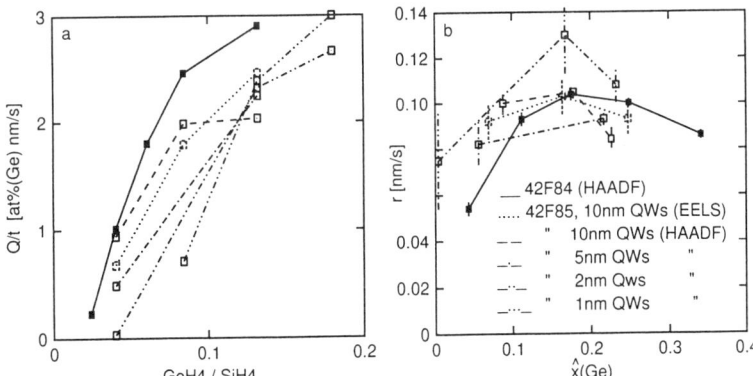

Fig. 5: Plot of the Ge deposition rate as a function of the germane/silane flux ratio for SiGe layers of various widths from different multi quantum well structures (a) and of the total growth rate as a function of the Ge peak composition in the quantum wells (b).

4. CONCLUSION

In conclusion, it has been shown that
1. the activation energy for interdiffusion is halved in thin SiGe/Si layers as compared with Si or SiGe bulk, due to strain;
2. diffusion lengths are still almost negligible at temperatures around 600 °C;
3. Ge segregation broadens the interfaces considerably, and the standard 2-state-exchange model is insufficient to explain the interface broadening;
4. the leading SiGe-on-Si interface is strongly broadened by an increase in the growth temperature and growth interrupts lead to local decreases in the Ge content. Both effects can be explained by a smaller Ge-H bond energy as compared with Si-H;
5. in the early deposition stage, the growth rate is enhanced by about a factor of 2 as compared with the steady-state value reached after about 10 nm SiGe deposition, but less Ge is incorporated in the beginning of SiGe deposition. This broadens the quantum wells and reduces the maximum peak composition of very thin SiGe layers.

ACKNOWLEDGEMENT

TW would like to thank the Engineering and Physical Sciences Research Council, Swindon, and the DRA for a CASE studentship during which course this work has been performed. The Institute of Physics, London, and the Royal Microscopical Society, Oxford, are gratefully acknowledged for travel assistance.

REFERENCES

Antoniadis D A 1982 J. Electrochem. Soc. **129**, 1093
Baribeau J-M, Pascual R and Saimoto S 1990 Appl. Phys. Lett. **57**, 1502
Boucaud P, Wu L, Guedj C, Julien F H, Sajnes I, Campidelli Y and Garchery L 1996 J. Appl. Phys. **80**, 1414
Butz R and Kampers S 1992 Appl. Phys. Lett. **61**, 1307
Copel M, Reuter M C, Kaxiras E and Tromp R M 1989 Phys. Rev. Lett. **63**, 632
Copel M and Tromp R M 1991 Appl. Phys. Lett. **58**, 2648
Dorner P, Gust W, Predel B, Roll U, Lodding A and Odelius H 1984 Philos. Mag. A **49**, 557
Egerton RF 1996 Electron Energy-Loss Spectroscopy (New York & London: Plenum, 2. ed)
Fahey P, Iyer S S and Scilla G J 1989 Appl. Phys. Lett. **54**, 843
Fujita K, Fukatsu S, Yaguchi H, Shiraki Y and Ito R 1991 Appl. Phys. Lett. **59**, 2240
Fukatsu S, Fujita K, Yaguchi H, Shiraki Y and Ito R 1991 Appl. Phys. Lett. 59, 2103
Gu S, Zheng Y, Zhang R, Wang R and Zhong P 1994 J. Appl. Phys. **75**, 5382
Harris J J, Ashenford D E, Foxon C T, Dobson P J and Joyce B A 1984 Appl. Phys. A **33**, 87
Hirth J P and Tiller W A 1984 J. Appl. Phys. **56**, 947
Larsson M I and Hansson G V 1993 Surf. Sci. **291**, 117
Li Y, Hembree G G and Venables J A 1995 Appl. Phys. Lett. **67**, 276
Lu Z H, Baribeau J-M and Lockwood D J 1994 J. Appl. Phys. **76**, 3911
McVay G L and DuCharme A R 1974 Phys. Rev. B **9**, 627
Ohtani N, Mokler S M, Xie M H, Zhang J, Joyce B A 1993 Surf. Sci. **284**, 305
Rez P 1982 Ultramicroscopy **9**, 283 (Hartree-Slater cross-sections, used in EL/P 3.0 software distributed by Gatan, Pleasanton, California, USA)
Robbins D J, Glasper J L, Cullis A G and Leong WY 1991 J. Appl. Phys. **69**, 3729
Sinniah K, Sherman M G, Lewis L B, Weinberg W H Jr, Yates J T and Janda K C 1989 Phys. Rev. Lett. **62**, 567
Sunamura H, Fukatsu S, Usami N and Shiraki Y 1993 Appl. Phys. Lett. **63**, 1651
Surnev L and Tikhov M 1984 Surf. Sci. **138**, 40
Walther T 1996 PhD Thesis, University of Cambridge
Walther T, Humphreys CJ and Robbins D J 1997 Defect and Diffusion Forum (Proc. Conf. DIMAT 96, Nordkirchen), in print by Trans Tech Publications

Inst. Phys. Conf. Ser. No 157
Paper presented at Microsc. Semicond. Mater. Conf., Oxford, 7–10 April 1997
© 1997 IOP Publishing Ltd

In-situ HREM irradiation study of point defect clustering in strained Ge_xSi_{1-x}/(001)Si heterostructure

L Fedina,* O Lebedev**, G Van Tendeloo and J Van Landuyt

University of Antwerp, EMAT, Groenenborgerlaan 171, B-2020 Antwerpen, Belgium
* on leave from Institute of Semiconductor Physics RAS, pr. Ac Lavrentyeva 13,
    630090 Novosibirsk , Russia
** on leave from Institute of Crystallography RAS, Leninsky pr. 59, 117333
    Moscow Russia

ABSTRACT: The process of point defect clustering in compressively strained Ge_xSi_{1-x}/(001)Si heterostructures is found to be very sensitive to the amount of misfit strain. At a low misfit (0.28%) extended defects of both interstitial and vacancy type with a density of about 10^{12} cm^{-2} are created inside the strained layer only. An increase of the misfit up to 0.46% leads to the formation of vacancy type defects inside the Ge-Si layer and defects of interstitial type are created in the Si matrix close to the interface. In material with a large misfit (0.7%) the complete transformation of the strained Ge-Si layer into Ge rich islands occurs by misfit dislocation introduction at the islands / Si-matrix interface during electron irradiation.

1. INTRODUCTION

Epitaxial Ge_xSi_{1-x} layers grown on a Si substrate are of major importance for the fabrication of advanced electronic and opto-electronic devices (Schaffer and Rodewald, 1996). One of the most important aspects from a scientific, as well as from a technological point of view is the possibility of relaxing strained heterostructures by accumulation of point defects. It has been known for more than two decades that self-interstitial atoms in Si and Ge crystals at intermediate temperatures cluster into extended defects, which are characterized by a very small displacement vector (for a review of this problem see e.g. Aseev et al,1994). Such defects seem to be suitable for the relaxation of a low mismatch. As recently established, intensive HVEM and HREM electron irradiation at room temperature leads to the clustering of both interstitials and vacancies in Si crystals (Fedina et al, 1995 and1997a,). It offers new possibilities to investigate the interaction of intrinsic point defects with the various interfaces having a different sign of deformation.
    This paper presents a study by in-situ electron irradiation in a HREM of the point defect clustering in a strained Ge_xSi_{1-x}/Si heterostructure.

2. EXPERIMENTAL DETAILS

The samples are p-type modulated Ge_xSi_{1-x} heterostructures containing 5, 12 and 17% of Ge, grown using solid source MBE VG Semicon equipment (Whall et al 1994). The Ge concentration in the strained Ge-Si layers has been measured by Rutherford backscattering analysis. Cross-section samples suitable for high resolution electron microscopy were prepared by Ar ion milling. In-situ HREM experiments were carried out in a JEM-4000EX microscope. HREM images of the defect clusters were simulated using the Mac Tempas multislice program.

3. RESULTS

Before intense irradiation in a HREM the strained layers are defect free (Fig. 1a). After several minutes of irradiation small clusters of point defects are created, but the process of point defect clustering in compressively strained $Ge_xSi_{1-x}/(001)Si$ heterostructures is found to be very sensitive to the amount of misfit strain. At a low misfit (0.28%) extended defects of both interstitial and vacancy type with a density of about 10^{12} cm^{-2} are created almost exclusively inside the strained GeSi layer (Fig.1b).

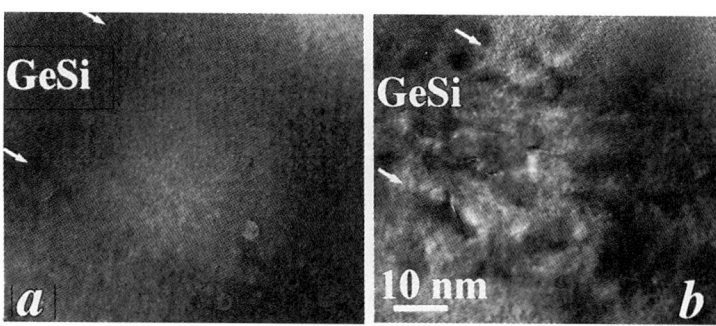

Fig.1. Low magnification TEM image of the GeSi/Si heterostructure with 0.28% misfit in the initial stage(a) and after intense irradiation at 400 keV at room temperature (b).

An enlarged image of a typical extended defect found within the strained GeSi layer is shown in Fig.2a. This image was recorded near Scherzer focus, so that the atomic columns are imaged as dark spots. The defect displays a larger contrast of separated columns which are located in the {113} plane. The atomic model for the defect, superimposed on the experimental HREM image, highlights the interstitial type defect consisting of several split <100> interstitial atoms in the (113) plane (Fig.2b). The displacement introduced by the defect found by direct measurement of the atom image position on the experimental HREM image of Fig.2a is about (0.13 ± 0.01) nm in the [001] direction which is approximately equal to a/4<001>. The small displacement vector creates a problem to position di-interstitial atoms in the (110) plane because of the strong compression of the Si-Si bonds which reaches values up to 46%. As a result, two split-interstitial atoms placed at the distance of the neighbouring split positions (a/2[110]) in the [110] direction perpendicular to the (110) surface move towards each other in order to form a relaxed configuration. The simulated HREM images of this defect using the above model are represented in Fig.2c,d. The best agreement between the experimental and the calculated images of a {113} defect consisting of split-<100> interstitials has been obtained for a 10% compression of the Si-Si bonds between the split atoms (instead of 2% for the fully relaxed structure). Although this type of defect is the most dominant one in the low misfit heterostructure, it is not the only one. Another type of the defect is shown in Fig.3 at higher magnification with an atomic model superimposed on the experimental image. The defect can be considered as an extended multi-vacancy cluster created by a succession of linked di-vacancies located within the {113} plane. Every di-vacancy in this model corresponds to a di-vacancy chain extended in the [110] direction, i.e. along the incident electron beam. HREM images have been simulated for a single di-vacancy chain (the unit cell is shown in the white rectangle in Fig.3b) with varying length of the chain and for a fixed crystal thickness of 10 nm. Good agreement between the experimental HREM images of the separated di-vacancy chain and the calculated ones has been obtained by assuming that 20 to 40% of the atomic positions in the vacancy chain are still occupied by Si (Fig. 3a and c,d,e).

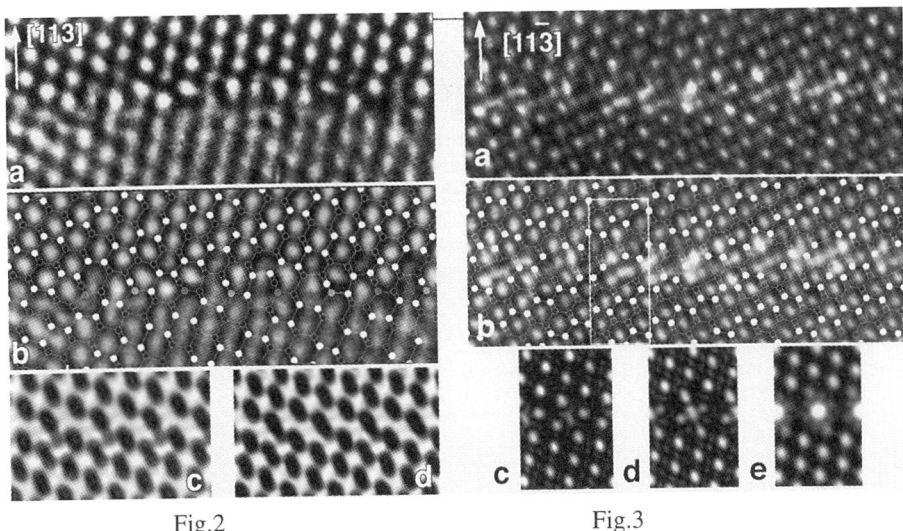

Fig.2   Fig.3

Fig.2. a) [110] HREM image of a {113}defect created inside the compressively strained GeSi layer with a 0.28% misfit; b) corresponding atomic model superimposed on the image. c), d) calculated images of the defect using the above model for 2% (c) and 10% (d) Si-Si bond compression (crystal thickness 5 nm and defocus value 55nm).

Fig.3. a) HREM image of the defect on a {113} plane created inside the compressively strained GeSi layer with a low misfit (0.28%); b) corresponding atomic model superimposed on the image; c),-e) Simulated HREM image of the defect on the basis of the above model (the unit cell is outlined by a white rectangle) for various lengths of the vacancy chain and for a crystal thickness of10nm and defocus value 55nm: (c)- 4nm; (d)-8nm; (e)-10nm.

It should be noticed that the defect in Fig.3a exhibits a very small distortion -hardly visible by the eye- of the {113} planes parallel to the defect plane and having a vacancy sign.

Electron irradiation of the strained GeSi/Si heterostructure with a 0.46% misfit reveals a large difference between the inside and outside the GeSi layer. A high density of point defect clusters is created outside the strained layer immediately after starting the irradiation at 3 to 5 nm distance from the interface(see Fig.4).

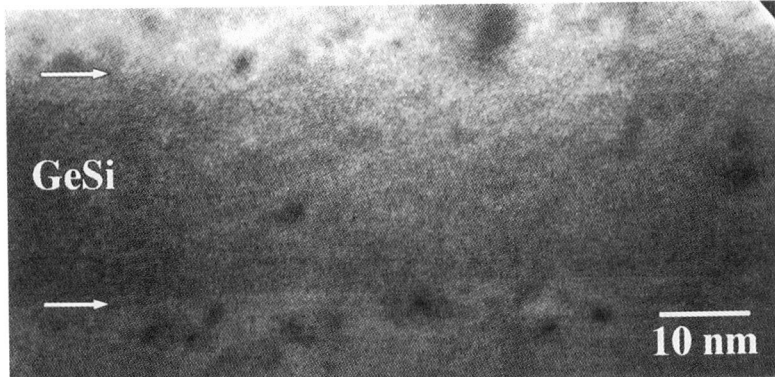

Fig.4. Point defect clustering in a compressively strained GeSi/Si heterostructure with a 0.46% misfit during in situ 400kV electron irradiation for 1 minute.

58

Most of the very extended defects found after a long time of irradiation appear to be concentrated outside the strained layer. Prolonged irradiation leads to the formation of other extended defects within the GeSi layer, located close to the compressively strained interface. Comparison of the experimental images with the calculated images for the di-vacancy chain allows to conclude that the present defects are extended defects of the vacancy type. On the other hand, analysis of the defects created outside the strained GeSi layer showed that they are {113} defects of interstitial type in the form of a split-<100> configuration. A more detailed analysis of the defects will be published elsewhere (Fedina et al 1997b).

Further increase of the misfit strain up to 0.7% in the GeSi heterostructure and prolonged irradiation leads to the complete transformation of the compressively strained layer (Fig.5). In the initial stages of irradiation extended defects of interstitial type are created outside the strained layer at a small distance from the interface. Continued irradiation however transforms the initially continuous GeSi layer into narrow islands which are growing along the interface during irradiation. The islands exhibit a visible distortion of the {111} planes in the initial stages of growth and there is strong diffraction contrast around the islands. Finally misfit dislocations are introduced at the interfaces between the islands and the silicon matrix at a distance of about 5-6 nm from each other. (see Fig.5b).

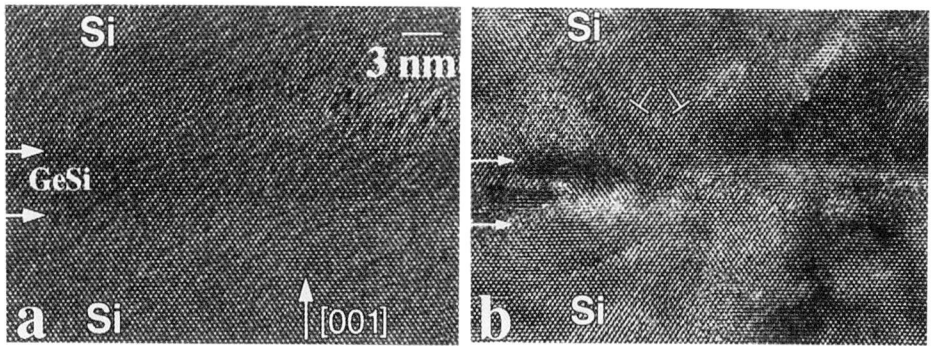

Fig.5. HREM image of the GeSi/Si heterostructure with a 0.7% misfit before (a) and after (b) intense electron irradiation .

ACKNOWLEDGEMENTS

This work was sponsored by NFWO. L.I. Fedina and O.I. Lebedev are indebted to the Belgian Government for their fellowships at the university of Antwerp (EMAT). The authors are grateful to Prof. E.H.C. Parker for the specimens supply and Prof. A.G. Vasiliev for RBS analysis.

REFERENCES

Aseev, A.L., Fedina, L.I., Hoehl, D. and Barsch H., 1994, Clusters of Interstitial Atoms in Silicon and Germanium, Academy Verlag, Berlin, 152P.

Fedina, L., Van Landuyt, J., Vanhellemont, J. and Aseev, A., 1995, Nucl. Instr. and Meth. in Phys. Res., B, **112**, 133.

Fedina, L., Gutakovskii, A., Aseev, A., Van Landuyt, J. and Vanhellemont, J., 1997a, Phil. Mag. A, (accepted for publication).

Fedina, L., Lebedev, O., Van Tendeloo, G., Van Landuyt, J., 1997b, Phil. Mag. A., (submitted for publication).

Schaffer C. and Rodewald M., 1996, Journal of Crystal Growth, **156**, 61.

Whall, T.E., Mattey, N.L., Plews, A.D., Phillips, P.J., Mironov, O.A., Nickolas, R.J. and Kearney M.J., 1994, Appl. Phys. Lett., **64**, 357.

Inst. Phys. Conf. Ser. No 157
Paper presented at Microsc. Semicond. Mater. Conf., Oxford, 7–10 April 1997
© *1997 IOP Publishing Ltd*

Structural characterisation of MnSb/GaAs and MnSb/Si heterostructures grown by Hot-Wall Epitaxy

H Tatsuoka, P D Brown[1], Y Xin[1], K Isaji, H Kuwabara, Y Nakanishi[2], T Nakamura, H Fujiyasu and C J Humphreys[1]

Faculty of Engineering, Shizuoka University, Hamamatsu 432, Japan.
[1]Department of Materials Science and Metallurgy, University of Cambridge, Pembroke Street, Cambridge CB2 3QZ, UK.
[2]Research Institute of Electronics, Shizuoka University, Hamamatsu 432, Japan.

ABSTRACT: MnSb layers have been grown on GaAs(001), {111}A, {$\bar{1}\bar{1}\bar{1}$}B and Si(111) substrates using hot-wall epitaxy. It is found that the Sb/Mn flux ratio affects the precise orientation of MnSb layers grown on GaAs(001). The crystalline quality of the layers on GaAs{$\bar{1}\bar{1}\bar{1}$}B is superior to that of layers grown on GaAs{111}A. MnSb(0001) layers are formed on Si(111) substrates with a thin interfacial layer of MnSi, rotated 30 degrees with respect to the Si substrate, and a predominant epitaxial relationship given by (0001),[$1\bar{2}10$]MnSb // (111),[$2\bar{1}\bar{1}$]MnSi // (111),[$1\bar{1}0$]Si.

1. INTRODUCTION

Magnetic epitaxial layers grown on semiconductor substrates have generated much recent interest because they offer wide ranging possibilities for the fabrication of new devices (Prinz 1990). For example, MnAs layers have been grown on GaAs and Si substrates by molecular beam epitaxy (Tanaka et al 1994; Akeura et al 1995).

MnSb is a well known ferromagnetic material, and its magnetic and magneto-optical properties have been studied. Epitaxial MnSb is one of the candidates for development within ferromagnetic-semiconductor hybrid structures for applications such as magnetic memories and switches. It has been reported that the magnetic properties of MnSb depend on the atomic composition (Okita and Makino 1986; Seshu Bai and Rama Rao 1984). MnSb adopts the hexagonal NiAs crystal structure in the composition range 50 to 58 at.%Mn (Okita and Makino 1986), and its lattice constants at 50 at.%Mn are a = 0.4128 and c = 0.5789 nm (Singh 1988). Epitaxial MnSb layers have been successfully grown by molecular beam epitaxy (MBE) on GaAs (Akinaga et al 1995), and by hot-wall epitaxy (HWE) on GaAs and Si substrates (Tatsuoka et al 1995, 1996a). The microstructure of the MnSb/GaAs interface has been investigated by transmission electron microscopy (Xin et al 1995). MnSb epitaxial layers can be grown in an Sb-excess environment with the growth rate being determined by the Mn flux. The epilayer structural integrity is of major importance for the application of MnSb within devices. Hence, a series of MnSb/GaAs and MnSb/Si depositions were performed for varying experimental conditions, and the structural properties of the epilayers were characterised using conventional and high resolution TEM combined with X-ray diffraction (XRD) measurements.

2. EXPERIMENTAL

The MnSb layers were grown by HWE on GaAs(001), GaAs{111}A, GaAs{$\bar{1}\bar{1}\bar{1}$}B and Si(111) substrates. Elemental Mn and Sb were used as the source materials, with the Sb and Mn flux being controlled by the source temperatures. The growth procedure has previously been described in detail (Tatsuoka et al 1996a, 1996b). The growth condition dependence of the heterostructures was investigated by conventional XRD, while the fine scale defect microstructure was characterised using a JEOL 4000EX-II. TEM sample foils were prepared using conventional mechanical polishing and argon ion milling procedures.

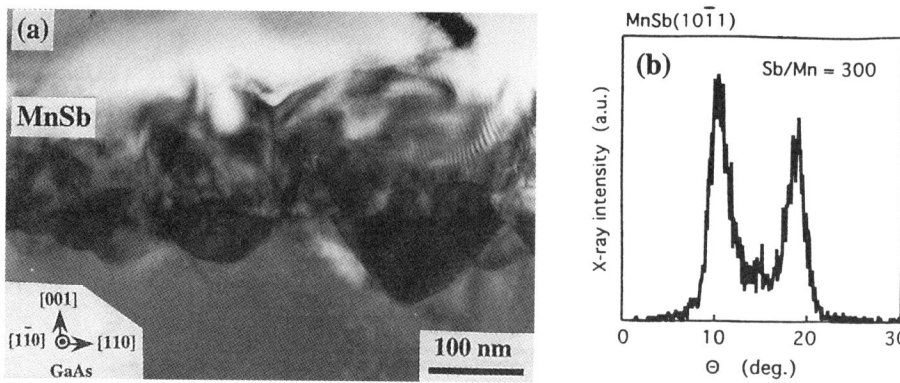

Fig. 1 (a) Cross-sectional TEM image and (b) X-ray rocking curve for a MnSb layer grown under conditions of high Sb/Mn flux ratio.

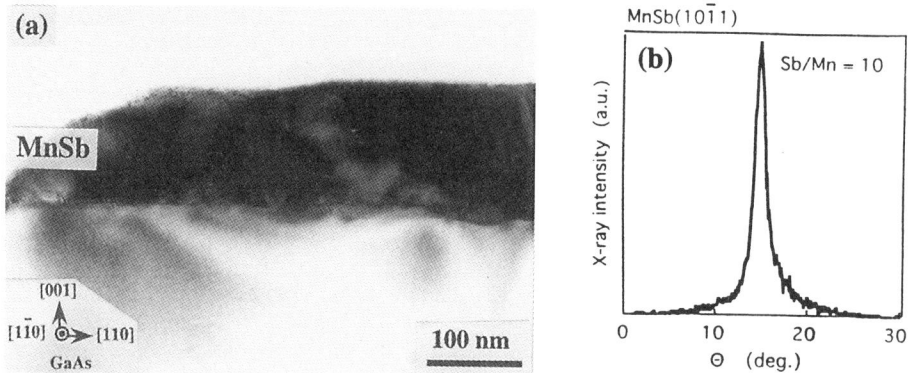

Fig. 2 (a) Cross-sectional TEM image and (b) X-ray rocking curve for a MnSb layer grown under conditions of low Sb/Mn flux ratio.

3. RESULTS AND DISCUSSION

3.1 MnSb($10\bar{1}1$) on GaAs(001)

It is found that the MnSb/GaAs(001) interface structural integrity is strongly influenced by the growth conditions used. While ($10\bar{1}1$)-oriented MnSb layers are formed on GaAs(001) substrates, the precise growth orientation is found to be affected by the Sb/Mn flux ratio. Figs 1(a) and 1(b) show a cross-sectional TEM image and an X-ray rocking curve for MnSb($10\bar{1}1$) diffraction peak of a layer obtained under growth conditions of high Sb/Mn flux ratio. The high flux ratio acts to produce etched grooves within the substrate during the initial stage of growth. MnSb c-planes tend to nucleate on exposed GaAs{111}A facets with the subsequent evolution of the domains in two distinct orientations. Two {$10\bar{1}1$} diffraction peaks, caused by the formation of the two growth variants, are shown in Fig. 1(b). The consequence of this is that MnSb($10\bar{1}1$) planes for each domain variant lay slightly inclined by about 4.5 degrees to the GaAs(001) surface.

Figures 2(a) and (b) show a cross-sectional TEM image and an X-ray rocking curve for MnSb($10\bar{1}1$) diffraction peak of a layer obtained under growth conditions of low Sb/Mn flux ratio. XRD measurements indicate a slight change in growth orientation such that the two {$10\bar{1}1$} diffraction peaks overlap as shown in Fig. 2(b). The corresponding micrograph demonstrates a smooth MnSb/GaAs(001) interface. Thus, with decreasing Sb/Mn flux ratio, domains in which the MnSb($10\bar{1}1$) planes lie parallel to GaAs(001) surface become dominant.

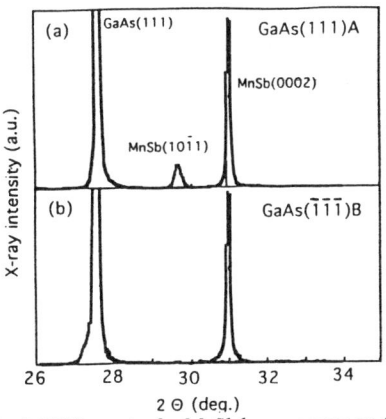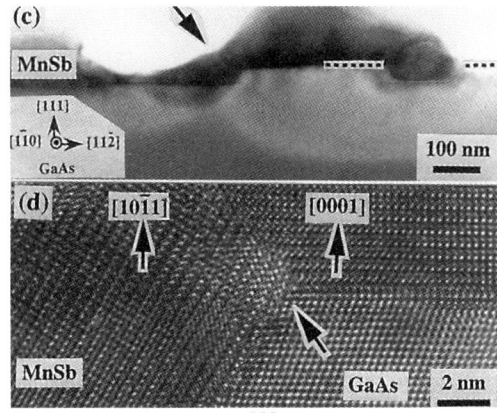

<u>Fig. 3</u> XRD spectra for MnSb layers grown on (a) GaAs{111}A and (b) GaAs{$\bar{1}\bar{1}\bar{1}$}B substrates. (c) and (d) show conventional and high resolution cross-sectional TEM images of the MnSb layer grown on GaAs{111}A. The arrow in (c) and (d) denotes a grain boundary between MnSb(0001) and MnSb(10$\bar{1}$1). The dotted line denotes the original substrate surface.

3.2 MnSb(0001) on GaAs{111}

Improved crystalline quality of MnSb(0001) occurs for growth on GaAs{$\bar{1}\bar{1}\bar{1}$}B rather than GaAs{111}A, as previously reported (Tatsuoka et al 1996b). Figures 3(a) and 3(b) show conventional X-ray spectra for layers grown under flux ratio of Sb/Mn = 10 on GaAs{111}A and {$\bar{1}\bar{1}\bar{1}$}B, respectively. The growth of MnSb(0001) is dominant for both cases. But the formation of MnSb(10$\bar{1}$1) oriented domains is additionally observed for the layer on GaAs{111}A. Figures 3(c) and 3(d) show conventional and high resolution TEM images for the layer grown on GaAs{111}A, respectively. It is evident that the MnSb(10$\bar{1}$1) domains originate from etched regions below the original substrate surface. Even though the etched regions tend to have irregular side walls, they are also characterised by a well defined flat base on {111}A in the same manner as the original growth surface. To explain the observation of preferential (10$\bar{1}$1) domain formation within etched regions of the substrate, consideration must be given both to lattice mismatch and to the chemistry of bonding of MnSb on opposite polar faces of GaAs. For the case of such domain formation on exposed etched facets, the lattice mismatch along the growth direction must be taken into account, i.e. the $d_{111} = 0.3264$ nm spacing of GaAs has a mismatch of 12.7% and 7.3% with the d_{0002} and $d_{10\bar{1}1}$ spacings of MnSb, respectively. Thus, the formation of (10$\bar{1}$1)-oriented material is preferable for growth within etched regions since this minimises the strain energy. Further, the MnSb crystal structure comprises a periodic layered structure of alternate Mn and Sb atoms along the (0001) growth direction. For the case of nucleation on GaAs{$\bar{1}\bar{1}\bar{1}$}B, As atoms on the surface can only bond to Mn atoms, thereby inducing a preferential stacking sequence along the MnSb c-axis. Conversely, for nucleation on the GaAs{111}A surface, Ga atoms can bond to both Mn and Sb atoms. The first atomic layer of MnSb deposited may consist of both Mn and Sb atoms, which when combined with the lattice mismatch considerations favours the formation of MnSb(10$\bar{1}$1) domains within etched regions of the substrate. As a result, the crystalline quality of the layers grown on GaAs{111}A is degraded as compared with that of the layers on GaAs{$\bar{1}\bar{1}\bar{1}$}B.

3.3 MnSb(0001) on Si(111)

A low magnification cross-sectional TEM image of a MnSb layer grown at a substrate temperature of 300° and Sb/Mn flux of 100 is shown in Fig.4(a). The deposit takes the form of discrete islands of typical size 500nm. The HRTEM image of Fig.4(b) reveals a distribution of particles of typical size 10nm of apparent random orientation between the original Si substrate surface and a MnSb grain. The presence of a well aligned uniform layer of 10nm thickness is also apparent extending below the line of the original substrate surface. Nevertheless, the MnSb has a strong epitaxial relationship with the Si substrate. A sequence of selected area diffraction patterns recorded by moving from the MnSb grain through the

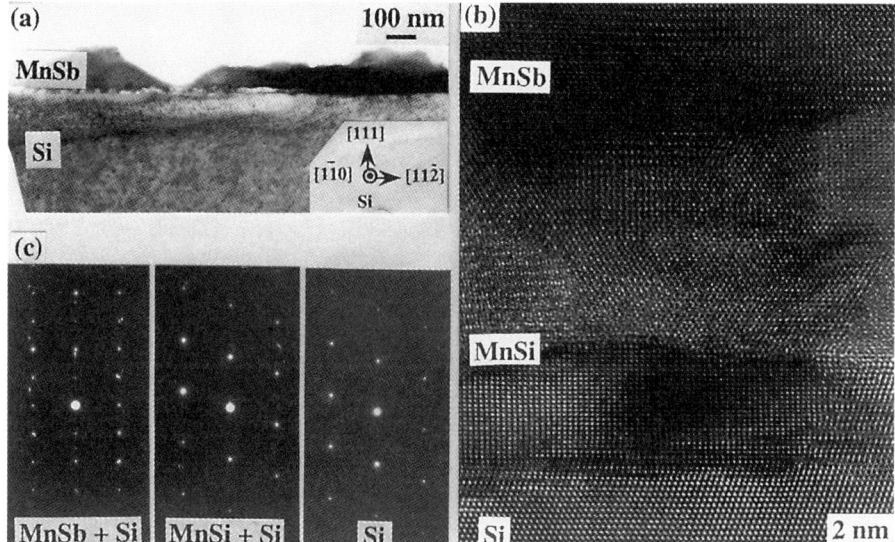

Fig. 4 (a) Conventional and (b) high resolution TEM images of epitaxial MnSb/Si(111) which, when combined with (c) diffraction data, confirmed the presence of MnSi in the hear interface region.

interface region into the Si substrate (e.g. Fig.4(c)) combined with the HRTEM data confirmed the well defined interfacial layer to be MnSi, rotated 30 degrees with respect to the Si substrate. Accordingly, the predominant epitaxial relationship within this sample is $(0001),[1\bar{2}10]$MnSb // $(111),[2\bar{1}\bar{1}]$MnSi // $(111),[1\bar{1}0]$Si. The MnSi crystal structure is cubic with a lattice parameter of 0.456nm. Only evidence for the presence of MnSi was found in the near interface region of this sample, with no evidence for any other MnSi phase.

4. CONCLUSION

The structural properties of MnSb/GaAs and MnSb/Si heterostructures grown using hot-wall epitaxy have been characterised. It is found that the integrity of the interfacial structure of these systems is strongly influenced by the growth conditions used. $(10\bar{1}1)$-oriented MnSb layers are predominantly formed on GaAs(001) substrates, although the precise growth orientation is affected by the Sb/Mn flux ratio. A smooth MnSb/GaAs(001) interface can be obtained under conditions of low Sb/Mn flux ratio, while a high flux ratio acts to produce etched grooves within the substrate leading to the introduction of domains with two distinct orientations. Improved crystalline quality of MnSb(0001) is demonstrated for growth on GaAs$\{\bar{1}\bar{1}\bar{1}\}$B rather than GaAs$\{111\}$A. MnSb(0001) layers have also been grown on Si(111) substrates, although a thin interfacial layer of MnSi is formed.

REFERENCES

Akeura K, Tanaka M, Ueki M and Nishinaga T 1995 Appl. Phys. Lett. **67** 3349
Akinaga H, Tanaka K, Ando K and Katayama T 1995 J. Crystal Growth **150** 1144
Okita T and Makino Y 1986 J. Phys. Soc. Jpn. **25** 120
Prinz G A 1990 Science **250** 1092
Seshu Bai V and Rama Rao K V S 1984 J. Appl. Phys. **55** 2167
Singh P 1988 Mater. Lett. **7** 293
Tanaka M, Harbison J P, Sands T, Cheeks T L, Keramidas V G and Rothberg G M 1994 J. Vac. Sci. Technol. **B12** 1091
Tatsuoka H, Kuwabara H, Oshita M, Nakanishi Y, Nakamura Y and Fujiyasu H 1995 J. Appl. Phys. **77** 2190
Tatsuoka H, Kuwabara H, Oshita M, Nakanishi Y, Fujiyasu H and Nakanishi Y 1996a Thin Solid Films **281/282** 499
Tatsuoka H, Isaji K, Kuwabara H, Nakanishi Y, Nakamura Y and Fujiyasu H 1996b, 8th Int. Conf. on Solid Films and Surfaces, in press.
Xin Y, Brown P D, Boothroyd C B, Humphreys C J, Tatsuoka H, Kuwabara H, Oshita M, Nakamura T, Fujiyasu H and Nakanishi Y 1995 J. Crystal Growth **166** 754

Inst. Phys. Conf. Ser. No 157
Paper presented at Microsc. Semicond. Mater. Conf., Oxford, 7–10 April 1997
© *1997 IOP Publishing Ltd*

Atomistic modelling and HREM-imaging of dislocations associated with steps at Si/Si(001) vicinal interfaces

A Y Belov, D Conrad, K Scheerschmidt and U Gösele

Max Planck Institute of Microstructure Physics, Weinberg 2, D-06120 Halle, Germany

ABSTRACT: Molecular dynamics simulations have been performed to predict stable configurations of interfacial planar and line defects resulting from bonding Si(001) wafers with single and double layer surface steps. A new type of interfacial line defect associated with a rebonded double layer step D_B at a vicinal Si(001) surface is proposed. The defect can be envisaged by placing a line of vacancies in the core of a 60° shuffle set dislocation with subsequent core reconstruction. Single layer steps adhere to a flat (001) surface producing planar defects with a 2×2 reconstruction. HREM images of the strip-like and rectangular planar defects have been simulated to analyze the defect shape influence on the image contrast.

1. INTRODUCTION

There is considerable interest in the atomic and electronic structures of synthetically made Si/Si(001) interfaces, since a (001) silicon wafer remains the most common material used in wafer bonding technology. Direct plan-view observations of interfaces obtained by hydrophobic Si(001) wafer bonding have revealed (Gafiteanu et al 1993, Benamara et al 1994) arrays of screw and 60° dislocations. The former are introduced to compensate the rotational misorientation of wafers. The latter result from the tilt of Si wafers from the (001) orientation and thereby can be directly related to atomic structures of bonded vicinal Si(001) surfaces, in particular, to the structure of surface steps accomodating the miscut angle. The structural disorder in the vicinity of the dislocations effects electrical properties of the interfaces and overall devices, which motivates atomistic modelling of dislocation cores. Previous atomistic studies of dislocations, using a wide range of approaches: from empirical interatomic potentials to ab initio electronic structure calculations, primarily addressed the problem of dislocation mobility (Marklund 1980, Heggie and Jones 1987, Nandedkar and Narayan 1990, Duesbery et al 1991, Bigger et al 1992, Öberg et al 1995). The dislocation core models employed in these studies are directly related with the glide process and stem from the geometrical modelling of Hornstra (1959) and Hirth and Lothe (1982). Here we report atomic structures of line and extended defects related to steps on the Si(001) surface and predicted by classical molecular dynamics (MD) simulations with the empirical many body potential of Tersoff (1989).

2. SIMULATION PROCEDURES

A vicinal silicon surface is known to consist of terrace-like domains of dimer rows and atomic steps (along [110] or/and [$\bar{1}$10]) separating the domains and accommodating the miscut angle. According to Chadi (1987), there is a considerable difference in energy of different types of steps. Among double layer steps, the rebonded step D_B (Fig. 1) has the lowest energy, whereas the absolute energy minimum is achieved in the single layer step S_A, as shown in Fig. 2. These two types of steps are of the main concern in the present study.

64

Using constant-energy-volume MD and rescaling the velocities to remove the kinetic energy, we performed the energy minimization simulation for the dislocation dipoles (along $[\bar{1}10]$), the cores of which reproduce the atomic structures of the steps. The computational cell contains a block of 40 single atomic layers with a dipole in the central plane. Periodic boundary conditions were imposed in the directions $[\bar{1}10]$ and $[110]$. The dimension of the cell in the $[001]$ direction was chosen such as to keep the upper and lower surfaces of the block free. To control the influence of the free surfaces, simulations with periodic boundary conditions in three directions were also performed.

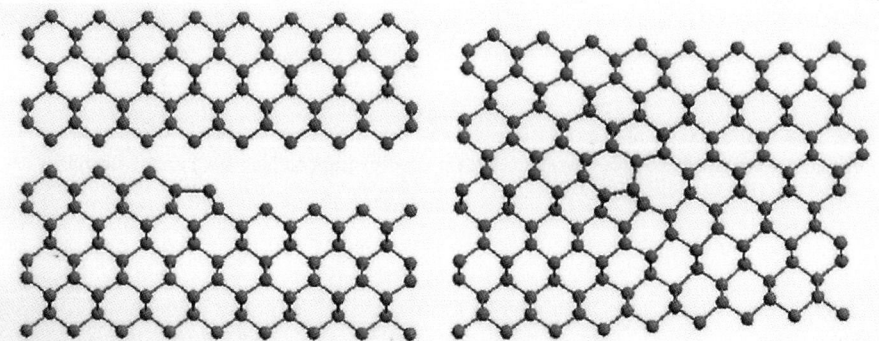

Fig. 1. A step D_B adheres to a surface, forming a dislocation retaining the step structure

2.1 Double layer steps

MD studies of defects associated with a double layer step demonstrated that a new stable reconstruction in the core of a 60° dislocation is possible. It can be envisaged by placing a line of vacancies in the core of a 60° shuffle set dislocation with a subsequent reconstruction to eliminate dangling bonds. There exist two ways to place the line of vacancies avoiding dangling bonds, one of which merely transforms shuffle set dislocations into glide set ones (Hirth and Lothe 1982) and gives rise to the undissociated glide set dislocation according to Amelinckx (1982). The other one, shown in Fig. 1, retains the atomic structure of the rebonded double layer step D_B. Like the glide set dislocation, the defect obtained contains five- and seven-membered rings in the core, however the bond arrangement is different in these two cases. MD simulations at temperatures $T = 300K$ and $1200K$ have shown that this atomic configuration is stable and does not experience a transformation into the configuration corresponding to the glide set dislocation core provided that the potential of Tersoff is used.

2.2 Single layer steps

Interaction of a single layer step S_A with a flat domain on the (001) surface, as illustrated in Fig. 2, inevitably entails a planar defect (either interstitial or vacancy type). Such a planar

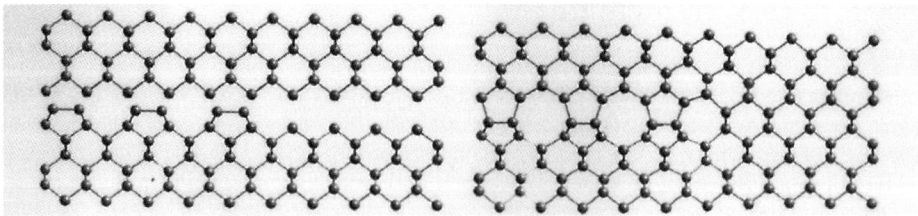

Fig. 2. A step S_A adheres to a surface, producing a planar defect with 2×2 reconstruction

defect can occur in Si (space symmetry group $Fd\bar{3}m$) due to the presence of a four-fold screw axis 4_1. Owing to this symmetry element, when a single atomic layer normal to the screw axis is removed, the two half-crystals adjacent to the layer turn out to be rotated by 90° about it. The subsequent joining of the two half-crystals by a translation towards each other creates a planar defect which is equivalent to the (001) 90° twist grain boundary. It can arise from a 90° rotation of one part of a crystal or from the aggregation of point defects (vacancies or interstitial atoms), that is without real rotation. MD examinations of the twist boundary revealed that energy minimum ($E = 0.85\ Jm^{-2}$) is achieved when its core is 2×2 reconstructed and consists of special structural units of twist (dreidls with the $\bar{4}2m$ point group symmetry) which retain fourfold coordinaton of all atoms (Belov et al 1997). The structural units fit two rotated half

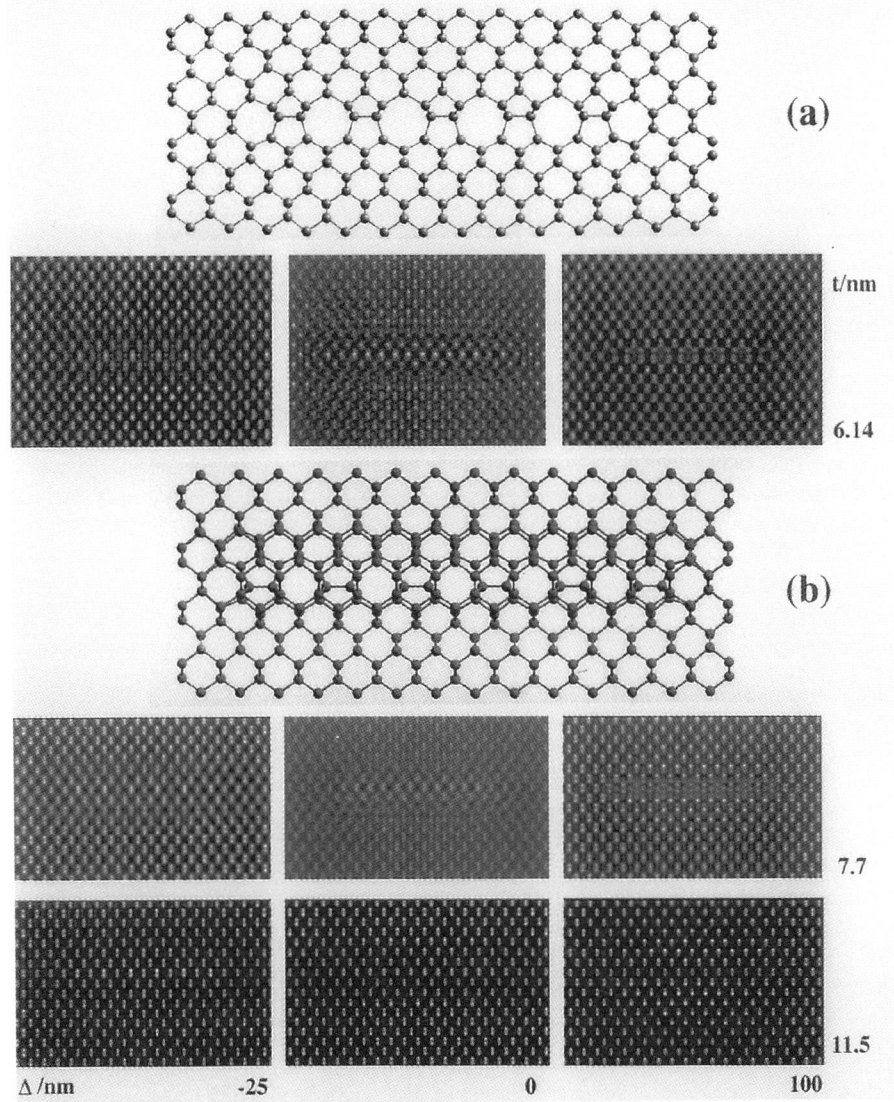

Fig. 3. Simulated HREM images for 2D (a) and 3D (b) interstitial defects

crystals with a minimal structural disorder. The bonds are stretched by -1.4% to 3.8% of the ideal bond length and bond angle distortions range from $-12.4°$ to $15.9°$.

MD studies of the twist boundary allows for the Burgers vector of partial dislocations bounding the planar defect to be evaluated. As has been shown by Belov et al (1997), the free volume (per unit area) associated with the twist boundary is equal to $t = 0.64$ Å. For the planar defect of the interstitial type the Burgers vector has only one component normal to the defect plane, which is calculated as $b_{[001]} = a/4 + t = 2.00$ Å, where a is the lattice parameter of Si. This value is in agreement with the experimental data of Muto and Takeda (1995) for self-interstitial precipitates with the same structure in Ge. In turn, for the vacancy type defect the Burgers vector has the component $b_{[001]} = a/4 - t = 0.72$ Å and, besides, the component $b_{[100]} = a/2$ in the defect plane.

3. RESULTS AND DISCUSSION

Fig. 3 shows the structure and the resulting HREM contrast of the (001) planar defects obtained by the energy minimization simulations with the potential of Tersoff. The defects have the form of a strip (Fig. 3a) and a rectangle (Fig. 3b), in the three-dimensional case the defect being also incorporated into the perfect lattice without forming dangling bonds. The imaging parameters assumed (electron beam energy $E = 200$ kV, spherical aberration $C_s = 1.2$ mm, the defocus value Δ (nm), and the specimen thickness t (nm)) yield the contrast variation with the change in the defect shape and defocus value. The effect of the 2×2 reconstruction on the image contrast (an additional double contrast periodicity with respect to the perfect lattice contrast) turns out to be very sensitive to the defect geometry. In the three-dimensional case (Fig. 3b) this effect is vanishing. However it can be enhanced by choosing the appropriate defocus value.

REFERENCES

Amelinckx S 1982 Dislocations in Crystals, ed F R N Nabarro (Amsterdam: North-Holland), Vol. 2, pp 67-460
Belov A Y, Conrad D, Scheerschmidt K and Gösele U 1997 (in press)
Benamara M, Rocher A, Laânab L, Claverie A, Laporte A, Sarrabayrousse G, Lescouzères L and Peyre-Lavigne A 1994 C.R. Acad. Sci. Paris **318**, Série II, 1459
Bigger J R K, McInnes D A, Sutton A P, Payne M C, Stich I, King-Smith R D, Bird D M and Clarke L J 1992 Phys. Rev. Lett. **69**, 2224
Duesbery M S, Joos B and Michel D J 1991 Phys. Rev. B **43**, 5143
Gafiteanu R, Chevacharoenkul S, Gösele U and Tan T Y 1993 Microscopy of Semiconducting Materials, eds A G Cullis, A E Staton-Bevan and J L Hutchinson (Bristol and Philadelphia: Institute of Physics) pp 87-90
Heggie M I and Jones R 1987 Microscopy of Semiconducting Materials, eds A G Cullis and P D Augustus (Bristol: Institute of Physics) pp 367-74
Hirth J P and Lothe J 1982 Theory of Dislocations (New York, McGraw-Hill)
Hornstra J 1959 Physica **25**, 409
Marklund S 1980 Phys. Stat. Sol. B **100**, 77
Muto S and Takeda S 1995 Phil. Mag. Lett. **72**, 99
Nandedkar A S and Narajan J 1990 Phil. Mag. A **61**, 873
Öberg S, Sitch P K, Jones R and Heggie M I Phys. Rev B 1995 **51**, 13138
Tersoff J 1989 Phys. Rev. B **39**, 5586

Inst. Phys. Conf. Ser. No 157
Paper presented at Microsc. Semicond. Mater. Conf., Oxford, 7–10 April 1997
© *1997 IOP Publishing Ltd*

The characterisation of ultrathin doping layers in semiconductors using high-angle annular dark-field imaging

C P Liu, C B Boothroyd, P D Brown and C J Humphreys

Department of Materials Science and Metallurgy, Pembroke St, Cambridge, CB2 3QZ

ABSTRACT: High-angle annular dark-field imaging in a scanning transmission electron microscope has been used extensively for providing atomic resolution, atomic number dependent images, but still little work has been done on obtaining quantitative information using this technique. Here, we consider a very simple method for characterising ultrathin doping layers in semiconductors and assess the validity of this method by comparing the results from this technique with the results from high resolution and Fresnel methods.

1. INTRODUCTION

It is widely believed that annular dark field images show highly localised scattering from each atomic column so in principle they can provide quantitative chemical information. Pennycook et al (1990) thought the annular dark field detector can be considered as a Bloch state filter, with only the electrons close to the atom sites (s-state) contributing to the high angle scattering, so the ADF images can be visualised as a consequence of incoherent Rutherford scattering independent of neighbouring atoms. If it is true, the ADF image intensity can be simply thought as a convolution of the probe intensity profile with an array of delta functions located on the atoms, whose strength is dependent on the scattering cross section of each atom, which is proportional to Z^2. However, a probe localised on a single atom column still interacts with neighbouring atoms, and hence produces a diffraction pattern from the crystal. In addition, the high angle scattering of incident electrons by a crystal is mainly due to phonon scattering (thermal diffuse scattering) which tends to Rutherford scattering (i.e. Z^2 scattering) at high angles. Further, strain in the crystal causes static displacements of atoms which can scatter to high angles, but the strain scattering does not vary as Z^2. The quantitative interpretation of HAADF images is therefore not always straightforward. Electron channelling effects must also be considered (Treacy et al, 1988). The aim of this paper is to see how ADF images can be used as a quantitative tool for the characterisation of ultrathin doping layers in semiconductors by comparing simple simulations with the results from this technique and others such as the Fresnel technique and high resolution imaging (Liu et al, 1997).

2. EXPERIMENTAL RESULTS

The material examined here is (001) InP grown in the presence of As_2 whose growth was interrupted for periods of 1, 2, 4, 8, 16 and 32s (for details see Liu et al, 1997). A VG HB501 scanning transmission electron microscope was used to record bright field and high angle annular dark field images of this material as shown in Fig.1. The inner collector angle for the annular dark field detector was 100mrad which is large enough to eliminate most higher order Laue zone diffraction (approximately 80mrad for the second order Laue zone diffraction ring for InP). From qualitative inspection of the HAADF images it can be seen that the layers are somewhat wavy. In order to measure the layer width of this kind of ultrathin layer from HAADF images, great care has been taken to ensure that the layers are

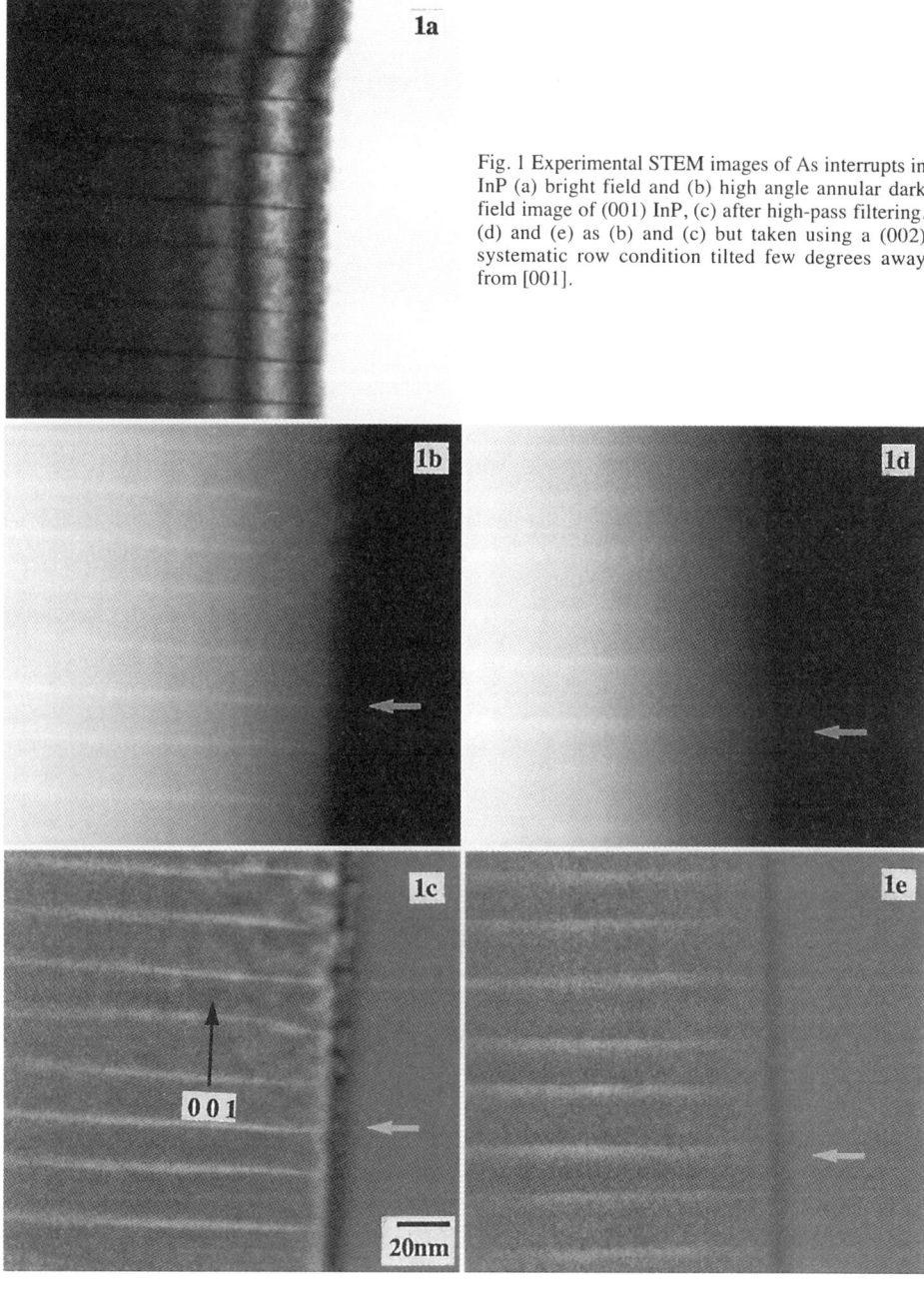

Fig. 1 Experimental STEM images of As interrupts in InP (a) bright field and (b) high angle annular dark field image of (001) InP, (c) after high-pass filtering, (d) and (e) as (b) and (c) but taken using a (002) systematic row condition tilted few degrees away from [001].

Table 1: Intensity ratio (r) calculated for 3 different layers from Fig.1(b) and Fig.1(d)

layer(interrupt time)	32sec	16sec	8sec
intensity ratio (r)	0.023 ± 0.005	0.0145 ± 0.005	0.0097 ± 0.005

both exactly edge-on and the microscope is very close to focus. The layer width was measured to be 2.1±0.2nm for the layer with the highest As concentration, arrowed in Fig.1(b). This is slightly wider than the layer width of 1.8nm measured using the Fresnel technique (Liu et al, 1997). The layer width measured from high angle dark field is expected to be broadened by the probe size.

From the appearance of Fig.1(b) and Fig.1(d), it is clear that the contrast level is much higher when the beam is on the zone axis (Fig.1(b)) than when tilted a few degrees off (Fig.1(d)), indicating that the HAADF image is very sensitive to channelling. To investigate the contrast quantitatively, we have extracted an area covering the thickness range from 10 to 40nm from Fig. 1d (the weakly diffracting case, and thus the most easily quantified) and projected along the layers to give the line trace. The intensity ratio (r) of the layer is calculated as the area under the layer (A in Fig. 2(c)) divided by the area under the InP for a width of 2nm (B in Fig. 2(c)) and is shown in table 1. This intensity ratio is proportional to the amount of As in the layer and is independent of how much the layer is spread due to the probe size.

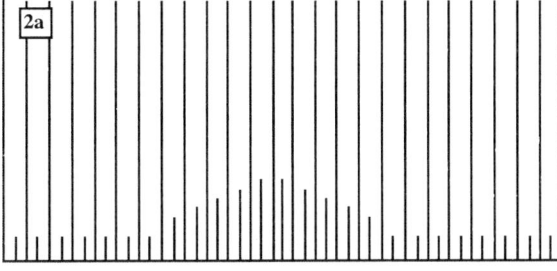

InP monolayer number

3. SIMULATIONS

The simulations were based on the simple convolution of the probe shape with atomic positions, in which the scattering cross section of each atom, shown in Fig. 2(a), was assumed to be proportional to Z^2. The total length of the model was 25 InP monolayers while the layer width of 1.8 nm and the dopant distribution were taken from the results of Fresnel analysis (Liu et al, 1997). This simple convolution will only be valid for thin specimens at a thickness where absorption is negligible and for a large inner collector angle. The tetragonal distortion due to the lattice mismatch is also crucial for ADF images since it will vary the scattering power

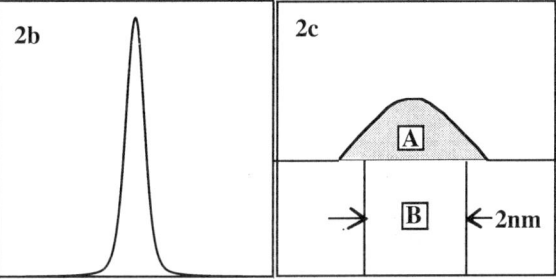

Fig. 2 (a) Atomistic model of layer constructed for the specimen examined, (b) STEM probe intensity profile (c) model convoluted with probe and include definition of r = area A/area B.

locally and so it has also been taken into account in this model using conventional elasticity theory. However, the elastic diffuse scattering has been neglected (see later). The incident probe intensity profile is shown in Fig.2(b) for an underfocus of 100nm, Cs of 3.1mm, an accelerating voltage of 100 keV and assuming a realistic broadening due to instabilities. This probe shape has about the right amount of broadening to broaden the layers of width

Table 2: Intensity ratio (r) calculated for various probe profiles and dopant distribution models. All models contain the same amount of As.

probe profile width (nm)	4.83	6.16	4.83	4.83
dopants distribution model	spread diffusely over 10 monolayers	spread diffusely over 10 monolayers	abrupt-4 monolayers of As	spread uniformly over 8 monolayers with 50% As for each monolayer
intensity ratio r	0.077	0.075	0.079	0.075

1.8nm obtained from Fresnel analysis to the width observed in the HAADF images.

The effect of different As dopant profiles and different probe shapes on the intensity ratio is shown in table 2. Here we see that the intensity ratio, r, is fairly constant as both the dopant profiles and probe shapes are changed, indicating that it is a good measure of the total amount of As present in the layer regardless of its distribution or the microscope conditions. Accordingly, the intensity ratio r was calculated as a function of As concentration for the model where the As is distributed evenly over 8 monolayers and is shown in Figure. 3. Not surprisingly, r increases linearly with As content (here measured as the number of monolayers it would occupy if all concentrated together). From Fig. 3, it can be seen that the ratio for the 32s layer (0.023) corresponds to an As content of 1.2 monolayers, which is much smaller than the value of 3.2 to 3.8 monolayers found earlier by the Fresnel method. This suggests the assumption that scattering is proportional to Z^2 is incorrect. Pennycook(1989) found experimentally that the Born approximation fails in this high angle regime and also pointed out that no reliable formula so far describes the cross section very well. He used the following formula which is close to experimental data for cross sections and corrects for the screening effect and multiple scattering

$$\sigma_x \propto Z_x(\frac{\theta_2 - \theta_1}{\theta_1 \theta_2 \theta_a'}) + \frac{1}{\theta_a'} \ln \frac{\theta_1(\theta_2 + \theta_a')}{\theta_2(\theta_1 + \theta_a')} \text{ where } \theta_a' = e\theta_0^2(1.13 + 3.76\alpha^2), \ \alpha = \frac{Z}{137\beta},$$

$\theta_0 = 1.13 \frac{Z^{1/3}}{137\beta}$, $\beta = \frac{v}{c}$ and θ_1 and θ_2 are the inner and outer detector angles. This cross-section gives the lower line in Fig. 3, and from this we find the 32s layer contains 2.3±0.5 monolayers of As. This compares with 3.2 to 3.8 monolayers from Fresnel method and 2.8 monolayers from high resolution (Liu et al, 1997).

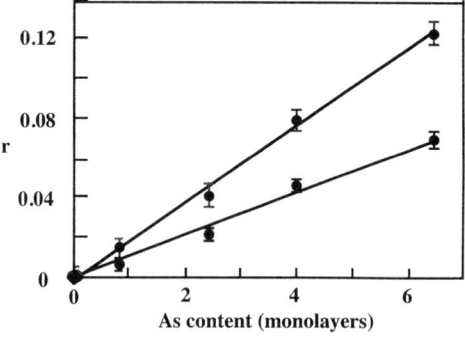

Fig. 3 Intensity ratio (r) plotted for scattering proportional to Z^2 (top line) and scattering from Pennycook's formula (bottom line), as a function of the As content of the layer.

4. CONCLUSION

We find that the width of the 32s As interrupt layer measured using high angle annular dark field (2.1±0.2nm) is consistent with that measured using Fresnel method (1.8nm) when the expected broadening due to the probe is taken into account. The amount of As measured from high angle annular dark field is lower than that measured using either high resolution or Fresnel method, even after an improved model is used for the high angle scattering cross sections. This suggests that a more complete treatment of scattering is needed before the As concentrations can be measured absolutely from high angle annular dark field. In particular, high angle thermal diffuse scattering due to phonons and high angle elastic diffuse scattering due to strain needs to be taken into account.

Acknowledgements

With thanks to Peter Skevington and Graham Davies of BT Labs., Martlesham Heath for provision of the δ-doped sample studied.

REFERENCES

Liu C P, Dunin-Borkowski R E, Boothroyd C B, Brown P D and Humphreys C J 1997 to be published in JMSA
Pennycook S J 1989 Ultramicroscopy **30**, 58
Pennycook S J,. Jesson D E and Chisholm M F 1990 SPIE **1284**, 182
Treacy M M J, Gibson J M, Short K T and Rice S B 1988 Ultramicroscopy, **26**, 132

Inst. Phys. Conf. Ser. No 157
Paper presented at Microsc. Semicond. Mater. Conf., Oxford, 7–10 April 1997
© *1997 IOP Publishing Ltd*

Electron diffraction from cross-sectional semiconductor heterointerfaces using subnanometer electron probes

A Radefeld and H Lakner

Werkstoffe der Elektrotechnik, Gerhard-Mercator-Universität
Duisburg, Bismarckstraße 81, 47048 Duisburg, Germany

ABSTRACT: We investigated cross-sectional heterointerfaces by electron diffraction using subnanometer electron probes. For interfaces in different material systems we observed unexpected asymmetries in diffraction patterns which are not present in patterns recorded from off-interface locations. With the assumption of ideally abrupt interfaces multislice simulations were made under consideration of thermal diffuse scattering (TDS) based on a frozen phonon model. With the multislice model and TDS it is possible to simulate the observed asymmetries. We compare simulated and experimental Convergent Beam Electron Diffraction (CBED) patterns.

1. INTRODUCTION AND EXPERIMENT

Epitaxially grown heterostructures of the ternary and quaternary material system (Al,Ga)As and (In,Ga)(P,As) on InP- or GaAs-substrate and the novel material system (In,Al,Ga)N/GaN are increasingly used for the fabrication of e.g. optoelectronic devices. One key parameter for the performance of such devices is the crystalline structure and especially the quality of the internal interfaces. Spatially resolved CBED (Convergent Beam Electron Diffraction) allows e.g. strain mapping across heterointerfaces. But the influence of the heterointerface on the pattern itself in cross-section orientation is not well known. Lakner et al (1996) has shown the existence of asymmetries in measured CBED-pattern caused by the heterointerface. Ungerechts et al (1996) has compared measurements from heterointerfaces qualitatively with simulated patterns and has demonstrated the existence of asymmetries in the simulated patterns.

We investigated cross-sectional heterointerfaces by electron diffraction using subnanometer electron probes. The experiments were performed using a field-emission STEM (VG HB 501) which is equipped with a high dynamic range CCD camera for CBED detection. At 100 keV the minimum probe size is less than 0.3nm. Additionally, coherent CBED patterns can be recorded because the field emission gun is a coherent source.

2. MULTISLICE-SIMULATIONS AND THERMAL DIFFUSE SCATTERING (TDS)

For simulations of CBED-patterns from interfaces Ungerechts et al (1996) used the elastic multislice approach which has been presented: Kirkland et al (1987) and Loane et al (1988). The consideration of inelastic effects for the simulation leads to a more quantitative interpretation of CBED-pattern. One major contribution to the inelastic diffraction is the TDS. The consideration of TDS in multislice simulations produces automatically a thermal diffuse background and Kikuchi bands. For the simulation results in this publication the modified frozen phonon approach of Loane et al (1991) with slight changes for the simulations of heterostructures has been used. Each simulated CBED-pattern with TDS is an average of patterns obtained from different discrete phonon configurations. Figure 1 shows the influence of TDS on CBED-patterns. Figure 1a shows the result of purely elastic multislice simulations. Figure 1b shows the average result of two different phonon configurations. The pattern appears more diffuse and a lot of Bragg disks disappear in comparison to Figure 1a. On the other hand weak Kikuchi bands begin to appear. Figure 1c shows the average result of 16 different phonon configurations. In Figure 1c the Kikuchi bands are clearly visible. To use more phonon configurations (e.g. 64) is very time consumptive and further improvements in the quality of the obtained patterns can be expected, but the profit is little.

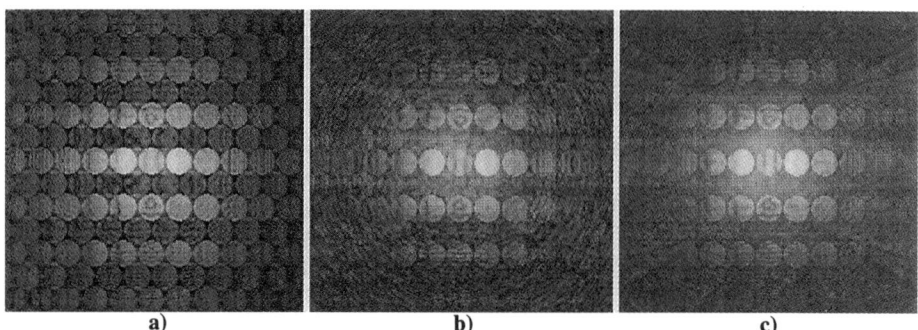

a)   b)   c)

Figure 1: CBED-Simulations for GaN in {120}-orientation and a thickness of 33nm, a) without TDS, b) with TDS (2 frozen phonon configurations), c) with TDS (16 frozen phonon configurations)

The model described here was used further on to simulate CBED-patterns from interface as well as from off-interface locations of the electron probe. Standard simulation parameters are: E=100 keV, C_S=1.3mm, Δf=-70nm.

3. ASYMMETRIES IN EXPERIMENTAL CBED-PATTERNS

As an example we observed unexpected asymmetries in experimental CBED-patterns from the interface of InGaN/GaN heterostructures. The experimental patterns and a Z-contrast image of the heterostructure are shown in Figure 2 which shows the central Bragg disks of the measured CBED-patterns from the InGaN/GaN heterostructure.

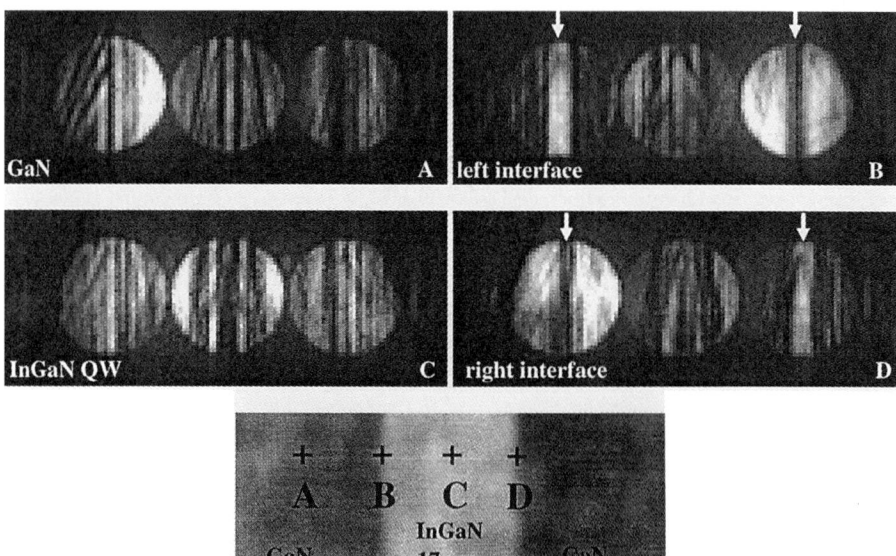

Figure 2: Central Bragg disks and Z-contrast image of the InGaN/GaN quantum well. The orientation of the specimen is 7° tilted of the {210} pole along the {200}-Kikuchi band.

In Figure 2 two types of asymmetries are visible. The first type is related to the non centro-symmetrical crystal structure of the InGaN/GaN material system and is visible in all CBED-patterns of Figure 2. In the CBED-patterns from the interface (Figure 2(B,D)) additional asymmetries (see arrows) along the {200}-Kikuchi band are visible. Also we have observed asymmetries in Bragg disks, Kikuchi lines and HOLZ (High Order Laue Zone) lines in other material systems like AlGaAs/GaAs, InGaAs/InP and InGaAs/InAlAs. These observations indicate that both, elastic and inelastic scattering processes contribute to the asymmetries. We conclude that for simulations inelastic contributions must be considered.

4. COMPARISON OF SIMULATED AND EXPERIMENTAL CBED PATTERNS

For the InGaAs/InP heterostructure inelastic multislice simulations were performed and the results are compared with experimental data in Figure 3. A good agreement exists between simulated and experimental CBED-patterns. The asymmetry in the {200} Bragg disc intensities in the interface patterns is clearly visible (see arrows). Much stronger asymmetries can be observed for the case that the specimen is tilt off a low index pole orientation.

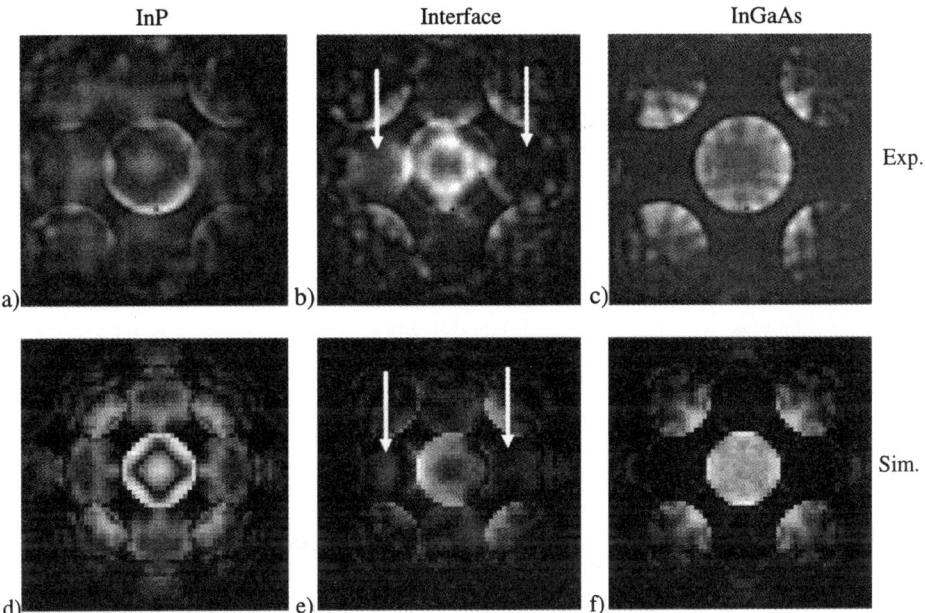

Figure 3: (a-c) are experimental and (d-e) are simulated CBED-pattern for InGaAs/InP in {100}-orientation with a thickness of 82nm

As an example Figure 4 e.g. shows CBED patterns from an AlGaAs/GaAs heterostructure near the {320}-orientation. The left (a,d) and the right (c,f) patterns are recorded from an AlGaAs/GaAs interface and show asymmetries (see white arrows) in the intensities near the {200}-Kikuchi lines. The fit between experimental and simulated data must still be improved. But, as a first step the simulations demonstrate, that the experimentally observed asymmetries are present in the simulations in principle. For the example shown here, the Bragg discs overlap and therefore coherence effects can be expected.

74

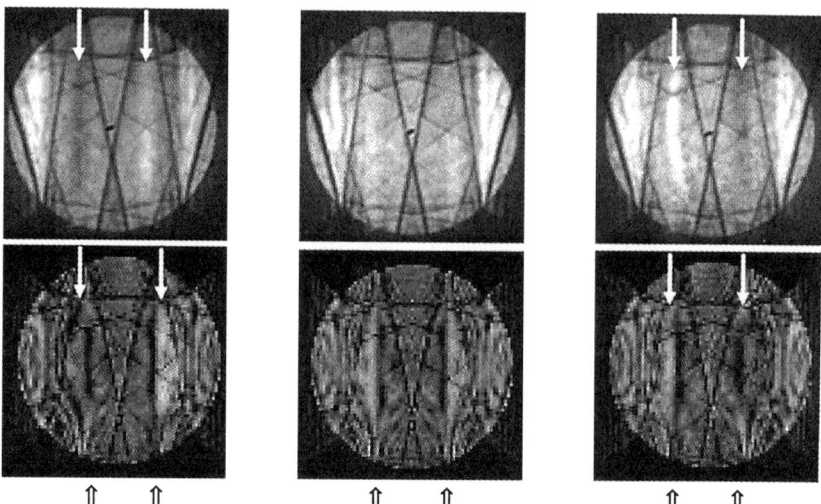

⇑ ⇑   ⇑ ⇑   ⇑ ⇑

Figure 4: (a-c) are measured and (d-e) are (with multislice and TDS) simulated CBED-pattern with an thickness of 120nm from an AlGaAs/GaAs heterostructure near the {320}-orientation. The probe position was in a/d) at the GaAs/AlGaAs interface, in b/e) in AlGaAs and in c/f) at the AlGaAs/GaAs interface. The ⇑ shows the position of the {200}-Kikuchi lines.

5. DISCUSSION

The presented results of experimental and simulated CBED-patterns from heterointerfaces show asymmetries in the investigated material systems. The use of multislice simulations allows simulation of the observed asymmetries in Bragg disc and HOLZ line intensities in principle. But there is still a lot of work to be done in order to improve the quality of fit. Especially the impact of the defocus on the asymmetries -which is present in experiments and simulations- will be studied in detail in the future. There are two possible explanations for the existance of the observed asymmetries: They may be caused either by refraction due to a change of inner potential in the crystal at the interface or by interface channeling effects. Additionally, more realistic models for non-perfect interfaces should be used as well. Nevertheless, we have demonstrated that CBED from interfaces delivers additional information, which may be utilised for interface characterisation in the near future.

REFERENCES and ACKNOWLEDGEMENTS

Kirkland E J, Loane R F and Slicox J 1987, Ultramicroscopy **23**, 77
Lakner H, Bollig B, Ungerechts S and Kubalek E, J. Phys. D: Appl. Phys. (1996) **29**, 1767
Loane R F, Kirkland E J and Silcox J, Acta Cryst. 1988 **A44**, 912
Loane R F, Xu P and Silcox J, Acta Cryst. 1991 **A47**, 267
Ungerechts S and Lakner H 1996, Proceedings of the 11th European Congress on Electron Microscopy (EUREM), Dublin (Ireland), 26-30. August 1996 (CD-ROM, vol 1, published by: EUREM 96, UCD, Belfield, Dublin 4, Ireland)

The software for the multislice simulations was developed at the School of Applied and Engineering Physics of the Cornell University (J. Silcox and E.J. Kirkland). This work was financially supported by the Deutsche Forschungsgemeinschaft.

Inst. Phys. Conf. Ser. No 157
Paper presented at Microsc. Semicond. Mater. Conf., Oxford, 7–10 April 1997
© 1997 IOP Publishing Ltd

The use of electron holography for composition profiling of semiconductor heterostructures

P A Midgley*, J Barnard and D Cherns

H H Wills Physics Laboratory, University of Bristol, Tyndall Avenue, Bristol, BS8 1TL.
*Now at: Department of Materials Science and Metallurgy, University of Cambridge, Pembroke Street, Cambridge, CB2 3QZ.

ABSTRACT: The possibility of applying off-axis electron holography in a FEGTEM to determine the composition profile of semiconductor heterostructures has been investigated. This paper reviews how to record a hologram, its subsequent processing and phase reconstruction. It is shown how a phase change can be related to a change in the mean inner potential and thus to a change in composition. An analysis of a $Si_{1-x}Ge_x$ quantum dot structure is given as an example of how quantitative composition profiles can be obtained at nanometre resolution.

1. INTRODUCTION

A number of micro-analytical electron microscopy techniques have been developed in recent years in order to study the composition near interfaces and boundaries in a variety of materials. The requirement for a reliable quantitative technique is driven largely by the need to determine the composition of semiconductor heterostructures at the nanometre scale in order to assess the likely performance characteristics of the device. In the STEM, high spatial resolution X-ray microanalysis and high angle annular dark field (HAADF) imaging coupled with point-by-point energy loss spectroscopy have been developed for this purpose and are powerful techniques. In the TEM, two techniques have emerged which look promising: core-loss imaging and electron holography. For these techniques to be truly quantitative, a number of systematic errors must be minimised and great care is needed in interpreting the final result. This paper concentrates on the last technique, electron holography, and describes how it can be used in a quantitative fashion to determine the composition of heterostructures with examples of results taken from typical semiconductor device structures.

2. OFF-AXIS ELECTRON HOLOGRAPHY

2.1 Theory

Although electron holography was first proposed nearly 50 years ago (Gabor 1949) initially as a means of improving the resolution of the TEM, it is only in the last few years with the introduction of field emission guns on commercial instruments that holography has been applied to a variety of materials science problems. The essence of holography is that it provides a means of converting phase information into an amplitude (and thus intensity) which can be recorded on a negative or a CCD camera. A number of variations of holography exist (Cowley 1992) both for the STEM and the TEM but here we will concentrate on the most widely-used, 'off-axis holography', as applied in the TEM. In this technique, an electron biprism, typically a fine (0.3μm diameter) quartz fibre coated in gold, is positively (or sometimes negatively) biased so that the electron wavefront is deflected to a form an interference region, see Fig. 1. If the electron source is coherent, as in a FEG microscope, this interference region will be filled with sinusoidal fringes whose origin can be traced back to two virtual sources in the back focal plane of the objective lens. As the biprism voltage is increased the virtual sources separate further, the interference region widens and the fringe spacing narrows. As the voltage increases, the fringe contrast also diminishes because the separation of the two sources becomes comparable with the spatial coherence length of the electron wavefront. If a sample is placed in the way of part of this wavefront then the electro-magnetic potential of that sample will alter the amplitude and phase of the wavefront which is seen as a reduction in the fringe intensity and a change in the fringe spacing relative to the vacuum, respectively. In the absence of magnetic fields, for an electron beam travelling parallel to the z-direction, the phase of the wavefront $\phi(\mathbf{R})$ is

$$\phi(\mathbf{R}) = C_E \int V(\mathbf{R},z) \, dz \qquad (1)$$

where C_E is a wavelength-dependent constant (equal to $0.731 \text{Å}^{-1} \text{keV}^{-1}$ for 200kV electrons). Thus the phase is sensitive to the electrostatic potential, $V(\mathbf{R},z)$, projected along the beam direction and away from strong diffraction conditions, and in the absence of any net electric fields (such as in ferroelectrics and at p-n junctions), the phase is given simply by

$$\phi(\mathbf{R}) = C_E V_0 t \qquad (2)$$

where V_0 is the mean inner potential of the sample and t is the thickness of the sample traversed by the beam. V_0 is essentially a volume-averaged atomic potential and is sensitive to bonding, structure and composition. Thus if the first two remain constant, any change in composition will be reflected in a change of V_0 and thus a change in the phase, ϕ, which can be measured from the hologram.

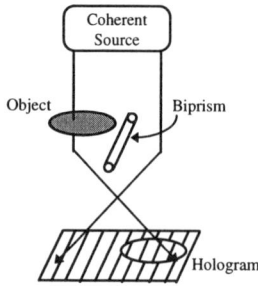

Fig. 1 Schematic representation of the formation of an off-axis electron hologram.

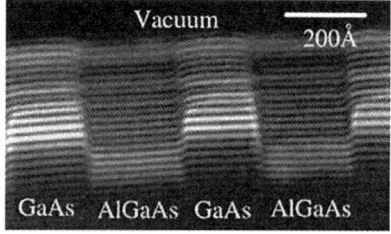

Fig. 2 Hologram taken at the edge of a cleaved AlGaAs/GaAs superlattice. The fringes are deliberately coarse to emphasise any displacement between the two regions and the vacuum.

2.2 Recording a Hologram

For off-axis holography, spatial coherence is needed only in a direction perpendicular to the biprism axis. Thus it is common to illuminate the sample with a highly astigmatic beam to increase the electron flux (to reduce exposure time and minimise drift) but preserve beam coherence in one direction. Fig. 2 shows a hologram taken at the edge of a cleaved AlGaAs/GaAs superlattice. The change in mean inner potential by the introduction of the Al can be seen by the displacement of the fringes. In this example part of the wavefront has passed through vacuum (constant phase) and thus the phase change measured is *absolute*. Where the region of interest is well away from an edge, it is often the case that the whole of the wavefront passes through the sample and then only a *relative* phase change is measured. For the latter in particular, energy filtering, to allow only elastically scattered electrons to contribute to the hologram, can improve the contrast of the fringes and thus improve the phase resolution.

2.3 Phase Reconstruction

Whilst there are a number of different approaches to phase reconstruction, using optical benches for real-time processing (Chen et al 1995) or alternative numerical methods (Lehmann et al 1994), by far the most popular is the Fourier transform method outlined in Fig. 3. A fast Fourier transform (FFT) of the hologram shows two sidebands corresponding to the original fringe frequency. A change in potential will give rise to a change in fringe spacing and this will displace the sideband in Fourier space. By masking a circular region around the side band and back-transforming a complex image is recovered from which a phase image can be determined. The size of the mask dictates the spatial resolution in the final image but its maximum radius is limited to about $^1/_3$ of the sideband frequency to avoid artefacts arising from the inclusion of part of the central self-correlation peak.

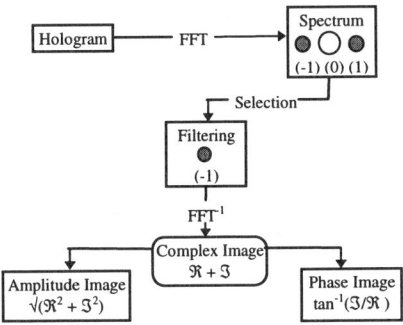

Fig. 3 The reconstruction process.

Fig. 4 Phase Image showing the change in phase across the CCD brought about primarily by 'pin-cushion' distortion in the GIF.

2.4 Phase Corrections

The first correction made to all the images obtained on the CCD is to deconvolute the point spread function (psf) which had been measured previously (Saunders et al 1996). Then, and by far the most important correction needed for the system used in Bristol, is to account for the geometric 'pin-cushion' distortion inherent in the Gatan Imaging Filter (GIF) used to record the holograms. This correction is achieved by recording a 'blank' or 'reference' hologram taken under identical conditions as the 'object' hologram but with the sample removed from the field of view. The two holograms are processed identically and the complex object image divided by the reference to correct for any unwanted phase distortions. Fig. 4 shows the phase image reconstructed from a reference hologram and reveals the effects of pin-cushion distortion at the corners. The sharp black-white jumps correspond to a phase change of 2π and are a consequence of the arctan function. For very high resolution (fringe spacing $\sim$ 30 pm) the phase changes introduced by the objective lens are important. However, for these studies at medium resolution, for images taken at or near focus the contrast transfer function (ctf) of the objective lens is relatively flat for the frequencies of interest and no contrast reversals are encountered.

2. 5 An Illustrative Example: $Si_{1-x}Ge_x$ Quantum Dots

To illustrate the potential of the holography technique, a quantum dot structure was studied which was composed of $Si_{1-x}Ge_x$ islands grown on a Si substrate with a Si capping layer (courtesy of W. Jäger, Kiel). Such a structure may lead to an optoelectronic device in the future. Preliminary results shown here illustrate how holography can be used to profile the dots' composition in a quantitative way at nanometre resolution.

Fig. 5(a) shows a low magnification phase image (fringe spacing $\sim$ 1nm) of a $Si_{1-x}Ge_x$ island. The bright stripe indicates an increase in the mean inner potential caused by the incorporation of Ge into the island. Making the assumption that any change in potential with respect to the Si matrix can be related linearly to a change in composition, then

$$\Delta\phi = C_E \Delta V_0 t \qquad \text{and} \qquad \Delta V_0 = x \, (V_0^{Ge} - V_0^{Si}) \qquad (3)$$

If part of the wavefront passes through vacuum then the thickness can be measured by the phase change encountered in the (pure) silicon layer leading up to the island. If not, as in this case, the thickness must be determined otherwise, either by EELS through a t/λ map or by 2-beam CBED with the probe focused on the substrate but next to the island. Any extra tilt achieving the 2-beam condition must be taken into account when calculating the thickness through which the beam passed for the hologram. In this case, using the 2-beam method, for the region studied, t = 1930Å ± 80Å and the peak concentration was determined to be 22% ± 4%. The major error in the calculation arises from uncertainties in the values of V_0 used for pure Si and Ge. The value for Si was determined recently (Gajdardziska-Josifovska 1993) and is thought to be accurate but that for Ge is far less so and work is currently in progress to measure a more accurate value.

Fig. 5(b) shows a much higher resolution phase picture from the $Si_{1-x}Ge_x$ islands region. The 'phase noise' in the image comes primarily from poor fringe contrast in the original hologram (fringe spacing ~ 1Å). Nevertheless, the asymmetry of the profile, seen better in the line trace of Fig. 5(c) is clearly revealed with the sharp rise corresponding to the substrate side and the slow fall indicating some diffusion of the Ge into the Si capping layer.

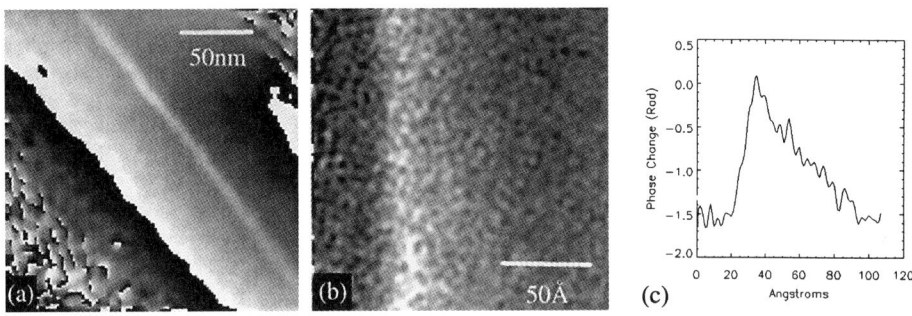

Fig. 5 (a) Reconstructed phase image showing the increase in phase at a Si/Ge island (bright stripe), (b) as (a) but at a much higher spatial resolution revealing the asymmetry of the composition profile, and (c) a line trace averaged over 20 pixels taken through the centre of (b).

3. CONCLUSIONS

It has been shown that electron holography can be used to determine the composition profile of semiconductor heterostructures. After corrections for artefacts arising from the recording or reconstruction of the hologram and careful interpretation of the final result, quantitative composition profiles at the nanometre scale can be obtained.

REFERENCES

Chen J, Hirayama T, Lai G, Tanji T, Ishikuza K and Tonomura A 1995 Electron Holography, ed A Tonomura, L F Allard, G Pozzi, D C Joy and Y A Ono (Amsterdam: Elsevier Science) pp 81-92
Cowley J M 1992 Ultramicroscopy **41** 335
Gabor D 1949 Proc. Roy. Soc. London A **197** 454
Gajdardziska-Josifovska M, McCartney M R, de Ruijter W J, Smith D J, Weiss J K and Zuo J M 1993 Ultramicroscopy **50** 285
Lehmann M, Völkl E and Lenz F 1994 Ultramicroscopy **54** 335
Saunders M, Midgley P A, Walsh T D, Menon E S K, Fox A G and Vincent R 1996 Proc. 15th Pfefferkorn Conference On Electron Image and Signal Processing (to be published)

Inst. Phys. Conf. Ser. No 157
Paper presented at Microsc. Semicond. Mater. Conf., Oxford, 7–10 April 1997

79

Imaging dislocation kinks, their motion and pinning in Si

J C H Spence, H R Kolar and H Alexander*

Dept. of Physics, Arizona State University, Tempe, AZ. 85287, U.S.A. (Spence@asu.edu)
*Universität Köln, Abteilung für Metallphysik im II Physikalishen Institut, Köln, Germany.

ABSTRACT: Lattice images have been obtained using "forbidden" reflections generated by (111) stacking faults in silicon lying normal to the beam at temperatures up to 600°C. Stationary and video images of 30°/90° partial dislocations relaxing toward equilibrium are studied. The images show directional fluctuations which are believed to be kinks, since, as expected from mobility measurements, a higher density is observed on 90° partials than on 30° partials, whereas artifacts contribute equally. Video difference images are used to estimate kink velocity. Observations of kink delay at obstacles, thought to be oxygen atoms at the core, yield unpinning energies and the parameters of the obstacle theory of kink motion. The kink formation energy is obtained from the distribution of kink pair separations in low-dose images. Kink migration rather than formation energy is thus found to control the velocity of unobstructed dislocations in silicon.

1. INTRODUCTION

One of the oldest questions in semiconductor dislocation theory remains to be answered - what limits dislocation velocity for given conditions of stress and temperature. Candidates for the rate-limiting process include the double kink formation energy $2F_k$, the migration energy W_m, and kink obstacles, possibly on an atomic scale. In an earlier paper (Alexander at al, 1986), we showed that TEM images formed using certain space-group-forbidden Bragg reflections in silicon may be used to directly reveal dislocation cores running normal to the electron beam at a resolution of about 0.3 nm. The aim of this work is to address this old question using new lattice images of moving and stationary kinks and to estimate the energy barriers of the processes (Kolar et al 1996).The images reveal the approximate position of kinks, not their structure. Kink movement was induced by warming, in the TEM, samples containing quenched-in stacking faults (SF) of non-equilibrium spacing.

First consider the information which could in principle be obtained from images of kinks, stationary and in motion, at the 0.3 nm resolution level. From recordings of kink velocity at known temperature and stress as discussed more quantitatively below), an Arhenius plot might yield kink migration activation energies (different for left and right kinks (Bulatov et al 1995)) and, from the pre-factor given by the intercept, the entropy term can be obtained. Very little is known about this term (Marklund 1985). These results could be obtained for both partials on dissociated dislocations, comparisons made and the stress dependance of the kink velocity determined. Measurements of kink density at low stress would yield directly the double kink formation energy. But perhaps more importantly, "movies" of kink motion would answer the question "Are kinks colliding" (Maeda and Yamashita 1993), and so provide a direct test of the Hirth-Lothe theory and its two regimes. In addition, the observation of kink delay at obstacles and the measurement of waiting times would yield unpinning energies and provide a test of the obstacle theory of dislocation motion (Celli et al 1963). The proposed correlation between kink nucleation events on different partials could be sought (Moller 1978). Finally, entirely new phenomena and mechanisims might be observed. All this information could be provided by an imaging method which allows the position of a kink to be localised to within about 0.5 nm, and does not require an image interpretable at the atomic resolution level. Earlier dynamical simulations of HREM images of kinks (Spence 1981) have shown that obstacles, such as foreign atoms, could not be identified in such images, and that many other processes (for example kink nucleation at impurites or solitons) would require much higher spatial resolution for structural analysis. The distinction between hetrogeneous and homogeneous kink nucleation could, however, probably be made using our 0.3 nm resolution video images under ideal conditions. Our original aim was to make a direct measurement of the nucleation energy barrier for comparison with the Seeger-Schiller model, from the measured kink pair distribution function, however this has not proven possible.

Two experimental difficulties must at once be confronted: the difficulty of distinguishing the effects of atomic-scale surface roughness from kinks in HREM images, and the effects of electron-beam induced damage or enhancement of glide (Werner et al 1995, Kusters and Alexander 1983). Our approach has been as follows. By subtracting successive images which are identical apart from the effects of kink motion we hoped to eliminate the effects of surface roughness. By working below the knock-on threshold for damage, using low-dose techniques and the new Fuji image plates and a CCD camera as detectors, and by turning off the beam during kink motion we planned also to make beam-induced effects negligible. A final difficulty is the accurate control of temperature needed to obtain a kink velocity for given stress which is measurable using video-rate imaging. Thickness variations in the sample also severely limit the field of view.

Related experimental work is summarized elsewhere (Alexander and Teichler 1991); a recent paper proposes several new mechanisms for kink motion (Bulatov et al 1995).

2. "FORBIDDEN" BRAGG REFLECTIONS.

We first review the forbidden reflection lattice image method. In 1971, Lynch performed three-dimensional multiple scattering calculations for gold [111] zone-axis electron diffraction patterns, and noted the occurance of additional reflections at the (-422)/3 positions in computed and experimental patterns if the crystal contained $p \neq 3m$ layers of atoms (m an interger) (Lynch 1971). Figure 1 indicates the location of these reflections. Such reflections are forbidden by the symmetry elements of the space group for a crystal of infinite dimensions. Calculations for similar "termination" reflections in MgO were subsequently reported (Goodman and Moodie 1974). TEM dark-field images were first formed with these reflections by D. Cherns in 1974 , who used them to image monatomic surface steps on (111) gold films (Cherns 1974). Dynamical calculations (Spence 1975) showed the optimum orientation and thickness to be used. Although there are no termination reflections in the wurtzite structure, similar contrast effects have been analysed at SF's in this structure (Glaisher 1987). With the development of ultra-high vacuum transmission electron microscopy, these same termination reflections could be identified in transmission patterns from thin (111) silicon crystals with (7X7) reconstructed surfaces (Tanishiro et al 1986) and analysed (Spence 1983). These reflections can also occur in f.c.c. and diamond structure materials due to twinning, or to SF's parallel to the surface. They were first observed as additional spots in microdiffraction patterns from stacking faults in 1986, using a field-emission STEM probe narrower than the ribbon of SF separating two partial dislocations (Alexander et al 1986). Figure 1 shows such a pattern, obtained recently using convergent-beam electron diffraction (CBED). Since the edges of the SF ribbon define partial dislocation cores, an image formed with the inner six of these "termination" or forbidden reflections in the (111) zone provides a lattice image of the SF alone, and its boundary at the dislocation core. The d-spacing for the "forbidden" planes is $d_{422} = h = 0.33$ nm., or one Peierls valley wide. These valleys run along the <011> tunnels in the diamond structure, orthogonal to (42-2)/3. Additional studies based on the use of termination reflections can be found elsewhere (Iijima 1981, Ourmazd et al 1983). Suzuki et al (1996) have considered their use for the observation of core reconstruction.

Termination reflections may be understood in several ways. A single (111) double layer of silicon atoms produces a much denser reciprocal lattice (which includes {-422}/3 reflections) than does an infinite crystal. This occurs because atoms in a single double layer are more sparsely packed that those in a [111] projection of three double layers, which overlap in projection, leading to a less dense reciprocal lattice without $g = (-422)/3$ type reflections. Dark field images formed with them show single atomic-height surface steps on thin foils. This can be understood as follows. Firstly we note that, if the [-1,-1,-1] beam direction is taken into the foil, then a (1-11) reflection lies in the first order Laue zone (FOLZ) directly above the (2,-4,2)/3 ZOLZ termination reflection shown in figure 1. We can thus consider the (2,-4,2)/3 ZOLZ spot to be the tail of the crystal shape-transform (or rocking curve) laid down around the

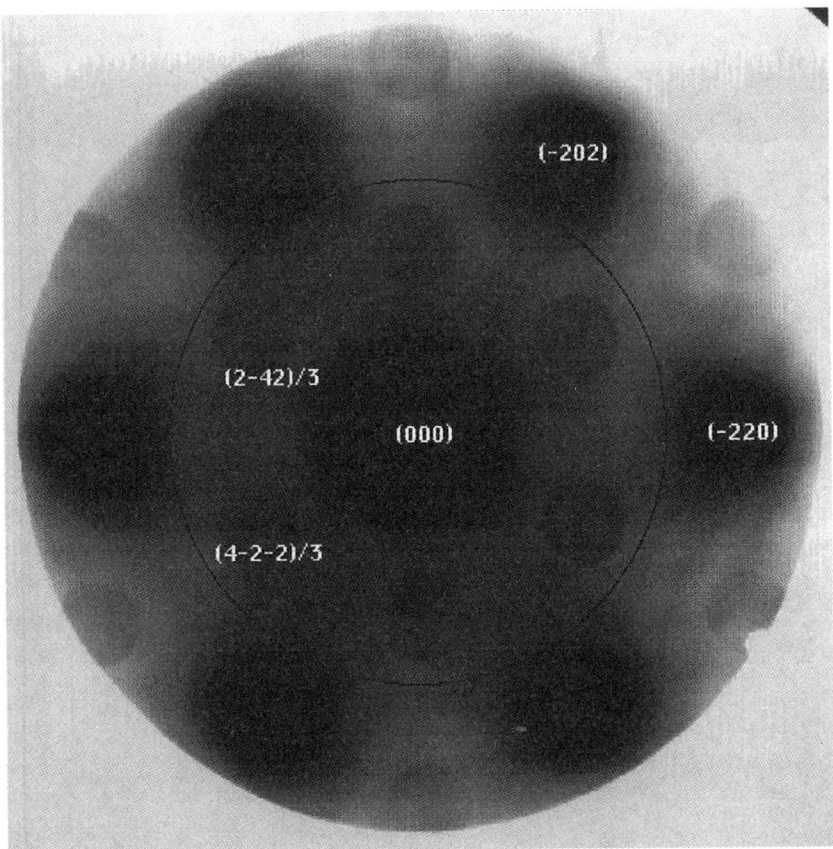

Figure 1. Experimental CBED pattern from an intrinsic stacking fault on (111) in silicon lying normal to the beam. The "termination" {-422}/3 type forbidden reflections can be clearly seen inside the bulk allowed {220} reflections. Circle shows aperture used for imaging (Philips FEG 400 ST, 120 kV. Probe size 2 nm.). The (1-11) point lies directly above (2,-4,2)/3.

(1-11) HOLZ spot, and extending down into the ZOLZ. Since the (1,-1, 1) reflection is weak in the ZOLZ, kinematic theory can be used to give the intensity of the (1,-1,1) beam as

$$\left|\phi_g(z)\right|^2 = z^2 \sigma^2 V_g^2 \frac{\sin^2(\pi \cdot z \cdot s_g)}{(\pi \cdot z \cdot s_g)^2} \qquad 1.$$

The period of this function in z (giving rise to thickness fringes) is $L = S_g^{-1}$. The excitation error S_g of the (1-11) FOLZ reflection evaluated in the ZOLZ (i.e. at the (2,-4,2)/3 position) is just equal to the height of the FOLZ, or $|g(111)|/3 = (\sqrt{3}/a)/3 = 1/(3d_{111}) = S_{111}$, where a is the silicon conventional cubic cell constant. Thus the period of thickness fringes is $t_0 = 3 d_{111} = 0.94$ nm, and we expect a sinusiodal intensity variation with thickness, with a period of three atomic double-layers. Termination reflections thus give "weak beam" thickness fringes with the period of the lattice in the beam direction.

This approximate treatment assumes that the (2,-4,2)/3 reflection is at the Bragg condition (rather than the [111] zone axis orientation used in these experiments), and it ignores atomic structure within the 0.94 nm spacing along the beam path. Alternatively, we may explicitly evaluate the (2,-4,2)/3 structure factor for silicon. If we choose an unconventional hexagonal unit cell for silicon with the c axis along the cubic [111] direction so that atoms have coordinates such as (1/3,1/3,1/3) etc, then a "forbidden" reflection **g** has hexagonal indicies such as (11.0), and its structure factor becomes

$$V_g = \frac{47.878}{\Omega} \sum_j f_j^e(g) \exp(-2\pi g.r_j) = \frac{47.878}{\Omega} 2 f^e(g) \sum_{n=0}^{N} \exp(-4n\pi i / 3) \qquad 2$$

for a crystal of N double layers, again at the Bragg condition for (2,-4,2)/3 (cubic indicies). Here V_g is in volts, Ω is the cell volume and $f^e(g)$ the electron structure factor. Each term in the final sum is proportional to the scattering from one double layer, and these terms may be represented on an Argand diagram, as shown in figure 2. The imaginary part of V_g has been plotted horizontally and real part vertical. One side of the triangle represents the scattering from a single (111) double layer of atoms in silicon - to analyse the effects of shuffle or glide termination, half the lengths of the sides should be used. The lateral shear of each double layer in the diamond structure introduces a 120° phase shift, making a closed triangle ABC every three layers with zero resultant scattering. A thin crystal containing 3m double layers above an intrinsic SF and 3n below it has stacking sequence (3m) AB/ABC (3n), as shown. Below the SF the total scattering vector runs from the origin to each of the corners on the upper triangle in turn, and the intensity is never zero. If the scattering amplitude from one double layer is F (with intensity F^2), then the change in intensity due to the addition of a single double layer in an unfaulted crystal is either zero (on adding a B layer to an A layer) or F^2 (for addition of an A or C layer). For a faulted crystal the results depend on the depth of the fault, however all the possible cases may be obtained by starting at one corner of the lower triangle and ending at one on the upper. In particular, for the kink images, we may compare the diffracted intensity produced by a column of crystal within a ribbon of SF with that generated outside it. In the most favorable case, an unfaulted region ABCABCABC produces zero intensity, but a faulted crystal ABCAB/ABCA of the same thickness generates intensity $3F^2$. Multiple scattering calculations (Alexander et al 1986) confirm these kinematic estimates. Figure 3 shows multislice calculations for the intensity of the (-2-24)/3 reflection as a function of thickness, and we see that the addition of a C layer (shown primed in figures 2 and 3) after the fault changes the intensity by about $4F^2$, in agreement with figure 2. These calculations (unlike figure 2) correctly take account of the excitation errors for the termination reflections, and when this is done we find that the six {-422}/3 reflections are not equivalent - there are two groups of three, reflecting the three-fold (not six-fold) symmetry of a (111) slab consisting of p ≠ 3m layers. At the [111] zone axis orientation the forbidden reflection intensity nevertheless still falls to zero every three double layers.

Real crystals have atomically rough surfaces, and these effects must also be considered. A full analysis (Alexander et al 1986) shows the following. TEM cross section lattice images of Si/SiO_2 interfaces suggest that the roughness will be one or two double-layers. Images were not recorded if contamination could be seen growing at the edges of our samples. Large atomically flat surface islands produce sharp forbidden reflections unless p =3m. As the island size becomes small compared with the coherence width of the electron beam these forbidden reflections broaden out into diffuse elastic scattering peaks. For a random distribution of surface vacancies there are no termination reflections. The width of the diffuse peaks increases with the depth of the surface roughness. We can understand this by recalling that the projected potential for a thin "perfect" crystal with atomically rough surfaces is not a periodic function, hence its diffraction pattern contains elastic diffuse scattering. Because we do not see surface islands in forbidden reflection lattice images from unfaulted crystal, we assume that the roughness can be modelled as random vacancies in the surface layer. Depending on the crystal thickness, these vacancies may or may not lie within the atomic column which contains a kink, thereby altering its contrast. The still images of dislocation cores, however, show a much higher density of kinks on the 90° partial dislocation than on the 30° partial, suggesting that surface roughness is not the dominant contrast effect. In addition, by digital subtraction of successive video frames we can isolate the moving kinks from the stationary surface noise. Figure 3 shows that the average image intensity from by a rough, faulted crystal is higher than that of a rough, unfaulted crystal, and kinks form the boundary between these regions. (The average scattering, rather than the image intensity, from a crystal

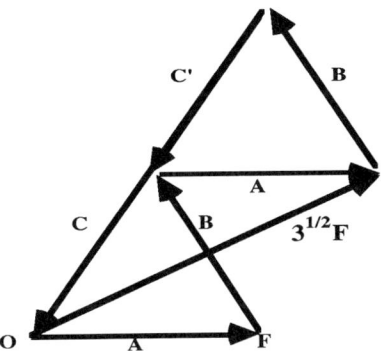

Figure 2. Argand diagram for a {-422}/3 reflection in silicon. Real part of structure factor F plotted vertically, imaginary part horizontal. The kinematic amplitude of Bragg scattering is proportional to a vector from the origin to one corner of the figure. A crystal with stacking sequence ABCAB/ABCA produces the amplitude √3 F shown.

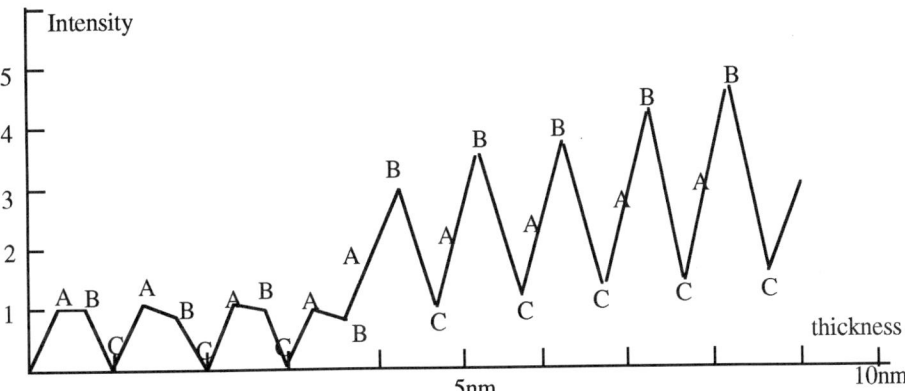

Figure 3. Multiple scattering calculations (using 1000 beams) giving the thickness dependance of the termination reflection (-2-24)/3 in silicon at 100 kV, with beam along [111]. Each letter represents the addition of one atomic layer. A stacking fault occurs at ABA. Three of the six beams are equivalent.

with a uniform distribution of terminations is represented by a point in the center of each triangle). Nevertheless, in still images this surface roughness contributes a large error to our estimates of kink concentration. (The error is calculated by tracing every reasonable boundary to the SF). The ability to introduce dislocations in-situ into a sample whose surfaces have been cleaned by heating in-situ would improve the quality of our images, but would also superimpose the silicon (7X7) reconstruction onto the images.

3. KINK MOTION AND PINNING.

Our first experiments consisted of TEM video recording of partials relaxing toward their equilibrium spacing at 600°C. Our silicon samples were formed by a two stage deformation process, ending with cooling under high stress (Wessel and Alexander 1977). This produces stacking faults on (111) of non-equilibrium width d, which relax back to equilibrium if the sample is warmed in the microscope. TEM samples were prepared by chemical etching at room temperature (no hot glues). In equilibrium, the elastic repulsive force between the partials is just balanced by the attractive force resulting from the work needed to create SF, and the SF width is d= 5.8 nm. Otherwise the shear stress σ in the direction of the partial Burgers vector b can be determined if d is known using $\sigma = (\gamma -(A/d))/b$, where the SF energy $\gamma = 0.058$ J m^{-2} for Si, and $A = 3.36 \times 10^{-10}$ N. Video recordings were obtained during this relaxation on the Akashi 002B TEM using the Gatan heating stage, at 0.25 nm resolution at temperatures of 600°C (above the kink nucleation temperature). The objective aperture shown in figure 1 was used. The beam remained on during the dislocation motion and video recording.Figure 4 shows a sketch of the experimental geometry used. In 4 the diagonal lines are the [011] "tunnels" , or Peierls valleys,through the diamond lattice, with separation 0.33 nm. Kinks are shown at K and K' where the dislocation throws a loop of line forward into the next [011] low energy valley by nucleating a kink pair. The line advances in direction V if the kinks move apart to the end of the segment, or until they collide with kinks of opposite sign. Figure 4(b) shows a side view. The SF plane generates six <42-2>/3 Bragg beams within the SF ribbon and not elsewhere, and only these are used to form the images. The d-spacing for the planes is $d_{422} = h = 0.33$ nm., or one Peierls valley wide. These valleys run along the <011> tunnels in the diamond structure, orthogonal to (42-2)/3.

Video rate images of several 60° dislocation segments dissociated into 30° and 90° partials were recorded at 600°C. Consistent with the earlier finding that 90° partials are more mobile than 30° partials (Alexander and Teichler 1991), motion (on the atomic scale) was confined to the 90° partial, and, accordingly, later stationary images of quenched samples showed a higher kink density on the 90° partial than on the 30° partial. (Mobility depends on which partial is leading or trailing). Figure 5(a) shows the difference between video frames recorded before and after motion of a 90°/30° dislocation at 600 $^{o} \pm$ 10° C. In principle, such a difference image should show zero intensity everywhere except in the region where SF has been created or elliminated. The high constrast outside this region is due to noise, which has been amplified in the process of increasing the contrast digitally in order to reveal the SF, which is the dark strip. Observations at the edge of the sample showed no growth of contamination. The overall SF is narrowing toward equilibrium by motion of the 90° partial alone. Cross-correlation between stationary regions was used to align successive video frames, which are inherently noisy. The dark region is a thin strip of SF on the 90° partial, whose width measured normal to the dislocation line is three Peierls valleys (3 d_{422} =0.99 nm). This strip, suggested by the crossed shading in Figure 4, has been eliminated by the passage of several kinks moving parallel to V_k shown. The kinks encounter obstacles at K and K'. Figure 5(b) shows the collapse of a segment of SF whose width is one Peierls valley (0.33nm) wide. A study of individual frames shows that the motion spans several 33ms frames, so that upper and lower limits on the kink transit time can be made with an error of one frame. A typical single-width segment L = 11.7 nm long gave a velocity of 205 $\pm$ 111 nm/sec , at 600°C. The stress $\sigma = 108.5 \pm$ 7.5 Mpa, was obtained from the total SF width of 9.95$\pm$ 0.5 nm. From the waiting times $\tau = v_D^{-1} \exp(E_u/kT)$ at obstacles such as K and K' in figure 4(a) unpinning energies E_u may also be obtained -the average of two single-width cases gave $E_u =$ 2.4 $\pm$ 0.04 eV. Applying the obstacle theory of kink motion (Celli et al 1963, Moller 1978, Rybin and Orlov 1970) we obtain a velocity $v'_k = L\, v_D \exp (-E_u/kT)$ = 2.4 nm/sec for the average velocity of kinks encountering many obstacles. A comparison with the instantaneous kink velocity of 205 nm/sec shows that the transit time is short compared with the waiting time, as assumed in obstacle theories. Several experimentally indistinguishable kink mechanisms may be responsible for these observations - in particular we cannot distinguish single kink unpinning at one end of the segment K from homogeneous double kink nucleation at mid-segment followed by outward propagation to obstacles K, K' at the ends of the segment. Since extended defects are not seen at cores in still images at atomic resolution, the obstacles are most likely to be dragging points which move forward with the line. We now consider the likely origin of the obstacles.

During the 800°C first stage of deformation, dislocations getter the P $(2 \times 10^{13}$ cm^{-3}), O $(<10^{16})$ and C $(<10^{16})$ impurities in our FZ sample. Carbon is known to be ineffective in dislocation pinning (Sumino 1983). Vacancies and interstitials have much larger values of L, as does P (although strongly pinning) due to its low diffusion rate (Sumino and Imai 1983).

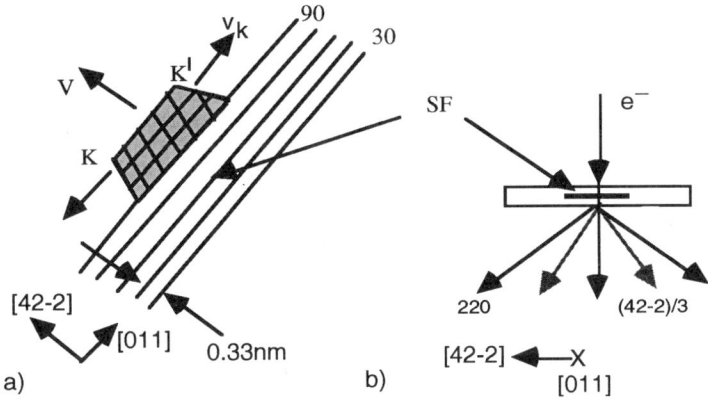

Figure 4. Stacking fault SF on (111) plane (parallel to page) separating 30⁰ and 90⁰ partial dislocation lines, with low-energy Peierls valleys along [1̄10] and kink pair K, K' shown. By running together these kinks advance the dislocation line in direction V. (b) shows side view, along [011̄], indicating "forbidden" (42̄2)/3 reflections from SF, and bulk (220) beams.

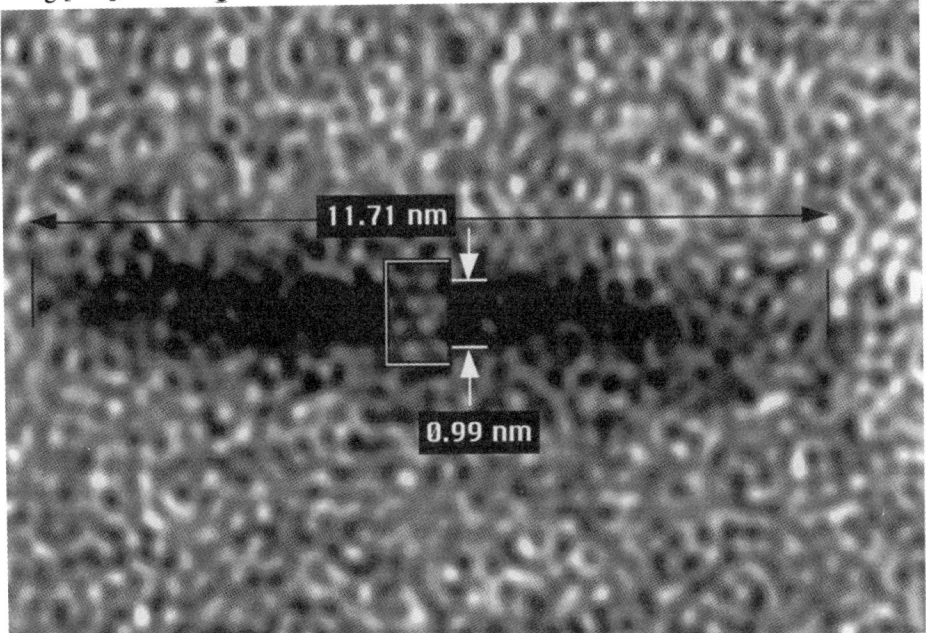

Figure 5(a). Difference between filtered video images of a moving 90⁰ partial dislocation in silicon at 600⁰C, viewed along [111]. Dark strip is SF three Peierls valleys wide (0.99nm), eliminated by passage of several kinks. Inset shows experimental SF image for scale. The dark patch is a portion of the shaded region in figure 4.

We find no evidence (at 0.27 nm resolution) for the O or P complexes previously proposed as pinning centers (Sumino 1989, Sumino 1983). We see no evidence of climb-induced pinning. The impurity with highest concentration is oxygen, whose pinning effect has been studied extensively by X-ray topography in samples with controlled impurity concentrations. This work suggests (Sato and Sumino 1985), at high temperature, an oxygen pinning center with L ≈ 14.0 nm and an unpinning energy of about 3 eV. Ab-initio electronic structure cluster calculations for a variety of likely structures suggest (Umerski and Jones 1993) that a single

86

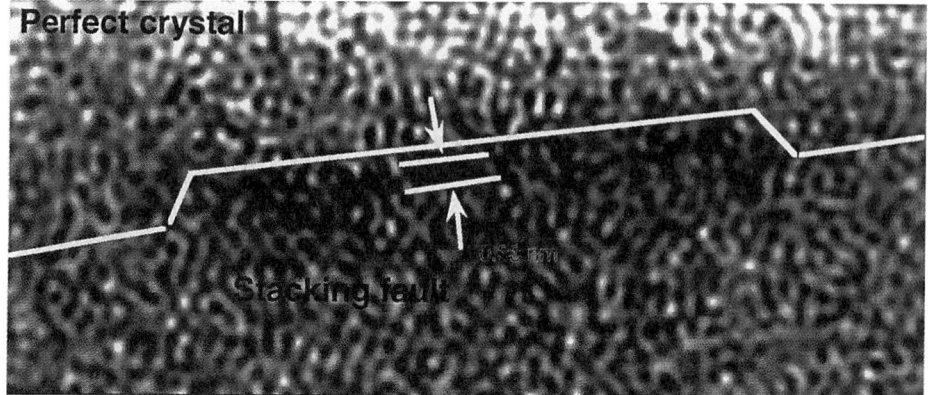

Figure 5(b). Similar for 90⁰ partial segment , one Peierls valley wide.

oxygen atom on the most stretched bond around the anti-phase (soliton) defect can explain this X-ray imaging result. The concentration and unpinning energy (release rate) of this defect are in rough agreement with our observations, however beam induced pinning effects must be considered. Previous work (Hirsch et al 1981, Louchet 1981, Gottschalk 1982) suggests that the beam has two effects a) At energies above the threshold for ballistic knock-on damage (about 140 kV) strong pinning centers are introduced, and b) Enhanced diffusion of kinks (REDG) and impurities by beam-induced electron-hole pair recombination at these defects occurs (Maeda and Takeuchi 1989). The first effect was minimized by conducting a series of experiments in which the beam energy was reduced until, at 130 kV (where all this work was done), the concentration of resolvable beam-induced defects was found to be negligible. (Knock-on damage is easily visible in lattice images at higher voltages). The second effect depends on the intensity and duration of tbe exposure. The spacing of these obstacles is consistent with the "garland" and cusp features seen on 90⁰ partials in previous work, thought to be due to recombination enhanced diffusion of impurities from the thin foil surfaces to dislocations (Gottschalk et al 1987), and which are not seen in unirradiated samples. The ability of impurities to create a soliton-antisoliton pair on arrival at a reconstructed core, and for these sites subsequently to nucleate oxide precipitates, has been noted (Heggie et al 1993). In summary, we speculate that the 2.4 eV unpinning energy we observe is due to oxygen at anti-phase defects, affected in some way by irradiation. Self interstitials may also be involved.

4. KINK MIGRATION ENERGY

By analysing the motion of a partial at a temperature low enough to ensure that no new kinks are created during the motion, the kink migration energy can be determined if the kink density is known. The velocity of a kink of height h, for a dislocation with Burgers vector b and core period a is

$$v_k = \frac{\sigma b h v_D a^2}{kT} \exp(-W_m / kT) = \frac{\sigma b h}{2kT} D_p \qquad 3$$

where v_D is the Debye frequency (1.3×10^{13} sec^{-1} - more strictly the phonon frequency in the direction of motion should be used) , D_p the double kink diffusion coefficient and $W_m = U_m - TS_m$ is the free energy (strictly enthalpy) of kink migration, U_m an enthalpy (experiments are performed at constant pressure). (Similarly $F_k = U_k - TS_k$). Theoreticians compute internal energies U'_m (the sum of elastic and core energies, usually computed at constant volume); experiments measure either W_m or U_m (depending on data analysis). Our U_m values assume (Marklund 1985) $S_k (90^0) = 0.5$ k and $S_m (90^0) = 5k$)

Stationary images were therefore recorded on video and film of a 30⁰/90⁰ dislocation under low dose conditions (to avoid the introduction of pinning centers), both before and after annealing at 130⁰C (below the kink nucleation temperature) for 15 mins. The electron beam

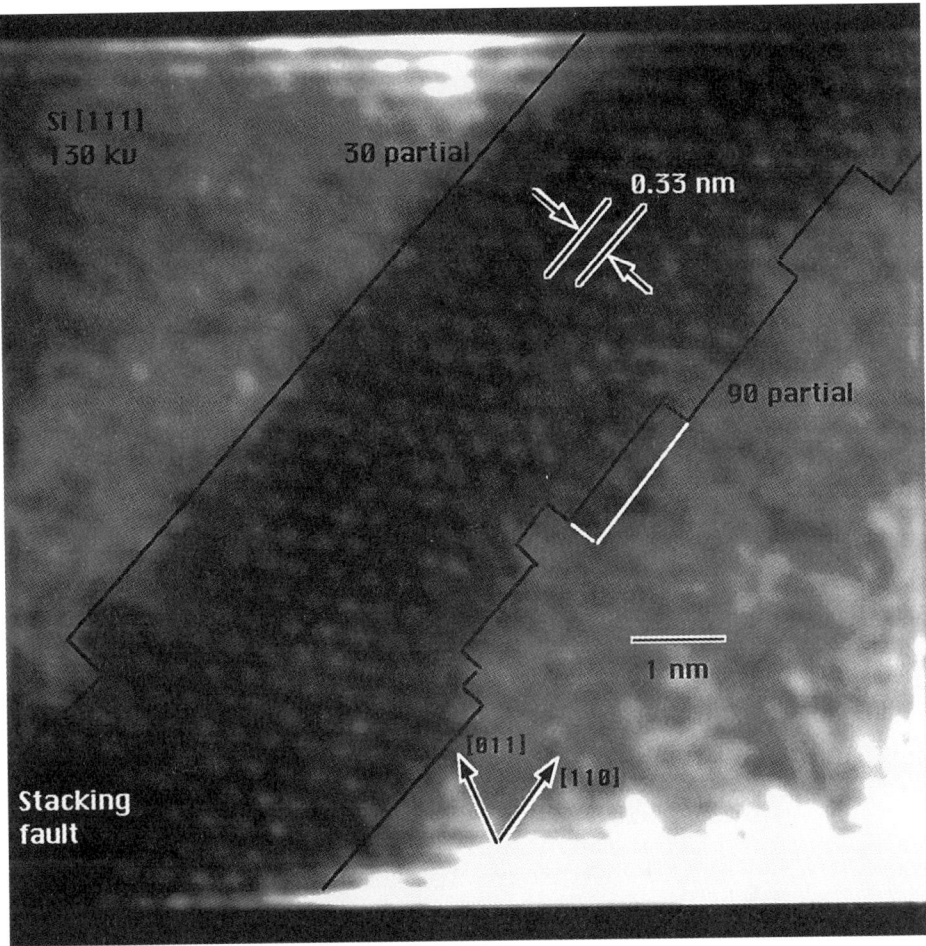

Figure 6. TEM image of dissociated 60⁰ dislocation in silicon after relaxation. The bright diagonal band of regular dots are six-membered rings in the ribbon of SF separating 30⁰ and 90⁰ partial dislocation lines. Black lines run along cores of the two partial dislocations. Fine white line shows typical alternative boundary used to estimate error in counting kinks.

was switched off during the dislocation motion. Figure 6 shows a typical lattice image formed from the inner six "forbidden" reflections in figure 1. The partials have moved apart in the TEM toward their equilibrium separation of 5.8 nm. (Unlike 90/30 dislocations and screws, the 30/90 contracts during the initial ex-situ deformation due to lattice friction (Alexander and Teichler 1991). Image calculations (Spence 1981) show that the bright diagonal band of regularly spaced dots is a lattice image of the double layer of atoms (spaced 0.33 nm apart) which form the SF plane. (Extrinsic and intrinsic faults produce similar images). Pairs of atoms appear as a single dark spot, bright spots are centered on the six-fold rings of a single double-layer. The borders of this band of regular dots forms the partial dislocation cores, as shown. The white scale lines indicate one Peierls valley, 0.33 nm wide. The average SF width corresponds to a stress on the partials of 275 MPa . We note that the density of kinks is greater on the 90⁰ partial than on the 30⁰ partial (see also figure 7). An accurate determination of kink density is complicated by the effects of surface roughness, however the higher density on one partial (seen also in larger fields of view such as figure 7 and in many different cases (Alexander et al 1986)) suggests that surface effects are not dominant. In addition, monatomic

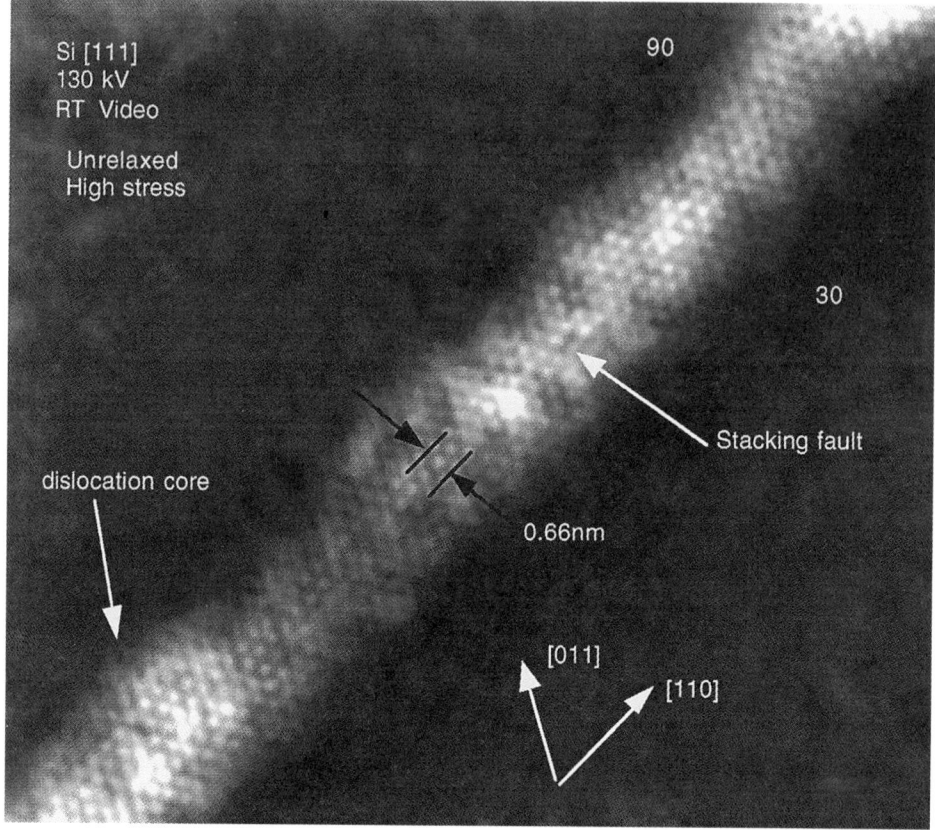

Figure 7. Similar to figure 6, but for an unrelaxed pair of partials spaced more closely than the equilibrium separation.

surface islands are not seen outside the SF - these would produce similar (but lower) contrast to the SF. "Kinks" smaller than the critical separation x* = 0.81 nm (defined below) are evidently due to surface roughness. Figure 6 shows how the error in kink density was estimated. These images may be used to estimate W_m from the kink density c and distance Δs the partial dislocation moves (with beam off), using $V_{dis} = 2 c h V_k = \Delta s/\Delta t$, where h = 0.33 nm, the kink height. Now $v_k \Delta t = \Delta y$, hence $\Delta s = 2ch\Delta y$, where Δy is the mean distance a kink moves in time Δt. Using $\Delta y/\Delta t = v_k$ in equation 3 yields $W_m = 1.24 \pm 0.07$ eV (U_m = 1.55 eV). (Within our limited field of view, less than the kink mean free path, obstacles are unimportant if present in unirradiated material). Figure 7 shows a still from a low-dose video frame for a different dislocation, in this case an unrelaxed pair of partials more narrowly spaced than their equilibrium separation.

5. KINK NUCLEATION ENERGY

By measuring the distribution of kink-pair separations in these images, the kink formation free energy F_k may be estimated. This could be most easily done under conditions of low stress, when the concentration of kinks is close to thermal equilibrium, and simply related to F_k by a Boltzmann factor. Our annealing experiments were unsuccessful is providing a sufficiently large field of view of fully relaxed partials within the acceptable range of sample thickeness for good imaging. It was therefore necessary to use the high stress theory of dislocation motion to analyse our images. The most successful theory of dislocation motion (Hirth and Lothe 1982) (but see Maeda and Yamashita (1993)) considers the nucleation of double kinks (with separation x), their diffusion and drift under an external stress σ by one-dimensional analogy with classical steady-state nucleation theory for particles of size x. Then the distribution of kink separations (constrained to zero kink current) per unit length of line is

$$c_c(x) = \frac{1}{a^2}\exp(-F(x)/kT) \qquad\qquad 4$$

with dims L^{-2} . Here the free energy of kink-pair formation in the presence of the large stresses used in our experiments is (Seeger and Schiller 1962)

$$F(x) = 2F_k - \frac{\mu b^2 h^2}{8\pi x} - \sigma bhx \qquad\qquad 5$$

where μ is the shear modulus. The second term represents the attractive kink-kink strain interaction (tending to annihilate by recombination kink pair embryos less then the critical separation x^*) while the last term describes the external stress, driving kinks apart. The net double kink nucleation rate is $J = D_p C_0(x^*)/2x'$, and the dislocation velocity $V = 2h(Jv_k)^{1/2}$ if the dislocation segment length is larger than the kink mean free path λ. Then V depends exponentially on $Q = F_k + W_m$. Here $x'=kT/(\sigma bh)$. Other theoretical treatments have assumed that kink velocity is limited by obstacles (such as anti-phase defects, impurities or point defects) (Celli 1963, Moller 1978, Rybin and Orlov 1970). Several theoretical calculations for W_m and F_k have been published (Gottschalk et al 1993, Huang et al 1995).

Table 1 shows the distribution of kink pair separations measured from images such as figure 6. The large error was obtained by drawing in all reasonable core boundaries. From the table, values of c(x) (with dims L^{-2}, not L^{-1}), the unconstrained number of kink pairs with separation between x and x+dx per unit length of dislocation may be obtained. The critical separation at the saddle point is $x^* = (\mu bh/8\pi\sigma)^{1/2} = 0.81$ nm. By extrapolation from Table 1 we obtain $c(x^*) = 8.9 \times 10^7$ m^{-2}, which, according to Zeldovich (Zeldovich 1943) is half the constrained value $c_0(x^*)$. Using eqn. 4 with a =0.384 nm, T = 420°C gives $F(x^*) \approx 1.455$ eV, so that, from eqn. 5, $F_k = 0.5 (F(x^*) + (\mu \sigma b^3 h^3/2\pi)^{1/2}) = 0.727 \pm 0.15$ eV ($U_k = 0.74$ eV). This may refer either to a reconstructed kink, or to a kink associated with a soliton dangling bond.

6. DISCUSSION AND SUMMARY

Measurements on the 30° partial give approximately one-third the kink density as the 90°. Thus we find $F_k(30) = 0.797 \pm 0.15$ eV. Our value of $W_m = 1.24$ eV may be compared with recent ab-initio LDA calculations, which give 1.8 eV (Oberg et al 1995) and is consistent with recent ab-initio calculations (Bigger et al 1992) favoring reconstruction of the 90° core, which hinders kink motion (by doubling the jump distance) and clears the band-gap of deep states . Our value is consistent with values of 1.0 - 1.2eV measured previously (Hirsch et al 1981, Gottschalk et al 1987). Our finding that $F_k = 0.727$ eV may be compared with measurements by other methods which fall in the range 0.4 to 1.1 eV (Gottschalk et al 1993, Hull and Bean 1993), and with calculations giving 0.1 eV (Oberg et al 1995), based on a hydrogen terminated cluster for the smallest kink pair, with elastic interactions dominating. (These authors comment that their method may underestimate F_k). For segments much longer than the kink mean free path we thus obtain $Q = F_k + W_m = 0.727 + 1.24 = 1.97 \pm 0.2$ eV. Since $F_k < W_m$, we find that, unlike metals, kink mobility rather than formation is the rate-limiting step controlling the motion of free dislocations in silicon. Since we cannot demonstrate the absence of widely spaced or unresolvable obstacles in unirradiated material, this work cannot distinguish the obstacle (Celli et al 1963) and secondary Peierls-valley (Hirth and Lothe 1982) theories of kink motion. In irradiated material, unpinning of kinks at obstacles is directly observed for the first time, yielding the parameters of the obstacle theory. In future work, we are experimenting with samples with reconstructed and hydrogen passivated surfaces to reduce surface noise.

N	m	x (nm)
3	4	1.5
2	5	1.9
2	6	2.3

TABLE 1. The measured number of kink pairs N with separations $x = ma$ (m integer, a=0.384 nm core period) for a 90° partial dislocation in silicon.

ACKNOWLEDGEMENTS

N.S.F. award DMR9526100. Drs. R. Jones, P. Pirouz and W. Schröter for discussions.

REFERENCES.

Alexander H, Spence J, Shindo D, Gottschalk H and Long N 1986, Phil. Mag **A53**, 627
Alexander H and Teichler H 1991 Dislocations (VCH, Weinheim)
Iijima S 1981 Ultramic. **6** , 41
Bigger J, McInnes D, Sutton A, Payne M, Stich I, King-Smith R, Bird D and Clarke L 1992 Phys. Rev. Letts. **69**, 2224
Bulatov V, Yip S and Argon A S 1995 Philos. Mag. **A72**, 453
Celli C, Kabler M, Ninomiya T and Thomson R 1963 Phys. Rev. **131**, 58
Cherns D 1974 Philos. Mag. **30**, 549
Glaisher R, Kuwabara M, Spence J and McKelvy M 1987 Inst. Phys. Conf. Ser. **87**, 349
Goodman P and Moodie A F 1974 Acta Cryst. **A30**, 280
Gottschalk H 1982 Electron Microscopy 1982 Vol **2**, 527
Gottschalk H, Alexander H and Dietz V 1987 Inst. Phys. Conf. Ser. **87**, 339
Gottschalk H, Sauerland N,Specht S and Alexander H 1993 Phys. Stat. Sol. **(a)138**, 547
Heggie M Jones R and Umerski A 1993 Phys. Stat. Sol. **(a)138**, 383
Hirsch P B, Ourmazd A and Pirouz P 1981 Inst. Phys. Conf. Ser. **60**, 29
Hirth J P and Lothe J 1982 Theory of dislocations (Wiley,New York)
Huang Y, Spence J and Sankey O 1995 Phys. Rev. Letts. **74**, 3392
Hull R and Bean J C 1993 Phys. Stat. Sol. **A138**, 533
Kolar H, Spence J and Alexander H 1996 Phys. Rev. Letts. 77, 4031
Kusters K and Alexander H 1983 Physica **116B**, 594
Louchet A 1981 Philos. Mag. **43**, 1289
Lynch D 1971 Acta Cryst. **A27**, 399
Maeda K and Yamashita Y 1993 Phys. Stat. Sol, **(a)138**, 523
Maeda N and Takeuchi S 1989 Inst. Phys. Conf Ser. **104**, 303
Marklund S 1985 Solid State Communications **54**, 555
Moller H J 1978 Acta Met. **26**, 963
Oberg S, Sitch P, Jones R and Heggie M 1995 Phys. Rev. **B51**, 13138
Ourmazd A, Anstis G and Hirsch P1983 Philos. Mag. **A48**, 139
Rybin V and Orlov A 1970 Sov. Phys. Solid State **11**, 2635
Sato M and Sumino K 1985 University of Tokyo Press p.391
Seeger A. and Schiller P 1962 Acta Met. **10**, 348
Spence J 1981 Proc.39th EMSA., Eds G Bailey. (Claitors, Baton Rouge), 120
Spence J 1975 Proc. EMAG 1975. Inst. of Phys. p. 257
Spence J 1983 Ultramicrosc. **11**, 117
Sumino K and Imai M 1983 Philos. Mag. **A47**, 753
Sumino K 1989 Inst. Phys. Conf. Ser., 104, (1989) , 245
Sumino K 1983a J. de Phys. **44**, C4
Sumino K 1983b Mat. Res. Soc. Symp. Proc., **14**, 307
Suzuki, K, Maeda,N., Takeuchi, S. 1996 Phil. Mag., **A73**, 431
Tanishiro Y, Takayanagi K and Yagi K 1986 J. Micros. **142**, 211
Umerski U and Jones R 1993 Philos. Mag. **A67**, 905
Werner M, Weber E, Bartsch M 1995 U.Messerschmidt, Phys. Stat. Sol. **(a)150**, 337
Wessel K and Alexander H 1977 Philos. Mag. **35**, 1523
Zeldovich J B 1943 Acta Physiochem USSR **18**, 1

Inst. Phys. Conf. Ser. No 157
Paper presented at Microsc. Semicond. Mater. Conf., Oxford, 7–10 April 1997

Analytical expression for the kink profile

M E Polyakov

The Institute of Physics, Academy of Sciences, Minsk 22072, Belarus

ABSTRACT: A simple expression for the kink profile situated between two adjacent Peierls peaks has been obtained from the Euler equation.

1. INTRODUCTION

An analytical expression for the kink profile, for the case of absence of action of an external force, has already been derived. The solution is given in real numbers. In the field of a homogeneous external force, in the case where kink ends taken at infinity are displaced from the Peierls valley by a finite value y_1, the expression for the kink profile has been obtained in the form of a meromorphic function (Polyakov 1989). In this solution, to each value x in one of the two alternating space regions, there are corresponding real and complex numbers. The regions with solutions in the form of complex numbers were interpreted as space portions where the dislocation profile exists in the case where the dislocation segment is fixed at the poles. Release of the segment at these points is accompanied by kink annihilation.

2. FORMER METHOD

The peculiarity of the approach of Polyakov (1989) to the problem of the kink profile is in the abandoning of the Pich-Keler formula (Fridel 1967) in which the expression for the force acting on the dislocation is equal to the product of tension τ by the Burgers vector amplitude b. This formula was obtained from the condition of equality of two energies: one that has been spent in moving the lower half of the crystal relative to the upper half along the cut surface l_o the distance equal to the modulus of the Burgers vector $\varphi = \tau l_o b$ and the other one resulting from the dislocation movement through the entire crystal with force $f_o - f_o l_o$. The expression $f_o = \tau b$ was obtained for distances exceeding the modulus of the Burgers vector. As applied to the movement of dislocations within the Peierls barrier period, it follows from this expression that the force acting on the dislocation is equal and equally directed both before and after the segment overcomes the dislocation barrier and means that the energy used in moving dislocations is spent also after the barrier is overcome. Since such an approach is strictly illogical however, the universally accepted expression for the force acting on a dislocation is unsuitable for infinitesimal increments within the limit for Burgers vector amplitudes and therefore cannot be formally applied to problems concerning kinks as for example, in Jones (1983) and Guyot and Dorn (1967) where the increments are considered to be smaller than the Burgess vector.

Beneath all criticism is the steady form for writing the expression for double kink energy U used at its minimization. For instance, the expression U for a double kink in the case of kink ends shifted by the value y_o is usually written in the form (Guyot and Dorn 1967)

$$U = \int_{-\infty}^{+\infty} \left\{ \Gamma(y) \left[1 + \left(\frac{dy}{dx} \right)^2 \right]^{1/2} - \Gamma_0 - \tau b (y - y_0) \right\} dx, \qquad (1)$$

where $\Gamma(y)$ is the coordinate y-dependent energy of the dislocation modulus unit, Γ_o being its minimum value.

In accordance with this formula, energy is spent in displacing dislocations also after the peak of the Peierls barrier is overcome, which is indicated by the term $\tau b(y - y_o)$. In essence, formula (1) contains the difference of two varying energy values representing a certain parameter which does not lend itself to physical interpretation. Any correct writing of the expression for kink energy in such a form is out of the question.

3. NEW METHOD

Why should one use the approximate Pich-Keler formula if for the determining force one can employ the derivative with respect to energy at each point of the Peierls relief? This principle is used in our approach: the force acting on a dislocation at each point of the Peierls relief is equal to the barrier reaction.

Besides obviating the need to use the approximate Pich-Keler formula, it is possible to extend the formalism of introduction of boundary conditions in the case of disposition of dislocations at the Peierls relief bottom to any point on the Peierls relief. The dislocation is balanced by the reaction force of the barrier at any point in the same way as at the bottom of the Peierls relief and the force is equal in absolute magnitude to the derivative with respect to the energy at a given point in the relief. In this connection the expression for the kink energy U_l is given by

$$U_l = \int_{-\infty}^{+\infty} \Gamma(y)\sqrt{1+\left(dy/dx\right)^2}\,dx, \tag{2}$$

which means that for each point of the kink relief the dislocation energy is equal to the energy at the preceding point, changed by an amount determined by the kink slope.

The key point of the variation principle in determining the kink profile is the calculation of the constant "C_o" (Guyot and Dorn 1967). In the absence of the action of an external force on the dislocation it is equal to the value of the energy of the Peierls valley bottom, Γ_o. At any displacement from the valley, which does not differ formally from a different disposition of the kink ends, the value of C_o, by analogy, should be equal to the quantity of the energy of the kink ends location. In the case of displacement by the value y_l it should be equal to $\Gamma(y_l)$, as shown by Polyakov (1989). In the case where the kink is situated between two adjacent peaks, the quantity "C_o" should be equal to Γ_m where Γ_m is the maximum value of dislocation energy. This assumption is supported below.

4. THE KINK PROFILE

In the present paper we suggest an investigation of a simple variant of a kink profile between the peaks of adjacent Peierls barriers. The extreme value of the kink energy is determined by the Euler equation (Arfken 1970):

$$\frac{d}{dx}\left(f - \frac{dy}{dx}\frac{df}{\partial(dy/dx)}\right) = 0, \tag{3}$$

from which at

$$f = \Gamma(y)\sqrt{1+\left(dy/dx\right)^2} \tag{4}$$

follows the relation

$$\Gamma(y) = C_0\sqrt{1 + (dy/dx)^2} \tag{5}$$

A sinusoidal Peierls relief is selected

$$\Gamma(y) = \Gamma_0 + (\Gamma_M - \Gamma_0)\sin^2\frac{\pi y}{a}, \tag{6}$$

where a is the Peierls relief period.

The value of C_0 is found from the condition $dy/dx = 0$ at $y_2 = a/2$, $x_2 = -\infty$ and at $y_3 = 3a/2$, $x_3 = -\infty$ and is equal to

$$C_0 = \Gamma_m. \tag{7}$$

The previous assumption about C_0 has been justified. In this case we use the periodic boundary conditions of the form $(dy/dx) = 0$ at points $y_2 = a/2$, $x_2 = +\infty$ and $y_3 = 3a/2$, $x_3 = -\infty$ (translational invariance with respect to y) to match the physical and mathematical models of the kink. If nonperiodic boundary conditions are set, the equation of a higher order should be employed. The main drawback of the previous models of the kink is the mismatching of physical and mathematical models that leads to a minimisation procedure of some physically meaningless parameter.

If for the case of the absence of the action of an external force the solution was given in real numbers and for the case of displacement of the kink ends from the valley by the value y_1, the solution is given in real and complex numbers, then in the case of the kink being positioned between two barriers the solution should be complex. We test the hypothesis:

Taking into account (6), from (5) follows the expression for the kink slope

$$\frac{dy}{dx} = \frac{\sqrt{\Gamma^2(y) - \Gamma_m^2}}{\Gamma_m}. \tag{8}$$

After integrating (8), taking into account (6)

$$\int \frac{dy}{\sqrt{\Gamma^2(y) - \Gamma_M^2}} = \frac{x}{\Gamma_M}$$

we obtain

$$\sqrt{\Gamma^2 - 1}xi = -\frac{\Gamma a}{2\pi\sqrt{1 - \Gamma_1}}\ln\frac{\sqrt{1 - \Gamma_1\sin^2\frac{\pi y}{a}} - \sqrt{1 - \Gamma_1}\sin\frac{\pi y}{a}}{\sqrt{1 - \Gamma_1\sin^2\frac{\pi y}{a}} + \sqrt{1 - \Gamma_1}\sin\frac{\pi y}{a}},$$

where Γ is the ratio between maximum and minimum energies of the Peierls relief,

$$\Gamma_1 = (1 - \Gamma)/(1 + \Gamma).$$

The solution is purely complex. It reflects the fact that the kink ends are fixed at infinity of adjacent peaks of the barrier, e.g. by locks. Release of one or both ends of the kink is accompanied by its annihilation. One released end jumps (fold catastrophe) onto the adjacent Peierls peak. In Frenkel-Kotorova's model the presence of breaks has become customary (Milchev et al 1988).

5. DISCUSSION

What other motives made us revise critically the conventional existing variation principle for the kink profile in the Peierls model? Firstly, it was as a result of the absence of an analytical solution as to the profile of the whole kink for barriers of any form in the force field. A solution only for the initial portion exists. A simple solution of this problem in a complex form representing a limiting case of the general solution for arbitrary displacement of the kink ends from the Peierls valley confirms the validity of the general and most complicated solution presented by Polyakov (1989).

The second motive is the complete disregard in the account of the deformation component in the expression for the dislocation segment.

Thirdly, the Peierls model, being more elaborate than the Frenkel-Kontorova model, has not been further developed within the last decade. The main disadvantage of the Frenkel-Kontorova model is that it fails to allow for crystal tension in the direction normal to the sliding plane. However it has received more attention because of the possibility of analytical treatment with the aid of the Sine-Gordon equation (Braiman et al 1993). In The Peierls model the analogous analytical generalisation is the work of Polyakov (1989), which may be unfamiliar to many researchers.

6. SUMMARY

This paper completes the solution of the problem of the kink relief shape in the field of a homogeneous external force in the Peierls model. In future, the problem of the kink profile should be solved for the case of the field of an inhomogeneous external force when the ends of the right and left branches of the kink are displaced different distances from the valley.

Thus, for the case where the dislocation kink is situated between two adjacent peaks of the Peierls relief, a simple expression for the kink profile has been obtained in analytical form. It has a complex nature.

REFERENCES

Arfken G 1970 Mathematical Methods for Physicists (Moscow: Atomizdat) 681
Braiman Y, Baumgarten J and Klafter J 1993 Phys. Rev. **B47** 11159
Fridel J 1967 Dislocations (Moscow: Mir) 49
Guyot P and Dorn J E 1967 Canad. J. Phys. **45** 983
Jones R 1983 J. de Physique **44** C4-61
Milchev A and Mazzuchelli D 1988 Phys. Rev. **B38** 2808
Polyakov M E 1989 Phys. Stat. Solidi (b) **153** 479

Inst. Phys. Conf. Ser. No 157
Paper presented at Microsc. Semicond. Mater. Conf., Oxford, 7–10 April 1997
© *1997 IOP Publishing Ltd*

Domain boundaries in epitaxial GaN grown on {1 1 1}B GaAs and GaP by molecular beam epitaxy

Y Xin, P D Brown, T S Cheng[1], C T Foxon[1] and C J Humphreys

Dept. of Mat. Sci. and Metallurgy, University of Cambridge, Pembroke St., CB2 3QZ, UK.
[1]Dept. of Physics, University of Nottingham, University Park, Nottingham, NG7 2RD, UK.

ABSTRACT: Epitaxial GaN/{$\bar{1}\bar{1}\bar{1}$}B GaAs grown by MBE at 700°C exhibits a mosaic cell structure with sub-grain boundaries delineated by predominantly mixed type threading dislocations. The main defect structure formed within epitaxial GaN/{$\bar{1}\bar{1}\bar{1}$}B GaP are domain boundaries on {$1\bar{2}10$} or {$1\bar{1}00$}, without a polarity inversion across them. The {$1\bar{2}10$} boundary extends a short distance along the c-axis, has a displacement of 1/2<$10\bar{1}1$> and is associated with single growth faults in the basal plane. Conversely, the {$1\bar{1}00$} boundary originates at the epilayer/substrate interface and runs through the whole epilayer.

1. INTRODUCTION

Recent progress in the fabrication of high brightness blue/green LEDs (Nakamura et al 1994) and the rapid development of continuous wave, room temperature blue lasers (Nakamura et al 1997) using GaN and its related alloys have stimulated renewed interest world-wide in III-V nitride research. Such device materials contain very high densities of threading defects, which is linked to the limited choice of lattice and thermally matched substrates. It is of interest to understand why extended defects within these devices do not deleteriously influence the light output. Thus, detailed atomic configurations of defect structures are required as a starting point for band-structure calculations. There is also need for defect reduction (certainly within laser structures) because of concerns over light scattering (Liau 1996). Hence, the characterisation of structural defects within epitaxial GaN is also required for the optimisation of growth procedures.

2. EXPERIMENTAL

Epitaxial layers of wurtzite GaN were grown on {$\bar{1}\bar{1}\bar{1}$}B GaAs and GaP substrates by MBE using an elemental solid source of Ga and an activated N plasma source at 700°C, as described previously (Cheng et al 1995). The substrates were desorbed under N and As flux at 620°C prior to growth. The active nitrogen arrival rate was adjusted to create a slightly N rich environment. The samples were examined in plan view and cross-section using Philips CM30 and Jeol 4000EX-II microscopes.

3. RESULTS AND DISCUSSION

3.1 GaN grown on {$\bar{1}\bar{1}\bar{1}$}B GaAs

Epitaxial GaN/{$\bar{1}\bar{1}\bar{1}$}B GaAs contains a very high density of threading dislocations (typically > 10^{11} cm^{-2}). The epitaxial relationship is given by [$000\bar{1}$]$_{GaN}$//[$\bar{1}\bar{1}\bar{1}$]$_{GaAs}$, [$1\bar{1}00$]$_{GaN}$//[$2\bar{1}\bar{1}$]$_{GaAs}$ and [$11\bar{2}0$]$_{GaN}$ // [$01\bar{1}$]$_{GaAs}$, therefore the mismatch between GaN and GaAs is 38.2% between {$1\bar{1}00$}$_{GaN}$ and {$2\bar{2}0$}$_{GaAs}$, with the epilayer in tensile strain. Figs. 1a and 1b are weak beam images of the epilayer viewed in cross section near the [$11\bar{2}0$] zone axis, using **g** = $000\bar{2}$ and **g** = $1\bar{1}00$. Most of the dislocations have a line direction of <0001>. Examples of perfect edge (**b** = 1/3<$1\bar{1}20$>) and screw type dislocations (**b** = <0001>) were identified, but most dislocations (typically ~70%) can be seen in both images, hence they are mixed-typed having Burgers vector components of **a** and **c** (i.e. 1/3<$2\bar{2}03$>).

By applying Ishida's rule (Ishida et al 1980), the Burgers vectors of some of the dislocations can be determined more precisely. For example, the screw dislocation S (Fig. 1a), imaged using **g** = $000\bar{2}$ has two thickness fringes terminating on the right-hand side

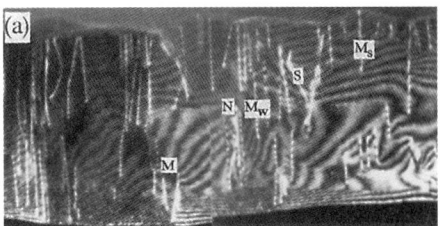
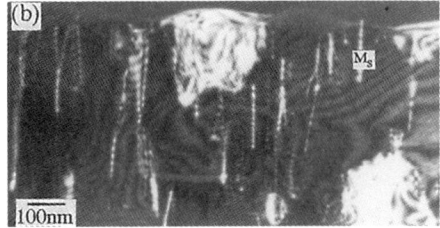

Fig. 1 Weak beam images of GaN epilayer viewed in cross section using (a) **g** = 0002 and (b) **g** = 1$\bar{1}$00.

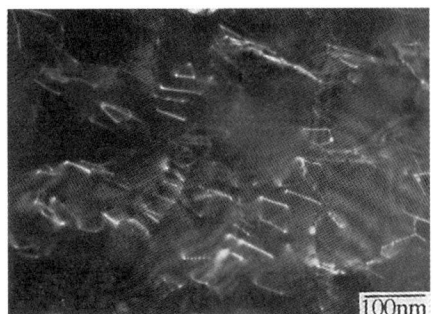
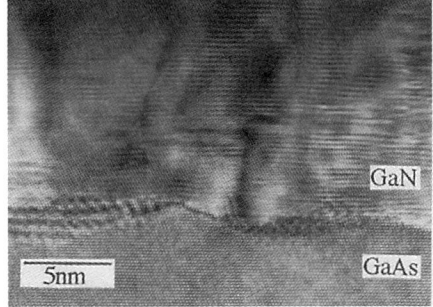

Fig. 2 Weak beam image of a tilted plan view sample showing a mosaic structure with sub-grain boundaries delineated by threading dislocations.

Fig. 3 HRTEM image showing a rough interfacial region between the GaN epilayer and the {$\bar{1}\bar{1}\bar{1}$}B GaAs substrate.

of the dislocation, with the dislocation line pointing towards the growth direction ([000$\bar{1}$]). Therefore **g.b** = -2L = 2, so **b** is [000$\bar{1}$] for this dislocation which is a parallel screw-type dislocation, i.e. **b** is parallel to the dislocation line direction. For dislocations of mixed-type, it is noted from Figs. 1a and 1b that some have strong contrast in both images, while others have strong contrast when imaged using **g** = 000$\bar{2}$ and weak contrast when **g** = 1$\bar{1}$00. For the former, for example the dislocation indicated M$_s$ in Fig. 1a, thickness fringes terminate at the left-hand of the dislocation with the dislocation line pointed downwards along [0001]. Hence, **g.b** = -2L = -2 for **g** = 000$\bar{2}$; **g.b** = x-y = -1 for **g** = 1$\bar{1}$00, thus its Burgers vector is either 1/3[$\bar{2}$113] or 1/3[$\bar{1}$2$\bar{1}$3]. For the case of strong contrast by **g** = 000$\bar{2}$ and weak contrast by **g** = 1$\bar{1}$00, e.g. dislocation labelled M$_w$ in Fig. 1b, such dislocations have **b** of ±1/3[11$\bar{2}$3] or ±1/3[11$\bar{2}\bar{3}$], and the weak contrast is due to **g.(bxu)** ≠ 0. Because the thickness fringes connected with this type dislocation are not clear in the images, a more detailed value of **b** cannot be determined for such dislocations. It is also noted that some dislocations have opposite values of **b** to others, e.g. those indicated as M and N in Fig. 1a. These two opposite types of dislocations delineate a sub-grain. The sub-grain shows bright contrast when imaged by **g** = 1$\bar{1}$00 (Fig. 1b), because the misoriented sub-grain is tilted close to the Bragg condition.

The associated weak beam image of a tilted plan view sample (Fig. 2) shows a mosaic structure with typical domain sizes of 100nm to 300nm, and further illustrates the aligned threading dislocations delineating the sub-grain boundaries. Fig. 3 is an HRTEM image showing the GaN/{$\bar{1}\bar{1}\bar{1}$}B GaAs interface region. It can be seen that the substrate surface is quite rough on the scale of several nanometres. It is suggested that the generation of the threading dislocations is due to the coalescence of misoriented islands during the initial growth (Ning et al 1996). Initial growth proceeds with the formation of 3D islands which become both tilted and twisted due to the large misfit (Schowalter et al 1990) and the rough substrate surface. The combination of island tilt and twist gives rise to the predominance of the mixed-type dislocations upon island coalescence (Ponce 1996).

3.2 GaN grown on {$\bar{1}\bar{1}\bar{1}$}B GaP

Two different kinds of domain boundaries are found to predominate within epitaxial GaN/{$\bar{1}\bar{1}\bar{1}$}B GaP, in addition to the presence of threading dislocations. These two boundary types on {11$\bar{2}$0} and {1$\bar{1}$00} are denoted DB-I plane and DB-II respectively, as

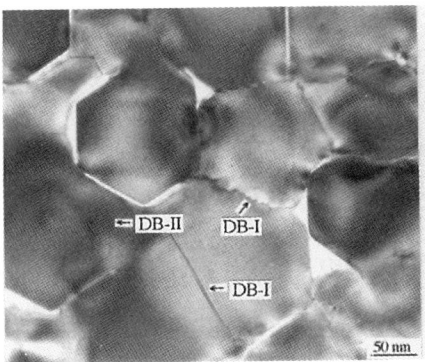

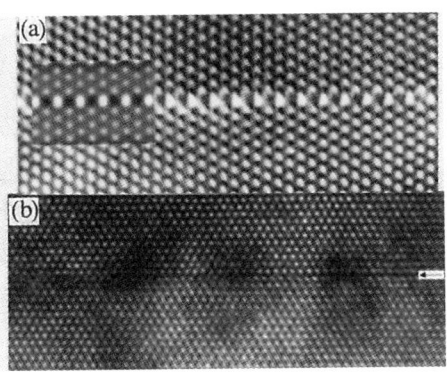

Fig. 4 Plan view image of GaN/{$\bar{1}\bar{1}\bar{1}$}B GaP showing two types of domain boundary on {11$\bar{2}$0} (DB-I) and {1$\bar{1}$00} DB-II.

Fig. 5 HRTEM images of boundaries viewed along <0001> on (a) {11$\bar{2}$0} (DB-I) and (b) {1$\bar{1}$00} DB-II.

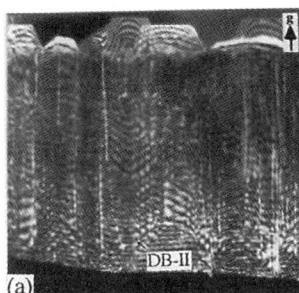
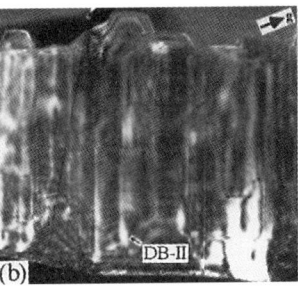
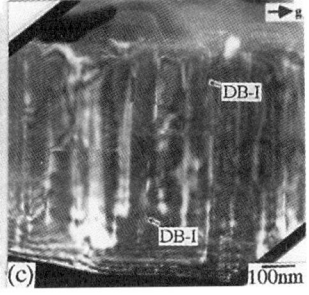

Fig. 6 Weak beam images of a sample close to [11$\bar{2}$0]; (a) $\mathbf{g} = 000\bar{2}$, (b) $\mathbf{g} = 1\bar{1}0\bar{1}$ and (c) $\mathbf{g} = 1\bar{1}00$.

seen from the plan view image of Fig. 4. Again, a mosaic type structure is formed with typical domains size of 100nm to 300nm. Figs. 5a and 5b show HRTEM images of the two boundaries. For such plan view imaging conditions the boundary DB-I exhibits a displacement of 1/2<10$\bar{1}$0> (Fig. 5a), while DB-II has no explicit displacement across the boundary (Fig. 5b).

Additional information about the displacement associated with each boundary type in the growth direction may be obtained using diffraction contrast (α-fringe) imaging. One important point, made by Yanfa Yan et al (1997), is worth emphasising when using this technique to determine the rigid body displacement associated with an inversion domain boundary (IDB). For an IDB, the measured phase shift (α) is the total phase shift which includes the phase shift due to the inversion operation across the boundary (δ) in addition to the rigid body displacement ($2\pi\mathbf{g.R}$). Therefore, it is inappropriate to obtain displacement information of an IDB using $\mathbf{g.R} = 0$ when the boundary is out of contrast, as the required condition should be $\delta + 2\pi\mathbf{g.R} = 0$. Figs. 6a to 6c are weak beam images of a sample cross-section close to [11$\bar{2}$0] obtained using $\mathbf{g} = 000\bar{2}$, $\mathbf{g} = 1\bar{1}0\bar{1}$ and $\mathbf{g} = 1\bar{1}00$. The sample exhibits a turreted growth surface. Since multiple beam dark field imaging using $\mathbf{g} = 000\bar{2}$ (Fig. 6a) shows no contrast changes between domains, and CBED data obtained from different surface islands (i.e. from different domains) confirmed no polarity inversion across the epilayer, the domain boundaries in this sample are not of the IDB type, and therefore the $\mathbf{g.R} = 0$ condition can be used to determine $\mathbf{R}$.

The boundary type DB-I lying on (1$\bar{2}$10) or (2$\bar{1}$10) shows fringe contrast for $\mathbf{g} = 1\bar{1}00$ and no contrast for both $\mathbf{g} = 000\bar{2}$ and $\mathbf{g} = 1\bar{1}0\bar{1}$. If the displacement of DB-I is $\mathbf{R} = [H, K, I, L]$, then from $\mathbf{g.R}$ analysis using the three reflections referred to above, the displacement of DB-I is [K+1/2, K, I, 1/2]. In particular there is a shift of c/2 along [0001]. Combined with the HRTEM data (Fig. 5a), the total displacement associated with DB-I is determined to be 1/2[10$\bar{1}$1] (Xin et al 1997). A model showing the projected atomic structure of this boundary is shown in Fig. 7 (a simulation based on this model is inset in Fig. 5a). Boundaries of this type emanate from stacking faults in the growth plane and it is considered that they arise as a consequence of coalescence of faulted and unfaulted growth islands.

98

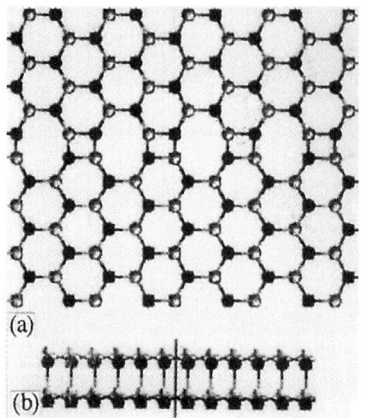

(a)

(b)

Fig. 7 Atomic model of the domain boundary on {11$\bar{2}$0} viewed along <0001> and (b) <1$\bar{1}$00>.

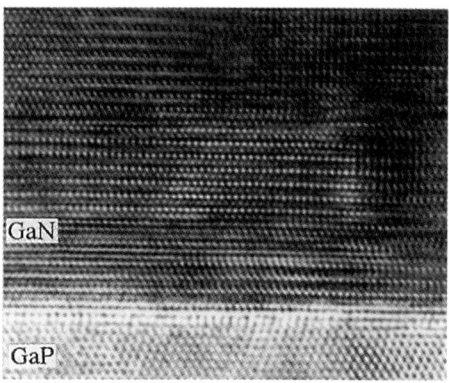

GaN

GaP

Fig. 8 HRTEM image showing a smooth interfacial region between the GaN epilayer and the {$\bar{1}\bar{1}\bar{1}$}B GaP substrate.

Strictly speaking, DB-I is a prismatic stacking fault which forms in the shape of a loop (i.e. mosaic cell). The displacement associated with an island growing above a stacking fault on the basal plane is 1/6<2$\bar{0}\bar{2}$3> (i.e. 1/3<10$\bar{1}$0> + 1/2<0001>). When displaced and unfaulted islands coalesce to leave a 1/2[10$\bar{1}$1] displacement across the boundary as observed, an extra 1/6<1$\bar{1}$00> displacement must be accommodated by a stair-rod screw dislocation at the basal plane stacking fault/prismatic stacking fault intersection (Drum 1964).

Determination of a precise displacement associated with boundaries of the type DB-II proved to be somewhat more difficult. Such boundaries originate at the epilayer/substrate interface, lie on {1$\bar{1}$00} and run along <0001> through the whole epilayer. They are out of contrast only when $\mathbf{g}$ = 1$\bar{1}$00, and show contrast when imaged with $\mathbf{g}$ = 000$\bar{2}$. The deduced displacement from $\mathbf{g}.\mathbf{R}$ is of the form [K, K, I, L] with L not being ±1/2 or integer. Thus, this domain boundary has a displacement along <0001> of 1/n<0001> (n>2) and a displacement of a fraction of 1/3[11$\bar{2}$0] within the boundary plane. Comparison between simulated and experimental HRTEM images indicates the displacement in the basal plane is 1/3n <11$\bar{2}$0> with n > 3 and a shift along the c-axis of 1/n <0001> with n > 3.

The HRTEM image of Fig. 8 shows a remarkably smooth interface at the GaN/{$\bar{1}\bar{1}\bar{1}$}B GaP interface. Thus, the reason for the growth of these very different defect structures on {$\bar{1}\bar{1}\bar{1}$}B GaAs and GaP substrates is partly attributed to the differing extent of epilayer/substrate interface roughness in the first instance, in that 3D growth and greater extent of island tilting leads to the dislocated mosaic structure on the rougher {$\bar{1}\bar{1}\bar{1}$}B GaAs substrate.

REFERENCES

Cheng T S, Jenkins L C, Hooper S E, Foxon C T, Orton J W and Lacklison D 1995 Appl. Phys. Lett. 66 1509
Drum C M 1964 Phil Mag 11 313
Ishida Y, Ishida H, Kohra K and Ichinose H 1980 Phil. Mag. A42 453
Liau Z L, Aggarwal RL, Maki PA, Molnar RJ, Walpole JN, Williamson RC and Melngailis I 1996 Appl. Phys. Lett. 69 1665
Nakamura S, Mukai T and Senoh M 1994 Appl. Phys. Lett. 64 1687
Nakamura S, Senoh M, Nagahama S, et al 1997 Appl. Phys. Lett. 70 868
Ning X J, Chien F R, Pirouz P, Yang J W and Asif Khan M 1996 J. Mater. Res. 11 580
Ponce F A, 1996 Proc. Int. Symp. on Blue Lasers and LEDs, Chiba, Japan , p225
Showalter L J, Hall E J, Lewis N and Hashimoto S 1990 Thin Solid Films 184 437
Xin Y, Brown P D, Humphreys C J, Cheng T S and Foxon C T 1997 Appl. Phys. Lett. 70 1308
Yanfa Yan, Terauchi M and Tanaka M 1997 Phil. Mag. A75 1005

Inst. Phys. Conf. Ser. No 157
Paper presented at Microsc. Semicond. Mater. Conf., Oxford, 7–10 April 1997
© *1997 IOP Publishing Ltd*

Basal and non-basal dislocations in deformed Aluminium Nitride

V Feregotto, A George and J P Michel

Laboratoire de Physique des Matériaux, CNRS URA n°155
Ecole des Mines, Parc de Saurupt, F-54042 Nancy, France

ABSTRACT : Several kinds of dislocations were identified by TEM in AlN polycrystals strained by 10 % in compression above 1820 K. **a** dislocations were seen to be (i) undissociated, (ii) dissociated by glide in two Shockley partials or (iii) dissociated by glide and climb in two Frank-Shockley partials. In the latter two cases, the stacking fault was in (0001). (**c+a**) dislocations were seldom observed. They could glide in $\{10\bar{1}0\}$ prismatic planes rather than in $\{11\bar{2}2\}$ second order pyramidal planes.

1. INTRODUCTION

Aluminium nitride (AlN) is a III-V compound with the wurtzite structure. Dislocations observed so far in this materials have **a** type Burgers vectors. Dissociated dislocations were observed in microcrystals grown from the vapour phase (Delavignette et al. 1961, Blank et al. 1962). In bulk material, extended triple nodes were seen in as-grown ceramics but no dissociation appeared after plastic deformation at low temperature, under confining pressure (Denanot and Rabier 1989 a and b, Audurier 1994). In this paper, we report on observations of dissociated basal dislocations and non-basal dislocations induced by a large plastic strain at high temperature.

2. EXPERIMENTAL

Sintered and hot-pressed ceramics were purchased from ESK. The grain size was about 5 μm. Samples with dimensions 3 x 3 x 8 mm³ were deformed in compression, at the constant imposed strain rate of $5 \times 10^{-6} s^{-1}$ or in creep at constant stress in the range 150 MPa $\leq \sigma \leq$ 250 MPa, under neutral (Ar or He) atmosphere. Testing temperatures ranged from 1820 K up to 1920 K. The maximum strain, ε, that could be reached previous to fracture was ~10 %. Samples were cooled down with the final load applied in order to freeze in dislocation configurations. Thin foils were obtained by ion-milling and observed in a Philips CM 200 TEM operating at 200 kV.

3. TEM OBSERVATIONS

In as-grown materials, only 10% of the grains contain dislocations, in density lower than 5×10^7 cm⁻². These are always basal dislocations, with **a** = $1/3<11\bar{2}0>$ Burgers vectors. They do not show preferred orientations and look undissociated in weak-beam conditions.

High temperature deformation induces two kinds of dislocations : many basal dislocations, and very few dislocations with **c+a** = 1/3<11$\overline{2}$3> Burgers vectors.

The density of **a** dislocations can reach 10^{10} cm^{-2} locally but is highly inhomogeneous. At ε = 0.1, nearly all grains contain dislocations. Most of dislocations lie in (0001) planes, at stresses higher than ~ 300 MPa. When stress is increased, a preferred screw character is observed. At σ < 300 MPa, one third of the dislocation lines are in the basal plane, about one third is on {10$\overline{1}$0} cross-slip planes and the rest out of any low indexes planes. Climb is most probably involved in such configurations.

3.1 Dissociation of basal dislocations

Wide dissociations in the basal plane were observed to be common after 10 % strain at stresses lower than 300 MPa.

Identifying the partial dislocations required the use of several kinds of diffraction vectors. With **g** of the kind 11$\overline{2}$0, stacking faults are out of contrast and the visibility of partials is easily assessed. Burgers vectors which are compatible with observations are either 1/3<10$\overline{1}$0>, in which case the partials would be Shockley dislocations and the dissociation done by slip, or 1/6<20$\overline{2}$3>, in which case the partials would be of the Frank-Shockley type, with climb involved in the dissociation process.

With **g** of the 10$\overline{1}$0 family, stacking fault contrast makes difficult to appreciate the visibility of partials. Unambiguous assessment could be obtained with diffraction vectors of the kinds 11$\overline{2}$2. With these **g**, stacking faults are expected to be out of contrast since the fault vector **R** can always be defined as the sum of the Burgers vector of a limiting partial plus one vector of the basal plane. This is not observed however, and two cases must be considered.

a) The contrast of the fault is weak and the visibility of partials can be clearly decided. In such cases, the contrast of partials is consistent with 1/3<10$\overline{1}$0> Burgers vectors, i.e. with Shockley dislocations.

b) The contrast of the fault is strong and the visibility of partials can only be appreciated, under the assumption that their contrast is an increasing function of |**g.b**| (equal to 0, 1 or 2). Then, the Burgers vectors can only be 1/6<2$\overline{2}$03>.

An example of each case is given.

Case a - Dissociation in Shockley partials

Figure 1 shows a dislocation with **b** = 1/3[2$\overline{1}$$\overline{1}$0], which lies in (0001) but contains a jog, so that the two arms on each side of the jog have crossed, forming an apparently closed

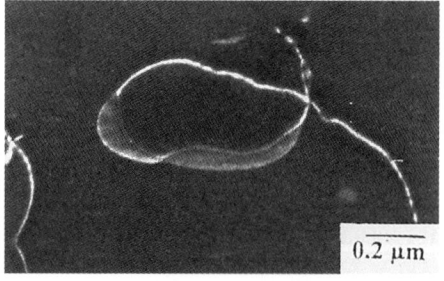
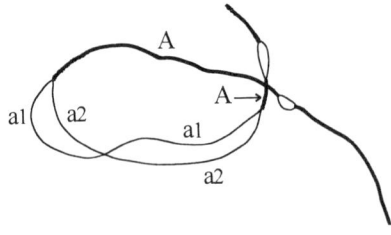

Figure 1   **a** dislocation. Glide dissociation in the basal plane with crossing of the Shockley partials at a jog. Sintered AlN deformed by ε = 10,5 % at $\dot{\varepsilon}$ = 5 x 10^{-6} s^{-1}. T = 1920 K. Beam direction [0001], weak beam (**g**, 3.5g), **g** = 10$\overline{1}$0.

loop. Except at the jog, the dislocation is dissociated in the basal plane according to :

$$\frac{1}{3}[2\bar{1}\bar{1}0] \rightarrow \frac{1}{3}[10\bar{1}0] + \frac{1}{3}[1\bar{1}00]$$

It appears that one partial is more rigid and that the two Shockleys cross each other at the jog.

In hcp structures, the only faults which are limited by Shockley partials are intrinsic and called I_2 (Hirth and Lothe1982). In wurtzite compounds, this fault is characterized by two three-layer wide sheets of sphalerite structure. The sequence of partial dislocations simply depends on the level along the c axis where the fault is situated. At a crossing point, the jog must be n x c/2 height, where n is an odd integer. The two faults on each side of the jog are strictly identical. Contrarily, in cubic III-V compounds, crossing of partials would mean a change from an intrinsic stacking fault to an extrinsic one (see for example the observations in InP by Azzaz et al. 1993). In AlN, crossing of partials are rather common.

Unusual stacking fault contrast was revealed by diffraction vectors with a component along the **c** axis. In dark field, with s ≠ 0, the contrast changed from bright to dark when **g** was changed to -**g**, without reversing the sign of **s**, whatever its modulus, and also when **s** was reversed for the same **g**. Such a contrast asymmetry first observed by Cullis and Booker (1972) and then several times, even in centrosymmetric materials, has never been really explained.

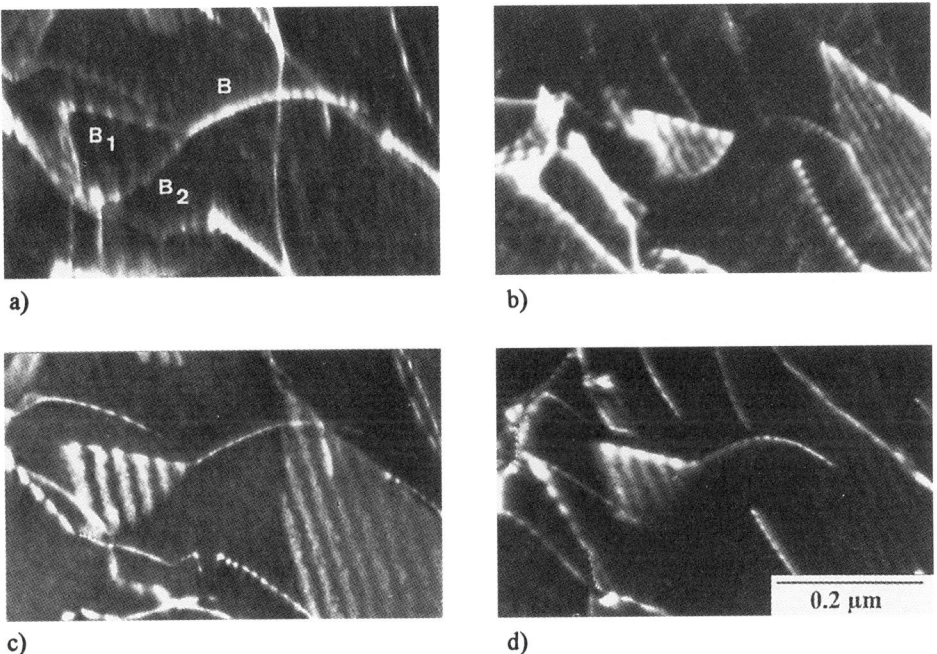

a)   b)

c)   d)

Figure 2   **a** dislocation. Dissociation into Frank-Shockley partials involving climb. Sintered AlN deformed by ε = 9 % in creep at T = 1820 K, σ = 150 MPa.
a) Beam direction : [0001], **g** = $2\bar{1}\bar{1}0$; b) Beam direction : [$4\bar{2}\bar{2}3$], **g** = $\bar{1}2\bar{1}2$
c) Beam direction : [0001], **g** = $\bar{1}100$; d) Beam direction : [$2\bar{1}\bar{1}3$], **g** = $\bar{2}112$
Weak beam (**g**, 2**g**) except c) (**g**, 3**g**).

Case b - Dissociation in Frank-Shockley partials

This kind of dissociation was observed for a minority of dislocations, but could be found in 75 % of the grains. In fact, it appeared at extended nodes formed in the basal plane by three sets of reacting basal dislocations. With **g** of $11\bar{2}2$ family, **g.b** is not simultaneously equal to zero for all partials and the observed contrast is zero, weak or strong as expected for Frank-Shockley partials with |**g.b**| = 0, 1, 2 respectively. The dissociation shown in figure 2 takes place according to :

$$\frac{1}{3}[2\bar{1}\bar{1}0] \rightarrow \frac{1}{6}[2\bar{2}0\bar{3}] + \frac{1}{6}[20\bar{2}3]$$

Of course, since the full dislocation is glissile in (0001), the climb or Frank components of the two partial Burgers vectors are equal and opposite.

We are not aware of such climb dissociation of glissile dislocations in other III-V compounds. It appears that for climb dissociation to occur a high stress level is not necessary. Our observations are all for $\sigma < 300$ MPa. (It must be said however that climb forces may vary from one grain to another and cannot be determined). The total strain, the duration and the temperature are not independent parameters in creep tests. The observation that either a long time or a high temperature are required for climb dissociation points to diffusion as a rate-controlling process. We have no ideas about diffusive species. It could be oxygen, the most important impurity in AlN ceramics (~ 2 weight per cent), nitrogen or aluminium ... Some amount of precipitation could explain the curious contrast of the faults, which never vanishes totally, especially when Frank-Shockley partials are formed.

3.2 (c+a) dislocations

In samples deformed at constant strain rate, full dislocations with a net component of the Burgers vector along the **c** axis were also observed. With the help of appropriate diffraction vectors, their Burgers vectors could be assessed to be of the **c+a** or $1/3<11\bar{2}3>$ type. Such dislocations are never seen isolated but always gathered in walls at small angle from the basal plane. In such a wall, dislocations with **b** = $1/3[\bar{1}2\bar{1}3]$ were seen to align along $[\bar{1}2\bar{1}0]$, which could suggest that they could glide in $(10\bar{1}0)$ prismatic planes, rather than in the so-called pyramidal planes of the second kind $\{11\bar{2}2\}$. To our best knowledge, no other observation of **c+a** dislocations in the wurtzite structure has been reported so far.

ACKNOWLEDGEMENTS

The authors would like to thank S. Choux, J.P. Feiereisen and J. Poirson for skilfull technical help.

REFERENCES

Audurier V, Thesis, Poitiers, 1994.
Azzaz M, Michel JP, Jacques A and George A 1993 Phys. Stat. Sol. **137**, 401
Blank H, Delavignette P and Amelinckx S 1962 Phys. Stat. Sol. 7, 747
Cullis A G and Booker G R 1972 Proc. Fifth European Congress on Electron Microscopy, Manchester pp 532-3
Delavignette P, Kirkpatrick H B and Amelinckx S 1961 J. Appl. Phys. **32**, 1098
Denanot M F and Rabier J 1989a J. Mater. Sci. **24**, 1594
Denanot M F and Rabier J 1989b Microscopy of Semiconducting Materials, eds A G Cullis and J L Hutchison (Bristol : Institute of Physics) pp 439-44
Hirth J P and Lothe J 1982 Theory of Dislocations (New York : Wiley) pp 354-6

Inst. Phys. Conf. Ser. No 157
Paper presented at Microsc. Semicond. Mater. Conf., Oxford, 7–10 April 1997
© *1997 IOP Publishing Ltd*

Structure of the GaAs/InP interface obtained by wafer fusion

G Patriarche, F Jeannès, J L Oudar and F Glas

France Télécom, Centre National d'Etudes des Télécommunications, DTD, Laboratoire de Bagneux, 196 avenue Henri Ravéra, BP 107, 92225 BAGNEUX Cedex, France

ABSTRACT: We investigate by TEM the structure of the GaAs/InP interface obtained by wafer fusion. We identify three interfacial dislocation networks. The first one accommodates both the lattice mismatch and the twist between the two crystals, the second one the tilt. The third one appears during cooling. The first network allows the mapping of the non-bonded areas.

1. INTRODUCTION

Wafer fusion (also known as epitaxial bonding) is increasingly used to fabricate heteroepitaxial structures, in particular those involving an interface between thick strongly mismatched layers (Liau and Mull 1990). These structures cannot be obtained by conventional growth techniques, which produce a high density of threading dislocations affecting adversely their electrical and optical properties. In the wafer fusion technique, two misfitting half structures are first grown conventionally, each on its lattice-matched substrate, and then put in contact, pressed and annealed until atomic bonds reconstruct across the interface. We demonstrated previously that, for epitaxial stackings grown respectively on GaAs and InP (lattice-mismatched by $\varepsilon_0 \approx 3.7\%$), no threading dislocation appears (Patriarche et al 1995). We have recently perfected the process (annealing temperature, surface deoxidisation, specimen holder) to fabricate a microcavity bistable optical switch operating at a 1.55 μm wavelength, which advantageously combines high reflectivity GaAs/AlAs Bragg reflectors with an InP-based active material (Jeannès et al 1996). Here, we study by TEM the structure of the GaAs/InP interface obtained by this technique. So far, only bonding of two identical crystals has been studied in depth (Benamara et al 1995). We examine zones of plan-view and cross-sectional specimens containing the bonded interface. Bonding was performed at 630°C for 30 minutes in a purified H_2 or N_2 flow on (001)-oriented samples carefully deoxidised.

2. MISFIT AND MISORIENTATIONS TO BE ACCOMMODATED

Let us first consider the geometrical constraints at the bonded GaAs/InP interface (Fig. 1). Accommodation of the misfit ε_0 requires dislocations whose Burgers vectors (BVs) **b** have an edge component in the bonding plane. This could be achieved simply by a square network of pure edge dislocations oriented along the <110> interface directions, with **b** of the $(a/2)<110>$ type, spaced by $b/\varepsilon_0 \approx 11$ nm. In addition, misorientations inevitably exist between the two crystals. The first one results from slight errors in positioning equivalent directions of the (001) crystal surfaces parallel to each other before annealing. This produces a twist θ around [001], which can be accommodated by dislocations with BVs having a screw component in the bonding plane, for instance a square network of pure screw dislocations along the <110> interface directions, with **b** of the $(a/2)<110>$ type. The second one results from the slight initial deviations of the surfaces to be bonded from their nominal (001) orientations. This produces a tilt ω around a certain direction of the interface (see section 4); its accommodation requires dislocations with an edge component normal to the interface.

3. EXPERIMENTAL RESULTS

3.1 The three networks of dislocations

The images reveal three networks of dislocations located in the interface or close to it. The first network, a square grid of dislocations with spacings of the order of 10 nm, is best studied in weak-beam (WB) images (Fig. 2(a)). The second network consists of a single set of roughly parallel wavy dislocations with a typical spacing of the order of 100 nm (Fig. 2). The third network is an orthogonal grid with irregular spacings of the order of 1 μm (Fig. 2(b)). Whereas the orientations of the first two networks vary between specimens, the dislocations of the third one are along the two interfacial <110> directions. The first two networks are in the interface. The third one is located close to it but in no well-defined plane; hence, its lines sometimes disappear from the micrographs.

3.2 Joint accommodation of misfit and twist by the first network

Contrast analysis shows that the dislocations of the first network have $(a/2)<110>$-type BVs normal to (001). They are locally interrupted by rectangular patches with <110>-oriented sides (Fig. 2(a)). The WB images thus reveal the distribution of the non-bonded interfacial areas, where the geometrical misfit between the two overlapping crystals remains, but not the dislocations. This allows a precise measurement of the non-bonded fraction of the interface, currently below 10 % in our samples. The high resolution electron microscopy (HREM) cross-sectional images confirm the location of the network at the interface and the nature of the BVs. By showing the alignment of the atomic columns across the interface, they also prove that, between the dislocation cores, a single crystal structure encompassing both GaAs and InP has been reconstructed (Patriarche et al 1995).

Both the BVs and the spacing of these dislocations call to mind the network of pure edge dislocations which could accommodate the misfit (section 2). They however differ from this ideal network since they are rotated by an angle α from the <110> directions and thus have a mixed edge-screw character; they are also more closely spaced (Figs. 2(a) and 3). We now demonstrate that they accommodate both misfit and twist. Half the dislocations of the ideal edge network accommodating ε_0 and half those of the ideal screw network accommodating θ (section 2) would have a BV $\mathbf{b} = (a/2)[110]$, the rest $\mathbf{b} = (a/2)[1\bar{1}0]$. However, accommodation of misfit and twist only prescribes the total lengths of edge and screw segments per unit area, $L_e = 2\varepsilon_0/b$ and $L_s = 2\theta/b$ respectively, and can also be achieved via a single square network of mixed dislocations incorporating edge and screw segments in the proportion L_e/L_s, thereby producing lines rotated on average with respect to the <110> directions by $\alpha = \tan^{-1}(\theta/\varepsilon_0)$, and spaced by $D = b/\sqrt{\varepsilon_0^2 + \theta^2}$. Fig. 3 shows the calculated variations of D and α with twist. Each measurement of these quantities yields an independent estimate of θ. For any given bonded sample, this method gives close values of θ (Fig. 3), which confirms our interpretation. Moreover, the plan-view diffraction patterns (Fig. 4) allow a precise measurement of θ through that of e.g. the angle β between $\mathbf{g}_2$ and $\mathbf{g}_1 - \mathbf{g}_2$, where $\mathbf{g}_1$ and $\mathbf{g}_2$ are reciprocal vectors with identical indices belonging respectively to GaAs and InP. Our measurements (Fig. 3) agree with the calculations which show that $\tan\beta = \theta/\varepsilon_0$, irrespective of $\mathbf{g}_1$.

3.3 Accommodation of tilt by the second network

The second network comprises dislocations of a single sample-dependent orientation, spaced typically by about 100 nm. It is best studied in WB 220 plan-view images, where the dislocations display no contrast but are nevertheless easily detected because upon crossing them, each visible line of the first network is shifted laterally by $D/2$ (Fig. 2(a)). The lines of the second network often deviate from their average orientation to run along lines of the first network, whose spacing is then increased to $3D/2$. A similar effect was observed at the Si/GaAs interface (Zhu and Carter 1990) and at the bonded Si/Si interface (Benamara et al 1995). In both cases, the BVs were identified as of the $(a/2)<101>$ type inclined to the interface. This is the case here again. Only a BV component in the (001) plane can explain the shift of the dislocations of the first network as a result of dislocation

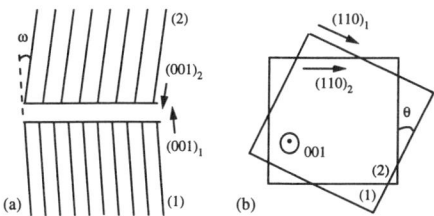

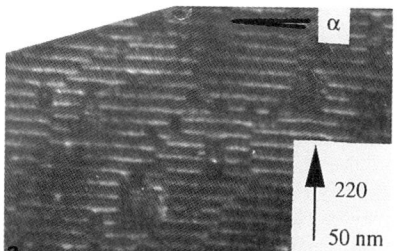

Fig. 1: Schematic side view (a) and top view (b) of the two halves of the structure (1: GaAs, 2: InP).

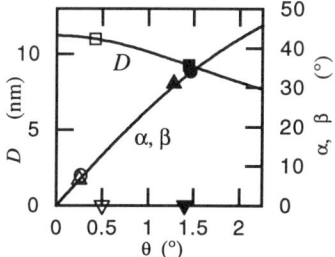

Fig. 3: Curves giving the calculated variations with twist of the spacing D between the dislocations of the first network, of their angle α with <110> and of the angle β measured in the diffraction patterns. Experimental data (D: squares, α: up triangles, β: circles) for two samples (empty and full symbols) are placed on the curves at ordinates given by the measurements so that the corresponding abscissae give the twist. Down triangles are direct twist measurements.

Fig. 2: (a) $\mathbf{g} = 220$ plan-view weak-beam image showing half the dislocations of the first network as white lines locally shifted by lines of the second network; the other half would be imaged with $\mathbf{g} = 2\bar{2}0$. (b) $\mathbf{g} = 220$ plan-view dark field image showing the second dislocation network (wavy lines) and lines of the third one (marked by segments).

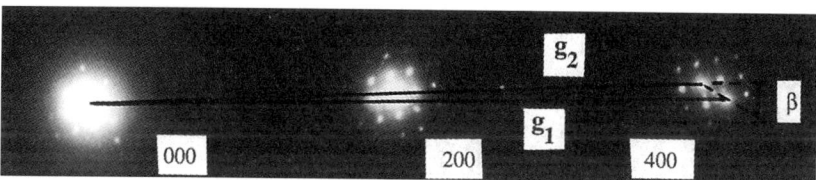

Fig. 4: Part of a diffraction pattern of a plan-view specimen containing the bonded area, showing double diffraction effects.

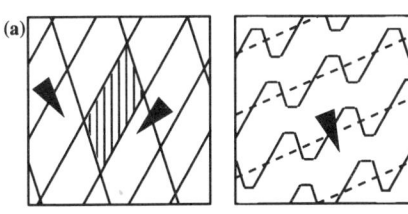

Fig. 5: (a) The two arrays of steps on the crystals facing each other before bonding. Arrows give the 'down' directions. The hatched zone is a terrace where bonding starts independently of its neighbours. (b) Same area after bonding. The dislocations of the second network (full lines) incorporate segments of the two initial step arrays. The arrow gives the down component of their BVs. The dislocations evolve towards the dashed lines, whose common direction is the tilt axis.

interaction and only a BV component normal to (001) can accommodate the tilt. This was confirmed by HREM and by direct measurement of the tilt (a few tenths of degree) by electron diffraction.

4. FORMATION OF THE BONDED INTERFACE AND COOLING OF THE STRUCTURE

On the basis of these observations, we propose that bonding between GaAs and InP occurs in the following way. The surfaces of the two crystals are initially slightly misoriented with respect to their respective (001) planes; this corresponds to two arrays of steps, with a single average orientation on each surface, separating terraces of height $a/2$. The crystals are put in contact with a slight twist error. Reconstruction of the atomic bonds between GaAs and InP starts simultaneously on many terraces, so that the relative twist of the two halves of the structure is now fixed forever. On portions of terraces facing each other (hatched in Fig. 5(a)), the reconstruction only requires the formation of the first network, which accommodates both lattice mismatch and twist. When however a terrace of *e.g.* InP starts bonding to two terraces of GaAs separated by a step, a dislocation with an $a/2$ BV component along [001] must appear to accommodate the difference of level. The various dislocation segments replicating the original step networks recombine to form the second network (Fig. 5(b)). The tilt ω is thus determined via this network by the misorientations of the initial surfaces. Some transport of matter occurs at the interface since the non bonded areas often consist of hollows located opposite each other on both sides of the interface, a configuration which did not exist at the start of the process.

Upon cooling from the bonding temperature, GaAs and InP contract differently so that their misfit increases by $\varepsilon_t \approx 0.1\%$. Our experiments show that this extra misfit is accommodated both elastically and plastically. Pure elastic relaxation would produce a global curvature of the structure with a radius R~0.6 m. Our bonded samples are indeed curved. In one case, we evaluated optically the actual radius of curvature to be about 1 m. Only part of the thermal misfit ($\varepsilon_e \approx 0.06\%$) is thus relieved elastically. However, the third network, constituted of 60° dislocations, accommodates a misfit $\varepsilon_p \approx 0.03\%$. Since $\varepsilon_t \approx \varepsilon_e + \varepsilon_p$, the third network very likely forms upon cooling. Because in the materials considered the dislocations are mobile enough to reach the interface only above about 400°C, we indeed expect only about a third of the thermal misfit to be accommodated plastically.

5. CONCLUSIONS

Wafer fusion is an efficient substitute to epitaxy for the fabrication of complex structures with a few strongly mismatched interfaces. Our study of fusion between GaAs and InP confirms the absence of threading dislocations and the reconstruction of the atomic planes across the interface, where we have identified three dislocation networks. The first one accommodates both the lattice misfit and the inevitable twist between the two crystals. By reducing this twist, the total length of these dislocations could be reduced towards the minimum prescribed by the misfit. The second network accommodates the tilt between the two crystals. Its density could be reduced by an even better control of the orientation of the bonded surfaces. The gaps of the first network map the non-bonded patches, whose total area might be further reduced by perfecting the deoxidisation and the planarity of the bonded surfaces. The third network forms upon cooling to relieve part of the difference of contraction between GaAs and InP, the rest of which is accommodated by bending.

REFERENCES

Benamara M, Rocher A, Laporte A, Sarrabayrousse G, Lescouzères L, PeyreLavigne A and Claverie A 1995 Micr. Semicond. Mat. 1995, IOP Conf. Ser. No 146, eds A G Cullis and A E Staton-Bevan (Bristol: Institute of Physics) pp 103-106

Jeannès F, Patriarche G, Azoulay R, Ougazzaden A, Landreau J and Oudar J L 1996 IEEE Photon. Technol. Lett. **8**, 539

Liau Z L and Mull D E 1990 Appl. Phys. Lett. **56**, 737

Patriarche G, Jeannès F, Glas F and Oudar J L 1995 Micr. Semicond. Mat. 1995, IOP Conf. Ser. No 146, eds A G Cullis and A E Staton-Bevan (Bristol: Institute of Physics) pp 409-412

Zhu J G and Carter C B 1990 Philos. Mag. **62**, 319

Inst. Phys. Conf. Ser. No 157
Paper presented at Microsc. Semicond. Mater. Conf., Oxford, 7–10 April 1997
© 1997 IOP Publishing Ltd

Influence of light illumination on the rosette microstructure in indented GaAs and the photoplastic effect

S Koubaïti+, J J Couderc*, C Levade and G Vanderschaeve

CEMES / CNRS, BP 4347, 31055 Toulouse cedex 4, France
* Physique de la Matière Condensée, INSA, 31077 Toulouse cedex, France
+ Present address: Faculté des Sciences, Kénitra, Morocco

ABSTRACT: Room temperature Vickers indentation tests have been made on the (001) face of GaAs crystal in darkness and under laser light illumination (λ = 940 nm). An illumination-induced decrease of the Vickers hardness is evidenced (negative photoplastic effect) provided low loads are applied to the indentor. The plastic zone around the indents consist of dislocations moving beneath the indentor, rosette dislocations and microtwins. Only the mobility of α dislocations is significantly increased by photonic excitations.

1. INTRODUCTION

Dislocations in large band gap semiconductors are very sensitive to either electronic excitation (cathodoplastic effect) or light illumination with a wavelength close to the band absorption edge (λ_e) of the crystal (photoplastic effect ; PPE). If the PPE is well documented in II-VI compounds, where it appears as an illumination-induced increase of mechanical parameters (positive PPE) such as flow stress and hardness (Osip'yan et al 1986, Koubaïti et al 1996), there are however only few studies of the influence of light illumination on the mechanical properties and related dislocation behaviour in III-V compounds. Mdivanyan and Shikhsaidov (1988) have reported a significant decrease in flow stress under illumination (negative PPE) in GaAs, which is evidenced in the wavelength (λ) range 700-1100 nm, with a maximum magnitude of the effect for λ = 900 nm, slightly higher than λ_e (875 nm).

Microindentation is also a convenient, though less sensitive, method to study the effect of illumination on the plastic behaviour of a semiconductor. Nevertheless any influence of light illumination on hardness remains questionable. The present study aims at a better understanding of the light illumination effect on the hardness number in GaAs, the only III-V compound for which the negative PPE, as observed in dynamical straining conditions, has been studied.

2. EXPERIMENTAL PROCEDURE

Room temperature Vickers indentations are performed on the (001) face of a n-type GaAs crystal in darkness and under laser light illumination (λ = 940 nm). The indentor diagonals are along the <110> directions. Relatively low loads (in the range 0.049-0.196 N) are applied to the indentor in order to prevent cracking of the samples. (A few experiments have been performed at higher load). The Vickers hardness is estimated from a statistical analysis on 20 indentation tests.

The plastic zone around the indents is investigated by transmission electron microscopy (200 kV and 1 MV) : Series of 20 x 20 indentations (applied load 0.049 N) are made in a square arrangement with distances of 50 μm in both directions. Samples are thinned from the backside of the indentations in a bromine methanol solution.

3. RESULTS

3.1 Vickers hardness

Table 1 reports the averaged impression diagonals (d) and the corresponding hardness numbers (H_v) for different loads (P) applied to the indentor.

	Darkness			Illumination (λ = 940 nm)		
P(N)	d(µm)	σ_n	H_v(MPa)	d(µm)	σ_n	H_v(MPa)
0.049	4.16	0.19	5250 ± 450	4.52	0.27	4450 ± 480
0.098	5.58	0.18	5850 ± 350	6.49	0.25	4300 ± 300
0.196	7.40	0.23	6650 ± 400	8.35	0.28	5200 ± 300

σ_n: mean root square, as determined from the dispersion of experimental results.

Table 1

It is seen that :
i) In darkness as well as under illumination, H_v increases with increasing P. Experiments performed with higher applied loads (up to 1.96 N) in darkness indicate that H_v reaches a plateau for P ≥ 0.196 N.
ii) Laser light illumination induces a significant decrease of H_v (about 20 % for P = 0.098 N and 0.196 N). This softening effect seems to vanish for higher applied loads.

3.2 Dislocation microstructure around the indents

Fig. 1 shows a typical view of an indented region in a crystal deformed in darkness.
The plastic zone consists in :
i) a highly strained region around the impression which results from the glide of dislocations with Burgers vector inclined to the surface ; they are submitted to a high lattice friction.
ii) rosette dislocations (noted r on Fig. 1) with Burgers vector parallel to the (001) plane. Most of them look like elongated hairpins ; their long arms are screw. Rosette dislocations extending along [110] (respectively [1$\bar{1}$0]) have their non screw parts of α (respectively β) type. The length of perpendicular rosettes are approximately equal.
iii) microtwins (noted m on Fig. 1) which are observed only on {111} planes parallel to the [110] direction. They are due to the propagation of identical Shockley partials on successive close-packed planes. The shape of the twinning dislocations suggest that they are nucleated on – or very close to – the indented surface. The segment parallel to the surface is 30° β, whereas the segment moving away from the indentation site (and emerging at the surface) is 30° α.

3.3 Influence of laser light illumination on the dislocation microstructure around an indent

Fig. 2 shows a typical view of a crystal deformed under illumination. The microstructure is rather similar to that observed in crystals indented in darkness, with the important exception that the length of the [110] rosette arms is increased by about 50 %. Neither the length of the [1$\bar{1}$0] rosette arms nor the length of twin bands are significantly affected by illumination : Illumination affects mainly the mobility of perfect α dislocations at room temperature.

4. DISCUSSION

4.1 Influence of light illumination on Vickers microhardness

The negative PPE can be evidenced by indentation tests provided that low loads are applied to the indentor and that the illumination wavelength is close to the fundamental absorption edge of the semiconductor. Similar conclusions have been given in the study of the positive PPE in the II-VI compound ZnS (Koubaïti et al 1996). It is suggested that under high applied load the resulting strong workhardening in the plastic zone could reduce or even conceal the PPE. Similar effects occur in semiconductors deformed in compression at high strain (Couderc et al 1990, Mdivanyan and Shikhsaidov 1988). Incidentally, in GaAs indented either in darkness or under illumination, the Vickers hardness depends on the applied load, and consequently on the extent of the plastic zone around the indent. These results are qualitatively in agreement with Yoshioka's (1994) model of indentation plasticity (Koubaïti et al 1997).

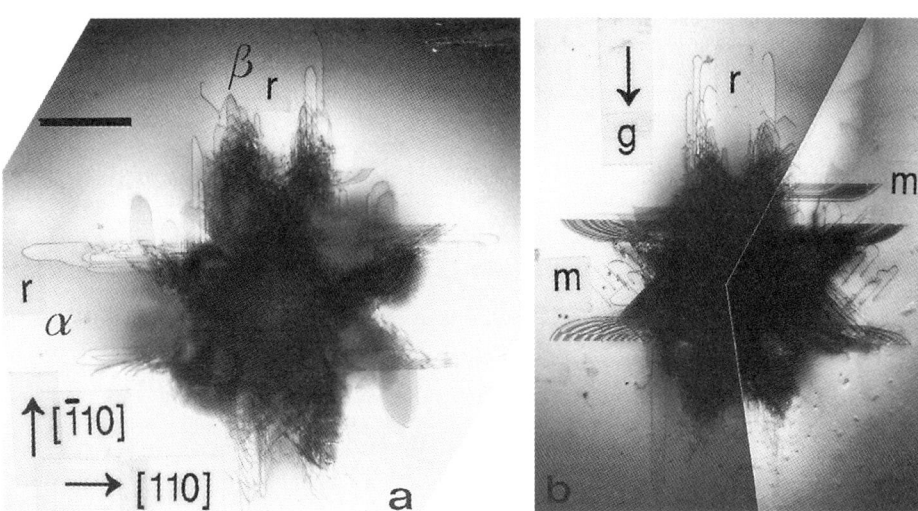

Fig. 1: General views of the plastic zone around 2 different indents. High voltage electron microscope images (a) multibeam conditions ; (b) $\mathbf{g} = [2\bar{2}0]$ the α rosette is out of contrast ; scale bar: 2 μm.

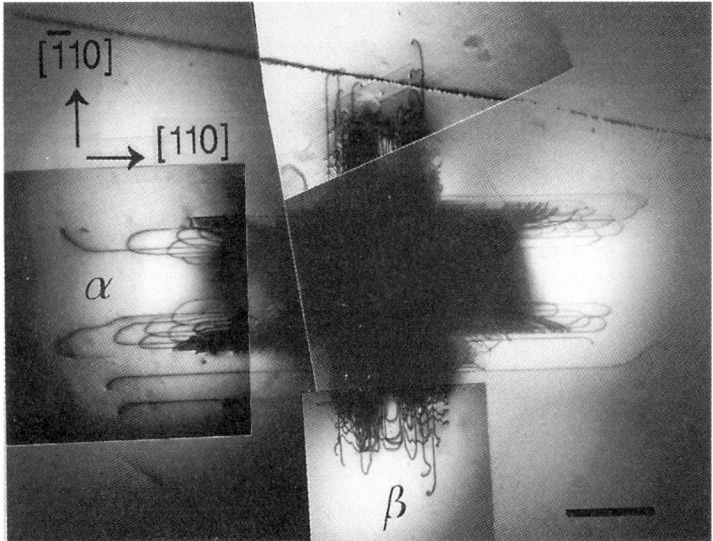

Fig. 2: Effect of light illumination. Note the increase of the length of the α rosette ; scale bar: 2 μm.

4.2 Structure of the plastic zone

The plastic zone consists of a highly deformed zone corresponding to dislocation glide on converging and diverging planes beneath the indentor and rosette dislocations, consistent with the predictions of Roberts et al (1986). Moreover twin bands in those {111} planes which are parallel to [110] are observed. Twinning is known as a low temperature - high stress deformation mode in III-V compounds (Höche and Schreiber 1984, Lefebvre et al 1988, Androussi et al 1989, Boivin et al 1990, Ning et al 1995). Our results suggest that twinning dislocations nucleate on the indentation surface in agreement with a model proposed by Pirouz (1994). At variance, the elongated hairpin shape of the perfect rosette dislocations indicate that they are nucleated deeper in the crystal, probably at indent facets (Ning et al 1995).

The well known α/β asymmetry is not evidenced in our experiments, since perpendicular rosette arms have similar lengths. Possibly this could be related to the low value of the applied load, resulting in a low value of the stress level in the region where rosette dislocations propagate. Nevertheless, the observation that only 30° partials are present in the twin bands is a strong indication that 90° partials have the highest mobility.

4.3 The effect of illumination : the negative PPE

The illumination-induced softening (negative PPE) of GaAs at room temperature is clearly connected to an enhanced mobility of α perfect dislocations, the mobility of β perfect dislocations being not significantly modified. In their investigation of the influence of photon excitation on the crack pattern of indented {111} GaAs crystals, Fujita et al (1988) arrived at the same conclusion.

The radiation enhanced dislocation glide of α dislocations, already observed by Mdivanyan and Shikhsaidov (1988) under light illumination and Maeda and Takeuchi (1983) under electron irradiation in a scanning electron microscope (both using bending tests), originates certainly from non radiative recombinations on electronic levels associated with dislocations (Maeda and Takeuchi 1983). In indentation tests performed at room temperature, β dislocations are not affected by illumination. However, Maeda and Takeuchi (1983) have shown that β dislocations move faster under electron excitation (but in the temperature range 500-650 K). An extrapolation of their results to room temperature suggests that β dislocations are not mobile enough, even in the excited state, to induce a significant effect.

REFERENCES

Androussi Y, Lefebvre A and Vanderschaeve G 1989 Philos. Mag. A **59**, 1189.
Boivin P, Rabier J and Garem H 1990 Philos. Mag. A, **61**, 647.
Couderc J J, Levade C and Kara A 1990 Rev. Phys. Appl. **25**, 1129.
Fujita S, Maeda K and Hyodo S 1988, Phys. Stat. Sol. (a), **109**, 383.
Höche H R and Schreiber J 1984 Phys. Stat. Sol. (a) **86**, 229.
Koubaïti S, Couderc J J, Levade C and Vanderschaeve G 1996 Scripta Mater. **34**, 869 ; 1997 Philos. Mag. submitted.
Lefebvre A, Androussi Y and Vanderschaeve G 1988 Phys. Stat. Sol. (a) **99**, 405.
Maeda K and Takeuchi S 1983 J. Phys. **44**, C4, 375.
Mdivanyan B E and Shikhsaidov M Sh 1988 Phys. Stat. Sol. (a) **107**, 131.
Ning X J, Perez T and Pirouz P 1995 Philos. Mag. A **72**, 837.
Osip'yan Y, Petrenko V F, Zaretskii M Sh and Whitworth R W 1986 Adv. in Physics **135**, 115.
Pirouz P 1994 Twinning in Advanced Materials, eds M H Yoo and M Wuttig (TMS) pp 275-295.
Roberts S G, Warren P and Hirsch P B 1986 J. Mater. Res. **1**, 162.
Yoshioka M 1994 J. Appl. Phys. **76**, 7790.

Inst. Phys. Conf. Ser. No 157
Paper presented at Microsc. Semicond. Mater. Conf., Oxford, 7–10 April 1997
© 1997 IOP Publishing Ltd

Dislocation behaviour in strained layer interfaces

P J Goodhew and G MacPherson

Department of Materials Science & Engineering, University of Liverpool, L69 3BX

ABSTRACT: The origin and energetics of misfit dislocations in interfaces between misfitting layers are reviewed. Experimental results on the distribution of misfit dislocation (MD) spacings in GaAs-based systems reveal the random nucleation behaviour of MDs and the gradual evolution of populations of 90° edge dislocations. Etch pit results indicate that dislocation multiplication becomes significant only when strained layers exceed a thickness of $4h_c$. The contribution of MDs to the formation of cross-hatch is clarified by observations of layers grown on vicinal substrates.

1. INTRODUCTION

For the purposes of this paper a strained layer will be considered to be an epitaxially-grown layer of similar cubic structure, but different in-plane lattice parameters, to its infinitely-thick substrate, which remains flat. Recent work on thin "compliant" substrates (Lo, 1991; Chua et al, 1994; Powell et al, 1994; Cartercoman et al, 1996; Freund & Nix, 1996) is therefore ignored, although many of the considerations detailed below still apply.

Strained epitaxial layers are grown for one or more of three reasons: To retain their strain in order to exploit the properties of strained material (e.g. resulting from the symmetry-breaking associated with distortion of the unit cell away from cubic); to relax their strain in order to provide a layer of different lattice parameter from its (presumably cheaper) substrate, or; by accident. In each of these three cases the behaviour (or potential behaviour) of dislocations is of central importance both to achieving the desired configuration and to ensuring its long term stability. Device structures containing strained layers are almost always not in their lowest energy state and are therefore intrinsically liable to be unstable. The extent of the regions of metastability has not been fully studied in most semiconductor systems, and receives little attention even in this conference.

It is necessary to distinguish between various dislocation geometries and terminologies: A misfit dislocation (MD) is any dislocation whose line direction lies approximately parallel to the epilayer surface and whose Burgers vector contains a component which acts to relieve the strain in the layer. An MD may lie in the interface between layers (typical behaviour when the composition changes abruptly, illustrated in Figure 1), or it may lie within the strained epilayer (typical behaviour in a compositionally graded layer). A threading dislocation (TD) has a line direction which takes it from the substrate, through the epilayer(s), to the free surface. The Burgers vector of a TD can be in any direction: It will not relieve significant strain because it acts only at a point in the interface. A third type of dislocation can be defined but is rarely considered in the context of semiconductor layers, although examples exist. This is the dislocation with Burgers vector perpendicular to the

112

strained layer interface. Such an interfacial dislocation provides a step in the interface, but relieves no strain.

A variety of essentially equivalent arguments all lead to the concept of critical thickness for a strained layer. One of the earliest, and still most frequently cited, is due to Matthews and Blakeslee (1974, 1975, 1976). Their original derivation, in terms of the

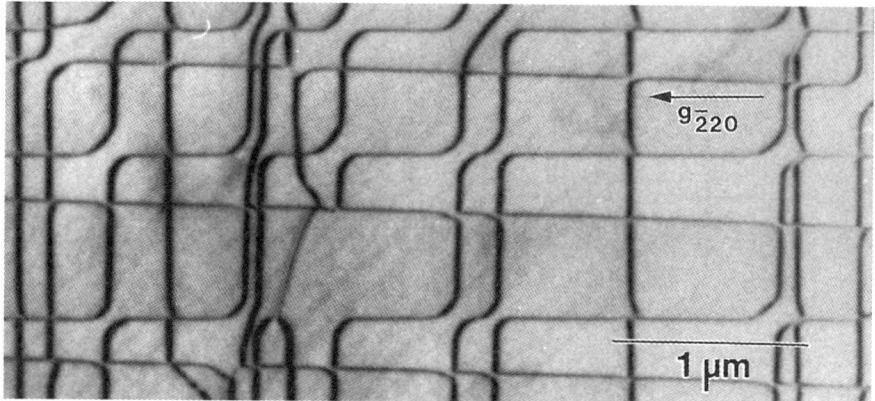

Figure 1 - An array of MDs in InGaAs on GaAs offcut 2° from (001)

energetics of misfit dislocation lines compared to the elastic strain energy of a misfitting layer, has been reworked and refined many times. Most analyses conclude that the critical thickness h_c, at which it becomes energetically favourable for MDs to exist in or near the interface depends on the lattice parameter misfit f (for a pair of cubic materials) via an equation something like this;

$$h_c = (x/f)\ln(h_c/y)$$

where x and y are constants based on the geometry and core parameter of the dislocations (see Fitzgerald, 1991, for a review). The details of the various critical thickness analyses do not in practice matter very much, for three main reasons:

• Kinetic effects (e.g. dislocation nucleation) or dislocation availability (e.g. the density of threading dislocations in the substrate or buffer layer) usually play a bigger role in determining the dislocation content of an interface:

• We do not in any case know what density of MDs will lead to the unacceptable degeneration of device performance. Indeed MDs themselves may be of no concern since they often lie outside the active region of the device. It is their propensity to increase the density of TDs which may be more damaging.

• MDs are almost always introduced gradually and do not reach their ultimate density until the layer is very many times its "critical thickness"

The whole concept of critical thickness has also been confused (and one of the present authors carries a small part of the blame) by debates relating to the ability of various experimental techniques to detect the arrival of the first misfit dislocations (and thus to determine the critical thickness) and the operation of various dislocation generation mechanisms (which might lead to the definition of conceptually different critical thicknesses e.g. Dixon & Goodhew (1990)). Our current view is that none of this matters very much since two more important factors, from the point of view of the users of strained layers, are whether the structure as grown will remain stable for the designed device lifetime and whether a flat surface can be maintained. If a structure is strained and contains few or no MDs, can we be sure that these will not be nucleated? If a structure is "fully relaxed" we have to recognise that it probably is not actually fully relaxed and ask whether the remaining strain will in time create further dislocations (Beanland et al, 1996). However since we have little information on these last important questions (e.g. Bennett and Delalamo, 1993) we will continue to explore what is known of the interfacial structures (metastable or otherwise) which we can observe, and will comment on the contribution of MDs to the surface morphology of the epilayer.

2. ORIGIN OF MISFIT DISLOCATIONS

Dislocations which have been observed lying in or parallel to interfaces are there for one or more of three reasons: They either relieve misfit strain, or they were introduced to accommodate the non-equivalent positioning of neighbouring islands during 3D growth or they have been unable to escape. We will assume that high quality growth is usually 2D and concentrate only on the first category, but it is wise to bear in mind that some of the dislocations we observe may be there for other reasons.

Misfit dislocations in semiconductor systems based on the cubic diamond or sphalerite structures usually have a Burgers vector of $1/2<110>$ and a $\{111\}$ slip plane. They usually adopt $<110>$ line directions and therefore the normal observation is of two orthogonal sets of MDs in a $\{100\}$ interface, three equivalent sets in a $\{111\}$ interface and more complex arrangements in other interfaces (Mitchell and Unal, 1991). Dislocations which reach a $\{100\}$ or a $\{111\}$ interface by glide will have a Burgers vector at 60° to their line direction and are usually known as 60° dislocations. Pure edge dislocations, with their Burgers vector perpendicular to their line direction can be formed from reactions between 60° dislocations and can form directly from the meeting of 3D growth islands. Occasionally, dislocations have been observed to adopt $<100>$ line directions (e.g. Bonar et al, 1992; Goodhew, 1994) and this may imply that other slip systems can operate under high stresses.

MDs may have come from a variety of sources. In many cases (especially for III-V systems) the original substrate wafer will have contained a significant density of threading dislocations, perhaps as high as 10^5 cm^{-2}. If they exist (e.g. in GaAs wafers) the glide of these dislocation, leaving an MD line behind them in the interface, is responsible for the formation of the earliest MDs as the epilayer grows past some critical thickness. A second source is the edge of the wafer, where the interface meets the surface (e.g. Tuppen et al, 1990; Albrecht et al, 1995a). A third source can, in appropriate circumstances, be the nucleation of loops at the top surface of the layer (e.g. Higgs et al, 1991; Cullis et al, 1995; Zou and Cockayne, 1996), while as a fourth possibility there are a plethora of mechanisms available for dislocation multiplication (see for example recent papers such as LeGoues et al, 1992; Lefebvre and Ulhaq-Boillet, 1994; Albrecht et al, 1995b; Beanland et al, 1996). These mechanisms are generally based on a Frank-Read type of process with either a single pinning point or two pinning points defined by interactions which lead to sessile segments of dislocation. As we will demonstrate below, most of these mechanisms can only operate if the layer thickness is well in excess of its critical thickness.

Our current understanding is that the first MDs originate from the slip of threading dislocations, starting shortly after the critical thickness has been exceeded; subsequent dislocations are generated by one or more multiplication mechanisms only when the layer thickness has increased substantially. In-situ observations using synchrotron radiation during MBE growth tend to confirm this sequence (e.g. Whitehouse et al, 1992; Barnett et al, 1995).

It is difficult to generalise the multitude of observations relating to MD formation but many aspects were reviewed in detail by Perovic and Houghton (1995) at the last MSM conference. The many variables which can affect the process include the material system(s), the growth technique, the growth temperature, the quality (e.g. dislocation density and cleanliness) of the substrate or buffer layer, the wafer orientation (major plane and vicinal angle), the misfit and its direction (leading to a tensile or compressive layer), the abruptness of the interface (stepped in composition or graded), the doping and the final thickness of the layer. Tensile layers may crack (e.g. Tsuchiya et al, 1994; Murray et al, 1993) but in the remainder of this paper we will concentrate on observations in the InGaAs/GaAs system, grown mainly by CBE, where the layer is in compression.

3. TEM AND ETCH PIT RESULTS

If MDs arise from TDs in the substrate or, later, from the operation of multiplication sources we would expect that they would not be uniformly distributed across the interfacial plane. Studies on whole wafers of GaAs have shown that the distribution of dislocations in the substrate is unlikely to be uniform, with clusters occurring in particular locations in each quadrant of the wafer (e.g. Barnett et al, 1995). Such a distribution should lead to arrays of MDs which are more closely spaced in some areas, although once each dislocation has slipped across a large fraction of the wafer the overall distribution might appear uniform. Once dislocation multiplication occurs most sources are expected to be capable of operating repeatedly and each active source should give rise to a closely spaced set of parallel MDs. This would certainly lead to bunches of dislocations superimposed on the more uniform background array arising from the original TD population. We therefore carried out a series of careful measurements of dislocation spacings in a series of InGaAs/GaAs layers (MacPherson et al, 1995a, 1995b). An example of the results is shown in Figure 2.

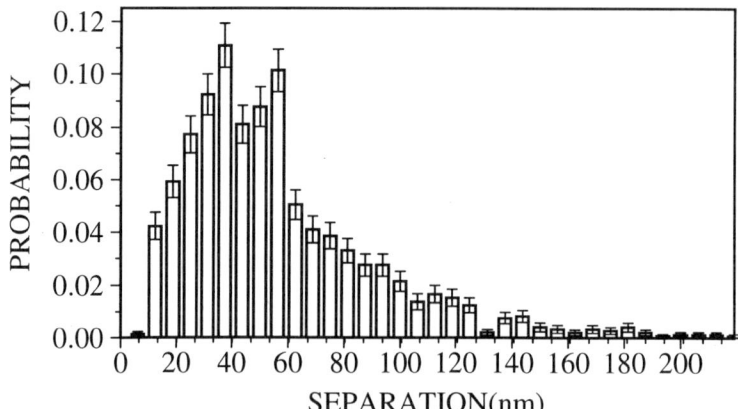

Figure 2 Histogram of interfacial dislocation spacings in a 250nm thick $In_{0.18}Ga_{0.72}As$ layer on a GaAs substrate.

Several interesting conclusions can be drawn from Figure 2 and similar data. The first peak in the distribution (at a spacing of about 17nm) arises from the basic array of 60° MDs. The sharp cut-off at smaller spacings occurs only because the microscope has a limited resolution for dislocation lines and pairs of dislocations with spacings below this point become hard to distinguish. There is however a very clear second peak (at about 33nm) which needs to be explained. This peak corresponds to the presence of edge dislocations in addition to the

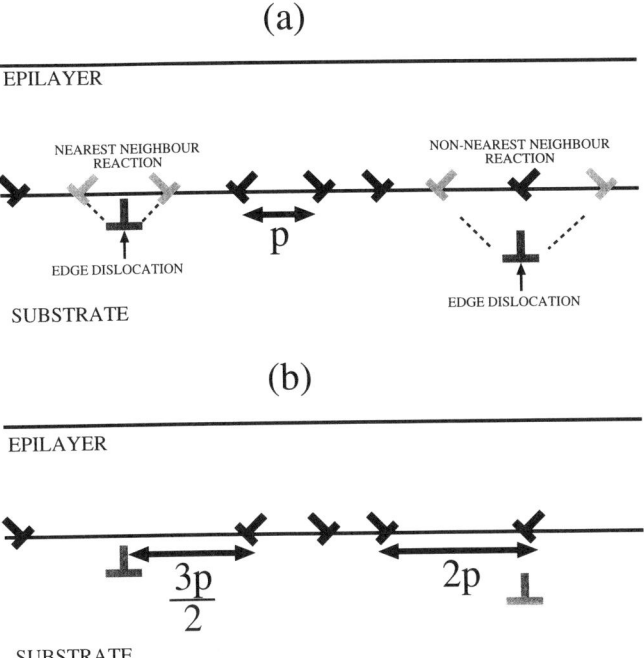

Figure 3 The interactions between 60° dislocations to create edges and change the mean dislocation spacing

expected array of 60° dislocations. These edge dislocations arise from interactions between pairs of 60° MDs . However only MDs which are originally quite close together can interact, since to do so they need to glide out of the interface on intersecting {111} planes. Figure 3 shows that if the original 60° spacing is p, then the interaction of two adjacent MDs will create a local spacing of 3p/2, while the interaction of next-nearest neighbours will likewise create an apparent spacing of 2p. Interactions between more-distant neighbours is very unlikely. The appearance of the second peak in Figure 2 at a spacing of almost exactly 2p seems to confirm this interpretation.

A further interesting conclusion can be drawn from the long tail at large spacings. We have modelled the distribution (MacPherson et al, 1995b) by generating an array of dislocations of average spacing p, with randomly generated Burgers vectors and at random locations. The shapes of the tails of the experimental and computed distributions fit precisely and this suggests strongly that the dislocations in our InGaAs/GaAs layers were generated randomly and no bunching could be detected. The layers on which we made these observations were grown by MBE and ALMBE with compositions from $In_{0.1}Ga_{0.9}As$ to $In_{0.2}Ga_{0.8}As$.

It is not until 60° MDs start to react to form edge dislocations, which must lie out of the interface, that the blocking of 60° dislocations on intersecting planes and the generation of further TDs becomes likely. We can use this observation to design layers which should reduce the TD density by allowing existing TDs to slip out of the wafer, but not allowing MD interactions to form new dislocations which lie away from the interface. If a layer is grown which is not thick enough to allow its average-spaced MDs to glide out of the interface to meet each other then no edges will be formed and all dislocations will be confined to interfaces. It has been shown that this approach works for a set of layers with ultimate composition $In_{0.3}Ga_{0.7}As$ in which each individual layer thickness was kept below 64nm (MacPherson et al, 1996a). A similar approach relies on the observation (Beanland, 1992) that a thickness of at least $4h_c$ is necessary before even a single-ended dislocation multiplication source can operate inside an epilayer. MacPherson and Goodhew (1996) found that the density of threading dislocations could be reduced by as much as a factor of ten

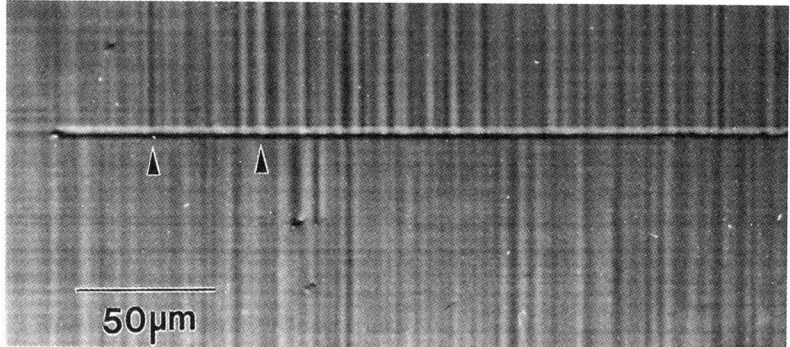

Figure 4. An etched InGaAs layer showing dislocation blocking

below that found in the substrate by keeping layer thicknesses below $10h_c$. They found the maximum effect at thicknesses of $4h_c$.

Further evidence of the role played by edge dislocations can be revealed using etch pits. Although etchants which lead to pits at dislocations are usually used on substrates and thick layers (e.g. van der Ven et al, 1994a and 1994b) we have found that aqueous $HF-CrO_3$ solutions reveal dislocations which lie parallel to the surface after the removal of only 100nm or so of layer, making it possible to reveal TDs, MDs and edge dislocations in fairly thin layers. The ability to etch a large area makes it possible to demonstrate that a single edge dislocation lying out of the interface in the epilayer can block many tens or hundreds of 60° dislocation trying to glide in the orthogonal direction (MacPherson and Goodhew, 1997). Figure 4 shows an example of this behaviour in a layer with a low density of TDs. If the threading dislocation density is higher (Figure 5), then alignment of the TDs can be seen, implying the existence of repeating dislocation sources.

In the above discussion we have neglected to mention the interaction between dislocations in the two orthogonal sets (for {001} surfaces), and we have not mentioned the effect of doping on the behaviour of MDs or TDs. There is a huge literature on the former effect (e.g. Dixon and Goodhew, 1990; Lefebvre et al, 1991) including a variety of possible dislocation multiplication sources, but little data on the effect of doping.

We have recently observed an unusual dislocation configuration in S-doped InGaAs. The gamma-shaped loop shown in figure 6 only occurs in doped layers and has not been seen in undoped layers of the same composition grown by the same CBE technique. It appears that cross-slip or climb must be involved in the development of such a configuration. It bears some resemblance to a single-ended spiral source (Beanland et al, 1996) but although the gamma configuration has been seen several times its development into a source has not been reported. It is possible that the effect of the dopant is to lower the barrier to cross-slip, allowing complex dislocation geometries to form more easily (e.g. as discussed by Ulhaq-Bouillet, Lefebvre and Di Persio, 1994). However it is more likely that climb is involved. Kamejima et al (1979) have observed a similar configuration in LEC-grown bulk GaAs and attributed it to climb associated with condensation of interstitials.

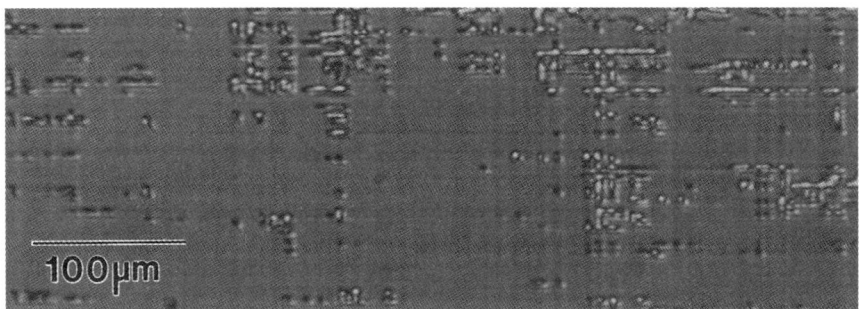

Figure 5. An etched layer with a high density of TDs

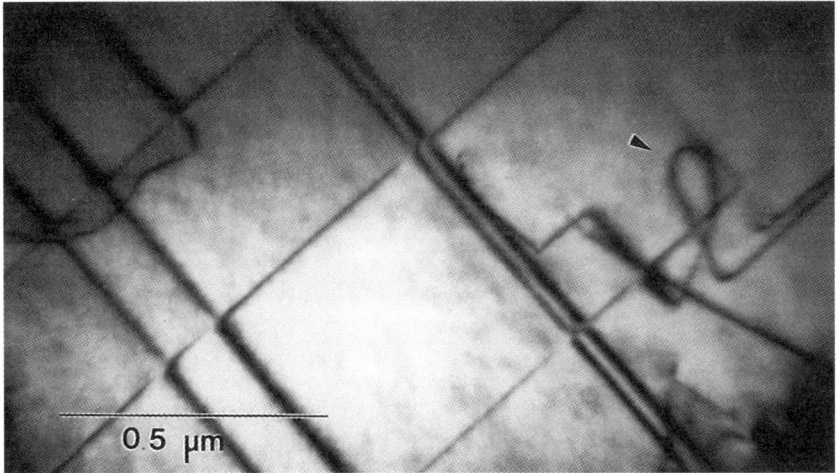

Figure 6. A loop of dislocation in an S-doped (8×10^{17} cm^{-3}) In$_{0.27}$Ga$_{0.73}$As layer

4. SURFACE MORPHOLOGY

The development of non-flat surface morphologies is commonplace in almost all epitaxial semiconductor growth. Cross-hatched surfaces have been reported, for example, in InGaAs/GaAs (Beanland et al, 1995), GaP/Si (Imaizumi et al, 1991), SiGe/Si (Fitzgerald et al, 1992; Cullis, 1994; Albrecht et al, 1995a), InGaAs/InP (Morishita et al, 1994), C-GaAs/GaAs (Kim, Kim and Min, 1996) and even mercury cadmium telluride/cadmium zinc telluride (111) growths (Tobin et al, 1995). The terminology is confusing: As well as the phrase cross-hatch, other workers have referred to hillocks, ridges and troughs, ripples, elliptical islanding, roughening, undulations and striations. These features appear to have in common that they involve alignment along first one, and then two prominent crystallographic directions (three on {111} surfaces). These surface features can be detected using Nomarski optical techniques, by AFM and, during growth, by laser light scattering (e.g. Boyd et al, 1996). There is still however considerable uncertainty over the origin of these effects. Even before the arrival of MDs strain-induced modulation of the surface can occur (e.g. Cullis et al, 1995; Cullis, 1996). The ridges of subsequent cross-hatch patterns have been related by various workers to the presence of edge dislocations (or bunches of such dislocations) in the epilayer above the interface (Beanland et al, 1995), to surface steps created by the slip of threading or misfit dislocations (Lutz et al, 1995), and to misfit dislocations lying in the interface (Albrecht et al, 1995a; Jonsdottir, 1995).

We have recently made some observations on InGaAs layers grown on vicinal plane substrates offcut 3 degrees from (001) towards [-110]. Figure 7 is a Nomarski micrograph showing the early stage of cross-hatch development on a layer of thickness $1\mu m$. The key feature is that in one of the directions there are two almost parallel sets of hatching lines. The angle between these pairs of lines exactly corresponds with the angle which is found between the two sets of $60°$ MDs which accommodate the strain at the interface, as described by Kightley et al (1991) and shown in Figure 1. When these two sets of dislocations interact (as they easily can because they intersect) the resultant edge dislocation usually lies in the epilayer and is aligned along a <110> in the layer, not at a small angle away. Only the MDs in the interface adopt the directions revealed in the cross-hatch and this demonstrates that MDs in the interface must play a significant part in controlling the surface morphology, as predicted by Jonsdottir (1995). It cannot be that edge dislocations within the epilayer are the dominant feature.

5. CONCLUSIONS

We have reviewed the concepts which control the occurrence and behaviour of dislocations in the interfaces between strained layers and have identified a number of features which are still unclear. In particular the effect of dislocation interactions which lead to an increase in the density of threading dislocations has been highlighted. Studies of dislocation spacings have revealed that TD sources at least operate to give a uniform distribution of MDs. Observations of the cross-hatch patterns on layers grown on vicinal plane substrates reveal that MDs in the interface play a large part in the development of this surface morphology. The potential effect of dopants on the behaviour of MDs is noted and an example has been shown.

Figure 7 The cross-hatch pattern on an InGaAs layer grown on a vicinal (001) GaAs substrate

ACKNOWLEDGEMENTS

The authors would like to thank Dr Tim Joyce for growing many of the layers, the EC ESPRIT programme and EPSRC for funding support, Georgios Petkos and Kostas Giannakopoulos for figures 6 and 7 and Dr Richard Beanland for useful discussions.

REFERENCES

Albrecht M, Christiansen S, Michler J, Dorsch W, Strunk H P, Hansson P O and Bauser E 1995a Appl. Phys. Lett. **67**, 1232

Albrecht M, Christiansen S and Strunk H P 1995b phys stat sol (a) **150**, 453

Barnett S J, Keir A M, Cullis A G, Johnson A D, Jefferson J, Smith G W, Martin T, Whitehouse C R, Lacey G, Clark G F, Tanner B K, Spirkl W, Lunn B, Hogg J C H, Ashu P, Hagston W E, Castelli C M 1995 J. Phys. D - Appl. Phys. **28**, A17

Beanland R, Aindow M, Joyce T B, Kidd P, Lourenco M and Goodhew P J 1995 J. Crystal Growth **149**, 1

Beanland R, Dunstan D J and Goodhew P J 1996 Advances in Physics **45**, 87

Bennett B R and Delalamo J A 1993 Appl. Phys. Lett. **63**, 1122

Bonar J M, Hull R, Walker J F and Malik R 1992 Appl. Phys. Lett. **60**, 1327

Boyd A R, Joyce T B and Beanland R 1996 J. Cryst. Growth **164**, 51

Cartercoman C, Brown A S, Jokerst N M, Dawson D E, Bicknelltassius R, Feng Z C, Rajkumar K C, Dagnall G 1996 J. Electronic Mats. **25**,1044

Chua C L, Hsu W Y, Lin C H, Christenson G, Lo Y H 1994 Appl. Phys. Lett. **64**, 3640

Cullis A G 1994 Scanning Microscopy **8**, 957

Cullis A G 1996 MRS Bulletin **21**, 21

Cullis A G, Pidduck A J and Emeny M T 1995 Phys. Rev. Lett. **75**, 2368

Dixon R H and Goodhew P J 1990 J. Appl. Phys. **68**, 3163

Fitzgerald E A 1991 Mater. Sci. Rep. **7**, 87

Fitzgerald E A, Xie Y H, Monroe D, Silverman P J, Kuo J M, Kortan A R, Thiel F A and Weir B E 1992 J. Vac. Sci. Tech. **B10**, 1807

Freund L B and Nix W D 1996 Appl. Phys. Lett. **69**, 173

Goodhew P J 1994 J. Phys. Chem. Solids **55**, 1107

Higgs V, Kightley P, Goodhew P J & Augustus P D 1991, Appl. Phys. Lett. **59** 829

Imaizumi M, Saka T, Jimbo T, Soga T and Umeno M 1991 Jap. J. Appl. Phys. 1, **30**, 451

Jonsdottir F 1995 Modelling Simul. Mater. Sci. Eng. **3**, 503

Kamajima T, Marsui J, Seki Y and Watanabe H 1979 J. Appl. Phys. **50**, 3312

Kightley P, Goodhew P J, Bradley R R and Augustus P D 1991 J. Crystal Growth **112**, 359

Kim S I, Kim M S and Min S K 1996 Solid State Comms. **97**, 875

Lefebvre A and Ulhaq-Bouillet C 1994 Phil. Mag. **A70**, 999

LeGoues F K, Meyerson B S, Morar J F and Kirchner P D 1992 J. Appl. Phys. **71**, 4230

Lo Y H 1991 Phys. Lett. **59**, 2311

Lutz M A, Feenstra R M, LeGoues F K, Mooney P M and Chu J O 1995 Appl. Phys. Lett. **66**, 724

MacPherson G and Goodhew P J 1996 J. Appl. Phys. **80**, 6706

MacPherson G and Goodhew P J 1997 Appl. Phys. Lett., in press

MacPherson G, Beanland R and Goodhew P J 1995a Scripta Met. & Mat. **33**, 123

MacPherson G, Goodhew P J and Beanland R 1995b Phil. Mag. **A72**, 1531

MacPherson G, Beanland R and Goodhew P J 1996, Phil. Mag. **A73**, 1439

Matthews J W and Blakeslee A E 1974, J. Cryst. Growth **27**, 118

Matthews J W and Blakeslee A E 1975, J. Cryst. Growth **29**, 273

Matthews J W and Blakeslee A E 1976, J. Cryst. Growth **32**, 265

Mitchell T E and Unal O 1991 J. Electronic Mats. **20**, 723

Morishita Y, Goto S, Nomura Y Tamura M, Isu T and Katayama Y 1994 Jap. J. Appl. Phys. 2, **33**, L9

Murray R T, Kiely C J, Goodhew P J and Hopkinson M 1993 Inst. Physics Conf. Series **138**, 309

Perovic D D and Houghton D C 1995 Inst. Physics Conf. Series **146**, 117

Powell A R, Iyer S S and LeGoues F K 1994 Appl. Phys. Lett. **64**, 1865

Tobin S P, Smith F T J, Norton P W, Wu J, Dudley M, Dimarzio D and Casagrande L G 1995 J. Electronic Mats. **24**, 1189

Tsuchiya T, Taniwatari T, Komori M, Tsuneta R and Kakibayashi H 1994 Jap. J. Appl. Phys. 1, **33**, 230

Tuppen C G, Gibbings C J, Hockly M and Halliwell M A G 1990 Appl. Phys. Lett. **56,** 140

Ulhaq-Bouillet C, Lefebvre A and Di Persio J 1994 Phil. Mag. **A69**, 995

van der Ven J, Weyker J L, van der Meerakter J E A M and Kelly J J 1986 J. Electrochem. Soc. **133**, 799 and **134**, 989

Whitehouse C R, Barnett S J, Soley D E J, Quarrell J, Aldridge S J, Cullis A G, Emeny M T, Lohnson A D, Clarke G F, Lamb W, Tanner B K, Cottrell S, Lunn B, Hogg C and Hagston W 1992 Rev. Sci. Instrum. **63**, 634

Zou J and Cockayne D J H 1996 J. Appl. Phys. **79,** 7632

Inst. Phys. Conf. Ser. No 157
Paper presented at Microsc. Semicond. Mater. Conf., Oxford, 7–10 April 1997

A mechanism for "double half dislocation loops" nucleation in low misfit epitaxial $Ge_x Si_{1-x}$ on Si

P B Hirsch

Department of Materials, University of Oxford, Parks Road, Oxford OX1 3PH

ABSTRACT: A three-stage mechanism is discussed for the formation of "double half loops" observed by Perovic and Houghton (1992) in low misfit epitaxial layers of $Ge_x Si_{1-x}$ on Si. First, prismatic loops are nucleated in Ge precipitates at the interface to relieve their strain. Second, these loops are transformed into two glide loops on different (111) planes. Third, these loops bow out by growth of the precipitates in the epitaxial layer and expand along the interface. On reaching a critical size, the loops expand in the epitaxial layer to form the "double half loops". Several steps in the process are controlled by vacancy concentration and diffusion and are consistent with the recently reported reduction in nucleation density by Si irradiation (Stirpe et al 1997).

1. INTRODUCTION

Perovic and Houghton (1992, 1995) have shown that in low misfit strain $Ge_x Si_{1-x}$ epitaxial layers on (001) Si, 60º misfit dislocations are generated from "double half loops", which are thought to be nucleated from small (~ 15 Å diameter) Ge platelets at the interface. The activation energy for this nucleation stage was found to be 2.5 ± 0.5 ev, independent of growth technique and heterostructure geometry, and attributed to a process of nucleation of a vacancy loop by the condensation of vacancies (Perovic and Houghton 1995). Figure 1 (taken from Perovic and Houghton 1992) is a TEM plan view micrograph of double half loops, and (in the inset) of the small interstitial type strain centres in the interface. Figure 2 shows the geometry of these half loops, which lie on two different (111) planes, which intersect the (001) interface in two orthogonal [110] directions, and whose Burgers vectors are at 60 º to one another.

This paper examines the conditions under which such platelets might act as sites for nucleation of "double half loops" which expand across the epitaxial layer and act as sources for the 60º misfit dislocations.

2. NUCLEATION OF DISLOCATION LOOPS IN PRECIPITATE

2.1 Steps in the nucleation process

We assume that Ge platelets occur at the (001) interface, in depressions in the Si. There will be large misfit stresses parallel to the platelet, and it is proposed that the first step in the nucleation process is the formation of a prismatic loop, Burgers vector $\frac{1}{2}[110]$ normal to the platelet (fig. 3a), by aggregation of vacancies. This relieves the misfit strain in the [110] direction, but not in the orthogonal [1$\bar{1}$0] direction. In order to produce relief in the [1$\bar{1}$0] direction the transformation shown in figs 3b, c, d in projection parallel to (001) is proposed: First, the loop lying originally on (110) tilts about [$\bar{1}$10] by prismatic glide until

122

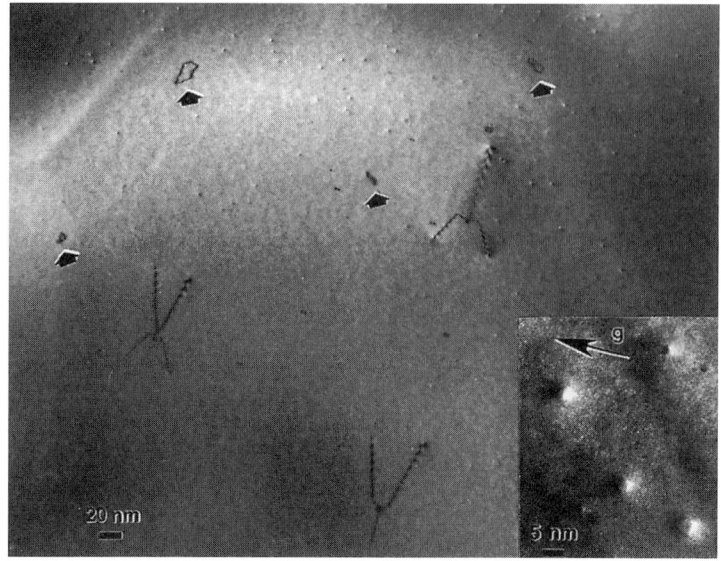

Fig.1. Plan view TEM image of interstitially strained platelets at the Ge$_{0.5}$Si$_{0.5}$/(001)Si interface (see insert), and "double half loops" in the epitaxial layer (from Perovic and Houghton (1992)).

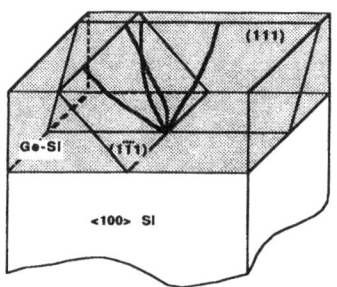

Fig.2. Geometry of "double half loops"; they lie on two different (111) planes which intersect (001) along orthogonal <110> direction (from Perovic and Houghton (1992))

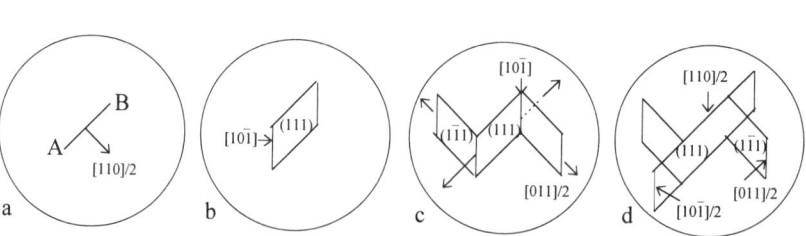

Fig. 3. Nucleation of prismatic loop $\frac{1}{2}$[110], and its dissociation into glide loops $\frac{1}{2}$[10$\bar{1}$] on (111) and $\frac{1}{2}$[011] on (1$\bar{1}$1), in Ge platelet at (001) Si/Ge$_x$Si$_{1-x}$ interface; projected onto (001).

its plane is parallel (111). The short sides of the loop A, B, are then assumed to be parallel to $[10\bar{1}]$ (fig. 3b). The dislocations along $[10\bar{1}]$ then dissociate into $\frac{1}{2}[10\bar{1}]$ on (111) and $\frac{1}{2}[011]$ on $(1\bar{1}1)$; the latter can then glide on $(1\bar{1}1)$ (fig. 3c), relieving the strain along $[1\bar{1}0]$, while the former glides on (111) relieving the strain along [110]. The final configuration is shown in fig. 3d; it consists of two glide loops, on (111) and $(1\bar{1}1)$ respectively, with Burgers vectors $\frac{1}{2}[10\bar{1}]$ and $\frac{1}{2}[011]$ at 60° to each other, which are the nuclei of the "double half loops" observed after their expansion in the alloy.

There is the possibility that the initial defect is a tetrahedron of stacking faults (Silcox and Hirsch (1959)), but the details of the transformation to shear loops remain to be worked out. Alternatively, the shear loop may be nucleated by formation of a Frank loop, followed by unfaulting by nucleation of a Shockley partial dislocation. This generates a dislocation loop in which the glide dislocation at the top and bottom of the platelet precipitate lie on different, parallel, (111) planes. The separation of these dislocations by the stress field of the epitaxial layer is likely to be easier than for the case considered here where these dislocations are coplanar. This mechanism looks promising, but the details need to be determined.

2.2 Nucleation of prismatic loop

The aspect ratio of the Ge platelet is not known. We have calculated the activation energy U_n, for formation of the original prismatic loop assuming a spherical precipitate, and circular dislocation nucleus, and assuming a linear misfit for the Ge precipitate of 4% in the Si matrix. With the dislocation energy given by Bacon and Crocker (1965) and the usual core parameter $\alpha = 1$, we find the critical radius $r_c \sim 1.45b$, and $U_n = 2.34$ ev. If the thickness of the platelet is much smaller than its diameter, the compressive stress parallel to the platelet is smaller than for the spherical precipitate and the critical radius is found to be $r_c \sim 2.2b$, and $U_n = 10$ ev which is not plausible. In either case the activation energy may be reduced if there is a supersaturation of vacancies present (Brown and Woolhouse 1970). Preliminary estimates suggest that the activation energy for nucleation of a tetrahedron of stacking faults is only ~ 1 ev for a spherical precipitate.

It appears from these estimates that if the thickness of the platelets is less than $\sim 3b = 12$ Å, prismatic loops would not be stable (unless a supersaturation of vacancies exists). The image contrast of the small strain centres in figure 1 is compatible with equal radial displacements in the defect in all directions parallel to the interface; it is not consistent with the presence of a prismatic vacancy loop of the type discussed above. This suggest that these precipitates are too thin or do not contain sufficient vacancies for loops to be nucleated.

The question arises why such precipitates are stable. Assuming that the alloy behaves as an ideal solution in thermodynamic equilibrium, the free energy ΔE of the precipitate (radius r, thickness t) relative to the state in which all the Ge in the precipitate is dissolved in the alloy, and replaced by Si is (neglecting differences in shear modulus between precipitate and epitaxial layer)

$$\Delta E = kT \frac{\pi r^2 t}{\Omega} \left\{ c_p \ell n c_p + (1 - c_p)\ell n (1 - c_p) + c_p \ell n \frac{(1 - c_0)}{c_0} \right\}$$
$$- \pi r^2 t 4\mu \left(\frac{(1 + v)}{(1 - v)} \right) \varepsilon^2 c_p \left(c_0 - \frac{1}{2} c_p \right) \tag{1}$$

where T is the temperature, k the Boltzmann constant, Ω the atomic volume, c_0, c_p the Ge concentration in the epitaxial alloy and precipitate respectively, μ the shear modulus, v Poisson's ratio, ε the linear misfit between Ge and Si (4×10^{-2}). The first two terms in the curly bracket comprise the configurational entropy of the precipitate, the third term that lost in the alloy when Ge atoms are transferred to the precipitate, the fourth term is the decrease in strain energy of the alloy, and the fifth term the strain energy of the precipitate. It is easy to show that for a constant r, t the energy is a minimum when $c_o = c_p$. The images show that this is clearly not the case, and we have to conclude that the precipitates are far from thermodynamic equilibrium, i.e. diffusion is too slow. On the other hand we may expect vacancies to be attracted and bound to the Ge platelets, thereby reducing their strain energy. It is worth noting that the number of vacancies required for the critical radius for nucleation of a loop in a spherical precipitate is approximately 19, for a tetrahedron it is only about 6. Such large numbers are plausible only in precipitates which are much larger than the critical loop radius. This suggests that loop nucleation is more likely in the largest precipitates within the distribution, and that the rate controlling process is diffusion of vacancies to the precipitates.

The nucleation of prismatic loops in the precipitates is the first step in the proposed mechanism for the formation of the "double half loops". The subsequent steps are discussed in the next section.

2.3 Nucleation of glide loops and their expansion in the epitaxial layer

No attempt has been made to estimate the activation energy for the dissociation process in figures 3b, c, d. It is possible that the dissociation takes place first at the periphery of the precipitate, where the shear stresses are very large (e.g. Weatherly 1968), i.e. when the initial prismatic loop has expanded across the platelet by climb, and A and B in figure 3a are at the periphery. The stresses for this particular dissociation route remain to be worked out. But for the glide dipoles to be stable their separation d, and therefore the precipitate thickness must exceed a critical value. For a thin platelet and a biaxial stress given by $2(1 + v) \mu\varepsilon/(1 - v)$, the stability condition becomes

i.e.
$$\frac{d}{b} > \frac{\left(1 - \frac{v}{4}\right)}{4\pi(1 + v)\varepsilon S} \ell n \frac{\alpha d}{b} \qquad (2)$$

where ε is the Ge misfit (4%), S the Schmid factor. Assuming $v = 0.215$, $\alpha = 1$, $S = 0.408$, we find a critical value of $d/b = 7.8$.

This value of d/b corresponds to a minimum platelet thickness of $d \sin 54°44' = 6.4b$, which is considerably greater than that required for nucleation of the initial prismatic loop (see section 2.2).

In order for the glide loop to expand in the epitaxial layer, the misfit stress has to be sufficient to overcome the line tension of the dislocation and the attraction of the dislocation remaining at the interface. A lower limit to the precipitate thickness can be estimated from the condition that the misfit stress must exceed the attraction between two infinitely long parallel 60° glide dislocations of opposite sign lying at the top and bottom of the precipitate on the same (111) plane; the minimum thickness of precipitate is found to be ~ $3.1b/c_o$ where c_o is the Ge concentration of the epitaxial layer. For $c_o = 0.25, 0.2, 0.15, 0.1$ this value is 12.4b, 19.5b, 20.7b, 31b respectively. The corresponding diameters of the platelets are expected to be larger than these thicknesses. [These values may need to be modified somewhat to take account of the platelet stresses just outside the precipitates.] When the critical thickness is exceeded, the dislocation at the precipitate/epitaxial layer interface begins to bow out; there is a driving force for the dislocation pinned at the interface between the Ge precipitate and the silicon substrate to lengthen by precipitate growth along the interface to enable the dislocation loop in the epitaxial layer to bow out further. At this stage in the process growth along the interface can occur at the expense of thickness by diffusion within the precipitate. There will be a configuration at which the loop in the epitaxial layer becomes unstable in the layer; this has not yet been determined. The expansion of the glide loop towards the surface is envisaged as shown in figure 4.

This mechanism therefore imposes minimum size conditions on the precipitates for it to operate. The three steps; the nucleation of the prismatic loop, the transformation to glide

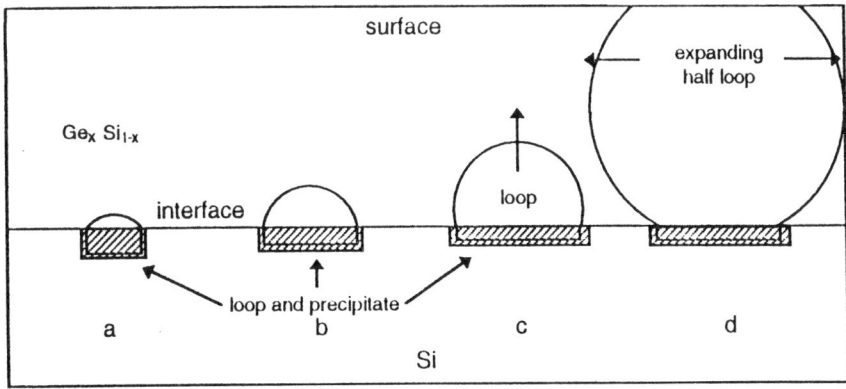

Fig. 4. Expansion of glide loop from a to d by lateral growth of precipitate along the interface, and by glide in the epitaxial layer (precipitates shown shaded).

loops within the precipitate, and the expansion of these glide loops in the epitaxial layer require Ge (or Ge rich) precipitates whose minimum thickness and/or diameter increases with each step.

It is of course possible that the "double half loops" are formed from a few original precipitates which have the required size for all three steps in the process. However, to transform the original loop nuclei, most of which are estimated to be ~ 4b in diameter, into loops of much larger diameter and thickness, we propose that the original precipitates grow either by combining with other precipitates, or by diffusion of Ge from the epitaxial alloy to the precipitates. With regard to the first of these mechanisms there is an elastic interaction between the dislocation loops and the interstitial type Ge precipitates without loops, which varies as Ad^{-4} where d is the distance between the precipitates and A is the loop area (Bastecká and Kroupa 1964). This interaction is attractive over a large range of angles between the Burgers vector and the line joining the loop and precipitate. The driving force for this amalgamation is the interaction energy between the dislocation and the strain field of the precipitate; the self strain energy of the precipitates remains unchanged. Similarly there is an attractive interaction between two suitably oriented vacancy or shear loops, which varies as $A_1A_2d^{-4}$, where $A_1 A_2$ are the areas of the two loops (Kroupa 1962). When precipitates with loops combine with other precipitates with or without loops, the areas of the loops increase and the attractive force increases, resulting in an acceleration of the accretion process.

With regard to the second mechanism, i.e. growth of the precipitates with dislocation loops by diffusion of Ge from the alloy to the precipitate, there is a driving force from the expansion of the dislocation loop, and from the reduction in strain energy of the epitaxial layer; the strain energy of the precipitate acts as a retarding force, and the configurational entropy causes the precipitate to be diluted and can provide a positive or negative driving force depending on the concentration of the precipitate relative to that in the alloy. In practice it is more likely that growth occurs by combining with other precipitates because the concentration of vacancies needed for diffusion will be greater in the precipitates than in the epitaxial layer because of the interaction with the larger compressive stress field, and because the activation energy for diffusion of Ge and Si in Ge is less than that in Si. The precipitates can come together by Ge atoms diffusing around the surface of the precipitates from one side to the other, or by Si atoms diffusing through the precipitates.

3. DISCUSSION

The three-step model presented for the formation of double half loop sources imposes conditions of minimum sizes on the precipitates, which are considerably larger than those of the majority of the precipitates observed by TEM, as judged by the image contrast. The nucleation process is likely only for any suitably large precipitates within the distribution, or for those for which the conditions for growth by combination with neighbouring precipitates are favourable. Thus only a very small proportion of the observed precipitates is likely to be successful in generating "double-half loops".

With regard to the kinetics of the nucleation process, the first step depends on the presence of sufficient numbers of vacancies in the precipitates, which is controlled by vacancy diffusion from the epitaxial layer to the precipitate. The activation energy for the second step, i.e. the nucleation of the shear loops from the prismatic loops (or perhaps tetrahedra) is not known; however precipitate growth involving diffusion may be needed before nucleation can take place.

For the third step, i.e. the expansion of the shear loops in the epitaxial layer, the driving forces make it possible in principle for the original shear loops to expand into the double half loop sources in the epitaxial layer by a combination of diffusion for precipitate growth along the interface, and loop expansion by glide. The experimentally observed activation energy for the nucleation stage of strain relaxation (2.5 $\pm$ 0.5 ev) (Perovic and Houghton 1995, Stirpe, Perovic, Lafontaine and Goldberg 1997) is consistent with dislocation glide and diffusion (particularly of Ge and Si in Ge). The first step in the proposed nucleation mechanism, which depends on vacancy concentration, and the subsequent steps which involve vacancy controlled diffusion, are consistent with the observed reduction of the nuclei density by Si irradiation (Stirpe et al 1997), since the concentration of vacancies will be reduced by the Si interstitials.

This paper should be regarded as a preliminary study. There may be easier pathways to produce the "double half loops". As mentioned above the mechanism involving the formation of a Frank loop followed by unfaulting appears promising, and is being explored at the present time. However, the conclusions concerning the importance of vacancies and diffusion will remain unchanged.

ACKNOWLEDGEMENTS

My thanks are due to Professor Perovic for helpful discussions, and for providing figures 1, 2, and to Dr A. Cerezo and Professor A. Cullis for valuable advice.

REFERENCES

Bacon D J and Crocker A G 1965 Phil. Mag. **12**, 195
Bastecká J and Kroupa F 1964 Czech. J. Phys. **B14**, 443
Brown L M and Woolhouse G R 1970 Phil. Mag. **21**, 329
Kroupa F 1962 Phil. Mag. **7**, 783
Perovic D D and Houghton D C 1992 Mat. Res. Soc. Symp. Proc. **263**, 391
Perovic D D and Houghton D C 1995 Inst. Phys. Conf. Ser. No **146**, 117
Silcox J and Hirsch P B 1959 Phil. Mag. **4**, 72
Stirpe M B, Perovic D D, Lafontaine H L and Goldberg R D 1997 Inst. Phys. Conf. Ser., in press
Weatherly G C 1968 Phil. Mag. **17**, 791

Inst. Phys. Conf. Ser. No 157
Paper presented at Microsc. Semicond. Mater. Conf., Oxford, 7–10 April 1997
© *1997 IOP Publishing Ltd*

Controlling misfit dislocation generation in strained layer epitaxy by point defect injection

M B Stirpe[†], D D Perovic[†], H L Lafontaine[*] and R D Goldberg[#]

† Department of Metallurgy and Materials Science, University of Toronto, Toronto, M5S 3E4, Canada
* Institute for Microstructural Sciences, National Research Council, Ottawa K1A 0R6, Canada
# Department of Physics, University of Western Ontario, London N6A 3K7, Canada

ABSTRACT: A detailed study on the evolution of dislocations has been conducted on Ge_xSi_{1-x}/Si heterostructures during rapid thermal anneals (RTA) of 540 keV self-irradiated Si versus as-grown samples. The implant fluence of 2×10^{14} ions/cm^2 and a substrate temperature of 25°C were used for these experiments. Nomarski interference microscopy has been used to quantitatively examine the behaviour of misfit dislocation generation. Dislocation densities were reduced by as much as an order of magnitude in the implanted samples relative to the as-grown material. Secondly, nucleation rates were reduced significantly following implantation. These results have been interpreted in terms of a point defect-controlled misfit dislocation nucleation mechanism.

1. INTRODUCTION

Considerable attention has been given to the development of strained layer heterostructures consisting of the group-IV alloys SiGe. The retention of structural perfection during epitaxial growth and thermal processing is crucial for the development of Ge_xSi_{1-x}/Si strained layer devices such as heterojunction bipolar transistors (HBT), resonant tunneling diodes, and light-emitting diodes (LED). Such heterostructures are, in general, metastable and can relax through the injection of misfit dislocations at the Ge_xSi_{1-x}/Si interfaces upon elevated temperature exposure (Houghton, 1991).

The sources of nucleation of misfit dislocations were extensively examined by the group of Perovic *et al* (1990). It was shown using transmission electron microscopy (TEM) that interfacial nucleation of 60° misfit dislocations was associated with sub-nanometre sized Ge-rich platelets that evolve from strain-induced surface segregation during growth (Perovic and Houghton 1992, 1993). This paper will further consider nucleation processes in low misfit heterostructures, which maintain planar interfaces during growth and subsequent annealing treatments.

Low misfit heterostructures were chosen for this work to maintain consistency with the nucleating mechanisms taking place in the "Stage I" regime (dislocation densities $\leq 10^5$ cm^{-2}). Perovic and Houghton (1995) defined this regime to be *nucleation-limited*, where misfit dislocation nucleation was observed to increase linearly with time, beyond some background density, N_o (see Houghton, 1991). The misfit dislocation velocity remains constant through this regime resulting in a linear increase in the overall strain relaxation rate with time. Activation energies were observed to be $Q_n = 2.5 \pm 0.5$ eV and $Q_v = 2.25 \pm 0.05$ eV for nucleation and glide, respectively (Houghton 1991). The activation energy for strain relaxation in the Stage I regime was found to be $Q_r = 4.2 \pm 0.5$ eV. This value is consistent with combining the nucleation and glide energies (i.e. $Q_r = Q_n + Q_v$) in series. The nucleation activation energy value found experimentally in this work (Q_n) is consistent with very recent work by Wickenhäuser *et al* (1997).

In this study, we present a novel method for controlling the generation of misfit dislocations in strained layers by point defect injection via ion-implantation in low misfit heterostructures of Ge_xSi_{1-x}/Si ($x < \sim 0.13$). Bulk measurements were made to determine the misfit nucleation rates and dislocation densities for various Ge_xSi_{1-x}/Si heterostructures. Two structures chosen for this study which possess varying effective stresses, (i.e. $\tau_{eff} = 103$ MPa and $\tau_{eff} = 57$ MPa) to drive misfit dislocation formation (see Houghton 1991). Ion implant energies and fluences were chosen following the previous study by Labrie *et al* (1996).

2. EXPERIMENTAL DETAILS

Ge_xSi_{1-x}/Si heterostructures were grown by ultra-high vacuum chemical vapour deposition (UHV-CVD) on <100>-oriented Si substrates using the procedure outlined by Lafontaine *et al* (1996). The UHV-CVD growth technique was chosen to ensure low values of the *initial* misfit dislocation source density, N_o, present in the as-grown structure at $t = 0$ (i.e. substrate threading dislocations, substrate-buffer precipitates) (Perovic and Houghton, 1995). N_o is $\leq 10^2$ cm^{-2} for UHV-CVD grown samples due to the thorough substrate cleaning prior to growth. Table 1 displays relevant data for the two heterostructures considered in this study.

Table 1: Structural parameters for UHV-CVD Ge_xSi_{1-x}/Si heterostructures

Layer	x	$h^{(SiGe)}$ nm	$H^{(Si)}$ nm	Periods	τ_{eff} (MPa)
CVD-9	0.10	15	21	10	103
CVD-61	0.13	7	29	10	57

The structures were subsequently implanted with Si ions at an energy of 540 keV with an implant fluence of 2×10^{14} ions/cm^2 at a substrate temperature of 25°C (see Goldberg, 1995 for implantation procedures). The as-grown and the implanted samples were rapid thermal annealed (RTA) in a nitrogen atmosphere using a Heatlpulse 410 furnace for times ranging from 5 to 100 s, at temperatures up to 1000°C.

Nomarski interference microscopy of chemically etched (Schimmel etch, 4 parts 48% HF and 5 parts 0.3M-CrO$_3$) samples was performed to quantitatively measure misfit dislocation nucleation rates and densities. Representative areas (several cm^2) away from edges and scratches were scanned optically.

3. RESULTS

Nomarski interference microscopy revealed large areas (several cm^2) of etched surfaces for quantitative measurement of misfit dislocation nucleation rates and densities following post-growth implantation and annealing. Misfit dislocation evolution can be calculated per unit time and per unit area to determine nucleation rates. Spacings between misfit segments can be measured to calculate the extent of misfit strain relaxation (readily obtained from micrographs such as in Figure 1).

Although nucleation rates vary between samples, due to layer geometry and effective stresses, the activation energy for nucleation is constant at a value of $Q_n = 2.5 \pm 0.5$ eV. This value is recognized as a universal value for misfit dislocation nucleation in low misfit GeSi/Si heterostructures, regardless of growth technique or heterostructure geometry (Perovic and Houghton 1995). The activation energy for the extent of misfit strain relaxation is measured to be 4.3 ± 0.4 eV. This is about double that for misfit glide and close to value of 4.75 eV reported by Houghton (1991).

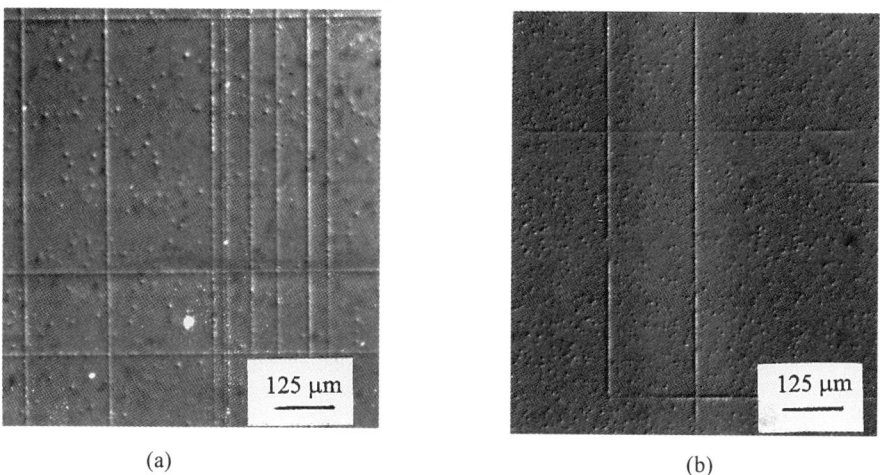

(a)   (b)

Figure 1: Nomarski images showing densities of misfit dislocations between (a) as-grown and (b) implanted CVD-9 (annealed at 850°C for 30s)

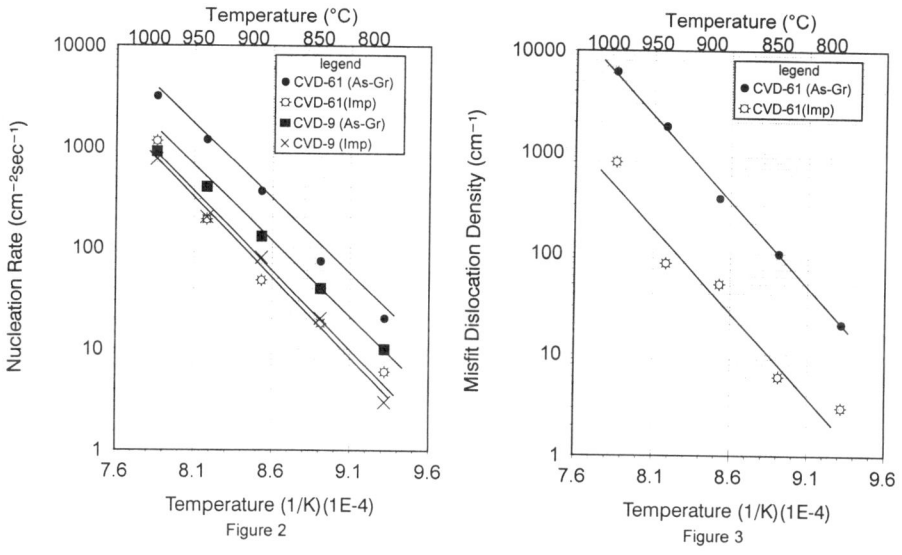

Figure 2: Misfit dislocation nucleation rate vs. anneal temperature for the Ge_xSi_{1-x}/Si heterostructures (see Table 1). (As-Gr: as-grown and RTA, Imp: implanted and RTA)

Figure 3: Comparison of misfit dislocation density vs. anneal temperature.

For the first time, it has been shown that the nucleation of misfit dislocations can be controlled and reduced by the ion implantation procedure outlined above. The dislocation densities of the as-grown and implanted samples differ by as much as a factor of eight for CVD-61 and a factor of five for CVD-9. The rate-limiting step in misfit dislocation nucleation has been attributed to vacancy formation and/or migration during the generation of the incipient loop at growth-induced interstitial perturbations (i.e Ge-rich platelets) (Perovic *et al*, 1995). The results shown here are consistent with a mechanism where Ge-rich platelets, which possess an interstitial elastic stress field, can be relieved by the injection of point defects (i.e. vacancies) and hence reduce the overall stress concentrations surrounding the perturbation which assist misfit dislocation generation. Therefore, an overall decrease in strain relaxation via misfit dislocations is expected, which is in agreement with our results.

4. SUMMARY

Based on a quantitative study conducted on Ge_xSi_{1-x}/Si heterostructures, it has been demonstrated that the onset of strain relaxation via misfit dislocations can be controlled by point defect injection. Point defects which are generated by ion implantation will diffuse during post-growth rapid thermal annealing and subsequently relieve the stress concentrations surrounding growth-induced Ge perturbations. The activation energy value of 2.5 ± 0.5 eV is universal for as-grown and implanted GeSi/Si samples indicating that misfit dislocation nucleation is the rate-limiting step in the strain relaxation process.

ACKNOWLEDGEMENTS

The authors would like to thank Dr. D C Houghton and Dr. J-M Baribeau for their contributions to this work. Financial support from the Natural Sciences and Engineering Research Council of Canada is gratefully appreciated.

REFERENCES

Goldberg R D, Simpson T W, Mitchell I V, Simpson P J, Prikryl M, and Weatherly G C 1995 Nucl. Inst. Meth. B **106**, 216
Houghton D C 1991 J. Appl. Phys. **70**, 2136
Labrie D, Aers C, Lafontaine H, Williams R L, and Charbonneau S 1996 Appl. Phys. Lett. **69**, 3866
Lafontaine H, Houghton D C, Elliot D, Rowell N L, Baribeau J-M, Laframboise G, Sproule G I, and Rolfe S J 1996 J. Vac. Sci. Technol. **B 14(3)**, 1675
Perovic D D and Houghton D C 1992 Mat. Res. Soc. Symp. Proc. **263**, 391
Perovic D D and Houghton D C 1993 Phys. Stat. Sol. (a) **138**, 425
Perovic D D and Houghton D C 1995 Inst. Phys. Conf. Ser. **146**, 117
Perovic D D, Weatherly G C and Houghton D C 1990 Mat. Res. Soc. Symp. Proc. **160**, 65
Wickenhäuser S, Vescan L, Schmidt K and Luth H 1997 Appl. Phys. Lett. **70** 324

Inst. Phys. Conf. Ser. No 157
Paper presented at Microsc. Semicond. Mater. Conf., Oxford, 7–10 April 1997
© 1997 IOP Publishing Ltd

Defect distribution in compositionally graded epitaxial SiGe layers on Si substrates

K Lyutovich, F Ernst*, F Banhart*, I Silier, A Gutjahr and M Konuma

Max-Planck-Institut für Festkörperforschung, Heisenbergstrasse 1, 70569 Stuttgart, Germany
*Max-Planck-Institut für Metallforschung, Heisenbergstrasse 1, 70569 Stuttgart, Germany

ABSTRACT: We have grown compositionally graded SiGe layers on Si with atmospheric pressure vapour phase epitaxy and liquid phase epitaxy. Ge composition profiles were determined using energy-dispersive X-ray spectroscopy. Atomic force microscopy revealed the surface morphology of the layers. Transmission electron microscopy of cross-sectional specimens showed differences in the dislocation distribution depending on composition profile and growth method.

1. INTRODUCTION

In recent years, strained SiGe/Si heteroepitaxial systems have received increasing attention. Besides that, however, strain-relaxed SiGe layers become more important (Whall and Parker 1995). SiGe layers are compatible with Si-based device technology and allow to control fundamental properties by changing Si/Ge ratio, but relaxed SiGe layers are more stable compared to strained systems. Particular interest has been given to compositionally graded layers because of the possibility to use such layers with an enhanced range of physical parameters for designing electronic devices as well as to accommodate the lattice mismatch in heteroepitaxial systems and, therefore, to reduce the strain (buffer layers).

None of the growth methods presently in use can provide the entire range of SiGe parameters needed for device application. Therefore, different epitaxy techniques are used to produce SiGe/Si heterostructures, e.g., molecular beam epitaxy (MBE), vapour phase epitaxy (VPE), liquid phase epitaxy (LPE). In this work comparative investigations of graded SiGe layers grown by two different methods, LPE and atmospheric pressure vapour phase epitaxy (APVPE), are carried out. By using transmission electron microscopy (TEM) and atomic force microscopy (AFM), the surface morphology and features of dislocation distribution in relaxed SiGe layers with various composition profiles deposited on (111) Si substrates are shown.

2. EXPERIMENTAL

$Si_{(1-x)}Ge_x$ epitaxial layers were grown on (111) Si substrates by APVPE and LPE. Both techniques allow the growth of thick compositionally graded layers at relatively high temperatures (given below), thus promoting the relaxation of lattice mismatch-induced strain.

For the growth of layers by APVPE, the $SiCl_4$–$GeCl_4$–H_2 system and a temperature range 1100 - 1000°C were used, which provide simultaneous growth and etching reactions during the heterogeneous hydrogen reduction of $SiCl_4$ and $GeCl_4$ on the Si surface. In order to produce layers with compositional grading, gas phase composition and growth temperature during the deposition were changed either continuously (Lyutovich et al 1990) or stepwise.

LPE was performed using indium as a solvent. A centrifugal system designed for epitaxial growth on 4" wafers (Konuma et al 1993) was applied to move the solution over the Si substrate. The composition, thickness, and the growth rate of the layers were varied by changing the composition of the In–Si–Ge solution, growth temperature interval, and cooling rate. Compositionally graded LPE layers were deposited within the temperature interval 920 - 750°C. The $Si_{(1-x)}Ge_x$ layers discussed in this paper had thicknesses between 2 and 20 μm. The final Ge concentration at the layer surface ranged between x=0.03 and 0.23. To measure the Ge concentration profiles from the layer/substrate interface to the surface, energy-dispersive X-ray analysis (EDX) was performed with a scanning electron microscope (SEM), JEOL JSM 6400, and with a dedicated scanning transmission electron microscope (STEM), VG HB-501. By means of an AFM, Park Scientific Instruments, the influence of the Ge content on the surface morphology of the layers was studied.

3. RESULTS AND DISCUSSION

Fig. 1a shows Ge composition profiles that were measured for APVPE graded layers by EDX of the STEM. The profiles show a mostly linear increase of Ge concentration with increasing distance from the substrate/layer interface. Fig. 1b presents Ge composition profiles determined for LPE layers. The composition profiles are typical for the LPE layers grown in the temperature and solution used in growth experiments performed.

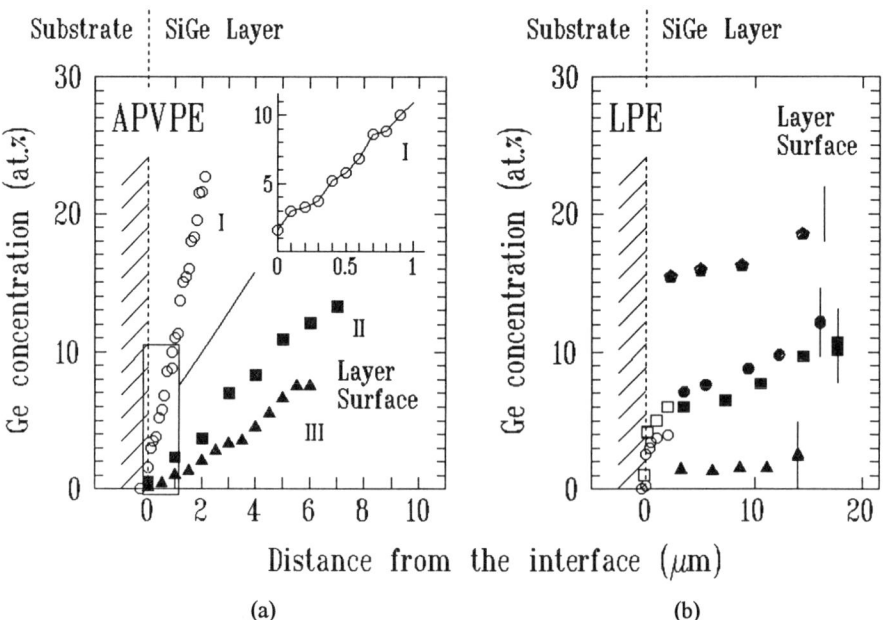

(a)   (b)

Fig.1. *Ge concentration profiles in $Si_{(1-x)}Ge_x$ layers grown by a) APVPE with various Ge compositional gradients, I - stepwise (see enlargement), II and III - continuous, measured by STEM, and b) LPE, measured by STEM (open symbols) and SEM-EDX (closed symbols).*

AFM studies showed that the graded APVPE layers with a Ge content up to x=0.03 have a planar surface. At higher Ge concentrations the surface became rough. LPE layers with comparable Ge concentration showed a similar surface morphology (Fig. 2). With increasing Ge content, the surface roughness became more pronounced on both, APVPE and LPE layers. AFM data allowed to estimate the scale of these surface undulations. Such modulations are commonly associated with the compressive strain originating from the lattice mismatch between layer and substrate (Cullis 1996). The ripples lie along <110> directions. The difference between the undulations on APVPE and LPE layers consists in their habitus and extension and may arise from the different growth mechanisms and kinetics. Growth rates of APVPE layers are 60-90 times higher than that of LPE layers within the ranges of growth parameters which we have used in these experiments. Layer growth in LPE occurs close to thermodynamical equilibrium, whereas in APVPE the layer growth takes place far from the equilibrium. This could explain the 20 - 40 times higher amplitude of the undulations in APVPE layers

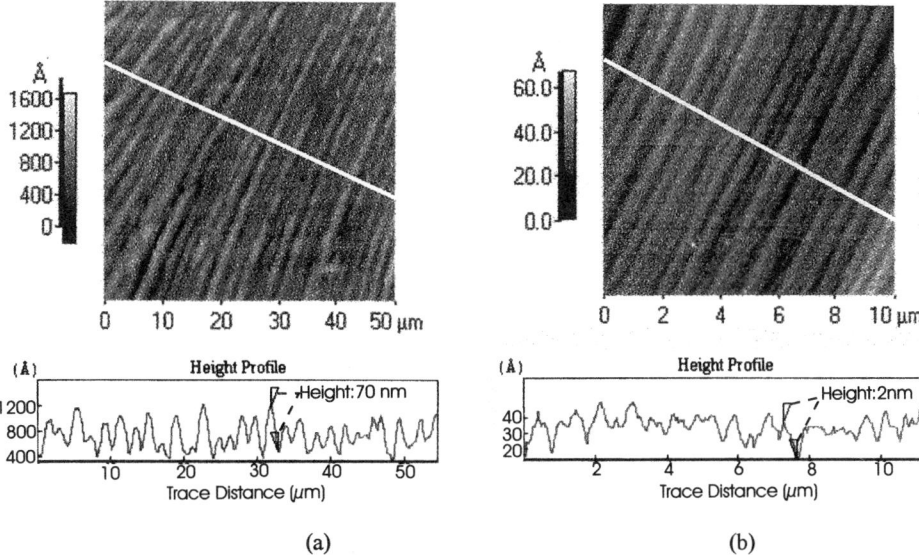

(a)   (b)

Fig. 2. AFM scans of the surface and surface profiles of $Si_{0.8}Ge_{0.2}$ layers grown by a) APVPE and b) LPE.

as compared with that of about 24Å in LPE layer. The wavelength of the undulations in APVPE layers is about 5 times larger than that of LPE-grown layers with similar composition.

Cross-sectional bright field TEM images from APVPE grown layers with Ge concentrations of 0.03 hardly show any contrast between layer and substrate. In these layers, a coarse-meshed misfit dislocation network lies parallel to the interface. In layers with Ge contents between 0.03 and 0.10, the interface appears with a distinct contrast. With increasing steepness of the composition profile the misfit dislocation density increases and the network extends into the layer, the overall dislocation density is higher than in layers with a lower Ge content, according to the increased lattice mismatch between layer and substrate. In the bulk of the layers with the continuous Ge gradient a complicated network of misfit dislocations, located partly parallel to the interface with different distances between them, is observed (Fig. 3a). In the layers with the stepwise Ge profiles, misfit dislocations are generated with a certain periodicity (Fig. 3b) to relax the strain in the corresponding depth of the layer. Their distribution conforms with the Ge profiles in the layers.

TEM was also used to investigate LPE layers. These layers with a Ge content of x=0.03 at the surface show a low density of misfit dislocations at the interface (Fig. 4a). Layers with Ge content of x=0.12 at the surface show a higher misfit dislocation density and a few threading dislocations (Fig. 4b). Relaxation in the layers occurs by lateral propagation of dislocation loops. However, these loops cannot spread over a large lateral distance because of the high density of obstacles such as threading dislocations. Layers with x=0.20 Ge show a higher density of threading dislocations which extend to the surface (Fig. 4c). The starting concentration of x=0.07 Ge at this interface results in a higher density of misfit and threading dislocations; thus the propagation of loops in the layers is effectively hindered (Banhart and Gutjahr 1996).

The characteristic difference between dislocation distributions in LPE and APVPE SiGe layers can be explained by the different growth mechanisms. The LPE layers have a higher dislocation density close to the interface because their growth starts with a higher Ge concentration (Fig. 1b). This fact also explains the formation of threading dislocations. These dislocations probably originate from the coalescence of islands which initially form on a substrate surface. The abruptness of the interface and the correspondingly high lattice mismatch between layer and substrate are assumed to lead to a three-dimensional growth mode (Chen et al 1992). The chemical interface in APVPE layers is more

134

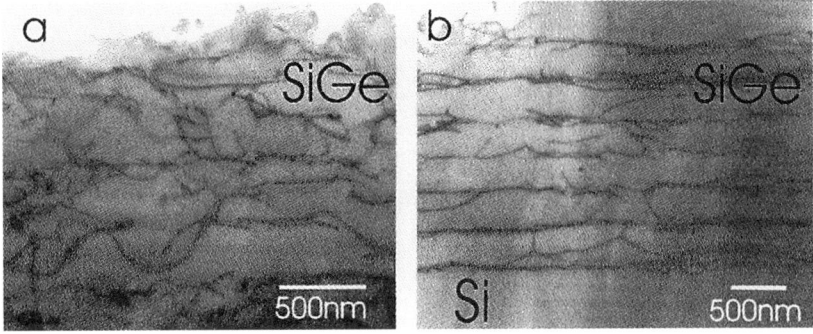

Fig. 3. *Cross-sectional TEM view of APVPE grown Si$_{(1-x)}$Ge$_x$ layers in which Ge concentration is continuously (a) and stepwise (b) graded.*

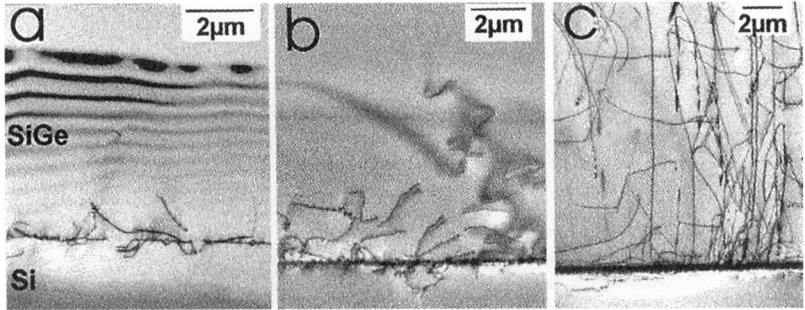

Fig. 4. *Cross-sectional TEM micrographs of LPE Si$_{(1-x)}$Ge$_x$ layers with x=0.003(a), 0.12(b), and 0.20(c) at the surface.*

diffuse because of the higher growth temperature (Bruley et al 1992). Interdiffusion of Si and Ge leads to a gradual accommodation of the lattice parameters during the initial stages of growth. This process promotes two-dimensional rather than three-dimensional growth and thus limits the density of threading dislocations. Discontinuous relaxation of misfit stress which is built up during subsequent growth of the layers leads to the observed nonuniform distribution of dislocations, whereas the stepwise Ge profile leads to a well organised sequence of two-dimensional misfit dislocation networks.

ACKNOWLEDGMENTS: The authors are thankful to H.-J. Queisser and E. Kasper for continuous interest and fruitful discussions. We gratefully acknowledge assistance of J.Thomas, K. Rombach, U. Bäder, S. Zehender, K. Davies. One of us (K.L.) thanks the Max-Planck-Gesellschaft for financial support.

REFERENCES

Banhart F and Gutjahr A 1996 J. Appl. Phys. **80**, 6223
Bruley J, Ernst F, and Lyutovich K 1992 Mat. Res. Soc. Symp. Proc. **238,** 469
Chen J X, Ernst F, Hansson P O, and Bauser E 1992 J Cryst. Growth **118**, 452
Cullis A G MRS Bulletin 1996 **21,** 21
Konuma M, Czech E, Silier I, and Bauser E 1993 Appl. Phys. Lett **63,** 205
Lyutovich K, Kulagina L, and Turaev E 1990 Phys. J. of Uzbek Academy of Sci **1,** 55
Whall T E and Parker E H C 1995 J. Mater. Sci. Mater. Electron. **6**, 249

Inst. Phys. Conf. Ser. No 157
Paper presented at Microsc. Semicond. Mater. Conf., Oxford, 7–10 April 1997
© *1997 British Crown Copyright*

On the growth of high quality relaxed $Si_{1-x}Ge_x$ layers on Si by vapour phase epitaxy

A J Pidduck, D J Robbins, D Wallis, G M Williams, A C Churchill, J P Newey, C Crumpton and P W Smith

DERA, St. Andrews Road, Malvern, WR14 3PS, UK

ABSTRACT: We present a direct comparison of the effects of growth temperature (over the range 560-800°C) and grading rate (over the range 4-50% Ge/μm) on the surface morphology and defect structure of relaxed $Si_{1-x}Ge_x$ layers (0.21<x<0.29) grown on Si substrates. Material quality is also compared with electron mobilities in strained Si quantum wells grown on these layers. We obtain best surface quality and highest electronic mobilities in the high temperature - low grade rate regime. The factors underlying optimisation of growth conditions are discussed with reference to the literature.

1. INTRODUCTION

Epitaxial growth of Si, under tensile strain, on relaxed $Si_{1-x}Ge_x$, is necessary to achieve a conduction band (CB) offset, and hence to confine electrons, in the Si-Ge materials system (Fitzgerald 1995). A high electron mobility channel can be formed in the strained Si layer, using remote doping to supply carriers. For $x \geq 0.2$ (-0.84% strain) the CB offset is large enough to be exploited in a modulation-doped field effect transistor (MODFET). The realisation of such devices with technologically useful performance and yield then depends on the availability of high quality relaxed $Si_{1-x}Ge_x$ substrates.

The usual approach to preparation of relaxed $Si_{1-x}Ge_x$ substrates has been, starting with commercially-available dislocation-free Si wafers, to deposit epitaxially a relaxation layer (RL), in which the Ge composition is increased (monotonically or step-wise) to the desired final value, followed by a constant Ge composition buffer layer (BL). The aim is to confine misfit dislocations (MD), which form and propagate to relieve the strain as it is introduced, within the RL. Ideally, the MD number density in the RL should be the minimum necessary to relieve the total misfit, so as to minimise the number of terminations able to thread upwards (threading dislocations, TD) into the active device layers. The roles of the BL are to separate the overlying active device layers from the MD strain fields, and to maximise the extent of strain relaxation at the BL surface.

Very high Si-channel electron mobilities, μ_e, in modulation-doped structures grown on such layers, were first reported by Mii et al (1991), Schaffler et al (1992) and Fitzgerald et al (1992) using molecular beam epitaxy (MBE). μ_e values of 96,000-180,000cm^{2}V^{-1}s^{-1} at 0.5-1.5K were achieved. The RL was grown at high temperature (T_g=700-900°C) with a slow Ge grading rate (GR=10%Ge/μm) in order to give existing MDs the maximum opportunity to increase in length by gliding outwards, along <110>, towards the edge of the wafer. The high T_g was thus chosen to maximise the MD glide velocity, and the low grade rate to reduce dislocation interactions (by increasing their vertical spacing) and to suppress the probability of nucleating new dislocations (by keeping the residual surface strain during growth, ε_s, small). The resultant TD density in such layers, graded to 24-32% Ge, was between $2 \times 10^5/\text{cm}^2$ and $5 \times 10^6/\text{cm}^2$ (Fitzgerald et al (1992), Hohnisch et al (1995)). This compares with TD levels down to near 10^3cm^{-2} reported by Dutartre et al (1994) in buffer layers grown by chemical vapour deposition (CVD) at 820°C on RLs graded to 32% Ge at rates $\leq 10\%\text{Ge}/\mu\text{m}$.

Subsequently, Ismail et al (1994, 1995, 1996) attained μ_e values at 0.4K of 300,000-400,000 $\text{cm}^2\text{V}^{-1}\text{s}^{-1}$ using ultrahigh-vacuum (UHV)-CVD at low temperatures (500-560°C) on relaxed $Si_{1-x}Ge_x$ layers step-graded to 25-40% Ge over 1μm. A BL thickness of about 1μm was necessary to ensure near-complete surface strain relaxation at these higher RL grade rates, and TD densities of about 10^6 cm^{-2} were measured in similar layers (Mooney et al (1995)). Earlier, TD densities as low as 10^4 cm^{-2} for a 20% Ge BL grown on a superlattice RL graded at $100\%\text{Ge}/\mu\text{m}$ had been reported (LeGoues et al 1992).

The above very high electron mobilities indicate the promising electronic quality of the strained Si channel. However, as already noted, mobility alone does not guarantee the suitability of the material for device applications. Indeed, as shown by Ismail 1996, μ_e values are insensitive to TD densities below about 10^8cm^{-2}, and cosmetic defects, which likewise only occupy a small fraction of the total surface area even if present at high levels, may also have little affect on μ_e values. Practical circuit microfabrication depends critically on minimising the densities of cosmetic defects and of TDs, and also the level of surface roughness, which invariably exhibits a pronounced <110> crosshatch pattern reflecting the underlying MD network. The present work aims, by use of a temperature-agile, UHV-background, vapour phase epitaxy (VPE) reactor (Robbins et al (1987, 1991), to make a direct comparison of the consequences of high vs low growth temperature (800°C and 560-610°C), and high vs low RL grade rate (<10 and ~50 %Ge/μm) on these wider aspects of relaxed $Si_{1-x}Ge_x$ layer quality. Recently, under high temperature low grade-rate conditions, Churchill et al (1997) have reported 0.3K electron mobilities of up to 270,000 $\text{cm}^2\text{V}^{-1}\text{s}^{-1}$ using this growth system.

2. EXPERIMENTAL

VPE was carried out on chemically cleaned 100mm dia. Si wafers after *in-situ* oxide desorption at 850-900°C in 130Pa H$_2$. $Si_{1-x}Ge_x$ layers were then grown at 560-800°C from SiH$_4$/GeH$_4$ mixtures in 15 Pa H$_2$. As noted previously, the layer structure comprised first a linearly- or stepwise- graded-Ge-composition RL followed by a constant composition BL. In some cases a further multilayer structure containing the modulation-doped strained-Si active layer was grown near 600°C.

Table 1 gives details of the layers grown for this study spanning the four regimes : (A) high T_g and GR, (B) low T_g / high GR, (C) low T_g and GR, (D) high T_g / low GR. The GR value listed is the average over the RL (i.e. final Ge x-value / RL thickness). Low grade rate (C and D) layers were grown using 4 times higher SiH$_4$/GeH$_4$ partial pressures than those used for the high grade rate (A and B) layers. The growth rates were approximately 0.4 and 0.1nm/s respectively. Epilayer surface morphology was monitored *in-situ* using laser light scattering (LLS) and examined *ex-situ* using Nomarski optical microscopy, differential phase contrast scanning optical microscopy (SOM-DPC; Bio-Rad Lasersharp SOM150), and atomic force microscopy (AFM; Digital Instruments Dimension 3000) in air with microfabricated Si cantilevers (Nanosensors). BL compositions and thicknesses were measured using secondary ion mass spectrometry (SIMS; Cameca 4f spectrometer) with 8keV O_2^+ incident ions. Thicknesses were calibrated from stylus profilometer measurements of SIMS crater depths, and compositions were calibrated against X-ray diffraction (XRD) measurements of selected BLs. XRD double-crystal rocking curves and triple-crystal reciprocal space maps of {004} and {044} reflections were obtained using a Bede D3 diffractometer. Where measured, the degree of BL relaxation was found to exceed 90%. Cross-sectional transmission electron microscopy (TEM) was carried out in a JEOL 4000EX microscope at 400KV under bright-field conditions after specimen thinning to electron transparency using a combination of mechanical, chemical and ion-assisted techniques.

Table 1 also lists the low temperature mobilities (μ_e) and corresponding 2-dimensional electron gas carrier concentrations n_e in specimens containing an active layer. There is a fixed relationship between μ_e and n_e for the group D layers which has been discussed by Churchill et al (1997). The final LLS intensities and root-mean-square (rms) roughness figures are included in table 1 as a measure of the surface quality. Rms roughnesses of the surface crosshatch pattern present on all specimens were calculated over 10µm x 10µm areas, after line-by-line tilting, from AFM images. The range of values given represents the position-to-position variation.

In the following we shall now consider in detail the surface and defect quality of layers corresponding with growth regimes A to D sequentially.

Table 1. Details of epitaxial layers

	Relaxation layer			Buffer layer		Active Layer		Surface quality	
Sample no.	T_g (°C)	GR$_{ave}$ (%Ge /µm)	form of grade	x	t (µm)		μ_e (1.5K) (cm^2/Vs) (n_e (cm^{-2}))	final LLS intensity (cps)	rms roughness (nm)
A1	800-650	45	6 steps	0.28	0.33 (650°C)	Yes	n.d.	3×10^6	20-25
A2	800	42	6 steps	0.27	0.33 (650°C)	Yes	n.d.	4×10^6	23-25
B1	560	36	10steps	0.29	0.09	Yes	24,000 (8×10^{11})	8×10^4	2-3
B2	560-605	52	9 steps	0.26	0.87	Yes	130,000 (6×10^{11})	5×10^5	3-4
C1	610	5	linear	0.22	1.84	No	-	$< 10^4$	6-9
D1	800	5	linear	0.24	0.88	Yes	260,000 (5×10^{11})	$< 10^4$	2-6
D2	800	5	linear	0.24	0.88	Yes	68,000 (3×10^{11})	$< 10^4$	3-6
D3	800	7	linear	0.23	0.68	Yes	202,000 (4×10^{11})	$< 10^4$	2-6
D4	800	4	linear	0.21	1.94	No	-	$< 10^4$	5-11

3. RESULTS AND DISCUSSION

3.1 Regime A : high T_g, high GR

A large rise in LLS intensity occurred at an early stage in the growth of step-graded A-type layers. Fig.1(a) shows the LLS trace obtained during growth of A1 in which the major intensity rise occurred as the Ge fraction was stepped from 10% to 14% at 800°C. Fig.1(b) is an AFM image from A1 after growth showing the surface crosshatch to comprise orthogonal <110> lines of 0.6-1μm width with a network of shallow flat-bottomed pits at their intersections. The final layer surface shows a high rms roughness level consistent with the high final LLS intensity.

In order to investigate the origin of the LLS rise, layers were prepared by terminating growth at different points in the 14% Ge step. Fig.2 compares the surface morphology (a) 13nm after incrementing the Ge concentration to 14% and (b) after completion of the 130nm-thick 14% layer. Fig.2(a) shows very clearly the origin of the LLS rise to be the development of <100> surface ripples associated with partial elastic relaxation of the layer strain (Pidduck et al 1992).

The ripple structure in fig.2(a) is locally perturbed by a few linear <110> features clearly associated with the first MDs. In some cases the perturbation is rather weak, whereas in most cases there is a deep central trough bordered by ridges and then by further weaker troughs. The systematic asymmetry in the ridge heights suggests a possible relationship with the Burgers vector of the underlying MD. By the time, in fig.2(b), that the 14% step was complete, <110> troughs have formed to such a density that the surface comprises a network of submicron-sized square or rectangular faceted pyramidal islands. Almost no <100> oriented ridges remain and the submicron lengthscale characteristic of the final layer surface (Fig.1(b)) has now set in.

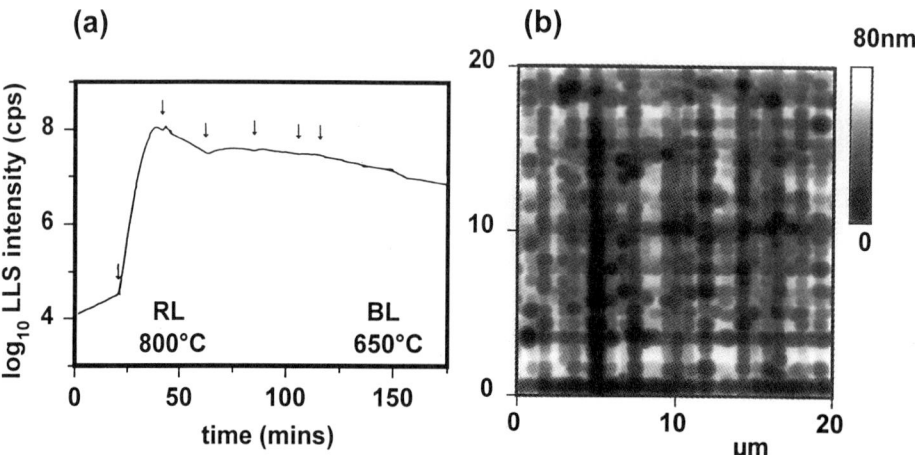

Figure 1. (a) logarithm of LLS intensity vs time during growth of specimen A1 RL and BL. Arrows denote stepwise increases in Ge concentration, from 10% (at t=0) to 30% in 6 steps, before finally reducing to 27% for BL growth (final arrow). T_g was reduced, also in steps, from 800°C to 650°C, the temperature used for BL growth, as the Ge fraction was increased. (b) 20μm AFM image of layer A1 surface. <110> directions lie parallel to the image axes.

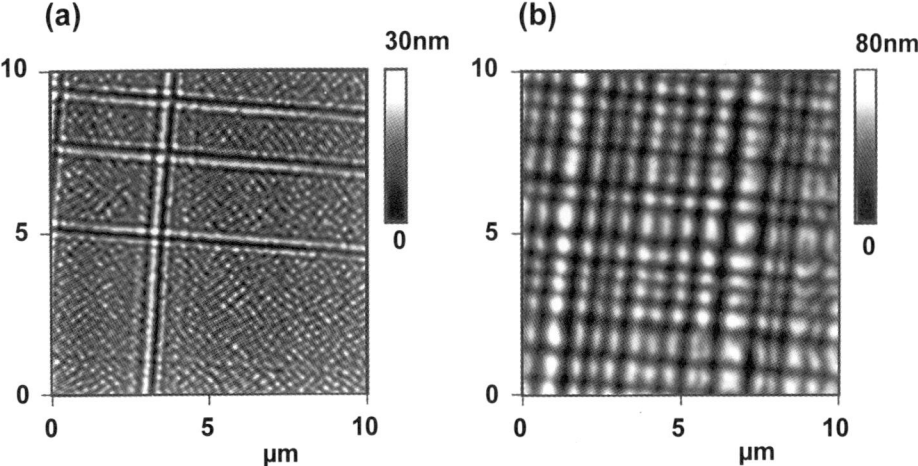

Figure 2. 10μm AFM images from two layers grown at 800°C and stopped (a) 13nm and (b) 130nm after increasing the Ge fraction from 10% to 14% (which is the point denoted by the first arrow in Fig.1(a)). <110> are tilted by about 5° from the figure axes.

These results confirm that the elastic relaxation mechanism is incompatible with high quality relaxed $Si_{1-x}Ge_x$ substrate formation, because of the inherently high roughness and the spatially fluctuating strain state. There is also an enhanced probability of dislocation nucleation at ripple troughs (Jesson et al 1993, Cullis et al 1995). The roughening is enabled by surface diffusion during layer growth and accelerates rapidly once initiated. This effectively results in a kinetic threshold primarily determined by the level of surface strain ε_s and T_g, but also dependent on growth rate and (since Ge-Ge bonds are weaker than Si-Si bonds) on surface Ge fraction. For VPE growth at a rate of about $0.1 nm.s^{-1}$ at 800°C in this work, since clearly no significant relaxation had occurred in the 10% Ge layer, this ε_s threshold must lie between 0.004 and 0.006. This compares with a threshold ε_s of 0.006-0.0075 observed by Dutartre et al (1994) at growth rates of about 2nm/min at 820°C. The phenomenon can clearly be avoided by growth at lower temperature (Cullis et al 1992) or, because ε_s depends directly on GR (Tersoff 1993), by using a lower GR. The dependence of TD density and roughening on GR has been studied in depth by Dutartre et al (1994). In the following we first explore the effect of reducing T_g, which should render the surface stable to higher strains (and thus higher GR) at higher Ge fractions.

3.2 Regime B : low T_g, high GR

These layers exhibited strong haze when viewed under bright light illumination. The haze was patchy on a scale of a few mm. The poor surface quality was consistent with a high final LLS intensity. The main LLS rise occurred at higher Ge fractions (15-21%) than during growth of A-type layers. Nomarski microscopy and SOM-DPC showed a distribution of surface pits, clustered in some areas into <110> linear or rectangular patterns. Fig.3 compares SOM-DPC images, and Fig.4 compares AFM images, from regions of high and low pit density. The pits were of various sizes and depths, typical dimensions being ~1μm dia. and 30-50nm depth. Their number density where densely packed was of order 10^6-$10^7 cm^{-2}$, and of

140

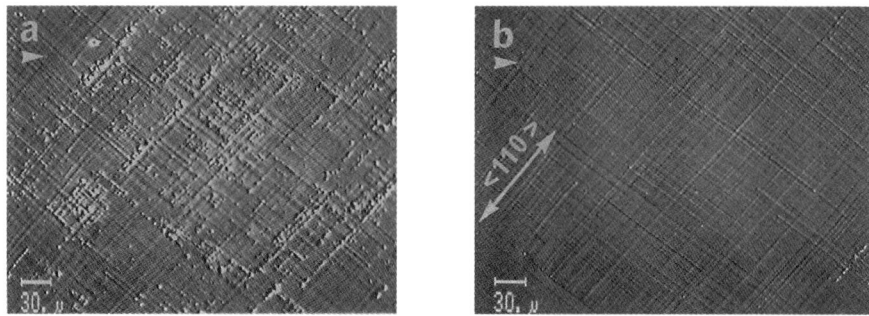

Figure 3. SOM-DPC images from (a) densely-pitted and (b) relatively smooth (pit-free) locations on layer B2. The arrow denotes the DPC differentiation direction (imagine oblique illumination from this direction).

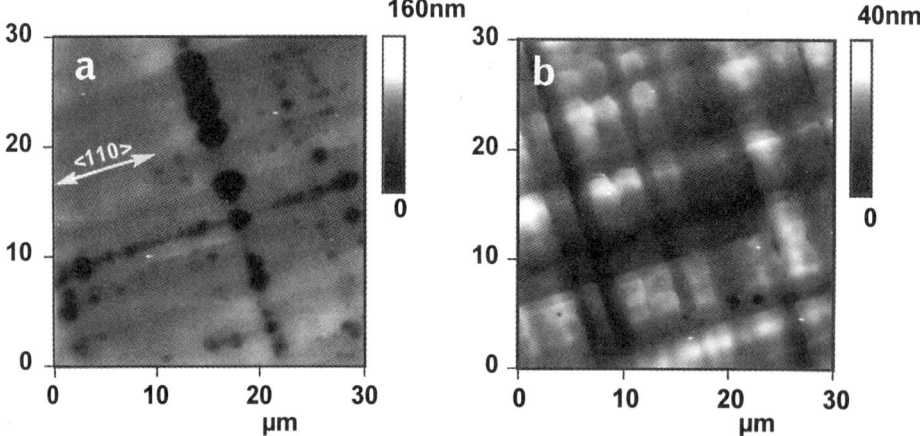

Figure 4. AFM images from (a) densely-pitted and (b) relatively smooth regions of layer B2.

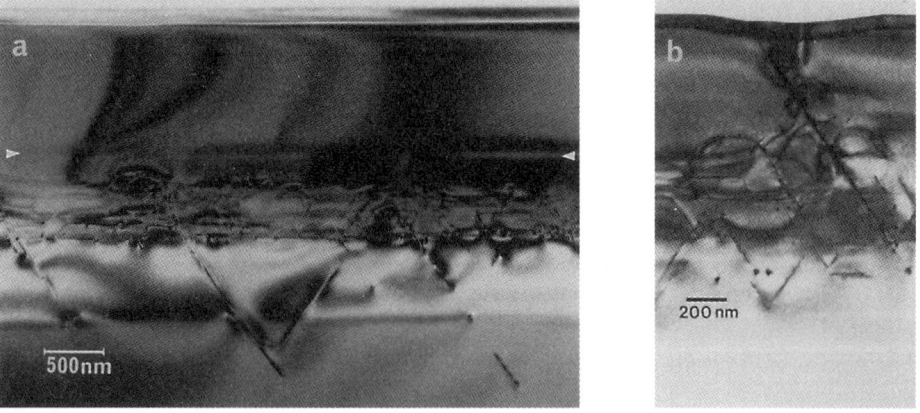

Figure 5. Bright-field many-beam cross-sectional TEM images taken from (a) the centre of layer B2, and (b) from a pitted region at the edge of layer B1.

order 10^5-10^6cm^{-2} elsewhere. There was a background crosshatched surface texture with a distinctive tiled appearance and a characteristic 2μm periodicity.

Some similar defect features have occasionally been referred to in the literature. Nelson et al (1993) remarked on the cloudy appearance of thicker layers grown in the low T_g high GR regime. Fitzgerald et al (1992) reported, in electron beam induced current (EBIC) images of MBE layers grown under high T_g low GR conditions, groups of threading dislocations forming <110> linear features and separated typically by 100s to 1000s of μm. These were assigned to heterogeneous sources able to continue punching out dislocations during RL growth, forming a dense pile-up able to pin MDs travelling along orthogonal directions. The density of these TD features (termed dark-line defects) was found to increase in CVD-grown layers as GR was increased (Watson et al 1994). Similar "bundles" of TDs, spaced by 10-40μm, were noted in low T_g high GR material by Mooney et al (1993).

Fig.5(a) shows a typical TEM cross-section from the centre of layer B2. The BL, active layers and top of the RL appear free of TDs. In this case the RL had been step-graded to a Ge fraction about 2% above that of the BL in order to encourage near-complete relaxation (Tersoff (1993)). This layer, which marks the top of the RL, is arrowed in fig.5(a). Dislocations are indeed confined within the RL except for a few deep pile-ups, notably spaced by about 2μm, penetrating into the Si substrate along {111} glide planes. Such dislocation pile-up into the substrate was first reported by LeGoues et al (1992). However, we observed no characteristic traces of the "modified Frank-Read" multiplication process noted by these workers in our layers.

Fig 5(b) shows a TEM cross-section containing a depression found near the edge of wafer B1. The depression, about 50nm depth and 1μm across, contained a tangle of threading dislocations associated with a {111} slip plane. Previously, Hsu et al (1992) observed pits in AFM images of MBE layers grown at higher T_g, and associated them with the emergence points of TDs. It was proposed that the cost in surface energy of forming the pit, due to the surface area increase, is compensated by the strain energy saved in reducing the TD length. If this is the case then the pits themselves may have a pinning action, once formed, on TD motion. In our case, at low T_g, we postulate that such pits are particularly likely to form due to the relatively slow glide velocity.

The effect of such pits on cosmetic surface quality, especially when clustered, renders the layers grown in this regime inherently unsuitable for microcircuit fabrication, even though (as shown in Table 1) very high μ_e values can be achieved in a strained Si channel grown on the BL. It is probable that the situation can be improved by employing (a) a lowered RL GR, to space out the MDs further and reduce their interactions (b) a linear RL grade to maintain a consistent surface strain, and (c) an increased T_g to increase the dislocation glide velocity.

3.3 Regime C : low T_g, low GR

Layer C1 was cosmetically much improved, with strong haze now limited to isolated mm-sized <110> cross-shaped features at very low number density (a few 10s per cm^2). The crosses comprised lines of pits as seen in regime B. Overall, the surface was covered by a crosshatch surface texture, which had both a finer-scale periodicity (0.7-1.0μm) and a larger amplitude than that on regime B layers, interspersed by occasional small pits. Some pits were clearly located at the termination of <110> crosshatch troughs, confirming their assignment to (single or multiple) TDs. Close inspection of the crosshatch, as shown by the AFM image in fig.6, revealed the presence of a higher density of very small pits, which from the faint effect they have on the surrounding surface topography, can confidently be assigned to the emergence of individual TDs. Their density was estimated from AFM images to be in the range 10^6-10^7 cm^{-2}.

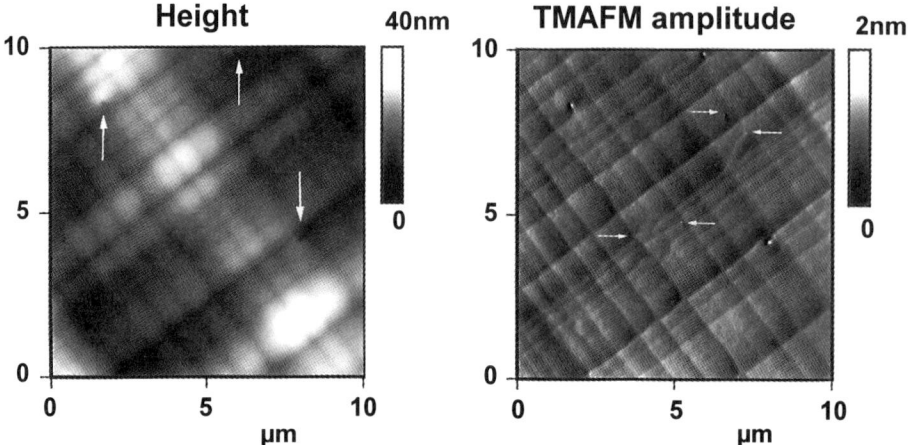

Figure 6. 10μm AFM height and tapping-amplitude (AFM error signal) images from layer C1. Amplitude image contrast is proportional to surface slope and emphasises finescale features. Three TDs are visible in the height image. In addition, four fainter features, associated with termination of shallow <110> troughs or threading segments and thus also assigned to the emergence points of TDs, are arrowed in the amplitude image.

3.4 Regime D : high T_g , low GR

Regime D layers exhibited the best cosmetic surface quality of all. No extended haze patches were observed. Instead, light scattering was confined to point defects, which generally turned out to be isolated pyramidal pits at a density of 10-50cm^{-2}, and the familiar near-specular diffraction from the crosshatch pattern. Fig. 7 compares SOM-DPC and AFM images of the surface topography. As seen in fig.7(a) the intensity of the crosshatch lines varies considerably over a wide lengthscale range. This is reflected by the range of AFM rms roughness values, calculated from 10x10μm areas at different positions, in table 1. The thickest BL has the greatest surface roughness. There are also, on the scale of 100s - 1000s of μm, domain boundaries visible in the crosshatch pattern. These are likely to be associated with the pinning of gliding dislocations. Evidence is accumulating that the heterogeneous defects, responsible for the light point defects noted above, may also be responsible for the patchiness of the surface haze noted in regimes B and C, and also, in this high T_g low GR regime, for the longer-range excursions in cross-hatch amplitude which may themselves define the locations of barriers to dislocation glide.

Fig. 8 shows a TEM cross-section from layer D3. As typical of group D layers, there are no dislocations observed in the BL or top of the RL, and those in the RL are much more widely spaced than in fig.5(a) due to the much increased RL thickness. Very few dislocations penetrate into the Si substrate, although there is still weak evidence of an approx. 2μm spacing between {111} pile-ups in the lower half of the RL. This correlates with a weak surface undulation, of 1.4-1.8μm period and 2-3nm amplitude, in AFM images from group D layers. The AFM images contained no surface pits. Wet chemical defect etching using a diluted Schimmel etchant (Gibbings et al 1989) was thus used to reveal the TDs. The etched surface features resembled those in fig.6 and had a number density of typically $2\text{-}9\times10^5\,\text{cm}^{-2}$.

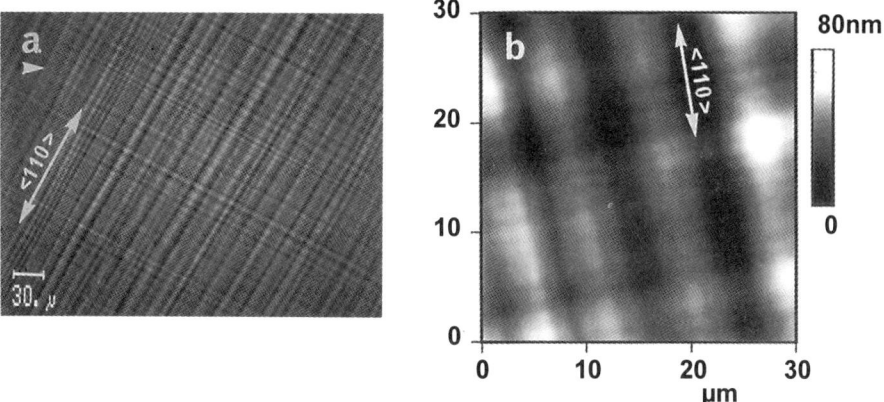

Figure 7. Representative (a) SOM-DPC and (b) AFM images from layer D1.

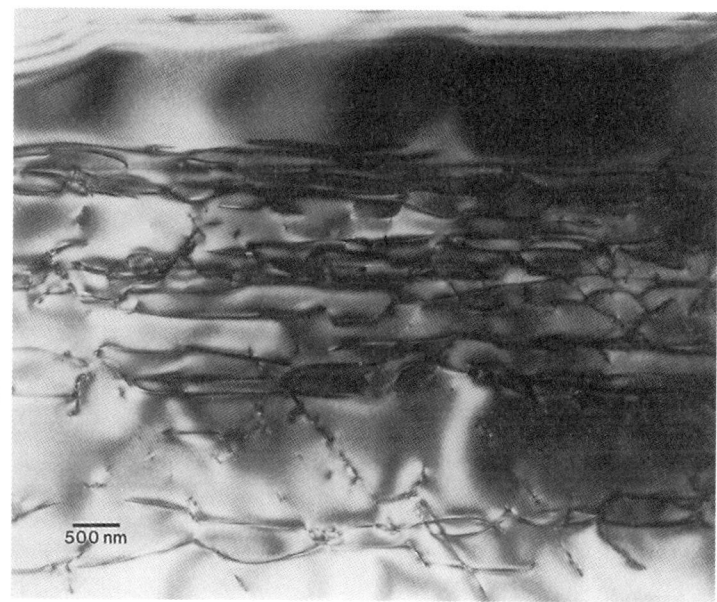

Figure 8.

Bright-field
many-beam TEM
cross-section
from layer D3.

4. CONCLUSIONS

We have employed a temperature-agile UHV-VPE reactor to make a direct comparison of the effects of high vs low growth temperature (800°C and 560-610°C) and high vs low RL grade rate (<10 and 40-50 %Ge/µm) on the quality of relaxed $Si_{1-x}Ge_x$ (0.21<x<0.29) epitaxial buffer layers. We obtained best results under high T_g / low GR growth conditions. This resulted in the virtual elimination of clustered surface pits, which led to poor cosmetic quality in other growth regimes. High T_g / low GR growth maintains a minimum level of surface strain, thus reducing the probability of nucleating dislocations, and also of elastic relaxation roughening, whilst at the same time maximising the opportunity for relaxation by glide of existing dislocations. BLs grown by this means contained 10-50cm^{-2} pyramidal pit defects, 10^5-10^6 threading dislocations cm^{-2}, and surface roughness levels of 2-10nm rms. Further reduction of these levels is necessary to meet the stringent demands of large-scale integrated circuit microfabrication.

REFERENCES

Churchill A C, Robbins D J, Wallis D J, Griffin N, Paul D J and Pidduck A J 1997 Semicond. Sci. Technol. **12**, 943

Cullis A G, Robbins D J, Pidduck A J and Smith P W 1992 J. Cryst. Growth **123**, 333

Cullis A G, Pidduck A J and Emeny M T 1995 Phys. Rev. Lett. **75**, 2368

Dutartre D, Warren P, Provenier F, Chollet F and Perio A 1994 J. Vac. Sci. Tech. **A12**, 1009

Fitzgerald E A 1991 Xie Y-H, Green M L, Brasen D, Kortan A R, Michel J, Mii Y-J and Weir B E Appl. Phys. Lett. **59**, 811

Fitzgerald E A, Xie Y-H, Monroe D, Silverman P J, Kuo J M, Kortan A R, Thiel F A and Weir B E 1992 J. Vac. Sci. Technol. **B10**, 1807

Fitzgerald E A 1995, Annu. Rev. Mater. Sci. **25**, 417

Hohnisch M, Herzog H-J and Schaffler F 1995 J. Cryst. Growth **157**, 126

Hsu J W P, Fitzgerald E A, Xie Y H, Silverman P J and Cardillo M J 1992 Appl. Phys. Lett. **61**, 1293

Ismail K, LeGoues F K, Saenger K L, Arafa M, Chu J O, Mooney P M and Meyerson B S 1994 Phys. Rev. Lett. **73**, 3447

Ismail K, Arafa M, Saenger K L, Chu J O and Meyerson B S, 1995 Appl.Phys.Lett. **66**, 1077

Ismail K 1996 J. Vac. Sci. Technol. **B14**, 2776

Jesson D E, Pennycook S J, Baribeau J-M and Houghton D C 1993 Phys. Rev. Lett. **71**, 1744

LeGoues F K, Meyerson B S, Morar J F and Kirchner P D 1992 J. Appl. Phys. **71**, 4230

Mii Y J, Xie Y H, Fitzgerald E A, Monroe D, Thiel F A, Weir B E and Feldman L C 1991 Appl. Phys. Lett. **59**, 1611

Mooney P M, LeGoues F K, Chu J O and Nelson S F 1993 Appl. Phys. Lett. **62**, 3464

Mooney P M, Jordan-Sweet J L, Ismail K, Chu J O, Feenstra R M and LeGoues F K 1995 Appl. Phys. Lett. **67**, 2373

Nelson S F, Ismail K, Jackson T N, Nocera J J, Chu J O and Meyerson B S, 1993 Appl. Phys. Lett. **63**, 794

Pidduck A J, Robbins D J, Cullis A G, Leong W Y and Pitt A D 1992 Thin Solid Films **78**, 222

Robbins D J and Young I M 1987 Appl. Phys. Lett. **50,** 1575

Robbins D J, Glasper J L, Cullis A G and Yeong W Y 1991 J. Appl. Phys. **69,** 3729

Schaffler F, Tobben D, Herzog H-J, Abstreiter G and Hollander B 1992 Semicond. Sci. Technol. **7**, 260

Tersoff J 1993 Appl. Phys. Lett. **62**, 693

Gibbings C J, Tuppen C G and Hockly M 1989 Appl. Phys. Lett. **54**, 148

Watson G P, Fitzgerald E A, Xie Y-H and Monroe D 1994 J. Appl. Phys. **75**, 263

© British Crown Copyright / DERA (1997)

Inst. Phys. Conf. Ser. No 157
Paper presented at Microsc. Semicond. Mater. Conf., Oxford, 7–10 April 1997
© *1997 IOP Publishing Ltd*

Relaxation of strained epitaxial layers by dislocation rotation, reaction and generation during annealing

R Beanland, M A Lourenço* and K P Homewood*

GEC-Marconi Materials Technology, Caswell, Towcester, Northants, NN12 8EQ
*Department of Electronic and Electrical Engineering, University of Surrey, Guildford, Surrey GU2 5XH, UK

ABSTRACT: We present an analysis of misfit dislocations in annealed $In_{0.1}Ga_{0.9}As$ layers on (001) GaAs. We show that $\sqrt{2}$ more misfit relief is produced by rotation of $a/2<101>$ dislocations into edge orientation. However, the lowest energy configuration consists of $a/2<110>$ edge dislocations, which can be formed by dislocation reactions. We propose that this lowest energy state cannot be reached due to the random nature of the original 60° dislocation array. The observation of $a/2<101>$ edge dislocations in annealed $In_{0.1}Ga_{0.9}As$ layers implies that generation of new dislocations is a relatively slow process.

1. INTRODUCTION

The stability of strained semiconductor epitaxial layers at high temperatures is potentially of interest for devices grown on top of relaxed or partially relaxed buffer layers (Beanland, Dunstan and Goodhew 1995). It has been found that the strain, ε, in high-quality epitaxial layers - which are the most promising candidates for such buffer layer structures - is always several times the equilibrium value (e.g. Dunstan et al. 1993). These layers are thus in a metastable state. In the Ge_xSi_{1-x}/Si system, this has been ascribed to the distinct lack of dislocation sources and the low velocity of threading dislocations (e.g. Dodson and Tsao 1987, 1988). However, in low-misfit $In_xGa_{1-x}As/GaAs$ (x < 0.25), the strain has been found to be considerably higher than equilibrium for a wide range of growth conditions, indicating that the cause of metastability is not the finite dislocation velocity in this system (Dunstan et al. 1993). This deviation from equilibrium primarily results from the high structural perfection of the material. High-quality single-crystal materials, such as $In_xGa_{1-x}As/GaAs$, have very few dislocations, and those which are present can only move in limited ways, i.e. by glide. This means that without dislocation multiplication and/or nucleation processes, the strain in highly perfect strained layers should stay virtually constant (Beanland and Boyd 1995).

The observation that strained layers relax at all indicates that new dislocations are formed even when the density of original dislocations is very low. Annealing a strained layer system can accelerate existing dislocation introduction mechanisms and produce new means of relaxation. In this paper we consider the latter processes, in particular the increase in relaxation that can occur once the restriction of dislocation glide is removed.

2. EXPERIMENTAL OBSERVATIONS

The dislocation structure of a 140nm thick layer of $In_{0.1}Ga_{0.9}As$ on (001) GaAs, grown by MBE at 500°C, is shown in the plan-view transmission electron microscope image of Fig. 1a. Long, straight 60° $a/2<101>$ dislocations are present, which have <110> line directions. These dislocations are introduced by glide on {111} planes, and their density is consistent with that expected from glide of pre-existing dislocations originating from the substrate (e.g. Fitzgerald 1991). The dislocation structure after an anneal at 950°C for 500 s is shown in Fig. 1b,c. Details of annealing conditions and the macroscopic strain relaxation behaviour of this sample has been described previously (Lourenço et al. 1994). The misfit dislocation array is now comprised of short edge dislocation segments with Burgers vectors of the type $\mathbf{b} = a/2<101>$ and line directions $\xi = <100>$, which combine to form $a/2<110>$ edge dislocation segments with <110> line directions, following the well-known reaction $a/2[101] + a/2[01\bar{1}] \leftrightarrow a/2[1\bar{1}0]$. In a two-beam $\mathbf{g} = 220$ PVTEM image (Fig. 1b), these latter dislocations have double images when lying parallel to [110] ($\mathbf{g}.\mathbf{b} = 2$) and are invisible when lying parallel to $[\bar{1}10]$, since $\mathbf{g}.\mathbf{b} = 0$ and $\mathbf{g}.(\mathbf{b}\times\xi) = 0$ (e.g. Hirsch et al. 1977). When viewed using $\mathbf{g} = \bar{2}20$, the previously invisible dislocations become visible and vice versa (Fig. 1c). Some of the a/2<110> dislocations rise out of the interface for some distance (typically a few μm) only to rejoin the array at a reaction point. The density of misfit dislocations in the annealed sample is considerably higher than in the as-grown sample, indicating that new dislocations have been formed. However, some progress towards equilibrium has also occurred due to the removal of the restriction of dislocation glide.

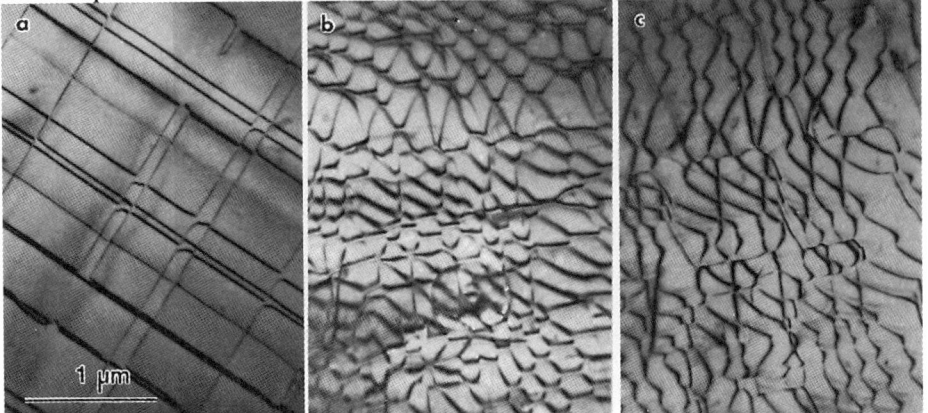

Figure 1. Bright field, g=220 type images of (a) an as-grown sample, and (b, c) a region of an annealed sample. The arrows indicate the direction of g.

3. THEORETICAL ANALYSIS

When dislocations arrive at the interface by glide, their Burgers vectors contain components which do not relieve the misfit strain. The relaxation of the misfit strain by the introduction of misfit dislocations is thus an imperfect process.

In as-grown $In_xGa_{1-x}As$ layers, the misfit dislocation array is comprised of 60° dislocations with $\mathbf{b} = a/2<101>$ and line directions $\xi = 1/\sqrt{2}<110>$. There are four possible Burgers vectors for the two <110> line directions, giving eight dislocation sets in total. The deformation produced by eight dislocation sets can be very complicated; for example, the dislocation set $\mathbf{b} = a/2[101]$ | $\xi = 1/\sqrt{2}[110]$ gives not only misfit relief, of magnitude $\rho b/2$, where ρ is the linear interfacial density and b is the magnitude of the Burgers vector of the dislocation set, but also a shear of the layer $\rho b/2$ due its screw component, and a tilt of the layer of magnitude $\sqrt{2}\rho b/2$ about [110] due to the edge component perpendicular to the interface. 'Pure' misfit relief only occurs if all eight a/2<101> sets have the same density; in this case, the tilts and shears cancel, leaving only misfit relief of magnitude $\rho b/2$, where ρ is now the density of all dislocations with the same line direction, irrespective of Burgers vector. This does not occur in real 60° dislocation arrays, even on a large scale; it is invariably found that the deformation of relaxed layers is complicated, with variations from place to place across a wafer and a net tilt of the layer. We can only justify the use of 'pure' misfit relief in the present analysis by the simplicity of the result.

When a dislocation set rotates by climb, the Burgers vector of the dislocations remains the same, but the line direction is changed. In the present case, an a/2[101] dislocation will tend to rotate from a $1/\sqrt{2}[110]$ line direction to an [010] line direction, eliminating its screw component by doing so. This eliminates the shear of the layer and increases the misfit relief by a factor of $\sqrt{2}$; the tilt of the layer is now about [010], but is of the same magnitude as before. Again, when there is no imbalance between the different types of Burgers vector, the net tilt of the layer is zero. Since the dislocation densities along <100> must be the same as the original dislocation densities along <110>, the dislocation array now relieves a factor of $\sqrt{2}$ more misfit.

In the actual structure (Fig. 1), rotated dislocation segments coexist with a/2<110> dislocation segments which have $1/\sqrt{2}<\bar{1}10>$ line directions. This is because the original 60° dislocations may also eliminate their screw components by reacting to produce 90° a/2<110> dislocations. It is straightforward to show that the homogeneous deformation produced by two arrays which can be transformed into one another by dislocation reactions is identical; the deformation produced by an a/2<110> | $1/\sqrt{2}<\bar{1}10>$ array is thus identical to that produced by an a/2<101> | $1/\sqrt{2}<110>$ array.

The presence of both a/2<101> | <010> and a/2<110> | $1/\sqrt{2}<\bar{1}10>$ dislocations in the array cannot be explained solely from the efficacy of misfit relief, since a rotated array relieves a factor of $\sqrt{2}$ more misfit than a reacted one. We therefore consider total energy of the three configurations and calculate the equilibrium strain expected for each. We follow the simple Matthews-Blakeslee approach (Matthews and Blakeslee 1974).

The self-energy per unit area of a square array of dislocations with similar Burgers vectors, a distance h from a free surface in elastically isotropic material is, according to Freund (1990),

$$E_{array} = \frac{\rho \mu b_e^2}{2\pi(1-v)}\left(\ln\left(\frac{2\alpha h}{b}\right) - \frac{\cos 2\theta}{2}\right) + \frac{\rho \mu b_s^2}{2\pi}\ln\left(\frac{2\alpha h}{b}\right), \tag{1}$$

where μ is the shear modulus, α is the core parameter, b_e is the edge component and b_s is the screw component of the Burgers vector, θ is the angle between the Burgers vector and the normal to the surface and ρ is the linear interfacial dislocation density. This is only a rough approximation since the energy due to the interaction of the stress fields within and between different dislocation sets has been neglected, and - as noted by Jain et al. (1992) - this interaction energy can be a significant fraction of the total energy of the array. The elastic strain energy of an elastically isotropic layer which experiences pure misfit relief is

$$E_{elastic} = \frac{2\mu(1+v)}{(1+v)}\left(\varepsilon_0 - \varepsilon_r\right)^2 h, \tag{2}$$

where ε_r is the strain relieved by the dislocation array and h is the thickness of the layer. Equilibrium in the Matthews-Blakeslee model occurs when the total energy, i.e.

$$E_{total} = E_{array} + E_{elastic} \tag{3}$$

is a minimum.

From eqns. (1), (2) and (3), the total energy per unit area of an array of a/2<101> dislocations with 1/√2<1̄10> line directions is

$$E_{total}\left\{\frac{a}{2}\langle 101\rangle | \frac{1}{\sqrt{2}}\langle 1\bar{1}0\rangle\right\} = \frac{2\mu(1+v)h}{(1-v)}\left(\varepsilon_0 - \frac{\rho b}{2}\right)^2 + \frac{\mu b^2 (4-v)\rho}{8\pi(1+v)}\ln\left[\frac{2\alpha h}{b}\right]. \tag{4}$$

The equilibrium strain may be obtained by minimising the total energy, i.e. by differentiating eqn (4) with respect to ρ. The result is

$$\varepsilon_{eq}\left\{\frac{a}{2}\langle 101\rangle | \frac{1}{\sqrt{2}}\langle 1\bar{1}0\rangle\right\} = \frac{(4-v)b}{16\pi(1+v)h}\ln\left[\frac{2\alpha h}{b}\right]; \tag{5}$$

and applying the same procedure to a/2<101> | <010> and a/2<110> | 1/√2<1̄10> arrays gives

$$\varepsilon_{eq}\left\{\frac{a}{2}\langle 101\rangle |\langle 010\rangle\right\} = \frac{b}{2\sqrt{2}\pi(1+v)h}\ln\left[\frac{2\alpha h}{b}\right] \tag{6}$$

and

$$\varepsilon_{eq}\left\{\frac{a}{2}\langle 110\rangle | \frac{1}{\sqrt{2}}\langle 1\bar{1}0\rangle\right\} = \frac{b}{8\pi(1+v)h}\left(\ln\left[\frac{2\alpha h}{b}\right] + \frac{1}{2}\right). \tag{7}$$

These are plotted in Fig. 2. It can be seen that the equilibrium strain in a layer with rotated dislocations is considerably higher than in the unrotated array, due to the higher self-energy of 90° a/2<101> dislocations. The release of strain energy in the layer due to the rotation of dislocations is thus a progression towards a local energy minimum, rather than a global one. The observation of this local energy minimum in the final configuration is a consequence of the random nature of the original 60° dislocation array. The four possible Burgers vectors associated with each 1/√2<1̄10> line direction in the original 60° dislocation array occur in a random sequence. This means that some dislocations would have to climb around others in order to pair up and produce a/2<110> edge dislocations. Since the misfit strain in the layer and the dislocation self energy tends to confine dislocations to the interfacial plane, the energy barrier to the lowest energy state will be prohibitively high. The next-lowest energy state is a mixture of 90° a/2<110> and 60° a/2<101> dislocations, both with 1/√2<1̄10> line directions, i.e. a mixture of the two lowest energy states. The observation that the array does not tend towards this structure implies that even at the high temperatures used in annealing, misfit relief by generation of new 60° dislocations does not occur as rapidly as misfit relief by dislocation rotation. This in turn implies that the density of new dislocation sources remains relatively low during the anneal, i.e. homogeneous surface half-loop nucleation does not occur. The

148

final structure is a mix of the lowest energy state ($a/2<110>$ edge dislocations) and a 'metastable' state ($a/2<101>$ edge dislocations) which is maintained by a prohibitively high energy barrier.

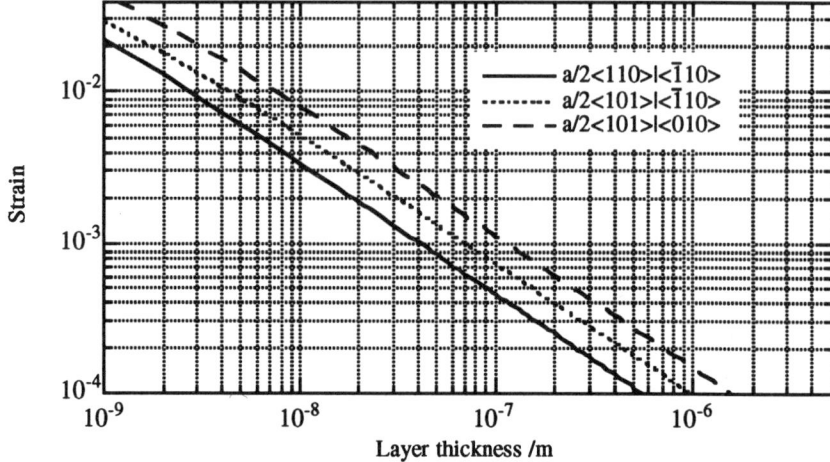

Figure 2. Equilibrium strain of layers relieved by the three different arrays as a function of thickness.

4. SUMMARY AND CONCLUSIONS

The misfit dislocation structure of an annealed $In_{0.1}Ga_{0.9}As$ layer on GaAs has been considered. The 60° array which is present in the as-grown sample does not relieve misfit efficiently; each dislocation set produces ancillary tilts and shears of the layer, which only cancel if the density of each set in the array is the same. Rotation of a 60° dislocation set into a 90° dislocation set produces greater misfit relief by the conversion of screw components into misfit-relieving edge components, whereas reaction of 60° $a/2<101>$ dislocations to produce 90° $a/2<110>$ dislocations eliminates the screw components of the Burgers vectors, but does not lead to greater misfit relief. If the supply of dislocations is unlimited, the lowest energy configuration is a 90° $a/2<110>$ array, and the highest energy configuration is a 90° $a/2<101>$ array. Due to the random sequence of Burgers vectors in the original 60° dislocation array and the large amount of energy required to change the sequence, it is very unlikely that the lowest energy configuration can form during annealing. The lowest energy configuration that can form solely by 60° dislocations gliding on {111} planes is a mixture of $a/2<110>$ edge dislocations and $a/2<101>$ 90° dislocations, with no rotated dislocations. The observation of 90° $a/2<101>$ dislocations with <010> line directions thus implies that dislocation rotation occurs much more readily than the generation of new 60° $a/2<101>$ dislocations. It thus appears that homogeneous generation of dislocation half-loops, which would give an unlimited supply of 60° dislocations, does not occur.

REFERENCES

Beanland R 1993 Phil. Mag. A **67**, 585
Beanland R and Boyd A R 1995 Microscopy of Semiconducting Materials 1995, eds A G Cullis and A Staton-Bevan (Bristol: Institute of Physics) 153-8
Beanland R, Dunstan D J and Goodhew P J 1995 Advances in Physics **45**, 87
Dodson B W and Tsao J Y 1987 Appl. Phys. Lett. **51**, 1325
Dodson B W and Tsao J Y 1988 Appl. Phys. Lett. **53**, 2498
Dunstan D J, Dixon R H, Kidd P, Howard L K, Wilkinson V A, Lambkin J D, Jeynes C, Halsall M P, Lancefield D, Emeny M T, Goodhew P J, Homewood K P, Sealy B J and Adams A R 1993 J. Cryst. Growth **126**, 589
Fitzgerald E A 1991 Mat. Sci. Reports **7**, 87
Freund L B 1990 J. Mech. Phys. Solids **35**, 657
Hirsch P B, Howie A, Nicholson R B, Pashley D W and Whelan M J 1977 Electron microscopy of thin crystals (Malabar FL: Krieger)
Jain S C, Gosling T J, Willis J R, Totterdell D H J and Bullough R 1992 Phil. Mag. A **65**, 1151
Lourenço M A, Homewood K P and Considine L 1994 Mat. Sci. and Eng. B**28**, 507
Matthews J W and Blakeslee A E 1974 J. Cryst. Growth **27**, 118

Inst. Phys. Conf. Ser. No 157
Paper presented at Microsc. Semicond. Mater. Conf., Oxford, 7–10 April 1997
© 1997 IOP Publishing Ltd

Correlation between defects, residual strain and morphology in continuously graded InGaAs/GaAs buffers

L Lazzarini, C Ferrari, S Gennari, A Bosacchi, S Franchi, M Berti*, A V Drigo*, F Romanato* and G Salviati

Maspec-CNR Institute, via Chiavari18/A, 43100-I Parma
* INFM at the Physics Dept., University of Padova, via Marzolo 8, 35131-I Padova

ABSTRACT: The influence of different types of grading (linear, parabolic and square-root) and growth conditions on residual strain, Threading Dislocation (TD) density, Misfit Dislocation (MD) confinement and surface morphology of MBE grown InGaAs/GaAs buffer layers has been studied by TEM, RBS, SIMS and AFM techniques. Non-linear buffers have wider MD-free surface regions and higher MD concentrations near the substrate where the compositional gradients are higher. Lowering the growth temperature and using an As$_2$ beam leads to symmetric and smoother cross-hatch morphology and removes asymmetries in the residual strain.

1. INTRODUCTION

The growth of opto-electronic grade quality, highly mismatched semiconductor heterostructures on commercially available substrates requires the interposition of suitable Buffer Layers (BL) (e.g. Fitzgerald 1991 and references therein; Beanland et al. 1996 and references therein). These buffers can be obtained by controlling the strain relaxation of ternary alloys. The seminal work of Tersoff (1993) points out the rules to control the MD multiplication and their concentration profile by appropriately designing the composition grading profile of BLs. Epilayers grown by continuosly grading the composition profile represent the state of the art to control the lattice constant by reducing at the same time the degrading effects of TDs. The available models of strain relaxation cannot take into account asymmetries in the strain relief or the influence of the growth parameters. Moreover the surface density of TDs and the achievement of the designed surface lattice constant cannot be considered as the unique parameters to evaluate the quality of a BL. In fact, the surface roughness and the efficiency of the buffer in confining MDs far from the surface must be considered as well. In this work three types of composition profiles for InGaAs BLs are considered (linear, parabolic and square-root), the non-linear profiles having slopes decreasing toward the top of buffers. Special attention will be paid to the relationships between grading profile, growth conditions, surface morphology and relaxation asymmetries.

2. EXPERIMENTAL PROCEDURE

The considered buffers have been designed as reported by Bosacchi et al. (1997) according to the predictions of the semiempirical model described by Salviati et al. (1995). Following the basic work of Tersoff (1993), our model is able to predict the surface residual strain, the MD depth distribution and the MD-free surface region. The calculations are based on the empirical relaxation law for InGaAs/GaAs single layers (Drigo et al.,1989), even if other models are available in the literature (e.g. Dunstan 1991). A further development of this model is based on the consideration of the excess strain energy density with respect to that of the free structure (Napolitani, 1996). This allows the extension of the model to any grading profile (including tensile and compressive misfits) and at the same time improves the agreement between predictions and experimental results. Since the MD concentration is proportional to the gradient of the composition profile, the effects of the MD interaction on the surface quality can be investigated by chosing composition gradings with increasing slope of composition. In$_x$Ga$_{1-x}$As BL with Linear (L), Square-Root (SR) and Parabolic (P) (with the maximum on the surface) composition profile were grown on (001) seminsulating GaAs substrates by Molecular Beam Epitaxy (MBE). For all the BL the composition was graded from x(In)=0 to a nominal value of x(In)=0.35 in a thickness of 2.3 μm. The layers were grown at 400°C and 500° C and by using As_2 or As_4 molecular beams.

TEM investigations to evaluate layer microstructure, MD distribution and TD density were performed with a JEOL2000FX microscope operated at 200 kV. (001) plan and (110) oriented cross sections were prepared by a conventional mechano-chemical thinning procedure followed by room temperature Ar ion milling. The surface composition and the residual strain at the surface, $\varepsilon_{//}$, were measured by Rutherford Backscattering Spectrometry (RBS) analysis in random and channeling conditions using 2-4 MeV $^4He^+$ beams of the VdG accelerator at LNL (Legnaro, Italy). The surface morphology of the samples was investigated by Atomic Force Microscopy (AFM) with a Park Scientific Instruments microscope in the "contact mode". Ultralever tips were used to minimize the tip-surface convolution.

3. RESULTS AND DISCUSSION

Table I summarizes the results obtained for the examined samples. The surface (maximum) composition of the layers measured by RBS is always slightly lower than the nominal value. However the shape of the In composition profiles, measured by SIMS, were found to be in perfect agreement with the designed ones.

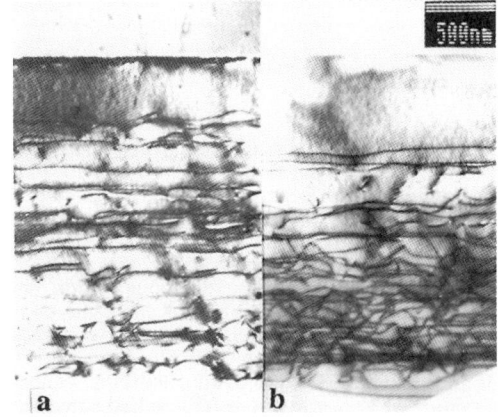

Fig. 1 (110) TEM cross section of linear (a) and parabolic (b) BL.

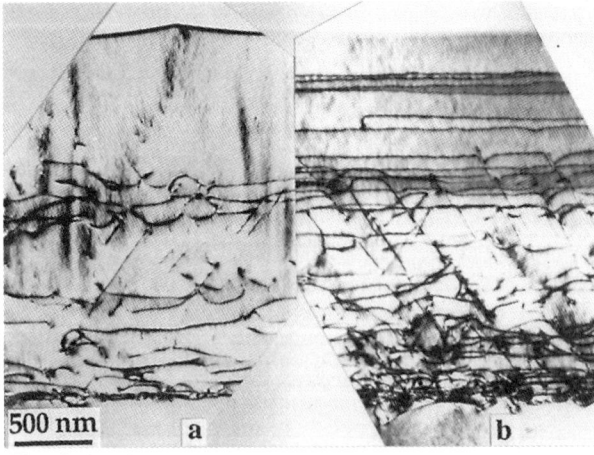

Fig. 2 (1-10) and (110) TEM cross sections (a and b respectively) of the SR BL.

Fig.1a and b shows the TEM cross sections of two BL (linear and parabolic) grown at 400°C with an As_2 beam. While in the linear BL the MDs are almost uniformly distributed from the substrate interface to the MD-free surface region, in the parabolic BL MDs mainly concentrate close to the substrate interface, according to the composition gradient, and a wider MD-free region. Fig.2a and b shows the (1-10) and (110) TEM cross sections of the SR BL grown at 500°C with an As_4 beam. The depth extension of the MD distribution appears to be very different in the two directions. This fact represents a general feature of the BLs grown under these conditions as shown by the results reported in Table I. This fact must be related to the well known zinc-blende lattice asymmetry leading to MD prefential nucleation along the [1-10] direction (Abrahams et al., 1972). The data of Table I also show that such an asymmetry is eliminated or strongly reduced by growing at 400°C with As_2 beams. In any case, the model predictions tend to over-estimate the actual MD-free region. The effect of growth conditions on the MD depth distribution is directly reflected on the misfit relaxation, as shown by the comparison of the surface residual strain, $\varepsilon_{//}$, in the [110] and [1-10] directions.

From the data in Table I it appears that, for BLs grown at 500°C and with As$_4$, $\varepsilon_{//}$ along the [110] direction is always lower than that in the [1-10] direction. This asymmetry almost completely disappears in the P and SR BL (according to the asymmetry decrease in the thickness of the MD-free region) by decreasing the growth temperature and by using As$_2$. However the average strain value shows nearly no dependence on the growth temperature and is in good agreement with the model predictions. Nevertheless the model predictions *overestimate* the strain relaxation in contradiction with the MD-free region overestimation. Of course no better agreement with the experimental results can be achieved by using relaxation models based on higher relaxation rates (Beanland et al., 1996). On the contrary the experimental data suggest that full strain relaxation is not achieved in the MD region, probably because of work hardening or insufficient driving force for the dislocation glide.

Fig.3a and b show the AFM micrographs of SR BL#1008 and #914. Fig.3b shows a strongly asymmetric cross-hatch pattern which is mainly directed along the [1-10] direction suggesting a correlation with the MD distribution asymmetry. In order to allow a rapid characterization of the morphology asymmetry the root mean square (rms) roughness, ρ, along the two <110> orthogonal directions has been measured and is reported in Table I. All the samples grown at 500° C present pronounced and asymmetric cross-hatch patterns, the main direction being parallel to the [1-10] direction, i.e. the preferential direction for MD nucleation. Lowering the temperature reduces both the rms roughness and its asymmetry. Changing the molecular beams from As$_2$ to As$_4$ further reduces the roughness.

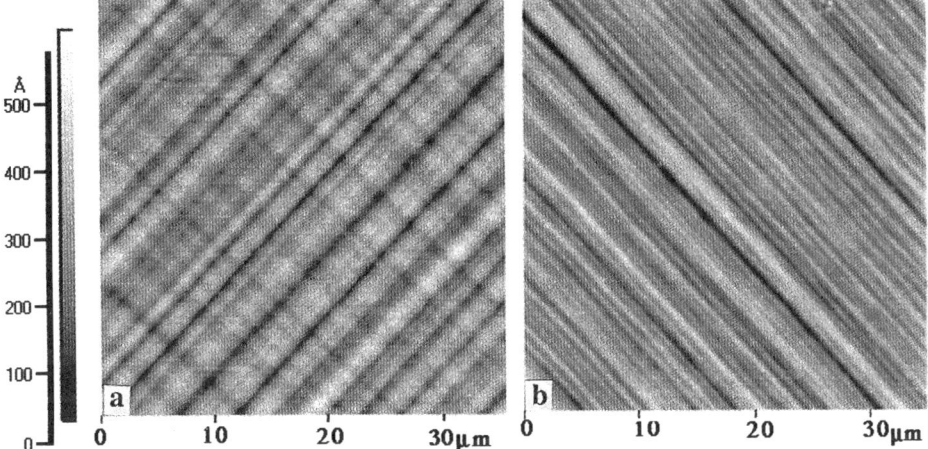

Fig. 3 AFM micrograph of square root BLs grown at 400°C As$_4$(a) and 500°C As$_2$ (b).

Let us begin the discussion on the correlation between morphology and strain release processes by comparing the results for different composition profiles. SR, P and L profiles have decreasing values of the maximum composition gradient and therefore of the maximum MD concentration. Correspondingly, decreasing values of ρ have been measured. The strain asymmetry is greater when the MD concentration is higher. This suggests a relationship between MD interaction, strain release and surface cross-hatch. However asymmetries in the MD distribution along the two <110> directions are not strongly related to strain and morphology asymmetries. For instance the data for the linear BLs indicate that the strain asymmetry, if any, is not related to the MD distribution asymmetry. On the other hand the SR BL grown under optimized conditions (symmetric strain and MD distribution) still shows asymmetric morphology.

In this context, the epitaxy kinetics seem to play a major role, as shown by the overall improvement of the surface morphology and symmetry in the strain relaxation using As$_2$ at 400°C. At 500°C the enhanced adatom surface mobility increases the probability of reaching surface equilibrium configurations by elastic strain relaxation developing surface morphology. The preferential orientation of the morphology indicates that this process is triggered or strongly influenced by the preferential nucleation of [1-10] MD. In turn the cross-hatch valleys, due to the stress concentration, become preferential sites for the nucleation of further MDs in the same direction. This mechanism explains the observed asymmetry in the MD distribution and strain relaxation. In fact the difference in the activation energy for [110] and [1-10] MDs alone should lead to *enhanced* asymmetry at lower growth temperature.

Table I: Summary of the experimental results obtained for linear (L), parabolic (P) and square root (SR) BLs and different growth conditions. t_{free} is the thickness of the MD-free surface region parallel to the indicated <110> direction; ρ and $\varepsilon_{//}$ represent respectively the rms roughness and the surface parallel residual strain measured along the indicated <110> direction.

Sample	Growth	x_{In} RBS (%)	t_{free} [1-10] (nm)	t_{free} [110] (nm)	t_{free} model (nm)	$\varepsilon_{//}$[110] (%)	$\varepsilon_{//}$[1-10] (%)	$\varepsilon_{//}$ [%] model	ρ[110] (nm)	ρ[1-10] (nm)
912 L	T=500° As$_4$	32.6±0.2	250	600	540	-0.49±0.02	-0.49±0.02	-0.52	9.5	3
998 L	T=400° As$_4$	33.3±0.3			540	-0.53 ±0.02	-0.53±0.02	-0.52	5	2
1000 L	T=400° As$_2$	32.5±0.3	430	430	540	-0.51±0.02	-0.51±0.04	-0.52	3	2
913 P	T=500° As$_4$	32.9±0.2	350	800	870	-0.27±0.06	-0.48±0.03	-0.31	17	6
1002 P	T=400° As$_4$	32.9±0.3			870	-0.37±0.02	-0.37±0.02	-0.31	10.5	3.5
1004 P	T=400° As$_2$	32.9±0.3	700	700	870	-0.41±0.02	-0.41±0.02	-0.31	4.5	4.5
914 SR	T=500° As$_4$	32.9±0.2	200	700	770	-0.23±0.06	-0.61±0.06	-0.42	19	7
1006 SR	T=400° As$_4$	34.1±0.3			770			-0.42	7	2.5
1008 SR	T=400° As$_2$	34.1±0.3	600	600	770	-0.50±0.01	-0.50±0.01	-0.42	6.5	3.5

4. CONCLUSIONS

The influence of different types of grading and growth conditions for InGaAs/GaAs layers on the residual strain has been studied by RBS, TEM and AFM. It has been found that lowering the growth temperature and using an As$_2$ beam results in a symmetric and smoother cross-hatch morphology and removes asymmetries in the residual strain.

Our results do not allow us to better define the correlation between morphology, strain relaxation and defects because they are coupled through the mechanisms of MD nucleation that are not yet well understood. However, the strong experimental correlations shown in this work suggest that the investigations of the mechanisms of strain relaxation must take into account the kinetics of epitaxy at the layer surface.

ACKNOWLEDGMENTS

The authors are indebted to Mr. M. Scaffardi and A. Sambo for their technical assistance.

REFERENCES

Abrahams M S, Blanc J and Buiocchi CJ, 1972 Appl. Phys. Lett. **21**, 185
Beanland R, Dunstan DJ and Goodhew PJ, 1996 Adv. Phys. **45**, 87
Bosacchi A, De Riccardis AC, Frigeri P, Franchi S, Ferrari C, Gennari S, Lazzarini L, Salviati G, Drigo AV and Romanato F, 1997 J. Cryst. Growth in press
Drigo AV, Aydinli A, Carnera A, Genova F, Rigo C, Ferrari C, Franzosi P and Salviati G, 1989 J. Appl. Phys. **66** 3334
Dunstan D J, Kidd P, Howard LK and Dixon RH, 1991 Appl. Phys. Lett. **59** 3390
Dunstan D J, 1991 Semicond. Sci. Technol. **6** A76
Fitzgerald EA, 1991 Mater. Sci. Rep. **7** 87
Napolitani E, 1996 'L'adattamento epitassiale di strutture ad alto disaccordo reticolare su substrati di GaAs', Degree Thesis, University of Padova
Salviati G, Ferrari C, Lazzarini L, Franchi S, Bosacchi A, Taiariol F, Mazzer M, Zanotti Fregonara C, Romanato F and Drigo AV, 1995, Inst. Phys.Conf. Ser.**146** 337
Tersoff J, 1993 Appl. Phys.Lett.**62** 693

Inst. Phys. Conf. Ser. No 157
Paper presented at Microsc. Semicond. Mater. Conf., Oxford, 7–10 April 1997
© 1997 IOP Publishing Ltd

The stacking faults in GaSb/(001)GaAs heterostructure

André M Rocher

CEMES-CNRS, BP 4347, F-31055 Toulouse, France.

ABSTRACT: GaSb/(001)GaAs is a free stress system owing to the perfect dislocation network obtained through the island growth of the GaSb. The relaxation mechanism of GaSb islands is apparent directly by the creation of 90° partial dislocations at the step edge of the GaAs substrate surface. The driving force comes from the local stress variation in the GaSb film caused by the substrate step edge.

1. INTRODUCTION

Metamorphic layers free of defects are the dream of most epitaxy laboratories. Numerous attempts have been made to obtain a high crystalline quality of semiconductor thin films when the lattice mismatch is large as it in GaAs/Si. The GaSb/(001)GaAs heterostructure is probably not the best system for infrared optical applications, but it is a very good one from the point of view of understanding the basic phenomena involved in the epitaxy of materials with large lattice mismatch. It appears to be ideally relaxed with a perfect grid of Lomer dislocations, as shown by Rocher & Kang (1995). Lomer dislocations are recognised to be the most efficient ones for accommodating the lattice mismatch for the GaSb/(001)GaAs interface. The perfection of the dislocation network is attributed to island growth limited by {111} facets, which allows us a periodic mechanism of creation of misfit dislocations as discussed by Rocher et al (1991).

Stacking faults are observed in most of the GaSb islands. These defects have already been discussed by Rocher & Atmani (1992), whose conclusions were partially wrong. The aim of this contribution is to discuss the origin of these stacking faults in terms of the relaxation of misfit strain.

2. EXPERIMENTS

The growth procedure has been described elsewhere by Raisin et al (1986). A 1 μm buffer layer of GaAs has been deposited by MBE at 580°C in order to improve the crystalline quality of the surface to be epitaxied. The temperature of the specimen is then reduced to 470°C in order to grow the GaSb layers with a nominal thickness from 6Å to 800Å.

Thin film morphology is investigated by Transmission Electron Microscopy (TEM) using plan view samples. Experimental observations have been made using both the moiré fringe and the weak beam techniques, discussed by Rocher et al (1991).

The GaSb films thinner than 100Å appear not to be continuous but consisted of isolated islands as shown in Fig. 1. Most of the islands contain at least one stacking fault.

The moiré pattern shown in Fig.2a consists of straight fringes with a mesh corresponding exactly to the value calculated from the bulk parameters: GaSb islands are perfectly relaxed. Fig. 2b shows the misfit dislocation network related to GaSb/(001)GaAs: the dislocations are Lomer type and organized on a square grid with a mesh length D_d of 54Å. In addition, a shift of the dislocation lines is observed, labelled S on Fig. 2c. This shift, perpendicular to the dislocation lines, is equal to half the distance between misfit dislocations $D_d/2$. The stacking faults and the dislocation shift appear to occur at the same location in the interface.

154

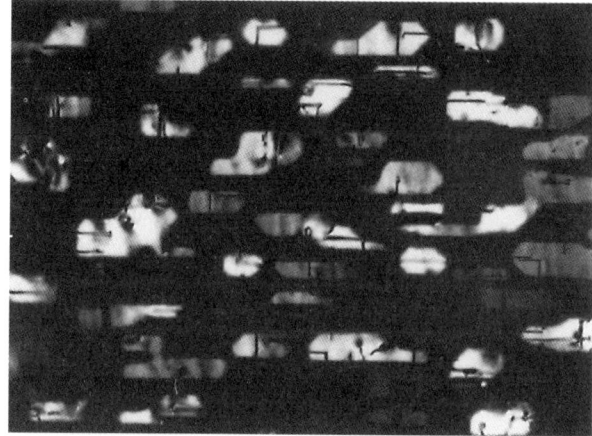

0.5μm .

Fig. 1 : Plane view of the 7nm
GaSb film deposited on GaAs;
each dark line inside the islands
is a stacking fault.

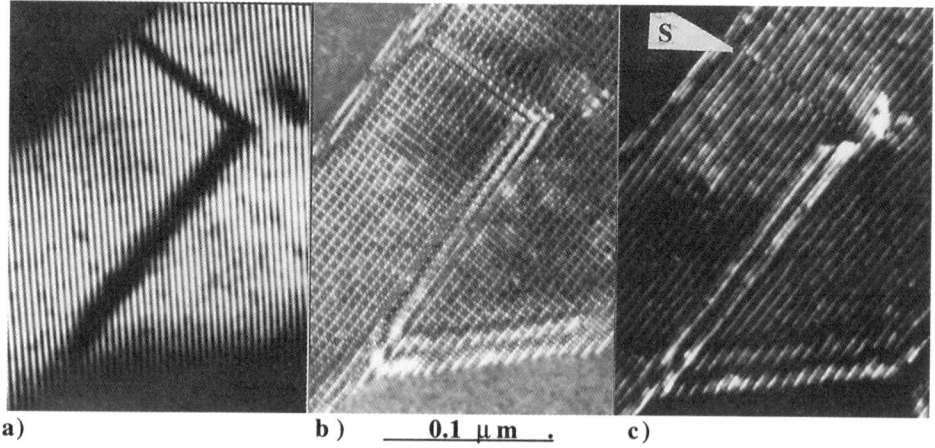

a)   b)   0.1 μm .   c)

Fig. 2 : Moiré pattern and weak beam image of a 270Å GaSb island: a) stacking fault contrast
observed as a dark line by a shift of (002) moiré fringes. b) weak beam: grid of Lomer
dislocations; c) weak beam image of one family of misfit dislocation: a $D_d/2$ shift on the misfit
dislocation lines is observed on S.

3. DISCUSSION

The high quality of the misfit dislocation network is due to a periodic relaxation of the
misfit stress by the direct creation of a Lomer dislocation network, a mechanism already
described by Rocher et al (1991). The object of this paper is to discuss where the first GaSb
relaxation occurs. The experimental result to be discussed here is the observation of a stacking
fault in most of the GaSb islands.

Stacking faults are always related to a shift of the dislocation line crossing the stacking
fault. The rigid body translation introduced by atomic steps gives a translation of $D_d/2$ of the
misfit dislocation network as shown by Zhu and Carter (1990). The two arrays of dislocations
separated by a stacking fault are not in the same plane of the substrate. They differ by an
atomic plane with a difference of level equal to a/2. This difference of level introduces a shift
$D_d/2$ in the coincidence site lattice of the {110} planes in the (GaSb, GaAs) system.

From this observation, we can assume that the surface step is at the origin of the stacking
fault. The stacking faults would thus be created at the surface step edges in the first stage of the
growth.

3.1 Accommodation efficiency of misfit dislocations

The dislocation efficiency for accommodating the lattice mismatch is a function of the type of its Burgers vector. In the {001}semiconductor interface, most of them are parallel to <$\bar{1}$10> and have a Burgers vector b with the components b_e, b_t and b_s defined in Fig. 3. The accommodation of the lattice mismatch is governed by the edge component in the interface b_e. The accommodation efficiency of a dislocation is measured by the parameter R, defined as the square of the ratio of the efficient part of the Burgers vector to its modulus. Thus:

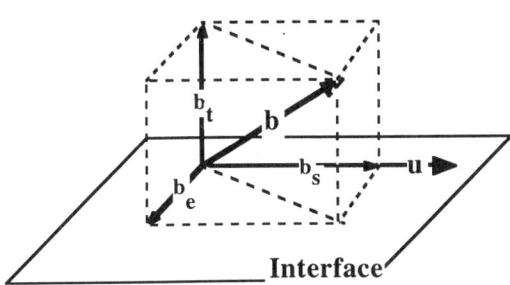

$$R = \frac{W_e}{W} = \frac{b_e^2}{b_e^2 + b_t^2 + b_s^2}$$

Fig. 3 : The components of the misfit dislocation Burgers vector.

This parameter corresponds to the ratio of the elastic energy of the dislocations, which is a function of b^2. The energy W_e is then the elastic energy transferred from the strained to the relaxed system. It needs to be compared with the total energy of the dislocation, W.

The efficiency R of a Lomer dislocation is equal to 100%: all the Burgers vector accommodates the lattice mismatch. R equal 25% for a 60° dislocation, owing to its screw component and its edge component perpendicular to the interface. 75% of its elastic energy is not furnished by the misfit strain and this dislocation is not the most likely one to be created by direct relaxation. Nevertheless, only 60° dislocations are able to accommodate lattice mismatch by a glide mechanism into a thick uniform epilayer. Screw dislocations, which are known to play no role in the misfit relaxation, have an efficiency R equal to zero.

The stacking faults observed in the GaSb islands are limited at the interface by a partial 90° dislocation with a Burgers vector defined by:

$$b_{1\bar{1}2} = \frac{1}{6}[1\bar{1}2] = b_e + b_t = \frac{1}{6}[1\bar{1}0] + \frac{1}{6}[002]$$

The accommodation efficiency of this dislocation is 33%. Its elastic energy is 3 times smaller than that of a Lomer dislocation. Using the energy criterion of critical thickness, this 90° partial dislocation would be created in the first monoatomic layer.

3.2 Interfacial Discontinuity at the level of the step edge

Owing to the pseudomorphic stress, a deformation ε_{001} occurs along the growth direction throughout the strained unit. ε_{001} is equal to 2f, with f = 8% . If we now consider the position of the epilayer atoms on both sides of a surface step as shown in Fig. 4, the Sb atom (B) is lower than the (A) since it lies atom above the additional surface plane related to the step. Because the difference of level a/2 related to the atomic step, the distance between A and B in the [001] direction is

$$\delta_z = \varepsilon_{001} * a/2 = f * a$$

3.3 Mechanism of creation of stacking fault at the interface

The surface step introduces a discontinuity in the atom organisation as shown in Fig. 4. An extra stress is created by the step on the deposited atoms. Now, the displacement vector related to a partial dislocation with the component b_t normal to the interface is a combination of the vectors in the GaSb and GaAs crystals. Its component along [001] is :

156

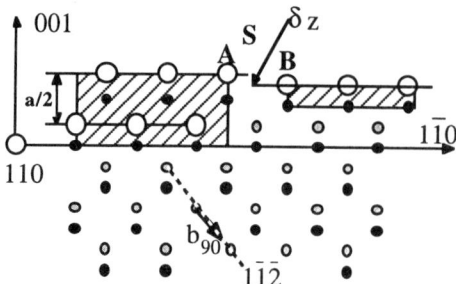

$$b'_t = b_t + \delta_z$$

$$= \frac{1}{6}[00\overline{2}] + f[001] \quad \text{with} \quad f \approx \frac{1}{12}$$

$$b'_t \approx \frac{1}{4}[00\overline{1}]$$

Fig. 4 : scheme of the atomic arrangment at a [1$\overline{1}$0] step of GaSb/GaAs interface

The efficiency ratio, applied to the 90° dislocation with the component b_t, becomes more favorable for relaxation after the introduction of the component related to the step. The strain energy considering b'_t is reduced by a factor 2/3: in GaSb b_t is equal to 1/3a while it is equal to 1/4a at the step. The efficiency ratio is hence modified: it is equal to 33% on a flat surface but equal to 50% at the level of the step edge. The stacking fault created in the very first stage of the GaSb growth, with an energy that is a function of the surface, is negligible compared with the strain energy, at least for the first layers. In addition, the Schmidt factor of this 90° partial dislocation, calculated to be 0.47 by Patriarche et al (1993), is greater than for any of the other misfit dislocation types. This extra stress controls the mechanism of creation of the partial dislocation accommodating the lattice mismatch.

The atoms deposited close to the step take a low energy position by creating directly a stacking fault, this is not the pseudomorphic position. More precisly, nucleation sites would be located at the edge of two adjacent {111} planes. Such stacking faults are also observed in other heterostructures such as GaAs/Si.

4. CONCLUSION

The relaxation of misfit stress by creation of a partial dislocation has been described. The mechanism involves a surface step introducing a discontinuity of the stress favorable to the direct creation of a partial dislocation. This result shows the importance of surface defects on the nucleation and the relaxation of the epilayer with large lattice mismatch. Work is in progress to establish a more exact model of the growth.

Acknowledgements: Many thanks to R. Bonnet, D. Caillard, P. Hawkes and M. Heggie for their fruitful discussions.

REFERENCES

Patriarche G, Rivière JP and Castaing J 1993 J Phys. III **3**, pp 1189-99
Raisin C, Saguintaah B, Tegmousse H, Lassabatère L, Girault B & Allibert C 1986 Annales des Télécommunications **41**, pp 1-9
Rocher A, Kang JM, Atmani H, Crestou J, Vandershaeve G, Lassabatère L and Bonnet R 1991 Inst Phys Conf ser **117**, pp 509-514
Rocher A and Atamani H 1993 Mat Science Forum, **126**, pp 559-62
Rocher A and Kang JM, 1995 Inst Phys Conf **146**, pp 135-142
Zhu G and Carter C B 1990 Phil Mag A **623**, pp 319-328

Inst. Phys. Conf. Ser. No 157
Paper presented at Microsc. Semicond. Mater. Conf., Oxford, 7–10 April 1997
© 1997 IOP Publishing Ltd

Stacking fault trapezoids, stacking fault tubes and stacking fault tetrahedra in ZnSe/GaAs(001) pseudomorphic epilayers

K K Fung, N Wang and I K Sou

Department of Physics, The Hong Kong University of Science and Technology, Clear Water Bay, Kowloon, Hong Kong

ABSTRACT: The fault planes and stair-rod dislocations in stacking fault trapezoid, stacking fault tube and stacking fault tetrahedron have been characterized with reference to the stacking fault pyramid in ZnSe/GaAs(001) pseudomorphic epilayers. All stacking fault configurations can be regarded as originating from an array of dimers on the (001) interface. The extended stair-rod dislocation dipoles in stacking fault trapezoids and tubes can act as diffusion channels for pipe diffusion of point defects during degradation.

1. INTRODUCTION

Device degradation leading to short lifetime is an important issue in the development of ZnSe-based II-VI semiconductor blue-green light emitters and laser diodes. Transmission electron microscopy (TEM) studies (Guha et al 1993, 1994, Hua et al 1994) have shown that degradation results from the formation of dark line defects from grown-in stacking faults originating at the ZnSe/GaAs interface and the diffusion of point defects. Neither the origin of the stacking faults nor the mechanism of diffusion of point defects are known. Two forms of stacking faults, single triangular faults and pairs of oppositely oriented triangular faults have been reported extensively. In a detailed TEM study of stacking faults in ZnSe/GaAs(001) pseudomorphic epilayers, we have found a variety of intersecting stacking fault configurations which include stacking fault pyramids, stacking fault trapezoids and stacking fault tubes (Fung et al 1996a, Wang et al 1996). We have shown that the stacking fault pairs can be regarded as incomplete stacking fault pyramids and single faults are probably stacking fault trapezoids. We report here the characterization of the fault planes and stair-rod dislocations in stacking fault trapezoids, stacking fault tubes and stacking fault tetrahedra.

2. EXPERIMENTAL RESULTS

The epilayers we have studied were grown by molecular beam epitaxy (MBE) in a VG MBE system dedicated to ZnSe-based II-VI growth in a single chamber. Details of the growth conditions and process have previously been reported (Sou et al 1995). These layers were studied by TEM in plan-view and cross-sections. Damage free plan-view specimens with millimetre size electron transparent areas have been prepared by back side etching with a 4:1 (NaOH: H_2O_2) solution (Wang and Fung 1995). TEM images were taken in a Philips CM-20 electron microscope at 200 kV.

158

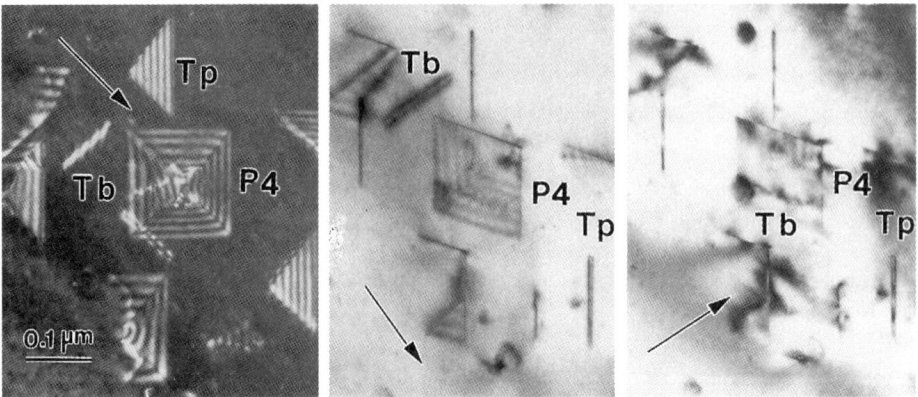

Fig. 1 (a) [001] 040 dark field image, (b) and (c) [101] 040 and $\overline{2}$02 bright field images

Stacking faults in pseudomorphic layers of various thicknesses have been studied in detail. Typical stacking fault configurations are shown in a [001] plan-view 040 dark-field image in Fig. 1(a). Complete and incomplete stacking fault pyramids (P4), stacking fault trapezoids (Tp) and stacking fault tubes (Tb) are clearly visible. Inclined {111} stacking faults running from the interface to the surface of the epilayer are delineated by triangular patches of parallel dark-bright fringes. The stacking fault pyramid (Booker 1964) will be used as an internal reference for the characterization of the stacking fault trapezoid and stacking fault tube. The {111} faults of the stacking fault pyramid intersect at about 109° along <101> giving rise to obtuse stair-rod dislocations (Fig. 2). Note that the contrast of the $(\overline{1}11)$ faults in Tp and P4 are different. The difference is due to the fact that contrast in P4 is due to a single fault whereas contrast in Tp is due to two closely spaced overlapping parallel faults. The closely spaced $(\overline{1}11)$ parallel faults are clearly seen connected by the $(\overline{1}\,\overline{1}1)$ fault edge-on when the specimen is tilted to the [101] zone axis (Fig. 1(b)). The $(\overline{1}\,\overline{1}1)$ fault intersects the $(\overline{1}11)$ faults at about 71° and 109°. The fringes in the defects (Tb) along $[\overline{1}01]$ and [011] are parallel to the fringes of the $(1\,\overline{1}1)$ and the $(\overline{1}\,\overline{1}1)$ faults of P4. As shown below, these defects are stacking fault tubes. The stacking fault tube is a rather complicated structure consisting of two pairs of closely spaced narrow faults intersecting in four parallel and closely spaced stair-rod edge dislocations. In order to characterize such a fault, it is important to separate the contrast of the faults from that of the stair-rods. This is to make sure that the contrast of the dislocation will not be obscured by that of the fault.

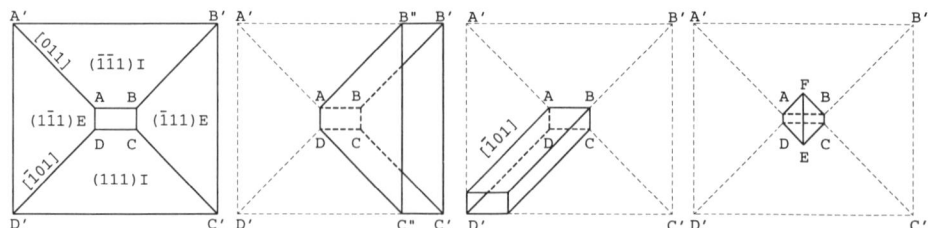

Fig. 2 Schematic of a stacking fault pyramid, trapezoid, tube and tetrahedron.

For an edge dislocation with Burgers vector **b** along **u**, the imaging invisibility criteria for a diffraction vector **g** are **g** · **b** = 0 and **g** · **b** × **u** = 0. These conditions for the $[\bar{1}01]$ stair-rod dislocation are satisfied by imaging with **g** = $\bar{2}02$ in the [101] zone axis. Images taken with **g** = 040 and $\bar{2}02$ in the [101] zone axis are shown in Fig. 1(b) and 1(c). The 040 image is dominated by stacking fault contrast. The $[\bar{1}01]$ stair-rod of the stacking fault pyramid is shadowed by the $(1\bar{1}1)$ and (111) faults of P4. Both Tb defects are in contrast. However, the $(1\bar{1}1)$ and (111) faults are out of contrast in the $\bar{2}02$ image. The $[\bar{1}01]$ stair-rod of P4 and the $[\bar{1}01]$ Tb fault are also out of contrast, but the [011] Tb fault is still in contrast. This implies that the Burgers vector of the stair-rods of the $[\bar{1}01]$ Tb fault is the same as that of the stacking fault pyramid, i.e. 1/6 [101], and the stacking fault planes are $(1\bar{1}1)$ and (111). Similarly, it can be shown that the fault planes of the [011] stacking fault tube Tb are $(1\bar{1}1)$ and $(\bar{1}\bar{1}1)$, and the Burgers vector of the stair-rod dislocations is 1/6 $[0\bar{1}1]$.

In addition to the extended stacking faults described above, defects showing point contrast can also be seen in Fig. 1(a). It turns out that these are stacking fault tetrahedra at the ZnSe/GaAs interface. Larger stacking fault tetrahedron can sometimes be observed (Fig. 3). There are two stacking fault tetrahedra in the weak beam dark-field images in Fig. 3. The large one (labelled T), about 50 nm, is clearly visible while the presence of the small one can be discerned. Stacking fault tetrahedra are in fact the most numerous of all the faults. The size of the tetrahedra can vary by a few times, with a typical value of about 10 nm. Unlike the other open faults described above which extend from the interface to the surface as the epilayer grows, the stacking fault tetrahedron closes on itself and terminates near the interface. The stacking fault trapezoid and stacking fault tetrahedron have also been observed in [110] HREM images. The separation between the intrinsic and extrinsic faults in a stacking fault trapezoid is typically about 10 nm.

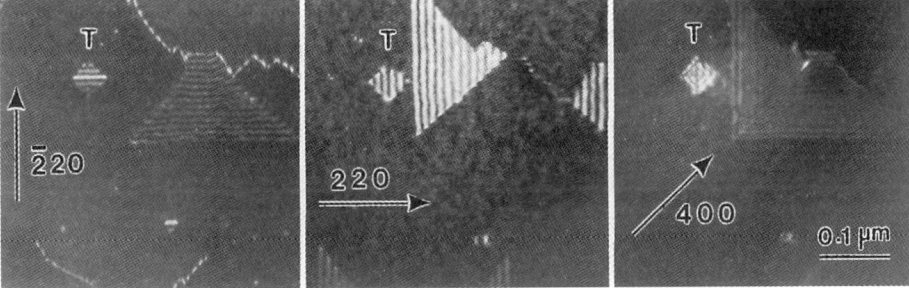

Fig. 3 (a) $\bar{2}20$, (b) 220 (c) 400 weak beam dark-field images of stacking fault tetrahedra.

3. ORIGIN OF STACKING FAULTS

It can be inferred from the cross-sections of the stacking fault pyramids, stacking fault trapezoids and stacking fault tubes that the faults initiate from a rectangular loop ABCD at the ZnSe/GaAs interface (Fig. 2). AB is is typically 10 nm long. The {111} faults from the loop ABCD in a pyramid fan out and intersect symmetrically at about 109° giving rise to obtuse stair-rod dislocations along <101>. When a pair of parallel intrinsic and extrinsic $(\bar{1}11)$ faults from AD and BC intersect the intrinsic $(\bar{1}\bar{1}1)$ and (111) faults from AB and DC, a stacking fault trapezoid is obtained. The "folded-in" AB"C"D fault intersect the $(\bar{1}\bar{1}1)$ and (111) faults at about 71° giving rise to acute stair-rod dislocations along [101] and $[0\bar{1}1]$

(Booker 1964). The closely spaced parallel acute and obtuse stair-rod pair form a dislocation dipole (Wang et al 1996). This is similar to the C dipole predicted by Kumar and Hirth (1992). Similarly, a stacking fault tube will be obtained when two pairs of parallel intrinsic and extrinsic faults from the initiating loop ABCD intersect, say along $[\bar{1}01]$. The stacking fault tube has a lozenge-shaped cross-section consisting of two intrinsic faults and two extrinsic faults. Faults of the same kind intersect in acute stair-rods while faults of opposite kind intersect in obtuse stair-rods. Adjacent acute and obtuse stair-rods form dislocation dipoles. The stacking fault tube is similar to but not the same as the extended dislocation dipole lozenge considered by Kumar and Hirth (1992). An intrinsic stacking fault tetrahedron can readily be obtained from an initiating loop ABCD elongated along $[\bar{1}10]$. In this case, the folded-in faults intersect the fanned-out faults in acute stair-rods before they intersect along [110] (EF) forming a tetrahedron. We have proposed that the initiating loop is the boundary of an array of dimers at the ZnSe/GaAs(001) interface (Fung et al 1996b, Wang et al 1996). Se/As dimers are formed on the oxide-free 2×1 reconstructed surface of the GaAs substrate (Pashley 1989). The dimer bonds are along $[\bar{1}10]$ while the dimers are aligned in rows along [110]. The bonding of atoms across the $[\bar{1}10]$ boundary is rather simple and leads to the formation of an intrinsic fault.

4. CONCLUSIONS

High densities of stacking fault tetrahedra, stacking fault trapezoids and stacking fault tubes have been observed in pseudomorphic ZnSe/GaAs(001) epilayers. These stacking faults originate from a rectangular array of dimers at the interface. It is not clear what role, if any, the stacking fault tetrahedra play in the degradation process. But it is expected that the stacking fault trapezoids and tubes will play an important role since their extended stair-rod dislocation dipoles can act as channels for pipe diffusion of point defects in the early stage of degradation.

ACKNOWLEDGEMENT

This work reported is supported by a grant from the Research Grants Council of Hong Kong, Grant No. HKUST621/95P.

REFERENCES

Booker G R 1964 Disc. Faraday Soc. **38**, 298
Fung K K, Wang N and Sou I K 1996a Proc. 6th Asia-Pacific Conf. Electr. Microsc. Hong Kong 1996, pp 163-4
Fung K K, Wang N and Sou I K 1996b J. Chin. Electr. Microsc. Soc. **15**, 191
Guha S, DePuydt, J M, Haase M A, Qiu J and Cheng H, 1993 Appl. Phys. Lett. **63**, 3107
Guha S, Cheng H, Haase M A, DePuydt J M, Qiu J, Wu B J and Hofler G E 1994 Appl. Phys. Lett. **65**, 801
Hua G C, Otsuka N, Grillo D C, Fan Y , Han J, Ringle M D, Gunshor R L, Hovinen M and Nurmikko A V 1994 Appl. Phys. Lett. **65**, 1331
Kumar A M and Hirth J P 1992 Phil. Mag. A **65**, 841
Pashley M D 1989 Phys. Rev. B **40**, 10841
Sou I K, Mou S M, Chan Y W, Xu G C and Wong G K L 1995 J. Cryst. Growth **147**, 39
Wang N and Fung K K 1995, Ultramicroscopy **60**, 427
Wang N, Sou I K and Fung K K 1996, J. Appl. Phys. **80**, 5506

Inst. Phys. Conf. Ser. No 157
Paper presented at Microsc. Semicond. Mater. Conf., Oxford, 7–10 April 1997
© *1997 IOP Publishing Ltd*

Global plastic relaxation of strained-layer superlattices with non-compensated strains

G Patriarche, E V K Rao, A Ougazzaden and F Glas

France Télécom, Centre National d'Etudes des Télécommunications, DTD, Laboratoire de Bagneux, 196 avenue Henri Ravéra, BP 107, 92225 BAGNEUX Cedex, France

ABSTRACT: We investigate by TEM the plastic relaxation of strained-layer-superlattices (SLSs) with non-compensated strains. Grown without dislocation beyond their critical thickness, the SLSs relax plastically during subsequent rapid thermal annealing. The dislocations appear in pairs located at the interfaces between the SLS and the buffer and the SLS and the capping layer. Their dissociation geometry is investigated. Depositing a silicon oxide mask on the structure prior to annealing strongly reduces the amount of plastic relaxation.

1. INTRODUCTION

Plastic relaxation of a single misfitting layer (L) epitaxially grown on a substrate (S) occurs through the introduction of dislocations at the S/L interface. This is energetically favourable only if the layer thickness exceeds some critical value $t^*(\varepsilon_0)$, which depends on its stress free (SF) strain ε_0 with respect to the substrate (Frank and van der Merve 1949). If a capping layer (C) lattice-matched to the substrate is grown on top of L, the critical thickness, defined from this energetical point of view (irrespective of the processes which might allow the actual introduction of dislocations), is increased. Moreover, the dislocations may be introduced either only at the S/L interface or as pairs located at both the S/L and L/C interfaces, depending on the thickness of C (Houghton 1991).

In this paper, we consider strained-layer superlattices (SLSs), consisting of n layers of thickness t_1 and SF strain ε_1 separated by $(n-1)$ layers of thickness t_2 and SF strain ε_2 buried under a capping layer of thickness t_c, for which the situation is more complicated since several possible modes of plastic relaxation exist. If any individual layer of the SLS has a thickness larger than a critical value (about twice $t^*(\varepsilon_i)$) determined by its particular SF strain ε_i and the fact that it is buried under the upper layers of the SLS and C, dislocations may be introduced at its interface with its two neighbouring layers in the SLS. In addition, even if each layer is kept below this critical thickness, the SLS may still relax globally via the introduction of either single dislocations at the S/SLS interface or pairs of dislocations at the S/SLS and SLS/C interfaces (Houghton et al 1994). The latter processes are characterized by their own critical thicknesses t_s^* and t_d^*, respectively, which depend on the average of the SF strain in the SLS, $\bar{\varepsilon} = \left[n\,t_1\varepsilon_1 + (n-1)\,t_2\,\varepsilon_2\right]/\left[n\,t_1 + (n-1)t_2\right]$, and on t_c. Only dislocation pairing allows C to remain in its SF state, because the strain introduced in C by each dislocation at the S/SLS is then exactly cancelled by the opposite strain introduced by the second dislocation at the SLS/C interface. Hence, one usually has $t_s^* < t_d^*$ for a given t_c.

When SLSs made of semiconducting materials are used to fabricate optoelectronic devices, care must be taken taken to avoid the introduction of extended defects which trap the charge carriers. Since t_s^* and t_d^* decrease when $\bar{\varepsilon}$ increases, it is desirable to keep $\bar{\varepsilon}$ as close to zero as possible; this is called strain compensation. Very efficient components have indeed been obtained (see for instance Ougazzaden et al 1995) by using strain-compensated SLSs with individual layers alternatively under large tension and large compression ($|\varepsilon_i|$ between 1 and 2 %). However, some devices require SLSs with non-compensated strains and are therefore particularly vulnerable to plastic relaxation. This is the type of SLS studied here.

2. SPECIMENS AND ANNEALING EXPERIMENTS

We consider non-strain-compensated SLSs grown on an (001)-oriented InP substrate by metalorganic vapour phase epitaxy (MOVPE) at a temperature $T_g = 650°C$, with layers made of two quaternary $In_xGa_{1-x}As_yP_{1-y}$ alloys ($n=10$ wells with $\varepsilon_1 = 0.9\%$, $t_1 = 10.4$ nm and $n-1=9$ barriers with $\varepsilon_2 = 0$, $t_2 = 6.9$ nm). Each SLS is grown on top of a buffer layer B made of another quaternary alloy lattice-matched to InP and capped by a stack of several layers also lattice-matched to InP, of total thickness $t_c = 144$ nm (Fig. 1). Because these layers are lattice-matched to InP, we may safely consider, as will indeed be confirmed below, that the S/SLS interface coincides with the top of B, and that the capping layers behave as a single layer.

No dislocation appeared either during growth or upon annealing for several hours at T_g in the MOVPE apparatus under a phosphine flux. The SLSs were subjected to rapid thermal annealings (RTAs) for 15 seconds at several temperatures $T_a \geq T_g$ in order to alter their characteristic emission wavelength by producing some interdiffusion of the elementary constituents (Hamoudi et al 1995). Some structures were covered by a silicon oxide mask. This mask was either undoped, or doped with various levels of phosphorus in the hope of minimizing the likely loss of this element from the structure during RTA and also to enhance interdiffusion. An InP wafer was placed directly above the samples during RTA to reduce further the loss of phosphorus. All RTAs produced a significant decrease in the photoluminescence efficiency of the SLSs. To understand its origin, a TEM study was carried out. We investigated both cross-sectional and plan-view specimens in a Philips CM20 microscope operated at 200 keV.

3. RESULTS

3.1 Dislocation geometry

Whatever the annealing temperature, dislocations appear, but never at the interfaces between the individual layers. Instead, they all lie at the two interfaces separating the buffer from the first strained well, and the last strained well from the capping layer (Fig. 1). With a few exceptions, all dislocations are dissociated, and thus appear as short segments when viewed edge-on (Fig. 1). Contrast analysis shows that the dislocations are of the 60° type. High resolution electron microscopy (HREM) images demonstrate that the 90° partial lies closest to the interface between the SLS and B or C and the 30° partial lies away from the interface (Fig. 2), in either B or C. The same dissociation geometry has been observed by Chen et al (1993) and Zou et al (1994) at the interface between a GaAs substrate and a single strained $In_xGa_{1-x}As$ layer. The dissociation length in this study is about 7 nm, which is also close to the values measured in the latter references.

Moreover, the 90° dislocations are paired, one lying at the base of the SLS, the other at the top, separated by a distance whose projection in the (001) growth plane varies slightly and is always larger than the projection in the same plane of their (111) glide plane. This pairing appears best in plan-view images of specimens with low dislocation densities (Fig. 3), which also reveal some degradation of the upper surface of the sample during RTA. The dislocation lines are oriented along the <110> direction (Fig. 3). In the plan-view images, we never observe any interruption which would indicate that a threading arm crosses the SLS. We thus estimate that each line extends for at least 100 μm in the interfaces. The pairs of dislocations are irregularly distributed and tend to occur in small clusters. Global relaxation of an uncapped SLS by formation of single dislocations at the B/SLS has been reported (Houghton et al 1990) but, to our knowledge, this is the first time that the previously predicted global plastic relaxation of a SLS by the formation of such pairs of dislocations has been observed.

3.2 Effects of RTA temperature and oxide mask

With respect to the situation pertaining before RTA, where the SLS relaxation is wholly elastic, the in-plane misfits of the wells and barriers with respect to InP become respectively $\varepsilon_1 - \varepsilon_r$ and $-\varepsilon_r$, where $\varepsilon_r > 0$ measures the amount of plastic relaxation. The observed relaxation by paired dislocations bordering the whole SLS thus produces a decrease of the elastic energy in the wells and its increase in the initially unstrained barriers. Given the type of the dislocations, ε_r is related to their measured average spacing D in each interface and along each <110> direction by $\varepsilon_r = a/(2\sqrt{2}D)$, where a is the lattice parameter of InP.

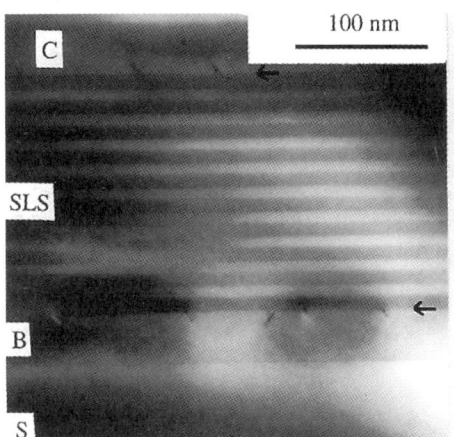

Fig. 1: 200 dark-field cross-sectional TEM image showing the SLS and the dissociated dislocations located at the interfaces (marked by arrows) between the SLS and the buffer and capping layer. In the SLS, the barriers appear darker than the wells.

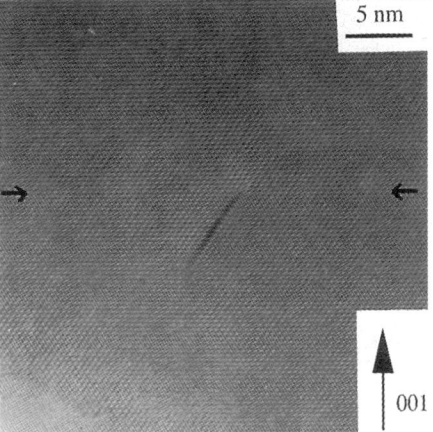

Fig. 2: HREM image showing a dissociated dislocation lying at the buffer/SLS interface, marked by arrows. The contrast between buffer and first SLS barrier is weak. The 90° partial (top of the fault) lies closer to the interface than the 30° partial (bottom of the fault).

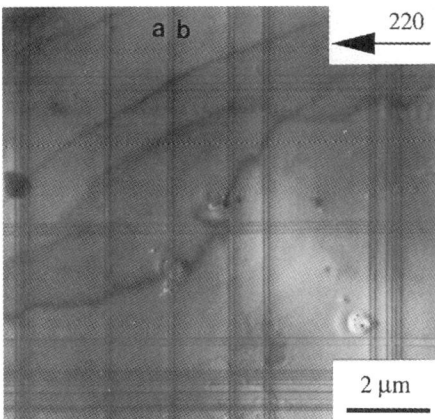

Fig. 3: 220 dark-field plan-view image showing pairs of 90°dislocations (such as ab), one of which lies at the buffer/SLS interface, the other at the SLS/capping layer interface. In this low magnification image, each dissociated dislocation appears as a single line.

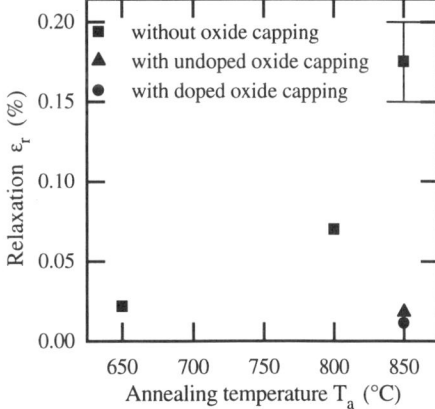

Fig. 4: Variation with annealing conditions of the amount of in-plane relaxation of the SLS with respect to the substrate. Each symbol gives the average for a given annealed sample. The error bar shows the variations measured on the particular sample considered.

Fig. 4 gives the observed values of ε_r for various RTA conditions. In the absence of a silicon oxide mask, ε_r increases rapidly with T_a. At the highest RTA temperature, the presence of a mask reduces ε_r dramatically. Phosphorus doping of the mask reduces ε_r further, but significant plastic relaxation remains. Fig. 4 gives the result for a concentration ratio P/Si $\sim 10^{-3}$; for a ratio $\sim 10^{-4}$, the dislocation distribution becomes very irregular, but the average relaxation is about the same.

4. DISCUSSION AND CONCLUSIONS

Applying the work of Houghton et al (1994) to a SLS with the same alternating layers as ours and a capping layer of the same thickness, we calculate a critical thickness $t_d* = 50$ nm for global relaxation by formation of dislocation pairs (corresponding to a total 'critical thickness' of strained wells $t_w* = 30$ nm). Moreover, we deduce from Houghton et al (1990) that, as long as the density of paired dislocations is low enough so that $\varepsilon_r \leq 0.34$ %, this mode of relaxation is energetically preferable to the introduction of dislocations at the B/SLS interface alone. This is consistent with all our observations regarding the occurrence of relaxation (since the total thickness of strained well is here 104 nm, well above t_w*) and its mode (dislocations at both upper and lower SLS interfaces), given that the amplitude of relaxation observed remains less than 0.34% (Fig. 4). Several points remain, however, to be discussed.

First, during and after growth, the SLS was already above its critical thickness (especially before the growth of C, when plastic relaxation would have required the introduction of only one set of dislocations at the B/SLS interface), but it did not relax plastically; neither did it relax during long annealings at T_g in the MOVPE apparatus. However, the SLS relaxed during RTA, even at T_g (Fig. 4). It is thus not only the elevated temperature which allows the SLS to relax during RTA. We suggest that plastic relaxation became possible during RTA because the change of atmosphere with respect to the situation during growth allowed the introduction of a larger concentration of vacancies or interstitial atoms in the sample. The observation of a degradation of its surface, as well as the large reduction of ε_r produced by the oxide mask (Fig. 4) which presumably reduces the loss of phosphorus, strengthen this hypothesis. The modification of the composition profile of the SLSs during RTA seems rather weak, however, although we did not attempt to measure it by TEM; due to the fall in photoluminescence efficiency induced by the dislocations, it was also impossible to evaluate its effect on the characteristic emission wavelength of the structure. The RTA temperature has nevertheless a large influence on ε_r, and it is striking that, at $T_a = 850°C$ and without mask, half of the optimal relaxation of 0.34% has already been achieved after only 15 seconds of RTA.

This study illustrates the difficulty of using SLSs with large non-compensated strains. Once the SLS has been grown beyond a critical thickness (which may depend in particular on the capping layer), even if it has not relaxed during growth, it may do so during subsequent processes. Depositing a silicon oxide mask on top of the structure prior to RTA reduces strongly the amount of plastic relaxation, without however making it disappear completely. If possible, it is thus advisable to use SLSs with compensated strains, since their critical thicknesses are in principle infinite, and in practice (considering the inevitable compensation errors) high enough for most applications.

REFERENCES

Chen Y, Zakharov N D, Werner P, Liliental-Weber Z and Washburn J 1993 Appl. Phys. Lett. **62** 1536

Frank F C and van der Merwe J H 1949 Proc. Roy. Soc. (London) A **198**, 216

Hamoudi A, Rao E V K, Krauz P, Ramdane A, Ougazzaden A, Robein D and Thibierge H 1995 J. Appl. Phys. **78**, 5638

Houghton D C 1991 J. Appl. Phys. **70**, 2136

Houghton D C, Davies M and Dion M 1994 Appl. Phys. Lett. **64**, 505

Houghton D C, Perovic D D, Baribeau J-M and Weatherly G C 1990 J. Appl. Phys. **67**, 1850

Ougazzaden A, Mircea A and Kazmierski A 1995 Electronic Lett. **31**, 803

Zou J, Cockayne D J H and Jiang S S 1994 J. Appl. Phys. **75**, 731

Inst. Phys. Conf. Ser. No 157
Paper presented at Microsc. Semicond. Mater. Conf., Oxford, 7–10 April 1997
© 1997 IOP Publishing Ltd

Critical thickness of quantum-well structures: Modified Matthews-Blakeslee formula and experimental support gathered by means of synchrotron X-ray reflection topography

P Möck[1*], B K Tanner[1], G Lacey[2], C R Whitehouse[2], and G W Smith[3]

[1] Department of Physics, University of Durham, Science Laboratories, South Road, Durham DH1 3LE
   ([1*] now at: Interdisciplinary Research Centre for Semiconductor Materials, Imperial College of Science, Technology and Medicine, (University of London), The Blackett Laboratory / Huxley Undercroft, Prince Consort Road, London SW7 2BZ; p.moeck@ic.ac.uk, see also http://www.dur.ac.uk/~dph0pm/me.htm)
[2] Department of Electronic and Electrical Engineering, University of Sheffield, Mappin Street, Sheffield S1 3JD
[3] Defence Research Agency, Malvern; St Andrews Road, Great Malvern, Worcestershire WR14 3PS

ABSTRACT: Qualitative experimental evidence that certain types of misfit dislocations shrink and eventually disappear when a capping layer is grown on an essentially unrelaxed strained epilayer has been gathered by means of synchrotron X-ray reflection topography. This evidence supports the modification of the line tension force term and the omission of a factor of two in the classical Matthews and Blakeslee critical thickness formula for a quantum-well structure.

1. INTRODUCTION

It was concluded in a thorough review by Fitzgerald (1991) that in disagreement with the classical Matthews and Blakeslee paper, the critical thickness of a quantum well structure cannot simply be double the value of a similar strained epilayer because there is no physical justification for a double line tension force term in Matthews' and Blakeslee's (1974-1976) equations. Accordingly, Matthews' and Blakeslee's claim that the critical thickness of a multi-quantum-well structure which is on average lattice matched to the buffer layer and consists of alternating layers of the same thickness should be just four times the value of a related strained layer has no physical justification either. An alternative critical thickness formula for the quantum-well case has not been given by Fitzgerald, but it is straightforward to derive as long as the critical thickness of the strained epilayer is defined as the one at which the *very first* misfit dislocations are introduced during the growth of the quantum-well structure, indicating that kinetic relaxation effects are being neglected.

The first aim of this paper is to derive a modified Matthews-Blakeslee-type formula for the critical thickness of a quantum-well structure. Secondly, we will present qualitative experimental evidence gathered by means of synchrotron X-ray reflection topography in order to support our modification.

2. THEORY

2.1 Modified Matthews-Blakeslee formula

Using the definition of the critical thickness given above, it could be simply argued that the capping layer on an essentially unrelaxed strained layer which itself is just at its critical thickness, i.e. which possesses a misfit dislocation density of as low as about 1-10 per cm^{-2} (see, e.g., Möck et al. 1996a) at the substrate - strained layer interface, experiences as its growth surface an almost identical lattice constant to that of the substrate. Hence, the capping layer will not relax to any significant amount and threading dislocations will essentially not be bent into the epilayer - capping layer interface. The critical thickness of this structure should be calculated by applying the Matthews (1975) formula (or even better by using modifications by Fitzgerald (1991), Chidambarrao *et al.* (1990), Jesser and Kui (1993), and Freund (1994) to name but a few) for a strained layer, taking into account the different distance of the misfit dislocations from the free surface. This means, in more precise terms, the cut-off length of the misfit dislocation strain field has to be regarded as the sum of the strained layer and the capping layer thicknesses and the expression of the line tension force has to be modified accordingly. Hence, in the Matthews framework the critical thickness of a quantum well which is reached during growth is given by

$$h_{str} = \frac{b(1 - \nu\cos^2\theta)\ln\{\alpha(h_{str} + h_{cap})/b\}}{8\pi\varepsilon(1+\nu)\cos\lambda}$$ (1),

where b is the magnitude of Burgers vector, ν is the Poisson's ratio, θ is the angle between **b** and dislocation line direction (**u**), α is the core-energy parameter, ε is the misfit strain, and λ is the angle between **b** and normal to **u** in the interface (consult Hull and Bean's (1992) review and Fitzgerald's (1991) paper). This modified relation differs from the well known Matthews-Blakeslee formula by a factor of ½ as well as by the dependence on the thickness of the capping layer.

The consideration above can be expressed in simple mathematical terms as well, since, according to the Matthews / Matthews and Blakeslee models, it should be the integral (Dunstan *et al.* 1994) of the product of the misfit strain, the magnitude of the Burgers vector, some material constants and angular factors over the epilayer thickness taken from the free surface down to any depth in the epitaxial structure which yields the force that tend to drive the threading arm of the dislocation through the layer by means of a single- or double-kink mechanism. Since the capping layer does not experience misfit strain to any significant amount, this force has to be regarded as almost independent of the thickness of the capping layer. Because this force is balanced with the line tension force of the dislocation acting against its elongation, which itself increases with the thickness of the capping layer, the critical thickness of the quantum well will be moderately higher than the critical thickness of a similar strained layer and will increase slowly with the capping layer thickness.

2.2 Prediction of a "critical thickness observation window"

The above consideration implies that a quantum-well structure will have a "critical thickness observation window" depending on both the thickness of the embedded strained layer and the capping layer due to the following processes. Misfit dislocations will nucleate in the strained layer at pre-existing threading dislocations as soon as its critical thickness is exceeded. During further growth of the strained layer, these misfit dislocations will grow in length and further misfit dislocations will nucleate at other sources and elongate as well. The subsequent growth of the capping layer will change the force balance acting on the threading arms of either some or all of the dislocations in favour of the force acting against the elongation of their misfit segment and, hence, this elongation will stop and these misfit dislocations will subsequently shrink in length. Above a certain thickness of the capping layer, the structure will eventually either become free of misfit dislocations or remain misfitted to a lower amount. In other words, the "critical thickness observation window" will either be closed or remain open.

Which one of these two possibilities will happen at this certain thickness of the capping layer will most likely depend on the amount by which the critical thickness of the strained layer was first exceeded during its growth, indicating that the "critical thickness observation window" will only exist between well defined boundaries. Whereas the lower thickness boundary of this window would be just above the critical thickness for misfit dislocation nucleation at pre-existing threading dislocations, the upper boundary may be just below the "second" critical thickness of misfit dislocation generation by other mechanisms which was discovered during our earlier *in-situ* synchrotron X-ray reflection topography experiments (see, e.g., Barnett *et al.*, 1995, Möck *et al.* 1996a). The propagation mechanism of the former dislocations will almost certainly be governed by the movements of single kinks, but the latter misfit dislocations might move by the double-kink mechanism, depositing misfit-relieving segments at both interfaces. It is straightforward to imagine that these "secondary" dislocations, which possess misfit-dislocation dipoles, will probably not encounter a significant force-balance shift due to the growing capping layer which could drive them into shrinkage and disappearance.

3. EXPERIMENTAL SUPPORT

3.1 Result of an *in-situ* synchrotron X-ray reflection topography study

Using purpose-built equipment (Whitehouse *et al.* 1992), a capped $In_{0.04}Ga_{0.96}As$ structure was grown by molecular beam epitaxy (MBE) at $555\,^\circ C$ on a (001) $\pm$ 0.1° oriented, Si-doped vertical gradient freeze Bridgman (VGFB) GaAs substrate. This epitaxial structure was topographed *in situ* at the synchrotron radiation source (SRS) in Daresbury. The topographic study was performed during a number of growth steps of typically 20 minutes while the sample was maintained at the growth temperature or annealing temperatures and subject to a constant flux of As. Grazing incidence 224 and -224 reflection topographs of the substrate reflection were taken with an X-ray beam of 0.1488 nm wavelength, emerging from a 333 channel-cut Si crystal. The study included a critical thickness measurement and annealing experiments of the capped structure which will be reported elsewhere.

The most interesting result of this study in the context of this paper is that misfit dislocations at the substrate - strained epilayer interface do indeed shrink when a GaAs capping layer is grown. This is clearly demonstrated in Figure 1, where one and the same set of misfit dislocations is depicted (at the growth temperature) for both structures: a nominally 140 nm thick strained epilayer prior to (Fig. 1a) and after (Fig. 1b) the deposition of a nominally 20 nm thick capping layer.

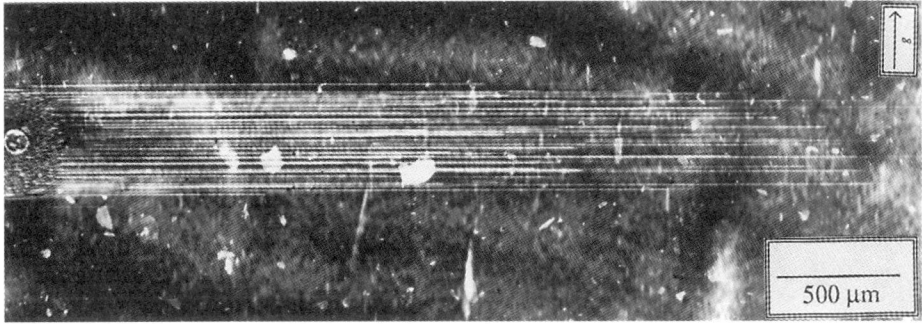

Fig. 1a) *in-situ* 224 reflection topograph of a set of misfit dislocations within a single strained epilayer

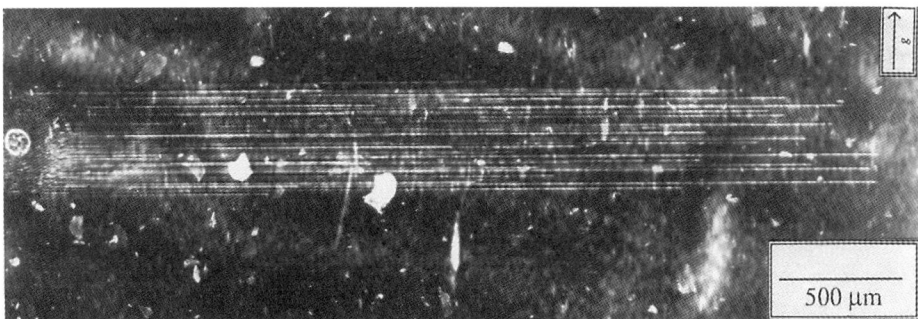

Fig. 1b) *in-situ* 224 reflection topograph of the same set of misfit dislocations within a quantum well

3.2 Result of an *ex-situ* synchrotron X-ray reflection topography study

An $In_{0.06}Ga_{0.94}As$ layer of nominally 60 nm thickness was grown by MBE at 517°C on a (001) ± 0.1° oriented, Si doped VGFB GaAs substrate. A GaAs cap of nominally 1μm thickness was grown on this strained epilayer subsequently.

From our earlier *in-situ* XRT experiments of single strained layers (Barnett *et al.* 1995, Möck *et al.* 1996b), we infer the critical thickness for bending of dislocations threading from the substrate into the substrate - strained layer interface to be around 50 nm and a "second" critical thickness due to other nucleation processes to be about 90 nm. Accordingly, this epilayer is expected to have relaxed prior to the deposition of the capping layer to an amount defined more or less by the average density of the pre-existing dislocations threading from the substrate (i.e. ≤ 20 cm^{-2}, Barnett *et al.* 1995).

This quantum well structure has been topographed *ex situ* at the SRS Daresbury employing a series of symmetrically equivalent 224 reflections of the substrate in grazing incidence at a wavelength of 0.1478 nm in a (+n,-m) double-crystal setting with the symmetric 333 reflection of Si as monochromator. Only three short segments of misfit dislocations were imaged within the whole 2 inch sample, i.e. the averaged density of misfit dislocations was two orders of magnitude lower than the average density of threading dislocations. This result illustrates the effectiveness of the capping layer in removing misfit dislocations originated by bending of threading dislocations.

4. CONCLUSION

Qualitative experimental evidence that certain types of misfit dislocations do indeed shrink and eventually disappear during the growth of a capping layer on an essentially unrelaxed strained epilayer has been gathered by means of synchrotron X-ray reflection topography. The classical Matthews and Blakeslee formula for the critical thickness of a quantum-well structure has been modified accordingly.

ACKNOWLEDGEMENTS

The authors acknowledge collaboration with all of the members (http://www.dur.ac.uk/~dph0pm/people. htm) of the "*In-situ* X-ray imaging of strained epilayer relaxation processes" project (http://www.dur.ac.uk/ ~dph0pm/in-situ.htm). Financial Support from the EPSRC and DRA is gratefully acknowledged.

REFERENCES

Barnett S J, Keir A M, Cullis A G, Johnson A D, Jefferson J, Smith G W, Martin T, Whitehouse C R, Lacey G, Clark G F, Tanner B K, Spirkl W, Lunn B, Hogg J C H, Ashu P, Hagston W E and Castelli C M 1995 J. Phys. D: Appl. Phys. **28**, A17
Chidambarrao D, Srinivasan G R , Cunningham B, and Murthy C S 1990 Appl. Phys. Lett. **57**, 1001.
Dunstan D J, Kidd P, Fewster P F, Andrew N L, Grey R, David J P R, González L, González Y, Sacedón A and González-Sanz F, 1994 Appl. Phys. Lett. **65**, 839
Fitzgerald E A 1991 Mater. Sci. Reports. **7**, 87
Freund L B 1994 Advances Appl. Mechanics **30**, 1
Hull R and Bean J C 1992 Critic. Rev. Solid State Mater. Sci **17**, 507
Jesser W A and Kui J 1993 Mater. Sci. Engin. **A164**, 101
Matthews J W and Blakeslee A E 1974 J. Cryst. Growth **27**, 118 ; 1975 ibid. **29**, 273 ; 1976 ibid. **32**, 265
Matthews J W 1975 J. Vac. Sci. Technol. **12**, 126
Möck P, Tanner B K, Whitehouse C R, Cullis A G, Lacey G, Clark G F, Lunn B, Hogg J C H, Keir A M, Johnson A D, Smith G W and Martin T 1996a The Physics of Semiconductors, eds. M Scheffler and R Zimmermann (Singapore, New Jersey, London, Hong Kong: World Scientific) Vol. 2, Sec. B, pp 927-30
Möck P, Tanner B K, Li C R, Keir A M, Johnson A D, Lacey G, Clark G F, Lunn B and Hogg J H C 1996b Semicond. Sci. Technol. **11** 1051, Erratum 1363
Whitehouse C R, Barnett S J, Soley D E J, Quarrell J, Aldridge S J, Cullis A G, Clark G F, Lamb W, Tanner B K, Cottrell S, Lunn B, Hogg C, Hagston W 1992 Rev. Sci. Instrum. **63** 634

Inst. Phys. Conf. Ser. No 157
Paper presented at Microsc. Semicond. Mater. Conf., Oxford, 7–10 April 1997
© *1997 IOP Publishing Ltd*

Crack interactions in tensile-strained epilayers

R T Murray, C J Kiely and M Hopkinson*

Department of Materials Science & Engineering, The University of Liverpool, Liverpool
L69 3BX, UK.
*Department of Electronic & Electrical Engineering, The University of Sheffield, Sheffield
S1 4DU, UK.

ABSTRACT: When epilayers are grown under tensile strain cracks are formed where the critical thickness, t_c, is exceeded. At thicknesses less than about $4t_c$ these form a unidirectional array on $(1\bar{1}0)$ planes but at a thickness of greater than $6t_c$ an orthogonal (110) array is introduced. By analysis of the intercrack spacing in $(1\bar{1}0)$ gratings a high degree of self ordering is demonstrated. When orthogonal cracks interact three outcomes have been identified, namely "T" junctions, straight intersections and jogs, and it is shown that these are in response to the relative depth of the two cracks at the junctions.

1. INTRODUCTION

If an epilayer is deposited pseudomorphically on a single crystal substrate, whose lattice constant is greater than that of the layer, elastic strain energy is stored in the structure, predominantly within the epilayer. As the layer thickens the stored energy per unit area increases and beyond a critical thickness t_c, cracks can form so as to relieve some of this energy as shown by Matthews and Kloleholm (1972) for magnetic garnets. To calculate a critical thickness for crack formation it is necessary to postulate that an incipient crack can only harvest energy within a finite range of itself. By equating the energy stored within a range t to the energy of the crack surfaces Murray et al (1995) deduced that $t_c = 8.6 \times 10^{-12} f^{-2}$ metres. Remarkable agreement between this equation and the observed conditions for crack growth have been reported by Murray et al (1996).

Where the epilayer thickness exceeds t_c an array of cracks forms. For InGaP grown on GaAs (Olsen et al 1974), GaAsP grown on GaAs (Aragon et al, 1993) and InGaAlAs on InP (Murray et al, 1995) this initial array has been found to be on $(1\bar{1}0)$ planes. When the thickness reaches about $4-6t_c$ an orthogonal family of cracks nucleates on the (110) plane. Whilst members of the $(1\bar{1}0)$ family of cracks are essentially straight and continuous across the wafer the path of the (110) cracks which formed after the $(1\bar{1}0)$ set are frequently disrupted by them. In this study we will report on features which arise through interactions between either parallel or orthogonal cracks.

2 EXPERIMENTAL

Three InGaAlAs layers were grown epitaxially on [001] InP substrates by MBE at temperatures of $500 \pm 20°C$ and their physical parameters are listed in table 1.

The crack spacings have been measured from TEM [110] cross sections at magnifications between 5K and 20K chosen so that 10 or more neighbouring cracks may be seen on a single negative. Orthogonal interactions have been observed by plan view TEM and Nomarski interference microscopy. Both cross section and plan view TEM specimens were prepared by Ar ion milling in a Gatan pips machine at an angle of 4° as this gave good preservation of crack morphology.

TABLE 1   Growth parameters and crack dimensions for the three layers

Wafer		Thickness t(nm)	tc	Composition	Array Type	$\bar{\Lambda}$ (μm)	$\delta\Lambda_{\frac{1}{2}}$ (μm)	f
A	M944	70	21	$In_{25}Ga_{75}As/$ InP	$(1\bar{1}0)$	1.15	.87	2×10^{-2}
B	M1048	75	21	$In_{25}Ga_{37}Al_{38}/$ InP	$(1\bar{1}0)$	1.8	1	2×10^{-2}
C	M530	150	21	$In_{25}Ga_{19}Al_{56}/$ InP	(110) & $(1\bar{1}0)$	-	-	2×10^{-2}

3.   UNIDIRECTIONAL ARRAYS

Layers A and B (Table 1) were grown to between t_c and $5t_c$ and display cracks only on $(1\bar{1}0)$ planes. C is thicker than $6t_c$ and is populated with cracks lying on both (110) and $(1\bar{1}0)$ planes. The majority of cracks in a unidirectional array are jog free and continuous across the 2" wafers used. However, Nomarski interference microscopy demonstrates that a few cracks either terminate or nucleate far from the perimeter of a wafer.

Numerous cross sections of A and B have been observed by TEM and over 100 crack spacings have been statistically analysed in the form of histograms for each - Fig 1 and the results summarised in Table 1.

The principal features which can be gleaned from these histograms are :
(i)   The full width at half maximum of the principle peak is less than half the average intercrack spacing $\bar{\Lambda}$.
(ii)   There is a minimum observed spacing which is about $0.5\bar{\Lambda}$.
(iii)   No crack spacings greater than $4\bar{\Lambda}$ have been observed.
(iv)   In the case of wafer B the distribution is biomodal with a weak peak at $2\bar{\Lambda}$.
(v)   The values of $\bar{\Lambda}$ and Λ, the most popular spacing, coincide within experimental error. Fig 2 is a typical cross section as used in the generation of Fig 1.

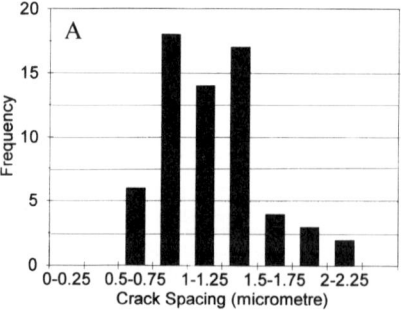
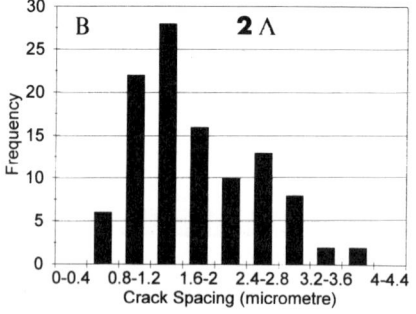

Fig 1 - Histograms of Crack Spacings in Wafers A and B. Note the secondary peak at $2\bar{\Lambda}$ in Wafer B. The histograms are obtained by analysing TEM cross sections.

Fig 2 - A cross section from wafer A,
as used for the statistical analysis
of crack spacings. Note that each crack
propagates down into the compressed
substrate.

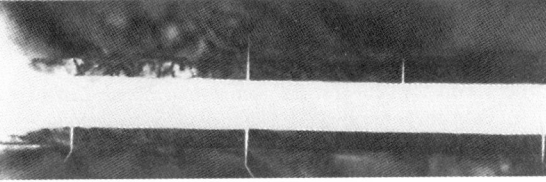

0·5 µm

Points (i) through (iv) strongly indicate that the crack positions are non-random and that one needs to invoke the concept of self ordering through intercrack interactions. In particular, if the positioning of the cracks was random as would be expected in the absence of interactions the probability function would peak close to the intercrack resolution falling off in exponential fashion according to $P(\Lambda) = m_c^{-(m\Lambda)}$ as shown for interfacial dislocations by MacPherson and Goodhew (1995). This is far from the nature of the distributions recorded in Fig 1.

Let us then consider the strain distribution in the MESA between the existing cracks separated by Λ and with the width W at their mouths. The average relaxation is then w/Λ and the average strain at the free surface is $(f - w/\Lambda)$. However, Harker et al (1995) have shown, by finite element modelling, that the relaxation is a maximum near the edge and that the residual strain rises to a flat maximum at $\Lambda/2$; thus the stored energy density and hence the energy which can be released by a new crack is a maximum at $\Lambda/2$. It is natural to associate a higher probability with such a high energy density location but the precise link between energy density and crack probability has not yet been defined. Nonetheless we propose that the occurrence of a subsidiary peak at $2\overline{\Lambda}$ in distribution B is support for the energy density mechanism.

4.   ORTHOGONAL CRACKS

During the growth of an orthogonal array of cracks the $(1\overline{1}0)$ family forms first and reaches a substantial fraction of its final population density before the generation of a second array is triggered at about $5t_c$. This (110) array is therefore faced with the need to interact with the pre-existing orthogonal cracks. Three situations may be envisaged depending upon the relative depth of penetrations of the two interacting cracks (Fig 3).

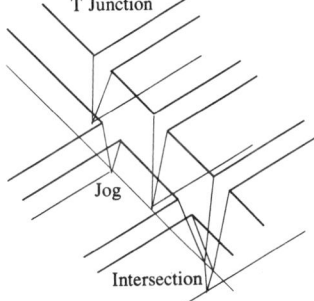

Fig 3 - Geometry of crack junctions leading to
continuous intersections, Joa and T junction,
through propagation of strain fields beneath the
earlier $(1\overline{1}0)$ crack

Fig 4 - Plain view micrographs of the three types of junctions observed.   Specimens have been prepared using low milling.

(a) A "T" junction   (b) A continuous intersection   (c) A jog   0·5 µm

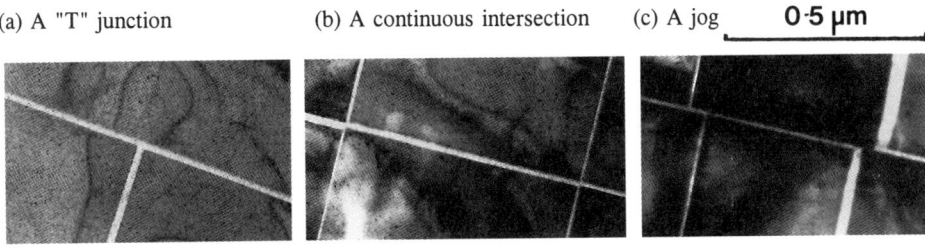

(i)   The new crack may be shallower than the $(1\bar{1}0)$ crack it encounters in which case virtually no penetration of its strain field through to the other side will occur. The crack will then terminate as a "T" junction, as illustrated in Fig 4a.

(ii) Where the new crack is deeper than the pre-existing $(1\bar{1}0)$ crack a strong penetration of its strain field under the shallower crack will occur leading to re-nucleation as a continuous crack (shown in Fig 4b). This is the most likely scenario since the (110) cracks are formed when this layer is thicker and a higher density of stored energy is available to drive their growth.

(iii) Where the two cracks are of the same depth there is only a weak linkage of the strain field into the virgin material on the other side. The crack may continue in a straight line or it may suffer a slight jog due to dynamic strain waves propagating from parallel cracks, as seen in Fig 4c.

Seventy interactions photographed at random from ion milled samples of specimen C have been sorted into three groups, dependent upon the geometry of the observed junction. Table 2 shows that the number of undeviated intersections slightly exceeds the number of "T" junctions. A jog occurs in about 20% of interactions but substantial jogs are rare. We, therefore, estimate that jogs are probable when this mismatch in crack depths is less than 10-20% of the observed range of depths.

TABLE 2   Statistical analysis of junction types observed by plan view TEM on Wafer C

Wafer		Number of Intersects	Number of "T" Junctions	Small Jogs	Large Jogs
C	M530	25	21	11	3

5.   CONCLUSIONS

The strain fields associated with individual cracks of an array interact and determine the geometry of further crack formation. Thus, for a grating of parallel cracks the most probable location for a new crack is at the mid point of an existing pair. At the junctions of two orthogonal cracks the ability of the strain field of the nascent crack to propagate under the pre-existing barrier into the empty zone on the other side will give rise to a distribution of "T" junctions, jogged junctions and undeviated intersections.

ACKNOWLEDGEMENTS

We thank EPSRC for their financial support for this work. Discussions with Dr T J Bullough and Mr S Westwater (Department of Materials Science & Engineering, The University of Liverpool) have enhanced our understanding of parallel interactions.

REFERENCES

Aragon G, DeCastro M J, Molina S I, Gonzalez Y, Gonzalez L, Briones F and Garcia R, 1993, MRS Fall Meeting
Matthews J W and Klokholme, 1972, Mat Res Bul, 7, 213
MacPherson G and Goodhew P J, 1995, Inst Phys Conf Series, 147, 381
Murray R T, Kiely C J, Hopkinson M and Goodhew P J, 1995, Inst Phys Conf Series, 146, 207
Murray R T, Kiely C J and Hopkinson M, 1996, Phil Mag, A74, 383
Murray R T, Kiely C J and Hopkinson M, 1997, to be published
Olsen G H, Abrahams M S and Zamerowski T J, 1974, J Electrochem Soc, 121, 1650
Westwater S P, Bullough T J, Joyce T B, Davidson B R and Hart L, 1997, App Phys Lett, 70, 60

Inst. Phys. Conf. Ser. No 157
Paper presented at Microsc. Semicond. Mater. Conf., Oxford, 7–10 April 1997
© 1997 IOP Publishing Ltd

Structural characterisation of GaN layers : influence of polarity and strain release

J-L Rouviere, M Arlery and A Bourret

CEA-Grenoble/ Département de Recherche Fondamentale sur la Matière Condensée /SP2M/ME
17 rue des Martyrs - 38054 Grenoble Cedex 9, France

ABSTRACT : We comment on the structural quality of wurtzite GaN layers grown by Metal-Organic Chemical Vapour Deposition (MOCVD) on $(0001)Al_2O_3$. We acknowledge that the nitridation of the sapphire substrate and the low temperature buffer layer are the key factors for selecting the polarity and releasing the strain of the films. The Ga-polarity appears to be the best solution to avoid threading {0,1,-1,0} Inversion Domains and to obtain a flat surface. The numerous threading <2,-1,-1,0> edge and [0001] screw dislocations present in the layer do not release strain. They are just necessary to compensate respectively the in-plane and out-plane rotations of adjacent grains that have coalesced during growth. Most of the -13.9% mismatch between the sapphire and the GaN layers is compensated in the highly defective buffer layer. We introduce the idea that the roughness of the buffer layer may be the fundamental property that allows the creation of a flat GaN surface : by producing periodic regions of good fit a wavy boundary condition at the buffer layer is helpful to create a flat surface at the top of the GaN layer. All this discussion is conducted by reviewing the different extended defects (Inversion Domain Boundaries, Translation Domain Boundaries, Dislocations, Nanopipes) present in these layers.

1. INTRODUCTION

Several companies in the world can now commercialised blue emitting diodes similar to the ones first realised by Nakamura et al (1991). With the recent realisations of several blue emitting lasers working at room temperature (Nakamura et al 1997, Akasaki et al 1996), but with a limited life time, there is no doubt that the III-V nitrides (InN, GaN and AlN) with GaN in the central position, will be important semiconductors in optoelectronic applications. These last years, great efforts have been realised to better understand the GaN material which can be used in devices in spite of a high density of dislocations (Lester et al 1995). Recently Rosner et al (1997) showed that dislocations in GaN are recombination centres but that the diffusion length of minority carriers is lower than the distances between dislocations (<250nm). Sigiura (1997) calculated that these dislocations are much less mobile than in other III-V material like GaAs. These two studies allow one to understand why diodes can work in a highly defective material. However, one would have certainly to improve the crystallographic quality of the GaN layers in order to extend the lifetime of working lasers.

We have already reported (Rouviere et al 1996) that the polarity of the layer is an important factor that greatly influences the surface morphology of the layer (Rouviere et al 1997). Polarity is controlled by the nitridation or/and the low temperature buffer layer. Figure 1 summarises the main results. All the flat samples, we have observed have a Ga-polarity. They were obtained with an adapted nitridation of the sapphire substrate and a low temperature GaN or AlN buffer layer (Briot et al. 1997, Nieburh et al 1995). The N-polar material has generally a surface formed of hexagonal flat tops or/and pyramidal steps. This polarity was obtained when no or a too thin buffer was used. As we will see, the defect contents of the GaN layer also depend critically on the polarity. So it is very important to determine the polarity of the layer when one wants to compare the crystallographic quality of different layers.

In this paper, we review the main extended defects present in GaN layers. We describe the origin of these defects (in which conditions these defects are created) and how they can be removed or at least reduced. Part 2 summarises the experimental procedures. Part 3 reviews the planar defects. Part 4 deals with dislocations. Part 5 is devoted to the holes often observed in GaN layers. Part 6 briefly discusses how the strain in the GaN layers can be relaxed.

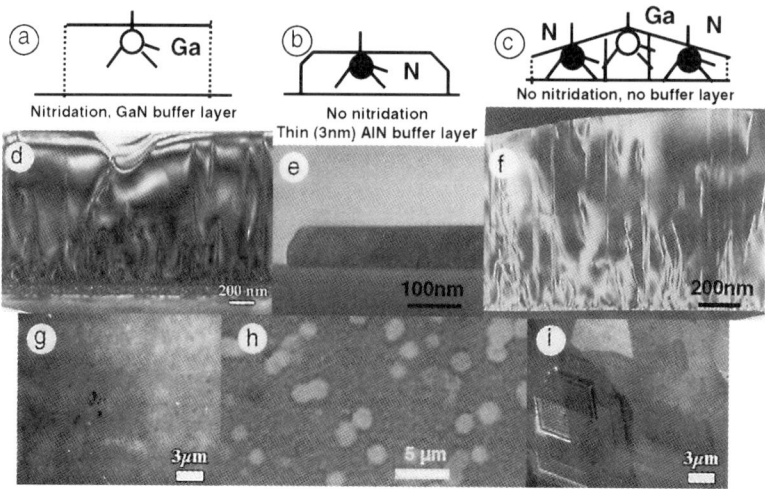

Fig 1. Relationship between the surface morphology and the polarity of three different MOCVD GaN layers . (a-b-c) Schemes outlining the polarity of the layers. The Ga-polarity is defined in fig. 2 (d-e-f) TEM cross-sections. (g-h-i) Optical images of the GaN surface. (a-d-g) A nearly flat (one pit is show here) Ga-polar GaN layer obtained by optimising the nitridation of the sapphire substrate and using a GaN buffer layer. (b-e-h) A thin N-polar GaN layer containing many hexagonal flat tops obtained without the nitridation of the sapphire substrate and with a thin AlN buffer layer. (c-f-i) A N-polar GaN layer presenting many hexagonal pyramids obtained with no nitridation and no buffer layer. The N-polar layer contains many Inversion Domains.

2. EXPERIMENTAL PROCEDURES

Most of the defect descriptions of this paper come from Transmission Electron Microscopy (TEM) observations realised on GaN layers grown by MOCVD on (0001) Al_2O_3 substrate. Conventional and High Resolution Electron Microscopy (HREM) studies were realised on a JEOL4000EX electron microscope, specially equipped for HREM (Scherzer resolution about 0.17nm, tilt ±20°). Convergent Beam Electron Diffraction (CBED) experiments were made on a JEOL3010FX microscope (Scherzer resolution about 0.17nm, tilt ±20°). Specimens for TEM were prepared using the standard techniques : mechanical polishing and Argon ion milling. The energy of the tetracoordinated models was computed using a Keating potential. The two parameters of this potential were taken from Kim et al (1996).

The investigated GaN layers have a wurtzite structure and grow epitaxially on the (0001) Al_2O_3 substrate, with [0001] and [2,-1,-1,0] GaN respectively parallel to [0001] and [1,0,-1,0] Al_2O_3 (Kuwano et al 1991).

3. PLANAR DEFECTS

Planar defects (PD) have been found on the following planes : (0001), {0,1,-1,0}, {2,-1,-1,0} and {0,1,-1,1}. The recent literature has been a bit confusing as far extended PD names are concerned. Several different names have been proposed for the same kind of defects. In general these names are too related to a particular defect (like staking fault (SF) is specific to the wrong stacking of the close-packed planes) to be really used as a generic term or they are too restrictive because they either incorporate the a-priori atomic structure of the defect (Anti Phase Boundary

(APB) says that there are wrong bonds on the model) or they give only one possible origin of the defect (Stacking Mismatch Boundary (SMB)). Following Pond (1989), the defect types should be classified according to the symmetry relating the two crystals on each side of the defect. Consequently, the term Inversion Domain Boundary (IDB) describes a defect where the two crystals on each side of the defect can be deduced from each other by a combination of an inversion and a translation. By analogy with the term IDB, we propose that Translation Domain Boundary (TDB) designs a defect where only a translation relates the two crystals. The SF is then only a special TDB that is defined on a close-packed plane.

So the first step in analysing a planar defect is to determine its type. CBED is the most straightforward and convincing technique to do so. A CBED pattern is taken on each side of the defect or inside and outside the domains when the defects formed closed domains (fig. 2). In cross-section, the <0,1,-1,0> direction is the most convenient direction of observation (Rouviere et al 1995). In plan view other directions have been used (Cherns et al, 1997, Vermaut et al 1997). Other experimental observations can indicate that the observed defect is an Inversion Domain Boundary, but, unless really mastered, they are generally not proofs. The HREM contrast argument (Rouviere et al. 1995) is very arguable and can only be considered as an indication. The reversal of contrast of thickness fringes when crossing the interface is more reliable (Westwood et al 1995) (fig. 2, fig. 4).

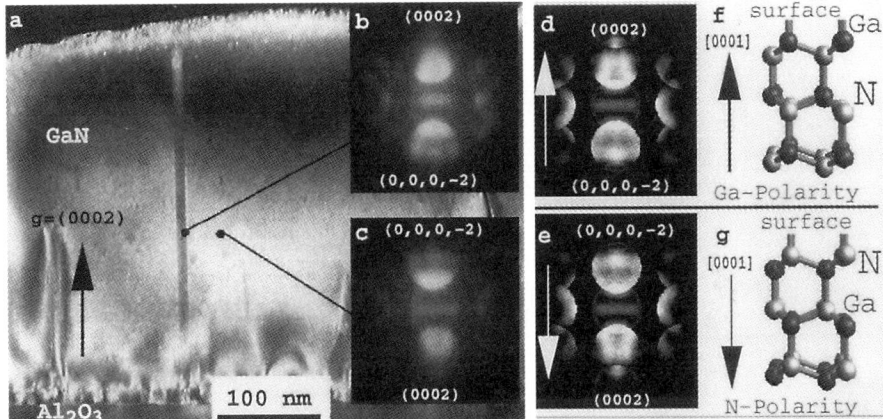

Figure 2. a) Low magnification [2,-1,-1,0] cross-section picture of a sample with pyramidal steps. Two vertical lines can be seen. They correspond to two of the six edges of an hexagonal column which forms an Inversion Domain (of Ga-polarity) in a matrix having a N-polarity. b-e) "[0,1,-1,0] CBED" patterns obtained inside (b) and outside (e) the ID. The two patterns, rotated by 180° from each other are a proof that the column is an ID. c-f) CBED simulations (crystal thickness : 47.5nm) for the two polarities. d-g) Atomic structure of the two possible GaN polarities. Polarity is defined with respect to the Ga-N bonds parallel to the c-axis. The positive c-direction starts on the Ga atom and points towards the N atom of these bonds. A Ga-polar (respectively N-polar) layer gives a natural Ga-terminated (respectively N-terminated) surface. The projection is along [2,-1,-1,0].

Following the knowledge acquired on tilt grain boundary in silicon (Thibault et al 1990), we think that the tetracoordinated bonds are sufficiently soft to allow the construction of tetracoordinated models without antisite Ga-Ga or N-N bonds at the interface. Recent ab-initio computations (Northrup et al 1996) have calculated that these wrong bonds have a high formation energy and are highly improbable without the introduction of impurities. We build and compute the energy of all the simpler structural models of the {0,1,-1,0}, {2,-1,-1,0} PDs that are tetracoordinated and without wrong bonds (fig. 3). We find that model IDB*1 and TDB1 have the lowest Keating energy for the respective a* and a-plane.

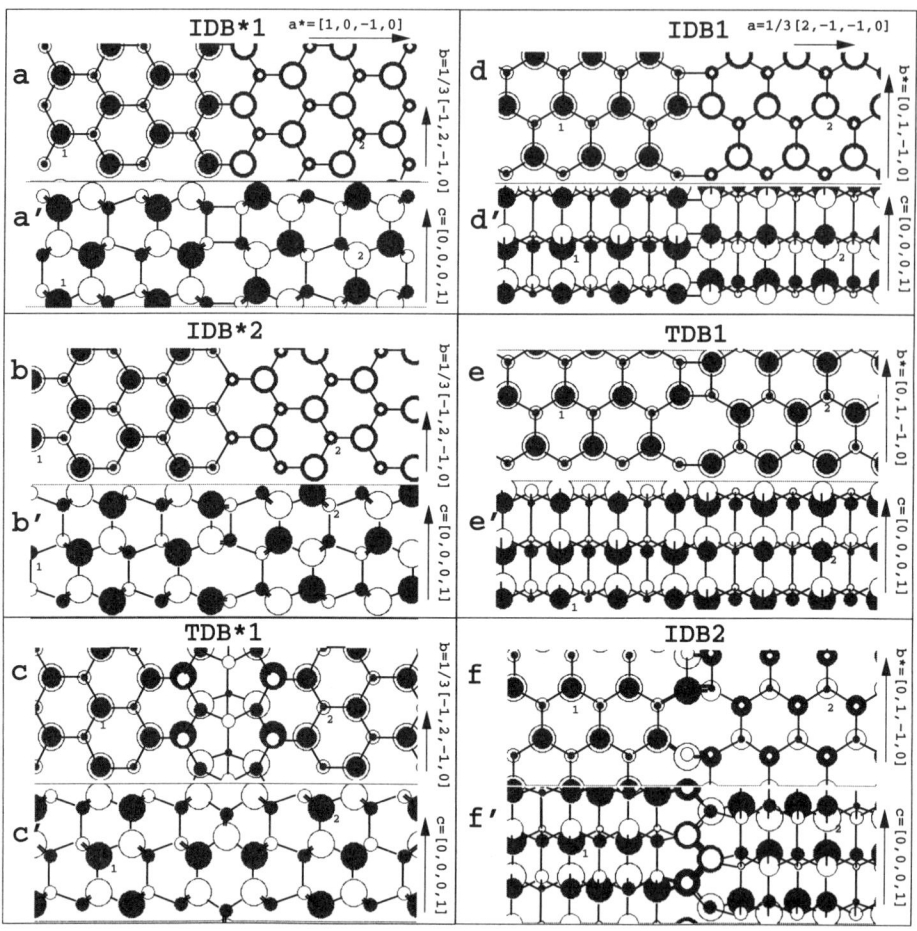

Figure 3 Projections of 6 atomic models of planar defects. (a-b-c-d-e-f) are projections on the c-planes. (a'-b'-c') are projections on the b=(-1,2,-1,0) plane. (d'-e'-f') are projections on the b*=(0,1,-1,0) plane. (a-a'-b-b'-c-c')are atomic models of PDs on the a*-plane. (d-d'-e-e'-f-f') are atomic models of PDs on the a-plane. The atoms have been represented by a circle whose radius is a function of the depth of the atoms along the viewing direction. N atoms are grey. Ga atoms are white. The bonds between atoms have been drawn. All atoms are tetracoordinated, i.e. have four neighbours. Only model f contains antisite Ga-Ga, N-N bonds (six per periods) that have been outlined (bold bonds).

model name	PD plane normal	Keating Energy (J/m2)	Ab-initio energy	Nature of symmetry	Ta* [1,0,-1,0]	Tb [-1,2,-1,0]/3	Tc [0001]
IDB*1	a*	0.73	0.4	T + I	0.0018	0	0.5
IDB*2	a*=[1,0,-1,0]	1.5		T + I	0.0287	0	0.5
TDB*1	a*	2.23	1.68	T	0.493	0	0.5
					Ta	Tb*	Tc
IDB1	a	2.52		T + I	0.554	0	0
IDB2	a=1/3[2,-1,-1,0]	antisite bonds		T + I	0.445	0	0.25
TDB1	a	1.71		T	0.53	0	0.5

Table I Characteristics of the six atomic models of figure 3. The translation operation between the two crystals on each side of the PD (for instance translation between atoms 1 and 2 of fig. 3) are given in crystallographic (a, a*, b, b* and c) units, modulo a perfect lattice translation. In the ab-initio calculations , model TDB*1 was not tetracoordinated, because not totally relaxed.

3.1 (0001) planar defects

The flat GaN layers (Ga-polar material) obtained by nitridating correctly the sapphire substrate and growing an adapted buffer layer do not contain any PDs on the (0001) planes. We only observe a high density of (0001) PD in some N-polar samples. Generally, threading defects, which are partial dislocations, terminate on these planar defects (Rouviere et al 1995, Liliental-Weber et al 1995, Xin et al 1997). Because of the well known low energy of the (0001) stacking faults and because they do not enclose a domain, these PD are always considered as SFs, that is to say as a special class of TDB. It is worth while noting that a high density of SP is a sign of a local cubic phase.

3.2 {2,-1,-1,0} planar defects

We observed these planar defects only in N-polar samples grown on a vicinal surface. They are equally distributed on the three equivalent {2,-1,-1,0}. They can be associated in pairs and formed elongated ribbons of a constant width or they can be more randomly distributed (Rouviere et al 1995). Since then several groups have reported them (Vermaut et al 1997, Xin et al 1997). These defects can be best viewed in plan view. However, the [0001]HREM plan view images do not allow one to distinguish between the three models of figure 3 (d-e-f). We have not yet proved the exact nature of the defect. It is tempting to say that the ones associated in pairs are IDBs (model IDB1) whereas the others can be TDBs (model TDB1). The IDB2 model would be least probable as it has antisite bonds. The energy calculation favours the TDB1 model. By analysing the extinction of these defects Xin et al (1997) determine the translation of the planar defect. In their case, the translation state corresponds to the TDB1 model.

3.3 {0,1,-1,0} planar defects

These PDs have been observed both in Ga and N-polar materials. Present on six planes, they enclose hexagonal domains. In N-polar materials these domains propagate from the buffer layer to the top GaN surface (fig. 2). In Ga-polar material, they rapidly close and form house-shaped domains (Rouviere et al 1997) (fig. 4). The average diameter (about 8nm) of these domains is relatively constant from sample to sample. It has been proved by CBED experiments that these domains in the N-polar material are IDs (Rouviere et al 1996). House shaped domains seem to be IDs due to their special two-beam contrast (fig. 4).

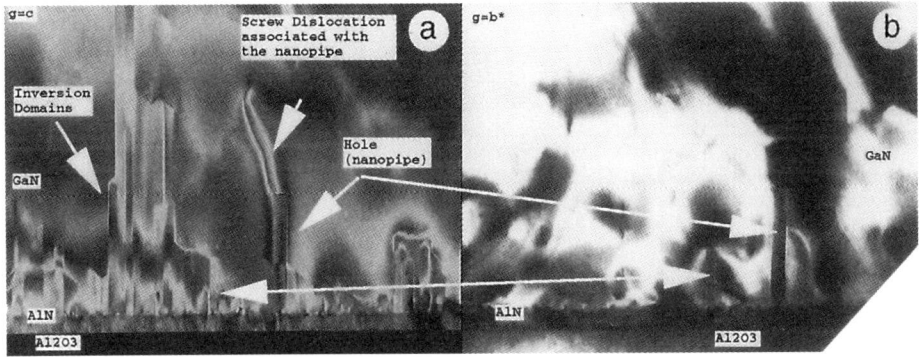

Fig. 4 Low magnification TEM images of a Ga-polar flat sample presenting many domains that do not propagate in the GaN layer. These domains can be terminated along the growth direction by a flat (0001) facet (they are empty holes-nanopipes- in the layer) or by roof-like {1,0,-1,1} facets (they are the house-shape Inversion Domains). a) and b) are image of the same area. In b) the sample has been oriented in order to see only the nanopipes.

There is still a controversy on the exact structure of these IDBs. On plan view images they can be confused with hexagonal holes (the so called nanopipes see fig.7) (Rouviere et al 1995) when in TEM preparation, the two polarities are thinned at a different rate. In HREM plan

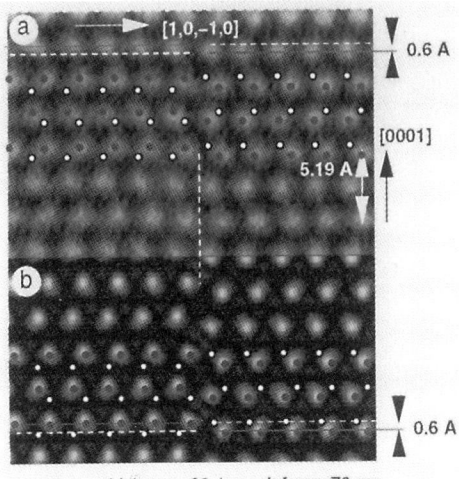

N ● Ga ○ thickness 20.4nm–defocus 70 nm

view images, these IDs can be moreover difficult to observe because when projected on the c-planes, their atomic models look like perfect crystal (fig. 3). On b=1/3[-1,2,-1,0]-cross-sections, they can occupy only a part of the TEM sample and the [-1,2,-1,0]-HREM images are very sensitive to a small amount of tilt and their contrast varies quickly with defocus. However, a rather good agreement can be obtained with the IDB*1 model (fig. 5) (Romano et al 1996) which is by far the lower energy model. These IDs are out of contrast (fig 1c) when g=(0,1,-1,0) indicating that the "displacement" (translation or rotation) associated with this defect is contained in the (0,1,-1,0) plane, which is consistent with model IDB*1 (table I).

Fig. 5 HREM image (a) and simulation (b) of an a*-IDBs. The model IDB*1 has been used in the simulation

3.4 {0,-1,-1,1} planar defects

{0,-1,-1,1} planar defects have only been observed terminating the house-shape IDs (Fig. 4). No atomic structure of these IDs have been proposed yet. We think that the house-shape inversion domains close because of the different speed of the two polarities :the surrounding Ga-polar material growing faster than the N-polar material would swallow the IDs.

4. LINEAR DEFECTS

All the GaN layers have a high density of dislocations. The best reported layers have a density of 10^{+8}/cm^{-3} (Heying et al 1996). Three types of perfect Burgers dislocations can occur in wurtzite GaN layers : (i) a=1/3<2,-1,-1,0>, (ii) c=<0001> and (iii) a+c=1/3<2,-1,-1,3> dislocations. As the dislocation lines are generally parallel to the c-axis this leads to (i) edge a-dislocations, (ii) screw dislocations and (iii) mixed a+c dislocations. Depending on the substrate where the layer is deposited, not all the types can be present. On a SiC substrate, no screw dislocations are observed (Vermaut et al 1995). On GaN and (0001)sapphire substrate the three types have been reported (Ponce et al 1996, Christiansen et al 1997). We restrict our presentation to the flat material, that is to say to the Ga-polar material. We have determined the Burgers vector by a classical two beam analysis. The densities of the three main types of dislocations were of the same order of magnitude. For instance, at a distance of 0.8μm from the buffer layer we find the following densities :a-edge (2.5 10^{+10}/cm^2), a+c mixed (2. 10^{+10}/cm^2) and c-screw (10^{+10}/cm^2). Nearer the BL, we observed a few interactions between these different types of dislocations. However, as the dislocation density in this region is very high, it was difficult to really identify these interactions.

4.1 Edge a-dislocations

These dislocations start in the buffer layer and extend to the GaN surface. The hexagonal flat tops of the N-polar material contain a high density of these edge-a-dislocations (10^{+11}/cm^2). They bound grains that are slightly disoriented from each other by in-plane rotations (fig. 6a). such samples exhibit a very good (0002) rocking curve widths as these edge dislocations do not disturb the (0002) planes (Heying et al 1997). Ga-polar samples have a smaller density of edge dislocations (<10^{+10}/cm^2). We suppose that this reduction comes from the improvement of the buffer quality. Once created, no mechanisms seem to annihilate these dislocations (Qian et al 1995).

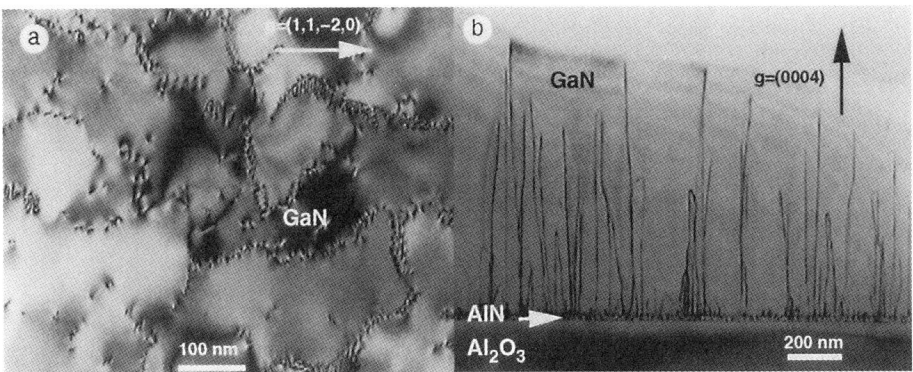

Fig 6 Two-beam images showing (a) a-edge (plan view) and (b) c-screw (cross-section) dislocations

4.2 Screw c-dislocations

These dislocations accommodate out-of-plane rotations. The magnitude of these out-of-plane rotations is generally 4 to 10 times smaller than the magnitude of the in-plane rotations. They are certainly introduced because the grains in the buffer layer are slightly rotated. The nice thing with these screw dislocations is that pairs of screw dislocations with opposite Burgers vectors tend to annihilate each other when more and more material is deposited (fig. 6b). After a deposited thickness of 0.8 μm, a reduction by a factor four can be obtained. The mechanism is the following. An array of two perpendicular families of screw dislocations is necessary to accommodate an out-of-plane rotation. At the beginning of the deposition, only the vertical family is formed. When more and more material is deposited, the horizontal screw dislocations would tend to appear. However, the material prefers to bend the vertical screw dislocations of opposite Burgers vectors in order to annihilate them : the crystallographic quality of the layer improves with the thickness of the layer.

These screw dislocations have a specific property : they can dissociate and recombine during the growth leaving empty hole in the structure (fig. 7). We have only observed these nanopipes in Ga-polar material.

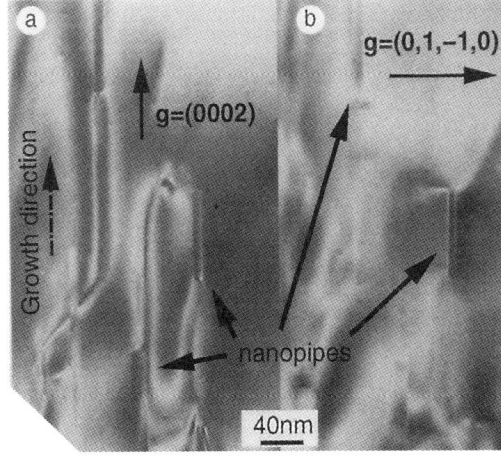

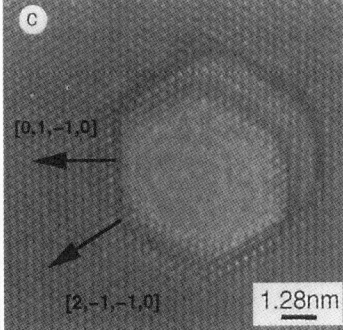

Fig. 7 (a-b) Two beam images of the same area (cross-section) taken with two g-vectors showing c-screw dislocations «opening and closing». c) Plan view image of a nanopipe.

4.3 Mixed a+c dislocations

The mixed a+c dislocations, which can be considered as the addition of one edge a-dislocation and one screw-c dislocation, accommodate a combination of in-plane and out-of-plane rotations. In MOCVD cross-section samples, we have not observed open core a+c dislocations. However in one MBE plan view sample, by making a Burgers circuit around the HREM plan view image of some voids, we observed a few voids containing an a-dislocation component like the one reported by Liliental et al (1996). Such studies would tend to prove that in some cases, the a+c dislocations could also have an empty void at their core.

5. HOLES OR/AND NANOPIPES

GaN has the surprising possibility of growing by leaving empty holes in its layers. These holes can have an irregular shape and be rather wide, or have an hexagonal columnar shape, being relatively narrow (8nm on the average). The first ones generally appear in poor quality materials that contain many grains highly disoriented from each other. The second ones have been first reported by Qian et al (1995) and have been named nanopipes. As said in the previous parts, it is now clear that these nanopipes are " dissociation of screw-dislocations " or may be " dissociation of mixed-(a+c) dislocations " and that they can open and close during growth. The closing of the nanopipes during growth is quick as the top of the nanopipes are flat (fig. 4, fig.7). As reported in 4.3, in MOCVD samples, only screw-dislocations had an empty core. It is not yet clear why these " dissociations " occur. In the next part, we indicate that holes near the BL could play an important role in the strain relaxation mechanisms.

6. STRAIN RELAXATION

We end this paper, by discussing the small residual strain of the GaN layer, which is generally of the order of 0.1% for thick samples (Hiramatsu et al 1993). All the 13% mismatch between the sapphire substrate and the GaN layer is thus compensated in the buffer layer or at the sapphire/buffer layer interface. By analysing HREM images, we deduce that the first GaN layers have nearly their bulk lattice parameter value : the relative strains in the c-direction and a*-direction are respectively lower than 0.2 and 1%; a result similar to Ning et al (1996). Like in the Si-Ge system, this residual strain is an important factor that influences the surface morphology of the layers and could be responsible for ripples at the surface of some GaN layers.

To release this residual strain, we think that the fundamental BL characteristic is its roughness. The exact crystallographic structure of the BL should not be so important as several groups have reported different crystallographic structure of the buffer layer : the AlN or GaN buffer layer can contain many cubic domains ((111)//(0001)) or be formed of tiny hexagonal grains or composed of grainlets (Albrecht et al 1997).

Two different arguments can be developed to justify this importance of the BL roughness. The first one involves the elastic theory and is a kind of reverse of the Grinfeld instability (Grinfeld 1993). In order to produce a flat GaN surface a wavy BL/GaN interface would be necessary ; a too flat BL surface would lead to a rough GaN surface like in the Grinfeld instability (fig. 8). The second one is based on the Coincident Site Lattice theory. One can define a Coincident Site Lattice (CSL) between a 1% strained GaN and an unstrained GaN lattice (period of this lattice $d_{cSL}=100a*=27.6nm$). This lattice defines points of good fit separated by areas of bad fit. The roughness of the BL would become critical when the lateral size d_i of the island produces a period equal to one of the CSL (fig. 8). For that critical value d_{CSL}, the top of the islands produce areas of good fit. In classical CSL theory, the area of bad fit contains dislocations. In the case of the GaN material, holes can be left in these areas of bad fit as is shown in fig. 8. This would explain why different GaN morphologies can be obtained by changing the BL thickness by a small amount (changing from 20nm to 25 and 30nm (Briot et al 1997)). Indeed, changing the BL thickness produces BL grains of different sizes. Only the grain size, and thus roughness, presenting the adapted period d_{CSL} of the CSL would release the residual strains. Of course such analyses have to be developed but could be one of the key points of GaN growth success on sapphire.

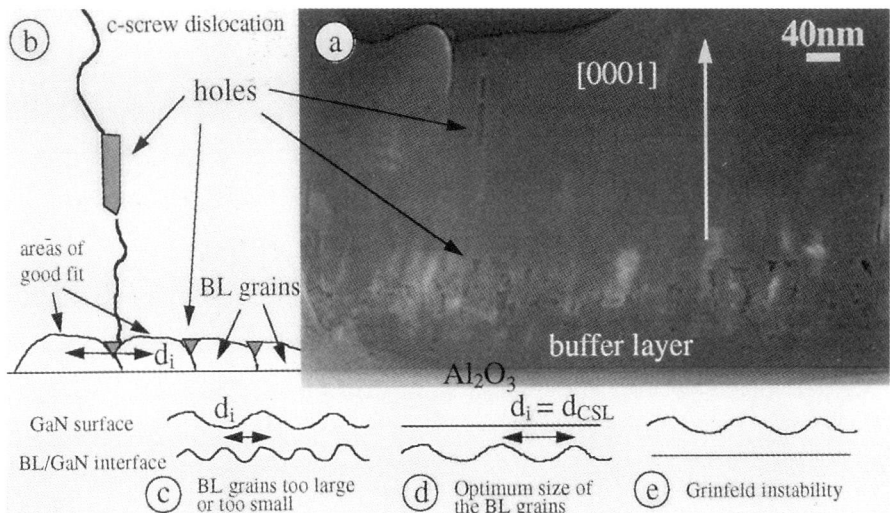

Fig. 8 a) Low magnification image showing voids near the rough buffer layer. Voids can be experimentally visualised by inclining the sample in order that no strong reflections are excited. b) Scheme of figure a outlining the voids and the BL grains that define regions of good fit. c-d-e) Schemes illustrating the influence of the BL grain size on the flatness of the GaN surface. Scheme e corresponds to the Grinfeld instability where a strained layer minimises its energy by roughening its surface.

7. CONCLUSION

We have tried to review the main extended defects that have been reported in the literature and that we have observed in our TEM samples.

The behaviour of the defects appears to be different in the two polarities. In Ga-polar material, IDBs propagate through the whole thickness of the sample, while they tend to disappear in N-polar material. (0001) stacking faults are not observed in Ga-polar material. The dislocation content also seems different. c-screw dislocations with an open-core (nanopipes) have only been observed in Ga-polar material. Tetracoordinated models without antisite Ga-Ga or N-N bonds can be constructed for all the main types of planar defects. We have computed the Keating energy of these models.

The threading dislocations do not release strain : they mainly accommodate in- and out-of-plane rotations of adjacent grains that have coalesced during growth. Most of the strain is released in the buffer layer. The roughness of the BL could be a critical parameter for releasing the residual strain of the GaN layer and achieving a flat GaN surface.

REFERENCES

Akasaki I, Sota S, Sakai H, Tanaka T, Koike M, Amano H 1996 Electronics Lett. **32**(12) 1105
Albrecht S, Christiansen S and Strunk HP 1997 Appl. Phys. Lett. **70**(8) 952.
Briot O, Alexis JP, Gil B and Aulombard RL 1996 Mat. Res. Soc. Symp. Proc. **395** 207.
Briot O, Alexis JP, Gil B and Aulombard RL 1997 Mater. Sci. Eng. B43 147.
Cherns D and Ponce FA this proceeding
Daudin B, Rouviere JL and Arlery M 1996 Appl. Phys. Lett. **69**(17) 2480.
Grinfeld MA 1993 J. Nonlin. Sci. **3** 35
Heying B, Wu XH, Keller S, Li Y, Kapolnek D, Keller BP, DenBaars SP and Speck JS 1996 Appl. Phys. Lett. **68**(5) 643.
Hiramatsu K, Detchprohm T and Akasaki I 1993 Jpn. J. Appl. Phys. **32** 1528.
Kuwano N Shiraishi T Koga A Oki K Hiramatsu K Amano H Itoh K and Akasaki I 1991 J. Cryst. Growth **115** 381
Kwiseon K, Walter RL, Lambrecht RL and Segall B 1996 Mat. Res. Soc. Symp. Proc. **395** 399.
Lester SD, Ponce FA, Craford MG and Steigerwald DA 1995 Appl. Phys. Lett. **66** (10) 1249.

182

Liliental-Weber Z, Ruvimov S, Kisielowski C, Chen Y, Swider W, Washburn J, Newman N, Gassmann A, Liu X, Schloss L, Weber ER, Grzegory I, Bockowski M, Jun J, Suski T, Pakula K, Baranowski J, Porowski S, Amano H and Akasaki I 1996 Mat. Res. Soc. Symp. Proc. **395** 351.

Nakamura S, Mukai T and Senoh M 1991 Jpn. J. Appl. Phys. **30** L1998

Nakamura S, Senoh M, Nagahoma S, Iwasa N, Yamada T, Matsushitat T, Sugimoto Y and Kiyoku H 1997 Appl. Phys. Lett. **70(7)** 868

Niebuhr R, Bachem KH, Dombrowski K, Maier M, Pletschen W and Kaufmann U 1995 J. Electronic. Mat. 24 1531.

Ning XJ, Chien FR, Pirouz P, Yang JW and Asif Khan M 1996 Journal of Materials Research **11(3)** 580.

Northrup JE, Neugebauer J and Romano LT 1996 Phys. Rev. Lett. **77(1)** 103.

Ponce FA, Bour DP, Gotz W and Wright PJ 1996 Appl. Phys. Lett. **68(1)** 57.

Ponce FA, Bour DP, Young WT, Saunders M and Steeds JW 1996 Appl. Phys. Lett. **69(3)** 337.

Ponce FA, Cherns D, Young W and Steeds J 1996 Appl. Phys. Lett. **69(6)** 770.

Pond RC 1989 Dislocations in Solids Ed. Nabarro FRN (Elseviers Science Publishers) Chap 38 pp5

Qian W, Rohrer GS, Skowronski M, Doverspike K, Rowland LB and Gaskill DK 1995 Appl. Phys. Lett. **67(16)** 2284.

Qian W, Skowronski M, De Graef M, Doverspike K, Rowland LB and Gaskill DK 1995 Appl. Phys. Lett. **66(10)** 1252.

Romano LT, Northrup JE, and O'Keefe MA (1996).Appl. Phys. Lett. **69(16)** 2394.

Rosner SJ, Carr EC, Ludowise MJ, Girolami G and Erikson HI 1997 Appl. Phys. Lett. **70 (4)** 420.

Rouviere J, Arlery M, Bourret A, Niebuhr R and Bachem K 1995 Microscopy of Semiconducting Materials. Ins. Phys. Conf. Ser. No **146**, edited by : AG Cullis, AE Staton-Bevan 285.

Rouviere JL, Arlery M, Bourret A, Niebuhr R and Bachem KH 1996 Mat. Res. Soc. Symp. Proc. **395** 393

Rouviere JL, Arlery M, Niebuhr R, Bachem KH and Briot O 1997 Mat. Sci. Eng. B43 161.

Rouviere JL, Arlery M, Niebuhr R, Bachem KH and Briot O 1996 Mat. Res. Soc. Internet J. Nitride Semicond. Res. **1**, 33

Sigiura L 1997 Appl. Phys. Lett. **81(4)** 1633.

Sverdlov BN, Martin GA,Morkoc, H.Smith, DJ 1995 Appl. Phys. Lett. **67(14)** 2063.

Thibault J, Rouviere JL and Bourret A 1991 Material Science and Technology Ed Cahn RW, Haasen P and Kramer EJ Vol 4 Electronic Structure and properties of semiconductors (VCH) 321

Vermaut P, Ruterana P, Nouet G, Salvador A and Morkoç H 1997 this proceeding.

Vermaut P, Ruterana P, Nouet G, Salvador A, Botchkarev A, Sverdlov B and Morkoç H K 1995 Microscopy of Semiconducting Materials. Ins. Phys. Conf. Ser. No **146**, edited by : AG Cullis, AE Staton-Bevan 289.

Westwood A.D. YRA, McCartney M.R., Cormack A.N., Notis M.R. 1995 J. Mater. Res. **10** 1270.

Xin Y, Brown P, Humphreys C, Cheng T and Foxon C 1997 Appl. Phys. Lett. **70(11)** 1308.

Yoshida S. MS, Gonda S. 1983 J. Vac. Sci. Technol. **B 1 (2)** 250.

We thank Dr. R. Niebuhr, Dr K-H Bachem and Dr. O. Briot for providing MOCVD grown GaN layers. Part of this work has been supported by the EU commission under the Brite-Euram II programme and we thank Dr U. Kaufmann the co-ordinator of this project.

Inst. Phys. Conf. Ser. No 157
Paper presented at Microsc. Semicond. Mater. Conf., Oxford, 7–10 April 1997
© *1997 IOP Publishing Ltd*

Polarity study by CBED of GaN films grown on (0001)$_{Si}$ 6H-SiC

P Vermaut, P Ruterana, G Nouet, A Salvador* and H Morkoç*

LERMAT, Unité associée CNRS N° 6004, ISMRA, 6 Blvd Maréchal Juin 14050 Caen
Cedex, France.
*University of Illinois-Urbana, Coordinated Science Laboratory, Urbana, Illinois, IL 61801
USA.

ABSTRACT: The polarity of GaN films grown on (0001)$_{Si}$ 6H-SiC has been investigated
by convergent beam electron diffraction. The free surfaces of the GaN layers are
determined to be Ga-terminated. Moreover, closed domains with $(\bar{1}2\bar{1}0)$ prismatic walls
observed in the layers are not inversion domains. The layers seem to be unipolar.

1. INTRODUCTION

Due to a lack of suitable substrates, GaN has to be heteroepitaxially grown. The more
commonly used substrate is Al$_2$O$_3$ which is non-polar. It has been determined that nitridation
of its (0001) surface before growth greatly influences the layer polarity and its surface
morphology (Rouvière et al. 1996), and Inversion Domains Boundaries (IDBs) have been
identified in such layers (Potin et al. 1997a and b). In contrast to sapphire substrates, SiC is
polar. Sasaki and Matsuoka (1988) have studied the influence of the chemical nature of the
substrate surface i.e. (0001)$_{Si}$ and $(000\bar{1})_C$ on the surface morphology and
photoluminescence properties of GaN layers. From X-ray photoelectron spectroscopy
measurements, they concluded that GaN films grown on (0001)$_{Si}$ and $(000\bar{1})_C$ of SiC are
terminated with N and Ga respectively.

In the present work, the polarity of GaN layers grown on (0001)$_{Si}$ 6H-SiC by
Molecular Beam Epitaxy has been studied by Convergent Beam Electron Diffraction
(CBED). Results on the polarity of the layers and on both sides of prismatic defects which
may be suspected to be IDBs or Stacking Faults (SFs) are presented. In the light of these
investigations, conditions for IDB formation in GaN layers grown over SiC are discussed.

2. EXPERIMENTAL DETAILS

6H-SiC wafers were cut 3.5° off the basal plane toward [11$\bar{2}$0]. The (0001)$_{Si}$ surface
was cleaned in the classical way followed by a hydrogen plasma step to reduce the amount of
oxygen-carbon bonds to below the X-ray photoemission detection limit. The details of this
procedure were reported by Lin et al. (1993). Growth was performed by electron cyclotron
resonance plasma enhanced molecular beam epitaxy at a rate of 40 nm/h with a substrate
temperature between 750 and 800° C.

TEM cross section samples were thinned down to 100 μm by mechanical grinding and dimpled down to 10 μm. Electron transparency was achieved by ion milling with a LN₂ cold stage at 5 kV. CBED experiments were carried out on a Jeol EM 2010 microscope operating at 200 kV with a probe diameter of 25 nm. CBED patterns were simulated using the EMS software (Stadelmann 1987).

3. GaN LAYER POLARITY

CBED is an efficient technique to underline the polar character of the structures and identify inversion domains. Along the [10$\bar{1}$0] zone axis, the contrast difference between the 0002 and 000$\bar{2}$ discs in CBED patterns allows determination of crystal polarity and identify inversion domains in GaN layers (Rouvière et al 1996).

CBED patterns are characteristic of the material but are also a function of the crystal thickness. The simulation of a series of zone axis patterns has been carried out for GaN and SiC using the Bloch wave method and for crystal thickness values ranging from 50 to 250 nm with a step of 5 nm. For comparison between the experimental and simulated patterns, the number and position of the fringes visible inside the discs were used.

In the present work, the absolute orientation of 6H-SiC substrates is known, determined before growth by polarity dependence of the oxidation rate of the (0001)$_{Si}$ and (000$\bar{1}$)$_C$ surfaces (Von Müench and Pfaffeneder 1975). This was confirmed by CBED in a first stage, using the difference between the 0006 and 000$\bar{6}$ discs.

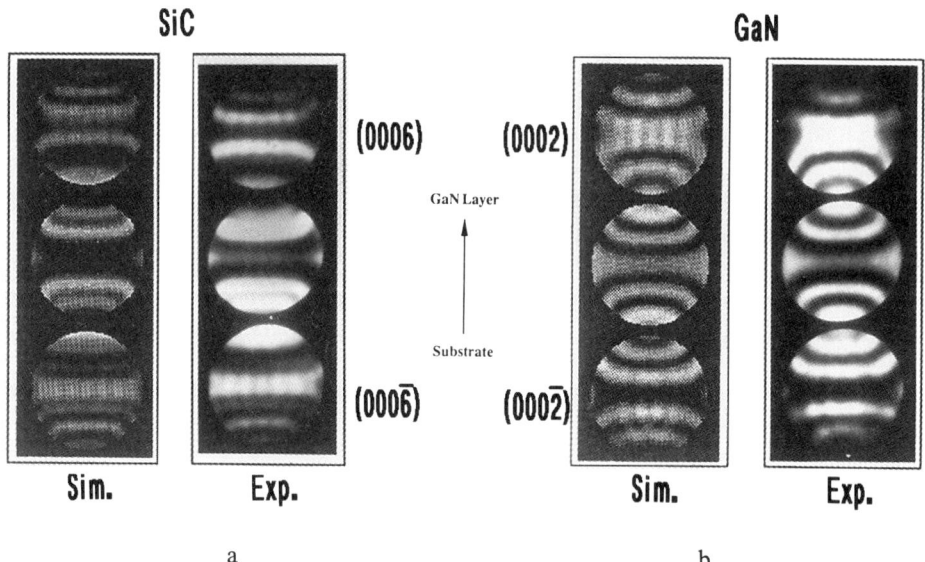

Fig. 1: [10$\bar{1}$0] zone axis CBED patterns of a) 6H-SiC for a crystal thickness value of 135 nm and b) 2H-GaN for a crystal thickness value of 150 nm, with their corresponding simulations. The SiC substrate → GaN layer direction is indicated.

Experimental [10$\bar{1}$0] zone axis CBED patterns recorded in the SiC substrate for a crystal thickness value of 135 nm and in the GaN layer at a 150 nm crystal thickness are shown in figure 1a and b respectively. The SiC substrate→GaN layer direction which is known to be [0001] is indicated. As expected, the black and white fringes observed inside the discs exhibit a $(11\bar{2}0)$ mirror symmetry and a good match is obtained between the experimental and simulated patterns. Then, CBED patterns allow us to determine that [0001] direction in the GaN films and in the 6H-SiC substrates are parallel. The same analysis has been carried out in different areas inside the GaN films, but also in different films deposited with and without an AlN buffer layer. Assuming that for the two materials, the surfaces are defined by cutting bonds parallel to the growth direction, the GaN films grown on the Si-terminated (0001) surface of SiC, with and without an AlN buffer layer, exhibit a Ga-terminated free surface.

4. $\{11\bar{2}0\}$ PRISMATIC DOMAIN POLARITY

$(11\bar{2}0)$ prismatic defects have been observed in GaN layers grown on sapphire (Rouvière et al 1995) and were suspected to be IDBs. Similar prismatic planar defects are present in the GaN layers grown on SiC. A careful study of the interface steps and the related prismatic defects has been reported by the authors. In this case, the defects were identified as SFs with a $1/2[\bar{1}101]$ fault vector and bounded by Frank partial dislocations (Vermaut et al 1997a). However, sometimes, these defects have been observed to form closed domains in the layers (Vermaut et al 1996) in which case, an inversion character can be suspected and may be looked for. A bright field planar view image of such a domain is shown in figure 2.

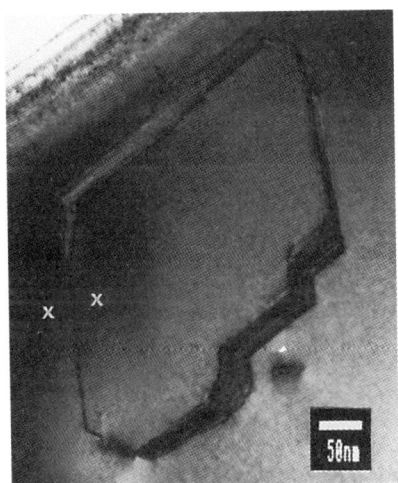

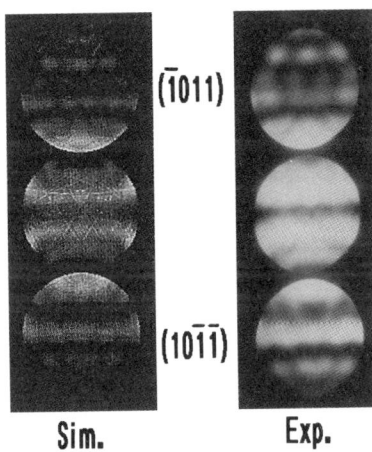

Fig. 2: Bright field planar view image of a typical $\{11\bar{2}0\}$ prismatic domain observed in the GaN layers. Crosses indicate the position of the probe during CBED experiments.

Fig. 3: $[10\bar{1}2]$ zone axis CBED pattern of GaN with its simulation for a crystal thickness value of 90 nm.

As the density of the closed domains is low, they are not easily observed in cross section samples. Thus, CBED experiments were carried out on planar view samples in order to determine if polarity changes when the domain wall is crossed. In that case, the $[10\bar{1}2]$ zone axis has been used, where $\bar{1}011$ and $10\bar{1}\bar{1}$ discs can be distinguished. An experimental CBED pattern within its simulation obtained for a crystal thickness of 90 nm is shown in figure 3. The probe has been positioned on both sides of the domain wall (indicated by crosses on fig. 2) and identical CBED patterns are obtained meaning that this defect is not an inversion domain.

5. DISCUSSION AND CONCLUSION

The results obtained by CBED show that the free surfaces of GaN layers grown on $(0001)_{Si}$ 6H-SiC are Ga-terminated in contradiction with that of Sasaki et al (1988).

The polarity of the deposited layers is determined by the structure and chemistry of the substrate surface. Geometrical considerations have shown that Si-N or C-Ga and Si-Ga or C-N bonds are needed to grow islands of opposite polarities, independently of the substrate surface structure (Vermaut et al 1997b). In the investigated samples, it is shown that the only prismatic defects present in the layers are SFs related to interface steps (Vermaut et al 1996) and that the layers are unipolar. Therefore, only one kind of bonds which can be Si-N or C-Ga is present at the interface. This conclusion is supported by an *ab initio* study of the GaN/SiC interface reported by Capaz et al (1995) who showed that Ga-terminated films with Si-N bonds at the interface are energetically favourable, in comparison to N-terminated ones having Si-Ga bonds.

It has been shown that oxygen and Al vacancies are needed to nucleate and stabilize the IDBs observed in sintered AlN (Westwood et al 1995). In that case, a single layer of octahedrally bonded aluminium and oxygen was observed. Similar defects could be generated in GaN since, as aluminium, gallium can also be bonded octahedrally to oxygen.

In summary, the polarity of GaN layers grown on $(0001)_{Si}$ surface of 6H-SiC has been identified in layers grown by electron cyclotron resonance plasma enhanced molecular beam epitaxy. The polarity of the substrate was shown to be transmitted to the deposited layers as if Si and C were species of group III and group V respectively.

REFERENCES

Capaz R B, Lim H and Joannopoulos J D 1995 *Phys. Rev. B* **51**, 17755.

Lin M E, Strite S, Agarwal A, Salvador A, Zhou G L, Teraguchi N, Rockett A and Morkoç H 1993 *Appl. Phys. Lett.* **62,** 702.

Potin V, Ruterana P and Nouet G 1997a submitted to *J. of Appl. Phys.*

Potin V, Ruterana P, Nouet G, Salvador A and Morkoç H 1997b *this proceedings*.

Rouvière J L, Arlery M, Bourret A, Niebuhr R and Bachem K 1995 *Inst. of Phys. Conf. Series N°* **146**, 285.

Rouvière J L, Arlery M, Bourret A, Niebuhr R and Bachem K 1996 *Mat. Res. Soc. Symp. Proc.* **395**, 393.

Sasaki T and Matsuoka T 1988 *J. Appl. Phys.* **64**, 4531.

Stadelmann P A 1987 *Ultramicroscopy* **21**, 131.

Vermaut P, Ruterana P, Nouet G, Salvador A and Morkoç H 1996 *Mat. Sci. Eng. B* (in press).

Vermaut P, Ruterana P, Nouet G and Morkoç H, 1997a *Phil. Mag. A* **75**,239.

Vermaut P, Ruterana P, Nouet G 1997b *Phil. Mag. A* (in press).

Von Müench W and Pfaffeneder I 1975 *J. Electrochem. Soc.* **122**, 642.

Westwood A D, Yougman R A, McCartney M R, Cormack A N and Notis M R 1995 *J. Mater. Res.* **10**, 1270.

Inst. Phys. Conf. Ser. No 157
Paper presented at Microsc. Semicond. Mater. Conf., Oxford, 7–10 April 1997
© *1997 IOP Publishing Ltd*

The analysis of nanopipes and inversion domains in GaN thin films

D Cherns[1], W T Young[1], M A Saunders[1], F A Ponce[2] and S Nakamura[3]

[1] H H Wills Physics Laboratory, University of Bristol, Tyndall Avenue, Bristol BS8 1TL
[2] Xerox Palo Alto Research Center, 3333 Coyote Hill Road, Palo Alto, CA94304, USA
3 Nichia Chemical Industries, 491 Oka, Kaminaka, Anan, Tokushima 774, Japan

ABSTRACT: Transmission electron microscopy is used to analyse defects in α-GaN thin films grown on (0001) sapphire by MOCVD. It is shown that hollow tubes 5-25nm across (nanopipes) lying along the c-axis ([0001]) are coreless screw dislocations with Burgers vectors $\pm\mathbf{c}$. The films also contain inversion domains typically 10-100nm across with boundaries mostly of $\{10\bar{1}0\}$ type. A quantitative analysis of displacement fringes at inclined domain boundaries is used to deduce the boundary structure. The origin and significance of these defects for understanding opto-electronic properties is discussed.

1. INTRODUCTION

In the last few years it has been shown that GaN thin films grown by metalorganic chemical vapour deposition (MOCVD) can be used for light-emitting diodes and lasers operating at the blue end of the visible spectrum. Surprisingly, such films have defect densities $\geq 10^{14}\text{m}^{-2}$, at least 5 orders of magnitude higher than the densities acceptable for GaAs light-emitting devices. In order to understand these differences, it is important to understand the structure of defects in GaN.

This paper uses transmission electron microscopy (TEM) to examine defects in GaN films grown by MOCVD on (0001) sapphire. The films were epitaxial with (0001) GaN//(0001) sapphire and $[11\bar{2}0]$GaN//$[1\bar{1}00]$ sapphire. Previous studies (Ponce et al, 1996) by convergent beam electron diffraction (CBED) have shown that the GaN has grown predominantly with a Ga-terminated growth surface; henceforth we will call this direction [0001]. Further studies have shown that the films contain **a**, **c** and **c** + **a** dislocations with line directions mostly close to [0001]. In addition to the dislocations, the films contain two further defects which we describe here.

2. EXPERIMENTAL

TEM studies were carried out on plan-view specimens produced by mechanical backthinning followed by ion milling at liquid N_2 temperatures with 4kV Ar^+ ions at an angle of incidence of 10°. Specimens were examined in a Philips EM430 microscope operating at 250kV. In the regions examined the sapphire substrate had been completely removed to leave only the GaN epilayer.

3. NANOPIPES

Fig. 1 shows bright field images taken from the same area with [0001] either normal to the beam (fig. 1(a)) or tilted by about 30° to the beam direction (fig. 1(b)). The micrographs show hollow tubes which are end-on in fig. 1(a) and obliquely inclined in fig. 1(b) where the tube contrast has been enhanced by imaging slightly out-of-focus. In general, observations

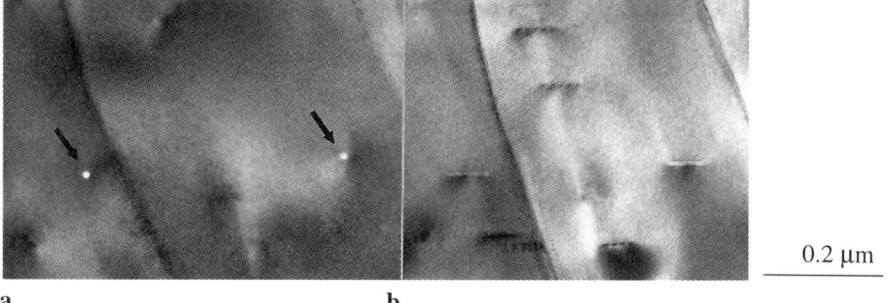

0.2 μm

a   **b**

Fig. 1. Bright field images showing nanopipes (a) end-on with [0001] parallel to the e-beam direction and (b) inclined at about 30° to the e-beam. Nanopipes are arrowed.

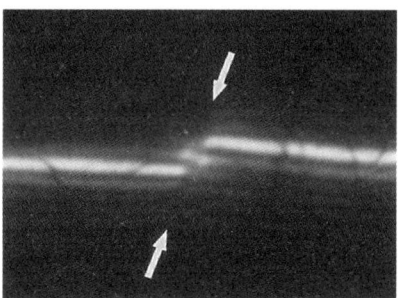

Fig. 2. A dark field LACBED pattern showing a 20$\bar{2}$2 reflection crossing the left hand nanopipe in fig. 1. The trace of the nanopipe is arrowed.

showed that these hollow tubes, which we term "nanopipes", varied in diameter in the approximate range 5-25nm and lay accurately parallel to [0001], mostly with {10$\bar{1}$0} facets. In the areas studied, nanopipes were present at densities ~10^{12}m^{-2}.

Examination of inclined nanopipes under strong diffracting conditions showed some evidence of long range strain fields similar to those associated with dislocations. To examine this further, large angle convergent beam electron diffraction (LACBED) studies were carried out to investigate the intersection of nanopipes with high order, quasi-kinematical two-beam diffraction contours. Fig. 2 shows a LACBED pattern in which a 20$\bar{2}$2 reflection intersects the nanopipe on the left hand side in fig. 1. The contour shows a splitting characteristic of a **g.b** = 2 dislocation (Cherns and Preston, 1989). In contrast LACBED observations of the same nanopipe in reflections of type {hki0} showed continuous contours. These results can be explained if we assume that the nanopipe is associated with a pure screw dislocation with a Burgers vector **c**. This gives **g.b** = 2 for g = 20$\bar{2}$2 and **g.b** = 0 for all basal plane reflections.

More extensive studies of nanopipes to be reported elsewhere (Cherns et al 1997(a)) have confirmed that more than 90% of nanopipes were associated with screw dislocations having Burgers vectors **c** or -**c**. The reason for this is not clear. One possibility is that the nanopipes are coreless dislocations which form to minimise the dislocation energy as Frank (1951) has suggested. The release of elastic strain energy is to be balanced by the energy of the surfaces thus created. Calculations suggest that the {10$\bar{1}$0} surface energy is too high for this to occur, even under MOCVD growth conditions (J. Northrup et al: manuscript in preparation). Instead we have proposed (Cherns et al, 1997(a)) that nanopipes grow from pinholes formed in the initial GaN growth in which threading dislocations become trapped. A screw dislocation can produce a potentially-stable growth spiral along **c** with growth steps restricted to the (0001) surface. In contrast pinholes with **a** and **c** + **a** dislocations must grow out since **a** ledges are generated as growth proceeds. However, although this model is consistent with the observations, further work is clearly needed to clarify the early stages of nanopipe formation.

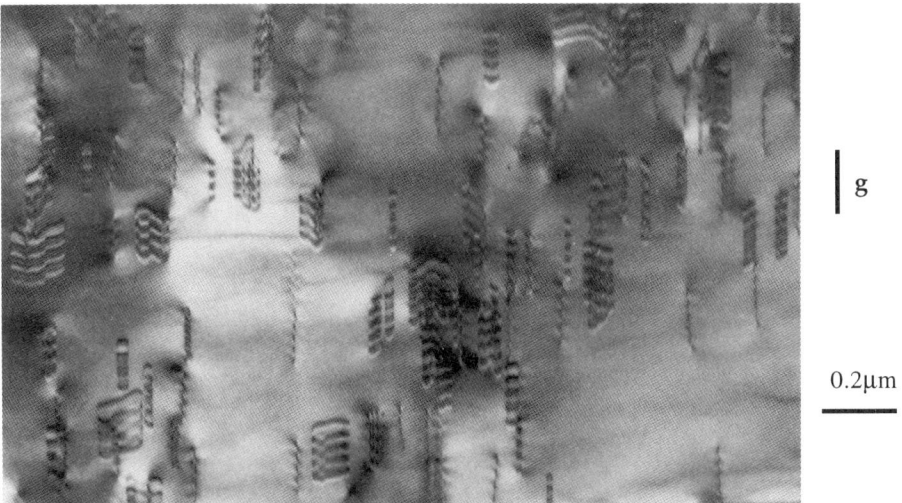

Fig. 3. Bright field image showing inversion domains in two-beam diffracting conditions (s≈0) with g = $\bar{1}$101. Intersections of inversion domains with the top surface of the foil are towards the top of the micrograph.

4. INVERSION DOMAINS

In addition to the nanopipes, the films contained a high density of crystal domains, also aligned parallel to **c** and mostly with {10$\bar{1}$0} sides. These domains showed strong fringe contrast when tilted and observed in g = hkil reflections where l ≠ 0, 8 (Fig. 3). However, the domains showed no evidence of either fringe contrast or strain contrast when viewed in {hki0} reflections.

CBED studies have confirmed that these domains were inversion domains with an outward normal [000$\bar{1}$] opposite to that of the surrounding matrix (Ponce et al, 1997). To investigate the structure of the inversion domain boundaries (IDBs), we have examined the displacement fringes observed under strong two-beam diffracting conditions (deviation parameter s~0). Under these conditions and for relatively thick foils (as in fig. 3), the terminating fringes were found to be either both dark or both bright under dark field conditions and of opposite sign under bright field conditions. To explain these observations we note that the outermost displacement fringes are due to single boundaries which are of opposite type, a matrix/domain boundary on one side and a domain/matrix boundary on the

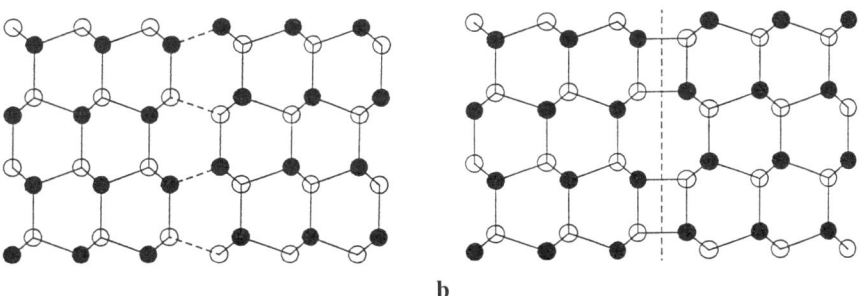

a   **b**

Fig. 4. {11$\bar{2}$0}projection of possible models for the {10$\bar{1}$0} IDB (a) a pure IDB with Ga and N sublattices reversed, (b) as (a) but with an additional relative shift of $c/_2$. Open and closed circles are Ga and N respectively.

190

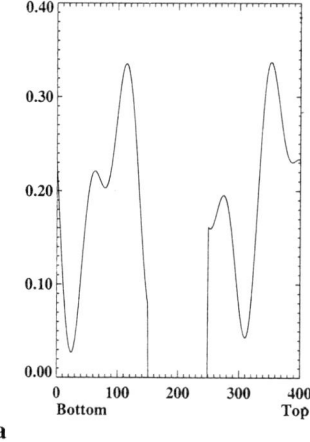
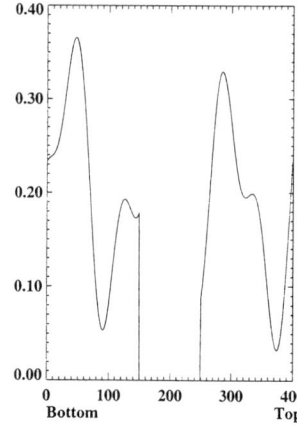

Fig. 5. Computed images for the IDBs in fig. 3. (a) corresponds to the model in fig. 4(a), (b) corresponds to fig. 4(b).

a   **b**

other. Since the boundaries are opposite types, we expect the fringe contrast to behave in an opposite fashion to that for a single boundary where the bright field image is symmetric and the dark field image is asymmetric (Gevers et al 1963).

To explain the contrast quantitatively we have compared the experimental images for reflections g = hkil for various l with dynamical calculations for IDBs. The details of these calculations will be presented elsewhere (Cherns et al, 1997(b)). Two possible IDB models which preserve the 4-fold bonding are illustrated in Fig. 4. Both models have a zero lattice displacement in the basal plane, consistent with the lack of strain contrast observed in {hki0} reflections. Fig. 4(a) shows a "pure IDB" in which Ga and N sublattices have been simply reversed. Fig. 4(b) shows the same boundary on which a relative shift of $c/2$ has been applied. Fig 5(a) and (b) show simulated images for the two IDBs corresponding to the image in Fig. 3. The asymmetry of the bright field image is confirmed. The results in Fig. 3 are consistent with those in Figs 4(b)/5(b) and the model in Figs. 4(a)/5(a) is clearly ruled out. Results for reflections with l = 2, 4 and 8 confirmed this result.

8. CONCLUSION

These studies establish the structure of nanopipes and IDBs in high quality GaN films grown on (0001) sapphire. It is interesting to note that our model for the IDBs has 4-fold bonding with Ga-N bonds throughout. This suggests that IDBs in GaN may have no intrinsic electronic states and may, therefore, be electrically inactive defects.

ACKNOWLEDGEMENTS

The Bristol Group gratefully acknowledges the use of NATO grant #CRG 960690 in pursuit of this work. The work was partially supported by the Department of Commerce Advanced Technology Program (70NANB2H1241) and by ARPA (Agreement #MDA 972-95-3-0008).

REFERENCES

Cherns D and Preston A R, 1989 J.Electron Microsope Technique **13**, 111.
Cherns D, Young W T, Steeds J W, Ponce F A and Nakamura S, 1997a J.Cryst.Growth: in press.
Cherns D, Young W T, Saunders M, Steeds J W, Ponce F A and Nakamura S 1997b, Phil.Mag:submitted.
Frank F C 1951 Acta Crys. **4**, 497.
Gevers R, Art A and Amelinckx S, 1963 Phys.Stat.Sol. **3**, 1563.
Ponce F A, Bour D P, Young W T, Saunders M and Steeds J W, 1996, Appl.Phys.Lett. **69**, 337.
Ponce F A, Cherns D, Young W T, Steeds J W & Nakamura S 1997, MRS Proc. **449**, 405.

Inst. Phys. Conf. Ser. No 157
Paper presented at Microsc. Semicond. Mater. Conf., Oxford, 7–10 April 1997
© *1997 IOP Publishing Ltd*

HREM study of the {10$\bar{1}$0} inversion domains in GaN grown on (0001) sapphire substrates

V Potin, P Ruterana, G Nouet, A Salvador* and H Morkoç*

Laboratoire d'Etudes et de Recherches sur les Matériaux, Unité associée CNRS 6004, ISMRA, 6 Bd du Maréchal Juin, 14050 Caen Cedex, France
*University of Illinois, Materials Research Laboratory and Coordinated Science Laboratory, 104 South Goodwin Avenue, Urbana, Illinois 61801, USA

ABSTRACT: In this work, the atomic structure of the {10$\bar{1}$0} inversion boundaries in nanometric domains inside GaN layers on top of sapphire has been studied using high resolution electron microscopy (HREM) and extensive image simulation. The experimental observations were found to agree with a pure inversion model in which the boundary plane contains two wrong bonds (N-N, Ga-Ga) per atom.

1. INTRODUCTION

Recently, the possibility to fabricate blue laser diodes (Nakamura et al 1996) based on GaN was demonstrated. However, many growth defects are still present inside the active layer. They are mainly dislocations, basal stacking faults and planar defects (Ponce et al 1995, Vermaut et al 1997). In this work, we focus on the atomic structure of the {10$\bar{1}$0} domain boundaries crossing the whole epitaxial layers. To study their structure, six models were constructed and used in extensive image simulation and comparison to experimental high resolution images.

2. EXPERIMENTAL DETAILS

The GaN layers were grown on the (0001) sapphire substrate by electron cyclotron resonance assisted molecular beam epitaxy (ECR-MBE). The deposition was performed at a 40 nm/h rate with a substrate temperature of 800°C. Details of layer growth can be found in Lin et al (1993).

The TEM samples were prepared by standard techniques of mechanical polishing and ion milling until electron transparency. HREM observations were carried out on a Topcon 002B electron microscope operating at 200 kV with a point to point resolution of 0.18 nm. The images were simulated using the electron microscopy software (EMS) of Stadelmann (1987).

3. RESULTS AND DISCUSSION

GaN is of wurtzite type and noncentrosymmetric whereas sapphire is centrosymmetric. Two parts of a crystal related by an inversion operation could present strong contrast in dark field conditions (Serneels et al 1973). This is due to the violation of the Friedel's law and true only in multibeam conditions along a zone axis that reveals the noncentrosymmetry of the crystal. We have imaged a cross section specimen along the $[10\overline{1}0]$ zone axis and recorded dark field images using g = 0002 and g = 000$\overline{2}$ (Fig. 1a and 1b). The observed contrast is complementary, so an inversion domain is present.

Afterwards, we have tried to determine if a translation exists besides the inversion. Fig. 2a and 2b present a cross section specimen in two beam conditions with respectivly g = 10$\overline{1}$0 and g = 0002. Whereas for g = 10$\overline{1}$0 the domains are out of contrast, for g = 0002 strong contrast is exhibited. Therefore, if any fault vector exists, it has to be along the c-axis.

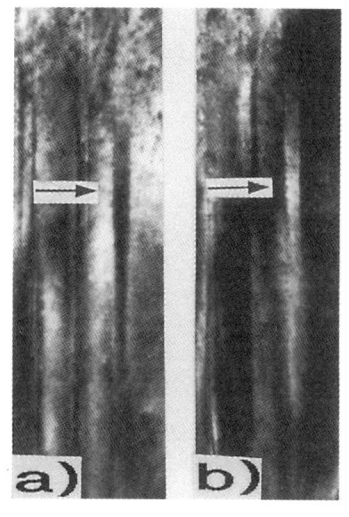

a) b)

40 nm

Fig. 1 Multiple dark field along a $<10\overline{1}0>$ zone axis with g = 0002 (a) and g = 000$\overline{2}$ (b).

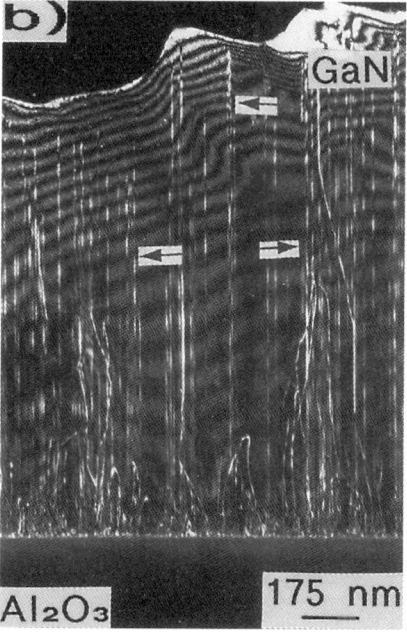

Fig. 2 Dark field images in two beam conditions with g = 10$\overline{1}$0 (a) and g = 0002 (b).

The domains are characterized by a very small size (5-20 nm) and a density of 10^{10} cm^{-2} (Fig. 2b). This small size has prevented us from using the convergent beam technique in order to determine their polarity (Vermaut et al 1997).

Considering the above results, six models have been constructed (Fig. 3). They all present an inversion operation and some of them a translation along the c-axis. The first type of model was proposed by Austerman and Gehman (1966). In this model, the anion sublattice is undeviated across the boundary, switching only the cation sublattice in the inversion operation (A1-A2). This corresponds to an inversion and a translation of 1/3 <0001>. Another type of model was proposed by Holt in 1969, in which only a pure inversion is applied (H1-H2). The last type of model consists on an inversion followed by a translation of 1/2 <0001> (V1-V2). Two locations can exist for the {10$\bar{1}$0} boundary: in the in-plane bonds (which are in the {11$\bar{2}$0} projection plane) or in the out-of-plane bonds (where two bonds per atom are cut) which results in six models.

Fig. 4a shows an experimental image at a defocus of 59 nm. The boundary plane is projected into an atomic plane (arrow) and formed by alternated weak and strong white

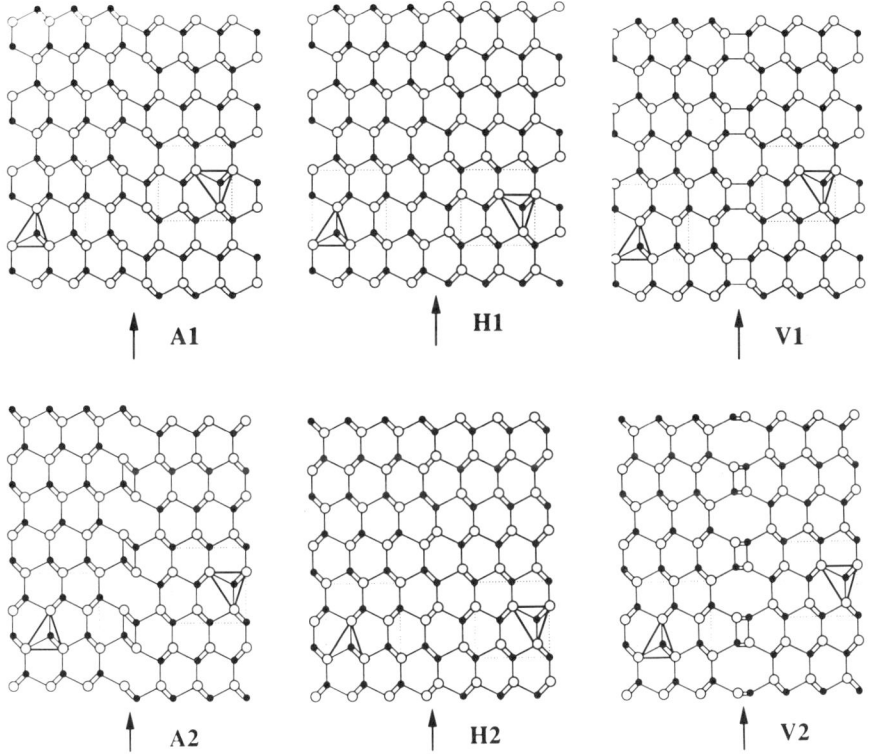

Fig. 3 Geometrical models for {10$\bar{1}$0} inversion boundaries in GaN shown in the <11$\bar{2}$0> projection . White circles represent nitrogen and black ones gallium. Single lines represent bonds which are in the {11$\bar{2}$0} projection plane and double lines correspond to a projection of two bonds, one pointing behind and the other pointing in front of the projection plane.

spots. Among the six proposed models, only the simulation of the H2 model presents the similar features. In Fig. 4a and 4b, the H2 model is fitted to the experimental images at a defocus of 59 and 79 nm for a thickness of 5 nm.

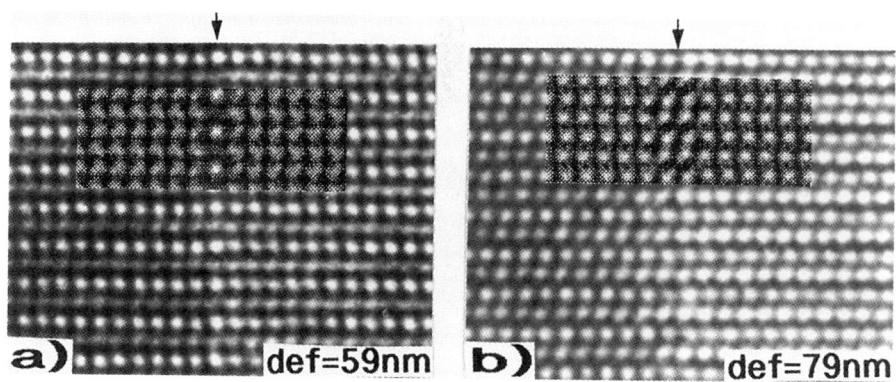

Fig. 4 HREM experimental images at 59 and 79 nm defocus. The insets show simulated images for the H2 model (thickness = 5 nm)

4. CONCLUSION

Use of the HREM technique combined with image simulation has allowed us to propose a model for the atomic structure of the $\{10\overline{1}0\}$ domain boundaries in GaN/ sapphire layers. This model is a Holt type one in which only a pure inversion is applied. The boundary is found to be made of N-N and Ga-Ga bonds in equal number and two such wrong bonds exist for each atom. This configuration is expected to be of high energy and may explain the small size of the domains (5-20 nm).

REFERENCES

Austerman S B and Gehman W G 1966 J. Mater. Sci **1**, 249
Holt D B 1969 J. Phys. Chem. Solids **30**, 1297
Lin M E, Sverdlov B N and Morkoç H 1993 J. Appl. Phys. **74**, 5038
Nakamura S, Senoh M, Nagahama S, Iwasa N, Yamada T, Matsushita T, Kiyoku K and Sugimoto Y 1996 Jpn. J. Appl. Phys. **35**, L74
Ponce F A, Major J S, Plano W E and Welch D F 1994 Appl. Phys. Lett. **65**, 2302
Serneels R, Snykers M, Delavignette P, Gevers R and Amelinckx S 1973 Phys. Stat. Sol. B **58**, 277
Stadelmann P A 1987 Ultramicroscopy **21**, 131
Vermaut P, Ruterana P, Nouet G and Morkoç H 1997 Phil. Mag. A **75**, 239
Vermaut P, Ruterana P, Nouet G 1997 Phil. Mag. A (in press)

Inst. Phys. Conf. Ser. No 157
Paper presented at Microsc. Semicond. Mater. Conf., Oxford, 7–10 April 1997
© *1997 IOP Publishing Ltd*

Growth of GaN layers on nitrided GaAs/Si and GaAs/SIMOX composite substrates by OMVPE

Chimin Hu[1], N T Nuhfer[1], S Mahajan[1], J W Yang[2], C J Sun[2], M Asif Khan[3] and H Temkin[4]

[1]Dept of Mats Sci and Engineering, Carnegie Mellon University, Pittsburgh, PA 15213, USA
[2]APA Optics Inc., Blaine, MN, USA
[3]Dept of Electrical Engineering, University of South Carolina, SC, USA
[4]Dept of Electrical Engineering, Texas Technical University, Lubbock, TX, USA

ABSTRACT: We have investigated the conversion of thin MBE-grown GaAs layers deposited on (111) Si and SIMOX substrates into GaN by conventional and high resolution transmission electron microscopy. Results show that the converted area is highly faulted, and many voids are observed close to the GaN/Si interface. In the GaN layer grown at 750°C, the zinc-blende phase is dominant, and the wurtzitic GaN is observed when the growth temperature is raised to 900°C. The major defects in the wurtzitic phase are dislocations, stacking faults and inversion domains.

1. INTRODUCTION

GaN is a promising material for the fabrication of light emitting diodes (LEDs) and lasers operating in the blue-green region of the spectrum. The layers are generally deposited on sapphire and SiC substrates, and are highly defective because of the large mismatch or the polarity differences between GaN and the substrates. The threading dislocation density in epitaxial layers is about 10^9 to $10^{10} cm^{-2}$. Numerous research groups have grown GaN layers on different substrates, including Si (Basu et al. 1994) and GaAs (Yang et al. 1995). In this study, we explored the possibility to deposit GaN layers on Si substrates since Si is inexpensive and provides large areas. We developed a unique dual-growth procedure to obtain good quality GaN layers on nitrided GaAs/Si and GaAs/SIMOX (Separation by IMplantation of OXygen) substrates. In the present paper, we have evaluated the microstructural characteristics of these epitaxial layers.

2. EXPERIMENTAL DETAILS

Since the basal plane of wurtzitic GaN is parallel to the {111} plane of silicon, we chose (111) silicon and SIMOX substrates for layer growth. After cleaning the substrates, a 15 nm layer of GaAs was first grown at 300°C using MBE. An additional 10 to 20 nm thick GaAs layer was next deposited at 550°C. The composite substrates were then transferred to a low pressure organometallic vapour phase epitaxy (LPOMVPE) system. The surface of GaAs was exposed to NH$_3$ at 750°C for 30 minutes. This was followed by the deposition of a 150 nm thick GaN layer grown at 750°C. This layer will be referred as the low temperature grown GaN (LT GaN). The subsequent growth of GaN layer was carried out at 900°C and will be denoted as the high temperature grown GaN (HT GaN).

The microstructures of as-grown layers were assessed by transmission electron microscopy (TEM). For this purpose, cross-sectional samples were prepared using the standard procedure. All electron micrographs were obtained using JEM 4000EX equipped with a Gatan imaging filter and Philips EM420T.

3. RESULTS

Figures 1(a) and (b) show, respectively, dark-field and high resolution electron micrographs obtained from an as-grown GaAs/Si sample. Four variants of {111} twins were observed, and the dominant twin plane is parallel to the growth surface as shown in Figure 1(a). No voids are seen in the layer.

After nitridation, EELS and EDS spectra from the converted layer show the presence of both nitrogen and arsenic. Figures 2(a) and (b) indicate that the converted area is highly faulted and many faceted voids are present near the GaAs/Si interface. We also found that there is no difference in the microstructures of the GaN layers grown on Si or SIMOX substrates.

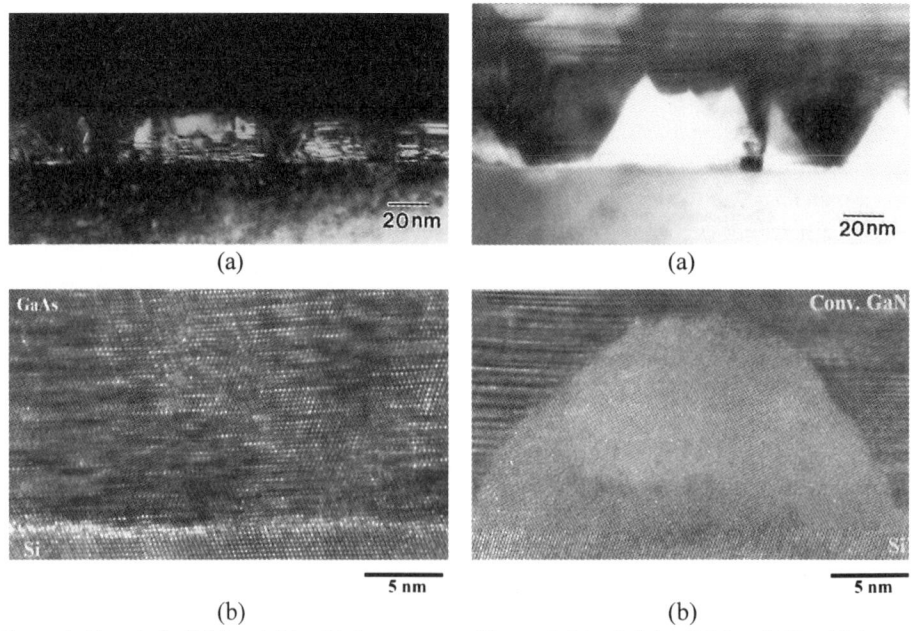

(a)   (a)

(b)   (b)

Figure 1 (a) a dark-field and (b) a high resolution micrographs obtained from an as-grown GaAs/Si sample showing twins. Voids are not observed at the interface.

Figure 2 (a) a bright-field and (b) a high resolution micrographs showing that the converted layer is highly faulted and contains many faceted voids .

As shown in Figure 3(a), the LT GaN has predominantly the zinc-blende structure, although wurtzitic GaN is also found in Figure 3(b). Twins, particularly those parallel to the (111) growth surface, are still the dominant defect. As apparent from Figure 2(a), the interface between the converted area and the LT GaN cannot be discerned, and the density of twins in the LT GaN decreases as its thickness increases.

We established by electron diffraction that the HT GaN has the wurtzite structure. The columnar structure, threading dislocations and planar defects are observed in this layer as shown in Figures 4(a) and (b). The average lateral size of the columns is about 100 nm. Threading dislocations are the primary defect and are mainly located at the interfaces between the columns. The dislocation density is $\sim 1 \times 10^{10}$ cm^{-2} on the surface of the samples. Stacking faults and inversion domains that are respectively parallel and normal to the growth surface are seen.

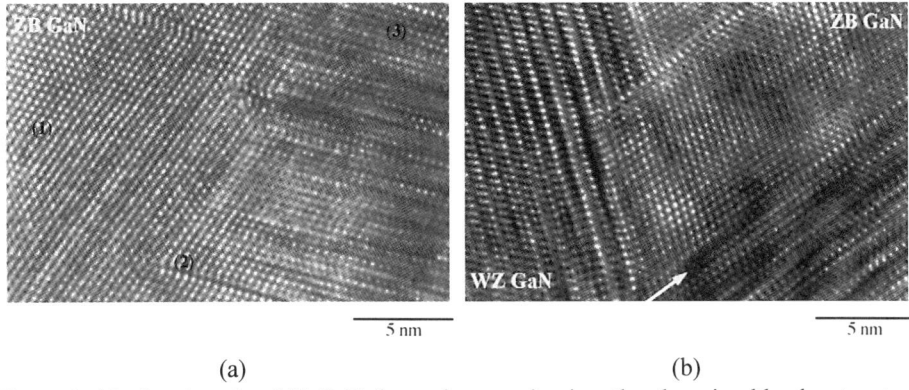

(a)   (b)

Figure 3 (a) showing the LT GaN layer has predominantly the zinc-blende structure, although as shown in (b) wurtzitic GaN is occasionally found. Note in (a), inclined twins present in region (3), and the white arrow in (b) indicating the direction parallel to (111) growth surface.

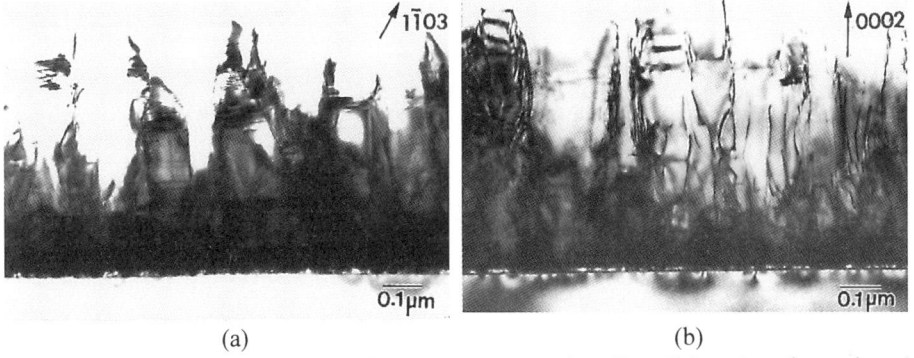

(a)   (b)

Figure 4 (a) and (b) showing the columnar structures, threading dislocations located at the interfaces between the columns, and some planar defects both parallel and perpendicular to the growth surface.

4. DISCUSSION

Our results indicate that on nitridation, GaAs layers are converted into GaN but the conversion is not complete. In addition, triangular voids are observed at the GaN/Si interface. These voids could arise from two sources. The first possibility is the volume decrease (49%) associated with the conversion. Second, tensile stresses can also develop at the GaN/Si interface because the lattice parameter of GaN is smaller than that of Si.

An important objective of this study was to grow the GaN layers on the converted GaN. Our observation shows that the interface between the converted area and the LT GaN is not discernible. This confirms that the converted GaN provides a better lattice-matched substrate for the subsequent growth of GaN.

Our observation indicates that the interface between HT and LT GaN layers is rough due to the presence of planar defects lying on inclined {111} planes. The existence of non-planar surfaces may affect the subsequent growth, leading to a columnar structure. Many threading dislocations are observed at the interfaces between the columns. They could arise due to the slight twist and/or tilt misalignment occurring between the adjoining columns. The threading dislocation density in our samples is comparable to that in the GaN devices grown on sapphire (Ning et al. 1996) and SiC substrates (Hu et al. 1996). In this study, GaN devices have surprisingly good optical and electrical characteristics but still contain a high density of defects. It is an interesting observation for which we cannot present a satisfactory explanation at the moment.

5. CONCLUSIONS

We have demonstrated a successful growth of GaN on (111) Si and SIMOX substrates by using GaAs as the initiation layer. GaAs is converted into GaN by nitridation and faceted voids form in the converted layer. GaN grown at 750°C has the zinc-blende structure and at 900°C GaN is wurtzitic. A variety of defects are observed in the as-grown layers and their origins are discussed.

REFERENCES

Basu S N, Lei T and Moustakas T D 1994 J. Mater. Res. **9**, 2370

Hu Chimin, Mahajan S, Dabkowski F P, Pendse D R, Barrett R J and Chin A K 1996 SPIE **2886**, 59

Ning X J, Chien F R, Pirouz P, Yang J W, Khan M Asif 1996 J. Mater. Res. **11**, 580

Romano L T, Northrup, J E, O'Keefe M A 1996 Appl. Phys. Lett. **69**, 2394

Sitar Z, Paisley M J, Yan B and Davis R F 1990 Mat. Res. Soc. Symp. Proc. **162**, 537

Yang J W, Kuania J N, Chen Q C, Khan M Asif, George T, De Graef M, and Mahajan S 1995 Appl. Phys. Lett. **67**, 3759

Yang J W, Sun C J, Chen Q, Anwar M Z, Khan M Asif, Nikishin S A, Seryogin G A, Osinsky A V, Chernyak L, Temkin H, Hu Chimin and Mahajan S 1996 Appl. Phys. Lett. **69**, 3566

Inst. Phys. Conf. Ser. No 157
Paper presented at Microsc. Semicond. Mater. Conf., Oxford, 7–10 April 1997
© 1997 IOP Publishing Ltd

Crystal defects and optical properties of GaN grown with different techniques: stacking fault related luminescence

G Salviati*, C Zanotti-Fregonara*, M Albrecht°, S Christiansen°, HP Strunk°, M Mayer^, A Pelzmann^, M Kamp^, KJ Ebeling^, MD Bremser†, RF Davis†, YG Shreter‡

* CNR-MASPEC Institute, Via Chiavari 18/A, I-43100 Parma, Italy
° Institut für Werkstoffwissenschaften, Mikrocharakterisierung, Universität Erlangen-Nurnberg, Cauerstrasse 6, 91058 Erlangen, F.R.G.
^ Abt. Optoelektronik, Universität Ulm, Albert Einstein Allee 45, 89069 Ulm, F.R.G.
† Department of Physics, North Carolina State University, Raleigh, NC27695-8202 USA
‡ AF Ioffe Physico-Technical Institute, Russian Academy of Sciences, Polytechnicheskaya 26, St Petersburg 194021, Russia

ABSTRACT: Low temperature spectrally resolved CL and TEM investigations have been carried out on GaN epilayers grown on (0001) oriented Al_2O_3 and $\alpha(6H)$-SiC $(0001)_{Si}$ substrates with different techniques. CL studies revealed an additional emission line at 3.425 eV whose intensity ratio with respect to the near emission band transition depended on the electron beam energy. This line, independently of the growth conditions, has been found to be present only in samples with a very high density of planar defects and has been ascribed to excitons bound to stacking faults.

1. INTRODUCTION

III-V nitride structures are promising for potential applications in the field of optical light emitters and detectors from 1.9 eV (red) to 6.2 eV (UV) and of high frequency and high power electronic devices working at high temperature (e.g. Morkoç 1995, Morkoç et al. 1994 and Pankove 1990 and references therein). However, due to the large lattice mismatch between GaN and commonly used substrates like sapphire (>13%), the successful development of the devices still requires a better understanding of the defect formation and recombination processes in GaN. After the pioneering work of Pankove et al. (1970) on the one hand and of Dingle et al. (1971) and Monemar (1974) on the other hand, several groups have carried out photoluminescence (PL) and cathodoluminescence (CL) studies of hexagonal and cubic GaN (for a detailed summary and discussion of the different results see for instance Ramirez-Flores et al. 1994; Menninger et al. 1995 and Tchounkeu et al. 1996 and references therein). On the other hand, only a few groups have correlated optical and structural properties of crystal defects (Weeks et al. 1995; Ponce et al. 1996a; Shreter and Rebane 1997; Ponce et al.1996b; Christiansen et al. 1996).

In this paper we correlate CL and TEM studies of nominally hexagonal GaN grown with different techniques on different substrates. We find an additional CL emission at about 3.41-3.42 eV only when a very high density of planar defects is present in the epilayers. We determine the nature of the structural defects and we ascribe the additional CL line to excitons bound to stacking faults (SFE).

2. EXPERIMENTAL

Four different epitaxial layers have been studied. The samples A and B, 1.4 µm thick, have been grown by gas source molecular beam epitaxy (GSMBE) on the top of nearly 50 nm thick GaN buffer grown on (0001) oriented Al_2O_3 substrates (Kamp et al. 1996). The GaN buffer of sample B has been grown by GSMBE at 550°C whereas the buffer of sample A has been grown by metal organic chemical vapour deposition (MOCVD) at high temperature and then transferred into the GSMBE chamber for the further growth of the epilayer.

Specimens C and D, 400 nm thick, have been grown by metal organic vapour phase epitaxy (MOVPE) onto a 100 nm thick AlN buffer deposited simultaneously on 6°off- and on-axis α(6H)-SiC $(0001)_{Si}$ substrates (Weeks et al 1996).

The optical emissions of the GaN epilayers have been mainly studied by low temperature spectrally resolved CL performed in a commercial Oxford MONOCL system fitted to a 360 Cambridge SEM. To account for optical inhomogeneities of the samples, CL investigations have been performed also to characterise the sample luminescence spatially as well as spectrally. Occasionally, also low temperature PL investigations have been carried out.

The microstructure of the layers has been characterised by conventional transmission electron microscopy in a Philips CM30 operating at 300 keV. High resolution transmission electron microscopy has been performed in a Philips CM 300UT with a point to point resolution of 0.17 nm. Cross sectional TEM samples have been prepared by standard techniques including mechanical grinding and polishing with diamond paste followed by ion milling.

3. RESULTS AND DISCUSSION

3.1 CL analyses of samples A and B

In Figure 1 a comparison between 10K CL spectra of the emissions of samples A and B is reported. In addition to a broad yellow-luminescence band centred around 2.25 eV(not shown) the two spectra present a strong emission line at about 3.473 eV.

Because of the instrumental resolution of the CL system (max 10 meV for the sample with the highest

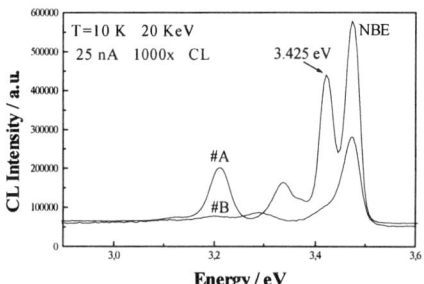

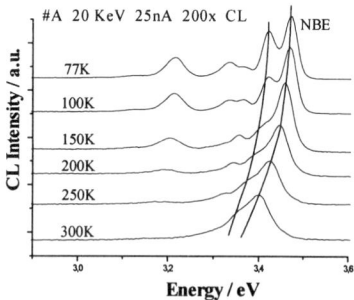

Fig. 1 CL spectra of samples A and B. Note the intense emission at 3.425 eV in sample A. FWHM of the NBE line of A=30 meV; B=50 meV

Fig. 2 Temperature dependence of the intrinsic emissions of sample A

luminescence efficiency), this band could not be resolved in separate emissions the CL spectra. As a consequence, the lines at about 3.47 eV in samples A and B are most likely due to a double transition (complementary PL spectra revealed the presence of a shoulder at about 3.460 eV).

The temperature dependence of the transitions at 3.47 eV in the two samples (e.g. for sample A in Figure 2) revealed that the emissions were still dominant at 300 K. From this result and on the basis of the comparison of our data with those reported in the literature for hexagonal GaN (e.g. Dingle et al. 1971 and Lagerstedt and Monemar 1974, Menninger et al. 1996), we ascribed those CL lines to near band edge (NBE) radiative transitions (e.g. bound excitons).

In Figure 1, sample A showed a much more pronounced and intense line around 3.425 eV, followed by its phonon replica at about 3.336 eV. Also the emission at 3.21 eV is followed by two weak phonon replicas at 3.122 and 3.033 respectively. Sample B also shows a broad peak at about 3.29 eV followed by two weaker emissions, most likely phonon replicas, at 3.20 and 3.11 eV.

From the energy positions of the peaks at 3.210 eV for sample A and at 3.292 for sample B and from the literature data (e.g. Glaser et al 1995 and references therein), we concluded that the two broad bands can be attributed to donor acceptor pair transitions (D°, A°).

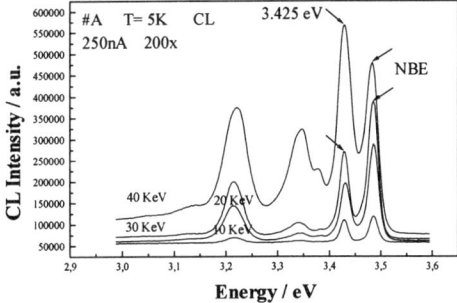

Fig. 3 Low temperature CL spectra of sample A at different beam energies. Note the increse of the 3.425 eV emission with respect to the NBE line at 40 KeV.

In the following we will focus on the intense additional CL emission at 3.425 eV of sample A. CL investigations performed at different beam energies showed an increase of the intensity of the 3.425 eV emission with respect to the NBE one. A typical example is reported in Figure 3 for sample A where the effect is clearly visible at 40 KeV. Such effects were not found in sample B. Corresponding results have been obtained from depth dependent PL experiments on the same samples (Christiansen et al. 1996). The results of Montecarlo simulations on the penetration depth at different accelerating voltages are also reported in Table 1. The values correspond to the penetration depth at which the maximum dose occurs (~25-30% of the Gruen range).

Table 1 Montecarlo simulations for samples A, B, C and D for different beam energies

Samples A and B on Al₂O₃	Maximum dose depth
E=10 KeV	150 nm
E=20 KeV	600 nm
E=30 KeV	1200 nm
E=40 KeV	1600 nm
Samples C and D on SiC	**Maximum dose depth**
E=10 KeV	150 nm
E=20 KeV	200-400 nm (constant)
E=30 KeV	400 nm
E=40 KeV	>600 nm

Spatially resolved CL studies carried out at constant beam energy on different areas of sample A also showed the onset of an anticorrelation between the NBE and the additional transitions.

The influence of the beam energy on the intensity of the additional emission in sample A, indicates a different role of the epilayer/substrate interface in the two samples. In this respect, the possible role of different structural properties will be discussed in the following on the basis of the accurate analysis of structural defects obtained by TEM studies on samples A and B.

3.2 TEM investigations of samples A and B

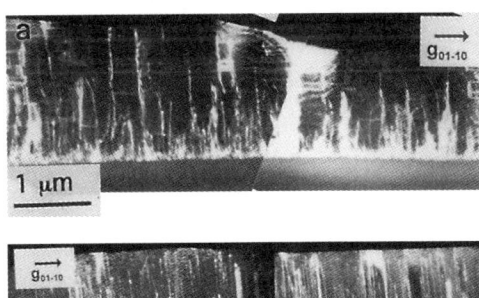

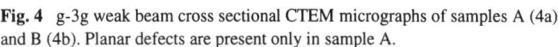

Fig. 4 g-3g weak beam cross sectional CTEM micrographs of samples A (4a) and B (4b). Planar defects are present only in sample A.

Fig 4 shows typical cross sectional TEM micrographs of samples A and B. In both samples a high density of threading dislocations extending through the layer from the interface to the top surface is present. The dislocation density in both samples is highest at the epilayer/substrate interface. It is apparent that the main structural difference between the two samples is in the high density of planar defects on (0001) planes in sample (A) which shows the intense emission at 3.425 eV. No such defects can be observed in sample B. These planar defects are statistically distributed throughout the whole layer volume of the sample A even if they increase in density near the interface (6.0 10[13] cm[-3] in the region extending 150 nm from the interface into the layer instead of 1.5 10[13] cm[-3] in the 700 nm thick top portion of the layer).

From high resolution TEM investigations (Figure 5) we have found four different planar defects: (i) stacking faults bound by Frank partial dislocations (**b**=1/6<2-203>, cf. Figure 5a) (ii) stacking faults bound by Frank partial dislocations (**b**=1/2<0001>, cf. Fig.5b), (iii) stacking faults bound by Shockley partial dislocations (**b**=1/3<1-100>, cf. Fig.5c) and inversion domain boundaries (cf. Fig. 5d) . Comparing the defect distribution in the layer volumes of samples A and B we can say that: (i) while in the two samples all types of dislocations are present, planar defects only occur in the epilayer which shows the additional line at 3.425 eV; (ii) the density of pure edge type threading dislocations (**b**=1/3>11-20> in sample A showing the additional line is one order of magnitude lower than in sample B; (iii) the number of screw dislocations that are present in sample A showing the additional line is 2 to 3 times lower than in sample B.

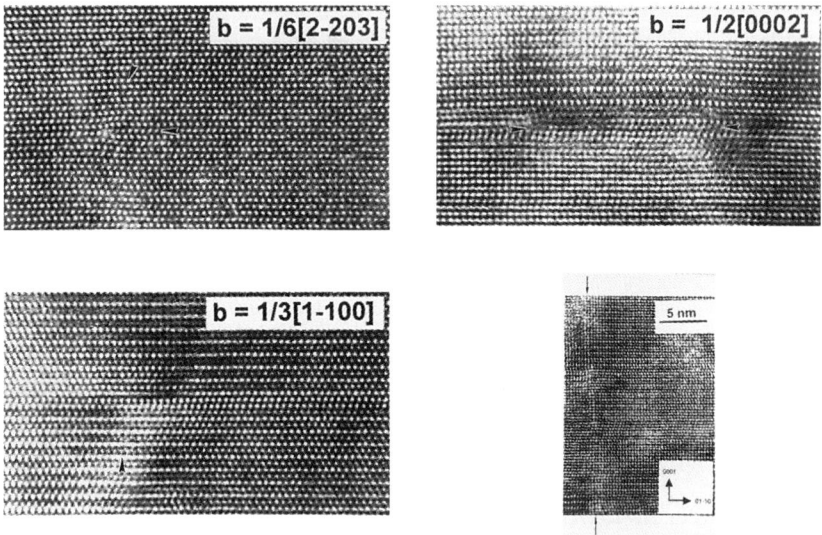

Fig. 5 Bright field zone axis HREM micrographs of the typical planar defects found in samples A and B.   a) Stacking faults (SFs) bound by Frank partial dislocations (**b**= 1/6<2-203>)   b) SFs   bound   by Frank   partial   dislocations (**b**=1/2<0001>)   c) SFs bound by Shockley partial dislocations (**b** =1/3<1-100>)   d) inversion domain boundaries.

As a first conclusion, the TEM and CL observations seem to indicate that the additional blue line in sample A can be correlated with the presence of SFs in the GaN epilayers.

3.3   CL and TEM studies of samples C and D

In order to support this hypothesis, two additional specimens grown 6° off- (C) and on-axis (D) on α(6H)-SiC (0001)$_{Si}$ substrates by a different technique (MOCVD) have been studied following the same approach. The CL spectra of samples C and D are shown in Figure 6. Focusing our attention only on the emission lines between approximately 3.4 and 3.5 eV, again in addition to a probable NBE transition at about 3.468 eV (also in this case supported by literature data and temperature dependent analyses), only sample C presents an intense additional CL emission line at ~ 3.410 eV.

Also in these samples, CTEM (cf. Figure 7) and HREM studies revealed that sample C showing the additional line contained a high density of SFs not present in the sample D. The nature of extended and planar defects was the same as in sample A.

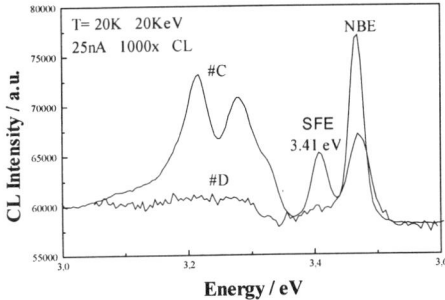

Fig. 6 Typical CL spectra of samples C (6° off axis) and D (on axis) grown on SiC by MOCVD. Only sample C shows the additional SFE line.

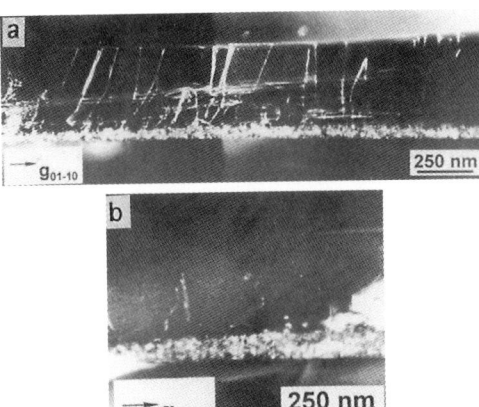

Fig. 7 g-3g weak beam cross sectional CTEM images of samples C (7a) and D (7b). Also in this case SFs are present only in sample C showing the additional CL line at 3.41 eV.

The results of TEM contrast analyses on the distribution of different types of threading dislocations and planar defects in all the samples are summarised in Table 2.

In principle, the additional lines in samples A and C might be ascribed to atomic defects present in different densities in the specimens. However, since the extra lines do not depend on the growth methods and on substrates, the previous hypothesis can be ruled out. Further, recent investigations (Shreter and Rebane 1997) on samples with an inversion domain boundaries density of 10^9 cm^{-2} did not show any additional transition in addition to the free exciton one. Also, recent theoretical work by Northrup et al (1996) showed that inversion domain boundaries do not induce gap states. According to Shreter and Rebane 1996; Schreter et al. 1996) and following Osypian et al. (1986), misfit dislocations in the (0001) GaN plane (60° type with **b**=1/3<11-20>, **b**=1/3<11-23>, 30° and 90° Schokley partials with **b**=1/3<1-100>, Frank partials with **b**=1/6<2-203>) are expected to be highly charged due to the strong ionic character of the bond.

The strong electric field of such dislocations destroys the dislocation excitons that are usually responsible for the dislocation luminescence (Shreter and Rebane 1997). The high concentration of SFs revealed by detailed TEM investigations of samples A and C, the anticorrelation of the additional line with respect to the NBE emission and its increase in intensity near the buffer/layer interface, suggest the emissions at ~ 3.41 and 3.42 eV are strictly related to excitons bound to SFs within the epilayer.

Table 2 Distribution of the different types of threading dislocations and planar defects in samples with and without the additional blue line as obtained from TEM contrast analyses. SFE line means: SF related excitonic emission

Sample	SFE line	b=1/3<11-20>	b=1/3<11-23>	b=<0001>	SFs	Frank loops	IDBs
# A	yes	0.4 10^9 cm^{-2}	0.4 10^9 cm^{-2}	0.5 10^9 cm^{-2}	2.3 10^{13} cm^{-3}	1.6 10^{13} cm^{-3}	1.3 10^9 cm^{-2}
# C	yes	2.5 10^9 cm^{-2}	< 10^6 cm^{-2}	0.2 10^9 cm^{-2}	11 10^{13} cm^{-3}	0.9 10^{13} cm^{-3}	-
# B	no	9.9 10^9 cm^{-2}	< 10^6 cm^{-2}	1.3 10^9 cm^{-2}	-	-	-
# D	no	1.8 10^9 cm^{-2}	< 10^6 cm^{-2}	1.1 10^9 cm^{-2}	-	-	-

The wurtzite GaN has an AaBbAaBbaAbB stacking sequence; the SFs bound by Shockley partials have an AaBbAaBbCcAaCcAa sequence (Osypian and Smirnova 1971). These SFs represent layers of cubic phase surrounded by the wurtzite phase. This cubic phase can be considered as uniaxially compressed; in this respect, they form potential wells for electrons. In the valence band, the deformation potential is negative and the SFs are potential barriers for holes. A theoretical model (Rebane et al. 1997, Albrecht et al. 1997), predicts for SFs a depth of about 120 meV in the conduction band and a height of the barrier of ~ 60 meV in the valence band. Thus, electrons are attracted to the SFs and the holes are repelled and the interface between wurtzite and cubic GaN is similar to a type-II heterojunction. The holes can be attracted to electrons bound to SFs via Coulomb force, forming excitons bound to SFs (Rebane et al. 1997).

4. CONCLUSIONS

GaN epilayers grown on (0001) oriented Al_2O_3 and α(6H)-SiC $(0001)_{Si}$ substrates with different techniques have been studied by TEM and CL in the SEM. The results have shown that an additional transition at ~3.42 eV has been found only in samples with a high density of SFs in the epilayers. The line has been ascribed to excitons bound to SFs.

REFERENCES

Albrecht M, Christiansen S, Strunk HP, Salviati G, Zanotti-Fregonara C, Shreter Y, Rebane YT, Mayer M, Pelzmann A, Kamp M, Ebeling KJ, Bremser MD, Davis RF, paper presented at the MRS 1997 Spring Meeting, St Francisco, CA-USA, March 31-April 4 1997

Christiansen S, Albrecht M, Dorsch W, Strunk HP, Pelzmann A, Mayer M, Kamp M, Ebeling KJ, Salviati G, Zanotti-Fregonara C 1996, Proc.E-MRS Strasbourg, Submitted toThin Solid Films

Dingle R, Sell DD, Stokowski SE, Illegems M 1971, Phys. Rev. **B4** 1211

Glaser ER, Kennedy TA, Doverspike K, Rowland LB, Gaskill DK 1995, Phys. Rev. **B51**, 13326

Kamp M, Mayer M, Pelzmann A, Thies A, Chung HY, Sternschulte H, Marti O, Ebeling KJ 1996, MRS Proceedings **395**, 135

Lagerstedt O and Monemar B 1974, J. Appl. Phys. **45**, 2266

Menninger J, Jahn U, Brandt O, Yang H, Ploog K 1996, Phys. Rev. **B53**, 1881

Monemar B 1974, Phys. Rev. **B10**, 676

Morkoç H, in Semiconductor Heteroepitaxy, Ed. by B Gil and RL Aulombard (World Scientific, Singapore, 1995), pp238-249

Morkoç H, Strite S, Gao GB, Lin ME, Sverdslov B, Burns M 1994, J. Appl. Phys. **73**, 1363

Northrup JE, Neugebauer J, Romano LT 1996, Phys. Rev. Lett. **77**, 103

Osipyan YuA, Petrenko VF, Zaretskii AV, Withworth R 1986, Adv. Phys. **35**, 115

Osipyan YuA and Smirnova IS 1971, J. Phys. Chem. Solids **32**, 1521

Pankove JI 1990, Mater. Res. Soc. Symp. Proc. **162**, 515

Pankove JI, Bergtreisser JE, Marushka HP, Wittke J 1970, Solid State Commun., **8**, 1051

Ponce FA, Bour DP, Gotz W, Jonhson NM, Helava HI, Grzegory I, Jun J, Porowsky S 1996a, Appl. Phys. Lett. **68**, 917

Ponce FA, Bour DP, Gotz W, Wright PJ 1996b, Appl. Phys. Lett. **68**, 57

Ramirez-Flores G, Navarro-Contreras H, Lastras-Martinez 1994, Phys. Rev. **B50**, 8433

Rebane YT, Shreter YG, Albrecht M 1997, paper presented at the MRS 1997 Spring Meeting, St Francisco, CA-USA, March 31-April 4 1997

Shreter YG and Rebane YT 1996, Proceedings of the 23rd International Conference on the Physics of Semiconductors, Berlin, in press

Shreter YG, Rebane YT, Bremser MD, Perry WG, Davis RF, Steeds JW 1996, MRS Fall Meeting, in press

Shreter YG and Rebane YT 1997, Proc. 2Conference on Extended Defects in Semiconductors", Giens, France, in press on J. Physique III

Tchounkeu M, Briot O, Gil B, Alexis JP, Aulombard RL 1996, J. Appl. Phys. **80**, 5352

Weeks TW, Bremser MD, Ailey KS, Carlsson E, Perry WG, Davis RF 1995, Appl. Phys. Lett. **67**, 401

Weeks TW, Bremser MD, Shown AK, Carlsson E, Perry WG, Piner EL, El-Masry NA, Davis RF 1996, J. Mater. Res. **11**, 1011

Acknowledgments

One of the authors (GS) is indebted to Prof. M. Guzzi of the Physics Dept. of the Univ. of Milano for very helpful discussions on CL data. Thanks are also due to Mr. M. Scaffardi for technical assistance.

Inst. Phys. Conf. Ser. No 157
Paper presented at Microsc. Semicond. Mater. Conf., Oxford, 7–10 April 1997
© *1997 IOP Publishing Ltd*

Heteroepitaxy of cubic GaN: influence of interface structure

A Trampert, O Brandt, H Yang*, B Yang and KH Ploog

Paul-Drude-Institut für Festkörperelektronik, Hausvogteiplatz 5–7, D-10117 Berlin, Germany
* Institut of Semiconductors, Chinese Academy of Sciences, Box 912, Beijing 100083, China

ABSTRACT: We report on the epitaxial growth and the microstructure of cubic GaN. The layers are deposited by plasma-assisted molecular beam epitaxy on GaAs and Si substrates. Despite the extreme lattice mismatch between these materials, GaN grows in the metastable cubic phase with a well-defined orientation-relationship to the GaAs substrate including a sharp heteroboundary. The preference of the metastable phase and its epitaxial orientation originates in the interface structure which is found to be governed by a coincidence site lattice.

1. INTRODUCTION

At ambient conditions, the thermodynamically stable modification of the group III-nitrides AlN, GaN and InN is the wurtzite structure. In contrast, the cubic zinc-blende modification is metastable and can be stabilized only by heteroepitaxial growth on topologically compatible substrates (Powell et al 1993). However, the growth on suitable cubic substrate materials, such as GaAs or Si, is scarcely payed attention to because of the very high lattice mismatch to the nitrides. This high mismatch might be expected to result in a breakdown of epitaxial growth leading to polycrystalline growth and eventually to a phase transformation. In the following, we discuss the role and the influence of the interface structure on these phenomena by investigating various GaN/GaAs as well as GaN/Si(001) heterostructures by transmission electron microscopy (TEM).

2. EXPERIMENTAL DETAILS

Cubic GaN films were grown on GaAs(001), (110), (311)A and Si(001) substrates by molecular beam epitaxy using a DC glow discharge nitrogen plasma source (Brandt et al 1995). In the case of the GaAs substrates, a buffer layer is grown to realize an atomically smooth surface consisting of a minimized step density. The growth temperature and growth rate for the GaN were typically at 600 - 650 °C and 0.3 monolayers (ML) per second, respectively. These growth conditions are optimized such as to obtain the highest quality interface and epitaxial film structure (Yang et al 1996, Brandt et al 1997). Prior to the growth on Si wafers, having a miscut of 3° toward the <110>, the substrates were annealed at 850 °C to remove the native oxide. In-situ reflection high energy electron diffraction is used to carefully control the selected growth conditions.

The preparation of transmission electron microscopy (TEM) specimen from the as-deposited GaN layers is done by mechanical pre-thinning and final argon ion milling using a specimen cooling unit. The TEM observations are carried out in Jeol JEM 4000EX/FX microscopes operating at 400 kV and the JEM ARM 1250 with an acceleration voltage of 1250 kV.

3. RESULTS AND DISCUSSION

An overview of the GaN/GaAs(001) heterostructure is shown in the cross-sectional TEM image in Fig. 1. The corresponding selected area diffraction pattern (inset) shows the cubic structure of the GaN and the epitaxial relationship to the GaAs substrate. The high density of stacking faults and microtwins contained in the GaN epilayer are characteristic for cubic GaN films. These planar defects originate at the heteroboundary as a result of island coalescence during the initial nucleation

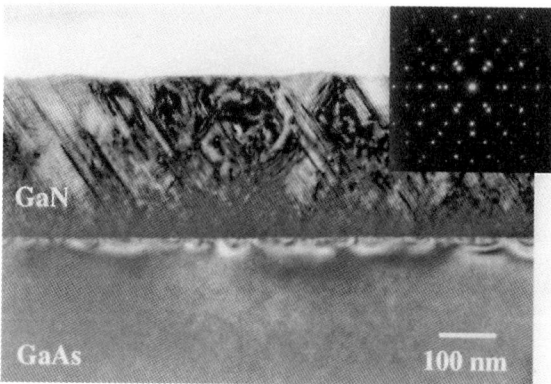

Fig. 1: Cross-sectional TEM bright field micrograph with the corresponding selected area electron diffraction pattern (inset) of the GaN/GaAs(001) heterostructure taken along the >110> direction.

stage (Trampert et al 1997). Their density decreases with the distance from the interface.

High-resolution TEM micrographs taken along the [110] and [100] directions are shown in Fig. 2a (left) and 2b (right), respectively. Closer inspection of the interface reveals an overall roughness of a few ML and the presence of strain releaving defects. Five lattice planes of GaN correspond to four planes of GaAs just at the interface. The periodicity of these misfit dislocations is most easily seen when tracing the {111} planes of GaAs and GaN in Fig. 2a. This 5/4-ratio directly corresponds to the lattice mismatch between GaN and GaAs of 20% (Strite et al 1991). Its high periodicity renders a simple explanation for the occurrence of epitaxy of the metastable cubic phase despite the large lattice mismatch.

Being confident that the strain energy represents the most important part of the total interfacial energy (the chemical portion is assumed to be comparable between the cubic and hexagonal phase and can thus be neglected), the occurrence of the "cube-on-cube" oriented interface can be explained by a near coincidence lattice model. Perfect coincidence sites between the epilayer lattice (a_e) and substrate lattice (a_s) would occur when $a_e/a_s = m/n$, where m and n are integers. If $m = n\pm1$, there is one extra lattice plane in each unit cell of the coincidence lattice, i. e., a geometrical edge dislocation

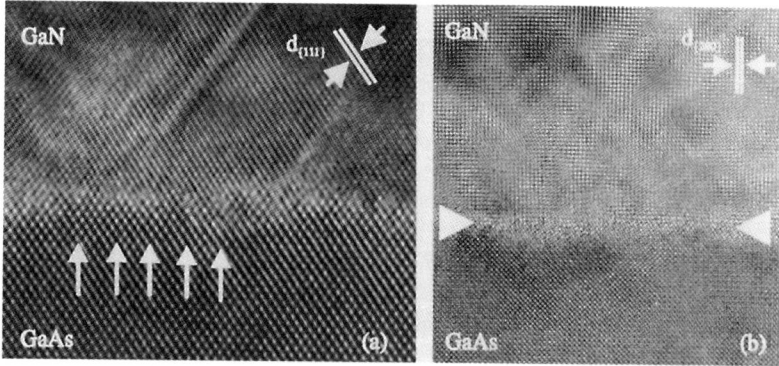

Fig. 2: Cross-sectional high-resolution TEM images of the GaN/GaAs(001) interface taken along the <110> (left) and the <100> (right) direction. The arrows indicate the misfit dislocations.

is generated at the interface. In general, however, the epitaxial system is not expected to be at true coincidence, and the *coincidence-lattice mismatch* f expresses this deviation from perfect coincidence as $f = (ma_s - na_c)/ma_s$. This deviation introduces elastic strain at the interface in addition to the strain accommodated by the geometrical misfit dislocations. Therefore, the energy of heteroboundaries is expected to be small and epitaxy can take place if and only if f does not deviate substantially from true coincidence. For the GaN/GaAs heterosystem, we obtain $f = -0.0002 \pm 0.0020$ by taking $m/n = 4/5$ and the most accurate values for the lattice constants available at growth temperature, namely, $a_{GaN} = 0.455 \pm 0.01$ nm and $a_{GaAs} = 0.568886$ nm. Thus, this system is close to true coincidence and a perfect periodic array of geometric misfit dislocations with a period of 5 GaN lattice planes will indeed account for the entire misfit. The occurrence of this perfect coincidence between cubic GaN and GaAs provides an explanation of the phenomenon of epitaxial growth for a strain at which epitaxy of covalently bonded materials is usually no longer achieved, particularly so for the case of a metastable phase. At this place we want to point out that we use the expression "geometrical" misfit dislocation in order to contrast these to conventional misfit dislocations as defined by Matthews. This latter type of misfit dislocation is, in general, of different physical nature because it corresponds to a bulk dislocation which moves to the interface.

The argument of almost perfect coincidence is generally valid and must be applicable to any interface orientation of the GaN/GaAs system. Figure 3 shows two cross-sectional high-resolution TEM images of a GaN/GaAs(311)A interface in [$\bar{2}33$] and [$\bar{1}21$] projection. The interface is significantly more abrupt compared to the (001) interface, suggesting that the GaN coverage during nucleation must be completed at a very early stage and thus inhibits nitridation. Again, the periodicity of perfectly matched planes, which are here the {220} planes running perpendicular to the interface in the [$\bar{2}33$] projection (Fig. 3a), and the {111} planes which are inclined to the interface in the [$\bar{1}21$] projection (Fig. 3b), is directly visible. The spacing of the matching sites corresponds to the expected 5/4 ratio. However, one should bear in mind that the (311) surface exhibits a lower symmetry compared to the (001) surface, and that the coincidence site lattice must share the symmetry of the adjacent planes of both materials at the interface. Since the [$\bar{2}33$] and the [$\bar{1}21$] axes are directions with a high density of coincident sites, one might assume that the principal directions of strain relaxation are perpendicular to these directions. However, we note that this simple consideration might not hold when taking into account the bonding configuration at the interface, which may lead to more complicated arrangements. In this case, the actual Burgers vector may not be observable in the TEM micrograph.

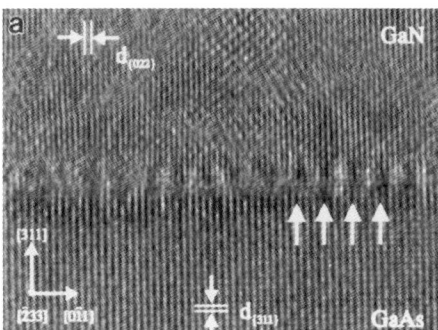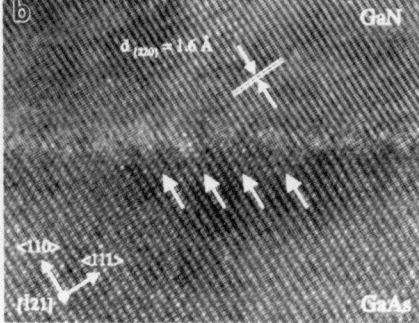

Fig. 3: Cross-sectional high-resolution TEM images of the GaN/GaAs(311) interface taken along the [$\bar{2}33$] (a) and the [$\bar{1}21$] (b) direction. The arrows indicate the misfit dislocations.

The interface structure between GaN and Si(001) appears completely different. While the mismatch between GaN and Si of about 16.8 % is close to the coincidence lattice periodicity of 6/5, the coincidence site lattice mismatch $f = -0.0013$ is about an order of magnitude higher than in the case of GaAs. Nevertheless, we find interfacial regions where the misfit dislocations are indeed in a distance corresponding to the 6/5 ratio (Fig. 4a). Frequently, however, an amorphous interlayer is observed between Si and GaN which presumably consists of Si_xN_{1-x} formed during the initial nucleation (Basu et al 1994, Yang et al 1997). Subsequent growth of GaN leads to the formation of domains of cubic GaN, separated from each other by hexagonal columns extending from the amorphous patches to the film surface (Fig. 4b).

4. CONCLUSION

In conclusion, we have demonstrated that the nucleation and the corresponding interface structure are important for the resulting microstructure of the GaN epilayer. The interface structure between cubic GaN and GaAs is determined by the strain state and can be explained by an extended coincidence site lattice model which takes into account the importance of the deviation from perfect coincidence. The microstructure of the GaN/Si system is determined by an amorphous interlayer which is formed during the initial nucleation stage.

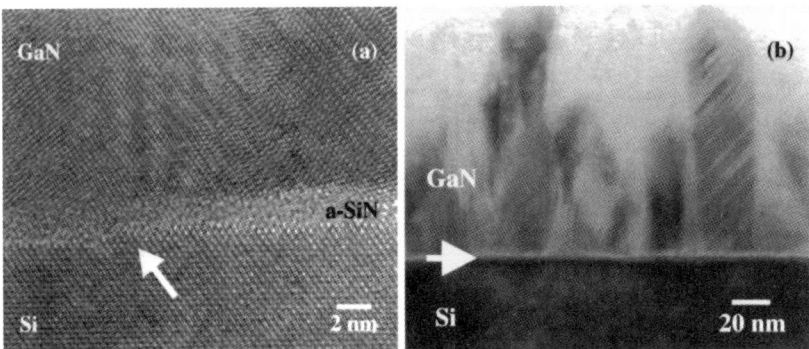

Fig. 4: (a) High-resolution TEM image of the GaN/Si(001) interface revealing the epitaxial oriented GaN directly connected to the Si substrate (arrow).
(b) TEM micrograph of the GaN/Si heterostructure showing the amorphous interlayer (see text for details).

Acknowledgements

Part of this work was sponsored by the Bundesministerium für Forschung und Bildung of the Federal Republic of Germany. The authors would like to thank the Max-Planck-Institut für Metallforschung and the Forschungszentrum Jülich for providing the microscope facilities.

REFERENCES

Basu S N, Lei T, Moustakas T D 1994 J. Mat. Res. **9**, 2370
Brandt O, Yang H, Jenichen B, Suzuki Y, Däweritz L, and Ploog K H 1995 Phys. Rev. B **52**, R2253
Brandt O, Yang H, Trampert A, Wassermeier M, and Ploog K H 1997 to be published
Powell R C, Lee N-E, Kim Y-W, Greene J E 1993 J. Appl. Phys. **73**, 189
Strite S, Ruan J, Li Z, Salvador A, Chen H, Smith D J, Choyke W J, and Morkoç H 1991 J. Vac. Sci. Technol. B **9**, 1924
Trampert A, Brandt O, Yang H, Ploog K H 1997 Appl. Phys. Lett. **70**, 583
Yang H, Brandt O, Trampert A, and Ploog K H 1996 Appl. Surf. Sci. **104/105**, 461
Yang B, Brandt O, Trampert A, Jenichen B, Ploog K H 1997 to be published

Inst. Phys. Conf. Ser. No 157
Paper presented at Microsc. Semicond. Mater. Conf., Oxford, 7–10 April 1997
© *1997 IOP Publishing Ltd*

Highly spatially resolved electron energy loss spectroscopy in the bandgap regime

U Bangert, A Harvey, R Keyse* and D Freundt**

Department of Physics, UMIST, Manchester M60 1QD, U.K.
*Department of Materials Science and Engineering, The University of Liverpool, Liverpool L69 3BX, U.K.
**Institut fuer Schicht- und Ionenforschung, Forschungszentrum Juelich, 52425 Juelich, Germany

ABSTRACT The possibility and application of electron energy loss spectroscopy in the loss regime below 6 eV is investigated. Requirements on experimental conditions and the modelling of the background are discussed. Residual scattering intensities obtained from GaAlInP and GaN films after subtraction of the model background gave the bandgap in both materials reasonably well, though the statistics in the measurement of the scattering intensity in the GaAlInP spectra masked the details of the shape of the loss spectrum with the current experimental set-up. The GaN spectra, obtained with improved statistics, reveal changes in the density of states, correlated to position and therefore to distinct features in the film.

1. INTRODUCTION

Energy loss spectroscopy in the eV regime was pioneered several years ago by Batson (1987). The energy losses in this regime are due to interband scattering and can in principle be used to extract the bandgap and the density of states function, which is observed by the onset energy and the general shape of the corrected low loss spectrum. However, these energy losses are not easily detected because of the proximity to the zero loss peak. High voltage instabilities also limit the energy resolution to 0.3 eV in the best case. Batson undertook substantial modifications to his STEM, in order to reduce the zero loss tail and to increase the energy resolution (Batson 1989). The approach adopted in this paper in order to separate the low loss spectrum from the measured spectrum is careful data processing, using knowledge of zero loss peak. The energy resolution is improved by summing over data taken with very short acquisition times. Though still inferior to optical absorption spectroscopy in terms of energy resolution, this method can give highly spatially resolved information about variations in the density of states function.

Materials with wide bandgaps are most appropriate for low EELS. Energy losses between 1 and 2 eV lie in the critical region. The zero loss tail depends very much on the day to day condition of the microscope (e.g. tip voltage, general stability), and can vary considerably in this energy region. Therefore bandgap studies of GaAs (1.42 eV) are difficult. The situation improves for GaAlInP (2.2 eV), as shown in paragraph 4. GaN (3.2 eV) is an ideal material, though the scattering intensity even in this regime does not usually exceed 50% of the zero loss tail and it therefore proved neccessary to obtain total counts in the zero loss peak of the order of 10^6, as discussed later.

The GaN film, which was used in this study, was grown on sapphire with a 10 nm buffer layer. It is suspected that Al-N-O as well as cubic GaN phases formed near the interface. The film subsequently grew in the hcp phase and is heavily twinned. It showed yellow emission in PL. The intention is to find out whether it is possible to obtain and interpret scattering information in the bandgap regime and furthermore, whether differences in spectra taken at distinctive points in the GaN film can be detected.

2. EXPERIMENTAL

The samples were MBE grown GaN layers on sapphire and MOCVD grown GaAlInP/GaInP heterostructures on GaAs. Cross-sections were investigated in a VG601UX FEGSTEM equipped with a parallel GATAN EELS spectrometer. EELS was carried out at 0.05 eV per channel. The probe size was of the order of 1 nm. In order to avoid low frequency instabilities in the high voltage supply, acquisition times were kept at 0.1 s and several spectra were recorded, aligned and summed for each particular point of interest in the specimen. The instrumental broadening function (PSF) was recorded and deconvolved from each experimental spectrum. With these procedures in place an energy resolution of 0.15 eV could be achieved. Data processing and analysis was undertaken with the EL/P3.0 software package. A zero loss peak model was subtracted from the experimental spectrum in order to obtain the residual spectrum, the onset of which should be at the bandgap of the respective material.

3. MODELLING AND DATA PROCESSING

An accurate model of the zero loss peak was obtained by convolving the Fowler-Nordheim field emission distribution, to which phonon resonance peaks (Lorentzian distributions) were added, with a Gaussian distribution, to account for some degree of energy smear-out due to high frequency high voltage instabilities. Fig.1 shows the dependence of the width and tail of the field emission distribution on the extraction field strength. This in turn depends on the work function. It becomes clear that reducing the work function from 4.5 to 3.5 eV (i.e. by using an appropriate tip material) halves the width and substantially reduces the tail of the distribution at 0.5 eV. Agreement between the modelled and the experimental zero loss peak, after deconvolution of the PSF, proved to be

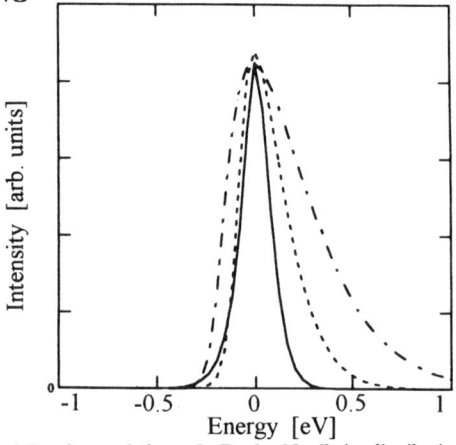

Fig.1 Zero loss peak due to the Fowler-Nordheim distribution without instrumental broadening. Solid curve: field strength F= 7x10⁶ V/cm, work function φ=3.5 eV; dotted and dash-dotted curves: F=2x10⁷ and 4x10⁷ V/cm and φ=4.5 eV. The distributions are displayed as energy loss spectra.

very good over the entire range of magnitudes in the intensity. Fig.2. shows a GaAs spectrum with superimposed tail model. It can be seen that the onset of interband scattering is obscured, and it is therefore essential to subtract the tail of the zero loss peak. Though in GaN the onset of the interband scattering is at higher energies, the tail still dominates the low scattering intensities and subtraction of the model tail is equally essential.

4. RESULTS AND DISCUSSION

Fig.3 shows spectra for $Ga_{0.15}Al_{0.35}In_{0.5}P$ and GaAs, after deconvolution of the PSF, together

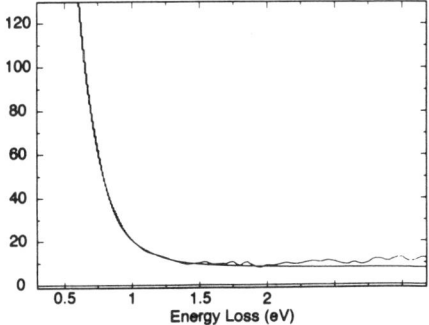

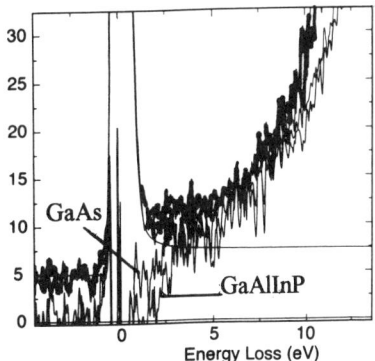

Fig.2 Tail of the modelled zero loss peak superimposed on experimental spectrum

Fig.3 Low loss spectra of GaAs, GaInAlP (thicker lines), model zero loss peak and residual GaAs and GaInAlP (thinner lines, bottom)

with the modelled zero loss peak. Further down the difference between the spectrum and the modelled zero loss peak is shown. The spikes around zero loss reflect differences in the intensities of the experimental peak and the modelled zero loss peak of approx. 1%. The only major (as yet unexplained) discrepancy is in the tail region between 1 and 1.5 eV. Unfortunately this is the region of the bandgap energy of GaAs and hence it will not be possible to detect the onset of the GaAs spectrum. The onset of the GaAlInP spectrum can be seen at approximately 2.1 eV, which agrees roughly with the bandgap of 2.2 eV for the nominal composition. Though the spectra are the sum of 16 acquisitions the statistics are poor. This, however, does not conceal the fact that there is a marked difference between the GaAs and the GaAlInP spectrum around bandgap energies, in all other parts of the spectra the differences between GaAs and GaAlInP are masked by the statistics. For more meaningful statements better statistics are therefore needed.

Fig.5a) shows a high angle dark field image of a cross-section of GaN on sapphire. It exhibits columnar grain growth and is heavily twinned. The grains are small in the buffer layer and become larger towards the layer surface. The buffer layer is approximately 10 nm wide and microdiffraction shows a mixed cubic/hexagonal phase in some places. The film above it is just hexagonal. We have obtained EELS spectra at several positions in the buffer, at the buffer/sapphire interface within large grains, at grain boundaries and at the layer surface. A selection is shown in fig. 4. There appear to be differences, in particular between the buffer layer

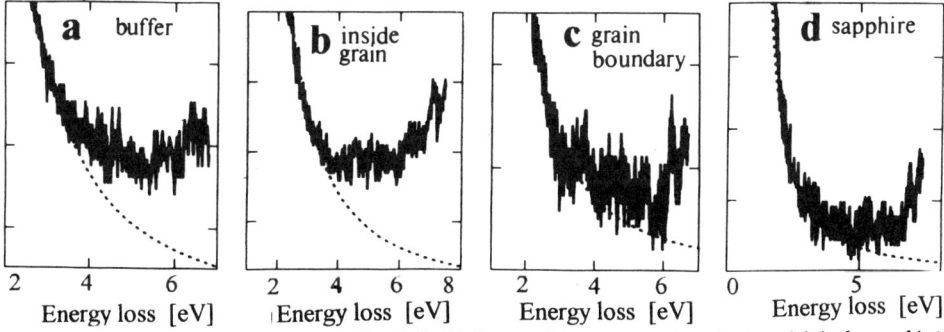

Fig.4 Low loss spectra, taken at points a (buffer), b (inside large grain), c (on grain boundary) and d (in the sapphire) marked by the respective letters in fig.5a). Sum of 8 spectra each, modelled background line is superimposed.

spectra and those taken within large grains higher up in the film, but the uncertainty due to the poor statistics prohibited conclusions to be drawn. We increased the number of acquisitions at point (a) in the buffer layer and at point (b), in a region of good crystal perfection higher up in the film (see fig.5a), to 100. The sum of the aligned spectra is shown in Fig.5b). The statistics have markedly improved and the differences in the two spectra can now be seen clearly: the 'large grain' spectrum (b) exhibits a peak at 2.5 eV and a subsequent increase in the intensity, which, when extrapolated backwards, goes to zero at about 3.1 eV. This is in good agreement with the bandgap in hcp GaN. The peak at 2.5 eV could be connected with the yellow luminescence, which was observed in PL of this sample. In the buffer layer spectrum (a) the intensity rises at about 1.7 eV. The spectrum is smeared out and does not show the distinct features of spectrum (b). This would suggest a wide distribution and large number of midgap states.

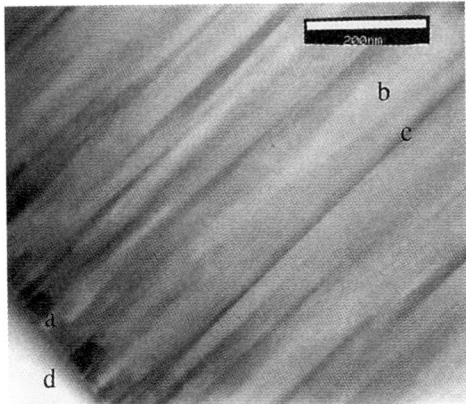

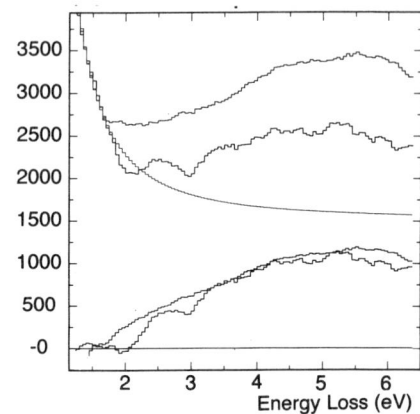

Fig.5a) Cross-sectional high angle dark field image of GaN on sapphire. The letters mark the region in the buffer layer and in the film, where the EELS in fig.4 and fig.5b) were taken.

Fig.5b) Sum of 100 aligned low loss spectra, in descending order, of the buffer layer region (a), of region (b) within a large 'grain' towards film surface, of the modelled tail of zero loss peak, of buffer layer and 'grain' spectrum after subtraction of modelled background.

5. CONCLUSIONS

Low loss spectra taken in a dedicated STEM can give valuable information about the local bandgap energy and density of conduction band states. This was demonstrated in GaN, which shows a distinct peak at 2.5 eV and an onset of the interband scattering at 3.1 eV in good crystalline regions and a much less defined spectrum with onset at 1.7 eV in the buffer layer. In order to obtain maximum energy resolution and sufficiently good statistics many spectra with short acqisition times were aligned and summed to give a maximum zero loss peak intensity of 10^6 counts, the instrumental broadening function was deconvolved and an accurately modelled background was subtracted from the original data.

REFERENCES

Bangert U, Harvey A J and Keyse R 1997 Ultramicroscopy, in press
Batson P E, Kavanagh K L, Wong C Y and Woodall J M 1987 Ultramicroscopy 22 89
Batson P E 1989 Ultramicroscopy 28 32

Inst. Phys. Conf. Ser. No 157
Paper presented at Microsc. Semicond. Mater. Conf., Oxford, 7–10 April 1997
© 1997 IOP Publishing Ltd

Developing a methodology for the electron energy-loss spectroscopy of defects in GaN

M K H Natusch, G A Botton and C J Humphreys

Department of Materials Science and Metallurgy, University of Cambridge, Pembroke Street, Cambridge, CB2 3QZ

ABSTRACT: The optical properties of wurtzite GaN are investigated using electron energy-loss spectroscopy (EELS). In this paper we develop a methodology to extract the information on the electronic structure from low-loss spectroscopy data and we apply this methodology to MgO as a test case. By considering the instrumental and methodological limitations of low-loss spectroscopy a decision whether it can be applied to a spatially resolved study of interband transitions and in particular differences in the band gap on and off crystal defects can be made. It is found that while the low-loss region provides the possibility of high-quality investigations of the optical properties of bulk materials, the limited spatial resolution due to delocalisation effects necessitates different tools for the study of nanometre scale defects, such as the near-edge fine structure of core excitations.

1. INTRODUCTION

Defects and surfaces in semiconductors can introduce defect states in the forbidden energy gap and therefore modify the macroscopic properties of these materials. Electron energy-loss spectroscopy (EELS) in the low-loss region can extract this information from the loss function via the Kramers-Kronig transformation. To decide whether such an approach can be successfully used to study defects one must first consider the spatial resolution and energy resolution and secondly whether other more fundamental effects allow these studies and goals to be achieved. In this study we first demonstrate the validity of the methodology on MgO smoke crystals and assess the dielectric data obtained and then apply the knowledge gained to GaN.

2. INSTRUMENTATION

Several instrumental parameters determine the success of a spatially resolved low-loss EELS experiment and the electronic structure retrieval. These are most importantly the spatial resolution attainable and the energy resolution of the electron gun, the spectrometer and the detector which will together limit the electronic structure details that can be observed and the lowest energy which can be reliably extracted from spectra. The measurements in this paper were performed on a VG STEM HB501, which is a dedicated STEM with a cold field emission gun, and on a Philips CM30 TEM with a LaB_6 filament. Both microscopes were operated at 100 kV. The spatial resolution and the energy resolution will be discussed separately.

2.1 Spatial Resolution

Besides the electron optical parameters defining the probe size (Kopf 1981) the broadening of the electron beam, which is geometric in EELS, and the delocalisation of inelastic scattering contribute to the ultimate spatial resolution of analysis. The geometric broadening in the specimen in EELS is linear with specimen thickness and also depends on the solid angle subtended by the collector aperture. Inelastic scattering is delocalised by a distance defined by the impact parameter. A good approximation for this parameter has been given by Pennycook (1982) and even though more detailed calculations are available, for instance by Muller and Silcox (1995), the simple approximation is sufficient for the present purpose.

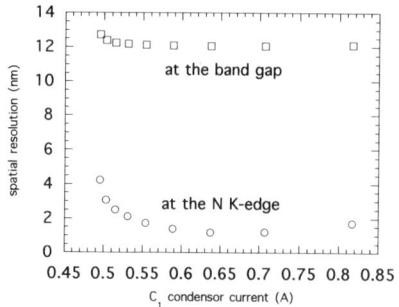

Fig. 1 Ultimate spatial resolution attainable in present STEM at 3.4 eV and 400 eV energy-loss.

Assuming the contributions to the spatial resolution are gaussian it is possible to add the electron optical resolution, the aperture function and the impact parameter in quadrature. The resulting spatial resolution in the experimental configuration used here as a function of the STEM C_1 condenser current is given in Fig. 1 for two energy losses, in the low-loss region at the GaN band gap of 3.4 eV and for higher losses at the nitrogen K-edge at about 400 eV. Muller and Silcox (1995) show how the spatial resolution can be improved by collecting the EELS signal at a high momentum transfer. However, this approach is not useful here since this procedure would yield the band gap for non-vertical transitions.

2.2 Energy Resolution

With a perfect spectrometer the electron source in the microscope provides the ultimate limit of the attainable energy resolution. Assuming a Maxwellian energy distribution the thermal spread for a LaB_6 tip is 0.4 eV whereas the energy spread for field emission is determined by the Fowler-Nordheim distribution which has a long high energy tail that stems from electron-electron scattering within the tip (Lea and Gomer 1970). Other effects limit the resolution further, mainly the modulation transfer function of the detector, the Boersch effect in TEM and instabilities in the STEM. At short acquisition times the energy resolution is approximately 0.4 eV for the STEM and 0.8 eV for a slightly undersaturated filament of the LaB_6 TEM.

In order to extract efficiently the low loss distribution from a measured spectrum the following approach was used. First a gaussian was deconvolved from the experimental zero-loss peak to remove the effect of instabilities during the acquisition of the zero-loss. A modified Fowler-Nordheim model was fitted to the result and this was then convolved with a gaussian to allow for instabilities during the acquisition of the spectrum. Finally the result from the previous step was deconvolved from the spectrum in a modified version of Egerton's (1996) Fourier-log deconvolution which also removes plural scattering.

3. OPTICAL PROPERTIES

3.1 Determination of Optical Properties using EELS

EELS measures the loss function which via Kramers-Kronig transformation provides the real and the imaginary parts of the dielectric function, ε_1 and ε_2. Assuming parabolic bands for the conduction and valence band, ε_2 is proportional to $(1/E)(E - E_g)^{1/2}$ for interband transitions to the conduction band, where E_g is the band gap and E the energy loss (Batson et al. 1986). Near the band gap ε_1 takes the form $1 + E_p^2/(E_g^2 - E^2)$, where E_p is the plasmon energy. The Kramers-Kronig analysis has been performed with the Fourier procedure described by Egerton (1996). The effects of the collection and convergence angles have been corrected for. It is essential that the loss function starts at zero and at the high energy end returns to zero. This means that the spectrum either needs to be acquired at a lower dispersion or that the loss function must be extrapolated beyond the range over which measurements were made assuming an inverse power law.

The band gap can in principle be obtained from the zero of the loss function or of ε_2 or by the same token from a sharp maximum in ε_1. Radiation losses, however, can be visible as Cerenkov peaks when $\varepsilon_1 v^2/c^2$ exceeds unity, which is the case in many insulators and III-V semiconductors close to the band gap (Festenberg 1971). Here v is the speed of the electron and c the speed of light. Alternatively the expected shape of ε_2 can be fitted to the spectrum. The ideal fitting region only contains the interband signal. As it happens, this region overlaps with the position of the surface plasmon which therefore must be removed. The method of recording the spectra off-axis to make use of the fact that surface scattering has a narrower angular distribution is not possible since this would only consider non-vertical transitions across the band gap. A possible solution is the iterative correction for surface losses described by Egerton (1996). The presence of the band gap shifts the free electron plasmon energy, E_{free}, to $\left(E_{free}^2 + E_g^2\right)^{1/2}$ which could be used to estimate E_g. A general source of errors in the determination of the band gap is the presence of strain in the specimen.

3.2 Optical Properties of Magnesium Oxide

To build up confidence in the methodology described above we have looked at a material with a wide band gap, MgO, and assessed whether problems arise in the comparison of the results obtained from the Kramers-Kronig analysis (e.g. the gap) and the direct observation from spectra. In addition, MgO is also a clean specimen, is simple to use and has been extremely well studied in the past. It also allows a check on the performance of the Kramers-Kronig analysis which by virtue of its sum rule yields the specimen thickness. This value can easily be compared with the dimensions of the single MgO cube visible in a bright field image.

The loss function and the real and imaginary parts of the dielectric function shown in Fig. 2 agree qualitatively and quantitatively very well with previously published EELS data, for instance by Venghaus (1971). Further to this the measured thickness from the bright field image agrees with the Kramers-Kronig sum rule. Clearly visible are a variety of interband transitions, the volume (22.3 eV) and surface (15.7 eV) plasmons and various core excitations such as O $2s_{1/2}$ (24.8 eV) and Mg $2p$ (53.5 eV). The zero of ε_1 agrees with the observed plasmon peak if damping is taken into account. Most surprising is the excellent agreement of the behaviour of ε_1 near the band gap (7.6 eV) with the prediction given above. Cerenkov losses can be expected up to about 10 eV and ε_2 and the loss function show counts below the band gap. This shows that a direct observation of the band gap from a spectrum or a fit would therefore be unreliable.

A comparison with optical data is given in Fig. 3 which shows the refractive index as calculated from the experimental ε_1 and ε_2 together with the refractive index as determined from optical data and tabulated by Roessler and Huffman (1991).

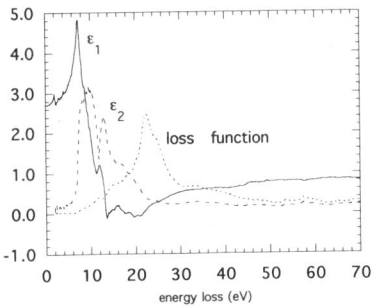

Fig. 2 Loss function and real and imaginary part of the dielectric function of MgO.

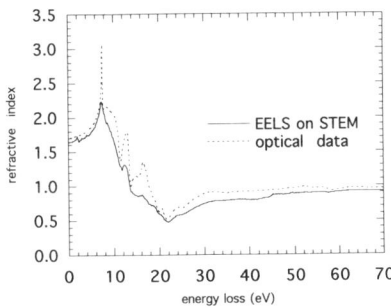

Fig. 3 Comparison of the refractive index of MgO determined from the measurement with optical data.

3.3 Optical properties of Gallium Nitride

Having verified the procedure for MgO, it was then applied to GaN. It was found that carbon contamination is an extremely critical factor and results comparable to optical measurements could only be obtained after the removal of the amorphous layer from the specimen using a plasma cleaner.

Fitting the region between 7 eV and 13 eV to the expected shape yields the band gap as 3.4 ± 0.3 eV. It is, however, questionable to what extent the assumption of parabolic bands is a realistic description of the conduction and valence bands in GaN. The volume plasmon in wurtzite GaN can be

216

calculated to be at 22.0 eV assuming no Ga $3d$ electrons are in the valence band. The presence of the band gap shifts this to 22.3 eV. ε_1 crosses zero at 18.2 eV as can be seen in Fig. 4. Hedman and Mårtensson (1980) report 19.7 eV for the onset of the Ga $3d$ excitation. It therefore seems that the lower one of the two peaks at 19.4 eV and 22.0 eV is the volume plasmon shifted by the Ga $3d$ and interband transitions which also cause this unexpected double peak. No clear estimate of the band gap can therefore be made from the position of the plasmon. The dielectric function and the loss function as given in Fig. 4 agree well with the calculation and synchrotron ellipsometry measurement by Logothetidis et al. (1994). Good agreement is also found for the effective number of valence electrons per atom calculated from the present measurement and shown in Fig. 5. The steep increase even after the volume plasmon can be explained by the onset of the Ga $3d$ transitions.

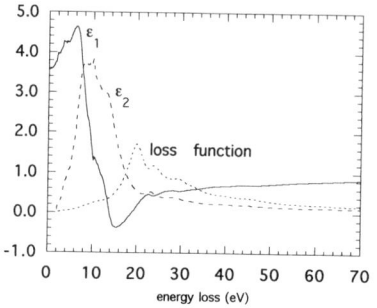

Fig. 4 Loss function and real and imaginary part of the dielectric function of wurtzite GaN.

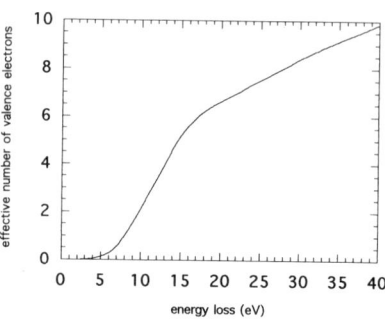

Fig. 5 Effective number of valence electrons per atom in wurtzite GaN.

4. CONCLUSION

Spectral features in low-loss EELS do not correspond directly to interband transitions and an attempt to directly interpret data from spectra can fail. Even though Kramers-Kronig analysis of electron energy-loss spectra provides the correct dielectric function, the application of low-loss spectroscopy to the study of defects is limited by the spatial resolution. The study of isolated structural defects to reveal possible defect states in the band gap therefore belongs to the realm of core-level excitations in order to learn about the unoccupied density of states from the near-edge fine structure.

Acknowledgements

The authors wish to thank PD Brown and Y Xin for the provision of the GaN specimen which has been grown by CT Foxon and TS Cheng. MKHN gratefully acknowledges financial support by Gatan Inc. and the EPSRC.

REFERENCES

Batson PE, Kavanagh KL, Woodall JM and Mayer JM 1986 Phys Rev Lett **57** 2729
Egerton RF 1996 Electron energy-loss spectroscopy in the electron microscope (New York: Plenum Press) 2nd ed.
Festenberg C 1969 Z Phys **227** 453
Hedman J and Mårtensson N 1980 Phys Scr **22** 176
Kopf DA 1981 Optik **59** 89
Lea C and Gomer R 1970 Phys Rev Lett **25** 804
Logothetidis S, Petalas J, Cardona M and Moustakas TD 1994 Phys Rev B **50** 18017
Muller DA and Silcox J 1995 Ultramicroscopy **59** 195
Pennycook SJ 1982 Contemp Phys **23** 371
Roessler DM and Huffman DR 1991 Handbook of optical constants of solids II (London: Academic Press) ed. Palik ED pp 919-55
Venghaus H 1971 Opt Commun **2** 447

Inst. Phys. Conf. Ser. No 157
Paper presented at Microsc. Semicond. Mater. Conf., Oxford, 7–10 April 1997
© *1997 IOP Publishing Ltd*

Probing the effect of defects on band structure in GaN

D M Tricker, M K H Natusch, C B Boothroyd, Y Xin, P D Brown, T S Cheng*, C T Foxon* and C J Humphreys

Department of Materials Science and Metallurgy, University of Cambridge, Pembroke Street, Cambridge CB2 3QZ, UK.
*Department of Physics, University of Nottingham, University Park, Nottingham, NG7 2RD.

ABSTRACT: Conventional and high resolution electron microscopy and high spatial resolution electron energy loss spectroscopy in a scanning transmission electron microscope are used to examine the effect of extended defects on the electronic band structure of MBE-grown GaN. The band gaps have been determined experimentally to be 3.4 ± 0.3 eV for bulk GaN and 2.8 ± 0.2 eV on a domain boundary, obtained by low loss electron energy loss spectroscopy, and the accuracy of these is limited by delocalisation of inelastic scattering which restricts the spatial resolution to 12nm. Fitting the shape of the nitrogen K-edge is considered to be the preferred route for comparing band structures.

1. INTRODUCTION

It is of interest to understand why high brightness GaN-based LEDs and lasers operate in the presence of extremely high defect densities, since this is contrary to experience gained from GaAs-based technologies. Our approach to address this fundamental question is to obtain accurate atomic coordinates of extended defects commonly found within epitaxial GaN and to use these as a basis for band structure calculations. In parallel, we seek to obtain high spatial resolution electron energy loss (EEL) data on and off defect structures to probe the effect of these defects on material electronic (and opto-electronic) properties. Prismatic stacking faults formed within MBE grown GaN are described and preliminary observations of the effect of these boundaries on low loss EEL spectra are presented. We then briefly consider nitrogen K-edges acquired on and off threading dislocations.

2. DEFECT CHARACTERISATION

Epitaxial GaN samples grown by MBE at 700°C on $\{\overline{111}\}$B GaAs and GaP substrates (Cheng et al 1995) were prepared in plan-view and cross-sectional geometries using conventional techniques (Newcomb et al 1985). Samples were examined using JEOL 4000EX-II and Philips CM30 electron microscopes, and in a VG HB501 scanning transmission electron microscope (STEM) equipped with a Gatan imaging filter.

A high density of threading dislocations is observed within epitaxial wurtzite (Mg-doped) GaN grown on $\{\overline{111}\}$B GaAs (Fig. 1), most of which are straight with line vector approximately parallel to the <0001> growth direction. These dislocations sometimes comprise very small angle grain boundaries separating regions of slightly differing orientation and lead to the formation of a mosaic cell structure. The majority of the threading dislocations are screw or mixed in character which contrasts with (0001) wurtzite GaN growth on the basal plane of sapphire, where the threading dislocations are mainly edge in character (Ning et al 1996, Wu et al 1996). Stacking faults are also observed on the basal plane of GaN (Fig. 2) leading to the introduction of vertical boundaries (prismatic stacking faults) which divide faulted and unfaulted columns of GaN. It is not uncommon for several parallel (0001) faults to lie above each other in a column. Some relationships between dislocations and fault boundaries can be discerned: it appears that some dislocations terminate at the plane of a fault.

218

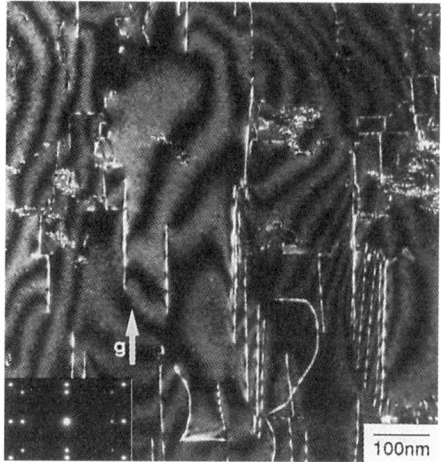

Fig. 1: Mg-doped GaN on {$\overline{1}\overline{1}\overline{1}$}B GaAs
Cross-sectional specimen, dark field, **g** = 000$\overline{2}$.
Inset: <10$\overline{1}$0>GaN // <11$\overline{2}$>GaAs

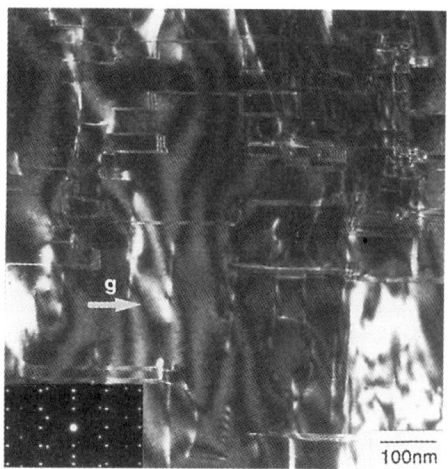

Fig. 2: Mg-doped GaN on {$\overline{1}\overline{1}\overline{1}$}B GaAs
Cross-sectional specimen, dark field, **g** = 10$\overline{1}$0
Inset: <11$\overline{2}$0>GaN // <1$\overline{1}$0>GaAs

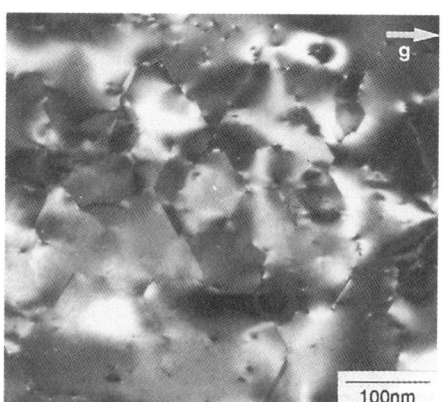

Fig. 3: (000$\overline{1}$) GaN on {$\overline{1}\overline{1}\overline{1}$}B GaP
Plan view specimen of near interface region
Dark field, **g** = 10$\overline{1}$0, taken near <0001>

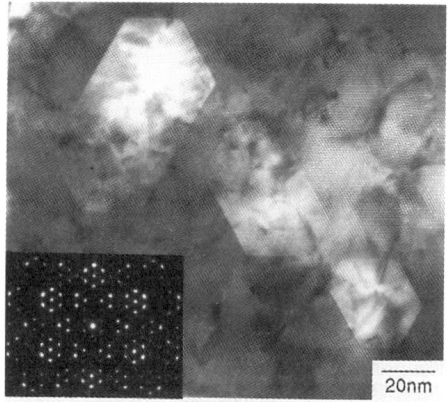

Fig. 4: (000$\overline{1}$) GaN on {$\overline{1}\overline{1}\overline{1}$}B GaP
Bright field at <0001> of interface region in plan
view, showing moiré fringes. Inset: SADP

Not all dislocations are straight: the two strikingly curved dislocation lines seen in Fig. 1 are shown in Fig. 2 to interact with (and maybe terminate at) basal plane faults. Thus, although it is our aim to study the effect of the different defects on band structure separately, in practice, different extended defect types may in fact be associated.

Fig. 3 shows the near interface region of a GaN film grown on {$\overline{1}\overline{1}\overline{1}$}B GaP. Two symmetry related sets of {1$\overline{2}$10} domain boundaries can be seen, lying near vertical in the foil. The third set, the trace of which lies parallel to the diffracting vector, is out of contrast. These boundaries delineate columns of GaN which has a turreted growth surface. This origin of this cell structure seems to relate to etching of the {$\overline{1}\overline{1}\overline{1}$}B surface, which is clearly seen in a plan view sample back thinned to intersect the GaN/GaP interface region (Fig. 4), suggesting that substrate etch features may be associated with the initial formation of these mosaic cells.

These {1$\overline{2}$10} boundaries are prismatic stacking faults (Xin et al 1997). Fig. 5 shows an <0001> lattice image of such a defect lying on {1$\overline{2}$10} that apparently terminates in perfect crystal at both ends. A 'burgers circuit' indicates a measured closure failure of $1/3[1\bar{2}10]$ (i.e. perpendicular to the defect). This is equivalent to an extra half plane at one end of the defect, or a displacement of $1/6[1\bar{2}10]$ across the boundary itself. This displacement is confirmed using moiré fringes where substrate and epilayer overlap in a plan view specimen (Fig. 6). Neither technique measures displacements parallel to the electron beam. An additional $c/2$ displacement parallel to the beam would give a fault vector of $1/6[1\bar{2}13]$, which is equivalent to $1/2[10\bar{1}1]$, the fault vector found by Drum (1964) in AlN. The total circuit around the defect would then have a closure of $1/3[1\bar{2}13]$, which is a perfect lattice vector. Thus, the boundary introduces a slight misorientation between adjacent columns of GaN.

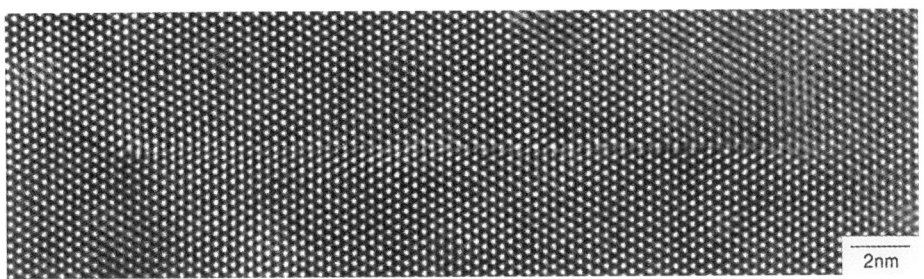

Fig. 5: $(000\bar{1})$ GaN on $\{\bar{1}\bar{1}\bar{1}\}$B GaP, plan view. <0001> lattice image showing $\{1\bar{2}10\}$ boundary in GaN.

Fig. 6: $(000\bar{1})$ GaN on $\{\bar{1}\bar{1}\bar{1}\}$B GaP. Dark field taken with both the $1\bar{2}10$ GaN and the $2\bar{2}0$ GaP spot within the objective aperture, showing moiré fringes parallel to a $\{1\bar{2}10\}$ boundary (lying horizontal). Note the extra half plane terminating near the right end of the boundary, and the fringe displacement across the boundary.

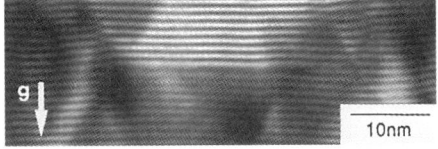

3. ELECTRON ENERGY LOSS SPECTROSCOPY ON AND OFF DEFECTS

Low-loss electron energy-loss spectroscopy has the potential to retrieve the optical properties of a material at high spatial resolution by probing the local joint density of states, which under the assumption of parabolic bands in turn is proportional to $(E - E_g)^{1/2}$, where E is the energy-loss and E_g is the band gap. Defects and surfaces can introduce additional states in the forbidden energy gap. The band gap will then appear to be lowered. Here we describe an attempt based on the work of Batson et al (1986) to find a change in the band gap of wurtzite GaN in the vicinity of the $\{1\bar{2}10\}$ boundaries seen in Figs. 3 and 5. Our intention is to extend this study to include isolated threading dislocations.

The field-emission gun in a VG HB501 STEM can provide an energy resolution of 0.4 eV at a spatial resolution of 1 nm. Inelastic scattering, however, is delocalised and at the band gap in GaN (about 3.4 eV) the spatial resolution decreases to 12 nm (Natusch et al 1997). The expected signal is superimposed on the intense tail of the zero-loss peak and this tail shows comparable intensity to the expected signal. A theoretical model such as that given by Lea and Gomer (1970) must be deconvoluted from the experimental spectrum. Once this contribution is removed the band gap could in principle be obtained from the zero of the loss function. A further complication is the occurrence of radiation losses (Festenberg 1969) and the presence of a large number of interband transitions (Bloom et al 1974) as can be seen in Fig. 7. It is therefore advisable to fit the theoretical model given above to the experimental spectrum using the Levenberg-Marquardt method as given by Press et al (1992).

Fig. 7 shows GaN spectra acquired from the bulk crystal and the $\{1\bar{2}10\}$ boundary for 32 ms from a 30 nm thick part of the sample at 0.02 eV per channel dispersion using a Gatan imaging filter. The spectra have been scaled to the respective loss functions. To obtain statistically meaningful values for the band gap, spectra have been acquired at various different

220

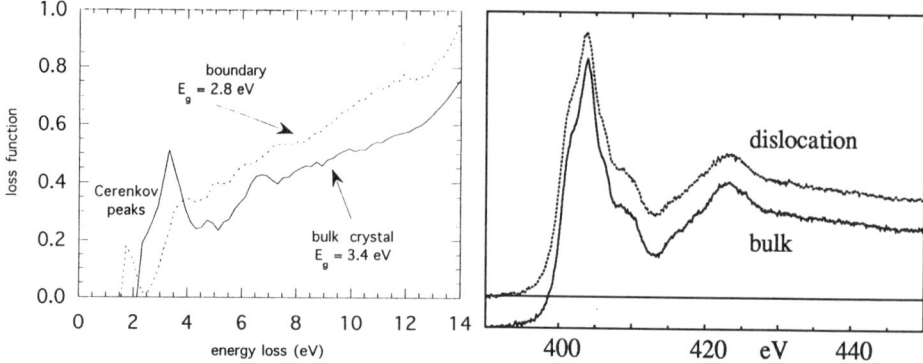

Fig. 7: Low loss EELS spectra acquired on and off a prismatic stacking fault in GaN on {1̄1̄1̄}B GaP

Fig. 8: Nitrogen K-edges acquired on and off dislocations in GaN grown on {1̄1̄1̄}B GaAs

regions in the bulk and on the boundary. Specimen drift has been corrected for and the spectrometer has carefully been kept in focus. No contamination effects were visible. The numerical fitting was performed in the region between 7 eV and 13 eV and yields a band gap of 3.4 ± 0.3 eV in the bulk and 2.8 ± 0.2 eV on the boundary. The limited accuracy in this measurement can be explained not only by the arguments given above but perhaps also by the presence of strain in the sample. The change in the band gap is likely to derive from a change in band structure. Possible segregation of phosphorus impurities present in the sample has not been found. In light of the limited spatial resolution in the low-loss region, a better approach would be to compare the shape of the nitrogen K-edge directly with the calculated band structure, as this is expected to give a resolution near 1nm.

A comparison of nitrogen K energy loss spectra acquired on and off dislocations in GaN grown on {1̄1̄1̄}B GaAs is shown in Fig. 8. There is no visible difference between the spectra, and this correlates with low loss spectra which also show no change between dislocations and bulk. This preliminary data thus suggests that the conduction band structure in GaN is altered in the vicinity of the {12̄10} planar boundaries but not in the vicinity of threading dislocations.

REFERENCES

Batson P E, Kavanagh K L, Woodall J M and Mayer J W 1986 Phys. Rev. Lett. **57** 2729

Bloom S, Harbeke G, Meier E, Ortenburger I B 1974 Phys. Stat. Sol. **66** 161

Cheng T S, Jenkins L C, Hooper S E, Foxon C T, Orton J W and Lacklison D 1995 Appl. Phys. Lett. **66** 1509

Drum C M 1964 Phil. Mag. **11** 313

Festenberg C 1969 Z. Phys. **227** 453

Lea C and Gomer R 1970 Phys. Rev. Lett. **25** 804

Natusch M K H, Botton G A and Humphreys C J 1997 these proceedings

Newcomb S B, Boothroyd C B and Stobbs W M 1985 J. Microsc. **140** 195

Ning X J, Chien F R, Pirouz P, Yang J W, Asif Khan M 1996 J. Mater. Res. **11** 580

Press W H, Teukolsky S A, Vetterling W T, Flannery B P 1992 Numerical Recipes 2nd ed. (Cambridge: Cambridge University Press) pp 683-7

Wu X H, Brown L M, Kapolnek D, Keller S, Keller B, DenBaars S P and Speck J S 1996 J. Appl. Phys. **80** 3228

Xin Y, Brown P D, Cheng T, Foxon C T and Humphreys C J 1997 these proceedings

Inst. Phys. Conf. Ser. No 157
Paper presented at Microsc. Semicond. Mater. Conf., Oxford, 7–10 April 1997
© 1997 IOP Publishing Ltd

STEM characterisation of MOVPE-grown (In, Ga) N quantum wells

G Brockt, C Mendorf, A Radefeld, F Scholz [+] and H Lakner

Werkstoffe der Elektrotechnik, Gerhard-Mercator-Universität Duisburg,
Bismarckstr. 81, 47048 Duisburg, Germany
[+]4.Physikalisches Institut, Universität Stuttgart, Pfaffenwaldring 57,
70569 Stuttgart, Germany

ABSTRACT: (Al, In, Ga) N heterostructures grown on sapphire substrate were characterised by scanning transmission electron microscopy (STEM). Here emphasis was put on quantitative analytical methods like Z-contrast, electron energy loss spectroscopy (EELS) and convergent beam electron diffraction (CBED). In addition to the characterisation of dislocations and interface properties by bright-field imaging fluctuations of the chemical composition were analysed by Z-contrast imaging. Strain or relaxation of the layers was quantified by CBED and corresponding simulations. EELS analysis in the low loss region was used for a spatially resolved investigation of band-gap properties.

1. INTRODUCTION

Recent improvements in metal organic vapour phase epitaxy (MOVPE) gave rise to major advances in the growth of group III nitrides. This material is of great interest for the development of high performance light emitting diodes and laser diode devices working in the blue and ultraviolet region. For the active region of laser structures ternary $In_xGa_{1-x}N$ quantum wells are suitable. Their bandgap and thus emitted wavelength is adjusted by the In-content in the layer. But there are problems regarding the growth of such layers. Especially the In incorporation seems to be hindered by the miscibility gap or the difference in the atomic radii of In and N. However the quality of these devices primary depends on the chemical homogeneity, interface sharpness and crystallographic properties of the individual layers. In this work these properties of MOVPE grown (Al, In, Ga) N heterostructures on sapphire substrates were investigated by analytical scanning transmission electron microscopy (STEM).

2. EXPERIMENTAL METHODS

Cross-sectional specimens of (In, Ga, Al), N heterostructures have been investigated by STEM. The heterostructures are grown by MOVPE on sapphire substrates at a temperature of 700^0 C (Scholz et al.). The microscope used for the described investigations is a cold field-emission STEM (VG Microscopes: HB501) equipped with a high-resolution pole piece. The minimum diameter of the electron probe is approximately 0.3 nm at an energy of 100 keV (Lakner[1] et al.). Bright-field imaging was used to analyse defects in the crystallographic structure of the specimen. Information on the variations of chemical composition and corresponding interface abruptness is provided by atomic number (Z)-contrast imaging. Here electrons scattered elastically and quasi-elastically in an angle range of 60-180 mrad are detected by a high-angle dark-field detector to form an image with contrast sensitive to the atomic number Z, for the image signal is proportional to Z^{α} ($2 \geq \alpha \geq 1,5$) (Lakner[2] et al). Information on the local crystalline structure is given by evaluation of CBED patterns. Especially the position of the High Order Laue Zone (HOLZ) lines in CBED patterns is sensitive to the local strain and can be used to determine the strain and relaxation quantitatively (Lakner[1] et al.). The analysis of the energy spectrum of electrons which have been scattered inelastically allows the analysis of the chemical composition quantitatively and applied to the low loss regime of the energy spectrum yields information about the electronic band structure of the specimen as well. For this purpose we made use of a parallel EELS system equipped with an Isaacson magnetic sector field followed by a magnetic three-quadrupole optic (McMullan). This device yields an energy resolution of less than 0.5 eV and thus allows effective applications in the bandgap region of the energy loss spectrum.

3. RESULTS AND DISCUSSION

3.1 Z-contrast imaging

The STEM analysis results of $In_{0.12}Ga_{0.88}N$ quantum wells of 2nm and 17nm width respectively are presented here. The quantum wells are embedded in GaN layers. To overcome lattice mismatch to a sapphire substrate the GaN is adapted by an AlN nucleation layer. This layer structure can be recognised in the bright field image of the 17nm quantum well in Fig.1a). Additionally, lattice mismatch induced defects can be observed here. They origin at the substrate interface and propagate in growth direction crossing every layer including the InGaN quantum well.

As well this (Al, In, Ga) N heterostructure was analysed by Z-contrast imaging using different specimen orientations as shown in Fig. 1. Corresponding to three different orientations bright-field images, Z-contrast images, Z-contrast linescans and in order to demonstrate the chosen orientation CBED patterns are shown.

Fig. 1a), b) and c) show the results recorded in {120} pole orientation. In the Z-contrast image of Fig. 1b) a striking contrast between the InGaN quantum well and GaN barriers is visible as well as in the case of the AlN nucleation layer and the sapphire substrate. Additionally, the dislocations can be observed here extensively.

When changing the specimen orientation to planar channelling conditions the behaviour of Z-contrast becomes quite different. Fig. 1f) points out the specimen tilt in the direction of the layers following the CBED pattern along the [200]-Kikuchi band. In this specimen orientation the Z-contrast image and the bright-field image show reduced contrast of the dislocations crossing the GaN barriers and the active region. But additional contrast of dislocations in the transition area of the AlN nucleation layer to the GaN becomes visible.

Furthermore in the Z-contrast of Fig. 1e) a lack of contrast between the GaN layer and the InGaN quantum well compared to that of the Z-contrast image recorded in on-pole conditions, Fig. 1b), is observable. This fact can be made clear by a comparison of the corresponding linescans in Fig. 1b) and e). Additionally, exclusive in the Z-contrast image recorded at planar channelling conditions a dark hue to both sides of the quantum well can be seen. This phenomenon,which is partly visible in the bright field-image as well, reveals an area of strained material where the lattice constant of the GaN has to adapt to that of InGaN.

Bright-field images   Z-contrast images

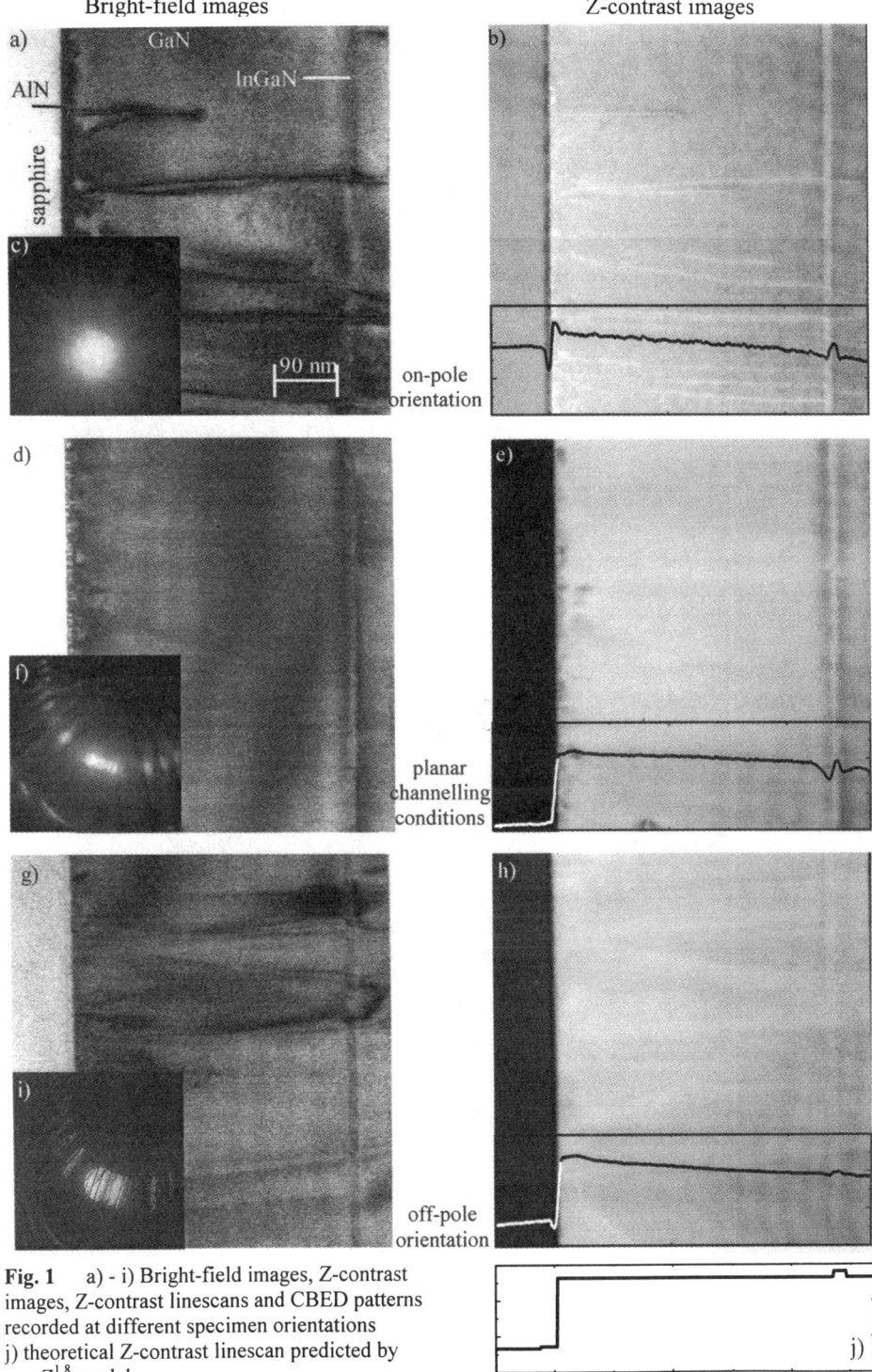

Fig. 1   a) - i) Bright-field images, Z-contrast images, Z-contrast linescans and CBED patterns recorded at different specimen orientations
j) theoretical Z-contrast linescan predicted by
a $Z^{1.8}$-model

The striking change in the Z-contrast brightness of the sapphire substrate when changing the orientation from on-pole to planar channelling conditions is due to the fact this substrate is rotated by 30^0 degree regarding to the other layers, which has been demonstrated before (Lakner[3] et al.).

The third series in Fig. 1g) - i) represents the results visible when additionally tilting the specimen off the [200] Kikuchi Band so that channelling effects are suppressed. Under these conditions the Z-contrast image no longer reveals dislocations in the AlN/GaN transition area and the contributions from the strained areas in the region of the InGaN quantum well vanish as well.

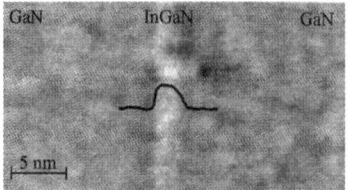

Fig.2 Z-contrast image and linescan of a 2nm InGaN quantum well

Furthermore this off-pole condition shows reduced Z-contrast between the InGaN quantum well and the surrounding GaN layers. Again this can be verified from the plotted linescans of Fig. 1. In addition Fig. 1j) shows a theoretical line scan predicted by the Z^α-model for the case of this specimen and $\alpha=1.8$. The comparison with the shown linescans from the Z-contrast images recorded at planar channelling conditions and on-pole conditions demonstrate that the Z-contrast at off-pole conditions yields the best agreement with the predictions of the Z^α-model. This result is what one would expect, for the Z^α-model is based on interactions of an electron beam with individual scattering centres. The interactions of an electron beam passing a specimen at conditions where channelling is suppressed are more similar to that assumption, whereas the transmission of an electron beam passing under channelling conditions is dominated by interactions of the beam electrons with the lattice structure of the specimen. These interactions are primary influenced by the heavy atoms of the lattice. Therefore the InGaN/GaN contrast at on-pole conditions might be enhanced over that at off-pole conditions, for these layers differ in the composition of their heavy atoms.

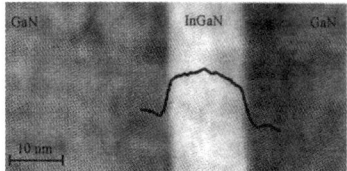

Fig.3 Z-contrast image and linescan of a 17nm InGaN quantum well

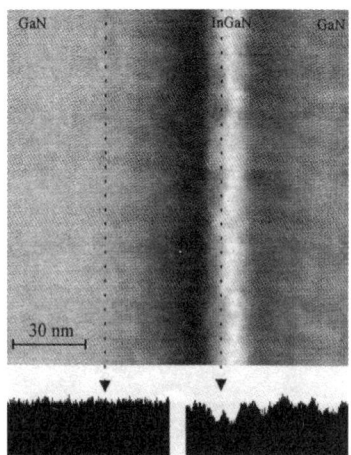

Fig.4 Z-contrast image and linescans of a 17nm In GaN quantum well and a GaN layer

At higher magnifications the Z-contrast images of Fig. 2 and 3 show material contrast within the quantum wells. In the case of this ternary material the contrast results from fluctuations in the In content of the active layer. Especially, the interface regions of the quantum wells appear to be asymmetric. In both cases the lower interface in growth direction is more abrupt than the upper one, where a grading in the In concentration is significant. The observed grading results from an evaporating of In when rising the temperature again after the growth of the InGaN quantum well for continuing with the GaN layer.

Additionally, Z-contrast imaging reveals fluctuations in the direction parallel to the interfaces. The example in Fig. 4 shows strong brightness variations within the active layer due to fluctuations of the In content. The magnitude of these fluctuations becomes clear by a comparison of linescans taken from the quantum well and the neighbouring GaN layer, Fig.

4b). From a quantitative analysis of these data the fluctuations of the nominal In-content of $x = 0.12$ within the quantum well can be deduced to be in the range of $\Delta x = 0.03$ which is in agreement with previous EDX results (Narukawa).

In summary it can be concluded that Z-contrast reveals information about the chemical composition corresponding to the $Z^{1.8}$ model on the one hand. But on the other hand experimental conditions which result in deviation from this model due to channelling effects can yield useful information for the analysis of the chrystallographic structure.

3.2 CBED analysis

In addition to the chemical composition we investigated the structural properties of the quantum wells by CBED. Fig.5 shows the CBED patterns of the 2nm quantum well and the neighbouring GaN layer recorded at planar channelling conditions. The patterns differ clearly in a shift of the HOLZ lines in the marked regions.

These CBED patterns have been simulated kinematically, as shown in Fig.6, in order to determine the lattice constants a and c of the wurzite structure. A comparison of the experimental and simulated patterns yields the local lattice constants a and c to be a = 0.319nm, c = 0.519nm for the GaN layer and a = 0.319nm, c = 0.523nm for the InGaN quantum well.

Theoretical values of the relaxed lattice constants can be calculated regarding to Vergard's law to be $a_{rel} = 0.319$nm, $c_{rel} = 0.519$nm for GaN and $a_{rel} = 0.322$nm, $c_{rel}= 0.524$nm for $In_{0.12}Ga_{0.88}N$. Out of this data it can be concluded that within the 2nm quantum well the cell unit shows a relative strain of $\Delta a/a_{rel} = 9.31*10^{-3}$.

For the case of the 17nm quantum well the lattice constants were measured to a = 0.323nm and c = 0.527nm. These data reveal no hints for strained material for the lattice constants are in good agreement

a)

b)

Fig. 5 Experimental CBED patterns
a) GaN layer and
b) 2nm InGaN quantum well

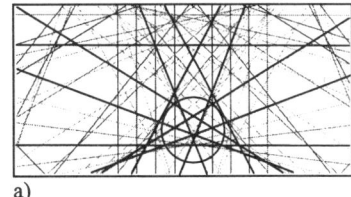
a)

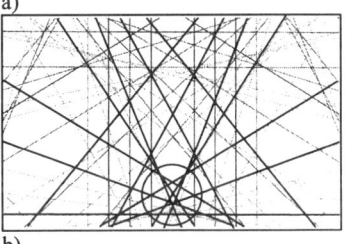
b)

Fig. 6 Simulated CBED pattern
a) GaN and
b) InGaN wurzite structure

to their theoretical values, but give evidence of a relaxed layer. This result was proved as well by the application of convergent beam imaging (CBIM) technique. In CBIM the defocused probe is used to obtain a diffraction pattern where a shadow image of the real structure is superimposed. In the recorded patterns, Fig. 7, we did not find any hints for relaxation in the case of the 2nm quantum well. But in Fig. 8, which shows detail of the pattern recorded from the 17nm quantum well, a bending of the HOLZ lines is visible demonstrating that the lattice of this quantum well is relaxed.

226

3.3 EELS analysis

Additionally EELS was applied to analyse the low loss region of the energy loss spectrum (0-30 eV). This region of the energy loss spectrum reveals information about the properties of the outer shells of the investigated atoms.

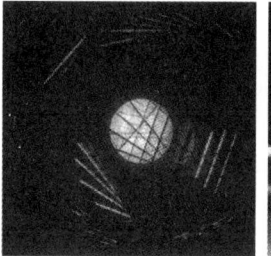 

Fig. 7 CBIM image of 2nm quantum well **Fig. 8** CBIM image of 17nm quantum well

Fig. 9 shows the low loss spectra of AlN and GaN. In comparison to the well knownplasmon-energy of sapphire (E_P = 26 eV) the plasmon-energy of AlN is located at 21 eV. The plasmon-energy of GaN could be determined as 19.5 eV. Furthermore the band-gap of these semiconducting materials could be measured spatially resolved. Thus the spectra in Fig 9 show additional features on the fading end of the zero-loss peak. These rises in the intensity of the energy loss spectrum result from interband transitions. Their starting points on the energy scale are in excellent agreement with the theoretical band-gap energies of AlN ($E_{g(AlN)} = 6.2$ eV) and GaN ($E_{g(GaN)} = 3.5$ eV).

These first results demonstrate the capability of EELS for a spatially resolved analysis of wide-band materials regarding their electronic and optical properties.

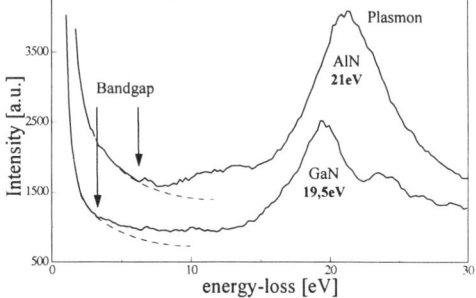

Fig. 9 EEL spectra of the low-loss region

In summary, the described STEM techniques are powerful inspection tools for the heterostructure and interface characterisation with respect to local crystal properties, spatial distribution of chemical composition and band-structural properties. Its application to (Al, In, Ga) N heterostructure revealed useful information regarding the results and necessary improvements of the growth process.

4. ACKNOWLEDGEMENTS

Parts of this work were financially supported by the Deutsche Forschungsgemeinschaft (Schwerpunkt Gruppe III Nitride) and the Volkswagen-Stiftung.

REFERENCES

Lakner[1] H, Maywald M, BalkL J, Kubalek E 1992 Surf. and Interf. Anal. **19**, 374
Lakner[2] H, Bollig B, Ungerechts S and Kubalek E 1996 J. Phys. D: Appl. Phys. **29,** 1767
Lakner[3] H, Brockt G, Mendorf C 1997 to be published in J. Electr. Mat.
McMullan D, Fallon P J, Ito Y and McGibbon A J 1992 Electron Microscopy **I** EUREM
Narukawa Y, Kawakami Y, Shizuo F, Proceedings of MRS Fall Meeting 1996 Boston
Scholz F,. HärleV, Streuber F, Sohnmer A, H,Bolay, V.Syganow, A.Dörnen, J.S. Im,
        A.Hangleiter, J.Y. Duboz, P.Galtier, E.Rosencher, O.Ambacher, D.Brunner,
        H.Lakner,Proc. MRS Fall Meeting 1996 Symposium N (III-V Nitrides)
I-hisu Jo, Stringfellow G B 1996 Appl. Phys. Lett. **69**, 2701

Inst. Phys. Conf. Ser. No 157
Paper presented at Microsc. Semicond. Mater. Conf., Oxford, 7–10 April 1997
© 1997 IOP Publishing Ltd

TEM characterisation of GaN grown on sapphire

B Pécz, MA di Forte-Poisson*, L Tóth and G Radnóczi

Research Institute for Technical Physics of the Hungarian Academy of Sciences, H-1325, Budapest, PO Box 76. Hungary
* THOMSON-CSF/LCR, Domaine de Corbeville 91404 Orsay Cedex France

ABSTRACT: The defect structure of GaN layers on (0001) sapphire substrates grown by MOCVD is studied in this paper by TEM. GaN layers were deposited onto exactly oriented and miscut substrates as well. HREM images revealed an AlN layer at the interface formed in the nitridization process of sapphire. The hexagonal GaN layers contain high density of defects like misfit dislocations, threading dislocations and planar defects. The role of the GaN nucleation layer deposited at lower temperature is discussed.

1. INTRODUCTION

GaN is extensively studied as a promising material for high temperature and optoelectronic devices. MOCVD (Metalorganic Chemical Vapour Deposition) growth of GaN onto sapphire is usually carried out by a two step procedure (Amano et al 1986 and Nakamura 1991). Very often the surface of sapphire is nitridated before the two step growth.

Although the misfit in this system is high (13.8%) it is possible to grow GaN layer onto sapphire with specular surfaces. In this paper the defect structure of flat GaN layers grown onto sapphire is discussed.

2. EXPERIMENTAL

1.4 μm and 0.88 μm thick GaN layers were grown onto (0001) sapphire by MOCVD using a two step procedure at 600 mbar pressure. At first a nucleation layer was deposited at 510°C onto both exact and miscut (3.5° off toward the $\overline{2}110$ axis) substrate. The growth temperature of the GaN layer varied between 960 and 1000°C. Before the two step growth procedure the nitridization of the sapphire substrate was carried out introducing ammonia into the reactor. Triethylgallium (TEG) and ammonia (NH_3) were used as Ga and N sources.

Cross sectional and plan view specimens were prepared for conventional and high resolution transmission electron microscopy (HREM) by ion milling thinning by 10 keV Ar^+ ions at low angle of incidence by the method described by Barna et al (1997). As a final step of preparation a new surface cleaning procedure with low energy Ar^+ ions was also applied as described in another paper in this volume (Barna et al 1997b).

High resolution electron microscopy was carried out in a JEOL 4000 EX TEM at the Electron Microscopy Department of University of Antwerp, Belgium. A Philips CM20 TEM was used at 200 kV for the conventional electron microscopy.

3. RESULTS AND DISCUSSION

GaN layers with specular surfaces were grown at 990°C onto exactly oriented (0001) sapphire surfaces. When the temperature of growth was decreased to 960°C the surface became rough. Mirror like surfaces could be obtained at 960°C on miscut substrates. The GaN layers are hexagonal and epitaxial to the sapphire substrate. The orientational relationships are the following: GaN [10$\overline{1}$0]//Al$_2$O$_3$ [11$\overline{2}$0]

GaN [0001]//Al$_2$O$_3$ [0001].

Although the FWHM data of the high resolution X-ray measurements show the increasing of defect density on the miscut substrates (typical FWHM data are: 450" for the (0002) peak and 230" for the (11$\overline{2}$4) peak) the range of the optimal deposition parameters can be extended by using miscut sapphire substrates.

A cross section of a typical GaN/sapphire sample grown at 960°C onto miscut substrate is shown in Fig. 1. The buffer layer of this sample was 22.5 nm thick. The GaN layer is homogeneous in thickness.

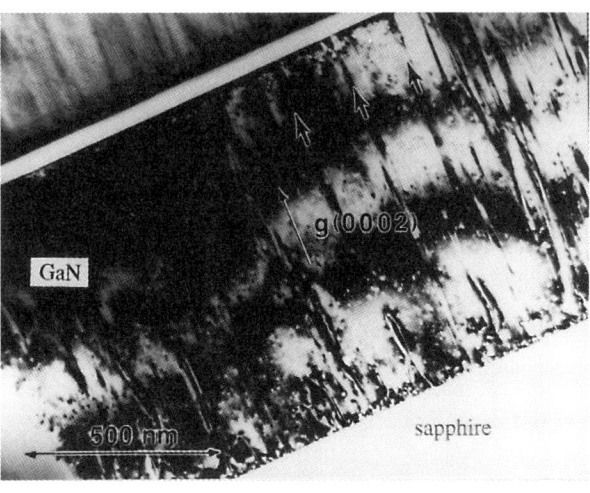

Fig.1. Cross section of GaN/sapphire sample grown at 960°C onto miscut substrate.

Fig. 2. Weak beam dark field image of the GaN layer.

Dislocation loops can be observed in the interface region. Lines of dislocations can be observed more clearly on the weak beam dark field image taken with the (0004) reflection of the GaN (Fig. 2). Beside the dislocations straight lines of planar defects can be observed in Fig. 1. (some of those are marked by arrows) which run perpendicular to the interface and extend to the surface. The latter type of defects was observed by Rouviére at al, 1995 and 1996 and

Smith et al 1997 and were described as inversion domain boundaries (IDB). In the low temperature grown GaN buffer layer we observed many steps, basal dislocations and planar defects (Fig. 3). Misfit dislocations can be seen in Fig. 3 clearly.

Fig. 4 shows a plan view bright field image of the sample surface with the threading dislocations. Dislocations are arranged in walls, separating slightly misoriented cells and originate from the nucleation layer. This observation is in agreement with the results of Ning et al 1996.

Fig. 5. shows regular shaped hexagonal domains on the top surface of GaN layer. The domain boundaries correspond to the planar defects marked by arrows in Fig.1 shown in cross section. Our HREM investigations show that those features represent hexagonal rods running through the whole layer and are formed always on steps in the GaN buffer layer.

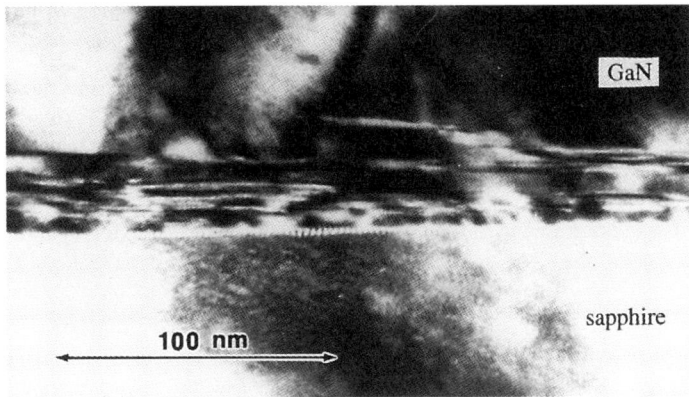

Fig. 3 Misfit, basal dislocations and planar defects at the interface region (g=10$\overline{1}$0).

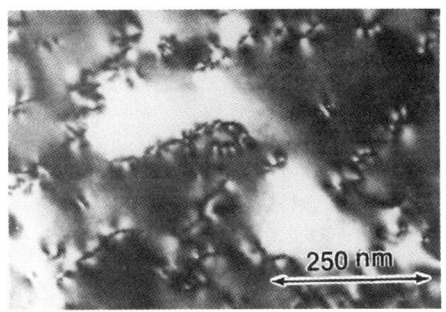
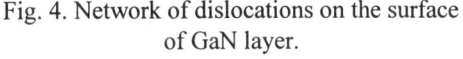
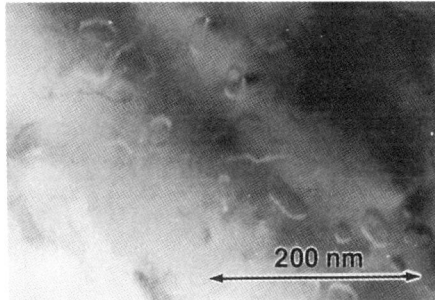

Fig. 4. Network of dislocations on the surface of GaN layer.

Fig. 5. Inversion domain boundaries on the surface of GaN layer.

High resolution images were taken from the GaN/sapphire interface region. The images show a region about 4 nm thick at the interface which is hexagonal AlN. Within this layer there are small (15-20 nm long) grains of cubic AlN. The observed AlN was formed probably by a reaction of the substrate with the nitrogen precursor. This observation agrees well with the findings of Yamamoto et al 1994 and Kawakami et al 1988, who have observed the formation of AlN in the nitridization process of sapphire, and disagrees with the results of Uchida et al 1996, 1997 who found the formed AlN layer to be amorphous. The oxygen content of the AlN layer (Uchida et al 1996, 1997) formed during the nitridization process might be an explanation of the above discrepancy.

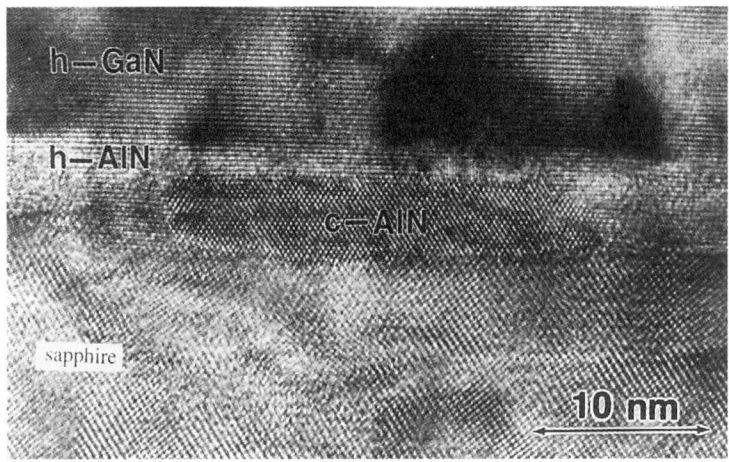

Fig. 6 High resolution image of the interface region showing the AlN layer formed in the nitridization process of the substrate.

4. CONCLUSION

The two step growth of GaN on nitridated sapphire surface always means the growth of GaN on an AlN layer formed in the nitridization process of the sapphire substrate.

ACKNOWLEDGEMENT

One of the authors (LT) expresses his gratitude to the Flamish and Hungarian Academies of Sciences whose cooperation made it possible to carry out the present HREM studies by sponsoring his stay at Antwerp University in the frame of the Inter-Academic Exchange Program. The work was partly supported by the PHARE TDQM project No.: HU-9305-02/1086.

REFERENCES

Amano H, Sawaki N, Akasaki I and Toyoda Y 1986 Appl. Phys. Lett. 48, 353
Barna Á, Radnóczi G and Pécz B 1997a Handbook of Microscopy ed. S. Amelinckx et al, (VCH Verlag, Weinheim, 1997) Vol. 3 pp 751-801
Barna Á, Tóth L, Pécz B, Radnóczi G 1997b *paper in this volume*
Kawakami H, Sakurai K, Tsubougchi K and Mikoshiba N 1988, Jap. J. Appl. Phys. L161
Nakamura S 1991 Japanese J. of Appl. Phys. 30, L1705
Ning XJ, Chien FR, Pirouz P, Yang JW and Khan MA 1996 J. Mater. Res. 11, 580
Rouviére J-L, Arlery M, Bourret A, Niebuhr R and Bachem K 1995 Inst. Phys. Conf. Ser. 146, 285
Rouviére J-L, Arlery M, Bourret A, Niebuhr R and Bachem K 1996 Mat. Res. Symp. Proc. 395, 393
Smith DJ, Tsen S-CY, Sverdlov BN, Martin G and Morkoc H 1997 Solid-State Electr., 41, 349
Uchida K, Watanabe A, Yano F, Kouguchi M, Tanaka T and Minigawa S 1996 J. Appl. Phys. 79, 3487
Uchida K, Watanabe A, Yano F, Kouguchi M, Tanaka T and Minigawa S 1997 Solid-State Electronics, 41, 135
Yamamoto A, Tsujino M, Ohkubo M, Hashimoto A 1994 J. Crystal Growth 137, 415

Inst. Phys. Conf. Ser. No 157
Paper presented at Microsc. Semicond. Mater. Conf., Oxford, 7–10 April 1997
© 1997 IOP Publishing Ltd

On the microstructure of GaN buffer layers grown at low temperatures on (0001) sapphire

H Selke, S Einfeldt[1], U Birkle[1], D Hommel[1] and P L Ryder

Institut für Werkstoffphysik und Strukturforschung, [1]Institut für Festkörperphysik,
Universität Bremen, D-28334 Bremen, Germany

ABSTRACT: A GaN buffer layer grown by molecular beam epitaxy at 500 °C on basal plane sapphire has been studied by reflection high-energy electron diffraction and high-resolution electron microscopy after subsequent growth of a GaN epilayer at 850 °C. Both the buffer layer and the epilayer showed mainly the wurtzite structure, with about 15% zinc blende. Most grains had the usual epitaxial orientation relationship – with a small scatter, leading to low-angle grain boundaries – but about 15% in the buffer showed random rotations about the growth direction. The growth of the rotated grains stopped in the epilayer.

1. INTRODUCTION

The optical and electrical properties of GaN epilayers grown on sapphire, such as band-gap photoluminescence, background carrier concentration and mobility, can be significantly improved by the deposition of AlN or GaN buffer (nucleation) layers at low temperatures prior to the epilayer growth (see Morkoç el al. (1994) for a review). Amano et al. (1986) and Akasaki et al. (1989) studied the surface morphology of AlN buffer layers and its influence on the GaN epilayer grown by metal organic vapour phase epitaxy (MOVPE). Akasaki et al. (1989) developed a model for the growth of GaN on buffer layers, which was later modified by Hiramatsu et al. (1991) on the basis of additional transmission electron microscopy (TEM) experiments. According to this model, the function of the buffer layer is to provide a high density of nucleation sites. The grains are initially randomly orientated, and the dominance of grains with [0001] parallel to the growth direction is produced by 'geometrical selection', since [0001] is the fastest growth direction. Qian et al. (1995) and Kapolnek et al. (1995) studied the structure of MOVPE grown GaN films in great detail using TEM. The films were grown on basal plane sapphire covered with AlN or GaN buffer layers. A fixed epitaxial relationship was always found between the sapphire (S) and the wurtzite (W) GaN film: $(0001)_S \parallel (0001)_W$ with $[2\bar{1}\bar{1}0]_S \parallel [\bar{1}010]_W$. They reported low-angle subgrain boundaries formed by pure edge dislocations parallel to $[0001]_W$ with Burgers vectors $\boldsymbol{b} = 1/3 \langle 2\bar{1}\bar{1}0 \rangle$. Kapolnek et al. (1995) suggested that these dislocations are formed by copying dislocations present in the buffer layer during the process of epitaxial growth. It seems that the structure of the buffer layer itself is strongly affected by the growth conditions, because sapphire [0001] is parallel to the [0001] wurtzite and [111] zinc blende stacking direction, respectively. Thus, sapphire (0001) is polytype neutral (Morkoç 1994). Akasaki et al. (1989) reported that the buffer layer is amorphous or hexagonal, whereas Kapolnek (1995) found a predominantly cubic structure.

A more detailed analysis of the microstructure of buffer layers and of its influence on the subsequent growth of GaN is required. Here we report on the results of a TEM study of the microstructure of a GaN buffer layer grown on (0001) sapphire by molecular beam epitaxy (MBE).

2. EXPERIMENTAL PROCEDURE

The sample was grown by MBE on a nitridated sapphire (0001) substrate using an electron cyclotron resonance plasma source. An approximately 15 nm thick GaN buffer layer was deposited at about 500 °C. The temperature was then raised to about 850 °C to grow an approximately 650 nm thick GaN epilayer. The growth was monitored using reflection high-energy electron diffraction (RHEED). Cross-sectional and plan-view specimens were prepared for TEM by ion milling using Argon at 4 kV. The TEM and high-resolution TEM (HRTEM) experiments were carried out on a Philips EM420 operating at 120 kV and a Philips CM20/UT operating at 200 kV, respectively.

3. RESULTS

RHEED experiments carried out during growth of the buffer layer showed sharp reflections, indicating a crystalline, three-dimensional buffer structure, with the usual epitaxial relationship mentioned above. Initially there were some grains with different orientations, but these disappeared during growth of the GaN epilayer at higher temperatures. Electron diffraction experiments carried out after growth in the TEM confirmed the usual epitaxial relationship between the substrate (S) and the wurtzite epilayer (W), i.e. $(0001)_S \parallel (0001)_W$ and $[2\bar{1}\bar{1}0]_S \parallel [\bar{1}010]_W$. However, the HRTEM image taken from the sapphire/buffer layer interface in Fig. 1a indicates the existence of regions with different orientations with respect to the substrate. The image of the region labelled A in Fig. 1a corresponds with the usual epitaxial relationship, as can be seen by comparison with the contrast simulation (inset) calculated for $[\bar{1}010]_W$ parallel to the electron beam and $[0001]_W$ perpendicular to the substrate interface. In this case only the $(0002)_W$ lattice fringes of the wurtzite structure are resolved. The structural image of the other region, labelled B in Fig. 1a, significantly differs from that of region A. It is very similar to the image calculated for the case that $[2\bar{1}\bar{1}0]_W$ is parallel to the electron beam, with the orientation of the c-axis unchanged, which is shown as an inset in that region. It is also evident from this figure that the growth of this rotated grain stops a few nanometres away from the substrate interface. This was also found for other crystallites deviating from the usual epitaxial relationship. To study the structure of the buffer layer in more detail, plan-view specimens of the buffer layer/sapphire interface were prepared. Fig. 1b shows a typical HRTEM image exhibiting crystallites in the buffer, which are typically 10 to 40 nm in diameter. In this case both the buffer layer and the sapphire substrate are penetrated by the electrons giving rise to the Moiré patterns observed in the image. Variations in the Moiré patterns indicate different orientations of the individual wurtzite grains in the buffer. The regions labelled A, which cover approximately 70 % of the sapphire, correspond to the usual epitaxial relationship, i.e. $(0001)_S \parallel (0001)_W$ with $[2\bar{1}\bar{1}0]_S \parallel [\bar{1}010]_W$. The region B has $(0001)_S \parallel (0001)_W$ and

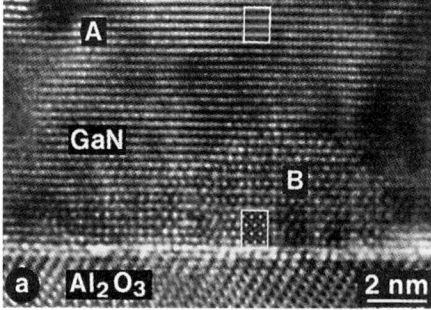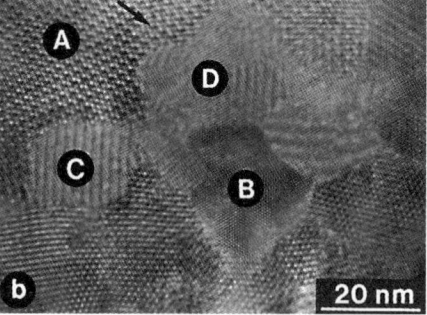

Fig. 1: HRTEM images of the sapphire/buffer layer interface: (a) in cross-section taken along $[2\bar{1}\bar{1}0]_S$ and (b) in plan-view taken along $[0001]_S$. Comparison with the calculated contrast simulations (insets) in (a) indicates different orientations of the wurtzite GaN with respect to the sapphire. This is also evident from variations in the Moiré patterns in (b). A 'dislocation' in the Moiré pattern in (b) is marked by an arrow. Labels A, B, C and D see text.

$[2\overline{1}\,\overline{1}0]_S \parallel [2\overline{1}\,\overline{1}0]_W$, as was also observed in Fig. 1a. There are other crystallites whose orientations could not be identified by using the Moiré patterns. Most probably this is due to a small tilt of $[0001]_W$ with respect to $[0001]_S$ (C) and for some other reason (D) pointed out below. Many 'dislocations' are observed mainly in the type A Moiré patterns. A close inspection using digital Fourier filters indicates that they are due to the presence of low-angle grain boundaries in the wurtzite GaN. These are formed by dislocations running nearly parallel to $[0001]_W$ with Burgers vectors of the type $\boldsymbol{b} = 1/3 \langle 2\overline{1}\,\overline{1}\,l \rangle$. The value of l could not be determined from this orientation, but the results are consistent with the usual Burgers vectors in close-packed hexagonal structures, i.e. $\boldsymbol{b} = 1/3 \langle 2\overline{1}\,\overline{1}0 \rangle$ or $\boldsymbol{b} = 1/3 \langle 2\overline{1}\,\overline{1}3 \rangle$.

The selected area diffraction pattern shown in Fig. 2, which was taken along $[0001]_S$ from a large area similar to that in Fig. 1, gives some insight in the possible orientations of the grains in the buffer layer. In addition to the diffraction spots of the $[0001]_S$-zone those of the $[0001]_W$-zone with $[2\overline{1}\,\overline{1}0]_S \parallel [\overline{1}010]_W$ are observed. The wurtzite spots are not sharp indicating a small scatter in this epitaxial relationship up to about ±3 ° due to the low-angle grain boundaries mentioned earlier. Moreover, a ring pattern is observed. Since all rings can be indexed as reflections of the $[0001]_W$-zone there must be some grains for which the c-axes are again parallel to that of the sapphire but there is no fixed orientation relationship in the $(0001)_S$ plane. From dark field experiments it was estimated that about 10 to 15 % of the sapphire is covered with these rotated grains.

Apart from the wurtzite GaN, there is also a appreciable proportion of zinc blende GaN present in the buffer layer. This may give rise to the Moiré patterns (D) in Fig. 1b. The crystallites with the zinc blende (ZB) structure exhibit in most cases the epitaxial relationship $(0001)_S \parallel (111)_{ZB}$ and $[2\overline{1}\,\overline{1}0]_S \parallel [11\overline{2}]_{ZB}$. About 10 to 15 % of the crystallites in the buffer possess the zinc blende structure as was estimated from dark field images.

In order to obtain information on the structure of the interface between the buffer layer and the epilayer, the TEM specimen was carefully thinned further from the sapphire side. In the GaN epilayer the proportion of the rotated grains decreases dramatically, but some grains can still be observed close to the buffer layer interface. This is exemplified by Fig. 3a which shows a grain rotated approximately 20 ° around $[0001]_W$. The Moiré pattern observed at the grain boundary indicates that this grain is partly covered by the surrounding GaN, which shows the usual epitaxial relationship. Other rotated grains, which were already completely overgrown could be observed in the plan-view specimen as well (compare also region B in Fig. 1a). It is noteworthy that this coverage generally proceeds without the formation of additional defects. However, low-angle grain boundaries as observed in the buffer layer are still present in the epilayer as can be seen in Fig. 3b. As in the buffer layer, the bounding dislocations run more or less perpendicular to the substrate interface. The Burgers vectors, determined directly from the HRTEM images, were again consistent with $\boldsymbol{b} = 1/3 \langle 2\overline{1}\,\overline{1}0 \rangle$ or $\boldsymbol{b} = 1/3 \langle 2\overline{1}\,\overline{1}3 \rangle$.

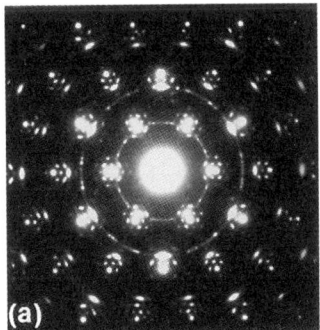 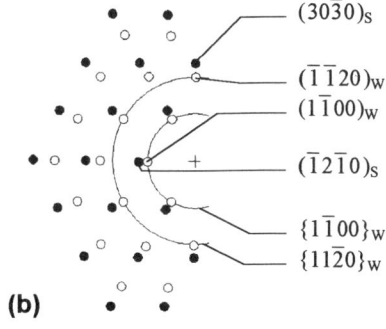

Fig. 2: (a) Selected area diffraction pattern of an area similar to Fig. 1b taken along $[0001]_S$ and (b) indexing. Filled and open circles represent sapphire and wurtzite GaN reflections with $[2\overline{1}\,\overline{1}0]_S \parallel [\overline{1}010]_W$, respectively. Ring patterns are also indicated. Reflections not shown in (b) are due to double diffraction.

Labels in (b): $(30\overline{3}0)_S$, $(\overline{1}\,\overline{1}20)_W$, $(1\overline{1}00)_W$, $(\overline{1}2\overline{1}0)_S$, $\{1\overline{1}00\}_W$, $\{11\overline{2}0\}_W$

234

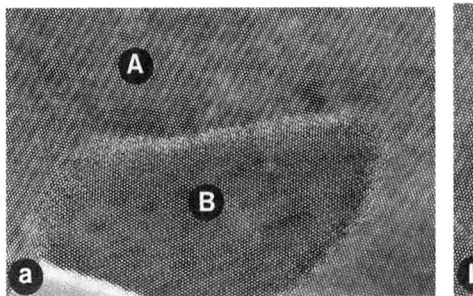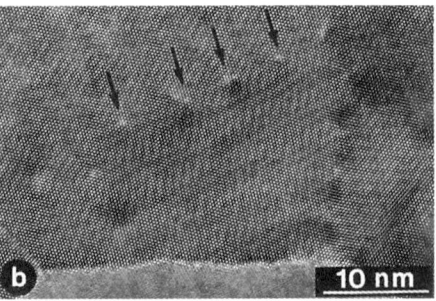

Fig. 3: HRTEM plan-view images taken along $[0001]_W$ of regions in the GaN epilayer located close to the buffer layer/epilayer interface: (a) Normal GaN orientation (A) surrounding a grain (B) rotated 20° about $[0001]_W$. (b) Low-angle grain boundary with associated dislocations (some marked by arrows) running parallel to $[0001]_W$. The Burgers vectors are of the type $\mathbf{b} = 1/3 \langle 2\bar{1}\bar{1}\,l \rangle$, $l = 0, 3$). Same scale in both figures.

Other defects running parallel the $\langle \bar{1}010 \rangle_W$-directions have not yet been identified. Most probably these are dislocations with $\mathbf{b} = 1/3 \langle 2\bar{1}\bar{1}\,l \rangle$, which may relax the lateral strain. Zinc blende GaN was also found in the GaN epilayer in a comparable proportion to that in the buffer layer.

4. SUMMARY AND CONCLUSIONS

The GaN was found to have mainly the wurtzite structure, but about 10-15% of the grains in the buffer layer and the epilayer had the zinc blende structure. All grains had $[0001]_W$ or $[111]_{ZB}$ parallel to the growth direction $[0001]_S$ with a little scatter. Most grains also showed a fixed epitaxial relationship in the $(0001)_S$ plane with $[2\bar{1}\bar{1}0]_S \parallel [\bar{1}010]_W$ or $[11\bar{2}]_{ZB}$, the remaining grains showing random rotations about the growth direction. The rotated grains make up about 15% of the buffer layer, but their proportion decreases rapidly with distance from the interface to the epilayer. The driving force for this process is thought to be the interfacial energy of the boundary between the rotated and the 'normal' grains. The epitaxial orientation relationships showed a scatter of around ±3 ° corresponding to the presence of subgrain boundaries formed by dislocations originating in the buffer layer and extending in the growth direction. In the wurtzite phase the Burgers vectors of the dislocations were $\mathbf{b} = 1/3 \langle 2\bar{1}\bar{1}0 \rangle$ or $\mathbf{b} = 1/3 \langle 2\bar{1}\bar{1}3 \rangle$. A more systematic study of the buffer layer structure as a function of growth parameters is in progress.

REFERENCES

Akasaki I, Amano H, Koide Y, Hiramatsu K and Sawaki N 1989 J. Crystal Growth **98**, 209
Amano H, Sawaki N, Akasaki I and Toyoda Y 1986 Appl. Phys. Lett. **48**, 353
Hiramatsu K, Itoh S, Amano H, Akasaki I, Kuwano N, Shiraishi T and Oki K 1991 J. Crystal Growth **115**, 628
Kapolnek D, Wu X H, Heying B, Keller S, Keller B P, Mishra U K, DenBaars S P and Speck J S 1995 Appl. Phys. Lett. **67**, 1541
Morkoç H, Strite S, Gao G B, Lin M E, Sverdlov B and Burns M 1994 J. Appl. Phys. **76**, 1363
Qian W, Skowronski M, De Graef M, Doverspike K, Rowland L B and Gaskill D K 1995 Appl. Phys. Lett. **66**, 1252

Inst. Phys. Conf. Ser. No 157
Paper presented at Microsc. Semicond. Mater. Conf., Oxford, 7–10 April 1997
© 1997 IOP Publishing Ltd

Hexagonal growth hillocks of MOCVD-grown GaN on (0001) sapphire

A Mohammed, C Trager-Cowan, P G Middleton, K P O'Donnell*
W Van Der Stricht, I Moerman and P Demeester**

*Dept. Physics and Applied Physics, University of Strathclyde, Glasgow G4 0NG, Scotland, UK.
**IMEC-INTEC, University of Gent, Gent 9000, Belgium.

ABSTRACT: Hexagonal growth hillocks in GaN epilayers grown by MOCVD on sapphire have been studied by scanning electron (SE), atomic force and optical microscopy. The SE micrographs obtained at incident electron energies ≥ 15 keV show pyramid-like and plate-like hillocks of hexagonal cross-section, while smaller hexagonal holes/pittings are observed at lower incident electron energies. Again at the lower electron energies, some of the plate-like hexagonal hillocks disappear from the image, leading us to surmise that these features are sub-surface. Monte-Carlo simulations of electron beam solid interactions provide a way of interpreting the SEM images.

1. INTRODUCTION

Gallium nitride (GaN) has attracted a great deal of interest due to its potential and actual device applications in the visible to ultraviolet wavelength region. Recently, high-performance violet, blue and blue-green light emitting diodes (LEDs) have been fabricated (Asaki & Amano 1994; Nakamura et al 1994a & b) and Nakamura 1997 has recently reported a blue injection laser with a life time of 35 hours. GaN also finds application in the fabrication of high-frequency, high-temperature transistors (Khan et al 1995). Research is ongoing on growing GaN on lattice-matched substrates (e.g., Nicholls et al 1996), however to date sapphire, mismatched by 13%, remains the favoured substrate for GaN epilayer growth.

The high quality gallium nitride films in the present work were prepared by a two-step growth technique (Nakamura 1991), in which growth of a low temperature buffer layer is followed by that of a high temperature film. The films exhibit a predominantly smooth morphology but small groups of hexagonal hillocks are found to populate the surface, particularly at the sample edges. In this paper, we report studies of these hexagonal hillocks using scanning electron (SE), atomic force and optical microscopy. In particular, we discuss the interpretation of the SE micrographs in light of the results obtained from Monte-Carlo electron trajectory simulations in gallium nitride epilayers on sapphire substrates.

2. EXPERIMENT

GaN films were grown on sapphire (0001) substrates by metalorganic chemical vapour deposition, in a vertical rotating disk reactor. A GaN buffer layer of thickness 20 nm was grown on the substrate at a temperature of 475 °C, followed by a 1μm epilayer at 1050 °C. A full account of the growth procedure has been presented elsewhere (Van Der Stricht et al 1996).

Surface morphology was studied using a Cambridge instruments scanning electron microscope (SEM) and a Burleigh personal atomic force microscope. Optical micrographs of the epilayers were taken using an Olympus microscope equipped with a video camera.

3. RESULTS

Optical microscopy of the GaN epilayers shows that small groups of hexagonal hillocks with a diameter of about 50 μm populate the smooth surface (Figure 1a). Closer inspection of the pyramids reveals that the sides are not smooth facets, but rather consist of several series of steps rising to an apex (figure 1b).

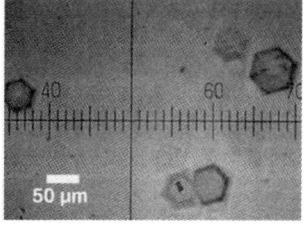

Figure 1. Optical micrograph of
a) a typical epilayer surface,
b) the stepped structure of a hexagonal pyramid.

With the surface of the sample kept normal to the incident beam in the SEM, several micrographs of a particular group of hexagonal hillocks were taken at different beam energies. An optical micrograph of the same area is provided for reference (Figure 2). The SEM images were obtained using a conventional Everhart-Thornley detector biased at +250V. Both secondary and backscattered electrons are detected. Figure 3 shows that, at high beam energy (25 keV) the hexagonal hillocks appear as bright, plate-like islands. As one lowers the beam energy the hexagonal hillocks apparently change shape from plate-like to pyramid-like and become less bright. A further decrease in the energy leads to the disappearance of hillocks C and D. At low beam energies some hexagonal holes/pittings and a ridge are observed on the surface. The number and depth of the pits seem to increase with decreasing electron beam energy. These pits have also been observed by Sasaki 1993 and Rouviere et al 1996.

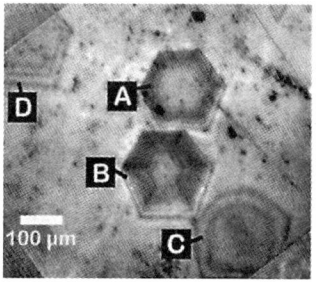

Figure 2. Optical micrograph of hillock region.

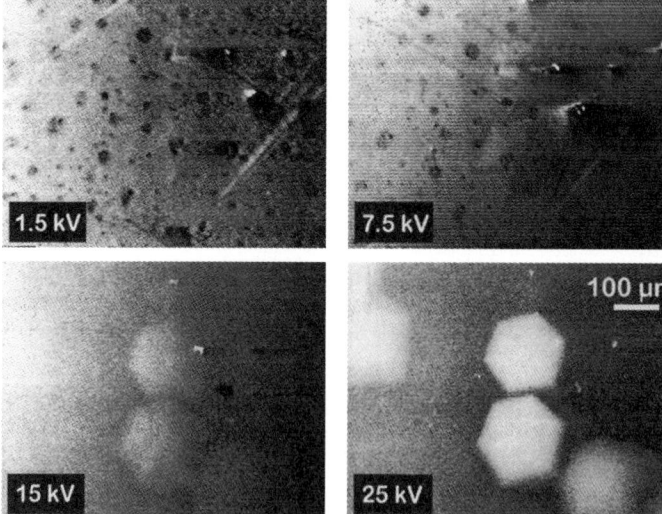

Figure 3. SEM images taken at different electron beam energies.

The surface of the film where Hexagon C is observed in the SEM and optical micrographs, was mapped using an atomic force microscope, in a series of overlapping 70μm x 70μm scans (Middleton et al 1996). None of the internal structure which is visible in the optical micrograph of Figure 2 is discernible.

4. MONTE-CARLO SIMULATIONS.

Monte-Carlo electron trajectory simulation techniques have proven to be of great utility in the study of electron sample interactions in scanning electron microscopy (see for example Joy, 1995). In the present work, the Monte-Carlo code PLOT2Z (Joy 1995) was used to simulate the interaction of an incident electron beam with our films. PLOT2Z considers the case where the target is made of two separate materials and generalizes a plural scattering code (in which every electron is considered to travel exactly the same total path length within the specimen before coming to rest). This code was run using 3000 trajectories to study the electron beam interactions in GaN epilayers on sapphire substrates at beam energies corresponding to the SE micrographs of Figure 3. Results are depicted in Figure 4. Note the different length scales. The variation of the Bethe range (the depth to which an electron can penetrate before giving up all its energy) and the interaction diameter (the horizontal spread of the beam as measured from Figure 4) with incident electron beam energy is summarised in Table 1.

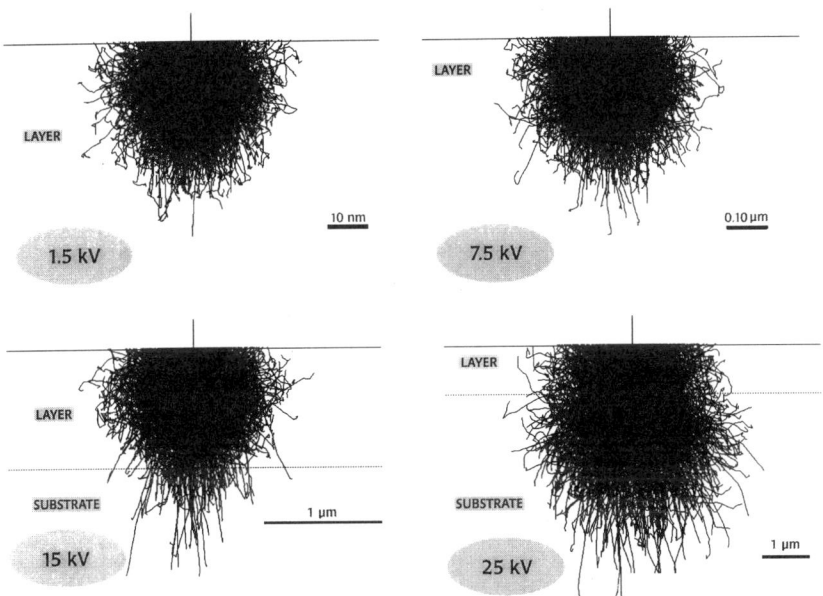

Figure 4. Monte-Carlo simulations of the electron trajectories

Electron Beam Voltage (kV)	Bethe Range (μm) (in GaN)	Interaction Diameter (μm)
1.5	0.03	0.03
7.5	0.42	0.30
15	1.36	1.12
25	3.33	2.79

Table 1. Bethe range and the interaction diameter as a function of electron beam energy.

5. DISCUSSION.

Figure 4 and Table 1 clearly illustrate that changing the electron beam energy has a dramatic effect on the Bethe range and interaction diameter. Both increase by nearly two orders of magnitude as the beam energy increases from 1.5 to 25 keV. At the lowest beam energies, 1.5 and 7.5 keV, the Bethe range and the interaction diameter are small, allowing the acquisition of high resolution images. The beam-specimen interaction is also more localised, which results in high image contrast (Pawley 1984 & Gauvin et al 1995). This partly explains why in Figure 3 the SE micrographs acquired at 1.5 keV and 7.5 keV show the most detail.

The secondary electrons which generate our image can be produced in a number of ways. SE1s are secondary electrons produced directly by the incident beam. Only those generated within the secondary escape depth (typically 5-15 nm) find their way to the surface and escape, thus contributing to the image. SE2s are secondary electrons generated by exiting backscattered electrons (incident electrons which penetrate deep into the target and are subsequently backscattered towards the surface). While SE1s are capable of producing high spatial resolution images, SE2s produced in a much greater volume, give lower spatial resolution (see for example Joy 1984 and 1995). As the number of SE2s increases with increasing incident electron beam energy, it is expected that images will become blurred as the electron beam energy is increased.

Finally, in Figure 3, in the micrographs obtained at 15 keV and 25 keV, we observe hillocks C and D which are not visible in the 1.5 and 7.5 keV micrographs. We postulate that these hillocks in fact lie beneath the surface and are therefore not imaged by low energy incident electrons. These hillocks can only be observed when the incident electron beam penetrates deep enough into the sample to generate backscattered and SE2 electrons from the buried crystallites.

6.CONCLUSION.

We studied hexagonal growth hillocks of GaN epilayers on sapphire. Optical microscopy reveals the hillocks to have a step-like structure. SEM and AFM studies show the presence of hexagonal pittings/holes on the surface and large hillocks of different heights some of which lie beneath the surface of the film. Monte-Carlo simulations provided an accurate and convenient way of interpreting the SEM images.

REFERENCES

Akasaki I and Amano H 1995 J. Electrochem. Soc. **141** 2266.
Gauvin R and Drouin D 1995 Scanning **17** (suppl. V) V12.
Joy D C 1995 Monte Carlo Modeling for Electron Microscopy and Microanalysis (Oxford Press).
Joy D C 1984 J. Microsc. **136** 241.
Khan M A, Shur M S, Kuznia J N, Chen Q, Burm J and Schaff W 1995 Appl. Phys. Lett. **66** 1083.
Middleton P G, Trager-Cowan C, Mohammed A, O'Donnell K P, Van Der Stricht W, Moerman I and Demeester P 1997 to be published in MRS Symposium Proceedings **449**.
Nakamura S 1997 MRS Internet J. Nitride Semicond. Res. **2**, 5.
Nakamura S, Mukai T and Senoh M 1994a Appl. Phys. Lett. **64** 1687.
Nakamura S, Mukai T and Senoh M 1994b J. Appl. Phys. **76** 8189.
Nakumara S 1991 Jpn. J. Appl. Phys. **30** L1705 - L1707.
Nicholls J F H, Gallagher H, Henderson B, Trager-Cowan C, Middleton P G, O'Donnell K P, Cheng T S, Foxon C T, and Chai B H T 1996, Mat. Res. Soc. Symp. Proc. **395**, 535.
Pawley James 1984 J. Microsc. **136** 45.
Rouviere J L, Arlery M, Bourret A, Niebuhr R and Bachem K H 1996 MRS. Sym. Pro. **395** 393.
Sasaki 1993 J. Cryst. Growth **129** 81.
Van Der Stricht W, Moerman I, Demeester P, Crawley J A, Thrush E J, Middleton P G, Trager-Cowan C, O'Donnell K P 1996 Mat. Res. Soc. Symp. Proc **395** 231.

Inst. Phys. Conf. Ser. No 157
Paper presented at Microsc. Semicond. Mater. Conf., Oxford, 7–10 April 1997
© 1997 IOP Publishing Ltd

Mechanisms of strain-induced surface ripple formation and dislocation multiplication in Si_xGe_{1-x} thin films

D E Jesson, K M Chen, and S J Pennycook

Solid State Division, Oak Ridge National Laboratory, TN 37831-6030, USA

T Thundat and R J Warmack

Health Sciences Research Division, Oak Ridge National Laboratory, TN 37831-6123, USA

ABSTRACT: We discuss the stress driven roughening transition of Si_xGe_{1-x} thin films. In the case of annealed films, nucleation effects dominate the nature of the surface ripple which formed by a cooperative nucleation mechanism. Facetting can however be suppressed at high supersaturations, resulting in a transition with characteristics of the Asaro-Tiller-Grinfeld instability. The relationship between morphological evolution and dislocation nucleation and multiplication is considered.

1. INTRODUCTION

In the fabrication of structures for device applications, the choice of electronic materials is usually limited to so-called lattice matched systems where the misfit strain is negligible. Ideally it would be advantageous to extend the range of available materials to other systems with the aim of obtaining new properties such as increased carrier mobilities or novel band gap alignments. However, it is well appreciated that misfit strain can present major problems in making this extension. For example, if we deposit Ge on Si, the larger lattice constant of Ge gives rise to a misfit strain of 4%. A line defect, or misfit dislocation will therefore occur every 25 planes or so to relieve the misfit strain which can be deleterious for many device applications.

One way to avoid line defects is to make use of the well known critical thickness for the introduction of misfit dislocations (van der Merwe 1963, Matthews 1975). A dislocation naturally costs a certain line energy but also relieves misfit strain in the film. If the film is sufficiently thin, not enough strain energy is released to justify the presence of the dislocation. So by growing thin enough layers it is possible to fabricate coherently strained pseudomorphic films with new properties.

More recently however it has been realized that, even without dislocations, misfit strain can also cause the film surface to roughen (Eaglesham and Cerullo 1990). Again, such roughening is highly deleterious for device applications which demand planar interfaces. One notable exception is the self assembly and self organization of quantum dots where the interest is to create regularly spaced 3D islands of uniform size (Service 1996, Krishnamurthy et al. 1991, Leonard et al. 1994, Solomon et al. 1995, Moison et al. 1994) In this case, the islands produced by the misfit strain are buried in a wider band gap material to fabricate quantum dot structures. Therefore, whether the interest is to grow planar interfaces for device applications or the self assembly of quantum dots, it is clearly important to identify and understand the kinetic pathways to island formation.

2. SURFACE RIPPLING

Often, the strain induced roughening of semiconductor films is manifested by the appearance of a continuous ripple morphology as shown in Fig. 1. This is generally interpreted in terms of the Asaro-Tiller-Grinfeld (ATG) instability (Asaro and Tiller 1972,

Grinfeld 1986) illustrated in Fig. 2. The Si_xGe_{1-x} film in Fig. 1 is under compressive stress so that an undulation in the surface allows lattice planes, which are compressed in the bulk, to relax towards the ripple peaks (Fig. 2). This lowers the elastic energy stored in the film but also increases the surface energy relative to the planar layer. A balance between the elastic energy released and the associated cost in surface energy gives a condition for the minimum wavelength λ_c for which the undulation is stable

$$\lambda_c = \frac{2\mu\pi\gamma}{(1-\upsilon)\sigma^2} \tag{1}$$

where γ is the surface energy density, σ is the misfit stress, μ is the shear modulus and υ Poisson's ratio of the layer. Beyond λ_c, the film will roughen naturally via surface diffusion. The ripple forms via gradients in the surface chemical potential such that atoms at the valley which experience a stress concentration will prefer to migrate towards more relaxed regions at the peaks. In this model there is no energy barrier to roughening except for mass transport along the surface. The planar surface is therefore unstable to strain induced roughening and one would anticipate a continuous evolution in ripple amplitude across extended regions of the film. At first sight, this is in agreement with typical experimental observations such as those contained in Fig. 1. However, in the following section we demonstrate that rippling occurs by a cooperative nucleation mechanism (Jesson et al. 1996a). This kinetic pathway for morphological evolution turns out to be useful for manipulating structures on the mesoscopic scale (Jesson et al. 1996b, Jesson et al. 1996c).

Fig. 1 Crossectional TEM image showing a typical rippled surface of a Si_xGe_{1-x} film grown on Si (100).

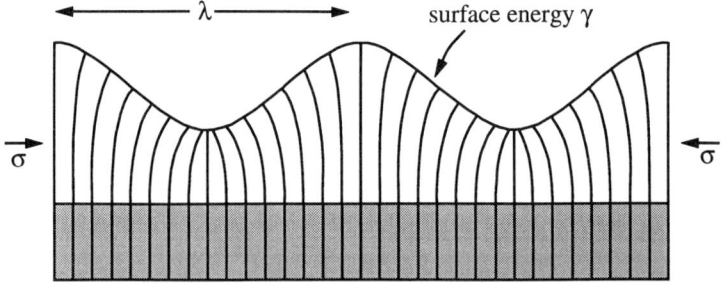

Fig. 2 Crossectional schematic illustrating the physical basis of the ATG instability.

3. COOPERATIVE NUCLEATION

To investigate the mechanism of surface ripple formation we have utilized a simple annealing experiment. Initially, a 5nm thick $Si_{0.5}Ge_{0.5}$ layer is grown on Si(100) at 400°C to produce a coherently strained layer with a nominally planar surface. The temperature of the film is then gently increased up to the anneal temperature of 590°C. Figure 3 is an Atomic Force Microscopy (AFM) image capturing the formation of the surface ripple morphology. In this image, the bright regions are islands and the dark regions correspond to pits in the surface. The islands and pits are bounded by {501} facets and are elongated along elastically soft ⟨100⟩ directions which is reminiscent of the hut cluster morphology observed by Mo et al. (1990). Most importantly, the image shows surface ripple domains consisting of alternating islands and pits which are separated by planar regions of the strained film. This indicates that the ripple is forming by the lateral growth of individual domains which involves the sequential nucleation of islands and pits (Jesson et al. 1996a).

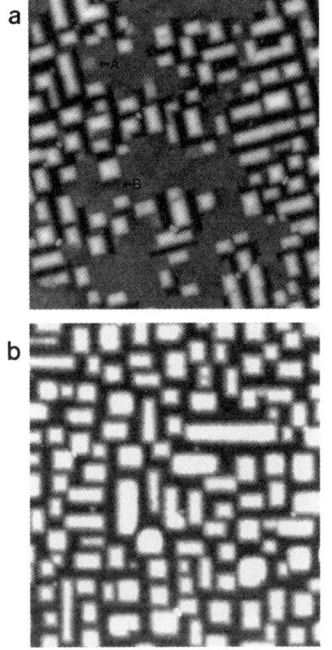

Fig. 3 (a) Atomic-force-microscopy image of a $Si_{0.5}Ge_{0.5}$ alloy layer grown on Si(001) and annealed for 5 min. at 590°C. This image reveals cooperative nucleation of surface ripple domains where islands and pits nucleate adjacent to each other as indicated. Eventually the domains merge to form the continuous ripple array shown in (b). The scanned area in both images is 1×1 μm², and the typical height (depth) of islands (pits) is 2.5 nm in (a) and 5.0 nm in (b). In both images, the islands and pits are bounded by {501} facets.

The physical origin of this cooperative nucleation mechanism reflects the fact that above a critical size, an isolated island or pit will lower the elastic energy of the strained layer. However, to nucleate an island or pit it is first necessary to overcome an energy barrier (Tersoff and LeGoues 1994). Since the elastic interaction between an island and a pit is negative (i.e. attractive) it is conceivable that cooperative effects may play an important role by lowering the activation barrier for domain growth. The essential physics of cooperative nucleation is captured in the following simple 2D model for surface ripple propagation. In Fig. 4 we represent a ripple domain in terms of rectangular islands and pits of fixed width L and height A. For the domain to propagate to the right as indicated by the arrow in Fig. 4(a), the shaded island of increasing height h must nucleate adjacent to the existing domain in the position shown in Fig. 4(b). The energy barrier to nucleate the island can be estimated by noting that the increase in surface energy per unit length is simply the strain energy per unit area γ_f associated with the vertical island facets multiplied by the height of the islands or depth of the pits. This is compensated by the elastic energy released per unit length which is found by summing over the elastic monopole interactions between vertical facets. The interaction energy between two facets a distance L apart is given by $C_m \sigma^{-} h_1 h_2 \ln(L/a)$ where

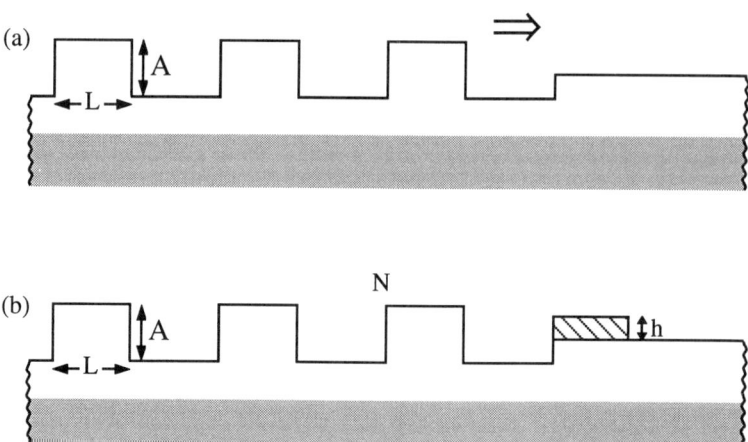

Fig. 4 A simple 2D model for domain propagation. (a) A preexisting domain consisting of pits and islands of fixed height A and fixed width L. The shaded island of varying height h nucleates adjacent to a preexisting domain in (b) causing the domain to propagate to the right. The total number of islands and pits is N.

h_1 and h_2 are the facet heights, σ is the misfit stress and C_m is an elastic constant. The microscopic cutoff length a is usually taken as a lattice constant. The interaction is positive (attractive) if the facets are the same sense (e.g., both down facets) or negative (repulsive) for opposite sense facets (e.g., up and down facets). The total energy is the sum of these two contributions and if we plot this as a function of h we can determine the energy barrier to form the island. The energy change associated with the nucleation of a single island is shown for example in Fig. 5.

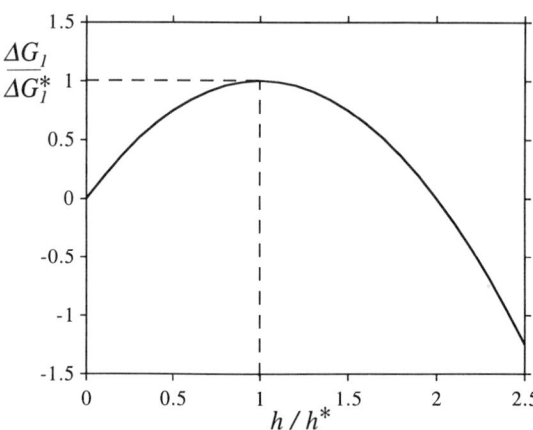

Fig. 5 Energy ΔG_1 for the nucleation of a single island ($N = 1$) as a function of height h. To attain a stable geometry the island must first overcome the energy barrier ΔG_1^* which occurs at height h^*.

The next event in the domain propagation is the nucleation of a pit adjacent to the freshly nucleated island followed by the nucleation of another island next to the pit and so on. We can evaluate the energy barriers for each nucleation event as a function of domain size N where N is the total number of islands and pits in the domain. This is plotted in Fig. 6 where the first point (N=1) corresponds to the activation barrier to nucleate an isolated island. However, the energy barrier falls on nucleating a pit next to the island because of the elastic attraction. The barrier then remains approximately constant as further islands and pits are added to the domain. This simple model is therefore very suggestive that an isolated island or pit can first act as a seed for domain growth and the domain then grows rapidly through a cooperative nucleation mechanism involving the sequential nucleation of islands and pits. This is rather different from the conventional view of ripple formation as an instability and emphasizes the importance of nucleation phenomena in the evolution of strained layer morphology.

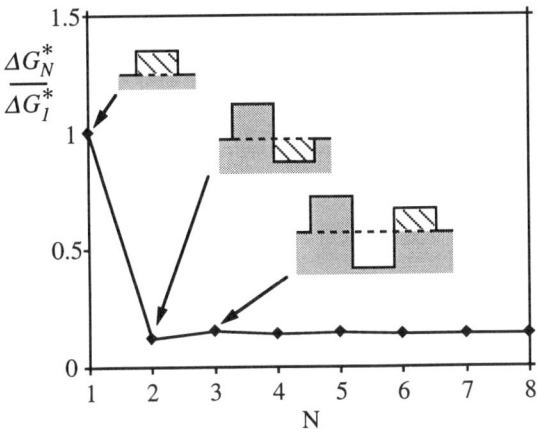

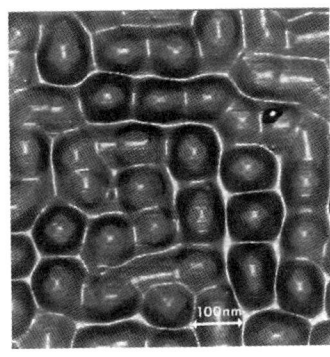

Fig. 6 Energy barrier for the growth of a domain as a function of the total number of islands and pits N. We assume $L = 45$nm, $A = 2.5$nm, $a = 2$Å. $\gamma_f = 1.0$ Jm^{-2}, with a shear modulus $\mu = 40$GNm^{-2} and Poisson's ratio $\nu = 0.3$ in the elastic constant $C_m = (1 - \nu)/\pi\mu$. The misfit $\varepsilon = 0.02$.

Fig. 7 Atomic-force-microscopy image of ridge splitting during annealing. The elongated ridges are fissioning into smaller islands.

4. FISSIONING OF ELONGATED RIDGES

Eventually, the domain structures merge, as shown in Fig. 3(b), leaving a continuous ripple array consisting of elongated islands and pits. At this stage, annealing the elongated ridges at a slightly greater temperature can give rise to island fissioning, as indicated in Fig. 7. The splitting process is reminiscent of a Rayleigh instability and can be explained by the destabilizing repulsive elastic interaction between the closely packed ridges. By forming separate islands, the system can relax along orthogonal [100] directions. This is optimized further by the increased aspect ratio of the islands which is achieved at the expense of low energy {501} facets. This might explain the 'dome-like' shape of the islands during the transition which corresponds to the appearance of higher order facets. Note that fissioning can be thought of as inverse Ostwald ripening since the mean particle size decreases with time.

The resulting morphology shown in Fig. 8 consists of a high density of self assembled islands with good size uniformity. They are also self-organized to some extent due to ridge fissioning, which tends to spatially align islands along [100] directions. In view of the interest in spontaneously generating quantum dots through self organization, cooperative nucleation phenomena coupled with island fissioning might provide a particularly attractive kinetic pathway for this purpose.

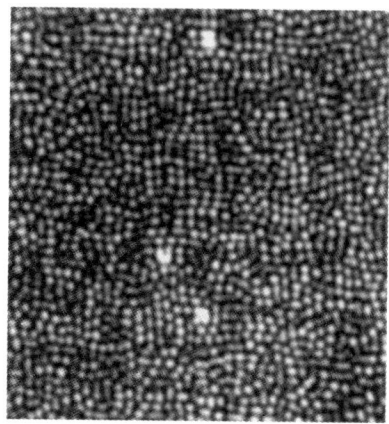

Fig. 8 Raft of quantum dots resulting from cooperative nucleation and island fissioning. The image scan is $5 \times 5 \ \mu m^2$.

5. DISLOCATIONS AND SURFACE MORPHOLOGY

A network of stress concentrations weaves in between the array of islands and eventually causes dislocations to nucleate and propagate in the film. In the AFM image in Fig. 9, a misfit dislocation has nucleated and propagated, regimenting the surface islands above it. This is a useful feature because it readily facilitates the study of dislocation nucleation and propagation by AFM. A most interesting feature of the image is the observation that a misfit dislocation can reproduce itself through interaction with the surface morphology (Jesson et al. 1995). This can be understood from the transmission electron microscope (TEM) images shown in Fig. 10. Large stress concentrations, manifested by the bright lines in the weak-beam image, develop between the islands which decorate the dislocation line. This is because the Burgers vector of the edge dislocation lies orthogonal to its line direction so that significant stress exists along the line. Such stresses offer preferred nucleation sites for dislocation nucleation (Jesson et al. 1995) although at present the detailed nucleation mechanisms are unclear. Dislocations could in principle occur by a mechanism of direct 'crack-blunting' (Jesson et al. 1995) or by a more gentle process of vacancy condensation at the stress concentration, followed by the reaction of dislocations to produce the observed edge dislocation (Cullis et al. 1995). In any event, it is clear from Figs. 9 and 10 that the interaction between surface morphology and dislocation nucleation can certainly play an important role in dislocation multiplication. (See also Albrecht et al. 1995.)

6. RELEVANCE OF THE ATG INSTABILITY

Given the obvious importance of nucleation effects in surface ripple formation, we now consider situations where the ATG instability might give a useful description of stress-induced roughening. The ATG analysis is relevant to situations where the surface energy anisotropy is very small. This might apply in some cases to chemical vapor deposition (with hydrogen terminated surfaces) or liquid phase epitaxy.

A further important effect which can reduce surface energy anisotropy is the dynamical roughening of the surface. This situation has been considered in detail by Nozières and Gallet (1987) who showed that above a critical supersaturation the surface will be rough, even below the roughening temperature. The reduced surface energy anisotropy associated with dynamical roughening has led Jesson et al. (1997) to propose that stress induced roughening under these conditions will assume the character of an ATG instability. This is supported by obsevations of the roughening of $Si_{0.5}Ge_{0.5}$ layers at high supersaturation which is consistent with an instability (Jesson et al. 1993). In particular, the non-linear behavior of the instability is manifested by the appearance of sharp cusp-like features, which are capable of injecting misfit dislocations into the film.

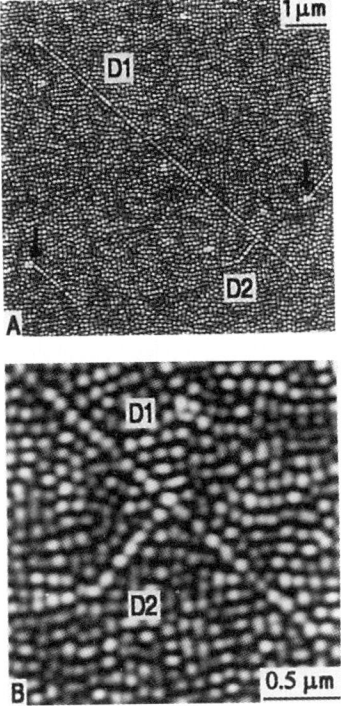

Fig. 9 (a) Atomic-force-microscopy image showing surface islands that have aligned along dislocation lines D1 and D2. An enlarged view of the intersection region between the dislocations is contained in (b)

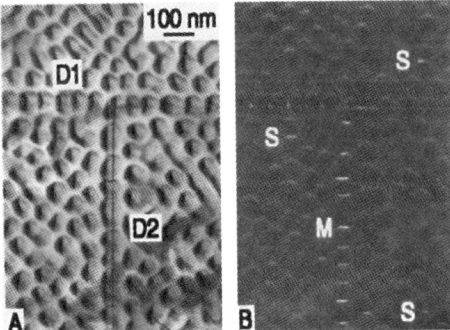

Fig. 10 (a) Bright-field (220) two-beam and (b) (2$\bar{2}$0) (g, 3g) weak beam TEM images of dislocation multiplication induced by morphological evolution. Large stress concentrations associated with the crack-like intersections between islands (S and M) are revealed as short bright segments in (b). The dislocation lines D1 and D2 define [110] and [1$\bar{1}$0] directions, respectively. The nature of the diffraction conditions means that dislocation D2 is in strong contrast in (a), and D1 is in contrast in (b). Similarly, stress lines (M) associated with D2 are in strong contrast in (b).

246

7. CONCLUSIONS

The stress driven roughening transition of semiconductor films exhibits a rich variety of interesting phenomena on the mesoscopic scale. In the case of annealed films, nucleation effects dominate the nature of the surface ripple which forms by a cooperative mechanism between islands and pits. High deposition rates however appear to suppress facetting resulting in a transition with characteristics of the ATG instability including a non-linear evolution to surface cusps. In the general case we might expect intermediate behavior between nucleation dominated and diffusive roughening regimes.

A fascinating consequence of stress-driven morphological evolution is the highly localized surface stresses which develop at the intersections between neighboring islands and at surface cusps. Such stresses can play an important role both in dislocation nucleation and multiplication, revealing an intimate link between surface morphology and defect introduction during strain relaxation.

8. ACKNOWLEDGMENTS

This research was sponsored by the Division of Materials Science, U.S. Department of Energy, under Contract No. DE-AC05-96OR22464 with Lockheed Martin Energy Research Corporation.

REFERENCES

Albrecht, M, Christiansen, S, Hansson, P O, Michler, J, Strunk, H P and Bauser, E (1995) Microscopy of Semiconducting Materials 1995, eds A G Cullis and A E Staton-Bevan, Institute of Physics Conference Series **146** (Institute of Physics Publishing, Bristol and Philadelphia) 65
Asaro, R J and Tiller, W A 1972 Metall. Trans. **3**, 1789
Cullis, A G, Pidduck, A J and Emeny, M T 1995 *Phys. Rev. Lett.* **75**, 2368
Eaglesham D J and Cerullo M 1990 Phys. Rev. Lett. **64**, 1943
Grinfeld, M A 1986 Sov. Phys. Dokl. **31**, 831
Jesson, D E, Pennycook, S J, Baribeau, J-M and Houghton, D C 1993 Phys. Rev. Lett. **71**, 1744
Jesson, D E, Chen, K M, Pennycook, S J, Thundat, T and Warmack, R J 1995 Science **268**, 1161
Jesson, D E, Chen, K M, Pennycook, S J, Thundat, T and Warmack, R J 1996a Phys. Rev. Lett. **77**, 1330
Jesson, D E, Chen, K M, Pennycook, S J, Thundat, T and Warmack, R J 1996b Proc. 23rd Int. Conf. Physics of Semiconductors **2**, 1253
Jesson, D E, Chen, K M and Pennycook, S J 1996c MRS Bulletin **21**, 31
Jesson, D E, Chen, K M, Pennycook, S J, Thundat, T and Warmack, R J (in press 1997) Journal of Electronic Materials
Krishnamurthy, M, Drucker, J S and Venables, J A 1991 J. Appl. Phys. **69**, 6461
Leonard, D, Pond, K and Petroff, P M 1994 Phys. Rev. B **50**, 11687
Matthews J 1990 J. Vac. Sci. Technol. **12**, 126
Mo, Y-W, Savage, D E, Swartzentruber, B S and Lagally, M G 1990 Phys. Rev. Lett. **65**, 1020
Moison, J M et al. 1994 Appl. Phys. Lett. **64**, 196
Nozières P and Gallet F 1987 J. Physique **48** 353
Service, R F 1996 Science **271**, 920
Solomon, G S, Trezza, J A and Harris, Jr., J S 1995 Appl. Phys. Lett. **66**, 991
Tersoff, J and LeGoues, F K 1994 Phys. Rev. Lett. **72**, 3570
van der Merwe J H 1963 J. Appl. Phys. **34**, 123

Inst. Phys. Conf. Ser. No 157
Paper presented at Microsc. Semicond. Mater. Conf., Oxford, 7–10 April 1997

247

Kinetic critical thickness for morphological instability in GeSi/Si strained layer epitaxy

B Bahierathan , D D Perovic and H Lafontaine[*]

Department of Metallurgy and Materials Science, Univ. of Toronto, Toronto M5S3E4 Canada
* Institute for Microstructural Sciences, National Research Council, Ottawa K1A 0R6 Canada

ABSTRACT: To predict the onset of surface wave formation in the GeSi/Si system, a morphological instability model has been refined by coupling continuum/atomistic treatments to include the existence of an energy barrier to the morphological instability. The model predicts accurately the kinetic critical thickness for surface wave formation as a function of temperature, growth rate and composition when compared to published experimental data.

1. INTRODUCTION

Several theoretical models have been developed to predict the evolution of surface morphological instability of strained films, based on either continuum or atomistic descriptions. The physical mechanism for the stress driven morphological instability can be described by considering the interplay between the stabilizing influence of surface energy and the destabilizing effect of misfit strain induced strain energy. Most theories employ thermodynamic or energy minimization treatments of semi-infinite, uniaxially strained solids, using either linear stability analyses of dynamic models for surface evolution (Asaro *et al.* 1972, Srolovitz 1989) or static energy minimization calculations (Grinfeld 1986 Gao 1994). Such models are neither capable of describing the evolution of a static film nor predicting the surface morphology or stability of a growing film.

Voorhees and coworkers (Spencer *et al.* 1993) have developed the most comprehensive continuum theory, which treats the dynamic case of a growing epitaxial film deposited on a substrate, taking into account the elastic constants, elastic moduli, Poisson's ratios and surface energies of the film and the substrate. Guyer *et al.* 1995 included the effect of surface stresses due to inhomogeneity of the composition profile along the surface coupled with the misfit stresses. Although a critical thickness for the onset of morphological instability can only be defined in the case of a perfectly rigid substrate (Spencer *et al.* 1993), an apparent "kinetic" critical thickness (h*) is defined for a growing film, when the growth rate of perturbation exceeds the instantaneous growth rate of the film (Snyder *et al.* 1992, Spencer *et al.* 1993, Guyer *et al.* 1995,).

To predict the onset of surface wave formation in the GeSi/Si system, the models of Voorhees and coworkers were modified by coupling continuum/atomistic considerations to incorporate the existence of an energy barrier to the morphological instability. The refined model predicts accurately the kinetic critical thickness for surface wave formation as a function of temperature, growth rate and composition.

2. EXPERIMENTAL

In order to accurately model the onset of surface wave formation, a range of $Ge_xSi_{1-x}/Si(100)$ single and multi layer heterostructures were grown by ultra high vacuum chemical vapor deposition (UHVCVD) to study the effects of increasing strained layer thickness at constant Ge fraction and increasing Ge fraction for constant layer thickness. Samples were examined using x-ray diffraction (XRD), atomic force microscopy (AFM) and transmission electron microscopy (TEM).

3. RESULTS AND DISCUSSION

Comparison of our experimental kinetic critical thickness data with the prediction of the models by Spencer *et al.* (1993) and Guyer *et al.* (1995) yielded poor agreement since continuum models do not incorporate an activation barrier for surface instability. Non-continuum or atomistic descriptions of strain-induced islanding predict the existence of an

Table1: Comparison of kinetic critical thickness calculated using the models of Spencer *et al.* (1993) [h*(S)] and Guyer *et al.* (1995) [h*(G)] with experimental data from the literature. LPCVD - low pressure chemical vapour deposition; UHVCVD - ultra high vacuum chemical vapour deposition; GSMBE - gas source molecular beam epitaxy; SSMBE - solid source molecular beam epitaxy; S - Film is stable (flat); U- Film is unstable (wavy).

Reference	Growth Technique	% Ge	T(°C)	R (nm/s)	h* (expt) (nm)		h*(S) (nm)	h*(G) (nm)
Cullis *et al.* 1992	LPCVD	18	610	0.083	66.7	S	167	254
Cullis *et al.* 1992	LPCVD	18	750	0.083	66.7	U	69	136
Cullis *et al.* 1992	LPCVD	19	750	0.083	50	U	44.5	92
Cullis *et al.* 1994	LPCVD	25	610	0.083	65	U	20.7	27.5
Pidduck *et al.* 1992	LPCVD	21	750	0.083	40	U	23.2	45.2
Sunamura *et al.* 1995	GSMBE	37	700	0.165	3.2	U	2.2	2.0
Sunamura *et al.* 1995	GSMBE	64	560	0.055	21	U	0.3	0.1
Jesson *et al.* 1993	SSMBE	50	400	0.2	9	U	19.6	4.5
Tersoff *et al.* 1994	UHVCVD	15	560	0.1	500	S	1059	1268
Tersoff *et al.* 1994	UHVCVD	30	560	0.1	20	U	13.4	13.3
Tersoff *et al.* 1994	UHVCVD	30	500	0.1	20	S	25	19.6
Tersoff *et al.* 1994	UHVCVD	40	500	0.1	20	U	6	3.7
Xie *et al.* 1994	CVD	50	650	0.1	5	U	0.51	0.24
Perovic 1986	SSMBE	25	550	0.6	360	S	244	124
Perovic 1986	SSMBE	50	550	0.6	23	U	6.8	2.7

activation energy barrier for the 2D to 3D growth transition (Tersoff *et al.* 1994, Xie *et al.* 1994, Chen *et al.* 1996). Incorporation of an activation energy for surface wave formation in the models of Spencer *et al.* (1993) and Guyer *et al.* (1995) resulted in very good agreement with our experimental data (Perovic *et al.* 1996). Using the Ge on Si (100) surface diffusion data of Lagally 1993, with an activation energy 0.5 eV, excellent agreement with our experimental data was obtained by introducing an additional activation energy component of 1.0 eV for the extreme case of pure Ge on Si, and linearly interpolating for intermediate Ge fractions (Perovic *et al.* 1996).

Table 1 compares the kinetic critical thickness calculated using the refined continuum/ atomistic models described above with experimental data from other workers. Generally there is very good agreement between the predicted and observed data irrespective of growth technique or growth conditions and the discrepancies quoted by Guyer and Voorhees (1996) are reconciled. Since the calculation of kinetic critical thickness is strongly dependent on temperature, growth rate (R) and strain in the film, any small changes in the measurements of these quantities will result in some deviation from the observed data.

The model of Spencer *et al.* (1993) appears to be more accurate than the model of Guyer *et al.* (1995). Guyer's model was expected to yield more accurate results since it considers both misfit and compositional stresses whereas Spencer's treatment considers only misfit stress. It has been shown that lateral Ge segregation during growth (i.e. lower Ge concentration at troughs and higher Ge concentration at peaks than the average composition of the film) is observed. This can be understood as the strain induced migration of larger Ge atoms to relaxed peak region (Walther *et al.* 1995, Perovic *et al.* 1997). On this basis the model of Guyer *et al.* (1995) should be more applicable. Further work is in progress to understand these differences.

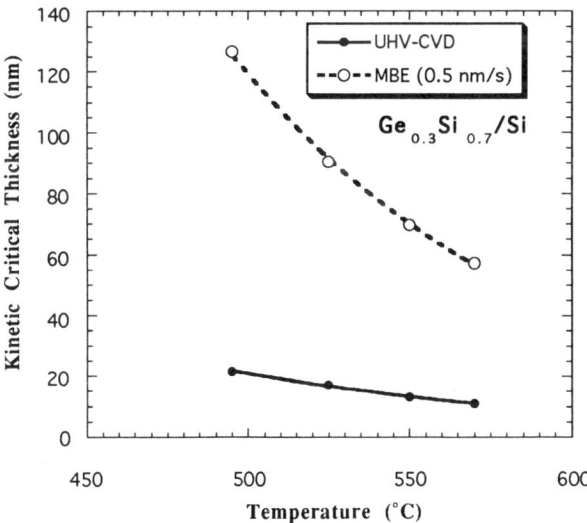

Fig. 1: Kinetic critical thickness for surface wave formation in $Ge_{0.3}Si_{0.7}/Si$ (100) heterostructures grown under UHVCVD vs. MBE conditions (see text for details).

The stability of a growing film is a sensitive function of the growth conditions. Fig. 1 shows a strain relaxation instability map for $Ge_{0.3}Si_{0.7}/Si$ (100) illustrating the effects of temperature (T) and deposition rate (R) for MBE and UHVCVD growth conditions. The effects of coupling deposition rate and growth rate have a profound effect on surface wave instability limits. For UHVCVD growth conditions the onset of surface wave formation occurs at much smaller thicknesses relative to typical MBE conditions (R= 0.5 nm/s) for the same composition. Unlike MBE, it is not possible to decrease the growth temperature to significantly enhance metastability in UHVCVD since the stabilizing effect of decreasing temperature is compensated by the concomitant exponential decrease in deposition rate which kinetically destabilizes the planar state.

4. SUMMARY

A theoretical model has been used to predict the surface morphological instability in the SiGe/Si (100) system using the continuum models by Spencer and Voorhees and Guyer and Voorhees modified to include an activation energy barrier to the onset of surface wave formation. The refined models provide accurate information about instability conditions for all growth parameters such as temperature, growth rate, composition and thickness. In addition, very good agreement is observed between our predicted data and experimental literature data regardless of growth technique employed.

REFERENCES

Asaro R J Tiller W A 1972 Met. Trans. **3**, 1789

Chen K M Jesson D E Pennycook S J Thundat T and Warmack R J 1996 Mat. Res. Symp. Proc. **395**

Cullis A G Robbins D J Pidduck A J and Smith P W 1992 J. Cryst. Growth **123**, 333

Cullis A G Robbins D J Barnett S J and Pidduck A J 1994 J. Vac. Sci. Technol. **A12**, 1924

Gao H 1994 J. Mech. Phys. Solids **42**, 741

Grinfeld M A 1986 Sov. Phys. Dokl. **31**, 831

Guyer J E and Voorhees P W 1995 Phys. Rev. Lett. **74**, 4031

Guyer J E and Voorhees P W 1996 Mat. Res. Soc. Symp. Proc. **395**

Jesson D E Pennycook S J Baribeau J-M and Houghton D C 1993 Phys. Rev. Lett **71**, 1744

Lagally M G 1993 Jpn. J. Appl. Phys. **32**, 1493

Perovic D D 1986 unpublished

Perovic D D Bahierathan B Houghton D C Lafontaine H and Baribeau J-M 1996 Mat. Res. Soc. Symp. Proc. **395**

Perovic D D Bahierathan B Lafontaine H Houghton D C and McComb D W 1997 Physica A **239**, 11

Pidduck A J Robbins D J Cullis A G Leong W Y and Pitt A M 1992 Thin Solid films **222**, 78

Snyder C W   Mansfield J F and Orr B G 1992 Phys. Rev. B **46**, 9551

Spencer B J Voorhees P W and Davis S H 1993 J. Appl. Phys. **73**, 4955

Srolovitz D J 1989 Acta Metall. **37**, 621

Sunamura H Shiraki Y and Fukatsu S 1995 Appl. Phys. Lett. **66**, 953

Tersoff J and LeGoues F K 1994 Phys. Rev. Lett. **72**, 3570

Walther T Humphreys C J Cullis A G Robbins D J 1995 Mater. Sci. Forum **196-201**, 505

Xie Y H *et al.* 1994 Phys. Rev. Lett. **73** 3006

Inst. Phys. Conf. Ser. No 157
Paper presented at Microsc. Semicond. Mater. Conf., Oxford, 7–10 April 1997
© *1997 IOP Publishing Ltd*

Decomposition analysis of $Ga_xIn_{1-x}As_yP_{1-y}$ heterostructures by STEM

C Mendorf, G Brockt, Q Liu, F Schulze, E Kubalek, I Rechenberg[1], A Knauer[1], A Behres[2], M Heuken[2], K Heime[2] and H Lakner

Werkstoffe der Elektrotechnik, Gerhard-Mercator-Universität Duisburg, Bismarckstr. 81, 47048 Duisburg, Germany
[1]Ferdinand-Braun-Institut für Höchstfrequenztechnik, Rudower Chaussee 5, 12489 Berlin
[2]Institut für Halbleitertechnik, RWTH Aachen, Templergraben 55, 52056 Aachen

ABSTRACT: $Ga_xIn_{1-x}As_zP_{1-z}/Ga_yIn_{1-y}As_zP_{1-z}$ multi quantum wells and $Ga_xIn_{1-x}As_yP_{1-y}$ single layers grown by MOVPE were investigated by STEM using bright-field and atomic number (Z-) contrast imaging, PEELS and CBED. $Ga_xIn_{1-x}As_yP_{1-y}$ layers on different substrate orientations and of different chemical compositions were compared in the context of their location in the plot of the miscibility gap. The use of misoriented substrates intensifies the decomposition phenomena. For the case of strong Z-contrast variations we determined the local chemical composition of the decomposed phases by PEELS using a 0.5 nm probe. The variations of x and y were $\Delta x=0.02$ and $\Delta y=0.04$, respectively. These results were confirmed by CBED strain measurements. The influence of pressure and temperature in MOVPE is discussed.

1. INTRODUCTION

The growth of $Ga_xIn_{1-x}As_yP_{1-y}$ heterostructures for the fabrication of optoelectronic devices has become more and more attractive because this alloy offers a great flexibility in selecting bandgap and lattice constant independently. For $Ga_xIn_{1-x}As_yP_{1-y}$ a miscibility gap is predicted. The existence of this miscibility gap restricts the opportunity of manufacturing tailor-made devices with these quaternary alloys. Therefore it is necessary to investigate the dependence of the properties of $Ga_xIn_{1-x}As_yP_{1-y}$ heterostructures on parameters such as chemical composition, growth temperature, reactor pressure and substrate orientation to obtain information about the real extension of the miscibility gap.

The periodicity of the observed contrast variations lies in the range of 10 - 100 nm, which is reported also for example by Henoc et al. (1982). Due to the need for spatial resolution in the nanometer range we used characterisation techniques like bright-field and Z-contrast imaging as well as parallel electron energy loss spectroscopy (PEELS) and convergent beam electron diffraction (CBED), performed in a cold field-emission scanning transmission electron microscope (STEM).

2. EXPERIMENTAL DETAILS

The $Ga_xIn_{1-x}As_yP_{1-y}$ heterostructures were grown by metalorganic vapour phase epitaxy (MOVPE) on GaAs (923 K) and by low pressure MOVPE on InP (913 K). In a field-emission STEM (VG Microscopes: HB 501) cross-sectional thin specimens of the quaternary heterostructures were investigated. The system has been modified and improved with respect to high-resolution capabilities (0.3 nm probe diameter at 100 keV). Bright-field imaging was used for the characterisation of the structural properties of the layers. Additionally, CBED strain measurements were carried out. The CBED patterns were recorded by a single crystal YAG scintillator which is coupled to a CCD camera (512x512 pixel, 18 bit dynamic range) by a mirror and lens system. For qualitative chemical composition analysis the heterostructures were analysed by Z-contrast images which are obtained by detecting elastically and quasi-elastically scattered electrons using a high-angle annular dark-field detector. Quantitative composition analysis of the specimens was performed by PEELS while the spectra were recorded by a CCD-camera. A detailed description of the applied characterisation tools can be found in Lakner et al (1996).

3. RESULTS

3.1 $Ga_xIn_{1-x}As_yP_{1-y}$ single layers grown on GaAs

Using GaAs as substrate material, $Ga_{0.74}In_{0.26}As_{0.5}P_{0.5}$ single layers were grown at a temperature of 923 K on substrates with different orientations. This composition is located deep inside the predicted miscibility gap (see Fig. 1, marker 3). Figure 1 exhibits the calculated miscibility gap for 910 K, following theoretical models of Stringfellow (1982) and Onabe (1982). For clarity, only the miscibiltity gap for 910 K is shown. The miscibility

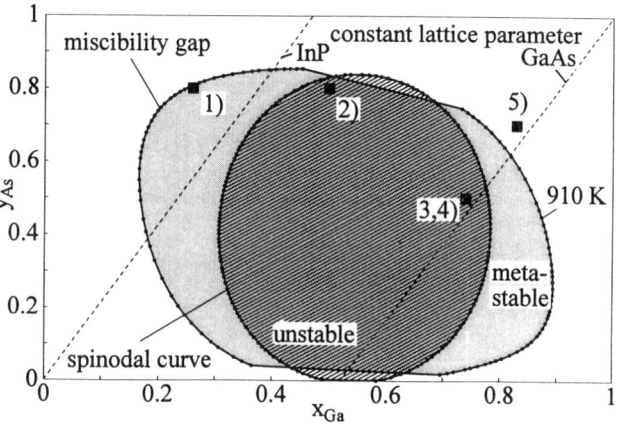

Fig. 1: Miscibility gap and spinodal curve of $Ga_xIn_{1-x}As_yP_{1-y}$ calculated for T=910 K.

gap for 923 K is slightly smaller. The following results refer to layers grown on a misorientated substrate (2 ° towards {111} B).

In Figure 2, a STEM bright-field image (referring to marker 3 in the plot of the miscibility gap) is presented showing both, the substrate and quaternary alloy. While GaAs is of excellent quality, a quasi-periodic pattern occurs in the $Ga_{0.74}In_{0.26}As_{0.5}P_{0.5}$ layer parallel to the growth direction. A Z-contrast image of the decomposed $Ga_{0.74}In_{0.26}As_{0.5}P_{0.5}$ layer is shown in Figure 3a.

The GaAs substrate shows no significant contrast variations whereas in the $Ga_{0.74}In_{0.26}As_{0.5}P_{0.5}$ the contrast of the pattern is alternately bright and dark. This indicates a chemical decomposition of the quaternary layer. A linescan extracted from the Z-contrast image (Fig. 3b) marked in Figure 3a reveals a quasi-periodicity which is determined to be about 25 nm.

To get quantitative information on decomposition PEELS was carried out. Figure 4 shows EEL spectra of the Ga-L- and As-L edge recorded from the spots marked in Figure 3a. It is clearly visible that the spectra recorded at a bright area (A) possess higher intensities of both, the Ga-L- and As-L edges compared with the spectra recorded at a dark area (B). This is in accordance to the contrast behaviour of the Z-contrast image, mainly because more As means less P and therefore a higher mean atomic number. The relative variation of Ga was determined as $\Delta x=0.02$ and that of As as $\Delta y=0.04$.

To confirm this result, CBED patterns were recorded at the same locations. The change of chemical composition leads to a change of the lattice constant due to Vegard's law, so that by means of the determination of the lattice constant with CBED conclusions can be drawn with respect to the chemical composition. The observed shifts in the HOLZ-lines indicate that changes in the lattice constants are small $(\Delta a/a \le 2*10^{-3})$. This is in accordance with the PEELS results. Additionally, in the CBED pattern recorded in the quaternary layers (see Fig. 5), a splitting of so-called HOLZ-lines can be observed.

Fig. 2: STEM Bright-field image of a $Ga_{0.74}In_{0.26}As_{0.5}P_{0.5}$ single layer lattice matched grown on GaAs.

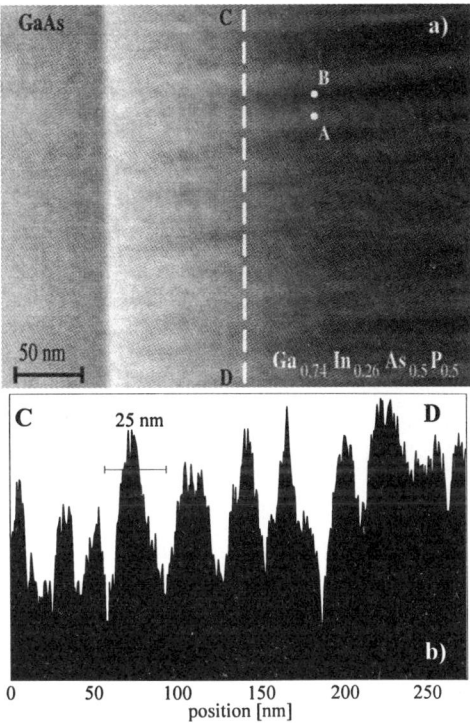

Fig. 3: STEM Z-contrast image a) and linescan b) of the $Ga_{0.74}In_{0.26}As_{0.5}P_{0.5}$ layer grown on GaAs.

254

Additionally, a $Ga_{0.74}In_{0.26}As_{0.5}P_{0.5}$ single layer grown on an exactly orientated substrate was investigated. The location in the plot of the miscibility gap is marked with number 4. In this case the decomposition effects are much weaker. The investigation of a specimen which was located outside the miscibility gap (marker 5 in figure 1) showed, as expected, no signs of decomposition.

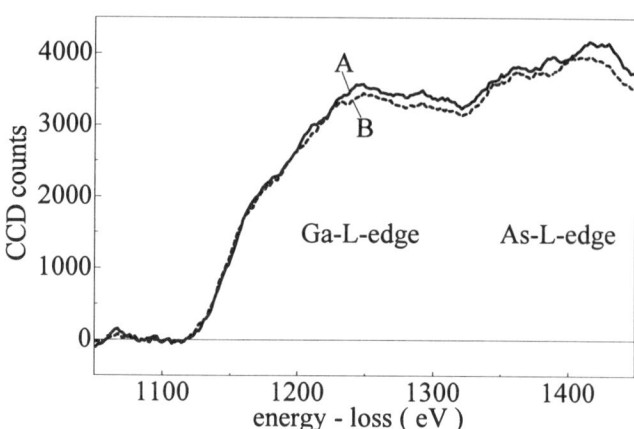

Fig. 4: EEL-Spectra recorded from marked points at Figure 3.

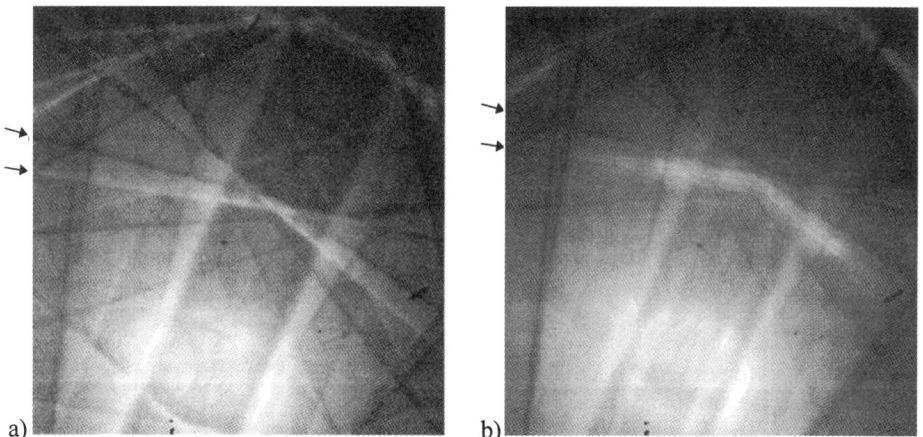

Fig. 5: CBED pattern recorded from a) GaAs substrate and from b) $Ga_{0.74}In_{0.26}As_{0.5}P_{0.5}$ layer, revealing splitting of HOLZ-lines.

3.2 $Ga_xIn_{1-x}As_zP_{1-z}/Ga_yIn_{1-y}As_zP_{1-z}$ multi quantum wells (MQW)

In addition to $Ga_xIn_{1-x}As_yP_{1-y}$ single layers, $Ga_xIn_{1-x}As_zP_{1-z}/Ga_yIn_{1-y}As_zP_{1-z}$ multi quantum well (MQW) structures were investigated. These MQWs were grown strain compensated on InP at a temperature of 913 K under optimised conditions (Behres et al, (1996)). The location of the layers used as wells and barriers is marked in the plot of the miscibility gap (Fig. 1, markers 1 and 2). The $Ga_{0.26}In_{0.74}As_{0.8}P_{0.2}$ wells are close to the margin of the miscibility gap whereas the $Ga_{0.5}In_{0.5}As_{0.8}P_{0.2}$ barriers are located clearly inside the gap.

Referring to recent examinations (Lakner et al, (1997)) the investigations carried out here were concentrated on Z-contrast imaging related to contrast variations in the wells and barriers, respectively. A Z-contrast image of the analysed superlattice is shown in Figure 6, where the $Ga_{0.26}In_{0.74}As_{0.8}P_{0.2}$ layer appears bright and the $Ga_{0.5}In_{0.5}As_{0.8}P_{0.2}$ layer appears dark, corresponding to their mean atomic number. Both wells and barriers reveal much less contrast variation than observed in the single layers. This weak contrast variation exceeds the detection limits of PEELS as well as CBED.

The excellent properties of the MQW are a result of series of tests varying the reactor pressure and the As-content. The result of the tests is shown in Figure 7. Decreasing the As-content, the pressure must be increased to prevent decomposition.

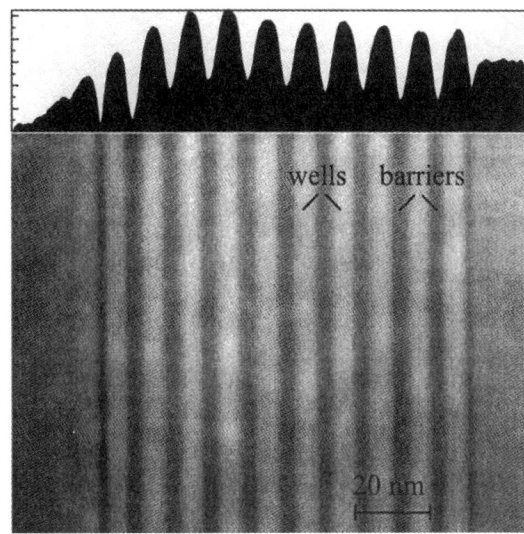

Fig. 6: Z-contrast image and linescan of the $Ga_{0.26}In_{0.74}As_{0.8}P_{0.2}/Ga_{0.5}In_{0.5}As_{0.8}P_{0.2}$ MQW

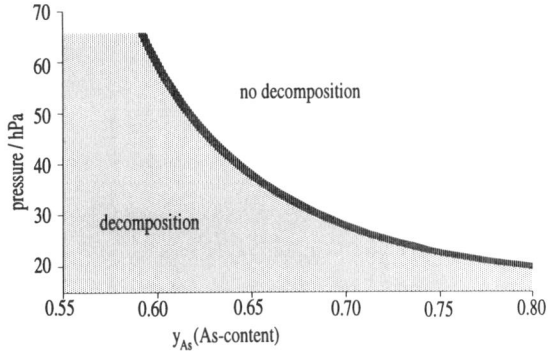

Fig. 7: Critical pressure in dependence on As-content.

4. DISCUSSION

By comparing quaternary layers with the same composition but different substrate orientations (markers 3 and 4 at Fig. 1), specimens grown on an exactly orientated substrate show a much less decomposed behaviour than layers grown on a misorientated substrate. Thus the use of misorientated substrates intensifies the decomposition process.

The $Ga_{0.26}In_{0.74}As_{0.8}P_{0.2}/Ga_{0.5}In_{0.5}As_{0.8}P_{0.2}$ multi quantum well, represented by markers 1 and 2 in the plot of the miscibility gap, shows only little chemical fluctuation despite the fact that at least the layer used as a barrier is located inside the spinodal curve which defines the instable region. This can be explained by the optimised growth conditions, changing the temperature, As-content and reactor pressure.

The shape and extension of the miscibility gap can be influenced e.g. by varying the growth temperature or the total pressure. Between 913 K and 933 K a minimum of spinodal decomposition can be found.

Reducing the As-content results in a location of the quaternary material deeper inside the miscibility gap. The increase of the pressure is a counterbalance for that. This is caused by the influence of the total pressure on the surface migration length. The influence of the miscibility gap can be reduced by increasing the total pressure (Behres et al, (1997)). This can be explained as follows. The calculated miscibility gap is valid for the thermodynamic equilibrium. The more the growth process is off the thermodynamic equilibrium, the smaller is the real gap. It follows that quaternary $Ga_xIn_{1-x}As_zP_{1-z}/Ga_yIn_{1-y}As_zP_{1-z}$ layers which are nominally located inside the theorectical miscibility gap can be grown homogeneously when the pressure is high enough.

The observed strong contrast variations of the quaternary single layers using Z-contrast imaging are difficult to quantify. In accordance with Treacy et al (1985) this quasi-periodic contrast variation is strongly enhanced by surface relaxation resulting from preparation of thin specimens. The surface relaxation causes a splitting of HOLZ-lines and ZOLZ-lines which has been proved by experimental CBED patterns. The effect of surface relaxation on Z-contrast images as well can not be neglected. Therefore, imaging techniques alone are not suitable for a fully quantitative understanding of decomposition. More astonishing is the small change of the chemical composition, determined as $\Delta x=0.02$, respectively as $\Delta y=0.04$ by PEELS.

In summary, the combination of the described STEM methods of Z-contrast, PEELS and CBED are highly efficient tools for a quantitative chemical and structural analysis of heterostructures. The application yields useful information about the behaviour of quaternary $Ga_xIn_{1-x}As_yP_{1-y}$ alloys grown within the predicted miscibility gap.

ACKNOWLEDGEMENTS

The authors like to thank Dr. B. Opitz, Robert Bosch GmbH, for the numerical calculation of the miscibility plot. The financial support of this work by the Volkswagen-Stiftung is gratefully acknowledged.

REFERENCES

Behres A, Opitz B, Werner H, Kohl A, Woitok J, Geurts J and Heime K 1996 Conf. Proc. of the 8th Int. Conf. on InP and Relat. Mater. IEEE 486
Behres A, Heuken M, Heime K, Mendorf C, Lakner H and Kubalek E 1997 :
    accepted for publication at the 7th European Workshop on MOVPE and Related Growth Techniques, Berlin
Henoc P, Izrael A, Quillec M and Launois H 1982 Appl. Phys. Letters **40** 963
Lakner H, Ungerechts S, Behres A, Kohl A, Opitz B, Heime K and Woitok J 1997 J. Crystal Growth **170** 732
Lakner H, Bollig B, Ungerechts S and Kubalek E 1996 J. Phys. D: Appl. Phys. **29** 1767
Onabe K 1982 J. Phys. Chem. Solids **43** No. 11 1071
Stringfellow G B 1982 J. Crystal Growth **58** 194
Treacy M M J, Gibson J M and Howie A 1985 Phil. Mag. A **51 (3)** 389

Inst. Phys. Conf. Ser. No 157
Paper presented at Microsc. Semicond. Mater. Conf., Oxford, 7–10 April 1997
© *1997 IOP Publishing Ltd*

Investigations of ordering in AlGaInP

A Dunbar, S Hall, M Halsall and U Bangert

Department of Physics, UMIST, Manchester M60 1QD

ABSTRACT Atomic ordering in GaAlInP and AlInP grown at different off axis substrate orientations is investigated by TEM and by Raman scattering. The tendency of ordering decreases with increase in the off axis orientation and is less for GaAlInP than for AlInP. The latter was found to order strongly even when grown at 10^0 off axis. Raman results show tendencies and relative strength of ordering in good agreement with TEM results.

1. INTRODUCTION

Long range ordering of atoms on {111} planes is known to occur in alloys of GaAs and InAs as well as GaP and InP. It is a feature of epitaxial, in particular MOCVD growth, and is influenced by the growth temperature, the growth rate, the substrate orientation and the surface reconstruction (Seong et al 1991, Baxter et al 1991, Augarde et al 1989, Seong et al 1995). Ordering has received attention from industry because the emission peaks of optical components incorporating ordered GaInAs and GaInP epitaxial layers show shifts of several tens of meV with respect to those expected for a random alloy. The GaAlInP system is employed in red lasers and diodes as cladding and confinement layer and the discussion arose, whether the phenomenon of step bunching in the active quantum well layer is related to ordering in the neighbouring layers. Ordering in GaAlInP has not been studied extensively, but the suggestion is that it is enhanced by the addition of Al. If the proposed model is applicable (e.g. see Norman et al 1993), the driving force for ordering is strain compensation. The smaller group III alloy atom segregates into compressive regions underneath surface dimers and the larger atom into the tensile regions underneath gaps between the dimers. However, the addition of Al should not affect the degree of the ordering, because the lattice constant of AlP is not much different from that of GaP. In this paper we study ordering in a $Ga_{0.15}Al_{0.35}In_{0.5}P$ and $Al_{0.52}In_{0.48}P$ layer, grown at different off axis orientations. We furthermore evaluate results obtained by Raman spectroscopy against TEM results to find out whether Raman spectroscopy, a fast, non destructive method, could find applications in industrial materials characterisation.

2. EXPERIMENTAL

The samples consisted of AlInP and GaAlInP layers grown by MOCVD to a nominal composition lattice-matched to GaAs, as shown in the table below. (01-1) and the (0-1-1) cross-sectional samples were made for investigations in a CM20 TEM. Raman scattering at room temperature in backscattering geometry was conducted on all samples. The 514 and 488 nm lines of an Argon-ion laser were used as excitation sources and the light was dispersed and detected by a Spex 1401 double spectrometer equipped with an intensified diode array detector at a

258

resolution < 0.5 cm^{-1}.

GaAs cap	0.05	μm	4 samples in total
Al$_{0.52}$In$_{0.42}$P	1.0		with substrate type
GaAs	0.1		(100) GaAs on axis,
Ga$_{0.15}$Al$_{0.35}$In$_{0.5}$P	1.0		2^0 off axis to <110> and
GaAs	0.1		10^0 and 15^0 off axis to
n-GaAs substrate			<111>

3. RESULTS AND DISCUSSION

The diffraction patterns of the various samples are shown in figs.1a)-e). Extra spots only show in one of the 2 perpendicular cross-sectional orientations (i.e. (01-1)), and two sets of superlattice spots can be seen. Fig.1a) shows distinct extra spots for the on axis AlInP layer, fig.1b) shows less intense but distinct extra spots for the on axis GaAlInP layer. This indicates ordering on the (111) and the (1-1-1) planes. The extent of ordering is larger in the AlInP than in the GaAlInP. Nearly identical diffraction patterns were obtained for the 2^0 off axis layers. Figs.1c) and d) show results for the 10^0 off axis layers. The diffraction pattern of the AlInP layer exhibits a wavy intensity pattern in the region around the extra spots originating from streaking of the 1/2<111> and the 1/2<1-1-1> spots in directions inclined slightly off <100> towards <111> and <1-1-1> respectively. In the GaAlInP layer in fig.1d) the extra spots and the streaking are barely visible. In the 15^0 off axis samples neither the AlInP nor the GaAlInP layer show any ordering and their diffraction patterns are identical to that of the substrate, as shown in fig.1e).

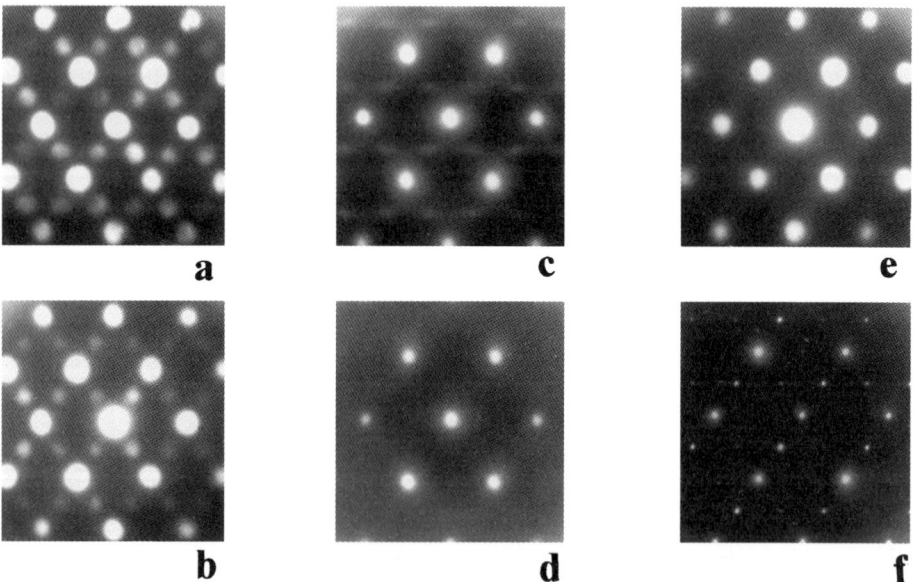

Fig.1 (0-11) diffraction patterns of a) and b): AlInP and GaAlInP grown on axis, c) and d): AlInP and GaAlInP grown 10^0 off axis towards (111), e) GaAs substrate, f) Ga$_{0.5}$In$_{0.5}$P grown on axis

For comparison in fig.1f) is shown the diffraction pattern of an MOCVD grown on axis $Ga_{0.5}In_{0.5}P$ layer showing extra spots not only at 1/2<111>, but also at 1/2<200> and 1/2<220>. This is an indication for extensively ordered, well defined, interwoven domains with the latter types of spots originating from interference of the 1/2<111> and the 1/2<1-1-1> beams. There are signs of weak streaking of the extra spots in this material in direction parallel to <100>, which could be indicative of planar defects such as antiphase boundaries (Baxter et al 1991) with orientation close to the growth direction.

The on axis and the 2^0 off axis AlInP layer shows very similar characteristics, except for the 1/2<200> and the 1/2<220> spots. This is possibly due to the fact that the domains are not so closely interwoven or less numerous. This appears to be increasingly the case when Ga is incorporated in the layer. The 10^0 off axis AlInP layer exhibits the characteristics of ordering in the presence of inclined antiphase boundaries or other planar defects (e.g. twins). Again the incorporation of Ga appears to upset the ordering. The higher misorientation of 15^0 has resulted in the disappearance of ordering altogether in the AlInP and the GaAlInP. The disappearance of one variant in GaInP with off axis misorientation is well documented (e.g. Bellon et al 1989). We observe the disappearance of both variants at off axis orientations of 15^0. This might be due to the fact that we approach a {113} surface with a different surface reconstruction. A new observation is that the incorporation of Al appears to reverse the tendancy to disorder and indeed AlInP layers show significant ordering at misorientations of 10^0. It is not obvious on the base of lattice distortion and strain compensation arguments, why this should be so, since AlP and GaP have much the same lattice parameter and hence no large differences between alloys incorporating one or the other element ought to be expected. A possible reason could be the difference in bond strength.

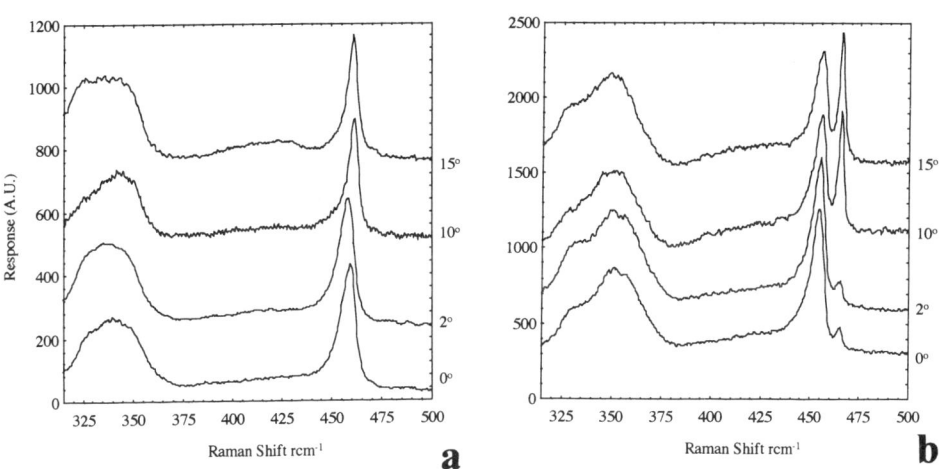

Fig.2 Raman spectra for a) AlInP and b) GaAlInP. The substrate misorientation is noted on the right hand side beside the respective spectrum

A fast and non-destructive assessment of ordering is often desired. Raman spectroscopy can offer such an assessment, though Raman spectra of AlInP are relatively unstudied . Fig.2a) and b) show Raman spectra of the 4 samples from the above table. Fig.2a) shows the spectrum measured with the 488 nm excitation. At this wavelength the light is strongly absorbed by the AlInP layer. The resulting spectra show strong resonant Raman scattering from this layer, such as occurs for wavelengths of excitation close to the materials band gap. Two main phonon features at 340 cm^{-1} and 460 cm^{-1} can be seen. Previous reports (Kondow et al 1987, Kondow et al 1988) have ascribed the peak at 340 cm^{-1} to an InP-like alloy mode and that at 460 cm^{-1} to an AlP-like alloy mode. It is known that the AlP-like frequency has a strong dependence on the alloy. It has been observed in the GaInP system (Alsina et al 1996) that the effect of ordering on the Raman spectrum is to 'zone fold' the L-point phonon producing a shoulder on the low energy side of the GaP-like phonon peak. A similar effect is expected for the AlInP, the phonon dispersion along the [111] direction is expected to be small due to the large difference in atomic masses. Thus we should expect a folded L point phonon to produce a broadening on the low energy side of the zone centre phonon peak in the case of an ordered alloy. Such a broadening is seen in fig.2a) for the on axis and the 2^0 off axis samples. A possible route to gain quantitative information from the Raman spectra may be via the FWHM of the Raman peaks. These are 10, 10, 6.5 and 6 cm^{-1} in order of increasing misorientation.

Fig.2b) shows the spectra taken with 514 nm light. At this photon energy the AlInP layer is relatively transparent and scattering from the GaAlInP is observed. The spectra are similar to those of the AlInP, though the AlP-like modes are broader. This time there is no pronounced narrowing of the AlP-like modes with increasing substrate misorientation, suggesting that the ordering is weak in all 4 samples.

Though the results from electron diffraction patterns give a more detailed picture, we conclude that the Raman results are in agreement with the electron diffraction results, showing the tendency of a decrease in the ordering as the substrate misorienation is increased and as Ga is incorporated into the alloy.

ACKNOWLEDGEMENTS

We greatly acknowledge the interest and support from S. Bland and G. Jones from Epitaxial Products Ltd, who stimulated the investigations of this materials system and supplied the samples.

REFERENCES

Alsina F, Mestres N, Pascual J, Eng C, Ernst P and Scholz F 1996 Phys. Rev. B 53 12994
Augarde E, Mpaskoutas M, Bellon P, Chevalier J P and Martin G P 1989 Inst. Phys. Conf. Ser. 100 155
Baxter C S, Stobbs W M and Wilkie J H 1991 Inst. Phys. Conf. Ser. 117 469
Bellon P, Chevalier J P, Augarde E, Andre J P and Martin G P 1989 J. Appl. Phys. 66 2388
Kondow M, Minagawa S and Satoh S 1987 Appl. Phys. Lett. 51 2001
Kondow M and Minagawa S 1988 Appl. Phys. Lett. 64 793
Norman A G, Seong T-Y, Philips B A, Booker G R and Mahajan S 1993 Inst. Phys. Conf. Ser. 134 279
Seong T-Y, Norman A G, Hutchison J L, Booker G R, Cullis A G, Bass S J and Taylor L L 1991 Inst. Phys. Conf. Ser. 117 463
Seong T-Y, Booker G R, Norman A G, Harris P J F and Cullis A G 1995 Inst. Phys. Conf. Ser. 146 241

Inst. Phys. Conf. Ser. No 157
Paper presented at Microsc. Semicond. Mater. Conf., Oxford, 7–10 April 1997
© 1997 IOP Publishing Ltd

Structural characteristics of highly ordered (GaIn)P

J C Jiang, A K Schaper [1], Z Spika, W Stolz [2], P Werner [3] and L Tóth [4]

Center for Materials Science, [1] Department of Geosciences; [2] Department of Physics, Philipps-University, 35032 Marburg, Germany; [3] Max-Planck-Institute of Microstructure Physics, 06120 Halle, Germany; [4] Institute for Technical Physics, Academy of Sciences, 1325 Budapest, Hungary

ABSTRACT: Depending on the conditions of growth, a new type of domain structure was observed in highly ordered $Ga_{0.5}In_{0.5}P$ epilayers. Rod-like shaped subdomains with characteristic tube-like boundaries could be distinguished from the usual antiphase domains by cross-sectional as well as plan-view transmission electron microscopy (TEM). The boundaries are characterized by a polarity reversal between the inner domain rods and the surrounding matrix domains.

1. INTRODUCTION

As is well-known, metalorganic vapor phase epitaxy (MOVPE)-grown (GaIn)P on (100) GaAs substrate exhibits spontaneous ordering on $\{111\}$ lattice planes (see, e.g., Su et al. 1994, Baxter et al. 1991, Suzuki et al. 1990). Although theoretical studies show two different types of CuPt structure to be possible (Zhang et al. 1995), only the $CuPt_B$ variant has been experimentally found until now. The degree of ordering is generally controlled by the misorientation of the substrate (Su et al. 1994), the growth temperature (Kurtz et al. 1990), and the growth rate (Su et al. 1994). Recently, the important role of surface reconstruction on ordering has been emphasized by Bernard et al. (1991), Suzuki et al. (1992) and Philips et al. (1994). As reported by Spika et al. (1997), enhancing the diffusion of the group III species at the reconstructed surface by a modulated growth (MG) technique, the degree of order can be markedly improved as compared to conventional growth (CG). Those highly ordered samples show a novel rod-like domain structure. The formation of these domains is considered a three-dimensional self-organization process inherent in the growth of the (GaIn)P layers, with no correlation to the morphology of the substrate surface.

2. EXPERIMENTAL

The $Ga_{0.5}In_{0.5}P$ epitaxial layers were grown on (100) GaAs substrates misorientated up to 4° in the [1$\bar{1}$1] direction using the modulated MOVPE technique described by Spika et al. (1997). The modulation consisted of an alternating interruption of the constituent trimethylgallium and trimethylindium gas streams. The growth temperature was 650°C. Samples for TEM investigations were prepared by grinding, polishing and Ar-ion milling at liquid nitrogen temperature. The TEM investigations were carried out using Philips CM20 (200kV) and JEOL JEM 4000EX (400kV) microscopes. Multislice image simulations were performed using the MacTempas software package.

3. RESULTS

3.1 Observation of Rod-Like Subdomains

The [011] cross-sectional electron diffraction pattern of an as-grown MG epilayer (Fig. 1a) shows one dominant and one weak set of superlattice reflections of the two $CuPt_B$ variants. There are only slight differences in spot shapes as compared to corresponding diffraction patterns of conventionally grown layers (see, e.g., Baxter et al. (1991)). However, dark-field imaging using the $1/2(1\bar{1}1)$ reflections (Fig. 1b) shows, in addition to the usual antiphase boundaries (APBs at "A"), a new type of subdomain and boundaries ("R") crossing the APBs. The new boundaries appear somewhat apart from the interface and pass through the epilayer up to top surface. They are also visible in dark-field images using the fundamental reflections except (h00) (h = even). Under the conditions applied, the boundaries are inclined from the normal to the surface by a few degrees towards the $[0\bar{1}1]$ direction. Fig. 1c is a plan-view that shows the subdomains and boundaries top-on. The rod-like shape of the domains becomes clearly apparent in this projection. The domains have rectangular up to circular 50-80 nm sized cross-sections, and are regularly arranged along the [010] and [001] directions.

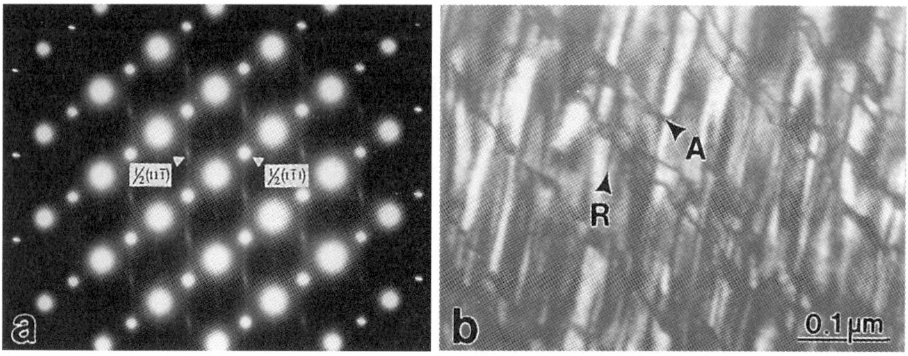

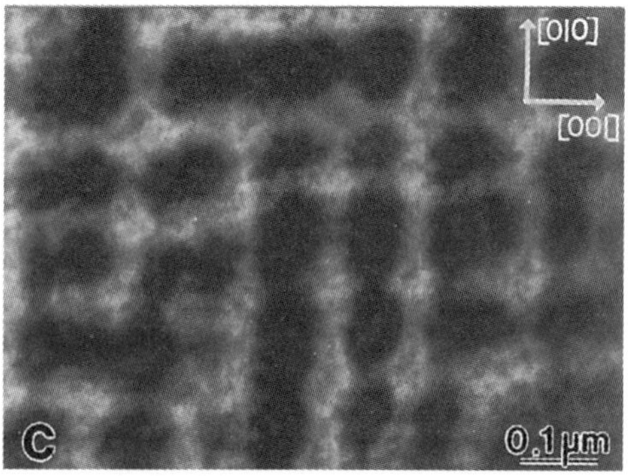

Figure 1: (a) [011] cross-sectional electron diffraction pattern of a modulation-grown (GaIn)P epilayer; (b) dark-field image using the $1/2(1\bar{1}1)$ superlattice reflection ("A": antiphase boundaries, "R": rotation boundaries, see below); (c) (100) plan-view of an epilayer showing the crystallographically arranged rod-like domains top-on.

Figure 2: Cross-sectional $1/2(1\bar{1}1)$ dark-field image showing strain-induced contrasts in the GaAs buffer layer (top) directly related to the rod-like domain structure in the (GaIn)P epilayer (bottom).

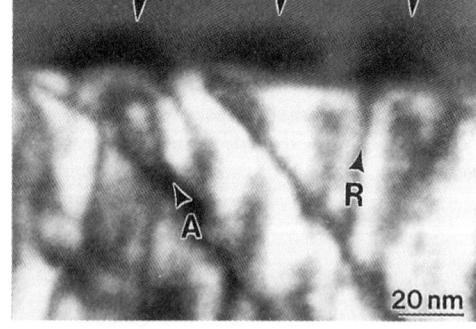

Another interesting feature connected with the existence of the new domain structure was observed in thin (GaIn)P epilayers covered by a GaAs buffer layer. As shown in the [011] cross-sectional image in Fig. 2 particular contrast figures are detected inside the buffer layer, which are directly related to the underlying rod domains and R-boundaries, respectively. Obviously, the domain growth is accompanied by the development of strains, which propagate into the top GaAs layer. Because the growth conditions were different for the two samples, the domain size in Fig. 2 is slightly reduced in comparison with Fig. 1.

3.2 Structural Characteristics of the Domain Boundaries

Fig. 3a is a high magnification electron micrograph directly revealing the superlattice ordering and two types of boundary structures. While antiphase domains are shifted against each other by $d(1\bar{1}1)$, the displacement of the new boundaries is characterized by $\frac{1}{2} d(1\bar{1}1)$. The Fourier filtered [011] close-up of a boundary region with adjoining domains in Fig. 3b is indicative of a polarity reversal of the Ga/In-P bond projections when crossing the boundary. In the (GaIn)P sphalerite structure there are two nonequivalent displacement vectors: $\frac{1}{4}a<111>$, corresponding to P and (Ga,In) sublattice origins at (000) and $\frac{1}{4}(111)$, respectively, and $-\frac{1}{4}a<111>$, corresponding to the lattice sites (000) and $\frac{1}{4}(\bar{1}\bar{1}\bar{1})$. The polarity is reversed between both lattices, which corresponds to a 90° rotation about the [100] axis. The situation is illustrated by a simplifying sketch of the [011] boundary projection in Fig. 3c. The [011] zone-axis image simulation in Fig. 3d is based on the assumption of a gradual transformation of one domain D_O into the other D_I as indicated in Fig. 3c (top). Due to this and other simplifications the agreement between the calculated and the experimental image is, however, not yet perfect.

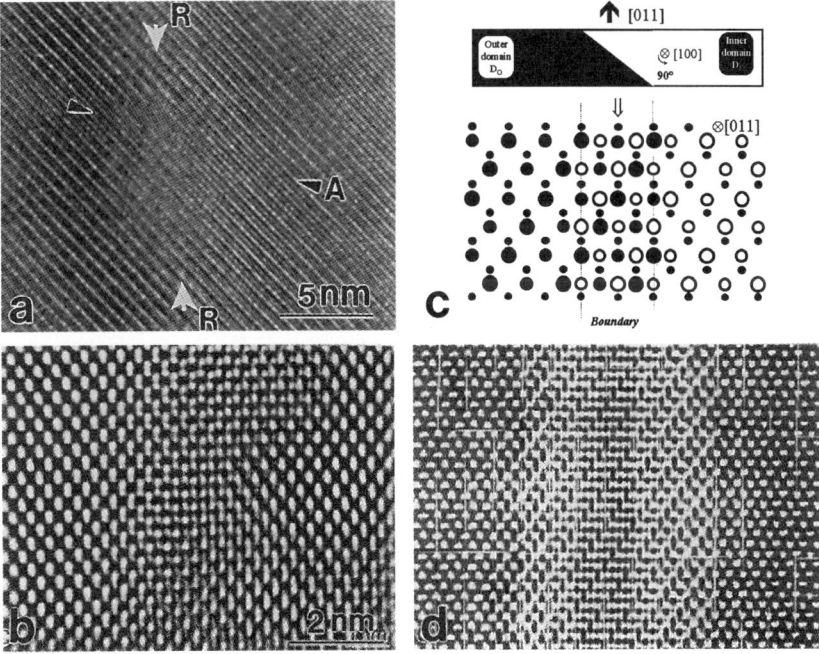

Figure 3: (a) Lattice fringe image using $\pm1/2(1\bar{1}1)$, $\pm(1\bar{1}1)$, and $\pm3/2(1\bar{1}1)$ beams showing superlattice ordering, an APB ("A") and a rotation boundary ("R"); (b) Fourier filtered [011] high-resolution micrograph of a R-boundary; (c) illustration of the [011] projection of a R-boundary and (d) corresponding simulated image (t = 5.6 nm, Δf = 50 nm, C_s = 1 mm)

3.3 Correlation between structure and ordering

Fig. 4 shows the results of room temperature photoluminescence measurements of a continuously grown and a modulated-grown sample. A considerable reduction of the band gap energy is revealed from 1.786 eV for the CG layer to 1.749 eV for the MG layer. This reduction is a clear indication of an increased ordering in the modulation-grown material with the new domain structure. The broadening of the PL linewidth of the MG sample is ascribed to a broadening of the size distribution of the domains with varying degree of ordering (Horner et al. 1994).

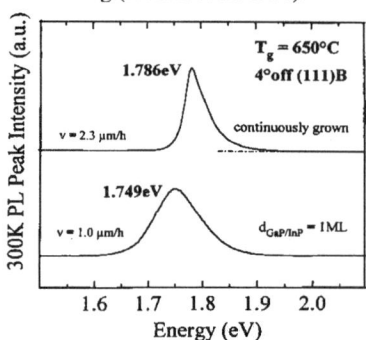

Figure 4: Room temperature photoluminescence (PL) measurements of CG and MG grown (GaIn)P.

4. DISCUSSION AND SUMMARY

In highly ordered $Ga_{0.5}In_{0.5}P$ epilayers novel rod-like domains were observed, which are suggested to be the result of an inherent self-organization process. The samples were prepared by a modulated MOVPE technique leading to exceptionally high degrees of CuPt superlattice ordering. The domains appear in addition to the well-known antiphase structure and are characterized by a reversed polarity with respect to the matrix domains, i.e., besides the $CuPt_B$ type of ordering also the $CuPt_A$ type seems possible. The existence of the new domain structure could be reproduced for various degrees of substrate misorientation between 0° and 4°. Obviously, the formation of the structure is not correlated with the morphology of the GaAs substrate surface. Future work is planned to elucidate the formation mechanism of the rod-like domains in detail. Moreover, further experiments are needed to clear-up the influence of the hierarchical domain structure onto the optoelectronic properties.

We thank Mrs. S. Hopfe, MPI of Microstructure Physics Halle, for her help during specimen preparation. This work has been supported by the Deutsche Forschungsgemeinschaft (SFB 383). L.T. gratefully acknowledges a grant from the Hungarian National Science Foundation (TO14091).

REFERENCES

Baxter C S, Stobbs W M and Wilkie J H 1991 Inst. Phys. Conf. Ser. **117**, 469.
Bernard J E, Froyen S and Zunger A 1991 Phys. Rev. **B 44**, 11178.
Horner G S, Mascarenhas A, Alonso R G, Froyen S, Bertness K A and Olson J M 1994 Phys. Rev. **B 49**, 1727.
Kurtz S R, Olson J M and Kibbler A 1990 Appl. Phys. Lett. **57**, 1922.
Philips B A, Norman A G, Seong T Y, Mahajan S, Booker G R, Skowronski M, Harbison J P and Kermidas V G 1994 J. Cryst. Growth **140**, 249.
Spika Z, Zimprich Z, Stolz W, Göbel E O, Jiang J, Schaper A and Werner P 1997 J. Cryst. Growth **170**, 257.
Su L C, Ho I H and Stringfellow G B 1994 J. Appl. Phys. **75**, 5135.
Suzuki T, Gomyo A and Iijima S 1990 J. Cryst. Growth **99**, 60.
Suzuki T, Gomyo A and Iijima S 1992 Springer Series in Materials Science **17**, 363.
Zhang S B, Froyen S, and Zunger A 1995 Appl. Phys. Lett. **67**, 3141.

Inst. Phys. Conf. Ser. No 157
Paper presented at Microsc. Semicond. Mater. Conf., Oxford, 7–10 April 1997
© *1997 IOP Publishing Ltd*

The effect of substrate misorientation on atomic ordering in $Ga_{0.52}In_{0.48}P$ epilayers grown on GaAs (001) substrates by Gas-Source MBE

C Meenakarn, A E Staton-Bevan, M D Dawson*, G Duggan*, A H Kean* and S P Najda*

Department of Materials, Imperial College of Science, Technology and Medicine, Prince Consort Road, London SW7 2BP, UK.
*Sharp Laboratories of Europe Ltd, Edmund Halley Road, Oxford Science Park, Oxford OX4 4GA, UK.

ABSTRACT: The ternary III-V semiconductor $Ga_{0.52}In_{0.48}P$ grown on GaAs substrate has been studied for visible wavelength light sources for information processing, laser printing and compact disk systems. In epilayers grown by MOCVD or Solid Source MBE, optical emission with reduced energy is known to originate from atomic ordering of the alloy and the degree of ordering of the group III elements has been found to be significantly influenced by the degree of substrate misorientation from (001). This paper reports a Transmission Electron Microscopy (TEM) study conducted on $Ga_{0.52}In_{0.48}P$ epilayers grown on misoriented (001) GaAs substrates by Gas-Source Molecular Beam Epitaxy. For a growth temperature of 530°C, substrate off-cut angles of 0°, 7°, 10° and 15° towards [111]A were investigated. Selected Area Diffraction Patterns obtained, indicated that the antiphase domain size decreases with increasing off-cut. TEM results have been correlated with band gap measurements obtained from PL and PLE spectra. The band gaps of $Ga_{0.52}In_{0.48}P$ epilayers grown by GS-MBE were found to be larger than those of the same composition grown by MOCVD or Solid Source MBE. This indicates potential for laser devices of shorter wavelengths.

1. INTRODUCTION

The ternary III-V semiconductor $Ga_xIn_{1-x}P$ grown on GaAs substrate has been the most widely studied material for visible wavelength light sources required for information processing, laser printing and compact disk systems (Kobayashi et al 1985, Ikeda et al 1985, Ishikawa et al 1996). Until recently, the majority of $Ga_xIn_{1-x}P$ epilayers have been deposited by Metalorganic Chemical Vapour Deposition (MOCVD) (Gomyo et al 1987, Kondow et al 1988a) or conventional solid source Molecular Beam Epitaxy (MBE) (Blood et al 1980, Kawamura et al 1981, Hafich et al 1989). For laser applications, the important parameter is the band gap energy since this determines the wavelength of the emitted light. This has been found to depend on the MOCVD or MBE growth conditions such as growth temperature and V/III ratio (Gomyo et al 1986, 1987, Y. Ohba et al 1996, Andre and Bellon 1992).

Many studies have shown that optical emission with reduced energy originates from atomic ordering of the alloy (Gomyo et al 1987, Bellon et al 1988). Photoluminescence (PL) and photoluminescence excitation spectroscopy (PLE) results can be correlated with the CuPt-type spontaneous ordering of the alloy (Honer et al 1993). A CuPt-type ordered phase of $Ga_xIn_{1-x}P$ is observed in (001) epitaxial layers and has been studied extensively (Kondow et al 1988b, 1988c). The ordered alloy consists of antiphase domains with alternating monolayers of GaP and InP stacked on the $(\bar{1}11)$ and $(1\bar{1}1)$ planes. Gomyo et al (1987) showed that two $Ga_{0.5}In_{0.5}P$ samples, one with complete order and the other with complete disorder, had different band gap energies (1.85 and 1.90 eV at 300K respectively). The situation for intermediate degrees of order has not been determined but it is generally believed that the band gap energy increases as the degree of order decreases.

For Ga$_x$In$_{1-x}$P layers grown by MOCVD and MBE, the degree of ordering of the group III elements has been found to be significantly influenced by the degree of substrate misorientation from (001). Both optical properties and surface morphology are also affected (Kikuchi et al. 1991, 1992). This paper reports the first study of substrate misorientation effects in Ga$_{0.52}$In$_{0.48}$P epilayers grown, lattice matched to GaAs (001), by Gas-Source MBE. This growth technique allows the advantages of conventional MBE to be retained without the normal difficulties of As:P flux ratio control normally experienced with P-bearing materials.

2. EXPERIMENTAL PROCEDURE

Ga$_{0.52}$In$_{0.48}$P epilayers, 2 μm thick, lattice matched to within 0.1%, were deposited on GaAs (001) substrates by Gas-Source Molecular Beam Epitaxy (GS-MBE) at Sharp Laboratories of Europe. Substrates were misoriented 0°, 7°, 10° or 15° towards [111]A and for each sample a 1 μm GaAs buffer layer was employed. The growth temperature and growth rates were 530°C and 1 μm/h respectively.

Samples for cross-sectional TEM were prepared by cleavage, mechanical polishing and Ar-ion thinning using standard procedures. Specimens were prepared parallel to the ($\bar{1}$ 10) and (110) crystal planes. Selected area diffraction patterns of cross-sectional samples were obtained using Jeol 120 CX and 2000 FX electron microscopes at Imperial College.

Photoluminescence emission (PL) and photoluminescence excitation (PLE) spectra were obtained at 5 K with the samples mounted in a continuous-flow He cryostat. A 543 nm green HeNe laser and a 100W tungsten lamp/0.22m monochromator combination was used as the excitation source with a fluence at the sample of ~0.01 nWcm^{-2}. The PL emission was detected by a 0.5m spectrometer, with a cooled GaAs photomultiplier and photon counting electronics. The overall spectral resolution of the lamp measurements was approximately 3.5 meV.

3. RESULTS AND DISCUSSION

3.1 Transmission Electron Microscopy Selected Area Diffraction Patterns (SADPs)

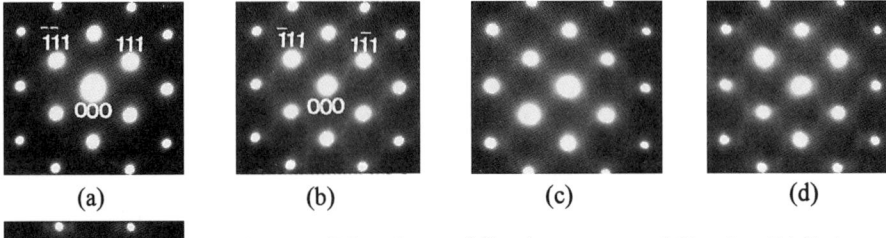

(a)   (b)   (c)   (d)

(e)

Figure 1: Selected area diffraction patterns of Ga$_{0.52}$In$_{0.48}$P/ GaAs cross-sections. (a) Selected area diffraction patterns at the ($\bar{1}$ 10) orientation of the sample grown with 15° off-cut, (b)-(e) selected area diffraction patterns at the (110) orientation, for samples grown with 0°, 7°, 10° and 15° off-cut, towards [111]A respectively

Since ordering in Ga$_{0.52}$In$_{0.48}$P occurs on ($\bar{1}$ 11) and (1$\bar{1}$ 1) but not (111) or (11$\bar{1}$) planes, no superlattice reflections were observed in any of the (1$\bar{1}$ 0) cross-sections, e.g. Figure 1(a) for 15° off-cut. However, Figures 1(b) - 1(e) show SADPs of the (110) cross-sections for the layers grown with 0°, 7°, 10° and 15° off-cut towards [111]A. All four patterns show ordering reflections which appear as wavy streaks exhibiting an intensity distribution centred around $\frac{1}{2}$(1$\bar{1}$ 1) and $\frac{1}{2}$($\bar{1}$ 11) positions. The wavy streaks consist of segments inclined at approximately ±10° to [001]. A single-domain crystal exhibiting CuPt-type ordering would exhibit superlattice spots at $\frac{1}{2}(1\bar{1}1)$ and $\frac{1}{2}(\bar{1}11)$, however,

Baxter et al. (1991) have pointed out that an array of equal-sized, plate-shaped antiphase domains of separation w would produce zero intensity at the position of the superlattice Bragg reflection $1/2\{\bar{1}11\}$ and satellite spots along the direction parallel to the APD boundary normal, at $1/2\{\bar{1}11\} \pm n/2w$, where $n = 1,2,3,...$ and the maximum intensity is for $n = \pm 1$. In the present case the presence of streaks rather than satellite spots indicates that the APD plates thicknesses are irregular. The position of the maximum intensity in the streak therefore corresponds to the predominant antiphase domain separation w_p at a distance $\pm 1/2w_p$ from $1/2\{\bar{1}11\}$. Since this distance increases with increasing off-cut, as may be seen in Figs. 1(b) to 1(e), this indicates that the APD thickness decreases with increasing misorientation to [111]A.

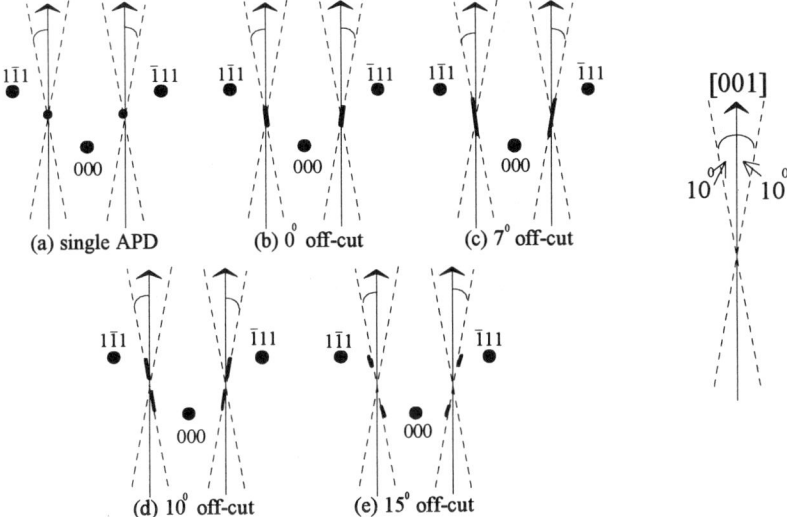

Figure 2: Schematic diagram illustrating the effect of degree of misorientation to [111]A on the $Ga_{0.52}In_{0.48}P$ epilayer APD size.

Additional streaking parallel to [001] results from the existence of very thin (~2nm), (001) laminae of the two ordered variants, $(1\bar{1}1)$ and $(\bar{1}11)$, (Baxter et al 1991). This is most visible where the intensity is concentrated, i.e. at positions distant from $1/2\{1\bar{1}1\}$ for a small APD size. This additional streaking leads to the APD streaks being broadened to produce the familiar "wavy" pattern.

3.2 Correlation of TEM diffraction data with optical measurements

Energy gap values measured by photoluminescence (PL) and photoluminescence excitation (PLE) spectroscopy are given in table I.

The band gap energies measured by PLE (more accurate than PL (Dawson and Duggan 1993)) suggest a very slight increase with increasing off-cut angle. This corresponds to a decrease in APD size indicated by TEM. The band gap energies are larger than those reported for MOCVD (Gomyo et al 1987, Kondow et al 1988a) and solid source MBE (Kushino et al. 1993), showing potential for shorter wavelength lasers.

It is unfortunate that the APD sizes are too small to allow the formation of HRDF images of adequate quality. This together with the diffuse nature of the superlattice reflections makes it difficult to make quantitative estimates of the degree of order using TEM. The presence of exciton peaks in the PLE spectra of the 10° and 15° offcut samples and their smaller Stokes shifts, $(E_{GAP}(PLE)-E_{GAP}(PL))$, indicate that these samples are less ordered than the 0° and 7° offcut samples. The broad PL peak FWHM for the 7° sample suggests that this is the most highly ordered sample.

Table I: 5K PL and PLE data for $Ga_{0.52}In_{0.48}P$ epilayers of Fig.1

Degree of Misorientation to [111]A	Band Gap Energy (eV)		Stokes Shift (eV)	PL FWHM (meV)	PLE Exciton peak
	PL	PLE			
0°	1.961±0.001	2.001±0.001	0.040±0.002	8.115±0.002	✗
7°	1.954±0.001	2.008±0.001	0.054±0.002	23.211±0.002	✗
10°	1.981±0.001	2.012±0.005	0.031±0.007	12.418±0.010	✓
15°	1.997±0.001	2.013±0.005	0.016±0.007	7.925±0.010	✓

4. CONCLUSIONS

For $Ga_{0.52}In_{0.48}P$ epilayers grown by GS-MBE at 530°C, on GaAs(001) substrates offcut 0°, 7°, 10° and 15° towards [111]A, TEM diffraction patterns show that the APD size decreases with increasing angle of offcut. This corresponds to a very slight increase in energy gap measured by PLE. The 15° sample had the highest degree of disorder. The band gap energies of GS-MBE-grown epilayers are higher than those reported for epilayers grown by MBE or SS-MBE showing potential for laser devices of shorter wavelengths.

The authors thank Prof. McLean for provision of research facilities in the Department of Materials.

REFERENCES

Andre J P and Bellon P 1992 Mat. Res. Soc. Symp. Proc. **262**, 835
Baxter C S, Stobbs W M and Wilkie J K 1991 J. Cryst. Growth **112**, 373.
Blood P, Bye K L and Roberts J S 1980 J. Appl. Phys. **51**, 1790
Dawson M D and Duggan G 1993 Phys. Rev. B **47**, 12598.
Gomyo A, Kobayashi K, Kawata S, Hino I, Suzuki T and Yuasa Y 1986 J. Cryst. Growth **77**, 367
Gomyo A, Suzuki T, Kobayashi K, Kawata S, Hino I, and Yuasa T 1987 Appl. Phys. Lett. **50**, 673
Hafich M J, Quigley J H, Owens R E, Robinson G Y, Li D and Otsuka N 1989 Appl. Phys. Lett. **54**, 2686
Honer G S, Mascarenhas A, Froyen S, Alonso R G, Bertness K and Olson J M 1993 Phys. Rev. B **47**, 4041
Ikeda M, Mori Y, Sato H, Kaneko K and Watanabe N 1985 Appl. Phys. Lett. **47**, 1027
Ishikawa M, Ohba Y, Sugawara H, Yamamoto M and Nakanisi T 1996 Appl. Phys. Lett. **48**, 207
Kawamura Y, Asahi H and Nagai H 1981 Jpn. J. Appl. Phys. **20**, L807
Kikuchi A, Kishino K and Kaneko Y 1991 Electron. Lett. **27**, 1391.
Kikuchi A and Kishino K 1992 Appl. Phys. Lett. **60**, 1046.
Kobayashi K, Kawata S, Gomyo A, Hino I and Suzuki T 1985 Electron. Lett. **21**, 931
Kondow M, Kakibayashi H and Minagawa S 1988a J. Cryst. Growth **88**, 291
Kondow M, Kakibayashi H, Minagawa S, InoueY, Nishino T and Hamakawa Y 1988b Appl. Phys. Lett. **53**, 2053
Kondow M, Kakibayashi H and Minagawa S 1988c J. Cryst. Growth **93**, 412
Kushino K, Kikuchi A, Nomura I and Kaneko Y 1993 Thin Film **231**, 173.
Ohba Y, Ishikawa M, Sugawara H, Yamamoto M and Nakanisi T 1996 J. Cryst. Growth **77**, 374

Inst. Phys. Conf. Ser. No 157
Paper presented at Microsc. Semicond. Mater. Conf., Oxford, 7–10 April 1997
© *1997 IOP Publishing Ltd*

Study of the structural and optical properties of ordered domains in GaInP alloys

L Nasi, F Fermi*, C Ferrari, L Francesio, L Lazzarini, C Zanotti-Fregonara, S Pellegrino° and G Salviati

CNR-MASPEC Institute, Via Chiavari 18/A, 43100 Parma, Italy
* INFM, Physics Department, University of Parma, Viale delle Scienze, 43100, Parma, Italy
° TELETTRA-ALCATEL, Via Trento 30, 20059 Vimercate (MI), Italy

ABSTRACT: GaInP layers, grown on differently misoriented (001) GaAs substrates, exhibit $CuPt_B$-type domains. The morphology of the ordered domains and their boundaries are strictly related to the surface step configuration. The increase in ordered domain size, with increasing misorientation angle, results in a significant decrease in the degree of order and in a sharper distribution of the ordering parameter. Finally, the spectrally resolved cathodoluminescence emission from the ordered samples is discussed on the basis of the inhomogeneity of the degree of order.

1. INTRODUCTION

The $CuPt_B$ type ordering that spontaneously occurs at the surface of GaInP alloys using typical growth conditions has important implications for optoelectronic devices since the bandgap energy is found to depend markedly on the degree of order. In addition, it has been recently demonstrated that increasing the number of antiphase boundaries (APBs) between ordered regions may result in a diminished performance of ordered lasers (Geng et al 1997). Therefore, knowledge about material parameters that depend on the degree of order and on the domain morphology is interesting both for basic research and for device applications. In this respect, a lot of work has been performed (e.g. Seong et al 1994, Friedman et al 1994, Ernst et al 1994 and references therein) in order to clarify the origin of APBs and their role in determining the optical properties of the layers.

In this paper the influence of substrate misorientation on the morphology and the degree of order of $CuPt_B$ domains in GaInP layers is discussed on the basis of transmission electron microscopy (TEM), X-ray diffraction (XRD), atomic force microscopy (AFM), spectrally resolved cathodoluminescence (SCL) and photoluminescence (PL) results.

2. EXPERIMENTAL

The GaInP layers were grown by metal organic chemical vapour deposition (MOCVD) on lattice matched (001) GaAs substrates misoriented by angles of 0°, 2° and 6° in the [100] direction. The temperature was fixed at 660 °C, while the growth rate was 2.5 μm/h at a V/III ratio of 120.

TEM observations were performed using a JEOL 2000FX working at 200 kV. For the XRD analyses, a high resolution four circle diffractometer at LURE synchrotron radiation centre, with a fixed wavelength λ=1.38 Å, was used. Optical investigations were made using a XBO 150W high pressure Xe lamp having an excitation power of less than 1 μW. Finally, the spectrally resolved cathodoluminescence measurements were performed in a Cambridge 360 stereoscan Scanning Electron Microscope, while AFM contact mode images were taken using a Nanoscope III Digital Microscope.

3. RESULTS AND DISCUSSION

The presence of CuPt$_B$ ordered structure in all the layers was directly revealed by the appearance of $\overline{1}/2,1/2,1/2$ (variant I) and $1/2,\overline{1}/2,1/2$ (variant II) extra-reflections in both transmission electron diffraction (TED) patterns and XRD curves. The crystal field split-off energy of the valence band evidenced in the polarized photoluminescence excitation spectra (PLE) also showed a direct relation between CuPt structure and electronic energy gap. Detailed descriptions of the above results are reported by Nasi et al (1995) and Francesio et al (1996).

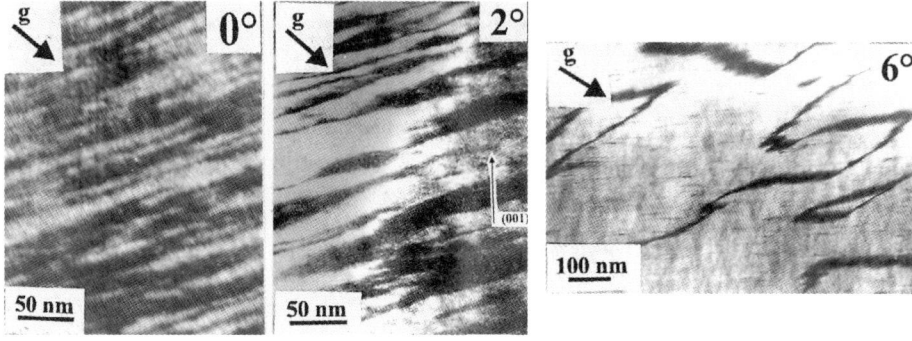

Fig. 1. (110) DF-TEM images ($g=1/2[\overline{3}31]$) showing $1/2,\overline{1}/2,1/2$ ordered domains.

Dark field (DF) TEM investigations of the size and morphology of the ordered domains confirmed the well-known fact (Zunger and Mahajan 1994) that an increase in the substrate misorientation towards a $[111]_B$ direction results in a single variant samples with an increasing domain size. In Fig. 1 the (110) cross section images show the enlargement of the variant II ordered domains as the substrate miscut angle increases from 0° to 6°. The ordered regions are distributed in plate-like domains laterally separated by dark bands that were confirmed to be APBs of the ordered superstructures from high resolution TEM observations (Nasi et al 1995).

Since atomic ordering formation and evolution are driven by surface mechanisms including surface reconstruction and step motion, the spatial distribution of the superstructure in the (001) projections was compared with the AFM surface images, in order to study the influence of the surface morphology on the distribution of ordered regions and their APBs. The comparison between the shape of the variant II ordered regions, obtained from TEM plan view analyses in the ($0\overline{1}3$) pole and AFM pictures, is shown in Fig. 2.

In the 0° off sample, the morphology of the superstructure observed in TEM plan view could be associated to the AFM surface configuration, consisting of monolayer steps whose misorientation component can give rise to either variant II or variant I domains. Since the steps are mainly inclined to the $[1\overline{1}0]$ direction, the presence of the "V" shaped APBs, observed in TEM plan view images between the variant II domains and variant I regions (not shown), may be due to the loss of coherence of the ordered structure at these step edges.

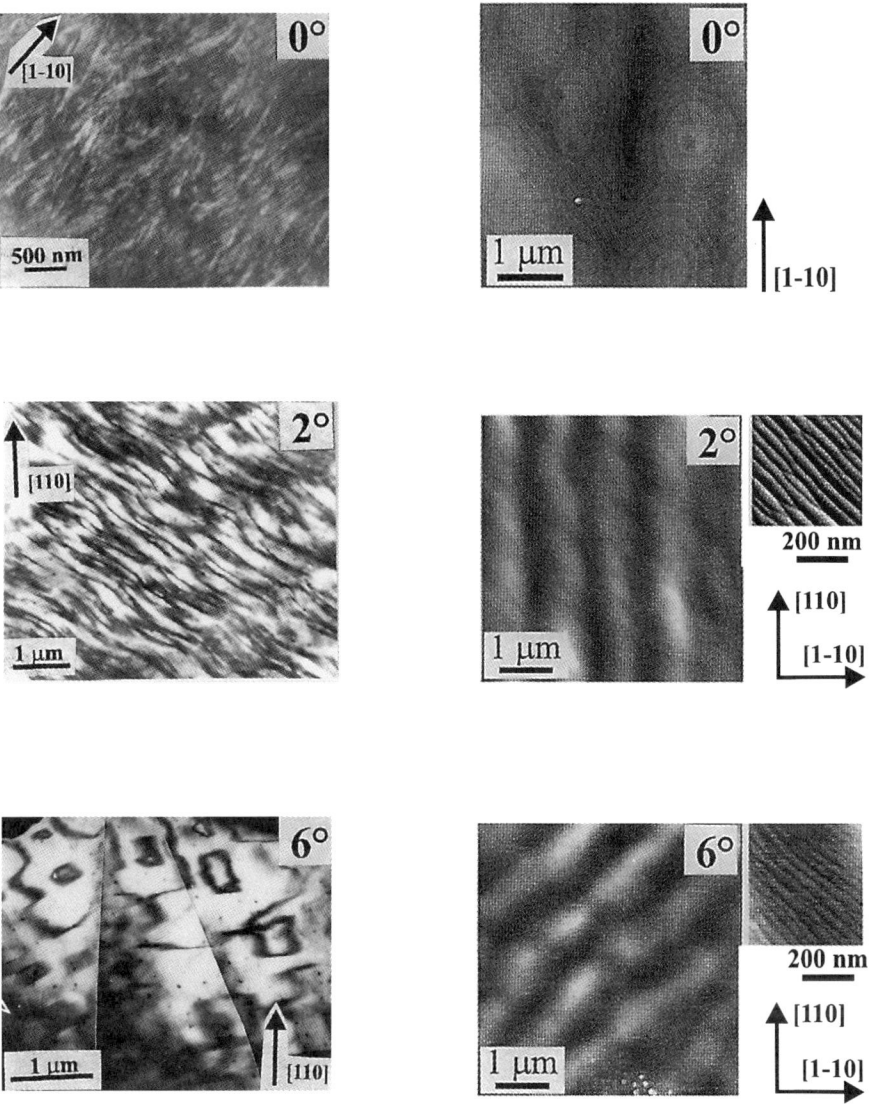

Fig. 2. Left: ($0\bar{1}3$) DF-TEM plan view images, taken with $\mathbf{g}=1/2[\bar{3}31]$, showing $1/2,\bar{1}/2,1/2$ ordered domains. Right: contact mode AFM images of the differently misoriented samples.

The analysis of the AFM pictures of the misoriented samples revealed the presence of a large-scale undulation (4-5 nm high) and a fine one (shown in the small images) related to the bunching of [100] steps required to accommodate the misorientation. These surface corrugations seem to be related to the distribution of the APBs of the superstructure. In particular, in the 2° off sample, the step edges of the fine modulation may represent the source of the wavy [100] oriented APBs shown in the TEM plan view image of Fig. 2. On the other hand, the APBs spacing and direction in the 6°

off sample, together with the frequent occurrence of square-shaped [100] boundaries, are consistent with the prediction that they originate from coarse modulation during growth. From the analysis of the TEM cross section of this sample, however, it is reasonable to suppose that the weak APBs running parallel to the interface (not seen in the plan view) come from fine modulation step edges.

XRD was used successfully for rapid estimation of the ordered domain size in the direction of the scattering vector. According to the kinematical diffraction theory, a crystal block of dimension D produces a diffraction peak having an approximate full width at half-maximum (FWHM) $\Delta\vartheta$ given by $\Delta\vartheta\cong\lambda/(2\cdot D\cdot\cos\vartheta_B)$, where λ is the radiation wavelength and ϑ_B the Bragg angle. In case of irregularly shaped blocks, D may be assumed to be the average dimension along the scattering vector direction. In Fig. 3 the size D of the variant II ordered domains, calculated from the measurements of the FWHM of the $3/2, \overline{3}/2, 3/2$ reflection, is reported as a function of the miscut angle. The values of 80 Å, 430 Å and 850 Å found for the 0°, 2° and 6° off samples respectively are in good agreement with the ordered domain size (limited by the APB spacing) in the $[1\overline{1}1]$ direction in the cross section TEM images (Fig. 1).

Additional information on the domain properties was obtained from the integrated intensities of the $3/2, \overline{3}/2, 3/2$ XRD variant II reflections, reported in Fig. 3. The decrease of the integrated intensity observed for the 6° off sample may only be explained by a decrease of the value of its ordering parameter. According to the kinematical theory, in fact, the other possible cause of this behaviour is ascribed to a decrease of the variant II domain volumes that can be ruled out by the above TEM and XRD findings. This result is confirmed by earlier optical investigations, summarised in Fig. 4.

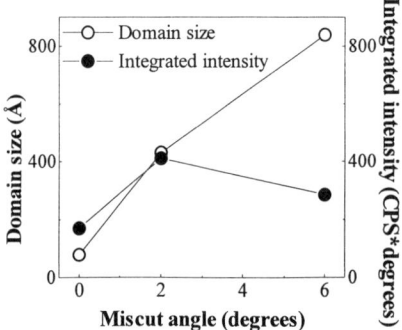

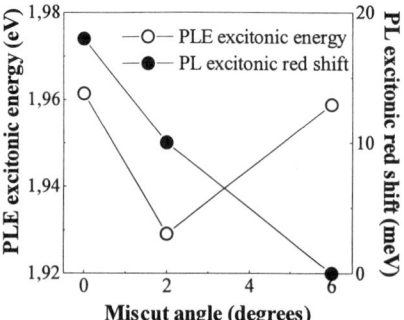

Fig. 3. Variant II domain size and integrated intensity of the $3/2, \overline{3}/2, 3/2$ reflection versus the substrate miscut angle.

Fig. 4. PLE excitonic peak energy and PL energy red shift versus the substrate miscut angle. T = 18 °K.

The higher excitonic peak energy of the 6° off sample, obtained from low-temperature PLE measurements, is indicative of a significantly lower degree of order in the more misoriented sample.

Moreover, as seen from the plot of Fig. 4, increasing the miscut angle results in a significant decrease in the red shift of the PL excitonic energy with respect to the PLE, that was interpreted in terms of an increase of the homogeneity of η inside the ordered domains. The overlapping of the PL and PLE excitonic peaks in the 6° off sample, together with the very narrow FWHM of the PL band (6 meV) and the insensitivity of the PL excitonic energy to a temperature increase at least up to 45 °K reported in Nasi et al 1995, led to the evidence that the distribution of its order parameter must be very sharp although its mean value is lower than for the 0° and 2° off specimens.

PL and CL investigations gave identical emission spectra (Nasi et al 1996). Taking advantage of the possibility of getting low-temperature monochromatic CL images, optical emission from sample areas with different degrees of ordering have been studied by scanning the excitonic peaks. However, due to the CL resolution, only a slight difference between monochromatic images taken at different excitonic peak position was found. Figure 5 shows SCL monochromatic pictures of the 0°, 2° and 6° off samples taken at the excitonic peak maximum wavelength.

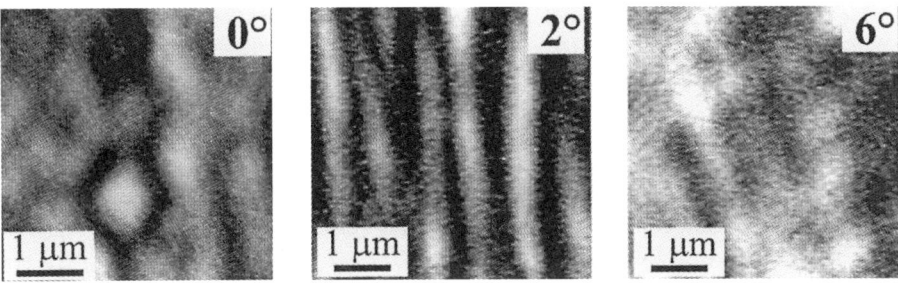

Fig. 5. SCL pictures of the differently misoriented samples taken at a wavelength corresponding to the excitonic peak maximum. T = 10 °K. The samples are oriented as in the AFM pictures of Fig. 2.

The unusual inhomogeneity in the CL emission efficiency, also observed by increasing the accelerating voltage, seems to reflect the AFM surface morphology shown in Fig. 2. Due to the preferentially growth of APBs at surface step edges or undulations, an attempt was made to correlate the low efficiency CL emission to the presence of boundaries between ordered domains. The higher band gap associated with the boundaries could act as a barrier to provide, at low sample temperature, spatial separation of the carriers. A possible alternative explanation would be correlated to refractive index fluctuations of the APBs (Geng et al 1997). Finally, a possible role of surface recombination velocity on CL emission cannot be ruled out and, in this respect, further investigations are in progress.

4. CONCLUSIONS

In summary, CuPt$_B$-type ordering was studied by both structural and optical techniques. The correlation between dark-field TEM images of the ordered domains and the AFM surface images of InGaP layers revealed that a different surface step rearrangement during growth results in a different morphology of the superstructure and their boundaries.

XRD was used successfully for estimation of the ordered domain size and the degree of ordering, in agreement with PL and PLE results. In particular, according to Su et al (1994), an increase in the misorientation angle from 0° to 6° lead to a decrease of the ordering degree, although only one variant is formed and the distance between APBs increases significantly. Moreover the single variant samples with a lower density of APBs show a very high homogeneity of the ordering degree.

Finally, an attempt was made to correlate the inhomogeneity in the CL emission efficiency from InGaP layers with the fluctuations of the electronic properties of the materials in proximity to APBs.

ACKNOWLEDGEMENTS: The authors would like to thank F. Vidimari of Telettra-Alcatel (Vimercate-MI, Italy) for growing the samples and M. Scaffardi for technical support.

REFERENCES

Ernst P, Geng C, Hahn G, Scholz F, Schweizer H, Phillipp F and Mascarenhas A 1995J. Appl. Phys. **79**, 2633

Francesio L, Alagna L, Capelle B, Ferrari C, Franzosi P and Sauvage M 1996 Proceed. 3rd European Symposium on X-Ray Topography and High Resolution X-Ray Diffraction

Friedman D J, Horner G S, Kurtz Sarah R, Bertness K A, Olson J M and Moreland J 1994 Appl. Phys. Lett. **65**, 878

Geng C, Moritz A, Heppel S, Mühe A, Kuhn J, Ernst P, Schweizer H, Phillipp F, Hangleiter A and Scholz F 1997 J. Crystal Growth **170**, 418

Nasi L, Lazzarini L, Salviati G, Fermi F and Lenzi G 1995 Microscopy of Semiconducting Materials, Inst. Phys. Conf. Ser. No. **146**, 249

Nasi L, Salviati G, Mazzer M and Zanotti-Fregonara C 1996 Appl. Phys. Lett.**68**, 3263

Seong Tae-Yeon, Norman A G, Booker G R and Cullis A G 1994 J. Appl. Phys. **75**, 7852

Su L C, Ho I H, Kobayashi N and Strinfellow G B 1994 J. Crystal Growth **145**, 140

Zunger A and Mahajan S 1994 Handbook of Semiconductor Vol **3** (19), 1399

Inst. Phys. Conf. Ser. No 157
Paper presented at Microsc. Semicond. Mater. Conf., Oxford, 7–10 April 1997
© 1997 IOP Publishing Ltd

TEM and TED studies of order-induced GaInP heterostructures grown by organometallic vapour phase epitaxy

J H Kim,[1] T-Y Seong,[1] Y S Chun,[2] and G B Stringfellow[2]

[1]Department of Materials Science and Engineering, Kwangju Institute of Science and Technology (K-JIST), Kwangju 506-712, Korea.
[2]Department of Materials Science and Engineering, University of Utah, Salt Lake City, Utah 84112, USA.

ABSTRACT: The effects of V/III ratio on ordering and antiphase boundaries (APBs) in $Ga_{0.5}In_{0.5}P$ layers grown on GaAs vicinal substrates at 670 °C have been investigated using TEM and TED. The degree of order was higher in the layer grown using a V/III ratio of 160 than in the layer grown using a V/III ratio of 40. The higher V/III ratio could be used to suppress APBs. In addition, the order-induced heterostructures could be used to block the propagation of APBs. Possible mechanisms are proposed to describe these phenomena.

1. INTRODUCTION

CuPt-type ordering in heteroepitaxial (001) layers of III-V compound semiconductors grown by organometallic vapour phase epitaxy (OMVPE) and molecular beam epitaxy (MBE) has been extensively investigated using TEM and TED techniques [Gomyo et al. (1994), Jun et al. (1996), Bellon et al. (1989), Stringfellow and Chen (1991), Baxter and Stobbs (1994), Jen et al. (1989), Seong et al. (1994a and b), Chen et al. (1991)]. It is well known that CuPt-type ordering is a surface related phenomenon occurring during epitaxial growth [Chen et al. (1991), Zunger and Mahajan (1994), Philips et al. (1994), Suzuki and Gomyo (1991)]. TED examination has shown that such ordering typically occurs on two of the four possible {111} planes, i.e., the $(\bar{1}11)$ and $(1\bar{1}1)$ planes. TEM dark field (DF) examination has revealed the individual domains for the $(\bar{1}11)$ and $(1\bar{1}1)$ ordered variants in MBE and OMVPE III-V alloy layers [Bellon et al. (1989), Baxter and Stobbs (1994), Seong et al. (1994a and b), Arent et al. (1993)]. The ordered domains commonly contained a number of antiphase boundaries (APBs). Such defects are known to adversely influence the optical and electrical properties of the layers. It is, therefore, very important to control APBs to enhance device performance.

In this paper, TEM and TED studies have been made of OMVPE GaInP layers grown onto (001) vicinal GaAs substrates to investigate effects of V/III flux ratio on the degree of order and the density of APBs in the ordered GaInP layers. Possible mechanisms are discussed.

2. EXPERIMENTAL

$Ga_{0.5}In_{0.5}P$ layers were grown in a horizontal, atmospheric pressure OMVPE reactor using trimethylindium, trimethylgallium and tertiarybutylphosphine (TBP). The substrates were Cr-doped semi-insulating (001) GaAs misoriented by 3° in the $[1\bar{1}0]$ direction. A GaAs buffer layer ~30nm thick was grown first and then followed by the GaInP layer grown at 670 °C with a rate of ~0.14nm/s. To investigate the effects of V/III flux ratio on ordering behaviour, the two types of samples with order-induced heterostructures were grown using a change in input V/III ratio: the type I sample was grown using an input V/III ratio of 160 (termed here '160 layer') for 48 min, after which the input V/III ratio was reduced to 40 (termed here '40 layer') and growth continued for 36 min. Growth conditions for the type II sample were the same as those of the type I sample, except that the order of the GaInP layers was inverted. The layers were all lattice-matched to the substrates, as determined from X-ray diffraction measurements. Pairs of orthogonal <110> cross-section (CS) thin foil films were prepared by mechanical polishing followed by Ar+ ion milling using a liquid N_2 cold stage. TEM and TED examination was performed in a JEM 2010 instrument operated at 200kV. The convergent beam electron diffraction technique was employed to determine the polarity between the [110] and [110] directions, and the thin foil thickness [Kelly et al. (1975)]. As for the type I sample, the thin foil thicknesses were measured to be ~186nm near the layer surface, ~234nm near the interface between the 40 and 160 layers, and ~363nm near the interface between the 160 layer and GaAs substrate.

3. RESULTS AND DISCUSSION

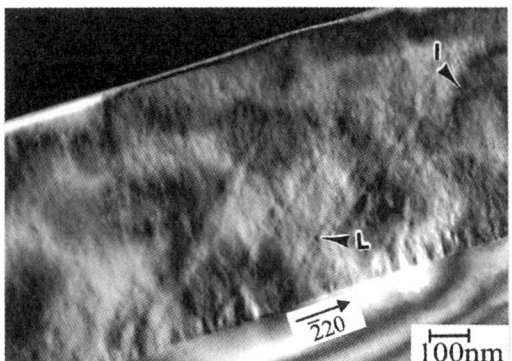

Fig.1 TEM ($\bar{2}$20) DF image of the type I sample.

A [110] CS TEM strain sensitive ($\bar{2}$20) DF image of the type I sample is shown in Fig. 1. The interface between the two layers grown using different V/III ratios is marked 'I'. Composition sensitive (002) DF results showed that although there are slight fluctuations in interfacial planarity of ±1 nm, the interface is relatively abrupt and sharp with 0.5nm resolution. The upper layer, ~300nm thick, was grown using an input V/III ratio of 40. The lower layer, ~400nm thick, was grown using an input V/III ratio of 160. For both layers, the image exhibits an ill-defined fine scale modulated contrast (8-15nm across) which is elongated along the [001] growth direction. The modulated contrast is a characteristic of alloy clustering occurring possibly as a result of the two dimensional spinodal decomposition [Seong et al. (1994c)]. There are line contrasts 1.0μm long inclined ~38° from the [110] direction (marked 'L'). TEM DF examination, to be discussed below, showed that the contrast is due to the strain associated with APBs. There is a strong modulated contrast (18-27nm across) near the 160 layer/GaAs interface. This contrast is also associated with APBs. These results indicate that the APB acts as a strong centre of strain fields leading to local thin foil bending, as observed even in the strain insensitive (002) DF images. Apart from these features, there are no other defects such as threading dislocations and stacking faults.

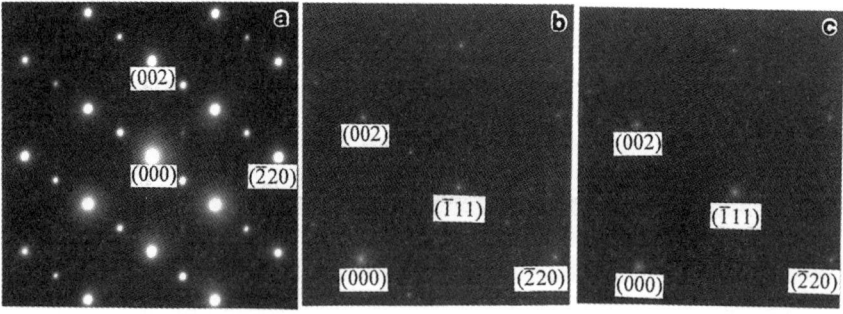

Fig.2 [110] TED patterns from the type I sample showing the main spots and ½{111}B extra spots.

Figure 2 shows [110] TED patterns taken from the type I sample. The pattern (Fig.2(a)) taken from the region including both 40 and 160 layers shows the fundamental spots and ½{111}B extra spots, indicating that long-range CuPt-type ordering occurs on the ($1\bar{1}1$) plane with much weaker ordering on the ($\bar{1}11$) plane. The observation of both variants is somewhat surprising, since misorientation toward [1$\bar{1}$0] typically produces only one variant [Bellon et al. (1989), Chen et al. (1991), Philips et al. (1994), Suzuki and Gomyo (1991)]. This could be attributed to the presence of different types of surface steps on the substrates, i.e., the steps introduced by the misorientation and the thermal steps [Pashley et al.(1988)] occurring during epitaxial growth. The ($1\bar{1}1$) variant, however, is overwhelmingly dominant over the ($\bar{1}11$) variant. This is because the number of steps introduced by the misorientation is far greater than that of thermal steps. The ½($1\bar{1}1$) superlattice spots are circular, indicating a low density of APBs in the ordered regions and/or the presence of large ordered domains [Seong et al.(1994a and b)]. The ½($\bar{1}11$) superlattice spots are elongated and inclined, indicating either a high density of planar defects in the ordered regions or the presence of small and narrow ordered domains [Bellon et al. (1989), Seong et al.(1994a and b)].

Figs. 2(b) and (c) show TED patterns obtained from the 160 and 40 layers, respectively, using the smallest selected area aperture. Comparison of the patterns shows that the degree of order is higher in the 160 layer than in the 40 layer. The results could be interpreted as follows. The higher input of phosphorus stabilises the (2x) surface reconstruction where the surface is covered with [$\bar{1}$10]-oriented phosphorus dimers. Murata et al.(1996), using surface photoabsorption (SPA), showed that the concentration of the [$\bar{1}$10]-oriented phosphorus dimers, which are characteristics of the (2x4) reconstruction, is increased with increasing the TBP partial pressure. Since the (2x) reconstructed surface is required to produce CuPt ordering [Chen et al. (1991), Zunger and Mahajan (1994), Philips et al. (1994), Suzuki and Gomyo (1991)], such a surface reconstruction leads to an increase in the degree of the order. This result is consistent with those obtained by Murata et al. (1996).

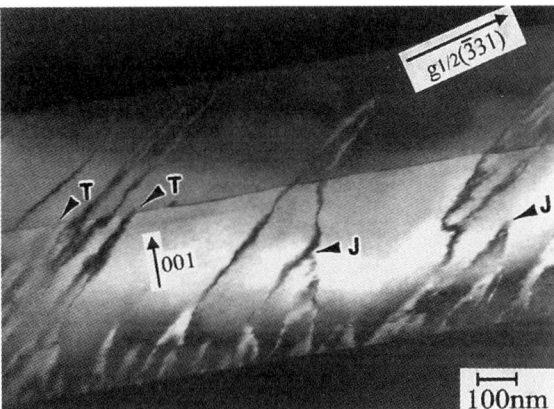

Fig 3. ½($\bar{3}$31) TEM DF image from the type I sample.

A ½($\bar{3}$31) TEM DF image from the type I sample (Fig. 3) reveals the ($1\bar{1}1$) ordered material as the bright regions. Within the ordered region there are wavy inclined dark lines 5~15nm across, corresponding to APBs, few of which propagate through the full layer thickness of ~0.7µm at an angle of ~38° to the (001) plane. The density of APBs decreases throughout the layer. It is ~8.1x10⁹/cm² near the interface, ~2.0x10⁹/cm² near the interface between the 40 and 160 layers, and ~1.5x10⁹/cm² in the middle of the 40 layer. Some of the APBs (marked 'J') within the 160 layer have joined together and annihilated one another. Other APBs (marked 'T') were terminated at the interface between the 40 and 160 layers, indicating that an increase in the V/III ratio retards the propagation of the APBs. The result suggests that multilayers or artificial superlattice layers consisting of individual ordered layers grown using different V/III ratios may be used to suppress (or eventually eliminate) the APBs. Note that a large decrease in the APBs' density occurs at the interface between the 40 layer and the 160 layer. This may be related to the stabilised [110] steps due to the high TBP partial pressure [Su and Stringfellow (1995)]. Su and Stringfellow (1995) suggested that the APBs arise from bunched steps. It was shown that an increase in the TBP partial pressure stabilises the bilayer steps, consequently preventing step bunching [Chun et al.(1996)].

Analogous TEM examination has been made of the type II sample to investigate APBs' behaviour. The results showed that APBs 5~15nm across propagate through the full layer thickness of ~0.7μm at an angle of ~37° to the (001) plane. The APB density again decreases throughout the layer, although very few APBs were terminated at the interface between the 160 and 40 layers as compared to the type I sample. This indicates that the inverted structure is less effective in suppressing APB propagation.

The effects of the V/III ratio on the degree of order were investigated using a new TEM technique demonstrated by Baxter et al.(1994) for obtaining quantitative information on the degree of order in GaInP layers. The ratio of the intensity profiles in the ½($\bar{3}$31) DF images (Fig. 3) and ($\bar{2}$04) DF images (not shown) obtained from both type I and II samples showed that the degree of order is higher in the 160 layer than in the 40 layer. This is in good agreement with TED results (Figs. 2(b) and 2(c)).

Examination of the interfaces corresponding to the change in TBP flow rate in Fig. 3 reveals the change in order parameter as a change in brightness, although the change is small because the change in order parameter is small, as indicated by the 25-30meV change in the bandgap energy. This preliminary data indicates that the change in order parameter is abrupt. The interfaces appear to be slightly wavy.

4. CONCLUSION

It was shown that the density of APBs can be reduced by producing a change in order parameter via a change in the input V/III ratio during growth. Changing to a high TBP partial pressure appears to hinder the propagation of APBs, resulting in fewer APBs in the upper layer. TEM DF results showed that the interfaces of the heterostructures were fairly sharp and abrupt although slightly wavy. TEM and TED results of both type I and II samples showed that the degree of order is higher in the 160 layer than in the 40 layer.

ACKNOWLEDGMENTS

The authors wish to thank Korea Science and Engineering Foundation (KOSEF) for financial support. GBS wishes to thank the Department of Energy for partial support of his work.

REFERENCES

Arent D J, Bode M, Bertness K A, Kurtz S R and Olson J M 1993 Appl. Phys. Lett. **62**, 1806
Baxter C S and Stobbs W M 1994 Phil. Mag. **69**, 615
Bellon P, Chevalier J P, Augarde E, André J P and Martin G P 1989 J. Appl. Phys. **66**, 2388
Chen G S, Jaw D H and Stringfellow G B 1991 J. Appl. Phys. **69**, 4263
Chun Y S, Murata H, Hsu T C, Ho I H, Su L C, Hosokawa Y and Stringfellow G B 1996 J. Appl. Phys. **79**, 6900
Gomyo A, Makita K, Hino I and Suzuki T 1994 Phys. Rev. Lett. **72**, 673
Jen H R, Ma K Y and Stringfellow G B 1989 Appl. Phys. Lett. **54**, 1154
Jun S W, Seong T-Y, Lee J H and Lee B 1996 Appl. Phys. Lett. **68**, 3447
Kelly P M, Jostons A, Blake R G and Napier J G 1975 Phys. Stat. Sol. **31**, 771
Murata H, Ho I H, Su L C, Hosokawa Y and Stringfellow G B 1996 J. Appl. Phys. **79**, 6895
Pashley M D, Haberern K W, Friday W, Woodall J M and Kirchner P D 1988 Phys. Rev. Lett. **60**, 2176
Philips B A, Norman A G, Seong T-Y, Mahajan S, Booker G R, Skowronski M, Harbison J P and Keramidas V G 1994 J. Cryst. Growth, **140**, 249
Seong T-Y, Booker G R, Norman A G and Ferguson I T 1994a Appl. Phys. Lett. **64**, 3593
Seong T-Y, Norman A G, Booker G R and Cullis A G 1994b J. Appl. Phys. **75**, 7852
Seong T-Y, Booker G R and Norman A G 1994c Inst. Phys. Conf. Ser. **134**, 301
Stringfellow G B and Chen G S 1991 J. Vac. Sci. Technol. **B 9**, 2182
Su L C and Stringfellow G B 1995 J. Appl. Phys. **78**, 6775
Suzuki T and Gomyo A 1991 J. Cryst. Growth **111**, 353
Zunger A and Mahajan S 1994 in Handbook on Semiconductors, edited by S. Mahajan, (Elsevier Science, Amsterdam), Vol.3, Chapt. 19.

Inst. Phys. Conf. Ser. No 157
Paper presented at Microsc. Semicond. Mater. Conf., Oxford, 7–10 April 1997
© 1997 IOP Publishing Ltd

TED, TEM and AFM studies comparing atomic ordering in InAs$_y$Sb$_{1-y}$ layers grown by MOVPE and MBE

T Y Seong[*], **G R Booker, A G Norman, P J F Harris**[1] **and G B Stringfellow**[2]

Department of Materials, University of Oxford, Parks Road, Oxford OX1 3PH, UK
[1]Department of Chemistry, University of Reading, Berks, UK
[2]Department of Materials Science & Engineering, University of Utah, Salt Lake City, Utah 84112, USA

ABSTRACT: CuPt-type atomic ordering in InAs$_y$Sb$_{1-y}$ layers grown by molecular beam epitaxy (MBE) and metal organic vapour phase epitaxy (MOVPE) on (001) substrates was investigated using TED and TEM. The degree of ordering was greater and occurred at higher temperatures, and the ordered domains were larger, in the MOVPE layers compared with the MBE layers. AFM showed surface ridges along the [110] direction for the MOVPE layers and along the [$\bar{1}$10] direction for the MBE layers. We suggest mechanisms to explain why the differences in ordering for the MOVPE and MBE layers occur.

1. INTRODUCTION

Previous work on ordering in MOVPE InAsSb has been reported by Jen et al (1989), Biefield et al (1994) and Follstaedt et al (1994), and in MBE InAsSb by Seong et al (1990 and 1994) and Kurtz et al (1990). In the present work we have examined further some of the MOVPE and MBE layers that we investigated previously (Jen et al 1989, Seong et al 1994). Our TED, TEM and AFM results show that the MOVPE and MBE layers exhibit different degrees of ordering, different ordered domain sizes and shapes, and different surface topographies. We suggest mechanisms to explain why the differences in ordering for the MOVPE and MBE layers occur.

Single MBE InAs$_y$Sb$_{1-y}$ layers of nominal compositions in the range y = 0.4 to 0.6 were grown at temperatures in the range 320 to 470°C and a rate of 0.27 nm/s on InAs buffer layers, which were grown on (001) GaAs substrates. Single MOVPE InAs$_y$Sb$_{1-y}$ layers of nominal compositions in the range y = 0.25 to 0.53 were grown at 450 and 480°C and rates in the range 0.15 to 0.33 nm/s on (001) InSb or InAs substrates. The layers were ~ 2 μm thick and were nominally undoped. For the TED/TEM examinations two orthogonal [110] and [$\bar{1}$10] cross-section thin foil specimens were prepared for each layer by mechanical polishing followed by Ar$^+$ ion thinning with the specimen in a liquid nitrogen cooled holder. Distinction between the [110] and [$\bar{1}$10] cross-sections was made using convergent beam electron diffraction techniques.

2. RESULTS

For the MBE growth in the range 320 to 400°C, TEM showed that each layer had spontaneously phase-separated during growth to give a series of thin plates approximately parallel to the (001) substrate surface (Seong et al 1993). The plates were of non-uniform thickness, alternated in composition from ~ InAs$_{0.3}$Sb$_{0.7}$ to ~ InAs$_{0.7}$Sb$_{0.3}$ and were elastically strained with tetragonal crystal lattices c/a > 1 and c/a < 1 respectively. Each layer comprising such plates was termed a natural strained layer superlattice (NSLS). For the MBE growth in the range 430 to 470°C and the MOVPE growth at 450 and 480°C, single composition layers occurred. RHEED patterns obtained from MBE layers grown at 340, 370, 400 and 430°C showed streaks corresponding to a (2x3) atomic surface reconstruction. InAsSb layers could not be grown by MOVPE at the low temperatures possible with MBE because there was insufficient energy for the organo-metallic incident

[*]Now at Department of Materials Science & Engineering, Kwangju Institute of Science and Technology, Kwangju 506-303, Korea.

molecules (available at that time) to dissociate.

For the $[\bar{1}10]$ cross-section TED patterns obtained from all of the layers, there were no $1/2(11\bar{1})$ or $1/2(111)$ superspots and hence no ordering on the $(11\bar{1})$ or (111) planes. For the $[110]$ cross-section TED patterns obtained from many of the layers, there were $1/2(\bar{1}11)$ and $1/2(1\bar{1}1)$ superspots and hence ordering on the $(\bar{1}11)$ and $(1\bar{1}1)$ planes. For example, for the 370°C MBE layer the $[110]$ TED patterns showed superspots that were elongated and tilted and of medium intensity (Fig. 1a), for the 430°C MBE layer the superspots were similar but of weak intensity, and for the 450°C MOVPE layer the superspots were circular and of strong intensity (Fig. 1b).

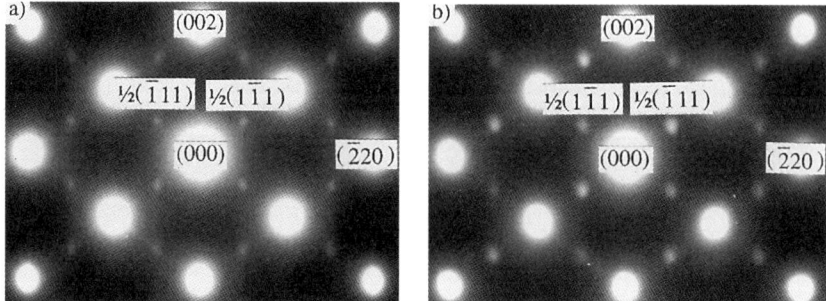

Fig. 1 $[110]$ cross-section TED patterns from (a) an MBE $InAs_{0.5}Sb_{0.5}$ layer grown at 370°C and 0.27 nm/s, and (b) an MOVPE $InAs_{0.39}Sb_{0.61}$ layer grown at 450°C and 0.22 nm/s, showing the main spots and $1/2\{111\}$ superspots.

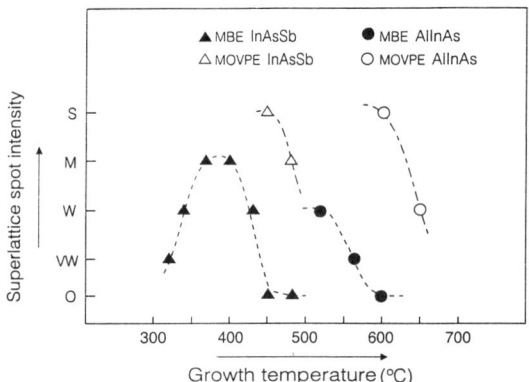

Fig. 2 Intensity of superlattice spots in $[110]$ cross-section TED patterns from InAsSb and AlInAs epitaxial layers grown by MOVPE and MBE as a function of growth temperature. S - strong, M - medium, W - weak, VW - very weak, O - zero.

The intensities of the superspots are plotted on an arbitrary scale against growth temperature in Fig. 2. For the MBE layers, there is a maximum in the degree of ordering at 370 to 400°C. There is no discontinuity in the ordering curve between 400 and 430°C, i.e. when the growth changes from a NSLS to a single composition layer, indicating that the change from severely strained to relatively unstrained material does not have a significant effect on the ordering. For the present MOVPE InAsSb layers, there were insufficient points to plot out the full ordering curve. Nevertheless the results indicate that for MOVPE, the degree of ordering is greater and occurs at higher growth temperatures than for MBE. Our earlier MBE and MOVPE AlInAs ordering results (Norman 1987, Norman et al 1987) showed similar trends (Fig. 2).

$[110]$ cross-section TEM DF images were obtained using individual superspots to observe directly the individual ordered domains. For the 370°C MBE layer, the $1/2(\bar{3}3\bar{1})$ spot showed the $(\bar{1}11)$ ordered material (variant I) as small bright blobs up to ~ 10 nm across (Fig. 3a) and these micro-domains were often elongated and tilted. The $1/2(\bar{3}31)$ spot showed the $(1\bar{1}1)$ ordered material (variant II) as similar bright blobs, and these micro-domains were also often elongated and titled. The micro-domains were more pronounced in the As-rich plates of the NSLS than the Sb-rich plates. The tilted micro-domains in the TEM images correlated with the tilted superspots in the TED patterns. Microtwin lamella on inclined $\{111\}$ planes often occurred. For the 430°C MBE layer, the superspots were not strong enough to give TEM DF images. However, it can be deduced from the elongated and tilted superspots for this layer that small elongated and tilted domains were present.

For the 450°C MOVPE layer, the $1/2(\bar{3}31)$ spot showed the variant II ordered material as bright columnar domains ~ 200 to 600 nm wide and running from the substrate to the layer surface. Two such domains labelled II are shown in Fig. 3b. The $1/2(\bar{3}3\bar{1})$ spot showed the variant I ordered material as similar bright columnar domains located between the variant II domains. One such domain obtained using the $1/2(\bar{3}31)$ spot appears dark and is labelled I in Fig. 3b. The boundary between adjacent domains was sometimes curved (C) and sometimes planar, the latter boundary corresponding to a micro-twin lamella. The variant I and II domains contained antiphase boundaries (APBs).

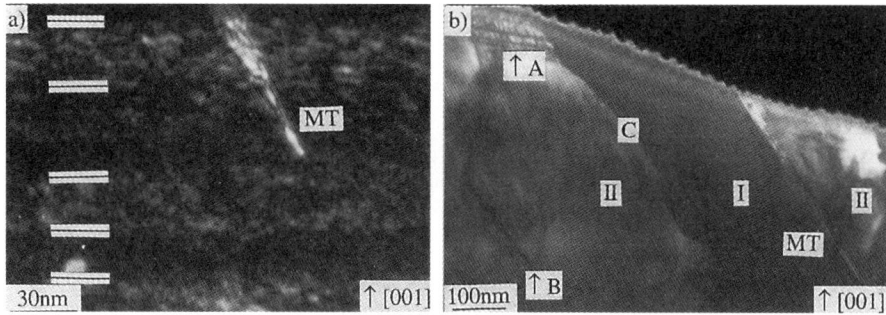

Fig. 3 [110] cross-section TEM DF images from (a) an MBE $InAs_{0.4}Sb_{0.6}$ layer grown at 370°C and 0.27 nm/s, and (b) an MOVPE $InAs_{0.39}Sb_{0.61}$ layer grown at 450°C and 0.22 nm/s. A and B - antiphase boundaries, C - curved domain boundary, MT - micro-twin lamella.

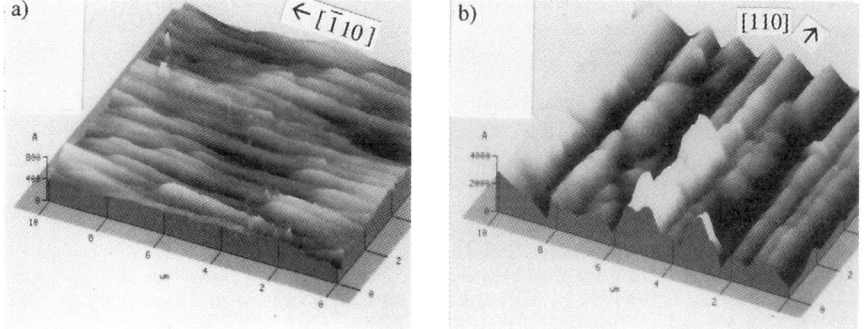

Fig. 4 AFM images. (a) MBE $InAs_{0.4}Sb_{0.6}$ layer grown at 370°C showing surface ridges along $[\bar{1}10]$ direction. (b) MOVPE $InAs_{0.39}Sb_{0.61}$ layer grown at 450°C showing surface ridges along [110] direction.

For the 370°C MBE layer, of Fig. 3a, AFM 3-D images showed surface ridges along the $[\bar{1}10]$ direction (Fig. 4a). The average ridge height (peak-to-valley) was 25 nm and the average width (valley-to-valley) was 0.7 μm. The average slope of the ridges facing in either the [110] or $[\bar{1}\bar{1}0]$ directions was 4°. For the 430°C MBE layer, AFM showed surface ridges also along the $[\bar{1}10]$ direction, the images being similar to those of Fig. 4a. The ridge height was 50 nm and the width was 0.7 μm. The slope of the ridges facing in either the [110] or $[\bar{1}\bar{1}0]$ directions was 7°. For the 450°C MOVPE layer, of Fig. 3b, AFM showed surface ridges along the [110] direction (Fig. 4b). The ridge height was 150 nm and the width was 1.0 μm. The slope of the ridges facing in either the $[\bar{1}10]$ or $[1\bar{1}0]$ directions was 10°.

3. DISCUSSION

Consideration by previous workers of atomic models suggested that the CuPt-type ordering for ternary and quaternary III-Vs occurred as follows. The ordering grows in at the (001) layer surface and arises from the atomic bond reconstruction that is present at the layer surface together with phase-locking that arises from the presence of surface atomic steps. The bond reconstruction produces regular arrays of atomic blocks at the surface, e.g. (2x3), the (2x) component corresponds to Group V surface dimers along the $[\bar{1}10]$ direction, and it is these dimers that initiate the ordering. A combination of dimers along $[\bar{1}10]$ and atomic steps along [110] is most favourable for ordering on {111}B planes, with steps facing $[\bar{1}10]$ giving ordering on $(\bar{1}11)$ planes (variant I domain), and steps facing $[1\bar{1}0]$ giving ordering on $(1\bar{1}1)$ planes (variant II domain). A combination of dimers along $[\bar{1}10]$ and steps along $[\bar{1}10]$ is least favourable for ordering on {111}B planes, little ordering occurring on the $(\bar{1}11)$ and $(1\bar{1}1)$ planes. For a review of CuPt-type ordering, see Norman et al (1993).

We suggest the following possible reasons to explain why the atomic ordering in the InAsSb layers occurs at higher growth temperatures for MOVPE then MBE. First, the (2x3) surface reconstruction required for the ordering occurs over a higher temperature range for MOVPE than MBE due to the effect of the different gaseous environments and molecular species present. Kamiya et al (1992) showed experimentally that the (2x4) surface reconstruction in GaAs occurs at a higher temperature for MOVPE than MBE and attributed this to the higher partial pressure of As-containing species present for MOVPE. An analogous behaviour may occur for InAsSb.

Second, the decrease in ordering that occurs for MBE on going from 400 to 450°C, while the ordering for MOVPE at 450°C is strong (Fig. 2), arises because annealing-out of the ordering occurs in the MBE layers faster than it does in the MOVPE layers at these temperatures. This annealing-out takes place in the bulk layer after the ordering has grown in at the surface and occurs because CuPt-type ordered material is less stable in the bulk semiconductor than unordered material. The reason it occurs faster in MBE is that the domains are small and poorly ordered, whereas in MOVPE the domains are large and well ordered.

We suggest that the main reason the domains are different for MOVPE and MBE is as follows. For MOVPE the surface ridges are along [110], this gives surface atomic steps along [110] and these are favourable for ordering. The sides of the ridges facing $[\bar{1}10]$ give well formed variant I domains and the sides facing $[1\bar{1}0]$ give well formed variant II domains. The mean width of individual domains is half the valley-to-valley separation, i.e. ~ 0.5 μm (from AFM). The ridges continue to propagate as the layer grows and so the individual domains propagate through the layer (Fig. 3b). For MBE the ridges are along $[\bar{1}10]$, the atomic steps are along $[\bar{1}10]$ and these are unfavourable for ordering. Consequently small growth islands of poorly ordered material on either the $(\bar{1}11)$ or $(1\bar{1}1)$ planes form in a random manner on both sides of the ridges, and so a 3-D distribution of variant I and II microdomains occurs throughout the layers (Fig. 3a).

REFERENCES

Biefield RM, Baucom KC and Kurtz SR, 1994 J. Cryst. Growth 137 231.
Jen HR, Ma KY and Stringfellow GB, 1989 J. Appl. Phys. 69 1154.
Follstaedt DM, Biefield RM, Kurtz SR, Dawson LR and Baucom KC, 1995 Inst. Phys. Conf. Ser. 144 224.
Kamiya I, Aspnes DE, Tanaka H, Florez LT, Harbison JP and Bhat R, 1992 Phys. Rev. Lett. 68 627.
Kurtz S, Olsen JM and Kibbler A, 1990 Appl. Phys. Lett. 57 1922.
Norman AG, 1987 D. Phil. thesis, Oxford University.
Norman AG, Mallard RE, Murgatroyd IJ, Booker GR, Moore AH and Scott MD, 1987 Inst. Phys. Conf. Ser. 87 77.
Norman AG, Seong TY, Philips BA, Mahajan S and Booker GR, 1993 Inst. Phys. Conf. Ser 134 279.
Seong TY, Norman AG, Booker GR, Droopad R, Williams RL, Parker SD, Wang PD and Stradling RA, 1990 MRS Symp. Proc. 163 907.
Seong TY, Norman AG, Ferguson IT and Booker GR, 1993 J. Appl. Phys. 73 8227.
Seong TY, Booker GR, Norman AG and Ferguson IT, 1994 Appl. Phys. Lett. 64 3593.

Inst. Phys. Conf. Ser. No 157
Paper presented at Microsc. Semicond. Mater. Conf., Oxford, 7–10 April 1997
© *1997 IOP Publishing Ltd*

Twinning of As precipitates in LT-GaAs

Ch Dieker,* S Ruvimov, H Sohn, J Washburn, and Z Liliental-Weber

Lawrence Berkeley National Laboratory, Berkeley, CA 94720 ms. 62/203
*on leave from: Institut für Schicht- und Ionentechnik, Forschungszentrum Jülich GmbH,
D-52425 Jülich, Germany

ABSTRACT: Twinning in As precipitates after 850°C annealing of LT-GaAs layers has been revealed by transmission electron microscopy. It was shown that twins are formed on $(10\bar{1}4)$ planes and that voids are often present in the same precipitates. Formation of such twins does not preserve the 3m orientation relationship as in the twinning observed in layers annealed at 600°C in larger precipitates. A large number of precipitates have two parallel twins.

1. INTRODUCTION

GaAs grown at temperature about 200°C (LT-GaAs) can be used to eliminate side-gating and back-gating in HEMT devices since the resistivity of these LT-layers can be higher than in the best semi-insulating bulk GaAs crystals (Smith et al 1988, Smith et al 1989). These as-grown layers incorporate about 1% excess As which results in an expansion of the lattice parameter (Liliental-Weber et al 1991a). The amount of incorporated As increases with decreasing growth temperature and with increasing beam equivalent pressure (BEP) during growth. Usually layers grown at 200°C are of good structural quality, however further lowering of the growth temperature can lead to the formation of structural defects, called pyramidal defects. As-grown layers are usually conductive, but annealing at 600°C leads to semi-insulating properties. Annealing also results in formation and coarsening of As precipitates. Heterogeneous nucleation takes place followed by growth of semi-spherical precipitates with a diameter range 2-6 nm. These precipitates have a specific (3m) orientation relationship with the GaAs matrix, $(0003)_{As}$ ‖ $<111>_{GaAs}$ (Claverie and Liliental-Weber 1992).

Since annealing of these LT-GaAs layers is necessary for most optoelectronic applications, it is important to understand the annealing dynamics. Previous work by Look et al (1993) of annealed LT-GaAs layers grown at 200°C showed strong anomalies in mobility and carrier concentration in the annealing temperature range 350-450°C. It was shown that the dominant acceptor strongly decreases and then increases again in this temperature range. Above 450°C, donors, acceptors, and As_{Ga} antisite defects all decrease in concentration.

For device applications ion implantation is often used together with LT-layers. To activate the dopant, annealing at high temperature is necessary. Based on previous reports (Warren et al 1992, Melloch et al 1992, Liliental-Weber et al 1995) it is expected that the amount of excess As remains essentially unchanged during annealing in this temperature range, but precipitate coarsening does take place. This coarsening decreases the total energy of the system due to reduction in total precipitate to matrix interface area. As a result small precipitates disappear and fewer larger precipitates remain. In this work we describe twinning in the precipitates which occurs in samples annealed at 850°C and compare it to twinning observed in the samples annealed at 600°C.

2. EXPERIMENTAL

A set of samples grown at 200°C with BEP of 20 has been studied using cross-section electron

284

microscopy. The 002B Topcon electron microscope with a resolution of 1.8 Å was used for this studies. Samples were annealed at 850°C by rapid thermal annealing (RTA). Proximity annealing was used for surface protection.

3. RESULTS AND DISCUSSION

As mentioned above a specific orientation relationship was observed for the layers annealed at 600°C (Fig. 1a) for heterogeneously nucleated precipitates $(0003)_{As} \parallel (111)_{GaAs}$. Nucleation of precipitates with this orientation relationship with the matrix is favored because it permits low energy coherent or semicoherent interfaces between matrix and precipitate whenever the interface is parallel to certain crystallographic planes of the matrix-precipitate crystals. Therefore, the nucleation barrier is smaller for certain orientation relationships. However, some deviations from this orientation relationship could be observed on some precipitates. During growth of the precipitate the coherent or semicoherent interface segments develop as facets. X-ray studies show that the diffraction peak from a LT-GaAs layer overlaps with the peak from the GaAs substrate and the FWHM is only slightly larger than for the substrate. This shows that the LT-layers are not highly strained (Liliental-Weber et al 1991b). For samples where dislocations are present in the layer (lower growth temperature or higher BEP) local strain near the dislocations leads to early nucleation and growth of much larger precipitates in their vicinity. Some of these also have 3m orientation relationship, and some others are nucleated with $(220)_{GaAs} \parallel (0003)_{As}$ (2mm) or $(200)_{GaAs} \parallel (0003)_{As}$ (m).

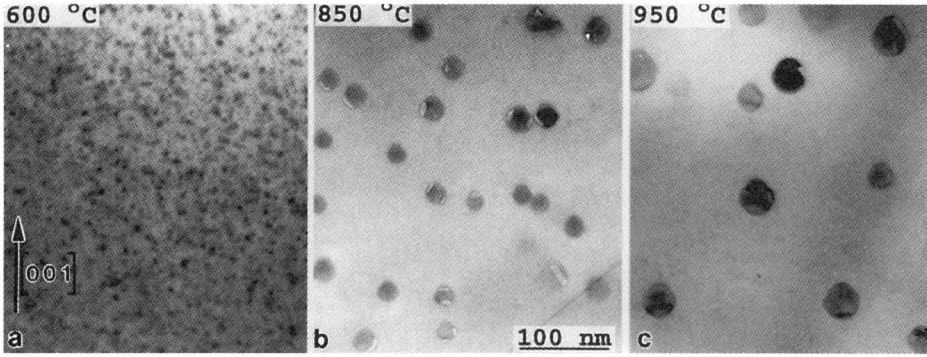

Fig. 1. Annealing of LT-GaAs layers for 10 min at: (a) 600°C, (b) 850°C, and (c) 950°C. Note formation of As precipitates and their coarsening with the increase of annealing temperatures. Note twinning and void formation at higher annealing temperature (b and c). The micrographs have been taken with g=(200).

The previous study showed (Liliental-Weber et al 1992) that these large precipitates are often twin crystals. In this case nucleation of a twin grain must occur during growth of the large precipitate and these twins always maintain the same 3m orientation relationship since the twinning occurs on (0003) basal plane of As with mirror $10\overline{1}2$ planes inclined 58° to the twin plane (Fig. 2).

For high annealing temperatures (850° and up) eg. above the melting point of As (810°) faceting of the precipitates on low-energy planes of GaAs was observed (Figs. 1b and c). As one can see from these figures the As precipitates have the shape of truncated tetrahedrons. In [110] projection two longer and two shorter facets are present. Convergent beam electron diffraction allowed determination of the polarity of these LT-GaAs layers. Based on these studies it was shown that the longer facets are parallel to a Ga surface of GaAs, and the shorter ones to the As surface of GaAs. On the longer facet normal Ga-As bonding is possible between precipitate and matrix, but on the shorter facet bonding is more distorted. It is interesting to note, that the ratio of the facet lengths remain unchanged for different annealing temperatures.

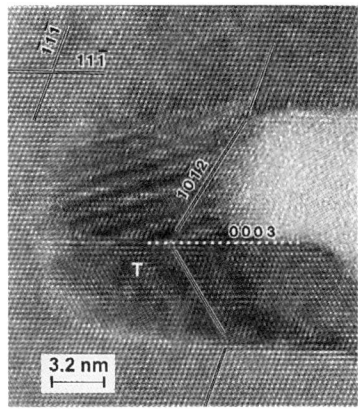

Fig. 2. Twinning of As precipitate formed in the vicinity of extended defect in the layer annealed at 600°C. This twinning occurred on $(0003)_{As}$ basal plane. 3m orientation relationship was preserved after twinning.

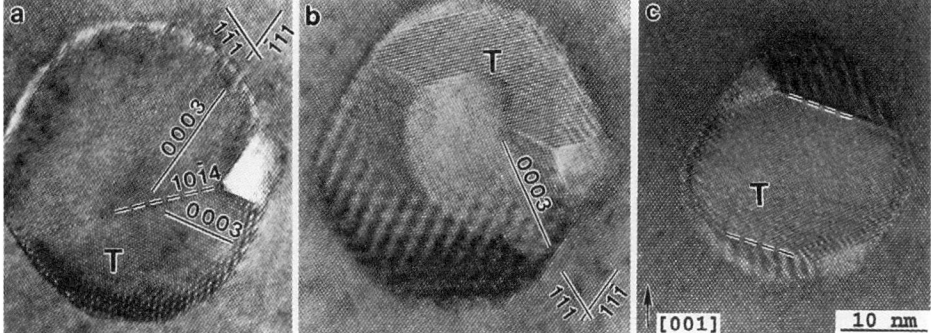

Fig. 3. Twinning of As precipitates along $(10\bar{1}4)$ plane shown for two different variants of 3m orientation relationship: (a) $(\bar{1}11)_{GaAs} \parallel (0003)_{As}$, (b) $(111)_{GaAs} \parallel (0003)_{As}$. Note that this type of twinning does not preserve orientation relationship and that each twin is associated with formation of void; (c) same variant of orientation relationship as (b), but with a few degree misorientation in the untwinned grain. Note formation of the second parallel twin. In this way on both sides of the precipitate 3m orientation relationship is preserved.

When a precipitate is melted by annealing above 810°C the GaAs matrix facets are probably determined by minimum energy of the GaAs surface. The shape that develops while the As is melted is probably largely preserved during rapid solidification. Small faceted voids were also observed in these resolidified precipitates due to the liquid to solid volume contraction of As. Extended x-ray absorption fine strucutre (EXAFS) measurements of amorphous As show that the As-As bond length is greater in the amorphous material than it is in crystalline As (Del Cueto and Shevchik 1987). Therefore, this void formation should be expected.

In a majority of cases this annealing above the melting point of As also leads to twinning. However, in this case the twinning plane is different, e.g. $(10\bar{1}4)$ compared to the (0003) twinning plane observed for annealing below the As melting temperature. Figs. 3a and b show examples of such twinned precipitates. Twinning along $\{10\bar{1}4\}$ planes does not preserve 3m orientation relationship in the twinned part. This twinning must also take place during solidification as a result of growth mistakes during rapid solidification, probably at the matrix/precipitate interface. In many cases it was noticed that during further growth of the twinned volume a second twinning (on a parallel plane) takes place that results in return to the original orientation (Fig. 3c). In this case the favorable matrix/precipitate orientation relationship exists on opposite sides of the precipitate.

Four equivalent variants of the 3m orientation relationship for As precipitates exist and two possible $(10\bar{1}4)$ twins can be formed for each variant. Therefore, 8 twins variants can be expected. In a particular $<110>_{GaAs}$ projection 4 of these twin variants should be observed. However, the experimental results show only one twin variant dominates each precipitate variant. This is probably related to the crystal polarity. It is observed that in a majority of cases the twin is nucleated at the longer facet which is terminated by Ga in the GaAs matrix. Measurement of the angles between a twin plane and a (002) plane of GaAs showed angles slightly different than expected from the determined orientation relationship for the annealing at 600°C. For the annealing temperatures above the melting point of As (e.g. 850°C) the measured deviations of $(0003)_{As}$ plane from $\{111\}_{GaAs}$ plane can be as large as 10° which will influence the twin plane position.

4. SUMMARY

This study shows that twinning on $\{10\bar{1}4\}$ planes is common within As precipitates after annealing of LT-GaAs above the melting point of As (810°C). It is believed that these twins are nucleated during the liquid/solid transformation within the precipitates by growth mistakes promoted by configuration of the matrix/precipitate boundary. This $\{10\bar{1}4\}$ twinning changes the matrix/precipitate orientation relationship for the twinned part of the precipitate. Lower temperature annealing also leads to twinning when some precipitates grow to large sizes during precipitate coarsening. However, this solid state precipitate growth always results in twinning on the $\{0003\}$ twinning plane which does not change the matrix/precipitate orientation relationship within the twinned part of the precipitate.

ACKNOWLEDGMENT

This research was supported by AFOSR-ISSA-90-0009. The use of the facility of the National Center for Electron Microscopy in Lawrence Berkeley National Laboratory supported by the U.S. Department of Energy under Contract No. DE-ACO3-76SF00098 is greatly appreciated.

REFERENCES

Claverie A and Liliental-Weber Z 1992 Phil. Mag. **A 65** 981.
Del Cueto J A and Shevchik NJ 1987 in „Gallium Arsenide" edts. by J.S. Blakemore (American Institute of Physics: New York) p 104.
Liliental-Weber Z, Swider W, Yu K M, Kortright J, Smith F W, and Calawa A R 1991a Appl. Phys. Lett. **58**, 2153.
Liliental-Weber Z, Caverie A, Washburn J, Smith F, and Calawa R 1991b Appl. Phys. **A53**, 141.
Liliental-Weber Z. Claverie A, Werner P, Schaff W, and Weber E R, 1992, in Materials Science Forum „Defects in Semiconductors 16" edts. Davis G, DeLeo G G and Stavola M (Trans Tech Publication) p 1045.
Liliental-Weber Z, Lin X W, Washburn J, and Schaff W, 1996 Appl. Phys. Lett. **66**, 2086.
Look D C, Walters D C, Robinson G D Sizelowe J R, Mier M G, and Stutz C E, 1993, J. Appl. Phys. **74**, 306.
Melloch M R, Otsuka N, Woodall J M, and Kirchner P D 1992, (Mater. Res. Society Proceedings: Pittsburg) **241**, 113.
Smith F W, Calawa A R, Chen C L, Manfra M J, and Mahoney L J, 1988 IEEE, **EDL-9**, 77.
Smith F W, Le H Q, Diadiuk V, Hollis M A, Calawa A R, Gupta S, Frankel M, Dykaar D R, Moureau G A, and Hsiang T Y, 1989, Appl. Phys. Lett. **54**, 890.
Warren A C, Woodall J M, Kirchner P D, Pollak F, Melloch M R, Otsuka N, Mahalingam K, 1992, Phys. Rev. **B 46**, 4617.

Inst. Phys. Conf. Ser. No 157
Paper presented at Microsc. Semicond. Mater. Conf., Oxford, 7–10 April 1997
© 1997 IOP Publishing Ltd

Features of excess arsenic precipitation in LT-GaAs delta-doped with indium

N A Bert[1], V V Chaldyshev[1], Yu G Musikhin[1,2] and P Werner[2]

1) Ioffe Physical-Technical Institute, 26 Polytekhnicheskaya, St.Petersburg 194021 Russia
2) Max-Planck-Institut für Mikrostrukturphysik, Weinberg 2, D-06120 Halle Germany

ABSTRACT: The isovalent indium δ-doping of LT-GaAs films, performed additionally to a uniform one with shallow donors (Si) and acceptors (Be), is shown by TEM to provide As cluster arrays ordered along the growth direction. At the appropriate growth and annealing conditions As/GaAs layered insulating-semiconducting structures have been obtained with GaAs spacers free of clusters between As cluster sheets as thick as 30 nm. In parallel, the indium diffusivity in LT-GaAs is estimated to be 5×10^{-18} cm^2/s in the temperature range from 500 to 600 °C.

1. INTRODUCTION

Gallium arsenide layers grown by molecular-beam epitaxy (MBE) at low substrate temperature (LT-GaAs) have attracted much attention over the last few years due to their interesting optical and electronic properties that make LT-GaAs useful for a range of solid-state devices. When the growth temperature is as low as 200°C up to 2 atomic % of excess arsenic is captured into growing layer. In the as-grown state the excess arsenic forms a large number of point defects of which the most important are arsenic antisites. Upon subsequent annealing the excess As affords nanoscale clusters built into the GaAs matrix, strongly reducing the concentration of nonstoichiometry-related defects. Conventional LT-GaAs grown at 200°C and annealed at 600°C exibits reduced charge carrier life-time at the level of 100 fs, suitable for ultrafast photodetectors and optoelectronic switches. It has also been used as a buffer layer in field-effect transistors to eliminate back-gating and side-gating due to its high resistivity.

Characteristic of the conventional LT-GaAs is a random distribution of the arsenic clusters over an epitaxial film. It seems to be of essential interest to produce the material with spatially ordered cluster ensembles. Arsenic clusters were found to accumulate within GaAs regions of LT GaAs/AlGaAs heterostructure (Mahalingam et al 1992) and within InGaAs regions of LT GaAs/InGaAs one (Cheng et al 1994a). Melloch et al (1992) and Cheng et al (1994b) also achived the accumulation of As clusters in undoped LT-GaAs matrix using δ-doping with Si or In. Recently we have succeeded (Bert et al 1996) in obtaining two-dimentional As cluster sheets in LT-GaAs uniformly doped with Si as a donor or Be as an acceptor inserting additional isovalent δ-doping with In. Here we present the results of TEM study on excess arsenic precipitation in differently doped LT-GaAs matrices containing In δ-layers.

2. EXPERIMENTAL

The LT-GaAs films were grown in a dual-chamber MBE system KATUN on undoped semi-insulating 2-inch GaAs(001) substrates which were prepared for the growth procedure in the conventional manner. A 85 nm thick buffer layer of undoped GaAs was grown on the substrate at 580 °C, after that the substrate temperature was lowered down to 200 °C, and an LT-GaAs film was deposited at the growth rate of 1 μm/h under As pressure of 7×10^4 Pa. During the growth the films were either undoped or uniformly doped with Si as a donor or Be as an aceptor, the impurity concentration being 7×10^{17} cm^{-3} in both cases. Indium δ-layers were inserted by interrupting the Ga beam and shutting on the In beam for 4 or 8 seconds that produced approximately 0.5 or 1 monolayer (ML) of InAs accordingly. The distance between In δ-layers was varied from 20 to 60 nm. The samples grown were cut into four parts of which one was kept as-grown, the others were subjected to annealing in MBE chamber under As overpressure for 10 min at 500, 600 or 700 °C. TEM specimens were prepared using wet etching for plan-view observation or the conventional route of mechanical treatment followed by Ar ion-beam milling for cross-sectional studies. Specimens were studied in Philips EM 420, and JEM 4000 instruments.

3. RESULTS AND DISCUSSION

The study performed using near infrared optical absorption showed the concentration of antisite-related defects of 1×10^{20} cm^{-3} in as-grown LT-GaAs. The photoluminescence spectra recorded at 4.2 K are qualitatively similar for all three kinds of the samples (n-, p-, and i-type) but contain important features indicating Si and Be incorporation in GaAs for low growth temperature as well as the fact of formation of shallow acceptor and donor states in the LT-GaAs matrix.

The indium concentration provided by 0.5 ML In δ-insertions was measured by electron-probe microanalysis to be 0.18±0.02 at.%. This value is close to one calculated by averaging In content over the volume accounting for 60 nm distance between In δ-layers in the upper part of the film. TEM and HREM studies of as-grown samples revealed InGaAs layers as thick as 2-3 ML at the positions of nominally 0.5 ML and 1 ML In δ-insertions.

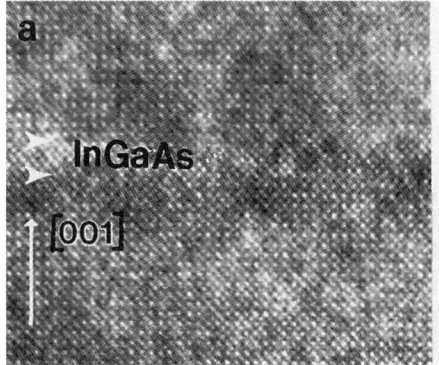 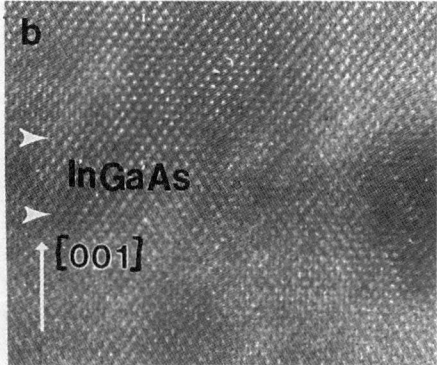

Fig.1. Cross-sectional HREM micrographs showing 0.5 ML In δ-insertions in (a) as-grown LT-GaAs and (b) after annealing at 500°C for 10 min.

Fig. 1a demonstrates HREM micrograph taken from LT-GaAs as-grown sample with nominally 0.5 ML indium δ-insertions. Observations of the samples annealed at the temperatures of 500, 600 or 700°C showed InGaAs layers to occupy 6-8 ML, 9-12 ML, and approximately 20 ML, accordingly. HREM micrograph from the sample annealed at 500°C is presented in Fig. 1b, showing a marked diffusion of In during annealing. Following Mallard et al. (1991) the diffusivity of indium in LT-GaAs is estimated to be 5×10^{-18} cm^2/s in the temperature range from 500 to 600 °C.

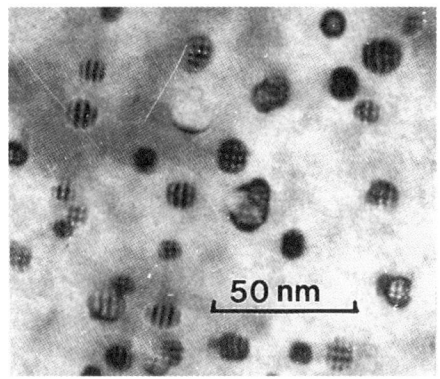

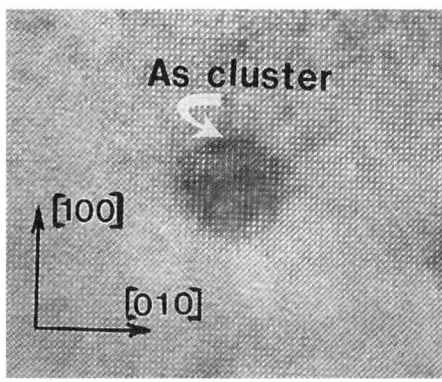

Fig.2. TEM micrograph along [001] zone axis exhibiting Moire fringes superimposed on the image of As clusters.

Fig.3. HREM image of a "small" As cluster along [001] GaAs matrix direction.

Annealing of the samples resulted in precipitation of excess As in small clusters of a size 2-15 nm depending on annealing conditions. At annealing temperature as high as 600°C almost all clusters are crystalline as one can see from Fig. 2 showing Moire patterns at precipitates. The cluster structure is hexagonal for "large" particles (5-15 nm in diameter) and tends to cubic with decreasing cluster size down to 2-3 nm. Fig. 3 shows a "small" As cluster of which the structure is cubic in all appeareance. This assumption is supported by our preliminary image simulations.

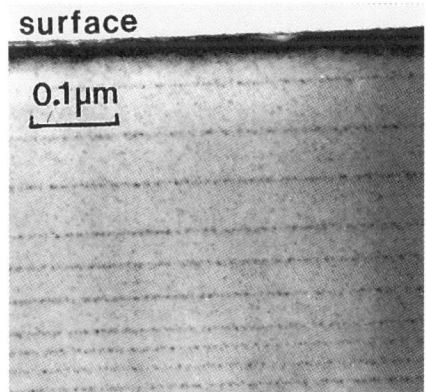

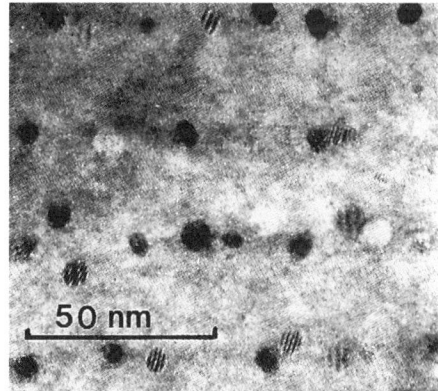

Fig.4. XTEM micrograph of LT-GaAs <Si, 7×10^{17} cm^{-3}> annealed at 600°C. The accumulation of As clusters at 0.5 ML In δ-insertions is clearly seen.

Fig.5. LT-GaAs with the incorporated 1 ML In δ-insertions and annealed at 600°C. The 30 nm GaAs spacers between As cluster sheets are almost free of clusters.

Observations of cross-sectional TEM specimens performed along [110] and [100] axes revealed a strong accumulation of As clusters at the positions corresponding exactly to In δ-insertions in all the samples, i.e. Si-doped, Be-doped, andundoped as well. An example of a regular As/GaAs structure is presented in Fig. 4 in which one can see vertically ordered As cluster sheets bordered with regions depleted of As clusters. When the nominal In content is 1 ML the spacers of the thickness up to 30 nm between As cluster sheets at the annealing temperature of 600°C are almost free of clusters. Such a layered insulating-semiconducting structure is demonstrated in Fig. 5.

TEM study of LT-GaAs samples incorporated with δ-insertions of various In content from 10^{12} cm^{-3} up to 10^{14} cm^{-3} allowed us to find out that the accumulation of As clusters starts when the nominal In δ-layer is as thick as 0.2 ML. In this case a weak accumulation of clusters was observed at annealing temperature of 500 °C. At elevated annealing temperatures the spatial distribution of As clusters becomes random. When 1 ML In δ-layer is inserted the artificial ordering in the As cluster array remains stable up to the highest annealing temperature used, i.e. 700°C

4. SUMMARY

Some features of excess As precipitation in LT-GaAs Si-doped, Be-doped, and as well in the precence of isovalent In δ-insertions are studied by TEM. It is shown that In δ-insertions serve as precursors for As precipitation despite some diffusion of In atoms during the annealing. It was confirmed for the first time that the diffusion coefficient of In atoms in LT-GaAs is close to that in conventional GaAs and by the order of magnitude is equal to 5×10^{-18} cm^{-3} in the temperature range 500-600°C. Layered As/GaAs structures have been grown for all three types of LT-GaAs under study. When In δ-insertions are nominally as thick as 1 ML the GaAs spacers of the thickness of up to 30 nm are almost free of As clusters.

This work is supported in its different parts by RFBR under the grants ## 95-02-05532, 96-02-17931 and by DFG under the cooperative grant DFG-RFBR. It is carried out under Russian National Programmes Physics of Solid State Nanostructures and Fullerens and Atomic Clusters.

REFERENCES

Bert N A, Chaldyshev V V, Faleev N N, Kunitsyn A E, Lubyshev D I, Preobrazhenskii V V, Semyagin B R, and Tret'yakov V V 1996 Semicond. Sci. Technol. **11**, 51

Cheng T M, Chin A, Chang C.Y., Huang M F, Hsieh K Y, and Huang J H 1994a Appl. Phys. Lett. **64**, 1546

Cheng T M, Chang C Y, Chin A, Huang M F, and Huang J H 1994b Appl. Phys. Lett. **64**, 2517

Mahalingam K, Otsuka N, Melloch M R, Woodal J M, and Warren A C 1992 J. Vac. Sci. Technol. **B 10**, 812

Mallard R E, Long N J, Booker G R, Scott E G, Hockly M, and Tailor M 1991 J. Appl. Phys. **70**, 162

Melloch M R, Otsuka N, Mahalingam K, Chang C L, Kirchner P D, Woodal J M, and Warren A C 1992 Appl. Phys. Lett. **61**, 177

Inst. Phys. Conf. Ser. No 157
Paper presented at Microsc. Semicond. Mater. Conf., Oxford, 7–10 April 1997
© 1997 IOP Publishing Ltd

Transmission electron microscopy, X-ray diffraction and photoluminescence study of InGaAs/GaAs heterostructures

J Kątcki, K Regiński, M Bugajski, J Adamczewska, W Lewandowski, J Ratajczak, W Rzodkiewicz, J A Kozubowski *

Institute of Electron Technology, Al. Lotników 32/46, 02-668 Warsaw, Poland
* Department of Materials Science and Engineering, Warsaw Universty of Technology, ul. Narbutta 85, 02-524 Warsaw, Poland

ABSTRACT: This paper reports the results of our investigation of InGaAs/GaAs heterostructures grown by molecular beam epitaxy (MBE) on a semi-insulating GaAs substrate. The heterostructures consist of layers of $In_{0.2}Ga_{0.8}As$ of different thickness embedded between thick GaAs layers. Cross-sectional transmission electron microscopy (XTEM) has been used to observe the perfection of layers and interfaces. Electron microscopic observations were combined with photoluminescence spectroscopy and x-ray diffraction studies of the heterostructures.

1. INTRODUCTION

InGaAs/GaAs heterostructures are potentially important in fabricating optoelectronics devices such as resonant tunneling diodes, infrared detectors and semiconductor lasers. A number of parameters are critical in controlling the properties of these devices, including the thickness of each layer, the indium content of the $In_xGa_{1-x}As$ compound layers, the roughness of interfaces between each layer and the crystallographic perfection of layers and interfaces. In order to evaluate the quality of the heterostructure combined techniques should be used. In this investigation we applied cross-sectional transmission electron microscopy (XTEM), X-ray diffractometry and photoluminescence (PL) spectroscopy to evaluate the crystallographic perfection of $In_xGa_{1-x}As$ layers and their interfaces in a heterostructure with layers of various thickness.

2. EXPERIMENTAL

A heterostructure containing $In_xGa_{1-x}As$ layers of various thickness was grown by means of molecular beam epitaxy (MBE) in a Riber 32P system. The heterostructure was grown on an (001) oriented semi-insulating GaAs substrate. First, approximately 500 nm of undoped GaAs was grown on the substrate. On this layer a heterostructure consisting of five $In_xGa_{1-x}As$ layers (each a few nanometer thick) separated by thick GaAs layers was grown. $In_xGa_{1-x}As$ layers have the same content of indium, x, but vary in thickness. The $In_xGa_{1-x}As$ layer were grown for 7, 14, 21, 28 and 35 seconds. On the top $In_xGa_{1-x}As$ layer a GaAs layer was epitaxially grown. In order to simplify the determination of In content, x, by means of X-ray diffractometry (XRD), approximately 1 micrometer of $In_xGa_{1-x}As$ of the same In content as other $In_xGa_{1-x}As$ layers was grown epitaxially on the top of the heterostructure.

Cross-sectional specimens for transmission electron microscopy (TEM) investigation were prepared in a <110> plane by the method described by Kątcki et al (1995). The specimens were studied in a JEM 200CX (200 kV, Inst. of Electron Technol.) and a JEM 3010 (300 kV, Warsaw University of Technology).

The indium content of the $In_xGa_{1-x}As$ layers was determined in an X-ray double crystal diffractometer and confirmed by PL. The heterostructure sample for PL study was prepared using a dimple grinder.

3. RESULTS AND DISCUSSION

A cross-sectional TEM micrograph of an $In_xGa_{1-x}As/GaAs$ heterostructure is shown in Fig. 1a. From this micrograph we measured the thickness of $In_xGa_{1-x}As$ and GaAs layers. Thicknesses of subsequent $In_xGa_{1-x}As$ layers $In_xGa_{1-x}As$ (counting from the substrate) were 2, 4 , 8, 12, 16 and 1000 nm. Thicknesses of subsequent GaAs layers (counting from the substrate) were 500, 90, 90, 180, 360 and 720 nm. Since an accurate growth time for each layer is known, this allowed the growth velocity for $In_xGa_{1-x}As$ and GaAs in MBE growth to be determined. The growth velocity for $In_xGa_{1-x}As$ was 0.46 nm/s and for GaAs 15 nm/s.

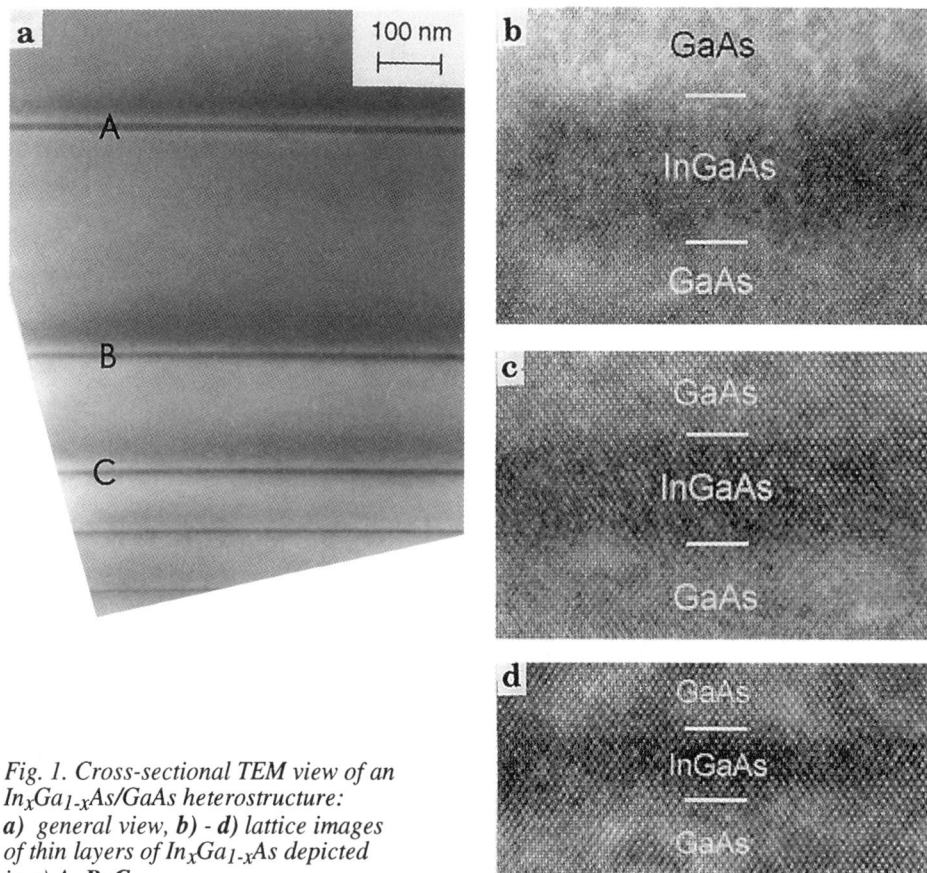

Fig. 1. Cross-sectional TEM view of an $In_xGa_{1-x}As/GaAs$ heterostructure:
a) general view, b) - d) lattice images
of thin layers of $In_xGa_{1-x}As$ depicted
in a) A, B, C

A heterostructure was observed in a high resolution transmission electron microscope. Figures 1b-d show lattice images $In_xGa_{1-x}As$ layers: 16, 12 and 8 nm thick, respectively. Layers containing In exhibit darker contrast than GaAs layers. Lattice defects were not observed at interfaces or in the layers. During observation the specimen should be well cooled since heating a thick $In_xGa_{1-x}As$ layer causes dislocations to form in the hetero-structure (Fig. 2).

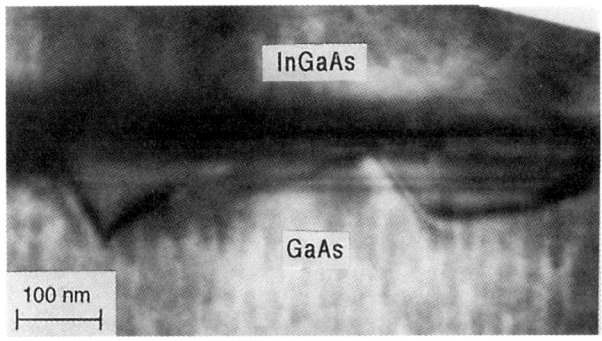

Fig. 2. Dislocations formed in a In$_x$Ga$_{1-x}$As/GaAs heterostructure

The indium content of In$_x$Ga$_{1-x}$As was measured in an X-ray double crystal diffracto-meter. Fig 3a shows an X-ray diffraction rocking curve obtained from the In$_x$Ga$_{1-x}$As/GaAs heterostructure. The peak on the right comes from GaAs which forms the bulk of the het-erostructure. The left peak corresponds to the top In$_x$Ga$_{1-x}$As layers. From the difference in angle between the two peak maxima the content of In in In$_x$Ga$_{1-x}$As was calculated. The in-dium content, x, was found to be 0.2. The same result was obtained from PL measurements. In Fig. 3b PL spectrum from the top InGaAs layer measured at 77K is shown. The position of the emission band indicated that the indium content x=0.2.

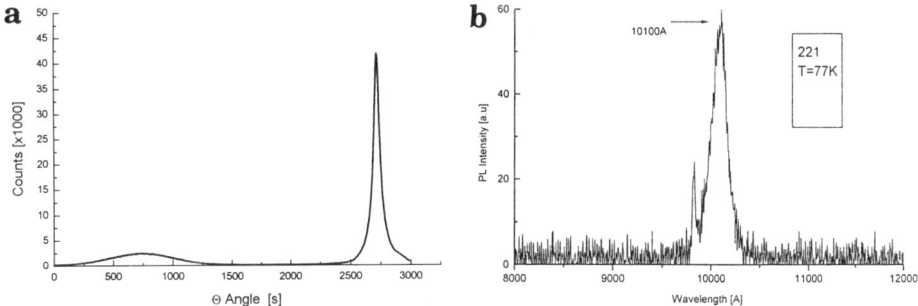

Fig. 3 a) X-ray rocking curve on the In$_x$Ga$_{1-x}$As/GaAs heterostructure; b) PL emission spectum of In$_x$Ga$_{1-x}$As/GaAs.

The full width at half maximum (FWHM) of the top In$_x$Ga$_{1-x}$As layer Bragg peak is about 10 times greater than FWHM of the GaAs layers. This is evidence for the presence of misfit dislocations in the layer. We observed these dislocations at the interfacial region In$_x$Ga$_{1-x}$As/GaAs.

In order to measure PL emission spectra from the layers located deeper in a heteros-tructure a dimple was ground in a sample. In Fig. 4 PL spectra at the edge of a dimple (the upper one, depicted as **A**) and slightly away from the edge (the lower one, depicted as **B**) are shown. Emission bands 9730Å, 9050Å and 8700Å visible on these spectra refer to In$_x$Ga$_{1-x}$As layers 16, 12 and 8 nm thick, respectively.

294

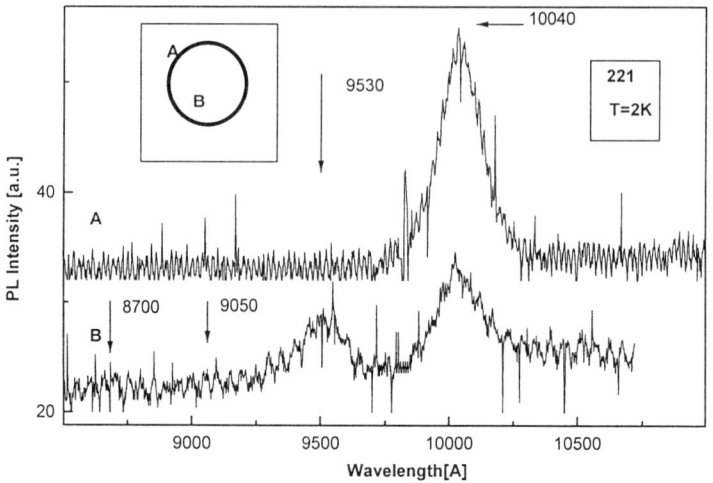

Fig. 4. PL spectra obtained from a dimple-ground heterostructure.

4. CONCLUSIONS

Cross-sectional TEM observations allowed the thickness of layers and growth rates for $In_xGa_{1-x}As$ and GaAs to be determined. No lattice defects were observed either in $In_xGa_{1-x}As$ layers or at $In_xGa_{1-x}As$/GaAs interfaces. Dislocations were observed in the top $In_xGa_{1-x}As$ layer.

Using both X-ray diffractometry and PL spectroscopy the content of In in $In_xGa_{1-x}As$ was determined and was found to be 0.2. From X-ray diffractometry investigation one can conclude that there are misfit dislocations in the top layer of a heterostructure. In PL spectra a shift of emission bands for subsequent $In_xGa_{1-x}As$ layers was observed.

ACKNOWLEDGMENT

This publication is based on work sponsored by the Polish Government under the project #8T11B.064.11. The authors are very much indebted to Mr M. Wesołowski for collaboration in the investigation, Ms D. Szczepańska for assistance in specimen preparation and Ms J Wiącek for careful preparation of micrographs. A transmission electron microscope, JEM 3010, used in this study was founded by the Foundation for Polish Science.

REFERENCES

Katcki J, Ratajczak J, Malag J A and Piskorski M 1995 Microscopy of Semiconducting Materials 1995, edited by A. G. Cullis and A. Staton-Bevan (Bristol, IOP), p. 273

Inst. Phys. Conf. Ser. No 157
Paper presented at Microsc. Semicond. Mater. Conf., Oxford, 7–10 April 1997
© 1997 IOP Publishing Ltd

Transmission electron microscopy investigation of FeAs precipitates in GaAs/AlGaAs heterostructures

J Kątcki, M Shiojiri [1], T Isshiki [1], K Nishio [1], Y Yabuuchi [2] and N Y Jin-Phillipp [3]

Institute of Electron Technology, Al. Lotnikόw 32/46, 02-668 Warsaw, Poland
[1] Kyoto Institute of Technology, Matsugasaki, Sakyoku, Kyoto 660, Japan
[2] Matsushita Technoresearch Inc., Moriguchi, Osaka 570, Japan
[3] Max-Planck Institut für Metallforschung, Institut für Physik, Heisenbergstrasse 1, 70569 Stuttgart, Germany

ABSTRACT: Furnace annealing at 850°C of GaAs/$Al_{0.4}Ga_{0.6}As$ heterostructures grown by liquid phase epitaxy (LPE) causes the formation of regularly shaped precipitates in a subsurface layer. Cross-sectional transmission electron microscopy (XTEM) observation of the heterostructures shows that the main boundaries of these precipitates are parallel to the low-index crystallographic planes.

A <110> lattice image of the GaAs layer containing the precipitates was digitized and processed on a computer. From a computer diffractogram of the image we have concluded that the precipitates are FeAs (orthorhombic) crystallites, with the orientation $[110](111)_{GaAs}//[101](010)_{FeAs}$. Another orientation found by TEM analysis was $[110](-111)_{GaAs}//[001](010)_{FeAs}$. The composition of precipitates was confirmed by energy-dispersive X-ray spectroscopy (EDX). On the EDX spectra the Fe peak is clearly visible. We believe that Fe atoms are deposited on the heterostructure surface during the final stage of the LPE growth or during later processing steps. The annealing causes the migration of Fe atoms from the surface and the formation of precipitates.

1. INTRODUCTION

Among the methods of epitaxial growth of III-V compounds liquid phase epitaxy (LPE) is still the cheapest technique for growing layers. This is why the LPE technique is used for growing heterostructures where the layer thickness is not critical. However, this method of growth can introduce some unwanted impurities into a semiconductor wafer. Annealing following epitaxy can cause the formation of precipitates in heterostructures. In this paper we report study of the precipitates formed in GaAs layers during furnace annealing accompanying the diffusion of dopant.

2. EXPERIMENTAL

In order to study the changes in crystal structure of GaAs/$Al_xGa_{1-x}As$ (x=0.4) heterostructures during the thermal processes accompanying diffusion and rediffusion of zinc the heterostructures GaAs:Sn; 1 μm/$Al_{0.4}Ga_{0.6}As$:Sn; 1.5 μm/GaAs:Sn; 10 μm were grown by LPE on a n-type (100) oriented GaAs substrate (Kątcki et al, 1994). After epitaxy samples were etched in HCl. Then a few nanometers thick surface layer of GaAs was removed from the samples by anodizing. The heterostructures were first annealed at a temperature of 670°C for 80 minutes and then at 850°C for 2 hours in a quartz ampoule (typical for diffusion and rediffusion of Zn, respectively). Before both thermal steps the surface was Ar^+ ion etched (250V) for 30s and covered by a 200 nm thick Si_3N_4 layer by sputtering (sputtron model Z-400, Leybold-Heraus). After both thermal steps the Si_3N_4 layer was removed from the surface by dry etching in CF_4/O_2 plasma.

Cross-sectional specimens for transmission electron microscopy (TEM) investigation were prepared in a <110> plane using the method described by Kątcki et al (1995). The specimens were studied in JEM 200CX (200 kV, Inst.Electron Technol.), JEM 4000FX (400 kV, Max-Planck Inst. für Metallforschung) and Topcon-EM002B (200 kV, Matsushita Technores., Inc.) transmission electron microscopes. TEM images were digitized in an image scanner Nikon LS-3510AF (spatial resolution on film - 16 μm; 12 bit depth) and processed in a NEC PC-H98 model U80 computer (Kyoto Institute of Technol.). An electron dispersive X-ray spectroscopy (EDX) analysis was carried out using a Kevex Delta Level 5 System attached to a Topcon-EM002B transmission electron microscope (Matsushita Technores., Inc.).

3. RESULTS AND DISCUSSION

Annealing of GaAs/AlGaAs heterostructures at a temperature of 850°C for 2 hours (typical for rediffusion of dopant) caused the formation of precipitates in the top GaAs layer. The distribution of the precipitates is shown in Fig. 1. The precipitates have a regular shape with boundaries parallel to low-index crystallographic planes. They formed at a depth of 100-400 nm. The main boundaries of these precipitates are parallel to the low-index crystallographic planes.

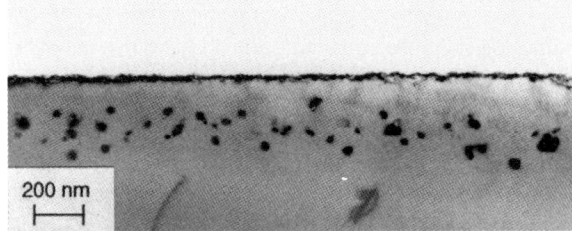

Fig. 1. Cross-sectional TEM view of a GaAs/AlGaAs heterostructure with a layer of precipitates.

200 nm

The precipitates were observed in JEM4000FX and Topcon-EM002B high resolution transmission electron microscopes. <110> lattice images of two typical precipitates are shown in Fig 2a and 2d. The images of the precipitates suggest the formation of another phase. In both cases the longest edge of the precipitate are extended along the $\{111\}_{GaAs}$ directions.

In order to identify the phase of the precipitates a <110> image of the GaAs layer containing the precipitates (shown in Fig 2a) was digitized and processed. Diffractograms were obtained both from an area containing only the matrix (Fig. 2b) and from an area containing the precipitate (Fig. 2c). Using the diffractogram obtained from the matrix (Fig 2b), the magnification of the image was corrected using the fact that a $(222)_{GaAs}$ spot corresponds to a spacing of 0.163 nm.

Diffraction spots due to the precipitate depicted by symbol **A-C** in Fig. 2c correspond to lattice spacing of 0.549 nm, 0.297 nm and 0.263 nm, respectively. An angle between spots **A** and **B** is 90.2°. These values agree with ones obtained from an <101> image of FeAs crystal (orthorhombic: a=0.6028 nm, b=0.5439 nm, c= 0.3373 nm)(Wyckoff, 1965). That is, the spots **A-C** are due to $(010)_{FeAs}$ (d=0.5434 nm), $(101)_{FeAs}$ (d=0.2944 nm) and $(111)_{FeAs}$ (d=0.259 nm), respectively. $(010)_{FeAs}$ meets $(101)_{FeAs}$ at a right angle. The precipitate is therefore considered to be an FeAs crystallite. The lattice relation between GaAs and FeAs is $[110](111)_{GaAs}//[101](010)_{FeAs}$.

In Fig. 2d another example of an FeAs precipitate is shown. In this case the orientation relationship between matrix and precipitate is $[110](-111)_{GaAs}//[001](010)_{FeAs}$. The electron beam is parallel to [110] GaAs and [001] FeAs. Fig 2e shows a diffraction pattern from the precipitate. Letters **F** and **G** represent FeAs and GaAs, respectively.

The composition of the precipitates was confirmed by energy dispersive X-ray spectroscopy (EDX). Typical spectra for a matrix and for a micro-area containing the

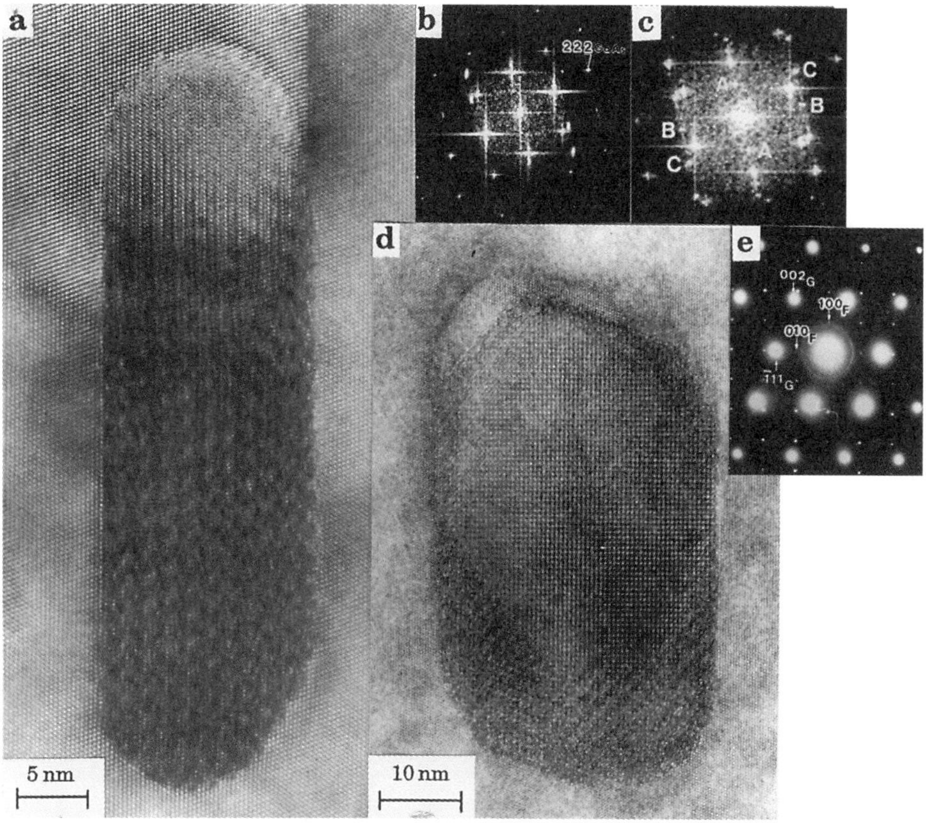

Fig. 2. <110> images of an FeAs precipitates formed in a GaAs/AlGaAs heterostructure:
a) [110](111)$_{GaAs}$//[101](010)$_{FeAs}$ (JEM4000FX); computer diffractogram of b) a matrix;
c) an area containing the precipitate; d) [110](-111))$_{GaAs}$//[001](010)$_{FeAs}$ (Topcon-EM002B);
e) electron diffraction of an area containing the precipitate.

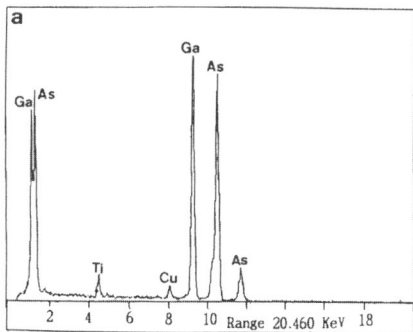

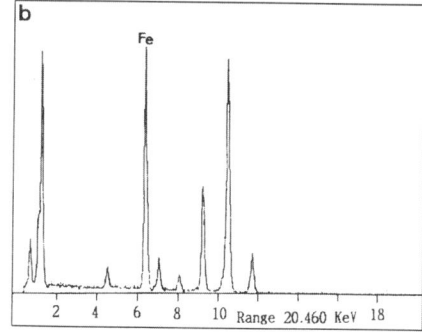

Fig. 3. EDX spectra a) of a GaAs matrix and b) of an area containing the precipitate.

precipitate are shown in Fig. 3a and b, respectively. The probe size of the microscope was about 8 nm. In Fig. 3b the peak due to Fe is clearly detected from the precipitates. The signals from Ga and As are also detected. On the spectra, Cu and Ti peaks can be also found both in the matrix and in the precipitate area. It is supposed that Cu is a background signal from the sample holder while Ti may come from a TEM sample. From these spectra one can conclude that the precipitates must be a compound of Fe-As or Fe-Ga-As.

Iron could contaminate the surface of GaAs/AlGaAs heterostructures during or directly after the LPE growth process. Surface contamination by Fe could also occur during photolithographic process and/or the first annealing step (at temperature of 670°C for 80 minutes). During the next annealing step (at 850°C for 2 hours) Fe atoms migrated from the sample surface to the bulk to form the precipitates at a depth of 100-400 nm. The mechanism of this phenomena can be explained as impurity gettering similar to that observed in silicon, where Fe atoms are often present.

4. CONCLUSIONS

Furnace annealing at 850°C (2 hours) of GaAs/Al$_{1-x}$Ga$_x$As heterostructures grown by LPE causes the formation of regularly shaped precipitates in a subsurface layer. Observation of cross-sections of the heterostructures in a transmission electron microscope revealed that precipitates formed at a depth of 100-400 nm. Precipitates were observed in JEM 4000FX and Topcon-EM002B high resolution transmission electron microscopes . A <110> image of a GaAs layer containing the precipitate was digitized and computer processed. From the computer diffractograms we have concluded that the precipitates are FeAs crystallites. We found two lattice relations between GaAs and FeAs. Those are [110](111)$_{GaAs}$//[101](010)$_{FeAs}$ and [110](-111)$_{GaAs}$//[001](010)$_{FeAs}$. The composition of the precipitates was confirmed by energy dispersive X-ray spectroscopy (EDX). On the EDX spectra the Fe peak is clearly visible.

We believe that the contamination of a heterostructure surface by the Fe atoms followed by their migration from the surface during annealing can be responsible for the formation of FeAs precipitates.

ACKNOWLEDGMENT

The authors are very much to Dr A. Maląg for collaboration in the investigations, Ms D. Szczepańska for assistance in specimen preparation and Ms J Wiącek for careful preparation of micrographs.

REFERENCES

Kątcki J, Malag A, Ornoch J and Ratajczak J 1994 *ITE Report*, no. 12, 2.
Kątcki J, Ratajczak J, Maląg J A and Piskorski M 1995 Microscopy of Semiconducting Materials 1995, edited by A. G. Cullis and A. Staton-Bevan (Bristol, IOP), p. 273
Wyckoff R.W.G 1965 Crystal Structure, *2nd ed*, p. 127

Inst. Phys. Conf. Ser. No 157
Paper presented at Microsc. Semicond. Mater. Conf., Oxford, 7–10 April 1997
© 1997 IOP Publishing Ltd

Characterisation of InPSb layers on different substrates (InAs or GaSb)

C Mendorf, G Brockt, A Behres[+], C von Eichel-Streiber[+], M Heuken[+], K Heime[+] and H Lakner

Werkstoffe der Elektrotechnik, Gerhard-Mercator-Universität Duisburg,
Bismarckstr. 81, 47048 Duisburg, Germany
[+]Institut für Halbleitertechnik, RWTH Aachen, Templergraben 55, 52056 Aachen, Germany

ABSTRACT: The structural quality of low-pressure MOVPE grown InPSb on InAs- and GaSb-substrates has been investigated by STEM, PL and XRD. InPSb layers on GaSb substrates are governed by defects on {1 1 1} whereas InPSb-layers on InAs substrates have less defects thus small scale contrast variations depending on growth parameters could be observed. This is in accordance with the miscibility gap of InPSb. Additionally, we investigated interface properties of InPSb/InAs superlattice (SL) structures grown under optimised conditions. The interface of a superlattice with a periodicity of 20 nm shows a minor asymmetric behaviour, caused by a Sb carry-over into the wells.

1. INTRODUCTION

InPSb/InAs heterostructures are a promising material system for laser diodes and detectors in the mid-infrared region (3 - 5 μm) for applications in communication, medicine and environmental analytical techniques. InPSb is requested as a barrier material providing suitable optical confinement. The ternary semiconductor InPSb may be an alternative material to Al containing alloys, because especially when grown by MOVPE, such alloys suffer from carbon and oxygen impurities. Nevertheless, only little is known of material quality with respect to defects and decomposition or ordering effects. Therefore we investigated the structural properties of InPSb grown on the substrates GaSb and InAs and applied methods like PL, X-ray measurements and brightfield as well as atomic number (Z-) contrast imaging performed in a field-emission scanning transmission electron microscope (STEM).

2. EXPERIMENTAL DETAILS

The heterostructures investigated here were grown by low pressure (20 mbar) metalorganic vapour phase epitaxy (MOVPE). The precursors were trimethylindium (TMIn), triethylantimony (TESb), arsine (AsH_3) and phosphine (PH_3). The growth temperatures were 480 °C, 520 °C and 570 °C. Photoluminescence was carried out at 10 K, respectively 4 K.

Cross-sectional specimens were investigated in a field-emission STEM (VG Microscopes: HB 501) operating at 100 keV. With respect to high-resolution capabilities (C_s=1.3 mm giving < 0.3 nm probe diameter) the system has been modified and improved (Lakner et al.). Brightfield imaging was used for the characterisation of structural defects. For chemical composition analysis as well as for the investigation of interface abruptness the heterostructures were analysed by atomic number (Z-) contrast imaging qualitatively.

The high-resolution X-ray measurements were carried out using a four crystal Ge(220) monochromator.

3. RESULTS AND DISCUSSION

In a first step the structural quality of InPSb layers grown on GaSb-substrates (T_{gr}=480 °C) was investigated. STEM brightfield images as well as Z-contrast micrographs show high dislocation densities in InPSb. Defects on the {111} planes due to misfit in InPSb are clearly visible (see fig. 1). Additionally, we observed microtwins which occur in the InPSb layers. The high-resolution brightfield image in figure 2 exhibits moiré fringes which are due to a superimposition of two crystals Σ=3 twinned with each other (e.g. Bender et al 1986; Shechtman et al 1993)

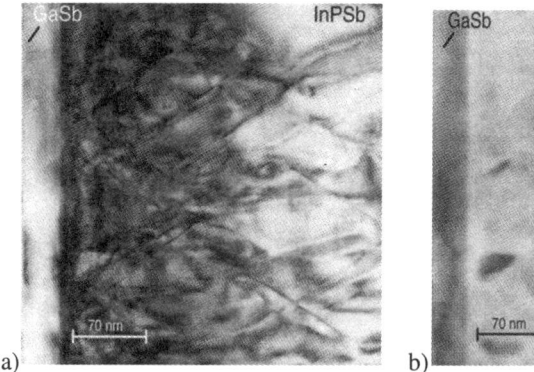 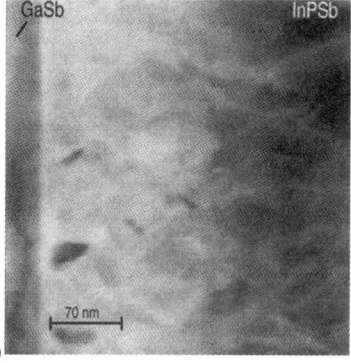

a)   b)

Fig. 1   (a) Brightfield and (b) Z-contrast image of InPSb on GaSb showing mismatch-induced defects.

The main reasons for the problems in epitaxial growth of InPSb on GaSb at the moment are on the one hand an insufficient cleaning of the substrates' surfaces and on the other hand chemical reactions at the GaSb surface during the heating process.

For further examinations therefore we used InAs as substrate material for the growth of InPSb. STEM brightfield micrographs of InPSb layers grown at 520 °C exhibit a good crystalline quality.

Nevertheless, the predicted miscibility gap of this alloy could limit the usefulness of this compound in a laser.

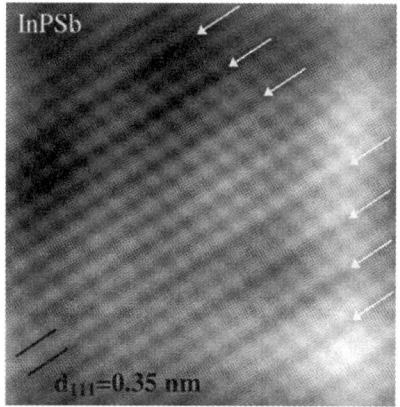

Fig. 2   High-resolution brightfield image showing moiré fringes.

Following theoretical models of Stringfellow (1982) and Onabe (1982), the extension of the miscibility gap of $InP_{1-x}Sb_x$ was determined numerically. Figure 3 describes the miscibility gap and spinodal curve in dependence on the growth temperature. For a typical temperature, e.g. 570 °C, all alloys with an Sb-content between $x_{Sb}=0.063$ and $x_{Sb}=0.961$ are located within the miscibility gap. Especially, the composition for lattice-matched growth on InAs ($x_{Sb}=0.31$) is located deep inside the gap.

To get additional information about the chemical composition in the thick InPSb layer, Z-contrast images were recorded by STEM (see fig. 4). The micrograph shows small scale contrast variations which indicate compositional fluctuations in the InPSb layer, whereas in InAs no significant contrast variation occurs. This result is in agreement with PL measurements carried out at 10 K on other samples. These PL spectra show a low energy tail which is a hint of compositional variations in the layer too.

After increasing the growth temperature to 570 °C and optimising the growth conditions, two 9 period InPSb/InAs superlattice structures were grown with a nominal periodicity of $\Lambda=20$ nm and $\Lambda=40$ nm, respectively, in order to investigate their interface properties.

The superlattice with $\Lambda=40$ nm is of very good structural quality. This is proved by X-ray diffractometry as well as by STEM data. Figure 5 shows a Z-contrast image and linescan of the superlattice. Due to Z-contrast the InPSb layers appear bright. The evaluation of the STEM results yields a periodicity of $\Lambda=39$ nm and thicknesses of 19 nm for wells (InAs) and 20 nm for barriers (InPSb). The linescan indicates a slightly asymmetric grading of the chemical composition at the interfaces. In comparison, Figure 6 shows the Z-contrast image and linescan of the superlattice with a nominally periodicity of $\Lambda=20$ nm.

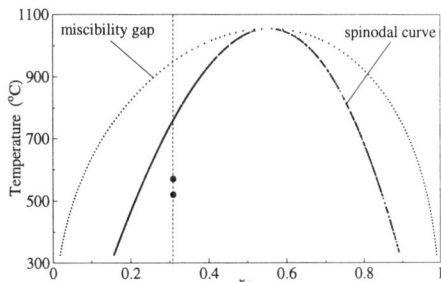

Fig. 3 Calculated miscibility gap for the $InP_{1-x}Sb_x$ system.

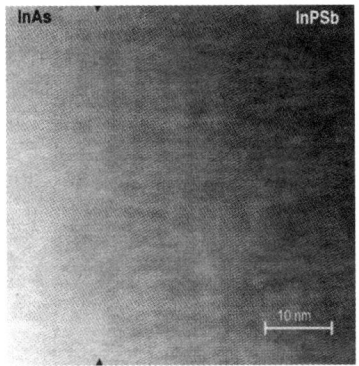

Fig. 4 STEM Z-contrast image showing small scale contrast variations in InPSb layer grown on InAs ($T_{gr}=520$ °C).

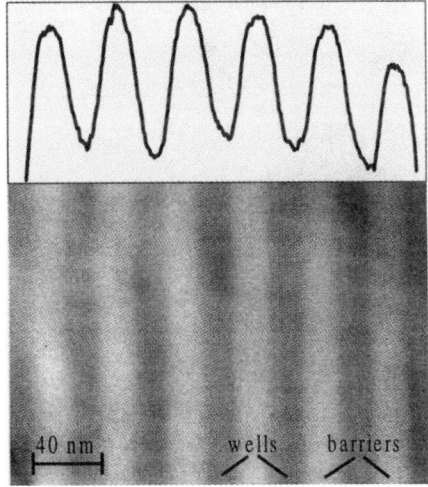

Fig. 5 Z-contrast image and linescan of the SL with a periodicity of $\Lambda=40$ nm ($T_{gr}=570$ °C).

302

Just as in the first superlattice the structural properties are excellent. We found neither hints for defects like misfit dislocations and ordering effects nor decomposition in the superlattice which can be attributed to the increased growth temperature. These results are proved by experimental and simulated X-ray diffraction patterns (see fig. 7). The experimental X-ray pattern shows well-defined satellites as well as distinct „Pendellösung" fringes between the first-order satellite peaks. The simulated curve shows a very good agreement related to the experimental data. Nevertheless, the relative intensity of the satellite peaks is a bit higher compared to the experimental spectra, which may indicate interfacial layers.

The periodicity was determined as $\Lambda=19.5$ nm with thicknesses of 9.5 nm for wells and 10 nm for barriers. The interface grading is much less compared with the superlattice investigated in figure 5. However, the linescan demonstrates an asymmetric behaviour which can be explained by an Sb carry-over into the wells (Heuken et al 1997).

The results demonstrate that despite the predicted miscibility gap at least for InAs substrates InPSb can be grown by MOVPE as a high quality lattice matched barrier material.

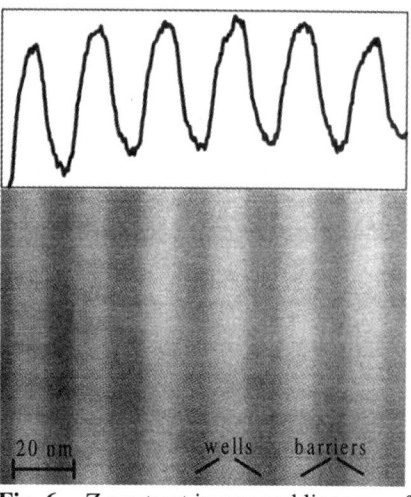

Fig. 6   Z-contrast image and linescan of the SL with a periodicity of $\Lambda=20$ nm ($T_{gr}=520$ °C).

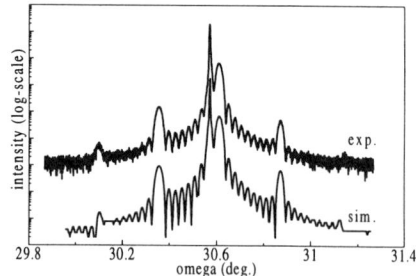

Fig. 7   Experimental and simulated X-ray diffraction pattern of a 9 period InPSb/InAs SL. The experimental curve is shifted for clarity.

ACKNOWLEDGEMENTS

The authors like to thank B. Opitz, Robert Bosch GmbH, for the numerical calculation of the miscibility plot. Parts of this work were financially supported by the Deutsche Forschungsgemeinschaft (within SFB 254) and the Volkswagen-Stiftung.

REFERENCES

Bender H, De Veirman A., Van Landuyt J and Amelinckx S 1986 Appl. Phys. **A 39** 83
Heuken M, v. Eichel-Streiber C, Behres A, Schineller B, Mendorf C, Brockt G and
    Lakner H 1997: submitted to J. Electr. Mat.
Lakner H, Bollig B, Ungerechts S and Kubalek E 1996 J. Phys. D: Appl. Phys. **29** 1767
Onabe K 1982 J. Phys. Chem. Solids **43** No. 11 1071
Shechtman D, Feldman A, Vaudin M D and Hutchison J L 1993 Appl. Phys. Lett. **62** 487
Stringfellow G B 1982 J. Crystal Growth **58** 194

Inst. Phys. Conf. Ser. No 157
Paper presented at Microsc. Semicond. Mater. Conf., Oxford, 7–10 April 1997
© 1997 IOP Publishing Ltd

TEM and HRXRD study of high strain InAlGaAs heterolayers

Yu G Musikhin, N A Bert and N N Faleev

Ioffe Physical-Technical Institute, 194021 St. Petersburg, Russia

ABSTRACT: A study of high strain $(Al_{0.3}Ga_{0.7})_{0.85}In_{0.15}As$ was carried out by using TEM and HRXRD techniques. A strained layer was grown on an (AlGa)As buffer layer. X-ray data showed that partial relaxation of the elastic layer strain occurred. Modulation of Al concentration was found within the buffer layer by TEM as well as by X-ray diffraction. TEM studies showed a non-uniform distribution of defects resulting from relaxation at the interface. A stacking fault and dislocation network, as well as non-uniform elastic strain, were found. This type of distribution of defects results in areas of high dislocation density within the strained layer. These areas have a dislocation density higher than $10^{10}cm^{-3}$ and are non-uniformly distributed.

1. INTRODUCTION

Recently, there has been considerable interest in the study of strained epitaxial layers, which are attractive materials for optoelectronic and microwave device fabrication. The accommodation of misfit strain and defect generation during the growth process are important for device applications. The AlInGaAs/GaAs system has been heavily studied (Androussi et al 1995): lattice mismatch (up to 7%) results in strain induced relaxation. It is well known, that the relaxation can occur through the formation of structural defects such as stacking faults and dislocation networks. Despite numerous investigations on this subject, there is not a full understanding of the process. On one hand, there are a lot of efforts to create pseudomorphic layers above their critical thicknesses, on the other hand, there is a great interest in 3D growth of quantum dots. In both cases, it is very important to obtain structures with low densities of defects. As is well known, the surface conditions have a strong effect on relaxation (Bert et al 1995). In this paper, we report on the influence of such conditions on the defect structure of AlGaInAs/GaAs heterolayers.

2. EXPERIMENTAL

Samples were grown by the MOCVD method and contained strained (AlGa)InAs layers and AlGaAs buffers grown at $650\,^{\circ}C$ on (001) GaAs semi-insulating substrates. The thickness of the buffer layers was about $1\,\mu m$: the thickness of the (AlGa)InAs layers varied from 0.2 to $0.4\,\mu m$. X-ray diffraction patterns have been obtained in the vicinity of the 004 GaAs reflection. Both ω-2Θ and ω-scan rocking curves have been measured. A narrow receiving slit has been used in front of the detector to diminish the intensity of inelastically scattered x-rays. Parameters of measured x-ray rocking curves allow us to detect deformation in epitaxial layers with extent of relaxation, and to estimate the type and density of structural defects, to evaluate the coherent length in epitaxial layersfor x-rays. The samples for TEM study were prepared by mechanical polishing with subsequent ion milling in Balzers IEC-101 and Gatan DuoMill 600 machines. The sputtering was performed with 4 kV Ar⁺-ions at glancing angles. TEM study was carried out using of Philips EM420 operating at 100 kV.

3. RESULTS

Both HRXRD and TEM show that the AlGaAs buffer layers have a modulation in Al concentration within the buffer layers: it is clearly seen in Fig.1. The modulation period is about 6 nm. This type of modulation very often occurs in Al containing layer grown by MOCVD. The double-crystal rocking curve presented in Fig.2. shows an additional ordering in AlGaAs layer. Combined analysis of this rocking curve and a number of ω-mode rocking curves measured nearby AlGaInAs layer position show that this layer is partially relaxed, as a result of generation of misfit dislocations. TEM study shows that the relaxation occurs by the formation of regular dislocation net with linear density equal to 5×10^5 cm^{-1} (Fig.3). Symbol 'D' marks the dislocations.

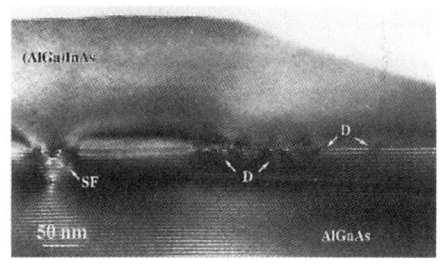

Fig.1 (220) BF cross-sectional image of strain and buffer layers.

These dislocations are usual 60⁰ dislocations lying along <110> directions. In addition, in Fig.3 one can see a particular contrast which may result from interface roughness. The roughness of the interface and some defects are shown in Fig 1. and Fig.4. In addition to

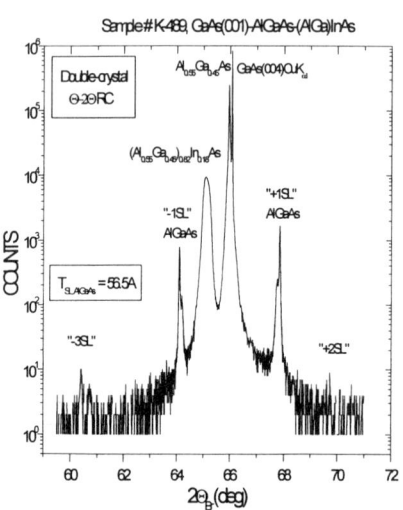

Fig.2. Double-crystal rocking curve in the vicinity of (004) GaAs, radiation CuK$_\alpha$. Receiving slit is 0.1mm.

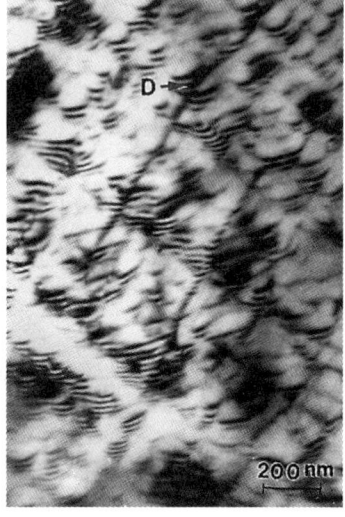

Fig.3 (220) BF plan view image. Symbol 'D ' is marked dislocation net.

regular dislocation net, the TEM study found non- uniform distribution of defects at the interface. They are indicated in Fig.1 as SF (stacking fault) and D (defect), which appears to be dislocations or agglomerate of point defects.

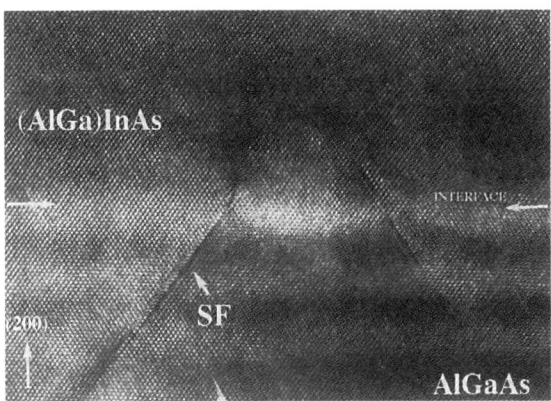

Fig.4 HREM image of interface beetwen buffer and strain layers.

We suggest that, this leads to strain concentration at local areas. The strain concentration is suggested to be the reason for highly elastically strained region within layer. Fig.5a shows the strained area, marked by arrows. Such regions are likely to relax with creation of high dislocation density observed in the layer (Fig.5b). The dislocation density was estimated to be more than $1 \times 10^{10} cm^{-3}$. The spatial distribution of such areas is non-uniform.

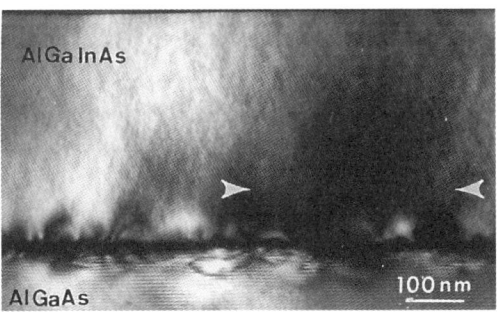

Fig.5a (220) BF image: arrows indicate an elastic strained region.

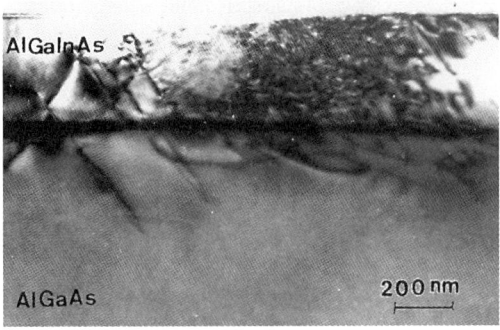

Fig 5b (220) BF image of high density dislocation area.

4. SUMMARY

The study of strained (AlGa)InAs layers by HRDXRD and TEM techniques shows that highly strained layer can lead to elastic strained areas. These areas can be relaxed through creation high dislocation regions. The dislocation density in these regions is more than 10^{10} cm^{-3}. We suggest that the reason of this relaxation is a local concentration of strain because of interface surface conditions.

ACKNOWLEDGMENTS

One of the authors is pleased to acknowledge the Royal Microscopical Society for financial support and Dr P. Werner from Max-Planck Institute, Halle, Germany for help with HREM studies. This work was supported in part by RFBR grant.

REFERENCES

N A Bert, N N Faleev and Yu G Musikhin, 1995, Proc. of IOP Conf. Series, 146, pp 173-5
Y Androussi, A Lefebver, C Dellamarre, L P Wang, A Dubon, B Courboules, C Deparise, and J Massies, 1995, Appl. Phys. Lett. 66(25), 3450.

Inst. Phys. Conf. Ser. No 157
Paper presented at Microsc. Semicond. Mater. Conf., Oxford, 7–10 April 1997
© 1997 IOP Publishing Ltd

A TEM study of Cu-In-Se thin films grown by molecular beam epitaxy

S B Lin, G L Gu and B H Tseng

Institute of Materials Science and Engineering, National Sun Yat-Sen University, Kaohsiung 80424, Taiwan-R.O.C.

ABSTRACT: Thin films of $CuInSe_2$ were grown on (001)GaAs substrates by the molecular beam epitaxy (MBE) technique. In this work, we demonstrate that a modification of the defect structure in the $CuInSe_2$ epitaxial film can be accomplished by a photo-assisted MBE (PAMBE) technique. Since this growth method is able to grow epitaxial films at a temperature as low as 300°C, the strain due to the lattice misfit is released much less than in films grown by conventional MBE at a substrate temperature of 500°C. TEM observations showed the formation of microtwins instead of dislocations in this film. In addition, thin films of a vacancy-ordered compound with a composition close to $Cu_2In_4Se_7$ were prepared and the structure was also characterized.

1. INTRODUCTION

$CuInSe_2$ has great potential for solar cell applications (Schock 1996). The growth of epitaxial films with good crystallinity may help in the development of new opto-electronic device applications. The films were grown on GaAs because of its reasonable cost and availability. The lattice mismatch of about 2.3% did produce a number of dislocations in the film but an epitaxial film could be obtained (Tseng and Lin 1994). Moreover, the difference in crystal structure of the film (tetragonal chalcopyrite structure) and the substrate (cubic zincblende structure) may cause the formation of orientation domains and antiphase domains in the epitaxial film. Our previous work showed that epitaxial $CuInSe_2$ film free of orientation domains could be grown on a (001)GaAs substrate and the antiphase domains in the films could be eliminated by an in-situ annealing process (Tseng et al. 1996). In this work, cross-sectional transmission electron microscopy is applied to explore the interface structures of $CuInSe_2$ epitaxial films grown by conventional MBE at 500°C and by a photo-assisted MBE (PAMBE) technique at 300°C. In addition, thin films of a vacancy-ordered compound with a composition close to $Cu_2In_4Se_7$ are prepared and the structure is also characterized.

2. EXPERIMENTAL PROCEDURES

Cu-In-Se thin films were grown on (001) GaAs substrates by the molecular beam epitaxy (MBE) technique. The background pressure of the MBE system was 4×10^{-9} torr. The temperatures of the Cu source and Se sources were kept at 1050°C and 210°C, respectively. The temperature of the In source was varied from 720°C to 770°C to control the Cu/In ratio of the films. The MBE system was also equipped with an Oriel Photomax 200W Hg lamp to conduct the photo-assisted growth.

A JEOL JEM3010 transmission electron microscope was used to reveal the interfacial structure and characterize the crystal structure. The TEM specimens were prepared by ion milling. A liquid-nitrogen-cooled stage was used to prevent damage caused by ion bombardment.

3. RESULTS AND DISCUSSION

3.1 Interfacial structure of $CuInSe_2$/GaAs

The temperature for the growth of $CuInSe_2$ epitaxial films on GaAs substrates by the conventional MBE technique was 500°C. For the growth at a lower substrate temperature, an UV light source was used to supply additional energy to the film surface and activate the surface processes such as surface diffusion and the dissociation of Se_2 and Se_4 molecules (Ohishi et al 1989). The interaction range of UV photons is limited to the near-surface region and the other parts of the film-substrate structure are only affected by thermal energy. Hence, the epitaxial growth temperature was dramatically reduced to 300°C.

Thin foils of cross-sectional specimens were examined in the TEM. Figure 1a shows a bright-field image of a $CuInSe_2$ film grown on (001)GaAs substrate. As can be seen, many threading dislocations were initiated at the interface. The microstructures of the films grown by PAMBE showed quite different features compared with the films grown by conventional MBE. Only microtwins and no dislocations were observed in the epitaxial films (see Fig. 1b).

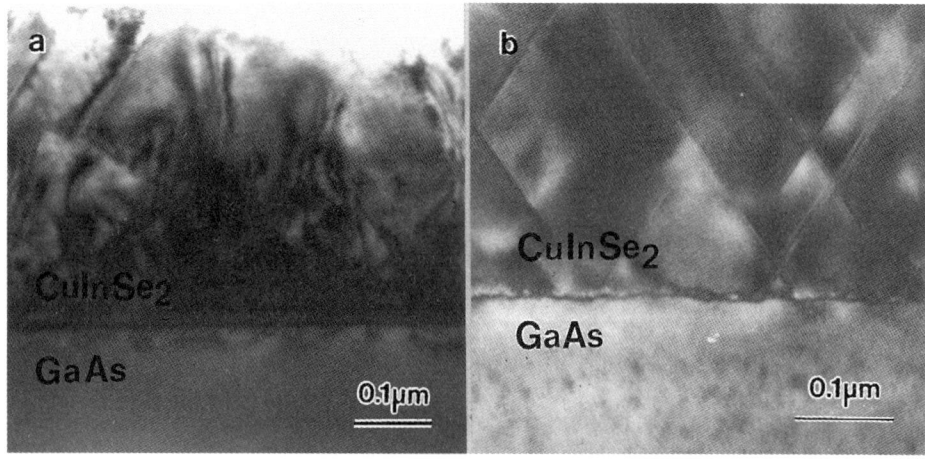

Fig. 1 Cross-sectional TEM micrographs of $CuInSe_2$ epitaxial films grown by (a) conventional MBE at 500°C and (b) PAMBE at 300°C.

Since the lattice misfit between $CuInSe_2$ and GaAs is about 2.3%, the strain in the film may be expected to build up as the film thickness increases and cause the formation of crystalline defects to relieve the strain when the film thickness exceeds a critical value. The thickness of the films was about 600 nm, which was well above the critical thickness. The differences in crystalline defects observed in the epitaxial films grown by the two techniques are thus attributed to different stress relief mechanisms in the films. It is known that the growth temperature determines the extent of the stress released (Tu et al 1992). A higher growth temperature may cause a significant part of stress to be released by the generation of dislocations in the film. In lower-temperature processes, only a small part of the stress is released by microtwins, having formation energies lower than those of dislocations, thus prevail in the film.

3.2   Structures of vacancy-ordered $Cu_2In_4Se_7$ thin film

An increase in the In content of the film growth resulted in the formation of a new phase with an ordering of Cu vacancies in the chalcopyrite structure. The chemical composition of this film was close to $Cu_2In_4Se_7$ as determined by energy dispersive X-ray spectroscopy. The atomic arrangement in a $Cu_2In_4Se_7$ unit cell may be constructed according to the measured chemical composition and the ordering sequence of Cu vacancies revealed by electron diffraction, see Fig. 2.

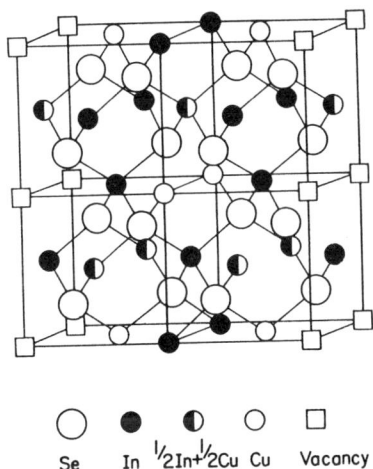

Fig. 2   Atomic arrangement in a $Cu_2In_4Se_7$ unit cell.

○ ● ◑ ○ □
Se   In   ½In+½Cu   Cu   Vacancy

An antiphase domain structure, which is similar to that found in $CuInSe_2$ epitaxial films, was also observed. Figure 3 shows the antiphase domain boundaries (APB's) revealed using a (101) superlattice reflection spot to form the image. An APB is actually a plane of antisite defects (Tseng et al 1996), which may detrimental to the film properties. An in-situ annealing technique is being developed to eliminate APB's from the film.

4.   CONCLUSIONS

Epitaxial films of $CuInSe_2$ have been successfully grown on (001)GaAs substrates by conventional and photo-assisted MBE technique. A reduction in epitaxial growth temperature from 500°C to 300°C was achieved by PAMBE. A high density of threading dislocations initiated from the interface was found in the epitaxial films grown by the conventional MBE technique, whereas microtwins were observed in the films grown by the low-temperature MBE process. The different types of crystalline defects observed in the films are attributed to the temperature dependence of the mechanisms for releasing misfit strain.

Another film with chemical composition close to $Cu_2In_4Se_7$ was also characterized by TEM. This film had many Cu vacancies that were arranged in an ordered sequence in the

310

chalcopyrite structure.   An antiphase domain structure was observed in this film.

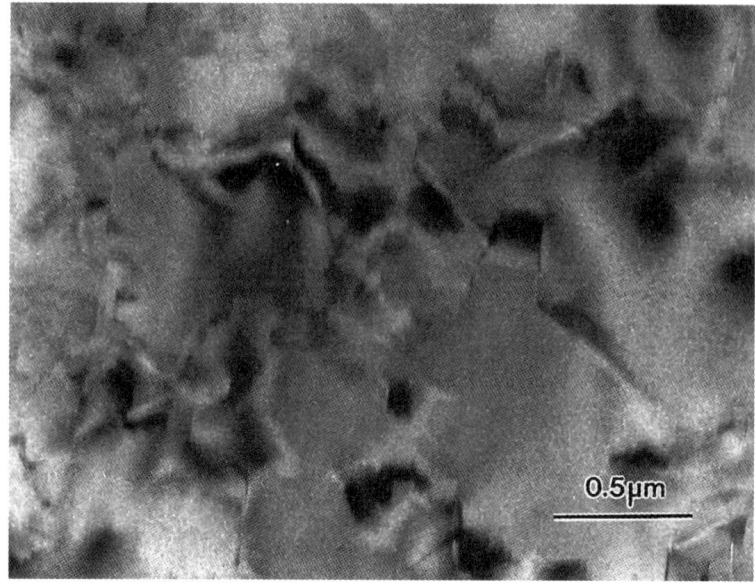

Fig. 3   A TEM micrograph of a plan-view $Cu_2In_4Se_7$ specimen shows APB's in the film.

ACKNOWLEDGEMENT

The authors gratefully acknowledge the support of the National Council of Science under Contract No. NSC 83-0208-M-110-009 and NSC 84-0208-M-110-001.

REFERENCES

Ohishi M, Saito H, Okano H and Ohmori K 1989 J. Crystal Growth **95**, 538
Schock H W 1996 Appl. Surf. Sci. **92**, 606.
Tseng B H and Lin S B 1994 MRS Symposium Proceedings Series **317**, 187.
Tseng B H, Lin S B, Gu G L and Chen W 1996 J. Appl. Phys. **79**, 1391.
Tu K N, Mayer J W and Feldman L C 1992 Electronic Thin Film Science p.174 (New York: Macmillan)

Inst. Phys. Conf. Ser. No 157
Paper presented at Microsc. Semicond. Mater. Conf., Oxford, 7–10 April 1997
© *1997 IOP Publishing Ltd*

Structural investigations of epitaxial CdMgSe/InAs(001) heterostructures

Th Walter, D Gerthsen, Th Litz*, A Waag* and G Landwehr*

Laboratorium für Elektronenmikroskopie, Universität Karlsruhe, Kaiserstraße 12,
D-76128 Karlsruhe, FRG
*Physikalisches Institut der Universität Würzburg, Am Hubland, D-97074 Würzburg, FRG

ABSTRACT: A microstructural study of an epitaxial $Mg_{0.33}Cd_{0.67}Se$ layer grown by molecular beam epitaxy on a InAs(001) substrate has been carried out by electron microscopy techniques. The epilayer grows predominantly in the zinc-blende structure with small volume fractions of the wurtzite structure. A high density of stacking faults with an anisotropy for the [110]- and [1$\bar{1}$0]-directions is found which induces additional reflections in the diffraction patterns. The surface shows pronounced facetting with ridges along one <110>-direction. The facetting and the stacking faults are correlated with the relaxation of the lattice parameter mismatch and the transformation of the wurtzite into the zinc-blende structure.

1. INTRODUCTION

Due to the high band gap energy, MgSe-containing II-VI semiconductors are an interesting source material for optoelectronic devices in the green and blue spectral region. In the thermodynamical equilibrium binary CdSe is stable in the hexagonal wurtzite structure. It exhibits a bandgap energy of E_{gap} = 1.67 eV at 300 K. By adding magnesium to CdSe, the bandgap energy of the resulting ternary compound can be shifted into the blue-green region of the spectrum. It is possible to stabilise epitaxially grown CdSe or ternary CdMgSe in the zinc-blende structure on (001)-oriented cubic substrates (Samarth et al. 1989, Phillips et al. 1992). The difference between the two phases is a different stacking sequence of the {111}-atomic plane with *AB AB AB* ... stacking for the wurtzite structure and *ABC ABC* ... for the zinc-blende structure. For the wurtzite CdSe phase a small stacking fault energy (γ = 14 ± 5 mJ/m^2 for the wurtzite structure Takeuchi et al. 1984) is observed which is the origin of the small transition energy between the wurtzite and the zinc-blende phase.

In the present study a structural characterisation of a $Mg_{0.33}Cd_{0.67}Se$/InAs(001) film is presented with special emphasis on the analysis of the defect structure which is induced by the misfit relaxation and the transition from the cubic zinc-blende into the hexagonal wurtzite structure.

2. EXPERIMENTAL TECHNIQUES

The $Mg_{0.33}Cd_{0.67}Se$ film with a thickness of 1.1 μm was grown by molecular beam epitaxy (MBE) on an InAs(001) substrate at about 250 °C. The surface morphology was imaged on a larger scale by a scanning electron microscope of the type JSM 6300F at 10 keV.

In a (001)-oriented sphalerite substrate the two orthogonal <110>-directions parallel to the surface are non-equivalent. The difference is caused by the tetrahedral bonding of the atoms of the two different face centred cubic sublattices. This leads to a different crystal polarity for the two orthogonal <110>-projections. Therefore cross-sectional samples were prepared along the two orthogonal <110>-orientations for the transmission electron microscopy (TEM) study. The determination of the crystal polarity was carried out using the convergent beam electron diffraction (CBED). The <110>-zone axis CBED pattern shows different patterns for the $\{hkl\}$- and the $\{hk\bar{l}\}$-reflections. To determine the crystal polarity the experimental CBED pattern must be compared with a numerically simulated CBED pattern. The simulation of the experimental pattern was carried out using the EMS software of Stadelmann (1987). The indexing of the crystal directions was adopted according to the convention of Gatos and Lavine (1960). For a III-V semiconductor the $\{\bar{1}\bar{1}\bar{1}\}$-surface is terminated with group V atoms. The specimen preparation by the standard mechanical preparation and Ar$^+$-ion milling followed the procedure described by Strecker et al. (1993). The TEM investigations were carried out using a Philips CM 200 FEG/ST transmission electron microscope which is equipped with an energy dispersive X-ray (EDX) system Noran Voyager 3000 with a germanium detector.

3. EXPERIMENTAL RESULTS

The scanning electron microscopy (SEM) image (Fig. 1) shows the surface of the $Mg_{0.33}Cd_{0.67}Se$ layer. The surface is characterised by ridges which are preferentially oriented along the $[1\bar{1}0]$-direction.

Fig. 1 *SEM image of the specimen surface with ridges along one the[$1\bar{1}0$]-direction.*

Conventional cross-sectional micrographs from samples oriented in the two different <110>-directions with $g = \{220\}$ are presented in Fig. 2. The first micrograph Fig. 2(a) shows the $[1\bar{1}0]$-projection parallel to the surface ridges. A high stacking fault density in the film can be recognized at the MgCdSe/InAs interface. The stacking fault density decreases with the distance from the interface. The surface exhibits a triangular facetting parallel to the $\{112\}$-planes with sharp edges which is magnified in the Fig. 2(a). The average base width of the ridges is 50 nm and the height is approximately 15 nm. A columnar structure with lamellae of the same width extends into the depth of the layer parallel to the growth direction. Each lamella shows a "herring bone pattern" consisting of 5 nm thick layers parallel to the facets with alternating bright and dark contrast. The film in the orthogonal [110]-projection

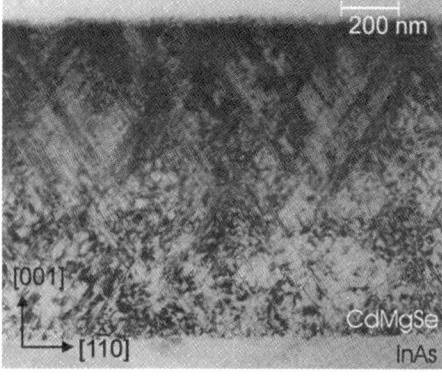

Fig. 2 *Cross-sectional TEM micrographs of CdMgSe/(001)InAs with a projection along (a) [$1\bar{1}0$] and (b) [110].*

shown in Fig. 2(b) contains a comparably high stacking fault density at the MgCdSe/InAs interface. For this [110]-projection the density still increases along the growth direction and becomes very high in the surface region. The surface is almost flat and the sample also shows an alternating bright and dark contrast parallel to the surface.

EDX analyses were carried out in the TEM nanoprobe-mode with a probe size smaller than 2 nm which showed a variation of the Cd concentration. Therefore a composition modulation of the ternary MgCdSe is the most reasonable origin of the bright and dark "herring bone pattern" contrast. Such a composition modulation can be caused by the rotation of the sample during MBE growth as shown by Wang et al. (1993).

HRTEM of the interface region shows that the misfit is mainly relaxed by Shockley Partial dislocations which terminate the stacking faults at the interface. The two perpendicular <110>-projections show a comparable stacking fault density near the interface. The density leads to a complete misfit relaxation even if 30 ° partial dislocations are assumed at the interface whose efficience for the mismatch relaxation is only 1/6 of the Lomer dislocations.

Diffraction patterns of the film in different zone-axis orientations show a large number of extra diffraction spots of weak intensity in addition to the fundamental reflections of the zinc-blende structure. The diffraction pattern of the <110>-zone axis (Fig. 3(a)) shows streaks along the <111>-directions. Some additional weak reflections are superimposed on the <111>-streaks. Fig. 3(b) shows a <111>-zone axis pattern with extra spots of the type 1/3{224} marked with arrowheads (all spot indices in brackets denote also the appropriate multiples). Additional reflections were also found in the zone axes of the type <224>, <013>, <112>, <123> and <233>. The indices of the extra diffraction spots are of the type 1/3{224}, 1/2{113}, 1/3{442}, 1/2{133}, 1/2{115}, 1/3{115} and 1/4{335}.

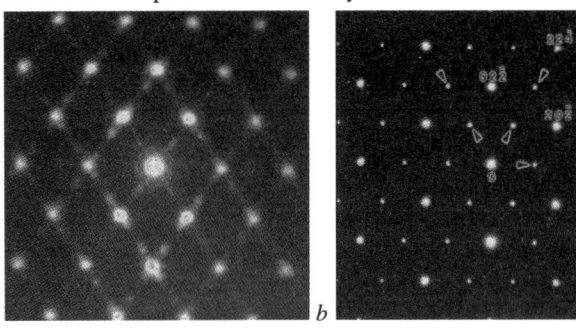

a *b*

Fig. 3 *Diffraction patterns from the MgCdSe film in (a) a <110>-zone axis and (b) a <111>-zone axis.*

4. DISCUSSION

The additional reflections, with the exception of the extra spots on the streaks (Fig. 3(a)) in the diffraction patterns, can be explained by the high density of stacking faults. A stacking fault on {111}-planes has a very small extension perpendicular to the stacking fault plane. This leads to an expansion of the reciprocal lattice points into reciprocal rods along this <111>-direction. The distance of the rods can be explained by the following consideration. The atomic positions in one {111}-plane of the zinc-blende structure are arranged in a way, that every third atomic {224}-plane perpendicular to this {111}-plane contains atoms of this particular plane. A stacking fault on this {111}-plane is caused by a shift of the atomic positions by a vector 1/6<112>. Therefore every third {224}-position which was originally occupied by an atom is now unoccupied and every third {224}-position belonging to the new positions is newly occupied which induces rods at the 1/3{224} positions in the reciprocal lattice.

Additional spots in the diffraction patterns always occur if the Ewald sphere for a special incident direction of the electron beam intersects the reciprocal rods (Xiao and Daykin (1994)). The Ewald sphere for an incident beam parallel to the <111>-direction intersects the reciprocal rods at lattice vectors $g = 1/3\{224\}$ perpendicular to this <111>-direction. This induces additional reflections at every $g = 1/3\{224\}$ in the <111>-zone axis. For an electron beam parallel to a <110>-direction, the Ewald sphere only touches the rods which results in streaking along the <111>-directions in the <110>-diffraction pattern (Fig. 3(a)). For any other incident beam direction [uvw], the reciprocal rods are intersected at different lattice vectors g_{hkl} which are in agreement with the experimental observations.

The additional <110>-zone axis reflections observed on the streaks in Fig. 3(a) result from small volume fractions of the wurtzite structure in the layer which is confirmed by HRTEM. The (0001)-basal planes of the hexagonal phase are oriented parallel to the {111}-planes of the zinc-blende structure. Therefore the weak extra spots in the diffraction pattern can be explained by the superposition of two wurtzite <$11\bar{2}0$>-diffraction patterns onto the zinc-blende <110>-zone axis pattern as shown schematically in Fig. 4.

The CdMgSe layer shows a complex microstructure which is correlated to the mismatch relaxation and the metastability of the zinc-blende structure. The $Mg_{0.33}Cd_{0.67}Se$ layer is in biaxial tensile strain on an InAs(001) substrate with a misfit of $f = 0.74$ %. The critical thickness for this layer composition is about 35 nm (calculated after Cohen-Solal et al. 1994). Therefore the critical thickness is largely exceeded for the film thickness of about 1.1 µm. It can be assumed that a considerable part of the tensile strain is already relieved by misfit dislocations at the interface

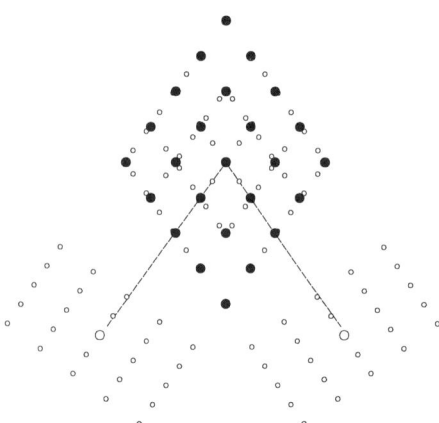

Fig. 4 *Schematic explanation of the diffraction pattern in Fig. 3a which is a combination of a zinc-blende <110>-diffraction pattern and two inclined wurtzite < $11\bar{2}0$ >-diffraction patterns.*

which is confirmed by the density of misfit dislocations at the interface.

A facetted surface similar to MgCdSe/InAs(001) was found by Brown et al. (1989) at the surface of a ZnSe/ZnS/GaAs(001) heterostructure. In both cases the surface ridges are preferentially oriented along the [$1\bar{1}0$]-direction. The facetting of the layer is likely to be related to the relaxation of the misfit which precedes the plastic relaxation. The presence of the facets at an early stage of the growth is confirmed by the triangulary shaped contrast modulation which starts at a layer thickness of less then 50 nm. The surface facet planes for the sample examined in this paper were found to be of the {112}-type. Brown et al. (1989) suggest that the orientation of surface facets is correlated with the different growth rates for varying crystal planes.

The high stacking fault density in the upper part of the layer is induced by the transition of the zinc-blende into the wurtzite structure which is exclusively present close to the layer surface.

REFERENCES

Brown P D, Russell G J, Woods J 1989 J. Appl. Phys. **66**, 129

Cohen-Solal G, Bailly F, Barbé M 1994 J. Cryst. Growth **138**, 68

Gatos H C, Lavine M C 1960 J. Electrochem. Soc. **107** , 427

Phillips MC, Wang MW, Swenberg JF, McCaldin JO, McGill TC 1992 Appl. Phys. Lett. **61**, 1992

Samarth N, Lao H, Furdyna J K, Qadri S B, Lee Y R, Ramadas A K, Otsuka N 1989 Appl. Phys Lett **54**, 2680

Stadelmann P A 1987 Ultramicroscopy **21**, 131

Strecker A, Salzberger U, Mayer J 1993 Prakt. Metallogr. **30**, 482

Takeuchi S, Suzuki K, Maeda K, Iwanga H 1984 Phil. Mag. A **50** , 171

Wang J, Steeds J W, Hopkinson M 1993 Semicond. Sci. Technol. **8**, 502

Xiao H Z, Daykin A C 1994 Ultramicroscopy **53**, 325

Inst. Phys. Conf. Ser. No 157
Paper presented at Microsc. Semicond. Mater. Conf., Oxford, 7–10 April 1997
© 1997 IOP Publishing Ltd

315

Transmission electron microscopy investigations of an epitaxial beryllium-chalcogenide-based superlattice

Th Walter, A Rosenauer, D Gerthsen, F Fischer*, R Gall*, Th Litz*, A Waag* and G Landwehr*

Laboratorium für Elektronenmikroskopie, Universität Karlsruhe, Kaiserstraße 12, D-76128 Karlsruhe, FRG
*Physikalisches Institut der Universität Würzburg, Am Hubland, D-97074 Würzburg, FRG

ABSTRACT: Electron microscopy techniques were used to investigate a BeSe containing II/VI-semiconductor superlattice grown by molecular beam epitaxy on a (001)-oriented GaAs substrate. The BeTe/ZnSe-superlattice was examined by scanning electron microscopy, conventional and high-resolution transmission electron microscopy. The superlattice shows cracks preferentially oriented along one particular <110>-direction to relax the tensile stress in the superlattice. High-resolution micrographs were evaluated by correspondence analysis (CA) to quantify the abruptness of the chemical transition at the interfaces and to determine accurately the layer thicknesses.

1. INTRODUCTION

Due to their high band gap energies, beryllium-chalcogenides are considered as promising materials for optoelectronic devices in the green and blue spectral region. Beryllium-chalcogenides are considerably harder than most wide gap II/VI-semiconductors which leads to significantly reduced defect densities – in particular during the operation of light emitting devices.

Binary BeTe and ZnSe are almost lattice matched to GaAs (a_{GaAs} = 5.654 Å). Compared to GaAs, the lattice constant of the BeTe a_{BeTe} = 5.626 Å is smaller whereas ZnSe has a slightly larger lattice parameter a_{ZnSe} = 5.667 Å. ZnSe and BeTe can be combined in BeTe/ZnSe-superlattices with alternating tensile and compressive strain on GaAs substrates. The average lattice parameter of the superlattice is given by $a_{SL}=(G_{BeTe}a_{BeTe}t_{BeTe}+G_{ZnSe}a_{ZnSe}t_{ZnSe})/(G_{BeTe}t_{BeTe}+G_{ZnSe}t_{ZnSe})$ where t denotes the layer thickness and G the bulk modulus. Lattice match with respect to the GaAs substrate is obtained if the BeTe/ZnSe layer thickness ratio is close to ½, assuming identical elastic properties which must be considered as a rough approximation. Values for the bulk modulus of the BeTe are, to our knowledge, not published.

In the present study a structural characterisation of BeTe/ZnSe-superlattices with the thickness of the ZnSe-layers being about twice of that of the BeTe-layers was carried out. Special emphasis was put on the determination of the strain state of the superlattice, the layer undulations, the roughness of the interface and the sharpness of the chemical transition.

2. EXPERIMENTAL TECHNIQUES

A BeTe/ZnSe superlattice with 75 periods each consisting nominally of 13 Å BeTe and 26 Å ZnSe was grown by molecular beam epitaxy (MBE) on GaAs(001) substrate at about 300 °C. The total thickness of the superlattice is 300 nm. The surface of this superlattice was etched for 3 min by a solution of Br_2-CH_3OH with 2.0 % Br_2 and investigated with a scanning electron microscope (SEM) of the type JSM 6300 F at 10 keV.

Since the superlattices show a crack anisotropy for the two orthogonal <110>-directions parallel to the surface the crystal polarity of the GaAs substrate was determined using the convergent beam electron diffraction (CBED), according to the technique described by Taftø and Spence (1982). The indexing of the crystal directions was adopted according to the convention of Gatos and Lavine (1960). For a III-V semiconductor the $\{\bar{1}\bar{1}\bar{1}\}$-surface is terminated with group V atoms. The BeTe and ZnSe layers of the superlattice are assumed to be of comparable polarity as the substrate, i.e. the positions of the group VI atoms correspond to the arsenic locations.

The transmission electron microscopy (TEM) investigations of the superlattice were carried out on cross-sectional specimens along the <110>-projections using a Philips CM 200 FEG/ST. To detect the abruptness of the chemical transition between the BeTe and the ZnSe layers, the evaluation program DALI (Digital Analysis of Lattice Images) was applied (Rosenauer et al. 1996) on digitized images.

3. EXPERIMENTAL RESULTS

The secondary electron SEM image (Fig. 1) shows the etched surface of the BeTe/ZnSe-superlattice. The surface is characterised by cracks preferentially oriented along the [110]-direction. The crack separation is 2 µm to 10 µm. The concave curvature of the superlattice between the cracks is deduced from bright/dark contrast transition at each crack which is also observed on the unetched surface.

In Fig. 2(a) a cross-sectional TEM micrograph with g = (004) from the sample oriented in the [110]-direction is presented. A crack with a V-shaped morphology is visible which is typical for all observed cracks. The

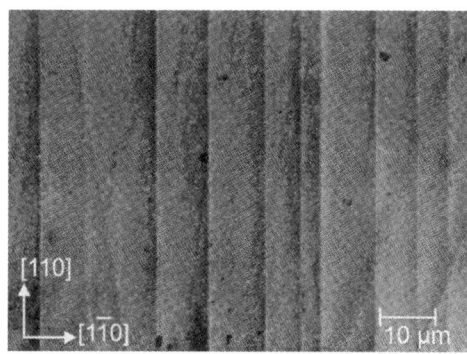

Fig. 1 *SEM image of the BeTe/ZnSe-superlattice surface etched by Br_2-CH_3OH with crack formation along the [110]-direction.*

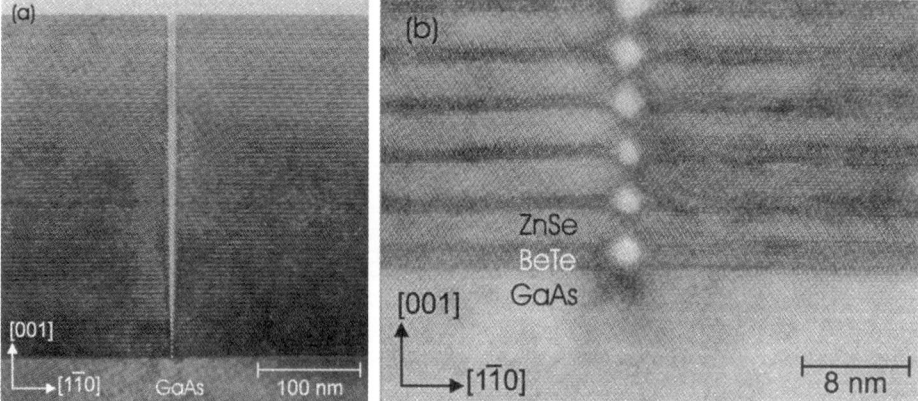

Fig. 2 *Cross-sectional images of a [110]-oriented specimen: (a) with g = (004) displaying a crack in the superlattice with a V-shaped profile (b) HRTEM image of the area close to the GaAs interface showing undulating crack edges.*

crack ends at the GaAs interface. Sometimes, cracks even enter the substrate up to 30 nm.

The HRTEM image (Fig. 2(b)) reveals undulating crack edges with a smaller crack width between the ZnSe layers. The wavy edges are attributed to artifacts of the specimen preparation procedure. The contact of the crack surface with oxygen after the MBE growth or with water during the sample preparation could induce the growth of oxides. Beryllium is known to have a large hydration enthalpy and a strong tendency to oxidize. The oxide is highly strained on the BeTe and could be preferentially removed during the Ar$^+$-ion milling.

The slight concave curvature of the superlattice layers is visible on larger sections of HRTEM images which indicates the presence of tensile strain. The GaAs interface does not show steps on the scale of the HRTEM micrographs. Misfit dislocations could not be observed which is confirmed by investigating planview specimens.

Fig. 3(a) shows a digitized HRTEM image of the superlattice. The dark areas correspond to the BeTe layers and the bright areas to the ZnSe. To determine the abruptness of the chemical

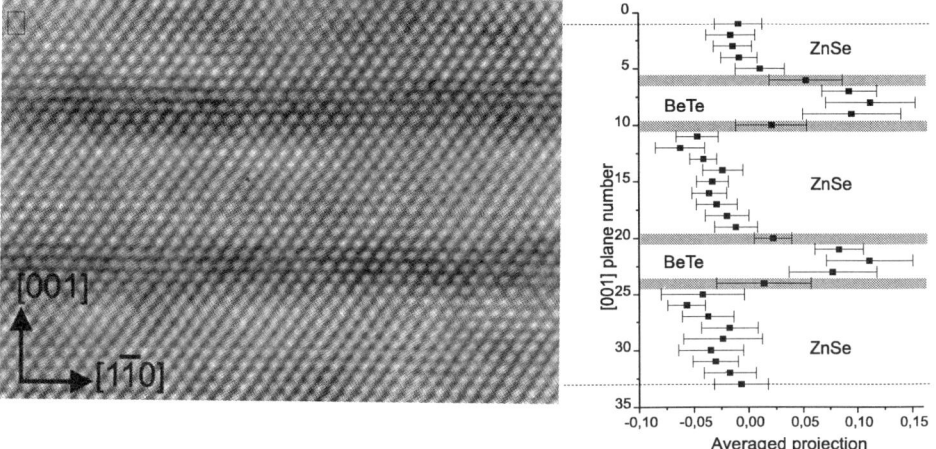

Fig. 3 (a) Digitized HRTEM image of the BeTe/ZnSe superlattice. (b) Projection of each image unit lattice cell onto the first eigencell, averaged along the [1 $\bar{1}$ 0]-direction.

transition the image was analysed using correspondence analysis (CA) (Aebersold et al. 1996) which was implemented in the DALI software package. The CA is a special variant of multivariate statistics (MS). The HRTEM image is subdivided into N image unit cells with an extension of the local (001)-plane along the growth direction and a width of the local (110)-distance (indicated close to the upper left corner of Fig.3(a)). The image unit cells are transformed into quadratic cells with $P = 2^n \times 2^n$ ($n = 5$) pixels. Each image unit cell is represented by a P-dimensional positive real vector r_k which contains the intensity of each pixel. The aim of MS is to investigate the shape of the cloud of points formed by the tips of the N image vectors. This problem is analogous to the well known task of finding the axis of least inertia of a distribution of N points of mass m_j, where m_j is given here by the integral intensity of cell j. The given problem is solved by calculating eigenvalues λ_j and eigenvectors e_j ($j = 1...M$) of the matrix of inertia. Each eigencell e_j represents a principal axis of the cloud of image points r_k. The main idea of this method is that only a small number M of principal axes is sufficient to describe the given distribution of points r_k. It can be shown that the corresponding M eigenvectors $e_{\alpha_1}...e_{\alpha_M}$ are eigenvectors to the M largest eigenvalues $\lambda_{\alpha_1} > \lambda_{\alpha_2} > ... > \lambda_{\alpha_M}$. In the present case it was found that the first eigenvector e_{α_1} contains characteristics of the contrast pattern of the BeTe region. This behaviour is shown in Fig. 3(b), where the value $r_k \cdot e_{\alpha_1}$ averaged along the [1 $\bar{1}$ 0]-direction is plotted versus the monolayer number in growth direction. The grey bars in Fig. 3(b) mark the estimated transition zone from BeTe to

ZnSe. This zone is found to be in the order of one atomic layer. The layer thicknesses can be accurately measured which typically yields 4 monolayers of BeTe and 10 monolayers of ZnSe.

4. DISCUSSION

The microstructure of the superlattice is significantly influenced by the stress distribution with respect to the GaAs substrate. The presence of stress is established by the cracks which were previously observed to relieve the misfit strain under tensile stress conditions (e.g. Murray et al. (1996)). However, there appears to be a discrepancy between the concave curvature of the superlattice layers between the cracks which is indicative of tensile stress and the measured BeTe/ZnSe layer thickness ratio of typically 0.4. Compressive stress is expected for the measured layer thickness ratio if, again, identical elastic properties of BeTe and ZnSe are assumed. The presence of tensile stress is confirmed by X-ray diffractometry which yields an average lattice parameter for the superlattice of $a_{SL} = 5.626$ Å.

The following reason could be responsible for the observed discrepancy. Small fractions of BeSe (lattice parameter $a_{BeSe} = 5.189$ Å) at the BeTe/ZnSe interfaces could reduce the average lattice parameter. Assuming the formation of the ternary BeSeTe compound at every BeTe/ZnSe interface with an average thickness of one atomic layer, this compound must contain 42 % BeSe to reduce the lattice parameter to a_{SL}. For quaternary ZnBeSeTe at the interfaces, the BeSe ratio must raised because the presence of Zn shifts the lattice parameter towards a higher value.

The critical thickness t_c of the superlattice can be approximately calculated by assuming a homogeneous layer with a_{SL}. The approximation is justified because the layer thicknesses of the BeTe and ZnSe layers is far below t_c for the relaxation between the individual layers (Matthews and Blakeslee 1976). A critical thickness of 60 nm is calculated according to Cohen-Solal et al. (1994) which is significantly lower than the total superlattice thickness of 300 nm. Although the critical thickness for the misfit dislocation generation is largely exceeded, the mismatch is only relaxed by cracks. Therefore, a strong barrier for the dislocation generation must exist. Since it is known that misfit dislocations are easily nucleated and mobile in mismatched ZnSe/GaAs-heterostructures (Rosenauer et al. 1996) the observed effect can be attributed to the BeTe.

The CA turned out to be a valuable analytical method to quantify the information contained in HRTEM images regarding the evaluation of the abruptness of the chemical transition and the layer thicknesses. The width of the transition region was determined to be in the order of one atomic layer which shows that abrupt interfaces can be grown in the ZnSe/BeTe system. This value for the transition from BeTe to ZnSe comprises effects of layer undulations along the direction of the electron beam and the sharpness of the chemical transition, which are difficult to separate.

REFERENCES

Aebersold J F, Stadelmann P A and Rouvière J-L 1996 Ultramicroscopy **62**, 171
Cohen-Solal G, Bailly F and Barbé M 1994 J. Cryst. Growth **138**, 68
Gatos H C and Lavine M C 1960 J. Electrochem. Soc. **107**, 427
Matthews J W and Blakeslee A E 1976 J.Cryst. Growth **32**, 265
Murray R T, Kiely C J and Hopkinson M 1996 Phil. Mag. A **74**, 383
Rosenauer A, Kaiser S, Reisinger T, Zweck J and Gebhardt W 1996 Optik **102**, 63
Rosenauer A, Reisinger T, Franzen F, Schütz G, Hahn B, Wolf K, Zweck J and Gebhardt W 1996 J. Appl. Phys. **79**, 4124
Taftø J and Spence J H C 1982 J. Appl. Cryst. **15**, 60

Inst. Phys. Conf. Ser. No 157
Paper presented at Microsc. Semicond. Mater. Conf., Oxford, 7–10 April 1997
© *1997 IOP Publishing Ltd*

Growth of SiC layers on off-axis 4H-SiC substrates

B Pécz, L Tóth, G Radnóczi, C Hallin* and E Janzén*

Research Institute for Technical Physics of the Hungarian Academy of Sciences, H-1325, Budapest, PO Box 76. Hungary
*Department of Physics and Measurement Technology, Linköping University, S-581 83 Linköping, Sweden

ABSTRACT: This paper deals with defect formation and in particular with various phase changes occurring during the CVD growth of homoepitaxial SiC films. The fine structure of growth defects at the phase boundaries of various SiC polytypes were characterised by means of cross sectional and high resolution TEM studies. In one of the examples shown the originally 4H epitaxial layer continued to grow in the cubic polytype. In another case by selecting proper deposition parameters the 4H phase was able to overgrow the locally formed cubic intermediate SiC phase.

1. INTRODUCTION

The growth of high quality homoepitaxial SiC layers is essential for the SiC electronics. CVD (Chemical Vapour Deposition) provides a tool for epitaxial layer growth at relatively high growth rates. The nucleation of 3C-SiC inclusions is a general problem during homoepitaxial growth onto hexagonal SiC substrates at lower temperatures. 3C-SiC has a lower surface energy than the hexagonal polytypes thus 3C nucleates on flat surfaces by 2D nucleation. To prevent the formation of 3C layers on 4H or 6H substrates miscut substrates are used. Once the density of surface steps is high (i.e. on miscut samples) the growth of SiC occurs by step flow growth mechanism (Kong et al 1988, Ueda et al 1990) which means that SiC nucleates at the step walls and the growing SiC layer follows the polytype of the substrate.

Beside the growth temperature and the silane flow rate the C/Si ratio in the growth chamber also influences the growth mode of the layer (Konstantinov et al 1995). It is found that at high C/Si ratio the density of 3C inclusions is high, while their density is low at low C/Si ratio. In this paper we present examples for the above phenomena and the defect structure of the samples is studied by TEM (Transmission Electron Microscopy).

2. EXPERIMENTAL

SiC epitaxial layers were grown by CVD using the silane-propane-hydrogen system at atmospheric pressure. The substrates were 4H-SiC (0001), n-type ($\approx 10^{18}$ cm^{-3}), Si-face, off-axis (3.5° toward the <11-20> direction). Samples were grown at 1450°C at growth rate of 1.5 μm/h, while the C/Si ratio in the gas was varied between 2 and 6.

Cross sectional specimens were prepared for conventional and high resolution transmission electron microscopy (HREM) investigation of the samples grown at different conditions. The specimens were thinned by 10 keV Ar⁺ ions at low angle of incidence by the method described by Barna et al (1997a). As a final step of preparation a new surface cleaning procedure with low energy Ar⁺ ions was also applied as described in another paper in this volume (Barna et al 1997b).

High resolution electron microscopy was carried out in a JEOL 4000 EX TEM at the Electron Microscopy Department of University of Antwerp, Belgium. A Philips CM20 TEM was used at 200 kV for the conventional electron microscopy.

3. HIGH C/Si RATIO - 3C CUBIC PHASE ON 4H HEXAGONAL PHASE

a.

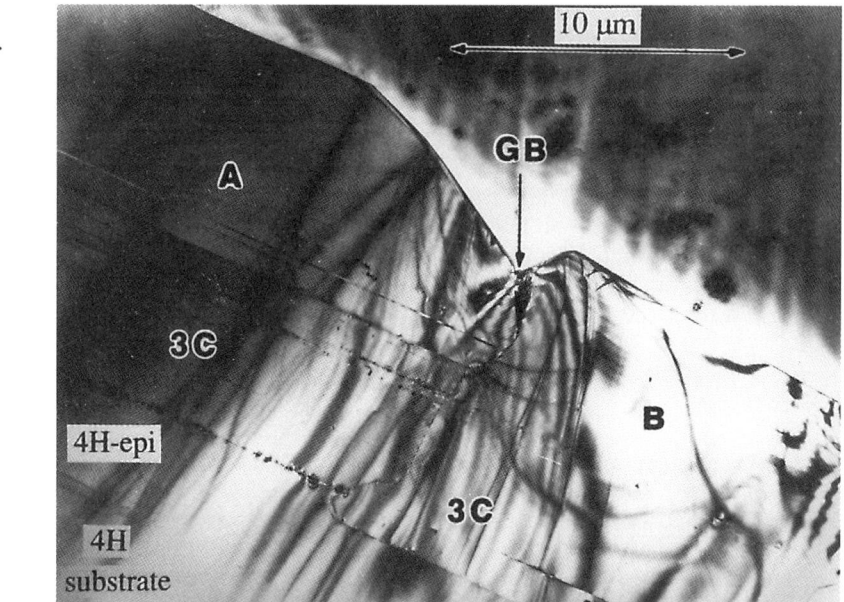

b.   c.

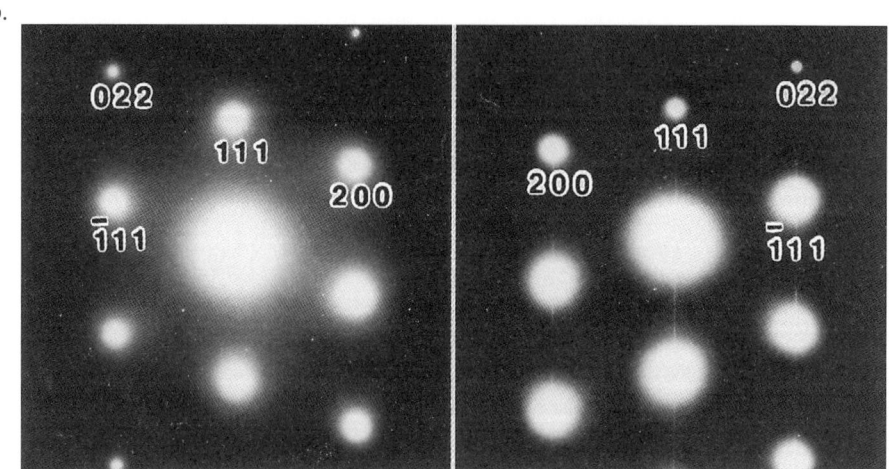

Fig. 1. (a) Cross sectional micrograph of the transition region in the 4H phase and the two cubic twin variants. (b) and (c) are the diffraction patterns of the twins A and B respectively.

Fig. 1. shows a cross sectional micrograph of the epitaxial layers on a 4H substrate in the [1-100]$_{4H}$ zone. The sample was grown with a C/Si atom ratio equal to 6 in the gas mixture. The 4H-SiC epitaxial layer is 2-3 μm and overgrown by a 10 μm thick 3C-SiC layer. The epitaxial orientation relationship between 4H and 3C allows the cubic phase to nucleate in two twin related variants (labelled with A and B in Fig. 1 a) which form twin boundaries upon coalescence (Figs. 1 b and 1 c respectively). A detailed analysis of the highly faceted twin boundary revealed (111) and (211) type boundary planes while the outer surface of the cubic layer exhibits (100), (211) and (311) facets on the twin boundary groove.

4. LOW C/Si RATIO - 4H HEXAGONAL PHASE OVER 3C CUBIC INCLUSION

The sample shown in Fig. 2 displays the opposite kind of phase change: in some places the epilayer starts to grow in the 3C cubic phase and this is later overgrown by the 4H hexagonal phase. This sample was grown with a C/Si ratio of 2 in the gas mixture. The 3C region was formed at the beginning of growth probably on a region where the density of surface steps was low.

In this case the XTEM specimen was cut parallel to the steps due to the surface misorientation thus the interfaces are virtually flat. The HREM observation revealed that the cubic intermediate layer itself is twinned relative to the (111) plane parallel to the original interface (Fig. 3). The first 3C twin was found to be about 2 - 5 nm thick while the second twin variant was 60 - 70 nm. The length of this cubic inclusion exceeded 10 μm. On the other hand the interface between the cubic phase and the overgrown 4H hexagonal region was always found flat and smooth without any intermediate layer (Fig. 4).

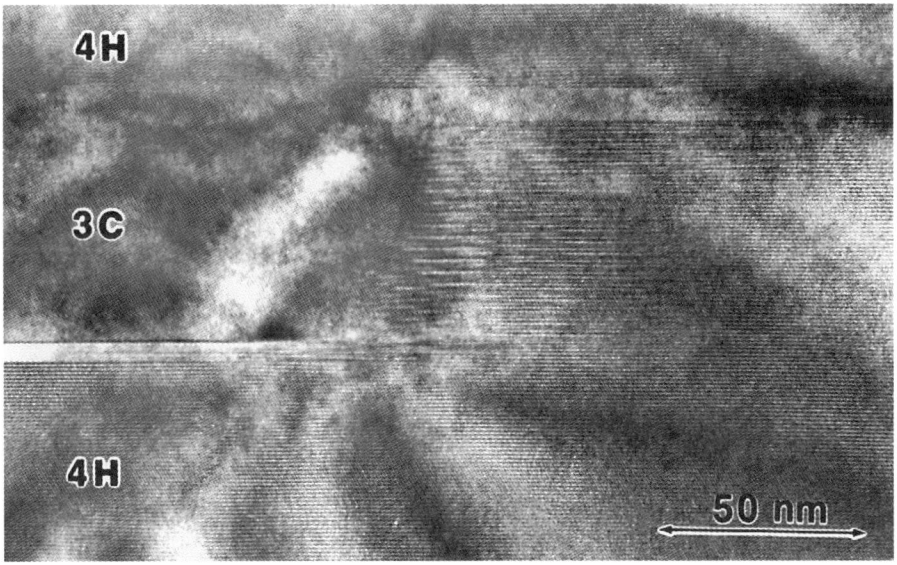

Fig. 2. Cross sectional (XTEM) micrograph of a twinned, cubic (3C) inclusion at the interface of 4H hexagonal SiC film epitaxially grown on 4H SiC (0001) substrate. The image shows a lateral transition region between the 3C and 4H phases as well where a virtual periodicity of 3 nm is enhanced due to overlapping polytypes.

322

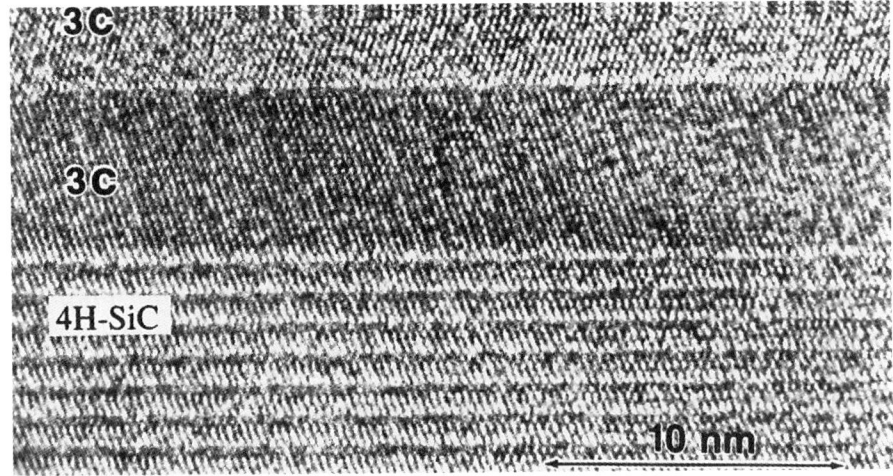

Fig. 3. High resolution XTEM micrograph of the two 3C twin variant near the 4H hexagonal SiC substrate.

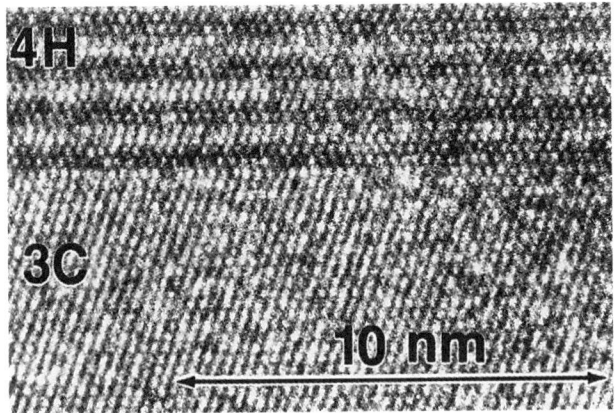

Fig. 4. High resolution XTEM micrograph of the transition region between the 3C cubic phase and the overgrown 4H region in SiC.

It has been shown that 3C region formed locally on 4H substrate can be overgrown by 4H at appropriate growth conditions.

ACKNOWLEDGEMENT

One of the authors (LT) wants to express his gratitude to the Flemish and Hungarian Academies of Sciences for sponsoring his stay at Antwerp University in the frame of the Inter-Academic Exchange Program which made it possible to carry out the above HREM studies. The work was partly supported by the National Science Research Foundation under OTKA Grant No. T14091.

REFERENCES

Barna Á, Radnóczi G and Pécz B 1997a Handbook of Microscopy ed. S. Amelinckx et al, (VCH Verlag, Weinheim, 1997) Vol. 3 pp 751-801

Barna Á, Tóth L, Pécz B, Radnóczi G 1997b *paper in this volume*

Kong H S, Glass J T and Davis R F 1988 J. Appl. Phys. **64** 2672-2678

Konstantinov A O, Hallin C, Kordina O and Janzén E 1995 IOP Conf.Ser. **142** (IOP, Bristol) pp. 249-252

Ueda T, Nishino H and Matsunami H 1990 J. Cryst. Growth **104** 695-700

Inst. Phys. Conf. Ser. No 157
Paper presented at Microsc. Semicond. Mater. Conf., Oxford, 7–10 April 1997
© *1997 IOP Publishing Ltd*

Self-organisation and defect mechanisms in heteroepitaxial growth

H P Strunk, M Albrecht, S Christiansen and W Dorsch

Institut für Werkstoffwissenschaften, Lehrstuhl Mikrocharakterisierung, Universität Erlangen-Nürnberg; Cauerstr. 6, 91058 Erlangen, Germany

ABSTRACT: During heteroepitaxial growth of single and multiple layers, misfit induced strain causes various growth and relaxation phenomena. These comprise i) the formation of regular ripple patterns and eventually islands (elastic relaxation of pseudomorphic misfit strain), ii) formation of dislocations preferentially at sites of high strain in the develloping three-dimensionally inhomogeneous strain distribution, iii) preferential growth at surface areas of low chemical potential as determined by the strain distribution and associated self limited growth with lateral and vertical alignment of ripples and islands, and iv) strain-induced concentration fluctuations. We present selected microscopic observations, suppported by finite element strain calculations, to discuss these phenomena and their interdependencies mediated by the misfit strain distribution.

1. INTRODUCTION

Heteroepitaxial growth of semiconductors occurs in most cases under the effects of strain that results from misfitting substrates. Only in a small number of cases is it possible to grow a lattice-matched layer by selecting its composition under the constraints of the intended electronic or optical properties. Growth under conditions of misfit has several consequences. One is well known: beyond a certain critical thickness misfit dislocations form to plastically relax the strain. For many applications dislocations cannot be tolerated in the epitaxial layer and special precautions are necessary to control dislocation formation. The alternative is to limit growth to below the critical thickness and to produce strained layers. Unfortunately, as shown theoretically by several authors (e.g. Asaro and Tiller 1972, Grinfeld 1986, Srolovitz 1989) and first verified experimentally by Cullis et al. (1991), planar pseudomorphically strained layers are unstable and develop an undulated growth surface. The driving force results from a concomitant redistribution of strain such that the total energy, ie. stored elastic strain energy plus surface energy, is reduced. Two main consequences follow from the rippling: i) the strain state varies along the surface and causes locally varying growth kinetics. One case is very interesting as it changes rippling into advantage: the local strain can be high enough to compensate the driving force for growth, given by the supersaturation. At these sites growth ceases whereas growth proceeds nearby which gives rise to self-terminated island formation. ii) Rippling and islands, both, determine the strain and stress distribution and thus the most likely sites of misfit dislocation nucleation.

In the following we shall consider these aspects of growth in some detail. We base our consideration on results obtained from the model system $Ge_x Si_{1-x}$ on Si grown from metallic solution (liquid phase epitaxy). Solution growth occurs with very small growth driving force and, due to the easy diffusion of the solute (atoms or molecules) in the metallic solvent, permits the incorporation of

324

atoms into the growing surface essentially according to the local energetic conditions. Thus the topology characteristic of the elastically strained state forms and impeding effects due to growth kinetics are minimized. We consider rippling, its development with growth time, and a very few aspects of misfit dislocation formation and their implications for further epitaxial growth. This analysis will be supported by finite element calculations of the respective three dimensional strain fields. These calculations also clarify the role of strain in the vertical alignment of ripples and islands in multilayered structures. The effect of the inhomogeneous strain distribution in the growing epitaxial layers can be formulated in terms of a rule, whose application will be exemplified by considering growth induced compositional variations and by analyzing an interesting growth mode of epitaxial GaN on sapphire.

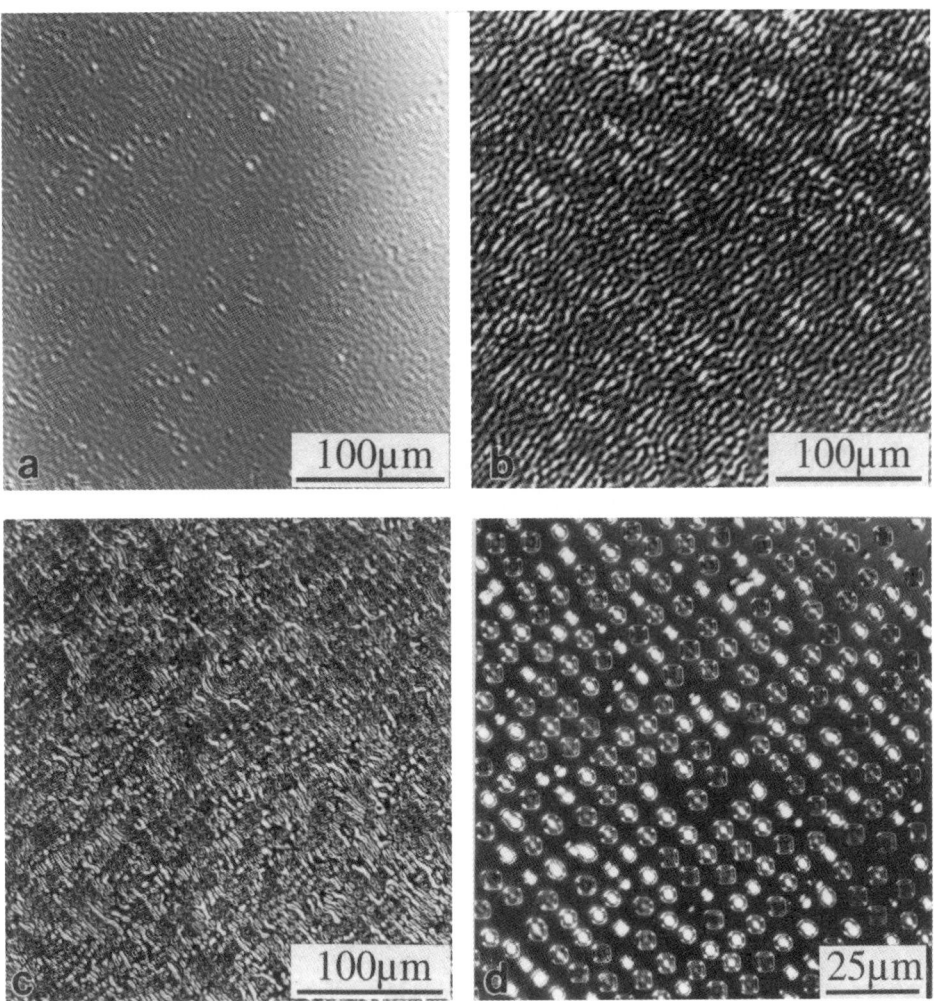

Fig. 1 a to d Successive growth stages of ripple formation until the eventual island formation (see text) $Ge_{0.05}Si$ / Si(001). Optical microscopy in NDIC.

2. EXPERIMENTAL ASPECTS

Our main considerations in Section 3 relate to epitaxial SiGe layers grown with various composition from the metallic solution onto (001) Si wafers. As regards the study of the developing surface topology (and defect structure), this liquid phase epitaxy has, in addition to the general advantages outlined in Section 1, the specific advantage to permit adjustment of the surface energy, (actual interface energy) γ. Its value enters the wavelength of the expected undulations which is $\lambda \sim \gamma/f^2$ with f: misfit parameter (eg. Srolovitz 1989). In our case, γ refers to the energy of the interface between solution and epitaxial layer and is, in the wetting case, lower than that of the free epitaxial layer, but can be selected by respective choice of the solvent, see Albrecht et al. (1994). The layers are mostly grown from In solution, starting with a solution at 900°C (supersaturated by 1°C cooling) that then is cooled at 10°C/h. The solvents contain the required amounts of Si and Ge to grow the epitaxial layer with the intended composition. The solution is transported on and off the substrate or epilayer by gravitational force using a combined tilting and sliding boat (for details, see Hansson et al. 1990). The Si substrates are cleaned by the RCA treatment, followed by a (2.5%) HF-dip and in-situ oxide desorption.

We characterise the surface morphology of the layers by optical microscopy with Nomarski differential interference contrast (NDIC), complemented by atomic force microscopy (AFM). We use various transmission electron microscopes, all Philips: a CM 20 microscope operated at 200 kV, an analytical CM 30 (300 kV) and a high resolution CM 300 (300 kV, resolution 0,17nm). Specimen preparation follows standard techniques, grinding, dimpling, and final thinning to electron transparency with Ar-ion beam etching.

We calculate the strain state of different morphological structures by applying the finite element method (FE) to appropriately defined model structures. The misfit is modelled within the framework of thermoelasticity. The temperature is taken as constant and island and substrate are assigned different thermal expansion coefficients such that the misfit f across the interface is met. The anisotropic elastic constants are derived from those of Si and Ge (Landolt-Börnstein 1982) by extrapolating linearly to the composition under consideration (Vegard's rule). The calculations yield the 3D strain and stress tensors as a function of the three coordinates x, y, z (and thus all hydrostatic and shear components). For details see eg. Christiansen et al. (1996a).

3. RIPPLES, ISLANDS AND DISLOCATION FORMATION

3.1 Development of ripple structure with epitaxial growth

Figure 1 shows a collection of growth topologies in different states of growth. In Fig. 1b one can easily see a well develloped, rather regular ripple structure, in which the crests and troughs lie at right angles along both <100> directions in the plane. (One set is rather pronounced because the line of no contrast in NDIC was not symmetrically set to both ripple sets.) The regular structure is very remarkable as it does not conform to the early state of rippling. Figure 1a shows such a state. Only patches of ripples are present, much less orderly arranged, and there are visible very large deviations from the <100> directions. Obviously the start of undulations does not occur homogeneously on the whole surface but occurs randomly at selected sites first. It is to be expected that in the course of further growth these patches grow in number and/or the initial patches extend until complete coverage of the growth surface is reached. In any case during this period the regular ripple structure forms exhibiting a pronounced long range order. This transformation requires an appreciable amount of matter transport along the interface over distances in the order of the wave length which is certainly aided by the easy diffusion in the adjacent solution. Consequently growth techniques that use an uncovered surface, such as molecular beam epitaxy or chemical vapour deposition, where surface diffusion is restricted tend to exhibit a less ordered ripple structure or even none (eg. Cullis 1994).

The development of the ripple structure during further growth yields an interesting result.

Further growth at this stage proceeds by an increase of the average layer thickness and an increase in ripple amplitude A at constant wavelength λ (Strunk 1996, Dorsch et al. 1995). Fig. 1c shows an advanced stage. Areas can be seen scattered across the surface of a few adjacent crests that are separated and surrounded by comparably deep troughs. At a few sites these structures have a geometry that suggests dissolution has occured in the troughs. In fact, a later stage of growth reveals an array of islands whose epitaxial material between them is removed down to the substrate surface, Fig. 1d. The most noteworthy result is that the islands have inherited from the ripple structure the wavelength λ and the long range order, and thus form a rather regular square array. Most of the islands are facetted because of surface energy minimization.

3.2 The strain state

It is evidently the strain (and stress) distribution caused by the ripples that determines the ripple development documented in Fig. 1. For a discussion, we refer now to some results on the strain field. The three dimensional strain tensor can be obtained comparably easy for any desired geometry by finite element calculations. Such calculations permit one to analyse the energetics and the various stress components and their variation within the considered geometry. Certainly, such an approach cannot describe the evolution of a structure or consider the stability against plastic relaxation. Analytical approaches such as performed eg. by Gao (1994), Guyer and Vorhees (1996) and Freund et al. (1996) are more appropriate. (The finite element code needs to be extended to yield the geometry as a result depending on growth parameters.)

Some experimental assessments of spatial strain distribution by different electron microscopy techniques exist (Cullis 1994 by measuring the relaxation induced tilt of lattice planes, Christiansen et al. 1994 by convergent beam electron diffraction). In view of the difficulties due to thin foil relaxation an independent method is advisable and micro-Raman experiments are being conducted (Groenen et al. 1997) for comparison and possibly gauging. Micro-Raman in addition can discriminate between strain and composition with comparably high spatial resolution (see Sect. 5).

Figure 2 shows finite element results for a two dimensional sinusoidal model undulation: the distribution of the strain energy density (and similarly the hydrostatic pressure) (Fig. 2a) and of the flow stress in a a/2<011>{111} glide system (Burgers vector inclined to the growth plane) (Fig. 2b).

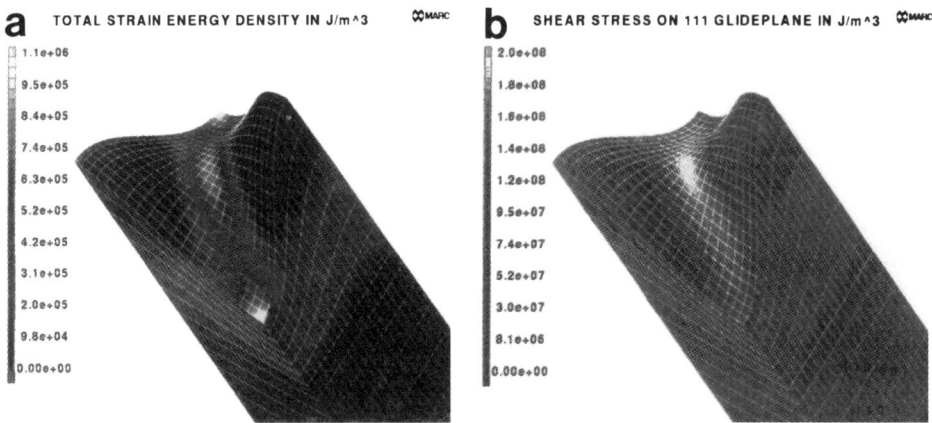

Fig. 2 Finite element calculations of a sinusoidal undulation. Aspect ratio amplitude/wavelength A/λ = 0.1, total height fluctuation 2A = 2 μm, on a scale of d = 1.25 μm. Scale perpendicular to surface exaggerated. Due to symmetry an array λ/2 x λ/2 was calculated. Gray scale values in J/m³.
a Strain energy density. **b** Flow stress in a 1/2 <110>{111} glide system.

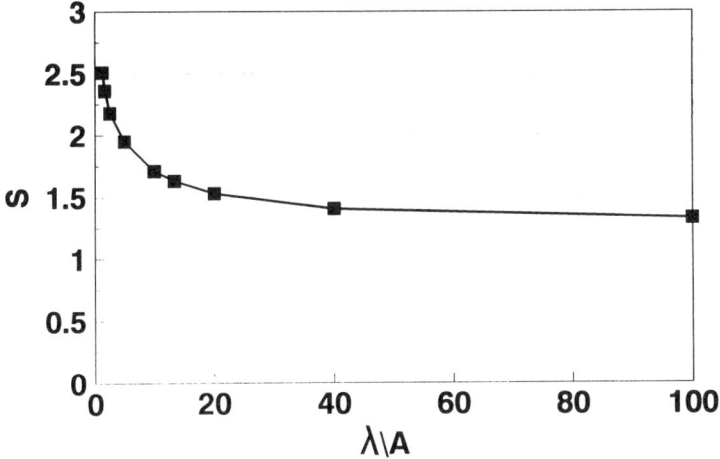

Fig. 3 Shear stress concentration for a glide system a/2 <110>{111} at troughs (normalized to the value in a planar layer of same misfit as a function of the inverse aspect ratio. Symbols represent FE calculations, line is a guide to the eye.

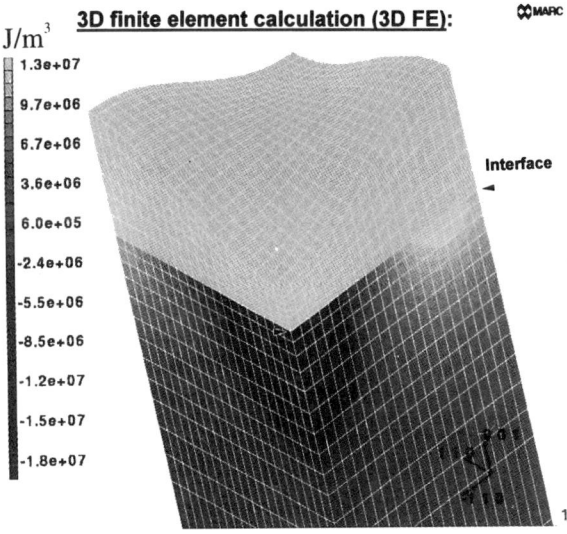

Fig. 4 Same as Fig. 2b, but with the scale changed to display strain distribution in the substrate.

We can summarize these results: i) in both cases large parts of the ridge tops are free of strain and stress (essentially those parts protruding beyond the bottom of the trough), ii) strain and stress are localized in the trough, the values are highest near/at the surface and decay rather rapidly into the interior to the pseudomorphic level, iii) the stress concentration factor is higher the larger the amplitude A (for constant wavelength λ), see Fig. 3, especially at large amplitudes, iv) strain redistribution and related relaxation are a characteristics of the ripples, the underlying epilayer, except for a certain thickness at its top, can be regarded pseudomorphic, v) in consequence, when this pseudomorphic layer is sufficiently thin, the modulated strain field penetrates into the substrate.

Figure 4 displays the calculation of Fig. 2b again but with a scale to highlight smaller strain values. The strain distribution within the substrate is such that below the crests the strain has the same sign as in the epitaxial layer but the opposite one below the trough area. The strain dristribution in the epilayer is sketched in Fig. 5a (for the case of compression, in the tension case the variation of the lattice parameter is inverted). At the crests the lattice is relaxed practically to the equilibrium bulk lattice parameter whereas in the troughs the compression is enhanced exceeding the pseudomorphic value. With this simple notion it is easy to understand the growth processes and especially also our growth experiment shown in Fig. 1. Also the strain induced compositional fluctuations can be treated with this sketch (Fig. 5b, Sect. 5). If we neglect for the moment compositional influences then the arriving atoms find their equilibrium lattice parameter at the crest and growth velocity there is correspondingly high, as given by the supersaturation. In the troughs, however, the enhanced strain presents an additional barrier to the incoming atoms and the growth velocity is reduced. Therefore, though growth occurs on the whole surface, the undulation amplitude increases permanently (initially exponentially in time, Srolovitz 1989, Strunk 1996. See also the respective experimental analysis by Jesson et al. 1993) and so does the strain in the trough, eg. Fig. 3. In our case, eventually, Fig 1c, an instant is reached where the stress concentration in the trough counterbalances the driving force and growth stops and, since growth at the crests proceeds, even overbalances it and dissolution sets in leading to the formation of islands, Fig. 1d. Thus the size of the island base is determined by the non-growth condition. It should be noted at this point that the self-organized growth of islands in Stranski-Krastanov growth mode with rather narrow base size distribution is frequently also explained by the growth limiting stress building-up at the base of the forming islands (LeGoues 1994, Dorsch 1995). Whether this type of growth mode is different from the above islanding process or can be regarded as limiting case, remains to be clarified.

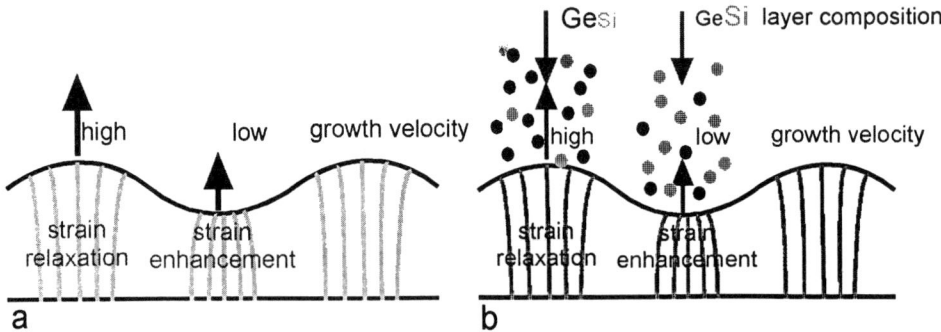

Fig. 5a Sketch to indicate, for a compressed epitaxial layer, the relaxation and strain enhancement due to ripples, also its action on growth velocity. **b** Additional influence of strain redistribution on the incorporation of the differently sized atoms during growth of a SiGe alloy (see text).

3.3 Dislocation formation

The calculated strain distributions in Fig. 2 directly indicate potential sites of (homogeneous) dislocation nucleation. Maximum hydrostatic pressure (Fig. 2a) exists in the surface-near layer at the trough and suggests dislocation formation by atomic defect condensation there. Maximum shear stresses (for the chosen type of glide system) exist also at the trough although generally slightly set off its center line. The maximum lies also off the saddle point but is nearer, the larger the ripple amplitude is (wave length λ constant) (Michler 1995). Therefore both conditions for dislocation formation, comparably low nucleation activation energy and high flow stress to extend a small dislocation loop segment by glide meet at around the same area in the trough.

Dislocation nucleation in troughs was analyzed in InGaAs/GaAs by Cullis et al. 1995. First a Frank partial dislocation is formed by atomic defect condensation. This is then converted by generation and glide of suitable partial dislocations into a partial stair rod or perfect Lomer dislocation. Androussi et al. (1995) analyzed in such a system an edge dislocation below a ripple trough and concluded its formation due to ripple coalescence. In our case more efficient dislocation sources situated at the rim of the substrate produce glissile threading dislocations that extend 60° dislocations at the interface. Above such misfit dislocations a band of relaxation exists which causes, according to our discussion before, rapid growth. Two such ridges are shown in Fig. 6a. The height profile in Fig. 6b shows that each ridge has an amplitude much larger than the ripples, a wavelength in the order of the ripple wavelength and is accompanied by a deep trough on either side. Consequently, very high stresses reside, see Fig. 3, there causing further dislocations to form and further ridges. Of course, the same happens in the perpendicular direction. Thus a cross hatch pattern of ridges forms. At a later stage of misfit dislocation formation the wavelength of the cross hatch pattern corresponds to the misfit strain not yet relaxed by misfit dislocations (Strunk 1996). According to Albrecht et al. (1995) the strain and stress fields this pattern creates within the substrate (analogous to Fig. 4) can operate misfit dislocation multiplication mechanisms within the substrate as observed by Lefebvre et al. (1991) and LeGoues et al. (1992). We do not consider the plastic relaxation processes any further here since dislocation processes have been studied extensively in continuous layers and we are interested in self organisation which occurs due to mediation by strain fields.

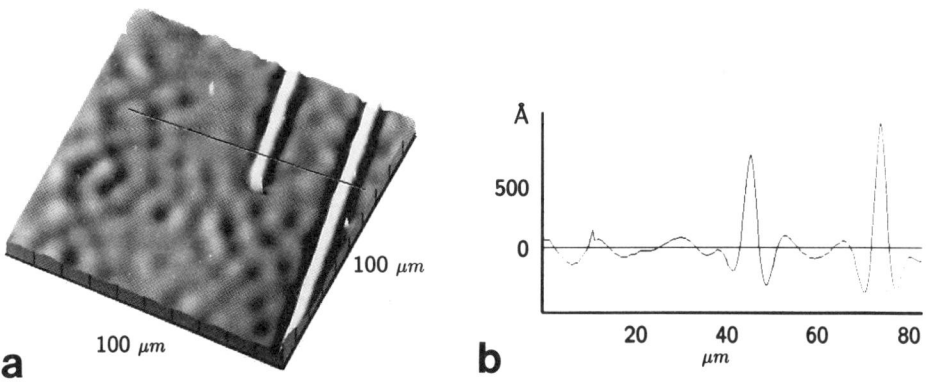

Fig. 6a Ridges formed by accelerated growth above an extended misfit dislocation. **b** Height profile along the indicated line.

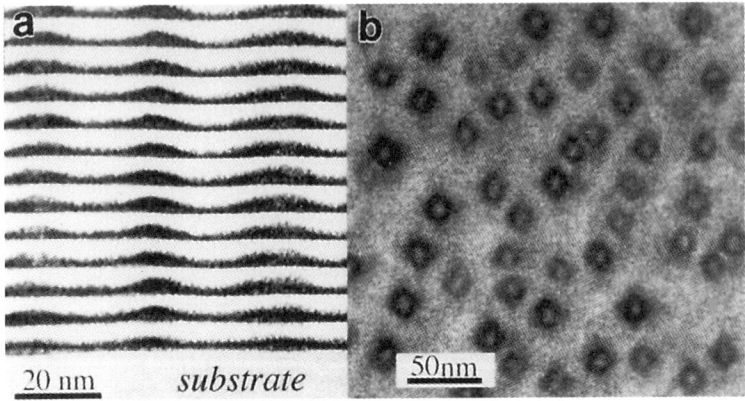

Fig. 7 $In_{0.50}$ $Ga_{0.50}$As quantum dots in GaAs. **a** Multilayer structure with excellent vertical alignment, XTEM. **b** Dot distribution as seen in plane view. Micrographs courtesy A. O. Kosogov and P. Werner, Max-Planck-Institut, Halle, Germany.

4. LATERAL AND VERTICAL ALIGNMENT

The growth mode present in Fig. 1 yields a pronounced long range order of islands. Conversely, the Stranski-Krastanov growth mode generally yields only a short range order if the arrangement of islands can be described as that at all. One example is shown in Fig. 7b and many other examples can be found in the literature including the one we shall refer to below (Vescan et al. 1992). Also, this type of arrangement appears practically not to depend on the growth technique and results also during solution growth, eg. Dorsch et al. (1995). Very probably the forming pseudomorphic islands have insufficient interaction since their strain fields are localized at their base lines and thus do not overlap until a very high density of islands is nucleated (Dorsch et al. 1995). In fact the to-be-islands of the ripple structure in Fig. 1b or c interact elastically very intensively and thus arrange for the long range order seen in Fig. 1d.

On the other hand, generally the vertical alignment in multilayer systems is strikingly good, be it isolated islands, (eg. Fig. 7a, Vescan et al. 1992) or undulations (eg. Ponchet et al. 1993, 1995). Such an alignment can be understood in terms of the inhomogeneous strain fields created by these structures. Fig. 8 shows a finite element simulation (Christiansen et al. 1996b) of a Ge-Si layer pair on the silicon substrate in the geometrical arrangement of the multilayer grown by Vescan et al. (1992). Because of the single-component character of each layer chemical misfit effects are absent. Arrowed lines identify the interfaces between substrate and Ge-layer and between Ge-layer and Si-layer. We know from above the strain state of the Ge-layer and can consider how growth of the Si-layer on top will proceed with the help of Fig. 5a. We will extend this consideration later in the sense of a "growth rule". The lattice parameter at the Ge crests is relaxed towards the bulk Ge value, whereas it is compressed in the Ge troughs to much smaller values, ie in the direction of the Si-value. Therefore Si will grow faster in the troughs than on the crests and eventually a planar Si surface is formed as shown in Fig. 8. However this planar surface exhibits regions of compression (positive hydrostatic pressure above the Ge troughs) and tension (above the Ge crests). Growth of another Ge-layer on top will proceed faster in the tension area where Ge finds a more appropriate lattice parameter and thus a new crest will form there. All other subsequent layers grow the same way and the initial lateral distribution of islands will be preserved as a vertically aligned arrangement.

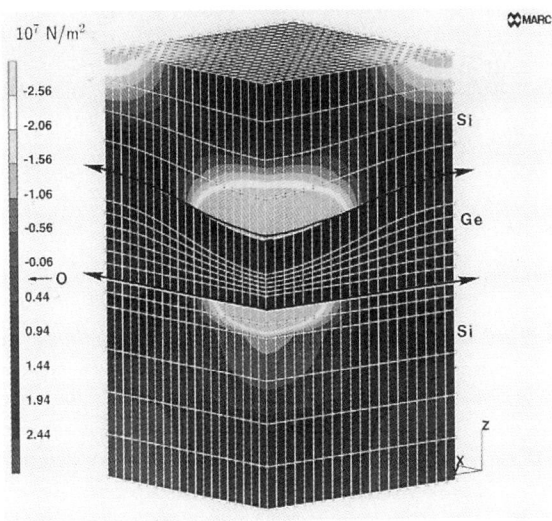

Fig. 8 Finite element calculation to show the strain induced origin of vertical alignment in multiquantum wells (see text).

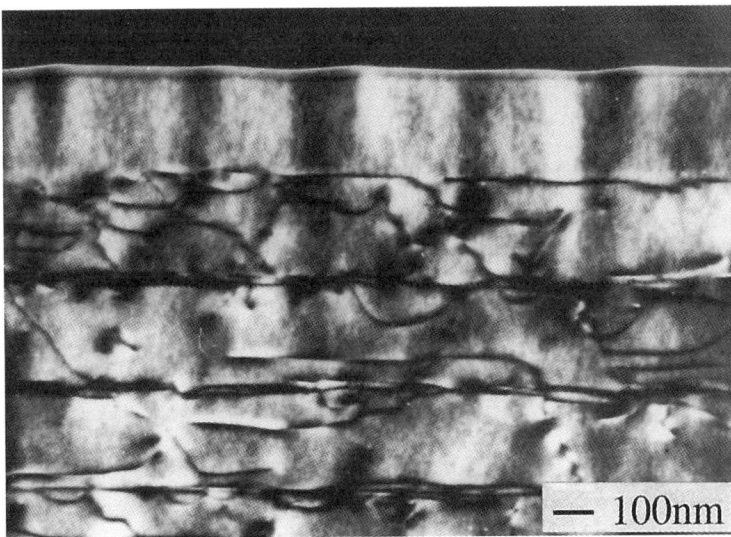

Fig. 9 In$_{0.50}$Ga$_{0.50}$As epitaxial layer on a step graded buffer layer. XTEM showing lateral contrast variations due to composition variation. These correlate with the undulations at the surface. Diffraction vector {220} along the surface. Courtesy G. Salviati.

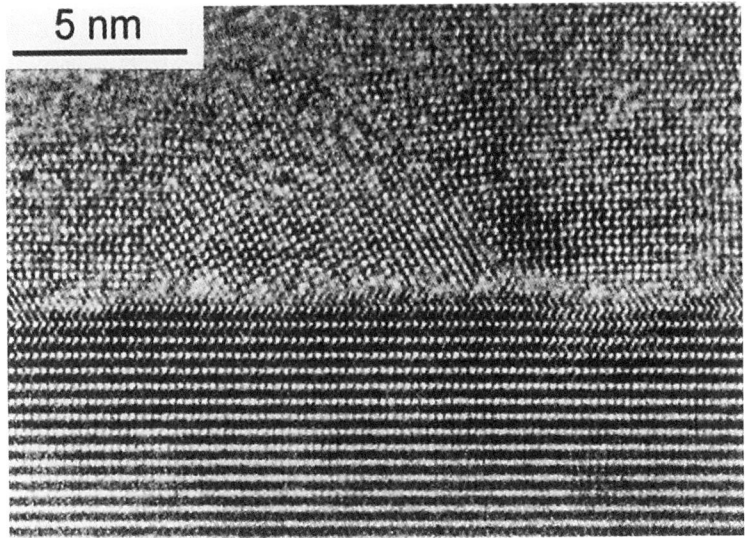

Fig. 10 Interface between sapphire (0001) substrate (bottom) and hexagonal GaN epitaxial layer (top) (gas source molecular beam epitaxy). In addition to the usual basal/basal plane orientation relationship (left and right) a grainlet with different orientation is seen. This grainlet is in tensile strain as opposed to the compressed surroundings. Its formation can be understood in terms of strain driven self organized growth.

5. STRAIN INDUCED COMPOSITION FLUCTUATIONS

We now consider misfitted growth of a binary compound. In many cases lateral contrast undulations can be detected in the electron microscope that indicate lattice parameter variations due to composition variations, ie composition induced strain waves along the growth surface. Fig. 9 shows an example; the contrast mechanism can be explained by thin foil relaxation of this compression-tension sequence (Treacy et al. 1985). In our context the observation that these contrasts are correlated with the ripples visible also in Fig. 9 is interesting. The suggested ripple and thus strain-induced composition modulation during growth can easily be described with the growth rule, see Fig. 5b. We neglect for the moment a varying growth velocity and describe the modulation development for the growth of SiGe alloy (which can be assumed to be completely miscible) on a rippled but homogeneous SiGe epilayer attached to a Si substrate. Due to the lattice parameter fluctuations a Ge rich alloy will grow on the crests whereas it will be Si rich in the troughs. The symbols introduced for Si and Ge atoms indicate the resulting lateral composition fluctuation. Of course to be compatible with real crystal growth, this very simplistic picture has to be modified by the additional modulation of the growth velocity. This is beyond the present scope and requires full account of growth kinetics (eg. Guyer and Vorhees 1996) and cannot be realized by our present finite element approach. In any case these composition modulations are generally energetically unfavorable, as compared to the homogeneously strained state, unless thermodynamically driven decomposition is involved (such as those in III-V compounds, eg. Treacy et al. 1985). The strain energy density averaged over a period of the modulation is proportional to $\overline{f^2}$ (f: local lattice misfit), whereas in the case of the homogeneous layer it is proportional to $(\overline{f})^2$ (f: misfit of homogeneous layer) and the inequality $\overline{f^2} \geq (\overline{f})^2$ holds. First evidence for such modulations in solution grown SiGe layers was recently obtained (Groenen et al. 1997).

6. STRAIN INDUCED EPITAXIAL GRAIN GROWTH

Fig. 10 shows the interface of a GaN epitaxial layer grown onto a sapphire (0001) substrate by gas source molecular beam epitaxy (eg. Kamp et al. 1996). Usually GaN grows with its basal plane parallel to the basal plane of the substrate, but rotated by 30° around the c-axis. This results in a misfit of almost 14 % and GaN is in compression. One would expect, therefore, that after nucleation the high strain would lead to growth of islands which can grow further only after dislocation formation. This certainly happened partly in the case of Fig. 10 also, but it indicates an alternative process by formation of a grainlet. The analysis shows that the visible grainlet is rotated around the direction of view such that its (01-11) plane is in contact with the sapphire. This orientation is correlated with a tensional state of the grainlet (f = -26%) which can compensate the compressional state of the usually oriented GaN. The arrangement in Fig. 10 can be understood in terms of a strain-induced self-organized growth process, a first account of which was recently given by Albrecht et al. (1997). In all brevity, when the usually orientated island ceases growing because of compressional strain accumulation at its base line (Fig. 11a), this base line is a preferential nucleation site for an island that compensates this strain (Fig. 11b). Repetitive nucleation may lead to a continuous layer that is strain free on average. Eventually on further growth one orientation, here the basal one, may overgrow the other for kinetic reasons and form a single crystalline epitaxial layer. Of course, in this case the geometric misfit to the substrate will be recovered and misfit dislocations will form at the overgrown grains. It is to be expected that a proper control of grainlet orientations, of which several are theoretically possible in this system, by growth condition variation can also control the population of threading dislocations. This is very desirable in the growth of III-nitride materials as long as homoepitaxy is not possible.

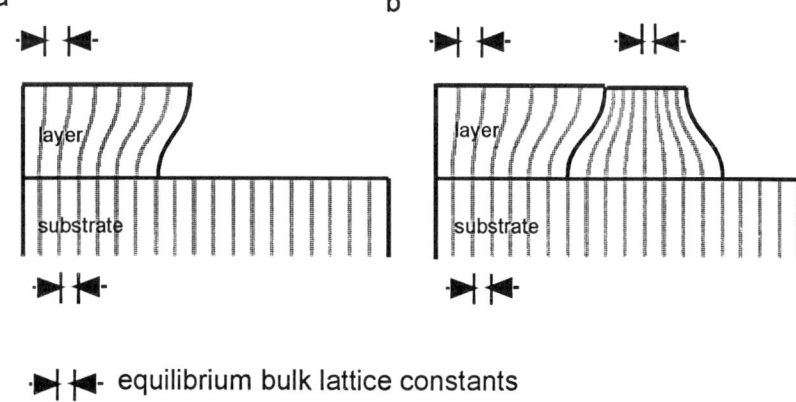

◀▶◀ equilibrium bulk lattice constants

Fig.11 Sketch to illustrate self organized formation of grainlets during the nucleation stage of a GaN layer on sapphire (0001).

334

ACKNOWLEDGMENTS

We would like to thank B. Steiner, H. Lorenz and T. Marek for crystal growth, technical assistance and discussion. We received in a longlasting fruitful collaboration many incentives and much support from Dr. E. Bauser, who deceased all to soon in 1996, and would like to dedicate this paper to the memory of her.

REFERENCES

Albrecht M, Hansson PO, Christiansen S, Dorsch W, Strunk H P and Bauser E 1994 Scanning Microscopy **8** 925
Albrecht M, Christiansen S and Strunk H P 1995 phys. stat. sol. (a) **150** 453
Albrecht M, Christiansen S and Strunk H P 1997 Appl. Phys. Lett. **70** 952
Androussi Y, Lefebvre A , Delamarre C, Wang L P , Dubon A, Courboulès B, Deparis C and Massies J 1995 Appl. Phys. Lett. **66** 3450
Asaro R J and Tiller W A 1972 Metall. Trans. **3** 1789
Bennabas T, François P, Androussi Y and Lefebvre A 1996 J. Appl. Phys. **80** 2763
Christiansen S, Albrecht M, Strunk H P and Maier H J 1994 Appl. Phys. Lett. **64** 3617
Christiansen S, Albrecht M, Michler J and Strunk H P 1996a phys. stat. sol. (a) **156** 129
Christiansen S, Albrecht M and Strunk H P 1996b Comput. Mat. Sci. **7** 213
Cullis A G, Robbins D J, Pidduck A J and Smith P W 1991 Microscopy of Semiconducting Materials 1991, eds. A G Cullis and N J Long (Bristol: IOP Publishing Ltd) 831
Cullis A G 1994 Scanning Microscopy **8** 957
Cullis A G, Pidduck A J and Emeny M T 1995 Phys. Rev. Lett. **75** 2368
Dorsch W, Christiansen S, Albrecht M, Hansson P O, Bauser E and Strunk H P 1995 Surf. Sci. **331-333** 896
Freund L B, Johnson H T and Kukta R V 1996 Mat. Res. Soc. Symp. Proc. **399** 359
Gao H 1994 J. Mech. Phys. Solids **42** 741
Grinfeld M A 1986 Sov. Physics Doklady **31** 831
Groenen J, Carles R, Albrecht M, Christiansen S, Dorsch W, Strunk HP, Wawra H and Wagner G 1997 submitted to Appl. Phys. Lett.
Gyer J E and Vorhees P W 1996 Mat. Res. Soc. Symp. Proc. **399** 351
Hansson P O, Werner J H, Tapfer L, Tilly L P and Bauser E 1990 J. Appl. Phys. **68** 2158
Jesson D E, Pennycook S J, Baribeau J-M and Houghton D C 1993 Phys. Rev. Lett. **71** 1744
Kamp M, Mayer M, Pelzmann A, Thies A, Chung H Y, Sternschulte H, Marti O and Ebeling K J 1996 Mat. Res. Soc. Symp. Proc. **395** 135
Landolt-Börnstein 1982 New Series Vol 17a to 17d (Berlin: Springer)
Lefebvre A, Herbeaux C, Bouillet C and DiPersio J 1991 Phil. Mag. Lett. **63** 23
LeGoues F K, Meyerson B S, Morar J F and Kirchner P D 1992 J. Appl. Phys. **71** 4230
LeGoues F K, Reuter M C, Tersoff J, Hammar M and Tromp R M 1994 Phys. Rev. Lett. **73** 300
Michler J 1995 Diploma Thesis University Erlangen-Nürnberg
Ponchet A, Rocher A, Emery J Y, Starck C and Goldstein L 1993 Microscopy of Semiconducting Materials 1993, eds.A G Cullis, A E Staton-Bevan and J L Hutchison (Bristol: IOP Publishing Ltd) p 485
Ponchet A, Rocher A, Ougazzadan A and Mircea A 1995 Microscopy of Semiconducting Materials 1995, eds. A G Cullis and A E Staton-Bevan (Bristol: IOP Publishing Ltd) p 199
Srolovitz D J 1989 Acta metall. **37** 621
Strunk H P 1996 Mat. Res. Soc. Symp. Proc. **399**, 313
Treacy M M J, Gibson J M and Howie A 1985 Phil. Mag. A **51** 389
Vescan L, Jäger W, Dieker C, Schmidt K, Hartmann A and Lüth H 1992 Mat. Res. Soc. Symp. Proc. **263** 23

Inst. Phys. Conf. Ser. No 157
Paper presented at Microsc. Semicond. Mater. Conf., Oxford, 7–10 April 1997
© 1997 IOP Publishing Ltd

Self-organisation processes in InSb quantum dots grown on InP(001) by ALMBE

J C Ferrer [a b], **F Peiró** [a], **A Cornet** [a], **J R Morante** [a], **T Utzmeier** [c], **G Armelles** [c] and **F Briones** [c]

[a] EME, Física Aplicada i Electrònica, Universitat de Barcelona, Diagonal 645-647, 08028, Barcelona, Spain
[b] Serveis Científico-Tècnics, Universitat de Barcelona, Lluís Solé i Sabarís 1-3, 08028 Barcelona, Spain
[c] CNM, Instituto de Microelectrónica de Madrid, Isaac Newton 8, Parque Tecnológico de Madrid, 28760, Madrid, Spain

ABSTRACT: The structural characterisation of uncapped InSb quantum dots grown on (001) InP substrates by ALMBE is reported. The analysis has been performed by AFM and TEM paying special attention to dot size, surface density and mechanism of strain relaxation. The amount of deposited InSb ranges between 2 and 7 ML's. Results show that good quality dots are obtained. For the first stages of growth, island density rises and sizes diminish. When the maximum value of density is attained, island size begins to increase. Dots have an anisotropic shape being elongated in [110], parallel to a surface undulation. Strain relaxation takes place by means of a misfit dislocation array.

1. INTRODUCTION

Recently it has been proved that a quite effectively method to obtain quantum structures is by epitaxially deposition of III-V materials. More precisely in those systems for which a large lattice mismatch is involved, we can take benefit from the self-organization properties of atoms on the growth surface. Since three-dimensional growth may occur spontaneously, this leads to the formation of nanometric structures such as quantum dots or wires. These nanostructures exhibit high quantum efficiency and narrow photoluminescence peaks, but their interesting properties may be lost because of dot inhomogeneities or the presence of crystalline defects.

In this paper, the growth of uncapped InSb quantum dots on InP substrates is reported. The morphology, size and evolution, depending on the deposited InSb amount, is studied by means of Transmission Electron Microscopy (TEM) and Atomic Force Microscopy (AFM).

2. EXPERIMENTAL

Four samples with an equivalent amount of 2, 2.8, 5 and 7 monolayers (ML's) of InSb were grown by Atomic Layer Molecular Beam Epitaxy (ALMBE) on semiinsulating (001) InP substrates at a temperature of 400°C, maintaining the V element flux pulsed in order to enhance the surface diffusion. After the formation of the InSb islands, the samples were annealed at a temperature of 440°C during 500 s to improve the dot quality maintaining a continuous Sb flux to avoid its loss from the surface. Prior to InSb deposition InP oxide was desorbed at 490°C and a 500 ML's thick InP buffer layer was grown at 400°C to obtain a good quality surface.

336

AFM images were obtained using a Digital Instruments Nanoscope III multimode AFM operating in tapping mode. Samples were prepared for TEM by conventional mechanical polishing and final Ar⁺ ion milling until perforation in a cooled stage in order to avoid preparation artifacts. TEM observations were performed in both [110] and [1$\bar{1}$0] in a Philips CM30 microscope operating at 300 KV.

Reflection High Energy Electron Diffraction (RHEED) patterns were used during the growth to assess the transition from two-dimensional to three-dimensional growth mode, which was found to take place at 1.2 ML's thickness.

AFM images allowed us to measure the density of dots on the surface of the substrate as well as to have a general knowledge of the morphology of the dots. On the other hand, TEM was used to measure more precisely the dot size, to analyze the dot faceting and to assess the crystalline quality and the relaxation state.

3. RESULTS AND DISCUSSION

Table I. Dot size and density depending on the amount of deposited InSb.

InSb amount (ML's)	size [110] (nm)	size [1$\bar{1}$0] (nm)	height (nm)	density (cm⁻²)
2	53	41	13	$3.4 \cdot 10^9$
2.8	50	33	13	$7.6 \cdot 10^9$
5	62	27	14	$8.8 \cdot 10^9$
7	90	33	18	$5.5 \cdot 10^9$

The first point to be addressed is the evolution of the dot size and density depending on the amount of InSb. The results of measurements are summarized in Table I in which the lateral sizes, heights and densities are presented. The most striking feature of these results is that the surface density of dots reach a maximum value between 2.8 and 5 ML's while the mean volume of single dots attains a minimum value in the same range. This behaviour may be explained taking into account the papers

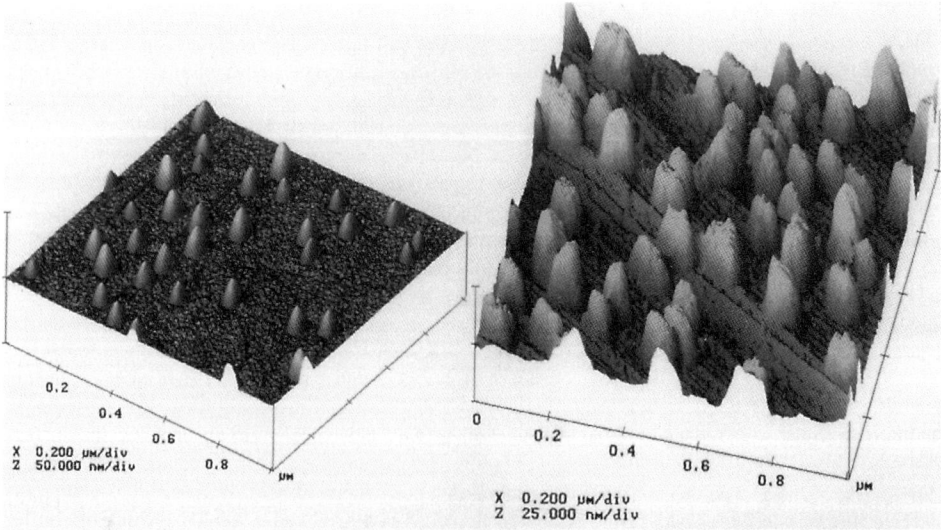

Fig. 1a. AFM image of a sample with 2 ML's. Dots have a uniform distribution and size.

Fig. 1b. Sample with 7 ML's. A high degree of anisotropy in dot size and in substrate surface morphology is observed.

of Tersoff (1993) and Stoop (1974). For the first stages of InSb deposition, beyond the transition from two-dimensional to three-dimensional growth mode, the minimum energy per unit volume is reached for high island densities, having an optimal size with length a_0. Thus, as growth proceeds, newer dots are generated and the interdot distance diminishes, activating material exchange among dots and reaching the optimal size which is lower than that of the first stages. If the inter-dot spacing diminishes they start to interact through the repulsive elastic strain field induced at the substrate vicinity. As the inter-dot distance has a minimum value because of the strain field interaction, further deposition of InSb results, not in the formation of new dots, but in the growth of larger islands. Furthermore, as the strain field increases with dot size, their density starts to diminish for higher InSb contents.

Figures 1a and 1b show AFM images of the samples with equivalent thicknesses of 2 and

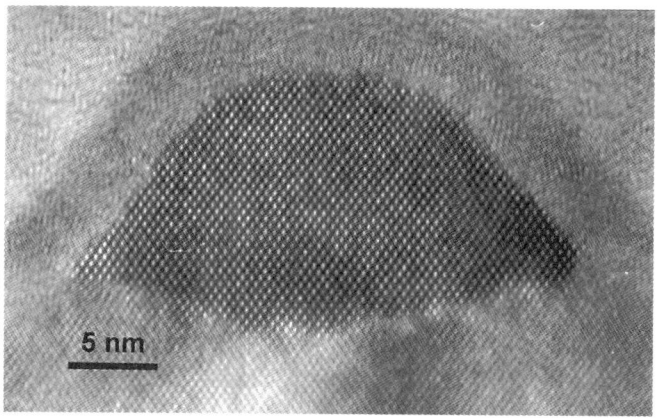

Fig. 2. High resolution TEM image from [110] direction of a sample with 5 ML's. Notice the dislocations at the InSb/InP interface and the facetting.

7 ML's of InSb respectively. We find a regular distribution of islands that arise as a result of the three-dimensional growth mode due to the lattice mismatch between InSb and InP that reaches 10.4%. The shape of the dots evolves from a rounded morphology for a total InSb content equivalent to 2 ML's to a dashed shape with dots elongated towards [110] for samples with 2.8, 5 and 7 monolayers.

TEM provides greater knowledge about dot structure. Figure 2 show a high resolution image of a single dot from [110] direction in which it can be observed that dots are faceted by low-index planes: a top (001) plane and {111} lateral facets. Moreover one may observe in [1$\bar{1}$0] images the presence of {113} and {114} planes.

On the other hand, from both TEM and AFM examinations we have noticed that the substrate surface has developed an anisotropic undulation with its ripples parallel to the dot elongation direction. This anisotropy also disappears for the sample with 2.0 ML's for which the dots are rounded.

Some alternatives may explain the presence of this anisotropy:

(i) The surface ripples have developed during the buffer layer growth as reported in the paper of Briones (1985). It may happen during the homoepitaxial growth of III-V materials that, in certain conditions a rugosity is developed. Thus, the different dot sizes in both <110> directions would be caused by the existing substrate anisotropy and, therefore, different diffusion coefficients for the atomic species in the directions parallel and perpendicular to the undulations modify the shape of the islands.

(ii) Due to different growth rates of {111}-A and {111}-B faces, dots are elongated in the [110] direction and the undulations would be the result of a strain field originated by the elongated dots. This option seems more likely because of the identical growth conditions for all samples.

338

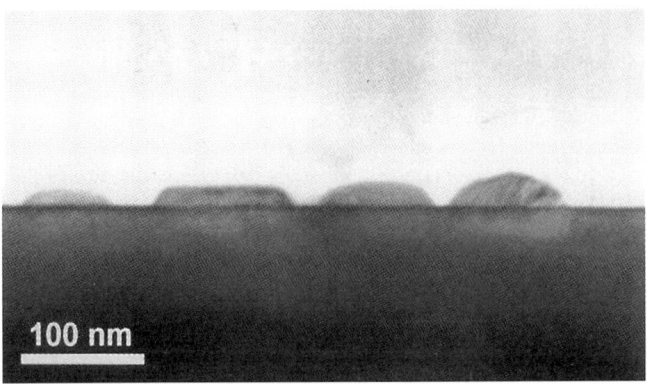

Fig. 3. TEM [1$\bar{1}$0] image of a sample with 7 ML's. Some planar defects are found in the islands. Notice the top (001) planes and the {111}, {113} and {114} facets..

If the later option is accepted, the enlargement of InSb islands in [110] direction could be explained following the work of Tersoff (1993). In this paper, it is demonstrated that, for sufficient atomic diffusion on the surface, strained islands tend to grow with a square shape until a critical size is reached. Beyond this limit, they become enlarged to minimize the total energy related to the extra dot surface and the island/substrate interface energy.

Finally, a careful examination of the interface reveals the presence of a misfit dislocation network that relaxes the strain due to the lattice mismatch. The majority of dislocations have a 90° Burgers vector (pure edge dislocations) that remains at the interface. It is known that these dislocations are those that relax most effectively the strain. The mean distance between two pure edge contiguous dislocations is 4.5 nm which agrees quite well with the value that would relax the structure completely.

4. CONCLUSIONS

In summary, InSb quantum dots have been grown by ALMBE on (001) InP substrates with InSb deposited amounts ranging from 2 to 7 ML's. AFM and TEM results show that dots elongated in the [011] direction are obtained, the elongation becoming greater as the InSb contents rises. This anisotropy affects also to the substrate surface which develops ripples parallel to the dot elongation direction. Dots are faceted by a (001) top plane, with {111}, {113} and {114} lateral facets. The strain due to the high lattice mismatch is relieved by an array of misfit dislocations placed at the island/substrate interface.

Island density reaches a maximum between 2.8 and 5 ML's and the dot volume attains a minimum at the same range. Both events are related to material exchange between dots and to interaction of dot strain fields.

ACKNOWLEDGEMENTS

This work has been funded by the Spanish CICYT Project MAT95-0966.

REFERENCES

Briones F, Golmayo D, González L, and De Miguel J L, Jpn. J. Appl. Phys. **24** L478 (1985)
Stoop L C A, Van der Merwe J H, J. Cryst. Growth **24/25**, 289 (1974)
Tersoff J, Tromp R M, Phys. Rev. Lett. **70**, 2782 (1993)

Inst. Phys. Conf. Ser. No 157
Paper presented at Microsc. Semicond. Mater. Conf., Oxford, 7–10 April 1997
© 1997 IOP Publishing Ltd

TEM and PL studies of self-assembling quantum dots

N Y Jin-Phillipp[1], M K Zundel[2], F Phillipp[1] and K Eberl[2]

[1] Max-Planck-Institut für Metallforschung, [2] Max-Planck-Institut für Festkörperforschung, Heisenbergstr.1, D-70569 Stuttgart, Germany

ABSTRACT: Self-assembling $In_{0.6}Ga_{0.4}As$ and InP quantum dots (QDs) grown by solid source molecular beam epitaxy (MBE) are characterised by transmission electron microscopy (TEM) and by photoluminescence spectroscopy (PL). The growth of the QDs is anisotropic. Strain relaxation takes place both in the QDs and in their neighbouring layers, which leads to defect injection when the nominal thickness becomes high. The structural properties of the quantum dots are correlated with the PL results. By optimising the growth conditions a pronounced alignment of the QDs along the growth direction is observed. PL from such multi-layers of QDs exhibits a single narrow peak.

1. INTRODUCTION

Quantum dots based on III-V semiconductor heterostructures are promising for future device applications, and, hence, interesting for fundamental research (Eberl et al 1995). Due to the Stranski-Krastanov growth mode in strained layer heteroepitaxy of materials with a large lattice mismatch three-dimensional (3D) islands assemble after a few monolayers (MLs) of pseudomorphic growth in a two-dimensional (2D) growth mode. This paper investigates the growth and structural properties of such self-assembling free standing $In_{0.6}Ga_{0.4}As$ QDs on (001) GaAs and InP QDs on $Ga_{0.51}In_{0.49}P$, and InP quantumn dots embedded in $Ga_{0.51}In_{0.49}P$ layers grown on (001) GaAs: the structures are correlated with the optical properties measured by PL.

2. EXPERIMENTAL

The samples were grown on GaAs (001) substrates by solid source MBE. For all samples a buffer layer of 200nm GaAs was first deposited. In the case of InP quantum dots, this was followed by a 45nm $Ga_{0.51}In_{0.49}P$ layer, lattice matched to GaAs. In the case of multilayer InP quantum dots three periods of alternate 3MLs of InP dots and a $Ga_{0.51}In_{0.49}P$ spacer layer and then a 45nm $Ga_{0.51}In_{0.49}P$ layer followed. The thickness of the spacer layer varied from 2 to 16 nm. The growth rate for InP and $In_{0.6}Ga_{0.4}As$ was 0.5ML/s, and that for $Ga_{0.51}In_{0.49}P$ 1ML/s. Plan-view as well as cross-sectional (X-) TEM specimens were prepared by the standard method. The lattice mismatch between InGaAs and GaAs is 4.3%, and that between InP and GaInP is 3.8%. TEM investigations were carried out on a JEOL 4000FX microscope operated at 400keV with a point resolution of 0.22nm. PL was measured at 8K using the 514.5nm line of a CW Ar^+ laser for the excitation.

3. RESULTS AND DISCUSSIONS

3.1 Free Standing Quantum Dots

In-situ reflection high energy electron diffraction (RHEED) shows that the transition from 2D to 3D growth mode takes place at less than 3MLs of InGaAs or 1.5MLs of InP. The distribution of the QDs in the sample with nominal 8MLs of InGaAs are shown in plan-view TEM (Fig. 1). The size and the shape of the dots may be well determined by the presence of moiré fringes. The QDs are statistically elongated along [$\bar{1}$10]. They appear wider on a (110) X-section than on a ($\bar{1}$10) X-section. This is clearly shown in the histogram of QD size, characterised as height (h) to width (w)

ratio (Fig. 2). For both samples with nominal 4 MLs and 8 MLs InGaAs the h/w spectrum lies on higher values on ($\bar{1}10$) X-sections in comparison to (110) sections. The same observations are also found in InP islands on GaInP. These indicate that the growth of the QDs is anisotropic, being faster along [$\bar{1}10$]. This growth anisotropy may be understood when structral difference of the surface steps along the two <110> orientations due to the surface reconstruction (Chadi 1987) is considered. One may also notice that the h/w spectrum is at markedly higher values in the case of 8MLs InGaAs, indicating the preferencial deposition of InGaAs on the top of preexisting islands.

Fig. 1 Plan-view micrograph of QDs of nominal 8MLs of InGaAs.

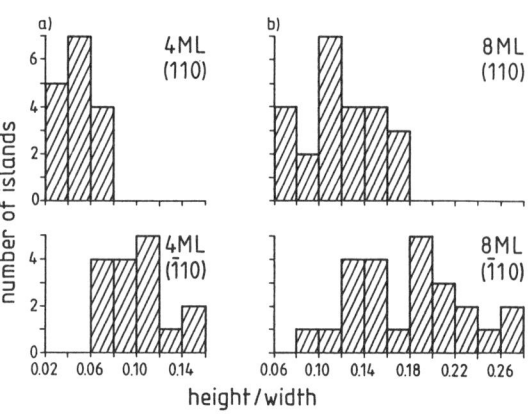

Fig. 2 Histogram of the h/w ratio of InGaAs QDs, measured on the two {110} X-sections of samples with a nominal thickness of (a) 4MLs and (b) 8MLs.

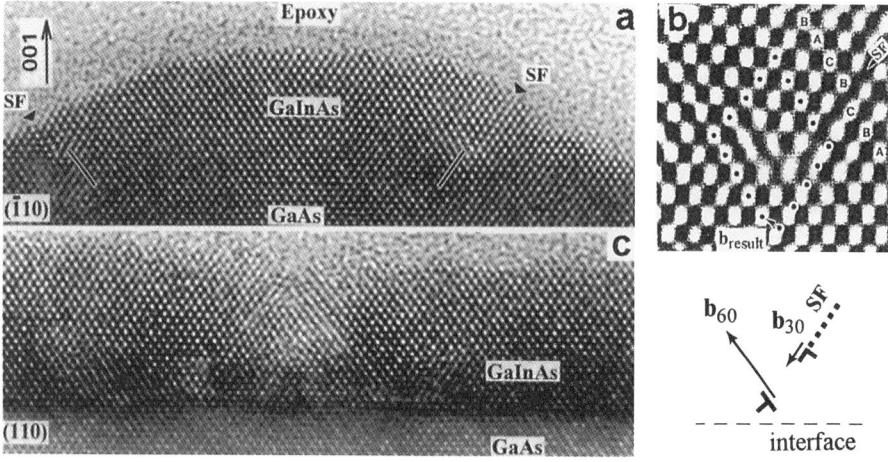

Fig. 3 HREM of defect structures in nominal 8MLs InGaAs. (a) Defect complexes introduced on both sides of a QD, (b) Burgers vector analyses of the defect complex, (c) Boundary-like defect formed after coalescence of two InGaAs QDs.

In the cases of 4MLs InGaAs/GaAs and 3MLs InP/GaInP no defects have been observed. On the contrary, introduction of defects, often observed on both sides of the QDs, takes place in the case of 8MLs InGaAs (Fig. 3a). Typically, defects are found to lie several MLs above the InGaAs/GaAs interface, and to be associated with intrinsic stacking faults (SFs). Burgers vector analyses of these defect complexes lead to a total defect vector of 1/3<111>. A sketch of the analysis on the defect complex on the right side of the QD in Fig. 3a is shown in Fig. 3b. Such a configuration appears to be very efficient for strain relaxation. In some cases, as the one shown in Fig. 3b, the defect vector is, in fact, the sum of the Burgers vectors of a perfect 60° dislocation and a 30° partial dislocation, which are very closely located to each other. We believe that the 60° dislocations have been first nucleated at the site of high stress concentration in the QDs (see below) and they draw further 30° partials, so that not only the lattice mismatch strain may be more effectively relaxed, but also the total elastic energy of the dislocations is reduced. When two QDs coalescence together, such defect complexes interact with each other and form a low-angle-boundary-like defect structure (Fig. 3c), consisting of 90° Lomer dislocations (as a result of the coalescence of two 60°dislocations), 30° partial dislocations and SFs.

The local lattice distortions of free-standing QDs are measured by quantitative analysis of HREM micrographs. Fig. 4 illustrate an InGaAs QD on GaAs. The lattice distortions in areas A and B of Fig. 4a are shown in Fig. 4b an 4c, respectively. The open points represent the real lattice image points and the bars show the deviation of the image pattern from an extrapolated pattern calculated from an undisturbed GaAs area far away from the InGaAs/GaAs interface (Bierwolf et al 1993). The InGaAs lattice in the QD is partially relaxed along both [001] and [110]. Close to the end of the QD (Fig. 4c) tensile distortions are observed in GaAs even at a position 25MLs below the interface/surface. The stress field around the InGaAs QD extends, therefore, more than 7nm.

Fig. 4 Lattice distortion in a free-standing InGaAs island in the sample of nominal 8 MLs InGaAs. (a) HREM of the island, (b) lattice distortion in the area A, (c) lattice distortions in the area of B with an enhancement factor of two.

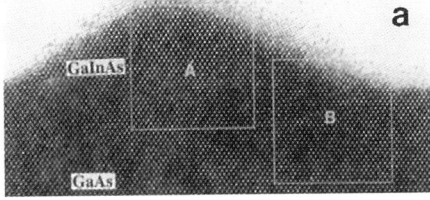

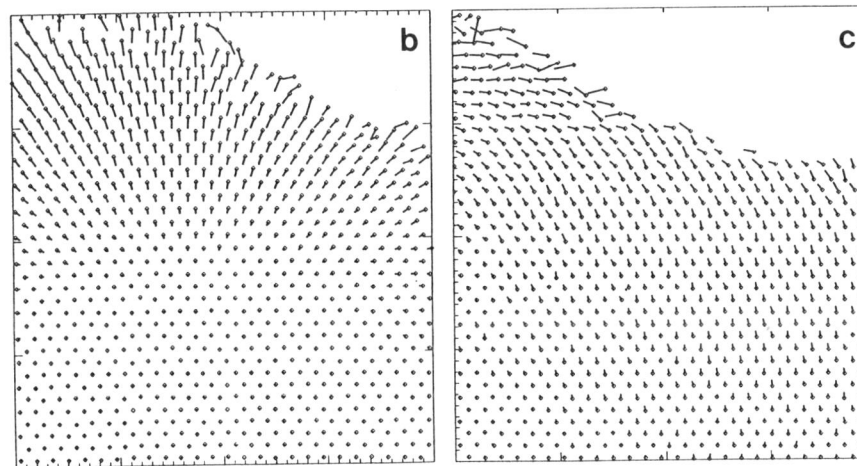

3.2 InP Quantum Dots Embedded in GaInP

TEM shows that in the case of nominal 3MLs of InP embedded in GaInP, the InP dots are about 20nm in diameter and the density is about $5 \times 10^{10} cm^{-2}$. PL spectra are measured from samples with one layer of InP islands embedded in $Ga_{0.51}In_{0.49}P$ on (001) GaAs (Kurtenbach et al 1996). The energy of InP dot PL is in the visible range (~1.80eV in the case of nominally 3 MLs of InP at 8K) and with very high intensity, being dominant even at room temperature. The strong blue shift of the InP dot PL against bulk InP is due to the confinement effect and the compressive strain within the QDs. It decreases with increasing nominal layer thickness, which is correlated with the increasing size of the islands. Only in samples with nominal layer thickness beyond 8MLs of InP does dot PL become broad and weak due to the nucleation of misfit dislocations.

In order to obtain a sharp luminescence peak, it is crucial to find a growth mode which minimises the size fluctuations of the islands, especially in the case of multi-layer quantum dots. The size of self-assembling QDs depends on many growth parameters, in particular the growth interruption after InP deposition and the thickness of the GaInP layers separating the layers of InP dots. An appropriate time of growth interruption improves markedly the homogeneity in island size in a single layer (not shown here). Fig. 5 shows TEM micrographs of three nominal 3MLs InP layers seperated with GaInP of various thicknesses d. A micrograph of a single layer of 3ML InP is also included (Fig. 5a) for comparison. In the case of $d = 16$nm (Fig. 5b) the InP QDs in the first layer vary in size. In the top two layers the dots become more uniform in size. It is often found that large QDs are correlated with two smaller QDs in the layers below. As a result, the PL shows a wide peak with a FWHM of 61meV (Fig. 6b), which is composed of two sub-peaks corresponding to the two groups of dots differentiated by their size. With decreasing d the FWHM of the PL becomes narrower. In the case of 4nm seperation of the InP layers, it is 26meV (Fig. 6d). TEM proves a pronounced alignment of the dots along [001] with good homogeneity in size in this sample (Fig. 5d). This vertical alignment is due to the strain field around the InP dots, as we demonstrated before (Fig. 4). One may also notice that the energy of InP dot PL shifts to lower values when the distance between the layers of InP dots decreases. This is presumably because of a higher degree of strain relaxation of the InP dots and/or higher degree of electron interaction between the dots when the spacer becomes thinner.

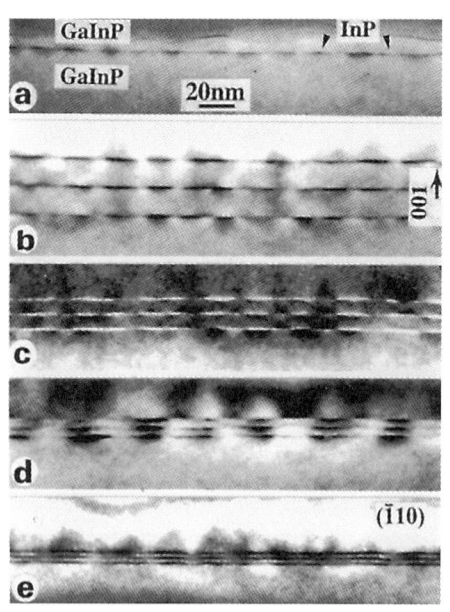

Fig. 5 Micrographs of three layers of 3MLs InP dots separated by GaInP layers with a thickness from (b) to (e): 16, 8, 4, and 2nm.

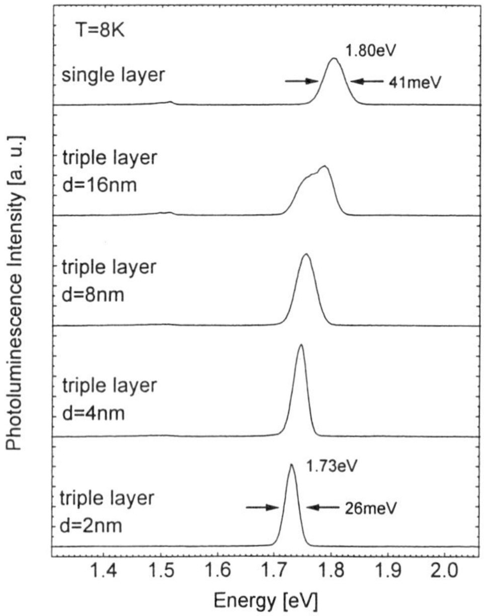

Fig. 6 PL spectra from samples with three layers of 3MLs InP dots separated by GaInP layers with various thickness d indicated in the spectra.

REFERENCES

Bierwolf R, Hohenstein M, Phillipp F, Brandt O, Crook G E, Ploog K 1993 Ultramicroscopy **49**, 273

Chadi D J 1987 J. Vac. Sci. Technol. A, **5**, 834

Eberl K, Petroff P M and Demeester P 1995 Low Dimensional Structures Prepared by Epitaxial Growth or Regrowth on Patterned Substrates, **298** NATO ASI Series E, Appl. Sci. (Kluver, Dordrecht)

Kurtenbach A, Ulrich C, Jin-Phillipp N Y, Noll F, Eberl K, Syassen K and Phillipp F 1996 J. Electronic Materials **25**, 395

Inst. Phys. Conf. Ser. No 157
Paper presented at Microsc. Semicond. Mater. Conf., Oxford, 7–10 April 1997
© *1997 IOP Publishing Ltd*

Strain-induced vertical ordering effects of islands in LPCVD-grown $Si_{1-x}Ge_x$/Si-bilayer structures on Si(001)

K Tillmann[†,*], B Rahmati[†], H Trinkaus[†], W Jäger[†,*], A Hartmann[‡], R Loo[‡], L Vescan[‡] and K Urban[†]

†Institut für Festkörperforschung and ‡Institut für Schicht- und Ionentechnik, Forschungszentrum Jülich GmbH, D-52425 Jülich, Federal Republic of Germany
*Present adress: Mikrostrukturanalytik, Technische Fakultät, Christian-Albrechts-Universität zu Kiel, Kaiserstraße 2, D-24143 Kiel, Federal Republic of Germany

ABSTRACT: The phenomenon of vertically ordered $Si_{1-x}Ge_x$-islands during LPCVD growth of $Si_{1-x}Ge_x$/Si-bilayers is examined by transmission electron microscopy. With decreasing thickness of a Si-interlayer separating two $Si_{1-x}Ge_x$-layers completely vertically ordered $Si_{1-x}Ge_x$-islands are observed. This correlation behaviour is quantitatively described in terms of a finite element method based model to predict local island nucleation probabilities. Likewise a lattice fringe analysis of HRTEM images substantiates strong compositional modifications compared to the aspired stoichiometry of the $Si_{1-x}Ge_x$-layers. Despite the observed geometrical uniformity of the $Si_{1-x}Ge_x$-islands the corresponding photoluminescence spectra show significant peak shifts dependent on the silicon interlayer thickness, which essentially result from compositional fluctuations of the islands.

1. INTRODUCTION

Over the past decade lattice mismatched semiconductor heterostructures epitaxially grown in the Stranski-Krastanow (1939) mode starting with layer-by-layer growth and followed by the nucleation of three-dimensional islands have gained considerable attention. Besides the development of new technological applications this interest results from the desire to understand the fundamental processes of nucleation and growth of pseudomorphic thin layers. Relatively recently regular arrays of islands in single layer structures have been observed and referred to anisotropic elastic material properties (Ruvimov et al 1995, Shchukin et al 1995). These self-assembled epitaxial islands are regarded as a promising possibility for the formation of quantum dots, whose electronic properties are determined by the island geometry, the layer composition and the mismatch induced internal elastic strains (Schittenhelm et al 1995).

In the present study we have focussed on highly lattice mismatched $Si_{1-x}Ge_x$/Si-bilayer structures with $x > 0.5$ grown in the Stranski-Krastanow mode by low pressure chemical vapour deposition (LPCVD). By varying the thickness of a silicon interlayer separating two $Si_{1-x}Ge_x$-layers we have examined the influence of buried $Si_{1-x}Ge_x$-islands on the spatial nucleation behaviour of subsequently grown $Si_{1-x}Ge_x$-islands on top of the interlayer. The general features of growth will be described based upon transmission and high-resolution transmission electron microscope as well as photoluminescence (PL) observations.

In addition to an earlier analytical approach (Rahmati et al 1996) to explain the vertical ordering of $Si_{1-x}Ge_x$-islands along the growth direction, also observed in periodical $Si_{1-x}Ge_x$/Si-multilayer structures (Jäger 1994), the present study uses finite element method (FEM) simulations to describe the spatially varying nucleation behaviour more realistically.

344

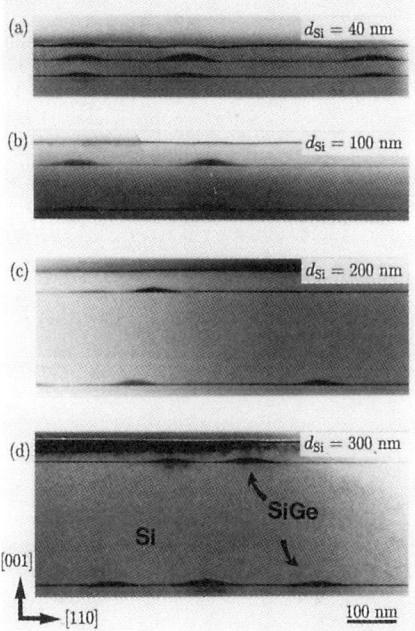

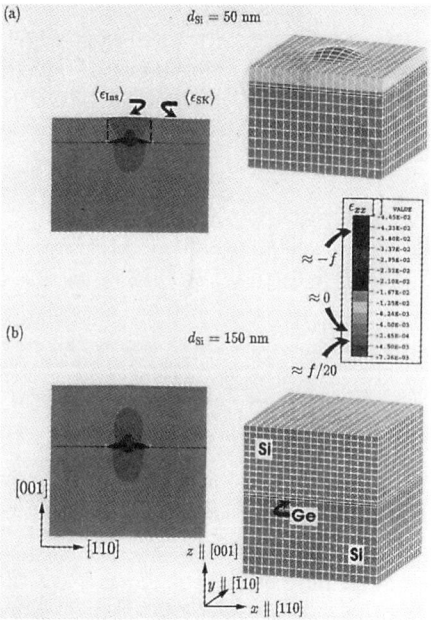

Fig. 1 Series of representative cross-sectional micrographs of the $Si_{1-x}Ge_x$/Si-bilayers
with two $Si_{1-x}Ge_x$-layers separated by a silicon interlayer with a thickness of $d_{Si} = 40$ nm
(a), 100 nm (b), 200 nm (c) and 300 nm (d)
taken along the [110]-direction under bright-
field conditions with $\mathbf{g} = [004]$. With decreasing d_{Si}-values and increased vertical ordering
of the $Si_{1-x}Ge_x$-islands is observed.

Fig. 2 Contour plot of the in-plane strain distribution ε_{xx} inside a Ge/Si-bilayer structure
after depostion of the silicon interlayer in case
of $d_{Si} = 50$ nm (a) and $d_{Si} = 150$ nm (b) as
gained by FEM simulations. The cross-sectional representations in the [110]-projection
(left in either case) illustrate that the mismatched buried Ge-islands (triangular shape)
strain the surrounding silicon in dilatation
(club-like contours).

2. EXPERIMENTAL DETAILS

The heteroepitaxial growth of the $Si_{1-x}Ge_x$/Si-bilayers was performed by LPCVD on
Si(001) substrates at a constant deposition rate of about 0.3 nm/min and a growth temperature of
$T = 700°C$. The structures consist of two $Si_{1-x}Ge_x$-layers with a nominal thickness of 0.8 nm and
an aspired composition of $x = 1$. These layers are separated by a silicon interlayer with a varying
thickness d_{Si} ranging between 40 nm and 300 nm. Each wafer has 10 nm of Si grown as a buffer
layer and a silicon capping layer of 50 nm in thickness.

The geometry and spatial arrangement of $Si_{1-x}Ge_x$-islands in these bilayers is analyzed by
conventional TEM of plan-view and cross-sectional samples. A lattice fringe analysis of digitized cross-sectional HRTEM images yields an estimation of the actual local $Si_{1-x}Ge_x$-layer
composition. Specimens were thinned to electron transparency by sequential mechanical polishing and Ar^+ ion milling.

Three-dimensional FEM simulations, taking into account the anisotropic elastic crystal
properties, were carried out on a CRAY Y-MP M94 computer by application of the ABAQUS
FEM code by Hibbitt, Karlsson and Sorensen (1994) to calculate the strain distributions in the bilayers and to predict strain induced variations in the local island nucleation behaviour.

3. RESULTS AND DISCUSSION

In the following we present our results concerning the topography of the island ensemble
inside the bilayers and the composition of the $Si_{1-x}Ge_x$-layers, focussing on strain induced vertical ordering phenomena of $Si_{1-x}Ge_x$-islands.

3.1 Vertical ordering of $Si_{1-x}Ge_x$-islands

The series of cross-section micrographs in Fig. 1 demonstrates that three-dimensional islands have been formed above a thin Stranski-Krastanow layer inside the $Si_{1-x}Ge_x$-periods of the bilayers. In highly lattice parameter mismatched systems this island formation is due to a lowering of the strain energy by an elastic relaxation mechanism (Eaglesham and Cerullo 1990, Vanderbilt and Wickham 1991, Christiansen et al 1995), i.e. by local displacements of lattice planes at the expense of an additional surface energy contribution compared to a two-dimensional layer. All bilayers are characterized by coherently strained interfaces between the $Si_{1-x}Ge_x$-islands and the surrounding silicon. Independent of the silicon interlayer thickness the islands are characterized by a regular pyramidal geometry with average lateral base extensions $l = 140$ nm and heights $h = 14$ nm, while the average mutual island distance as obtained from plan-view images amounts to $2\lambda = 430$ nm.

However, the interlayer thickness significantly influences the topographical arrangement of the islands nucleated in the subsequently grown $Si_{1-x}Ge_x$-layers. As demonstrated by Figs. 1(a) and (b) islands of the second $Si_{1-x}Ge_x$-layer are located exactly above positions of buried islands of the first $Si_{1-x}Ge_x$-layer along the growth direction for d_{Si}-values smaller than 100 nm. Similar observations are reported on InAs/GaAs multilayers (Xie et al 1995) and $Si_{1-x}Ge_x$-multilayers with an increased number of periods (Vescan et al 1992, Carlino et al 1996). This vertical correlation decreases with increasing interlayer extensions and results in a completely random arrangement at $d_{Si} = 300$ nm (Fig. 1 (d)). Analysing a large number of cross-section micrographs we obtain data on the fraction K of islands in the upper layer which is correlated to those of the lower layer as a function of the interlayer thickness (cf. ($\bullet$) in Fig. 4 later). In detail islands are denoted as vertically correlated if they overlap with half their lateral base extension l at least, differently from the definition used by Rahmati et al 1996.

In order to get a more quantitative understanding of the preferential vertical island ordering we discuss the experimental results in the framework of a nucleation model for islands of the second $Si_{1-x}Ge_x$-layer taking into account an inhomogeneously strained growth surface due to the strain fields of buried islands of the first $Si_{1-x}Ge_x$-layer. To simplify matters we assume pure germanium islands ($x = 1$) of the geometry given above. The implications of compositional modifications are discussed in section 3.2 later. Three-dimensional FEM simulations were applied to obtain quantitative information about the strain state on the growth surface before deposition of a second germanium layer in dependence on the silicon interlayer thickness. In the simulations periodic boundary conditions with a periodicity of the average island distance 2λ were applied.

For illustration, the in plane strain component ε_{xx} along the <110>-directions as obtained from these simulations is given by the contour plots in Figs. 2 (a) and (b) for two interlayer extensions of $d_{Si} = 50$ nm and $d_{Si} = 150$ nm in form of a cross-sectional and a perspective representation. The germanium island is concentrically buried inside the cube at the height of the Stranski-Krastanow layer. The darker greyish club-like contours around the trianglar germanium islands

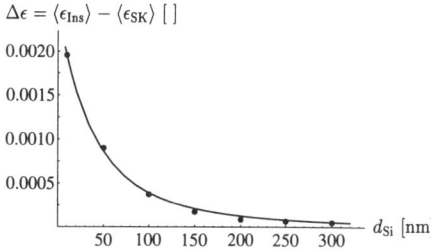

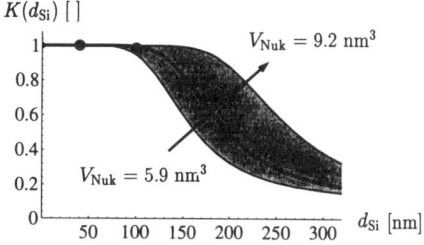

Fig. 3 Difference of the averaged strain values near the growth surface directly before deposition of the second germanium layer above a buried germanium island $\langle\varepsilon_{Ins}\rangle$ and above the two-dimensional Stranski-Krastanow layer $\langle\varepsilon_{SK}\rangle$ in dependence on the silicon interlayer thickness d_{Si}.

Fig. 4 Diagram illustrating the behaviour of the correlation function K in dependence on the silicon interlayer thickness d_{Si}. The three calculated curves base on island nucleation volumes V_{Nuc} amounting to 5.9 nm³, 7.0 nm³ and 9.2 nm³. The best fit to the experimental data ($\bullet$) is found for $V_{Nuc} = 7.0$ nm³.

(left in both cases) indicate that the surrounding silicon substrates and interlayers are strained in dilation with a maximum in the order of $f/20$, while the islands and the Stranski-Krastanow layer are nearly completely strained in compression in the order of the lattice-parameter mismatch amounting to $\varepsilon_{xx} = -f = -0.040$ for Ge/Si-heterostructures. Contrasting to $d_{Si} = 50$ nm the long-range dilatation-clubs do not reach through the surface at $d_{Si} = 150$ nm resulting in a nearly homogeneous strain distribution at the growth front for the subsequent deposition of the second germanium layer.

Fig. 3 illustrates the behaviour of the difference $\Delta\varepsilon$ of the averaged strains above the buried germanium island $<\varepsilon_{Ins}>$ and above the two-dimensional Stranski-Krastanow layer $<\varepsilon_{SK}>$ in dependence on the silicon interlayer thickness. For decreasing d_{Si}-values an increasing dilatation of the silicon interlayers lattice-parameter is predicted ($\Delta\varepsilon > 0$), resulting in an effectively reduced mismatch for the growth of the second germanium layer. Compared to a random arrangement a vertical ordering of islands is therefore energetically preferred, because of a reduced elastic interaction potential between buried germanium islands and newly formed nuclei of the second layer.

Assuming that the strain induced change in the local island nucleation probability can be described by a Boltzmann distribution characterized by an activation barrier dependent on the elastic interaction between subsequently grown germanium islands, the measured fraction K of vertically correlated islands can be calculated as a function of the interlayer thickness d_{Si} (for details see Tillmann 1997). For v, G and V_{Nuc} denoting the Poisson ratio, the shear modulus and the volume of a newly formed islands nucleus of the second germanium layer as well as a geometry parameter $\delta = l/2\lambda$, this correlation $K(d_{Si})$ is written as:

$$K(d_{Si}) = \{1 + ((1-\delta^2)/\delta^2)\exp[-4((1+v)/(1-v))(GfV_{Nuc}\,\Delta\varepsilon(\,d_{Si}))/(k/T))]\}^{-1} \quad (1)$$

All variables in this equation are either experimentally measurable (K, d_{Si}, T, δ), material properties (f, G, v) or determinable by FEM simulations ($\Delta\varepsilon(d_{Si})$) except the island nuclei volume V_{Nuc}, which is not accessible by ex situ TEM investigations. Thus V_{Nuc} can be used as a fitting parameter to adapt $K(d_{Si})$ on the experimental data obtained by cross-section micrographs. According-ing to Fig. 4 the optimal adjustment is achieved for $V_{Nuc} = 7.0$ nm^3 equivalent to an island nucleus formed by approximately 320 germanium atoms, which is somewhat larger than the value obtained by Rahmati et al 1996.

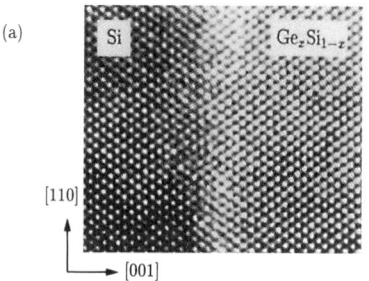

(a)

To test if this number is reasonable, we give a rough assumption on V_{Nuc} basing on energetical considerations. As mentioned previously island formation in mismatched heterostructures is caused by a lowering of the strain energy by an elastic relaxation at the expense of an additional surface energy contribution. Therefore an island nucleus is energetically inherently unstable below a certain critical volume, which is in a rather complicated manner dependent on the nucleus geometry, the amount of elastic relaxation and the specific surface energy γ of the material forming the nucleus (Tillmann 1997). Based on typical γ-values for various silicon surfaces (Eaglesham et al 1993 and reference therein) this critical nucleus volume amounts $V_{Nuc} = 5.9$ nm^3 ($\gamma = 1.3$ J m^{-2}) up to $V_{Nuc} = 9.2$ nm^3 ($\gamma = 1.5$ J m^{-2}). Taking these V_{Nuc}-values as a basis, the grey hatched area in Fig. 4 represents the whole range of reasonable $K(d_{Si})$-branches also including the experimental data. In this way it is demonstrated that the FEM-based island nucleation model successfully describes the vertical ordering of islands within the margins of the limitted

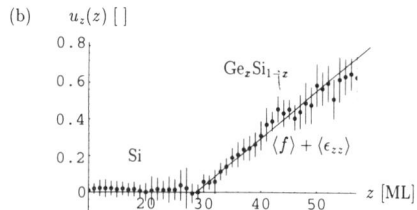

(b)

Fig. 5 Cross-section HRTEM image of the coherent interface between the Si(001)-substrate and a buried Si$_{1-x}$Ge$_x$-island in the [110]-projection (a). The gradient of the cumulative sum of deviations $\Delta u_z(z)/\Delta z$ represents the sum of the average mismatch $<f>$ and strain $<\varepsilon_{zz}>$ along the z-direction, i.e. the crystallographic [001]-direction.

accurancy of material parameters. Compared to the analytical approximation (Rahmati et al 1996) using an elastic dipole approximation ("equivalent sphere approximation") for the strain field of a buried island the V_{Nuc}-values obtained by FEM are of the same order of magnitude.

3.2 Composition of the $Si_{1-x}Ge_x$-islands

The coherency of the $Si_{1-x}Ge_x$-islands inside the bilayers motivates further investigations concerning the layer composition because the critical dimensions for misfit dislocation nucleation are largely exceeded. For example the Matthews (1972) approximation neglecting any elastic relaxation predicts a critical length of $l_c = 27$ nm for pure germanium islands of the observed geometry. Even if elastic relaxation is taken into account by an effectively reduced misfit in the order of $f_{red} = 0.65 f$ (Tillmann 1997) the experimentally observed island base extensions amounting to $l = 140$ nm are distincly overcritical when assuming $x = 1$.

To get an estimation on the real composition of the $Si_{1-x}Ge_x$-layers we performed a digital analysis of HRTEM images to extract the mismatch and mismatch induced strains from micrographs of the $Si_{1-x}Ge_x$/Si-interface such as Fig. 5 (a). The procedure (for details see Tillmann 1996a and references therein) involves the detection of intensity maxima positions by a centre-of-mass analysis inside the bright spots of HRTEM images, the determination of averaged (002)-monolayer plane distances $a(z')$ enumberated z' and subsequently the calculation of a cumulative sum of deviations $u_z(z)$ in relation to a generated reference lattice in the undisturbed silicon substrate characterized by the lattice parameter a_{Si}. According to the definition of the cumulative sum of deviations $u_z(z) = [\int_0^z dz'\ (a(z')/a_{Si})] - z \approx (f + \varepsilon_{zz})\ z$ the gradient of the $Si_{1-x}Ge_x$-island related branch in Fig. 5 (b) represents the average sum of the lattice parameter misfit and the strain component along the [001]-direction, at which the latter can be calculated by FEM simulations also considering thin foil relaxation in HRTEM-specimen. The straight line in Fig. 5 (b) yields $(f + \varepsilon_{zz}) = 0.02571$ which is equivilant to a germanium content $x_{min} = 0.37$ assuming an infinitely thick HRTEM specimen or to $x_{max} = 0.47$ if a vanishing specimen thickness is assumed, respectively. Hence clear evidence is given that the actual layer composition of the $Si_{1-x}Ge_x$-periods strongly differs from the yearned stoichiometry striving for $x = 1$. Similar observations have been reported for comparable $Si_{1-x}Ge_x$/Si-heterostructures (Carlino et al 1996, Walther et al 1996) and referred to an interdiffusion mechanism (Stenkamp 1993) causing a broadening of the nominal thickness of the germanium layer by producing a $Si_{1-x}Ge_x$-alloy.

However, the substantiated compositional modifications are not in contradiction to the island nucleation model presented in section 3.1 explaining the vertical ordering in case of pure germanium islands. For a fixd volume of the islands of the first $Si_{1-x}Ge_x$/Si-layer a decreased germanium content diminishes the difference of strains $\Delta\varepsilon$ because of a nominally reduced deformation of the silicon interlayer by a buried $Si_{1-x}Ge_x$-island. On the basis of Eqn. (1), when assuming $x = 0.5$, i.e. half of the values for f and $\Delta\varepsilon(d_{Si})$, the optimal fit to the experimental data is simply given by the fourfold nucleus volume of $V_{Nuc} = 28.0$ nm^3.

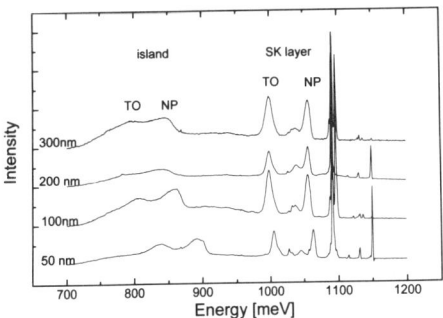

Fig. 6 The $Si_{1-x}Ge_x$-island related peaks (NP and TO) of the $Si_{1-x}Ge_x$/Si-bilayers show a significant blueshift in the PL spectra with decreasing silicon interlayer thickness d_{Si}.

3.3 Photoluminecence measurements

In addition to the TEM and HRTEM observations photoluminescence experiments give information on compositional changes of the $Si_{1-x}Ge_x$-islands subject to the silicon interlayer thickness. As demonstrated by Fig. 6 the PL spectra show a strong blue shift in the order of 80 meV of the island related non-phonon (NP) and transversal optical (TO) peaks with decreasing d_{Si}-values. Generally, with decreasing silicon interlayer thickness this observation may be explained by (i) an increased band gap of the $Si_{1-x}Ge_x$-alloy (Si: 1.2 eV, Ge: 0.7 eV) equivalent to a decreasing germanium content, (ii) an increased quantum confinement equiva-

348

lent to smaller island heights or (iii) strongly increased elastic strains inside the islands (Apetz et al 1995, Abstreiter et al 1996).

The latter possibility (iii) can be excluded because the FEM simulations (Fig. 2) show no significant differences of the $Si_{1-x}Ge_x$-islands strain values dependent on d_{Si}. In detail the micrographs in Fig. 1 demonstrate that the islands of the second $Si_{1-x}Ge_x$-layer in comparison to those of the first layer are characterized by an enhanced height with decreasing d_{Si}-values, which should cause a red shift of the island-related peaks (ii). On the other hand because of a constant nominal layer thickness and density of islands in all samples an increased island height with decreasing interlayer thickness is equivalent to a reduced average germanium content x. Therefore the experimentally observed blue shift is explained by the assumption that the mechanism (i) is dominant with decreasing d_{Si}-values.

4. CONCLUSIONS

In summary, it has been demonstrated that self-organization of coherently strained islands in form of a vertical ordering is observed in $Si_{1-x}Ge_x$/Si-bilayer structures grown by LPCVD on Si(001) under the chosen growth conditions. This phenomenon is attributed to a long-range elastic interaction between buried $Si_{1-x}Ge_x$-islands and newly formed $Si_{1-x}Ge_x$-nuclei separated by the silicon interlayer. An island nucleation model based on finite element simulations has been proposed successfully describing the vertical island correlation behaviour in dependence on the silicon interlayer thickness $K(d_{Si})$. By a digital analysis of lattice fringe images a strong modification of the $Si_{1-x}Ge_x$-layer composition is observed compared to the chosen stoichiometry, which is attributed to an interdiffusion during epitaxial growth. Photoluminescence measurements substantiate a silicon germanium interdiffusion which is enhanced at decreasing interlayer extensions resulting in a blue shift of the $Si_{1-x}Ge_x$-island related non-phonon peaks.

REFERENCES

Abstreiter G, Schittenhelm P, Engel C, Silveira E, Zrenner A, Meertens D and Jäger W 1996 Semicond. Sci. Technol. **11** 1521.
Apetz R, Vescan L, Hartmann A, Dieker C and Lüth H 1995 Appl. Phys. Lett. **66** 445.
Carlino E, Giannini, Geradi C, Tapfer L, Mäder KA and Känel H 1996 J. Appl. Phys. **79** 1441.
Christiansen S, Albrecht M, Strunk HP, Hansson PO and Bauser E 1995 Appl. Phys. Lett. **66** 574.
Eaglesham DJ and Cerullo M 1990 Phys. Rev. Lett. **64** 1943.
Eaglesham DJ, White AE, Feldman LC, Moriya N and Jacobsen DC 1993 Phys. Rev. Lett. **70** 1643.
Hibbitt, Karlsson and Sorensen Inc. 1994 ABAQUS 5.4 User's Manual (Pawtucket RI)
Jäger W 1994 Electron Microscopy of Boundaries and Interfaces in Materials Science, eds. J Heydenreich and W Neumann (Halle/Saale: MPI for Microstructure Physics) pp 221-234.
Matthews JW 1972 Surf. Sci. **31** 241.
Rahmati B, Jäger W, Trinkaus H, Loo R, Vescan L and Lüth H 1996 Appl. Phys. **A62** 575.
Roland C and Gilmer GH 1993 Phys. Rev. B **47** 16286.
Ruvimov S, Werner P, Scheerschmidt K, Gösele U, Heydenreich, Richter U, Ledentsov NN, Grundmann M, Bimberg D, Ustinov VM, Egrov AY, Kop'ev PS and Alferov ZI 1995 Phys. Rev. B **51** 14766.
Schittenhelm P, Gail M and Abstreiter G 1995 J. Cryst. Growth **157** 260.
Stranski IN and Krastanow L 1939 Akad. Wiss. Wien, Mathe.-Naturwiss. Kl IIb **146** 797.
Shchukin VA, Ledentsov NN, Kop'ev PS and Bimberg D 1995 Phys. Rev. Lett. **75** 2968.
Stenkamp D 1993 Reports of the Research Centre Jülich GmbH, JÜL-**2839**.
Tillmann K, Thust A, Lentzen M, Swiatek P, Förster A, Urban K, Laufs W, Gerthsen D, Remmele T and Rosenauer A 1996a Phil. Mag. Lett. **74** 309.
Tillmann K, Rahmati B, Trinkaus H, Jäger W, Loo R, Vescan L and Urban K 1996b Proc. XIth European Congress on Electron Microscopy, Dublin (CD-ROM).
Tillmann K (1997) Reports of the Research Centre Jülich GmbH, JÜL-**3360** and to be published.
Vanderbilt D and Wickham LK 1991 Mat. Res. Soc. Symph. Proc. **202** 555.
Vescan L, Jäger W, Dieker C, Schmidt K, Hartmann A and Lüth H 1992 Mat. Res. Soc. Symp. Proc. **263** 23.
Walther T, Boothroyd CB and Humphreys CJ 1996 Proc. XIth European Congress on Electron Microscopy, Dublin (CD-ROM).
Xie Q, Madhukar A, Chen P and Kobayashi NP 1995 Phys. Rev. Lett. **75** 2542.

Inst. Phys. Conf. Ser. No 157
Paper presented at Microsc. Semicond. Mater. Conf., Oxford, 7–10 April 1997
© 1997 IOP Publishing Ltd

TEM assessment of the growth mode and strain state of capped InSb dots grown on InP (001) substrates

J C Ferrer [a b], **F Peiró** [a], **A Cornet** [a], **J R Morante** [a], **T Utzmeier** [c], **G Armelles** [c] and **F Briones** [c]

[a] EME, Física Aplicada i Electrònica, Universitat de Barcelona, Diagonal 645-647, 08028, Barcelona, Spain
[b] Serveis Científico-Tècnics, Universitat de Barcelona, Lluís Solé i Sabarís 1-3, 08028 Barcelona, Spain
[c] CNM, Instituto de Microelectrónica de Madrid, Isaac Newton 8, Parque Tecnológico de Madrid, 28760, Madrid, Spain

ABSTRACT: Capped InSb islands have been deposited on InP substrates by ALMBE. An amount of InSb ranging from 0.7 to 15 monolayers was deposited in order to assess the transition from two-dimensional to three-dimensional growth mode as well as the beginning of island coalescence. TEM observations show uniform distributions of dots in a range between 1.7 and 5 monolayers. Below the former value no dot is observed probably due to aninitial two-dimensional growth mode. Above 5 monolayers of InSb, island coalescence occurs. It is possible to recognise both strained dots, by the typical lobe contrast around them, and relaxed dots, in which moiré fringes are evident.

1. INTRODUCTION

The epitaxial growth of different III-V alloys has been widely studied for a whole range of lattice mismatch: from the combinations where the small mismatch allows layer-by-layer growth up to thicker layers, to those that the raising mismatch leads to three-dimensional growth modes even at low thickness. Thus we can find pure two-dimensional mode (Franck van der Merwe), two-dimensional followed by island formation (Stransky-Krastanov) or directly three-dimensional growth (Volmer-Weber). The interest of three-dimensional growth comes from the possibility of synthesizing quantum structures, eg quantum dots or wires which exhibit carrier confinement and have potential applications in optoelectronics. See the paper of Bimberg (1996) for more details. In this framework, electron microscopy is revealed as a powerful tool to assess the quality of the grown nanostructures, and to determine which is the best choice of technological parameters to achieve good quality devices.

The aim of this work is to study the first stages of InSb growth on InP substrates in growth conditions for which island formation occurs. Transmission Electron Microscopy (TEM) is used to examine the dot morphology and distribution on the substrate surface as well as to assess the relaxation mechanisms and the range within which good quality dots are developed depending on the total InSb amount.

2. EXPERIMENTAL

Samples were grown by Atomic Layer Molecular Beam Epitaxy (ALMBE) in which the group V element flux is pulsed to enhance the surface migration. An InP buffer layer was deposited prior to dot growth to obtain a good quality surface. InSb was deposited obtaining eight samples with an equivalent amount of 0.7, 1.7, 2.2, 2.8, 3.2, 5, 10 and 15 monolayers (ML's); afterwards an InP cap layer was grown to protect the dots. The thickness of this layer was about 60 nm for the first four samples and about 45 nm for the last four.

TEM observations in both plan view and cross-section configurations were performed with a Philips CM30 microscope operating at 300 KV. Samples were prepared by conventional methods of mechanical polishing and final ion milling in a cooled stage in order to avoid preparation artifacts such as indium precipitates.

3. RESULTS AND DISCUSSION

The first point to be addressed is the evolution of the island growth and their morphology depending on the amount of deposited InSb. Table I. summarizes the approximate size and density of quantum dots.

Table I. Approximate mean size and density of InSb quantum dots.

InSb	Cap thickness	diameter	density (cm
0.6	60	-	-
1.7	60	13	$5.0 \cdot 10^9$
2.2	60	20	$2.7 \cdot 10^9$
2.8	60	14	$4.0 \cdot 10^{10}$
3.2	45	30	$1.1 \cdot 10^{10}$
5	45	30	$1.2 \cdot 10^{10}$
10	45	$\simeq 50$	coalescence
15	45	$\simeq 50$	coalescence

The sample with the lowest amount of InSb (0.6ML's) presents no contrast related to the presence of dots. Thus it may be suggested that there is an initial layer-by-layer growth and that the Volmer-Weber mode characterized by direct island nucleation may be discarded for the studied conditions. The next sample, with an equivalent amount of 1.7 ML's, is the first after the transition to three-dimensional growth. From this value to the sample with 5 ML's the dots seem to be uniformly distributed and isolated from each other. Notice that the density reaches a maximum that may be indicative of the beginning of dot joining. Samples with 10 and 15 ML's show large InSb aggregates as well as some isolated islands.

Above the onset of island nucleation, i.e. within the range between 1.7 to 5 ML's, we can distinguish mainly two types of contrast related to strained and relaxed dots. In the former case two contiguous lobes can be observed at the place where strained dots are placed, as can be seen in Fig. 1 corresponding to a planar view of the sample containing 3.2 ML's. These lobes are usually symmetric, with the plane of symmetry perpendicular to the g vector used to obtain the image in two-beam conditions. Such contrast comes from atomic plane distortion at the vicinity of islands, due to the high lattice mismatch between InP and InSb, which reaches   10.4%. On the other hand, relaxed dots can be identified by the presence of moiré fringes inside them (Fig. 2). The distance between contiguous fringes (about 4.5 nm) indicates that the islands are completely relaxed. The fraction of relaxed dots increases with the InSb contents as bigger dots have higher energy related to strain. Around some relaxed islands   can be seen a contrast very similar to that enveloping the strained dots. This leads us to suspect that the strain distribution over the dot is not uniform, and that there may be some residual strain at the dot border.

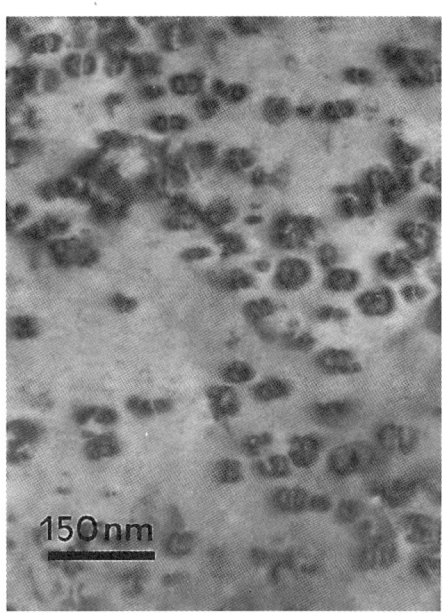

Fig. 1. Plan view image of a sample with 3.2 ML's. The line of no contrast between lobes is perpendicular to the g=220 vector.

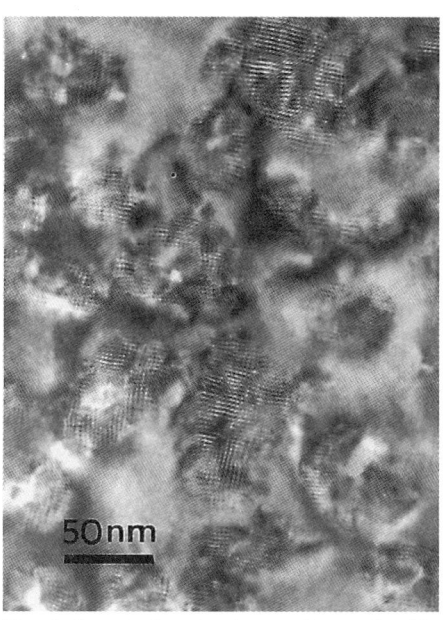

Fig. 2. On-axis plan view image of a sample with a total amount of 10 ML's. Dot coalescence has started. The moiré fringes indicate the relaxed state of InSb islands.

Another way to confirm the relaxation state is by means of selected area diffraction patterns (SADP's). Figure 3 shows the (001) on-axis SADP of an area that overlaps some relaxed dots, in which the spot splitting due to double diffraction is observed: each diffracted beam at the InP cap layer becomes a new incident beam for the InSb island that has different interplanar distances, and thus, leads to multiple (001) diffraction patterns. The distance between split spots compared to main spots can also indicate the relaxation degree of InSb islands. In our case the splitting is about the 10% of the distance between main spots, confirming the complete strain relie of relaxed dots. In our case, SADP's confirm the complete relaxation of dots. Similar procedures are used by Kiely (1989) to assess the strain state of InSb layers on GaAs with a larger lattice mismatch (14 %).

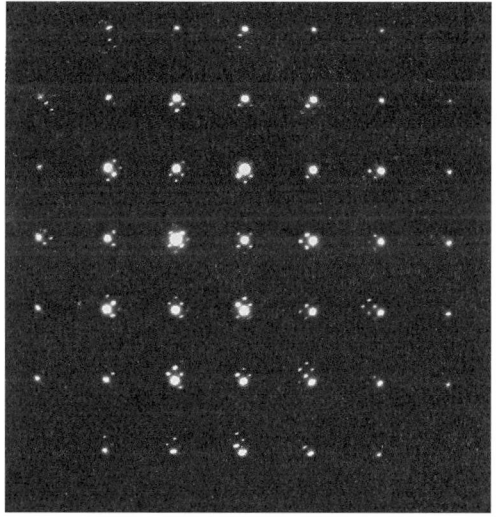

Fig.3. Selected area diffraction pattern of an area overlapping some relaxed dots.

In order to obtain further information about the structure of dots and understand the relaxation mechanism, high resolution studies in cross section configuration have been performed. Figure 4 shows a dot section of a sample with 5 ML's where the alternatively clear and dark contrast inside the dot suggest that the island is relaxed. This is confirmed by a careful examination of the interface between the dot and the substrate where there is an array of misfit dislocations. Different mechanisms of dislocation formation are discussed in the papers of Kiely (1989) and Chen (1996).

In view of the HRTEM images, it is also likely that the relaxation mechanism between dots and substrate is different from that between dots and the cap layer because of the different morphology of both interfaces. Whereas a network of misfit dislocations is placed at the bottom interface, the interface structure between the dot and the enveloping cap layer remains unclear. As it is not a planar surface, it is more difficult to assess the relaxation mechanism between both materials in its vicinity. Despite this, the generation of planar defects and the presence of dislocations in the proximity of dots seems to be the means of strain accommodation.

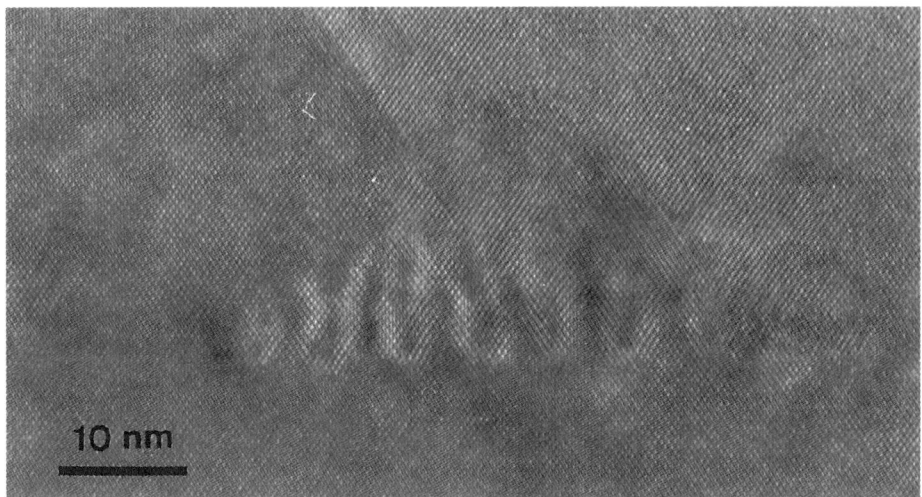

Fig. 4. High resolution cross-section image of a relaxed dot. Notice the dislocations placed at the dot/substrate interface.

4. CONCLUSIONS

In summary, the formation of capped InSb quantum dots on InP substrates has been studied by TEM. The analyzed range has allowed us to establish the onset of the transition from two to three-dimensional growth, as well as the coalescence limit. In the range between these two transitions, an homogenous distribution of similar size dots is obtained. TEM images have allowed us to assess the dot morphology and the relaxation state as it is possible to recognise the relaxation state of dots by the contrast related to plane distortion in strained dots or moiré fringes for relaxed islands.

ACKNOWLEDGEMENTS

This work has been funded by the Spanish CICYT Project MAT95-0966.

REFERENCES

Bimberg D, Ledentsov N N, Grundmann M, Kirstaedter N, Schmidt O G, Mao M H, Ustinov V M, Egorov A Yu, Zhukov A E, Kop'ev P S, Alferov Zh I, Ruminov S S, Gösele U and Heydenreich J, 1996 Phys. Status Solidi **194**, 159.
Chen Y, Lin X W, Liliental-Weber Z, Washburn J, Klem J F and Tsao J Y, 1996. Appl. Phys. Lett. **68**, 111.
Kiely C J, Chyi J I, Rockett A and Morkoç H, 1989 Phil. Mag. **A60**, 321.

Inst. Phys. Conf. Ser. No 157
Paper presented at Microsc. Semicond. Mater. Conf., Oxford, 7–10 April 1997
© *1997 IOP Publishing Ltd*

Structural and optical characterisation of MOVPE self-assembled InSb quantum dots in InAs and GaSb matrices

A G Norman, N J Mason*, M J Fisher[+], J Richardson[+], A Krier[+], P J Walker* and G R Booker

Department of Materials, University of Oxford, Parks Road, Oxford OX1 3PH, UK
*Clarendon Laboratory, University of Oxford, Parks Road, Oxford OX1 3PU, UK
[+]Advanced Materials and Photonics Group, School of Physics and Chemistry, Lancaster University, Lancaster LA1 4YB, UK

ABSTRACT: Transmission electron microscopy (TEM) and photoluminescence (PL) were used to study metal organic vapour phase epitaxy (MOVPE) self-assembled InSb quantum dots (QDs) grown for the first time in InAs and GaSb matrices. Coherently strained InSb QDs were observed in the InAs matrix sample with a number density $\approx 3 \times 10^9$ cm^{-2} and diameter 5-10 nm. PL emission was obtained at 3.5 μm, the longest wavelength yet reported for self-assembled semiconductor QDs. Coherently strained InSb QDs were also observed in the GaSb matrix sample together with larger relaxed InSb islands.

1. INTRODUCTION

There has been increasing interest over the last few years in the formation of assemblies of semiconductor quantum dots (QDs) for use in optoelectronic devices. The 3D carrier confinement of QDs is expected to lead to increased absorption and emission probabilities and improved thermal stability associated with the increased density of states and reduction in Auger processes when there are totally quantised levels. This should result in lower threshold current density and higher temperature stability operation semiconductor lasers (Arakawa and Sakaki 1982). One of the most promising routes towards the fabrication of such arrays of quantum dots is by the Stranski-Krastanow strained layer growth mechanism (Stranski and Krastanow 1939, Nötzel 1996) which can lead to an array of coherently strained QDs with a relatively small distribution of sizes ($\approx 10\%$) and high number density without the need for any patterning or photolithography. Such arrays are termed self-assembled. The majority of work has so far been on In(Ga)As self-assembled QDs, grown in a GaAs matrix by both molecular beam epitaxy (MBE) and metal organic vapour phase epitaxy (MOVPE), which luminesce in the near infrared (IR). In(Ga)As QD lasers demonstrating some of the improvements expected in QD devices have recently been reported (e.g. Kirstaedter et al 1994). Other QD/matrix systems have also been investigated e.g. GaSb/GaAs (Hatami et al 1995, Glaser et al 1996), InSb/GaAs (Glaser et al 1996) and InSb/InP (Utzmeier et al 1996) which luminesce in the near IR or visible regions.

There is a growing interest in optoelectronic devices operating in the mid-IR region (3-5 μm) for gas sensing, pollution monitoring, defence and medical applications. Present semiconductor lasers emitting in this region are based on quantum wells (QWs) and have disadvantages, e.g. poor temperature stability, and these could be improved by using QDs. However, no work has yet been reported for self-assembled QDs that emit in the mid-IR. Consideration shows that in order to achieve this, both the QD and the matrix materials need to have narrow band gaps. In this work we report results on the MOVPE growth and properties of InSb QDs, the III-V binary alloy with the narrowest band gap, formed in InAs and GaSb matrices, the two III-V binary alloys with the next narrowest band gaps. InAs and GaSb both have lattice parameters $\approx 7\%$ less than InSb and so Stranski-Krastanow strained layer growth should occur for both systems.

2. EXPERIMENTAL DETAILS

The two QD samples for this study were grown at 480°C in an atmospheric pressure MOVPE reactor described previously by Booker et al (1997). The first sample consisted of 1s deposition of InSb on a (001) InAs substrate with an InAs buffer layer, followed by a thin InAs capping layer. The second sample consisted of a 1s deposition of InSb on a (001) GaSb substrate with a GaSb buffer layer, followed by a thin GaSb capping layer. Five second growth pauses under flowing hydrogen were inserted in the growth sequence before and after deposition of the InSb QDs. Plan-view TEM specimens were prepared by chemical thinning using Cl_2/methanol solution and cross-section specimens by Ar^+ ion milling with the sample cooled to liquid nitrogen temperatures. Transmission electron microscopy (TEM) samples were examined in a Philips CM20 operated at 200 kV and an ultrahigh resolution JEOL 4000 EX operated at 300 kV. PL measurements were made at 4K with the samples excited using a 5W Ar^+ ion laser. The PL spectra were recorded using a 0.3 m monochromator and the light was detected with a cooled InSb detector using conventional phase-sensitive detection techniques.

3. RESULTS

For the InSb/InAs sample, plan-view **g**[220] dark field (DF) images (Fig. 1) showed small strain contrast features (QD) each consisting of a pair of dark and bright lobes separated by a line of no contrast perpendicular to the **g** vector. This contrast is consistent with coherently strained InSb QDs embedded in an InAs matrix. Plan-view [001] pole bright field (BF) images (Fig. 2) showed dark 'rings' of strain contrast (QD) and this contrast is also consistent with coherently strained InSb QDs. The number density of the QDs was $\approx$ 3 x 10^9 cm^{-2}. The [001] pole BF images showed that the QDs were 5-10 nm across and that in general they were irregular in shape, i.e. did not possess regular facets. {100} cross-section lattice images (Fig. 3) of individual QDs present at the InAs buffer/cap layer interface showed pairs of dark lobes also consistent with coherently strained QDs. {100} cross-section **g**[002]DF images (no micrograph included) revealed the presence of a continuous InSb wetting layer $\approx$ 0.5 nm thick between the InAs buffer layer and the InAs capping layer. PL spectra (Fig. 4) showed a weak peak at 3.10 μm (0.400 eV) with a shoulder at 3.02 μm (0.411 eV) corresponding to InAs excitonic and donor acceptor pair transitions (Fang et al 1990, Fisher and Krier 1997) and a strong peak at 3.5 μm (0.354 eV, FWHM $\approx$ 26 meV) well away from previously reported InAs transitions and concluded here to arise from the coherently strained InSb QDs. Further work is in progress to confirm this result. The 3.5 μm emission wavelength observed for this sample lies in the important mid-IR region which is currently of great interest for a number of applications.

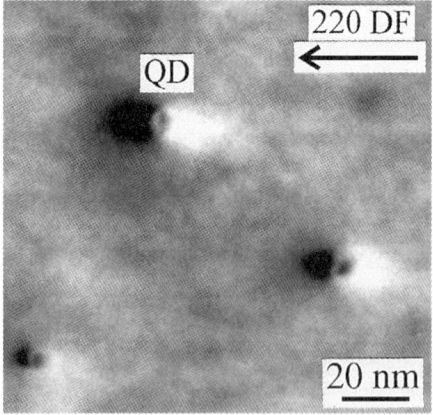

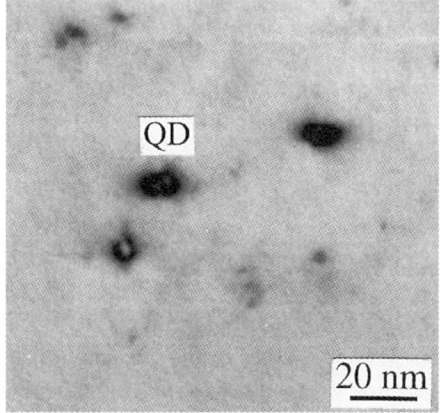

Fig. 1. Plan-view **g**[220]DF micrograph showing dark-bright strain contrast lobes, e.g. QD, arising from individual coherently strained InSb QDs in the InAs matrix.

Fig. 2. Plan-view [001] pole BF showing 'rings' of strain contrast, e.g. QD, arising from individual coherently strained InSb QDs in the InAs matrix.

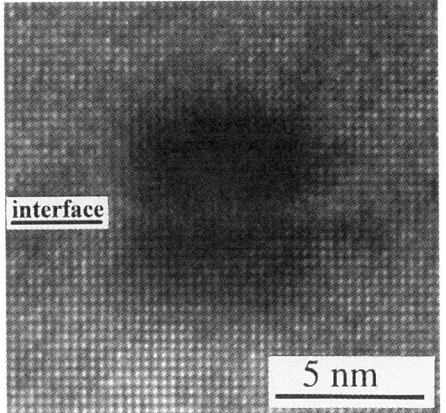

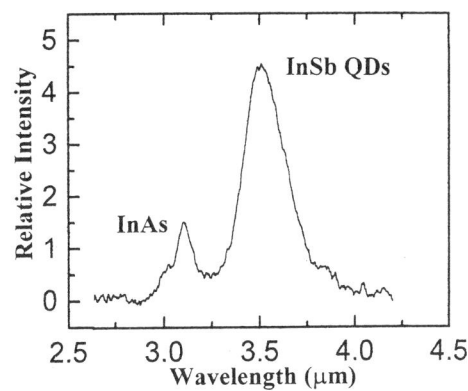

Fig. 3. {100} cross-section lattice image showing strain contrast associated with a coherently strained InSb QD at the InAs buffer/cap layer interface.

Fig. 4. 4K PL spectra showing weak peak at 3.1 μm arising from InAs matrix and strong peak at ≈ 3.5 μm from coherently strained InSb QDs.

For the InSb/GaSb sample, plan-view **g**[220]DF images (Fig. 5) showed large InSb islands which were approximately rectangular and elongated along one of the <110> directions. They were typically ≈ 180 by 85 nm across with a number density ≈ 10^8 cm^{-2} and exhibited moiré fringes of spacing ≈ 4.4 nm. Calculation shows that moiré fringes between 100% relaxed InSb and GaSb for **g**[220] would be of spacing 3.6 nm and so it is deduced that the islands were ≈ 80% relaxed. These images (Fig. 5) also showed features with dark and bright lobes typically 10 nm across and with a number density ≈ 10^9 cm^{-2}. These features are similar to those of Fig. 1 and are attributed to coherently strained InSb QDs in the GaSb matrix. {110} cross-section **g**[002]DF images (Fig. 6) showed inclined stacking faults and micro-twins, and dislocations originating at the relaxed InSb islands and which thread through the GaSb cap layer. A continuous InSb wetting layer ≈ 0.5 nm thick was also present. PL spectra (not included) showed only a broad peak at 1.58 μm (0.786 eV) probably corresponding to GaSb.

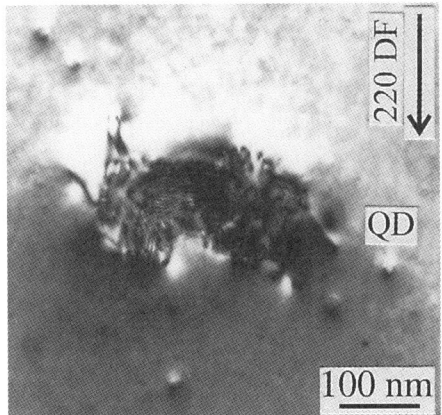

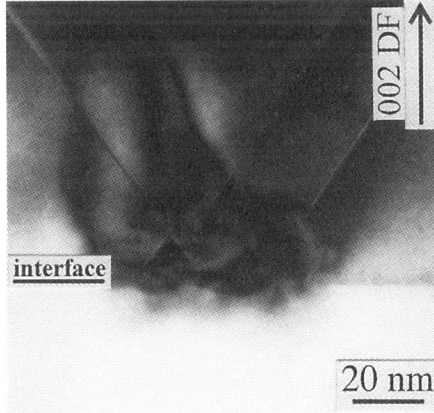

Fig. 5. Plan-view **g**[220]DF micrograph showing a large relaxed InSb island with associated defects, and small coherently strained InSb QDs, in the GaSb matrix.

Fig. 6. {110} cross-section **g**[002]DF micrograph showing large relaxed InSb island at the GaSb buffer/cap layer interface and associated defects.

4. DISCUSSION AND CONCLUSIONS

Our interpretation of the above results is as follows. For the InSb/InAs sample, all of the InSb islands were coherently strained and acted as QDs. For the PL, the recombination of the injected carriers was dominated by these QDs, radiative recombination occurred and this gave the strong peak at ≈ 3.5 μm. For the InSb/GaSb sample, excessive InSb had been deposited and some of the islands relaxed by the formation of interface dislocations. These islands then grew faster than the other islands present because less strain energy was involved. For the PL, the carrier recombination was dominated by these large islands and mainly occurred at the dislocations and planar defects present in these islands. Such recombinations were non-radiative and so there was insignificant QD emission in the mid-IR range.

The emission at ≈ 3.5 μm from the coherently strained InSb QDs in the InAs is to our knowledge the longest wavelength so far obtained from any epitaxial self-assembled semiconductor QD system. Furthermore, use of different growth conditions would give different mean QD sizes and hence different wavelengths in the mid-IR range, and increased QD number densities would give higher emission intensities. For example growth at higher temperatures should result in larger QDs and hence in emission at longer wavelengths due to reduced quantum confinement. Thus the InSb/InAs QD system shows great promise for applications in the important 3-5 μm mid-IR region. For the InSb/GaSb system, a significant decrease in the amount of InSb deposited would give InSb QDs that were all coherently strained and probably also strong QD emission in the mid-IR.

In conclusion we have successfully grown coherently strained self-assembled InSb QDs by MOVPE in an InAs matrix which emit at ≈ 3.5 μm (0.354 eV) in the important 3-5 μm mid-infrared region. Deposition of coherently strained InSb QDs in a GaSb matrix was also demonstrated although larger relaxed InSb islands were also present in this sample.

ACKNOWLEDGEMENTS

We wish to thank the EPSRC for financial support and K Belcher and S Moulder for technical help.

REFERENCES

Arakawa Y and Sakaki H 1982 Appl. Phys. Lett. **40** 939
Booker G R, Daly M, Klipstein P C, Lakrimi M, Kuech T F, Jiang Li, Lyapin S G, Mason N J, Murgatroyd I J, Portal J C, Nicholas R J, Symons S G, Vicente P and Walker P J 1997 J. Crystal Growth **170** 777
Fang Z M, Ma K Y, Jaw D H, Cohen R M and Stringfellow G B 1990 J. Appl. Phys. **67** 7034
Fisher M and Krier A 1997 accepted for publication in J. Infrared Phys. and Technol.
Glaser E R, Bennett B R, Shanabrook B V and Magno R 1996 Appl. Phys. Lett. **68** 3614
Hatami F, Ledentsov N N, Grundmann M, Böhrer J, Heinrichsdorff F, Beer M, Bimberg D, Ruvimov S S, Werner P, Gösele U, Heydenreich J, Richter U, Ivanov S V, Meltser B Ya, Kop'ev P S and Alferov Zh. I 1995 Appl. Phys. Lett. **67** 656
Kirstaedter N, Ledentsov N N, Grundmann M, Bimberg D, Ustinov V M, Ruvimov S S, Maximov M V, Kop'ev P S, Alferov Zh I, Richter U, Werner P, Gösele U and Heydenreich J 1994 Electronics Lett. **30** 1416
Nötzel R 1996 Semicond. Sci. Technol. **11** 1365
Stranski I N and Krastanow L 1939 Akad. Wiss. Lit. Mainz Math.-Natur Kl. 11b **146** 797
Utzmeier T, Postigo P A, Tamayo J, García R and F Briones 1996 Appl. Phys. Lett. **69** 2674

Inst. Phys. Conf. Ser. No 157
Paper presented at Microsc. Semicond. Mater. Conf., Oxford, 7–10 April 1997
© *1997 IOP Publishing Ltd*

Microstructural characterisation of CdSe quantum dots prepared by various routes

R R Nayak*, J R Galsworthy[#], P J Dobson[+] and J L Hutchison*

Departments of Materials* and Engineering[+] Parks Road, Oxford and Department of Inorganic Chemistry[#], South Parks Road, Oxford, UK.

ABSTRACT: From the many available synthesis routes for producing CdSe quantum dots we concentrate on the reaction of molecular species at high temperatures in co-ordinating organic solvents as first reported by Murray et al. This route yields monodisperse quantum dots that are stable, isolable in powder form with a robust surface passivation and good optical properties. A potential drawback is that a toxic metal alkyl is used directly as the cadmium source. An alternative approach considered by O'Brien et al, uses the metal alkyl to form a single molecule precursor providing both the metal and chalcogenide species that is then injected into the hot solvent. We characterise and compare the microstructure of CdSe quantum dots prepared using both routes using HRTEM and correlate this with the optical properties.

1. INTRODUCTION

Quantum size effects in nanocrystallite semiconductors make them interesting model systems to study the evolution of material properties with size and earns them the name 'Quantum Dots' (Brus 1991) (Steigerwald and Brus 1989) (Weller 1993). In quantum dots the size dependent effective bandgap can give photoluminescence that is tunable according to the dot size. This and other properties of quantum dots such as the higher non-linear optical effects per unit mass have potentially novel applications. Due to the small sizes involved, a significant proportion of atoms reside on the surface. Dangling bonds present on the surface give rise to surface states which are deleterious to the optical properties, particularly the luminescence. It is therefore preferable to devise some means of surface passivation which eliminates the effects of surface states.

A synthesis route which incorporates the production of semiconductor nanocrystallites with a narrow size distribution and the chemical passivation of the dangling bonds on the crystallite surfaces is one which is based on the pyrolysis of organometallic reagents by injection into a hot co-ordinating solvent and pioneered by Murray et al (1993). From High Resolution Transmission Electron Microscopy (HRTEM) studies we report in detail the fine scale microstructure of CdSe crystallites such as the shape of the crystallites, nature of the surfaces and defects in the crystallites.

The organometallic reagent used in the synthesis of CdSe quantum dots as reported by Murray et al (1993) is dimethylcadmium $(CH_3)_2Cd$ which is toxic and needs specialist handling using airless techniques normally compatible with larger scale high vacuum production routes such as Molecular Beam Epitaxy or Organometallic Vapour Phase Epitaxy. A further development in quantum dot synthesis based on this route has been to avoid directly injecting the $(CH_3)_2Cd$ into the hot reaction solvent. This is done by using the $(CH_3)_2Cd$ at an earlier stage to make a precursor molecule containing both the Cd and Se sources with the idea of avoiding repeated use of $(CH_3)_2Cd$ for every batch of quantum dots produced. This approach, referred to as the Single Molecule Precursor (SMP) route, has been first reported by Trindade and O'Brien (1996).

We compare here the microstructure of dots produced using $(CH_3)_2Cd$ as the direct cadmium source with that of the single molecule precursor approach and correlate the microstructure to the optical properties.

2. EXPERIMENTAL

CdSe crystallites prepared using both routes were studied using a JEOL 4000EX HRTEM by placing a drop of a dilute dispersion of the crystallites in toluene onto a holey carbon coated copper grid. The synthesis using $(CH_3)_2Cd$ adapted from Murray et al (1993) is described in brief. Standard airless techniques are used.

$(CH_3)_2Cd$ is mixed with Selenium dissolved in Trioctylphosphine (TOP) a co-ordinating solvent with a high boiling point. This mixture cooled to about -50°C is injected rapidly into Trioctylphosphineoxide TOPO) another co-ordinating solvent with a high boiling temperature which is maintained at 350°C. The rapid injection of reagents causes a temperature drop which results in a saturation of reagents leading to a temporally discrete nucleation of CdSe crystallites. This gives a narrow size distribution of CdSe crystallites which can then be grown to a desired average size by varying the temperature and time. Here the TOP and TOPO not only act as reaction solvents, they also bind or ligate to the CdSe surface passivating the surface and allow controlled growth of the crystallites under these conditions. After synthesis, a size selective precipitation enables further narrowing of size distributions, if required.

3. RESULTS AND DISCUSSION

Figs. 1-3 shows CdSe quantum dots prepared according to Murray et al (1993). We can see the crystallites that are on or near zone axis are highly facetted (Fig. 1). Although the wurtzite structure predominates, a mixture of the hexagonal close-packed (wurtzite) showing ABAB stacking and the closely related cubic close-packed (spharelite) structure showing ABCABC stacking is often seen. In these very small regions with mixed stacking it is not relevant to distinguish between the two structures. The preferred growth direction for the larger crystallites is in the direction of the 'c' axis. The coalescence of particles forming a 'bi-particle' characterises one of the processes by which larger particles form in the reaction vessel. Fig. 3 shows particles suspended over the edge of the holey carbon with some distortion in the lattice fringes particularly at the surfaces, possibly due to the stress associated with the shape of the crystallites. By studying crystallites in particular projections within a fraction having a narrow size distribution, information on the shape can be obtained as was previously shown for $CdS_xSe_{(1-x)}$ nanocrystallites in silicate glasses by Allais et al (1992). Their proposal that the crystallites grow following a hexagonal prism model with preferred growth in the direction of the 'c' axis has also been observed in our case. Such detailed information can be used in simulating X-ray diffraction data of the dots as has been carried out by Murray et al (1993) particularly in the refining stage, to provide a better fit to the experimental data.

Fig. 4 shows CdSe quantum dots prepared according to the alternative route by Trindade et al (1996) using the single molecule precursor. The crystallites are monodisperse with a good degree of crystallinity. The surfaces are mostly, but not consistently facetted, on crystallites near zone axes suggesting the surfaces are sometimes irregular. Also Cd metal nanocrystallites are deposited, Fig. 5, indicating that on decomposition of the single molecule precursor, the selenium source does not always react with the Cd source. The optical absorption spectra from samples obtained using both approaches are shown in Fig. 6.

The spectra are staggered for comparison. The quantum dots clearly show quantum size effects by the sharp excitonic peaks present. The sharpness indicates the presence of narrow size distributions in agreement with the HRTEM micrographs. On comparison, the spectrum from the sample using the single molecule precursor route shows some absorption in the infra-red region (700-900 nm) which arises most likely due to the presence of the Cd metal particles.

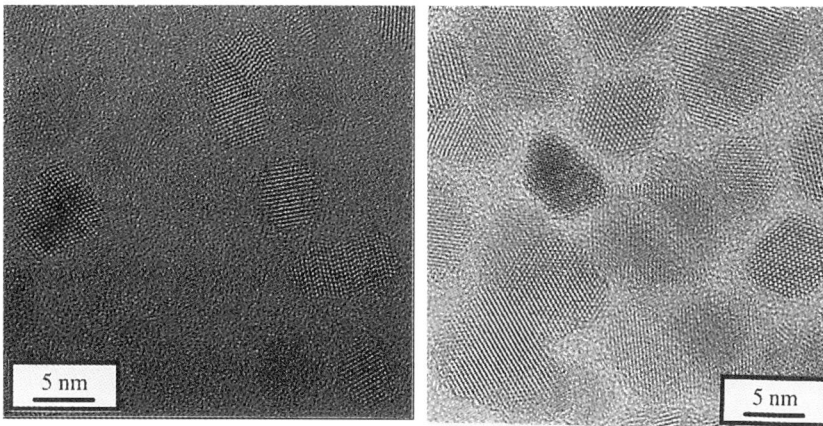

Fig. 1 Facetted CdSe Dots - [010] $_{HCP}$ Fig. 2 Coalescence of dots during growth

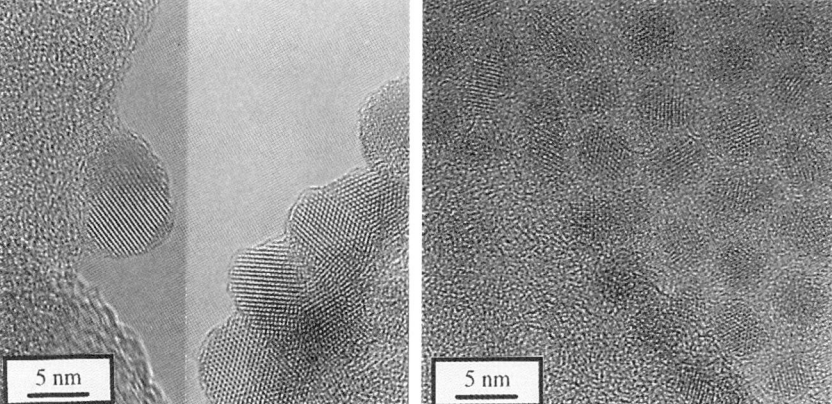

Fig.3 Lattice distortion in CdSe nanoparticles   Fig. 4 CdSe dots from the SMP route.

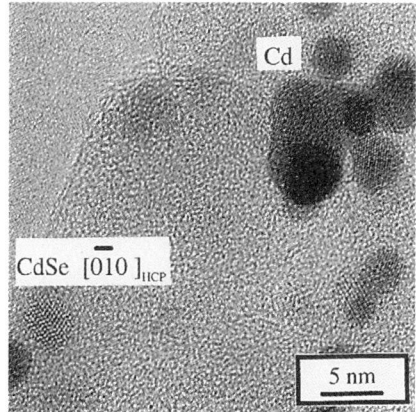

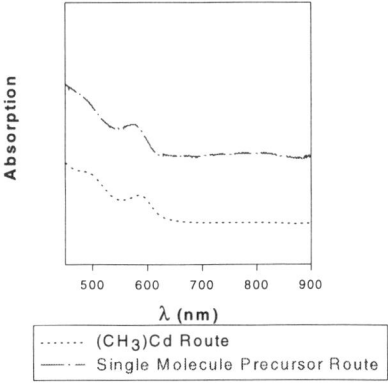

Fig. 5 Cd particles with a SMP- CdSe Dot   Fig. 6 Optical Absorption Spectra - CdSe

4. CONCLUSIONS

We show the quantum crystallites prepared using dimethylcadmium as the direct cadmium source and via the single molecule precursor route to have narrow size distributions and sharp excitonic peaks in the optical spectra. In the former method, we show that formation of 'bi-particles' and preferential growth of the crystallites parallel to the 'c' axis to be two processes leading to the formation of the larger particles. CdSe quantum dots prepared using single molecular precursor route show absorption in the infra-red most likely due to the presence of Cd metal particles formed as a result of the incomplete reaction of the cadmium and selenium from the precursor.

A combination of HRTEM studies in conjunction with optical measurements are important in understanding the relationship between the structure and optical properties of the crystallites and also contributes to a better understanding of the nucleation and growth processes of the crystallites.

REFERENCES

Allais M and Gandais M 1992 Philosophical Magazine Letters **65**(5): 243-248.

Brus L 1991 Applied Physics A **53**: 465-474.

Murray CB, Norris DJ and Bawendi MG 1993 Journal of the American Chemical Society **115**: 8706-8715.

Steigerwald M L and Brus LE 1989 Annual Review of Materials Science **19**: 471-95.

Trindade T and O'Brien P 1996 Advanced Materials **8**(2): 161-163.

Weller H 1993 Angewandte Chemie International Edition in English **32**: 41-53.

ACKNOWLEDGEMENTS

Help and encouragement with the synthesis of CdSe quantum dots using $(CH_3)_2Cd$ as the direct cadmium source, by Ken Kuno from Prof. M. Bawendi's group at the Chemistry Department, MIT is gratefully acknowledged.

We thank Tito Trindade from Prof. P O'Brien's group at the Inorganic Chemistry Department, Imperial College for providing the CdSe crystallites prepared according to the Single Molecule Precursor route.

R. Nayak is on an EPSRC studentship. Financial support towards conference presentations has been provided by the Institute of Physics, Department of Materials, St Cross College and the Royal Microscopical Society.

Thanks are due to Epichem (UK) for the provision of $(CH_3)_2Cd$.

Inst. Phys. Conf. Ser. No 157
Paper presented at Microsc. Semicond. Mater. Conf., Oxford, 7–10 April 1997
© *1997 IOP Publishing Ltd*

Application of the 113 weak beam imaging technique to the investigation of strain-induced InAs islands grown on InP and GaAs(001) by MBE

A Ponchet and D Lacombe

CEMES-CNRS, 29 rue Jeanne Marvig, BP 4347, 31055 Toulouse, FRANCE

ABSTRACT: InAs islands obtained by the strain-induced 2D-3D transition on (001) InP substrates are examined by the ($1\bar{1}3$) Weak Beam Imaging Technique in cross-sectional configuration. It is shown that some of the most useful of the strained island characteristics (density, width) can be obtained directly by this method with a good agreement with the values obtained by plan-views. This method is particularly suitable when plan-views are not easily available (islands buried by a thick or disturbed structure).

1. INTRODUCTION

The strain-induced formation of islands has raised a great interest over the past years. The island characteristics are highly dependant on the deposition parameters (lattice mismatch, thickness...) and on growth conditions (growth technique, growth temperature, growth rate, interface procedures etc...). The island characteristics should be carefully identified, for fundamental understanding of the island formation and for applications (the optical properties being dependant on the island shape, island size, size homogeneity etc...). Only TEM allows the local investigation of buried islands. Two configurations are generally used : (a) plan-view observations which allow to estimate the island spatial distribution (surface coverage, inter-island distances...) and (b) cross-sectional observations, generally performed with the interface being vertical, which allow to determine the island shape and size.

In some cases, observation in plan-view configuation is difficult, if the islands are buried under thick or disturbed structures. It is therefore necessary to remove precisely the part grown above the islands (Lacombe et al 1997a). Another solution consists of analysing a cross-sectional sample with the interface being inclined. With the [001] growth direction, the sample can be tilted around the [$1\bar{1}0$] direction. By a rotation of 25° from the [$1\bar{1}0$] zone axis, the [$3\bar{3}2$] zone axis is reached. As shown in detail by Yao et al. (1991), the use of the g=$1\bar{1}3$ diffraction vector in a weak beam condition is a powerful method to image the islands : the strained layers appear as fringes perpendicular to g, while the strained islands appear as bright disklike contrasts. In the following this will be referred as the weak beam imaging technique (WBIT).

2. COMPARISON OF PLAN-VIEW AND (113)WBIT

In a first stage, the characteristics obtained by this method have been compared to those obtained by plan-view, using one sample contained InAs islands grown by Molecular Beam Epitaxy on InP(001) (lattice mismatch 3.2%). The growth procedure consisted of a deposit of 4 InAs monolayers (ML) at a low growth rate (0.1 ML/sec) followed by a growth interruption of 30 sec. resulting in the formation of InAs islands. Then the islands were covered by 25 nm of InP. A more complete study of this sample is reported elsewhere (Ponchet et al 1995).

TEM was performed at 200 kV on a Philips CM20 microscope. Figure 1 shows the islands observed by plan-view with g=220 (fig.1a) and by the (113)WBIT (fig.1b). The same features were found by plan-view and (113)WBIT : no defects for more than 95% of the islands, inhomogeneous spatial distribution of islands (with large areas without islands), and formation of some wires along [1$\overline{1}$0]. There also exists a good agreement between the characteristics obtained by the two methods, as shown in fig.2. Note that the (113)WBIT allows the observation of the islands in thin specimen with thickness as large as 0.3 μm, while when the interface is vertical, the thickness is limited by superposition of the islands along the direction of observation (typically 50 nm in this sample).

In another study concerning InAs islands embedded in GaAs (Lacombe et al 1997b, Lacombe et al 1997a), we found a small discrepancy of 20% for the island densities obtained by plan-view (g=220) and by (113)WBIT.

Figure 1 : Observation of the InAs islands

0.5 μm

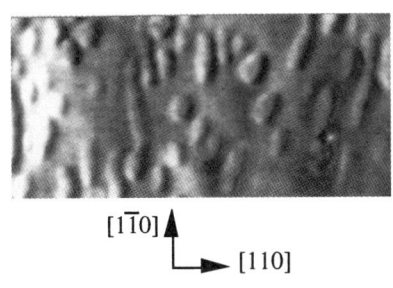

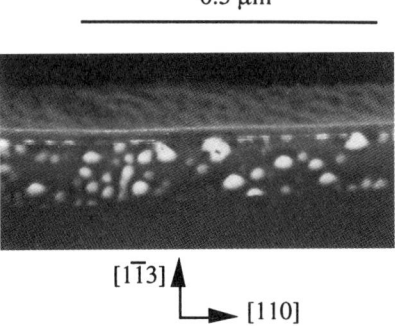

[1$\overline{1}$0] [110]

[1$\overline{1}$3] [110]

(a) in plan-view, with the direction of observation close to [1$\overline{1}$0] (g=220)

(b) in cross-section, with the direction of observation close to [3$\overline{3}$2] (g=1$\overline{1}$3, WBIT)

3. ISLANDS BURIED UNDER A THICK AND DISTURBED STRUCTURE

We show an application of this method on a sample for which plan-view was impossible to perform. This sample contained InAs islands grown by MBE on InP(001). The total deposit was 2 ML of InAs, deposited in four steps : (1) deposit of 1.5 ML (growth rate of 0.1 ML/sec), (2) growth interruption of 30 sec., (3) deposit of 0.5 ML, (4) growth interruption of 30 sec. The islands were covered by 0.1 μm of InP and then by a thick multi-quantum-well (lattice matched to InP).

As shown in figure 3 (which is typical of the whole sample), several different cross-sectional observations were performed using :

* g=002, the interfaces being vertical (fig.3a)
* g=1$\overline{1}$3, the interfaces being inclined of 25° around [110] (fig.3b and 3c)
* g=220, the interfaces being more or less inclined around [110] (not shown here)

Unlike in the previous sample, two kinds of islands were identified :

(1) strained islands, typically of 50 nm in width and 8 nm in height. They are characterised in the g=1$\overline{1}$3 weak beam conditions by a bright disklike contrast, as shown by Yao et al (1991). This contrast is very sensitive to the exact diffraction deviation as shown in fig.3b and 3c.

(2) larger islands, characterised in the g=1$\overline{1}$3 weak beam conditions by fringes. These islands are found to be at the origin of threading dislocations and it can be inferred that these islands are plastically relaxed.

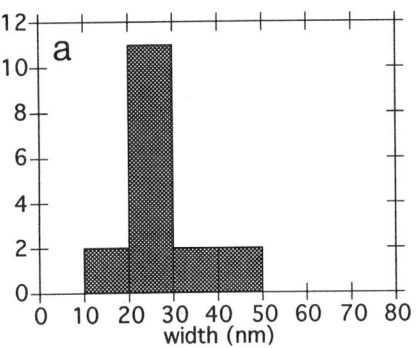

Figure 2 : histograms of island width

(a) from cross-sectional observations, the interface being vertical (g=002)

(b) from cross-sectional observations, the interface being inclined (g=1$\bar{1}$3)

(c) from plan-view observations (g=220)

The sample is that shown in fig.1.
The y axis represents the number of islands.

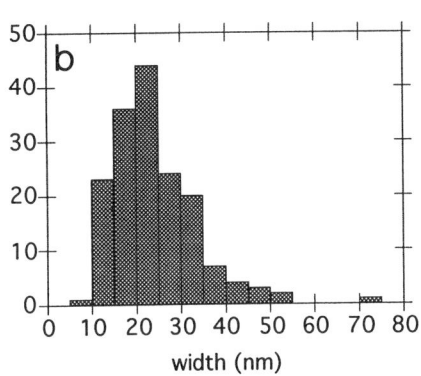

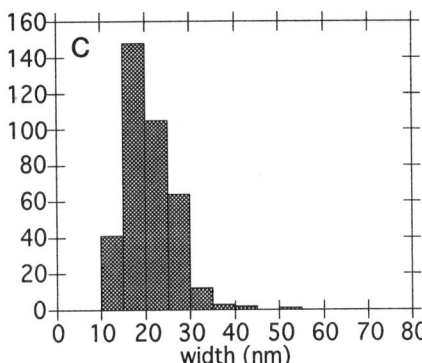

4. CONCLUSION

Using the (113)WBIT, some of the most useful characteristics of the strained-induced islands (density and lateral size) could have been obtained directly by cross-sectional observations. Comparison of the values obtained by plan-view and by (113)WBIT on some particular samples shows a good agreement, with a discrepancy no greater than 20%. This method can therefore be successfully applied to samples for which plan-view specimens are difficult to achieve (i.e. islands buried under thick or disturbed strutures).

Strained and relaxed islands observed with (113)WBIT present bright disklike contrast and fringes within the island, respectively. Combination of (113) WBIT and other diffraction vectors is particularly suitable for the observation of the defects originating from the islands (this work), or originating from the substrate when these defects meet the islands (Lacombe et al 1997b, Lacombe et al 1997a).

Finally, it is seen that different procedures of deposition can result in different relaxation processes (elastic and plastic) : although the deposit was only 2 ML, it is seen that the procedure used in the second case (§3) has resulted in the formation of some large islands which have plastically relaxed, unlike using the procedure previously presented (§2 of this article and Ponchet et al 1995)).

364

(a)

[1 $\bar{1}$ 0] [110]

(b)

[1 $\bar{1}$ 3] [110]

(c)

[1 $\bar{1}$ 3] [110]

0.1 μm

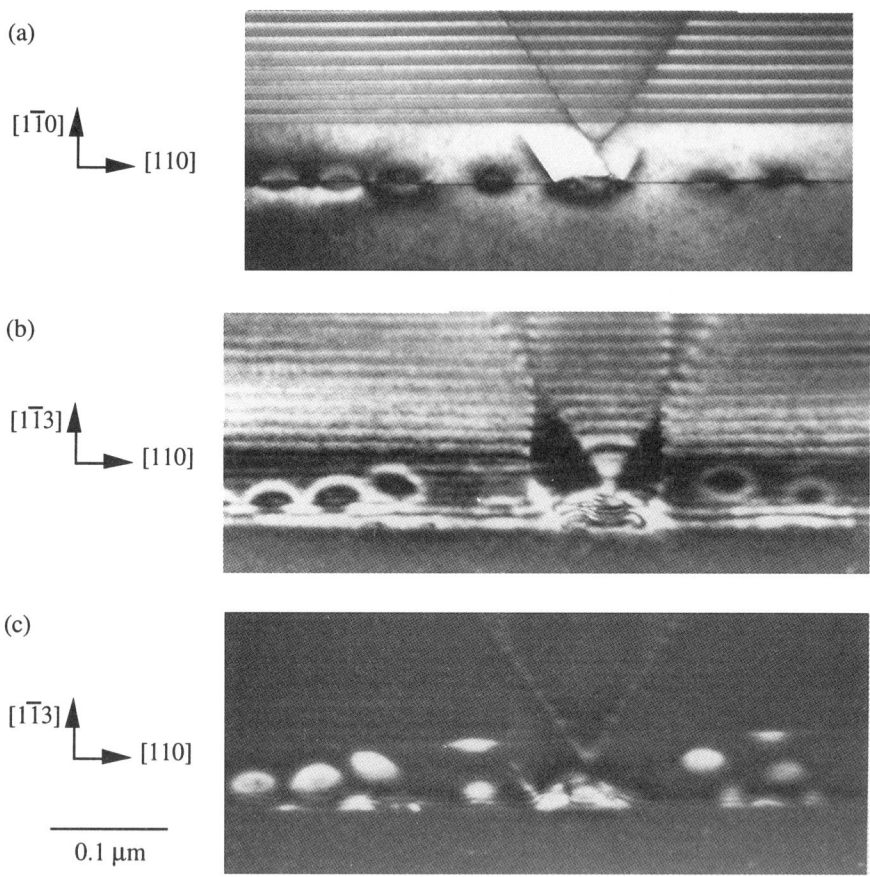

Figure 3 : strained and relaxed InAs islands (disklike contrasts and fringes, respectively)
(a) the interfaces are vertical, g=002
(b) the interfaces are inclined of 25° around [110], g=1$\bar{1}$3 in weak beam conditions **g/2g**
(c) the interfaces are inclined of 25° around [110], g=1$\bar{1}$3 in weak beam conditions **g/3g**

ACKNOWLEDGEMENT

We are grateful to A. LeCorre, S. Salaün, H. L'Haridon, B. Lambert for the growth of the samples and for their interest, and to France-Télécom for financial support.

REFERENCES

Lacombe D, Ponchet A and Gérard J-M 1997a Inst. Phys. Conf. Ser. (this volume).
Lacombe D, Ponchet A, Gérard J-M and Cabrol O 1997b Appl. Phys. Lett. 70 **in press**.
Ponchet A, LeCorre A, L'Haridon H, Lambert B and Salaün S 1995 Appl. Phys. Lett. **67,** 1850.
Yao J Y, Andersson T G and Dunlop G L 1991 Philosophical Magazine A **64,** 173.

Inst. Phys. Conf. Ser. No 157
Paper presented at Microsc. Semicond. Mater. Conf., Oxford, 7–10 April 1997
© 1997 IOP Publishing Ltd

Impact of the threading dislocations and residual stress on InAs islands grown on a (001) GaAs-on-Si pseudo-substrate relative to growth on standard GaAs

D Lacombe, A Ponchet and J M Gérard[1]

CEMES-CNRS, 29 rue Jeanne Marvig, BP 4347, 31055 Toulouse, FRANCE
[1] France Télécom, CNET-PAB, 196 av. Ravéra, BP 107, 92225 Bagneux, FRANCE

ABSTRACT : Strain-induced InAs islands have been grown by MBE simultaneously on a GaAs-on-Si substrate (containing a density of 10^7 threading dislocations per cm^2) and on a standard GaAs substrate (dislocation free). The impact of the substrate on the island growth mode has been examined by cross-sectional and plan-view TEM observations. Using the GaAs-on-Si substrate has resulting in the formation of a greater number of smaller islands. On the other hand, it is seen that several threading dislocations have been bent in the island plane, although the average thickness is lower than the critical thickness of plastic relaxation. Finally it is estimated that only a few percentage of the islands (less than 2%) is affected by the threading dislocations.

1. INTRODUCTION

Quantum boxes (QB) formed by the strain-induced 2D-3D transition have generated a great interest over the past years (Moison et al. 1994, Leonard et al. 1994, Kirstaedter et al. 1994 and references therein). However, QB size fluctuation, inherent in the growth process, results in inhomogeneous broadening of the QB array emission. So far, this has limited the efficiency of such QB based lasers compared with those based on 2D quantum well (QW) (Gérard et al 1996b). Recently the potential of such QBs compared with 2D-QW for the integration of emitting optical devices on Silicon substrates has been investigated by Gérard et al (1996a). Indeed, the threading dislocations in GaAs-on-Si substrates, with of an order of magnitude of 10^6 per cm^2 in the state of the art, behave as non-radiative recombination centers and supress the quantum efficiency of QWs. This is due to the 2D geometry, which cannot quench laterally the carrier diffusion towards the dislocations. One potential advantage of the QBs is that their geometry should quench laterally the carrier diffusion, preserving the properties of the QB array. This has been recently demonstated by Gérard et al (1996a) who compared the optical properties of an InAs QB array and a GaInAs 2D-QW, grown on a GaAs-on-Si substrate (10^7 threading dislocations per cm^2) or on a GaAs substrate (dislocation free) : as expected, the 2D-QW presented a dramatic decrease (of one order of magnitude) in photoluminescence (PL) intensity when grown on the GaAs-on-Si substrate relative to grown on a GaAs substrate. On the other hand, the quantum efficiency of the InAs QB array remained unaffected by the substrate : its integrated emission was similar to that of the 2D-QW grown on GaAs, but much larger than that of the 2D-QW grown on GaAs-on-Si (Gérard et al 1996a).

The aims of the TEM structural study are :
(a) to compare the island characteristics on the GaAs-on-Si and standard GaAs substrates
(b) to observe what happens when the threading dislocations cross the island plane.

2. EXPERIMENTS

The growth was carried out by Molecular Beam Epitaxy at 520°C. Samples A and B were grown simultaneously on the GaAs-on-Si substrate and on the GaAs substrate, respectively. The QB array was obtained by deposition of 2 monolayers of InAs at a low growth rate (within 25 seconds), resulting in the formation of InAs islands. In order to perform a PL study, which is reported elsewhere (Gérard et al 1996a, Lacombe et al 1997), the deposited heterostructure contained also a 2D quantum well and GaAlAs barriers, above the QBs, but we will focus here on the QB array.

TEM was performed at 200 kV on a Philips CM20 microscope. As the islands were buried under a thick structure (0.3 μm), a particular procedure was necessary to prepare plan-view samples : using a well-established etching rate, reactive ion etching was performed to remove exactly 0.25 μm of the wafer surface. Then the back side was classically thinned by mechanical polishing and ion milling, allowing the observation of the island plane. [1$\overline{1}$0] cross-sectional specimens were also thinned by mechanical polishing and ion milling. To determine the island density we used plan-view and also cross-sectional specimens with the (113) weak beam imaging technique (WBIT) previously proposed by Yao et al (1991). Some other applications of this method are given elsewhere (Ponchet and Lacombe 1997).

3. RESULTS

Plan-view observations are shown in fig.1. Cross-sectional observations using the (113)WBIT are presented elsewhere in detail (Lacombe et al 1997). Island characteristics (far from the dislocations) are reported in Table I. Whatever the method, the density in sample A is found 30% higher than in sample B. However, there exists a small discrepancy (20%) between the values obtained by plan-view or by cross-section. Its origin is not well known, but the diffraction vector being different could have different sensitivity to strained islands.

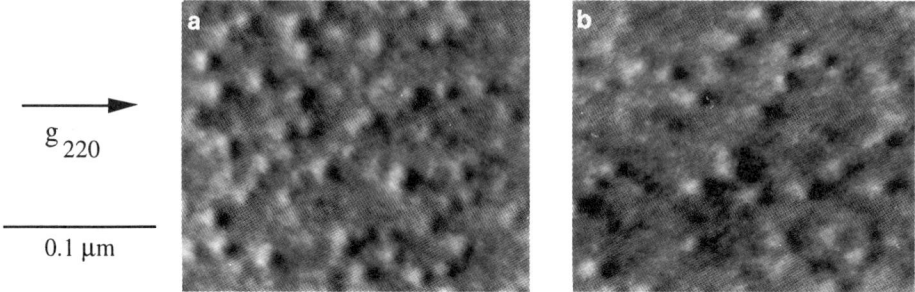

g$_{220}$

0.1 μm

Figure 1 : Islands observed in plan-view (dark field, g=220)
(a) GaAs-on-Si substrate, far from threading dislocations (b) standard GaAs substrate

sample	substrate	island width (nm) (a)	island height (nm) (a)	island density (10^{10} cm^{-2}) (b)	(c)
A	GaAs-on-Si	9 +/- 4	2	14	12
B	GaAs	11 +/- 3	2-3	11	8

Table 1 : Island characteristics **(a)** from cross-sectional observations (g=002) **(b)** from cross-sectional observations using the (113)WBIT **(c)** from plan-view observations (g=220)

The crossing of the island plane by the threading dislocations has resulted in several different behaviors identified by cross-sectional observations (Lacombe et al 1997) :
(a) no change in the dislocation direction
(b) the dislocation straightens up towards the [001] direction
(c) the dislocation is bent in the island plane.

Plan-views (fig.2) showed that an important number of threading dislocations are bent in the island plane, resulting in the formation of an irregular and asymetrical network along the <110> directions. The length per surface unit of the segments bent in the interface was estimated of $\ell \approx 1.2$ μm per μm^2 (the length along one of the <110> directions being two times higher than along the other). Finally, as shown in fig.3 and in cross-sectional observations (Lacombe et al 1997), the presence of threading dislocations has not altered the island growth mode close to the dislocations, even when they are bent in the interface.

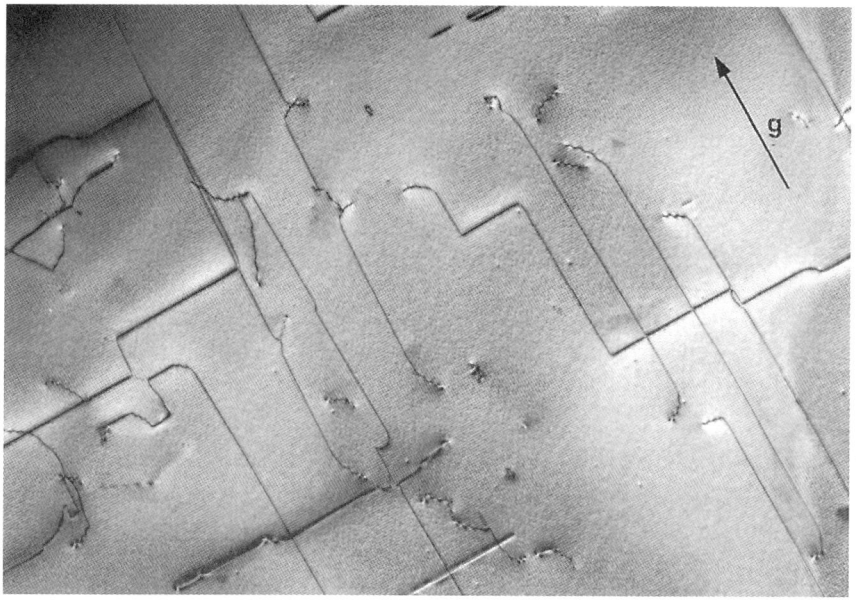

1 μm

Figure 2 : Plan-view of the island plane, showing threading dislocations originating from the Si-GaAs interface (dark field, g=220). The granular aspect of background is due to islands.

Figure 3 : Detail of the plan-view (dark field, g=220) showing a dislocation bent in the island plane. The island growth mode is not disturbed by the dislocation.

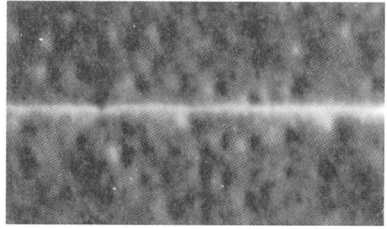

0.1 μm   g 220

4. DISCUSSION

TEM and PL experiments have shown that the use of a GaAs-on-Si substrate has resulting in the formation of a greater number of smaller islands than on standard GaAs (under same growth conditions). The formation of smaller islands could be due to an higher lattice mismatch (Tersoff and Tromp 1993). The origin of an higher lattice-mismatch in our case is that a thick GaAs buffer layer grown on Si is never completly relaxed and exhibits a small compressive residual stress (Landa et al 1989). Therefore, the actual in-plane lattice mismatch $\Delta a/a_{//}$ between InAs and the GaA buffer layer should be corrected by the residual in-plane strain in the GaAs buffer layer : $\Delta a/a_{//} = \Delta a/a_{InAs/GaAs} - \varepsilon_{//\,residual}$, with $\Delta a/a_{InAs/GaAs} = 7\%$ and $\varepsilon_{//\,residual}$ estimated as -0.1 to -0.2% (Lacombe et al 1997, Landa et al 1989).

A second important feature is that a certain number of threading dislocations originating from the substrate are bent in the island plane. It can be inferred that this mechanism occurs under the influence of the island stress. Indeed, the bending of dislocations originating from the substrate is one way to introduce misfit dislocations at the interfaces of mismatched heterostructures. This mechanism is efficient when the thickness of the strained layer exceeds the critical thickness of plastic relaxation (Matthews et al 1970, Matthews et al 1975). In our case however, the bending occurs for an average thickness of the strained layer (InAs) lower than the critical thickness. The usual model of plastic relaxation in the 2D growth mode is not sufficient, and a more realistic model should take into account the elastic relaxation resulting from the 3D growth mode, including a wide inhomogeneity of stress within the islands (the island edges being overstrained).

Finally, we considered that the optical properties of the islands which stand at a distance from a dislocation equal or lower than their width ($w \approx 0.01$ µm) can be detrimentally affected by this dislocation. Using the length of the segments bent in the interface ℓ, it is found that the surface occupied by these islands is $2w\ell \approx 0.02$ µm^2 per µm^2 which means that about 2% of islands are concerned. This explains why the optical properties of the InAs QB array are not significantly affected by the threading dislocations.

ACKNOWLEDGEMENT : We are grateful to L. Monin and to the L2M-CNRS for the reactive ion etching preparation of our plan-view samples.

REFERENCES

Gérard J M, Cabrol O and Sermage B 1996a Appl. Phys. Lett. 68, 3123.

Gérard J M, Marzin J Y, Zimmermann G, Ponchet A, Cabrol O, Barrier D, Jusserand B and Sernage B 1996b Solid-State Electronics 40, 807.

Kirstaedter N, Ledentsov N N, Grundmann M, Bimberg D, Ustinov V M, Rumivov S S, Maximov M V, Kop'ev P S, Alferov Z I, Richter U, Werner P, Gösele U and Heydenrich J 1994 Elect. Lett. 30, 1416.

Lacombe D, Ponchet A, Gérard J M and Cabrol O 1997 Appl. Phys. Lett. 70 in press.

Landa G, Carles R, Fontaine C, Bedel E and Munoz-Yagüe A 1989 J. Appl. Phys. 66, 196.

Leonard D, Pond K and Petroff P M 1994 Phys. Rev. B 50, 11687.

Matthews J W, Jackson D C and Chambers A 1975 Thin Solid Films 26, 129.

Matthews J W, Malder S and Light T B 1970 J. Appl.Phys. 41, 3800.

Moison J M, Houzay F, Barthe F, Leprince L, André E and Vatel O 1994 Appl. Phys. Lett. 64, 196.

Ponchet A and Lacombe D 1997 Inst. Phys. Conf. Ser. (this volume).

Tersoff J and Tromp R M 1993 Phys. Rev. Lett. 70, 2782.

Yao J Y, Andersson T G and Dunlop G L 1991 Philosophical Magazine A 64, 173.

Inst. Phys. Conf. Ser. No 157
Paper presented at Microsc. Semicond. Mater. Conf., Oxford, 7–10 April 1997
© *1997 IOP Publishing Ltd*

Microstructure study of GaAs quantum wire superlattice

H Matsuhata, X-L Wang and M Ogura

Electrotechnical Laboratory, 1-1-4 Umezono, Tsukuba, Ibaraki, 305 Japan.

ABSTRACT: Crescent-shaped GaAs quantum wire superlattices were fabricated on a V-grooved GaAs(001) substrate using a flow rate modulation epitaxy technique in metalorganic chemical vapor phase deposition under various growth conditions. The microstructures were investigated using electron microscopy. The growth temperature and barrier thickness between quantum wires were found to be important factors in fabricating quantum wires with excellent uniformity in size and shape. The optical properties of the fabricated quantum wires were also investigated. The relationship between the microstructures of the quantum wires and observed optical properties is discussed.

1. INTRODUCTION

Low dimensional structures such as quantum wires (QWRs) and quantum dots have been expected to show various distinct quantum effects (see for example: Waschke *et al.*, 1993 and Citrin, 1993) due to the discrete nature of the density of states. In order to utilize such low dimensional quantum structures in practical electron devices, superlattice structures with excellent size and shape uniformity are primarily required. Structures on the order of several tens of nanometers or less are necessary to obtain subbands separated by an energy greater than k_BT at room temperature. The tolerance of nonuniformity in such superlattice structures is considered to be at the nanometer level (for example: Zarem *et al.*, 1989). Thus, fabrication technologies for controlling the structures with an accuracy at the atomic layer level have been attracting attention.

A very promising technique for producing such QWRs is to grow a GaAs epitaxial thin film on a V-grooved AlGaAs layer. At the bottom of the grooves crescent-shaped GaAs QWRs can be formed by a self organization mechanism (Kapon *et al.*, 1989). To fabricate such structures by metalorganic chemical vapor deposition (MOCVD), a lower growth temperature is preferable because of lower residual impurity concentration. Thus, to grow GaAs QWRs at lower temperature, the flow rate modulation epitaxy (FME) technique, in which Ga atoms can migrate with high-mobility on the growing surface at relatively low growing temperature, was adopted and developed (Wang *et al.*, 1995).

In this paper we report on the fabrication of crescent-shaped GaAs quantum wire superlattices on a V-grooved (001) GaAs substrate using the FME technique and on microstructure observations of the resulting QWRs by electron microscopy. Appropriate growth conditions for uniform size and shape of GaAs QWR fabrication are investigated. The relationship between the measured optical properties and observed microstructures of GaAs QWR superlattices with uniform size and shape is discussed.

2. EXPERIMENTAL PROCEDURE

The epitaxial growth was carried out using a horizontal MOCVD system at 76 Torr using the FME technique (for details see Wang *et al.*, 1995) at various temperatures from 1003K to 873K. (001) GaAs substrates with periodically arrayed V-grooves formed by two {111}A planes along the [110] direction produced by photolithography with wet chemical etching were used. Then vertically stacked superlattice structures of GaAs QWRs were grown between AlGaAs barrier layers. The microstructures were observed using a high-resolution SEM Hitachi S900. Also cross-sectional TEM observations were performed with a JEOL 4000FX and 4000EX operated at 200kV. Optical properties of the GaAs QWRs were investigated by photo-luminescence spectroscope Atago PL-S/HR640.

3. RESULTS AND DISCUSSION

3.1 Size of GaAs QWRs

Figure 1 shows cross-sectional views of epitaxially grown GaAs/AlGaAs multilayers observed by a SEM at regions between two V-grooves. Figures 1a, b and c are the results of growth at 873K, 903K and 1073K, respectively. In Fig. 1a , the width of the (111)A side wall planes of grooves became narrower, and the (001) plane became larger, as the multilayer growth proceeded. Figure 1b shows that the width of the (111)A and (001) planes are almost constant except at the beginning of the growth. At higher temperature, as shown in Fig. 1c, the side-wall planes of grooves became wider as the growth proceeded. These images show that the diffusion of Al and Ga atoms on the growing (111)A surfaces towards the (001) surface depends on the temperature during the growth of the AlGaAs layers. Thus, at high-temperature the growth of the AlGaAs layer on the (001) plane is encouraged more than on the (111)A plane.

During growth of the thin GaAs layers, Ga atoms on the surface of the growing (111)A plane migrate to the (001) plane at the bottom of the grooves or up to the (001) plane located between two grooves. Thus, the size of the GaAs QWRs which grow at bottoms of grooves is considered to depend on the width of the (111)A side wall plane of the grooves.

Therefore, at 873K the QWRs formed at the bottom of V-grooves became smaller as growth proceeded, and at 1073K the formed QWRs became larger as the growth proceeded. At 903K, the size of the QWRs is almost invariable. Thus to choose an appropriate growth temperature is important for the fabrication of the QWRs superlattice with excellent size uniformity.

3.2 Shape of GaAs QWRs

Figure 2 shows GaAs QWRs with 4.5nm thickness. The AlGaAs barrier layers are 4.5 and 9 nm for Fig. 2a and 2b, respectively. These structures were fabricated at 903K to give excellent size uniformity. In Fig. 2a, the upper, crescent-shaped QWRs have larger radii of curvature than do the lower QWRs, indicating that the radius of curvature increased gradually as growth proceeded.

A high-resolution image of the lowest part of the specimen of Fig. 2a is shown in Fig. 3. The heterointerface around GaAs QWRs is shown clearly. For the first QWR, the upper and lower heterointerfaces are curved and crystallographic planes are not distinct. However, at the heterointerfaces of upper QWRs the (001) plane and {113} planes gradually became dominant.

During the growth of GaAs QWRs the curvature of the growing surface at the bottom of the V-groove becomes larger. However, during growth of the AlGaAs barrier layers, recovery to small curvature takes place because of the lower mobility of AlGaAs on the growing surface. Thus, if the AlGaAs barrier layer is not thick enough, the recovery of curvature is incomplete, as seen in Fig. 2a. With a sufficiently thick AlGaAs barrier layer, QWRs with an invariable curvature can be formed. The curvature of crescent shaped QWRs is almost invariable with 9nm AlGaAs barrier layers as seen in Fig. 2b. Thus, an appropriate barrier layer thickness between QWRs is important for achieving shape uniformity of QWRs.

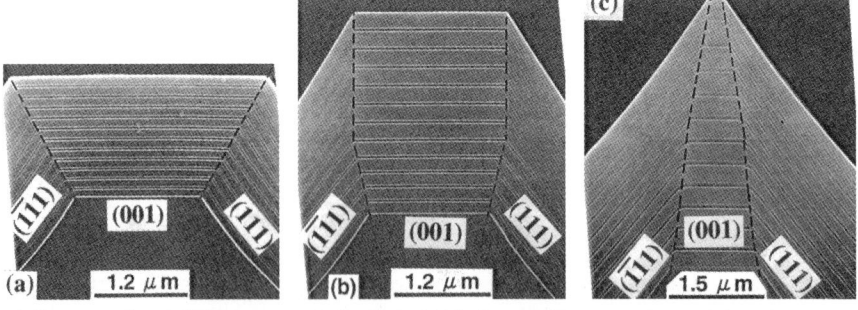

Fig. 1 Cross-sectional SEM images of the GaAs/AlGaAs multilayers at the positions between V-grooves. a), b) and c) are grown at 873K, 903K and 1073K, respectively.

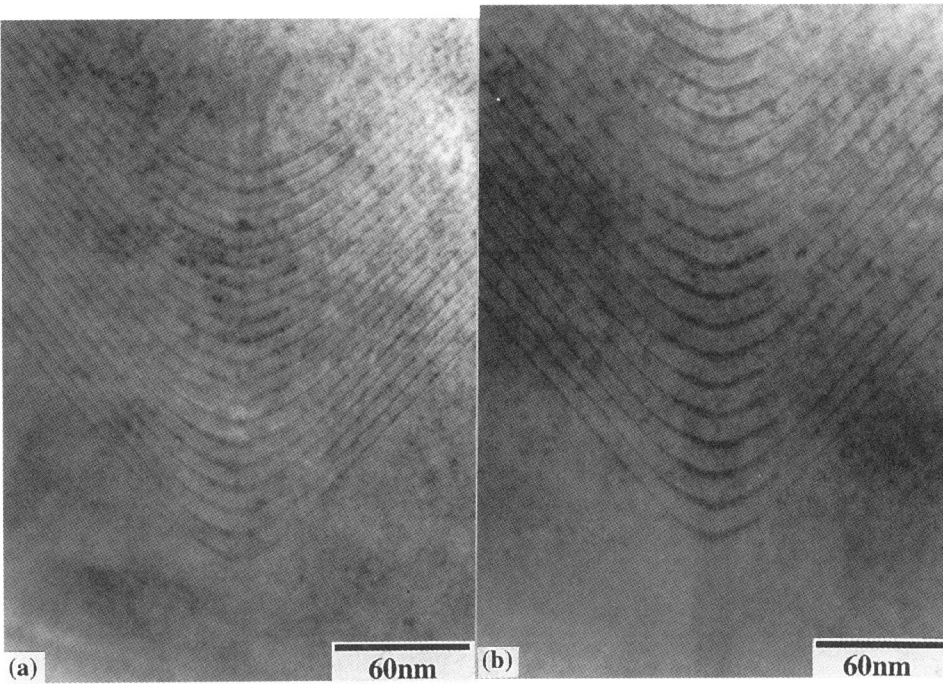

Fig. 2 GaAs QWRs superlattice structure. a) GaAs 4.5nm QWRs/AlGaAs 4.5nm barrier, b) AlGaAs barrier thickness is 9nm.

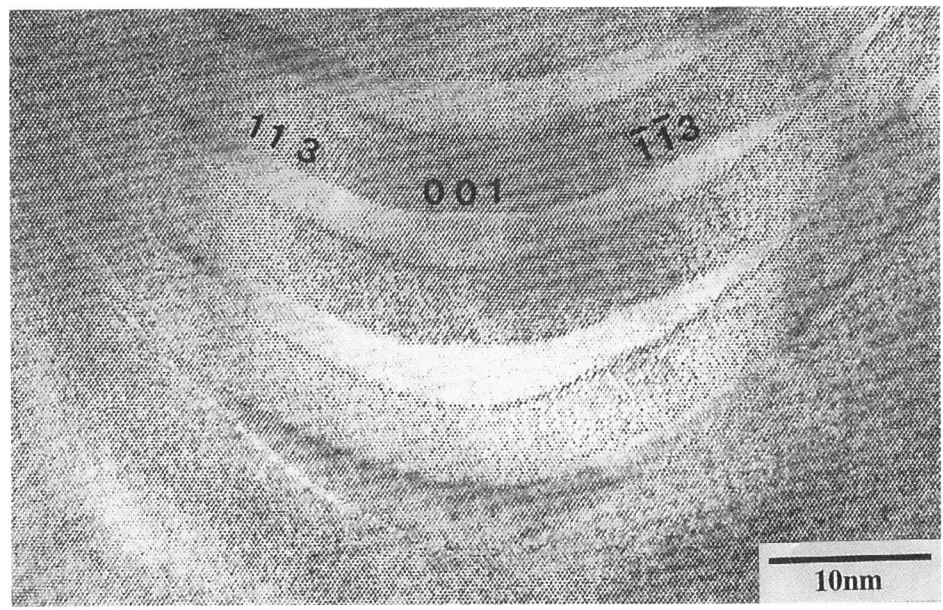

Fig. 3 High-resolution image of the lower part of the specimen of Fig. 2a.

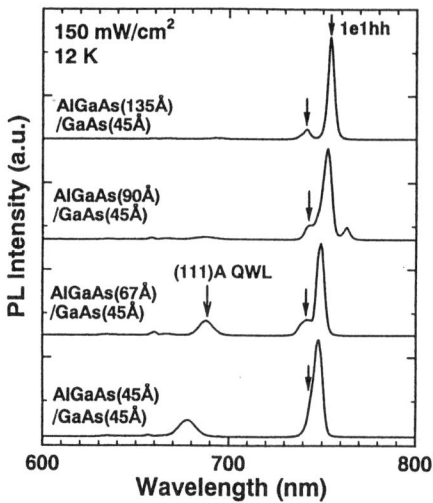

Fig.4 Photoluminescence spectra from various QWR superlattice structures.

3.3 Optical Properties

Figure 4 shows the measured photoluminescence spectra from various QWR superlattice structures. Sharp peaks observed around 755nm are considered to be due to the transition between the two energy levels newly created by a QWR. Side peaks appear around this strong peak in the QWRs superlattice specimens with thicker barrier layers. Because the QWR superlattice structures with thicker AlGaAs layers are more uniformly shaped, as we observed in the TEM image, these side peaks are not considered to be due to irregularity in size and shape of QWRs. The spectrum from the GaAs(4.5nm)/AlGaAs (4.5nm) specimen, which is nonunifrom in shape according to the TEM image, shows a broader peak without side peaks. Thus, we consider that the side peaks observed in Fig. 4 may be due to the QWRs superlattice structure with highly uniform size and shape. More detailed research on the causes of the side peaks due to the superlattice effects is now under way.

4. CONCLUSIONS

It was confirmed that growth temperature influences size uniformity of QWRs and that the shape uniformity of QWR depends on AlGaAs barrier layer thickness. The relationship between optical properties of QWR superlattices and their microstructure as investigated by electron microscopy was discussed. In the case of quantum wells, various techniques were utilized to investigate microstructure. However, in the lower dimensional quantum structure, the electron microscopy is the most useful technique for structure analysis at the moment.

ACKNOWLEDGMENT

Prof. Tomokiyo and Mr. Manabe are gratefully acknowledged for the opportunity to use the high-resolution electron microscope at Kyushu University.

REFERENCES

Citrin D S 1994 Phys. Rev. B49 1943.
Kapon E, Hwang D M and Bhat R 1989 Phys. Rev. Lett. 63 430.
Wang X-L, Matsuhata H and Ogura M 1995 Appl. Phys. Lett. 66 1506.
Waschke C, Roskos H G, Schwedler R, Leo K, Kurz H and Köhler 1993 Phys. Rev. Lett. 70 3319.
Zarem H, Vahala K and Yariv A 1989 IEEE J. Quantum Electro. 25 705.

Inst. Phys. Conf. Ser. No 157
Paper presented at Microsc. Semicond. Mater. Conf., Oxford, 7–10 April 1997
© 1997 IOP Publishing Ltd

Electron microscopy characterization of low-dimensional semiconductor structures grown on V-grooved substrates

A Gustafsson*, G Biasiol, B Dwir, F Reinhardt and E Kapon

Institute of Micro and Optoelectronics, Swiss Federal Institute of Technology (EPFL)
CH-1015 Lausanne, Switzerland.
*Present address: Division of Solid State Physics, Lund University, Box 118, S-221 00 Lund, Sweden. E-mail: anders.gustafsson@ftf.lth.se

ABSTRACT: We present an overview of various low-dimensional structures that can be produced using the approach of growth on V-grooved substrates in the GaAs/AlGaAs and InGaAs/AlGaAs material systems. We discuss both conventional quantum wire and vertical quantum well structures, as well as novel superlattice structures. Structural information derived from transmission electron microscopy and cathodoluminescence spectra of these structures are presented.

1. INTRODUCTION

Low-dimensional semiconductor structures, especially quantum wires (QWRs), have many properties that make them interesting from a fundamental point of view, as well as for potential applications in electronics and optoelectronics (Kapon et al 1992). To form the basis for useful optoelectronic devices, the QWRs must be of high quality in terms of luminescence intensity and emission linewidth, and should exhibit sufficiently large energy spacing of the one dimensional (1D) subbands (Kapon et al 1992). One of the most promising ways to fabricate such QWRs is by organometallic chemical vapour deposition (OMCVD) on V-grooved substrates (Kapon et al 1989). The V-grooved substrates can be fabricated by photolithography, laser holography or electron-beam (e-beam) lithography, as we will show below.

The success of QWR growth on V-grooved substrates relies on the differences in growth rates on the various facets which define the V-groove. In the first stages of GaAs growth on a self-limiting, sharp AlGaAs V-groove, the growth rate at the bottom of the groove is higher than in the surroundings sidewalls. The resulting expansion of the facets at the bottom of the groove, together with the thickening in the growth direction, leads to the formation of a GaAs QWR, connected to thinner quantum wells (QWs) on the side walls. A fundamental aspect of this approach is that the profile of the AlGaAs is defined only by the growth conditions and not by the initial shape of the V-groove. This self-limiting growth front is reached typically after less than 100 nm, depending on the profile of the initial groove. This implies that with proper control of the growth conditions, the reproducibility of the structures is very high. It also means that it is possible to grow a stack of identical QWRs, provided that the barriers are thick enough to let the self-limiting profile recover between successive QWRs.

The growth of a tertiary compound on a substrate containing several facets can lead to differences in the composition across the groove (Hersee et al 1986, Vermeire et al 1992). This is due to different incorporation rates of the growth species on different crystallographic planes, or at the intersection between two facets. The most prominent effect is Ga segregation in AlGaAs at the bottom of the groove during growth. This results in a sheet of Ga-rich material extended in the growth direction and along the groove. This difference in composition in a sheet typically narrower than ~20 nm forms a *vertical* QW (VQW) (Walther et al 1992).

For the growth of the structures, we have used low pressure (20 mbar) OMCVD. The kinetically limited nature of low pressure growth is essential, as it promotes better defined interfaces, as compared with conventional atmospheric pressure growth (Gustafsson et al 1995a). This allows the formation of well defined nm-sized facets at the bottom of the groove. Growth was performed in a standard horizontal reactor, with a rotating susceptor plate. The sources used were trimethylaluminium, trimethylgallium, trimethylindium and arsine, using hydrogen as carrier gas. The growth temperature range used was 550 to 765°C. All the structures presented here were nominally undoped, which results in low background doping of carbon (p-type) in the range $1-5*10^{16}$ cm^{-3}. The V-groove arrays were prepared by optical lithography for the 3 μm-pitch gratings and laser holography for the sub-micron pitch gratings. Variable pitch gratings were also fabricated, using e-beam lithography. After the mask patterning, the substrates were wet etched to produce the V-shaped grooves. All substrates used were (100) GaAs, with the grooves oriented along the [01$\overline{1}$] direction.

The transmission electron microscopy (TEM) was carried out along the [01$\overline{1}$] zone axis, i.e. along the direction of the grooves. The images presented here are either multi-beam bright field (BF) images, with the sample nominally aligned (g=(000)) using the central (000) beam, two (200) beams and four (111) beams, or dark field (DF) images, with the sample tilted (g=(200)), using one (200) beam. The tilt around the [100] growth direction was about 1-2°, kept low in order to minimize the effect of tilting the grooves away from beam direction and the smearing of the features that can result from the projection through the tilted sample (Gustafsson et al 1996). GaAs appears dark in all the images. Starting from GaAs, the appearance gets lighter with Al content and darker with In content. The cathodoluminescence (CL) studies were carried out in an SEM operated at 5-10 keV, using a cold stage, operated either with liquid helium or with liquid nitrogen, in the temperature range of 5-300 K.

2. VERTICAL QUANTUM WELLS

The VQW is interesting both for understanding the segregation on the different crystalline facets and for possible optoelectronics applications. It has been demonstrated that it is possible to fabricate lasers using the VQW as the active layer, yielding a device that exhibits a polarization of the light that is rotated by 90°, compared to conventional QW lasers (Kapon et al 1995a). Since the QWs are oriented perpendicularly to the surface with a high filling factor, they should be useful in top illuminated infrared optical detectors (Berger et al 1995). VQWs are also important for QWR applications, since the effect of pre-capture of carriers

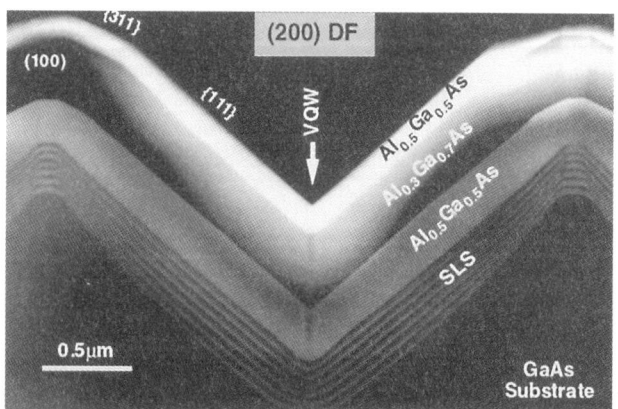

Figure 1. (200) DF image showing a full period of a VQW structure. From the bottom: GaAs substrate, long period SL, high-Al lower cladding, low-Al active VQW region and finally high-Al top cladding.

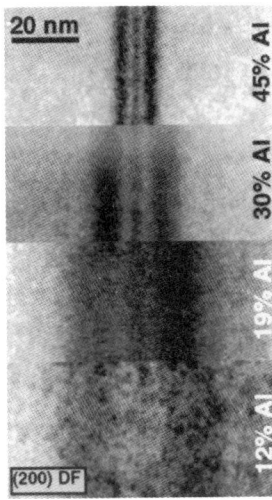

Figure 2. (200) DF images of VQWs with different nominal compositions $Al_xGa_{1-x}As$, ranging from X=0.12 to X=0.45. For all compositions, the VQW shows the same structure, with three branches.

into the VQW has been demonstrated to be the largest contribution to the carrier transfer into the QWR (Gustafsson et al 1995b).

Figure 1 is a TEM image of a 3 µm-groove on which a multilayer $Al_xGa_{1-x}As$ VQW structure was grown. It consists of a recombination layer surrounded by cladding layers with a higher Al content. A superlattice (SL) was grown directly on the etched substrate for obtaining a reproducible surface profile. The SL has a large period (25 nm AlGaAs+20 nm GaAs), so that it does not interfere with the emission from the VQW itself. This image also illustrates the various growth planes involved. The ridges between the grooves are (100) planes, connected to the sidewalls of the grooves by {311}A planes. The growth on the sides of the groove is generally not defined by any low-index plane; the angle between them varies between 70 and 110°, corresponding to {111}A and {211}A planes, depending on growth temperature and Al content. We will designate these planes as quasi-{111} planes. In this image, the VQW appears as a dark stripe that stems from the bottom of the groove and continues up to the surface. A closer inspection of the area around the ridges reveals variations in the grey level around the (100)-{311} and the {311}-quasi-{111} facets, related to variations in the Ga content in these regions.

At higher magnification and better resolution, the self-limiting profile of the VQW can be seen to consist of three branches, with one narrow central branch, surrounded by two wider side-branches. Figure 2 shows DF images of four VQWs of different Al content. As the Al content increases, the separation and width of the two outer branches decrease, whereas the central branch only decreases slightly in width. At a growth temperature of 700°C, the separation between the outer branches ranges from 28.7±0.2 nm for X=0.12 to 4.6±0.2 nm for X=0.75. This is much smaller than the 20-30 nm generally obtained for growth at higher pressures (Pan et al 1995, Gustafsson et al 1996). The branch separation is also influenced by the growth temperature (Biasiol et al 1996). For a given composition, e.g. X=0.42, the separation increases from 4.4±0.1 nm at 600°C to 12.1±0.3 nm at T=765°C. Furthermore, we have performed DF and high resolution imaging from the top, which exhibit the same three branches, indicating a high degree of homogeneity of the VQW structure.

The well defined structure with three separate branches is only found for growth at low pressures. We have observed this in samples grown at 20 and 40 mbar, where growth is limited by surface kinetics. In samples grown at 150 mbar or higher pressures, i.e. above the transition between kinetically and mass transport-limited regimes (Stringfellow 1989), the faceting is not as well defined and the VQW generally appears as a homogeneous layer (Gustafsson et al 1995c). In some cases, there is a hint of a Ga-rich central region, but no indication of three separate branches. The origin of the three branches at low

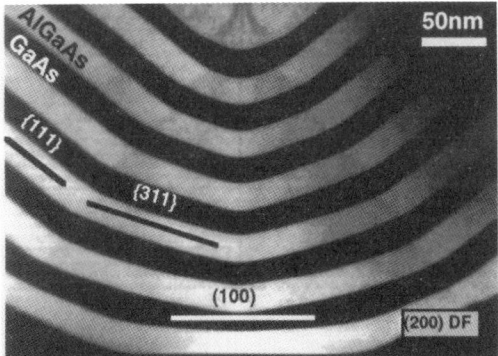

Figure 3. (200) DF image of the bottom of the groove revealing the origin of the three branches of the VQW.

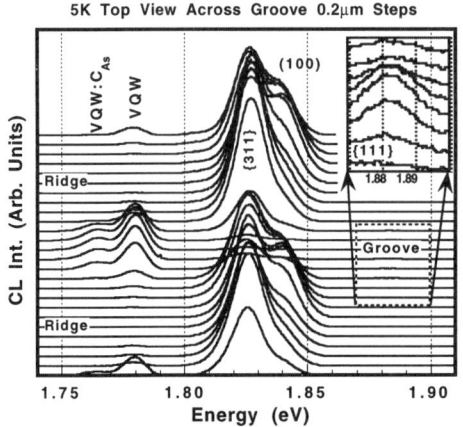

5K Top View Across Groove 0.2μm Steps

Figure 4. Series of CL spectra recorded from a line scan across a groove in top view. From this line scan the various peaks can be labeled.

growth pressures can be studied in the transition region between different materials, where the self-limiting profile has not yet been established. Figure 3 shows the lower part of a groove, which had a flat bottom before growth. In the region of the large period SL, unusually extended {311} facets develop before the self-limiting profile is established. The origin of the three branches becomes clear: the central branch originates in the short (100) facet, and the side branches originate in the {311} facets. The latter is asymmetrically placed towards the quasi-{111} facets.

With the many facets and boundaries between facets, this type of samples has a high number of possible emission lines. To establish the origin of the different emission peaks, we have performed a series of CL measurements, both in top view and side view. We have also varied the temperature from 5-200 K. Figure 4 shows a series of line scan spectra obtained in top view, over one period of the structure, with a step of 0.2 μm. From the scans, we can identify all the peaks: VQW at 1.779 eV, with a carbon acceptor related peak at 1.764 eV; a peak related to the {311} facet on the ridges at 1.827 eV; a peak at 1.840 eV from the (100) ridges and finally a peak at 1.882 eV, from the quasi-{111} sides. It is worth mentioning that there are drastic variations in the intensity distribution with sample temperature, reflecting carrier transfer between the different regions (Martinet et al 1997).

3. (In)GaAs QUANTUM WIRES BETWEEN AlGaAs BARRIERS

Most of the work in growth on V-grooved substrates has been focused so far on GaAs QWRs in AlGaAs. The quest has mainly been in the direction of increasing the spacing of the 1D subbands. A subband spacing of up to 45 meV has been reported (Kapon et al 1996). Because of the geometry, the confinement energy is smaller for the more extended direction across the groove than in the growth direction. To increase the subband spacing, the width of the crescent needs to be reduced, still without reducing the thickness in the centre. Figure 5 shows how this can be achieved. As the temperature is reduced from 765 to 600°C, the width of the QWR is significantly decreased. However, at lower temperatures, a higher surface roughness results in larger QWR non-uniformities, reflected in larger photoluminescence (PL) line widths and larger Stokes shifts in PL excitation (PLE) spectra (Vouilloz et al 1997). A useful criterion for the QWR quality is the ratio

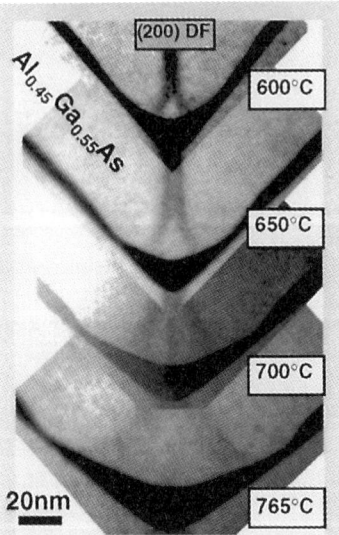

Figure 5. A series of (200) DF images of GaAs QWRs grown at various temperatures. The barrier material is Al$_x$Ga$_{1-x}$As with X≈0.45. As the temperature is lowered, the width of the QWR is reduced significantly.

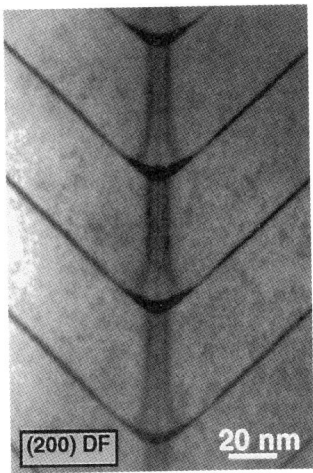

Figure 6. A (200) DF image of a stack of four identical QWRs in a stack of ten.

of the subband spacing and the emission linewidth or Stokes shift, which we found to be maximum for growth temperatures around 650°C.

An important factor for the luminescence intensity is the density of QWRs. A high density can be achieved by using a short-period (typically ≤ 0.5μm) grating and/or stacking several QWRs in the growth direction. In both cases, it is crucial that the size fluctuations of the QWRs are minimal (below ~5%) (Kapon et al 1992), from groove to groove as well as within the stack. Figure 6 exemplifies the stacking of identical GaAs/AlGaAs QWRs. The VQW gives very important information of the minimum possible spacing of the QWRs. As the GaAs QWR perturbs the self-limiting profile of the AlGaAs barrier, the VQW just above the QWR is much wider. When the profile recovers, the width is reduced until the self-limiting width is achieved. The point where the VQW is recovered indicates the minimum spacing required between identical QWRs. In figure 6, this minimum spacing of about 15 nm is well exceeded to ensure that the QWRs are identical.

One of the signatures of a QWR is the 1D subbands, discussed above. There are several ways to observe these subbands, one being PLE (Vouilloz et al 1997). Another way is to increase the excitation density to a point where the emission from the lowest lying subband is saturated and emission from the second subband can be seen. As the excitation density is increased further, even higher lying subbands can be filled, resulting in emission from more and more subbands. Figure 7 shows a series of CL spectra from InGaAs/AlGaAs QWRs, where the probe current was increased from 100 pA to 25 nA by doubling it from step to step. At the lowest current, only emission from the lowest subband is seen. At the highest probe current, at least four subbands can be seen, with a spacing of about 25 meV.

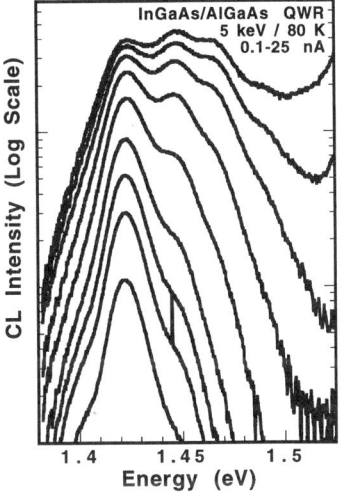

Figure 7. A series of CL spectra of an InGaAs QWR in AlGaAs barriers. For each spectrum, the probe current is increased by a factor of two, from 0.1 to 25 nA.

4. QUANTUM WIRES GROWN ON E-BEAM DEFINED GRATINGS

The optical lithography techniques are good for covering large areas of identical structures, but with little possibility of modulating the structure (e.g., along the grooves). An alternative way is to use e-beam lithography to write the grating. Before this tool can be used, a fundamental question must be answered: Does the spacing between the V-grooves play any role in determining the QWR size and shape? To answer this question, we produced a sample with grating of different pitch: 0.5 1.0, 1.5 and 2.0 μm. Here we display low magnification images of the 1.0 and 2.0 μm pitch structures (figure 8). The structure of the QWRs appears to be identical in the TEM analysis of the four areas, and thus the period plays no significant role in the formation of the QWRs. This indicates that it is possible to make modulated QWR structures with

378

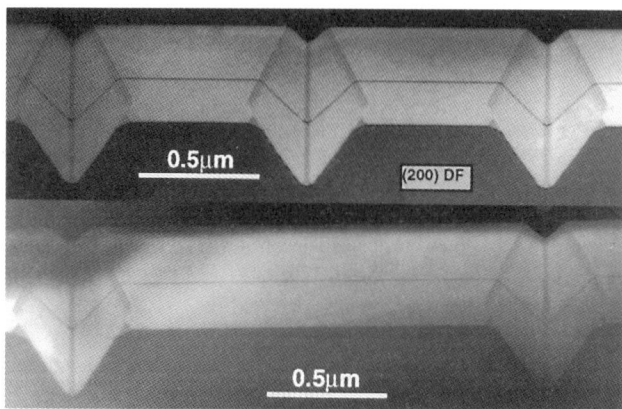

Figure 8. DF images of a single QWR structure grown on e-beam written gratings with a 1.0 and 2.0 μm pitch, respectively.

arbitrary periodicities (at least in the range mentioned above), without significant change in the QWR intrinsic properties.

One important feature of the e-beam written QWRs is that, for a given width of the groove the spacing can be made large enough for the spatial origin of the various luminescence peaks to be identified. Figure 9 shows three CL spot mode spectra of the structure grown on the 2.0 μm grating. From these, it is possible to identify four different regions of the structure: the QWR at 1.585 eV; the {311} facet at 1.615 eV and the (100) and the quasi-{111} facets at 1.685 eV. The high energy shoulder of the latter is consistent with the energy difference of a thickness variation of one monolayer in the QW. In a series of spot mode spectra, the ratio between these peaks varies, which is also consistent with thickness fluctuations. From this series, we can also conclude that the quasi-{111} related peak varies much more in its spectral position than the other peaks, caused by thickness variations due to an uneven upper interface of the quasi-{111} QW. The only difference between the 0.5 and 2.0 μm gratings is that the {311} QW is shifted towards higher energy in the 0.5 μm grating. This may suggest that the (100) area in the 2.0 μm grating supplies more material for the {311} QW than in the 0.5 μm pitch structures.

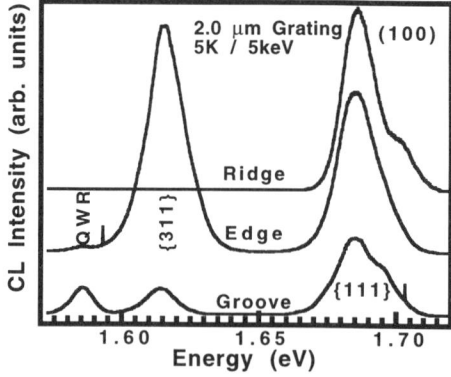

Figure 9. Spot mode CL spectra of a 2.0 μm period, e-beam written grating. The emission from four different areas can be identified.

5. InGaAs QUANTUM WIRES BETWEEN GaAs BARRIERS

The (In)GaAs/AlGaAs QWR structures are more difficult to incorporate in device structures requiring regrowth. The problem is the oxidation of the AlGaAs region between growth steps (Walther et al 1993). It would therefore be invaluable to find growth conditions, where GaAs can be used as a barrier material, with, e.g., an InGaAs QWR. The growth of InGaAs QWRs in AlGaAs looks structurally like the conventional GaAs QWR, but the challenge is to find growth conditions, where an InGaAs QWR is formed *in a GaAs matrix*, characterized by a self-limiting growth profile. Here we demonstrate for the first time to our knowledge that it is possible to grow, not only one, but a stack of identical InGaAs QWRs in GaAs, where the GaAs recovers its self-limiting profile after the perturbation of the InGaAs layer. The InGaAs QWR in GaAs is crescent shaped, similar to the conventional GaAs QWR in AlGaAs. A stack of 4 InGaAs QWRs in GaAs is shown in figure 10. At higher magnification, figure 11, a dark stripe can be

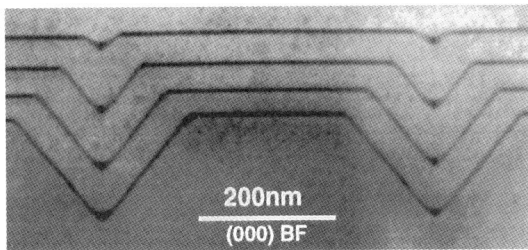

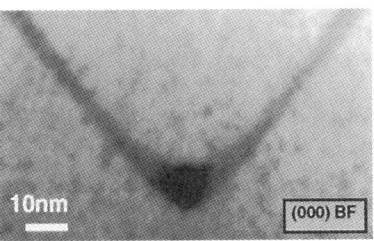

Figure 10. BF image of a stack of 4 InGaAs QWRs in GaAs. This image displays that it is possible to produce a stack of identical and self-limiting InGaAs QWRs in GaAs barriers.

Figure 11. BF image of one of the InGaAs QWRs in the stack of figure 10. The QWR is crescent shaped. The dark stripe in the centre is the equivalent of the VQW.

observed in the centre of the QWR, indicating a higher In content. This stripe is the equivalent of the AlGaAs VQW structure discussed above. CL data suggest the formation of a QWR with subbands separated by about 15 meV for a similar structure.

6. SUPERLATTICES OF QUANTUM WIRES

A different type of structure that can be grown on V-grooved substrates is a stack of QWRs, spaced so close that they are electronically coupled. This would lead to a QWR superlattice (QWR-SL) structure. Similar structures were demonstrated as serpentine superlattices (Miller et al 1992), but the V-groove approach offers much better control of the structures. The electronic coupling of the individual QWRs requires a spacing of a few nm, compared to the 10-30 nm of AlGaAs needed to recover the self-limiting profile (see figure 6). The QWR-SL will reach a profile somewhere between the self-limiting profiles of GaAs and AlGaAs. Figure 12 shows a DF image of such a QWR-SL. There is a transitional region of 2-3 periods before the QWR-SL reaches a stable state. This is a metastable state, as a minor shift in the period will alter the profile of the QWR-SL. This can be observed as a slight wiggling of the VQW in the AlGaAs in the QWR-SL. The properties of these QWR-SL structures will be discussed elsewhere.

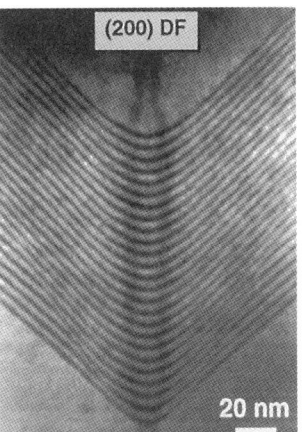

Figure 12. (200) DF image of a 1.8/3.9 nm period GaAs/$Al_{0.45}Ga_{0.55}As$ QWR-SL.

7. SUMMARY

Transmission electron microscopy and low temperature cathodoluminescence were employed to characterize the structure and luminescence properties of nanostructures grown by low pressure OMCVD on V-grooved substrates.

We have demonstrated that self-limiting growth of AlGaAs, GaAs and GaAs/AlGaAs SLs on these V-grooved substrates can be used for fabricating GaAs/AlGaAs, InGaAs/AlGaAs and InGaAs/GaAs QWRs, as well as QWR-SLs of high optical quality and with high spatial density.

Acknowledgments: This work was partly financed by Fonds National Suisse de la Recherche Scientifique. The TEM work was s carried out on a Phillips EM430-ST and the CL work was carried out on a Cambridge S360, both instruments belonging to CIME-EPFL.

REFERENCES

Berger V, Vermeire G, Demeester P and Weisbuch C 1995 Appl. Phys. Lett. $\underline{66}$ 218
Biasiol G, Reinhardt F, Gustafsson A, Martinet E and Kapon E 1996 Appl. Phys. Lett. $\underline{69}$ 2710
Gustafsson A, Reinhardt F, Biasiol G and Kapon E 1995a Appl. Phys. Lett. $\underline{67}$ 3673
Gustafsson A, Samuelson L, Hessman D, Malm J-O, Vermeire G and Demeester P 1995b J. Vac. Sci. Technol. B $\underline{13}$ 308
Gustafsson A, Dwir B, Reinhardt F, Biasiol G, Bonard J M and Kapon E 1995c Proc. Microscopy of Semiconducting Materials, Inst. Phys. Conf. Ser. $\underline{146}$ eds A G Cullis and A E Staton-Bevan (Bristol: Institute of Physics) pp 375-8
Gustafsson A, Malm J-O, Carlsson A and Vermeire G 1996 Semicond. Sci. Technol. $\underline{11}$ 1745
Hersee S D, Barbier E and Blondeau R 1986 J. Cryst. Growth $\underline{77}$ 310
Kapon E, Hwang D M and Bhat R 1989 Phys. Rev. Lett. $\underline{63}$ 430
Kapon E, Walther M, Christen J, Grundmann M, Caneau C, Hwang D M, Kolas E, Bhat R, Song G H and Bimberg D 1992 Superlattice Microstruct. $\underline{12}$ 491
Kapon E, Dwir B, Pier H, Gustafsson A and Bonard J M 1995a Technical Digest Vol. 16 / Quantum Electronics and Laser Science Conference 235
Kapon E, Biasiol G, Hwang D M and Colas E 1995b Microelectronics Journal $\underline{26}$ 881
Kapon E, Biasiol G, Hwang D M, Walther M and Kolas E 1996 Solid State Electron. $\underline{40}$ 815
Martinet E, Gustafsson A, Biasiol G, Reinhardt F, Kapon E and Leifer 1997 K Phys. Rev. B (submitted)
Miller M S, Weman H, Pryor C E, Krishnamurthy M, Petroff P M, Kroemer H and Merz J L 1992 Phys. Rev. Lett. $\underline{68}$ 3464
Pan W, Yaguchi H, Onabe K, Ito R and Shiraki Y 1995 Appl. Phys. Lett. $\underline{67}$ 959
Reinhardt F, Dwir B and Kapon E 1996 Appl. Phys. Lett. $\underline{68}$ 3168
Stringfellow G B 1989 Organometallic Vapor-Phase Epitaxy: Theory and Practice (Boston: Academic Press) Chap. 6
Vermeire G, Yu Z Q, Vermaerke F, Buydens L, Van Daele P and Demeester P 1992 J. Cryst. Growth $\underline{123}$ 513
Vouilloz F, Oberli D Y, Dupertuis M A, Gustafsson A, Reinhardt F and Kapon E 1997 Phys. Rev. Lett. $\underline{78}$ 1580
Walther M, Kapon E, Christen J, Hwang D M and Bhat R 1992 Appl. Phys. Lett. $\underline{60}$ 521
Walther M, Kapon E, Caneau C, Hwang D M and Schiavone L M 1992 Appl. Phys. Lett. $\underline{62}$ 2170

Inst. Phys. Conf. Ser. No 157
Paper presented at Microsc. Semicond. Mater. Conf., Oxford, 7–10 April 1997
© 1997 IOP Publishing Ltd

Quantitative Analysis of Al$_{1-x}$Ga$_x$As heterostructures using EELS

K Leifer [*,1] and P A Buffat[*]

[*] Centre Interdepartemental de Microscopie Electronique, EPFL, Lausanne, Switzerland
[1] Present address: Institut de Micro- et Optoelectronique, EPFL, CH-1015 Lausanne

ABSTRACT: The optimization of the acquisition and the evaluation conditions for the EELS spectra of Al$_x$Ga$_{1-x}$As heterostructures permits one to find the absolute concentration x with a precision of ±0.02 and to detect changes in concentrations of ±0.01-0.02 for x=0-0.5. The central element of this optimization is a Bloch wave calculation of electron channeling patterns in order to find a specimen orientation where electron channeling is minimized without introducing an unacceptable loss of spatial resolution due to sample tilt. Vertical quantum wells in MOCVD grown structures could be quantified and their detailed structure could be found.

1. INTRODUCTION

When the chemical composition profile of a specimen exhibits two dimensional variations on both the micrometer and nanometer scale the only techniques that may give precise concentrations on the two scales are energy dispersive X-ray spectroscopy (EDS) and electron energy loss spectroscopy (EELS) in a transmission electron microscope. When we acquire parallel EELS (PEELS) spectra with acquisition times of 1.5s and a collection angle of 50mrad, the standard deviation of the Ga concentration over several tens of spectra is about 4 times lower than in the case of EDS spectra that were acquired for 15s each. Therefore when several hundred spectra have to be acquired, it is more favourable to use PEELS for the acquisition of the spectra in order to reduce the influence of specimen drift, contamination and electron beam damage on the analysis.

2. DEPENDENCE OF ELECTRON CHANNELLING ON THE TILT ANGLE

In order to obtain the highest spatial resolution of the chemical analysis, the interfaces have to be parallel to the electron beam. In crystalline materials this orientation corresponds often to the excitation of a systematic row. This orientation leads to electron channelling effects which result in an uneven distribution of the electron intensity on the different atomic sites as shown by Long (1989) and Spence et al. (1988) for the case of III/V semiconductors. The number of ionizations of a core level of an element, and thus the EELS edge intensity is proportional to the electron intensity on its atomic sites. Close to an orientation of high symmetry or the excitation of a systematic row, the channelling effect varies rapidly with the orientation as well as with the specimen thickness. Although one might apply correction factors that take into account these experimental parameters for a few experimental points (Long 1989), both thickness and orientation of the specimen may significantly change along a linescan of several 100nm and make thus the determination of correction factors difficult if not impossible.

First we carry out a Bloch wave simulation of the electron propagation in the crystal for different electron beam tilts and specimen thicknesses in order to find a specimen orientation where channelling effects are reduced to an acceptable level and are stable for a wide range of orientation and thickness. Then we show that under such conditions for uncontrolled changes of orientation and thickness the correction factors become negligible.

The Bloch wave calculations were carried out within the EMS software package (Stadelmann

382

1987) including the Laue zones 0 to 2 for an incident plane electron wave (E=200keV) on a [110] zone axis of GaAs and AlAs crystals. The inclusion of more than 120 Bloch waves changes the computed intensities by less than ±2%. At every coordinate (x,y) the averaged electron intensity $\bar{I}(x,y)$, i.e. the electron intensity averaged over the intensities at all specimen thicknesses, is calculated. The probability of the ejection of a core electron depends furthermore on the impact parameter b_{rms} which is known as the delocalization of the interaction (Pennycook (1982) and Kohl and Rose (1985)). The formula given by Pennycook (1982) results in impact parameters which are in good quantitative agreement with the formula given by Kohl and Rose (1985) for the AlGaAs system:

$$b_{rms} = \frac{h \times v \times \theta_0}{\Delta E} \times \left[(\theta_0^2 + \theta_E^2) \times \ln\left(1 + \frac{\theta_0^2}{\theta_E^2}\right) \right]^{-1/2}$$, where h is the Planck constant, v the speed of the incident

electron, θ_0 the maximum scattering angle which is accepted by the objective aperture, ΔE the energy loss and $\theta_E = \Delta E/E$. The values for the L edges are $b_{rms}^{Ga\,L} = 0.51$ Å, $b_{rms}^{As\,L} = 0.44$Å and $b_{rms}^{Al\,K} = 0.37$Å which are comparable to the interatomic distances. Therefore the delocalization has to be included when we calculate the changes of the L edge intensities due to channeling.

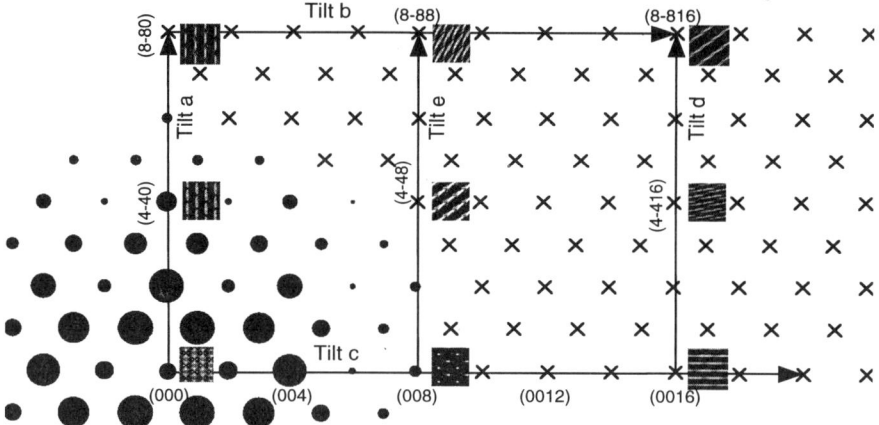

Figure 1: [110] diffraction pattern and map of the specimen tilts used to calculate the channelling effects. The diameter of the spots is proportional to their kinematic intensity, the crosses indicate the position of weak reflections. The solid lines correspond to the trajectory of the projection of the centre of the Laue circle during the tilt series. For some selected tilts the $\bar{I}_{el}(x,y)$ map is given in the inserts.

We use a Gaussian function to characterize the dependence of the transition probability on the impact parameter: $h^A(x, y) = (1/(\sqrt{2\pi} \times b_{rms}^A))\exp\left[-\left\{(x-x_0^A)^2 + (y-y_0^A)^2\right\}/(2b_{rms}^{A\,2})\right]$, where (x_0, y_0) is the position of the column of the A atoms as previously done by Spence et al. (1988). The number of ionizations of the atom A is proportional to the integral over the thickness-averaged electron intensity $I_{el}(x,y)$ and h(x): $H^A = \iint I_{el}(x, y) \cdot h^A(x, y)dxdy$. We use the ratios $R^{Al/As} = H^{Al}/H^{As}$ and $R^{Ga/As} = H^{Ga}/H^{As}$ to describe the deviation of the intensity of the As-L, Al-K, and Ga-K edges due to channelling. Therefore for R<1, the electron intensity is shifted on the As sites.

The tilt series that are carried out here, are indicated in Fig. 1 by the projection of the centre of the Laue zone (CLZ) on the diffraction pattern of the [110] zone axis. The values which we obtain for the GaAs system are in good agreement with those obtained by Spence (1988) and Roussow and Maslen (1987). Here we only discuss the case of the AlAs crystal and it can be shown that the $R^{Al/As}$ are an upper limit for the intensity ratios of the $Al_xGa_{1-x}As$ system (Leifer 1997).

In tilt series a) the CLZ moves along the $(2\bar{2}0)$ systematic row and therefore the III element planes and the As planes remain parallel to the electron beam. After a strong decrease at small tilt angles (Fig. 2a) $R^{Al/As}$ stabilizes at a value of about 0.5, i.e. the excitation of the As is stronger than the excitation of the Al due to channelling between the As and Al (002) planes. In the tilt b) the (002) planes are then inclined with respect to the electron beam and after a change from R<1 to R>1

(Fig. 2b) due to the inversion of the excitation error the electron intensity stabilizes at values around R≈1 until {111} planes are parallel to the electron beam (i.e. CLZ at (8$\overline{8}$16)).

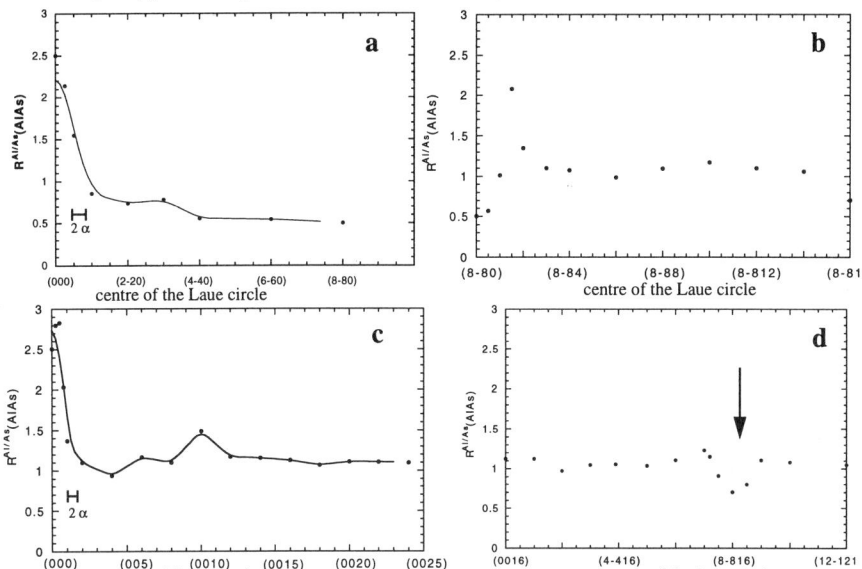

Figure 2: Relative intensities $R^{Al/As}$(AlAs) for a specimen thickness of 60.3nm as a function of tilt of the incident electron beam, denoted here by the CLC for tilt series a, b, c and d. The orientation with the (1$\overline{11}$) planes parallel to the electron beam in d) is indicated by an arrow. The lines in a and c represent the same ratio $R^{Al/As}$(AlAs) integrated over a semiconvergence angle α=2mrad.

The series c) starts again at the zone axis and the (2$\overline{2}$0) planes remain parallel to the electron beam (Fig. 2c). R stabilizes at higher tilt angles at $R^{Al/As}$≈1. This is due to the fact that the overlap of Al and As columns increases with the tilt angle. An additional tilt d) that starts from the CLZ at (0 0 16) to the CLZ at (10 $\overline{10}$ 16) results in a zone of nearly constant R≈1 up to tilt angles corresponding to the CLZ at (6 $\overline{6}$ 16). Therefore when we tilt the specimen such that the CLZ is at (3 $\overline{3}$ 16), the ratio $R^{Al/As}$ will fluctuate only by less than ±0.07 when the illumination angle relative to the specimen orientation varies by about ±1°. The orientation with the CLZ at (3 $\overline{3}$ 16) is convenient for the analysis since the interfaces of the observed structures are parallel to the (2$\overline{2}$0) planes.

When the illumination is changed from parallel (points in Fig. 2) to convergent illumination (solid line in Fig. 2a,c), $R^{Al/As}$ shows small deviations only for orientations close to the zone axis.

The ratio $R^{Al/As}$ stabilizes after a strong decrease close to the zone axis orientation for specimen higher than 5mrad. Therefore in microanalysis one often tries to reduce the effect of electron channelling by increasing the convergence angle to several ten mrad. This may be suitable for a comparison of concentrations from different specimen regions even if the orientation in these regions varies slightly. But since the calculation of the relative change of the edge intensities $R^{Al/As}$ is not straight forward, a priori the integration of $R^{Al/As}$ over the illumination cone does not result in a total $R_{tot}{}^{Al/As}$≈1. Therefore even for high convergence angles correction factors have to be introduced when specimen is close to an orientation with strong electron channeling.

3. ACQUISITION AND EVALUATIONS OF PEELS LINESCANS

The analysis is carried out on a Hitachi HF2000 FEG microscope at 200keV holding the cross sectional specimens at LN_2 temperature. The electron probe with a FWHM of 1.0-1.5nm (as measured with a CCD camera) is scanned across the interfaces and at each dwell point a PEELS spectrum with an acquisition time of 1.5s is acquired with a GATAN PEELS model 666 (collection angle 50mrad).

First we characterized a standard specimen containing various $Al_xGa_{1-x}As$ layers with x=0.05, 0.10,0.20, 0.30, 0.40, 0.50, which also have been quantified using cathodoluminescence and using the

384

thickness fringe method by Bonard and Ganière (1996). We used two methods for the extraction of the intensities: 1) fit of power law model to the pre-edge region and extrapolation of this model to the edge region as described in Egerton (1996) and 2) a Fourier ratio deconvolution of the edges with a spectrum of the low loss region (Egerton 1996) followed by the fit of theoretical Hartree-Slater cross sections to the Ga and As edges.We verified that for specimen thicknesses of more than 60nm $c_{Ga}+c_{Al}=c_{As}$ (Leifer 1997) and thus we only work with the Ga and As edges. This decreases the statistical fluctuations of the signal by a factor 3 since the Al K edge cross section is small compared to the cross sections of Ga and As. Both quantification methods 1 and 2 lead to errors in the absolute concentration which are smaller than $\Delta x=0.01-0.03$ when the specimen thickness is between 60nm-120nm. But the method 1 systematically underestimates the Ga concentration whereas the second method systematically overestimates the c_{Ga}. The mean value of the concentrations of both methods differs by less than $\Delta x=0.02$ from the values found with cathodoluminescence for all of the different layers.

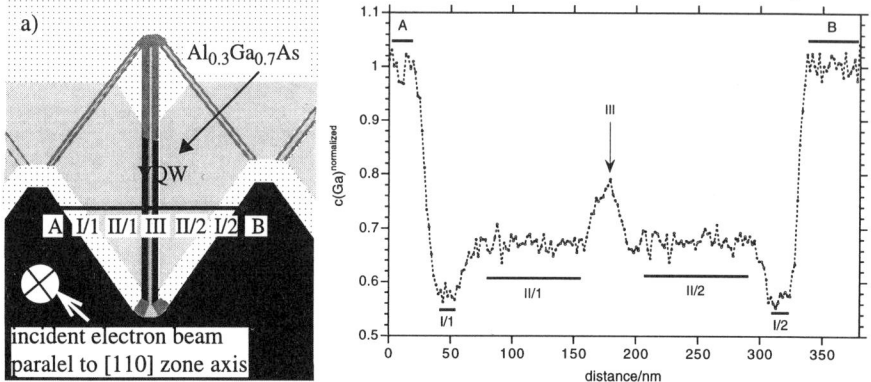

Figure 3: a) Schematic representation of $Al_xGa_{1-x}As$ structures that are grown on linearly patterned GaAs substrates containing a vertical quantum well (VQW). b) PEELS linescan through the structure as indicated in a) (here: c(Ga)=1-x). The edge intensities were extracted using method 1.

Second, we characterized $Al_xGa_{1-x}As$ structures which were grown by MOCVD on patterned GaAs substrates (Fig. 3a). The concentrations that were evaluated from the PEELS linescans not only show the regions of constant concentration A, B, I and II but prove the existence of a vertical Ga rich layer III in the centre of the groove. From PEELS linescans through these vertical quantum wells the detailed structure of the concentration profile in the region of these vertical wells can be found and together with photoluminescence measurements the confinement of carriers in these layers could be proved (Martinet et al.). When the same data as presented in Fig. 3b are evaluated using method 2, the standard deviation in regions II/1 and II/2 is reduced by about a factor of 2 to $\sigma=0.006$. When we take 3σ as a criterion for a significant deviation of the concentration, relative deviations of less than $\Delta x=0.02$ can be found with this technique. The standard error of a series of spectra i.e. in region II is typically 0.02, so that the relative error of the mean value is smaller than ±0.01.

We would like to thank P. Stadelmann for many interesting discussions and J.M. Bonard and E. Kapon and for having offered their excellent specimens for the EELS analysis.

REFERENCES

Bonard J M, Ganière J D 1996, Ultramicroscopy **62**, 249

Egerton R F 1996, EELS in the Electron Microscope (New York: Plenum Press)

Kohl H and Rose R 1985, Adv Electr. Electr. Physics **65**, 173

Leifer K 1997, Thesis, (Lausanne, EPFL)

Long N J, 1989, Inst. Phys. Conf. Series **100**, 59

Long N J, Norman G, Petford-Long A K, Butler B R, Cureton C G, Booker G R and Trush E J 1991, Inst. Phys. Conf. Series **117**, 69

Martinet E , Gustafsson A, Biasol G, Reinhardt F, Kapon E, Phys. Rev. Rap. Com., submitted

Pennycook S J 1982, Contemp. Phys. **23**, 371

Roussow C J, Maslen V W 1987, Ultramicroscopy **21**, 277

Spence J C H, Kuwabara M and Kim Y 1988, Ultramicroscopy **26**, 103

Stadelmann P 1987, Ultramicroscopy **58**, 35

Inst. Phys. Conf. Ser. No 157

Paper presented at Microsc. Semicond. Mater. Conf., Oxford, 7–10 April 1997

© 1997 IOP Publishing Ltd

Orientation dependent growth of TmAs wires in GaAs grown by MBE

A C Wright, M R Bennett[1] and K E Singer[1]

Advanced Materials Research Laboratory, NEWI Plas Coch, Mold Road, Wrexham
LL11 2AW
[1] Dept of EE+E and Centre for Electronic Materials, UMIST PO Box 88, Manchester,
M60 1QD

ABSTRACT: Thulium doped gallium arsenide has been grown by Molecular Beam Epitaxy
(MBE) on both (100), (311)B and (111)A oriented GaAs substrates. Above the solubility limit
for Tm in GaAs, free precipitation of TmAs occurs and this can be either as dots, wires or
bifurcated trees, depending on growth conditions. We show that the substrate orientation has a
marked effect on the form of the precipitates and provides a means to control the
microstructure of TmAs in its wire-like form.

1. INTRODUCTION

Rare-earth doping of compound semiconductors such as GaAs has been of interest because of
the possibilities for well defined luminescence through the internal transitions of the 4f shell of the
rare-earth ion. Erbium doped GaAs has been investigated because the 1.54 micron luminescence
matches the transmission window of silica optical fibres. However, previous work by Poole et al.
(1992) has shown that the solubility of Er in GaAs is rather low at around 10^{17} cm^{-3} and that above
this limit, precipitation of free erbium arsenide occurs. These precipitates were shown by
transmission electron microscopy to be very small in size (~2.5nm in diameter) with a narrow size
distribution and that the precipitation mechanism offers a way to directly synthesise so-called self-
organised quantum dot structures. Unfortunately, ErAs is not a semiconductor, being semi-metallic
in nature. Subsequent work by Singer at al. (1994) showed that further control over the shape of
the ErAs precipitates could be effected by altering the arsenic to gallium ratio (V/III) of the growth
flux. Both tree-like structures and wires could be grown when the V/III flux ratio was low. These
were shown to align only approximately in the growth direction. All of these structures were
grown on nominally on-axis (100) oriented substrates (±0.5 degrees). Clearly, one variable worth
investigating is the orientation of the substrate. In this study we have used thulium as the rare earth
dopant rather than erbium but find the same precipitation phenomena and it is with thulium that we
have investigated the effect of substrate orientation on the form of the precipitates.

2. EXPERIMENTAL

All layers were grown in a Riber 2300 MBE system using elemental Ga and As (As$_4$) sources
for the matrix elements. A small quantity of high purity (> 99.95 %) Tm sublimate was obtained
from Ames Laboratory. Ga and As$_4$ fluxes, and hence growth rates, were measured using RHEED
oscillations. Thulium concentrations were calibrated using SIMS and RBS analysis. Sample
temperature was measured using an Ircon Modline Series V infrared pyrometer. Layers were
grown on nominally on-axis semi-insulating (100) , n$^+$ (311)B and n$^+$ (111)A GaAs substrates
after degreasing, etching in H$_2$SO$_4$/H$_2$O$_2$ and thermal annealing to remove the native oxide.
Specimens for TEM were prepared in both cross-sectional and plan-view form using ion-thinning.
TEM was performed on a Philips EM430T at 300kV. Precipitate visibility is best observed in
centred dark field using the (002) reflection as described previously by Poole at al. (1992)
although in plan view bright field imaging was able to observe precipitates during the orientation
studies albeit at lower contrast.

3. RESULTS

Layers grown on (100)-oriented substrates with Tm concentrations above $5*10^{19}$ cm^{-3} and with V/III ratios of 2:1 or above were observed to contain a fine distribution of spherical precipitates of TmAs as shown in figure 1. As before, diffraction contrast experiments indicated that the structure of these precipitates was of rocksalt type in line with the known structure of bulk TmAs. The absence of misfit dislocations indicates that these precipitates are coherent with the GaAs matrix and the lattice mismatch between TmAs (0.5711 nm) and GaAs is +1.02 % (i.e. the precipitates are in compression). The range of particle sizes was determined from high resolution dark field images to be 1.2nm to 2nm and showed a strong Arrhenius dependence on the substrate temperature during growth, giving rise to an estimated activation energy of ~0.48eV. At temperatures above 620°C, however, the precipitates were shown to be elongated into wires oriented approximately in the growth direction ([100]). Figure 2 shows the size dependence of the precipitates on the growth temperature.

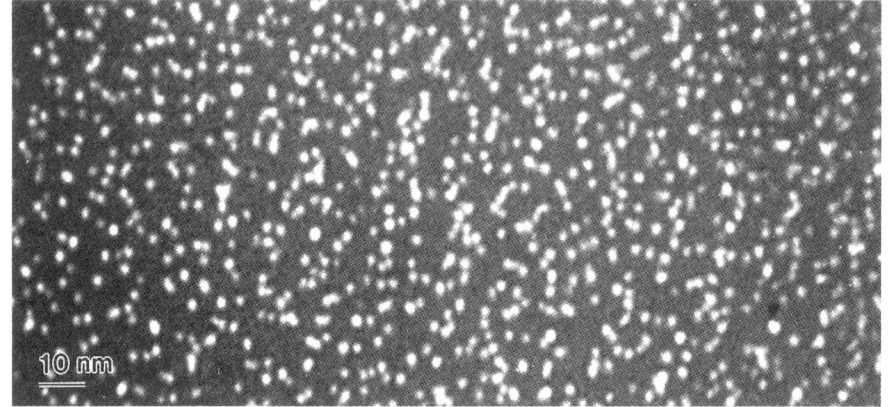

Figure 1. Dark field (002) beam image of spherical TmAs precipitates in GaAs grown onto GaAs(100) substrates under As/Ga flux ratio of 4:1.

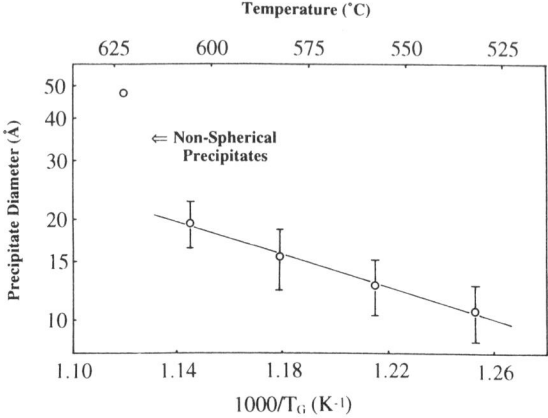

Figure 2. Size of spherical precipitates versus growth temperature under a As/Ga flux ratio of 4:1.

Further experiments with (100)-oriented substrates showed that reducing the As/Ga flux ratio to close to unity (<1.3:1), at growth temperatures which would normally be expected to give rise to spherical precipitates, resulted in the formation of bifurcated precipitate structures ('trees') which appeared to inhabit particular (110) planes. Figure 3(a) and (b) show orthogonal views of these tree-like structures along the [110] and [1$\bar{1}$0] projections respectively.

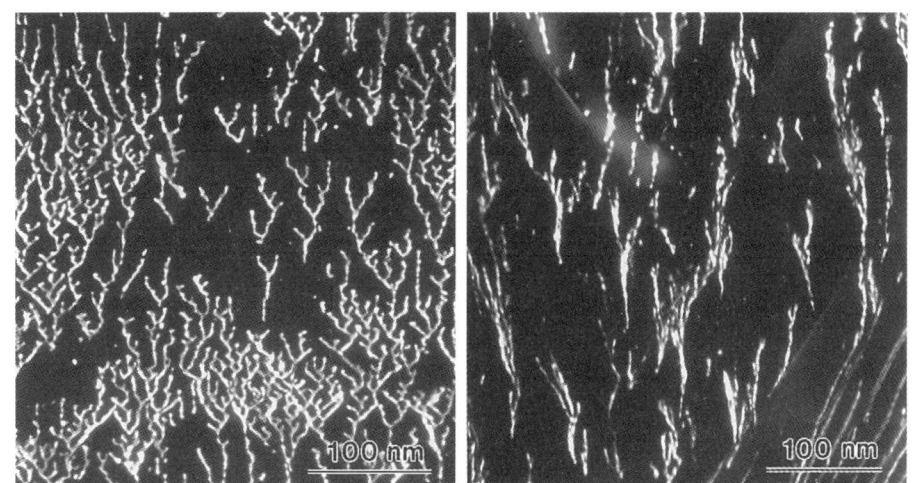

Figure 3. Tree-like structures of TmAs in GaAs as grown onto GaAs(100) as viewed along [110], in (a) and along [1$\bar{1}$0] in (b). Flux ratio As/Ga was close to unity.

When grown on substrates with orientations of (311)B and (111)A under the same conditions as those found to produce the tree-like structures on (100) substrates, the TmAs was seen to form wires or rods in the GaAs matrix which were consistently oriented in specific directions with respect to the growth direction. On (311)B substrates, these rods were highly aligned in one orientation as shown in the cross-sectional image of figure 4(a). No branching of the rods was seen in these images. Plan view TEM images taken in bright field mode, enabled the preferred orientation of these rods to be defined. For growth on these (311)B GaAs substrates, this direction was found to be very close to the [211] direction as shown in figure 4(b).

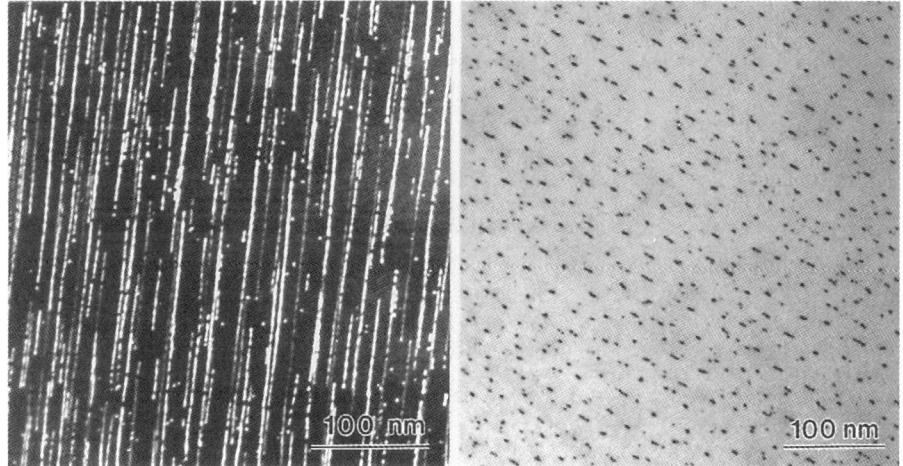

Figure 4. Rod-like TmAs precipitates in GaAs as grown onto (311)B GaAs, cross-section (a) and in plan view, (b) which is taken near the [211] pole showing the rods near end-on

For the (111)A growth, the pin-pointing of the exact direction was hampered by the poor morphology of the GaAs grown under such metal rich conditions for this substrate. The RHEED patterns were also spotty because of the rough morphology. Figure 5 shows a typical cross-section of these layer revealing that the orientation of the rods varies considerably from area to area. This is confirmed by bright field plan view images taken along the surface normal which clearly show that the alignment changes rapidly from one area to another in a random fashion rather like wheat in a field. In this sample, the angle the rods make with the surface normal is typically a maximum of four degrees. The poor morphology of the layers grown on (111)A substrates is also reflected in the large number of stacking fault defects seen in plan view images. It is probable that better quality material would result if (111)B oriented substrates (arsenic face) were available to this study.

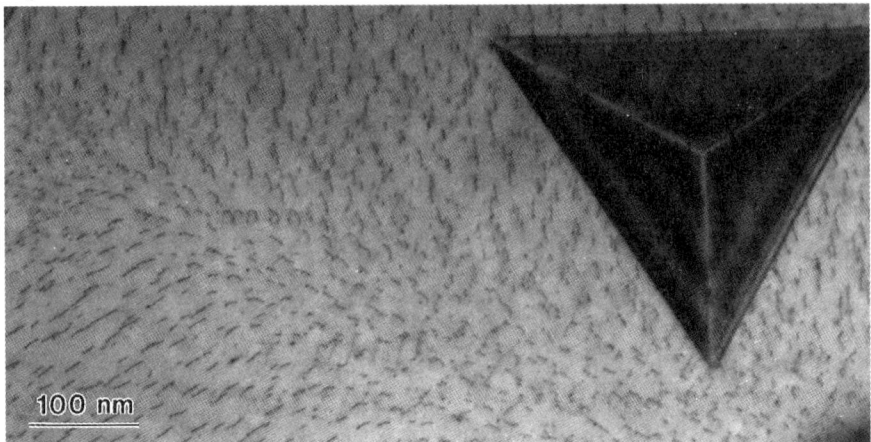

Figure 5. TmAs rod precipitates viewed near end-on in a plan view of growth on (111)A GaAs. View is along [111] pole and shows variation in rod alignment. Triangular features are stacking faults.

4. DISCUSSION

It is clear from these results that the orientation of the substrate has a marked effect on the morphology of the precipitate. Why this is so is less obvious and further work will be required to understand the nature of the surface during growth. A simple model for wire formation in the Er:GaAs system was described by Singer et al. in terms of the increased surface mobility of the rare-earth at low V/III ratios but does not explain why the precipitates are formed as straight rods as seen here. The mechanism which 'locks' the thulium diffusing around on the surface onto an existing TmAs site may involve surface steps which would have to remain in the same place as growth proceeds. The (100) orientation must lack this locking mechanism, possibly because of step flow. It is possible that surface steps on both the (311)B and (111)A orientations are similar as a very rough nominally (111) surface can have local orientations of (311) type as facets at an angle of 29.5 degrees to the surface plane. The possibility for reconstruction of the surface (and steps) and its effect on how the rare-earth is incorporated must also be considered in any model.

5. CONCLUSIONS

The substrate orientation has a marked effect on the form and morphology of self-organising rare-earth arsenides as grown by MBE in a GaAs matrix. While (100) oriented GaAs substrates yield wire-like structures under low V/III ratios, the development of the wire structure is random, unlike that occurring, for example, on (311)B substrates where the formation of straight rod-like wires is observed closely aligned to the [211] pole.

REFERENCES

Poole I, Singer K E, Peaker A R and Wright A C 1992 J. Crystal Growth **121**, 121
Singer K E, Rutter P, Peaker A R and Wright A C 1994 Appl. Phys. Lett. **64**, 707

Inst. Phys. Conf. Ser. No 157
Paper presented at Microsc. Semicond. Mater. Conf., Oxford, 7–10 April 1997
© *1997 IOP Publishing Ltd*

TEM and HREM structural studies of non-lithographically-produced CdS nanowires

J L Hutchison, D Routkevitch*, A Albu-Yaron# , M Moskovitz* and R R Nayak

Department of Materials, University of Oxford, Parks Road, Oxford OX1 3PH U.K.
*Department of Chemistry, University of Toronto, 80 St. George Street, Ontario M5S
1A1, Canada and #ARO, The Volcani Center, P.O. Box 6, 50250 Bet Dagan, Israel

ABSTRACT: CdS nanowires (~ 1 μm length by ~ 10 nm diameter) have been grown non-lithographically by an ac electrodeposition technique directly into highly oriented, very uniform and nearly parallel nanometer-sized pores in an anodic aluminium oxide (AAO) template. Detailed TEM and HREM investigations of the morphology and nanostructure of individual nanowires have been carried out. Here we report preliminary results of the HREM study which reveal the nanostructure of individual nano-wire arrays at close to atomic resolution. Our results constitute direct evidence of the anisotropy and crystallinity of these CdS nano-wire arrays, confirming that the axis of the wires corresponds to the c-axis of remarkably well ordered hexagonal CdS single crystals.

1. INTRODUCTION

There is considerable interest in growing spataially constrained semiconductors by new techniques which do not involve the expense, complexity or ultimate limitation of lithographic methods. Some of the present authors (Routkevitch et al, 1996a, 1996b, 1996c) have devised methods involving the creation of appropriately sized uniform pores in anodic aluminium oxide (AAO) layers. Under appropriate conditions, very regular arrays of pores can be created to provide AAO templates within which semiconductor material may be deposited directly. The morphology, size, structure, atomic composition and band-gap energy of CdS nanowire arrays grown within such pores by ac electrodeposition were previously investigated using SEM, EDX, TEM, elemental microanalysis, reflectance spectroscopy and resonance Raman spectroscopy (RRS). These studies indicated that the CdS nanowire arrays generated by a single step ac electrodeposition in an electrolyte containing Cd^{2+} and S in DMSO, directly into an AAO template, are not randomly oriented, but rather have a highly anisotropic structure of aligned ~ 1 - 3 μm long by ~ 9 - 35 nm diameter nanowires, a Cd:S stoichiometry of $1:1$ and that they appear to form the hexagonal phase of the semiconductor. Strong but indirect evidence based on previous XRD measurements suggests that after annealing, the CdS nano-wires produced in larger pores (i.e. 35 nm diameter) consist of several crystallites with average dimensions ~ 21 nm in the axial direction and ~ 10 nm in the radial direction; i.e. on average the thicker wires consist of a large number of crystallites in the axial direction and rather few in the radial direction. In contrast, nano-wires produced in pores with diameters < 12 nm supposedly consist essentially of a string of single, continuous crystallites along their length. In order to determine the details of the nanowire morphology with a greater degree of confidence and especially to determine the nature of the structural anisotropy, a HREM investigation has now been carried out.

2. EXPERIMENTAL

Specimens from nano-wire arrays were prepared for HREM studies by carefully dissolving away the AAO template. The self-supporting "forests" of nanowires were then suspended in alcohol and mounted on copper grids coated with either carbon or "lacey" carbon. The specimens were examined in a JEOL 4000EX(II) electron microscope, operated at 400 kV. Conventional TEM at moderate magnifications was employed to gain an overview of the morphology of the wires; high resolution imaging was then used on selected individual wires to reveal the extent of local ordering.

3. RESULTS

3.1 Morphology of nanowires

Preliminary examination of the specimens confirmed that the CdS was indeed in the form of long wires, up to several fm in length, with diameters in the 9 - 40 nm range, depending on the exact growth conditions. These dimensions were in accord with those deduced from XRD peak width measurements (Routkevich et al, 1996). A segment of typical nanowire of nominal diameter 10 nm and 1.3 μm in length, is shown in Fig. 1. Note the remarkably uniform width.

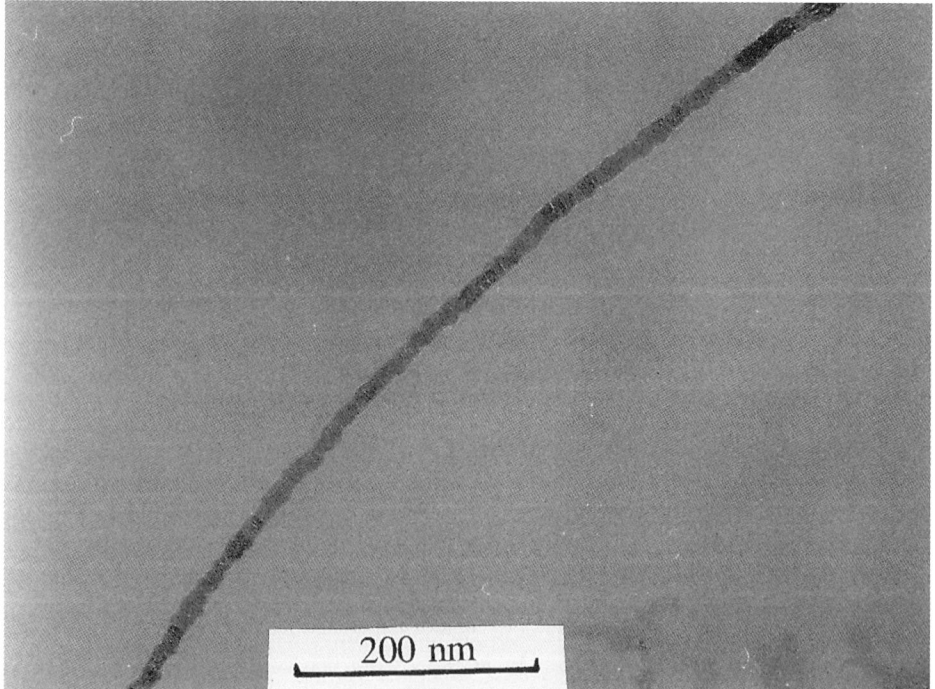

Figure 1. Example of a CdS nanowire of nominal diameter 10 nm.

3.2. Local ordering revealed by high resolution imaging

Individual wires were selected which could be oriented with either [1120] of the hexagonal or <110> of the cubic variant parallel to the optic axis. Images were recorded at Scherzer defocus, corresponding to dark "atom contrast". A typical section of a wire (nominal diameter 10 nm) is shown in Fig. 2. This image reveals several interesting features: firstly, it was particularly evident that the wire axis was normal to the close packed planes, i.e. {111} or (0001). Micrographs recorded sequentially along this wire confirmed that this orientation was maintained along its entire length. Large numbers of stacking faults and twinned segments are present. These are best visualised by viewing the micrograph at a glancing angle along the wire axis. The structure appears to be mainly the hexagonal form, interleaved with domains of cubic structure. This is broadly in agreement with the previous X-ray data.

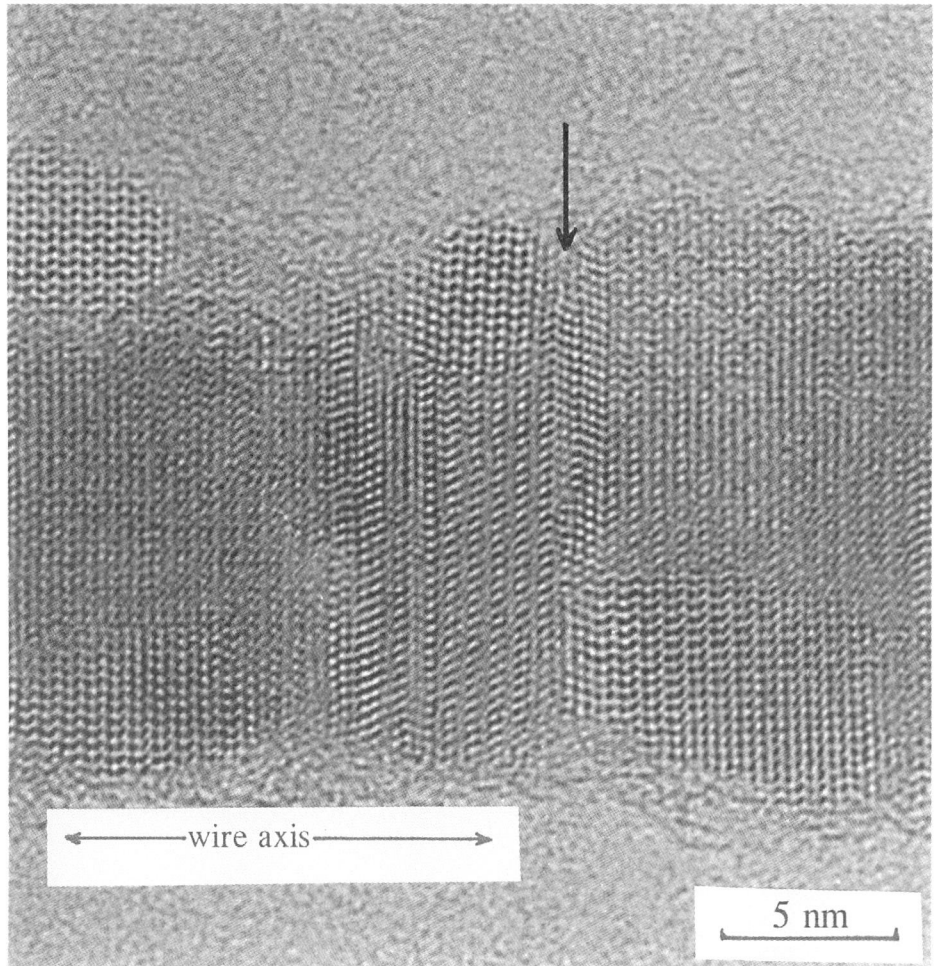

Figure 2. Enlarged section of a CdS nano-wire showing local ordering of {111 planes with <111> parallel to the wire axis.

Another feature was the presence of numerous planar features, parallel to $(1000)_{hex}$, which showed diffuse contrast. We interpret these as planes of rotational disorder in which local registry is lost. A further notable feature was the occasional appearance of edge dislocations, again parallel to the basal plane. An example of such a defect is arrowed in Fig. 2.

5. CONCLUSION

These preliminary HREM studies have confirmed that crystalline nano-wires have been grown in uniform pores in anodic aluminium oxide. Conventional TEM showed them to be remarkably uniform in width. HREM studies combined with electron diffraction revealed that many of the wires were well ordered with the wire axis normal to the close-packed {111} or (0001) planes. There was also evidence for rotational disorder, along with occasional edge dislocations. Further studies to investigate the effects of annealing on the ordering are in progress and will be reported elsewhere.

References

Routkevitch D, Bigioni T, Moskovits M and Xu J M 1996 J. Phys. Chem. **100** 14037
Routkevitch D, Tager A A, Haruyama J, Almawlawi D and Moskovits M 1996 IEEE Trans. on Electron Devices **43** 1646
Routkevitch D, Haslett T L, Ryan L, Bigioni T, Douketis C and Moskovits M 1996 Chem. Phys. **210** (1996) 343

Inst. Phys. Conf. Ser. No 157
Paper presented at Microsc. Semicond. Mater. Conf., Oxford, 7–10 April 1997
© 1997 IOP Publishing Ltd

TEM studies of processed Si device materials

J Vanhellemont, H Bender* and J Van Landuyt**

Wacker Siltronic AG, P.O. Box 1140, D-84479 Burghausen, Germany
*IMEC, Kapeldreef 75, B-3001 Leuven, Belgium
**University of Antwerp, EMAT, Groenenborgerlaan 171, B-2020 Antwerpen, Belgium

ABSTRACT: Recent developments in the field of TEM characterisation of Si device materials are discussed and illustrated by a few case studies of material in different stages of various kinds of processing. Important challenges are the ever decreasing defect densities and device feature sizes. Defect delineation techniques using large area inspection tools yielding accurate coordinates of the defects to be studied have therefore become an essential part of the TEM analysis procedure. The possibility to transfer these defect coordinates without loss of accuracy to tools for local TEM specimen preparation is also a conditio sine qua non for a successful analysis. In-situ TEM remains important as dynamic processes can be observed and analysed under well defined experimental conditions. As case studies illustrating new developments, results are presented on defects in as-grown Cz silicon, on in-situ studies in processed silicon, on problem sites in advanced integrated circuit structures and on assessment of localised strain fields in the nm size scale.

1. INTRODUCTION

During the last two decades transmission electron microscopy has become a well established technique widely spread not only in academic but also in industrial environments. At the same time however commercial single crystal silicon material has reached unprecedented quality with extremely low densities of grown-in crystal defects. Silicon processing induced defect densities have also decreased drastically while the feature sizes have dropped well below the micrometer level. Important challenges for TEM in the micro-electronics industry are thus the study of very low defect density material and the investigation of localised areas in device structures where a material problem is expected or detected by other measurements. Luckily there has been a parallel development of specimen preparation techniques and especially the use of tripod polishing (Anderson 1992) and focused ion beams (Young et al 1990, Pantel et al 1994) has helped to solve some of the above mentioned experimental problems.

Another trend is to use the potentiality of TEM as a microscopic laboratory where important physical processes can be followed in-situ under well controlled experimental conditions (e.g. Mat. Res. Soc. Symp. Proc. Vol. 404 1996 and references therein). Thin foils of the material can be heated, strained, irradiated,... Especially attractive is the possibility to use also microelectronic structures even under operation inside the microscope (Ross et al 1993).

In the present paper case studies are presented illustrating the study of ultra low defect density material, of in-situ stress induced dislocation nucleation and point defect studies using in-situ electron irradiation and heating, and localised defect studies and strain measurements in device structures.

2. ADVANCED SPECIMEN PREPARATION

The quality of a TEM image and thus of the analysis depends to a large extent on the quality of the prepared thin foil. While in the early days chemical thinning, ion milling and more recently also large area thinning using advanced polishing techniques satisfied most of the needs, the increased semiconductor and device quality has led to the need for localised thinning in those spots where other techniques have detected a problem which might be understood by local TEM analysis.

In the case of device structures the actual part of the device which is failing can often be determined quite accurately from the analysis of the electrical device characteristics in correlation with the layout of the actual circuit. In such a way the to be thinned area can relatively easily be found in the FIB and a thin foil can be prepared in which the problem site is incorporated.

For bulk Si materials with low defect densities the situation is somewhat more complicated. Some surface treatment can be used which delineates the defects which can then be observed with visible light scattering tools. With these surface inspection tools the coordinates of the defects on the wafer surface can be determined with an accuracy on the order of 10 μm (Schmolke 1997). Transferring the coordinates to the FIB specimen stage then allows to find the defect with the help of the imaging system and to put additional markers around it using the focused ion beam (Fig. 1a,b). Then the FIB tool is used to deposit a protecting metal layer over the area of interest (Fig. 1c) after which a thin foil can be prepared without damaging the defect area (Fig. 1c).

Fig. 1: FIB preparation of a cross-section TEM sample through a COP : a) FIB image of the defect (45° tilted), b) additional markers are introduced with FIB (45° tilted), c) TEM specimen prepared by FIB : a Pt protective layer is deposited over the region of interest and an additional marker is introduced. Diabolo-shaped thinning is used to allow large angle tilting in TEM, d) XTEM image of the double COP shown in a).

One approach to delineate substrate defects is to use a Cu decoration technique whereby Cu islands are preferentially deposited on weak spots in the gate oxide. A second approach is to use visible light scattering tools (or tools like AFM) to localise so called COP's (Crystal Originated Particles) which are observed on the wafer surface after prolonged cleaning treatments (Gräf 1996). A drawback of both the Cu decoration technique and the etching treatment to delineate defects is that the defect might be altered by the delineation procedure itself. Recently also infrared light scattering has been used to determine coordinates of defects inside the bulk of the material and to transfer these coordinates to the FIB (Kato et al 1997). In that case the risk of altering the defects is much smaller.

3. DEFECTS IN AS-GROWN SILICON

Most of the commercially available silicon materials have been pulled under conditions leading to vacancy-rich material. Although this material has particular advantages in particular with respect to controlling of interstitial oxygen precipitation, the quality of gate oxides grown on them is lower than that of those grown on epitaxial or on FZ (Floating Zone) wafers. The nature of the defects responsible for the reduced gate oxide integrity has remained unclear for many years also due to their low density which corresponds with values on the order of 10^5 to 10^6 cm^{-3} in the silicon substrate. Only recently it was shown that large octahedral voids are present in commercial Cz silicon with a similar density and that these voids are causing the midfield gate oxide breakdowns (Itsumi 1996). These voids should be considered as vacancy precipitates and their nucleation and growth during crystal pulling can be modelled as such.

Fig. 1d shows a cross-section TEM image of the double COP of Fig. 1a. The shape of the COP reflects the octahedral void bounded by {111}-planes which creates a square pit on (001) silicon surfaces, detected as a particle by surface inspection tools based on light scattering. This octahedral shape is further evidenced by the image in Fig. 2 of a buried D-defect which was observed adjacent to one of the COP. By an iterative thinning procedure with intermediate TEM inspection it was possible to thin the specimen further until high resolution images of the buried defect could be obtained. The observations are in agreement with a void nature of the defect although a mottled contrast is observed inside the defect which is typical of an amorphous phase. Most probably this is sputtered material which is redeposited and partly covered with a native oxide. The amorphous layer observed at the top of the cross-section image in Fig. 2a and also in Fig. 1d, is due to the inspection of the wafer surface with the Ga ion beam of the FIB in order to localise the COP with markers. On the internal {111} surfaces of the void a 4-6 nm amorphous layer is observed with a slightly darker contrast which could be a silicon oxide layer formed during the cooling of the crystal during the pulling process by indiffusion of interstitial oxygen, or a native oxide layer formed in the void after TEM specimen preparation. At the {100}-edges of the defect a gradual transition of crystalline silicon to the amorphous layer on the inner surface can be observed which is due to the inclination of the interface with respect to the image plane. A more detailed analysis will be published elsewhere (Bender et al 1997).

4. IN-SITU ANALYSES

4.1 Dislocation generation in patterned structures

Heating of plan view samples of silicon wafers covered with patterned thin films allows to study in situ dislocation nucleation and multiplication in a high voltage microscope (Vanhellemont et al 1995b). Typical examples are shown in Fig. 3 for Si$_3$N$_4$ film patterns along [110] and [100] directions on a (001) silicon surface. 60° dislocations are formed at a scratch close to the film edge and glide easily several 100 μm along the [110] film edge under the influence of the local strain field. Dislocation propagation along [100] edges is much more difficult and triangular half loops are frequently observed often formed by cross-glide mechanisms.

396

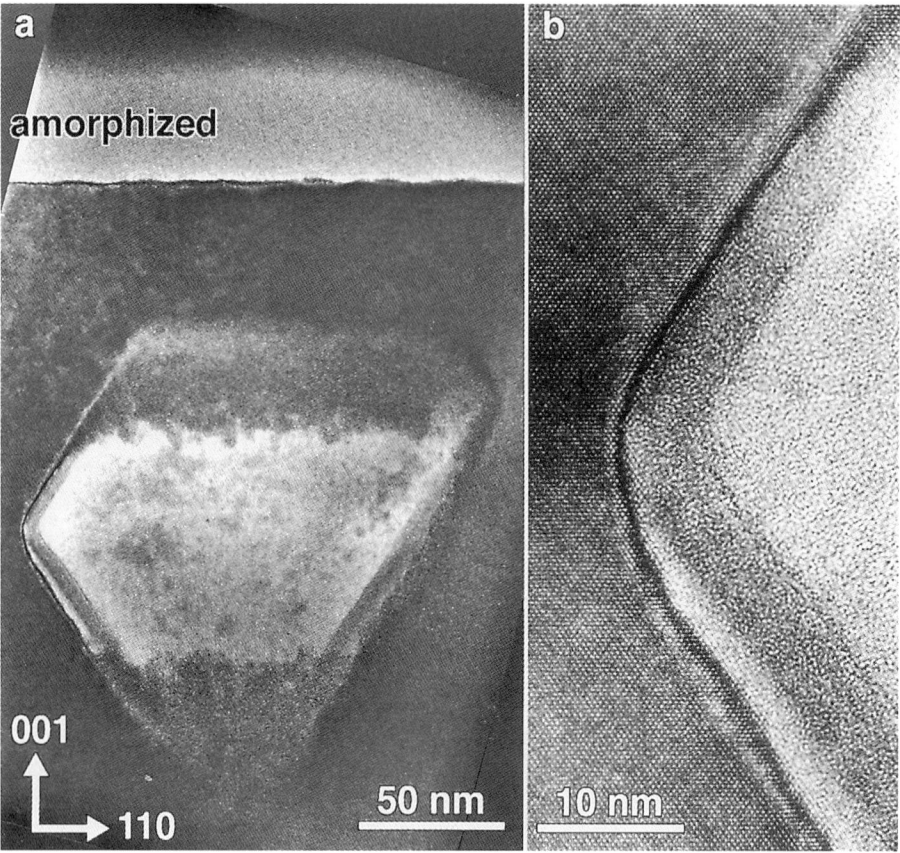

Fig. 2: a) 'Low' magnification image showing a buried D-defect which was found close to a COP. b) HREM image of the left part of the defect revealing the rounded corner formed by two {111}-facets. The 4-6 nm amorphous layer on the inner surface is also clearly visible.

When a scratch is present in between the film edges the formed dislocations loops have the same habit plane and Burgers vector as those created by a punching mechanism at silicon oxide precipitates. In another experiment the film pattern was etched off after dislocation nucleation. Heating these samples in the microscope reveals that single dislocations glide back to the wafer surface and disappear. Such in-situ studies help to understand the basic mechanisms of dislocation nucleation and multiplication in patterned film structures important for device manufacturing.

4.2 Irradiation induced defects in device structures

Understanding of intrinsic point defect reactions is the key to control not only grown-in defects during crystal pulling but also extended defect formation during processing steps like ion implantation, oxidation, ... By in-situ monitoring of the formation of electron irradiation induced extended defect clusters in layered structures, information is obtained on the role of surfaces, interfaces and extrinsic point defects as sinks for intrinsic point defects. Heating experiments allow also to estimate activation energies for the processes observed.

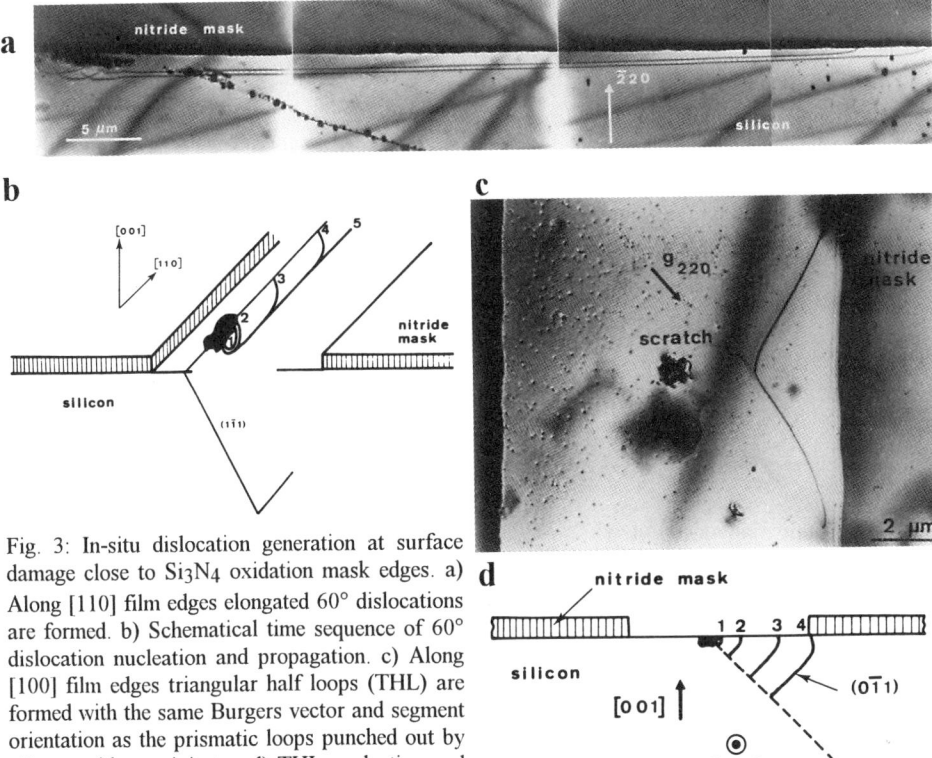

Fig. 3: In-situ dislocation generation at surface damage close to Si₃N₄ oxidation mask edges. a) Along [110] film edges elongated 60° dislocations are formed. b) Schematical time sequence of 60° dislocation nucleation and propagation. c) Along [100] film edges triangular half loops (THL) are formed with the same Burgers vector and segment orientation as the prismatic loops punched out by silicon oxide precipitates. d) THL nucleation and growth mechanism (Vanhellemont et al 1995b).

Irradiation of thin silicon foils using high fluxes of electrons which are accelerated with energies higher than 300 keV leads to the in-situ formation of extended point defect clusters which can be observed and characterised in the microscope (Aseev et al 1994). Well known are the so called {113}-defects which are also observed in ion implanted silicon and are considered as self-interstitial clusters and precursors of dislocations (Bender and Vanhellemont 1988). Detailed structural analyses of irradiation induced defects have been performed by Takeda et al 1994.

Recently it has become clear however that also extended vacancy clusters can be formed especially when the foil is very thin so that self-interstitials are captured efficiently by the two specimen surfaces (Fedina et al 1995). The vacancy concentration can then build up to much higher levels also because the sinks for vacancies in the „bulk" material are depleted fast. Fig. 4 shows an experimental image of intrinsic point defect aggregates in a 10 nm thin FZ specimen area. The defects are created by in situ 400 keV electron irradiation at room temperature in a JEOL 4000EX microscope. Both defects with {111}- and with {113}-habit planes are observed. The nucleation and growth of these defects can be studied in-situ in the HREM. A detailed structural analysis of the various intermediate stages of the defects by comparison of the experimental images with calculated ones shows that the {111}-defects originate from vacancy agglomerates and can gradually be filled with self-interstitials without full recombination of both types of intrinsic point defects (Fedina et al 1997). In such thin foils vacancy type extended defects are thus created simultaneously with interstitial type ones which is probably one of the first stages of amorphisation of the material.

Studying {113}-defect formation in processed wafers with device structures allows to identify the influence of local sinks for vacancies or self-interstitials such as highly doped areas or localised strain fields on point defect behaviour (Vanhellemont et al 1995).

Fig. 5 shows the evolution of the {113}-defect distribution near the tip of the bird´s beak of a local isolation structure. This experiment illustrates that the silicon oxide/silicon interface is a stronger sink for self-interstitials than for vacancies and that the localised strain field at the BB tip causes a higher local self-interstitial concentration by influencing the diffusion of vacancies (Vanhellemont and Romano-Rodríguez 1993).

Fig. 6 illustrates the impact of dopant profiles on {113}-defect generation. Irradiation of a silicon sample with implanted 2D boron profile results in the preferential formation of extended defects near the junction (Vanhellemont and Romano-Rodríguez 1994). The two dimensional boron concentration profile is easily recognised. In the higher doped layer the substitutional boron atoms form a strong sink for self-interstitials suppressing initially the formation of {113}-defects. During prolonged irradiation a very high density of small defects develops also in the highly doped area (arrow).

By varying the temperature at which the irradiation is performed the impact of dopants changes significantly as illustrated in Fig. 6(bottom) for a sample with a highly phosphorus doped surface layer created by ion implantation. At room temperature the phosphorus atoms form a strong sink for vacancies by the formation of so called E-centres (V-P pairs). Due to this a high supersaturation of self-interstitials is formed in the highly doped area leading to high densities of {113}-defects. At 520K the E-centre is no longer stable and defect formation is suppressed. The phosphorus concentration levels are superimposed on the TEM image.

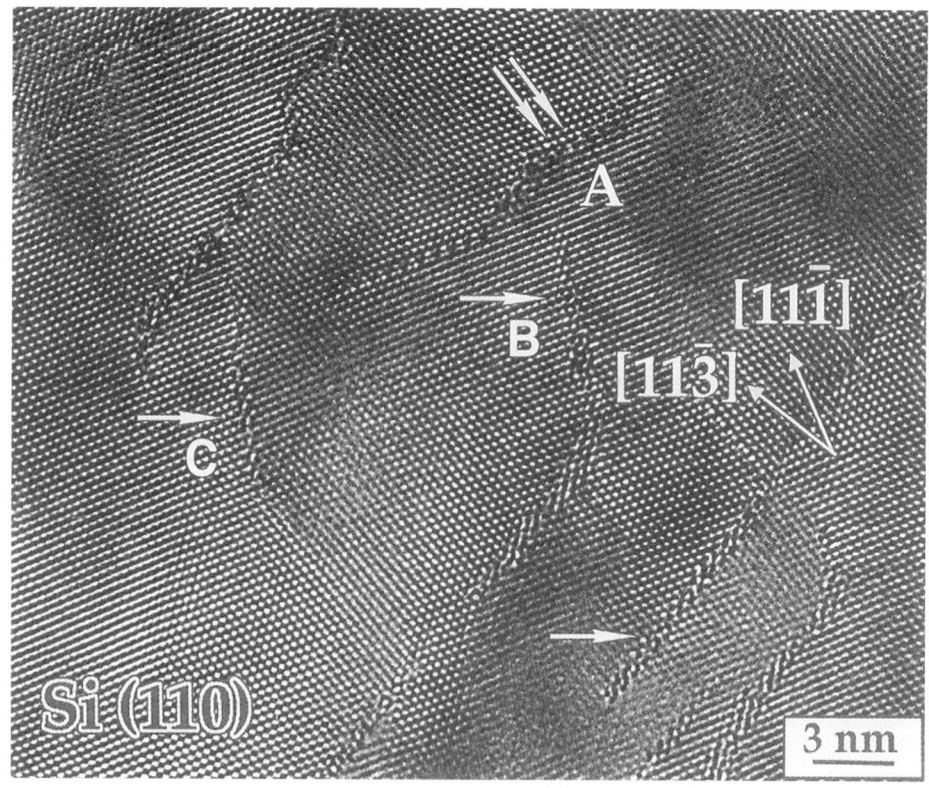

Fig. 4: Self-interstitial/vacancy agglomerates observed in an in-situ irradiated 10 nm FZ silicon foil after 35 minutes of irradiation. {113}- and {111}-defects are marked with single and double arrows, respectively (Fedina et al 1997).

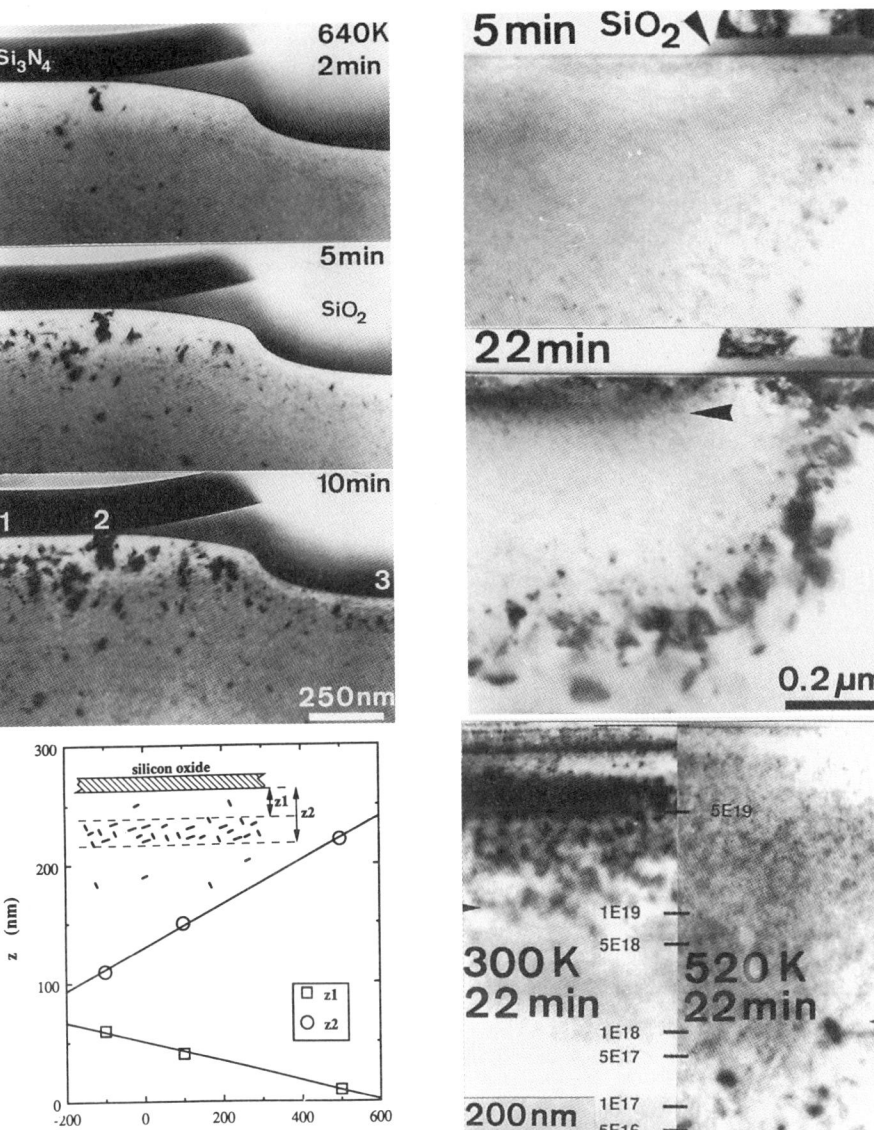

Fig. 5: Top: Time sequence of {113}-defect formation in a local isolation structure. Bottom: Dependence of {113}-defect distribution on the interface force per unit length.

Fig. 6: Top: {113}-defect formation at 640K in a B implanted and annealed Si wafer. Bottom: Irradiation temperature dependent {113}-defect formation in a P implanted and annealed Si wafer.

5. DEVICE STRUCTURES

Several reviews on TEM investigation of integrated circuit structures have been published in previous proceedings of this conference. In this section only a few examples will be given of work performed by some of the authors.

5.1 Geometry and defect problems

An important application of TEM in the microelectronics industry is the inspection of submicron device structures thanks to the excellent spatial resolution which can be obtained in combination with FIB specimen preparation. In some cases lattice images of the relevant device area can even be obtained (Bender and Roussel 1997). Some typical examples are shown in figures 7 and 8. The first figure shows a PELOX (poly encapsulated local oxidation) isolation structure after the pattern etching and deposition of amorphous silicon. The cavity size and filling need tight control. The poly etching of a 0.35 µm poly line with 0.35 µm spacing is illustrated in the second figure.

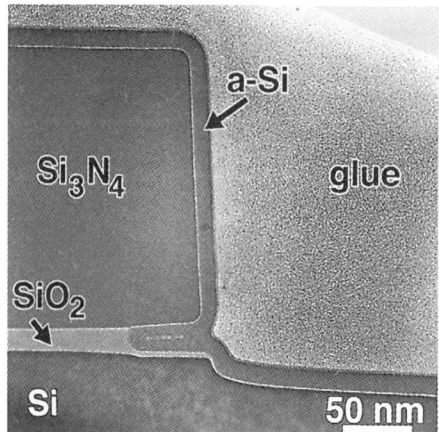
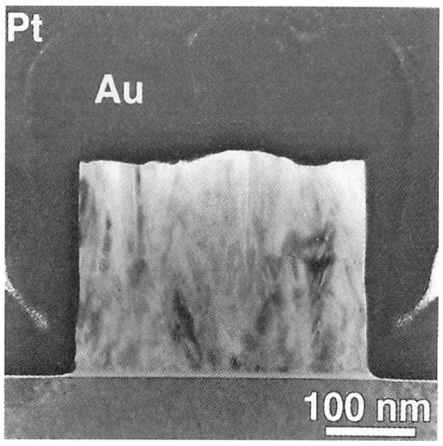

Fig. 7: FIB prepared XTEM of PELOX structure showing a 58 nm cavity filled with a-silicon.

Fig. 8: FIB prepared XTEM of 0.35 µm poly line after the poly etch.

5.2 Local strain determination

A new development of the last years is an effort to use of electron diffraction techniques to assess local strain distribution in device structures. The reason for this effort is that with the ever decreasing device sizes there is a trend to have increased local strain fields at device edges. These strain fields have a strong impact not only on physical processes during device processing but also on the electrical characteristics of the devices themselves due to the induced bandgap changes. During the last years well established techniques like convergent beam electron diffraction (CBED) and electron diffraction contrast imaging (EDCI) have been used with success to investigate localised strained areas in device structures (Armigliato et al 1995, Janssens et al 1995, Vanhellemont et al 1996).

CBED techniques have several advantages and allow to address the strain either in a point to point manner or also in a more global manner where strain contours are superimposed on the image (large angle CBED). A drawback of CBED is that rather thin foils have to be used and that the specimen has to be cooled in order to improve the quality of the diffraction pattern. The sensitivity which can be obtained is however similar to that of micro-Raman spectroscopy (µ-RS) with the advantage of the superior spatial resolution. The problem of strain relaxation due to the preparation of the thin foil can adequately be solved (Armigliato et al 1996).

The "old" technique of electron diffraction contrast imaging (EDCI) has been reinvented to assess local strain fields in a global way. A versatile software package, i.e. SIMCON (Janssens et al 1993 and Janssens 1996) has been developed which allows to calculate EDCI images of arbitrary displacement fields. The advantage of EDCI is that it is not a point to point technique like CBED but the images yield an overall view of the displacement field. The technique has been used with success to determine the strain distribution in local isolation structures and obtained results in good agreement with those of CBED and μ-RS. The main results obtained on a poly buffered LOCOS (PBL) structure have been summarised in Fig. 9. CBED and EDCI results are compared with those of a sophisticated SUPREM IV simulation and are also in excellent agreement with those of μ-RS analyses.

6. CONCLUSIONS

A lot of progress has been made the last years with respect to the analysis of local features in processed silicon device materials. Especially the advent of focused ion beam tools allowing to prepare thin foils of interesting specimen areas with unprecedented accuracy has led to the further introduction of TEM techniques in the microelectronics industry. As before TEM remains an indispensable tool for further improvement of processing steps and device structures due to its extremely good spatial resolution and imaging and the possibility to identify also the structure and nature of crystalline defects even present in extremely low densities. For a successful assessment of localised features the assistance of other defect delineation and localisation tools is essential and an integration of these tools in the TEM specimen preparation procedure remains one of the challenges for further progress.

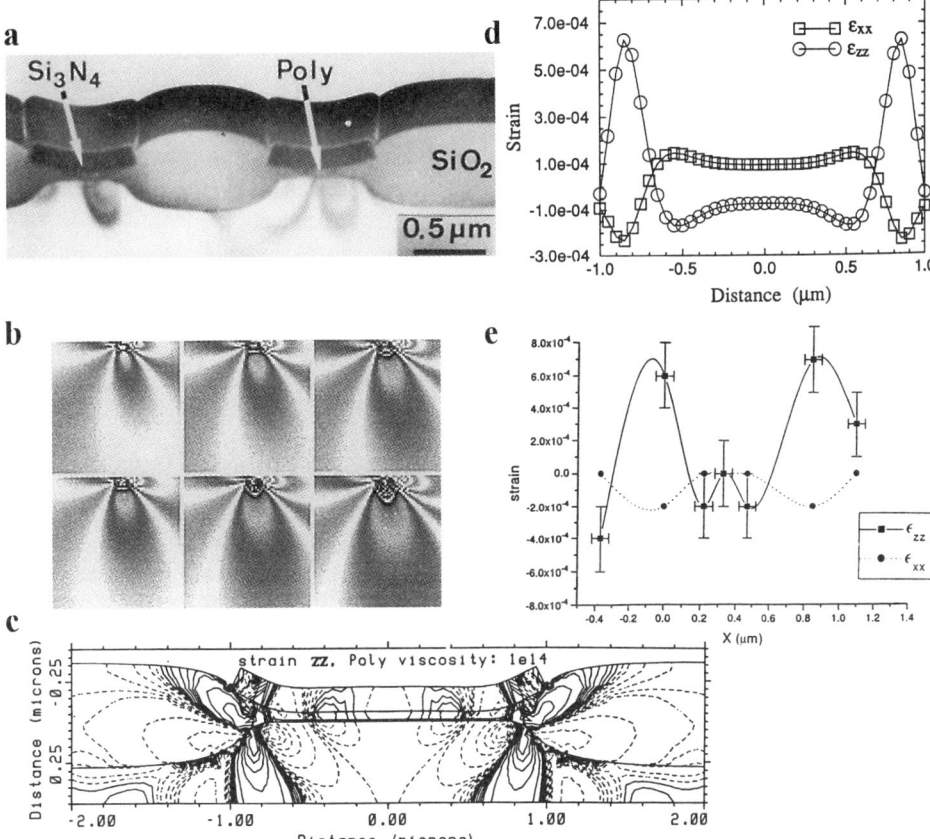

Fig. 9: a) Experimental EDCI image revealing strain concentration at local isolation structure edges. b) Simulated strain contrast images using a simple analytical model and different film edge forces (300 and 400 Nm^{-1}). c)-d) Strain (ε_{zz}) distribution in a 2 μm local isolation structure and strain components ε_{xx} and ε_{zz} at 300 nm below the silicon surface simulated with SUPREM IV. e) Strain components at 300 nm depth determined with CBED in a similar structure.

In-situ TEM studies offer unique possibilities to investigate important physical processes in real device structures and to study in-situ defect nucleation and growth processes. Although not discussed in this paper tremendous progress has also been made with respect to local chemical analyses with spatial resolution in the nm range. A new development is to use TEM also for local strain analysis in device structures which is a unique application where no other techniques are available at the moment.

As a general conclusion it can be stated that TEM has become an indispensable tool for the further development not only of advanced devices and processes but also of the new generation of large diameter silicon substrates.

ACKNOWLEDGEMENT

The authors would like to thank A Armigliato, A Aseev, R Balboni, C Claeys, I De Wolf, L Fedina, S Frabboni, K G F Janssens, H E Maes, A Romano-Rodríguez, P Smeys and O Van der Biest for collaboration on various topics and for the use of co-authored results. R Schmolke and M Suhren are acknowledged for supplying the wafer with COP coordinates and for stimulating discussions.

REFERENCES

Anderson R 1992 Mater. Res. Soc. Proc. 254, 141

Armigliato A, Balboni R, De Wolf I, Frabboni S, Janssens K G F and Vanhellemont J 1993 Inst. Phys. Conf. Ser. 134, 229

Armigliato A, Balboni R and Frabboni S 1996 Mat. Res. Soc. Symp. Proc.Vol. 405, 467

Aseev A L, Fedina L, Höhl D and Bartsch H 1994 Clusters of Interstitial Atoms in Silicon and Germanium, Academy Verlag, Berlin, 152 p

Bender H and Roussel P 1997 these proceedings

Bender H and Vanhellemont J 1988 phys. stat. sol. (a) 107, 455

Bender H et al 1997, to be published.

Fedina L, Van Landuyt J, Vanhellemont J and Aseev A L 1996 Nuclear Instruments and Methods in Physics. Research B 112, 133

Fedina L, Gutakovskii A, Aseev A, Van Landuyt J and Vanhellemont J 1997 these proceedings

Gräf D, Suhren M, Lambert U, Schmolke R, Ehlert A, von Ammon W and Wagner P 1996 Electrochemical Society Proceedings Volume 91-13, 117

Itsumi M, Akiya H, Ueki T, Tomita M, Yamawaki M 1996 Jpn. J. Appl. Phys. 35, 812

Janssens K G F, 1996 http://www.mtm.kuleuven.ac.be/~simcon/PhD thesis.html

Janssens K G F, Vanhellemont J, Maes H and Van der Biest O 1993 Inst. Phys. Conf. Ser. 134, 225

Janssens K G F, Van der Biest O, Vanhellemont J, Maes H E and Hull R 1995 Appl. Phys. Lett. 67, 1530

Kato M, Takeno H and Kitagawara Y 1997 proc. Symposium E 'Defects in Electronic Materials' of the 1996 MRS Fall Meeting, Boston, December 2-6, in press

Pantel R, Auvert G, Mascarin G and Gonchond J P 1994 Proc ICEM 13, 1007

Ross F, Hull R, Bahnck D, Bean J, Peticolas L and Kola R 1993 Inst. Phys. Conf. Ser. 134, 245

Schmolke R, Gräf D, Suhren M, Kirchner R, Piontek H and Wagner P 1997 proc. Symposium E 'Defects in Electronic Materials' of the 1996 MRS Fall Meeting, Boston, Dec. 2-6, in press

Takeda S, Kohyama M and Ibe K 1994 Phil. Mag. A 4, 287

Takeda S and Horiuchi S 1994 Ultramicroscopy 54, 144

Vanhellemont J and Romano-Rodríguez A 1994 Appl. Phys. A58, 541

Vanhellemont J and Romano-Rodríguez A 1993 Appl. Phys. A57, 521

Vanhellemont J, Romano-Rodríguez A, Fedina L, Van Landuyt J and Aseev A 1995a Materials Science and Technology 11, 1194

Vanhellemont J, Claeys C and Van Landuyt J 1995b phys. stat. sol. (a) 150, 497

Vanhellemont J, Janssens K G F, Frabboni S, Smeys P, Balboni R and Armigliato A 1996 Mat. Res. Soc. Symp. Proc. 406, 479

Young R J, Kirk E C G, Williams D A and Ahmed H 1990 Mat. Res. Soc. Symp. Proc. 199, 205

Inst. Phys. Conf. Ser. No 157
Paper presented at Microsc. Semicond. Mater. Conf., Oxford, 7–10 April 1997
© *1997 IOP Publishing Ltd*

Two- and three-dimensional characterisation of advanced LOCOS isolation using transmission electron microscopy

R Beanland, D J Bazley, S K Jones and B Scaife*

GEC-Marconi Materials Technology Ltd., Caswell, Towcester, Northants, NN12 8EQ, UK
*GEC Plessey Semiconductors, Tamerton Rd., Roborough, Plymouth PL6 7BQ, UK

ABSTRACT: Local oxidation of silicon (LOCOS) isolation processes are severely challenged by deep sub-micron CMOS design rules in two aspects; controlling active area encroachment and lifting at minimum active area line widths and corners; and preventing field oxide thinning at minimum active area spacing. This paper describes the use of dark-field weak-beam transmission electron microscopy to obtain a three-dimensional quantitative measurement of the topography of the Si/SiO$_2$ interface. We find that standard LOCOS can be optimised for isolation down to 0.3 micron geometries.

1. INTRODUCTION

Device isolation using local oxidation of silicon (LOCOS) is a widely used process. A thermal (pad) oxide of thickness 20-50 nm is first grown on the silicon surface. The active area of the device is then defined by depositing and lithographically patterning Si$_3$N$_4$. A second (field) oxidation is then performed. This consumes silicon not covered by Si$_3$N$_4$, and under the edges of the nitride some oxidation occurs which forces the edge of the nitride upwards and produces stress. This stress reduces the oxidation rate under the Si$_3$N$_4$ edge. The encroachment of oxide under the Si$_3$N$_4$ edge is known as a 'bird's beak', and reduces the size of the active area. Finally, the nitride and the pad oxide are removed, and processing continues to fabricate transistors in the active area regions, isolated from one another by the thick field oxide. Deep sub-micron CMOS design rules require strict control of active area encroachment both at edges and corners. Furthermore, the close packing of active areas can lead to field oxide thinning and so less effective isolation. Quantitative measurement of active area encroachment is thus essential to the development of deep sub-micron LOCOS.

Here, we describe a measurement of active area topography at the end of Si$_3$N$_4$ stripes using dark field, weak-beam transmission electron microscopy. The principle of the technique is simple; in a plan-view TEM sample, surface topography gives rise to thickness fringes which can be used to reconstruct the sample surface (Bazley et al. 1996). With the advent of digital image processing, this is a relatively straightforward process and can be applied readily to such images; measurements of surface topography can be made directly from the digital image and used in process modelling.

2. EXPERIMENTAL

Two versions of a LOCOS process were compared: (a) a standard version, applicable for 0.5 micron CMOS, using a nitride thickness of 130nm with oxidation performed in steam to achieve a field oxide thickness of 500 nm; (b) an optimised version, suitable for 0.35/0.25 micron CMOS, using a thicker nitride and oxidation at high temperature in dry O$_2$ to achieve a field oxide thickness of 400 nm. The TEM analysis described here was primarily used to assess their performance under 0.35 micron CMOS design rules. Using data obtained from the TEM measurements, simulations were used to predict performance down to 0.25/0.18 micron isolation pitch.

The Si$_3$N$_4$ stripes defining the active area were removed from the surface by etching the underlying SiO$_2$ in HF. The samples were then polished from the substrate side to a thickness of approx. 100 μm, followed by jet etching from the substrate side in 5:1 HNO$_3$:HF to produce a plan-view TEM sample. The thickness contours in the electron transparent sample show a combination of the surface topography of the silicon and the gradual increase in sample thickness from the edge of the etched hole. A dark-field, weak-beam imaging condition was used to reduce the effective extinction distance (e.g. Hirsch et al. 1977), and provides many contours over the height of the structure. The resulting images were then digitised and the position of the contours obtained by

404

image processing. Quadri-linear interpolation between the thickness fringes was then used to create an image of the surface topography. The 'background' due to the variation of sample thickness was then subtracted from the image by taking three points of known height on the sample and subtracting a wedge from the image with its top surface passing through these three points. Cross-section TEM images of the active area, far from the end of the stripes, were used to calibrate the vertical scale of the image.

3. RESULTS

The oxidation kinetics in a steam ambient produce a rather long (0.5 μm) and narrow bird's beak in the default LOCOS process. It is possible to compensate for this during the oxide etch after Si_3N_4 removal; however, for nitride line widths below 0.4 μm complete lifting of the nitride occurs (Fig. 1). A plan view TEM image (dark field, weak beam g, 3g, 220) of the end of the stripe is shown in Fig. 2, and an orthogonal projection of the silicon surface obtained from this image in Fig. 3. It is clear that the end of the stripe is even more susceptible to active area encroachment. This is not surprising given the larger source for oxygen diffusion at the ends of stripes and the reduced mechanical stress which occurs due to bending of the nitride.

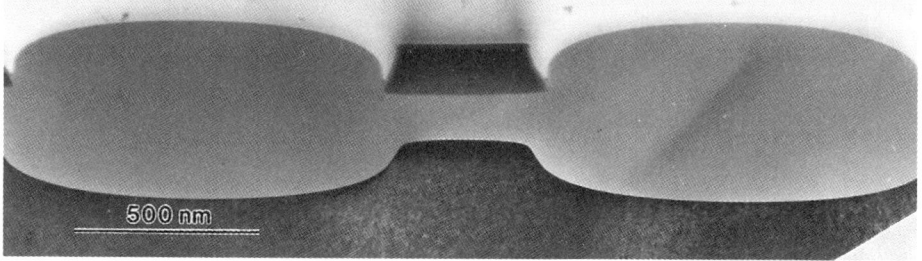

Figure 1. Cross-sectional TEM image (bright field, [110] axis), of the standard LOCOS process for line/space widths of 0.4/0.6 μm.

Figure 2. Plan view TEM image (dark field, weak beam g, 3g, 220) of the standard LOCOS process for line/space widths of 0.4/0.8 μm. Thickness between fringes is 13.5 nm.

The high temperature dry O_2 ambient used in the optimised LOCOS process gives both more favourable oxidation kinetics and a reduction in stress due to the high thermal budget. Oxidation becomes diffusion limited very quickly, and so oxide penetration under the active area nitride is minimised. The bird's beak is more abrupt, with length 0.12 μm (Fig. 4) and active area lifting does not occur at 0.4 μm line width. In addition, the amount of active area encroachment is considerably reduced at the ends of the stripes, as shown in Figs. 5 and 6.

The measured active area lifting for Si_3N_4 line widths from 0.4 to 1.0 μm is compared for the two processes in Fig. 7. LOCOS in steam shows a large thinning effect, but the optimised LOCOS process shows a reverse effect, with an increase in field oxide thickness at small active area spacing (Belutti et al. 1995). This is due to an increased out-flow of oxide from under the Si_3N_4 mask during the long high temperature oxidation. Simulations predict that this reverse thinning effect will be sustained down to a spacing of 0.2 μm, which may allow the nitride line width to be reduced and a more aggressive isolation pitch to be attained.

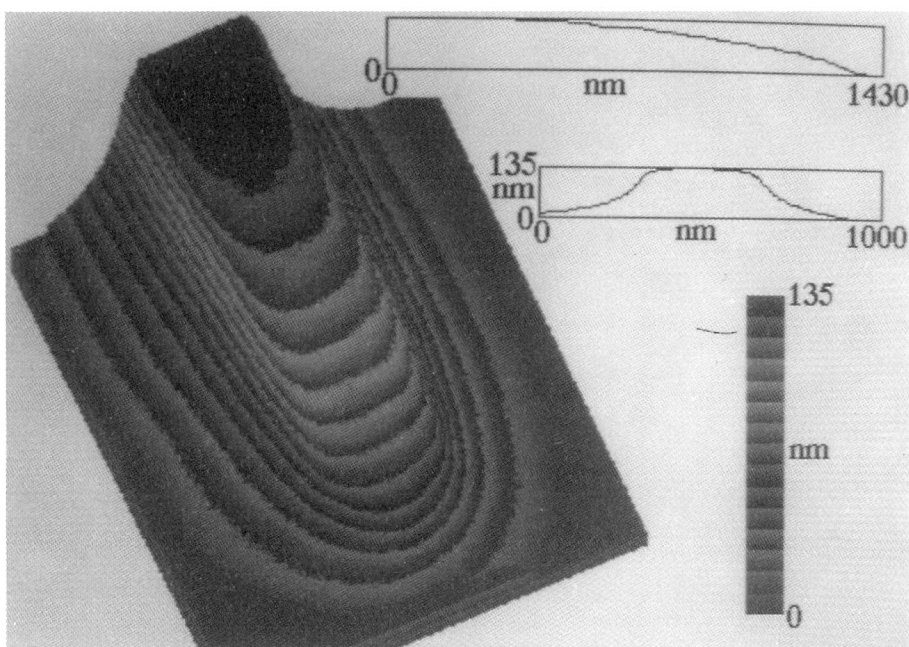

Figure 3. Orthogonal projection of the silicon surface, obtained from the thickness fringes in Fig. 2.

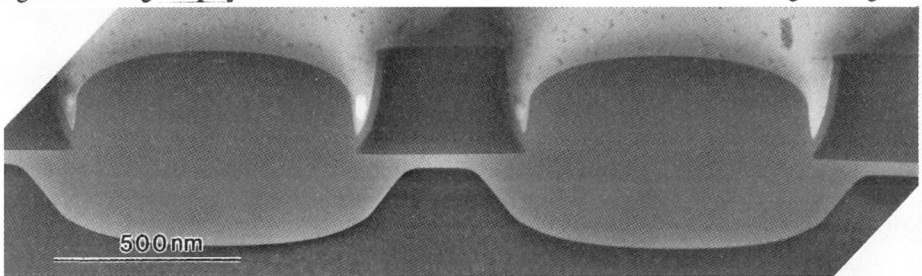

Figure 4. Cross-sectional TEM image (bright field, [110] axis), of the optimised LOCOS process for line/space widths of 0.4/0.6 μm.

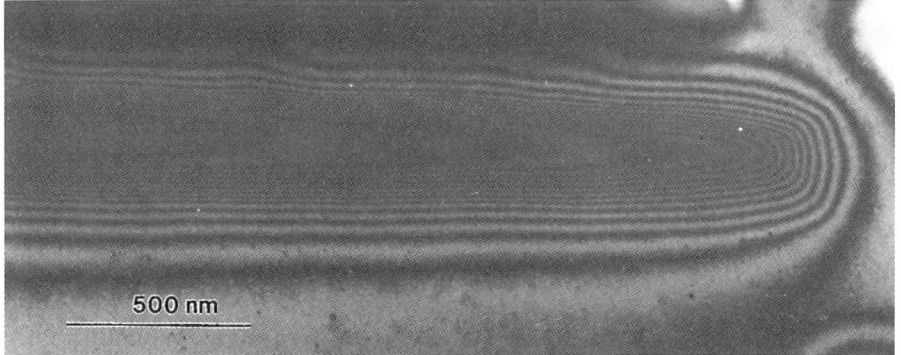

Figure 5. Plan view TEM image (dark field, weak beam **g**, 3**g**, 220) of the optimised LOCOS process for line/space widths of 0.4/0.6 μm. The end of line encroachment is much reduced.

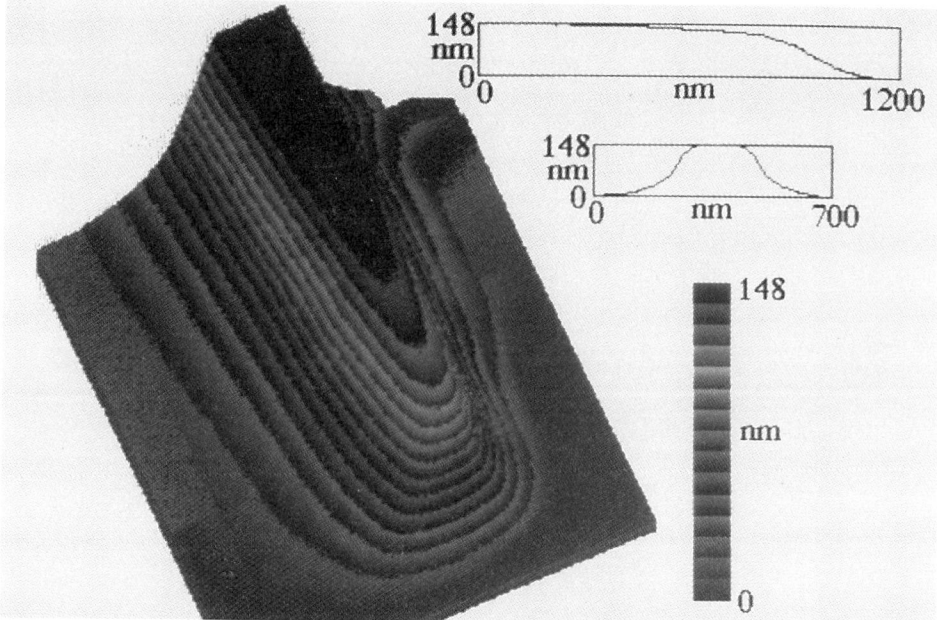

Figure 6. Orthogonal projection of the silicon surface, reconstructed from the thickness fringes in Fig. 5.

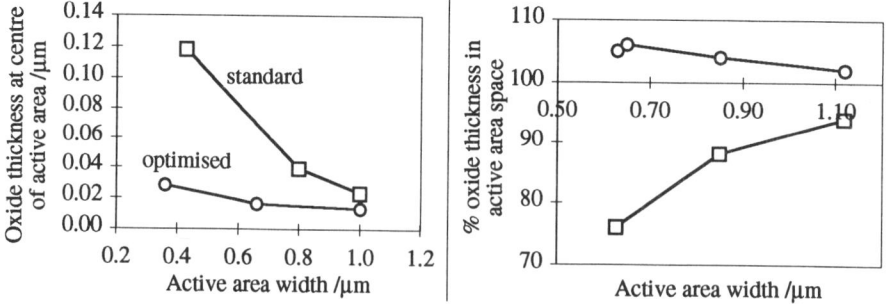

Figure 7. Measured active area lifting (left) and field oxide thinning (right).

4. CONCLUSIONS

We have developed a TEM technique for quantitatively measuring the active area encroachment in two dimensions in the LOCOS process. High temperature field oxidation in dry O_2 allows the use of LOCOS to be extended to sub-0.5 micron CMOS design rules. There is a potential penalty of dopant up-diffusion from the increased field oxidation budget, which may limit this approach to a non-epitaxial CMOS process. However, the benefits of using a high temperature dry O_2 oxidation ambient should be transferable to other advanced LOCOS schemes.

ACKNOWLEDGEMENTS
This work was partly funded by the EC ESPRIT 8002 ADEQUAT2 Project, ESPRIT 8150 PROMPT Project and by GEC Plessey Semiconductors.

REFERENCES
Bazley D J, Jones S K, Beanland R and Scaife B, Proc. ESSDERC 1996, eds E Baccarani and M Rudan (Editions Frontieres) p 623
Bell89utti P, Boscardin M, Soncini G, Zen M and Zorzi N, 1995 Semicond. Sci. Technol. **10**, 1700
Hirsch P, Howie A, Nicolson R B, Pashley D W and Whelan M J, 1977 Transmission Electron Microscopy of Thin Crystals (Malabar FL: Krieger)

Inst. Phys. Conf. Ser. No 157
Paper presented at Microsc. Semicond. Mater. Conf., Oxford, 7–10 April 1997
© 1997 IOP Publishing Ltd

Improved epitaxial quality following etch damage removal on plasma etched silicon surfaces

J M Bonar, J Schiz and P Ashburn

Department of Electronics and Computer Science, The University of Southampton, Southampton SO17 1BJ, UK.

ABSTRACT The quality of epitaxial growth on surfaces subject to plasma etching can be degraded due to damage to the silicon surface. The aim of this study is to obtain improvements in epitaxial quality by the use of an Etch Damage Removal (EDR) treatment following plasma etch or Reactive Ion Etch (RIE) stages in wafer processing. Optical and electron microscopy observations suggest increased defect density in epitaxial layers grown following some RIE processes, while wet etching or RIE + EDR will allow the growth of high quality epitaxial layers.

1. INTRODUCTION

Fabrication of silicon devices generally involves removing SiO_2 layers in defined regions from the surface of silicon wafers. After standard lithographic techniques produce a pattern on the surface of an oxide layer, the exposed regions of oxide may be removed by wet or dry etching techniques. Wet etching using buffered hydrofluoric acid (BHF) is fast and highly selective, that is it will remove oxide very quickly but silicon only very slowly. However, as it is an isotropic etch it will produce oxide features with sloping sidewalls, and the minimum feature size is limited. RIE can be made anisotropic by the choice of process gases and conditions, producing vertical sidewalls with good selectivity. However, surface damage is sometimes associated with the RIE process, which is significant when epitaxial layers are to be grown in the etched areas.

Particulates created during the RIE process, which would cause defects in the epitaxial layers, can be effectively removed by wet chemical cleaning. The surface damage produced by the RIE process may be structural or in the form of impurities remaining on the surface (Tseng et al 1995, Lou et al 1992). Epitaxial growth over damaged surfaces is expected to contain defects (Tseng et al 1995, Lou et al 1992), which would be expected to adversely affect device characteristics. Thus, elimination of the damage prior to epitaxy is an important goal in silicon device fabrication.

2. EXPERIMENTAL

Sample preparation began with the growth of a 400 nm thermal oxide on p-type CZ <100> silicon wafers at 1100°C in wet oxygen. A pattern in resist was then produced on the oxide using standard optical lithographic techniques. The oxide was removed from the exposed regions using an anisotropic SiO_2 etch using CHF_3 and Ar. The time to remove 400nm SiO_2 is calculated and the wafers were etched for 15% or 30% longer than necessary to produce 15% or 30% Over Etch (OE). The use of OE is standard practice in RIE in order to ensure oxide removal, as the material removal process is non-uniform both across features and across wafers. Some wafers also received an Etch Damage Removal (EDR) treatment. This consisted of 20 minutes at 950°C in dry oxygen, which produced ≈20 nm oxide. The EDR oxide was then removed by a brief etch

408

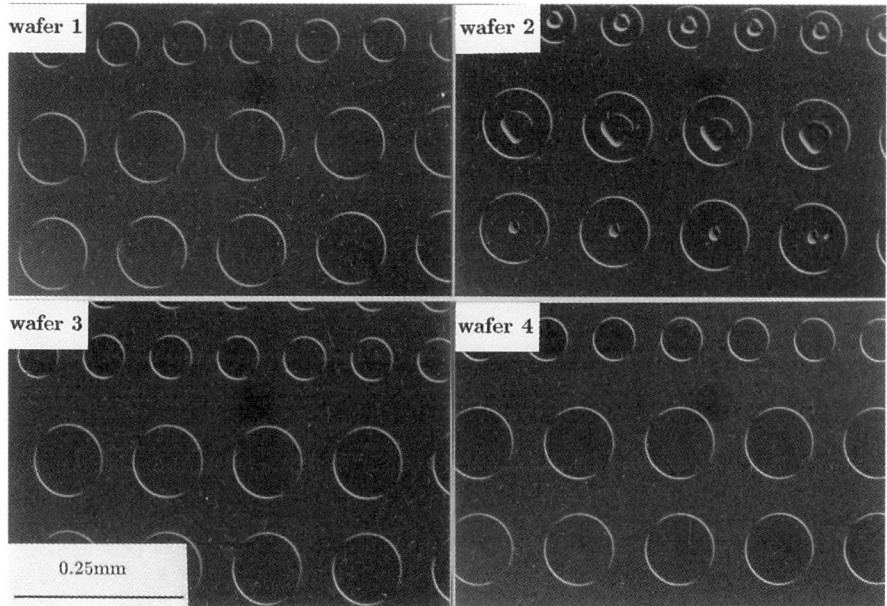

Figure 1: Nomarski contrast optical micrographs of wafers 1 through 4 following epitaxy.

in BHF. As a control, the oxide was removed from the exposed regions of one wafer using only BHF. This process produced wafers with an oxide mask with various geometry windows opened to the underlying silicon. Nominally identical epitaxial Si-SiGe-Si layers were then grown on the wafers by LPCVD (Bonar and Parker 1995). Over silicon areas epitaxial growth is expected to occur, while over oxide covered regions polycrystalline material will deposit as there is no crystal structure to act as a template for epitaxial growth. The sample set consists of wafer 1 (wet etched), wafer 2 (15% OE), wafer 3 (15% OE + EDR), wafer 4 (30% OE + EDR), and wafer 5 (30% OE). Wafers 1 through 4 received epitaxy, while wafer 5 did not.

After the epitaxy stage, the wafers were examined by Nomarski contrast optical microscopy. Sections of each wafer were Secco etched (Schimmel 1976) and the crystalline defect level counted. Cleaved sections were examined at an oblique angle in the Scanning Electron Microscope (SEM). Cross-section TEM observations were made of selected wafers in order to examine the crystallinity of the deposited layers. TEM samples were made by standard mechanical thinning and polishing followed by thinning to electron beam transparency by ion beam milling.

3. RESULTS AND DISCUSSION

Figure 1 is a Nomarski contrast optical micrograph of the surface of wafers 1 through 4. On each wafer are circular windows in which epitaxial layers have been grown as well as regions of oxide covered with polysilicon. Polysilicon has a rough surface morphology while epitaxial layers should appear smooth and featureless. The epitaxial areas in wafers 1, 3 and 4 appear smooth, although there is some faint roughness visible in the centre of some features on wafer 4. The silicon surfaces of wafer 5 also appear smooth and featureless. By contrast, the surface of the epitaxial layer on wafer 2 appears rough in the centre of the epitaxy in all the windows observed. The clear difference in surface morphology suggests wet etching or RIE with 15% OE + EDR will allow the growth of epitaxial layers with smooth surface morphology, suggesting high

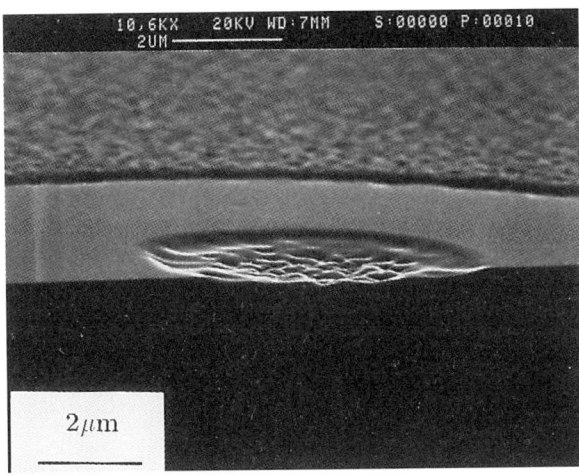

Figure 2: Oblique view SEM micrograph of a circular feature on wafer 2.

quality crystalline layers may be grown. The slight roughness present in some features in wafer 4 suggests that not all damage produced by a 30% OE is removed by the EDR treatment used in this case. However, the defect density following Secco etching confirmed that the epitaxial quality on wafers 1, 3 and 4 is suitable for device fabrication.

Following Nomarski contrast microscopy, the wafers were then cleaved and loaded into the SEM. The SEM observations confirmed that the surface of wafers 1, 3, 4 and 5 appeared smooth and featureless. Figure 2 is an SEM micrograph of wafer 2 which has been cleaved through a circular window in which epitaxy has been grown. The surface of the layer appears smooth at the feature edge with a region of gradually increasing roughness part way to the centre. The centre of the feature appears quite rough, with surface morphology similar to that of polysilicon. Thus the deposition in the features on wafer 2 is suspected to be polycrystalline rather than monocrystalline in nature.

Cross-section TEM observations were made of wafer 1 (wet etched), wafer 2 (15% OE), and wafer 5 (30% OE but no epitaxy). The epitaxial layer of wafer 1 appears to be quite perfect, with no twin or other defects visible in the field of view. This was confirmed by the diffraction pattern of the epitaxial layer, which showed no evidence of extra spots or streaking. The high crystalline quality of this sample, deduced from the smooth surface morphology, is thus confirmed by the TEM observations. TEM observation of wafer 5 was not conclusive. The silicon surface adjacent to the oxide sidewall was examined, and no defects extending below the surface were found. Thus although epitaxy on a surface which received RIE with OE produced poor quality growth, surfaces receiving RIE with OE do not demonstrate non-smooth surface morphology or defects extending below the surface.

Figure 3 is a cross-section image of an area of epitaxy on wafer 2, with an inset of the selected area diffraction pattern. The silicon substrate is visible in the lower area of the image, with the line of the substrate-deposition interface clearly visible. On the right hand side of the image the surface of the deposited layer appears fairly smooth and the layer is uniform in thickness although defects are visible. The material on the left hand side of the image appears to have a higher density of defects, and the layer is not of uniform thickness. Thus some areas of defective

410

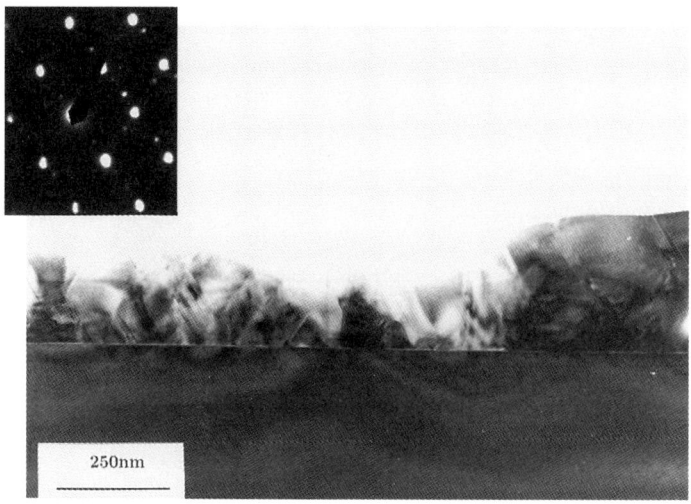

Figure 3: Cross-section TEM image of area of rough epitaxy on wafer 2.

material are rougher and more defective than others, which correlates with the SEM observation that the transition from smooth to rough is not abrupt. Although the material in the feature centre appears fairly defective, the diffraction pattern from this region is that of single crystal silicon with extra spots, possibly due to twin defects. Thus the determination of the nature of the deposit as defective epitaxy, rather than polycrystalline in nature, is quite definite.

4. CONCLUSIONS

In conclusion, oxide etching by RIE incorporating 15% or 30% OE plus EDR or wet etching will produce a silicon surface suitable for the epitaxial growth of high quality Si-SiGe-Si layers. However, surface damage is produced by RIE incorporating 15% OE without an EDR oxidation stage, and heavily defective but epitaxial layers will be produced. The present results do not allow determination of the nature of the surface damage, but further experiments are planned in an attempt to elucidate the mechanism. The EDR process developed has been demonstrated to be an effective means of removing the surface damage produced by RIE. Its incorporation into the silicon device fabrication process will ensure the quality of the epitaxial layers will not be affected by the RIE stages necessary to produce advanced device structures.

REFERENCES

Tseng H-C, Chang C Y, Pan F M and Chen L P 1995 J. Appl. Phys **78**, 4710
Lou J-C, Oldham W G, Kawayoshi H and Ling P 1992 J. Appl Phys **71**, 3225
Bonar J M and Parker G J 1995 Mat Sci and Tech **11**, 31
Schimmel D G 1976 J Electrochem Soc **123** p734

The authors would like to acknowledge the assistance of Dr. BA Cressey of The Science and Engineering Electron Microscopy Centre, University of Southampton.

Inst. Phys. Conf. Ser. No 157
Paper presented at Microsc. Semicond. Mater. Conf., Oxford, 7–10 April 1997
ⓒ *1997 IOP Publishing Ltd*

The effects of fluorine on the epitaxial regrowth of arsenic-doped amorphous silicon and polysilicon and of chlorine on the epitaxial regrowth of arsenic-doped polysilicon

C D Marsh, N E Moiseiwitsch*, G R Booker and P Ashburn*

Dept of Materials, University of Oxford, Parks Rd, Oxford OX1 3PH, UK.
*Dept of Electronics & Computer Science, University of Southampton, Southampton SO17 1BJ, UK.

ABSTRACT: The role of fluorine (F), chlorine (Cl) and arsenic (As) in the break-up of the native oxide and the regrowth of poly-Si or α-Si at anneal temperatures of 1015°C and 1065°C is investigated by TEM and SIMS. The results show that α-Si layers with F regrow slower than poly-Si layers with F, but faster than the poly-Si layers without F. The concentrations of F and As at the interface during the early stages of annealing were lower for the α-Si layer than for the poly-Si layer and it is considered that this causes a slower break-up of the native oxide and hence a slower regrowth. The poly-Si layers with Cl regrew slower than the poly-Si layers with F. At 1015°C they regrew slower while at 1065°C they regrew faster than layers without Cl and F.

1. INTRODUCTION

Arsenic doped poly-Si is used for the fabrication of emitters for high speed bipolar devices. The poly-Si is deposited onto single crystal Si and as a result a thin oxide layer is present at the poly-Si/single crystal Si interface. The presence of the oxide has the advantage of increased transistor gain, but the disadvantages of higher emitter resistance and non-uniform transistor gains across a wafer. Hence processing conditions which break up the oxide and regrow the poly-Si to single crystal Si to give lower emitter resistance and more uniform transistor gains are important. As device dimensions decrease it is important to reduce the thermal budgets for such processes.

Hoyt *et al* (1987) have shown that with a high As concentration (5×10^{20}/cm^3) complete epitaxial regrowth of poly-Si occurs for an emitter drive-in of 1150°C for 20s. Williams and Ashburn (1992) and Wu *et al* (1994) have shown that the break-up of the oxide and the regrowth of poly-Si can occur more rapidly if fluorine (F) is present. Marsh *et al* (1995) and Moiseiwitsch *et al* (1995) have shown that the presence of F enables the near complete regrowth (>92%) of poly-Si after annealing as low as 850°C (for 120mins). In the present paper we investigate the role of F on the oxide break-up and the regrowth of amorphous(α)-Si layers and poly-Si layers and the role of Cl on the oxide break-up and the regrowth of poly-Si layers.

2. EXPERIMENTAL

Cz Si (100) wafers, p-type resistivity 5-35Ωcm, were given a RCA pre-clean which resulted in a native oxide layer of 14Å. This was immediately followed by a low pressure chemical vapour deposition (LPCVD), at either 560°C or 610°C, of 400nm of undoped amorphous(α)-Si or poly-Si respectively. Some of the poly-Si layers were implanted (1×10^{16}/cm^2, 50keV) with either F or Cl and some of the α-Si layers were implanted (1×10^{16}/cm^2, 50keV) with F. All the samples were then implanted with As (1×10^{16}/cm^2, 70keV). The layers were then capped with 600nm LPCVD of oxide to prevent subsequent As loss and annealed at 1015°C or 1065°C for times between 4s and 180s.

3. RESULTS AND DISCUSSION

Table 1 summarises the regrowth of the poly-Si in the samples with and without F, annealed at 1015°C or 1065°C, as a function of the anneal time. In the samples without F, after 60s at 1015°C and 10s at 1065°C, the oxide layer has only partially broken up while in samples with F given similar anneals, after 60s at 1015°C and 7s at 1065°C, the oxide layer has completely broken up (fig. 1). In the samples without F, near complete regrowth, 90% and 98%, has occurred after 120s at 1015°C and 75s at 1065°C, respectively. In the samples with F, near complete regrowth, 98% and 95%, has occurred after 60s at 1015°C, half the time without F, and after 7s at 1065°C (fig. 1), one tenth the time without F, respectively. Hence the presence of F enables the complete break up of the oxide layer and near complete regrowth of the poly-Si layer to occur more quickly.

Deposited silicon	F dose (/cm^2)	Anneal temp(°C)	Anneal (s)	Oxide break-up	Regrowth pinned to oxide	Regrowth to surface	Amount of regrowth (%)
poly	none	1015	10	✓	✓	✗	<0.5
"	"	"	60	✓	✓	✗	8
"	"	"	90	✓	✓	✓	80
"	"	"	120	✓	✗	✓	90
"	"	1065	10	✓	✓	✗	0.7
"	"	"	30	✓	✓	✗	2
"	"	"	45	✓	✓	✗	8
"	"	"	75	✓	✗	✓	98
poly	10^{16}	1015	4	✓	✓	✗	0.5
"	"	"	10	✓	✓	✗	2
"	"	"	60	✓	✗	✓	98
"	"	"	90	✓	✗	✓	98
"	"	"	120	✓	✗	✓	98
"	"	1065	2	✓	✓	✗	0.5
"	"	"	7	✓	✗	✓	95
"	"	"	15	✓	✗	✓	98
"	"	"	30	✓	✗	✓	98

Table 1 : Summary of the regrowth of the poly-Si in the samples with and without F.

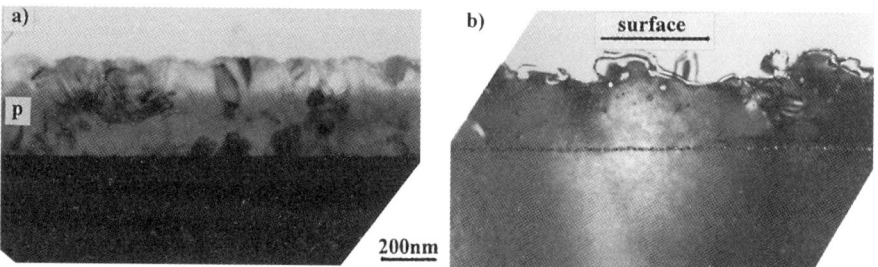

Fig. 1 : Samples a) without F annealed for 10s and b) with F annealed for 7s at 1065°C showing the complete break-up of the oxide layer and the greater regrowth in the sample with F (p=poly-Si).

Table 2 summarises the regrowth of the α-Si in the samples implanted with F and annealed at 1015°C as a function of the anneal time. During annealing the α-Si regrows to become poly-Si or single crystal Si. In the α-Si sample annealed for 60s the oxide layer has partially broken up, while in the corresponding poly-Si sample the oxide layer has completely broken up (fig. 2). In the α-Si samples near complete regrowth (99%) occurs after 120s, while for the poly-Si samples near complete regrowth (98%) occurs after 60s (table 1 & fig. 2). Hence both the break up of the oxide layer and near complete regrowth of the Si layer occur more quickly for a poly-Si than for an α-Si layer. The TEM results indicate that for the same anneals the α-Si layers result in slightly larger average grain sizes than the poly-Si layers. SIMS analysis has shown that both the integrated doses and peak concentrations of both F and As in the oxide layer are greater in the samples with the poly-Si layers. This suggests that it is likely the oxide layer break-up occurs more quickly for the poly-Si than for α-Si layers because the smaller grain size, and hence higher grain boundary density, enable a higher concentration of As and F to reach the oxide which enhances the break-up and regrowth.

Deposited silicon	F dose (/cm^2)	Anneal temp(°C)	Anneal time (s)	Oxide break-up	Regrowth pinned to oxide	Regrowth to surface	Amount of regrowth (%)
amorphous	10^{16}	1015	4	✗	-	✗	0
"	"	"	10	✓	✓	✗	0.5
"	"	"	60	✓	✓	✗	5
"	"	"	90	✓	✓	✓	60
"	"	"	120	✓	✗	✓	99

Table 2 : Summary of the regrowth of the α-Si layer in the samples with F and annealed at 1015°C.

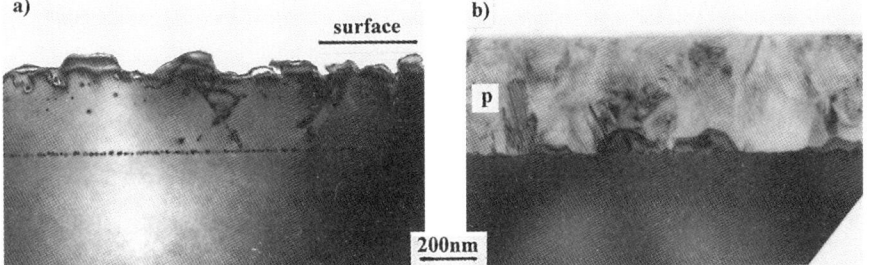

Fig. 2 : Samples with F a) poly-Si layer and b) α-Si layer annealed for 60s at 1015°C showing the complete break-up of the oxide layer and the greater regrowth in the poly-Si sample (p=poly-Si).

Table 3 summarises the regrowth of the poly-Si in the samples implanted with Cl and annealed at 1015°C or 1065°C as a function of the anneal time. In the Cl samples annealed for 60s at 1015°C and for 10s at 1065°C (fig. 3) the oxide layers have partly broken up and only a small amount of poly-Si regrowth has occurred, 2% and 0.5% respectively. This compares with the poly-Si samples with F (table 1) where the oxide has completely broken up and near complete regrowth (>90%) has occurred after 60s at 1015°C and after 7s at 1065°C (fig. 1b). Hence at both temperatures the oxide layer completely breaks up and near complete regrowth of the poly-Si occurs more quickly for poly-Si layers with F than with Cl. Hence Cl is not as effective as F in breaking up the oxide layer and enhancing the poly-Si regrowth. This may be because F is more electro-negative than Cl and hence F more readily breaks and forms new bonds.

Deposited silicon	Cl dose (/cm^2)	Anneal temp(°C)	Anneal (s)	Oxide break-up	Regrowth pinned to oxide	Regrowth to surface	Amount of regrowth (%)
poly	10^{16}	1015	10	✗	-	-	0
"	"	"	60	✓	✓	✗	2
"	"	"	90	✓	✓	✗	4
"	"	"	120	✓	✓	✓	25
"	"	"	180	✓	✗	✓	98
"	"	1065	10	✓	✓	✗	0.5
"	"	"	30	✓	✗	✓	95
"	"	"	45	✓	✗	✓	98
"	"	"	75	✓	✗	✓	98

Table 3 : Summary of the regrowth of the poly-Si layer in the samples with Cl.

414

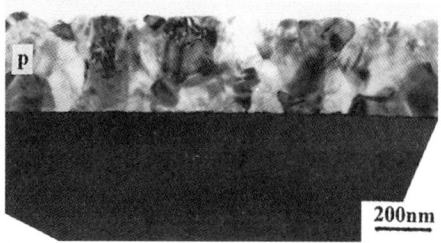

Fig. 3 : Poly-Si sample with Cl annealed for 10s at 1065°C showing less regrowth than the similar sample with F shown in fig 1b (p=poly-Si).

In the poly-Si samples with Cl annealed at 1015°C the oxide layer is only partially broken up after 120s, while in the sample without Cl or F the oxide layer is completely broken up. In the poly-Si samples with Cl annealed at 1065°C the oxide layer is completely broken up after 30s, while in the sample without Cl or F the oxide is only partially broken up (fig. 4). In the poly-Si samples with Cl (table 3) near complete regrowth (>90%) occurs after 180s and 45s, at 1015°C and 1065°C respectively, while in the samples with no F and Cl (table 1) near complete regrowth occurs after 120s and 75s, at 1015°C and 1065°C respectively. Hence at 1015°C break up of the oxide and near complete regrowth occurs more quickly in samples without Cl than with Cl, suggesting that Cl has an effect suppressing the break up of the oxide, but at 1065°C oxide break up and regrowth occurs more quickly in samples with Cl than without Cl or F suggesting that the precise role of Cl is a thermally activated process.

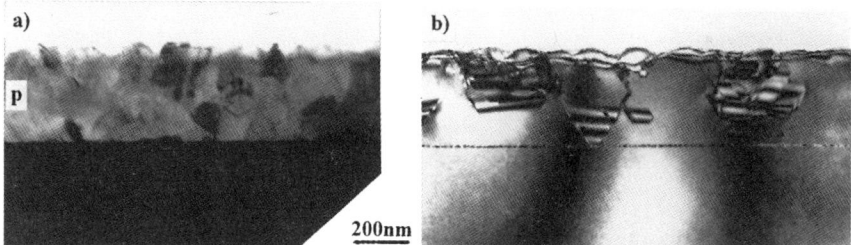

Fig. 4 : Samples a) without Cl or F and b) with Cl annealed for 30s at 1065°C showing at this temperature the greater regrowth of the poly-Si in samples with Cl (p=poly-Si).

4. CONCLUSIONS

At anneal temperatures of 1015°C and 1065°C both the oxide layer at the Si substrate/Si layer interface breaks up and the Si layer regrows to single crystal Si more quickly for i) poly-Si layers with F than for poly-Si layers without F, ii) poly-Si layers with F than for α-Si layers with F and iii) poly-Si layers with F than for poly-Si layers with Cl. The effect of Cl on the oxide break up and poly-Si regrowth is temperature dependent.

5. ACKNOWLEDGEMENTS

This authors would like to thank the EPSRC for financial support and the staff of the University of Southampton clean room for fabrication work.

REFERENCES

Hoyt J L, Crabbe, Gibbons J F and Pease R, 1987 Appl. Phys. Lett. 50 751.
Marsh C D, Moiseiwitsch N E, Ashburn P and Booker G R, 1995 Inst. Phys. Conf. Ser. 146 457.
Moiseiwitsch N E, Marsh C D, Ashburn P and Booker G R, 1995 Appl. Phys. Lett. 66 1918.
Williams J D and Ashburn P, J. 1992 J. Appl. Phys. 72 3169.
Wu S L, Lee C L, Lei T F, Chen C F, Chen L, Ho K and Ling Y, 1994 IEEE Elec. Dev. Lett. 15 120.

Inst. Phys. Conf. Ser. No 157
Paper presented at Microsc. Semicond. Mater. Conf., Oxford, 7–10 April 1997
© *1997 IOP Publishing Ltd*

Effects of static disorder on LACBED patterns of single crystal silicon implanted with hydrogen

S Frabboni[1], F Gambetta[1], R Tonini[1], R Balboni[2] and A Armigliato[2]

[1]) Istituto Nazionale Fisica della Materia (INFM) and Dipartimento di Fisica
  Università di Modena-Via G.Campi 213/A 41100 Modena, Italy
[2]) CNR-Istituto LAMEL, Via P. Gobetti 101, 40129 Bologna, Italy

ABSTRACT: The LACBED technique has been applied to <110> TEM cross-sections of Si sample implanted with H to study the effect of the static disorder on the Rocking Curves (RC) of high angle diffracted beams in presence of platelets and cavities. This effect has been revealed by comparing the RC originating from the perfect crystal and from the implanted layer, respectively. The results are consistent with an atomic root mean square displacement of the order of 10^{-2} nm which compares quite well with x-ray measurements performed in the Bragg-symmetric geometry.

1. INTRODUCTION

Static disorder in crystal structures may arise from an irregular arrangement of atoms in the lattice sites or from the presence of vacancies, interstitials or impurity atoms. The determination of this quantity is of importance in implanted materials as it increases with the radiation damage, which eventually results in its amorphisation. The most widely used diffraction technique for measuring static disorder is Double-Crystal x-ray Diffraction (DCD) (Balboni et al 1995). The same quantity has never been studied by transmission electron diffraction techniques although the use of electrons would provide some advantages: first, the probe can be focused to nanometric dimensions, then micro-crystalline areas can be analysed; next, high order reflections with different Miller indices can be easily excited. The aim of this paper is to apply the Large Angle Convergent Beam Electron Diffraction (LACBED) technique to the determination of the static disorder in H implanted, furnace annealed silicon samples. The method we propose is based on the quantitative comparison between the area under the Rocking Curve (RC) of high order diffracted beams ($15nm^{-1} < g < 21nm^{-1}$) originating from the perfect crystal and from the implanted layer, respectively. The results will be compared with the ones obtained by the DCD technique.

2. RELATIONSHIP BETWEEN ROCKING CURVE AND STATIC DISORDER

For crystals with defects it is not possible to use the formulas for diffraction intensities developed for ideal crystals, in particular it is not possible to insert into the expression of the structure factors F_g the atomic scattering factors and position of all the atoms of the specimen. The intensity must be evaluated in terms of the statistical relationship between these quantities (Cowley 1990). When an average periodic lattice can be defined (in a first approximation this is the case for defects with dimensions lower than the thickness of the sample), we can write the total scattered intensity distribution as a function of the scattering vector $\mathbf{s}$ ($|\mathbf{s}| = \sin(\Theta)/\lambda$), $I(\mathbf{s}) = |\langle F(\mathbf{s}) \rangle|^2 + |\langle \Delta F(\mathbf{s}) \rangle|^2$. $|\langle F(\mathbf{s}) \rangle|^2$ is the square of the Fourier transform of the electrostatic potential $\langle V(\mathbf{r}) \rangle$ of the average lattice and $|<\Delta F(s)>|^2$ represents the square of the Fourier transform of the deviation from the average lattice $<\Delta V(r)>$ which is essentially non-periodic (Diffuse Scattering). In the case of defects with the same scattering power as the matrix, if the relaxation of the lattice surrounding the defects is taken into account, each term of $<F(s)>$ can be expressed by the structure factor of the perfect crystal F_g multiplied by an exponential factor which depends on the square of reciprocal lattice vector $\mathbf{g}$ and on the atomic mean square static displacement $<u^2_x>_s$ along a direction x, normal to the crystallographic plane. This factor has the same form as the thermal Debye-Waller factor and is expressed by: $\exp[-2\pi^2 g^2 <u^2_x>_s] = \exp[-L_H]$. The corresponding diffuse scattering tends to have an approximately Gaussian shape with local maxima near to the reciprocal lattice point positions (Cowley, 1990). The most accurate method for measuring L_H is the quantitative analysis of RC of high order diffracted beams (Taftø and Metzeger, 1985). In fact it is well known that these reflections are most sensitive to the atomic positions and, in addition, their variations with F_g (or the extinction distance ξ_g) and thickness of the sample, t, are quite well described within a kinematical approach. In particular, the area under the RC is proportional to the quantity $|F_g|^2 \cdot t$. The evaluation of the ratio between the areas under the RCs taken in the perfect crystal (A_{Si}) and in the Implanted Layer (A_{IL}), multiplied by the inverse ratio of the thickness of the sample in the two areas, allows determination of L_H as:

$$2L_H = \ln\left[\frac{A_{Si}}{A_{IL}}\frac{t_{IL}}{t_{Si}}\right] \quad (1)$$

From an experimental point of view the best method to obtain RCs of high order diffracted beams is LACBED in the Tanaka mode [Tanaka, 1986]. With this technique, owing to the very large angle of convergence of the probe impinging onto the specimen, it is possible to obtain Dark Field (DFLACBED) patterns of high angle reflections. A line scan across the DFLACBED pattern reveals the local RC originating from an area of the specimen of the order of the spot size.

3. EXPERIMENTAL

Silicon <100> wafers were implanted with 1.6 x 10^{16} H$^+$ cm^{-2} 15.5 keV and subsequently annealed in vacuum at 600 C for 2 h. <110> TEM cross-sections were first prepared by dimpling (residual thickness of about 10 µm) and then ion milled with Ar$^+$ ions of 4 keV at glancing angle (14 degrees). A TEM Jeol 2010 was used for HREM analyses . LACBED patterns were taken with a Philips EM400 operating at 100kV in nanoprobe mode (nominal spot size 10 nm, selected area aperture of 10 µm) equipped with a 694 Gatan Slow Scan Camera.

4. RESULTS AND DISCUSSION

The defects detected in the implanted layer by TEM are located at a mean depth of about 210 nm and form a band 100 nm wide. In Fig. 1 two HREM images of the typical defects observed are reported: extrinsic {100} platelets and small cavities, formed by the coalescence of self-interstitials and vacancies, respectively (Cerofolini et al 1992, Muto and Takeda 1995). The DCD analysis in the Bragg symmetric geometry for the (400) reflection reveals a profile of static disorder where TEM detects extended defects. A maximum value of L_H=0.25 corresponding to $<u^2_x>_s$ =1.5 x 10^{-2} nm is detected at a depth of about 200 nm. Then this sample represents a good test for static disorder measurements by LACBED as the defective layer is formed by small defects which are clearly detectable in TEM and HREM images. In order to analyse the static disorder in the implanted layer with an electron diffraction technique, we have performed LACBED analysis following the method proposed in section 2. The sample was oriented near the [4$\bar{4}$1] zone axis where the {hh0} planes are in systematic orientation and the corresponding Bragg contours cross the defective layer at right angles. For the area-ratio measurements of (1) we have focused ourselves on the (660) and (880)

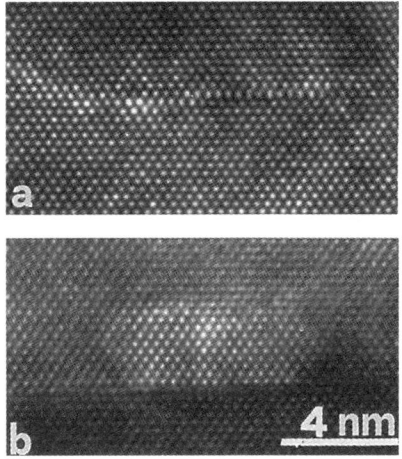

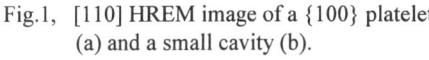

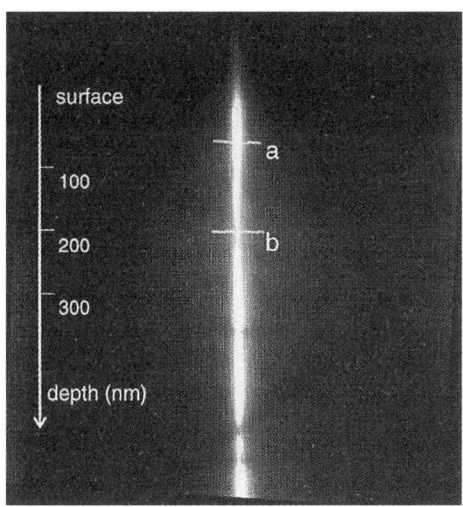

Fig.1, [110] HREM image of a {100} platelet (a) and a small cavity (b).

Fig.2, **g**= 880, DFLACBED pattern. **a** and **b** mark line scan position at a depth 70 and 210 nm respectively.

reflections. These reflections are characterised by extinction distance of 473 and 918 nm, respectively, and therefore "kinematical" for sample thicknesses lower than 150 nm. In addition, the corresponding crystallographic planes are not sensitive to the stress relaxation along the [100] and [1$\bar{1}$0] directions which are the free surfaces of the cross-sections. In Fig.2 is presented a DFLACBED pattern of **g**=880 with two line scans superimposed, where RC's reported in Fig 3a have been registered. It is evident that the intensity of the reflection decreases when it crosses the defective layer; meanwhile an increase in the diffuse scattering is observed. The baseline of Fig 3a is the sum of a linear background and the Gaussian curve which best fits the tails of the RC, where the oscillations are less visible as the intensity is mainly due to the diffuse scattering. These regions can be more easily detected in a differential plot of the RC. This was found to be the most reasonable and reproducible method of background subtraction. However the assumptions of a linear baseline only, versus a more stringent background evaluation lead to a variation of $4 \cdot 10^{-3}$ nm in the final value of $[<u_x^2>_s]^{1/2}$, which is twice the absolute error reported in Table 1. The RCs of **g**=880 after background subtraction are reported in Fig 3b; from these plots we have calculated the ratio of the areas under the RCs (see Table 1). The ratio of the local thickness was determined by measuring the periodicity of the oscillations in the tail of the RC of the (220) reflection available in the same zone axis. The thickness of the sample was found to be 120 and 150 nm for the line scan labelled a and b in Fig.2, respectively. The results, converted in atomic root mean square displacements, are reported in Table 1 together with the ones obtained applying the same method to the (660) reflection.

5. CONCLUSIONS

The LACBED technique has been applied to the determination of static disorder in H implanted Si after thermal treatment at 600°C. The defects present in the implanted layer produce a static disorder, with a peak value of 0.01 nm. which is of the same order of magnitude of the one measured by DCD. However a small difference seems to exists between the two measurements. A possible explanation could be the effect of the anisotropy of the crystal structure as the two displacements are measured along different crystallographic directions. Work is in progress to study this discrepancy. In particular the method could be applied to other reflections to sample the displacement field along different crystallographic directions.

418

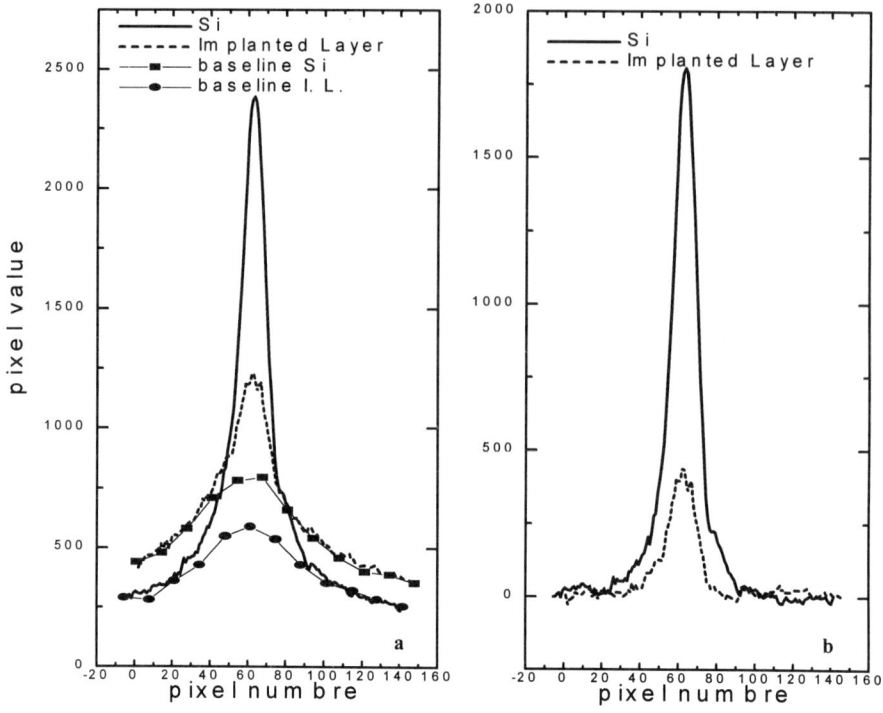

Fig 3: a) **g**= 880, line scans across the DFLACBED (Rocking Curve) in perfect Si and Implanted Layer (see Fig.2), raw data and baseline; b) **g**=880, RC after background subtraction.

	g=880 (21 nm^{-1})	g=660 (15 nm^{-1})
extinction distance (nm)	918	473
Ratio of the Area under RC	5.01	3.30
$[<u_x^2>_s]^{1/2}$ (nm) (LACBED)	0.010_2	0.011_3

Table1: Summary of the analysis of the RC's of **g**=880 and **g**=660

REFERENCES

Balboni R, Milita S, and Servidori M,1995 Phys.Stat. Sol (a) **148**, 95
Cerfolini G F, Meda L, Balboni R, Corni F, Frabboni S, Ottaviani G, Tonini R, M Anderle and Canteri R, 1992 Phys Rev B **46**, 2061
Cowley J M, 1990 Diffraction Physics pp 139-261
Muto S and Takeda S ,1995 Phil. Mag. Lett. **72**, 99
Taftø J, Metzeger T H, 1985 J. Appl. Cryst. **18**, 110
Tanaka M, 1986 J. Electron Microsc. **35**, 314

ACKNOWLEDGEMENTS: The authors are indebted to Prof. G. Ottaviani (Dip. Fisica Modena) and M. Servidori (LAMEL) for useful discussions. This work was partially supported by MURST and INFM.

Inst. Phys. Conf. Ser. No 157
Paper presented at Microsc. Semicond. Mater. Conf., Oxford, 7–10 April 1997
© *1997 IOP Publishing Ltd*

TEM characterisation of carbon ion implantation into epitaxial $Si_{1-x}Ge_x$

A Romano-Rodríguez[1], A Pérez-Rodríguez[1], C Serre[1], L Calvo-Barrio[2,1], A Bachrouri[1], O González-Varona[1], J R Morante[1], R Kögler[3] and W Skorupa[3]

[1]EME, Dept. Física Aplicada i Electrònica, Universitat de Barcelona, Avda. Diagonal 645-647, E-08028 Barcelona, Spain
[2]Serveis Científico-Tècnics, Universitat de Barcelona, Solé i Sabarís 1-3, E-08028 Barcelona, Spain
[3]Forschungszentrum Rossendorf e.V., Institut für Ionenstrahlphysik und Materialforschung, Postfach 510119, D-01314 Dresden, Germany

ABSTRACT: Strained and relaxed epitaxial $Si_{1-x}Ge_x$ layers (x=0.265-0.546) have been implanted with C ions at 500°C and doses corresponding to peak concentrations of x and 0.1x. The results show that C introduced precipitates in the form of β-SiC, causing a Ge enrichment of the $Si_{1-x}Ge_x$ layers. For the lower doses no degradation of the crystal quality and no relaxation occurs; for the higher doses, full relaxation and strong Ge migration are observed. This is likely to be related to segregation of Ge from the $Si_{1-x}Ge_x$ phase and the consequent strong Ge enrichment in this layer.

1. INTRODUCTION

There is an increased interest in the growth and characterisation of group IV semiconductor heterostructures, due to the possibility to control their electronic properties and, thus, their possible application in Si technology. Strong efforts have been dedicated to the study of the $Si_{1-x}Ge_x$ alloy system and its technology is mature enough to be used in modern devices. The possibility of incorporating substitutional C in the $Si_{1-x}Ge_x$ without loosing the high crystalline quality of the layers has opened the opportunity of band-gap engineering, something which up to now was restricted to III-V compounds.

High dose implantation is a well known technique for the processing and synthesis of heterostructures and multilayers such as silicon-on-insulator (Cristoloveanu 1994) and silicide contacts (Mantl 1994). High quality SiC crystals in silicon can be synthesised by high temperature implantation (Romano-Rodríguez *et al* 1996). The study of these processes in $Si_{1-x}Ge_x$ is useful to evaluate the incorporation of C into substitutional positions in the layer, the synthesis of new phases, such as SiC, and the stability of the $Si_{1-x}Ge_x$ layers under further treatments. In this work high dose, high temperature (500°C) carbon implantation into epitaxial $Si_{1-x}Ge_x$ layers is investigated.

2. EXPERIMENTAL

Two samples with different Ge compositions and relaxation states, grown on Si substrates, have been used as starting materials. Sample A had a composition of x=0.265, a thickness of 73 nm and was fully strained, while sample B had a composition of x=0.546, a thickness of 180 nm and was fully relaxed. C ions have been implanted at a temperature of 500°C in order to avoid the amorphization of the substrate, similar to the experiments performed previously on Si substrates (Romano-Rodríguez *et al* 1996). The implantation energies have been chosen to obtain a peak carbon concentration located almost in the centre of the $Si_{1-x}Ge_x$ layer. Furthermore two different doses have been employed for each sample: one dose which gives rise to a peak carbon concentration about 1/10th of the Ge content, which, when incorporated into substitutional position in the $Si_{1-x}Ge_x$ lattice (hereafter called "lower dose"), should give a strain-free ternary layer, and a dose giving rise to a peak concentration equal to the Ge content (hereafter called "higher dose"). According to TRIM simulations (Ziegler *et al* 1985), the energies and doses required in this study for the two samples are 15 keV and doses of 6×10^{15} and 6×10^{16} cm^{-2} for sample A and 40 keV and doses of 2×10^{16} and 2×10^{17} cm^{-2} for sample B.

The different samples have been analysed by Transmission Electron Microscopy (TEM), X-Ray Diffraction (XRD), both in low and high resolution (HRXRD), X-ray photoelectron spectroscopy (XPS) and Raman scattering.

3. RESULTS AND DISCUSSION

Fig. 1 shows the TEM images and corresponding SAD patterns of sample B (relaxed) before and after implantation at the higher dose. In the as-grown sample a high density of defects is visible at the interface, which extend both towards the surface and the bulk silicon (Fig. 1a). This is caused by the relaxation of the layer through the emission of dislocations, because its thickness is beyond the critical value for pseudomorphic growth (Matthews 1975). In the SAD pattern (Fig. 1b), the splitting of the diffraction spots corresponds to the epitaxial layer and to the substrate. After implantation it is clear from the image (Fig. 1c) that a high density of precipitates have been formed, which are identified as β-SiC from the SAD pattern (Fig. 1d). From the distribution of the β-SiC spots and from the Moiré fringes in Fig. 1c, a well defined orientation relationship between the precipitates and the Si and $Si_{1-x}Ge_x$ can be deduced (Pérez-Rodríguez *et*

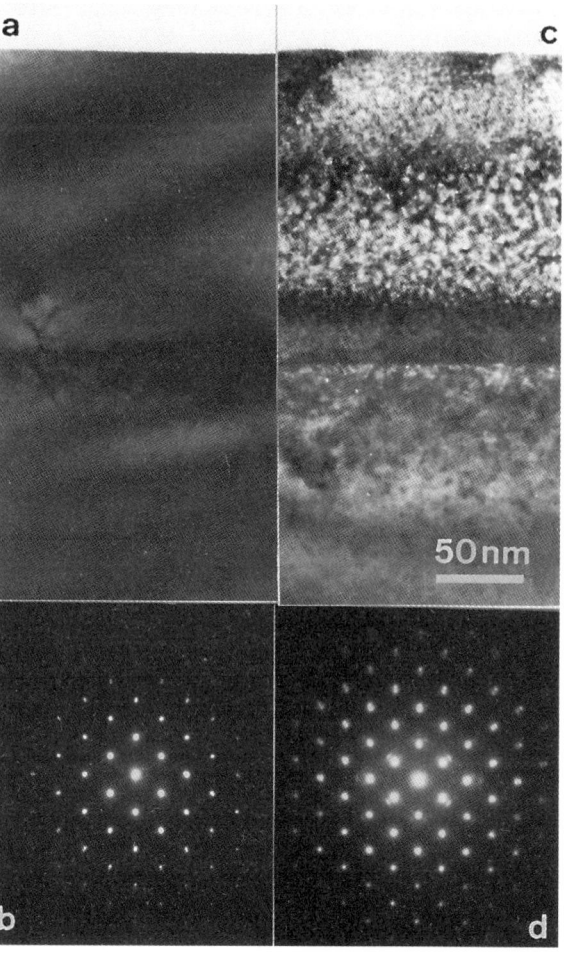

Fig. 1: a) TEM and b) SAD of sample B as-grown; c) TEM and d) SAD of sample B implanted at the higher dose.

al 1996, Romano-Rodríguez *et al* 1996). This result proves that phase separation occurs during very high dose implantation: the carbon atoms tend to precipitate in the form of β-SiC, and for this, Si from the $Si_{1-x}Ge_x$ is consumed, which causes a Ge enrichment of the layer. Similar results have been observed in $Si/Si_{1-x-y}Ge_xC_y/Si$ heterostructures grown by RTCVD, where SiC precipitation is induced by thermal annealing (Warren *et al* 1995). Confirmation of this mechanism is given by the increased separation of the Si and the $Si_{1-x}Ge_x$ spots in the SAD pattern of Fig. 1d, as compared with the as-grown sample (Fig. 1b). Furthermore all these results are confirmed by XRD measurements,

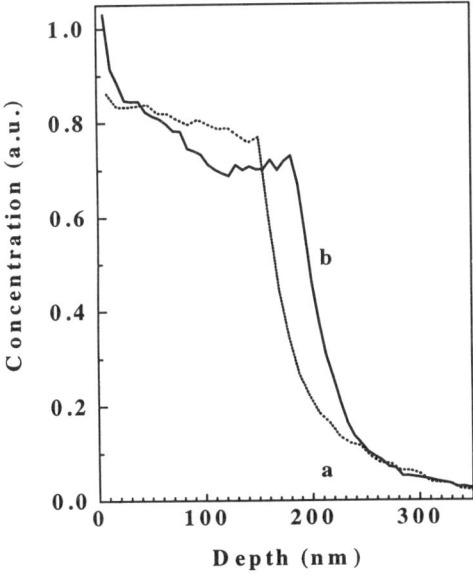

Fig. 2: Ge concentration in sample B as-grown (a) and after higher dose implantation (b). The outdiffusion of Ge in (b) is clear.

both in low and high resolution modes. From low resolution, the presence of β-SiC in the layer is confirmed and the orientation relationship observed by TEM is corroborated. HRXRD allows us to follow the position of the $Si_{1-x}Ge_x$ and Si peaks, from which the Ge enrichment is confirmed. For sample B, the Ge content is about x=0.75 after implantation. SiC precipitation also gives rise to an outdiffusion of Ge towards the regions with lower carbon concentration, as deduced from the XPS spectra of Fig. 2.

Fig. 3 shows the TEM images of sample A as-grown (a) and after implantation at the higher dose (b). In this case the splitting of the diffraction spots cannot be seen in the as-grown sample, because of the lower Ge content. Again the presence of β-SiC precipitates can be seen in the TEM image of Fig. 3b and the orientation relationship is the same as for sample B. In sample A, the interface $Si/Si_{1-x}Ge_x$ appears to be strongly damaged by the implantation, although it can still be identified. From HRXRD measurements it is deduced that, similar to sample B, Ge enrichment occurs in the layer. In this case the Ge content is about x=0.43 and the sample is fully relaxed, which is consistent with the strong Ge enrichment of the lattice. This relaxation can be seen in the HRXRD spectra of Fig. 4. In opposition, no outdiffusion has been observed in sample A.

These results contrast with the behaviour observed for the samples implanted at lower doses, where few differences are seen as compared with the as-grown samples. From TEM only the presence of some precipitates can be detected, whose nature is not yet clear. It is believed that, as already observed for the higher implantation dose, they could be β-SiC. Further observations are still ongoing to clarify this point.

On the other hand, XRD gives more insight into these samples. First, conventional XRD measurements show no β-SiC peak, probably because of the low implantation dose. Furthermore, Raman spectroscopy excludes the existence of amorphous or crystalline C precipitates. However from HRXRD measurements the composition and degree of relaxation of the epitaxial layer can be deduced, provided the stiffness coefficients of the alloy are known. The elastic constants of $Si_{1-x}Ge_x$ can be deduced for each alloy composition by linearly

Fig. 3: TEM image of the sample A as-grown (a) and implanted at the higher dose (b).

422

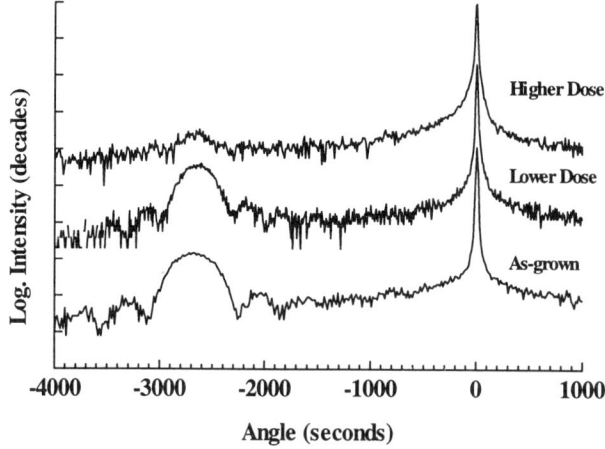

Fig. 4: (004) Rocking curve from sample A as-grown and implanted at the lower and higher doses. The presence of Pendellösung fringes after the lower implantation dose is visible.

interpolating the values between those of Si and Ge (Baker and Arzt 1995). In order to estimate the C incorporation into the $Si_{1-x}Ge_x$ layer, linear interpolation between the values for Si, Ge and C (diamond) has been assumed for the ternary compound. Furthermore, it has been assumed that the fraction of carbon which is not incorporated forms β-SiC, thus consuming Si from the layer. The results obtained confirm, within the resolution of the XRD technique, that in neither of the two samples carbon has been incorporated into the layer and, resulting in Ge enrichment, giving the values of x=0.269 for sample A and x=0.563 for sample B. On the other hand, in spite of the implantation conditions and of the precipitation of β-SiC, sample A only presents a relaxation of about 4%, i.e., remains essentially strained, as can be seen in Fig. 4, where this sample shows Pendellösung fringes. This result is due to little changes in the Ge content in the epitaxial layer, which is highly stable.

4. CONCLUSIONS

In this work the characterisation of high temperature carbon implantation into strained and relaxed $Si_{1-x}Ge_x$ layers has been performed. The high temperature has been used to avoid the amorphization of the $Si_{1-x}Ge_x$ layer during implantation. For the employed implantation conditions no carbon is detected in substitutional positions in the layer, observing phase segregation and formation of β-SiC. This consumes Si atoms from the $Si_{1-x}Ge_x$ layer, leaving Ge enriched regions on both sides of the implantation peak. Strain relaxation only occurs for the highest values of Ge enrichment, which correspond to the highest implanted dose. These data point out the high stability of the epitaxial layers for medium implantation doses.

REFERENCES

Baker S P and Arzt E 1995 Properties of strained and relaxed silicon germanium, ed E Kasper (London: INSPEC).
Cristoloveanu S 1994 Silicon-on-insulator technology and devices (Pennington: The Electrochemical Society).
Mantl S 1994 Nucl. Inst. Meth. Phys. Res. B **84**, 127.
Matthews 1975 J. Vac. Sci. Technol. **12**, 126.
Pérez-Rodríguez A, Romano-Rodríguez A, Serre C, Calvo-Barrio L, Cabezas R, González-Varona O, Morante J R, Kögler R, Skorupa W and Rodríguez A. 1996 Nucl. Inst. Meth. Phys. Res. B **120**, 173.
Romano-Rodríguez A, Serre C, Calvo-Barrio L, Pérez-Rodríguez A, Morante J R, Kögler R and Skorupa W 1996 Mater. Sci. Eng. B **36**, 282.
Warren P, Mi J, Overney F and Dutoit M 1995 J. Cryst. Growth **157**, 414.
Ziegler J F, Biersack J P and Littmark U 1985 The Stopping and Range of Ions in Solids (New York: Pergamon).

Inst. Phys. Conf. Ser. No 157
Paper presented at Microsc. Semicond. Mater. Conf., Oxford, 7–10 April 1997
© *1997 IOP Publishing Ltd*

Polycrystalline silicon grain structure in VLSI devices

R Lindsay, J N Chapman, A J Craven and D McBain[1]

Department of Physics & Astronomy, University of Glasgow, Glasgow, UK
[1]Motorola Ltd, Kelvin Industrial estate, East Kilbride, UK

ABSTRACT: Two TEM diffraction techniques are used to perform orientational analysis on polycrystalline silicon films. One method involves the analysis of Kikuchi patterns and the other a new selected area scanning technique. Cross-sections of fully processed silicon wafers, prepared using a tripod polishing technique, are used to reveal the degree of texturing in the polySi films. Both methods were successful in demonstrating the absence of texturing in the films studied.

1. INTRODUCTION

As devices shrink, it is becoming more and more important to understand fully the structure of films used in the devices. Electrical performance and surface roughness are significantly affected by the level of texturing (the existence of preferred orientations) in a film (Kakinuma 1995) and so there is a need to provide a quick and reliable technique to perform texture analysis on it. Conventional selected area diffraction using mechanical circular apertures often fails to fit the size and shape of the area studied. This paper uses polysilicon (polySi) films on a fully processed EEPROM silicon device to illustrate two cross-sectional TEM (XTEM) diffraction techniques: individual grain orientation using Kikuchi patterns and a new selected area scanning diffraction technique, the emphasis being on the latter.

2. EXPERIMENT

An EEPROM silicon device is prepared for XTEM analysis using the tripod polishing technique as described by Anderson (1992) Using a high magnification (x400) optical microscope to locate the exact (±1μm) area of interest on the chip, it is possible to produce a wedge shaped specimen with the area of interest lying along the electron transparent apex of the wedge. In this way, a specimen is prepared with an electron transparent region 1-2mm long, allowing many polySi areas to be investigated. Fig. 1 shows an EEPROM capacitor which

Fig. 1: XTEM image showing 2 polySi films forming an EEPROM capacitor.

contains two polySi films, defined as 'poly 1' and 'poly 2', which are ≈270nm and ≈470nm thick respectively. Shown here is an example of where the surface roughness of the poly 1 becomes a limiting factor on how close subsequent polySi layers in a capacitor can be deposited. According to Hasegawa (1990), surface roughness can be improved if the polySi is deposited using plasma enhanced CVD and is <110> textured.

3. SURFACE NORMAL GRAIN ORIENTATIONS USING KIKUCHI PATTERNS

The first diffraction technique involves probing individual polySi grains (50-400nm across) using the Philips CM20 FEG (S)TEM and examining the Kikuchi pattern for each grain to determine the surface normal grain orientation, n_s. In this microscope setup, the spot sizes can be reduced to less than a nanometer if required. An example of such a Kikuchi pattern is shown in fig. 2 with the surface normal direction obtained from a substrate diffraction pattern. On the schematic diagram, P is the pole out of the page (e.g. [345]) obtained from a Si Kikuchi map, V is a vector from the pattern (e.g. [0$\bar{4}$4]) and n_s is the surface normal direction. Note that P and V are not necessarily perpendicular. The xy-plane is in the plane of the diffraction pattern and x is defined by the projection of V onto the pattern. To calculate n_s, $n_s = \sin \psi. (P^\wedge V) + \cos \psi. (P^\wedge V)^\wedge P$, where unit vectors are used throughout.

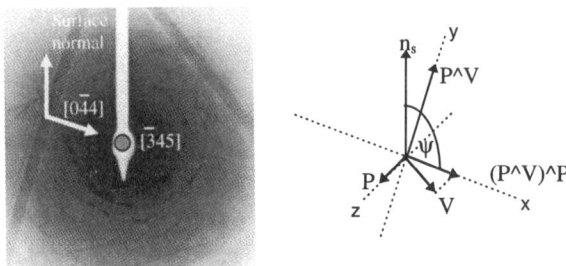

Fig. 2: An example of the analysis done on each Kikuchi pattern for a polySi grain.

The diagram in fig. 2 demonstrates how the surface normal grain orientation can be obtained from a Kikuchi pattern. The orientations are presented on a square θ–ϕ grid (spherical polar co-ordinates) shown in fig. 3. The cubic Si structure allows us to confine all possible orientations onto a section of the grid as shown. Looking at the orientation projections for both thicknesses of polySi in fig. 3 it is clear that although some orientations, e.g. <011>, <012> and <234> are slightly more abundant in poly 2; the results suggest that preferred orientations, or texturing, is highly unlikely for both cases.

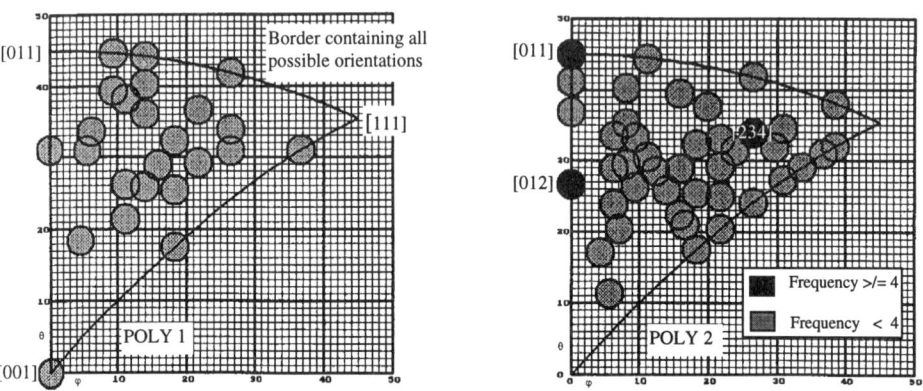

Fig. 3: Grain orientation projections showing the distribution of orientations for poly 1 and poly 2.

4. ORIENTATIONAL ANALYSIS USING SELECTED AREA STEM

If there is a large number of grains to be analysed the Kikuchi analysis mentioned before can be quite time consuming. The following method is performed in STEM mode whereby a selected rectangular area is scanned producing a pattern consisting of the individual grain diffraction patterns superimposed. The rectangular shapes of the polySi films as seen in fig. 1 show that this technique is well-suited for studying texture in the films. Once the microscope is set up for this analysis it is likely to be the quickest and most reliable way to highlight the presence of crystallographic texturing.

With many (>10) grains in each area studied it will be necessary to avoid diffraction spots overlapping, e.g. spots from the 311 and 220 planes. The angular resolution has to be greater than that produced by normal lens strengths on the STEM. Therefore once the microscope is aligned in normal STEM mode, further changes are made to decrease the convergence angle.

Using the smallest C2 aperture, the strength of the objective lens is decreased and the C2 lens strength is increased, as shown in fig. 4.

The scanning pivot points are positioned to be conjugate with the objective aperture plane which in turn is conjugate with the detector plane. In this way a stationary diffraction pattern is produced with all the possible silicon diffraction spots resolved.

Fig. 5 shows selected area STEM diffraction patterns of the two polySi films. Areas like the poly 1 region shown in fig. 1 are scanned across. Again the surface normal direction is obtained from a substrate diffraction pattern and is in the plane of the pattern.

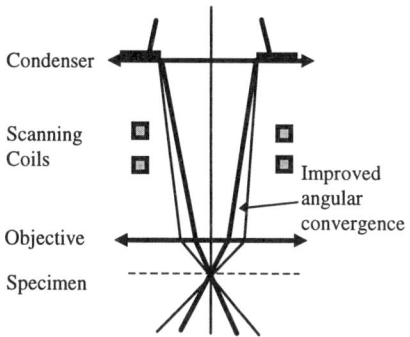

Fig. 4: Lens configuration for improved angular resolution.

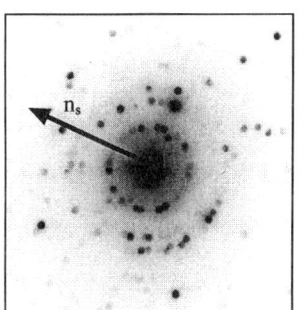 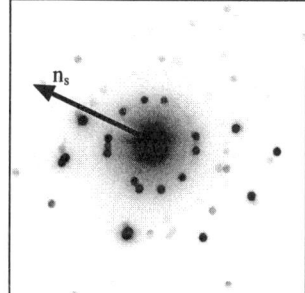 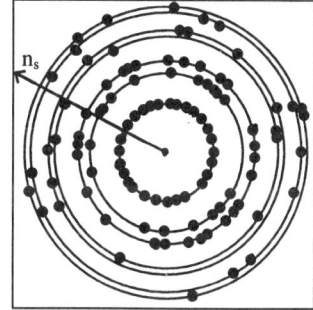

Fig. 5: Scanning diffraction patterns for poly 1 (left) and poly 2 (centre). 4 superimposed poly 2 diffraction patterns (right).

In this cross-section, the polySi layers are discontinuous so there is a limited length available for each scan with the areas scanned being similar in size for both poly 1 and poly 2. The grains in poly 1 are smaller leading to a higher spot density in the diffraction patterns. The relatively small number of grains present in the rectangular areas of poly 2 puts a restriction on how conclusive the results can be, so 4 diffraction patterns are superimposed (fig. 5) with the spots clearly approaching ring patterns.

Calculations done in Microsoft Excel allow for the display of any cubic theoretical textured diffraction pattern with a fixed axis and random azimuthal angle (here for the silicon structure) so they can be compared to the patterns from the areas scanned. The angle between the textured direction (always considered upwards) and each possible reflection for the diamond structure is calculated and this gives the horizontal and vertical component of the reflection. Texturing in the <110>, <210> and <234> directions are considered (fig. 6) since these were seen to be slightly more abundant in the analysis done using the Kikuchi patterns.

426

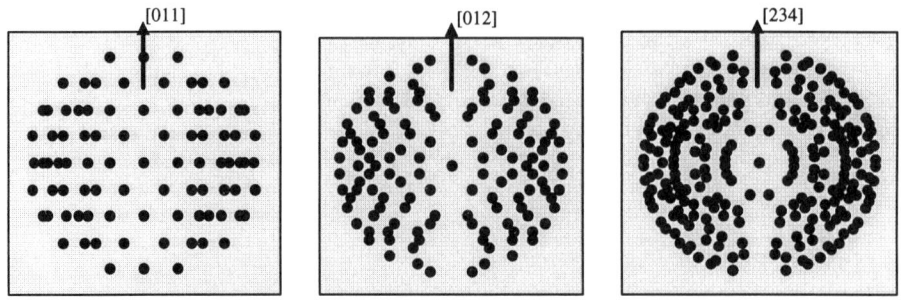

Fig. 6: Theoretical diffraction patterns for texturing in the <011>, <012>, and <234> directions respectively.

Texture around a given direction leads to a band through the pattern with a low density of spots as shown in fig. 6. This band lies parallel to the preferred direction. Examining the STEM diffraction patterns in fig. 5 for both polySi films, no evidence of texturing is found. Over 30 diffraction patterns are acquired from around 10 EEPROM cells and there is no sign of a low spot density band in any of the patterns, implying a randomly orientated grain structure.

The results from both diffraction analysis techniques agree very well in illustrating the random nature of the grain orientations. There may be a slight preference for orientations in the <011>, <012> and <234> directions but, even if so, it is insignificant.

5. CONCLUSIONS

The scanning technique itself seems very promising as a quick and reliable way to investigate the level of texturing in a given rectangular area. Normal selected area diffraction using circular apertures would be completely inadequate in performing the analysis required here because of the size and shape of the area selected. With such small film thicknesses in modern devices, mechanical apertures are insufficient for cross-sectional TEM diffraction analysis.

ACKNOWLEDGEMENTS

We should like to thank EPSRC and Motorola Ltd for their financial support.

REFERENCES

Anderson R M 1992 Mater. Res. Soc. Proc. **254**, 141
Hasegawa S, Yamamoto S and Kurata Y 1990 J. Electrochemical Soc. **137**, **11**, 3666
Kakinuma H 1995 J. Vac. Sci. & Tech. A-Vac. Surfaces & Films **13**, **5**, 2310

Inst. Phys. Conf. Ser. No 157
Paper presented at Microsc. Semicond. Mater. Conf., Oxford, 7–10 April 1997
© 1997 IOP Publishing Ltd

Structural and electronic properties of partially crystallised silicon

P D Brown, J P Smith[1], W Eccleston[1] and C J Humphreys

Dept of Mats Sci & Metallurgy, Univ of Cambridge, Pembroke St, Cambridge CB2 3QZ
[1]Dept of Elect Eng & Electronics, Univ of Liverpool, Brownlow Hill, Liverpool L69 3BX

ABSTRACT: The structural and electronic properties of polycrystalline silicon produced by the solid phase crystallisation of low pressure chemical vapour deposited amorphous films are compared. The amorphous to crystalline transition under annealing is characterised on the structural level by dendritic crystal growth and grain coalescence, and on the electronic level by a gradual change in the mode of conduction from hopping via localised states to an extended state mode of conduction. Grain size and the extent of layer crystallisation depend both on the film thickness and the time of annealing. A high density of stacking faults is found to develop during the very first stages of grain formation.

1. INTRODUCTION

In thin film transistors the high generation rate which occurs when a drain voltage is applied to polysilicon leads to a high leakage current which is undesirable for maintaining charge on the pixel. The trade off between film thickness and time of annealing is important to establish improved conductivity while maintaining relatively high activation energies. It is recognised that the deposition conditions affect the crystallisation behaviour of amorphous silicon (a-Si) films (Guillemet et al 1993) and this will have consequence for the mode of conduction. In this study, the structural and electronic properties of annealed amorphous silicon films deposited on $SiO_2/Si(001)$ substrates have been compared.

2. EXPERIMENTAL

Layers of a-Si were deposited by low pressure chemical vapour deposition (LPCVD) at 550°C from silane onto a silicon dioxide buffer layer on (001) oriented n-type silicon. The a-Si layers examined here, which did not receive external doping or hydrogenation, were nominally 300nm, 600nm and 900nm respectively in thickness. The silicon dioxide buffer layer was grown to a thickness of ~ 500nm using a wet thermal process at a temperature of 1100°C. Samples were annealed at 550°C for up to 16 hours to provide material covering different extents of crystallinity. Current - voltage measurements were carried out between surface electrodes on each sample at temperatures ranging from 30°C to 250°C in order to calculate the activation energy. Microstructural characterisation of the as-deposited and partially crystallised material was performed using a JEOL 4000 EX-II electron microscope.

3. RESULTS AND DISCUSSION

Reciprocal temperature vs current measurements comparing as-grown samples with those annealed for 16 hours are shown in Fig. 1. Line curvature is apparent for all the as-grown samples and this is attributed to two parallel modes of conduction, i.e. hopping conduction via localised states near the Fermi level (Nakashita et al 1984) and extended state conduction (Spear and Le Comber 1977). Solid phase crystallisation has the effect of reducing this curvature as a sample approaches full crystallisation due to increased electron mobility and the dominance of extended state conduction. Analysis of the high and low

428

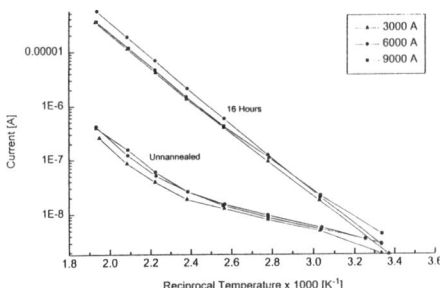

Fig. 1 Reciprocal temperature measurements for 300, 600 and 900nm samples of as-deposited a-Si compared with material annealed for 16 hours at 550°C in a nitrogen ambient.

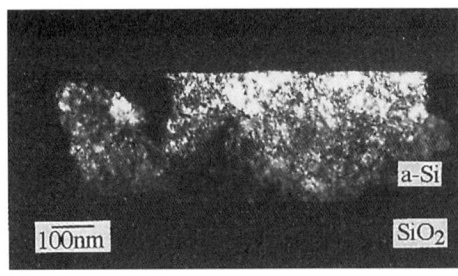

Fig. 2 Dark field cross-sectional image showing extended grain formation within a 300nm sample annealed for 16 hours.

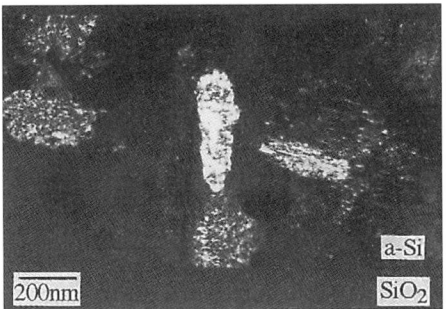

Fig. 3 Dark field cross-sectional image showing distribution of grains formed within a 900nm sample annealed for 16 hours.

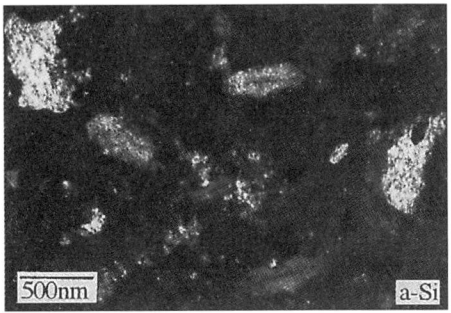

Fig. 4 Dark field plan view image showing grain formation within a 600nm sample annealed for 16 hours.

temperature regions of the curves for the as-grown samples yields activation energies in the range 0.78 to 0.82eV for extended state conduction and 0.16 to 0.18eV for hopping, respectively, as compared with the activation energy of 0.56eV (half the band-gap) associated with band-edge conduction in intrinsic crystalline material. Similarly, activation energies for extended state conduction in the samples annealed for 16 hours are determined to be in the range 0.60 to 0.64eV which is indicative of a wider effective band-gap and hence some remnant degree of disorder. In particular, there is also slight curvature associated with the 900nm sample which gives an activation energy for hopping of 0.41eV which is consistent with the removal of localised states from the band-gap during annealing.

Varying degrees of crystallinity are observed depending on sample thickness and time of annealing. No grain formation was evident within as-deposited unannealed layers, while dark field cross-sectional imaging of a 300nm sample annealed for 16 hours (e.g. Fig. 2) demonstrates the formation of well defined grains extending through the whole thickness of the film and up to ~ 700nm in length. Conversely, the dark field cross-sectional image of Fig. 3, recorded from the 900nm sample annealed for 16 hours, exhibits a more random mixture of grains, some of which propagate through the full film thickness and show dendritic character. Also, the lateral extent of grain growth in the plane of the a-Si film is noticeably less than for the 300nm sample annealed for the same time. Thus, even though thicker samples contain grains of comparable size to those within thinner films annealed for the same time, the fact that the lateral extent of grain growth is lower suggests that the mode of lateral electronic conduction through the film will differ depending on film thickness (for a fixed anneal time), i.e. thicker films will exhibit a greater extent of hopping mode conduction in agreement with the electrical measurements.

The dark field plan view image of Fig. 4 is recorded from a 600nm sample annealed for

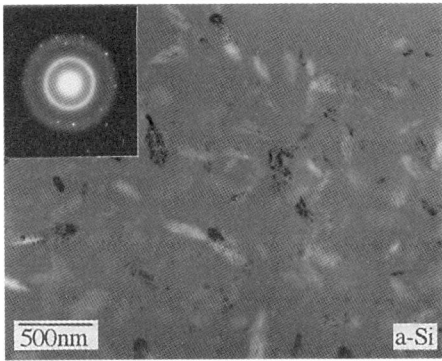

Fig. 5 Bright field plan view image showing formation of isolated grains within a 300nm sample annealed for 4 hours.

Fig. 6 Dark field plan view image showing dendritic grain growth within a 600nm sample annealed for 16 hours (lower grain <110> SAD pattern inset).

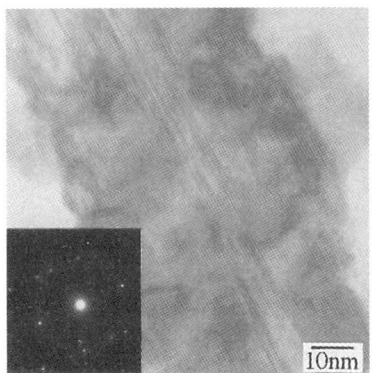

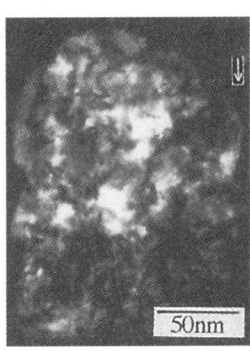

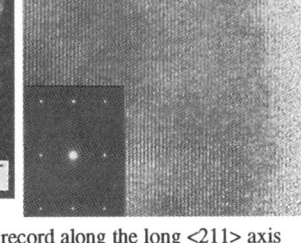

Fig. 7 <110> HREM image showing detailed core structure of a dendrite (twinned SAD pattern inset)

Fig. 8 (a) Dark field image record along the long <211> axis of a grain and (b) HREM image of top right hand corner (arrowed) showing {111} planes at grain surface.

16 hours. A mixture of dendritic grains and irregular grains are present up to ~ 1μm in size. Comparison of Figs. 3 and 4 demonstrates that dendritic grains can form both parallel and perpendicular to the substrate surface, indicating that grain nucleation can occur throughout the layer and is not just confined to the a-Si/SiO$_2$ interface and the a-Si layer free surface.

Samples annealed for intermediate times were also examined to characterise the early stages of grain growth. Fig. 5 shows a bright field plan view image recorded from a 300nm sample annealed for 4 hours. Isolated grains of typical size 200nm distributed randomly throughout the amorphous matrix are present. The majority are elongated in shape and again show dendritic character on closer inspection.

The dark field image of Fig. 6 shows the twin bands comprising the core of a well defined dendritic grain. The high resolution image of Fig. 7a also shows the structure of such a faulted dendritic crystal core. Such grains have long axis along <211> and exhibit a high density of thin planar defects propagating into the bulk of the grain from the {111} twin bands. This is in accordance with the in-situ annealing studies of Batstone and Hayzelden (1993), whereby the fastest growth front of dendritic grains occurs at the emergent point of these multiple twin bands. It is also suggested that the likelihood of such twinning increases in the presence of impurities (Batstone 1993).

To give a more complete view of the crystal dendrite geometry, Figs. 8a and 8b show dark field and high resolution images of a grain viewed along the long <211> axis. The grain is bounded by {111} and {110} planes.

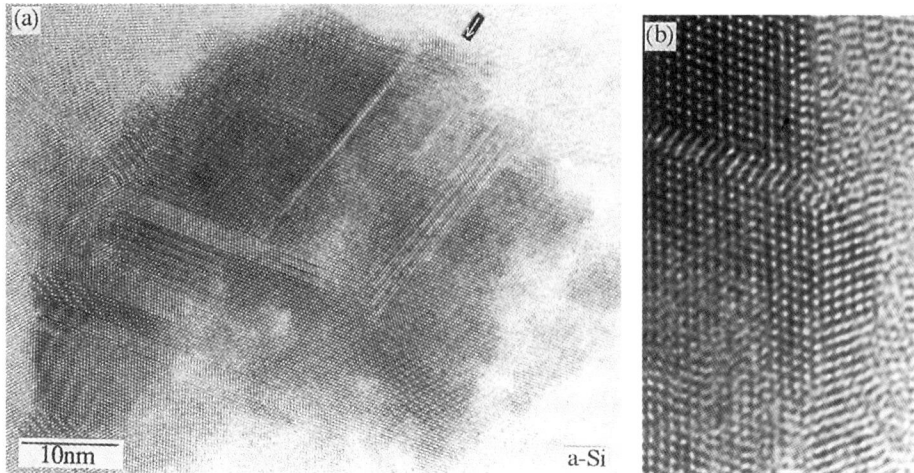

Fig. 9 (a) <110> HREM image showing a small grain in the initial stages of formation and (b) surface twin.

Fig. 9a shows a <110> projection of a small grain formed within a 300nm sample annealed for 16 hours. The grain contains a number of stacking faults and microtwins, either within the grain or nucleating on the {111} facets of the grain surface (Fig. 9b). It is likely that a combination of both growth and deformation induced twinning processes are operative during the very first stages of grain nucleation and growth, in a manner similar to that found for the very first stages of microtwin formation within sphalerite heteroepitaxial thin films (Brown et al 1995).

In summary, while electrical activation energy measurements demonstrate that increased time of annealing leads to a transition from hopping to extended state conduction, film thickness also plays an important role in determining the dominant conduction path within partially crystalline material, since thicker films exhibit higher levels of hopping via localised states than thinner films annealed under the same conditions. The defect microstructure of annealed layers of a-Si takes the form of evolving dendrite crystals which coalesce with increasing time of annealing. Thinner films show increasing extent of lateral crystal growth and extended state mode of conduction. Thicker films exhibit comparable grain sizes but a reduced lateral extent of crystallisation. It is this balance between grain size and film thickness for increasing times of annealing which will define the optimum electronic properties of the material. Dendritic grains are characterised by twin band core structures, and a high density of stacking faults is generally associated with grains from the first stages of nucleation.

Acknowledgements

With thanks to John Alderman of the National Microelectronics Research Centre in Ireland for the provision of amorphous silicon samples.

REFERENCES

Batstone J L 1993 Phil. Mag. A67 51
Batstone J L and Hayzelden C 1993 Inst Phys Conf Ser No. 134 165
Brown PD, Loginov YY, Stobbs WM and Humphreys CJ 1995 Phil. Mag. A72 39
Guillemet J P, De Maudit B, Pieraggi B, Campo E and Scheid E 1993 J. Mat. Sci. Lett. 12 910
Nakshita T, Hirose M and Osaka Y 1984 Japan J. Appl. Phys. 23 146
Spear WE and Le Comber PG 1977 Proc 7th Int. Conf. Amorphous and Liquid Semiconductors p309

Inst. Phys. Conf. Ser. No 157
Paper presented at Microsc. Semicond. Mater. Conf., Oxford, 7–10 April 1997
© 1997 IOP Publishing Ltd

431

Microstructure study of pure hydrogen RF-sputtered microcrystallized silicon thin films

F Gourbilleau, A Achiq, P Vermaut, P Voivenel and R Rizk

LERMAT, UPRES-A 6004, ISMRA, 6 Bd Maréchal Juin 14050 Caen Cedex France

ABSTRACT: Microcrystalline silicon (μc-Si) films were prepared by RF sputtering with pure hydrogen plasma without any post-treatment at various substrate temperatures, (T_s). The microstructure of the deposits was investigated by Transmission Electron Microscopy for T_s ranging between 50°C and 250°C. Upon T_s increase, the average grain size gradually evolves from 13 nm to 40 nm with a concomitant increase of the crystalline fraction. For the lowest T_s, the crystallized films grow on an amorphous buffer layer whose thickness is T_s dependent. The results are interpreted in terms involving the role of hydrogen species, and particularly that of SiH_2 in the crystallization mechanism.

1. INTRODUCTION

The grain size control in microcrystallized silicon (μc-Si) is of great interest for a variety of applications in optoelectronic and microelectronic industry. The μc-Si films can generally be obtained by solid phase crystallization through thermal annealing of the basic amorphous layer or by direct crystallization occurring during the deposition process. In such cases, and in almost all the techniques used, hydrogen constitutes the dominant element of the plasma or of the gas vector (Veprek 1990, Matsuda 1983, Nada et al 1983). There, the hydrogen species interact with the growing surface leading to the relaxation of the network towards the crystalline phase. Among the deposition techniques, sputtering is a promising one in as much as it allows, on the one hand, the control of hydrogen content (Rizk et al 1996, Achiq et al 1997), and, on the other hand, the deposition of large areas.

In previous studies dealing with sputtered films, we have examined the influence of hydrogen content and substrate temperature on the optical and electrical properties of the μc-Si films (Achiq et al 1997, Blas et al 1997). In order to complement these studies, we propose in this contribution, a transmission electron microscopy (TEM) investigation of the microstructure of μc-Si films obtained by sputtering without any post processing in a pure hydrogen plasma for different substrate temperatures.

2. EXPERIMENTAL PROCEDURE

The μc-Si thin films were deposited by RF sputtering in pure hydrogen plasma (99.9999%) on single-crystal silicon substrates whose temperature (T_s) was varied from 50°C to 250°C. The depositions were performed at a total pressure of 0.4 Torr with a fixed power density of 0.6 W/cm^2. The film thicknesses were found to range between 550 nm to 700 nm. More details about the deposition techniques are given elsewhere (Rizk et al 1996).

For TEM observations, the samples were glued with deposited films facing each other using the well known cross-section technique. The TEM observations were carried out in a Jeol 200 CX operating at 200 kV.

3. RESULTS AND DISCUSSION

The influence of the substrate temperature on the microstructure of the deposited films has been investigated for T_s varying between 50°C and 250°C. On the left-hand side of Fig. 1 are shown the bright-field images and their electron diffraction patterns (EDPs) obtained for T_s values at 50, 100, 150 and 250°C, whereas the corresponding (111) dark-field micrographs are displayed on the right hand-side. The EDP rings that are originating from the {111}, {220} and {113} lattice planes are due to a superposition of diffraction spots originating from the numerous Si micrograins. They are characteristic of a random distribution of the crystallographic orientations. On the other hand, one can notice a continuing increase of the density of diffraction spots with an increase of T_s. In this respect, the EDPs display fine rings for the lowest temperature (50°C, Fig. 1a), whereas for T_s=250°C some reinforced diffraction spots can be clearly distinguished (Fig. 1d). This evolution might be attributed to an enhancement of the crystalline fraction, f_c, in the film that could be associated with a rise in the average grain size, in addition to some change in the crystallization mechanism. Considering the difficult determination of f_c from TEM observations, the grain size was estimated from the measurement of the high contrast dots in the dark-field mode. The evolution of the average grain size values as a function of the substrate temperature T_s is illustrated in Fig. 2. In the same way, the bright-field images (left-hand side of Fig. 1) reveal the appearance of a columnar growth from the T_s value of 150°C. This columnar structure becomes more pronounced for higher T_s values, but becomes much less evident for $T_s < 150$°C. These features are consistent with the above-mentioned suggestions of an increasingly important crystalline fraction for increasingly higher T_s values. This has been recently confirmed by Raman spectroscopy measurements performed on the same samples, which report an evolution of the crystalline fraction from 56% for T_s=50°C to about 72% for the highest T_s (250°C) (Blas et al 1997). It is worth noticing that, according to previous studies, the growth of a columnar morphology is thought to be induced by SiH$_2$ radicals (Tsai et al 1986), because of the low surface diffusivity and the high sticking coefficient of these species, as compared to their counterparts for SiH$_3$ (Moon et al 1994). Thus, on the basis of the evidence of SiH$_2$ incorporation in our samples, as reported elsewhere, (Achiq et al 1997), one can similarly explain the growth of a columnar structure for $T_s \geq 150$°C, where the "chemical annealing" effects (Shimizu et al 1990) and diffusivity of SiH$_2$ are expected to be notably enhanced. This is supported by the quasi-absence of such columnar growth for low T_s (50°C and 100°C). In these latter cases, instead, an amorphous buffer layer appears at the interface between the substrate and the growing film whose origin seems to lie in the limited interactions of SiH$_2$ with the surrounding network at these low substrate temperatures. This is corroborated by the decrease of both thickness and roughness of this intermediate layer as T_s increases. The growth of such an amorphous buffer layer has been already reported by many authors (Kaneko et al 1994, Yang and Abelson 1995) and was attributed in these studies to the limited nucleation rate at low temperatures. In this connection, it is quite plausible that this nucleation rate would be governed by the above-mentioned effects of the hydrogen species.

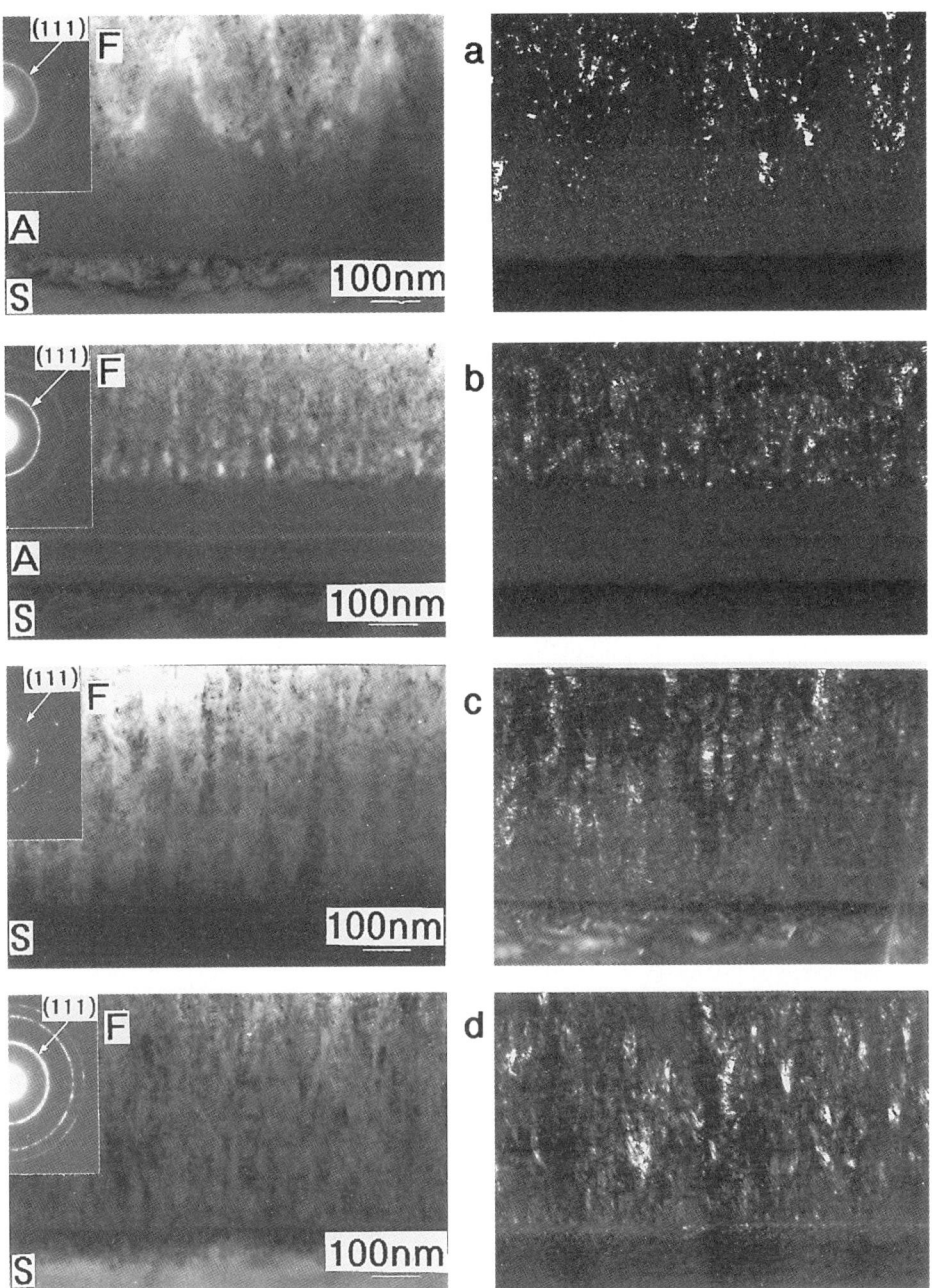

Figure 1: Microstructure evolution of the μc-Si layer with the substrate temperature T_s,
a) 50°C, b) 100°C, c) 150°C, d) 250°C
S : substrate ; A : amorphous buffer layer ; F : μc-Si film
On the left-hand side : bright-field images with the corresponding EDPs
On the right-hand side : (111) dark-field images

434

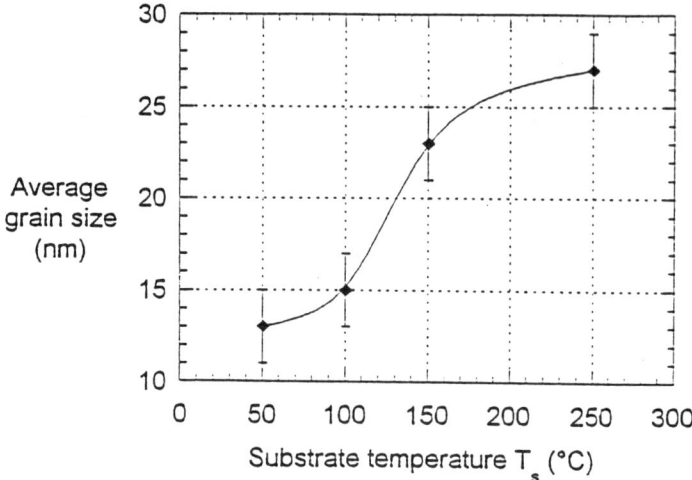

Figure 2 : Average grain size as a function of the substrate temperature T_s

4. CONCLUSION

This work attempts to control the grain size and the crystalline fraction by means of the substrate temperature variation of microcrystalline films obtained by sputtering in pure hydrogen plasma. The important role played by the hydrogen species in the crystallization mechanism deserves further work while the investigations of their influence on the formation of the amorphous buffer layer are in progress.

REFERENCES

Achiq A, Rizk R, Madelon R, Gourbilleau F and Voivenel P 1997 Thin Solid Film (In press).

Blas G, Achiq A, Macia J, Morante J R, Perez-Rodriguez A, Ruterana P and Rizk R 1997 Mater. Res. Soc. Symp. Proc. (In press).

Kaneko T, Wakagi M, Onisawa K I and Minemara T 1994 Appl. Phys. Lett. **64** 1865.

Matsuda A 1983 J. Non-Cryst. Solids **59&60** 767.

Nada M, Shimizu H, Kohnoh H and Ishida H 1983 J. Non-Cryst. Solids **59&60** 823.

Moon D G, Jung B H, Lee J N, Ahn B T, Im H B, Nam K S and Kang S W 1994 J. Mater. Sci.: Mater. In Electron. **5** 364.

Rizk R, Achiq A, Madelon R, Gourbilleau F and Cruège F 1996 Polycrystalline Semiconductors IV - Physics, Chemistry and Technolgy Solid State Phenomena, **51&52**, p. 243.

Shimizu I, Hanna J and Shirai H 1990 Mater. Res. Soc. Symp. **164** 195.

Tsai C C, Knights J C, Chang G. and Wacker B. 1986 J. Appl. Phys.**59** 2998.

Veprek S 1990 Mater. Res. Soc. Symp. Proc **164** 39.

Yang Y H and Abelson J R 1995 Appl. Phys. Lett. 67 3623.

Inst. Phys. Conf. Ser. No 157
Paper presented at Microsc. Semicond. Mater. Conf., Oxford, 7–10 April 1997
© 1997 IOP Publishing Ltd

The effect of doping and formation conditions on the microstructure of porous silicon

G Wakefield, JL Hutchison* and PJ Dobson

Department of Engineering Science, University of Oxford, Parks Road, Oxford OX1 3PJ
*Department of Materials, University of Oxford, Parks Road, Oxford OX1 3PH

ABSTRACT: The microstructure of luminescent porous silicon has been studied by high resolution transmission electron microscopy. The silicon substrates examined are non-degenerate and degenerate p-type and non-degenerate n-type.

In the case of p-type silicon the structure has been determined to be quantum dot like in non-degenerate materials, with the porous structure consisting of 3-5nm silicon dots, surrounded by a thin oxide layer. For degenerate p-type material the microstructure consists of interconnected quantum wires, with the degree of interconnection dependent on the anodisation current density. The higher the current density the more open the microstructure.

For n-type material formed under low intensity white or UV light the microstructure consists of macropores of width 50-150nm extending many microns into the substrate the surface and pore walls are covered with a thin layer of nanoporous material. As the light intensity increases the pores widen leaving behind a nanoporous quantum dot layer. The luminescent intensity of this layer increases with the intensity of illumination. A blue shift in the luminescence peak is also observed.

1. INTRODUCTION

Since the demonstration of efficient room temperature photoluminescence from porous silicon by Canham (1990), there has been wide interest in understanding the microstructure of this potentially important material. Various studies have shown that porous silicon can best be described by a quantum confinement model (Brus 1994), with the nanostructures shown as either undulating quantum wires (Cullis 1991), or dots (Cole 1992). These studies have been hampered by the fact that the porous silicon microstructure is highly dependent upon the doping of the silicon substrate, and the formation conditions used.

In this paper three porous silicon systems are studied. Degenerate p-type, non-degenerate p-type, and non-degenerate n-type. We will show that both the quantum wire and quantum dot models are valid in some cases, and that the most highly luminescent material can best be described by a quantum dot model.

2. EXPERIMENTAL DETAILS

All the porous silicon samples described are formed in a single cell with a platinum cathode parallel to a single crystal (100) silicon wafer. The wafer is backed with aluminium and baked in a nitrogen atmosphere for fifteen minutes in order to create an back ohmic contact. The cell fluid is HF:H_2O:C_2H_5OH=1:1:2 (ethanoic HF)

The doping of the silicon substrates is as follows. 1) Degenerate p-type: Boron doped, ρ=0.005 Ωcm. 2) Non-degenerate p-type: Boron doped, ρ=5-8 Ωcm. 3) Non-degenerate n-type: Phosphorous doped, ρ=10-20 Ωcm.

In the case of n-type material illumination is supplied from either a UV(254nm) or a tungsten halogen white light via a cold mirror, in order to eliminate heating of the cell fluid and any associated etching effects. Current densities and anodisation times for the various samples are stated by the relevant micrographs.

TEM samples are prepared by either scraping a fresh sample of porous silicon on to a copper grid, or polishing and ion-beam milling. Ion beam milled specimens are prepared in a Technoorg-Linda thinner, which allows the use of grazing incidence and a retarding potential on the specimen stage. In this way preparation artefacts are avoided.

3. RESULTS

3.1 Degenerate p-type porous silicon

Anodisation of degenerate p-type silicon results in a porous lattice. The structure is quite open, with pore sizes of 20-30nm common. A plan view of a sample formed at a current density of 250mAcm^{-2} is shown in Figure 1. The micrograph is defocused to emphasise the pore structure. The size of the interconnected features is 5-10nm. As the sample has been scraped onto a grid and placed directly into vacuum after formation, it is considered that the features are predominantly silicon. A diffraction pattern from this sample is indistinguishable from that of bulk silicon.

After preparation of a cross-section the majority of the pores are filled with oxide, although the crystal structure remains intact. As 250mAcm^{-2} is quite a high formation density, the structure is sufficiently 'open' to allow individual quantum wire structures to be observed. One such example, a single quantum wire of thickness 1-5nm, is shown in Figure 2.

A decrease in formation current density results in the wire structures becoming more interconnected. Figure 3 shows a multilayer specimen, formed by alternating the current between 250mAcm^{-2} (high porosity) and 150mAcm^{-2} (low porosity) periodically during anodisation. The structure remains wire-like over all formation conditions, however.

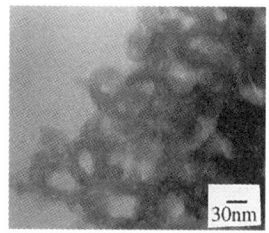

Figure 1. Plan view at 250mAcm^{-2}

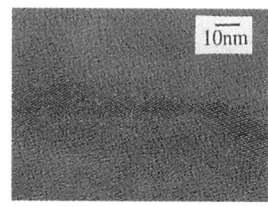

Figure 2. Undulating quantum wire.

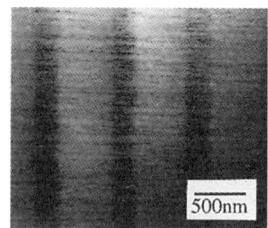

Figure 3. Multilayer structure

3.2 Non-degenerate p-type porous silicon

Luminescence studies show that the photoluminescence efficiency of porous silicon is greatest for non-degenerate systems. In the case of quantum confinement in an indirect semiconductor such as silicon we expect that the radiative transition is still principally vibronic, although enhanced by the confinement (Brus 1994). This means that 3D confinement in porous silicon would improve the luminescence over the pseudo-2D wires seen in section 3.1.

This is the situation for non-degenerate p-type material. In figure 4 a plan view specimen is shown. The formation current density being 200mAcm^{-2} in this case. Clearly the structure is much more compact then for the degenerate material. Even in this fresh sample the pores are hardly visible. The diffraction pattern inset shows that the crystallinity of the sample is quite poor. Diffuse scattering suggests an amorphous component to the sample, and the streaking of the points indicates some misorientation of the silicon lattice is occurring. Some lattice fringes can also be seen, and they appear to show isolated quantum dots. In figure 5, a cross-section of the sample, dots of 3-4nm are clearly seen slightly misaligned with respect to each other.

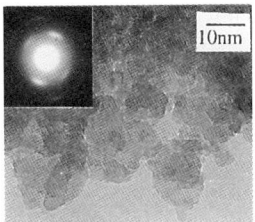

Figure 4. Scraped sample (inset:- diffraction pattern.

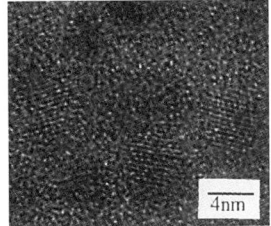

Figure 5. HREM cross-section

As a result of their dot like nature non-degenerate p-type samples are extremely fragile and prone to cracking when dried in air. To avoid this the samples shown are formed with alternating high and low porosity layers, the low porosity layers strengthen the structure to allow a cross-section to be prepared (Wakefield 1995).

3.3 Non-degenerate n-type porous silicon

The formation of porous silicon in an electrochemical cell requires holes to be present for dissolution (Lehmann 1993). Therefore illumination is required during anodisation. Illumination with low intensity white or UV light results in the formation of macropores. There is also a thin layer of luminescent nanoporous material on the surface and the pore walls of the sample. These macropores, shown in Figure 6, extend many microns into the bulk silicon. This example has undergone a 2.5mAcm^{-2} etch for 5mins under 254nm UV illumination. The

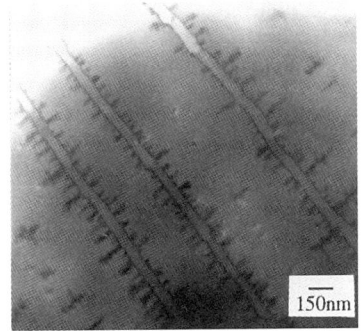

Figure 6. Macroporous n-type porous silicon

pores are lobed, although there is no interconnection between them. The microstructure is

438

the same under low intensity white light. In this case however, an increase in the light intensity results in an increase in the nanoporous layer. This occurs initially at the bottom of the macropore, which expands as shown in Figure 7. Nanoporous material remains in the structure. The flow lines of the cell fluid can also be seen in the resulting microstructure.

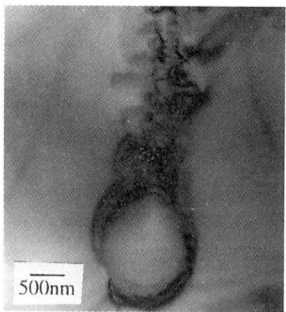

Figure 7. Increased light intensity

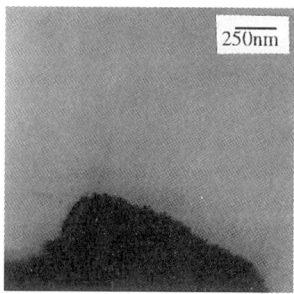

Figure 8. Fully nanoporous layer

Further increase in the light intensity results in a completely nanoporous and highly luminescent layer being formed, as in Figure 8. Again this layer is quantum dot like, having luminescent material identical to that of Figure 5. There is a size gradient throughout the cross-section, with the smallest dots being at the surface. Cracking effects are therefore not usually a problem in n-type material. As light intensity increases, the luminescence peak position blue shifts down to a minimum at 600nm.

4. CONCLUSIONS

It has been shown that porous silicon microstructures can vary between quantum wires and quantum dots due to the dopant density in p-type material. The high luminescence efficiency in non-degenerate material has been shown to be associated with the presence of quantum dots.

In n-type material the structure depends on light intensity during anodisation. As the intensity increases the structure becomes quantum dot like after expansion of the original macropores.

REFERENCES

Brus L 1994 J. Phys. Chem. **98**, 3575
Canham LT 1990 App. Phys. Lett. **57**, 1046
Cole MW 1992 App. Phys. Lett. **60**, 2800
Cullis AG and Canham LT 1991 Nature. **353**, 335
Lehmann V 1993 NATO ASI Series E. **244**, 1
Wakefield G, Hutchison JL and Dobson PJ 1995 EMAG Inst. Phys. Conf. Ser no 147, IOP Publishing Ltd. 377

Inst. Phys. Conf. Ser. No 157
Paper presented at Microsc. Semicond. Mater. Conf., Oxford, 7–10 April 1997
© *1997 IOP Publishing Ltd*

Temperature mapping of polysilicon microheaters using Raman micro-spectroscopy

M Bowden[1], D J Gardiner[1], A A Parr[1], R T Carline[2], R J Bozeat[2], R J T Bunyan[2] and M Ward[2]

(1) Department of Chemical and Life Sciences, University of Northumbria at Newcastle, Ellison Place, Newcastle upon Tyne, UK, NE1 8ST.
(2) DERA, St Andrews Road, Malvern, Worcestershire, UK, WR14 3PS

ABSTRACT: Raman micro-spectroscopy has been used to establish the relationship between the polysilicon optical phonon shift and temperature. This has enabled the temperature of polysilicon microheaters to be monitored as a function of electrical power. Raman Line focus methods were used to generate a temperature map of an active microheater approximately 100μm square with 1μm spatial resolution. The results will contribute to improved thermal modelling and device design along with the development of methods to study degradation of device performance.

1. INTRODUCTION

There is a great deal of interest in the use of micro-machining to form thermally isolated micro-structures. Such structures can be used in a wide range of applications including flow sensors, micro-catalytic gas sensors, infra-red scene generation and micro-Pirani gauges.

Raman spectroscopy measures molecular vibrations by analysing laser radiation scattered from a sample. Typically a cw gas laser operating in the visible region of the spectrum is used to excite the spectra. Scattered photons arise from both elastic (Rayleigh) and inelastic (Stokes and anti-Stokes) scattering, the latter resulting from excitation of the sample molecular vibrations by the photon. It is the red shifted, Stokes Raman scattering which is normally analysed and a plot of intensity against Raman shift expressed in wavenumbers (cm^{-1}) is used to display the Raman spectra. In the solid state, molecular vibrations are treated as vibrations of atom against atom, with the unit cells acting in phase within the crystalline structure. This is the wave vector $k = 0$ condition. Such vibrations resolve into acoustic and optical phonon vibrations. The frequency of the optical phonons is determined by the kinetic and potential energies of the structure which in turn relate to atomic masses and force fields. Increases in temperature allow mixing of the optical and acoustic vibrations resulting in a broadening of the phonon band and a shift to lower frequency.

2. EXPERIMENTAL

A Raman Microline Focus Spectrometer was employed which is capable of generating image maps of parameters calculated from the Raman spectra at points throughout an area. The line-focus approach has distinct advantages in the study of samples in which high laser power densities can modify the sample material or, as in the present example, directly alter the parameters being measured.. Initial studies were undertaken to determine the response of the polysilicon phonon band to temperature and to laser power density and to establish a reliable temperature calibration. Additionally changes in the phonon band resulting from annealing were investigated.

The microline Focus Spectrometer (MiFS) which we have developed has been described in detail elsewhere (Bowden et al. 1992). The instrument is capable of obtaining Raman spectra from a surface with sub-micron resolution and can generate profiles and images representing intensity (species concentration), frequency (stress) and bandwidth (crystallinity).

Fig. 1 shows a schematic of the instrument which uses cylindrical optics to produce a line focus of the incident laser beam on the surface of the sample.

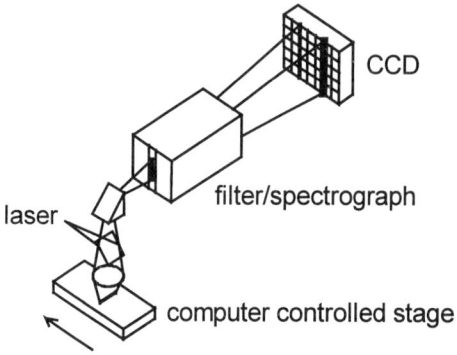

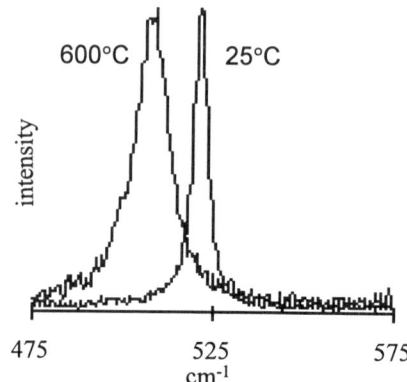

Fig. 1: Schematic of the Raman Microline Focus Spectrometer (MiFS).

Fig. 2: Raman spectra of polysilicon at 25 and 600°C

Light scattered from the focused line is then imaged through a subtractive filter stage and a spectrograph stage to form a spectrally dispersed image on a cooled ccd detector. The non-astigmatic imaging optics employed ensure that the spatial integrity of the origin of the scattered light is retained thus generating Raman data from the length of the focused laser line at a single exposure. Raman images are obtained by collecting data from a series of adjacent lines on the sample surface. At each line position the required spectral parameters can be extracted and then written as a greyscale image to a monitor screen before moving on to the next line position. In this way the Raman image is generated in near real time whilst at the same time the full dynamic range of the image data is retained for 3 dimensional plotting or for further analysis. Mapping and imaging software has been developed which controls the stage movement, removes background counts, integrates and carries out arithmetic on any number of Raman band intensities or determines band positions and bandwidths to generate the image data.

This approach is an objective compromise between the time consuming and high positional stability demands of point by point sampling of the surface and the spectroscopically limited global illumination method. In addition the method does not rely upon moving optical components and thus is inherently stable. In the case of this application the band positions were determined and converted to temperature values with a pixel determined spatial resolution of 1μm.

The heater structures were fabricated by silicon bulk micro-machining methods and typically comprised a meandering 10 μm wide strip of polysilicon covering an area 100 x 100 μm.

3. TEMPERATURE CALIBRATION

A Linkam TH1500 hot stage was used to heat the samples. Preliminary Raman studies on the polysilicon material of the micro-heaters revealed that there was no change in the state of anneal of the samples following heating to 600°C. Temperature calibration was carried out by measuring the frequency of the Raman phonon band for a micro-heater held at constant temperatures in the range from 25 to 600 ± 1°C temperature. Fig. 2 shows the changes in Raman shift and bandwidth observed in spectra from polysilicon obtained at 25°C and at 600°C. Two sets of measurements were made at

5mW and 10 mW of laser power at the sample, in order to determine the effect of heating from the focused laser. The results are shown in Fig. 3.

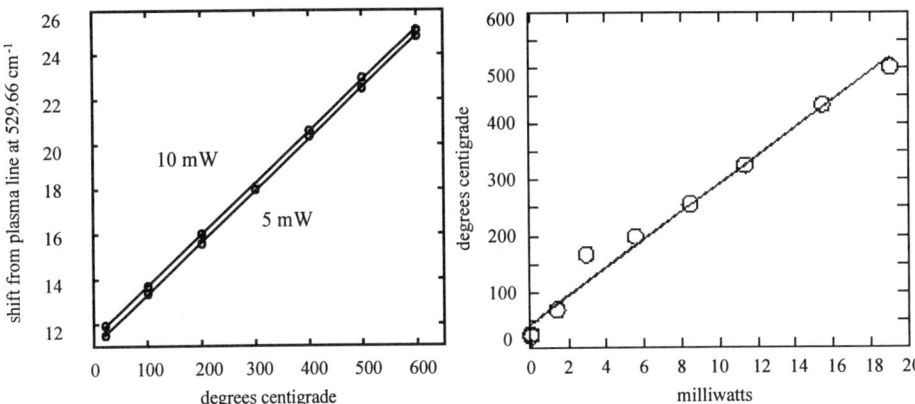

Fig. 3: Graph of the relationship between the wavenumber shift of the polysilicon phonon and temperature over the range 25 - 600°C.

Fig. 4: Graph of temperature against electrical power for a polysilicon microheater.

The data clearly indicate a linear relationship, for the two laser powers used, between temperature, in degrees centigrade (T) and the wavenumber shift of the phonon frequency (f), such that $\Delta T = 43.48 \Delta f$ which is within 5% of the value determined by Hart et al. (1970) for crystalline silicon

4. TEMPERATURE / POWER MEASUREMENTS

Positioning the focused laser line across the centre of the heaters and summing all of the resultant spectra, provides a method for determining the representative temperature of the heater at any applied bias. With applied voltages from 2 to 5 volts, the current was measured and the temperature calculated from the Raman shift for a number of heater elements. The temperature values were corrected for the known temperature increase caused by the laser power used. All the heaters measured showed a straight line relation between electrical power input and measured temperature rise over a range of applied voltage. A typical example is shown in Fig. 4.

5. TEMPERATURE MAPPING

An approximately 100 x 100 μm micro-heater held at 5.19 volts (1.54mA) was used for temperature mapping. The focused laser line was positioned across the heater parallel to one of the meander directions using a X40 microscope objective which resulted in a spatial resolution of 1μm per pixel. After each exposure a greyscale image representing temperature for the line of data was generated. The sample was then moved 1μm in a direction orthogonal to the focused laser line and the process repeated until the temperature of the whole heater had been mapped. The image was then corrected to allow for the effect of laser heating and the resulting temperature map is shown in Fig. 5.

Examination of the map reveals that the temperatures vary from ambient at the contact pads, rapidly increasing over about 40 μm to around 300°C which is then maintained to the end of the contact arm. There then follows a cooler ~200°C arm followed by a uniformly hot region ~325°C at the central arms of the meander.

442

40

100

175

240

275

325

300

300

300

300

Fig. 5: Temperature image of a 100 x 100 μm
polysilicon microheater. The temperatutes (°C) are
indicated for the right hand arm and for the centre of the
structure.

6. CONCLUSION

This study has demonstrated the possibilities for temperature mapping of silicon devices. The technique has distinct advantages over infrared thermal imaging which is restricted in spatial resolution and requires emissivity data to generate temperatures. Raman imaging provides a spatial resolution of 0.5-1.0 μm and following temperature calibration of the material will enable temperature maps to be generated. The micro-heaters studied here were free standing and thus had a small thermal mass. This in turn resulted in significant laser heating which was minimised by using low laser powers but which consequently led to long exposure times. Laser heating is significantly affected by the local geometry of these structures and consequently only an approximate correction for laser heating could be made. Consequently the accuracy of the temperature values may vary as much as ± 20°C. For more massive samples, for example electronic devices, higher laser powers could be used resulting in faster data collection times.

REFERENCES

Bowden M, Gardiner D J and Southall J M 1992 J. Appl. Phys., **71**, 521

Hart T R, Aggarwal R L and Lax B 1970 Phys. Rev. B, **1,** 638

Inst. Phys. Conf. Ser. No 157
Paper presented at Microsc. Semicond. Mater. Conf., Oxford, 7–10 April 1997
© *1997 IOP Publishing Ltd*

Development of a mechanical polysilicon layer for surface machined microelectromechanical systems using TEM, SEM, and Raman spectroscopy

D O King, M C Ward, A G Cullis *, **D Gardiner** ** and **M Bowden** **

DERA Malvern, St Andrews Road, Malvern, Worcs WR14 3PS
* now at University of Sheffield
** University of Northumbria

ABSTRACT : Low stress polysilicon is required to replace the standard gate polysilicon for the mechanical layer in a microelectromechanical (MEMs) process. A uniform structure through the thickness of the layer is required to stop freed beams curling up. In this paper electron microscopy techniques have been used to identify such a material. TEM was used to analyse the structure of polysilicon layers, and SEM to show up any residual stress gradient large enough to curl beams. A suitable process has been identified and used to fabricate accelerometers. Raman spectroscopy is currently being investigated to map stress in these devices.

1. INTRODUCTION

Polysilicon surface micromachining offers the possibility of producing mechanical components, such as accelerometers and gyroscopes, using standard CMOS batch processing techniques to deliver small, light, and potentially cheap devices with integrated electronics (Linder 1992, Petersen 1982, Yun 1992).

A simplified schematic of the process is shown in Fig. 1. Wherever possible, existing CMOS process layers are used: gate polysilicon is used for the bottom electrode and the phosphorous-doped oxide employed for the first metal isolation is used as the sacrificial layer. This enables micromachined structures to be inserted into a CMOS process.

Standard gate polysilicon is unsuitable for the mechanical layer as it is has a high stress gradient through the layer and therefore cantilever beams when released curl up (see Fig. 2a). The same set of beams fabricated in the mechanical polysilicon described in this paper are also shown (Fig. 2b).

2. SAMPLES

The samples investigated consisted of 2μm of LPCVD polysilicon sandwiched between two layers of phosphorous doped low temperature oxide (LTO). The LTO served as a sacrificial layer, a dopant source and a hard mask for plasma etching of the polysilicon. The polysilicon was grown at various temperatures from 570°C to 620°C and then annealed in argon at 1050°C for 45 minutes.

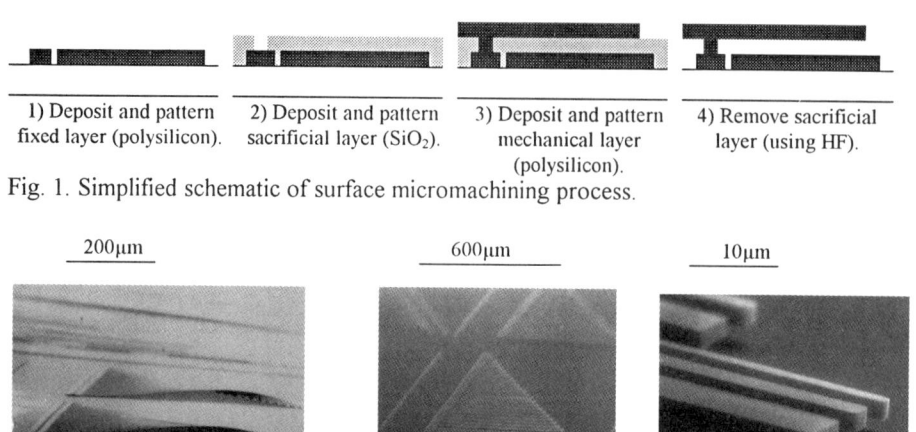

1) Deposit and pattern fixed layer (polysilicon).	2) Deposit and pattern sacrificial layer (SiO₂).	3) Deposit and pattern mechanical layer (polysilicon).	4) Remove sacrificial layer (using HF).

Fig. 1. Simplified schematic of surface micromachining process.

Fig. 2. SEMs of beams fabricated in a) standard gate polysilicon and b) mechanical polysilicon, with close-up of ends of longest set of beams.

3. ANALYSIS

3.1 Transmission Electron Microscopy (TEM)

The grain structure of the polysilicon layers before and after annealing were studied using TEM. There was a very distinct change in the structure of the polysilicon as the deposition temperature was increased.

Layers deposited at 5˝ ͻ°C and 580°C have about half the layer amorphous and half the layer polycrystalline (Fig. 3a.): it can be seen why structures fabricated in polysilicon deposited at 580°C curl (Fig. 2a.) when released. Once annealed both layers become fully polycrystalline with random size and orientation (Fig 3b).

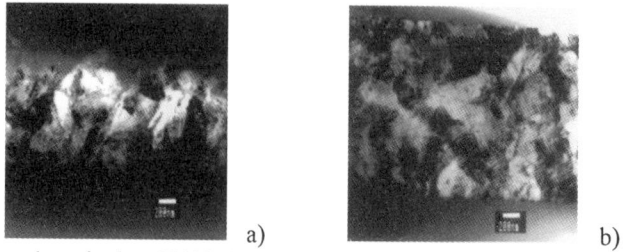

a)   b)

Fig. 3. Polysilicon deposited at 580°C (a) and annealed at 1050°C (b).

Layers deposited at 590°C are polycrystalline, with a large number of small randomly orientated crystals. After annealing the crystal size has increased and gives a structure very similar to that of annealed polysilicon deposited at 580°C.

Layers deposited at 600°C are polycrystalline, with a highly columnar structure (Fig. 4a.), after annealing the columnar structure remains with the crystal size increasing (Fig. 4b.). A seeding layer of randomly oriented crystals can be seen at the bottom of the layer, the thickness of this compared to the rest of the layer is small and therefore has little effect on the final structure.

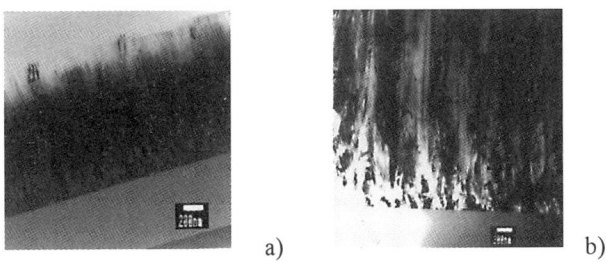

Fig. 4. Polysilicon deposited at 600°C (a) and annealed at 1050°C (b).

3.2 Scanning Electron Microscopy (SEM)

Test structures consisting of a cantilever beam 300µm long and a short (10µm long) reference beam (Fig. 5) were fabricated in polysilicon deposited at 580°C, 590°C, and 600°C and annealed at 1050°C. Any residual stress gradient would cause the tip of the long beam to be deflected up or down with respect to the reference beam. The results are shown in Fig. 6. The beams fabricated from 580°C and 590°C show a distinct deflection down, indicating there is still some residual stress gradient. The cantilever beam fabricated from 600°C polysilicon shows no discernible deflection. Long beams (>500µm) fixed at both ends fabricated with this polysilicon show a slight bowing due to a small amount of compressive stress still present in the layer. As this is uniform, structures can be designed to release this and so it is not a problem.

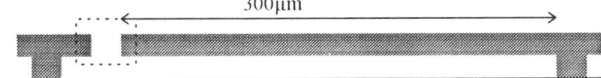

Fig. 5. Test structure consisting of a 300µm long cantilever beam and a 10µm reference beam.

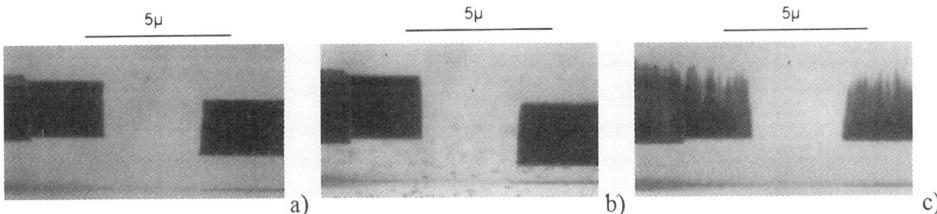

Fig. 6. SEMs of test structures fabricated in a) 580°C, b) 590°C, c) 600°C polysilicon and annealed at 1050°C, tips bending down indicate that a stress gradient is still present through the layer.

3.3 Raman Spectroscopy

Raman spectroscopy is currently being investigated to measure the stress in polysilicon. Spot measurements can be used to monitor layers during fabrication and scanning measurements to map the stress in devices. This will enable high stress points in structures to be identified and, where possible, designed out.

4. ACCELEROMETER

A series of accelerometers were fabricated with different suspensions to prove the process. These are open loop capacitative devices, consisting of a 300μm-square proof mass suspended from four suspension arms. A device with 10μm wide arms is shown in Fig. 7. CV tests carried out on the devices showed the characteristic latching action of an electrostatically controlled sprung mass. Fig. 8a shows CV curves for two devices. Devices with 10μm wide legs were bonded up as accelerometers and the tumble test results shown in Fig. 8b. The devices can clearly

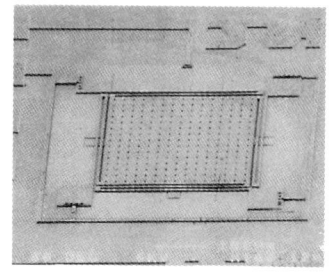

100μm

Fig. 7. SEM of an accelerometer with a 300μm square proof mass and 10μm wide suspension arms.

distinguish between up and down, performance can be improved by integrating the electronics on the same die as the device.

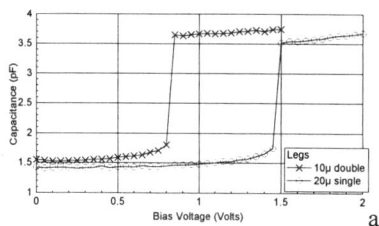

a)

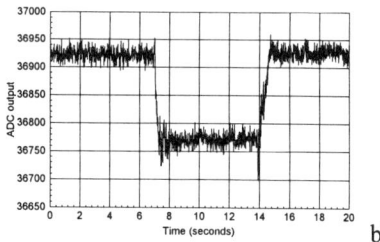

b)

Fig. 8. a) CV curves for two accelerometers with different suspensions; b) tumble test results for an accelerometer with 10μm wide suspension.

5. CONCLUSIONS

A suitable deposition/anneal process for polysilicon mechanical layers has been identified using TEM and SEM. The columnar structure of polysilicon deposited at 600°C and annealed at 1050°C in argon for 45 mins produces a layer with a negligible stress gradient through the thickness of the layer and a low overall compressive stress. Accelerometers have been successfully fabricated using this process. The techniques used so far have only given a qualitative indication of the stress, Raman is currently being used to measure and map the stress in devices.

REFERENCES

Howe R T and Muller R S 1983 J. Appl Phys. **54**, 4674
Linder C, Paratte L, Gretillat M-A, Jaecklin V P, and de Rooij N F 1992 J.Micromech.
    Microeng. **2**, 122
Petersen K E 1982 Proc. IEEE **70**, 420
Yun W, Howe R T and Gray P R 1992 Proc IEEE , 126

© British Crown Copyright 1997 DERA
Published with the permission of the Controller of Her Britannic Majesty's Stationary Office

Inst. Phys. Conf. Ser. No 157
Paper presented at Microsc. Semicond. Mater. Conf., Oxford, 7–10 April 1997
© *1997 IOP Publishing Ltd*

Dislocation structure in interfaces of bonded hydrophobic silicon wafers: experiment and molecular dynamics

M Reiche [1)], K Scheerschmidt [1)], D Conrad [1)], R Scholz [1)], A Plößl [1)], U Gösele [1)] and K N Tu [2)]

[1)] Max-Planck-Institut für Mikrostrukturphysik, Weinberg 2, D - 06120 Halle, F R Germany
[2)] Department of Materials Science and Engineering, University of California, Los Angeles, CA 90095-1595, USA

ABSTRACT: The defect structure in interfaces of directly bonded hydrophobic silicon wafers is studied by HVEM and HREM. It is shown that the twist component of the boundary causes a screw dislocation network, while the tilt component or a miscut are compensated by arrays of 60° dislocations. Interactions occur between both especially during annealing at higher temperatures. For annealing at temperatures below 1000°C the structure of the interface is more or less analogous to that of wafer pairs bonded under UHV conditions at room temperature without any subsequent heat treatment. Molecular dynamics simulations of the latter revealed that the screw dislocation network is caused by the relaxation of atoms into a mosaic-like structure with the formation of covalent bonds via the interface.

1. INTRODUCTION

Silicon wafer direct bonding has become of increasing interest for silicon-on-insulator (SOI) and power devices as well as for micromechanics and sensors. Here, wafers used for wafer bonding are covered with either a thin native oxide or a purposely grown thermal oxide which has been rendered hydrophilic by cleaning procedures. At room temperature, the initial bonds are hydrogen bonds caused by the presence of OH-groups at the interface, while at higher temperatures they are transformed into stable Si-O-Si bonds. The interfaces are atomically flat without any defects. Alternatively, silicon surfaces may be treated with diluted HF, which removes the native oxide layers and directly covers the silicon surface with hydrogen, thus rendering the surfaces hydrophobic. Here, van der Waals forces are generally assumed to be the origin of the attractive force at room temperature. At higher temperatures, they are transformed into Si-Si bonds. Depending on the misorientation between both wafers, dislocation networks are generated in the interface: the twist component causes a network of pure screw dislocations, while the tilt component is compensated by a periodic array of 60° dislocations. Both dislocation fractions were extensively investigated after annealing at T ≥ 1100°C (e.g. Gafiteanu et al 1993, Benamara et al 1994, 1995).

The present paper analyzes the interfacial defects of bonded hydrophobic wafer pairs occurring at lower temperatures (> 800°C). Furthermore, these interfacial defects are compared to the interfaces of wafer pairs bonded under ultra-high vacuum (UHV) conditions at room temperature without any subsequent heat treatment.

2. EXPERIMENTAL

Czochralski-grown silicon wafers (diameter 4 in., (100) orientation, p-type) were used for the experiments. The tilt component (cut-off) of all wafers is below ± 1°. After cleaning in standard

RCA 1,2 solutions they were dipped into ~ 1% HF for 2 minutes in order to make the surfaces perfectly hydrophobic. Without subsequent water rinsing the wafers were bonded in a micro-cleanroom set-up (Stengl et al 1988) under normal atmospheric conditions. After bonding one group of wafer pairs was annealed at temperatures between 800°C and 1200°C under inert or oxidizing conditions. Another group was transferred in a UHV chamber, separated, and bonded again at room temperature after annealing at 600°C < T < 800°C in order to desorb the hydrogen from the wafer surfaces before bonding (Gösele et al 1995). These UHV room temperature bonded wafers did not receive any subequent heat treatment.

After thinning down one of the wafers of a pair, samples for electron microscopy were prepared by chemical thinning in HF/HNO$_3$ solutions (plan-view or oblique cut) and by ion beam milling (cross-sectional samples). Both, high-voltage and high resolution electron microscopy were applied. The formation of interfacial defects for the case of UHV bonding at room temperature was simulated by molecular dynamics (Scheerschmidt et al 1996).

3. INTERFACES OF UHV-BONDED WAFER PAIRS

If two clean silicon surfaces meet (as under UHV conditions) strong covalent bonds develop already at room-temperature between the surfaces (Gösele et al 1995). HREM images show that the interface is not perfect. Instead, defects occur caused by the unavoidable misorientations between the two wafers. Analyzing such samples without any additional heat treatment reveals an analogous interface defect structure generally known from interfaces of wafer pairs bonded under atmospheric conditions (Benamara et al 1994, 1995). There is a 2-dimensional dislocation network having square-like meshes (Fig. 1). The dislocations forming the network are screw dislocations ($\mathbf{b}$ = a/2<110>), with spacings of about 66 nm, corresponding to a twist angle Θ ~ 0.3°. Furthermore, there are irregularities in the network suggesting a superposition of 60° dislocations. Molecular dynamics (MD) simulations were carried out starting with two perfect Si blocks

Fig.1: Formation of a screw dislocation network in the interface of bonded silicon wafers. The wafers were bonded under UHV conditions. TEM image of a bonded {100} interface (a). MD-simulation of rotationally misoriented wafers (b,c). Starting configuration (dimer rows at interface) with 4.58° misorientation (b) and after relaxation (annealing at 900 K for 1.25 ps (c)). Simulated TEM bright field images from (c) with $\mathbf{g}_{220}$ excited (thickness 4.3nm (d),12.9nm (e)).

of 2 x 1 reconstructed (100) surfaces. A small twist angle is assumed as a rotational misorientation without other defects (Fig. 1b). After sufficient relaxation a mosaic-like interface structure (size 7 nm) occurs where almost all atoms have a bulk-like environment separated by screw dislocations (Fig. 1c). The dislocations show irregularities due to randomly started bonding processes. Image simulations (TEM bright field contrast) based on the MD-model showed the square-like arrangement of dislocations as shown in the experimental images (Figs. 1d,e). The dark contrast effects between the dislocation lines are caused by a lattice plane deformation. Recently, MD-models on the basis of double-layer steps on the surfaces have also been developed for simulating the nature of 60° dislocations (Belov et al 1997).

4.INTERFACES OF WAFER PAIRS BONDED UNDER ATMOSPHERIC CONDITIONS

The hydrogen passivation of the wafer surfaces causes only van der Waals forces as the attractive force at room temperature. Therefore the bonding strength is low, requiring additional annealing at higher temperatures. Here wafers were bonded with a rotational misfit up to 5°. Interfaces of wafer pairs annealed at temperatures between 800°C and 1200°C show 4 different features:

i) There is a more or less periodical network of 60° dislocations originating from the tilt component. In most samples analyzed the spacing of these dislocations is of the order of several micrometers corresponding to tilt angles of about 0.02°.

ii) At higher magnifications an array of screw dislocations occurs respresenting the twist component or rotational misfit (Fig. 2a). At lower temperatures (T < 1000°C) the network is more regular than at higher temperatures. Here, the screw dislocations are displaced along almost straight lines running at an angle to the <110> directions. The displacements are due to interactions with the 60° dislocations described by Gafiteanu et al (1993) and Benamara et al (1995) for bonded wafers, and by Föll and Ast (1979) for grain boundaries in sintered silicon. The latter showed that the interaction causes *extraneous* dislocations lying in their {111} glide plane and resulting in a step in the boundary plane. This may be the reason for the high dislocation density (Fig. 2b).

iii) In some areas the dislocation network abruptly discontinues, with only Moiré fringes appearing in the images (Fig. 2c). They may represent bonded areas in which the misorientation is not adapted by misfit dislocations but by an adaption arrangement distributed over all interface atoms via slightly changed bond lengths and angles, or are due to interfacial defects filled with residual oxide (SiO₂) not removed by the HF dip. Such areas would correspond to the oxide-welded

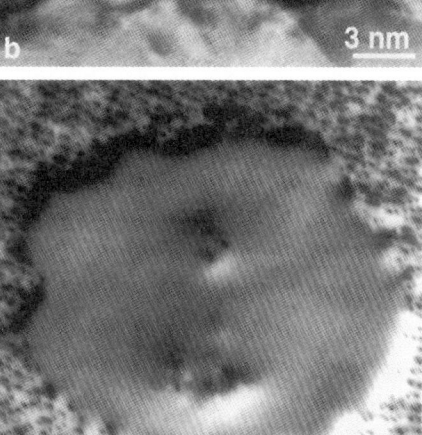

Fig. 2: Interfacial defects in wafer pairs bonded under atmospheric conditions.
a) Screw dislocation network superimposed with 60° dislocations (plan-view image).
b) Formation of *extraneous* dislocations due to interactions (cross-sectional image).
c) Interfacial area without the dislocation network (plan-view image).

boundaries (Perreault et al 1992).

iv) In the plan-view images some of the samples showed linear structures running at an angle of about 30° to a <110> direction. Their spacing is about 50 nm. Diffraction contrast experiments proved that these structures are not related to 60° dislocations. Cross-sectional samples prepared from different {110}- and {100} planes showed a zigzag-like structure in one {100} plane with the same periodicity (50 nm). Samples of all the other planes only show the dislocation network ({110} samples) or dot-like contrasts referring to oxide islands (the other {100} plane). As shown in Fig. 3, a terrace-like structure is reconstructed from all these images. Their nature is not yet known. Terraces may be caused by stress relaxation of the rotational misorientation.

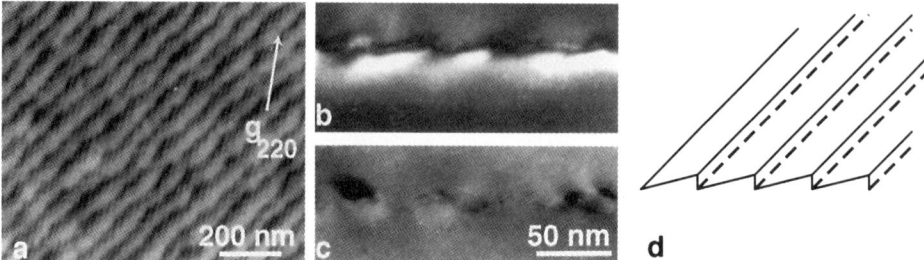

Fig. 3: Terrace-like structure in the interface. TEM-plan-view (a) and cross-sectional images of 2 different {100} planes (b,c). Schema of the terrace structure (d).

CONCLUSIONS

It is shown that the interfaces of bonded hydrophobic wafer pairs and such pairs bonded under UHV conditions correspond to grain boundaries having a twist and a tilt component. MD-simulations for clean and reconstructed surfaces (as obtained under UHV conditions) explain the generation of dislocation networks during bonding by the relaxation of interfacial atoms and the formation of covalent bonds. Furthermore, the defect structure in interfaces of wafers bonded under atmospheric conditions is more complex. Besides the screw dislocation network, interactions of screw dislocations with 60° dislocations, resulting from the miscut of the samples used, are observed more frequently. These interactions may be caused by the high annealing temperatures applied. In some cases, instead of the screw dislocation network, a terrace-like structure formswhich may constitute a form of adapting the rotational misorientation,at least partially.

REFERENCES

Belov A, Conrad D, Scheerschmidt K and Gösele U 1997 these proceedings

Benamara M, Rocher A, Laanab L, Claverie A, Laporte A, Sarrabayrousse G, Lescouzeres L and Peyre-Lavigne A 1994 C. R. Acad. Sci. Paris **318**, 1459

Benamara M, Rocher A, Laporte A, Sarrabayrousse G, Lescouzeres L, Peyre-Lavigne A, Fnaiech M and Claverie A 1995 Mat. Res. Soc. Symp. Proc. **378**, 863

Föll H and Ast D 1979 Phil. Mag. **A40**, 589

Gafiteanu R, Chevacharoenkul S, Gösele U and Tan T Y 1993 Inst. Phys. Conf. Ser. **134**, 87

Gösele U, Stenzel H, Martini T, Steinkirchner J, Conrad D and Scheerschmidt K 1995 Appl. Phys. Lett. **67**, 3614

Perreault G C, Hyland S L and Ast D G 1992 Proceedings of the 1st Internat. Symp. on Semiconductor Wafer Bonding: Science Technology, and Application, eds U Gösle, T Abe, J Haisma and M A Schmidt (Pennington: The Electrochemical Society) Proc. Vol. 92-7, pp365-374

Stengl R, Ahn K Y and Gösele U 1988 Jpn. J. Appl. Phys. **27**, L2364

Scheerschmidt K, Conrad D and Gösele U 1996 Comput. Mat. Sci. **7**, 40

Inst. Phys. Conf. Ser. No 157
Paper presented at Microsc. Semicond. Mater. Conf., Oxford, 7–10 April 1997
© *1997 IOP Publishing Ltd*

Interfaces of CVD diamond films on silicon (001)

D Wittorf, W Jäger[1], C L Jia, K Urban, A Flöter[2], H Güttler[2] and R Zachai[2]

Institut für Festkörperforschung, Forschungszentrum Jülich, 52425 Jülich, Germany
[1]Mikrostrukturanalytik, Technische Fakultät, Universität Kiel, 24243 Kiel, Germany
[2]Daimler-Benz AG, Forschungszentrum Ulm, Wilhelm-Runge Strasse 11, 89013 Ulm, Germany

ABSTRACT: Interfaces and orientation relationships between diamond films, Si and β-SiC interlayers were analyzed by conventional and high-resolution transmission electron microscopy for the early stages of film deposition on Si (001) substrates under optimized conditions using the microwave-assisted chemical vapour deposition process. Epitaxially oriented diamond nuclei form particularly on {111} facets of a nanocrystalline β-SiC interlayer. A model describing the early stages of diamond film formation on (001)-oriented Si substrates via formation of a β-SiC interlayer is presented

1. INTRODUCTION

Thin diamond films of high structural quality are candidates for future applications in high power or high temperature electronic devices. The heteroepitaxial growth of diamond films by microwave-assisted chemical vapour deposition (MWCVD) has been successfully performed on Si substrates. High densities of oriented diamond grains could be obtained for the first time by introducing a bias-supported nucleation stage (Jiang et al. 1993). Similar results were also recently obtained by growth on β-SiC substrates (Kawarada et al. 1995). High-resolution transmission electron microscopy of the interfaces performed for the diamond films after a growth stage on Si(001) substrates has revealed that direct epitaxy occurs for growth on Si (Jia et al. 1995) and on Si-C alloy buffer layers (Wittorf et al. 1996), with frequent inclusions of β-SiC nanocrystals at the interfaces. Independently, growth of diamond grains on β-SiC interlayers has been reported for the nucleation stage of film growth (Wurzinger et al. 1996, Stammler et al. 1996).

In this paper we report the results of interface studies during the early stages of diamond film deposition on Si(001) under optimized conditions. The studies suggest that formation of diamond nuclei on a thin interconnected interlayer of β-SiC nanocrystals governs the orientation relationships observed for the later stages of diamond film growth. Frequently, epitaxially oriented diamond nuclei form on {111}-facets of the β-SiC nanocrystals. A model describing these observations during the early stages of diamond film formation on Si(001) substrates will be presented.

2. EXPERIMENTAL DETAILS

Bias-enhanced nucleation and growth of diamond films were performed by MWCVD. Mirror-polished (001)-oriented Si wafers were first etched by a hydrogen plasma for 5 min and a hydrogen gas flow of 1500 standard ccm. The hydrogen flow was kept constant during the whole deposition process. During nucleation, a bias-voltage of -200 V was applied between the plasma and the substrate to accelerate positively charged ions to the substrate and to enhance the nucleation of diamond. Small amounts of N_2 (50 ppm) and O_2 (25 ppm) were added to the hydrogen gas flow during the nucleation stage. The power of the microwavesin the nucleation and in the growth stage was 2500 and 2000 W, respectively. Deposition conditions are summarized in table 1.

(110)-oriented cross-section and (001)-oriented plan-view samples were characterized by conventional (TEM) and high-resolution transmission electron microscopy (HRTEM) at an operating voltage of 400 kV (JEOL 4000EX). Film morphology and densities of oriented diamond grains were determined by scanning electron microscopy.

Sample	Nucleation stage	Growth stage
	time [min] / T [°C] / CH_4 [%]	time [min] / T [°C] / CH_4 [%]
A	4 / 875 / 0.4	---
B	11 / 875 / 0.4...0.9	---
C	15 / 875 / 0.4...0.9	60 / 810 / 2.3

Table 1: Optimized conditions for MWCVD diamond deposition used in present studies.

3. RESULTS

Fig. 1(a) shows a cross-section micrograph of the substrate surface region for a film deposition during the early nucleation stage (sample A). Small Si islands with heights of

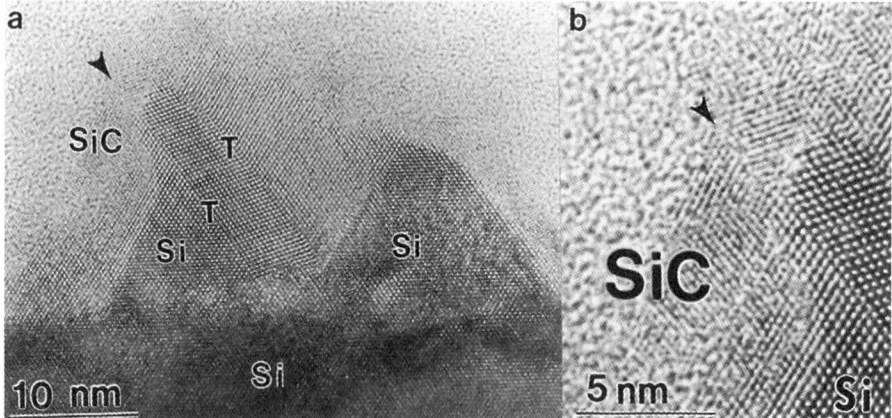

Fig. 1. (a) Faceted Si islands with microtwins (T) on (111)-planes. The Si is covered by small β-SiC nanocrystals. (b) SiC/Si(111) interface. A twin (arrow) in SiC is connected to the twin in Si. HRTEM micrograph taken in the [110]-orientation, sample A.

about 10 nm which are faceted parallel to the {111}-lattice planes were typically seen. The moiré patterns inside the Si islands originate from double diffraction of the electron beam from both {111}-Si and {111}-SiC lattice planes. Occasionally, coherent twins are observed. The Si substrate is covered with β-SiC nanocrystals of about 5 nm thickness. In fig. 1(b) the interface between SiC and Si is shown at a higher magnification. The SiC nanocrystal is epitaxially oriented, and the SiC lattice planes are in direct contact with those of Si at the interface. Particularly, the SiC nanocrystal is twinned, with a (111)-habit plane as is also observed for the Si (marked by an arrow). Parts of the interface plane between SiC and Si are faceted parallel to the $(\bar{1}1\bar{1})$ - and the (111)-lattice planes.

The β-SiC nanocrystals are either epitaxially oriented or slightly rotated with respect to the Si substrate (sample B, late nucleation stage). Figure 2(a) shows an example of β-SiC nanocrystals with small rotations with respect to the Si-substrate. The terminating {111}-SiC lattice planes are not regularly arranged along the interface. The two β-SiC nanocrystals, SiC1 and SiC2, are rotated about the [110]-zone axes by 7° with respect to each other. A (111)-SiC lattice plane is seen to terminate at the corresponding grain boundary between these grains (marked by an arrow). Rotation angles of β-SiC nanocrystals with respect to the Si-substrate deduced from selected area diffraction patterns are up to 9° about the [110]-zone axis (Fig. 2b) and about the [001]-zone axis.

Figure 3 shows an epitaxially oriented diamond grain on top of the β-SiC interlayer with an average thickness of 12 nm (sample B, late nucleation stage). The diamond grain has been nucleated 8 nm above the Si substrate on the SiC interlayer. The diamond grain is faceted parallel to the {111}-lattice planes and is twinned. The lower part of the diamond grain overlaps with the β-SiC interlayer in the projected image.

Deposition for one hour under growth conditions as shown in table 1, sample C, resulted in a 600 nm thick film with grain-sizes of about 300×300 nm^2 and with a density of oriented diamond grains (≤5° misorientation) of about 5×10^8 cm^{-2}. The average thick-

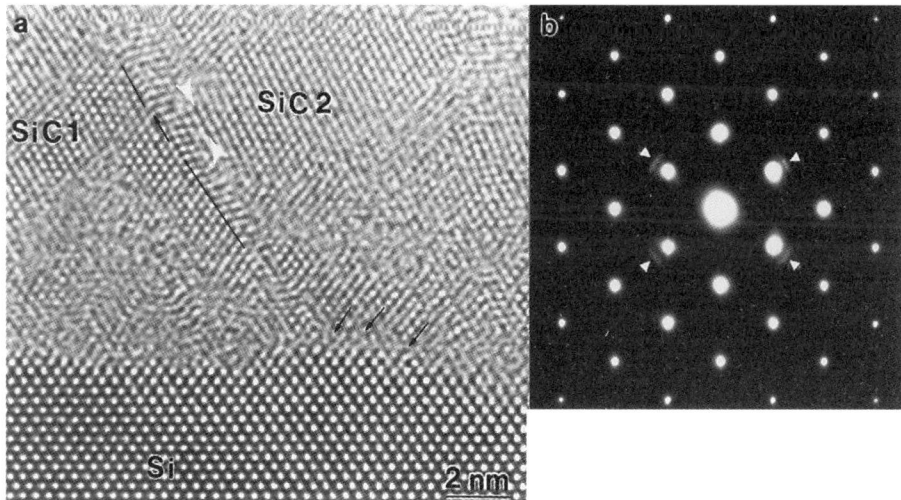

Fig. 2: (a) Misalignment between β-SiC nanocrystals and Si substrate. Terminating $(\bar{1}1\bar{1})$-SiC lattice planes at the interface and a terminating (111)-SiC plane at the grain boundary are marked by arrows. The nanocrystals SiC1 and SiC2 are tilted with respect to each other by 7° about the [110]-orientation. [110]-zone axis HRTEM micrograph. (b) [110]-zone axis selected area diffraction pattern with {111}-SiC reflections (arrows). Sample B.

454

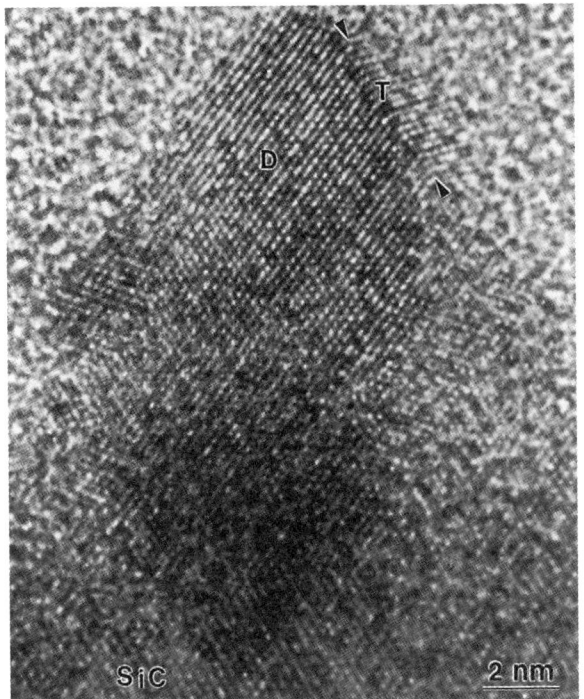

Fig. 3: Epitaxially oriented diamond grain on the β-SiC interlayer. The diamond grain is faceted parallel to the {111}-lattice planes. Twins (T) are formed in the diamond grain. HRTEM micrograph in the [110]-zone axis, sample B.

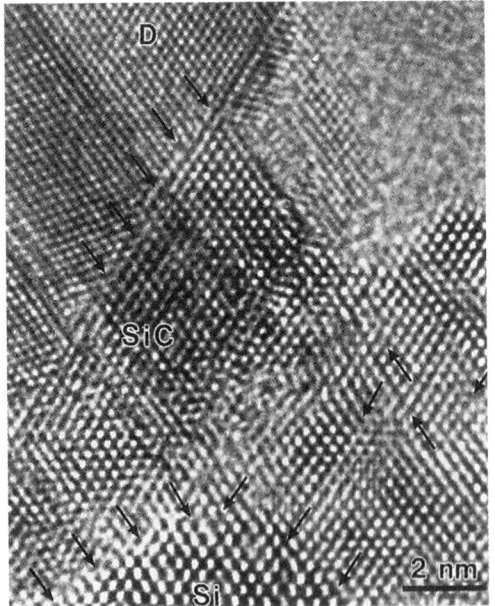

Fig. 4: Epitaxially oriented diamond grain on top of the SiC interlayer, with a matching of five (111)-diamond lattice planes with four (111)-SiC lattice planes along the faceted D/SiC interface. At this interface, and at the SiC/Si interface, terminating {111}-lattice planes are indicated by arrows. HRTEM micrograph in the [110]-zone axis, sample C.

ness of the β-SiC interlayer is 12 nm. A high fraction of the β-SiC nanocrystals show facets parallel to the {111}-lattice planes. The Si/SiC interface is roughened with amplitudes of up to 10 nm. An interface between diamond and a β-SiC nanocrystal is depicted in fig. 4 (sample C, growth stage). The interface is faceted parallel to the $(\bar{1}1\bar{1})$ - planes and is smooth on an atomic scale. The (111)-lattice planes of diamond and SiC are directly connected along the interface, with a regular matching of five lattice planes of diamond with four SiC lattice planes ('five-to-four'-registry). The corresponding interface between SiC and Si is also faceted parallel to the $(\bar{1}1\bar{1})$-planes and shows a registry between the (111)-SiC planes and the (111)-Si planes of 'five-to-four'. As a result, a registry of 'twenty-five-to-sixteen' between the (111)-diamond lattice planes and the (111)-Si lattice planes via the SiC interlayer is obtained.

4. DISCUSSION

4.1 Early stages of diamond film formation on (100) Si substrates

According to our studies different stages and processes have to be considered during the formation of diamond films on Si(001) which is briefly sketched in the following model:

(i) Substrate roughening by etching and redeposition of Si: The cleaning procedure of the Si substrate by the hydrogen plasma leads to an enrichment of the plasma with Si. At the beginning of the nucleation stage, subsequent redeposition of Si under the applied bias voltage may lead to the formation of small Si islands which frequently exhibit {111}-facets (fig. 1). Twins may be generated by the formation of stacking faults on {111}-planes during growth (Ernst et al. 1988), which are probably assisted by the continuous ion bombardment from the plasma.

(ii) Formation of a thin nanocrystalline β-SiC interlayer: Formation of β-SiC nanocrystals and an almost interconnected interlayer of about 12 nm thickness during the nucleation stage (figs. 1,2), require Si and C in sufficiently large amounts. On the basis of our results we propose that redeposition of etched Si from the plasma is a predominant source of Si for SiC formation. Under similar process conditions the substrate thickness etched away during these deposition stages was determined to be about 23 nm, leading to the conclusion that strong enrichment of the plasma by Si takes place (Wittorf et al. 1996). On the other hand, formation of SiC resulting from diffusion of Si through SiC or from C implantation are unlikely to account for the formation of the observed SiC interlayer of 12 nm thickness. The diffusion length of Si in β-SiC is below 1 nm at temperatures below 875° (Hon et al. 1980), and the average penetration depth of C^+-ions for ion energies below 200 eV is also smaller than 2 nm (Yugo et al. 1992).

(iii) Formation of {111}-SiC facets: Both the etching and the redeposition of Si under the applied bias have to be taken into account. The observed average SiC interlayer thickness increases with increasing nucleation time from 5 nm (sample A) to 12 nm (sample B), indicating growth of the SiC nanocrystals. If the resulting growth rate of the {001}-facets is larger than those of the {111}-facets, the nanocrystals will develop preferentially {111}-facets (fig. 3).

(iv) Nucleation of diamond on the {111}-facets of SiC nanocrystals (fig. 4): At the later nucleation stage (sample B) less Si is etched, and the CH_4 fraction in the plasma increases. Therefore, more and more CH_3^+ and C^+ are implanted due to the applied bias voltage into the SiC. This may lead to the formation of small oriented diamond nuclei within the near-surface region of the SiC lattice by ion-implantation enhanced transformation of sp^2-bonded into sp^3-bonded carbon near the surface (Yugo et al. 1992, Gerber et al. 1995) or by ion-induced transformation of sp^2- into sp^3-bonded carbon in amorphous Si-C phases on top of the surface (Wurzinger et al. 1996).

4.2 NCSL description of orientation relationships: diamond / SiC and diamond / Si

The $(\bar{1}\,1\,\bar{1})$ - D/SiC interface plane (fig. 4) is characterized by a regular arrangement of terminating (111)-diamond lattice planes resulting in a in a 'five-to-four'-registry of lattice planes. This seems to reduce the original misfit between diamond and SiC from 22.2% to $f = (5 \cdot a_D - 4 \cdot a_{SiC}) / (5 \cdot a_D) = 2.3\%$. Considering energetically favoured lattice sites and assuming an isotropic expansion of the SiC lattice by 2.3% a near-coincidence site lattice (NCSL) can be constructed (Balluffi et al. 1982). The basic structure for this diamond-SiC NCSL is a face-centered cubic lattice. This description of the orientation relationship between diamond and SiC by a NCSL is comparable to that between diamond and Si where a 'three-to-two'-registry between epitaxially oriented diamond and Si with a reduced misfit of 1.5% was observed (Jia et al. 1995).

5. SUMMARY

Studies of interfaces and orientation relationships between diamond films, β-SiC interlayers and Si were done with transmission electron microscopy for the early stages of film deposition on Si(001) substrates under optimized MWCVD process conditions. These studies show that diamond films form particularly via nucleation on the {111}-facets of β-SiC nanocrystals of a thin interlayer. A model for the early stages of diamond film formation is proposed which describes the observed phenomena. The orientation relationship between epitaxially oriented diamond and β-SiC can be described using a NCSL model assuming a strained SiC lattice. This is similar to the description of the orientation relationship between epitaxially oriented diamond and Si(001).

ACKNOWLEDGEMENTS

This work was financially supported by the Federal Minister of Research and Technology (BMBF Project 'Diamond as electronic material', Contract Nr. 03 N 1001 BO).

REFERENCES

Baluffi R W, Brokman A and King A H 1982 Acta metall. **30**, 1453
Ernst F and Pirouz P 1988 J. Appl. Phys. **64**, 4526
Gerber J, Sattel S, Jung K, Ehrhardt H and Robertson J 1995 Diamond Rel. Mater. **4**, 559
Hon H M, Davis R F and Newbury D E 1980 J. Mater. Sci. **15**, 2073
Jia C L, Urban K and Jiang X 1995 Phys. Rev. B **52**, 5164
Jiang X, Klages C-P, Zachai R, Hartweg M and Füßer H J 1993 Diamond Rel. Mater. **2**, 407
Kawarada H, Suesada T and Nagasawa 1995 Appl. Phys. Lett. **66**, 583
Stammler M, Stöckel R, Ley L, Albrecht M and Strunk H P 1996, Proceedings of
    DIAMOND '96, Tours '96 [Diamond Rel. Mater. , in press]
Wittorf D, Jäger W, Urban K, Gutheit T, Güttler H and Zachai R 1996, Proceedings of
    DIAMOND '96, Tours '96 [Diamond Rel. Mater. , in press]
Wurzinger P, Pongratz P, Gerber J and Ehrhardt H 1996 Diamond Relat. Mat. **5**, 345
Yugo S, Kimura T and Kanai T 1993 Diamond Rel. Mater. **2**, 328

Inst. Phys. Conf. Ser. No 157
Paper presented at Microsc. Semicond. Mater. Conf., Oxford, 7–10 April 1997
© 1997 IOP Publishing Ltd

Synchrotron X-ray reticulography: a versatile new technique for mapping misorientations in single crystals

A R Lang and A P W Makepeace

H H Wills Physics Laboratory, University of Bristol, Tyndall Avenue, Bristol BS8 1TL

ABSTRACT: Reticulography is an offshoot of white-radiation X-ray topography. In reticulography a fine-scale X-ray absorbing mesh is placed between a Laue-diffracting crystal specimen and the topograph-recording photographic plate. The mesh splits the diffracted beam into an array of individually identifiable microbeams. Direction differences between microbeams, which give the orientation differences between the crystal elements reflecting them, are measured from their relative shifts within the array when mesh-to-plate distance is changed.

1. INTRODUCTION

X-ray topography performed at synchrotron radiation sources conventionally employs either simple white-radiation (Laue) imaging of single crystals, or the slower and more complicated double-crystal topographic technique. In either case the commonest aim is detection and measurement of lattice misorientations in the specimen, whether due to grown-in or process-induced defects. Reticulography is nearly as simple to perform as single-crystal Laue topography, yet it yields quantitative data on misorientations that would demand long sequences of images if the double-crystal technique were applied. Moreover, when performed in the back-reflection mode the direction differences between the microbeam elements into which the mesh splits the diffracted beam are twice the orientation differences between the crystal elements reflecting them in both radial and tangential directions. Hence a single back-reflection reticulograph is directly transformable into a map of misorientation **vectors**. The angular sensitivity of reticulography depends upon the angular size of the X-ray source. At Station 7.6 at the SRS, Daresbury, 80m from the tangent point, and with source size FWHM = 0.23mm vertically, the incidence angular range in the vertical plane is only 0.6 arc second, and misorientations down to this magnitude are measurable.

2. EXPERIMENTAL ARRANGEMENTS

The general layout is shown in Fig. 1(a), where 'position 1' and 'position 2' represent photographic plate settings at short and long distances, M, from the mesh. Fig. 1(b) schematically illustrates the mesh action in splitting the diffracted beam into filaments that diverge or converge according to the orientation difference between the crystal elements producing them. Precision electrodeposited Au and Ag meshes are used: a coarser mesh period (68 or 64μm) and a finer period (34μm) together cover most needs. The plate carrier rides on a long optical track tilted to the required $2\theta_B$ (135° in all images shown here). The mesh carrier is mounted on the track close to the diffracting specimen. Versatility of the technique derives from its wide sensitivity range, which is simply controlled by choice of M. With imperfect crystals, such as melt-grown Cu and Australian natural diamonds, maximum useful values of M may be only a few cm. On the other hand, with unstrained, nearly perfect InP and GaAs the mean reticulograph pattern remains beautifully regular even with M = 1070mm; and at this high M local lattice tilts at outcrops of individual dislocations become detectable.

458

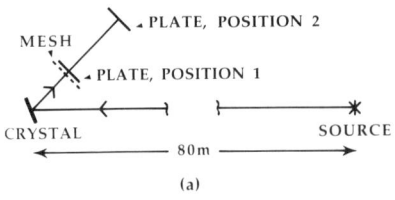

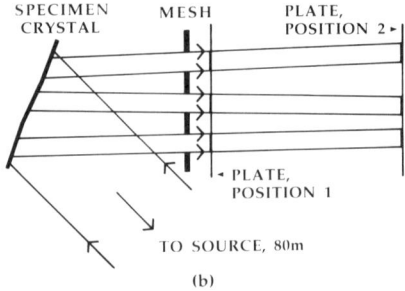

Fig. 1(a) Set-up for reticulography. (b) Production of mesh image distortions by specimen lattice tilts.

3. SOME APPLICATIONS OF RETICULOGRAPHY

Fig. 2 compares white radiation images of a surface cut and polished on a large synthetic diamond: 2(a) is a standard Laue topograph and 2(b) a representative reticulograph, one of a set shown in Lang and Makepeace (1996). This surface (indexed $(00\overline{1})$) is that from which the seed crystal, originally nearer the observer, was cut away. Evidence from colouration, IR absorption and CL shows that the near volume of the (001) growth sector occupying the centre of the image contains less substitutional nitrogen impurity, and hence has a smaller lattice parameter, than its flanking $\pm(100)$, $\pm(010)$ and octahedral growth sectors. Consequently, lattice planes parallel to the cut surface have an overall deformation, concave

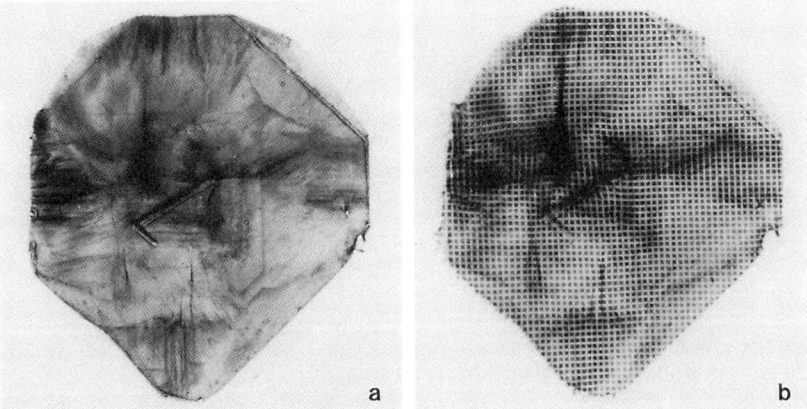

Fig. 2 Back-reflection images ($2\theta_B = 135°$) of a cube-cut surface on a cubo-octahedral synthetic diamond. Dominant reflection 004-type, $\lambda = 0.165$nm. Overall crystal width 3.9mm. (a) Standard-technique topograph, no mesh in diffracted beam. Specimen-to-plate distance (P) = 65mm. (b) Reticulograph, 68μm-period mesh. Mesh-to-plate distance (M) = 260mm, P = 315mm. (Reproduced with permission of IUCr).

towards the observer. The bending is largely localized at arrays of dislocations. Concavity towards the observer causes contraction of reticulograph mesh spacings, and, as M increases, first a focussing and then a folding of the mesh image. Orientation difference between 2 points on a crystal, A and B, say, is found from the closure vector between the step-wise path AB on a reticulograph recorded at appropriate M and the corresponding step-wise path on a perfect reference mesh. If the direct path A to B on the reticulograph concerned passes through focussing or folding then it is usually possible to detour step-wise around such regions to avoid ambiguity in counting mesh steps. This is a valuable feature of the 2D-lattice character of reticulographs.

Digital reticulographic analysis via TV microscopy is shown in Fig. 3, applied to a region straddling the left-hand vertical band of mesh image contraction seen in the lower left of Fig. 2(b). Misorientation vectors are measured relative to a block of nearly undistorted crystal

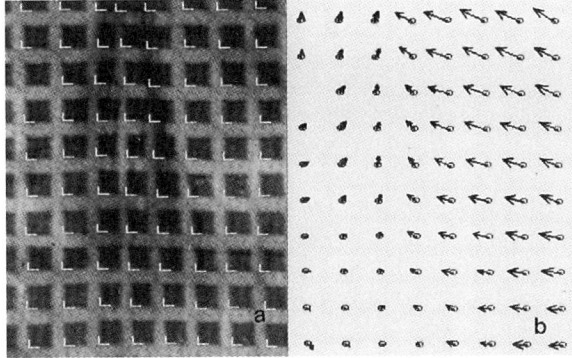

Fig. 3 Digital analysis of part of lower left quadrant in Fig. 2(b). Field width on crystal = 0.51mm. (a) CCD-camera-recorded micrograph of reticulograph with markers set on mesh cell corners. (b) Vectors from reference-mesh cell corners (small circles) to corresponding corners in (a).

on the left of Fig. 3. Identity of the fields in Figs. 3(a) and (b) is confirmed by corresponding locations of a missing lattice point in both (a) and (b) (upper left).

Lattice tilts associated with growth sector boundaries are directly measurable by reticulography. Fig. 4 shows part of a (110)-orientation polished plate of synthetic diamond, of interest because it had one unusually large {110}-type growth sector. Substitutional nitrogen impurity was only ~1ppm atomic in this sector but ~100ppm atomic in the adjacent {111} sectors (recently revised value). Synchrotron X-ray double-crystal diffractometry (Lang et al 1991) found a 1.2×10^{-5} fractional increase in lattice parameter of the {111} sectors relative to the centre of the $(1\bar{1}0)$ sector. X-ray topography shows the growth sector boundaries to be coherent and dislocation-free. To accommodate the lattice parameter difference across these boundaries, the surface lattice planes are convex outwards (increased mesh image spacing) on the {111}-sector side and concave outwards on the $(1\bar{1}0)$-sector side, as clearly evident in Figs. 4(b) and (c).

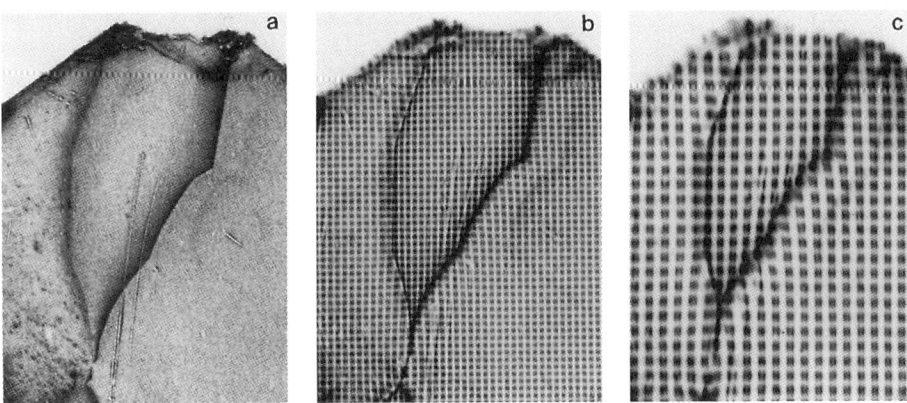

Fig. 4 Part of a (110)-polished synthetic diamond plate containing a $(1\bar{1}0)$ growth sector expanding towards the crystal's external surface (upper boundary of image) and abutted by the $(1\bar{1}\bar{1})$ sector and the $(1\bar{1}1)$ sector on its left and right, respectively. Dominant reflection 440, λ = 0.12mm. Specimen thickness 0.7mm. Field width 1.4mm. (a) Conventional topograph showing growth-sector-boundary diffraction contrast. (b) and (c) Reticulographs: (b) 34μm-period mesh, M = 500mm; (c) 68μm-period mesh, M = 1070mm.

Diffraction images of an ion implant are shown in Fig.5. A (110)-polished, dislocation-free natural diamond plate was implanted locally with 17MeV-energy fluorine ions, penetration depth ~5μm. In the implant shown the dosage was ~8×10^{14} ions mm^{-2} over a roughly circular patch, diameter ~0.9mm. Sharp lattice bending around the patch perimeter causes a ring of intense diffraction contrast. A standard reflection topograph (Fig. 5(a)) is uninformative on lattice tilts within the periphery, whereas double-crystal topographs show complex patterns, difficult to analyze. A single reticulograph is immediately informative:

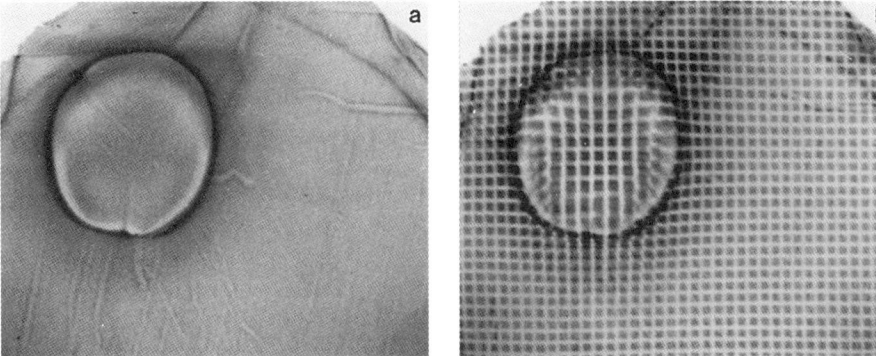

Fig. 5 Lattice distortion resulting from ion implantation. Seen upper left in each image is a ~0.9mm-diameter patch where fluorine ions have been implanted. Dominant reflection 440, λ = 0.12nm, P = 550mm. (a) No mesh. (b) Mesh period 64μm, M = 500mm.

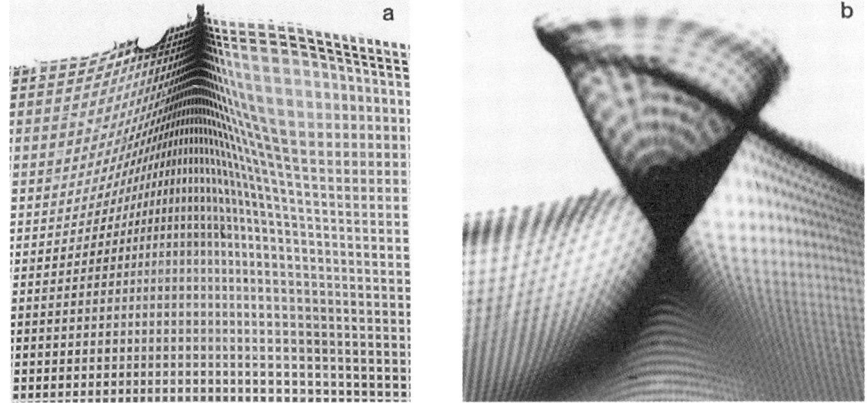

Fig. 6 Elastically deformed InP single-crystal plate, thickness 350μm, orientation (100). Field width 3.5mm. Top boundary of images is the cleaved plate edge (placed lower in (b) than in (a)). Mesh period 68μm. (a) M = 250mm. (b) M = 1070mm.

Fig.5(b) reveals that the central part of the patch is spherically domed outwards, radius 5m.

The reticulographs in Fig. 6 instructively reveal the misorientation field generated by local stress applied close to an edge of a plate of perfect InP. A pull on the plate's back surface, caused by a drop of adhesive, gives rise to a depression in the front, X-ray reflecting surface, deepening towards the top edge of the plate. A focus close to this edge in Fig. 6(a) moves to a less deformed area in Fig. 6(b). Continuity of the folded image of the top edge in (b) testifies to absence of slope discontinuities. The maximum departures from mean orientation seen are only ±0.5 milliradian.

ACKNOWLEDGMENTS

The authors thank the Director and Staff, Synchrotron Radiation Department, Daresbury Laboratory, for use of experimental facilities.

REFERENCES

Lang A R and Makepeace A P W 1996 J. Synchrotron Rad. **3**, 313
Lang A R, Moore M, Makepeace A P W, Wierzchowski W and Welbourn C M 1991 Phil, Trans. R. Soc. Lond. A**337**, 497.

Inst. Phys. Conf. Ser. No 157
Paper presented at Microsc. Semicond. Mater. Conf., Oxford, 7–10 April 1997
© 1997 Government of Canada

The use of transmitted color and interference fringes for TEM sample preparation of silicon

J P McCaffrey

National Research Council of Canada, Institute for Microstructural Sciences, M-50 Montreal Rd., Ottawa, Canada K1A 0R6

ABSTRACT: Silicon TEM samples backlit with an optical light source display a series of colors in regions of less than 10 microns thickness and a series of interference fringes in the regions of less than 2 microns thickness. These colors and fringes result from the transmission, reflection, absorption and interference of light within the sample and depend upon the light source and the sample thickness. We present direct measurements of silicon color versus thickness, the factors that influence this relationship, and the origin and significance of the interference fringes.

1. INTRODUCTION

The optical colors of silicon viewed in transmission have been used for many years as a technique for monitoring the thickness of silicon samples prepared for transmission electron microscopy (TEM) - (c.f. Ellington, 1988). However, the observation and demonstration of optical interference fringes as a subjective figure of merit for the quality of TEM samples is more recent (c.f. Gatan, 1995). These and other observations are a testament to the continual improvement in TEM sample preparation equipment and techniques, largely driven by the silicon-based semiconductor industry. In a previous study, the colors of silicon were related to thickness through the use of an optically translucent wedge of silicon where the thickness profile could be directly measured (McCaffrey, 1997). This work expands upon the previous study and investigates the observation of interference fringes at the thinnest edge of silicon TEM samples.

2. EXPERIMENTAL

Two silicon samples were used for these measurements and observations, a cleaved 7° wedge and a tripod-polished 2° wedge. The cleaved wedge of silicon was prepared by a variation of the small-angle cleavage technique and is described in detail elsewhere (McCaffrey, 1997). The tripod-polished wedge was prepared by epoxying two pieces of a silicon wafer together such that one piece would produce a thinned wedge viewed down the <100> zone axis, and the adjacent piece would be viewed down the <011> axis. The samples were then polished optically flat on one face with standard tripod polishing techniques (c.f. Klepeis, 1988), then turned over and polished on the other face at a 2° angle to the first face. This resulted in adjacent 2° wedges of <100> and <011> oriented silicon. The 7° angle of the cleaved wedge was determined from a series of SEM micrographs taken from different sample orientations (Fig. 1), and the 2° angle of the tripod polished sample was determined using simple trigonometry.

To observe the colors of silicon in transmission, both wedges were back-illuminated and photographed in an optical microscope with ISO 200 Kodak Gold daylight color film. The light source was a 150W quartz halogen tungsten lamp which approximates the CIE standard illuminant A (gas-filled tungsten lamp with a color temperature of 2854 K - Bouma, 1971) The addition of a Kodak 80B filter in series with this same lamp approximates average daylight; i.e., the CIE standard illuminant C (color temperature of 6770 K (Bouma, 1971)).

3. RESULTS

In order to quantitatively describe the observed transmitted color of the Si wedge, it is useful to employ the internationally adopted Commission Internationale de l'Eclairage (CIE) 1976 color coordinates U' and V' (cf. Bouma, 1971). These color coordinates form the axes of a chromaticity diagram and together can be used to represent the gamut of colors seen by the human eye. The color coordinates for the transmitted light through a Si wedge depend on the radiant energy of the spectral source, the transmittance of the Si wedge and the wavelength response of the human eye, specified by spectral tristimulus values (Bouma, 1971). For this study, the CIE color coordinates were calculated using formulae that integrate the transmittance through a single crystal silicon sample with a specified thickness over the wavelength region from 380 to 780 nm, weighted by the spectral source radiant energy and the tristimulus values. In addition, the luminous transmittance L, which is the integrated transmittance weighted with respect to the photopic curve of the eye, was also calculated. All calculations are presented in Table 1. As it was not possible to include the color diagrams here, a color description is included.

Observations of the samples showed that colors ranged from deep red for thicknesses of approximately 5 to 10 μm, to orange and yellow in the thinner regions, to colorless at

Fig. 1: Scanning electron micrographs of three views of the cleaved silicon wedge used in this study. The features identified by numbered arrows were used to help determine the thickness profile.

Table 1: 1976 CIE color coordinates are calculated for two light sources, CIE standard illuminants A and C. Note that the silicon transmission color for light source A at 1μm thickness is approximately equal to the silicon transmission color at 2 μm using light source C. Also note how rapidly the luminous transmittance L falls off with thickness; for CIE illuminant C at 5 μm thickness, only 1% of the illumination is transmitted.

TABLE 1	Si thickness	0 μm	1 μm	2 μm	3 μm	4 μm	5 μm
CIE	U'	0.256	0.333	0.388	0.433	0.468	0.495
standard	V'	0.524	0.546	0.540	0.535	0.530	0.526
illuminant	L	1.000	0.181	0.087	0.046	0.027	0.016
A	color description	no color	orange	orange/ red	reddish orange	red	red
CIE	U'	0.201	0.275	0.337	0.390	0.433	0.467
standard	V'	0.461	0.547	0.547	0.541	0.535	0.530
illuminant	L	1.000	0.155	0.068	0.033	0.018	0.010
C	color description	no color	yellow	orange	orange/ red	reddish orange	red

the thinnest regions, in agreement with the calculated values. For the tripod-polished 2° wedge, no difference in colors was seen between the two sample orientations <100> and <011> for regions of equal thickness. The presence of interference fringes (Fig. 2) were clearly noted for the tripod-polished sample but were not observed for the cleaved wedge. The explanation and appearance of these fringes is similar to that for Newton's rings - internal reflections and interference.

Interference fringes at the thinnest regions of the silicon wedges are produced by destructive interference between transmitted light and light that has undergone multiple internal reflections, as illustrated in Fig. 3. Many rays enter the silicon wedge through the first air-silicon surface and while some rays pass completely through the wedge, others undergo multiple internal reflections, undergoing a 180° phase shift and a reversal of direction at each silicon-air interface. When the addition of all phases of all exiting rays of a particular wavelength cancel each other out (interfere destructively), a dark band will appear.

Silicon has a refractive index that varies with wavelength (Table 2) and refraction within the silicon wedge can be calculated using Snell's law; $n_1 * \sin \phi_1 = n_2 * \sin \phi_2$, where n_1 is taken as 1.0 and n_2 is the refractive index of silicon. The calculation shows that a silicon wedge with an angle greater than approximately 5° will produce total internal reflection of internally reflected rays at all visible wavelengths, i.e., these samples will not display interference fringes. This explains the lack of interference fringes observed in out 7° sample, with its atomically flat (cleaved) faces. Note that surface roughening caused by mechanical polishing will enhance internal reflections and affect the internal reflectance angles, resulting in an increase in the number and intensity of fringes.

The dispersion of light in the silicon produces another interesting and observable effect - the dark fringes observed in the thin silicon sample appear to be green at higher magnification. Also visible are red fringes, but with a shorter period between adjacent fringes. Positions of the dark (green) fringes correspond to

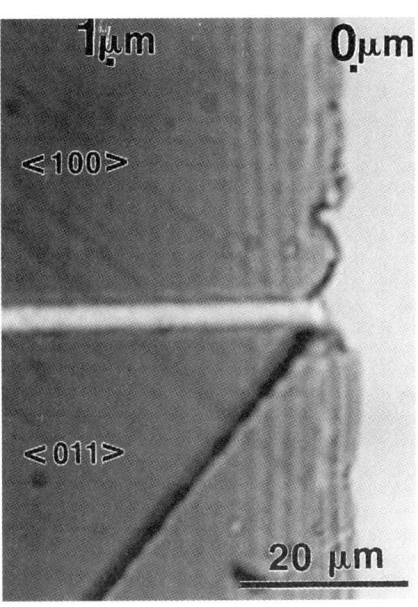

Fig. 2. Interference fringes in silicon for two orientations of silicon single crystal; <100> and <011>. The large numbers at the top of the figure indicate the thickness of the sample at that position.

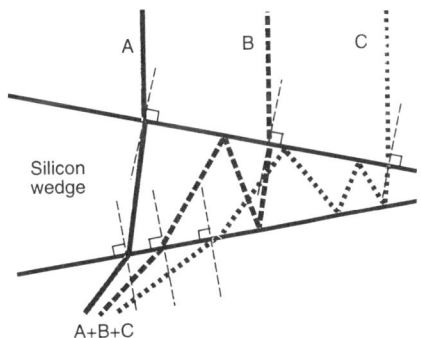

Fig. 3. Schematic drawing of coherent light rays interfering with each other after following different paths through a silicon wedge sample. The thin dotted lines represent the surface normals, and the dotted and solid lines labeled A, B and C represent the light rays.

the thicknesses where one expects destructive interference to occur for red (approximately 600 nm) rays. As red light is the dominant transmitted color, destructive interference will result in a darker region, colored by the remaining (predominately green) rays that do not interfere destructively at that position. Similarly, in the thinnest region of the sample (<1 μm in thickness), faint red fringes can be observed at the positions where 500 nm (green) light would be expected to interfere destructively. The fringe positions for green and red light were calculated and are shown in Fig. 4.

Wavelength	Si <100>	Si <011>
300 nm	5.06	5.10
400	5.59	5.57
500	4.29	4.29
600	3.94	3.94
700	3.77	3.77

Table 2. Refractive indices of <100> and <011> silicon at various wavelengths. The slight differences between <100.> and <011> are to be attributed to differences in surface roughness , as silicon is isotropic.

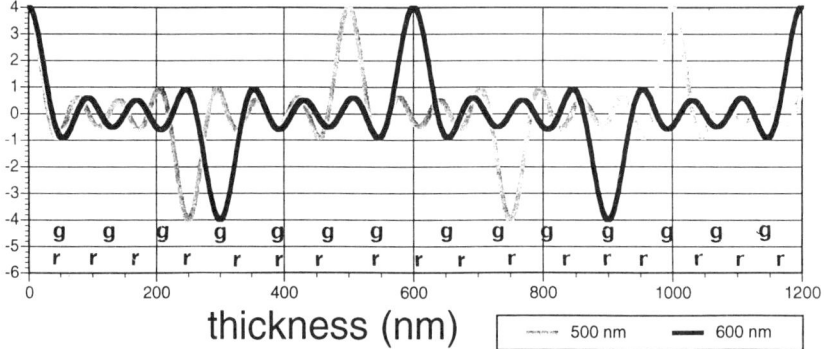

thickness (nm)

······ 500 nm   ▬▬ 600 nm

Fig. 4: Interference patterns for green (500 nm) and red (600 nm)light in a silicon crystal for multiply-reflected rays. Note that the higher wavelength (600 nm) produces fewer regions of destructive interference (dark bands, marked "g" for their dark green color) than the lower wavelength destructive interference (marked "r" for red), in agreement with the observations. Changing the geometry of the incident light relative to the silicon sample can slightly shift the position and period of the fringes.

4. CONCLUSION

The color of light transmitted through a silicon sample can provide useful information on the thickness of silicon samples during TEM sample preparation and other thinning and etching processes. Colors can vary with the light source, but in general, samples that appear deep red in color are within the 5-10 μm thickness region, and are progressively thinner as orange, yellow and clear regions are observed. Interference fringes are observable if the sample is thinner than approximately 2 μm, and are the result of destructive interference of the internally reflected and transmitted rays.

REFERENCES

Bouma, P.J. 1971, Physical Aspects of Color, (Macmillan and Co. Ltd, The Netherlands).
Ellington, M.B.1988, Mat. Res. Soc. Symp. Proc. Vol. 115, (Pittsburgh, Penn.: Materials Research Society) pp. 265-270.
Gatan, Inc. 1995, Gatan Technical Note 9204, pp. 1-2.
Klepeis, S.J., Benedict, J.P. and Anderson, R.M. 1988, Mat. Res. Soc. Symp. Proc. Vol. 115, (Pittsburgh, Penn.: Materials Research Society) pp. 179-184.
McCaffrey, J.P., Sullivan, B.T., Fraser, J.W., and Callahan, D.L. 1997, Micron, in press.

Inst. Phys. Conf. Ser. No 157
Paper presented at Microsc. Semicond. Mater. Conf., Oxford, 7–10 April 1997
© *1997 IOP Publishing Ltd*

Cross-sectional transmission electron microscopy and focused ion beam study of advanced silicon devices

H Bender and P Roussel

IMEC, Kapeldreef 75, B-3001 Leuven, Belgium, bender@imec.be

ABSTRACT: The use of Focused Ion Beam (FIB) for cross-section imaging and TEM specimen preparation for advanced devices is discussed. Attention is given to techniques to avoid surface damage by the Ga ion beam when the top surface layers have to be studied in unpassivated or partially processed devices. The possibilities for additional thinning of FIB-specimens after TEM observation will be illustrated. It is shown that high resolution imaging on these specimens with comparable quality as for conventional ion milling is possible. Comparison with the information which can be deduced from FIB cross-section images is made.

1. INTRODUCTION

The advanced silicon device technologies of 0.35 μm and below are characterised by an increasing number of layers, a decreasing layer thickness and shrinking feature sizes. The development and control of these processes requires very site specific cross-sectioning as well as high resolution imaging techniques. In many cases scanning electron microscopy (SEM) imaging on cleavage planes does not allow the necessary site specification. This problem can be overcome with Focused Ion Beam (FIB) imaging. On the other hand, even with the best FEG-SEM's and with FIB is the image resolution often inadequate. Therefore an increasing demand exists for cross-sectional transmission electron microscopy on processed devices.

The combination of FIB preparation and TEM observation fulfils both the needs for site localisation and high resolution imaging.

2. FIB IMAGING AND SPECIMEN PREPARATION

The principle of imaging with a Focused Ion Beam is to a large extent similar to the imaging with a scanning electron microscope, with the major difference that a Ga^+ ion beam is used instead of an electron beam (Stevie et al 1995). Therefore during imaging also some sputtering of material occurs. Modern FIB systems allow image resolutions on the order of 5-7 nm.

The procedures for TEM specimen preparation have already been discussed in detail in the literature (Pantel et al 1994, Walker et al 1995, Young et al 1990). Thin strips are sawed from a silicon wafer at the region of interest with a thickness of directly 50-70 μm so that no further polishing is necessary or with thickness of approximately 250 μm requiring further mechanical polishing to 50-70 μm before loading in the FIB. These strips are glued on a standard large slot copper TEM-grid and thinned stepwise by milling the top surface layers with decreasing beam currents until a thin slice remains at the selected spot. Typical thinned areas are 15-20 μm wide and 5-10 μm deep. Before the FIB thinning a protective metal layer (typically Pt) is deposited in-situ at the location of the final thinned region. In cases that large angle tilting or EDX analysis is required in the TEM, diabolo shaped craters have to be made (see e.g. Vanhellemont et al 1997).

3. RESULTS

3.1 Surface layers

In a development phase and for production control, non-passivated or only partially processed devices have to be analysed. In such cases often the top surface layers are of major interest. As one has to localise the region of interest first from a top view image, damage by the

Ga$^+$ ion beam cannot be avoided even not by the use of as low currents as possible and by the fast deposition of a protective Pt-layer. This problem is illustrated by the study of a selective epitaxial growth (SEG) structure.

Fig. 1 shows FIB cross-section images through the structure. To avoid surface damage, a Pt layer is deposited first. The isolating layers (SiO$_2$ and Si$_3$N$_4$) initially show up dark, while the poly-Si and substrate have the same contrast. In the top of the nitride an additional layer seems to be present. Longer imaging (Fig.1b) with the FIB results in a contrast change of the nitride due to conductance induced by the implanted ions, the top region of the nitride layer still shows a different contrast. Additional features appear in the image near the edge of the oxide layer and on the bottom of the nitride. Further interpretation of the images is possible by comparison with XTEM.

In Fig. 2 XTEM images of similar samples prepared by FIB are shown. In the regions with an initially uncovered silicon surface, the top layer is amorphised to a depth of approximately 60 nm (Fig. 2a) with a dark layer occurring in the top of this amorphised layer. Auger electron spectroscopy depth profiling through similarly deposited Pt layers shows that this is a knocked-in Pt layer. Also in the top of the nitride layer, a dark layer due to knocked-in Pt can be seen. This layer corresponds to the additional layer seen on the FIB image (Fig. 1). Near the poly silicon/SEG-Si interface two holes occur (over-exposed on the pictures) which are actually partially filled with polycrystalline silicon (see below). These porous holes correspond with the additional features appearing in the FIB images after longer imaging (Fig. 1b).

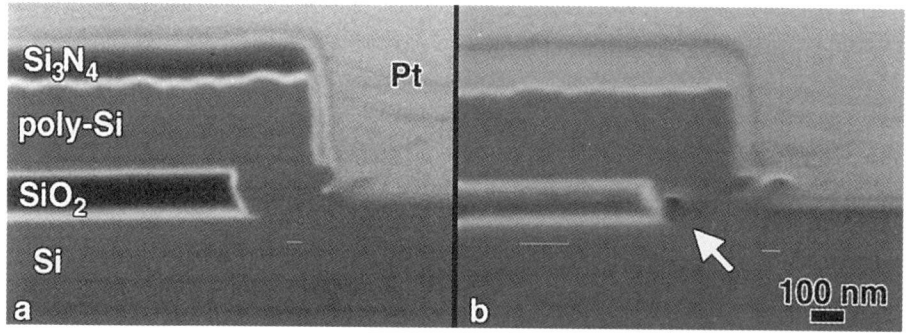

Fig. 1 : FIB cross-section images of a selective epitaxial growth structure : a) initial image and b) after longer imaging. The arrow indicates an additional feature becoming visible after longer imaging.

As the aim is to study the SEG layer, the amorphisation is an unacceptable artefact, and also some uncontrolled sputtering of the layer might have occurred. The depth of the amorphisation can be reduced by lowering the ion beam energy during the localisation of the region and the deposition of the Pt. Fig. 2b, indeed shows that the amorphous layer in that case is reduced to approximately 35 nm with still a knocked-in Pt layer near the surface and in the nitride (some non-amorphised regions in Fig 2a and b are due to some particles protecting the surface which are introduced by the sawing and polishing). Clearly an ex-situ deposited protective layer is needed. In case of Fig 2c, Au is sputtered before the FIB preparation with standard 30 keV Pt deposition. The thickness of the Au layer on the bottom of topography with a high aspect ratio is much less than on the planar surfaces. Whereas the Au on top of the nitride is 100-120 nm, it is only 60-70 nm on top of the SEG silicon. This effect is seen to be even worse in structures with much higher aspect ratios than shown in Fig. 2 so that the bottoms of the structures are then not or only partially covered. Furthermore it is observed that the Au layer grows in a porous dendritic-like structure. As a result of this the Au layer often does not sufficiently protect the surface, as can be seen in Fig. 2c where damage still occurs below the Au layer in the silicon. An alternative and efficient protecting layer is found to be the remains of the glue used for the polishing of the strips. Unfortunately, the glue coverage is difficult to control so that sometimes a layer over the whole structures is present which smooths all topography and makes the localisation of the region of interest difficult. Nevertheless it has successfully been applied to study for example deep open contact holes. A protective layer deposited during the device processing is to be preferred whenever possible from a processing point of view.

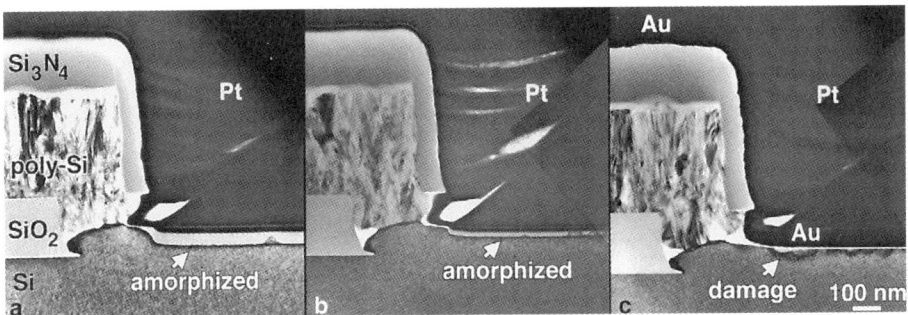

Fig. 2 : XTEM images of FIB prepared specimens of a SEG structure with initially uncovered Si surface : a) standard 30 keV Ga beam, b) 10 keV Ga beam for localisation and Pt deposition and c) 30 keV Ga beam with ex-situ sputtered Au layer before the Pt deposition.

3.2 Thickness control

The final thickness control of the XTEM specimens in the FIB is not straightforward and depends strongly on the experience of the operator. In the top view FIB image the surface of the thinned lamellae will be slightly rounded so that an accurate measurement of the real thickness at the depth of interest is hampered. Too thin samples are not advantageous for studying e.g. substrate defects and also result in low contrast between different (amorphous) dielectrica. However, to study e.g. 0.3 μm W plugs in cross-section, the specimens have to be as thin as possible in order to minimise the projection artefact of the layers which are not seen edge on.

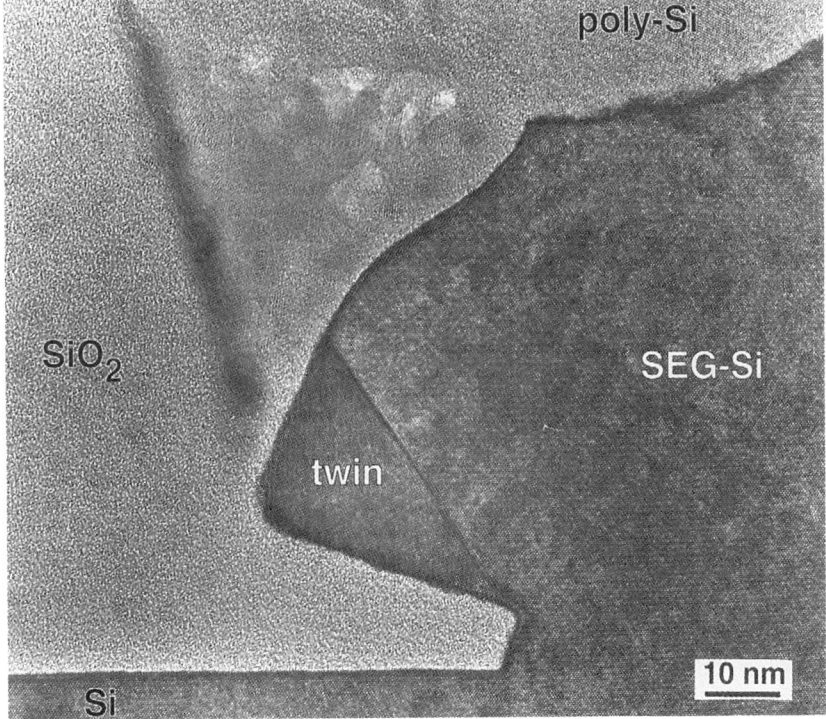

Fig. 3: HREM image of the foot of the SEG structure showing the partial filling of the pores with polycrystalline silicon.

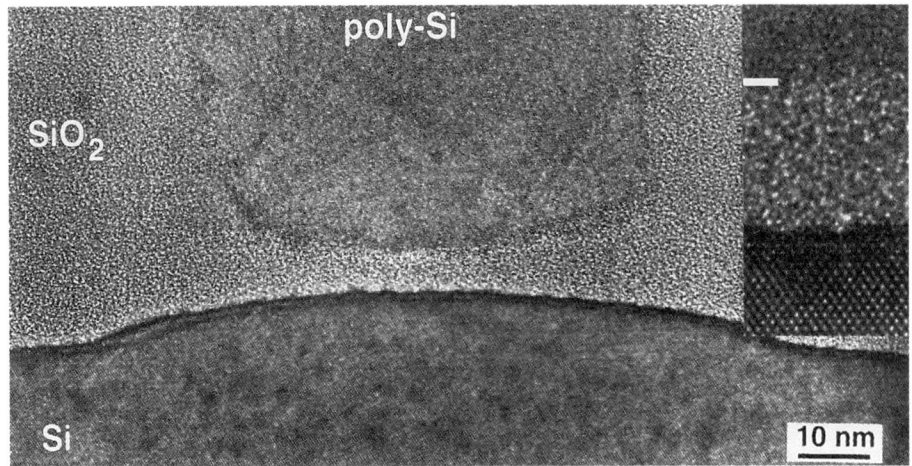

Fig. 4 : HREM image of a 80 nm gate region obtained after a third further FIB thinning of the specimen. The inset shows a higher magnification of the gate oxide.

Thinning the specimens to a thickness allowing high quality HREM can generally not be done in a predictive way. Whereas a thickness allowing vague high resolution imaging in the silicon at 200 keV can easily be reached ($\leq$ 50 nm), thinning to smaller thickness often needs additional thinning in the FIB after a first TEM inspection. After remounting the specimens in the FIB, the thinned lamellae has to be positioned again perfectly vertical. Whereas this is not very critical for initially thick specimens as discussed by Vanhellemont et al (1997) for studying extremely low density defects, the remounting becomes critical at the thicknesses where high resolution imaging is the aim. Total loss of the protective layer and eventually also of some top layer might occur at these thicknesses. This can be avoided by slight tilting of the thinned foil.

Fig. 3 shows a high resolution image of the foot of the SEG structure discussed before. It can be seen that the porous hole is indeed partially filled with polycrystalline silicon and that a twinned region is present on the edge of the SEG-layer. Fig. 4 shows a 80 nm wide poly-gate of a PMOS transistor with a 5.5 nm gate oxide. This high resolution image is obtained after a third FIB thinning step with the full structure still intact in the specimen. The absence of a wedge shaped specimen edge as in conventionally ion milled specimens does not allow the thickness to be measured from the thickness fringes. In both Fig. 3 and 4, however, the quality of the high resolution imaging allows estimation of the specimen thickness to be less than 20 nm.

4. CONCLUSIONS

FIB preparation of TEM specimens of samples for which the top surface layers are of interest need an ex-situ deposited protective layer. Sputtered gold is not suitable for structures with high aspect ratio. Remounting of the specimens for further thinning is possible. High quality HREM images can be obtained from FIB prepared specimens.

ACKNOWLEDGEMENT

The authors are indebted to the processing groups at IMEC especially Dr D Howard and Dr S Biesemans for the research material and to Dr A De Veirman (Philips Semiconductors) and Dr J Walker (FEI) for stimulating discussions. Mrs P Van Marcke and C Drijbooms are acknowledged for skilful specimen preparation and FIB imaging. The TEM analysis is done with the microscopes at EMAT, University of Antwerpen (RUCA).

REFERENCES

Pantel R, Auvert G, Mascarin G and Gonchond J P 1994 Proc. ICEM 13, 1007
Stevie F A, Shane T C, Kahora P M, Hull R, Bahnck D, Kannan V C and David E 1995 Surf. Interface Anal. **23**, 61
Vanhellemont J, Bender H and Van Landuyt J 1997 these proceedings.
Walker J F, Reiner J C and Solenthaler C 1995 Inst Phys Conf Ser **146**, 629
Young R J, Kirk E C G, Williams D A and Ahmed H 1990 Mat. Res. Soc. Symp. Proc. **199**, 205

Inst. Phys. Conf. Ser. No 157
Paper presented at Microsc. Semicond. Mater. Conf., Oxford, 7–10 April 1997
© 1997 IOP Publishing Ltd

Preparing TEM sections by FIB: stress relief to straighten warping membranes

J F Walker

FEI Europe Ltd, Brookfield Business Centre, Cottenham, CAMBS CB4 4PS

ABSTRACT: Stresses in a TEM section can cause considerable problems, both in the preparation of the section and in the subsequent microscopy. As the FIB thins a stressed specimen, warping occurs as the section offers less resistance because of reduced rigidity. Compressive stress is relieved as the membrane finds its own length through excessive local bending. Cutting the membrane at one end allows the thinned region to expand into the cut without the necessity to bend out of the original plane. The membrane straightens for further thinning or TEM observation. Under exceptional circumstances, residual strain can prevent complete straightening of the membrane. In this case both sides of the membrane can be freed to allow complete relaxation. This has always been sufficient to allow completion of the thinning and a full TEM examination.

1. INTRODUCTION

The preparation of transmission electron microscopy (TEM) sections of many materials by focused ion beam (FIB) milling has been of considerable interest over the past few years (Overwijk 1993, Young 1993, Hull 1995, Walker 1995 and Robinson 1997) as it offers many advantages over more conventional techniques. It can be site specific to the order of tens of nanometers, it is largely material independent and can offer very fast (typically 1-2 hours) preparation times. Circumstances arise whereby the membrane warps as it approaches the thickness required for TEM analysis. Clearly, stress in the membrane is responsible for this warping and this study identifies the origins of the stress and suggests methods of reducing the incidence of warping and for straightening an already warped membrane to assist further thinning. Whatever the source of the stress, the nature of the FIB preparation technique tends to concentrate the stresses in the region of interest. The thinner the section is made, the greater the likelihood of warping; to such an extent that misshapen membranes are rarely experienced in sections at 200nm thickness or more, but are often observed if sufficient electron transparency is required for lattice resolution.

2. SOURCES OF STRESS

Two principal sources of stress have been identified. An external stress can be caused by poor mounting techniques when applying the ground sample to the grid prior to FIB milling. The most commonly applied technique for preparing the sample prior to FIB is to cleave out a piece of the material to be examined and then to grind it down to 2mm by 0.05-0.1mm using a succession of diamond lapping films. This can then be attached to a slot grid from which a section has been cut away. This cut-away section allows access for the ion beam to the surface of the sample. Adhesives used to attach this piece of specimen to the grid sometimes require heat to cure them and usually experience some shrinkage. If a stiff, rigid grid is used then much of the stress is transferred to the specimen. A soft, annealed copper grid will itself deform rather than transfer stress to the specimen, making this type of grid ideal.

(a)   (b)

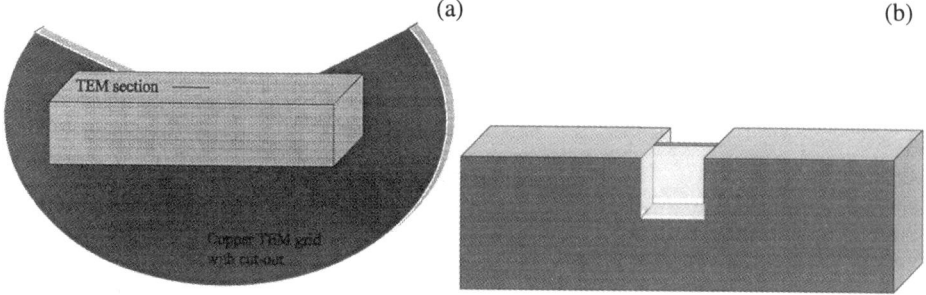

Figures 1(a) and 1(b). A 50-100mm thick piece of material is glued to a grid with a cut-out section (a) ready for FIB milling. Milling, with the ion beam coming in from above, removes channels leaving a thin membrane ready for TEM analysis (b).

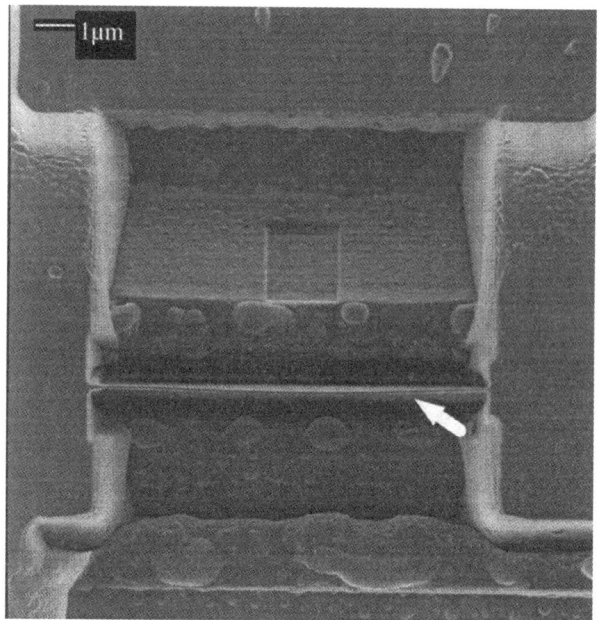

Figure 2. An almost completed TEM section as seen from the top. The membrane is seen end-on by the ion beam (used to generate this image) and the TEM electron beam would pass through the membrane from top to bottom or *vice versa*. The membrane (arrowed) is warped because of stress and no further thinning can be done without first straightening this.

Internal stresses are present as a result of the fabrication process of the material. Surface layers such as metallisation and inter-layer dielectrics deposited on a semiconductor are often under a degree of stress with silicon nitride and tungsten plugs being particularly susceptible.

3. STRESS RELIEF

The membrane shown in figure 2 shows a downward bowing due to compressive stress in the membrane. It is possible to relieve this stress by providing space for local expansion and a relief cut can be added, using the FIB, at one end of the membrane. Figure 3 shows this cut on the right edge of the membrane shown in figure 2. With most of the stress now gone, thinning of the membrane can continue.

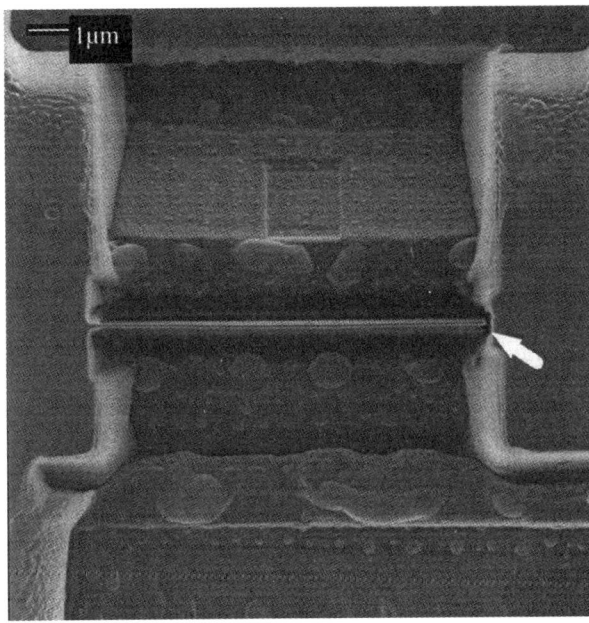

Figure 3. A cut (indicated) has been made on the right edge of the membrane and the resultant stress relief has straightened the membrane.

Some residual stress can remain even with the end cut and is often apparent when samples must be thinned for lattice imaging. As the membrane thins further under the milling ion beam, rigidity is lost and further warping prevents even milling. In this case stress can be relieved further by cutting the other side of the membrane, as in figure 4, such that it is only held at the bottom. This has almost always been sufficient to allow a membrane to be thinned to completion.

4. CONCLUSIONS

One of the limitations of TEM sample preparation by focused ion beams has been the tendency of thin membranes to warp if the sample has some internal or external stress. While good mounting practice can reduce or eliminate externally induced stress, another method has been described herein to resolve problems of internal stress. Cutting the almost finished membrane, either on one side or both, can straighten warping which will allow further thinning. The cut membranes, being held by only one edge may be inherently less robust than membranes held on three sides, yet experience in sending such samples through the post has shown no failures attributed to increased fragility.

472

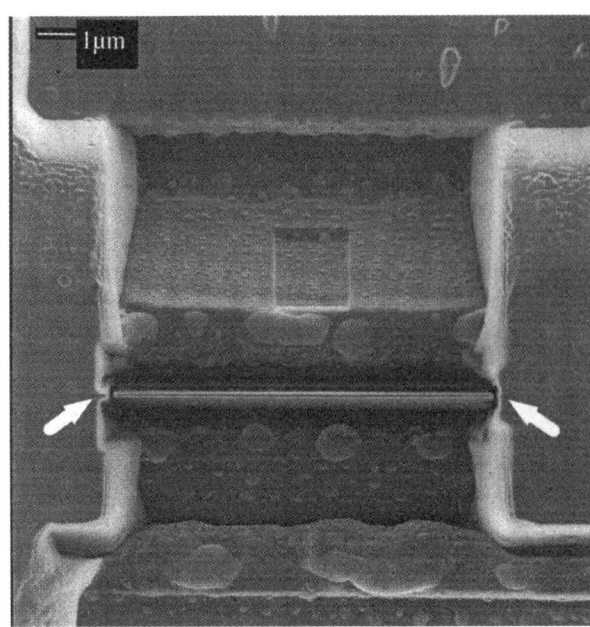

Figure 4. The membrane has been cut on both the left and right sides, as indicated. The membrane is straight and was thinned sufficiently for lattice imaging.

REFERENCES

Hull R, Moore M, Bahnck D, Geva M, Karlicek R F Jr., Stevie F A, Walker J F 1995 ECS, Reno, USA

Overwijk M H F, van der Heuvel F C, Bulle-Lieuwma C W T 1993 J. Vac. Sci. Tech. B **11** 2021

Robinson K, Pugh P J A, Walker J F, Newcomb S B 1997 Microscopy and Analysis **45**

Walker J F, Reiner J C, Solenthaler C 1995 Inst. Phys. Conf. Ser. No 146

Young R J 1993 Vacuum **44**, 353

Inst. Phys. Conf. Ser. No 157
Paper presented at Microsc. Semicond. Mater. Conf., Oxford, 7–10 April 1997
© *1997 IOP Publishing Ltd*

Surface damage of semiconductor TEM samples prepared by focused ion beams

J F Walker and R F Broom*

FEI Europe Ltd, Cottenham, CAMBS CB4 4PS
*Dept of Materials Science and Metallurgy, University of Cambridge, Cambridge CB2 3QZ

ABSTRACT: Presented here is information related to the damage caused by energetic gallium ions, used in FIB-TEM sample preparation, incident on the semiconductors frequently used in micro- and opto-electronics. The membranes were examined by TEM, using bright and dark field microscopy with the electron beam aligned a few degrees off the [110] zone axis, perpendicular to the plane of the membrane and parallel to the FIB-damaged surface. By this means the amorphous region was observed between the surface and the underlying crystal. The results for silicon, gallium arsenide and indium phosphide at energies of 10 and 30keV are compared with theoretical expectations from Monte-Carlo calculations (SRIM). The effects, dependent on ion energy, ion mass, incident angle and the properties of the target semiconductor, result in disruption of the semiconductor lattice mainly through the production of interstitial-vacancy pairs. Subsequent movement and interaction of these point defects give rise to damage observable in TEM.

1. INTRODUCTION

There are many methods of producing TEM samples and the criteria of quality are generally the sample thinness and the degree to which damage is minimised. Ion beams have been used in this context for some time and can produce thin samples with reasonable ease and little damage. However, the advent of focused ion beams (FIB) has facilitated the preparation of previously difficult or impossible samples as the section can be prepared about a defined position with submicron accuracy (Young, 1993). This is particularly important when a unique feature, such as a defect or a particular junction must be viewed. The FIB workstation can find the site and prepare the section in the correct place and in a short time. The production of highly focused ion beams requires 30keV gallium ions in highly developed systems and the question of ion damage inevitably arises.

When an ion penetrates the surface of a crystalline semiconductor, several processes take place. At the surface, secondary ions and electrons of low energy are produced which are typically used to form images in FIB systems. Neutrals are produced which is the sputtering responsible for the milling capabilities of these systems. Ions penetrating beyond the first few monolayers can impact with the constituent atoms and knock them out of their nominal position. This event will usually create a vacancy in the position of the original atom and an interstitial where that atom ends between the normal lattice points. The Monte-Carlo (Ziegler) calculation can simulate the impact of the ion and generate a file of data describing a possible route for the ion and the effects of its passing. The ion will typically leave a trail of Frenkel pairs (vacancy-interstitial pairs), ending in a gallium interstitial as the primary particle comes to rest. With many such impacts, data can be accumulated to give a reasonable approximation to the actual effects of many ion impacts. Such data, while intrinsically instructive, gains credence when compared to experimental studies of the damage. Several TEM sections were prepared to look at the relative surface damage produced by gallium ions. The materials studied were silicon, gallium arsenide and indium phosphide - the three most commonly used semiconductors. Damage was created by 30keV and 10keV gallium ions with sufficient dose to create an equilibrium level whereby as much damage is removed by sputtering as is created. The damaged layers were then coated with one micron of ion beam deposited platinum to protect the surfaces. While TEM is an excellent tool to study damage, it cannot be used to determine low levels of point defects and is used in this case to determine the amorphous-crystalline boundary. Previous work (Cerva, 1993) has shown that a crystalline semiconductor becomes amorphous at point defect densities greater than $10^{22} cm^{-3}$.

2. GALLIUM IONS IMPINGING ON SILICON

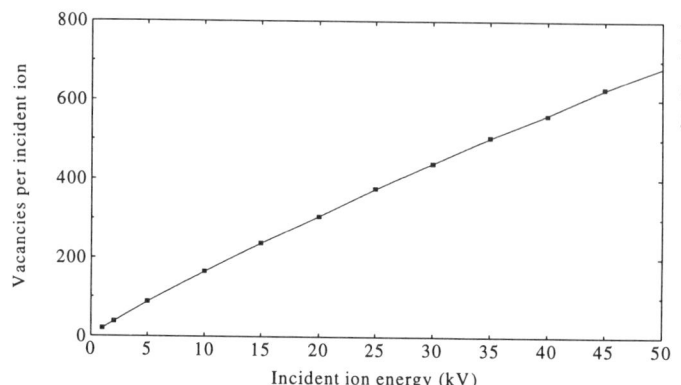

Figure 1. Number of vacancies produced in silicon per incident gallium ion.

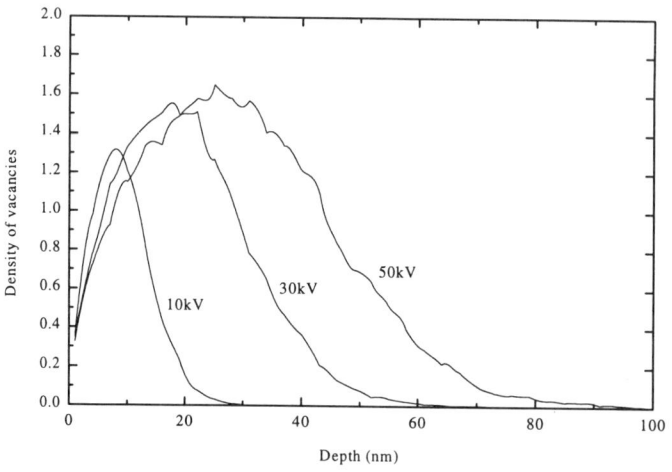

Figure 2. Silicon vacancy density ($Å^{-1}$ ion^{-1}) as a function of depth at 10, 30 and 50keV ion energy.

The total vacancy formation, as calculated from SRIM and seen in figure 1, shows neither the depth to which the damage penetrates nor the vacancy density. The depth is clearly important in defining the thickness of a damaged surface layer on the material. The damage will only be visible, however, if the vacancy (or interstitial) density reaches a critical quantity. This density defines the likelihood of their coalescing into dislocation loops and, at some critical density, turning the material amorphous. Figure 2 shows the vacancy density as a function of depth. While, at 30 kV, the tail of the distribution reaches over 70nm, the maximum occurs at about 18nm and the vast majority, 95%, is contained within 40nm. It is interesting to note that at the peak of the distribution, the number of vacancies per incident ion is largely invariant, but the damage depth and total damage increases with beam energy. It is important to consider that these distributions represent a snap-shot of the damage at the instant of creation and the real distribution of damage within a sample will be modified by several factors: primarily sputtering, whereby the surface is constantly being eroded and which will skew the real peak to the surface; and the defect mobility whereby vacancies and interstitials will spread from their original sites or could annihilate each other.

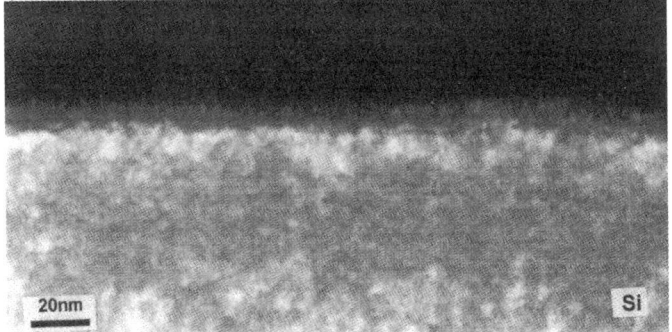

Figure 3. Dark field TEM image of a damaged (amorphous) layer of silicon above the single crystal substrate. Above the damage, in each case, is a protective platinum coating. The damage measures 6nm and is the result of impact by 10keV gallium ions.

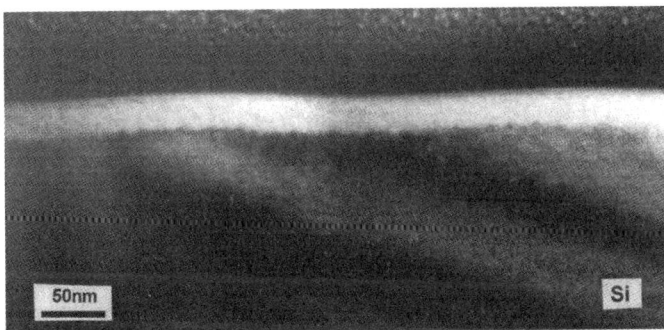

Figure 4. Bright field TEM image of silicon amorphised by 30keV gallium ions. The damage penetrates to 28nm.

3. GALLIUM IONS IMPINGING ON GALLIUM ARSENIDE AND INDIUM PHOSPHIDE

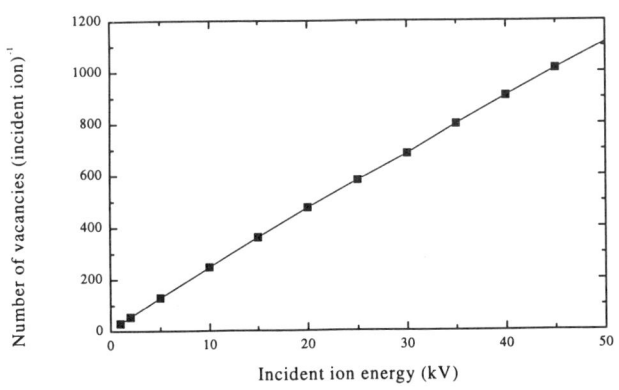

Figure 5. Number of gallium and arsenic vacancies produced per incident gallium ion.

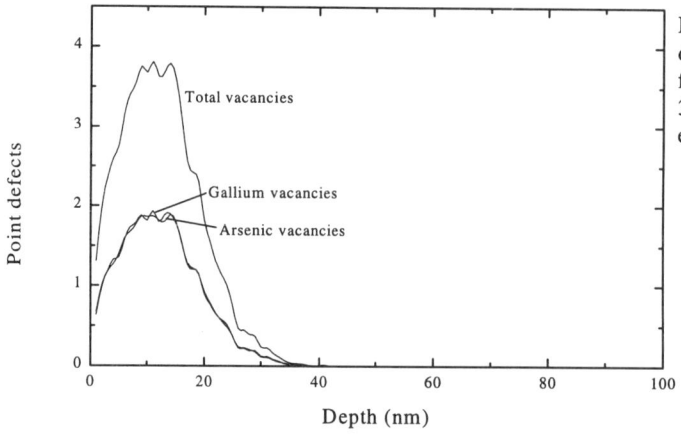

Figure 6. Vacancy density (Å^{-1} ion^{-1}) as a function of depth at 30keV gallium ion energy.

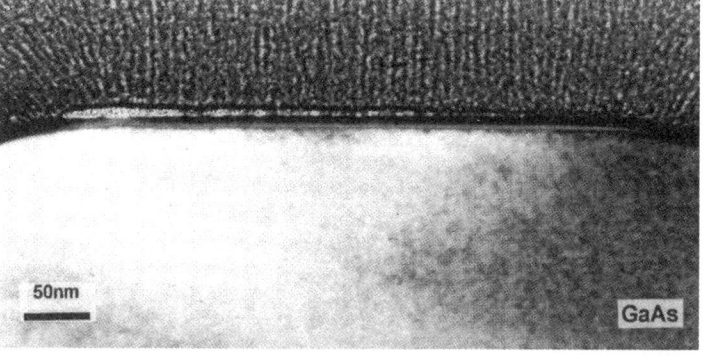

Figure 7. The thin damage layer for 10keV gallium ions in gallium arsenide is seen in this bright field image as the mid-grey layer below the dark platinum protection layer and above the brighter single crystal gallium arsenide. The thickness measures 4nm.

Figure 8. At 30keV gallium ion beam energy, the amorphous damaged region in gallium arsenide measures 24nm.

The analog in gallium arsenide of figure 1 is the graph shown in figure 5. Clearly the total point defect numbers produced is greater by nearly a factor of two. Similarly, figure 11 shows the point defect numbers in indium phosphide showing a slight increase over the silicon numbers. However, a truer picture is seen in the vacancy density graphs, figure 6 for gallium arsenide and figure 12 for indium phosphide, showing considerably greater densities over a shallower depth.

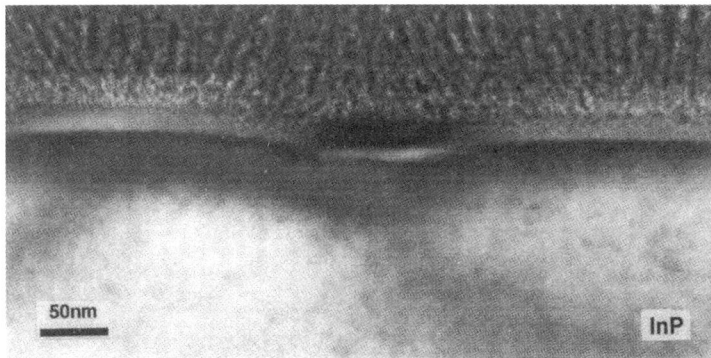

Figure 9. The amorphous damage region in indium phosphide from 10keV gallium ions is 15nm, clearly greater than for either silicon or gallium arsenide.

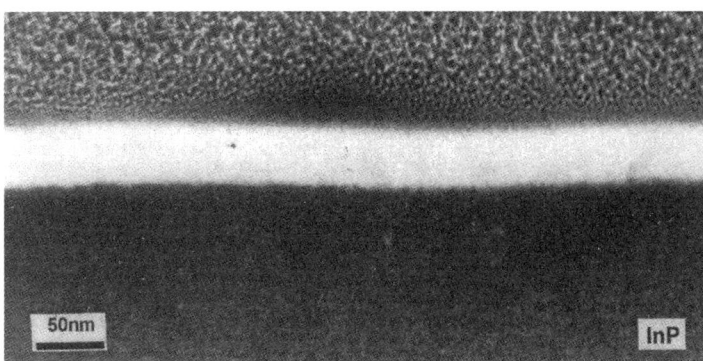

Figure 10. 30keV gallium ion damage in indium phosphide is shown in this bright field TEM image. The damage layer thickness measures 40nm.

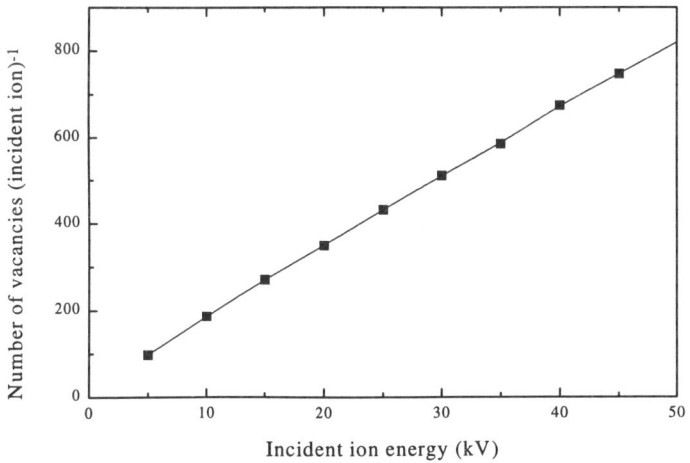

Figure 11. Number of vacancies produced in indium phosphide per incident gallium ion.

Confirmation of the reduced depth of the vacancies can be seen in figures 7 and 8 for gallium arsenide and figures 9 and 10 for indium phosphide.

478

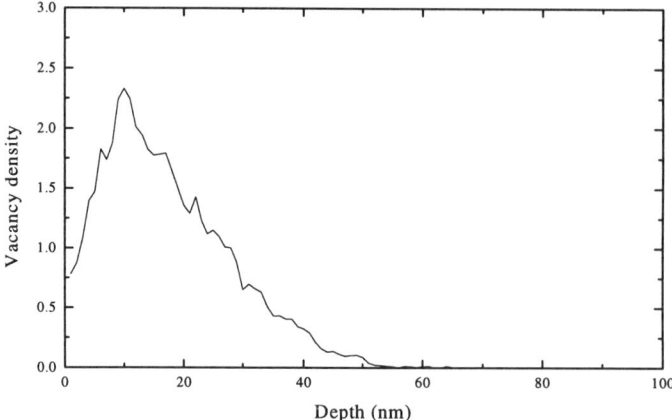

Figure 12. Vacancy (phosphorous and indium vacancies) density (Å^{-1} ion^{-1}) as a function of depth at 30keV incident ion energy.

4. CONCLUSIONS

While it is clear that there are differences between the three materials studied in terms of the depth of damage, it is difficult to definitively state why. Most likely, the gallium arsenide shows less damage depth because the incident gallium ion will interact more strongly, through momentum transfer, with target atoms of similar mass. The gallium ions will distribute their momentum more frugally in silicon, with its lighter mass, and indium phosphide, with its more disparate elemental masses. This is clearly illustrated by the high vacancy density maximum of almost 4 vacancies per incident ion per Ångstrom depth in gallium arsenide, despite the relative shallowness of the damage. With 30keV silicon the equivalent number is only 1.4 vacancies per incident ion per Ångstrom depth. At any non-zero temperature, some vacancies will be annihilated by interstitials at a rate dependent on both sample temperature and the density of vacancies or interstitials (generally, the density of the least common, though in this case the numbers are approximately equal). So for silicon and indium phosphide, the absolute number of point defects remaining, independently of the number created, will be greater than for gallium arsenide and the depth of damage will also be greater. It is satisfying that the data in table 1 reflects these expectations.

However, the situation described in this paper relates to normal incidence ions arriving at a surface. Ion damage is most often considered a detriment in TEM sample preparation by FIB where milling, or at least the final polishing step, uses glancing angle milling for which the ion penetration depths are significantly less. Therefore, this work should be seen as a worst case, with amorphisation damage of TEM membranes typically seen to be less than that described in this work.

	10keV	30keV
silicon	6nm	28nm
gallium arsenide	4nm	24nm
indium phosphide	15nm	40nm

Table 1. Listing measured amorphisation depth as a function of material and incident beam energy (data from figures 3, 4, 7, 8, 9 and 10).

REFERENCES

Cerva H and Hobler G 1993 Inst. Phys. Conf. Ser. No 134
Young R J 1993 Vacuum **44**, 353
Ziegler J F   SRIM program (for details contact: Internet: Ziegler @ Watson.IBM.Com)

Inst. Phys. Conf. Ser. No 157
Paper presented at Microsc. Semicond. Mater. Conf., Oxford, 7–10 April 1997
© *1997 IOP Publishing Ltd*

Ion energy effect on surface amorphisation of semiconductor crystals

Á Barna, L Tóth, B Pécz and G Radnóczi

Research Institute for Technical Physics of the Hungarian Academy of Sciences, H-1325, Budapest, PO Box 76. HUNGARY

ABSTRACT: A new, low energy ion gun has been developed for use as a final step of TEM sample preparation by ion milling. The goal was to reduce the thickness of the amorphised layer formed on the bombarded surfaces. By decreasing the ion energy from 10 keV to 250 eV no amorphous surface layer was detectable on GaAs and only a 1 nm thick amorphous layer was found on Si, where the oxidation process complicates the phenomenon. The cross sectional technique was capable of revealing the surface roughening effect of the ion beam as well.

1. INTRODUCTION

Ion beam milling is one of the most widespread techniques recently used for thin sample preparation for TEM and HREM in both cross section and plan view samples. It is known that ion bombardment, usually carried out by 3-10 keV energy Ar^+ ions, causes damage in the thinned crystals. In the case of semiconductor crystals (Si, GaAs) this damage appears mostly in the form of an amorphised layer on the milled surfaces of the thinned sample. Decreasing the ion energy and/or the angle of incidence (measured from the surface) of the ion beam is known to decrease this kind of damage. The angular dependence is generally accepted as obeying the cosine law while it is still not clear whether the ion energy dependence of amorphization follows a linear or square-root relationship (Ishiguro et al 1987 Schuhrke et al 1992 and Bulle Lieuwma 1987).

The amorphous layer thickness on Si and GaAs (001) single crystal surfaces have been measured on cross sections of ion milled surfaces as a function of the ion energy, and its effect on the HREM image quality was studied in a Philips CM20 electron microscope with TWIN objective lens (nominal point resolution 0.27 nm).

2. EXPERIMENTAL

A low energy ion gun with suitable power supply has been developed for use in the ion milling equipment for TEM sample preparation (Barna 1994). The new, hot cathode gun emits Ar^+ ions with a maximum energy controllable between 100 eV and 2 keV. Its ion current and sputtering speed (measured on Si single crystal at an angle of 30°) as a function of anode voltage are shown in Table 1. In the following, anode voltage is used for characterisation of ion gun power, although the majority of ions possess an energy of about

90 % of the nominal energy according to measurements (i.e. 1 kV gun voltage results in a mean value of 900 eV for the Ar^+ ions).

For the quantitative determination of the thickness of an amorphised surface layer on the (001) faces of Si and GaAs single crystals due to ion bombardment a new technique was applied. A thin film of aluminium (about 30 nm thick) was deposited onto the freshly amorphised surface immediately after bombarding the surface with Ar^+ ions of a certain energy in the same vacuum system. This single-crystalline/amorphous/polycrystalline layer system was then prepared as a cross section specimen by our standard procedure (Barna 1992) including low energy ion bombardment as a final step. These further steps in the preparation are supposed not to change the layer configuration and thickness, although some lateral oxidation cannot be fully excluded in the case of silicon. Temperatures during the cross sectional preparation have not exceeded 170 °C. Since no metal induced crystallisation of the amorphous Si film due the Al layer has been found, it is estimated that the sample temperature was also sufficiently low (Radnóczi et al 1991) during the ion milling.

Anode potential [V]	Ion current [μA]	Sputtering rate [μm/h]
200	17.5	1
500	20	2.5
800	50	9
1000	55	15
1400	80	20
2000	110	25

Table 1. Some parameters of the new, low energy ion gun

3. RESULTS AND DISCUSSION

The amorphised layer thickness has been determined by means of the HREM technique. The point resolution of our electron microscope allowed us to resolve the substrate crystal along the [011] zone axis, but for the Al film, only one-dimensional fringe contrast could be observed. Reliable measurement of the amorphous layer thickness was performed only when both the substrate crystal and some Al grains were resolved simultaneously. Fig. 1 shows a characteristic cross sectional HREM micrograph of a GaAs

Al

a-GaAs

c-GaAs

Fig. 1. Cross sectional HREM image of GaAs (001) surface bombarded with 1600 eV Ar^+ ions at an angle of 5° and covered in situ with an Al film.

(001) surface after bombardment with 1600 eV Ar⁺ ions at an angle of 5°, covered in situ with Al film. The images of both Si and GaAs (001) surfaces treated with 250 eV ions reveal much less amorphisation (Fig. 2).

The surface roughening, as well as the amorphised layer thickness on the bombarded surface of the Si and GaAs crystals, has been determined from the high resolution micrographs for various bombarding ion energies. The results are shown in Table 2.

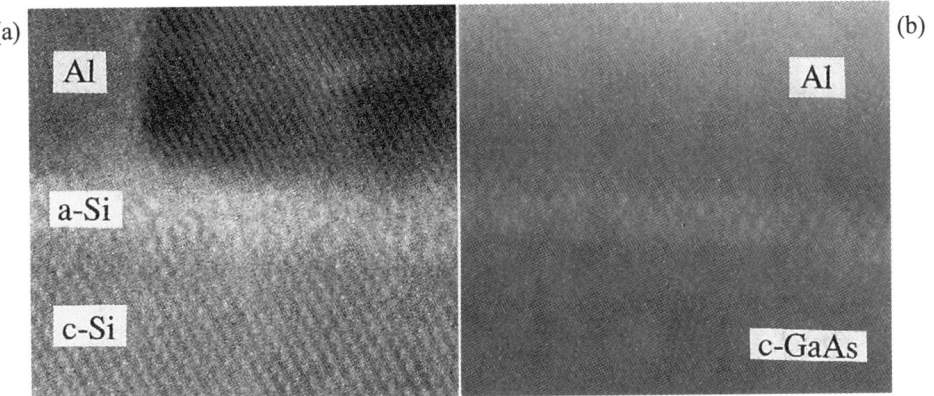

Fig. 2. Cross sectional HREM image of Si (a) and GaAs (b) bombarded with 250 eV Ar⁺ ions at an angle of 5° and covered in situ with Al film.

Material	Anode potential [V]	Mean thickness of the amorphous layer [nm]	Surface roughness
GaAs (001)	1600 V	2.1 nm (7.5 ML)	3-4 ML
GaAs (001)	250 V	0 (less than 1 ML)	0-1 ML
Si (001)	3500 V	5.1 nm (19 ML)	6-8 ML
Si (001)	250 V	1.0 nm (3.5 ML)	1-2 ML

Table 2. Amorphous layer thickness vs. bombarding energy (ML = monolayer)

The effect of low energy bombardment with the new ion gun on the quality of HREM images is illustrated on Fig. 3. It is obvious that the grainy contrast due to the amorphous surface layer has been significantly reduced in the case of both Si and GaAs (011) surfaces.

4. CONCLUSION

It has been shown that the reduction of the ion energy to 250 eV in the final stage of ion milling results in an effective improvement of the sample quality for high resolution electron microscopy. The effect is clearly and easily observable, comparing HREM images of samples prepared by the traditional low angle thinning method and samples thinned at energies as low as 250 eV. The new cross sectional technique, involving an in situ coating of the bombarded surface with a thin protective film, revealed that practically no amorphous surface layer can be detected on GaAs (001) after treatment with 250 eV Ar⁺ ions. In the case of Si (001) treated under similar conditions, the disordered layer thickness has been reduced to 1 nm, possibly influenced by an oxidation process. The surface roughening due to ion bombardment was also found to decrease with lowering the ion energy, although its quantification needs further investigation.

482

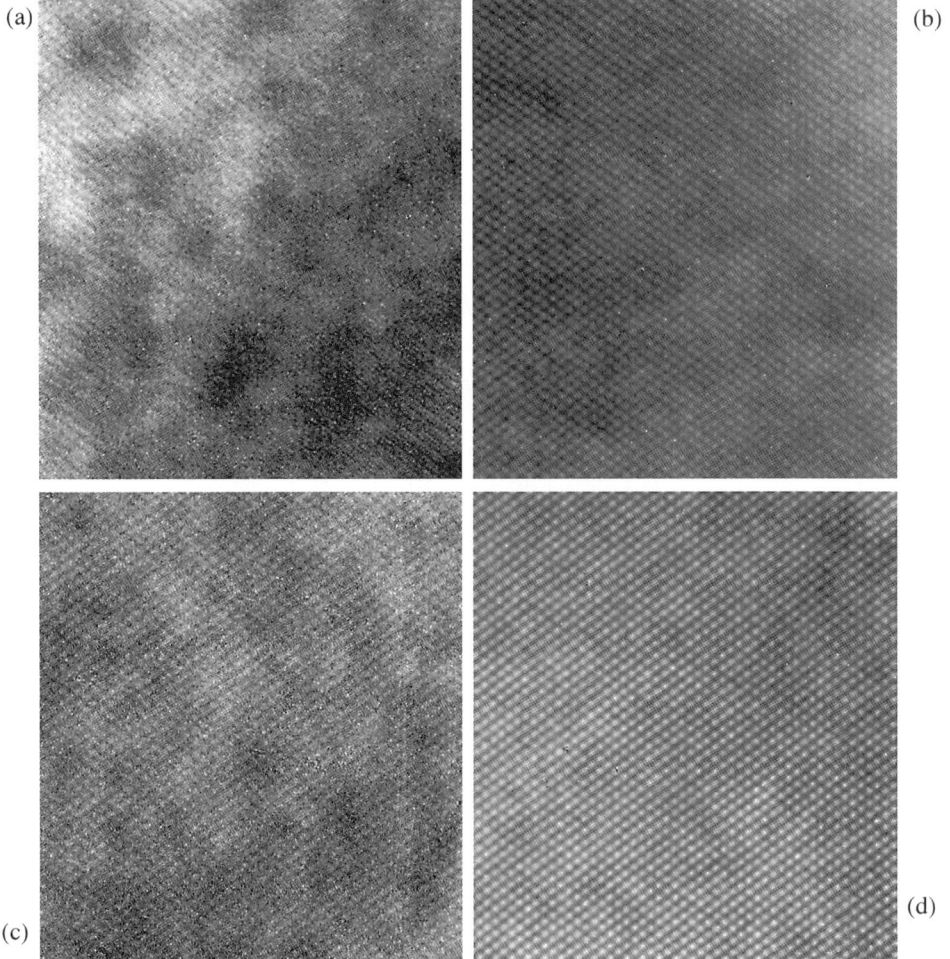

Fig. 3. [011] zone HREM images of Si (a,b), and GaAs (c,d) taken with a CM20 TWIN electron microscope. The samples in (a) and (c) were prepared with 10 keV and finished with 3keV Ar$^+$ ions, while those in (b) and (d) received an additional treatment at 250 eV.

REFERENCES

Barna Á 1992 Mater. Res. Soc. Proc. Ser, **254**, 3
Barna Á and Menyhárd M 1994 phys. stat. sol. (a) **145**, 263
Bulle-Lieuwma C W T and Zalm P C 1987 Surface and Interface Analysis **10**, 240
Ishiguro T, Suzuki T, Suzuki N and Ozawa M 1987 J. Electron Microsc. **36**, 163
Radnóczi G, Robertsson A, Hentzell H T G, Gong S F and Hasan M A 1991 JAP, **69**, 6394
Schuhrke T, Mändl M, Zweck J and Hoffmann H 1992 Ultramicroscopy **41** 429

The work was supported by the National Science Research Foundation under OTKA Grant No. T7592.

Inst. Phys. Conf. Ser. No 157
Paper presented at Microsc. Semicond. Mater. Conf., Oxford, 7–10 April 1997
© *1997 IOP Publishing Ltd*

The effects of surface relaxation and ion thinning on δ-doped semiconductor cross-sections

C P Liu, P D Brown, C B Boothroyd and C J Humphreys

Department of Materials Science and Metallurgy, University of Cambridge, Pembroke Street, Cambridge, CB2 3QZ, UK.

ABSTRACT: The measured distributions of As δ-doped layers within InP are found to vary depending on the technique used to prepare the TEM sample foil. Comparison of conventional cross-sectional specimens prepared by ion milling with cleaved-wedge samples is made using high resolution electron microscopy and Fresnel analysis. The observed width of the δ-doped layer is influenced by surface amorphisation due to the ion beam damage and surface relaxation due to the thin specimen. The effect of free surface relaxation is found to dominate HREM images acquired from cross-sectional specimens, although surface amorphisation should also be considered when data is analysed quantitatively.

1. INTRODUCTION

The quantitative analysis of modern semiconductor heterostructures, particularly those with very fine δ-doped layers, demands that every factor which can potentially contribute to the experimental measurement and subsequent image interpretation be considered. Of these factors, the cross-sectional specimen itself is particularly significant, and yet it is one which has received relatively little attention in the past. Ion beam thinning has become the standard technique for making cross-sectional TEM semiconductor specimens. Although artefacts are inevitably introduced during conventional ion beam milling procedures, their contribution to quantitative image interpretation is often ignored. Two problems associated with the preparation and the form of cross-sectional specimens can potentially contribute to the final image. The first one is that of free surface relaxation which will not only affect the accuracy of lattice spacing measurement, but also vary the contrast of conventional bright field and dark field TEM images (Treacy and Gibson 1986). The second problem involves the nature of the semiconductor material itself and the ease of damage under ion bombardment. For example, Vaseashta et al (1991) report that a GaAs surface under Ar+ ion milling can be influenced to a depth of a few hundred angstroms. The aim of this paper is to investigate the extent to which these problems influence the quantitative analysis of δ-doped layers within InP. Comparison is made between iodine milled cross-sectional specimens and cleaved-wedge specimens using the combined techniques of HREM and Fresnel contrast analysis.

2. EXPERIMENTAL

Very thin (δ-doped) layers of As doped InP were introduced into epitaxial InP/InP(001) grown by chemical beam epitaxy by interrupting the growth in the presence of As_2 for periods of 1, 2, 4, 8, 16 and 32 seconds, with spacings of about 20nm between each δ-doped layer (see Brown et al 1993 for further details). Cross-sectional specimens were prepared using conventional mechanical polishing and argon ion-milling procedures (i.e. 5kV at an angle of 15°) until foil perforation. Iodine milling (2kV, 20μA) was then used to eliminate the indium islands remnant on the sample foil surface after argon ion milling (Chew and Cullis, 1987). The 90°-wedge specimens were prepared by cleavage on {110} planes. HREM images along [100] were obtained using a JEOL 4000EX-II microscope operated at 400kV (C_S = 0.9mm, C_C = 1.3mm), and an objective aperture of semi-angle 20mrad (corresponding to an Airy disc radius of 0.05nm). Energy-filtered Fresnel contrast images

484

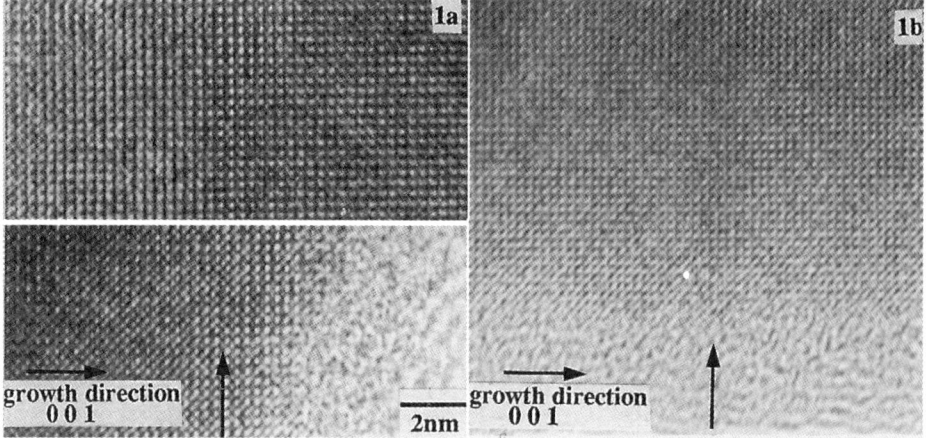

Fig. 1 HREM images from 32 second As growth interrupts in InP for (a) cross-sectional (thicker at top, thinner at bottom) and (b) cleaved-wedge specimens. The δ-layer and sample growth directions are indicated.

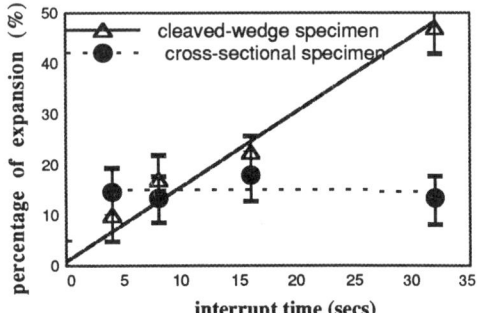

Fig. 2 Plot of lattice expansion percentage as a function of layer interrupt time for both specimens geometries.

of the As δ-doped layers were obtained using a JEOL 4000FX microscope, equipped with a post-column Gatan imaging filter (GIF), operated at 400kV (C_S = 2.0mm, C_C = 1.4mm), and an objective aperture of semi-angle 3.4 mrad (corresponding to an Airy disc radius of 0.29nm). The layers were tilted to an 020 systematic row condition a few degrees from the [100] zone axis and a 10eV energy window was used for energy-filtered imaging. The Fresnel series comprised 512 × 512 pixel digitally-acquired images which were obtained at a sampling density of 0.173nm/pixel and with an acquisition time of 2s per image. The point spread function of the detector was deconvoluted from each image. Special care was taken to ensure that knock-on damage was minimised when imaging with 400keV electrons. Microscope alignment was always performed away from the regions of interest, while image acquisition occurred as soon as possible for each experiment using a fresh specimen. Most of the detailed results reported on here will be concerned with the highest As concentration layer (an interrupt time of 32s).

3. RESULTS AND DISCUSSION

Figures 1a and 1b show HREM images of the cross-sectional and cleaved-wedge specimens, respectively, each recorded in the vicinity of the 32 second growth interrupt and digitised at a resolution of 7.2 pixels / 002 fringe (the layers are indicated). By inspection, there is always a greater extent of amorphisation at the edge of cross-sectional specimens than for cleaved-wedge specimens. The origin of this wider layer (typically by a factor of 1.5 to 2) for cross-sectional specimens is generally attributed to amorphisation introduced by the process of ion bombardment. From the direct image contrast of the original micrograph of full size it can also be seen that the δ-layer width becomes wider close to the edge of the cross-sectional foil, gradually becoming constant with increasing thickness, while the layer width is almost constant throughout the thickness range of the cleaved-wedge specimen.

While HREM images in general can be complicated by many factors such as lens aberrations and Fresnel contrast, the principal contribution to these essentially $InAs_xP_{1-x}$ HREM images at the [100] pole is the set of four {002} reflections which give rise to a high degree of chemical contrast.

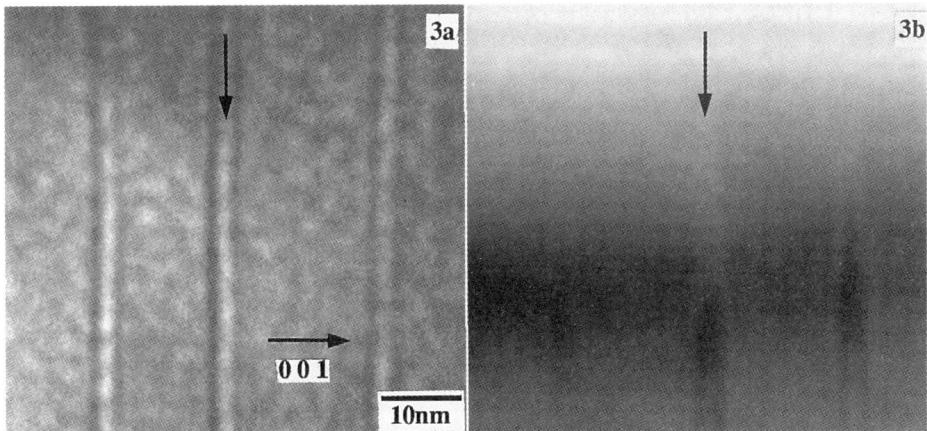

Fig. 3 Examples of zero-loss energy-filtered Fresnel contrast images of (a) cross-sectional (thickness of 60-70nm) and (b) cleaved-wedge specimens (thickness of 0-70nm) at a defocus value of +3500nm.

Mallard et al (1991) also pointed out the effect of surface relaxation on HREM images of CdTe/ZnTe multilayer materials in the form of a displacement of lattice fringes rather than a variation of local image intensity.

Before trying to measure the lattice distortions in our images (the lattice mismatch between InAs and InP substrate is about 3.23% and this will help to drive the local thin foil lattice distortions), correction must first be made for the distortions introduced both by the microscope and by the densitometer used for digitising the negatives. Such distortions were found to be

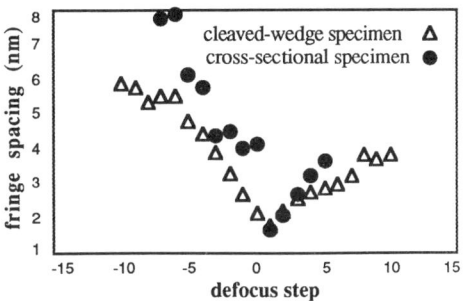

Fig. 4 Plot of Fresnel spacing as a function of defocus step for both specimens geometries at the thickness of 70nm.

large enough to affect the measured rigid lattice shifts appreciably and so were digitally removed from these InP HREM images (Liu et al, 1997) in advance of further image analysis. The technique of regressional analysis was then applied to determine the rigid lattice shift across the layers. This approach involves extrapolating the lattice fringes from regions on either side of each layer towards it and then measuring the mismatch between them to an accuracy that can approach 1% of the lattice fringe spacing (Stobbs et al., 1986). One dimensional lattice fringe profiles were obtained by projecting the lattice fringes within each region parallel to the direction of the layer and polynomial fits were used to determine the position of each peak and trough. Subsequently, the total rigid lattice shift was determined using an algorithm that has been described by Dunin-Borkowski and Stobbs (1997). The reliability of the results was established by examining the measured lattice shift as a function of both the widths of the regions on either side of the layer analysed and their distances from the layer, ensuring that the effects of diffraction and phase contrast at the layer were discounted. The final results are summarised in Fig. 2 which shows the lattice expansion percentage as a function of interrupt time at thickness of about 10nm for cross-sectional specimen and of 30nm for cleaved-wedge specimen. A linear relationship was obtained for the cleaved-wedge specimen where the measured lattice expansion was 0.0139 ± 0.02nm for the highest As concentration layer. Conversely, the lattice expansion is approximately invariable against interrupt time for the cross-sectional specimen. This result of an even distribution of strain state over the δ-layers formed by increasing interrupt times must partly be due to the surface relaxation phenomenon for such thin sample foils. Indeed, from simulation, Treacy and Gibson (1986) found that the lattice spacing deviates most strongly from the mean value for strained layer superlattice for the range $0.1 \leq t/\lambda \leq 3$, where t is the foil thickness and λ is the compositional modulation wavelength. Hence, significant relaxation is expected to occur up to a thickness of about 60nm for our materials system, which covers the range available for HREM analysis of typical cross-sectional specimens and also of cleaved-wedge specimens. However, possible ion beam thinning effects leading to randomisation of the growth interrupt profiles due to amorphisation, displacement damage and point defect creation cannot as yet be discounted.

The Fresnel contrast technique can facilitate layer width measurement to very high accuracy and thereby provide insight into the effect of the ion beam milling on these δ-doped specimens. Fig. 3 shows one of a through focal series of images at a defocus of +3500nm for both the cross-sectional and cleaved-wedge specimens. Because we only observe very faint Fresnel contrast from very thin regions of δ-doped material, the Fresnel series for the cross-sectional samples were recorded from thicker areas of about 70nm, though the accuracy of defocus and thickness values is not critical for the determination of layer width. A plot of Fresnel spacing as a function of defocus step is shown in Fig. 4 for both specimen geometries. The minimum spacing extrapolated from best fitting lines at either side of focus provide a good indication of the layer width, and this analysis yields an almost identical value of 1.8 ± 0.2nm for the longest growth interrupt layer for both specimen geometries. However, it is noted that slight speckle contrast is remnant in the image of the cross-sectional specimen (Fig. 3a), while such contrast is absent for the cleaved-wedge specimen (both samples were exposed to the imaging electron beam for the same time period of perhaps less than 1 minute before recording the Fresnel series). This confirms that some ion beam damage is present prior to data acquisition, as distinct from damage induced by the imaging electron beam. While Vaseashta et al (1991) showed that the amorphous to crystalline transition is gradual rather than abrupt, such an intermediate layer would appear to be of little significance from our Fresnel results acquired from thicker regions. However, it is possible that damage resulting from the process of ion milling acts to promote the relaxation of thin foils leading to uniformity of the residual strain distribution.

Finally, it is apparent that the curve for the cross-sectional specimen in Fig. 4 is more asymmetric. Such asymmetric behaviour of the Fresnel contrast or spacing with defocus is characteristic of materials showing strong diffraction effects, but a full explanation for the extra asymmetry must include the absorption and increase in beam divergence caused by the thin amorphous surface layers. These will influence the image contrast and quantitatively affect compositional analysis to some extent. Therefore, the influence of amorphous layers on sample foil relaxation and consequent image contrast from bulk materials is worth further investigation.

In summary, two complementary TEM-based techniques have been applied to assess the consequences of careful ion milling on a cross-sectional specimen comprising As δ-doped layers within InP. Free surface relaxation is found to dominate HREM images for the very thin areas of a conventional cross-sectional specimen prepared by iodine milling, with perhaps interdiffusion induced by the ion beam thinning process contributing to the homogenisation of the strain distribution. Conversely, layer width determinations are consistent within relatively thick areas of a sample foil for both cross-section and cleaved wedge geometries, as determined by the energy-filtered Fresnel technique.

Acknowledgements

With thanks to Peter Skevington and Graham Davies of BT Labs., Martlesham Heath for provision of the δ-doped sample studied.

REFERENCES

Brown P D, Bithell E G, Humphreys C J, Skevington P J, Cannard P J and Davies G J 1993 Inst Phys. Conf. Ser. **No. 134** 373
Chew N G and Cullis A G 1987 Ultramicroscopy **23** 175
Dunin-Borkowski R E and Stobbs W M 1997 submitted to Ultramicroscopy
Treacy M M J and Gibson J M 1986 J. Vac. Sci. Tech. **B4** 1458
Liu C P, Dunin-Borkowski R E, Boothroyd C B, Brown P D and Humphreys C J 1997 to be published in JMSA
Mallard R E, Feuillet G and Jouneau P-H 1991 Inst. Phys. Conf. Ser. **No. 117** 17
Stobbs W M, Wood G J and Smith D J 1986 Ultramicrosc. **14** 145
Vaseashta A, Elshabini-Riad A and Burton L C 1991 Mat. Sci. Eng. **B9** 489

Inst. Phys. Conf. Ser. No 157
Paper presented at Microsc. Semicond. Mater. Conf., Oxford, 7–10 April 1997
© *1997 IOP Publishing Ltd*

Practical epitaxial silicide technologies for ULSI applications

R T Tung and K Inoue[a]

Bell Labs. Lucent Technologies, Murray Hill, N.J. 07974 U.S.A.

[a]NEC Corp., ULSI Device Development Labs., Sagamihara, Kanagawa 229, JAPAN

ABSTRACT: Possible advantages of using epitaxial silicides in Si-based microelectronic devices were recognized some time ago, but until the recent discovery of a few "practical" technologies the application of epitaxial silicide in the manufacturing of ultra-large scale integration (ULSI) devices did not seem likely. This paper examines the phenomena underlying three epitaxial silicide technologies recently developed: high temperature sputtering (HTS), Ti-interlayer mediated epitaxy (TIME), and oxide mediated epitaxy (OME). Merits and drawbacks of these technologies from an applicational standpoint are also discussed.

1. INTRODUCTION

The discovery of the epitaxial growth of some silicides on Si in the 1970's (Tu et al 1974, Ishiwara et al 1978, Chiu et al 1981) generated much scientific interest in the various properties of these epitaxial silicides (Tung 1993). The high perfection of properly prepared epitaxial silicide interfaces seemed to offer, on one hand, an ideal testing ground for various concepts concerning metal-semiconductor contacts which, until then, could only be speculated, and, on the other hand, plenty of opportunities for novel applications. The first half of the expectations has indeed been fulfilled. Notably, the correlation of the Schottky barrier height (SBH) with the atomic structure at epitaxial silicide interfaces has significantly elevated our level of understanding of the SBH formation mechanism (Tung 1992). But on the practical side, attempts to find suitable applications for epitaxial silicides have not been substantially rewarded. The problem seems to be two-fold: epitaxial silicides are tedious to fabricate and they don't seem to offer a real advantage over polycrystalline silicides, as far as electrical resistances are concerned. The often-discussed metal base transistor (MBT) has proved to be unrealistic to implement with thin epitaxial silicide layers (Hensel et al 1985). Isolated uses of epitaxial silicides in novel devices, such as the permeable base transistor (Tung et al 1986), have thus far attracted only scientific interest. The plight of applying epitaxial silicides in real Si devices may be changing, however. As the dimensions of ULSI devices continue to shrink, there is renewed interest in using epitaxial silicides. Very thin and yet uniform epitaxial silicide layers are ideally suited for the formation of ultra-shallow junctions in ULSI devices. Thanks to some recent discoveries, the procedures to fabricate epitaxial silicides have also become simpler. With these recent developments, it seems plausible that we will see the introduction of epitaxial silicides into ULSI devices in the not too distant future.

Silicides are in wide use in Si ULSI devices as contacts, gates, and interconnects. The main advantage of employing silicides in devices is to reduce the parasitic resistances and, in that respect, it makes little difference whether the silicide is epitaxial or, as in most cases, polycrystalline. Ti silicide is presently the most popular self-aligned silicide (salicide) for high-performance ULSI devices. It is, however, difficult to process $TiSi_2$ devices with a gate length shorter than 0.25μm, because of the well-known problem with the polymorphic transformation, $C49\text{-}TiSi_2 \rightarrow C54\text{-}TiSi_2$, in narrow silicide lines (Lasky et al 1991). Therefore, as the device dimensions continue to shrink, a replacement for $TiSi_2$, preferably one which forms thin and uniform layer easily in narrow geometries, is being sought. The uniformity and the thermal stability of the silicide layer are critical for shallow junctions, as layer non-uniformity weakens the junction integrity and increases the effective junction depth. For ≤ 0.18μm devices, uniform silicide layers less than 30nm thickness are desired. Because of a lower interface free energy and a lack of grain boundaries, an epitaxial silicide layer is usually more uniform and has higher thermal stability than a comparable polycrsytalline layer, especially when the layer thickness is very small. The most suitable candidate for ULSI applications, among all the known epitaxial silicides, is $CoSi_2$ which has a cubic fluorite lattice structure, a small lattice mismatch with Si, ~ 1.2% at room temperature, and a low electrical resistivity, ~ 14μΩ-cm. When the formation of shallow junction is also taken into consideration, the fact that $CoSi_2$ can be used as a doping source (Liu et al 1988), thus avoiding the problem of transient enhanced diffusion of dopants (Stolk et al 1995), is an additional advantage of $CoSi_2$.

The epitaxial growth of $CoSi_2$ on Si(100) is by no means automatic. Routine salicide processing procedures generally do not lead to a significant fraction of epitaxially oriented grains in $CoSi_2$ films. Growth under ultra-high vacuum (UHV) conditions, by molecular beam epitaxy (MBE) (Tung 1993, Mantl and Bay 1992) can generate epitaxial $CoSi_2$ layers of any desired thickness. However, MBE growth of single crystal $CoSi_2$ on Si(100) requires the (co-)deposition of Si (Yalisove et al 1989, Jimenez et al 1990) and, hence, it is not compatible with self-aligned formation. Alternatively, epitaxial $CoSi_2$ can be fabricated by mesotaxy, using high-energy, high-dose, Co implantations (White et al 1987), or by pulsed laser melting (Tung et al 1983a). These techniques, in addition to a cost issue and a compatibility problem with conventional processing environments, cannot generate uniform $CoSi_2$ layers with thicknesses of less than 40nm. Recently, three methods have been reported to be able to fabricate epitaxial $CoSi_2$ layers, employing tools and practices which are compatible with standard clean-room settings. In this paper, we will describe these potentially useful technologies and briefly assess the merits and shortcomings of each method.

2. EPITAXIAL SILICIDE TECHNOLOGIES

2.1 High Temperature Sputtering (HTS)

One of the major concerns of Co salicide processes is the suppression of oxidation of as-deposited Co before and during the first rapid thermal anneal (RTA). The well known method of in-situ capping the Co with Ti or TiN was designed largely to contain the oxidation problem. Recently, it was shown that sputtering and in-situ annealing of Co film at an elevated temperature, ~ 400°C, could lead to an improvement in the electrical characteristics of $CoSi_2$ films subsequently annealed (Inoue et al 1995). The sheet resistances of Co silicide films, after various anneals, grown from 5nm Co sputtered at either room temperature (RT) or $400°C$ are shown in Fig. 1. The phase and orientation of precursor Co silicide formed during the HTS and the in-situ annealing processes are found to depend on the doping species. This precursor silicide reaction prevents the oxidation of Co in subsequent ex-situ anneals, which presumably is one major reason for the improved sheet resistance. Another improvement of the HTS process is in the thermal stability of the $CoSi_2$

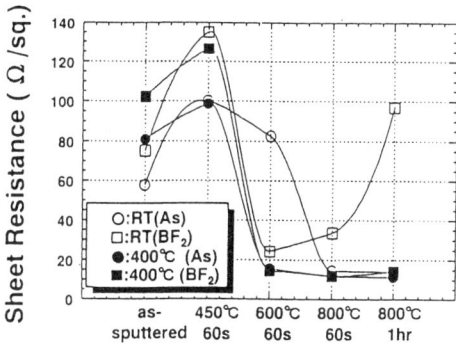

Fig. 1. *Sheet resistance of Co silicide layer grown from the sputtering of 5nm Co on Si(100) at either RT or $400°C$. The implantation received by the wafer prior to Co sputtering and the anneal performed after Co sputtering are also indicated.*

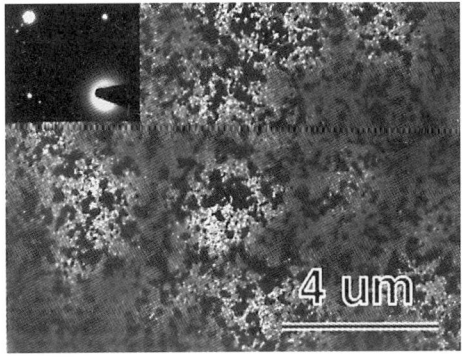

Fig. 2. *Planview, (200) dark-field, TEM image of a $CoSi_2$ layer grown on BF_2-implanted Si by HTS. The majority of this film is occupied by (100)-oriented $CoSi_2$.*

layer, especially for layers grown on BF_2 implanted Si, as shown in Fig. 1. $CoSi_2$ films grown by the HTS technique have predominantly epitaxial orientations on heavily doped Si(100) (Inoue et al 1997), likely accounting for the marked increase in their thermal stability. A dependence of the epitaxial orientation of the HTS grown $CoSi_2$ layers on the doping species is also found. As shown in Fig. 2, the majority of $CoSi_2$ layers has the regular (100) epitaxial orientation on BF_2-implanted Si substrate, with non-epitaxial $CoSi_2$ making up the remainder of the layers. On arsenic implanted Si(100), HTS processing leads to a complete coverage by oriented $CoSi_2$, but with two epitaxial orientations. The regular (100) orientation is found to be adopted by only a small fraction of the films on Si(As), while the majority of the films is found to have the (221) orientation. This orientation is equivalent to type B orientation on an inclined {111} interface. Imaging with (311)-related beams originating from these grains allows the four variants of the (221) orientation to be individually illuminated. Dark field images of (100)-oriented $CoSi_2$ and one of the four {221} orientations are shown in Figs. 3(a) and 3(b), respectively. One notes that the epitaxial orientations observed for HTS-grown $CoSi_2$ layers on differently doped Si are similar to those observed for $CoSi_2$ layer grown by the TIME technique.

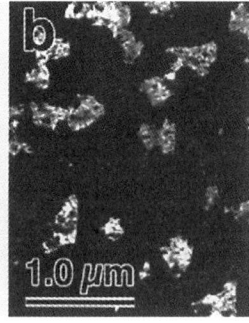

Fig. 3. *Planview TEM images of a $CoSi_2$ layer grown by HTS on arsenic-implanted Si(100). (a) (200) dark field image of (100)-oriented $CoSi_2$ grains and (b) $CoSi_2$ grains with one of the {221} orientations are imaged with a type B related (311) beam.*

Only slight modifications to a routine Co salicide process are involved in the HTS technique, making it a very attractive technique for device processing. The HTS process, however, is not well matched to the strategy to use silicide as the doping source (SADS) (Liu et al 1988), because $CoSi_2$ layers grown by HTS on lightly doped Si contain only a small fraction of epitaxial grains. Further improvements in the HTS process may be possible by exploring the apparent chemical effect giving rise to the observed dependence of the epitaxial orientation of $CoSi_2$ on the doping species.

2.2 Ti-Interlayer Mediated Epitaxy

The TIME technique makes use of the observation that a thin Ti "interlayer" may be used to promote the epitaxial growth of $CoSi_2$ (Dass et al 1991). When a layer of Co, with a thickness of > 4nm, is deposited at room temperature onto a Si(100) surface, a subsequent anneal at > 550°C invariably leads to the growth of a polycrystalline $CoSi_2$ layer. However, if a Ti layer (2-10nm) were first deposited on Si before the Co deposition (10-20nm), then a subsequent anneal would lead to the growth of an epitaxial layer of $CoSi_2$ (Dass et al 1991). After the reaction, Ti is found near the surface in the forms of TiN and other Ti-containing phases. It has been suggested that the Ti-interlayer serves as a diffusion barrier, which limits the Co-Si reaction until temperature exceeds 500°C, allowing one (Co_2Si) or both (Co_2Si and $CoSi$) precursor phases in the ordinary Co-Si reaction sequence to be skipped (Hsia et al 1992, Ogawa et al 1993). Skipping the precursor silicide phase(s) is then somehow related to the epitaxial growth of the $CoSi_2$ phase. The formation of an off-stoichiometric cubic silicide phase is sometimes discussed in connection with the nucleation of epitaxial $CoSi_2$ (Goncalves-Conto et al 1996).

Epitaxial $CoSi_2$ layers grown by TIME are often non-uniform in thickness. To improve on the lateral uniformity and the epitaxial quality of the TIME grown $CoSi_2$ layers, the thickness of the Ti interlayer need to be optimized. For Co layers of a common thickness of ~ 12-15nm, high quality epitaxial $CoSi_2$ layers are usually grown with a Ti interlayer of ~4-7nm thickness. When co-deposited $TiSi_x$ is used as the interlayer, the amount of Ti can be reduced (Tung and Schrey 1996). A thin Ti cap, 1-3nm thick, deposited over the Co layer to form a Ti/Co/Ti/Si starting configuration, was found to improve the degree of epitaxy and the homogeneity of the silicide reaction (Tung and Schrey 1995). With a proper interlayer (and cap), the silicide reaction proceeds such that after a RTA at 650-800°C, an essentially uniform, single crystal, $CoSi_2$ layer is grown next to the Si, while a $Ti_x Co_y Si_z$ layer, or a $Ti_x Co_y Si_z/TiN$ bilayer when nitrogen anneals are used, is formed at the surface. When a temperature higher than ~ 850°C is used to react the TIME layer, the surface $Ti_x Co_y Si_z$ decomposes and disrupts the $CoSi_2$ film, as shown in the planview TEM micrographs of Fig. 4. The majority of this film is occupied by epitaxial $CoSi_2$, which shows up bright in the 200 dark field image of Fig. 4(a), with Ti-containing, non-epitaxial regions, which show up dark in Fig. 4(a) and also in the bright field image of

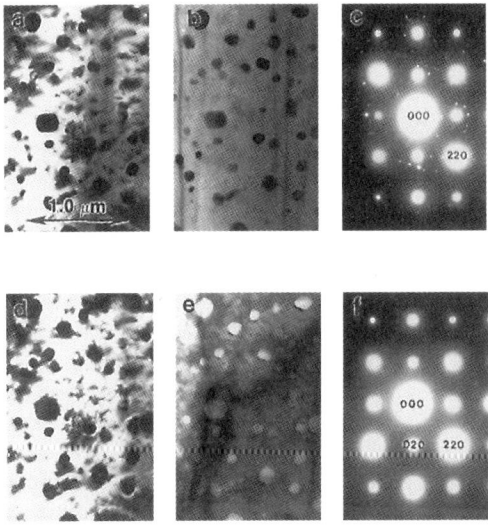

Fig. 4. *Planview micrographs and diffraction patterns of a $CoSi_2$ layer grown on $Si(100)$ by TIME using a 900°C anneal. (a), (b) and (c) are 200 df, bf, and TED, respectively, of the as-grown layer. The TEM sample was given a 2 min etch in a buffered oxide etching solution and reimaged to give (d), (e), and (f), which are, respectively, 200df, bf, and TED. Ti-rich columns in the as-grown epitaxial $CoSi_2$ layer were removed by the etch.*

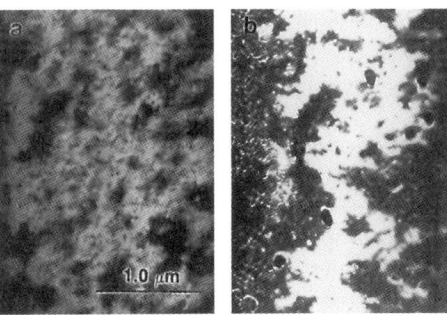

Fig. 5. *Planview, 200 dark-field, TEM images of TIME $CoSi_2$ layers. A uniform layer (a) is grown at 750°C, chemically etched, and annealed at 850°C for 20 min, while layer (b), grown at 750°C, annealed at 850°C, and then chemically etched, contains pinholes.*

Fig. 4(b), making up the rest of the film. These non-epitaxial grains can be removed selectively using a buffered oxide etch, as shown in Fig. 4(d-f). Since the inhomogeneity of a high-temperature-annealed epitaxial $CoSi_2$ layer is caused by the metastability of the ternary $Ti_x Co_y Si_z$ phase, the removal of the surface $Ti_x Co_y Si_z$ layer (Hsia et al 1992, Tung and Schrey 1996) prior to high temperature annealing can lead to the growth of a uniform layer of epitaxial $CoSi_2$, as shown in Fig. 5.

The TIME effect is an intriguing phenomenon and $CoSi_2$ layers grown by TIME on lightly-doped, blanket Si wafers are of very high quality. An obvious advantage of the TIME technique is its compatibility with standard Si fabrication environment. However, the TIME process has some problems, one of which is the growth of multi-oriented $CoSi_2$ on heavily arsenic doped Si. Significant fractions of $CoSi_2$ layers grown on As-implanted Si(100) are found to occupy {221} orientations, in addition to the regular {100} epitaxial orientation. There are four variants to the {221} orientation, which is likely the result of twinning (type B orientation) along the four inclined {111} planes. Another, more serious, problem of the TIME process is the formation of voids near the edges of $CoSi_2$ layers grown on Si surface with an oxide pattern (Tung and Schrey 1996, Byun et al 1996). This problem, likely caused by diffusion of silicon and the Ti-Si reaction, is serious enough that presently the TIME process is not deemed a viable technique for the processing of ULSI devices.

2.3 Oxide Mediated Epitaxy

The oxide mediated epitaxy (OME) technique (Tung 1996a 1996b) uses a phenomenon similar to that responsible for the TIME effect, but is able to avoid most of TIME's problems. A thin SiO_x layer, grown on Si substrates in an aqueous bath containing H_2O_2, is found to promote the epitaxial growth of $CoSi_2$. On atomically clean Si(100), deposition of 2nm Co at room temperature and annealing at 600°C lead to the growth of polycrystalline $CoSi_2$, as shown in Fig. 6(a). When the same deposition and

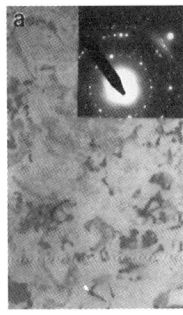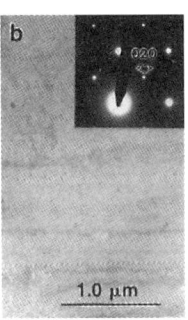

1.0 μm

Fig. 6. *Planview, bright field, TEM images of 5.4 nm thick $CoSi_2$ layers on Si(100). (a) a polycrystalline layer grown on atomically clean Si, and (b) an epitaxial layer grown on oxidized Si.*

annealing procedures are carried out on a thin SiO_x covered Si(100) surface, however, a single crystal $CoSi_2$ layer is grown, as shown in Fig. 6(b). Layers grown by OME have the ordinary epitaxial orientation, namely, $CoSi_2[100]//Si[100]$ and $CoSi_2[011]//Si[011]$. Microfacetting at the $CoSi_2/Si(100)$ interface is also indicated by the presence of some streaked diffraction spots in TED. The morphology and the epitaxial formation of the $CoSi_2$ layer depend on the thickness of the deposited Co. When less than 1nm Co is deposited, annealing leads to the growth of discontinuous, yet epitaxial, $CoSi_2$ islands, and when more than 3nm Co is deposited, the $CoSi_2$ layers contain non-epitaxial grains. Only in the thickness range of 1-3nm Co are $CoSi_2$ layers grown continuous and single crystalline. Essentially the same OME procedures can be applied to grow continuous and singly oriented $CoSi_2$ layers on any Si surface with an orientation far from the {111} (Tung 1996b). Furthermore, OME growth is not significantly affected by substrate doping at least up to levels commonly used for shallow junction formation (Tung 1996b).

After an epitaxial CoSi$_2$ layer is grown by OME, Auger electron spectroscopy (AES) detects oxygen and Si signals essentially unchanged from those detected prior to the cobalt deposition. The fine structure of the Si LVV transitions indicates that the majority of Si within the electron escape depth of the surface is still in an oxide-like bonding environment. Low energy electron diffraction (LEED) detects no ordered structure from the surface of epitaxial CoSi$_2$ layers grown on oxide-covered Si(100). These results suggest that the thin oxide layer originally residing on the Si surface remains largely on the surface of the epitaxial CoSi$_2$ layer. The success of the OME process depends on the cleanliness of the deposited cobalt film, as air exposure is known to impede the epitaxial growth. However, oxygen from the SiO$_x$ layer, which comes in contact with the deposited cobalt layer, does not impede the OME growth. These results are suggestive that Co does not react chemically with the SiO$_x$ layer, but rather diffuses through the thin oxide layer and reacts directly with the Si crystal. Such a behavior is in contrast to the intermixing of Co and Ti in the initial stage of the TIME process. Despite possible differences in detail, the overall role played by the SiO$_x$ layer in inducing the OME effect is likely analogous to that played by Ti interlayer in TIME: as a kinetics modifier.

OME grown, ultrathin CoSi$_2$ layers were used as template layers (Tung et al 1983b) for the growth of thicker CoSi$_2$ layers. Due to the presence of a thin SiO$_x$ layer on the surface of OME grown CoSi$_2$ layers, deposition of Co at elevated temperatures cannot be used to increase the thickness of the CoSi$_2$ layer (Tung 1996b). Instead, room-temperature deposition of cobalt and annealing at 600-800°C is used to increase the thickness of OME CoSi$_2$ layers. In the planview TEM image of a $\sim$ 18nm thick CoSi$_2$ layer grown by repeated deposition and annealing sequences, shown in Fig. 7, a network of misfit dislocations is observed. However, the silicide morphology and/or thickness appear not to be affected by the presence of a 100 nm thick field oxide pattern.

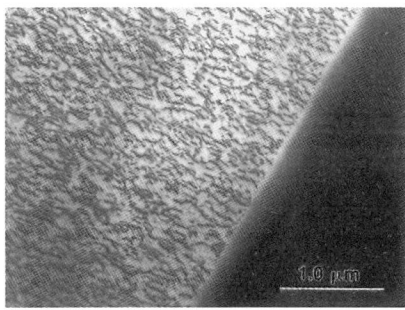

Fig. 7. *Planview, bright-field, TEM image of a 18 nm thick* CoSi$_2$ *layer grown by OME and template methods on Si(100), near the edges of a 100 nm thick field oxide layer.*

OME grown CoSi$_2$ template layers have a significant fraction of type B orientated grains on Si(111) and Si(211) surfaces. However, as these layers are made thicker with subsequent deposition and annealing steps, the orientation of the layers changes. The poor mobility of type B interfaces, compared with that of type A CoSi$_2$ interfaces, hinders the grain growth of type B CoSi$_2$ regions during the template growth process. Type A grains grow at a faster rate and eventually take over the type B oriented regions. This difference in the kinetics of the two interfaces have previously been explained for NiSi$_2$ (Tung et al 1983c, Hesse and Mattheis 1989). CoSi$_2$ layers can be converted to entirely type A orientation on Si(211) by the template technique. On Si(111), films with overwhelming (>99%) type A orientation can also be grown, as shown in Fig. 8. Previously observed formation of pinholes in type B CoSi$_2$ layers (Tung and Batstone 1988) on Si(111) is absent in the present experiment, likely because the present CoSi$_2$ surface is covered with a thin oxide cap which removes the thermodynamic driving force for pinhole formation, namely, a transition in surface structures (Hellman and Tung 1988). The increased interfacial mobility (of type A CoSi$_2$) also reduces the tendency for pinhole formation.

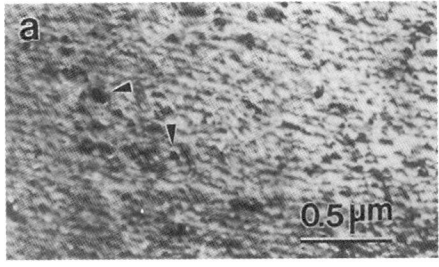

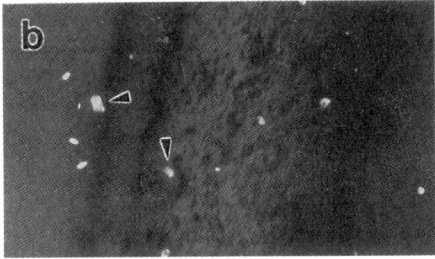

Fig. 8. *Planview TEM images of a 19.8nm thick* CoSi$_2$ *layer grown on Si(111) by OME and the template technique. (a), (200) dark field, showing type A oriented* CoSi$_2$, *and (b), 1/3(511) dark field, which is a type B related (111) diffraction.*

The OME technique is capable of generating high quality single crystal CoSi$_2$ layers on Si surfaces with any crystallographic orientation, of any common doping level, and without the formation of voids. Its ability to generate epitaxial CoSi$_2$ layers is also known to be independent of the linewidth of Si (Tung 1996b). Even though this technique is versatile and produces some very impressive results, a few important issues still need to be resolved before it can be applied to ULSI device fabrication. One problem concerns interfacial faceting of CoSi$_2$ layers on Si(100). Although the degree of faceting can be dramatically reduced by high temperature annealing (Tung 1996b), the added thermal budget is an undesirable burden for processes involving pre-existing doping profiles. In addition, the homogenization process during high temperature anneals is driven by the minimization of interface energy. Such a mechanism works well for blanket CoSi$_2$ layers, but does not need to work for narrow CoSi$_2$ lines. Other issues concerning the OME technique include the complexity of its procedures, the feasibility of applying this technique with sputtering tools, the quality of CoSi$_2$ layers grown on polycrystalline Si, and the prevention of oxidation during the OME process.

3. CONCLUSIONS

The morphology and structure of silicide thin films are often determined by the reaction kinetics. The formation of epitaxial silicide, even though favored by thermodynamics, is still largely dependent on kinetics. Of the three techniques discussed in this article, TIME and OME likely promote epitaxial growth by altering the kinetic path of the Co silicide reaction. The HTS technique and the doping-dependence of its epitaxial orientation are not well understood at this time. High quality epitaxial $CoSi_2$ structures, created by these techniques, are desirable for certain ULSI devices. Even though all three techniques are "practical" in the sense that they can be implemented using standard clean-room tools, they are limited by different sets of constraints. Because of various problems with $CoSi_2$ layers grown by the TIME technique, the likelihood of using TIME in future device manufacturing is small. The formation of thermally stable epitaxial $CoSi_2$ layers on pre-existing junctions by HTS involves only slight modifications to routine Co salicide processes. It is thus expected that HTS will be implemented in processing ULSI devices in the near future. However, the much discussed scheme of SADS for the formation of ultra-shallow junctions is not immediately compatible with the HTS technique. The OME technique, on the other hand, is very compatible with the SADS scheme, but this silicide process is complicated and it is still under development. In summary, the basic phenomena leading to the development of three epitaxial silicide technologies have been described. It appears that we are now closer than ever to witnessing the application epitaxial silicide in mainstream Si ULSI devices.

ACKNOWLEDGEMENT

We thank F. Schrey for his help with the experiments.

REFERENCES

Byun J S, Seon J M, Park J W, Hwang H and Kim J J 1996 MRS Symp. Proc. **402**, 167

Chiu K C R, Poate J M, Rowe J E, Sheng T T and Cullis A G 1981 Appl. Phys. Lett. **38**, 988

Dass M L A, Fraser D B and Wei C -S 1991 Appl. Phys. Lett. **58**, 1308

Goncalves-Conto S, Müller E, Schmidt K and von Känel H 1996 MRS Symp. Proc. **402**, 493

Hellman F and Tung R T 1988 Phys. Rev. **B 37**, 10786

Hensel J C, Levi A F J, Tung R T and Gibson J M 1985 Appl. Phys. Lett. **47**, 151

Hesse D and Mattheis R 1989 Phys. Stat. Sol. (a) 116, 67

Hsia S L, Tan T Y, Smith P and McGuire G E 1992 J. Appl. Phys. **70**, 1864

Ionue K, Mikagi K, Abiko H and Kikkawa T 1995 IEDM Tech. Dig. 445

Inoue K, Tung R T, Mikagi K, Chikaki S and Kikkawa T 1997 MRS Symp. Proc. **441**

Jimenez J, Hsiung L M, Thompson R D, Hashimoto S, Ramanathan K V, Arndt R, Rajan K, Iyer S S and Schowalter L J 1990 Mater. Res. Soc. Symp. Proc. **60**, 237

Ishiwara H, Nagatomo M and Furukawa S 1978 Nucl. Instr. Methods **149**, 417

Lasky J B, Nakos J S, Cain O J and Geiss P J 1991 IEEE Trans. Electron Dev. **ED-38**, 262

Liu R, Williams D S and Lynch W T 1988 J. Appl. Phys. **63**, 1990

Mantl S and Bay H L 1992 Appl. Phys. Lett. **61**, 267

Ogawa S, Fair J A, Dass M L A, Jones E C, Kouzaki T, Cheung N W and Fraser D B 1993 Ext. Abs. SSDM'93, p. 195

Stolk P A, Gossmann H -J, Eaglesham D J and Poate J M 1995 Nucl. Instr. Meth. B **96**, 187

Tu K N, Alessandrini E I, Chu W K, Krautle H and Mayer J W 1974 Japan. J. Appl. Phys., Suppl. 2, Part 1, 669

Tung R T, Gibson J M, Jacobson D C and Poate J M 1983a Appl. Phys. Lett. **43**, 476

Tung R T, Gibson J M and Poate J M 1983b Appl. Phys. Lett. **42**, 888

Tung R T, Gibson J M and Poate J M 1983c Phys. Rev. Lett. **50**, 429

Tung R T, Levi A F J and Gibson J M 1986 Appl. Phys. Lett. **48**, 635

Tung R T and Batstone J L 1988 Appl. Phys. Lett. **52**, 648

Tung R T 1992 in *Materials Interfaces: Atomic-level structure and properties*, Eds. D. Wolf and S. Yip (Chapman and Hall, London), p. 550

Tung R T 1993 Mater. Chem. Phys. **32**, 107

Tung R T and Schrey F 1995 Appl. Phys. Lett. **67**, 2164

Tung R T 1996a Appl. Phys. Lett. **61**, 3461

Tung R T 1996b Mat. Res. Soc. **427**, 481

Tung R T and Schrey F 1996 Mat. Res. Soc. **402**, 173

White A E, Short K T, Dynes R C, Garno J P and Gibson J M 1987 Appl. Phys. Lett. **50**, 95

Yalisove S M, Tung R T and Loretto D 1989 J. Vac. Sci. Technol. A **7**, 599

Inst. Phys. Conf. Ser. No 157
Paper presented at Microsc. Semicond. Mater. Conf., Oxford, 7–10 April 1997
© 1997 IOP Publishing Ltd

Micro-characterisation of Pt-silicides prepared on (100) silicon

S Jin, H Bender, R A Donaton and K Maex

IMEC, Kapeldreef 75, B-3001 Leuven, Belgium, sing@imec.be

ABSTRACT: Ultra-thin and uniform Pt-silicide layers are prepared by a one-step RTP process and by a two-step RTP process with a selective etch in between. In the case of the two-step RTP process, the first silicidation step is found to be crucial for the roughness of the final PtSi layer.

1. INTRODUCTION

Schottky barrier detectors based on metal-silicide/silicon structures are widely used in infrared detector applications. Platinum silicide has been reported as a potential material system for detector applications by Chin et al (1993) and Kosonocky (1987). Up to now, several Pt-silicide phases have been reported, and Pt-monosilicide (PtSi) is considered to have many advantages due to its Schottky barrier height of 0.23 eV to p-type silicon (Maex and Van Rossum 1995, Pellegrini 1994). The tight control of the PtSi layer uniformity and thickness in the range of 2-8nm is essential for optimal detector behaviour.

Pt and PtSi have a cubic (a=0.3923nm) and orthorhombic (a=0.5932nm, b=0.5595nm, c=0.3603nm) structure, respectively. The Pt_2Si phase has two crystal forms: tetragonal α-Pt_2Si (a=0.3933nm, c=0.591nm) and hexagonal β-Pt_2Si (a=0.654nm, c=0.3606nm). The $Pt_{12}Si_5$ phase is a high temperature form and has a tetragonal structure (a=1.3404nm, c=0.5451nm).

The rapid thermal process (RTP) technique has been successfully used for silicidation of Pt films on Si substrates (Dimitriadis, 1991). In this work, one-step RTP and two-step RTP processes are used.

In order to optimise the processing conditions for the fabrication of an ultra-thin and uniform PtSi film, transmission electron microscopy (TEM) is used in this paper for phase identification and to study the thickness and layer roughness.

2. EXPERIMENTAL

Pt is deposited on (100) silicon wafers by sputtering of 3nm ~ 100nm thick layers. Deposition is performed at room temperature. Two different silicidation processes are used: 1-step process: deposition followed by annealing at 450°C for 60s; 2-step process: the first annealing is performed by RTP at 200°C and 250°C for different annealing times followed by selective etch step (S.E.), and then the second RTP is carried out at 450°C for 60s. All annealings are done in a N_2 atmosphere. The etch is performed in $H_2O/HNO_3/HCl$ solution.

The samples are characterised with JEOL 200CX electron microscope. Cross-sectional specimens for TEM studies are prepared by mechanical thinning and subsequent ion milling. The normal is parallel to [011] direction of silicon. In order to get plan-view membranes, the wafer is thinned from the substrate side by grinding and milled from the backside with a single ion beam.

Platinum reacts with Si at a low temperature, so the temperature during TEM specimen preparation must be controlled to be low (below 80°C in our studies) in order to avoid Pt/Si reaction.

3. RESULTS AND DISCUSSION

Fig. 1 shows a cross-sectional high resolution electron microscopy (XHREM) image of a thin Pt-silicide layer after a single RTP silicidation at 450°C for 60s. A continuous Pt-silicide layer with thickness of 6nm is formed. In the case of the one-step process, the roughness of the interface between the silicide film and Si substrate is found to be dependent on the thickness of the initial Pt layer. Furthermore, electron diffraction patterns taken from plan-view samples, e.g. Fig. 2, show that the $Pt_{12}Si_5$ phase co-exists with PtSi grains. The distribution of the $Pt_{12}Si_5$ phase is very inhomogeneous over the sample and occurs only in certain regions.

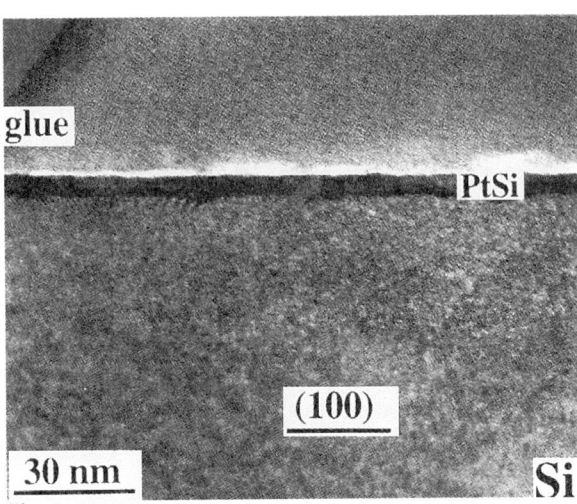

Fig. 1: XHREM image from the sample after the one-step silicidation at 450°C for 60s.

It was the initial thought that the two-step silicidation process follows the phase growth sequence of Pt-silicides, and that the first annealing (RTP1) is performed at a lower temperature to initiate the formation of the Pt_2Si phase. After the removal of the unreacted Pt, the second annealing (RTP2) is performed at a higher temperature so that a full transformation of Pt_2Si into PtSi is carried out (Torres et al 1995). However, a more detailed analysis of this process showed that a $Pt/Pt_2Si/PtSi/Si$ layered structure has been formed after the RTP1 step without the selective etch (not shown here, see Bender et al 1996 and Donaton et al 1997). Hiraki et al (1971) and Naem (1988) revealed that Pt reacts with Si initially forming Pt_2Si, and that the growth of Pt_2Si inhibits the formation of PtSi, therefore, only when all Pt is reacted to form Pt_2Si does PtSi start growing from the

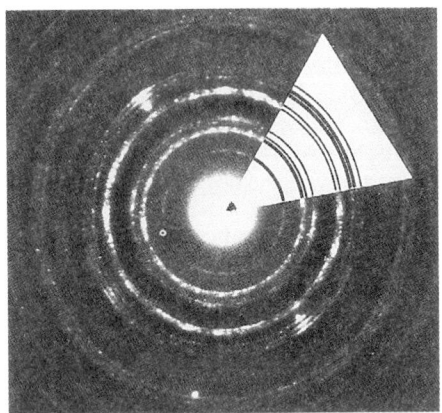

Fig. 2: Electron diffraction rings, taken from the $Pt_{12}Si_5$-rich area in the plan-view specimen, showing co-existence of PtSi (thick lines) and $Pt_{12}Si_5$ phases (thin lines).

Pt_2Si-Si interface, so Pt, Pt_2Si and PtSi are not simultaneously present. However, it has been shown that when oxygen is introduced in the film during the deposition or the annealing process, it will pile up at Pt/Pt_2Si interface and cause a halt in Pt_2Si growth before completion.

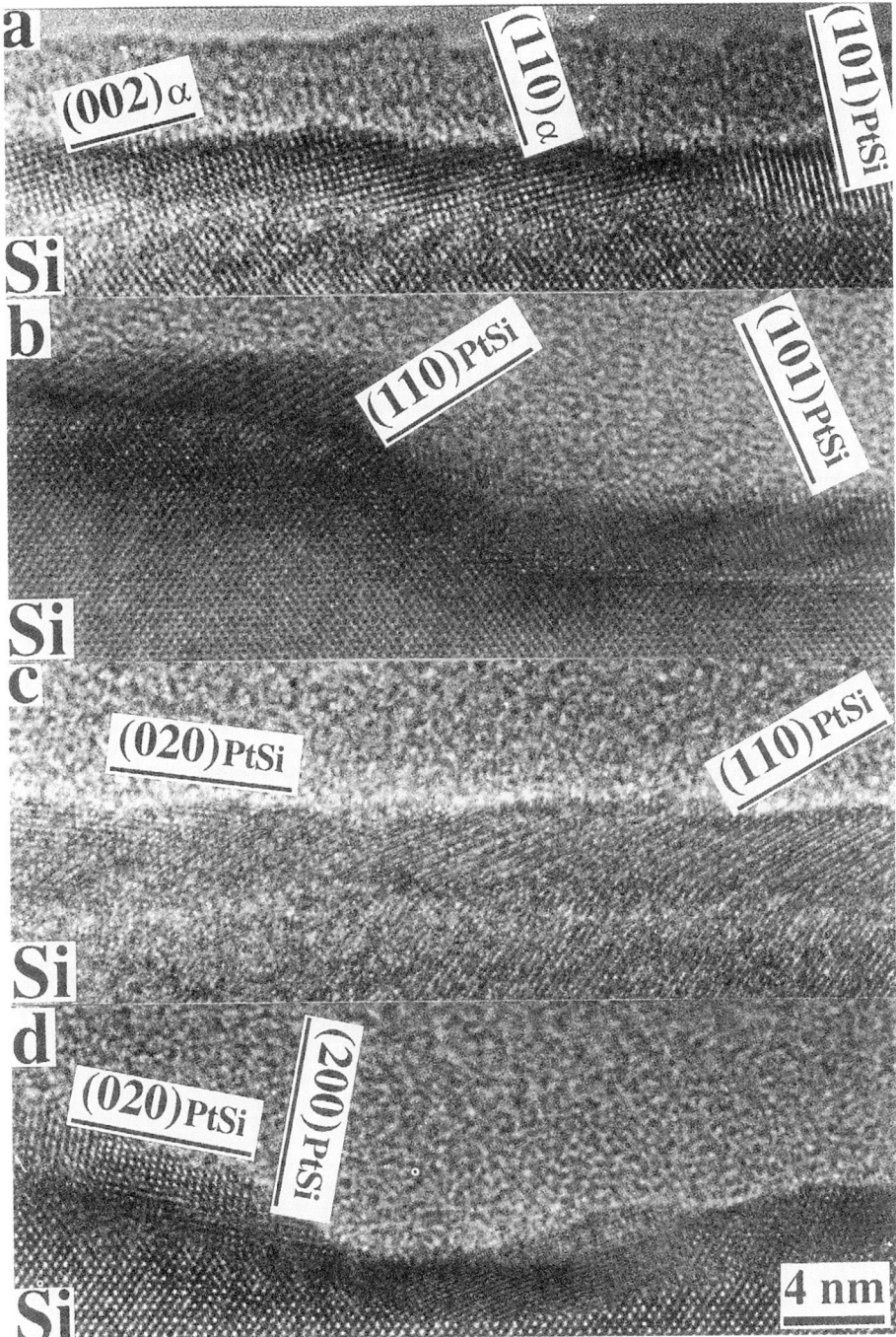

Fig. 3: XHREM images. a: RTP1 at 200˚C for 10s + S.E.; b: RTP1 at 250˚C for 10s + S.E.; c: RTP1 at 200˚C for 10s + S.E. + RTP2 at 450˚C for 60s; d: RTP1 at 250˚C for 10s + S.E. + RTP2 at 450˚C for 60s.

As a result, PtSi starts growing and Pt, Pt_2Si, PtSi and Si can exist simultaneously (Crider et al 1981, Poate et al 1974 and Muta et al 1972). In our work, Auger electron spectroscopy analysis gives no indication of the presence of oxygen in the layer. Therefore, another factor, e.g. softening of the Si bonds at the Pt/Si interface due to metal/silicon mixing during deposition, has to be invoked for the coexistence of the Pt films and the PtSi layers (Donaton et al 1997).

The first silicidation step is found to be crucial for the roughness of the final PtSi layer. Fig. 3 (a) and (b) give XTEM pictures of thin silicide layers after the first annealing at 200°C for 10s and 250°C for 60s, respectively, followed by the selective etch step. Fig. 3 (a) shows a very uniform silicide layer of 2-3 nm thickness and a smooth interface, whereas the layer is non-uniform and the interface is rough in Fig. 3(b). As shown in Fig. 3(c) and (d), the morphology after the RTP2 step is the same as after the first RTP. PtSi grains are seen along several low index zone axes, $(101)_{PtSi}$, $(020)_{PtSi}$ and $(200)_{PtSi}$ planes are often observed. In the case of the two-step silicidation process, the uniformity of the PtSi film and the roughness of the interface between the PtSi layer and the Si substrate are independent of the thickness of the initial Pt layer.

4. CONCLUSION

Ultra-thin and uniform Pt-silicide layers are prepared by a one-step RTP process and a two-step RTP process, respectively. In the case of the one-step RTP process, PtSi is the dominant phase, meanwhile, a small fraction of $Pt_{12}Si_5$ phase is inhomogeneously distributed. For the layer prepared by the two-step RTP process, the first silicidation step is found to be crucial for the roughness of the final PtSi layer.

ACKNOWLEDGEMENT

The transmission electron microscopy investigations are performed with the microscopes at EMAT, University Antwerpen (RUCA). K.M. is a research associate of the National Fund for Scientific Research, Flanders.

REFERENCES

Bender H, Roussel P, Torres A, Kolodinski S, Donaton R A, Maex K and Van der Sluis P 1996 Mat. Res. Soc. Symp. Proc. Vol. 402, 449
Crider C A, Poate J M, Rowe J E and Sheng T T 1981 J. Appl. Phys. 52, 2860
Chin V W L, Green M A and Storey J W V, Solid State Electron 1993 Vol 36, 1107
Dimitriadis C A, Polychroniadis E K, Evangelou E K and Giakoumakis G E 1991 J. Appl. Phys. 70, 3109
Donaton R A, Jin S, Bender H, Zagrebnov M, Baert K, Maex K, Vantomme A and Langouche G 1997 European Workshop on Materials for Advanced Metallization, Villard de Lans, France
Hiraki A, Nicolet M A and Mayer J W 1971 Appl. Phys. Lett. 18, 178
Kosonocky W F 1987 Proc. SPIE Vol 869, 90
Maex K and Van Rossum M 1995 EMIS Datareviews Series Vol. 14
Muta H and Shinoda D 1972 J. Appl. Phys. 43, 2913
Naem A A 1988 J. Appl. Phys. 64, 4161
Pellegrini P P 1994 Mat. Res. Soc. Symp. Proc., Vol 320, 27
Poate J M and Tisone T C 1974 Appl. Phys. Lett. 24, 391
Torres A, Kolodinski S, Donaton R A, Maex K, Roussel P and Bender H 1995 Proc. SPIE Vol 2554, 185

Inst. Phys. Conf. Ser. No 157
Paper presented at Microsc. Semicond. Mater. Conf., Oxford, 7–10 April 1997
© 1997 IOP Publishing Ltd

In situ TEM study of the evolution of CoSi$_2$ precipitates during annealing and ion irradiation

M Palard, M-O Ruault, H Bernas, M Strobel°, and K-H Heinig°

CSNSM, Bât. 108, F-91405 Orsay Campus France °Research Centre Rossendorf, PO Box
510119, D-01314 Dresden Germany

ABSTRACT : Cobalt implantation into Si at ca. 350°C forms epitaxial nanoprecipitates
which are either coherent with Si (A-type) or twinned (B-type). We compare the effect of
thermal annealing and 100 keV Si$^+$ ion beam irradiation at temperatures 600-750°C, via
transmission electron microscopy on line with an ion implanter. Identical behaviour is
observed, within a scaling factor : increasing annealing time or ion fluence leads to
increases in the size of A-type (at the expense of the B-type) precipitates. Since Co
diffusion is extremely rapid, the effect is ascribed to the precipitates' total free energy
difference (which increases with temperature), due to their surface energy difference.
Simulations confirm the latter, and agree with experimental results.

1. INTRODUCTION

The understanding of nucleation and growth mechanisms in realistic conditions is of
considerable basic interest, as well as vital to the design and control of nanocluster sizes and
densities for various applications. Ion beam synthesis can in principle be a solution, since it is
possible to control process parameters such as temperature, ion flux, fluence etc., but the
mechanisms of nucleation, growth and ripening must be studied under irradiation conditions.
The aim of this paper is to compare the effect of ion irradiation with that of pure thermal
annealing on diffusion and dissolution processes taking place at temperatures where high
species mobility occurs in Si containing CoSi$_2$ precipitates. We demonstrate that the results are
dominated by the effect of the surface energy difference between precipitates of identical
compositions, oriented differently relative to the host.

2. EXPERIMENTAL DETAILS

Thinned (001) Silicon samples are pre-implanted with 50 keV Co$^+$ ions, at 350°C with a
fluence of 10^{15} Co cm^{-2} and a flux of 1.3×10^{12} Co cm^{-2} s^{-1}. Sample preparation for
Transmission Electron Microscopy (TEM) and the implantation conditions were described in
Palard et al (1996). According to TRIM calculations, the peak   Co profile concentration
(depth about 45 nm, profile width ca. 35 nm) lies at half the TEM sample thickness. After Co$^+$
implantation, two types of CoSi$_2$ epitaxial precipitates (Bulle-Lieuwma 1991) are formed.
Octahedral (A-type) precipitates are in perfect epitaxy with the matrix and could only be
imaged by selecting the <200>CoSi$_2$ diffracting vector on the <100>CoSi$_2$ // <100>Si zone
axis. Platelet shaped twinned (B-type) precipitates are 180° rotated around a <111>Si
crystallographic axis and were imaged using a <111>CoSi$_2$ diffracting vector on the
<110>CoSi$_2$ // <114>Si zone axis. All experiments were performed on a CM12 Philips TEM
with a point by point resolution ≤ 0.4 nm. Due to contrast limitations, the precipitate visibility

limit was determined to be > 1 nm. Our measurements were done on a constant thickness area sample ($\approx$ 80 nm).

The initial measured B-type density is $\approx$ 7.10^{11} cm^{-2} and the A-type to B-type density ratio was estimated to be 10% from former experiments (Ruault et al 1994). The size of visible A-type and B-type ranges from 1 to 4 nm with an average of $\approx$ 2.5 nm. At each temperature (600, 650 and 750°C), the irradiation experiments were compared to the corresponding isothermal annealing experiments. Irradiation was performed with 100 keV Si$^+$ ions (fluences from 5x10^{14} to 2.5x10^{16} Si cm^{-2} at a constant flux of 1.3x10^{12} Si cm^{-2} s^{-1}). Because of their energy, the Si$^+$ ions go through the sample so that they loss their energy in the sample without concentration modification.

3. RESULTS

Because of the large proportion of precipitates with sizes near the visibility limit, especially in the as implanted samples, the precipitate density is underestimated. Thus the increase of the observed density at the beginning of the experiments is probably due to the growth of the smallest precipitates. Hence, relevant information could only be obtained after precipitate growth to mean sizes well above the visibility limit (around 4 nm).

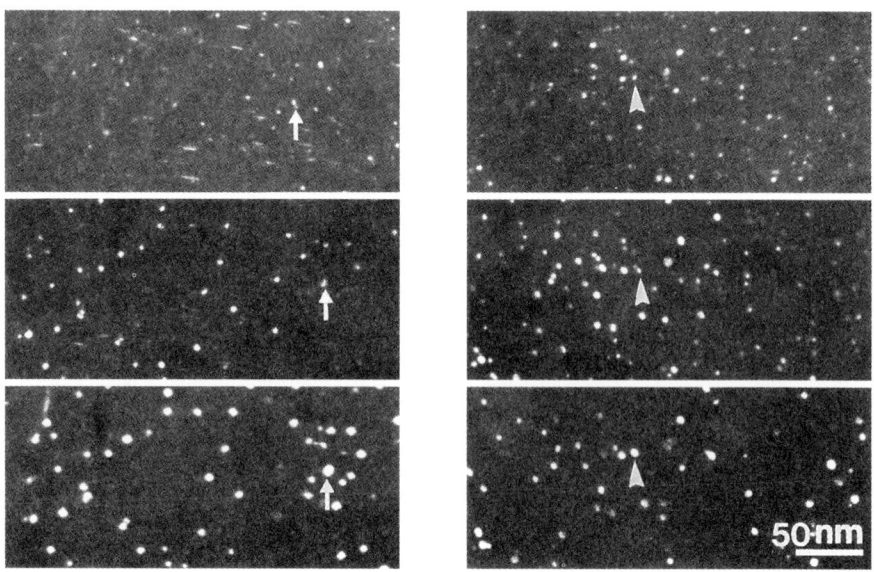

Figure 1 : Dark-field TEM pictures of a Si (100) thinned sample irradiated with different fluences 1-5-15x10^{15} Si cm^{-2} with a constant flux of 1.3x10^{12} Si cm^{-2} s^{-1} at 600°C. The A (on the left column) and B-type (on the right column) are imaged on the same area.

Precipitate density : The density evolution obtained during thermal annealing and Si$^+$ irradiation experiments show the same general trend (Fig. 1). In all cases the B-type precipitate density decreases (Fig. 2). The higher the substrate temperature, the smaller the remaining B-type density. At every temperature we observe an enhanced dissolution effect during irradiation (Fig. 2). Since the sample thickness is roughly constant, the area density evolution is similar to the volume density evolution. Thus in temperature range used for our experiments, kinetics in between $1/t^{1/3}$ and $1/t^{2/3}$ are deduced for the B-type density evolution. The kinetics of coarsening are slower than predicted by the LSW theory (Lifshitz et al 1961), (Wagner 1961). This theoretical approach and its extensions (Jayanth et al 1989) describes coarsening of second-phase particles in a supersaturated matrix considering the minimisation of the surface energy as a driving force. Within this theory, the density evolution follows a $1/t$ law in the diffusion controlled case, whereas in the reaction controlled case it follows a $1/t^{3/2}$ law. The reason for this discrepancy remains an open question.

The A-type density remains roughly equal to the as-implanted value ($\approx 7 \times 10^{10}$ cm^{-2}) during pure annealing. During irradiation assisted annealing, the density remains roughly constant within the error bars after a transient time in which the density evolution depends on the temperature. During the transient regime, the A-type density increases by a factor ≈ 2 at temperatures below 750°C and decreases by a factor ≈ 2 at 750°C. In situ TEM observation under irradiation shows that a constant A-type density actually corresponds to a balance between nucleation and dissolution due to Ostwald ripening.

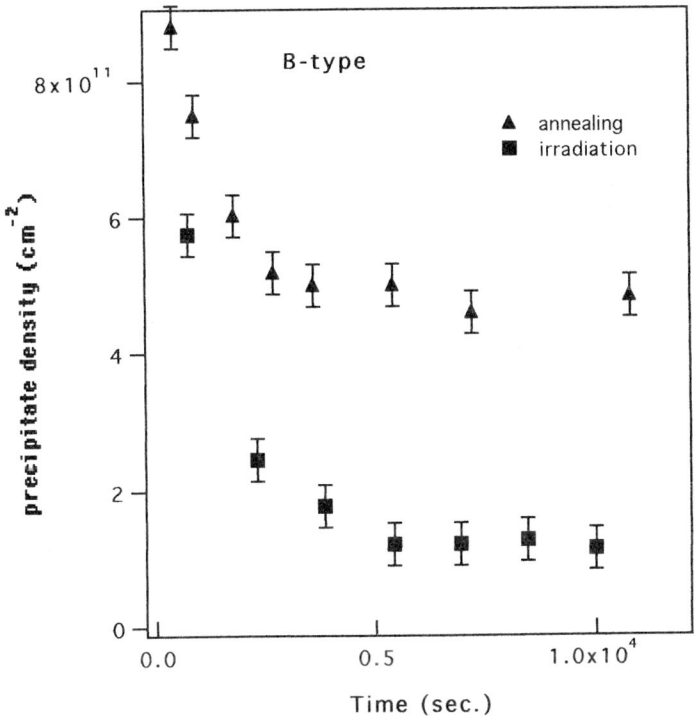

Figure 2 : B-type densities versus time during thermal annealing and Si$^+$ ion irradiation at a temperature of 750°C.

Precipitate size : The B-type size is defined as the diameter of the platelets observed along their main surface in the (114)Si // (110)CoSi$_2$ plane. The octahedral A-type size is defined as the diagonal of the double tetrahedron base observed in the (100) plane. As shown in Fig. 3, the average size for B-type increases and saturates whereas it increases for A-type. The higher the substrate temperature, the larger the average size. Because of the different shapes of A and B-type precipitates, the mean Co atom number in the former is larger than that of the latter. At a given temperature, no significant difference in the precipitate size evolution is observed between pure thermal annealing and irradiation assisted annealing.

504

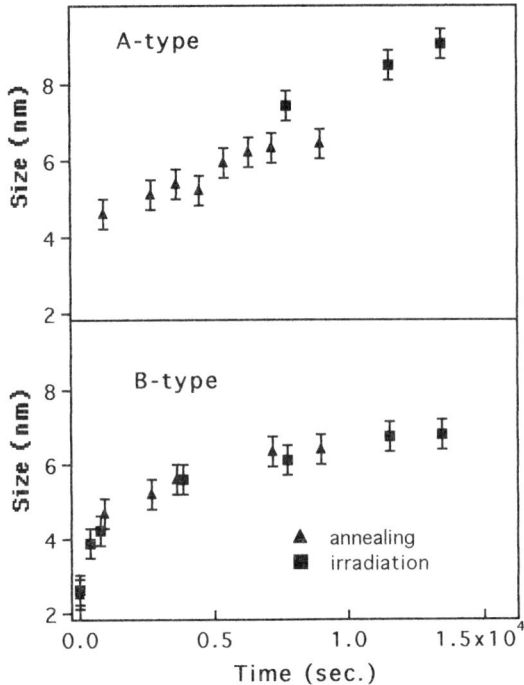

Figure 3 : Time evolution of the average size of the A and B-types during annealing and Si$^+$ ion irradiation at 650°C.

Co atom distribution : From the results on the size and density of the precipitates (Figs. 2 and 3) one observes a redistribution of Co atoms from the B-type to the A-type. This Co flow is irradiation enhanced.

4. DISCUSSION

The Co atom redistribution proceeds by a size increase and a decrease in the density of the B-type correlated to an A-type size increase. Due to the presence of both coherent and incoherent twin boundaries on B-type, while the A-type are coherent interfaces, the mean surface energy is larger for B-type than for A-type (Aaronson et al 1978). Thus the more stable A-type act as sinks for diffusing Co atoms. Here, the precipitates of both types interact via one common concentration field of Co monomers. Thus we have not two isolated subsystems with independent Ostwald ripening but rather one complex coarsening system. The competitive Ostwald ripening growth process (Mantl 1992) mainly occurs at the expense of smaller B-type precipitates by the growth of larger B and A-type precipitates.

Ion irradiation compared to thermal annealing increases the instability of the B-type. Above a critical temperature (in our case 650°C) an irradiation-enhanced dissolution effect is clearly observed. The A-type precipitate nucleation is enhanced and leads to an increase of the equilibrium precipitate density except at 750°C where significant Co redistribution occurs. This material redistribution is mainly due to a noticeable increase of Ostwald ripening which take place at this temperature. As shown above, in fact, A-type nucleation occurs all along but this effect is hidden by Ostwald ripening which increases with increasing temperature.

A computer simulation based on a numerical integration of the reaction-diffusion equations (Strobel et al 1996) of an ensemble of A-type and B-type precipitates has been performed. In the beginning the ratio of A-type and B-type precipitates ensembles is that of the experimental as-implanted state (Fig. 4). The initial (arbitrary) Particle Radius Distribution (PRD) for both A-type and B-type precipitate ensembles obeys the stationary PRD of a diffusion limited case of the LSW theory. The volume fraction of the precipitate material is 0.03 with a mean radius of 5 nm. In this model we assume a spherical shape for all precipitates. By ascribing different mean surface energies to the two types of precipitates, we try to describe their different ripening behaviour. The evolution of the whole precipitate ensemble was studied for various surface energy ratios : γ_B/γ_A

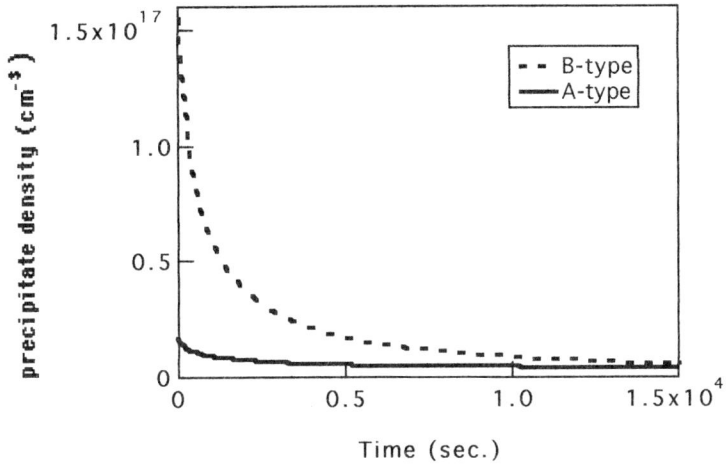

Figure 4 : The evolution of the A and B-type precipitate densities obtained by a computer simulation based on rate equation with $\gamma_B/\gamma_A = 1.05$.

The time evolution of these ensembles mainly depends on this ratio rather than on the characteristic of the coarsening process. A good agreement is obtained for a surface energy ratio $\gamma_B/\gamma_A = 1.05$. For this ratio, the Co atoms flow from the B to the A-type follows the experimental measurements (not shown here). This result is in good agreement with the previous TEM results obtained by Bulle-Lieuwma (1991) which shows that A-type precipitates are more stable than B-type precipitates during the growth stages. This agreement between the simulation and the TEM experimental results leads us to conclude that the surface energy minimisation is the driving force for the B-type precipitate dissolution relative to the A-type growth in a closed system.

This situation is drastically different from that observed during nucleation and growth of CoSi$_2$ precipitates by implantation i.e. with an external source, (Ruault et al 1995) where the whole system is far off the thermodynamic equilibrium. In that case, the A-type precipitates do not grow. During Co implantation the precipitate evolution is supposed to be controlled by kinetic rather than by energetic processes. Thus differences in interface reaction constants of the two types could be an explanation of this observed behaviour i.e. growth in size of B-type favoured. Moreover the small difference in mean surface energies of the two types is the reason of the different behaviour during the experiments with or without external source term.

In summary, we have shown, that pure thermal or irradiation assisted annealing of a system of A and B-type precipitates leads to a major redistribution of Co atoms from B to A-type.

This general effect may be well explained by ascribing different mean surface energies to both types and shows that A-type are energetically favoured.

ACKNOWLEDGEMENTS

We wish to thank O Kaïtasov for the implantation experiments. We would like to thank the Research Centre Rossendorf for financial support during student visit there. This work was partially supported by the PROCOPE project.

REFERENCES

Aaronson H I, Lee J K and Russell K C 1978 Precipitation Processes in Solids, eds. Aaronson, and Russell, TMS-AIME (New-York) p 31
Bulle-Lieuwma C W T 1991 thesis
Jayanth C S and Nash P 1989 J. Mater. Sci. **24** pp 3041-3052
Lifshitz I M and Slyosov V V 1961 J. Phys. Chem. Solids **19** pp 35-50
Mantl S 1992 Mater. Sci. Rep. **8**
Palard M, Ruault M-O, Kaïtasov O, Heinig K-H and Bernas H 1996 Nucl. Instr. & Meth. **B 120** p 212
Ruault M-O, Fortuna F, Bernas H and Kaïtasov O 1994 Nucl. Instr. & Meth. **B 84** p 135
Ruault M-O, Clayton J, Hardit C, Kaïtasov O, Bernas H, Reiss S and Heinig K-H 1995, IBMM Int. Conf. (Caberra, Australia) unpublished
Strobel M, Reiss S and Heinig K-H 1996 Nucl. Instr. & Meth. **B 120** p 216
Wagner C 1961 Z. Elektrochem. **65** p 581

Inst. Phys. Conf. Ser. No 157
Paper presented at Microsc. Semicond. Mater. Conf., Oxford, 7–10 April 1997
© *1997 IOP Publishing Ltd*

Heteroepitaxial Si/ErSi₂/Si structures grown in high vacuum

A Travlos, E Flouda, P Aloupogiannis and N Salamouras

Institute of Materials Science, National Centre for Scientific Research "Demokritos" GR 15310 Agia Paraskevi, Attiki, POB 60228, Greece

ABSTRACT: Buried $ErSi_2$ layers were grown in high vacuum (10^{-8} Torr) on both (111) and (100) Si substrates. $ErSi_2$ layers were grown by evaporation of Er and Si onto the Si substrate at 450°C and subsequent annealing for 30 min at 800°C. The top Si layer was formed by evaporation of Si onto the $ErSi_2$ layer at 450°C and subsequent annealing at 550°C for 30 min. The Si/$ErSi_2$/Si structures were characterized by X-ray diffraction and electron microscopy. Transmission electron microscopy characterisation revealed that the $ErSi_2$ and top Si layers were grown epitaxially on (111) Si, were continuous and of high crystalline quality. Layers grown on (100) Si were epitaxial as well, but of inferior crystalline quality.

1. INTRODUCTION

Silicide-silicon heterostructures are of technological interest because of their applications in micro- and optoelectronics and infrared detection (Shepherd and Yang 1973, Kosonocky et al 1985). The use of erbium silicide in such heterostructures is particularly attractive because of its low Schottky barrier height to n-type silicon (Duboz et al 1989).

Applications of $ErSi_{2-x}$ in tuneable internal photoemission sensors have already been reported in heterostructures of the form Pt/Si/$ErSi_{2-x}$/Si (Pahun et al 1991) and Ir/Si/$ErSi_{2-x}$/Si (Sagnes et al 1993).

Despite the technological interest, little work has been reported on the electron microscopy characterization of such heterostructures on (111) Si (D' Avitaya et al 1990, Veuillen et al 1994, Frangis et al 1996), and virtually none on heterostructures grown on (100) Si, which is compatible with existing C-MOS technology.

In this work the conditions of growth of Si/$ErSi_{2-x}$/Si heterostructures on (111) and (100) Si are presented and their structural characteristics are investigated by electron microscopy.

2. EXPERIMENTAL RESULTS AND DISCUSSION

2.1 Thin Film Growth

$ErSi_{2-x}$ and Si layers are grown in a high vacuum chamber ($\sim 10^{-8}$ Torr) equipped with two electron guns and a substrate heating system. Silicon (100) and (111) substrates are heated in vacuum at 850°C for 30 min in order to clean them. Subsequently a thin 20 Å layer of Er metal is evaporated onto the substrate held at 450°C. Annealing of this thin layer for 5 min at 800°C resulted in the formation of a thin layer of epitaxial $ErSi_{2-x}$ which served as a template for the growth of the thicker silicide layer. This thick silicide layer is grown by coevaporation of Si and Er at a ratio of two, onto the template held at 450°C, with subsequent annealing for 30 min at 800°C. Details and discussion of this method of growth of $ErSi_{2-x}$ layers are presented elsewhere:

Travlos et al. (1997). In the final stage of growth a single layer of Si is evaporated onto the formed ErSi$_{2-x}$ at 450 °C and then is annealed at 550 °C for 30 min.

2.2 X-ray Diffraction Results

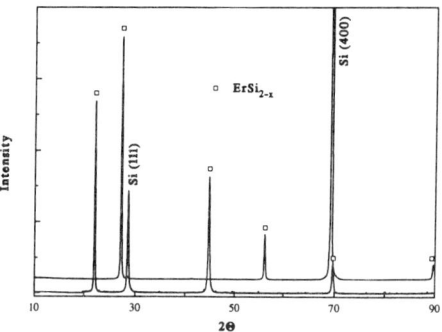

Fig. 1 shows the X-ray diffraction results for the Si/ErSi$_{2-x}$ /Si heterostructures grown on (111) and (100) Si, revealing lines corresponding to (0001), (0002), (0003) planes and (004), (008), (0012) planes respectively.

2.3 Electron Microscopy Characterization

Electron microscopy observations were performed on cross-section specimens of Si/ErSi$_{2-x}$ /Si heterostructures grown on (111) and (100) Si with a Philips CM20 TEM.

Fig. 2a shows a bright field micrograph of Si/ErSi$_{2-x}$ /Si heterostructure grown on (111) Si. The thickness of ErSi$_{2-x}$ layer is 75 nm while that of top Si layer is 68 nm. Both surface and interfaces of the layers are smooth. However, the substrate appears heavily stressed and the top silicon layer contains a large number of defects.

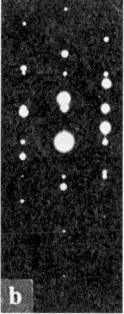
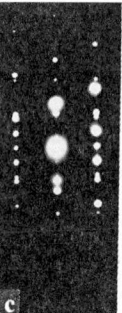

Fig. 2 a) Bright field micrograph of (111)Si/ ErSi$_{2-x}$ /Si heterostructure. b,c) SAED patterns along the [0$\bar{1}$1]sub Si from two different areas of the heterostructure.

Fig. 2b and 2c are selected area diffraction (SAED) patterns taken from two different areas of the heterostructure. From these SAED patterns one can see that the ErSi$_{2-x}$ layer is epitaxially grown on the (111) Si substrate and its crystalline structure is the AlB$_2$ hexagonal one with lattice parameters a=b=3.80 ± 0.03 Å and c=4.080 ± 0.03 Å, which is in agreement with the structure determined by Mayer et al. (1962) for bulk ErSi$_{2-x}$. The epitaxial orientation of ErSi$_{2-x}$ with respect to the (111) Si substrate is determined to be (0001) ErSi$_{2-x}$ // (111)Si and [1$\bar{2}$10]ErSi$_{2-x}$ // [0$\bar{1}$1]Si. Several electron diffraction experiments along the silicide layer revealed that it is almost single crystalline, consisting of large "grains" up to 5μm slightly misoriented with respect to each other. However, planar crystalline defects parallel to the $(01\bar{1}0)$ planes of the silicide are not uncommon.

Fig. 2b and 2c show also that the silicon overlayer grows epitaxially onto the $ErSi_{2-x}$ layer but in two orientation variants:

$(0001)\ ErSi_{2-x}\ //\ (111)\ Si,\qquad [\bar{1}2\bar{1}0]\ ErSi_{2-x}\ //\ [0\bar{1}1]\ Si\qquad\qquad$ and

$(0001)\ ErSi_{2-x}\ //\ (1\bar{1}1)\ Si,\qquad [\bar{1}2\bar{1}0]\ ErSi_{2-x}\ //\ [0\bar{1}1]\ Si$

The existence of these two orientation variants is not unexpected since the silicon overlayer is growing on the (0001) plane of hexagonal silicide where the two orientations are not distinguishable. We note that similar orientation variants have been observed by Veuillen et al. (1994), on 20 Å thin overlayers of silicon grown on $ErSi_{2-x}$.

Fig. 3 is a high resolution image of the $ErSi_{2-x}$ /epiSi interface taken along the $[\bar{1}2\bar{1}0]$ axis of the silicide, revealing the two orientation variants of the silicon epi-layer. Note also the two planar defects in the silicide layer which are perpendicular to the interface and parallel to $(01\bar{1}0)$ planes.

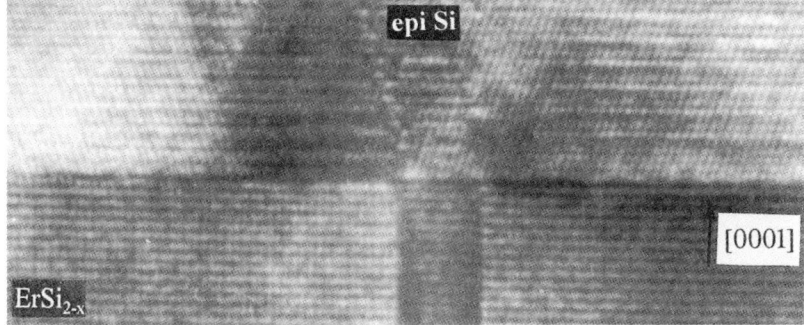

Fig. 3 HREM image of the $ErSi_{2-x}$ /epiSi interface of (111)/$ErSi_{2-x}$ /Si heterostructure revealing the two orientation variants of the silicon epilayer

This type of defect has been identified by Frangis et al (1996) as "twin" planes separating orientation variants of silicon vacancies superstructure in $ErSi_{2-x}$ thin films. Evidence of ordering of the silicon vacancies can also be seen in Fig. 2b and 2c in the form of weak spots in the direction parallel to the (0001) planes.

Fig 4a is a bright field micrograph of a Si/$ErSi_{2-x}$ /Si heterostructure grown on (100) Si and the corresponding SAED pattern.

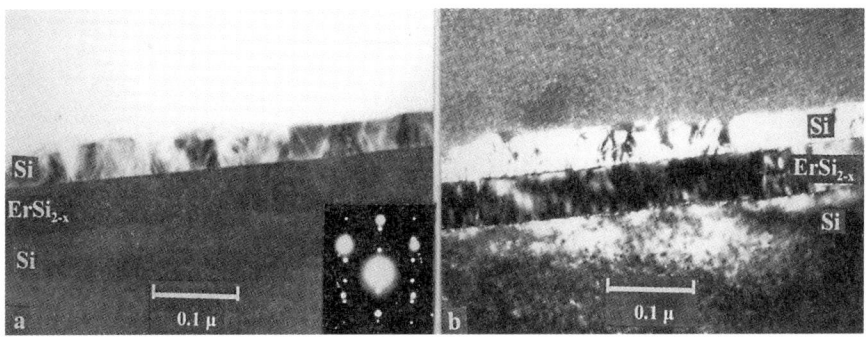

Fig. 4: a) Bright field micrograph of (100)/$ErSi_{2-x}$ /Si heterostructure and b) dark field micrograph of the same heterostructure

The ErSi$_{2-x}$ layer is 60 nm thick with smooth interfaces. It consists of "grains", up to 1μm, epitaxially grown on the (100) Si substrate slightly misoriented with respect to each other. The crystalline structure of the ErSi$_{2-x}$ layer is the tetragonal ThSi$_2$-type with a=b=3.96 ± 0.02 Å and c=13.26 ± 0.04 Å. The ThSi$_2$-type structure of ErSi$_{2-x}$ is a structural phase induced by the substrate orientation and was found to grow on (100) Si under specific conditions of growth by Kaltsas et al (1996). Details of the conditions of growth of this phase are discussed in detail elsewhere (Travlos et al 1997). The epitaxial orientation of ErSi$_{2-x}$ to the (100) Si substrate was found to be:

$$(100)Si//(001) ErSi_{2-x} \text{ and } [011] Si//[100] ErSi_{2-x}$$

The Si overlayer has a thickness of 50 to 60 nm and a rather rough surface. It consists of grains up to 100 nm in size, most of which are epitaxially grown on the ErSi$_{2-x}$ layer with the same orientation as the silicon substrate, i.e. (100)Si//(001)ErSi$_{2-x}$ and [011]Si//[100] ErSi$_{2-x}$. There are, however, some smaller grains, ~ 20 nm, with orientations other than those described above. Fig. 4b is a dark field micrograph of the heterostructure grown on (100) Si taken with the 200 spot of silicon revealing the smaller misoriented grains in the silicon overlayer.

The inferior crystalline quality of the heterostructures grown on (100) Si compared to those grown on (111) Si can be attributed to the difference in mismatch between Si and ErSi$_{2-x}$ in the two cases. The misfit in the case of (111)Si/ErSi$_{2-x}$/(111)Si is -1.1% while in the case of (100)Si/ErSi$_{2-x}$/(100)Si it is -3.1%. This larger misfit, in the later case, gives rise to larger stress in the heterostructure inhibiting the growth of good crystalline quality layers.

3. CONCLUSIONS

We have demonstrated the feasibility of growing thick silicon epilayers on ErSi$_{2-x}$ on both (111) Si and (100) Si substrates. The Si epilayers grown on ErSi$_{2-x}$/(111)Si heterostructures are of very good crystallinity while those on ErSi$_{2-x}$/(100)Si are of inferior crystalline quality because of the large misfit between Si and tetragonal ErSi$_{2-x}$.

ACKNOWLEDGEMENT

This work has been supported by the PENED subprogramme of the Greek General Secretariat for Research and Technology.

REFERENCES

D'Avitaya F A, Badoz P A, Campidelli Y, Chroboczek J A, Duboz J Y and Perio A 1990 Thin Solid Films **184**, 283
Duboz J Y, Badoz P A, D'Avitaya F A and Chroboczek J A 1989 Appl. Phys. Lett. **55**, 84
Frangis N, Van Tendeloo G, Van Landuyt J, Muret P and Nguyen T T A 1996 J. Alloys and Compounds **234**, 244
Kaltsas G, Travlos A, Nassiopoulos A G, Frangis N and Van Landuyt J 1996 Appl. Surf. Sci. **102**, 151
Kosonocky W F, Shallcross F V, Villani T S and Groppe J V 1985 IEEE Trans. Electron. Devices **ED22**, 1564
Mayer I P, Banks E and Post B 1996 J. Phys. Chem. **66**, 693
Pahun L, Campidelli Y, D'Avitaya F A and Badoz P A 1992 Appl. Phys. Lett. **60**, 1166
Sagnes I, Campidelli Y, Vincent G and Badoz P A 1993 Mat. Science and Eng. **B21**, 312
Shepherd F D and Yang A C 1973 I EDM Tech. Dig. p. 310
Travlos A, Salamouras N and Flouda E 1997 Appl. Surf. Sci. submitted
Veuillen J Y, D'Anterroches C and Nguyen T T A 1994 J. Appl. Phys. **75**, 223

Inst. Phys. Conf. Ser. No 157
Paper presented at Microsc. Semicond. Mater. Conf., Oxford, 7–10 April 1997
© *1997 IOP Publishing Ltd*

Radiation enhanced diffusion of ion implanted Fe in Si (100) observed in ion beam synthesis of β-FeSi₂

Y Maeda, T Fujita, K Umezawa and K Miyake*

Department of Materials Sciences, Osaka Prefecture University, Sakai, Osaka 593 Japan
*Power & Industrial Systems R & D Division, Hitachi Ltd., Hitachi, Ibaraki 316 Japan

ABSTRACT: Radiation enhanced diffusion (RED) of implanted Fe atoms to the surface was observed in Si (100) implanted by $^{56}Fe^+$ ion at room temperature. The RED takes place in a damaged surface layer containing large amounts of Si vacancies ($\sim 10^{18}$ atoms/cm²). The RED is related to remaining amounts of Si vacancies after annealing. In the highly damaged surface, formation of nucleation centers, *i.e.* $FeSi_8$ blocks, for β-FeSi₂ is found to be also enhanced (radiation enhanced nucleation: REN). These results show that RED and REN enhance effectively formation of orthorhombic β-FeSi₂ near the surface of implanted Si (100).

1. INTRODUCTION

The orthorhombic FeSi₂ (β-FeSi₂) compound is a semiconductor which has a direct band-gap of 0.8-0.85 eV(Bost and Mahan 1985). The semiconducting property of β-FeSi₂ comes from a Jahn-Teller effect because of its strong electron-lattice interaction. The β-FeSi₂ phase also shows a large Seebeck coefficient at high temperature (>600°C). These promising properties can be applied to optoelectric devices at near infrared wavelengths, for optical telecommunication, and thermoelectric power devices.

Ion beam synthesis (IBS) using $^{56}Fe^+$ ion implantation has been widely employed to form β-FeSi₂/Si heterostuructures (White *et al* 1987, Leong *et al* 1996, Katumata *et al* 1996). In the IBS of β-FeSi₂ the profile control of ion implanted Fe atoms and heavy implantation of $^{56}Fe^+$ with a dose of $\sim 10^{17}$ ions/cm² are crucial to a utilization of β-FeSi₂/Si heterojunctions.

The high dense implantation damage has been known to cause some radiation enhanced phenomena for diffusion, precipitation, segregation and surface melting because of its nonequilibrium state. In the case that heavy ion and high dose implantation such as $^{56}Fe^+$ into Si, large amount of damage is produced near the surface layer. In the implantation energy of less than 200 keV, the nuclear stopping power is larger than the electron stopping power, so that most of the energy is paid to make many displaced Si atoms in the lattice sites; for instance, in $^{56}Fe^+$ random implantation into Si (100), a simple calculation suggests that one Fe atom makes 14-15 displaced Si atoms. The $^{56}Fe^+$ implantation with a dose of 10^{16}-10^{17} ions/cm² which is required to make β-FeSi₂ produces over $\sim 10^{17}$-10^{18} atoms/cm² of displaced Si atoms in the damaged layer. This situation in the damaged layer is the same as that in the liquid state. Therefore, we expect various nonequilibrium phenomena related to the liquid-like damaged layer.

In this study, we examine the effects of damage (radiation effects), which are induced by room temperature implantation of $^{56}Fe^+$ ions into Si (100), on the redistribution of Fe atoms and formation (nucleation and growth) of β-FeSi₂ after annealing by X-ray photoelectron spectroscopy (XPS), Rutherford backscattering spectroscopy (RBS) and Raman spectroscopy.

2. EXPERIMENTAL

Mass-separated $^{56}Fe^+$ ion implantation into *n*-type-floating zone (FZ) grown-Si (100) was carried out at room temperature with a dose of 5x10¹⁶ ions/cm² at energies of 50, 80 and 100 keV at the incident angle of 7° (random implantation). The ion source of dehydrated FeCl₃ powder was used to achieve large ion beam current of ~1 mA/cm². The implanted Si (100) was lamp-annealed at 600-900°C for 2 h in an argon atmosphere (3 mTorr) to precipitate β-FeSi₂. The 2MeV-⁴He⁺ ion RBS data were measured at backscattering angle of 150°. The RBS data

512

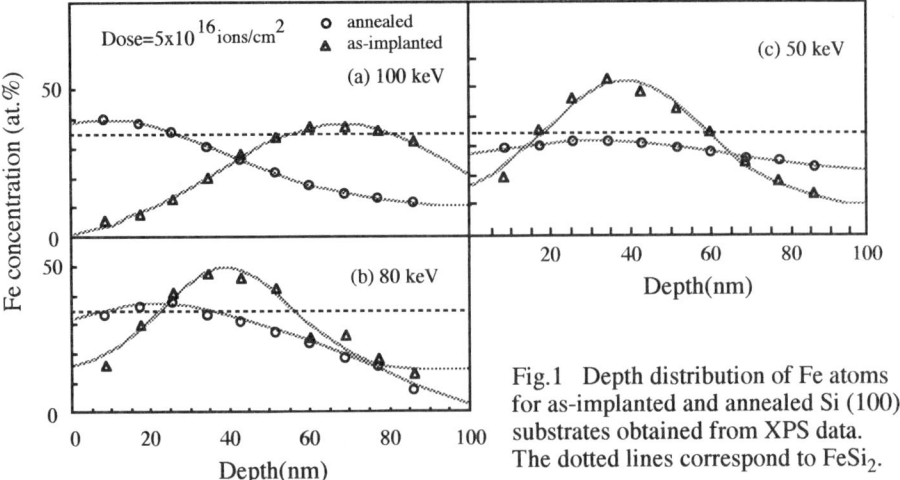

Fig.1 Depth distribution of Fe atoms
for as-implanted and annealed Si (100)
substrates obtained from XPS data.
The dotted lines correspond to FeSi$_2$.

were analyzed by RAMPU. The depth-resolved XPS were excited by Mg Kα for Si 2p$_{1/2,3/2}$ (99-100 eV) and Fe 2p$_{3/2}$ (707-708 eV) and measured   after Ar ion etching. Raman spectra were excited by a 514.5 nm Ar ion laser (~200 mW) and measured at room temperature by a 1000 mm focus-length double monochromator (Jobin Yvon U1000), a cooled photomultiplier and a photoncounter unit.

3. RESULTS AND DISCUSSIONS

3.1 Implantation Energy Dependence of RED

Figure 1 shows depth-distribution of Fe atoms derived from the XPS data of Si 2p$_{1/2,3/2}$ and Fe 2p$_{3/2}$ for as-implanted and annealed Si (100) substrates. The dose was 5x10^{16} ions/cm^2. The distribution for as-implanted Si (100) can be described by a simple Gaussian function. Higher energy implantation made a deeper and broader Fe atomic distribution in Si (100). However, after annealing at 800°C, it was observed that all the distributions shifted to the substrate surface. Significant change of the Fe distribution, shifted by ~60 nm, took place in the 100 keV implanted Si (100). With decreasing implantation energy, the shifts decreased as shown in Figs.1(b) and (c). The 50 keV-implanted Si (100) showed a uniform depth-distribution of Fe atoms. This redistribution observed in the case of 50 keV was different from those in the cases of 80 and 100 keV. Surface segregation of Fe atoms observed in 100 and 80 keV implanted Si (100) can be considered to be related to radiation enhanced diffusion (RED) of Fe atoms associated with lattice vacancies, *i.e.* Frenkel defects (Dearnaley *et al* 1973). All the cases achieved the composition near β-FeSi$_2$ (33 at.% Fe) as shown by the dotted lines in Figs. 1(a)-(c). In fact, the depth at which β-FeSi$_2$ was formed corresponds to that of about 30-35 at.%Fe in Fig.1 in the depth-analysis of Raman spectroscopy as shown in Figs.4 (a) and (b).

3.2 Annealing Temperature Dependence of RED

Figure 2 shows redistribution behavior of implanted Fe atoms after annealing at 600-900 °C. A big difference in the redistribution of Fe atoms was observed in the annealed Si (100) at 800°C. For annealing below 700°C, the redistribution was formed near the surface, *i.e.* surface segregation, while for annealing above 800°C a redistribution was formed also in the deep part (200-400 nm in depth). This observed difference in the redistribution can be considered to come from competition between RED in the damaged surface and normal Fick's diffusion driven by a difference in Fe concentration. Above 800°C, the process of crystallinity recovery and formation of the β-FeSi$_2$ lattice can be thermally activated, so that vacancies near surface (<~100 nm in depth) rapidly decrease. Therefore, above 800°C, no effect of RED on the redistribution after annealing can be observed. From these results in Figs.1 and 2, we found that the redistribution of

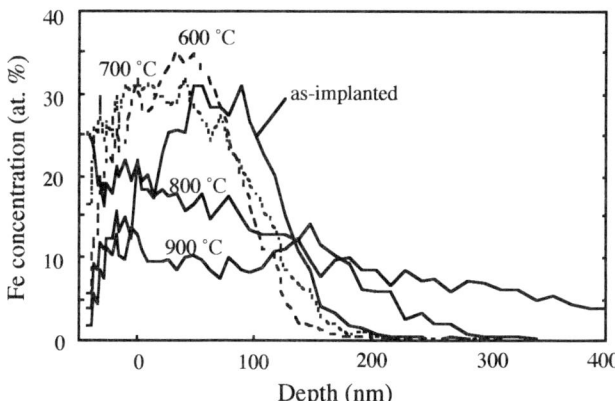

Fig.2 Depth-distribution of Fe atoms derived from the RBS data of the 50 keV-^{56}Fe$^+$ ion beam implanted Si (100) substrates after annealing at 600-900°C. The doses were 5x10^{16} ions/cm^2.

implanted Fe depends on radiation enhanced diffusion in the damaged surface through the recovery process and the formation of the β-FeSi$_2$ lattice, which are thermal activated processes.

Figure 3 shows the change in density of displaced Si atoms in as-implanted and annealed Si (100). The annealing at 800°C decreased the density of displaced Si atoms for all energies. The densities for 50 and 100 keV were respectively 1.6 and 1.7 (x10^{18} atoms/cm^2), while these decreased by 20% and 60%, respectively after annealing at 800°C for 2 hours. The vacancies for 100 keV decreased more rapidly than those for 50 keV. However, a small change in the density took place below 700°C at which RED was observed for 50 keV. These results suggest that the vacancies remaining during annealing determine whether RED takes place or not.

3.3 Raman spectrum for β-FeSi$_2$

Raman spectra were shown in Figs. 4 (a) and (b). After annealing at 600°C, for 100 keV Raman lines (A$_g$) at 195 and 248 cm^{-1} of β-FeSi$_2$ (Lefki *et al* 1991) and recovery of crystallinity of Si (TO phonon) were observed, but for 50 keV no lines were observed. Above 800°C, for both energies clear Raman lines of β-FeSi$_2$ and Si were observed. This difference observed at 600°C

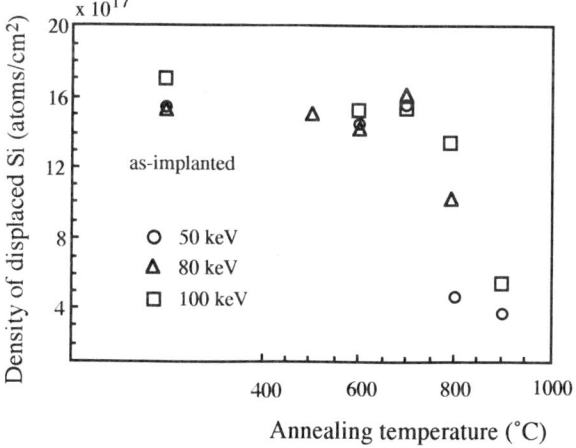

Fig.3 Change in density of displaced Si atoms derived from the RBS data.

514

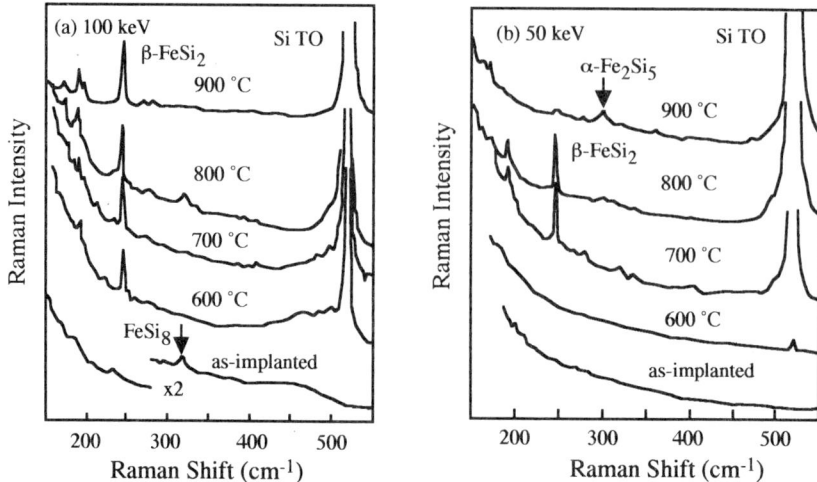

Fig.4 Raman spectra for (a) 100 keV and (b) 50 keV-implanted Si (100).

comes from nucleation centers: tetragonal $FeSi_8$ clusters for β-$FeSi_2$ reported by Tan *et al* (1992). For 100 keV, a Raman line at 320 cm^{-1} was observed in the as-implanted sample. The Raman line of α-Fe_2Si_5 (tetragonal) is close to the line at 320 cm^{-1}. Therefore, the line at 320 cm^{-1} may come from the $FeSi_8$ cluster. The formation of $FeSi_8$ cluster may be enhanced in the much damaged surface, *i.e.* radiation enhanced nucleation (REN). The big difference of the structure between Si and β-$FeSi_2$ makes direct nucleation very difficult. However, nucleation *via* $FeSi_8$ at the damaged surface is easy and accelerated by vacancies. We found nucleation enhancement due to REN.

4. CONCLUSIONS

We observed surface segregation of Fe atoms due to RED in highly damaged surface. The RED depends on the implantation energy and annealing temperature. For the much damaged surface close to liquid-like state, formation of nucleation centers ($FeSi_8$ clusters) is enhanced. It is found that both the nucleation and growth of β-$FeSi_2$ in the damaged surface are enhanced by an induced high density of vacancies. The damage can play an important role to enhance the formation of β-$FeSi_2$.

ACKNOWLEDGMENTS

We would like to acknowledge Dr. M. Sugiyama of Nippon Steel Co. for valuable discussions. The author (Y.M) acknowledges support from the Sumitomo Foundation (Contract No. 950055).

REFERENCES

Bost M C and Mahan J E 1985 J Appl. Phys. **58**, 2696.
Dearnaley G, Freeman J H, Nelson R S, Stephen J S 1973, "*Ion Implantation*" (North Holland Publishing) p.227.
Katumata H, Makita Y, Kobayashi N, Shibata H, Hasegawa M, Akesenov I, Kimura S and Obara A 1996 J. Appl. Phys. **80**, 5955.
Lefki K, Muret P, Bustarret E, Boutarek N, Madar R, Chevrier J, Derrien J and Burnel M 1991 Solid State Comm. **80**, 791.
Leong D N, Harry M A, Reeson K J, Homewood K P 1996, Appl Phys Lett. **68** 1649
Tan Z, Namavor F, Heald S M, Bouldin C E and Woicik J C.1992 Phys. Rev. **46**, 4077.
White A E, Short K T, Dynes R C, Gibson J M 1987 Appl Phys. Lett. **50**, 95.

Inst. Phys. Conf. Ser. No 157
Paper presented at Microsc. Semicond. Mater. Conf., Oxford, 7–10 April 1997
© 1997 IOP Publishing Ltd

Application of image filtering to semiconductor structures

P L Flaitz and A Domenicucci

IBM Analytical Services, 1580 Route 52, Hopewell Junction, NY 12533, USA

ABSTRACT: The development of filtered imaging, and, in particular, the Gatan Imaging Filter, has greatly enhanced the ability to visualize, and therefore help understand, complex structures. We have applied a variety of filtered imaging techniques to representative structures typical of semiconductor devices. With plasmon imaging, we will examine the shape of SiO_2 precipitates in Si. Single element mapping will be discussed in the context of detrimental layers at the polysilicon/silicon interface. Multiple element mapping with overlays will be applied to the Ti/TiN barrier structure of Al metallization to demonstrate the ability to highlight reacted layers of the order of 2 nm thick which might otherwise be overlooked.

1. INTRODUCTION

Semiconductor devices are continuing to decrease in all physical dimensions. With the very small structures associated with such devices, the defects which cause problems have also continued to decrease in size. Any techniques which help to visualize these defects greatly aid the understanding of the role these defects play in device failure. As the dimensions associated with devices has decreased, the demand for TEM analysis has substantially increased due to the ready ability of the TEM to image structures at sub-nanometer resolution[e.g., Cunningham et al , 1995].

When defects occur, however, imaging them is not always sufficient to identify and understand the defects. High spatial resolution analysis has increasingly been the feature of most interest when TEM analysis is required. While high spatial resolution analysis is becoming commonplace with the wide availability of Field Emission TEM, quite often the information can be most effectively conveyed by an image rather than through a ream of spectra. In the last few years, the ability to image chemical information at high spatial resolution has come to the fore through the availability of energy filtered imaging. This technique has been commercialized both by in-column filtering and post-column filtering[Krivanek, 1992]. We have applied a post-column filter, the Gatan Imaging Filter to several examples of semiconductor structures where chemical imaging enhances the visualization of these structures, and in some cases, reveals information which was not obvious before application of the chemical imaging.

2. EXPERIMENTAL

Analyses were performed with a Gatan Imaging Filter interfaced to a Hitachi HF-2000 TEM operating at 200 kV with a cold field emitter, and Digital Micrograph software on Macintosh computers. Images 512 x 512 pixels were obtained by 2x binning of 1024 x 1024 pixel slow scan camera, using 30 eV energy windows (for pre- and post-edge images), at TEM magnifications of ~2000 - 5000X (~40 kX - 100 kX at the slow scan camera). Typical exposures ranged 2-60 seconds, depending on edge intensity, specimen thickness and beam current. Core-loss maps computed from two pre-edge windows and one post-edge window using AE^{-r} background extrapolation and subtraction. Image registration was corrected using manual or automated cross-correlation. All specimens imaged were prepared by mechanical polishing using the tripod wedge technique[Benedict et.al., 1990]. Some specimens were additionally thinned for 2-10 minutes by Ar ion milling.

3. APPLICATIONS

3.1 Oxygen Precipitates in Silicon

Oxygen, left in solution in the silicon in a highly supersaturated condition, is a major impurity remaining in silicon grown by the Czochralski technique. As soon as there is sufficient oxygen mobility, the oxygen begins to aggregate with several possible detrimental effects, including the formation of oxygen precipitates which can be sources for interstitial silicon atoms or dislocations. The thermal treatment given to the silicon can affect the size and shape of the oxygen precipitates.

The determination of size and shape is usually done by high resolution electron microscopy. In Fig. 1a, we have imaged an oxygen precipitate at zero energy loss in lattice imaging mode. Note that the amorphous contrast can be seen in the middle of the defect, and that lattice defects have originated at the precipitate. Although an amorphous region can been seen, the strong lattice contrast makes it difficult to determine if that is the complete extent of the oxygen precipitate. In Fig. 1b and c, we have imaged the same defect using a 6 eV window in the energy filter at 17 eV and 27 eV loss, respectively. These values represent the plasmon energies for Si and SiO_2. In both cases, but particularly for the 27 eV image, the shape of the oxygen precipitate is more clearly seen than in the lattice image alone. With the visual clues obtained from the plasmon images, it can be seen that there is contrast in the lattice image outside the amorphous region which is suggestive of a greater extent of amorphous material. Even so, it is not as clear as the delineation obtain by a filtered plasmon image.

3.2 Polysilicon/Silicon Interface

A more typical example of the routine application of energy filtering is illustrated by the polysilicon/silicon interface. This interface is pervasive in semiconductor devices. Typically, the Si surface will be covered with a deposited insulator such as SiO_2. Openings through the insulator will be made and then filled with a conductor. Quite often this conductor will be polysilicon. If the surface of the opened Si substrate is very clean, the deposited polysilicon may grow epitaxially on the substrate silicon, leading to device problems. It is usual, therefore, to have a thin native oxide layer developed before continuing with the polysilicon deposition. If the process is not carefully controlled, however, the oxide that develops can itself cause an increase in contact resistance and a degradation in device performance.

In Fig. 2a, we show a contact which was reported to have an increase in resistance by a factor of 2-3. Imaging clearly showed an amorphous layer at the polysilicon/silicon interface, with a thickness of 1.8 nm. EDS analysis showed the presence of oxygen at the interface, but there was also

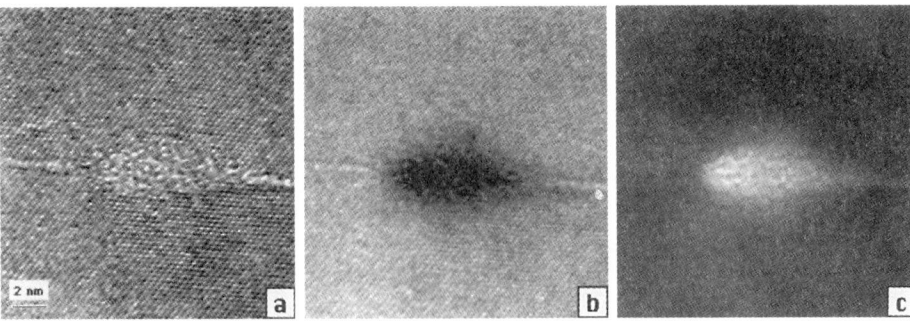

Figure 1. Cross-section TEM images of an oxygen precipitate. Images obtained through the Gatan Imaging Filter at a) zero eV loss, b) 17 eV loss and c) 27 eV loss.

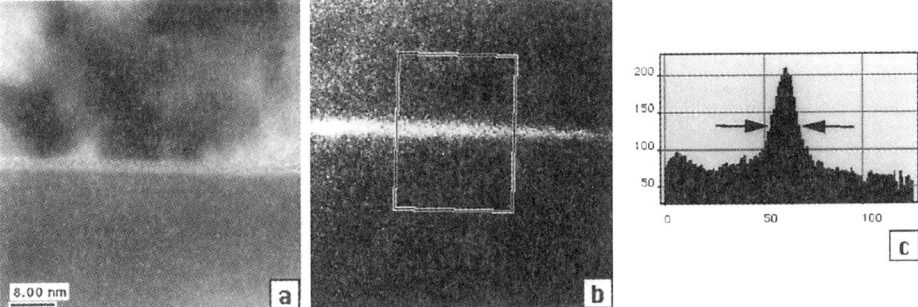

Figure 2. TEM images of a polysilicon/silicon interface. a) Image at zero, conventional contrast, showing an amorphous layer at the interface. b) Elemental map for oxygen. c) Histogram of pixel intensities for area highlighted in (b).

oxygen detected in the adjoining poly and silicon substrate due to specimen exposure to air after specimen preparation. To more clearly demonstrate the presence of oxygen at the interface, an elemental map for oxygen was obtained using the GIF. As can be seen in Fig. 2b, the oxygen layer at the interface is clearly detected. A line profile of the interface, integrated over 100 pixels, is shown in Fig. 2c. From the line profile, an estimate of the width of the layer was obtained from the half-maximum of the peak. The thickness of 1.5 nm is in excellent agreement with the thickness measured by lattice imaging.

3.3 Aluminum Metallization Barrier Layer

Aluminum is the most frequently used metallization in silicon devices. Reaction between the two metals, however, has led to the development of a variety of metal structures which act as a barrier to reaction[Nakamura, 1996]. In Fig. 3, we show an example of an aluminum metallization barrier structure over SiO_2. In this image, the specimen has been thinned to the point where all of the Al has been ion milled away. The remaining structure is a Ti/TiN barrier and a thin reacted $TiAl_3$ layer. Except for the $TiAl_3$ intermetallic, the structure appears to be unreacted. In Fig. 3b-d, we show the elemental maps for Ti, N and O. The Ti map clearly shows the two layer structure of the barrier, as well as mapping the intermetallic. The N clearly highlights the TiN layer, but also shows that there is some migration of the N to the SiO_2 interface. This would not have been expected without the N map clearly highlighting this interface. The O map shows some indication of O being present at the Ti and $TiAl_3$ layers. Since the specimen is quite thin, surface oxygen can present a significant signal for elemental detection. Note that the O is not seen for the TiN layer, lending credence to the argument that it is surface O.

In Fig. 3e, we have constructed an RGB (Red-Green-Blue) overlay of the three elements. Two features stand out which were not clearly evident in the individual maps. At the SiO_2/Ti interface, a bright band has been generated which is yellow in the RGB construct. This would indicate the presence of Ti (Red) and O (Green). When the bright field image is examined more closely, it can be seen that there is a distinct layer at this interface, most likely TiO_2. The other notable feature is a dark band at the $TiN/TiAl_3$ interface, which appears as blue in the RGB construct. This indicates the presence of N without Ti also being present. The most likely alternate species to have reacted with the N is Al. EDS analysis confirms the presence of AlN at this layer. This result has reported elsewhere [Inoue, 1994] and has been associated with increased resistance for an Al-Al contact with TiN barrier layer. Neither the TiO_2 nor the AlN layers are immediately obvious in the bright field image, or the individual elemental maps. The composite construct, however, makes these two layers very clear.

518

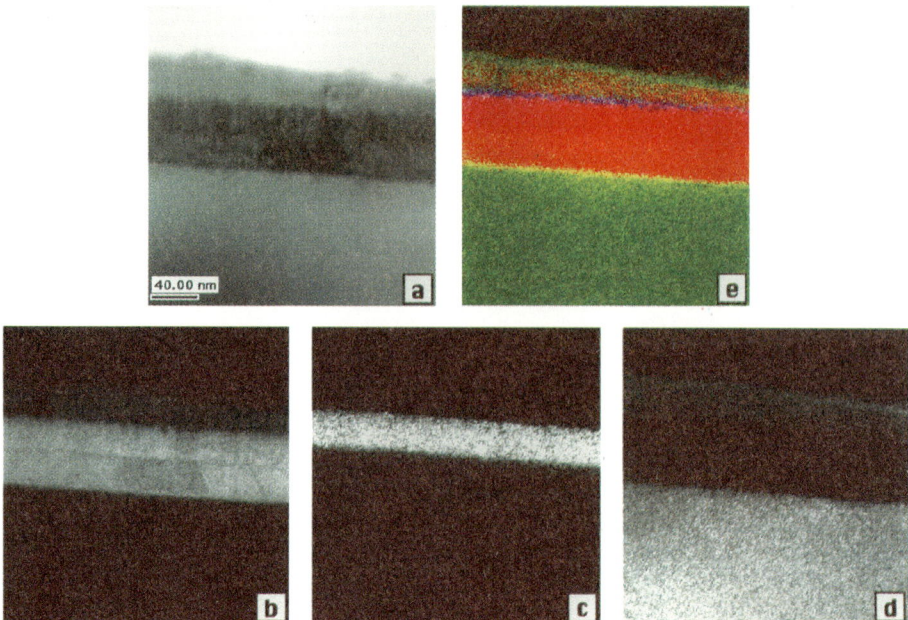

Figure 3. Barrier metal structure. a) TEM image of barrier layer over oxide. b-d) Elemental maps for b) Ti, c) N and d) O. e) RGB construct from b-d.

CONCLUSIONS

We have applied a Gatan Imaging Filter to a variety of structures typical for semiconductors. The ability to utilize chemical characteristics to image these structures has improved our ability to understand their structure and the reactions associated with those structures. The Imaging Filter has also greatly improved our ability to convey that information. What formerly required multitudes of spectra and lengthy explanations, can now be illustrated quite often with two or three images.

ACKNOWLEDGEMENTS

The authors wish to acknowledge the DRAM Development Alliance (IBM, Siemens, and Toshiba) for permission to use some of their structures in this paper.

REFERENCES

Benedict J P, Anderson R, Klepeis S J and Chaker M 1990 Proc. Mat. Res. Soc. **199** 189
Cunningham B, Joseph T W, Gignac L and Domenicucci A 1995 Inst. Phys. Conf. Ser. **146** 565
Inoue Y, Tanimoto S, Tsujimura K, Yamashita T, and Yoneda K 1994 J. Electochem. Soc. **141** 1056
Krivanek O L, Gubbens A J, Dellby N, and Meyer C E 1992 Microsc. Microanal. Microstruct. **3** 187
Nakamura H and Fushimi K 1996 Electr. Comm. Jpn, Part II: Electr. **79** 101

Inst. Phys. Conf. Ser. No 157
Paper presented at Microsc. Semicond. Mater. Conf., Oxford, 7–10 April 1997
© *1997 IOP Publishing Ltd*

Tungsten and tungsten nitride Schottky contacts to 4H-SiC

B Pécz, A Sulyok, G Radnóczi, O Noblanc*, C Arnodo*, S Cassette* and C Brylinski*

Research Institute for Technical Physics of the Hungarian Academy of Sciences, H-1325, Budapest, PO Box 76. Hungary
* THOMSON-CSF/LCR, Domaine de Corbeville 91404 Orsay Cedex France

ABSTRACT: Tungsten and tungsten nitride layers were deposited onto 4H-SiC by magnetron sputtering to obtain Schottky contacts stable at high temperature. The samples were annealed at 1200°C for 10 minutes. The solid phase reactions taking place during annealing were studied by cross-sectional TEM. Despite the formation of W_5Si_3 and W_2C phases, the contacts retained Schottky character after annealing at 1200°C.

1. INTRODUCTION

SiC is the most promising material of the high temperature devices. 4H polytype of SiC having a wide energy bandgap of 3.26 eV is the best candidate for high power, high temperature applications among the available substrates. Tungsten contacts on 3C SiC proved to be unreacted up to 850°C (Baud et al 1995). However, high frequency applications of SiC will require even higher annealing temperatures to anneal out the defects caused by ion implantation process. This means that contacts must withstand even higher temperature. This paper deals with the properties of W and WN_x contacts to 4H SiC annealed at 1200°C.

2. EXPERIMENTAL

W and WN_x layers were deposited onto 4H-SiC (n-type, 10^{18} cm^{-3}) by magnetron sputtering at room temperature to obtain high temperature stable Schottky contacts. WN_x was prepared in a mixed gas of Ar and N (18 at%) with a total pressure of 6×10^{-3} mbar. The thickness of the as-deposited layers was about 300 nm. The samples were annealed at 1200°C for 10 minutes.

The as-deposited samples as well as the annealed ones were investigated by cross sectional TEM (Transmission Electron Microscopy) and by Auger depth profiling. Auger depth profiling was carried out by alternating the Auger measurement with ion sputtering. The sample was rotated during sputtering by ion beam bombardment of 1.5 keV energy at 8 degrees angle of incidence measured from the surface (Barna et al 1992). TEM samples have been prepared by ion milling (Barna et al 1997). Electrical measurements were also carried out on the samples.

3. W/4H-SiC

The as-deposited W layer on SiC is 360 nm thick. The layer is polycrystalline with a grain size of 100-150 nm and shows columnar structure (Fig. 1). The layer is uniform although the surface of the tungsten layer is not flat. (Note the defects in the SiC substrate in the surface region, which shows that commercial substrates are not perfect.)

Fig. 1 Cross section of the as-deposited W/SiC sample.

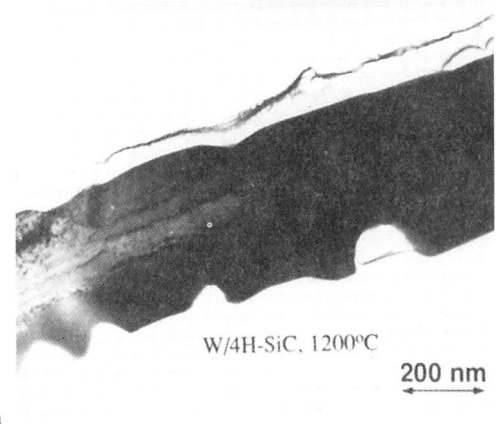

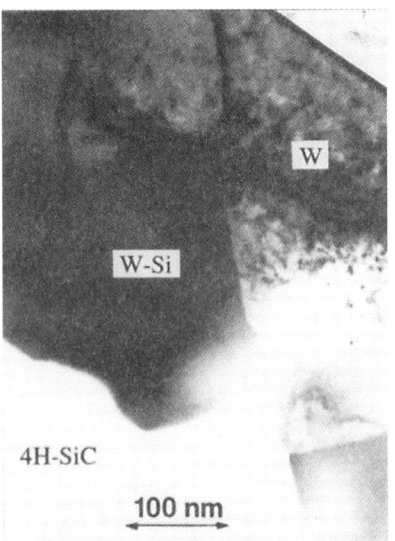

a   b

Fig. 2. a: Cross section of a W/SiC sample annealed at 1200°C. Large voids formed at the interface. b: tungsten silicide grain at a void.

Strong reaction took place in the sample during annealing at 1200°C and the columnar structure of the layer disappeared (Fig. 2). The formation of a tungsten-silicide has been observed in the annealed layer by EDS (Energy Dispersive System) microprobe analysis. In addition large voids were formed at the contact/SiC interface (Fig.2 a). Two phases can be distinguished in the layer: tungsten in the upper region and tungsten-silicide (W_5Si_3) grains at the interface with the grain size of 100-200 nm (Fig. 2b). The grains show quite different contrast in the TEM. (Tungsten shows always characteristic black dotted contrast due to artefacts (dislocations) of ion milling which are not present in tungsten silicide.)

4. WN$_x$/4H-SiC

The cross section of the as-deposited WN$_x$ layer (Fig. 3) shows an amorphous matrix with some small crystalline grains. The small crystallites have been identified as cubic W$_2$N grains by electron diffraction. Yamagishi et al (1987) identified the same phase in a WN$_x$ layer on GaAs at high temperature by X-ray diffraction. The AES (Auger Electron Spectroscopy) depth profile of the as-deposited sample shows a uniform distribution of nitrogen in the WN$_x$ layer.

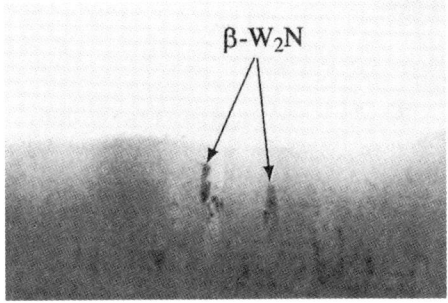

β-W$_2$N

Fig. 3. Bright field cross sectional image of the bottom part of a 300 nm thick as-deposited WN$_x$ layer on SiC. W$_2$N grains are shown by arrows in the amorphous matrix.

SiC   **100 nm**

The annealed contact layer consists of W$_5$Si$_3$ and W$_2$C phases (Fig. 4), which were identified by selected area electron diffraction. The same phases have been identified by Baud at al (1995) in W/3C-SiC contacts. Nitrogen evaporated from the sample during annealing as shown by the AES depth profile in Fig. 5. Small voids have been observed at grain boundaries within the contact layer as well as at the SiC surface. The size of the voids is smaller then in the case of W/SiC contacts.

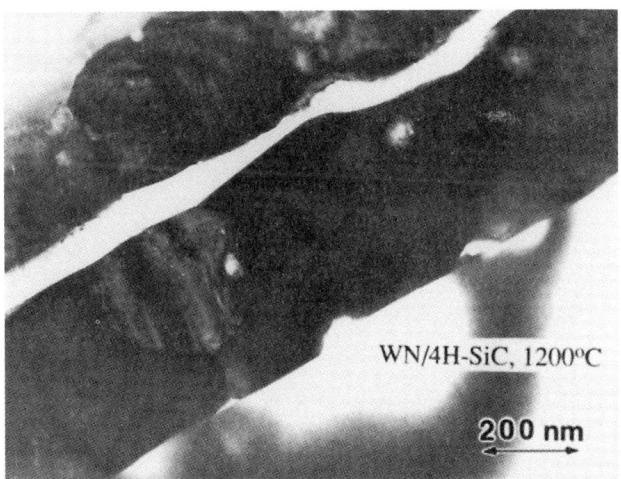

WN/4H-SiC, 1200°C

200 nm

Fig. 4. Cross section of WN$_x$/SiC sample annealed at 1200°C. (Two samples are shown embedded in face to face position.)

The AES depth profiles and the phase analysis carried out by selected area electron diffraction gave consistent results, i.e. the W$_2$C grains are closer to the surface.

Despite the above solid phase reactions, the contact remained Schottky at 1200°C according to the I-V measurements carried out on test circular diodes.

522

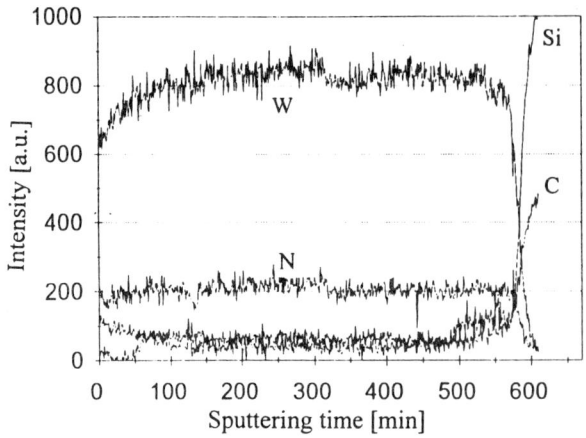

a

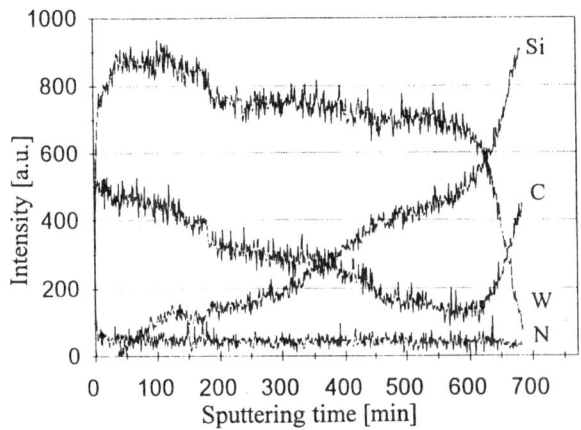

b

Fig. 5. AES depth profiles of the WN$_x$/SiC samples a: as-deposited, b: annealed at 1200°C.

5. CONCLUSION

WN$_x$ contacts to 4H-SiC gave better contacts - the formed voids are smaller -at high temperature than pure W layers, despite the nitrogen content evaporated from the sample during annealing. The contacts remained rectifying at 1200°C, which means that especially the WNx contacts are promising for high temperature applications.

ACKNOWLEDGEMENT
This work was supported by the COPERNICUS project No. CP940603.

REFERENCES
Barna Á, Sulyok A and Menyhárd M 1992 Surf. Interface Anal. **19**, 77
Barna Á, Radnóczi G and Pécz B 1997 Handbook of Microscopy ed. S. Amelinckx et al, (VCH Verlag, Weinheim, 1997) Vol. 3 pp 751-801
Baud L, Jaussaud C, Madar R, Bernard C, Chen JS and Nicolet MA 1995 Materials Science and Engineering **B29**, 126
Yamagishi H and Yamamoto Y 1987 Jap. J. Appl. Phys. **26**, 122

Inst. Phys. Conf. Ser. No 157
Paper presented at Microsc. Semicond. Mater. Conf., Oxford, 7–10 April 1997
© 1997 IOP Publishing Ltd

Distribution of Fe and extended defects in Fe-implanted InP

C Frigeri*, A Carnera^, B Fraboni^, A Gasparotto^, F Priolo°,
A Camporese+ and G Rossetto+

*CNR-MASPEC Institute, via Chiavari 18/A, 43100 Parma (Italy)
^INFM and University of Padova, Physics Department, via Marzolo 8, 35131 Padova (Italy)
°INFM and University of Catania, Physics Department, C.so Italia 57, 95129 Catania (Italy)
+CNR-ICTIMA Institute, C.so Stati Uniti 4, 35127 Padova (Italy)

ABSTRACT: The anomalous diffusion of Fe implanted into InP has been studied by SIMS and TEM. From the close correspondence, observed in the annealed samples, between the positions of the two Fe accumulation peaks around $R_p + \Delta R_p$ and the positions of the end of range loops and of the bottom interface of the twin band, respectively, it is suggested that the Fe distribution can be affected by gettering at the mentioned crystal defects. The origin of the Fe peak at $0.8\,R_p$ is also discussed.

1. INTRODUCTION

The anomalous diffusion of the implanted dopant is typical of Si (Sadana et al 1982, Eaglesham et al 1994, Claverie et al 1996) as well as of III-V compounds (Gauneau et al 1982, Ullrich et al 1991, Vellanki et al 1993). In particular, for InP implanted with transition elements with doses above the amorphization threshold the dopant distribution after annealing is characterized by the accumulation of the dopant at some specific positions, namely $0.8R_p$ and $R_p + \Delta R_p$, R_p and ΔR_p being the projected range and straggle, respectively, as shown by SIMS (secondary ion mass spectroscopy) (Gauneau et al 1982, Ullrich et al 1991, Vellanki et al 1993). When Fe was implanted the commonly accepted interpretation of the peak at $R_p + \Delta R_p$ was that it was due to the interaction between the Fe and the excess self-interstitials, typically P (Gauneau et al 1982, Ullrich et al 1991, Vellanki et al 1993). Here we present a combined study by TEM and SIMS of Fe-implanted InP whose results would suggest that the Fe distribution in the annealed material is mostly affected by its gettering at the extended crystal defects left after annealing.

2. EXPERIMENTAL

Fe-implantation into (100) InP substrates was carried out at room temperature with the energy of 2 MeV and fluence of $2 \cdot 10^{14}$ cm^{-2}. The InP wafers were either undoped (n < $1 \cdot 10^{16}$ cm^{-3}) or Sn-doped to $1.5 \cdot 10^{18}$ cm^{-3}. Annealings were done at 650 °C and 750 °C for 1.5 h under a P overpressure. The Fe distribution was measured by SIMS by monitoring ^{56}Fe$^+$ while sputtering with an 8 keV O$_2^+$ primary ion beam. TEM specimens were prepared by bombardment with Ar ions at liquid nitrogen temperature. TEM observations were performed in the bright field, weak beam dark field and high resolution (HREM) modes.

3. RESULTS

The distribution of Fe in the as-implanted and annealed samples is given in the SIMS spectra of Fig. 1 a) and b). The changes in the Fe distribution are more clearly visible after annealing at 750 °C and are greater in the undoped substrate than in the doped substrate. From the SIMS spectra R_p = 1260 nm and ΔR_p = 350 nm, so that 0.8 R_p = 1010 nm and $R_p + \Delta R_p$ = 1610 nm. Fig. 1a) shows that for the undoped substrate annealed at 750 °C the SIMS spectrum exhibits one accumulation peak for Fe at ~0.8 R_p and two peaks around ~$R_p + \Delta R_p$ (Table I). For annealing at 650 °C only the peak at the lowest depth around $R_p + \Delta R_p$ is present, the other ones are missing. For the doped substrate only one peak at ~$R_p + \Delta R_p$ is well pronounced (Fig. 1b). It exhibits a small shoulder towards the substrate for the annealing at 750 °C.

For both undoped and n-doped substrates, the as-implanted sample turned out to be fully amorphized down to a depth of ~1650 nm, as shown by HREM and selected area diffraction. The defect structure after annealing is the same for both type of substrates and consists of a band of twins followed by a band of end of range (Jones et al 1988) dislocation loops (Fig. 2a-c). The band of twins has replaced the amorphous layer and extends down to a depth of 1500-1440 nm, depending on the annealing temperature (Table I). A top surface layer, of thickness ranging between ~90 and ~150 nm, was instead free of twins indicating that the solid phase epitaxial growth had also started from the top surface as well as from the a/c interface. At the bottom and top regions of the twin band the twins have partially evolved into a tangled network of dislocations.

Beyond the twin band, a band of end of range (EOR) dislocation loops was detected in the region corresponding to the a/c transition region in the unannealed sample (Fig. 2a and b). The majority of the EOR loops were faulted and of the Frank type with Burgers vector **b** = 1/3<111>, as determined by the **g·b** = 0 analysis at the <011> zone axis and by tilting the cross sectional specimen to the <100> zone axis. The rest of the EOR loops were of the perfect prismatic type with **b** = 1/2<110> but still lying on the {111} planes. All loops were extrinsic. The maximum loop size increased from ~50 to ~70 nm by increasing the annealing temperature. Determination of the loop density in cross sectional samples was uncertain.

Table I compares the depths, measured from the top surface, of the the EOR bands and of the bottom edges of the twin band (error ±30 nm) with the positions of the Fe peaks in the SIMS spectra. It is seen that there is nearly an exact correspondence between the first Fe peak around $R_p + \Delta R_p$ (at 1500 or 1455 nm) and the bottom edge of the twin band as well as between the other Fe peak at 1700 (or 1580) nm and the position of the EOR band.

TABLE I - *Positions of the bottom edges of the twin band, of the top and bottom edges of the band of the EOR loops and of the SIMS Fe peaks around $R_p + \Delta R_p$ for the investigated samples.*

Substrate type	Annealing Temperature (°C)	Bottom edge of twin band (nm)	Range of band of EOR loops (nm)	SIMS peaks (nm)
Undoped	650	1500	1600-1800	1500
Undoped	750	1460	1600-1800	a) 1500 b) 1700
n+ Sn-doped	650	1460	1600-1750	1455
n+ Sn-doped	750	1440	1600-1750	1580

4. DISCUSSION

The presence of the twins in the annealed samples indicates that the solid phase epitaxial growth from the amorphous state was not complete. This point has been discussed elsewhere (Frigeri et al 1997). The EOR dislocation loops are due to the condensation of the recoil self-interstitials that have their maximum density at, and fairly beyond, the a/c interface (Jones et al 1988). We restrict our discussion here to the anomalous diffusion of Fe, as it is detected by SIMS. In this respect the existence of the twin band can be useful to understand the mechanisms leading to the formation of the implanted Fe accumulation peaks.

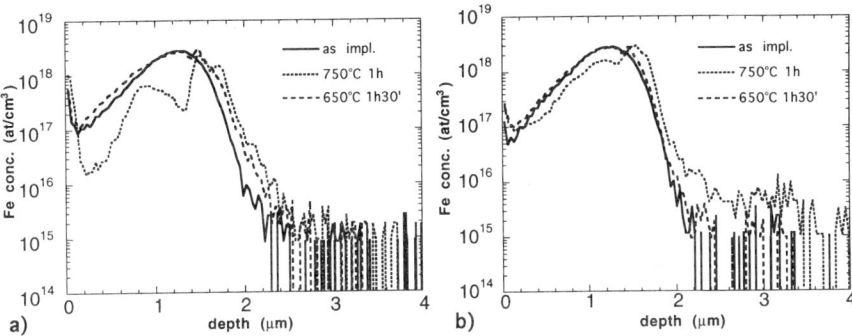

Fig. 1 SIMS spectra of the Fe distribution in InP substrates implanted with Fe.
a) Undoped substrate, b) n+ Sn-doped substrate.

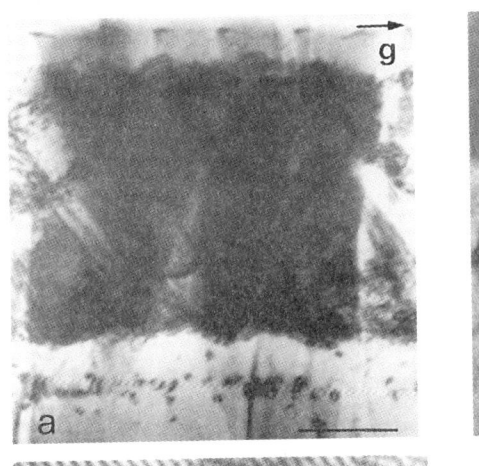

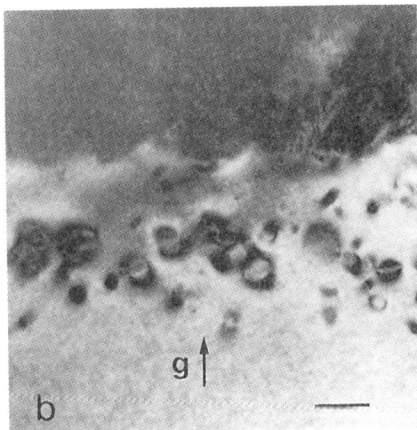

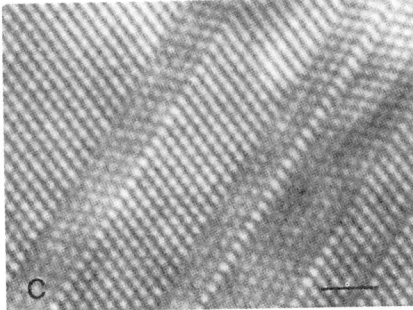

Fig. 2. Cross-sectional TEM images of the annealed undoped InP samples. a) Overview of the whole implanted layer, $\mathbf{g} = [220]$, bar = 0.5 μm; b) region of the EOR loops, $\mathbf{g} = [400]$, bar = 0.1 μm; c) HREM image of twins, bar = 2 nm.

The multiple peak structure of the SIMS spectra is typical of annealed InP substrates implanted with Fe, as well as other dopants such as Co and Ti, above the amorphization threshold (Gauneau et al 1982, Ullrich et al 1991, Vellanki et al 1993). The dopant peaks are always located at the depths of $0.8R_p$ and around $R_p + \Delta R_p$ (Gauneau et al 1982, Vellanki et al 1993). A peak at $2R_p$ was also reported (Ullrich et al 1991). The peak at $0.8R_p$ was ascribed to the enhanced gettering of the Fe by the implant damage that has its maximum at such depth. Although in the published SIMS spectra a double peak appears around $R_p + \Delta R_p$, as also shown in our spectra, this double peak was always considered as a single one and as such it was ascribed to the interaction of Fe with the excess P self-interstitials, which have their maximum concentration at this depth, according to Christel and Gibbons (1981), thus forming Fe-P complexes (Gauneau et al 1982, Vellanki et al 1993).

Our combined SIMS/TEM analysis seems to suggest another interpretation. By looking especially at the undoped substrate sample (Table I) it is apparent that the two SIMS peaks around $R_p + \Delta R_p$ are located at depths corresponding to the bottom edge of the twin band and to the band of the EOR loops, respectively. The accumulation of Fe at such positions is therefore believed to be due to the trapping of Fe at the twin band/perfect crystal interface and at the EOR loops, respectively. Trapping of Fe can be due to its gettering by the partial dislocations at the EOR loops and by the tangled dislocations at the bottom edge of the twin band, respectively, and possibly also by the strain fields associated with such dislocations. The dopant gettering by the EOR dislocation loops was previously suggested by Claverie et al (1996) for B implanted Si preamorphised with Ge and containing a band of EOR loops, to explain the anomalous diffusion of the dopant. If the two SIMS peaks at $R_p + \Delta R_p$ had been due solely to the interaction of Fe with the recoil interstitials, only one accumulation peak for Fe would have formed instead of two. On the other hand, as the recoil self-interstitials are concentrated in the region of the EOR defects, it is not likely that the accumulation peaks are only due to the interaction with the recoil interstitials themselves, since this would not explain the peak at the bottom edge of the twin band. Our results show that after annealing at 650 °C, Fe is first accumulated at the bottom edge of the twin band. The second peak at the greater depth of the EOR loops band forms at higher temperature (750 °C), in agreement with the expected higher amount of migrated Fe and longer diffused distance.

In our SIMS spectra, the peak at $0.8\,R_p$ does not really correspond to an increase of Fe concentration to levels higher than that of the as-implanted sample (Fig. 1a-b). It should not be due, therefore, to an enhanced accumulation of Fe caused by Fe diffusion to this position. This peak seems rather to appear because of the depletion of Fe at greater depths, i.e. roughly between $0.8\,R_p$ and $R_p + \Delta R_p$, associated with the gettering of the Fe atoms at the bottom edge of the twin band and at the EOR loops. SIMS results, not shown here, showed that the height of the peak at $0.8\,R_p$ further decreases by annealing at 800 °C. It cannot, however, be fully ruled out that the $0.8\,R_p$ peak can also be due, to a certain extent, to the expected large amount of implant damage that can trap Fe at such depth and prevent its outdiffusion towards the substrate and top surface.

In the n^+ Sn-doped substrate, the redistribution of Fe after annealing is less pronounced than for the undoped substrate. The presence of the shoulder in the peak at $R_p + \Delta R_p$ in the 750 °C annealed sample (Fig. 1b) would suggest that the same mechanisms discussed for the undoped substrate also hold for the doped substrate, for which, however, the amount of migrated Fe is smaller, as seen from the SIMS spectrum. This should be due to a smaller diffusivity of Fe in the n^+ Sn-doped substrate. Sn occupies In sites as Fe does. The n^+ Sn-doped substrate should contain a smaller density of V_{In}. Substitutional impurity diffusion by the interstitial-substitutional model is less effective when the vacancy density is high, whereas the opposite is true for the simple vacancy exchange mechanism (Gösele and Tan 1991). The latter mechanism might therefore be the one operating in our samples.

REFERENCES

Christel L A and Gibbons J F 1981 J. Appl. Phys. **52**, 5050
Claverie A, Bonafos C, Martinez A and Alquier D 1996 Solid State Phenomena **47-48**, 195
Eaglesham D J, Stolk P A, Gossmann H-J and Poate J M 1994 Appl. Phys. Lett. **65**, 2305
Frigeri C, Carnera A, Fraboni B, Gasparotto A, Cassa A, Priolo F, Camporese A and Rossetto G 1997 Mater. Sci. Eng. B, in press
Gauneau M, L'Haridon H, Rupert A and Salvi M 1982 J. Appl. Phys. **53**, 6823
Gösele U M and Tan T Y 1991 in Materials Science and Technology, ed R W Cahn, P Haasen and E J Kramer (New York: VCH) vol. 4, p. 197
Jones K S, Prussin S and Weber E R 1988 Appl. Phys. **A45**, 1
Sadana D K, Washburn J and Booker G R 1982 Phil. Mag. **B46**, 611
Ullrich H, Knecht A, Bimberg D, Krautle H and Schlaak W 1991 J. Appl. Phys. **70**, 2604
Vellanki J, Nadella R K, Rao M P, Holland O W, Simmons D S and Chi P H 1993 J. Appl. Phys. **73**, 1126

Inst. Phys. Conf. Ser. No 157
Paper presented at Microsc. Semicond. Mater. Conf., Oxford, 7–10 April 1997
© *1997 IOP Publishing Ltd*

Influence of doping on the native acceptors of gallium antimonide

P Hidalgo[1], B Méndez[1], J Piqueras[1], P S Dutta[2] and E Dieguez[2]

1) Dpt. Física de Materiales, Facultad de Físicas, Universidad Complutense, 28040 Madrid, Spain
2) Dpt. Física de Materiales, Facultad de Ciencias, Universidad Autónoma, Cantoblanco, 28049 Madrid, Spain

ABSTRACT : Cathodoluminescence (CL) in the scanning electron microscope (SEM) has been used to investigate the effect of doping with chromium and vanadium on the native acceptors and on the general structure of extended defects of gallium antimonide single crystals. Both impurities cause the reduction of the CL band at 775 meV (band A) attributed to a native acceptor. This effect is explained by the formation of a complex involving the dopant. CL images reveal the presence of precipitates and inhomogeneities in impurity distribution.

1. INTRODUCTION

Characterization of defects in GaSb is necessary for the development of practical applications of this material and its related ternary and quaternary compounds for micro and optoelectronic devices. In particular, these compounds have received attention for their applications to optical devices in the 0.3-1.6 eV range and for different tunnel structures. The interest of defect studies refers to undoped crystals and to crystals doped with different electrically active impurities. Undoped GaSb is p-type with a high concentration of intrinsic acceptors formed by a gallium vacancy (V_{Ga}) and a gallium antisite (Ga_{Sb}). The acceptor V_{Ga}-Ga_{Sb} is responsible of a luminescence band at 775 meV called band A (Jakowetz et al. 1972, Wu and Chen 1992) which enables the use of luminescence techniques to study the behaviour of acceptors under different treatments. CL in the SEM, which is known to provide information on the nature and spatial distribution of defects, has been to our knowledge only in the last years applied to characterize GaSb crystals, as reported in our previous works (Mendez et al. 1995, Panin et al. 1995, Dutta et al. 1996 a, b, Méndez et al. 1996). In this work CL has been used to study the effect of doping with Cr and V on the defect structure of GaSb single crystals. Doping with transition metals has been previously found to cause a reduction of the native acceptor related luminescence band (Georgitsé et al. 1991,1992, Méndez et al. 1996) which makes the CL a suitable technique to control the doping induced changes in the samples. Besides possible precipitation processes, the introduction of transition metal impurities in other well known III-V compounds usually generates deep levels involving the doping impurity and causes the increase in the electrical resistivity. Typical example of this situation is the chromium doping in GaAs which involves deep levels and can cause a semiinsulating behaviour. Similar effects in GaSb single crystals have not been previously investigated in detail and with the spatial resolution techniques used here.

2. EXPERIMENTAL METHOD

The samples used in this work were GaSb:Cr and GaSb:V single crystals grown by the vertical Bridgman method with a concentration of impurities in the starting melt of $10^{17} \, cm^{-3}$. The as-grown crystals had a length of about 3 cm and a diameter of approximately 1 cm. In a previous work on GaSb:Cr (Méndez et al. 1996) it has been observed that this growth procedure causes a segregation of Cr along the growth axis leading to different CL spectra at the Cr-poor and Cr-rich ends of the ingot. For this reason, samples obtained at different positions along the growth axis of the ingots were studied. In particular, four disks were cut perpendicular to the growth axis of each ingot. Disk labelled 1 corresponds to the end with higher Cr (or V) concentration, disks 2 and 3 at increasing distances from the end and disk 4 corresponds to the Cr (or V) poor end. For comparison CL emission of undoped samples was measured. The samples were prepared by conventional chemo-mechanical polishing, and rinsing in HCl for 3 minutes to remove any oxide layer.

The samples were observed in the emissive and CL modes in a Hitachi S-2500 scanning electron microscope at 77 K at accelerating voltages of 20-30 kV. For CL measurements an optical lens in the SEM chamber was used to concentrate the light to a window of the microscope on which a light guide was attached to feed the light either directly or through a computer controlled monochromator to an ADC germanium infrared detector. In order to study the composition of some features observed in the secondary electron micrographs, mapping of the elements Ga, Sb, Cr and V were obtained by wavelength dispersive x-ray microanalysis in a Jeol JXA - 8900 M superprobe.

3. RESULTS AND DISCUSSION

CL images of the undoped GaSb crystal show a contrast due to subgrain boundaries (Fig.1). Spectra of undoped crystals contain the band A at 775 meV of the V_{Ga}-Ga_{Sb} acceptor in addition to the near band edge emission (Fig.2). CL images of the V-doped samples have a more complex appearance which depends somewhat on the area considered. Most of the crystal surface appears well polished without features in the secondary electron image. The corresponding CL images are as Fig. 3-a with dark features at points of the subgrain boundaries. Comparison among the different V-doped samples shows that the size and number of the dark features increase for the V-rich samples suggesting a segregation process at the boundaries.

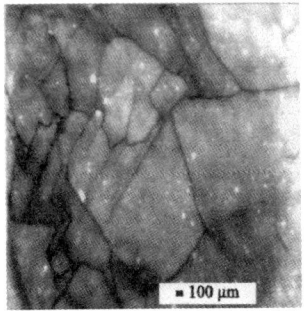

Fig. 1. CL image of the undoped GaSb crystal.

Other regions of the V-doped samples show topographic features (Fig.3c) associated with dark contrast in the CL images (Fig.3b). Wavelength dispersive analysis of those regions show that the protrusions are vanadium and, to a lesser extent, antimony rich which indicates that they correspond to precipitates with vanadium as main component. The effect of V is also observable in the CL spectra. Fig.4 shows the CL spectra of the V doped samples labelled 2, 3 and 4. It can be observed that for higher V content (sample 2) the emission in the region of band A is weaker. This is explained by the formation during growth of defects involving V instead of the above mentioned native acceptor. In particular, the acceptor is formed by the scheme:

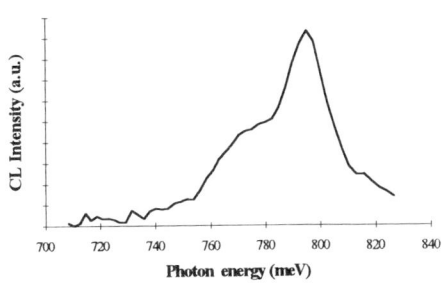

$Ga_{Ga} + V_{Sb} \leftrightarrow Ga_{Sb} + V_{Ga}$ but in the presence of Vanadium the complex (Vanadium)-V_{Sb} can be formed. Such situation has been observed in the case of GaSb:Mn by photoluminescence and electron spin resonance (Georgitsé et al. 1992).

Fig. 2. *CL spectrum of the undoped GaSb crystal.*

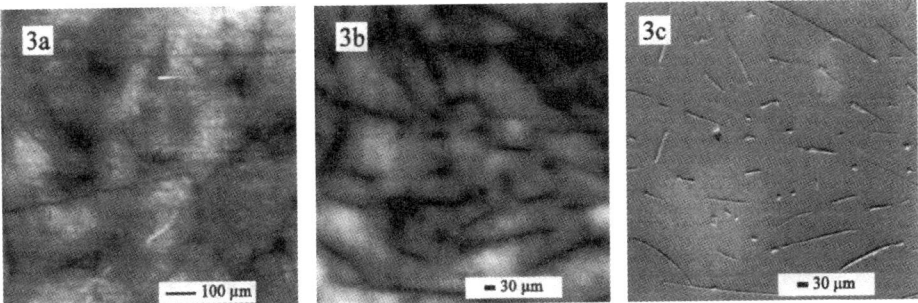

Fig. 3. *GaSb:V, a) Representative CL image of the samples , b) and c) CL and SE images of an area containing V-rich features.*

The CL images of the Cr doped samples show very clearly the differences among the samples obtained from different parts of the ingot (Fig. 5). In sample 3 (Fig. 5a) the CL image is similar to that of the undoped sample, while in sample 1 (Fig. 5b) a high number of precipitates are observed.

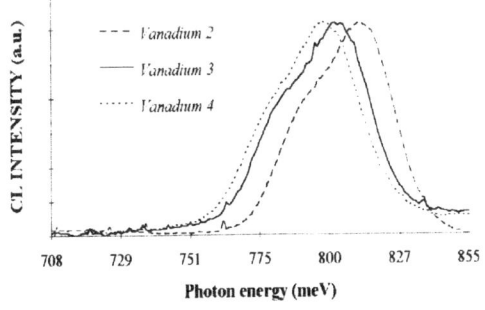

Fig. 4 . *CL Spectra of V doped samples from the different position in the ingot.*

Fig.6 shows the CL spectra of samples 1 and 4 of the Cr doped ingot. As in the case of V, Cr causes the reduction of the band A which by analogy with the case of Mn and V can be interpreted by the formation of the V_{Ga}-Cr complex.

530

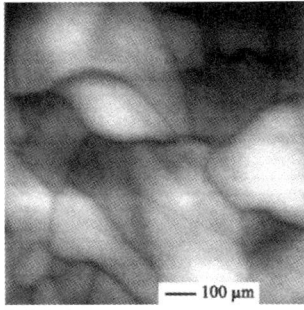

Fig. 5a. *CL image of GaSb:Cr. Sample with low Cr content.*

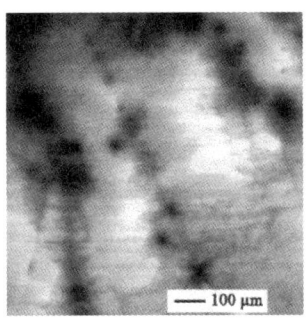

Fig. 5b. *CL image of GaSb:Cr. Sample with higher Cr content.*

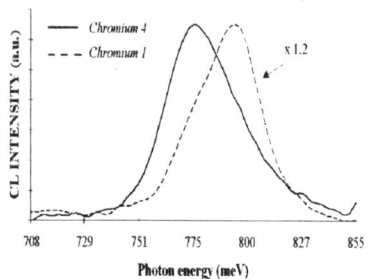

Fig.6. *CL spectra of GaSb:Cr samples with different Cr content.*

4. CONCLUSION

During growth of Cr or V doped GaSb by the vertical Bridgman technique a gradient of the impurity concentration along the growth axis appears which is detectable by CL microscopy. Both impurities have been found to cause a reduction of the native acceptors responsible for the luminescence band A. CL and X-ray microanalysis images reveal a distribution of precipitates mainly related to the sub-boundaries whose size and number depends on the position of the sample along the growth axis.

ACKNOWLEDGEMENTS

This work has been supported by DGICYT (Project PB 93-1256).

REFERENCES

Georgitsé E I, Gutsulyak L M, Ivanov-Omsky V Y, Smirnov V A and Yuldashev Sh U 1991, *Sov. Phys. Semicond.* **25** 1180

Georgitsé E I, Gutsulyak L M, Ivanov-Omsky V Y, Smirnov V A and Yuldashev Sh U 1992 , *Sov. Phys. Semicond.* **26** 50

Jakowetz W et al. 1972 , *Phys. Stat. Sol. (a)* **12** 169

Méndez B, Piqueras J, Dutta P S and Dieguez E 1995 *Appl. Phys. Lett.* **67** 2648

Méndez B, Dutta P S, Piqueras J and Dieguez E 1996 *Mater. Sci. Eng.* **B42** 38

Meng-Chyi Wu and Chi-Ching Chen 1992 *J. Appl. Phys.* **72** 4275

Panin G N, Dutta P S, Piqueras J, Dieguez E 1995 *Appl. Phys. Lett.* **67** 3584

Dutta P S, Marín C, Dieguez E and Bhat H L 1996 *Journal of Crystal Growth* **160** 207

Dutta P S, Prasad V and Bhat H L 1996 *J. Appl. Phys.* **80** 2847

Inst. Phys. Conf. Ser. No 157
Paper presented at Microsc. Semicond. Mater. Conf., Oxford, 7–10 April 1997
© 1997 IOP Publishing Ltd

Effect of high implantation temperatures on defect formation in 6H-SiC

A A Suvorova, O I Lebedev*, A V Suvorov** and I O Usov

Ioffe Physical-Technical Institute, St.-Petersburg 194021, Russia
*Institute of Crystallography, Moscow 117333, Russia
**Cree Research Inc., Durham, NC 27713, USA

ABSTRACT: This paper describes a TEM study of 6H-SiC after implantation with 40 keV aluminum ions within the range of implantation temperatures of 1100°C to 1800°C. At high implantation temperatures, TEM showed that aluminum precipitates associated with interstitial half-loops are present in a damaged layer. Defects referred to as end of range (EOR) damage form below the region containing the precipitates. With increasing temperature, the EOR defects evolve from loop nuclei exhibiting strain-contrast to elongated dislocation loops lying on the (0001) basal plane.

1. INTRODUCTION

In the present paper, the impact of increased implantation temperature upon the structure of 6H-SiC is studied and possible defect formation mechanisms which relate to the experimental results are discussed. The study is of interest due to the fact that the formation of p-type regions in SiC by aluminum implantation at high temperatures yields a reduction of lattice damage and an increased electrical activation of implanted aluminum, as reported by Edmond et al (1987), Suttrop et al (1992) and Suvorov et al (1995).

2. EXPERIMENTAL DETAILS

6H-SiC specimens were implanted within the range of temperatures of 1100°C to 1800°C with 40keV Al^+ ions to a high dose of $2 \cdot 10^{16}$ ions·cm^{-2} . These were examined using Philips EM420 and JEOL 4000EX microscopes operated at 100 and 400keV, respectively.

3. RESULTS AND DISCUSSIONS

3.1 Microstructure

After implantation, TEM revealed the presence of two distinct regions in a damaged layer in all implanted specimens: close to the surface a region contained aluminum precipitates and below there was a region without precipitates. Defects below the precipitates are referred to as end of range (EOR) damage. Figure 1 gives an example of the two regions of a damaged layer resulting from implantation at a temperature of 1500°C. Figure 2 shows the size distributions of Al precipitates and EOR defects for specimens implanted at different temperatures.

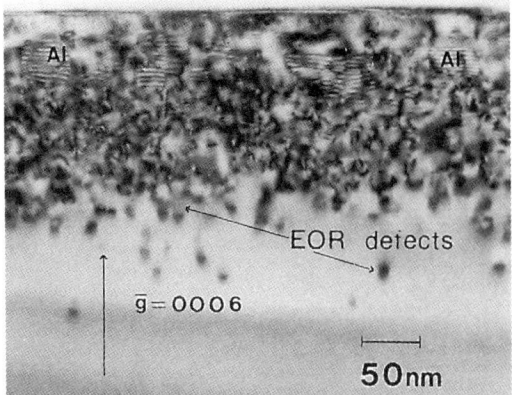

Fig. 1. TEM cross-sectional image of 6H-SiC implanted at 1500°C

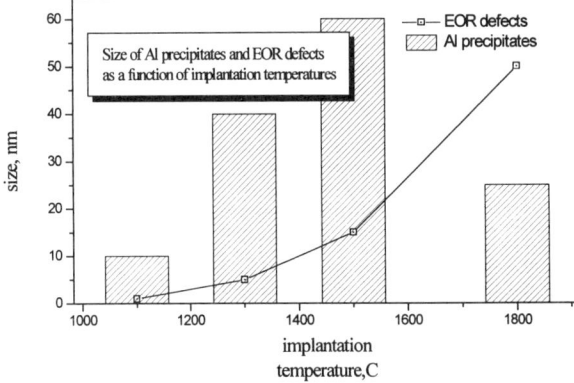

Fig. 2. Size of aluminum precipitates and EOR defects observed in 6H-SiC implanted with 40 keV aluminum ions as a function of the implantation temperature.

3.2 Al precipitate/ stacking fault configuration

Aluminum precipitates exhibiting moiré fringes were identified by their selected area diffraction patterns (SADP). The orientation relationship of aluminum precipitates with the matrix was derived from SADP and HREM analysis and was found to be $(111)_{Al}//(0001)_{6H-SiC}$ and $(110)_{Al}//(-2110)_{6H-SiC}$. Figure 3 shows a high-resolution image of an aluminum precipitate formed in the matrix. Detailed HREM studies of the precipitate/matrix interface have been reported in our work (Suvorov et al 1997) where it has been found that extended stacking faults originate at the precipitate/matrix interfaces and terminate in the matrix by the formation of partial dislocations. The structural model of the stacking fault was suggested to be ABCACB ACBCBCACB (Fig. 4). Similar faulted loops have been observed by Yano and Iseki (1990) in neutron-irradiated 6H-SiC. Such a configuration of Al precipitate associated with a stacking fault in implanted 6H-SiC was assumed to be an interstitial half-loop lying on the (0001) basal plane with Burgers vector $b=1/6[0001]$, generating at the precipitate/matrix interface. The evidence of this implication was obtained by tilting experiments resulting in the image of Fig. 5. Loops, similar to that seen in Fig. 5, surrounding stacking faults on basal (0001) planes were reported by Maeda et al (1988) in 6H-SiC plastically deformed at elevated temperatures. These workers suggested that the loops are formed by partial dislocation motion.

Fig. 3. HREM image of an
Al precipitate in the 6H-SiC
matrix viewed along [11-20]
direction

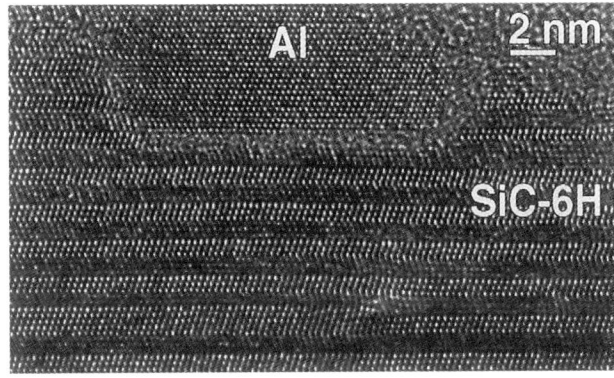

3.3 End of range defects

From Fig. 2, which shows the size distribution of EOR defects for specimens implanted at the different temperatures, it is evident that EOR defects grow in size with increasing implantation temperature. At temperatures of 1100°-1300°C, TEM revealed that the defects appeared as "black dots" exhibiting strain-contrast. These were assumed to be nuclei of small loops. Loops exhibiting marked black/white contrast were found to be generated at higher (1500°C) temperature. At the temperature of 1800°C, EOR defects were imaged as elongated stacking faults when viewing along [2-110] directions and were found to be interstitial dislocation loops lying on the basal (0001) plane as seen by tilting experiments (Fig. 6). Basal dislocations were also observed and were likely to have been formed due to the glide mechanism occurring in SiC at high temperatures (Maeda et al 1988, Pirouz et al 1991).

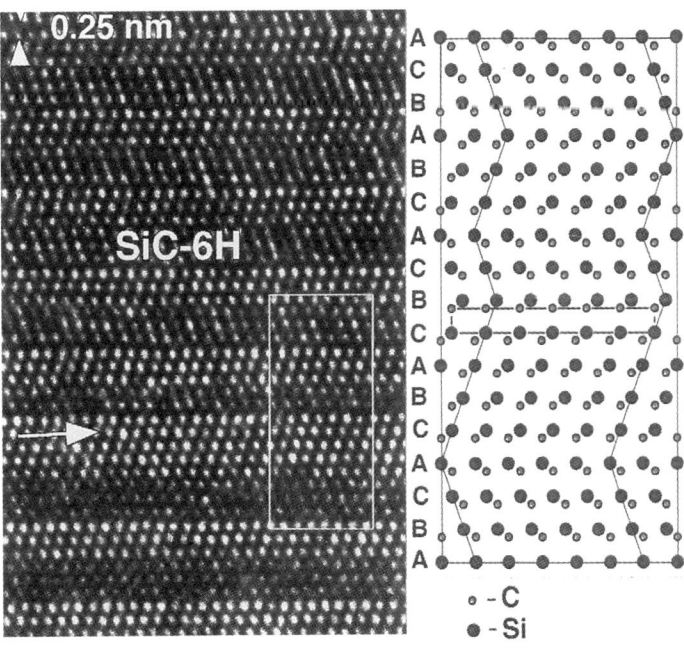

Fig. 4. HREM image of stacking fault sequence in a damaged layer and corresponding structural model along [11-20] direction

We propose that increased temperatures favor the agglomeration of interstitials induced by implantation into the basal dislocation loops. At high temperatures, thermally activated dislocation reactions seem to be responsible for most extended EOR defects seen after implantation.

534

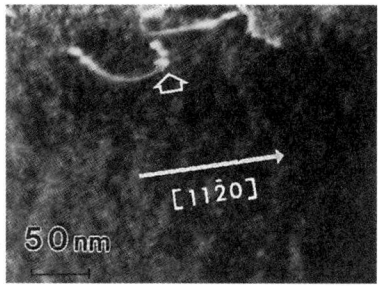

Fig. 5. Weak-beam (g,2g) image of a faulted loop taken by tilting about the [01-10] axis to the [4401] pole; g=11-20.

The edge of the loop associated with Al precipitate is marked with an arrow

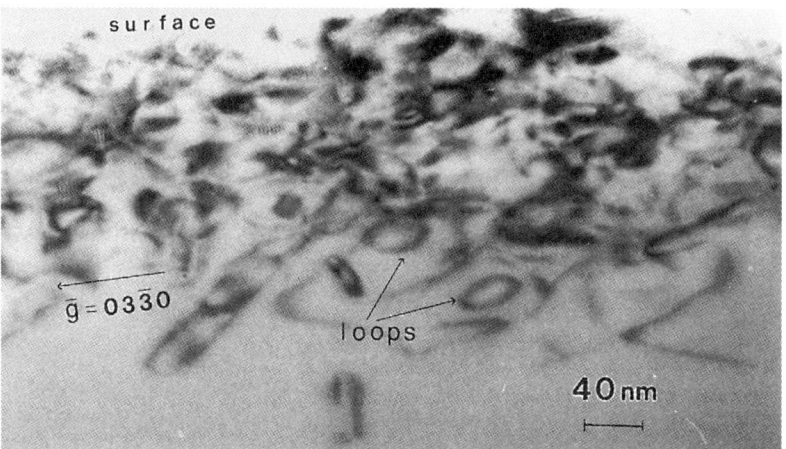

Fig. 6. Cross-sectional image of the sample implanted at 1800°C tilted to 30° from [11-20] axis

4. CONCLUSIONS

In summary, we have demonstrated that high temperature implantation of 40 keV Al+ ions into 6H-SiC results in the formation of aluminum precipitates associated with interstitial dislocation half-loops. End of range defects evolve from small basal loops to elongated ones of interstitial type.

ACKNOWLEDGEMENTS

AA Suvorova wishes to acknowledge the Royal Society for financial support.

REFERENCES

Edmond J A, Withrow S P, Wadlin W and Davis R F 1987 MRS Symp Proc. **77**, 193

Maeda K, Suzuki K, Fujita S, Ichihara H and Hyodo S 1988 Phil. Mag. A **57**, 573

Pirouz P, Yang J W, Powell J A and Ernst F 1991 Inst. Phys. Conf. Ser.**117**, 149

Suttrop W, Zhang H, Schadt M, Pensl G, Dohuke K and Leibenzeder S 1992 Springer Proceeding in Physics **71** (Springer-Verlag, Berlin) p 143.

Suvorov A V, Usov I O, Sokolov V V and Suvorova A A 1995 MRS Symp Proc **396**, 83

Suvorov A V, Lebedev O I, Suvorova A A, Van Landuyt J and Usov I O 1997 Nucl. Instrum. Meth. B **127/128** (to be published)

Yano T and Iseki T 1990 Phil. Mag. A **62**, 421

Inst. Phys. Conf. Ser. No 157
Paper presented at Microsc. Semicond. Mater. Conf., Oxford, 7–10 April 1997

© 1997 British Crown Copyright

X-ray topography of single crystal zinc germanium phosphide

M K Saker, A M Keir, A W Vere and L L Taylor

DERA Malvern, St. Andrews Road, Great Malvern, Worcestershire, WR14 3PS.

ABSTRACT: The ternary semiconductor $ZnGeP_2$ shows considerable potential for non-linear optical applications. In this paper we present the double crystal x-ray topography of a large single crystal $ZnGeP_2$ sample grown using the horizontal Bridgman technique. To our knowledge this is the first report of x-ray topography on this material system.

1. INTRODUCTION

The uniaxial birefringence and very high non-linear optical coefficient of 80pm/V of $ZnGeP_2$ make this material of great interest for the production of optical parametric oscillators (OPO's) (see Shay and Wernick 1975). Such a device can be made to cover the majority of the 3.9μm to 10μm range (Vodopyanov and Voevodin 1995).

$ZnGeP_2$ is one of many II-IV-V semiconductors which are a very close analogue of the III-V zincblende semiconductor compounds, see Fig.1. The II-IV-V crystal structure may be derived from the III-V zincblende semiconductor by substituting half the group III atoms by a group II atom, in this case the element zinc, and the other half by a group IV atom, in this case germanium. If this is done in an ordered manner the chalcopyrite structure is formed. Hence $ZnGeP_2$ has a double unit cell with tetragonal crystal class with a=b=5.467Å and c = 10.715Å.

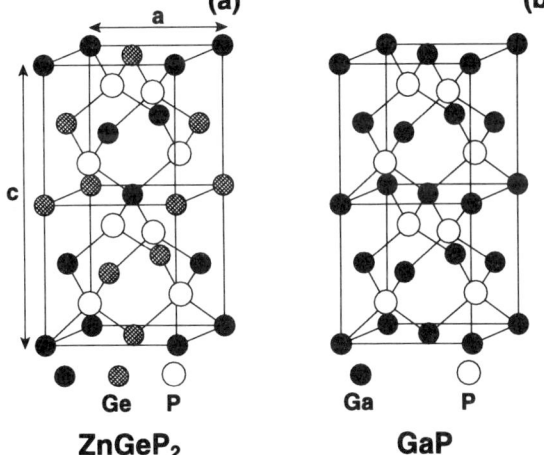

Fig 1. Crystal structure of
(a) $ZnGeP_2$ in which the lattice
constants are a=b=5.467Å and
c=10.715Å and (b) GaP the closest
analogue of $ZnGeP_2$.

There are two main problems in obtaining high quality crystals of $ZnGeP_2$ that are large enough for OPO devices. Firstly crystals of $ZnGeP_2$ grown by any method with the exception of vapour transport (see Xing et al 1990), exhibit broad near band edge absorption which extends to 2 to 3µm at room temperature. This absorption at these wavelengths is strong enough to make the samples sufficiently opaque that optical pumping at the required 1 to 3µm is not possible. Some success in reducing the near band edge absorption has been achieved by post growth annealing (Rud and Masagutova 1981) and by electron irradiation (Brudnyi et al 1978) who accounted for their results as due to movement of the Fermi level.

Two models for the origin of the absorption have been proposed. The first based on photoluminescence results by Dietz et al 1994 attributes the absorption to transitions between the broad donor - acceptor bands in the $ZnGeP_2$ band gap. Dietz et al 1994 explain the effects of annealing as being due to the reduction of Zn_{Ge} anti-site defects and the electron population of acceptor-like Ge_{Zn} defects. The second is based on EPR the results of Rakowsky et al 1994, photoinduced EPR results of Giles et al 1995 and the ENDOR results of Halliburton et al 1995. Absorption is attributed to V_{Zn} and V_P.

The second problem in obtaining high quality large single crystals is the high phosphorous vapour pressure during growth leading to cracking of the holder and hence failure of the growth run and, due to the anisotropic thermal expansion of $ZnGeP_2$, cracking of the ingot. Thus samples are often small crystallites, or larger crystals which are heavily cracked.

2. EXPERIMENTAL RESULTS AND DISCUSSION

The $ZnGeP_2$ OPO, Fig.2, was cut with the long axis parallel to the phase-matching direction, from the single-crystal region of an ingot produced by the horizontal Bridgman process. The equipment used was a Crystalox six-zone resistance-heated system.

Fig 2. The single crystal $ZnGeP_2$ grown by horizontal Bridgman technique. The sample has been aligned, cut and antireflection coated at each end for further assessment

In this process, the charge with 0.13% excess Zn is melted in a boron nitride coated quartz boat (22mm in diameter 20cm long) in a temperature gradient of $6°C$ cm^{-1}. A single crystal seed of (001) silicon, oriented along the growth direction, was used to nucleate growth. The material was moved from the hot zone at 1050°C to the cold zone at 950°C at a rate of 6mm/hr, the melting point of $ZnGeP_2$ is 1027°C, and then cooled to room temperature at an average rate of 4°C/hr.

After growth the OPO specimen was cut from the ingot and annealed at 600°C for 300 hours. During the annealing process 0.218g of Zn and 0.095g of P_2 were added to the ampoule with a volume of 9.88×10^{-5} m^3. Annealing the OPO reduced the absorption coefficient at 2.1µm from 3.3cm^{-1} to 0.6cm^{-1}. Before laser testing the OPO was antireflection coated using proprietary technology.

The broad sides of the OPO sample were polished and the sample mounted in the diffractometer such that the c-axis was in the diffraction plane and the (0 0 12) reflection used for the topography.

Double crystal x-ray topography was carried out on the OPO sample using a Bede D3 diffractometer and a Cu source rotating anode generator. The effective height of the source was 50μm and the total source-to-sample distance was 1m. The incident beam was conditioned using the (400) reflection from a (100) silicon crystal proven suitable for topography. An Ilford L4 nuclear emulsion plate, with an emulsion thickness of 25μm, placed 30mm from the sample was used to collect the image over a period of 40 hours. Fig. 3 shows the result of the topograph.

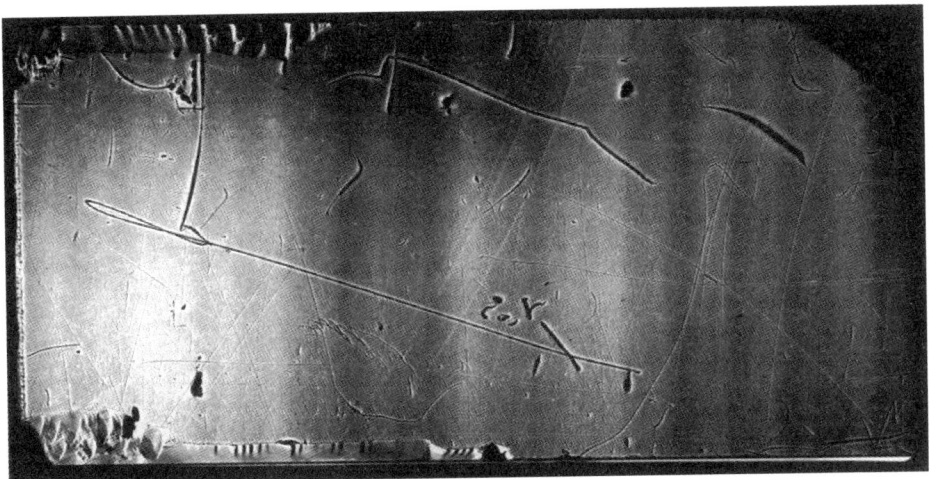

Fig 3. Double crystal x-ray topograph of the sample in Fig. 2 using the (0 0 12) sample reflection. The banding originates from the source used in the rotating anode generator (see text).

As can be seen from the topograph in Fig. 3 the image shows only surface scratches and the sample to be good quality single crystal with no sign of dislocations or twinning. The banding on the image originates from the non-uniformity along the 10mm wide x-ray source in the rotating anode generator at the time of the experiment. This has been confirmed by carrying out double crystal topography on a (111) silicon beam conditioning crystal using the (224) reflection. This too showed the same banding effect, and hence that these bands do not originate from defects (e.g., growth striae) in the OPO sample. Optical measurements are currently in progress on this OPO sample.

Hence we have shown how double crystal topography plays a vitally important role in the assessment of solid state optical device material. It has also been demonstrated by using this powerful technique that the $ZnGeP_2$ OPO sample, both grown and assessed at DRA Malvern, is of high crystallographic quality. As far as we are aware this is the first x-ray topography result reported for this material system.

© British Crown Copyright 1997/DERA

Published with permission of controller of Her Majesty's Stationary Office

REFERENCES

Brudnyi V N, Budnitskii D L, Krivov M A, Masagutova R V, Prochukhan V D and Rud Yu V 1978 phys. stat. sol. (a) **50** 379

Dietz N, Tsveybak I, Ruderman W, Wood G and Bachmann K J 1994 Appl. Phys. Lett. **65** 2759

Giles N C, Halliburton L E, Schunemann P G and Pollak T M 1995 Appl. Phys. Lett. **66** 1758

Halliburton L E, Edwards G J, Scripsick M P, Rakowsky M H, Schunemann P G and Pollak T M 1995 Appl. Phys. Lett. **66** 2670

Rakowsky M H, Kuhn W K, Lauderdale W J, Halliburton L E, Edwards G J, Scripsick M P, Schunemann P G, Pollak T M, Ohmer M C and Hopkins F K 1994 Appl. Phys. Lett. **64** 1615

Rud Yu V and Masagutova R V 1981 Sov. Tech. Phys. Lett. **7** 72

Shay J L and Wernick J H 1975 Ternary Chalcopyrite Semiconductors: Growth, Electronic Properties and Applications (Pergamon Press)

Vodopyanov K L and Voevodin V G 1995 Optics Comm. **117** 27

Xing G C, Bachmann K J, Posthill J B 1990 Appl. Phys. Lett. **56** 271

Inst. Phys. Conf. Ser. No 157
Paper presented at Microsc. Semicond. Mater. Conf., Oxford, 7–10 April 1997

© 1997 IOP Publishing Ltd

Mechanisms of breakdown in semi-insulating GaAs detectors under high reverse bias conditions studied by EBIC and OBIC

M Mazzer, A Cola, L Vasanelli
CNR-IME - Campus Universitario, via Arnesano, 73100 Lecce - Italy

M De Vittorio
INFM - Unità di Lecce - Campus Universitario, via Arnesano, 73100 Lecce - Italy

C Pennetta, L Reggiani
Dipartimento di Fisica e Dipartimento di Scienza dei Materiali Università di Lecce, Campus Universitario, via Arnesano - 73100 Lecce - Italy

ABSTRACT: Optical and electron beam induced current have been used to characterise a semi-insulating-GaAs Schottky diode for x-ray detection. The anomalies in the electrical response of the device under high reverse bias conditions are explained by showing that the non uniformity of the Schottky barrier is amplified by the applied voltage leading to the formation of low resistivity channels across the device which anticipate the electrical breakdown.

1. INTRODUCTION

Schottky barrier diodes based on semi-insulating (SI)-GaAs are currently used for fabricating high sensitivity x-ray detectors for medical applications (Cola et al. 1997). The detection of x-rays requires the active region of the device to be thick enough to allow the absorption of most of the impinging photons . For this purpose, a Schottky barrier is deposited on a 150 mm thick SI-GaAs wafer grown by liquid encapsulating Czochralski method. The back contact of the device is ohmic. Reverse bias voltages of the order of a few hundred volts are applied between the Schottky barrier and the ohmic contact in order to set up an electric field throughout most of the SI-GaAs. The main issue with these devices is the possibility of controlling the uniformity and the extension of the electric field to get the maximum detection efficiency, to keep the dark current at the lowest possible level and to prevent the early breakdown of the device (Castaldini et al. 1994). In the present paper, Optical Beam Induced Current (OBIC) and Electron Beam Induced Current (EBIC) were used to characterise the devices under their normal operating conditions. OBIC, in particular, is used to study the device in cross section configuration (the light beam is parallel to the metal to semiconductor interface) avoiding any charging effect of the sample which is unavoidable when an electron beam impinges on a semi-insulating material.

2. EXPERIMENTAL

The OBIC measurements were performed by using an home-made apparatus. The sample is placed on a high-resolution X-Y scanning stage with a precision 0.20 mm over a 25×25 mm² area. The central body of the apparatus consists of a large aperture objective (30×) focusing the laser beam to the diffraction limit. The 514.5 nm line of an Argon laser was used for these experiments, the spot size under optimised alignment conditions and beam expansion being of the order of

0.7 mm. The photocurrent generated by the optical beam is read-out by a femto-Amperometer. The system is driven by a computer, which controls both the X-Y motion of the sample and the acquisition two-dimensional chromatic maps in which higher currents are associated with lighter grey tones. The typical input power for the OBIC experiments is of the order of few tens of μW. The EBIC analyses were performed by using a JEOL JSM-840A Scanning Electron Microscope equipped with a Matelect ISM-5A unit.

3. RESULTS AND DISCUSSION

The ideal dark I-V characteristic of the SI-GaAs detector can be written (Luo et al. 1996):

$$V_{dark} = V_{bulk} + V_{barrier} = I_{dark}R_{bulk} - \frac{kT}{e} \log(1 - \frac{I_{dark}}{I_s}) \qquad (1)$$

where R_{bulk} is the resistance of the SI-GaAs and I_s is the inverse saturation current of the Schottky barrier.

Eq.1 shows that the voltage drop across the device is the sum of the voltage drop across the bulk GaAs plus the contribution of the Schottky barrier. At low applied voltages the barrier voltage is small and the ohmic term dominates as shown by the curves in Fig.1.

Fig. 1: I-V characteristic of the device. The solid line refers to the device before being cleaved, the dashed and the dotted line to the smaller and to the larger of the two pieces obtained after the cleavage.

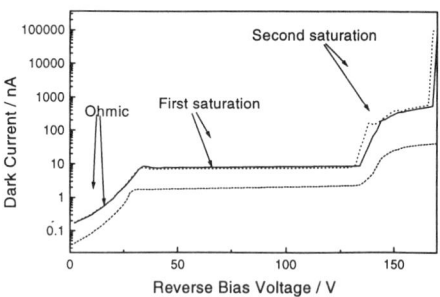

When the dark current approaches the saturation value, I_s, the logarithmic term becomes dominant and the current is almost independent of the applied voltage as expected for a reverse biased Schottky junction. In the case of an ideal Schottky barrier, the depletion region should propagate uniformly from the barrier towards the back ohmic contact with increasing applied voltage. The breakdown is expected to occur after the whole of the SI-GaAs layer is depleted and the electric field ceases to be negligible near the ohmic contact. At low applied voltages (<100 V), the experimental dark I-V curve exhibits the trend predicted by eq.1. In fact, the observed saturation current is given by the reverse dark current, I_s, plus a contribution due to leakage currents flowing along the perimeter of the sample. These leakage currents can easily be reduced or eliminated by using guard-rings to confine the electric field far from the lateral surfaces of the sample. However, this is not essential in our experiment since the perimetral currents are not significant compared to the bulk currents and the configuration without a guard-ring is more convenient for OBIC experiments.

The extension of the electric field in the bulk GaAs depends on the density of charged centres which are generated by the ionisation of shallow impurities or other defects. The state of occupation of the shallow levels is the result of competing processes. In particular, the higher the current density, the higher is the probability of charge neutralisation. In SI-GaAs, a mechanism of charge neutralisation of the EL2 centres at high electric fields has been suggested by McGregor et al. (1994) According to this model, the capture cross-section of the EL2 centres increases dramatically at electric fields above ~10 kV/cm. This causes a dynamic compensation of the charge in the depletion region leading to the formation of quasi-neutral regions and, consequently, the distortion of the depletion region in shape and dimensions. Now, if the Schottky barrier is uniform, so should be the depth profile of the electric field, that is, the intensity of the electric field should depend only on the distance from the barrier. When this not the case, the current density may vary greatly along the metal-to-semiconductor interface. As a consequence, the boundary of the depletion region can be highly irregular and, in some regions, the electric field can reach the ohmic contact at a relatively low applied voltage.

Fig.1 shows the dark I-V characteristics of a SI-GaAs detector before and after cleaving the device for OBIC analysis in cross section configuration. The solid line refers to the sample before being cleaved, the dashed line and the dotted line refer to the smaller (1/10) and to the larger (9/10) of the two pieces, respectively. The sum of the dark currents of the two pieces exceeds slightly the value read before the cleavage because of the additional perimetral currents. The graph shows that the reverse I-V curve exhibits a clear step at high voltages, well before the breakdown and the dark current reaches a higher but relatively stable value which is more than one order of magnitude higher than the low voltage saturation current.

The OBIC analysis makes it possible to understand the mechanism responsible for this behaviour. Fig.2 shows a sequence of OBIC images of sample A at increasing bias voltages in cross section configuration.

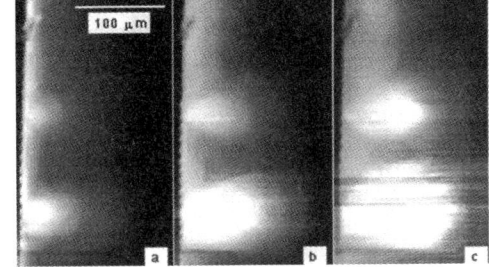

Fig 2: Sequence of cross-section OBIC images at increasing bias voltage (a: 60V, b: 100V, c: 120V) showing the development of regions of anomalous high gain. To the left is the Schottky barrier, to the right is the ohmic contact.

To the left is the Schottky barrier where the collection efficiency is higher. At a reverse bias voltage of 60 volts, that is just inside the saturation region of the I-V curve when the electric field builds up at the barrier, irregularities can be seen along the metal to semiconductor interface. Regions of very high collection efficiency appear at the interface and propagate towards the ohmic contact as the voltage increases. At V = 120V one of these high-efficiency domains reaches the back contact and, at the same time, the OBIC signal becomes quite unstable as the corresponding image shows. This voltage is just below the knee in the dark I-V curve which is at about 135V. The instability of the OBIC signal in the 120V map is the indication that the dark current is approaching the transition to what, in fig.1, is described as second saturation level. In fact, for higher values of the bias voltage, the illumination causes an irreversible increase of the dark current which persists, at zero bias, for a few minutes. By repeating the OBIC mapping at lower values of the optical-beam power, the current transition occurs at higher values of the reverse bias which tend to approach the dark critical-value of 135V. This suggests that the OBIC observation is a genuine picture of the mechanisms responsible for the anomalous behaviour of the device in the dark, the illumination causing, at most, a slight change of the critical parameters.

What OBIC shows is that the effects of a local non-uniformity of the Schottky barrier is amplified as the bias voltage increases. Where the barrier is lower, the current density is higher and, consequently, the rate of charge neutralisation is enhanced. This causes the limit of the depletion region to extend more and more towards the ohmic contact so as to keep the voltage drop across the sample constant. The very high collection efficiency in the anomalous regions is due to a photoconductivity effect which causes the gain, in this case the ratio between beam induced current and the photon flux, to be greater than one. By measuring the power of the light beam impinging on the sample we estimated the gain in the anomalous region to reach values as high as 10^5 at bias voltages of 100V. Moreover, it appears that the photoconductivity contribution to the total current in these anomalous regions becomes larger than one inside the first saturation region of the I-V

542

characteristics when a non negligible voltage begins to build up along most of the metal to semiconductor interface. This means that the regions of high collection efficiency originate along the metal to semiconductor interface where the barrier is locally reduced by the presence of defects or by other factors affecting the interfacial dipole distribution. As a matter of fact, the non uniformity of the Schottky barrier is an important issue in this kind of device operating at very high reverse bias. Fig.3 shows the EBIC image of one of the detectors in plan-view (the electron beam direction is perpendicular to the metal to semiconductor interface).

Fig.3 (left): plan EBIC image at 60 V.

Fig.4 (right): cross-section map of the resistivity resulting from the biased percolation calculation

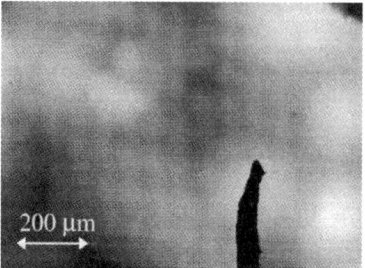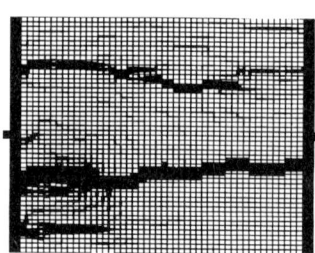

The image was collected at a reverse bias of 60V with a beam accelerating voltage of 30 keV which is enough to penetrate the metal layer and to excite electron-hole pairs in the underlying GaAs. The black strip at the bottom is the metal wire. It is quite evident that the collection efficiency is far from being uniform, the contrast of the signal being a non negligible fraction of the background signal. This confirms that the anomalous behaviour observed in these devices is due to the non-uniformities of the Schottky barrier. To complete the explanation of the dark I-V curve, the origin of what we called "second saturation region", whose onset has already been correlated with the observation of the high OBIC-gain regions reaching the back contact, is easily understood in terms of the development of ohmic channels connecting the two contacts. When the channels form, i.e. at a reverse voltage of 135V in the dark, the shunting effect causes the barrier voltage to drop thus giving a second ohmic branch in the I-V curve. The second saturation level of the total current is reached if the barrier voltage can build up again and a new equilibrium is attained, before any breakdown event occurs. To study the mechanism responsible for this behaviour a numerical model of biased percolation has been set up to simulate the occurrence and the development of the high gain channels observed by OBIC. The model describes the device in terms of a bidimensional network of resistors whose magnitude is controlled by the local current flowing through. When defects are introduced at the Schottky barrier, low resistivity domains develop and propagate towards the ohmic contact as shown in Fig.4. The shape and dimensions of these domains depend on the applied voltage and the system evolves towards a stationary configuration which is determined by the density and the nature of the defects at the barrier. More details will be presented in a forthcoming paper.

4. CONCLUSIONS

OBIC and EBIC analysis made it possible to explain the anomalous I-V characteristic of SI-GaAs detectors at high reverse bias voltages. High OBIC-gain regions develop at the metal to semiconductor interface leading to the formation of low resistivity channels connecting the Schottky barrier to the ohmic contact.

REFERENCES

Castaldini A, Cavallini A, Del Papa C, Alietti M, Canali C, Nava F, Lanzieri C, 1994 Scanning Microscopy **8**, 969
Cola A, Reggiani L, Vasanelli L, 1997 J. Appl. Phys. **81**, 999
Luo Y L, Chen T P, Fung S, Beling C D, 1997 Solid State Comm. **9**, 715
McGregor D S, Rojesky R A, Knoll G F, Terry F L, East J, Eisen Y, 1994 J. Appl. Phys. **75**, 7910

Inst. Phys. Conf. Ser. No 157
Paper presented at Microsc. Semicond. Mater. Conf., Oxford, 7–10 April 1997
© 1997 IOP Publishing Ltd

The impact of structural non-uniformity on the operation of $(Al_yGa_{1-y})_xIn_{1-x}P$ quantum well lasers at high strain

P C Mogensen[a], S A Hall[b], U Bangert[b], P Dawson[b], P M Smowton[a] and P Blood[a]

[a]Dept. of Physics and Astronomy, University of Wales at Cardiff, PO Box 913, Cardiff CF2 3YB, UK.
[b]Condensed Matter Physics Group, Dept. of Physics, UMIST, PO Box 88, Manchester, M60 1QD, UK.

ABSTRACT: Increased optical losses in the laser waveguide region of compressively-strained visible-emitting AlGaInP quantum well lasers are largely responsible for the measured increases in threshold current as the nominal mismatch is increased from 1% to 2%. TEM and X-ray microanalysis confirm compositional and structural non-uniformities in the strained GaInP layers and the formation of 3D indium-rich islands through Stranski-Krastanov-like growth, as a mechanism for the partial relaxation of the nominal strain in these structures. Significantly, this occurs before the Matthews-Blakeslee critical strain-thickness product is reached.

1. INTRODUCTION

The use of strained quantum wells in semiconductor lasers was first suggested by Adams (1986) and Yablonovich and Kane (1986) as a method of improving laser device performance through a reduction in the effective densities of states. The motivation for the development of strained AlGaInP lasers is the large direct bandgap which may be achieved, leading to emission in the ~610-690nm range for applications such as optical data storage, plastic fibre communications systems and 633nm HeNe replacements. The use of both biaxial tensile and compressive strains below about 1% in the quantum well regions is widespread (Bour et al 1994). However at higher strains there is evidence of poorer laser performance (Valster et al 1992).

The motivation of this work has been to study the physical mechanisms that lead to a degradation in laser performance at higher values of strain, as the Matthews and Blakeslee (1974) strain-critical thickness limit is approached. In the following section we describe the sample structures and present results from laser devices which show the deterioration in laser performance as the nominal compressive mismatch exceeds 1%.

2. LASER STRUCTURES AND DEVICE CHARACTERISATION

The lasers used in this work are separate confinement heterostructure lasers with a single $Ga_xIn_{1-x}P$ quantum well in a waveguide core and cladding region of $(Al_yGa_{1-y})_{0.52}In_{0.48}P$. The samples were grown by Metal-Organic Chemical Vapour Deposition on n-type GaAs substrates, mis-oriented by 10° from the [100] direction towards the [111] to limit the tendency of the alloys to spontaneously order. The laser structures have been designed so that only the Ga fraction in the well region differs between samples, changing the mismatch and hence the

Table 1 Laser structure, giving the nominal layer compositions, thicknesses and doping values. The strain in the quantum well depends on the Ga fraction (x). Four samples have been grown: Sample 0(1.0% strain) [x=0.38], Sample 1(1.4% strain) [x=0.32], Sample 2(1.7% strain) [x=0.28] and Sample 3(2.0% strain) [x=0.24].

Layer number and description	Material	Thickness (μm)	Doping:(type), element, density (cm^{-3})
8) Capping layer	GaAs	0.25	(p), Zn, >1x10^{19}
7) Cladding	(Al$_{0.7}$Ga$_{0.3}$)$_{0.52}$In$_{0.48}$P	1.5	(p), Zn, 5x10^{17}
6) Waveguide core	(Al$_{0.4}$Ga$_{0.6}$)$_{0.52}$In$_{0.48}$P	0.1	undoped
5) Quantum well	Ga$_x$In$_{1-x}$P	0.01	undoped
4) Waveguide core	(Al$_{0.4}$Ga$_{0.6}$)$_{0.52}$In$_{0.48}$P	0.1	undoped
3) Cladding	(Al$_{0.7}$Ga$_{0.3}$)$_{0.52}$In$_{0.48}$P	1.5	(n), Si, 1x10^{18}
2) Buffer layer	GaAs		(n), Si, >1x10^{18}
1) Substrate	GaAs		(n), Si, >1x10^{18}

strain between the well and the lattice matched quaternary alloy. The generic structure is shown in Table 1.

The wafers were processed to form 50μm wide oxide isolated stripe lasers and these were operated in a pulsed mode at a repetition rate of 1kHz and a pulse width of 300 ns. The threshold current density has been measured as a function of temperature and the results are shown in Fig. 1. The data is for 750μm long lasers, the least strained sample 0(1%) has the lowest threshold current density at all temperatures, indicating that an increase in the nominal strain beyond 1.4% has a detrimental effect on the threshold current. The structure with the largest nominal mismatch of 2%, sample 3, fails to operate as a laser.

The threshold current density depends directly on the round trip threshold gain which is itself dependent on several factors, these include: the optical confinement factor, which describes the overlap of the gain generating region of the structure (the quantum well) with the optical field profile in the laser waveguide, the optical scattering loss (light scattered out of the waveguide) and the mirror loss which provides the laser output. In addition, increasing the gain leads to increased carrier leakage from the well exacerbating the temperature dependant exponential increase in threshold current at high temperatures (see Fig. 1).

We have measured the optical scattering loss in the laser waveguide from a determination of the attenuation of photoluminescence intensity emitted from a cleaved facet as a function of the length travelled by the light along the waveguide, this has been adjusted by

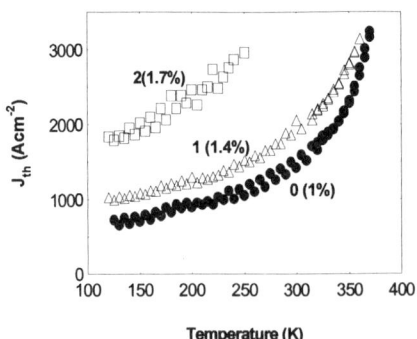

Fig. 1 Monotonic increase in threshold current density (J_{th}) is observed as a function of temperature with increasing nominal mismatch (numbers in brackets).

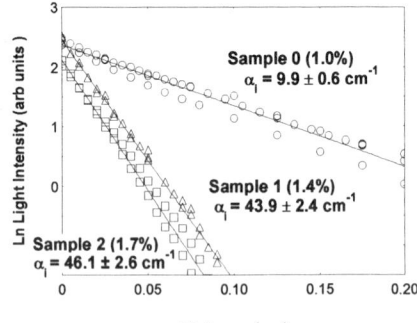

Fig. 2 Experimentally measured values of the waveguide scattering loss (α_i) for the three working laser structures.

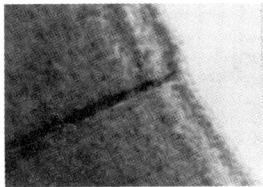

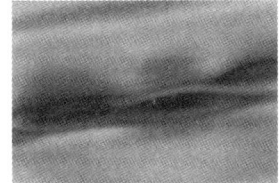

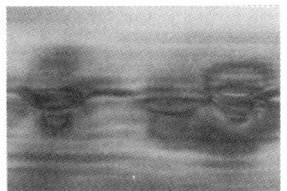

Fig. 3a Sample 0 (1.0%),
(290nm by 205nm).

Fig. 3b Sample 1 (1.4%),
(300nm by 205nm).

Fig. 3c Sample 2 (1.7%),
(550nm by 365nm).

traversing the PL pump beam across the sample. Fig 2. shows a dramatic increase in the scattering loss (from $10cm^{-1}$ to $>40cm^{-1}$) determined from the gradient of the graph, as the nominal mismatch increases. A loss of the order of $10cm^{-1}$ represents a typical scattering loss for a visible laser This increase alone will push the threshold gain up by a factor of ~2.4, if all other losses were to remain constant. A high scattering loss is suggestive of poor material quality within the waveguide and in the next section this is discussed in light of TEM and STEM measurements.

3. ELECTRON MICROSCOPY AND X-RAY ANALYSIS

TEM studies were performed on a Philips CM200 series microscope at the Manchester Materials Science Centre and STEM analysis on a VG HB601 system at the Northwest STEM facility. The bright field cross-sectional TEM images of samples 0,1 and 2 are shown in Figures. 3a, 3b and 3c respectively. In Fig. 3a the quantum well (the dark band running from right to left) is seen to be intact with a width of approximately 10nm. However in figures 3b and 3c the quantum well region, running from right to left, is seen to be 'wavy' and irregular with a fluctuating thickness, these irregularities in the well region are more pronounced in the nominally higher strain sample (Fig. 3c) which show distinct 'thick' and 'thin' regions where the well should be. The intended increase in the mismatch between the well region and the surrounding material has resulted in non-planar growth, with strong strain fields propagating into the material surrounding the well region.

The breakdown of uniform growth is confirmed in Fig. 4, which shows the presence of islands in the well region in the plan view TEM image of sample 2 (1.7%). These islands have a mean diameter of 150nm, which is comparable to the pitch of the undulations seen in Fig. 3c and an approximate density of $3x10^9cm^{-2}$, representing a coverage of ~40%. It appears that the increase in mismatch has led to the transformation of 2D layer Franck Van-der-Merwe growth in the well region to 3D Stranski-Krastanov (S-K) growth as a probable strain relieving mechanism. S-K growth is well documented as a preferred growth mode in high mismatch

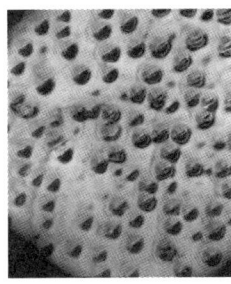

systems, including Si\Ge (Mo et al 1990, Eaglesham and Cerullo 1990) and InGaAs\GaAs (Snyder et al 1991). Therefore in these lasers the intended strain is accommodated through the formation of islands in what should be the well region. In addition to the increased scattering loss, the islanding of the well region will affect the optical confinement factor which is calculated assuming a uniform 2D gain generating layer.

In Fig. 5, the cross section STEM image for sample 2 shows 'thick' and 'thin' regions of contrast in the centre of the waveguide instead of a uniform quantum well. X-ray linescans were taken across these 'thick' and 'thin' regions at the positions marked in Fig. 5 and the resulting relative X-ray intensities of the Group III elements along the length of the scan are shown in Fig. 6. In the thick well region (the top row of scans), the waveguide core (width ~180nm) is visible as a change in the relative intensities of the Al and Ga

Fig. 4 Plan view micrograph of sample 2(1.7%) shows islands in the well region. (1.9μm by 2.2μm)

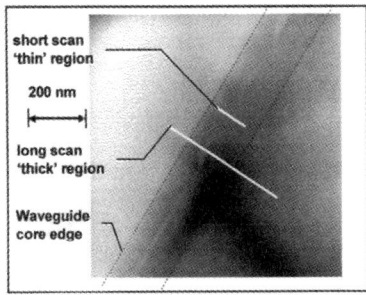

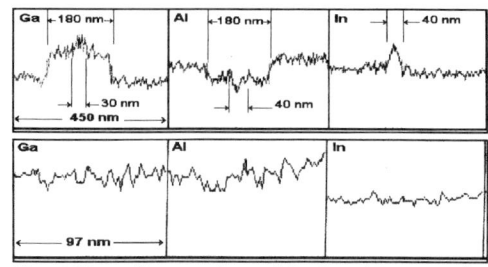

Fig. 5 STEM image of sample 2(1.7%) in cross section showing linescan positions across 'thick' and 'thin' well regions .

Fig. 6 X-ray linescan intensities in arbitrary units showing the difference in the relative ratios of the Group III elements across a 'thick' section of the well region (top row) and a 'thin' section (bottom row). Scan positions as marked in Fig. 5.

linescans. The position of the 'thick' well (width ~35nm) is shown as a central dip in the Al intensity and a corresponding peak in the Ga and In intensities. In the lower half of Fig.6, no such distinct changes in X-ray intensity are seen for the thin region of contrast, implying that there is no discernible quantum well in the region surrounding the islands.

Plan view X-ray microanalysis of sample 2, confirms the islands (which correspond to the 'thick' regions seen in Fig. 5) are In rich with a 25% In composition difference between the islands and their surroundings, the islands therefore constitute the lowest bandgap material in the laser.

4. DISCUSSION AND CONCLUSIONS

The measured increase in the threshold current of these lasers with increasing mismatch arises from growth non-uniformities in the quantum well region as seen in both TEM and X-ray analysis. This causes an increase in the optical scattering loss which may be enhanced by the similarity in the length scale of the island growth and the wavelength of the laser light. In addition the optical confinement factor is likely to be affected, since this depends on the overlap of the optical mode in the laser waveguide with the gain generating region in the quantum well which is no longer uniform. The observation of S-K type growth in these laser structures implies that it is the relative mismatch between the quantum well and substrate and not the final strain-thickness product that is the overriding factor in determining whether high quality 2D layers are produced.

ACKNOWLEDGEMENTS

P C Mogensen and S A Hall have been supported during this work by EPSRC CASE awards with Epitaxial Products International Ltd.

REFERENCES

Adams A R 1986 Elec. Lett. **22**,249

Bour D P, Geels R S, Treat D W, Paoli T L, Ponce F, Thornton R L, Krusor B S, Brigans R D and Welch D F 1994 J. Quant. Electron. **30**, 593

Eaglesham D J and Cerullo M 1990 Phys. Rev Lett. **64**,1943

Matthews J W and Blakeslee A E 1974 J. Cryst. Growth **27**, 118

Mo Y W, Savage D E, Swartzentruber B S and Lagally M G 1990 Phys. Rev. Lett. **65**,1020

Snyder C W, Orr B G, Kessler D and Sander L M 1991 Phys. Rev. Lett. **66**, 3032

Valster A, Van der Poel C J, Finke M N and Boermans M J B 1992 Digest 13th Int. Semicond. Laser Conf. **G-1**, 152

Yablonovich E and Kane E O 1986 J. Lightwave Technol. **4**, 504

Inst. Phys. Conf. Ser. No 157
Paper presented at Microsc. Semicond. Mater. Conf., Oxford, 7–10 April 1997
© *1997 IOP Publishing Ltd*

EBIC and TEM investigations of laser heterostructures grown on linearly-graded and step-graded buffer layers

M J Romero, F J Pacheco, D González, T C Rojas, D Araújo, S I Molina and R García

Departamento de Ciencia de los Materiales e I.M. y Q.I., Facultad de Ciencias, Universidad de Cádiz, Apartado 40, 11510 Puerto Real (Cádiz),Spain.

ABSTRACT : The transport properties and defect recombination activity of 1.3μm and 1.55μm laser heterostructures grown on different step- and linearly-graded InGaAs buffer layers are evaluated by means of electron-beam-induced-current. These transport parameters are related to the threading dislocation density and interface roughness obtained by means of transmission electron microscopy (TEM). The defect strength γ of dislocations, estimated from EBIC measurements, gives $\gamma \approx 10^3$ - 10^4 μm^{-2}. The EBIC data allow us to conclude that the interface roughness plays a fundamental role in laser degradation. The step-graded buffers investigated in this contribution exhibit an enhanced planarity and, hence, the lasers grown on them show better transport properties.

1. INTRODUCTION

High speed telecommunication technology requires the integration of 1.3μm and 1.55μm lasers, corresponding to minimum optical fiber absorption, into the high speed integrated circuits, that is, the availability of opto-electronic integrated circuits (OEICs), that combine the InP and GaAs technologies. Nevertheless, from the integration of these mismatched materials results a quality degradation of the grown devices. To reduce the level of defects in InP-based lasers, buffer layers must be introduced between the active layers and the GaAs substrate. The lower the level of the dislocation interaction achieved, the lower will be the density of dislocations threading across the active region. Therefore, a graded buffer layer (step- : Krishnamoorthy et al. 1992, Araújo et al. 1995 or linearly- : Chang et al. 1992, Molina et al. 1994) seems to be a solution to relax progressively the strain. In this contribution, the transport properties - minority carrier diffusion length and surface recombination velocity - of laser heterostructures grown on different step- and linearly-graded InGaAs buffer layers are evaluated by means of electron-beam-induced-current (EBIC). From EBIC measurements we also assess the recombination activity of threading dislocations (TD). The structural distribution of defects and interface roughness are studied by plan-view and cross-section transmission electron microscopy (PVTEM and XTEM, respectively). Thus, the transport parameters are related with the structural quality of the lasers.

2. EXPERIMENTAL

The 1.3μm double bulk heterostructure (DBH) lasers and 1.55μm multiquantum well (MQW) lasers were grown by metal-organic vapor phase epitaxy (MOVPE) on InGaAs buffer layers (Fig.1.) at 900K and nominal atmospheric pressure (III-V ratio : 100). Previously, the InGaAs buffers were grown by molecular beam epitaxy (MBE). For 1.3μm DBH-laser, the undoped active region consists of a 100 nm InGaAsP (λ = 1.3μm) layer sandwiched between to InGaAsP (λ = 1.15μm) layers of 50 nm. The cladding layers are composed of 1.5 μm p-doped

(Zn, 8×10^{17} cm^{-3}) and 1.0 μm of n-doped (Si, 2×10^{18} cm^{-3}) InP. A 100 nm capping layer of 4×10^{18} cm^{-3} p-doped In$_{0.53}$Ga$_{0.47}$As acts as contact. As for the 1.55μm MQW-laser, the cladding layers are essentially the same and the undoped active region consists in three InGaAs(λ = 1.55μm) QWs of 80 Å width separated by InGaAsP(λ = 1.15μm) barriers of 120 Å (both InGaAs and InGaAsP are matched to InP). This structure is sandwiched between two InGaAsP layers of 75 nm.

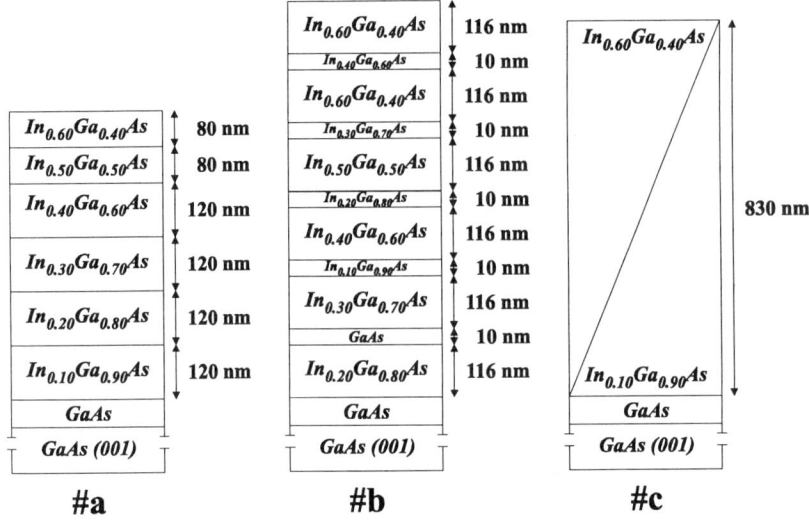

Fig. 1. Buffer design scheme

The TEM observations were performed in JEOL 1200EX and 2000EX transmission electron microscopes. PVTEM specimens were prepared by chemical etching methods (Br$_2$+CH$_3$OH and H$_2$SO$_4$+H$_2$O$_2$+H$_2$O) and XTEM specimens by Ar$^+$ ion milling. For transport property investigations, the EBIC linescans are performed on freshly cleaved (110) faces perpendicular to the epilayers in the [001] direction, labelled as normal-collector configuration. The planar-collector configuration is used for defect electrical activity assessment. The current induced (I$_{cc}$) by electron beam excitation is measured using a Matelect ISM-5A amplifier and Matelect IU-1 interface. The EBIC experimental set-up is installed on a JEOL-820SM scanning electron microscope.

3. RESULTS AND DISCUSSION

Fig. 2 displays the EBIC gain against electron beam energy of laser heterostructures grown on step (#a), inverse-step (#b) and linearly-graded (#c) buffer layers, and also on InP (#d), InGaAs/InP (#e) and GaAs(#f) substrates taken as references. From quantitative evaluation of EBIC measurements, applying Monte Carlo methods of electron-hole(e-h) pair generation and minority carrier transport phenomena involved in EBIC experiments (Romero et al. 1997a), the hole diffusion length L_h and surface recombination velocity S/D at the n- side of the laser are estimated. The accuracy of the minority carrier diffusion length

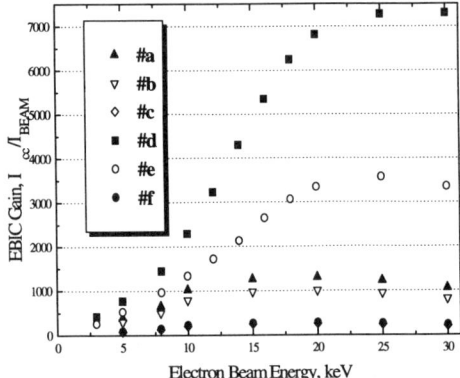

Fig.2. EBIC gain for laser heterostructures grown on different buffer layers and substrates.

determinations is estimated to be ± 0.02 μm. These results are presented in Table I together with the density of threading dislocations crossing the active layers, measured from PVTEM micrographs (Fig. 3).

Laser	ρ_{TD} (10^8 cm^{-2})	L_h(μm)	S/D (cm^{-1})
#a	2.20	0.15	$> 1 \times 10^6$
#b	4.20	$< \approx 0.1$	$> 1 \times 10^6$
#c	4.30	$<< 0.1$	$> 1 \times 10^6$
#d	< 0.01	0.56	1×10^5
#e	0.24	0.24	$> 1 \times 10^6$
#f	4.25	$<< 0.1$	$> 1 \times 10^6$

Table I. Threading dislocation density (ρ_{TD}), hole diffusion length (L_h) and surface recombination velocity (S/D) from the lasers diodes grown on different buffer layers and substrates.

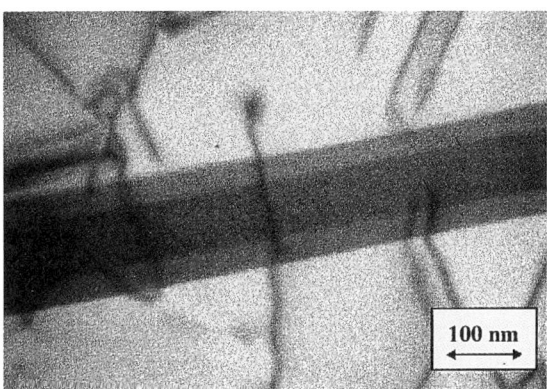

Fig. 3. BF (002) XTEM image of the active region of a double bulk heterostructure (DBH) laser grown on #c.

100 nm

As expected, the enhanced transport properties and structural quality corresponds to the laser on the InP substrate, they deteriorate with introduction of a layer of InGaAs between the active layers and the InP substrate, and dramatically worsen for the laser grown directly on the GaAs substrate.

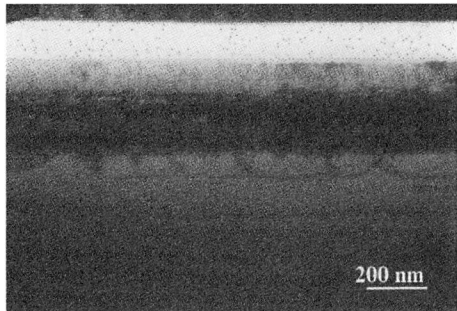

200 nm   200 nm

Fig. 4. BF (220) XTEM micrographs of the linearly-graded buffer structures for growth temperatures of 773 K (left) and 723 K (right).

We observe that the best EBIC gain and minority carrier diffusion length for a laser on buffer correspond to the linearly step-graded one (#a), while the introduction of the linearly-graded buffer (#c) implies considerable damage in the active layers. The reduction of the value of L for the laser grown on inverse-step layers (c#) *versus* the laser on #a results from its lower efficiency as a threading dislocation filter, revealed from measured TD density data and expected from the Dunstan et al.(1991) model of relaxation. Nevertheless, it has been demonstrated (Pacheco et al. 1995) for InGaAs buffer layers (up to Indium mole fraction of 0.30), that linearly-graded structures offer better capability for relaxation and TD filtering than step-graded ones. These unexpected results are due to the presence of a cross-groove pattern along <110> directions on top of the buffer, that supposes a high interface roughness which results in laser quality degradation. Thus, although an initial Indium mole fraction different from zero is not inconvenient for relaxation efficiency (Tersoff 1993, Dunstan et al. 1991), it originates a cross-groove pattern (the depth of the grooves is around of 150 nm) instead of a smooth cross-hatch pattern. Since these observations, efforts have been focused to optimize the growth conditions. A high planarity has been obtained decreasing the temperature from 773 K to 723 K, keeping the initial Indium mole fraction (0.10) and the composition gradient (60%In/μm) (Fig. 4). Further studies on lasers grow on these new linearly-graded buffers, produced at 723 K, are in progress, and we expect that their transport properties will improve strongly.

The assessment of defect activity has been carried out through Monte Carlo methods applied to EBIC data (Romero et al. 1997b). From evaluation of electron-beam-induced-current linescan profiles across threading dislocations (planar-collector configuration), at different electron beam energies and currents, the defect strength γ, defined as $\gamma = 1/L_{TD}^2 - 1/L^2$, is estimated to be between 10^3 and 10^4 μm^{-2}. This strength value does not explain the observed hole diffusion length decrease, and therefore, the recombination at the interface between active layers and buffer must play a fundamental role in laser degradation.

ACKNOWLEDGEMENTS

This work was supported by the CICYT (Comisión Interministerial de Ciencia y Tecnología), under MAT94-0538 Project and by the Junta de Andalucía through group TEP-0120. The authors would like to thank J.D.Lambkin of the National Microelectronics Research Centre of Cork (Ireland) for laser growth. This work was carried out at the División de Microscopía Electrónica de la Universidad de Cádiz.

REFERENCES

Araújo D, González D, Sacedón A, Calleja E and García R 1995 Appl. Phys. Lett. **67**, 3632
Chang J C P, Chen J, Fernández J M, Wieder H H and Kavanagh K L 1992 Appl. Phys. Lett. **60**, 1129
Dunstan D J, Young S and Dixon R H 1991 J. Appl. Phys. **70**, 3038
Hardingham C and Holt D B 1995 Inst. Phys. Conf. Ser. **146**, 621
Krishnamoorthy V, Lin Y W and Park R M 1992 J. Appl. Phys. **72**, 1752
Molina S I, Pacheco F J, Araújo D, García R, Sacedón A, Calleja E, Yang Z and Kidd P 1994 Appl. Phys. Lett **65**, 2460
Pacheco F J, González D, Molina S I, Villar M P, Sacedón A, Calleja E and García R 1995 Inst. Phys. Conf. Ser. **146**, 211
Pacheco F J 1997 PhD. Thesis, Universidad de Cádiz
Romero M J, Araújo D, Lambkin J D and García R 1997a Mat. Sci. Eng. B**44**, 57
Romero M J, Araújo D and García R 1997b Dependence of electron-hole generation function on EBIC contrast of defects (at this conference)
Tersoff J 1993 Appl. Phys. Lett. **62**, 693

Inst. Phys. Conf. Ser. No 157
Paper presented at Microsc. Semicond. Mater. Conf., Oxford, 7–10 April 1997
© 1997 IOP Publishing Ltd

TEM observation of degraded InGaAsP MQW laser diodes

T Matsuda, T Namegaya, A Kasukawa, Y Ikegami, N Tsukiji, T Ijichi and F Iwase

Yokohama R&D Laboratories, The Furukawa Electric Co., Ltd., 2-4-3, Okano, Nishi-ku, Yokohama 220, Japan

ABSTRACT: We observed three types of degradation in InGaAsP graded-index separate-confinement-heterostructure multiple quantum well laser diodes using cross-sectional and plan-view transmission electron microscopy. The first case is rapid degradation of ridge waveguide lasers due to oxidation of an exposed active region. The second case is rapid degradation of buried heterostructure lasers due to a climb motion of perfect dislocation half loops. The third is gradual degradation of buried heterostructure lasers due to generation of dislocation microloops at confinement/blocking interfaces for a long operation time.

1. INTRODUCTION

Degradation mechanisms in III-V compound semiconductor laser diodes (LDs) have been characterized by microstructual studies using transmission electron microscopy (TEM). The major degradation modes are rapid degradation caused by recombination enhanced dislocation climb or glide, gradual degradation caused by formation of defect clusters, and catastrophic damage caused by strong optical absorption at mirror surface defects, respectively (Ueda 1988).

High reliability has been reported for long wavelength InGaAsP/InP LDs as compared to short wavelength AlGaAs/GaAs or InGaP/GaAs LDs. However, the InGaAsP/InP LDs are not always faultless. So far, a few microstructual studies of degraded InGaAsP/InP LDs have been carried out. In rapidly degraded InGaAsP/InP LDs, climbed dislocation loops and plate-like precipitates or slip dislocations were observed (Wakita et al 1982; Ishida et al 1982). In gradually degraded buried structure InGaAsP/InP LDs, dislocation loops were observed at regrown interfaces (Chu et al 1988; Sim et al 1988; Cooman et al 1989; Bangert et al 1991). Catastrophic optical damage (COD) or COD-like degradation in InGaAsP/InP LDs was also observed (Snyder et al 1995; Chu et al 1988).

In this paper, we report on a TEM study of a few degradation modes in graded-index separate-confinement-heterostructure multiple quantum well (GRIN-SCH-MQW) LDs.

2. EXPERIMENTAL

The GRIN-SCH-MQW LDs are fabricated by metalorganic chemical vapour deposition (MOCVD) and wet etching techniques (Kasukawa et al 1992). Various compositional InGaAsP layers are lattice matched to a InP substrate in the GRIN-SCH-MQW. Fig.1 displays (a) a composition analysis by thickness fringes (CAT) image and (b) a (200) dark field image of the multilayer structure. These images show a step-like index (composition) change at the GRIN-SCH region and good flatness at each interface. Fig.2 illustrates schematic diagrams of fabricated (a) a ridge waveguide LD and (b) a buried heterostructure (BH) LD. The ridge mesa is formed by wet etching, and mesa sides are buried by polyimide. The BH-LD is fabricated by repeated wet etching and buried regrowth, respectively. The fabricated LDs are subjected to an accelerated lifetest under automatic power control (APC) at 5mW/85°C.

TEM samples of degraded LD chips for the lifetest are prepared by ion-milling. For plan-view preparation, we measure the ion-milling rate for upper layers of an active layer in a dummy LD chip or wafer using an in-lens type scanning electron microscope (SEM). After mechanical thinning of a substrate, an upper clad layer thickness is reduced until it is submicron (this value is changed with view position) by measured ion-milling. Finally, only the substrate side is ion-milled. For cross-sectional preparation, a LD chip is surrounded by four small bars of InP and bonded with epoxy glue, two of them are trimmed to the same thickness as the chip. This compound sample is thinned by conventional cross-section techniques.

552

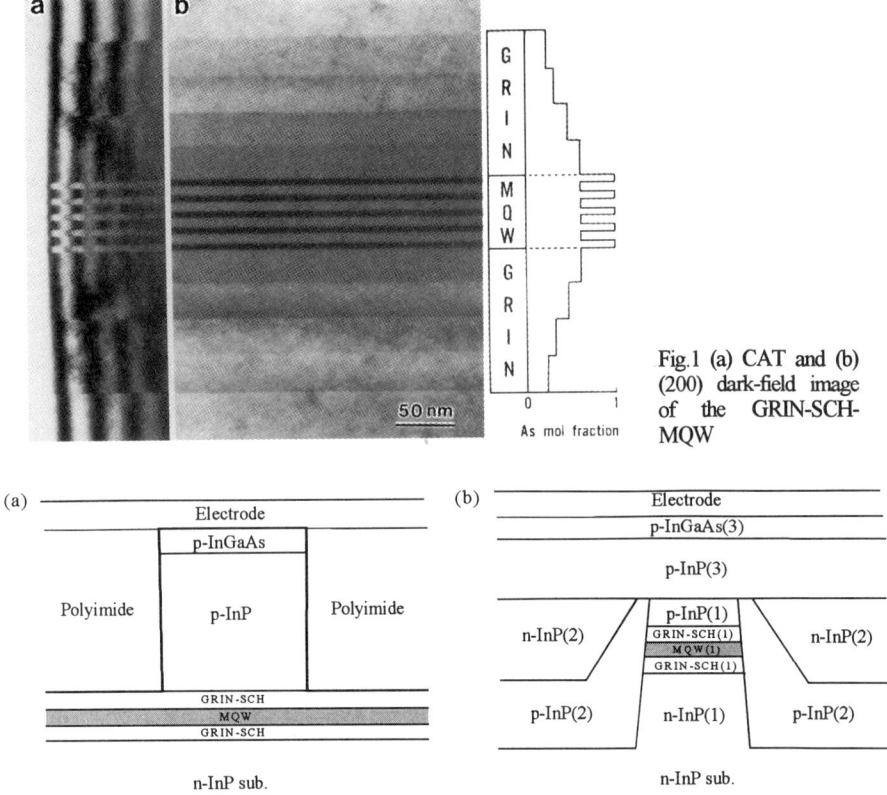

Fig.1 (a) CAT and (b) (200) dark-field image of the GRIN-SCH-MQW

Fig.2 Schematic diagram of (a) the ridge waveguide and (b) the BH-LD Parenthesized numbers in fig.(b) correspond to growth order.

3. RESULTS AND DISCUSSION

3.1 Rapid degradation of ridge waveguide lasers

In general, ridge waveguide LDs are believed not to degrad, however, we observed rapid degradation for particular devices. A driving current in the degraded devices doubled after 100h under the APC condition. Fig.3 shows cross-sectional TEM images of the failure device, which reveal that both sides of MQW active layers are exposed to polyimide or air in the opening, resulting from over etching of the mesa. Much oxygen was detected with In, Ga, As, P in amorphous regions at the exposed MQW surfaces by electron energy dispersive X-ray spectroscopy (EDX). Dislocations were not found around the active region. In addition, we observed no increase of driving current in devices which were fabricated from the wafer remaining same in the upper SCH region as in Fig.2(a). These results suggest that this degradation mode corresponds to oxidation of the quantum wells associated with recombination of carriers. In ordinary correctly fabricated devices, whose active layers are not exposed to the side, surfaces hardly degrade. Actually, the degradation mode of oxidation has almost been neglected in the InGaAsP/InP systems. However, oxidation of an active layer is inevitable at both cleaved mirror facets. From these experiments, the passivation of surfaces including mirror facets plays an important role not only in the GaAs systems but also in the InGaAsP/InP systems under a high power for a long operation time.

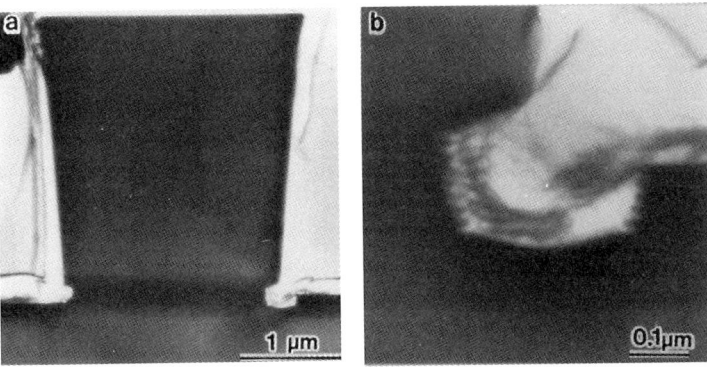

Fig.3 Cross-sectional bright field images of the rapidly degraded ridge waveguide LD

3.2 Rapid degradation of BH lasers

The lifetime of the InGaAsP/InP buried heterostructure (BH) lasers is more than ten million hours at room temperature. However, the LDs rapidly degraded under the APC condition in some circumstances. We observed many dislocations around the active regions in the failed devices whose driving currents tripled after 80-100h operation. Fig.4 shows plan-view bright field images of the dislocations, six different reflections with diffraction vectors **g** (indicated by big arrows) are excited. Each dislocation imaged in Fig.4 corresponds to an edge dislocation with Burgers vector $\mathbf{b}=a/2[01\bar{1}]$.

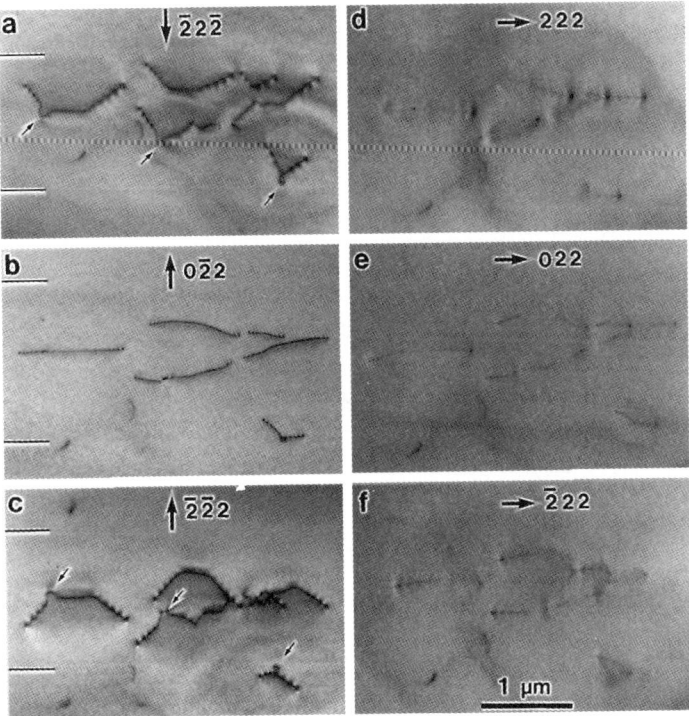

Fig.4 Plan-view bright field images of the rapidly degraded BH-LD

554

Residual contrasts of the dislocations are attributed to $\mathbf{g} \cdot \mathbf{b} \times \mathbf{u}$ ($\mathbf{u}$ is the direction of a dislocation segment). Shapes and locations of these dislocations were obtained by observing the location shift from the active stripe for tilt, and by stereo viewing. The dislocations form half loops on the $(01\bar{1})$ plane which is vertical to the active layer and parallel to the waveguide direction. A typical half loop in this failed device is shown in Fig.5. The half loop is assembled with three lines of dislocation, each of them elongates in the $[\bar{2}11]$, $[011]$, $[211]$ direction and lies on (111), (100), $(1\bar{1}\bar{1})$ plane, respectively. The inside-outside contrast method showed that each half loop was an interstitial type. Some dislocation half loops have small dipoles or loops at their own corners (small arrows in Fig.4). These dipoles and loops exist in the first growth region and in the first/ second growth interface. Each of them was also of interstitial type.

Similar large half loops exist only in the third growth region over active layers in an as-buried wafer from which the failed devices were fabricated. In addition, no small dipole or loop was found at the corner. Therefore, we consider that the large half loops are misfit dislocations induced by the third growth, the small dipoles and loops are climb dislocations assisted by nonradiative recombination during the laser operation. A few reports regarding dislocation climb in InGaAsP/InP LDs have been published (Wakita et al 1982; Ishida et al 1982), their climb direction is $\pm[010]$ or $\pm[001]$ that is parallel to an active layer and rotating $45°$ to a waveguide direction. In this investigation, we found dislocation climb from misfit dislocation half loops in the $[\bar{1}00]$ direction which was perpendicular to the active layer and the waveguide direction. The climb velocity is estimated to be about 5×10^{-11} cm/s under the APC condition (85°C/5mW), if the length of dipoles equals the climb distance. The value is close to the reported growth speed of $\pm[010]$ or $\pm[001]$ dipole (Ishida et al 1982).

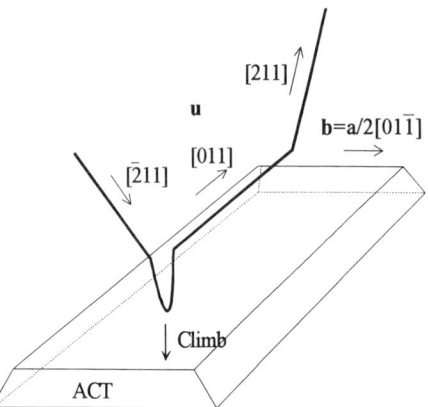

Fig.5 Schematic diagram of the dislocation half loop in the rapidly degraded BH-LD

3.3 Gradual degradation of BH lasers

If there are no dislocations, the BH-LDs gradually degrade during long operation time. Fig.6 illustrates cross-sectional bright field images around an active region in a gradually degraded device after the accelerated lifetest, showing small dislocation loops at buried regrown interfaces. The small loops were generated during a long aging time, because no dislocation was observed in a virgin device which was made from the same wafer and process. By the $\mathbf{g} \cdot \mathbf{b} \times \mathbf{u}$ contrast (Fig.7) and the inside-outside analysis, the nature of the dislocation loops was found to be an interstitial Frank type lying on the {111} plane and an interstitial prismatic type lying on the {100} plane. The former is a comparatively common type on gradual degradation (Ueda 1988; Cooman et al 1989; Bangert et al 1991). The latter type is rare in the zinc-blende structure, only Chu and Nakahara (1990) reported a 1/2<100>{100} type extrinsic loop in the degraded InGaAsP active region of 1.3μm wavelength CSBH LDs. It seems that this type is specific only in degraded InGaAsP/InP optical devices.

It is characteristic that almost dislocation loops exist at InGaAsP-SCH/buried-InP interfaces and a few exist at InP/InP regrown interfaces near the SCH region, while there is nothing at the MQW active region sides in the gradually degraded GRIN-SCH-MQW LDs. In particular, many dislocation loops are formed at p/n interfaces between n-InGaAsP GRIN-SCH (lowest two layers) and both p-InP buried blocking regions. The other SCH and the MQW regions are non-doped. The same result was obtained in another system of BH-LD. Fig.8 shows cross-sectional bright field images in a gradually degraded p-substrate liquid phase epitaxy (LPE) buried BH-LD that has a p-InGaAsP bulk type active layer and a n-InGaAsP optical guide layer. A schematic diagram of the structure is illustrated in Fig.9. Fig.8 shows that small dislocation loops are predominantly formed at interfaces between the n-InGaAsP guide layer and the p-InP buried blocking regions. Some of them are formed at n-InP/p-InP interfaces close to the guide layer, but are not found at p-InGaAsP/p-InP interfaces of the bulk active layer sides. The nature of the dislocation loops was also of interstitial type. Thus the small dislocation loops tend to form at a regrown interface between n-InGaAsP with a wider bandgap and p-InP buried region near an active layer.

These results suggest that this degradation is connected with the following : 1) generation of point defects at regrown interfaces; 2) electromigration and accumulation of charged interstitials to p/n junctions; 3) nucleation and growth of interstitial dislocation loops assisted by nonradiative recombination. In addition, we consider that the loops are difficult to generate in InGaAsP with its narrower bandgap because there are less surface defects formed by escape of phosphorus atoms at regrown surfaces and/or difficulty of dislocation climb by absorption of interstitials.

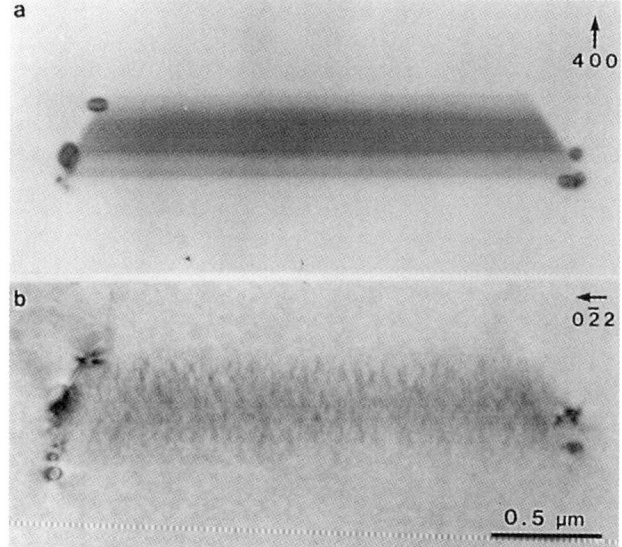

Fig.6   Cross-sectional bright field images of the gradually degraded BH-LD

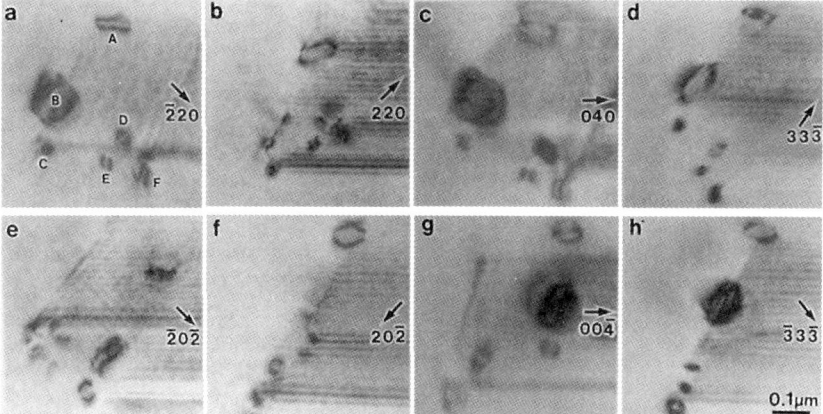

Fig.7 The $\mathbf{g \cdot b} \times \mathbf{u}$ analysis of dislocation loops in the gradually degraded BH-LD (bright field images) Each loop is lying on following plane: A: (100), B: (1$\bar{1}$1), C: (111), D: ($\bar{1}$11), E,F: (010), respectively.

556

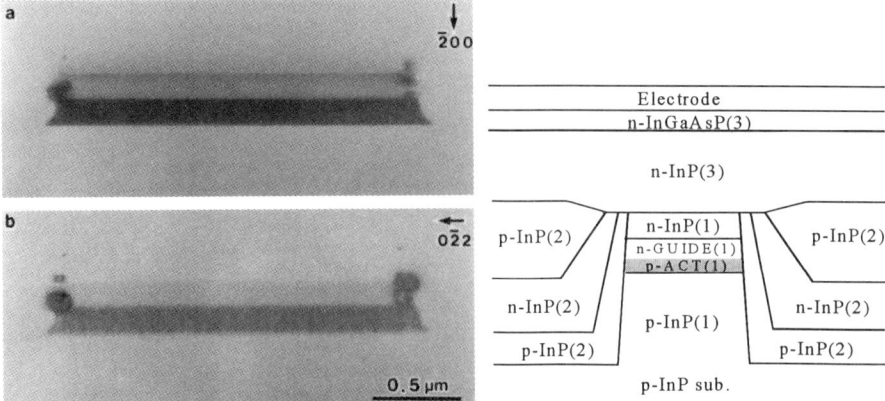

Fig.8 Cross-sectional bright field images of the gradually degraded p-substrate BH-LD

Fig.9 Schematic diagram of the p-sub. BH-LD Parenthesized numbers correspond to growth order.

4. CONCLUSIONS

A Study of degradation in InGaAsP GRIN-SCH-MQW LDs was carried out. The results of our investigation are summarized below.

(1) Exposure of the active region sides causes rapid degradation by recombination assisted oxidation even in the InGaAsP/InP system.

(2) Climb of the misfit dislocation half loops formed at buried regrowth causes rapid degradation of the BH-LDs. The climb direction is [$\bar{1}$00] which is perpendicular to the active layer and waveguide direction.

(3) Generation of interstitial dislocation microloops (Frank and lying on {100} type) during laser operation causes gradual degradation of the BH-LDs. The microloops are predominantly formed at the n-InGaAsP SCH/ p-InP blocking regrowth interfaces and not formed in the MQW active region.

REFERENCES

Bangert U, Charsley P, Davies I G A, Goodwin A R 1991 J Matt. Sci. Lett. **10**, 1185

Chu S N G, Nakahara S, Twigg M E, Koszi L A, Flynn E J, Chin A K, Segner B P, Johnston,Jr W D 1988 J. Appl. Phys. **63**, 611

Chu S N G, Nakahara S 1990 Appl. Phys. Lett. **56**, 434

de Cooman B C ,Bulle-Lieuwma C W T, de Poorter J A, Nijman W 1990 J. Appl. Phys. **67**, 3919

Ishida K, Kamejima T, Matsumoto Y 1982 Appl. Phys. Lett. **40**, 16

Kasukawa A, Matsumoto N, Namegaya T, Imajo Y 1992 IEICE Trans. Electron. E75-C, **12**, pp 1541-1554

Sim S P, Skeats A P, Taylor M R, Hockly M, Cooper D M, Nelson A W, Devlin W J, Regnault J C 1988 IEE Conf. Publ. No.292 Pt.1 pp 396-399

Snyder C W, Lee J W, Hull R, Logan A 1995 Appl. Phys. Lett. **67**, 488

Ueda O 1988 J. Electrochem. Soc. 14C

Wakita K, Takaoka H, Seki M, Fukuda M 1982 Appl. Phys. Lett. **40**, 525

Inst. Phys. Conf. Ser. No 157
Paper presented at Microsc. Semicond. Mater. Conf., Oxford, 7–10 April 1997
© 1997 IOP Publishing Ltd

Atomic processes at the laser front facet during laser operation

I Rechenberg[1], U Richter[2], A Klein[1], W Höppner[3], J Maege[1], G Beister[1] and M Weyers[1]

1) Ferdinand-Braun-Institut für Höchstfrequenztechnik, Rudower Chaussee 5, D-12489 Berlin, Germany, 2) Labor für Elektronenmikroskopie in Naturwissenschaft und Medizin, Weinberg 23, D-06120 Halle, Germany, 3) Institut für Halbleiterphysik GmbH Frankfurt/O., Walter-Korsing-Str. 2, D-15230 Frankfurt/Oder, Germany

ABSTRACT: With the aim to identify the initial processes of facet degradation, 980 nm-pump laser diodes were stressed and studied using transmission electron microscopy (TEM). From the changes of the TEM contrast observed in the light emission region there is evidence for interdiffusion of indium and aluminum.

1. INTRODUCTION

Facet degradation is the most important factor limiting the reliability of InGaAs/AlGaAs high power laser diodes. The present knowledge concerning the mechanism of catastrophic optical damage (COD) of laser facets is related to the role of nonradiative recombination and its influence on defect formation and propagation. However, although it was shown that the threshold level of catastrophic optical damage can be considerably enhanced by facet coating, further increase of laser diode output power requires a more detailed analysis of the atomic processes responsible for facet damage. While the phenomenon of the defect formation leading to COD was studied by different methods, only little is known of the underlying atomic processes. The square-root time behaviour of the degradation rate points to a diffusion process (Okayasu 1992). During laser operation a progressive change of the III/V ratio and an arsenic accumulation on the uncoated laser facet was detected (Houle et al 1992, Epperlein et al 1993). We have investigated the atomic processes responsible for the defect formation at the laser facet of 980 nm-pumplaser diodes.

2. EXPERIMENTAL

Ridge waveguide (RW) lasers with separate confinement heterostructure were prepared on (100) n-type GaAs substrates by MOVPE. The layer structures consist of an $In_xGa_{1-x}As$ (x=0.18) single (QW) or a double quantum well (DQW) sandwiched between GaAs spacer layers, $Al_yGa_{1-y}As$ waveguide (y=0.28) and cladding layers with y=0.3 or 0.5. Dopant elements are zinc for p-type and silicon for n-type. The laser facets were coated with antireflecting Al_2O_3 films on the front facet and highly reflective dielectric mirrors (pairs of Al_2O_3/TiO_2) on the rear facet.

For the analysis of the structural changes at the laser facet during operation transmission electron microscopy (TEM) investigations were carried out. The lasers were selected according to their degradation behaviour dependent on output power and stress time. As selection criteria for the lasers we have additionally used the results of P-V-I characteristics (Beister et al 1996) and facet inspections in a scanning electron microscope (SEM) as well as measurements of the increase in the facet temperature. For the latter investigations the laser diodes were driven in a SEM. The temperature increase of the laser facet was estimated from the peak shift of cathodoluminescence spectra dependent on the driving current/output power (Rechenberg et al 1995). For TEM investigations of selected laser diodes, focussed ion beam etching (FIB) was used to prepare crossections across the laser stripe near the front facet as well as along the laser stripe. The TEM investigations were done in a high voltage electron microscope JEOL JEM 1000 operating at 1000 kV. An energy dispersive analysis (EDX) was carried out in a TEM CM 20 FEG at 120 kV using the Voyager 2 analysing system. The diameter of the probe of the analysing system was 0.6 nm.

3. RESULTS AND DICUSSION

Fig. 1 shows a TEM image taken from a laser diode after an aging test over 1035 h at 90 mW/40°C. Optical microscopy showed that the laser facet remained undamaged. However, the temperature increase of the waveguide (by 90° C) and the cladding (by 60° C) layers during operation of this laser (8mW/µm) after aging is higher than that obtained from lasers of the same wafer before stress testing (50°C/30°C).

For lasers driven at a high output power, different stages of facet degradation can be distinguished. In an early stage (Fig. 2) a bright line encloses a lenticular region which resembles the near field intensity distribution calculated for this structure. In this region the contrast of the GaAs layers to the AlGaAs confinement layers has disappeared. Also the InGaAs QW shows displacement towards the p-side. In the region of the two corners of the lenslike area the contrast of the QW has vanished and dislocations have formed. After further facet degradation the QW is destroyed and the material contrast vanishes (Fig. 3). The dissolution of the QW structure is confirmed by EDS of the composition in the light output region. The In-peak has vanished indicating the interdiffusion of In. Furthermore, this stage of facet degradation is accompanied by defect formation in the active region at the facet. The dislocations expand laterally into the waveguide and cladding layers and in <110> direction into the interior of the laser diode (up to several tens of micrometres dependent on the operation conditions). In most cases the Burgers vectors are of the $a/2$ <110> type. The dislocations propagating in the <110> direction are 60°-dislocations, dislocations with nonstraight propagation direction are of mixed type. Also dislocation loops and dipoles were found. In the final stage of COD the dissolution of the QW structure can be observed up to several tens of micrometres in the depth from the facet (Fig. 4).

The disappearence of the GaAs spacer contrast in Fig. 2 and the dissolution of the InGaAs QW structure in Fig. 3 indicate an interdiffusion process of Al and In. From the Al and In diffusion lengths and the operation time of the laser diodes showing facet damage we can estimate the interdiffusion coefficients to be $D_{Al-Ga} \approx 10^{-21}$ cm^2/s and $D_{In-Ga} \approx 10^{-22}$ cm^2/s. These values indicate that the facet temperature was lower than 650° C in this early stage of COD. The basic processes underlying the interdiffusion appear to be similar to those during the annealing of GaAs/InGaAs quantum wells coated with dielectric cap layers (Bürkner 1995).

The temperature increase caused by nonradiative recombination and light absorption at the semiconductor/dielectric interface leads to diffusion of Ga into the dielectric coating film. The Ga vacancies generated in this way trigger the interdiffusion of In and Al and destroy the QW structure of the active region. Furthermore, As accumulates at the facet and enhances absorption and nonradiative recombination, resulting in a further temperature increase. The condensation of point defects results in the generation of dislocations according to the well-known REDM/REDR (recombination enhanced defect motion/reaction) mechanism (Kondo 1983). Our investigations indicate that the facet degradation is initiated by the interdiffusion of Al and In induced by nonradiative recombination and enhanced by reaction of the semiconductor with the coating material. The investigations show that efforts to enhance the COD level also must consider the ability of the antireflecting coating layers to dissolve Ga to a low degree. The positive effect of Si deposited on the laser facet to enhance the COD level (Tu et al 1996) supports this conclusion.

FIG. 1 - 3 TEM IMAGES FROM THE FACETS OF RW LASER DIODES
FIG. 4   TEM IMAGE OF A SECTION PREPARED PERPENDICULAR TO FACET

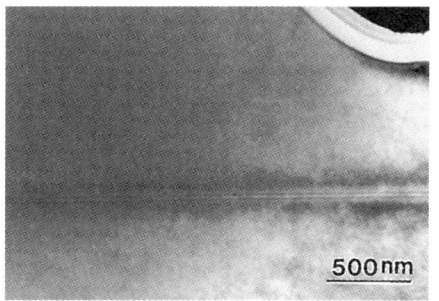

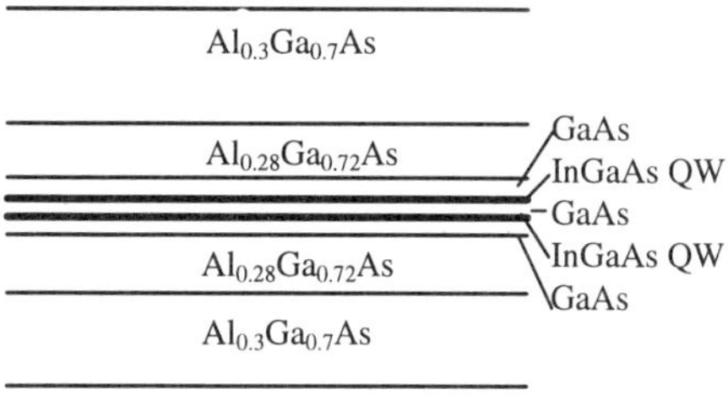

$Al_{0.3}Ga_{0.7}As$

$Al_{0.28}Ga_{0.72}As$ — GaAs / InGaAs QW

— GaAs

$Al_{0.28}Ga_{0.72}As$ — InGaAs QW — GaAs

$Al_{0.3}Ga_{0.7}As$

Fig. 1 a   Fig. 1 b
Undamaged laser facet after 1035 h operation   Scheme of the layer structure in Fig. 1 a
at 90 mW/40°C

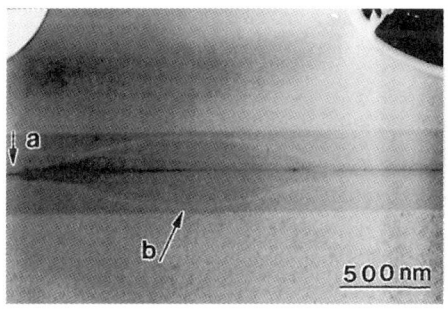

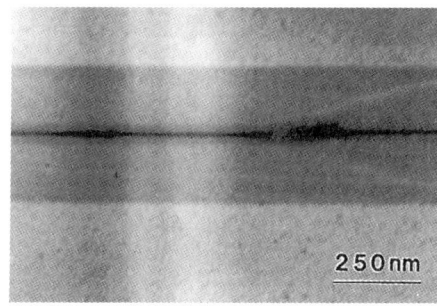

Fig. 2 a   Fig. 2 b
Early stage of facet degradation   Magnified detail from Fig. 2a
a) QW with spacer layers
b) region with interdiffusion

560

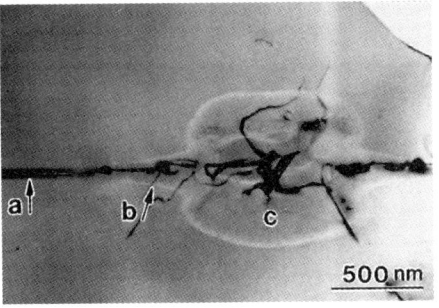

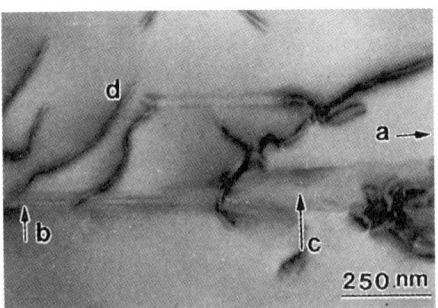

Fig. 3
Advanced stage of facet degradation
a) undamaged DQW structure
b) peripheral zone of the damaged region
c) damaged region of the DQW structure
   with dislocations

Fig. 4
Final stage of facet degradation
a) front facet
b) InGaAs-DQW
c) damaged region of the DQW
d) dark lines represent dislocations

ACKNOWLEDGMENT

The authors are thankful for the opportunity to carry out these investigations using the electron microscopes running at the Max-Planck-Institut für Mikrostrukturphysik Halle. We thank Dr. H. Stenzel and Dr. W. Erfurth for performing the EDX analysis.

REFERENCES

Beister G, Maege J, Erbert G, Rechenberg I, Sebastian J, Weyers M and Würfl J 1996, Int. Symp. on Compound Semiconductors, St Petersburg, Russia, Sept. 23-27
Bürkner S, Larkins E C, Baeumler M, Wagner J, Rothemund W, Flemig, G and Ralston J D 1995 Mater. Sci. and Technol. **11,** 840-843
Epperlein P W, Buchmann P, Jakubowicz A 1993 Appl. Phys. Lett. **62,** 455-7
Houle F A, Neimann D L, Tang W C, Rosen H J 1992 J. Appl. Phys. **72,** 3884-96
Kondo K, Ueda O, Isozumi S, Yamakoshi S, Akita K and Kotani T 1983 IEEE Trans. Electron. Devices **ED-30** 321-6
Okayasu M and Fukuda M 1992 J. Appl. Phys. **72,** 2119-24
Rechenberg I, Höpner A, Maege J, Klein A, Beister G and Weyers M 1995 Inst. Phys. Conf. Ser. **146,** 587-90, Proceed. of the Microsc. Semicond. Mater. Conf., Oxford 20-3 March 1995
Tu L W, Schubert E F, Hong M and Zydzik G J 1996 J. Appl. Phys. **80,** 6448-51

Inst. Phys. Conf. Ser. No 157
Paper presented at Microsc. Semicond. Mater. Conf., Oxford, 7–10 April 1997
© 1997 IOP Publishing Ltd

Application of secondary electron dopant contrast imaging to InP/InGaAsP laser structures

CP Sealy, MR Castell, CL Reynolds[*] and PR Wilshaw

Department of Materials, University of Oxford, Parks Road, Oxford OX1 3PH, UK
[*]Lucent Technologies Bell Laboratories, 9999 Hamilton Avenue, Breinigsville PA 18031-9359 USA

ABSTRACT: Electron microscopy techniques have great potential for dopant profiling because of their high spatial resolution in two-dimensions (2-D). Previous work has shown that contrast arises between *p*-, *n*- and *i*-doped material when observed in the secondary electron (SE) mode of a scanning electron microscope (SEM). This allows the direct 2-D mapping of the dopant distribution in a semiconductor structure. The results presented here apply the SE-technique to real fabrication issues that affect the operational characteristics and lifetimes of InP/InGaAsP capped mesa buried heterostructure (CMBH) laser devices.

1. INTRODUCTION

The use of modern fabrication techniques, enabling the reduction in semiconductor device dimensions, has demanded a corresponding improvement in characterisation techniques. Effective 2-D dopant profiling is crucial for the optimisation of all aspects of device fabrication and processing, as well as providing calibration for computer-aided-design simulation. Commonly used techniques such as secondary ion mass spectrometry (SIMS), capacitance-voltage and spreading resistance profiling essentially provide 1-D information only, whilst techniques such as transmission or scanning electron microscopy which can readily achieve 2-D spatial resolution usually require some sort of contrast enhancement treatment, such as staining or etching, prior to analysis (Subrahmanyan, 1992). However, recent work has drawn attention to dopant contrast which arises during SE-imaging and can produce direct qualitative (Perovic *et al*, 1995) and quantitative (Venables and Maher, 1996) dopant profiles in 2-D without complex sample preparation. The origin of the contrast observed between differently doped regions is believed to be electronic and is currently understood in terms of differences in ionisation energy. The ionisation energy is defined as the difference between the valence band edge and the vacuum level. Contrast arises because the ionisation energy is less in *p*-type than in *n*-type, which is reflected in their relative SE-yields and accounts for the bright appearance of *p*-doped regions with respect to *n*-doped areas.

2. EXPERIMENTAL METHODS

In this work, the SE-dopant imaging technique is applied to real InP/InGaAsP CMBH laser devices (wafers A, B, C and D) with substantially different lasing characteristics. Cross-sectional specimens were prepared simply by cleaving each wafer through the active layer stripe in air. SE-imaging was performed using two high resolution field emission gun SEMs offering an in-lens specimen position and SE-collection facility (a Hitachi S900 and a DSM 982 Gemini). In addition, a detailed quantitative analysis of the active layer of wafers C and D was achieved using a high

562

resolution standard SE-detector FE-SEM. The contrast values were determined from absolute secondary electron signal levels according to,

$$C_{pn} = (S_p - S_n)/S_n \qquad (1)$$

where S_p and S_n are the SE-signals from p- and n-type material respectively. The values obtained in this way are independent of the "contrast" and "brightness" settings on the microscope. In each case, working conditions of low accelerating voltage (1-2.5kV) and, where possible, low beam current were used (6 x 10^{-12} A).

3. DEVICE FABRICATION AND DEGRADATION MECHANISMS

The lasing characteristics and degradation behaviour of InP/InGaAsP CMBH lasers are determined by their constituent materials, method of operation and fabrication. This type of laser has a complex structure which requires three main fabrication steps (Figure 1): the first epi-growth to fabricate the buffer and InGaAsP active layers, followed by an etch delineation of the mesa structure; the second epi-growth over the mesa structure to fabricate a series of p-, n- and i-doped InP current blocking layers which serve to confine the injected current to the active region; and the third epi-growth to fabricate InP and InGaAsP capping and contact layers. The fabrication process is most

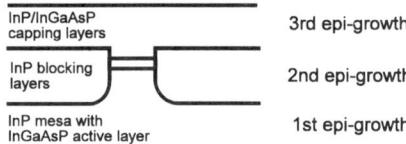

Figure 1 CMBH laser structure.

commonly performed by metal organic chemical vapour deposition (MOCVD) which is able to produce very sharp material and dopant interfaces. However, there are inherent weaknesses in this process because this type of device is very sensitive to the diffusion of zinc, the most commonly used p-type dopant, in or around the active and blocking layers during regrowth. This can displace pn-junctions, increase the threshold current and generate defects (Fukuda and Iwane, 1985). In particular, paths for the injected current to bypass the active layer can be created during fabrication of the blocking layers or subsequently by zinc diffusion from the capping layers which affect lasing efficiency. In the latter case, zinc dopant atoms diffuse from the p-type capping layers into the n-type blocking layers so that the region at their cusp is transformed from n- into p-type producing local "type conversion". In addition, a low zinc concentration in the active layer is essential for a low threshold current density and good lasing efficiency for this type of device. These effects can be minimised to a certain extent by a higher temperature and ratio of Group V:III species during low pressure MOCVD growth (Reynolds *et al*, 1994), and by maintaining an n-dopant level that exceeds that of zinc (Swaminathan *et al*, 1995).

4. RESULTS

The work presented here concentrates on two factors: (i) zinc diffusion at the cusp of the blocking layers and mesa; and (ii) zinc concentration in the active layer. The zinc concentration in the active layer is usually determined by SIMS analysis of a test structure, but this technique cannot generally provide dopant distribution information about the crucial region at the cusp of the mesa and blocking layers. SE-imaging in cross-section is ideal to investigate both these factors as it can directly detect dopant type and position, and can indicate relative concentrations.

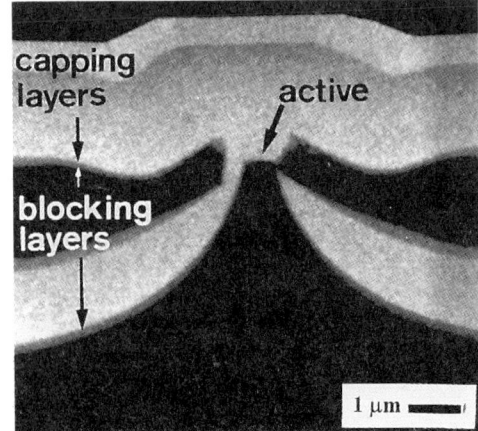

Figure 2 Wafer A on an in-lens SEM at 2kV.

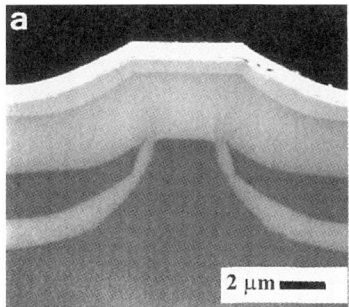
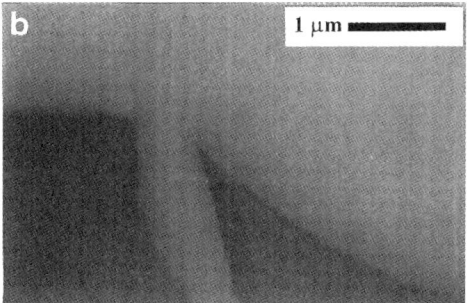

Figure 3 SE-image of wafer B (a) in overview and (b) in close-up of blocking layer - mesa cusp taken on an in-lens FE-SEM at 1kV.

The SE-technique can identify gross irregularities in structure, for example Figure 2 of wafer A, which shows an asymmetric laser device. It is apparent on the left hand side of the device structure that the n-type blocking layers do not make contact with the mesa, whilst the converse is true on the right hand side. This asymmetric structure is likely to give rise to poor lasing behaviour.

Under higher resolution imaging conditions these differences in structure are clearly evident and can be correlated with lasing characteristics. Figure 3a shows a device (wafer B) with poor lasing characteristics: (i) high threshold current of 13.6 mA; and (ii) low slope efficiency of 0.36 mWmA^{-1}. The SE-image of its dopant distribution shows that, as before, the n-type blocking layers do not reach the mesa and there is a wide p-type bypass path for injected current from the p-type capping layers into the bottom p-type blocking layer adjacent to the mesa. This region is shown in more detail in Figure 3b. It is likely that this p-type bypass path was produced by type conversion during the 3rd epi-growth stage.

Another example of a laser device with poor lasing characteristics, is shown in Figure 4 (wafer C), which has a threshold current of 10.9 mA and a slope efficiency 0.32 mWmA^{-1}. Again this poor behaviour can be accounted for by the presence of an injected current bypass route, but in this case provided by n-type material. During the 2nd epi-growth of the blocking region, the p-type layer has not extended as far as the mesa region so that a path is formed from the top n-type blocking layer directly into the n-type mesa which avoids the active region. A better device (wafer D) is shown in Figure 5 for comparison which has a substantially lower threshold current of 6.9 mA and greater slope efficiency of 0.41 mWmA^{-1}. It is evident from the SE-image that although a p-type bypass path is present at the top edge of the mesa linking the capping and blocking layers, it is significantly narrower. In addition, the active layer of D also has a relatively lower zinc concentration, see below.

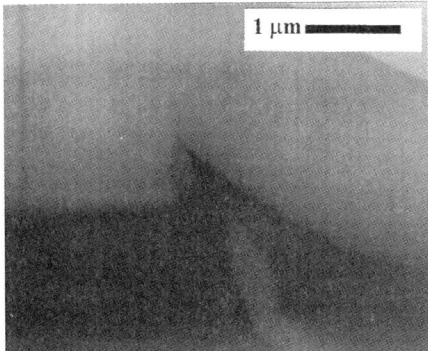
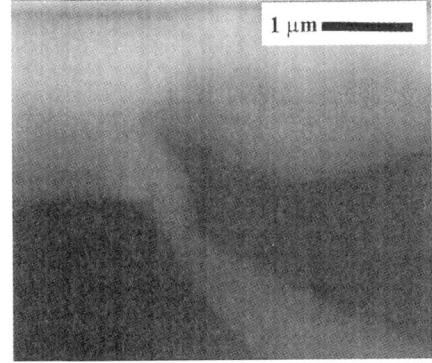

Figure 4 SE-image of wafer C taken at 1kV on an in-lens FE-SEM.

Figure 5 SE-image of wafer D taken at 1kV on an in-lens FE-SEM.

A second important factor that can affect the behaviour and lasing characteristics of this type of device is the zinc concentration in the active region. Laser samples B, C and D have active layer zinc concentrations of 8×10^{17} cm^{-3}, 2×10^{18} cm^{-3} and 4×10^{17} cm^{-3} respectively, as determined by SIMS on associated test structures, but these differences are not evident in the SE-images as reproduced here. However, when the SE-signals from wafers C and D were compared quantitatively, as described in Experimental Methods, it was possible to measure a substantial difference in contrast between their active layers. The active layer of C was found to have a contrast value relatively ~20% greater than D, in agreement with the corresponding zinc concentration determined by SIMS.

A summary of these results is shown below in Table 6 which display a good correlation between the known lasing characteristics and the dopant distribution determined by SE-analysis.

		Device specifications			
Wafer	Zinc concentration in active layer (from SIMS) /cm^{-3}	Slope efficiency /mWmA^{-1}	Threshold current /mA	Active layer width /μm	Dopant distribution (from SE-analysis)
A	-	low	high	0.5	poor
B	8×10^{17}	0.36	13.6	2.7	poor
C	2×10^{18}	0.32	10.9	4.1	poor
D	4×10^{17}	0.41	6.9	5.4	good

Table 6

5. CONCLUSION

The SE-imaging technique directly produces dopant distribution maps of cross-sectional semiconductor specimens very readily without the need for complex chemical preparation, such as staining or etching. The results presented, both qualitative and quantitative, have demonstrated that the technique can identify test laser structures with different dopant distributions in their active and blocking layers. The relative concentration of these dopant distributions, which are crucial to the characteristic behaviour and lifetime of InP/InGaAsP devices, can also be indicated. This work has shown that the SE-imaging technique is able to address successfully some current fabrication issues that have previously been difficult to tackle.

ACKNOWLEDGEMENTS

The authors would like to thank Dr. David Williams of The Cavendish Laboratory (University of Cambridge) and Dr. Sydney Nigrin of GEC-Plessey Semiconductors Ltd for the use of the S900 and DSM 982 Gemini SEMs respectively, and Dr. Robert Hull of the University of Virginia (formerly AT&T Bell Laboratories) for the provision of wafer A.

REFERENCES

Fukuda M and Iwane G 1985 J. Appl. Phys. **58** (8) 2932
Perovic DD, Castell MR, Howie A, Lavoie C, Tiedje T and Cole JS (1995) Ultramicroscopy **58** 104
Reynolds CL, Swaminathan V, Geva M, Smith LE, Luther LC 1995 J. Electron. Mater. **24** (6) 747
Seiler H 1983 J. Appl. Phys. **54** R1
Subrahmanyan R 1992 J. Vac. Sci. Technol. **B10** (1) 358
Swaminathan V, Reynolds CL, Geva M 1995 Appl. Phys. Lett. **66** (20) 2685
Venables and Maher 1996 J. Vac. Sci. Technol. **B14** (1) 421

Inst. Phys. Conf. Ser. No 157
Paper presented at Microsc. Semicond. Mater. Conf., Oxford, 7–10 April 1997
© *1997 IOP Publishing Ltd*

Degradation of electron-beam-pumped Zn₁₋ₓCdₓSe/ZnSe GRINSCH blue-green lasers

J-M Bonard, J-D Ganière, D Hervé[1], L Vanzetti[2], J J Paggel[2], L Sorba[2], E Molva[1] and A Franciosi[2]

Institut de Micro- et Optoélectronique, Ecole Polytechnique Fédérale, CH-1015 Lausanne, Switzerland
[1] LETI Département Optronique, 17 rue des Martyrs, F-38054 Grenoble Cédex 9, France
[2] Laboratorio Nationale TASC-INFM, Area di Ricerca, Padriciano 99, I-34012 Trieste, Italy

ABSTRACT: We explored the degradation in electron beam pumped $Zn_{1-x}Cd_xSe/ZnSe$ laser structures by combining cathodoluminescence measurements in a scanning electron microscope with transmission electron microscopy. We found that degradation occurred via the formation of dark spot and dark line defects, and that it involved the formation of a characteristic type of defect, namely dislocation loops within the ZnCdSe quantum well, which grew to form a characteristic network as degradation progressed.

1. INTRODUCTION

An intense effort has been devoted in the last few years to the realisation of incoherent and coherent light emitters based on II-VI semiconductors and operating in the blue-green range of the visible spectrum. Although important progress has been made, II-VI blue-green laser diodes operating at room temperature and in cw mode are still far from showing reliability and lifetime comparable to those of their III-V infrared counterparts.

Recently, microgun-pumped blue-green lasers have been demonstrated by Hervé et al (1995). Such electron beam (e-beam) pumped devices use $Zn_{1-x}Cd_xSe/ZnSe$ graded-index, separate confinement hetero-structures (GRINSCH) grown by molecular beam epitaxy (MBE), in combination with a lithographically patterned microtip cathode, which acts as a miniaturized electron source for injection. Because e-beam pumping does not require doping or contact fabrication, this approach allows one to circumvent some of the major limitations of current II-VI laser diodes, and to evaluate the remaining intrinsic material limitations.

We studied the degradation (i.e. the decrease of the luminescence yield) of e-beam pumped $Zn_{1-x}Cd_xSe/ZnSe$ GRINSCH structures by combining cathodoluminescence (CL) measurements in a scanning electron microscope (SEM) with transmission electron microscopy (TEM) studies of selected microgun-pumped structures. In the former type of measurements, the SEM directly supplied e-beam pumping for electron-hole pair excitation with an energy comparable to the lasing experiments, the degradation, in real time, being monitored by the CL signal with submicron spatial resolution. If this type of degradation study is performed on thin-foil samples prepared for TEM measurements, a direct correlation between the degradation behaviour and the formation of defects is possible.

2. EXPERIMENTAL DETAILS

All laser structures were grown by molecular beam epitaxy on $In_{0.01}Ga_{0.99}As$ (001) wafers. Two buffer layers, comprised of $In_{0.04}Ga_{0.96}As$ (1μm thick, lattice matched to ZnSe) and ZnSe (1μm thick), were grown sequentially on the substrate prior to the deposition of the GRINSCH. The GRINSCH itself included a $Zn_{1-x}Cd_xSe$ quantum well (QW) embedded between two 500nm thick $Zn_{1-x}Cd_xSe$ graded layers, where x varied continuously from x=0.05 (at the well boundaries) to x=0. Typical well thicknesses examined were in the 50-100Å range, and typical Cd concentration in the well were in the 0.15<x<0.25 range.

Plan-views, as well as cross-sections, were prepared from all structures by mechanical thinning down to a thickness of 50μm followed by 5keV-Ar$^+$ ion bombardment. The TEM observations were carried out at 300keV.

The CL measurements were done with a modified stage allowing measurements between 10K and room temperature. The sample CL was collected by an ellipsoidal mirror and focused on a Si-photodiode for polychromatic imaging.

3. ELECTRON-BEAM INDUCED DEGRADATION

All examined structures showed strong cathodoluminescence (with typical experimental conditions of E_b=20 KeV and I_b=10 nA) up to 300K. Polychromatic maps revealed non-uniform CL, and the defects affecting the CL were visible mainly as dark spot defects (DSD), with a few dark-line defects (DLD) oriented along [100] and [010].

Two kinds of degradation experiments were undertaken in the SEM. Scan mode degradation studies were performed by scanning the beam over a limited area of the sample surface at a fixed magnification of 5000 × with TV scan rate. This scan mode emulated the operating mode of the laser under electron-beam pumping. Spot mode degradation studies were performed by placing the beam at a point of the sample surface, and recording the behaviour of bright and dark spots under bombardment.

The degradations performed in scan mode showed three phases. At first, the intensity increased, up to a factor of 2 under certain injection conditions. The second phase was a near exponential decrease. A degradation half-time, τ, could be deduced by fitting an exponential function $I(t) = b\exp(-t/\tau)$ to the intensity profile. The third phase corresponded to intensities below 5 % of the initial intensity, and was characterized by a slow decrease of the degradation rate.

The degradation half-time on a structure depended strongly on the injection conditions. It decreased rapidly with increasing current, and varied with the beam energy at constant beam power. In fact, τ decreased when the initial CL intensity increased. The degradation behaviour thus did not depend only on the injected power, but also on the emitted luminescence intensity.

The influence of the structural parameters was evaluated by comparing the degradation half-times of different structures at constant beam current and beam energy. We found that the degradation half-time increased with increasing threading dislocation density (stacking faults did not play a significant role) and with increasing Cd content, and thus compressive strain, in the quantum well.

Degradation experiments in spot mode showed that there were two types of degradation behaviour. The first one was typical of the majority of bright spots. The intensity increased at first (or decreased slowly), prior to a catastrophic degradation after a certain lapse of time. CL maps acquired before and after this rapid decrease showed that it coincided with the creation of a DSD. The second type was encountered on DSD as well as on some bright spots. The intensity decreased in abrupt steps that were correlated to the extension of the DSD in the CL maps, followed by regions of slower degradation.

CL maps taken during experiments in scan mode on samples with high defect densities indicated furthermore that the bombardment resulted in the random extension of existing DSD and the creation of new DSD. On samples with low defect densities however, the DSD grew and extended along [100] or [010] directions, for a part of them actually along existing DLDs.

4. DEGRADATION EXPERIMENTS ON THIN-FOIL SAMPLES

In order to investigate in more detail the mechanisms leading to degradation, we performed subsequent CL and TEM observations on the same areas of TEM thin foils (plan-views as well as cross-sections). This procedure permitted a direct correlation between the CL contrasts and the presence of structural defects, as well as between the degradation behaviour and the presence and/or formation of extended defects.

The observation procedure is depicted in Fig. 1. In a first phase (figure 1a), the thin foil was characterized in the TEM, and a region with salient topographical features (such as pyramids, other 3D growth defects and/or cleavages) was selected. The thin foil was then characterized by SEM, the region surveyed in TEM being identified in secondary electron (SE) mode with the topographical features detected in TEM and then observed in CL. In a second phase (Fig. 1b), degradation experiments were performed in the SEM on the thin foil. Finally, the degraded region was reobserved in the TEM and compared to its nondegraded state.

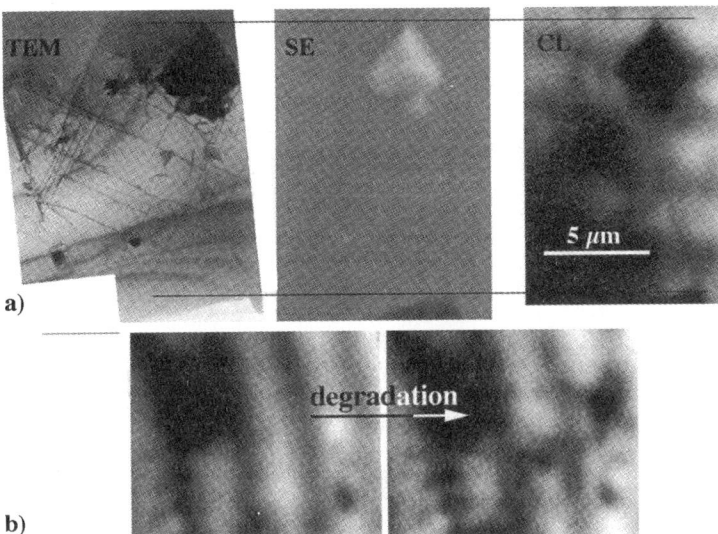

Fig. 1 (a) combined CL and TEM observation procedure: characterization of the thin foil by TEM, identification of the region in SEM in SE mode and CL mapping; (b) degradation of the thin foil in the SEM. The alternating dark and bright fringes parallel to the rim of the foil in the CL images are equal thickness fringes provoked by constructive and destructive interferences of the emitted CL in the thin foil.

Our observations showed that degradation inducing a significant decrease of the emitted luminescence (≥ 10 %) resulted in the nucleation of degradation defects. A typical defect is shown in Fig. 2a.

The defects started in most cases at a threading dislocation, as in figure 2a where the 60°-type dislocation is clearly visible and has bowed out during the formation of the defects. The defect in itself was constituted of little segments, each approximately 20-60 nm in length, that were arranged in V-shaped tree-like structures. The "V" opened in a $\pm[100]$ or $\pm[010]$ direction, and the branches of the "V" were nearly symmetrically disposed on each side of the mean direction. The direction of the segment lines varied from one defect to another and showed an angle of 29-38° with the mean direction. These variations probably reflected changes in the local strain around the QW as pointed out by Nakano et al (1996).

To determine the location of the defects, a cross-section was degraded in the SEM and observed along a [111] zone axis, as in Fig. 2b. The sample was viewed at an angle of 45° with respect to the plane of the interfaces. The two black line contrasts (indicated by two white lines on the TEM image) corresponded to the intersection of the quantum well with the foil surfaces. A threading dislocation could be seen just below the well, as well as three degradation defects. The extension of their contrast was strictly contained between the two lines bounding the well, which indicates clearly that they were located in the QW or at the QW interfaces.

The Burgers vector and the type of the defect were determined from observations in two-beam diffraction conditions. The Burgers vector was $\underline{b} = a/2[01\bar{1}]$ for degradation defects opening in $\pm[100]$ directions, and $\underline{b} = a/2[10\bar{1}]$ for defects opening in $\pm[010]$ directions. $\underline{b}$ showed therefore a 45° angle with the interface plane, and was perpendicular to the orientation of the defect. Under certain diffraction conditions, bright/dark lobe contrasts were observed, which indicated that the defects were constituted of small dislocation loops with diameter smaller than 10 nm. Careful analysis of the lobe shape allowed us furthermore to conclude that the degradation defects were formed of interstitial and vacancy dislocation loops in nearly equal proportions. It is worth noting that the defects visible in Fig. 2 have also been observed in II-VI laser structures designed for optical and/or electrical pumping by Gua et al (1993), Hovinen et al (1995) and Hua et al (1994). The Burgers vectors identified here also agreed with studies performed on III-V heterojunction lasers by Petroff et al (1974) and O'Hara et al (1977), as well as with the findings of Nakano et al (1996) Salamanca-Riba and Kuo (1996) on II-VI structures,

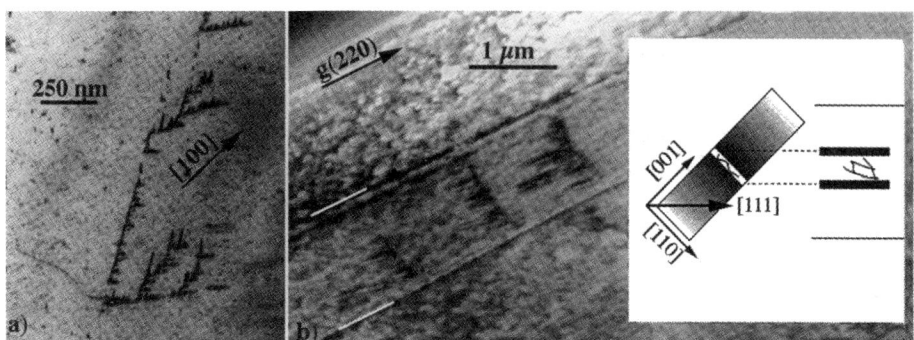

Fig. 2   (a) Degradation defect opening in the [100] direction viewed in plan-view near the [001] zone axis; (b) degraded cross section observed in two-beam diffraction conditions with g = (220) in dark field near the [111] zone axis. The two lines mark the intersection of the well with the surfaces of the thin foil.

5. CONCLUSIONS

We found that degradation in microgun-pumped blue-green GRINSCH lasers occurred via the formation of DSD and/or <100>-oriented dark line defects, in a manner compellingly similar to what has been observed in blue-green diode lasers. Our combined CL-TEM studies clearly indicated that the degradation involved the formation of a characteristic type of defect, namely small dislocation loops within the ZnCdSe quantum well, which grew to form a characteristic network as degradation progressed.

We would like to thank G. Peter, B. Garoni and B. Senior at the Centre Interdépartemental de Microscopie Électronique (CIME) for the expert technical support with the microscopes.

REFERENCES

Guha S DePuydt J M, Haase M A Qui J and Cheng H 1993 Appl. Phys. Lett. 63 3107
Hervé D Accomo R, Molva E, Vanzetti L, Sorba L and Franciosi A 1995 Appl. Phys. Lett. 67 2144
Hovinen M, Ding J, Salokatve A, Nurmikko A V, Hua G C, Grillo D C Li He Han J Ringle M and Gunshor R L 1995 Appl. Phys. Lett. 66, 2013
Hua G C Otsuka N Grillo D, Fan , Han J Ringle M D Gunshor R L Hovinen M and Nurmikko A V 1994 Appl. Phys. Lett. 65 1331
Nakano K, Tomiya S, Ukita M, Yoshida H, Itoh S, Morita E, Ikeda M, and Ishibashi A 1996 J. Electron. Mater. 25, 213
O'Hara S Hutchinson PW and Dobson PS 1977 Appl. Phys. Lett. 30, 368
Petroff P and Hartman RLJ 1974 Appl. Phys. 45, 3899
Salamanca-Riba L and Kuo LH 1996 J. Electron. Mater. 25, 239

Inst. Phys. Conf. Ser. No 157
Paper presented at Microsc. Semicond. Mater. Conf., Oxford, 7–10 April 1997
© 1997 IOP Publishing Ltd

Degradation dynamics of II-VI (ZnCdSe) quantum well materials using confocal photoluminescence microscopy

D T Fewer, C Jordan, S J Hewlett, E M McCabe, F P Logue, J F Donegan, J Hegarty, S Taniguchi*, T Hino*, K Nakano* and A Ishibashi*

Department of Physics, Trinity College Dublin, Dublin 2, Ireland
*Sony Research Centre, Fujitsuka-cho, Hodogaya-ku, Yokohama-shi 240, Japan

ABSTRACT: A confocal microscope set-up is used for the first time to image degradation processes in ZnCdSe-based quantum well heterostructures. As well as being non-destructive, this technique permits the degradation to be monitored in real time and over a large area in samples with low defect densities. Moreover, this set-up offers advantages over conventional optical microscopes, such as a higher lateral resolution and a very narrow depth-of-field. High-contrast images are presented that show the defect formation and propagation within the quantum well region of the device.

1. INTRODUCTION

Wide-band-gap semiconductor lasers are expected to provide the next generation of laser diodes (LD's) for high-density optical storage and optical displays operating in the blue and green regions of the visible spectrum. III-V GaN-based heterostructures (Nakamura *et al.* 1996) and II-VI ZnSe-based heterostructures (Taniguchi *et al.* 1996) are the most likely candidates. However, the short operating lifetime of these devices remains a major technical problem which must be overcome before commercialization.

Previous studies have shown (Honiven *et al.* 1995, Tomiya *et al.* 1995, Hua *et al.* 1994, Guha *et al.* 1994, Guha *et al.* 1993) that degradation in II-VI-based devices originates from pre-existing defects such as stacking faults and threading dislocations. These defects are formed at the GaAs/ZnSe hetero-interface and extend throughout the different II-VI layers. They are the precursors for the formation of dark spots in the active region that can be observed by imaging the luminescence during degradation.

Considerable effort has been put into reducing the dark spot density to less than 10^4 cm^{-2} in order to realise a LD that has no stacking faults in its stripe area. At present, a 100 hour lifetime II-VI LD operating at room-temperature and under cw excitation has been obtained with a stacking fault density of less than 3×10^3 cm^{-2} (Taniguchi *et al.* 1996).

In this paper, we show that confocal photoluminescence microscopy (Cox 1984) is a powerful tool for the real-time imaging of degradation. This method does not require any sample preparation and can provide high-contrast photoluminescence images of events that take place in the active layer of the LD. In addition, the lateral resolution is improved relative to that of the conventional microscope (Wilson and Sheppard 1984).

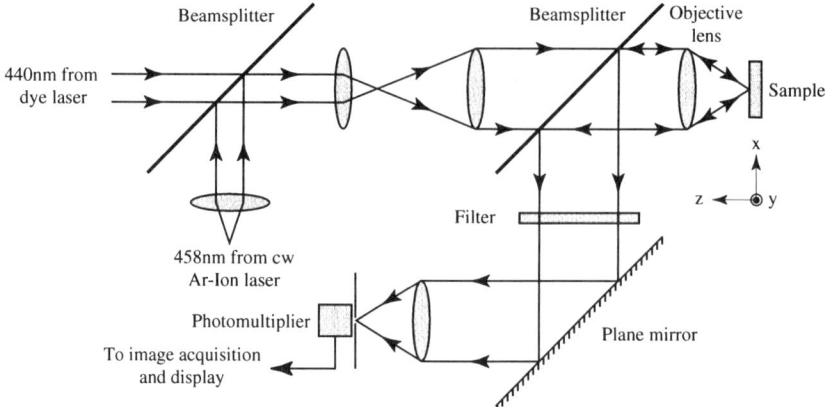

Figure 1: Optical set-up of the confocal photoluminescence microscope.

2. EXPERIMENT

The sample studied was a single-quantum-well separate-confinement heterostructure which was grown by molecular beam epitaxy (MBE) at the Sony Research Center. It consists of a GaAs buffer layer, a 30 nm ZnSe layer, a 150 nm ZnSSe buffer layer, a 500 nm ZnMgSSe cladding layer and a 100 nm ZnSSe barrier. The active region was a 6 nm $Zn_{0.75}Cd_{0.25}Se$ single quantum well and was followed by a further 100 nm ZnSSe barrier layer and a 200 nm ZnMgSSe capping layer. Although undoped, this is a typical laser structure in which the wide-band-gap ZnMgSSe layers provide good optical confinement in the active region.

Fig. 1 shows a schematic diagram of the confocal microscope set-up used for the degradation measurements. The sample was degraded at room temperature by optical injection using a synchronously pumped mode-locked dye laser operating at 440 nm and providing 5 ps pulses at a repetition rate of 76 MHz. A 140–μm–diameter excitation spot was formed at the quantum well by displacing the stationary sample axially from the focal plane of the objective lens. The degraded region was then probed using the 458 nm line of a cw Ar-Ion laser. Confocal imaging was achieved by placing the quantum well in the focal plane of the objective lens, raster scanning the sample in x and y and imaging the photoluminescence through a pinhole in the detection plane. In order to collect luminescence from the quantum well, a colour glass filter with an optical density ≈ 5 at the excitation wavelength was placed in the detection path. The images obtained are high-resolution (submicron), high-contrast maps of the photoluminescence emission from the quantum well. Defects involving non-radiative centers are detected by the absence of photoluminescence.

3. RESULTS

Fig. 2 shows the progressive stages of degradation during high intensity optical excitation of 1 μJ/cm^2 at 440 nm, corresponding to a carrier density of $\approx 10^{12}$ cm^{-2}. At the early stages of degradation, we can see the formation of dark spots ($t = 1$ min.). As these grow in size and become triangular in shape, dark line defects emanate from them

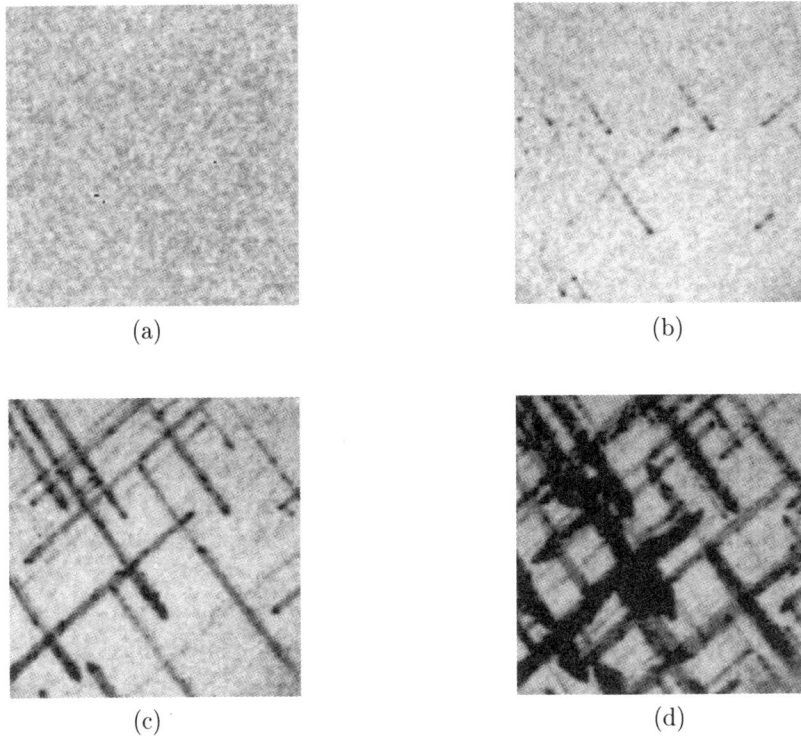

(a)

(b)

(c)

(d)

Figure 2: 90×90 μm photoluminescence images of the ZnCdSe quantum well showing the evolution of degradation over time with high optical excitation: (a) $t = 0$, (b) $t = 1$, (c) $t = 10$ and (d) $t = 30$ mins.

and propagate along oriented directions (t=10 min.). The existence of a high initial density of dark spots leads to a dense network of dark line defects which dramatically reduces the luminescence intensity after 30 minutes of optical excitation. We believe the initial dark spots are related to pre-existing defects as they are homogeneously distributed throughout the sample with a density as high as 2×10^5 cm^{-2}, independent of the excitation intensity.

Guha *et al.* (1993) and Hua *et al.* (1994) have shown that the initial dark spots are related to stacking faults present at the GaAs/ZnSe interface. The progressive darkening of the active layer is thought to be caused by the diffusion and aggregation of non-radiative centers of recombination present in the well and in the barrier material and not directly by photon or electron-hole non-radiative recombination (Chuang *et al.* 1996). The presence of oriented dark lines suggests that non-radiative centers of recombination can diffuse only along privileged directions of the crystal lattice (Haugen *et al.* 1995).

In general, photoluminescence confocal scanning microscopy has two important advantages over conventional microscopy for the imaging of defect formation in the active region of II-VI laser structures. Firstly, it offers superior lateral resolution due to the effect that employing a pinhole in the detection plane has on the systems optical transfer function (Wilson and Sheppard 1984). Secondly, the narrow depth of field permits depth discrimination which improves image contrast because light scatter from regions outside the range of the optical probe is not present in the final image.

4. CONCLUSION

We have shown that confocal photoluminescence microscopy is a powerful tool for imaging optically induced degradation in I I-VI semiconductor materials. This technique permits high contrast, sub-micron imaging of the degradation dynamics in the active region below the surface of the device. Degradation is initiated by dark spot defects whose density is found to be homogeneous throughout the sample. Dark line defects are observed by imaging the photoluminescence during degradation.

REFERENCES

Chuang S L, Ukita M, Kijima S, Taniguchi S and Ishibashi A 1996 Appl. Phys. Lett. **69**, 1588

Cox I J 1984 J. Microsc. **133**, 149

Guha S, Cheng H, Haase M A, Depuydt J M, Qiu J, Wu B J and Höfler G E 1994 Appl. Phys. Lett. **65**, 801

Guha S, Depuydt J M, Haase M A, Qiu J and Cheng H 1993 Appl. Phys. Lett. **63**, 3107

Haugen G M, Guha S, Cheng H, Depuydt J M, Haase M A, Höfler G E, Qiu J and Wu B J 1995 Appl. Phys. Lett. **66**, 358

Honiven M, Ding J, Nurmikko A V, Hua G C, Grillo D C, Li He, Han J, Gunshor R L 1995 Appl. Phys. Lett. **66**, 2013

Hua G C, Otsuka N, Grillo D C, Fan Y, Han J, Ringle M D, Gunshor R L, Honiven M and Nurmikko A V 1994 Appl. Phys. Lett. **65**, 1331

Nakamura S, Senoh M, Nagahama S, Iwasa N, Yamada T, Matsushita T, Sugimoto Y, Kiyoku H 1996 Appl. Phys. Lett. **69**, 4056

Taniguchi S, Hino T, Itoh S, Nakano K, Nakayama N, Ishibashi A and Ikeda M 1996 Electron. Lett. **32**, 552

Tomiya S, Morita E, Ukita M, Okuyama H, Itoh S, Nakano K and Ishibashi A 1995 Appl. Phys. Lett. **66**, 1208

Wilson T and Sheppard C J R 1984 Theory and practice of scanning optical microscopy Academic Press London

Inst. Phys. Conf. Ser. No 157
Paper presented at Microsc. Semicond. Mater. Conf., Oxford, 7–10 April 1997
© 1997 IOP Publishing Ltd

Antiphase Boundaries in GaAs/Ge Solar Cells

C Hardingham, D B Holt, L Lazzarini*, M Mazzer⁺, L Nasi*, B Raza and C Zanotti-Fregonara*

Department of Materials, Imperial College of Science, Technology and Medicine, London SW7 2BP
*CNR - MASPEC Institute, Via Chaivari 18/A, 43100-I Parma, Italy
⁺CNR - IME, Dipartimento di Scienza dei Materiali, Universita di Lecce, Via Anresand, 93100 Lecce, Italy

ABSTRACT: The interfacial recombination velocity parameter was measured for many antiphase boundaries (APBs) in GaAs/Ge solar cells using the Donolato theory for the EBIC contrast. The values covered a wide range. Low temperature cathodoluminescence imaging and spectral results are also reported. The APBs occured only in narrow bands round the wafer periphery. Cross-sectional TEM showed many small antiphase domains at the interface were soon grown over except in these bands, where some APBs extended up through the GaAs layer. These results are discussed.

1. INTRODUCTION

Antiphase boundaries (APBs) tend to occur in profusion in epitaxial films of polar semiconductor compounds grown on (100) Ge or Si substrates. They have received much attention in recent years because of the interest in GaAs/Si for integrated optoelectronics and in GaAs/Ge which is now the most economical material for solar cells for space satellites. It was found that APBs occurred in MOCVD (metal organic CVD) grown (100) GaAs/Ge only in narrow bands round the periphery of the wafers and that these defects gave strong dark contrast allowing both SEM EBIC and CL imaging (Holt et al 1995). The Donolato (1985) phenomenological theory of dark EBIC contrast was applied to the linescans to obtain values for the recombination velocity (Holt et al 1995, 1996). In this paper are presented new TEM and CL observations and many new determinations of the recombination velocities of APBs and the results of these measurements are discussed.

2. EXPERIMENTAL METHODS

The material was all grown by MOCVD at EEV Ltd. The EBIC measurements were made at Imperial College in a JEOL JSM-840A SEM fitted with a Matelect ISM-5 EBIC detection system and a Kontron image analysis system. The APB dark contrast linescan profiles were analysed using Donolato's phenomenological theory equations using the Mathematica software, at the University of Lecce.

The TEM work was carried out at MASPEC on a JEOL 2000 FX microscope at 130 kV. The CL observations were also carried out at MASPEC on a Cambridge 150 SEM fitted with an Oxford Instruments monoCL detection system.

The microprobe optical beam induced current (μOBIC) observations were made at the University of Lecce. For a brief account of this finely-focussed, laser-excited technique see DeVittoria et al (1997).

3. RESULTS

As in previous studies (Holt et al 1995, 1996) it was found in SEM EBIC images that the APBs occurred only in a narrow band around the periphery of the wafers. The antiphase domains (APDs) formed an interlocked jig-saw puzzle pattern in the centre of the band but became fewer toward the edges so only scattered individual APDs were seen and none occurred over most of the wafer area. EBIC images of these types were published previously (Holt et al 1995, 1996) and look just like that recorded with the new μOBIC technique (Figure 1). It was found to be best for quantitative contrast studies to make EBIC linescan recordings along lines, like that in Figure 1, across individual, isolated APDs near the edges of the band of APBs (Figure 2). Otherwise overlapping of the APB contrast dips in the adjacent APDs would have made it difficult to determine the real EBIC signal level in defect free material, as required for the analysis.

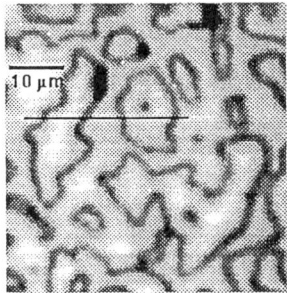

Figure 1. μOBIC micrograph showing an area in the band of APBs in a GaAs/Ge solar cell. The horizontal line shown would be a good one along which to record four reasonably resolved APB contrast signatures.

Figure 2. Typical quantitative EBIC linescan profile recorded across an isolated APD in an area like that in Figure 1. The apparent peaks actually represent reductions in the negative EBIC current (plotted downward).

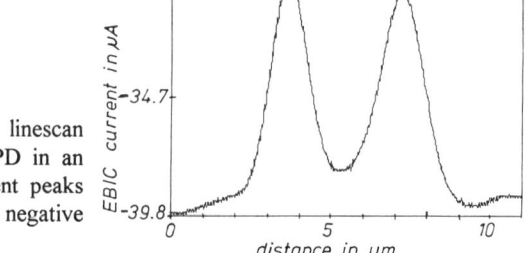

A few linescans were recorded with the sample cooled to liquid nitrogen temperature. Little change in contrast occurred on lowering the temperature, although there appeared to be a weak trend of s (the recombination velocity parameter with dimensions μm^{-1}) to decrease with temperature. The results of all the new APB measurements are given in Table 1 and these plus all the previously determined s values (Holt et al 1995, 1996) are plotted in the histogram of Figure 3. It can be seen that most APBs had small values, around the peak value of s = 10 - 20 μm^{-1}. However individual APBs had values over a wide range of larger magnitudes with some indication of the possible occurrence of additional peaks.

The cross-sectional TEM micrographs (Figure 4) showed that outside the band of APBs large numbers of very small antiphase domains (APDs) occurred to a height of about 200 nm from the interface and a few larger APDs extended up 1 or 2 μm but eventually all were

Table 1. APB Dimensionless Recombination Velocities s and Minority Carrier Diffusion Lengths derived from EBIC Linescan Analyses of linescans across APBs in specimens from wafer MR546-16-2

LSP & APB	s in μm^{-1}	L in μm	LSP & APB	s in μm^{-1}	L in μm	LSP & APB	s in μm^{-1}	L in μm
1a	36.8	0.518	12a	18.8	0.632	7'a	13.4	0.896
1b	101	0.44	14a	49.4	0.453	7'b	22.7	1.02
2b	227	0.484	14b	33.6	0.467	9'a	6.68	1.07
3b	87.9	0.52	15a	164	0.531	9'b	72.6	2.26
4a	141	0.493	15b	25.7	0.97	10'a	11.1	0.766
4b	49.6	0.509	16a	7.61	0.616	10'b	12.3	0.943
5a	54.4	0.469	16b	29.8	0.791	11'a	4.31	1.42
6a	218	1.58	17a	8.23	0.726	11'b	3.51	1.88
6b	28.6	1.67	17b	31.5	0.85	12'a	14.4	0.926
7a	57.9	0.84	18a	7.51	0.867	12'b	13.6	1.12
7b	50	0.721	19a	23.2	0.675	13'a	5.52	1.75
8a	87.5	0.93	19b	9.12	0.589	13'b	4.05	1.66
9a	34.6	0.778	2'a	5.49	1.14	14'a	15	0.896
9b	40.2	0.735	3'a	0.925	2.01	14'b	13.1	1.11
10a	29.2	0.636	3'b	0.953	1.49	16'a	12.5	1.06
10b	117	0.942	6'a	4.96	1.27	16'b	8.38	1.16
11a	17.5	0.675	6'b	8.34	1.63	17'a	11.6	0.974
11b	32.3	0.837				17'b	11.7	0.937

The LSPs are numbered and the letters denote two APBs that appeared in most profiles (cf. Figures 1, 2). Primed numbers are measurements made at different dates on different areas of the wafer.

576

overgrown during the deposition of the GaAs. In the band of macroscopically visible APBs (Figure 5) the 200 nm layer of small overgrown APDs again appeared but a few APBs grew up to the final free surface where steps 700 - 800 nm high occurred.

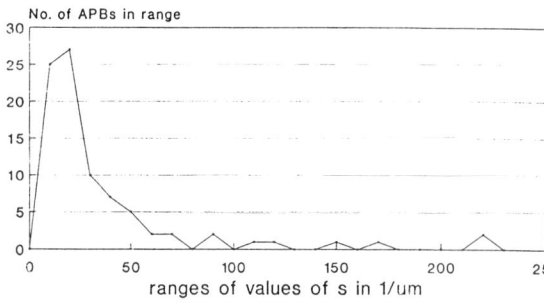

Figure 3. Histogram of interfacial recombination velocities of all the APBs measured in two samples from MOCVD (100) GaAs/Ge solar cell MR546-16-2 plus all the previously published values for APBs in samples from cells 3929-3 and 5423-3. At each value of s are plotted the number of APBs with recombination parameter values between s and s - 10.

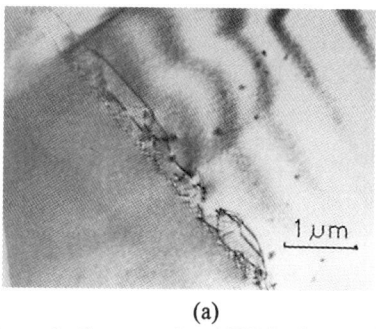
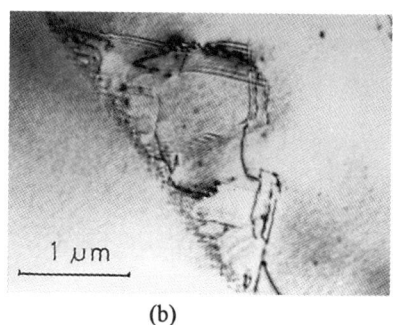

(a)   (b)

Figure 4. Cross-sectional TEM micrographs of areas outside the macroscopically visible band of APBs on solar cell specimen MR553-28. In such areas at the interface there is (a) a high density of dislocations and of small antiphase domains extending up no more than about 200 nm into the GaAs layer plus, sometimes, (b) a few larger APDs extending up 1 or 2 μm before being overgrown.

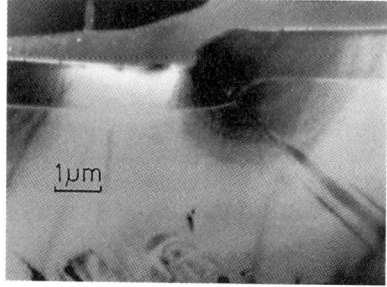

Figure 5. In the macroscopically visible APB band the first 200 nm above the interface again contains numerous overgrown APDs but very few dislocations. In addition there are a few APBs, spaced about 10 μm apart, that extend up to the final free surface where steps 700 - 800 nm high occur.

Liquid nitrogen temperature monochromatic (825 nm i.e. intrinsic) CL micrographs (Figure 6a) revealed the presence of the APBs in the bands just as EBIC was previously found to do. CL spectroscopy at this temperature showed that the large domains of opposite polarity, i.e. the single crystal material outside the APB band near the periphery of the wafer (region X) and that inside the band (region Y), gave near band gap, intrinsic emission bands

intrinsic emission band similar to that of region X (as marked for the near band-gap emission in Figure 6) plus an extrinsic, longer wavelength band presumably associated with APB radiative recombination.

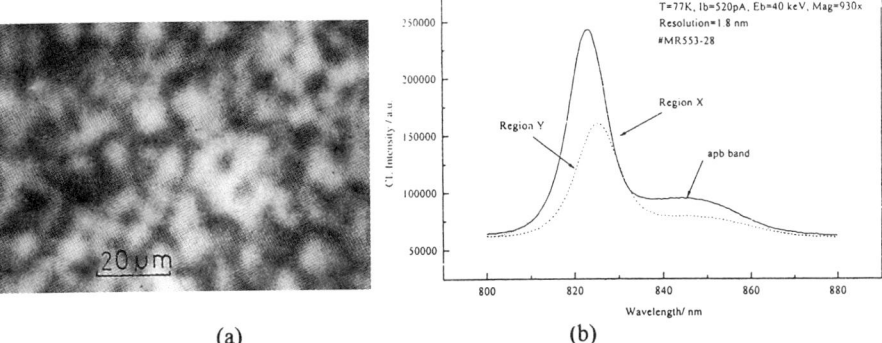

(a)   (b)

Figure 6. Liquid-nitrogen temperature (a) monochromatic (825 nm) CL image of an area in the band of APBs in a GaAs/Ge solar cell and (b) spectra of the regions inside (Y) and outside (X) the band of APBs.

4. DISCUSSION

The observation that μOBIC can also produce high dark contrast images of the APB structure of the material studied here extends the range of techniques giving approximately equally good results in such microcharacterization. Liquid nitrogen temperature CL images (e.g. Figure 3a) showed the APBs as fuzzier defects than they appear in EBIC and μOBIC pictures. Whether this is due to the recombination strength of the defects being less and/or their width greater in CL than for contrast in these other techniques is not yet known. The 3nm blue shift of the intrinsic emission band from one polarity domain (region Y) to the other (region X), although small, is new and may be significant. The observation of an apparent APB band is also new. This could be due to recombination via wrong bonds or impurities segregated to the APBs.

The recombination parameters s of APBs extend over a wide range but the most numerous APBs have s values near the peak around 15 μm^{-1}. This must be due to different APBs having different recombination centres or different area densities of recombination centres whether intrinsic (wrong bond like) or impurity related. It is known that crystallographically there are planes on which APBs would be polar (all one atom type i.e. all one wrong bond type: Ga-Ga or As-As) and others that are apolar (equal numbers of both kinds of wrong bonds) (Holt 1969). The occurrence of A(Ga), B(As) and apolar APBs would give three possible levels of recombination strength. Partially polar orientations can also occur. Recent etching studies (Li and Giling 1996) showed that, macroscopically, APBs tend to lie in {110} (apolar) planes. These APBs are of minimum energy but the wrong bonds could still give rise to states at two levels in the band gap to act as recombination centres and so reduce the EBIC signal locally. This gives rise to the dark contrast dealt with by the Donolato theory. Hence it seems likely that the numerous APBs with low recombination strengths relate to those of the apolar {110} type. Boundaries even when macroscopically planar can contain step facets of various orientations including polar ones. The spread of observed recombination strengths thus may arise from (i) the presence of varying percentages of the area in atomic scale steps in polar orientations or, of course, (ii) varying types and amounts of impurity decoration.

Fairly large, isolated APDs near the edges of the bands of APBs were selected for analysis to optimise contrast resolution. It is possible that this resulted in certain types of APBs being preferentialy analysed. It appeared from the micrographs that the range of contrast from weaker to stronger was wider in the centre of the APB bands. This is probably a misleading impression due to near overlap effects. The impression of a wider range of contrast and hence of recombination strengths among band-centre APBs was not supported by the results of quantitative analyses, where these could be carried out on linescans from central areas in the band.

A few low temperature EBIC observations were carried out to check whether any major changes occurred (as in the case of twin boundaries in polycrystalline Si solar cells which only become visible in EBIC at liquid nitrogen temperatures - Raza and Holt 1995). No such effect of the movement of the Fermi level with temperature on EBIC contrast was observed in the case of APBs. This means that the movement of the Fermi level from liquid nitrogen to room temperature did not carry it through the energy level or levels of the APBs, so it did not change the charge state and hence the recombination strength of these defects significantly.

The TEM results show that APDs are nucleated in high densities in the initial layers of GaAs deposited on Ge. The final production of APB free material allowing the successful production of high efficiency solar cells is due to the overgrowth of these nuclei over most of the area of the wafer areas by material of a single "phase".

REFERENCES

DeVittorio, M, Cingolani, R, Mazzer, M and Holt, DB 1997 Appl. Phys Lett. (to be published)

Donolato, C 1985 in Polycrystalline Semiconductors, ed. G Harbeke (Berlin: Springer-Verlag) pp. 138 - 154

Holt, DB 1969 J. Phys. Chem. Solids **30**, 1297 - 1308

Holt, DB, Mazzer, M, Zanotti-Fregonara, C, Hardingham, C, Salviati, G, Lazzaraini, L and Nasi, L 1995 in Inst. Phys. Conf. Ser. No.146 (Bristol: Inst. Phys.) pp. 713 -718

Holt, D B, Hardingham, C, Lazzarini, L , Nasi, L, Zanotti-Fregonara, C, Salviati, G, and Mazzer, M 1996 Mat. Sci. Eng. **B42**, 204 -207

Li, Y and Giling, L J 1996 J. Cryst. Growth **163**, 203 - 211

Raza B and Holt DB 1995 in Inst. Phys. Conf. Ser. No.146 (Bristol: Inst. Phys.) pp. 107 - 112

Inst. Phys. Conf. Ser. No 157
Paper presented at Microsc. Semicond. Mater. Conf., Oxford, 7–10 April 1997
© *1997 IOP Publishing Ltd*

EBIC and cathodoluminescence studies of grain boundary and interface phenomena in CdTe/CdS solar cells

S A Galloway*, P R Edwards and K Durose

Department of Physics, University of Durham, South Road, Durham, DH1 3LE, UK
* Now at Oxford Instruments, Old Station Way, Eynsham, Witney, Oxford, OX8 1TL

ABSTRACT: SEM/EBIC and cathodoluminescence (CL) have been used to investigate grain boundary and interface effects in CdTe/CdS/ITO/glass solar cells. EBIC images of grain boundaries in the CdTe near to the electrical junction were made by injection through the window (ITO/CdS) side of the cell. The contrast was injection dependent and indicated that the grain boundaries are passivated. Comparison of this with rear wall and cross section EBIC suggested a carrier distribution model for the material. CL microscopy and spectroscopy of CdTe in a bevelled solar cell was used to investigate grain boundary segregation and CdTe-CdS interdiffusion which may be related to the passivation.

1. INTRODUCTION

Thin film heterojunction p-CdTe/n-CdS solar cells are an emerging technology for cheap large area PV power generation. Since the cells are polycrystalline and yet have conversion efficiencies of up to 15.8% it is apparent that the grain boundaries are electrically passivated, perhaps by preferential segregation or an intergranular phase (Chu and Chu 1995). However no convincing microscopic observations of passivation with sufficient resolution to observe grain boundaries i.e. OBIC or EBIC, have been reported to date. For this material it is well known that high efficiencies are only achieved after processing the material by adding chloride ions and baking in an oxygen containing ambient. It has been speculated that this may introduce CdO (Chu and Chu 1995) to the grain boundaries, and indeed a preliminary TEM study by Loginov et al (1995) showed moiré fringes at grain boundaries in the CdTe which could be CdO. An alternative suggestion is that processing promotes formation of the alloy CdS_xTe_{1-x} at the interface and grain boundaries, the latter by a process of preferential interdiffusion (Dhere et al 1996). Interdiffusion itself has been measured by Dhere et al (1996) using SIMS, and is known to be responsible for influencing the spectral response and diode factors of the devices. CdTe and CdS are only miscible to ~14-16% at 650°C (Nonue et al 1990). Bowing of the band-gap/composition diagram gives the Te-rich compositions band gaps which are *lower* than that of CdTe, and conform approximately to the relation $E_g(CdS_xTe_{1-x}) = 1.45-0.5x$ (eV) for $x < 0.2$ (Jensen et al 1996). In this work EBIC was used to study recombination at grain boundaries in the cells and CL microscopy and spectroscopy were used to monitor grain boundary segregation and interdiffusion in bevelled samples. For similar cells Galloway et al (1995) have shown that EBIC images made by back wall injection display dark grain boundary contrast which is sharpened very considerably when the samples are cooled. The work lists the possible causes of contrast, but did not identify the mechanisms responsible. It was noted however that since the junction was only just accessed by the e-beam (after passing through 8μm of CdTe), small variations in the junction position would be significant.

2. EXPERIMENTAL

Cells comprising CdTe(12μm)/CdS(100nm)/TCO/glass were provided by ANTEC GmbH.

Both the CdTe and CdS layers were deposited by close space sublimation, the CdTe at 500°C. They had been treated with the standard chloride/air-bake treatment and had efficiencies of 10.1%. Quantitative EBIC was done in a JSM IC848 SEM with a nitrogen cold stage and Matelect ISM5 specimen current amplifier. Three injection geometries are discussed:

1. In back wall plan view mode - with injection through the CdTe and back contact - which is described in Galloway et al (1995) and is discussed here with reference to the current results.
2. In cross-section, as a function of temperature.
3. In front wall plan view mode - with injection through the ITO and CdS. The glass was removed by mechanical polishing and HF, the ITO acting as an etch stop layer.

CL microscopy and spectroscopy was done on similar samples using a JSM 840 SEM at Oxford Instruments equipped with a MonoCL2 system and He stage. (Temperatures reported here are those indicated by the stage thermocouple). The samples were bevelled by controlled immersion in Br_2/methanol/ethelene glycol to give a bevel angle of 0.1° i.e. a raster size of 100µm represented a depth variation of 0.17µm. Spectra were taken at 15kV at intervals along the bevel to investigate depth dependent effects in the CdTe and monochromatic imaging was possible.

3. RESULTS AND DISCUSSION

3.1 EBIC

Fig. 1. Cross section EBIC showing a reduction and shift in the peak collection with cooling.

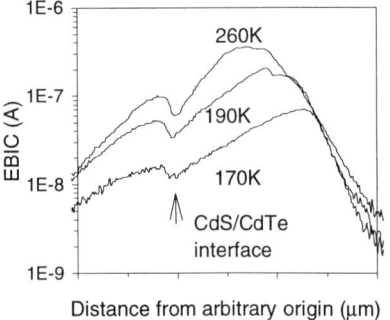

Distance from arbitrary origin (µm)

Cross section EBIC linescans were made as a function of temperature, as shown in Fig. 1. Care was taken to use a damage-free part of the sample. Upon cooling the peak of EBIC collection decreased and shifted towards the back wall of the cell, presumable due to a carrier depletion effect. This goes some way toward explaining the sharpening of dark grain boundary contrast in back wall injection EBIC as reported by Galloway et al (1995). Front-wall injection EBIC i.e. through the ITO and CdS with the glass removed, gives the surprising result that grains are delineated by bright boundaries under certain imaging conditions, as shown in Fig. 2. This indicates that there is enhanced EBIC collection efficiency at the grain boundaries. The magnitude of the contrast varies as a function of injection current as shown in

Fig. 3. Variable electron beam voltage (not presented here) was used to show that the contrast effect is due to the change in carrier injection density rather than beam current value *per se*. It was also shown that the EBIC collection efficiency fell off with increasing injection density. This is due to the onset of high injection conditions in which the carrier density in the material (p) is exceeded by the injected carrier density. Under such conditions the bands are flattened and collection efficiency is degraded. Fig. 3. shows three contrast regimes for excitation at 11kV, namely:- (1) low invariant contrast at low beam currents, (2) bright grain boundaries at intermediate currents and (3) reduced contrast at high beam currents. This variation, and the back wall EBIC behaviour, are consistent with a carrier distribution model which will now be presented. The model presumes that there is a variable p-type carrier concentration in the CdTe with high carrier concentration at the grain boundaries, implying that there is trapped negative charge at them. A second presumption is that there is a gradient in p from front to back of the cell, the higher carrier concentration being near the back wall. This is consistent with the doping method used, and the observation (by spectral response) that the p-n junction is buried in the CdTe. The three contrast regimes for front wall injection shown in Fig. 3. may be accounted for as follows:-

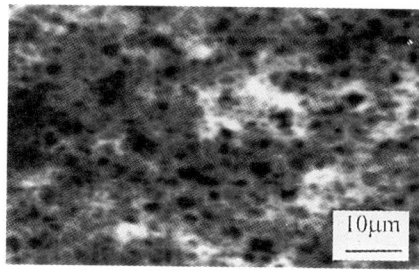

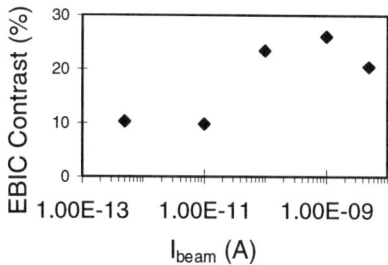

Fig. 2. Front wall injection EBIC image taken at 11kV with $I_b = 10^{-10}$A showing bright grain boundary contrast.

Fig. 3. Front wall injection EBIC contrast as a function of beam current.

(1) Low injection density regime - low contrast ($I_b < 10^{-11}$A). p is not exceeded anywhere by the injected carrier density and minority carriers in the CdTe are repelled by the grain boundaries. Normal photovoltaic operation is in this regime and corresponds to $I_b < 10^{-12}$A.

(2) Mid injection density regime - bright grain boundary contrast (10^{-11}A$<I_b<10^{-9}$A). p is exceeded by injected carrier density in the grains, but not at the grain boundaries. The grains are in high injection conditions and hence have reduced EBIC efficiency (they have low p carrier density), while the grain boundaries remain under low injection conditions and retain their collection efficiency (they have higher p). Hence the grain boundaries appear bright as shown in Fig. 3.

(3) High injection density regime - reduced contrast ($I_b > 10^{-9}$A). p is exceeded by the injected carrier density everywhere. There is reduced EBIC efficiency at the grains *and* at the grain boundaries, hence the contrast is reduced.

The back wall injection EBIC observation of dark grain boundaries is consistent with a junction position effect caused by higher carrier concentration at the grain boundaries; this would have the effect of moving the collecting junction toward the metallurgical interface hence taking it further from the excited volume.

3.2 Cathodoluminescence

CL spectra taken from the CdTe at a mid point on the bevel at 117K showed exciton related luminescence at 1.576eV (785 nm) and a lower intensity broad donor acceptor band in the range 1.491-1.406 eV (830-880nm). Monochromatic images formed with the spectrometer centred on 1.567eV (785nm) and 1.439eV (860nm) taken from the area are shown in Fig. 4. The excitonic luminescence is clearly confined to the grains, with the grain boundaries appearing as dark lines. The donor-acceptor luminescence is very diffuse and only the interiors of large grains are free from D-A luminescence, as shown in the figure. Association of the D-A luminescence with near grain boundary areas and with small grains is consistent with it being related to impurities (i.e. those impurities involved in processing) which diffuse from grain boundaries.

CL spectroscopy at 4.5K was used on a bevel in an attempt to evaluate interdiffusion by monitoring the position of the exciton peak as a function of position on the bevel. The exciton line energy for the alloy CdS_xTe_{1-x} is approximately $E_g(CdS_xTe_{1-x}) - E_{EX}(CdS_xTe_{1-x}) = E_g(CdTe) - E_{EX}(CdTe) - 0.5x$ for $x < 0.2$, i.e. over the miscibility range. This expression allows the relative variation in x to be assessed providing that it is assumed that the exciton binding energy $E_{EX}(CdS_xTe_{1-x})$ is not a strong function of x and that the slope of 0.5 is not different at the measurement temperature. In this work $E_g(CdTe) - E_{EX}(CdTe) = 1.6$eV, and the measurement of x by line shift should be sensitive at a level of 2.6×10^{-3}/nm. CL spectra were recorded from the bevel with a vertical interval of 0.2μm, this being decreased to 0.1μm near the interface. Excitonic lines were recorded at 1.599eV (777nm) and 1.602eV (775.5nm). However no measurable shift in the line position was observed. This indicates

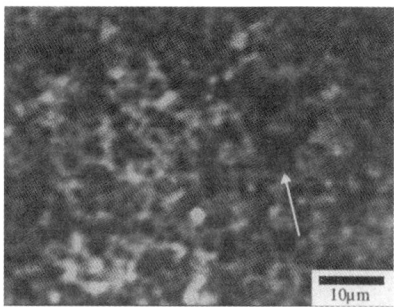

Fig. 4. Monochromatic CL images taken at 117K of the same area of CdTe on a bevel (3 ±0.08µm thick). Left image 785nm (exciton region), right image 860nm (DAP region). A prominent large grain is indicated by the arrow.

that either diffusion is insignificant in this sample or that the sampling by the electron beam is inadequate to probe near interface effects on the bevel. Further studies using samples processed under conditions thought to enhance interdiffusion (e.g. as reported by Dhere et al 1996) should be undertaken to test this.

4. CONCLUSIONS

An EBIC and CL study of polycrystalline CdTe/CdS/ITO solar cells (treated by chloride ion introduction and air baking) has been undertaken. EBIC imaging by carrier injection through the front wall shows (for the first time on a microscopic level) that the grain boundaries in the CdTe near the junction are passivated. This is likely to be a result of the introduction of impurity related dopant centres during post growth processing - which is known to be necessary in order to achieve high conversion efficiencies (Al-Allak et al1995). Comparison of front-wall, rear-wall and cross-section EBIC suggests a model of impurity distribution which would lead to grain boundary passivation. Indeed, CL shows that excitonic luminescence is excluded from the grain boundaries in the CdTe, while diffuse DA luminescence originates from small grains and grain boundaries; This supports the idea that impurity segregation and diffusion at grain boundaries is important. Although it is expected that a luminescence/bevel technique is capable of measuring interdiffusion of the CdTe and CdS, no such interdiffusion was measured in the sample examined. This indicates that interdiffusion is probably not responsible for the grain boundary observations in these samples. However it cannot be ruled out that the technique is insufficiently sensitive for measurement of thin layers if the luminescence is excited by an electron beam, in which case PL (which is more surface selective) may be more appropriate.

REFERENCES

Al-Allak H M, Galloway S A, Brinkman A W, Durose K, Richter H and Bonnet D, 1995 Proc. 13th EPVSEC, Nice, France Vol II 2135

Chu L C and Chu S S 1995 Solid-State Electronics **38**, 3 533-549

Dhere R D, Albin D S, Rose D H, Asher S E, Jones K M, Al-Jassim M M, Moutinho H R and Sheldon P, 1996 Mat. Res. Soc. Symp. Proc. **426** 361-366

Galloway S A, Holland A J, Wilshaw P R, Brinkman A W and Durose K, 1995b Proc. 13th EPVSEC, Nice, France Vol II 2072

Jensen D G, McCandless B E and Birkmire R W 1996 Mat. Res. Soc. Symp. Proc. **426** 325

Loginov YY, Durose K, Al-Allak HM, Galloway S A, Oktik S, Brinkman A W, Richter H, Bonnet D, 1995 J. Cryst. Growth **161**, 159-163

Nonue S, Hemni T, and Kato E 1990 J. Electrochem. Soc. **137** 1248

Inst. Phys. Conf. Ser. No 157
Paper presented at Microsc. Semicond. Mater. Conf., Oxford, 7–10 April 1997
© 1997 IOP Publishing Ltd

583

A study of the activation of CdTe/CdS thin film solar cells using OBIC

P R Edwards, S A Galloway[1], P R Wilshaw[2] and K Durose

Department of Physics, University of Durham, South Road, Durham DH1 3LE
[1]Now at Oxford Instruments Ltd, Station Way, Eynsham, Witney, Oxford OX8 1TL
[2]Department of Materials, University of Oxford, Parks Road, Oxford OX1 3PE

ABSTRACT: High resolution Optical Beam Induced Current (OBIC) has been used to investigate the effect of a post-deposition chloride treatment on thin film polycrystalline p-CdTe/n-CdS/ITO/glass solar cells. The light beam was focused through the glass front of the cells, and the induced current maps recorded as a function of bias voltage, hence generating illuminated I-V curves for each pixel investigated. The results were interpreted in terms of a distribution of diode responses. Trends in behaviour upon varying the chloride treatment and errors are discussed.

1. INTRODUCTION

Thin film (polycrystalline) cadmium telluride/cadmium sulphide is a leading candidate for use in low cost, large area photovoltaic power generation (Chu and Chu 1995), and conversion efficiencies of ~10% are routinely obtained from material deposited using a variety of techniques. Nevertheless, the efficiencies obtained fall short of the 29% figure expected from theory, and attention has focused on the role of grain boundaries and interfaces in loss. In electro-deposited cells the grains are ~300Å in size (Chu and Chu 1995), while those in close space sublimation (CSS) grown cells are typically 8μm. It has become evident that for all types of material the standard method of chloride treatment to form a homojunction in the CdTe also affords some degree of grain boundary passivation. The manner of applying the chloride varies with the deposition technology, but common features are that chloride is introduced into the CdTe layer before an air bake at 400°C for approximately 30 minutes, oxygen being considered to be essential (Basol 1988). Such post-growth treatment has the effect of enhancing the short circuit current (J_{SC}), open circuit voltage (V_{OC}), fill factor (FF) and efficiency (η), the latter typically increasing from 2 to 10%. In the case of the CSS cells in this study, chloride was introduced by evaporation of a layer of $CdCl_2$ onto the upper surface of the CdTe prior to air baking. The purpose of the work was to use high resolution OBIC microscopy in conjunction with J-V measurements to assess the optimum thickness of $CdCl_2$ to be applied in this process. Since the modal grain size is ~8μm, high resolution OBIC has sufficient spatial resolution to distinguish high and low performance areas on the scale of grains. The same technique has also previously been used by Galloway et al (1996a,b) to probe the differences between chloride treated and untreated CdTe/CdS cells.

2. EXPERIMENTAL

Solar cells provided by ANTEC GmbH for this study comprised Au(60nm)/CdTe(15μm)/ CdS(100nm)/TCO/glass, the TCO being supplied by Donnelley Applied Labs. The active layers were deposited by CSS, both at 500°C. In order to allow investigation of the influence of the post-growth processing, layers of $CdCl_2$ of thickness 15, 30, 60 and 120nm were deposited on the top surface of the CdTe using a set of solar cells from the same batch. These were then baked in air at 400°C for 25 minutes. Prior to contacting with gold, the CdTe was etched with a mixture of HNO_3 and H_3PO_4 for 30 sec to remove residuals.

J-V characteristics were measured under standard AM1.5 ($100mWcm^{-2}$) illumination. OBIC was carried out using a He-Ne laser ($\lambda = 632.8nm$) with apparatus previously described by Galloway (1996a,b). The light was focused onto the junction through the glass using a long working distance microscope objective. Sample scanning was achieved using a piezo-driven x-y stage. This allowed quantitative current maps of the sample to be generated as a function of bias voltage, with a spatial resolution in the order of 5µm. Since the current output was in the nA range, the light source was chopped and lock-in amplification used. The resultant induced current maps were recorded on a PC, from which linescans and histograms could be extracted.

3. RESULTS

Table 1 shows a summary of the J-V measurements of the cells. It should be realised that cells processed without $CdCl_2$ have efficiencies of ~2%. Use of 15-60nm of $CdCl_2$ enhances V_{OC}, J_{SC}, FF and η up to a plateau between 60 and 120nm, where no improvement is seen. Quantitative OBIC maps of the

$CdCl_2$ thickness [nm]	V_{OC} [mV]	J_{SC} [$mAcm^{-2}$]	FF [%]	η [%]
15	720	17.2	57.3	7.1
30	775	19.0	65.4	9.6
60	794	19.3	66.1	10.1
120	793	19.7	64.3	10.1

Table 1. Measured solar cell parameters of the samples

four cells at different bias voltages are shown in Fig.1. For clarity, the current corresponding to mid-grey has been adjusted to the mean current value for each image, while the contrast was adjusted to 20% (i.e. black and white correspond to 80% and 120% of the mean, respectively). Increasing reverse bias increased the overall brightness of each image (as expected from the J-V measurements), but also resulted in a marked reduction in contrast.

4. DISCUSSION

The aim of varying the $CdCl_2$ thickness was to optimise the solar cell performance (Table 1) without introducing second phases or inclusions at a level deleterious to operation. Performance is enhanced for $CdCl_2$ layers up to 60nm but is not further improved for thicker layers. OBIC images show contrast on the scale of ~8µm which is comparable to the larger grain sizes as observed by SEM. The influence of grains can be seen most clearly for the +0.5V biased images in Fig. 1 which show that areas of high grain response appear clumped for the 15nm $CdCl_2$ treated sample but less so for the 60nm sample. Some evidence of longer range behaviour (~100µm) reappears in the image of the 120nm sample. While the J-V characteristic of a cell represents the collective response of all grains within it, the bias dependent OBIC images in Fig. 1 contain the individual responses of each pixel interrogated by the beam. The material may be described as a collection of diodes (grains) acting in parallel. The distribution of diode behaviours was plotted as histograms which allowed calculation of the mean OBIC response as a function of bias as shown in Fig. 2. In principle the bias dependence of mean OBIC for each cell should scale as the J-V response but although OBIC was enhanced by increasing the $CdCl_2$ thickness from 15 to 60nm as expected, it did not increase in proportion with J_{SC}. Moreover, the 120nm sample showed unexpectedly low OBIC response. Since some months had elapsed between the measurements it seemed likely that contact degradation as reported by Singh et al (1995) was responsible, and repeat J-V measurements confirmed this. OBIC histograms were also used to generate Fig. 3 which represents the contrast in the images as the standard deviation of the distribution. This variation may be described in terms of the J-V behaviour of the individual grains. Diode behaviour is often represented by the diode ideality factor in the equation $J=J_0[\exp(qV/nkT)-1]$, with n=1 or 2 for diffusion and recombination limited transport respectively. However, since other factors can influence the J-V curve, fitting to this equation can yield values of n >2 for CdTe/CdS (Stollwerk and Sites 1995). Contrast due to recombination might be expected to decrease with

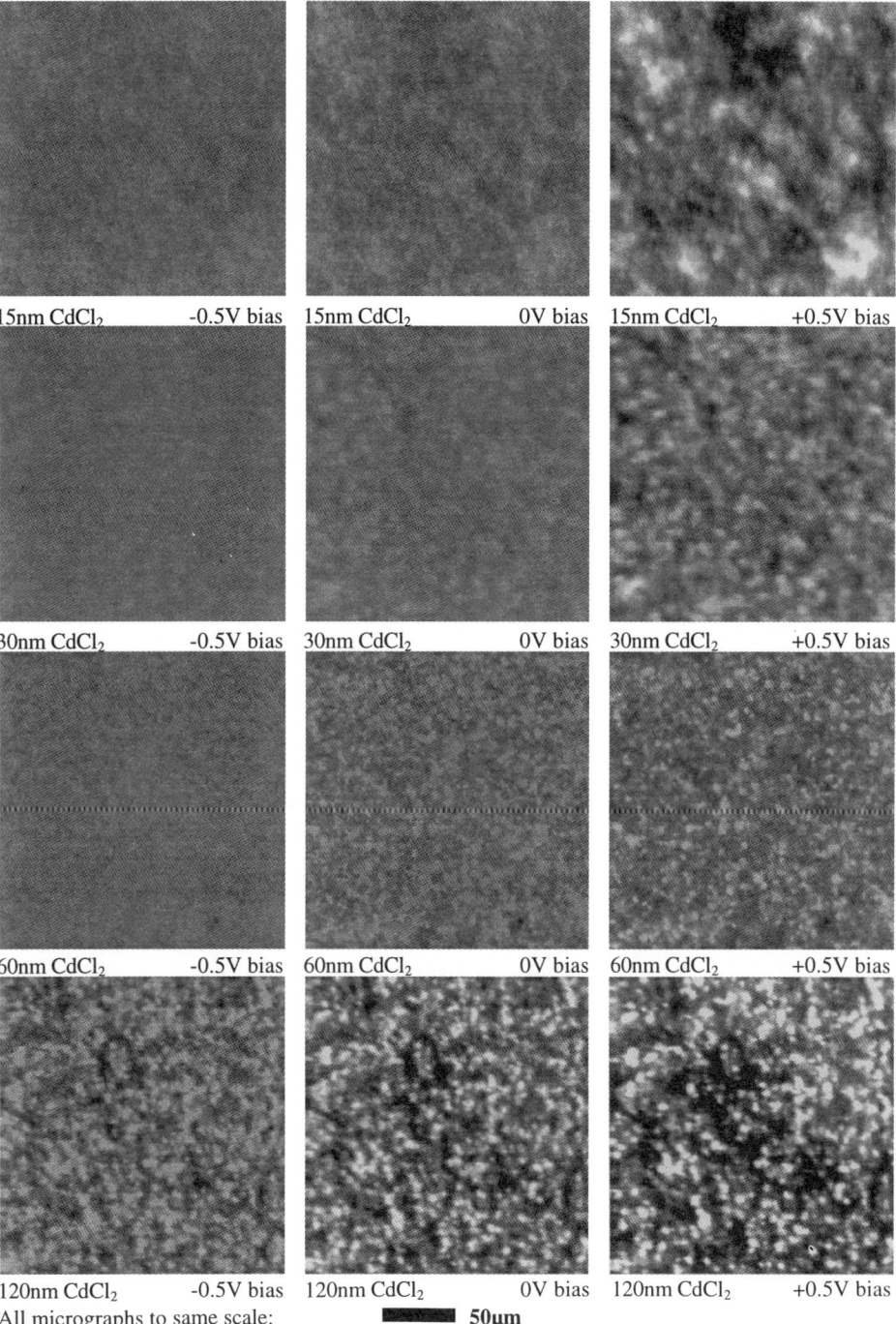

15nm CdCl$_2$ -0.5V bias   15nm CdCl$_2$ 0V bias   15nm CdCl$_2$ +0.5V bias

30nm CdCl$_2$ -0.5V bias   30nm CdCl$_2$ 0V bias   30nm CdCl$_2$ +0.5V bias

60nm CdCl$_2$ -0.5V bias   60nm CdCl$_2$ 0V bias   60nm CdCl$_2$ +0.5V bias

120nm CdCl$_2$ -0.5V bias   120nm CdCl$_2$ 0V bias   120nm CdCl$_2$ +0.5V bias

All micrographs to same scale: ▬▬ **50μm**

Fig.1. OBIC images of solar cells coated with varying thickness of CdCl$_2$ prior to heat treatment. All images are reproduced with the same level of contrast enhancement.

586

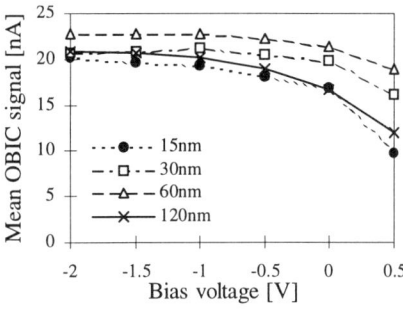

Fig.2. Mean OBIC signal vs. bias

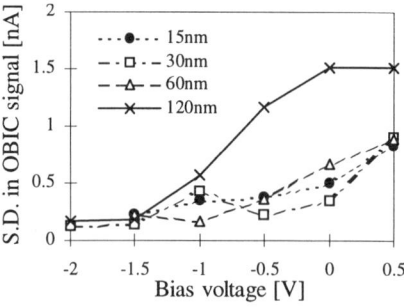

Fig.3. Standard deviation of OBIC vs. bias

increasing field since this would reduce loss in those grains with high densities of recombination centres. An alternative mechanism of contrast reduction results from variation in the position of the homojunction and the gradient of the change from n to p-type CdTe: in some grains carriers may be generated at a depth in the absorber at which collection is unlikely - unless the depletion region is extended in reverse bias.

Both the recombination and collection contrast mechanisms could be influenced on the scale of grains by the post-growth processing since supply of chloride, oxide and native species to the material at the electrical junction is probably by grain boundary and defect mediated diffusion. Since the grain structure varies over the cell, and the density of dislocations, stacking faults and twins varies (Loginov 1995), the impurity distribution and hence doping and grain boundary passivation is also likely to differ from grain to grain.

5. CONCLUSIONS

The reverse bias dependence of quantitative high resolution OBIC images provides a means of characterising the distribution of diodes which collectively represent the behaviour of polycrystalline solar cell devices. While the mean pixel response should in principle be related to the currents in J-V measurements (subject to errors caused by the choice of wavelength), the distribution of diode behaviours may be represented by its standard deviation. Contrast in CdTe/CdS cells arises on the scale of grains and may be due to either recombination or 'collection' loss; both mechanisms are likely to be influenced by local impurity distribution at the grain level. In this study OBIC-bias measurements did not reveal quantitatively the influence of varying the chloride treatment on microscopic cell performance due to degradation of the contacts used.

ACKNOWLEDGEMENTS:
The authors thank H Richter and D Bonnet at ANTEC GmbH for the samples, and A W Brinkman for valuable discussions. OBIC measurements were carried out with an SOM built by A J Holland. P R Edwards gratefully acknowledges the financial support of EPSRC.

REFERENCES

Basol B M 1988 Solar Cells **23**, 69-88

Chu L C and Chu S S 1995 Solid-State Electronics **38**, 3 533-549

Galloway S A, Brinkman A W and Durose K 1996a Appl. Phys. Lett. **68**, 26 3725-3727

Galloway S A, Holland A J and Durose K, 1996b J. Cryst. Growth **159**, 925-929

Loginov YY, Durose K, Al-Allak HM, Galloway S A, Oktik S, Brinkman A W, Richter H, Bonnet D, 1995 J. Cryst. Growth **161**, 159-163

Singh V P, Erickson O M and Chao J H, 1995 J. Appl. Phys. **78**, 7 4538-4542

Stollwerk G and Sites J R, 1995 Proc. 13th European Photovoltaic Solar Energy Conf. Nice, 2020

Inst. Phys. Conf. Ser. No 157
Paper presented at Microsc. Semicond. Mater. Conf., Oxford, 7–10 April 1997
© 1997 IOP Publishing Ltd

REBIC studies of electrical barriers in varistor ZnO

D Halls, D B Holt, C Leach* and J D Russell*

Department of Materials, Imperial College of Science, Technology and Medicine, London
SW7 2BP, U.K.
*Manchester Materials Science Centre, University of Manchester and UMIST, Manchester
M1 7HS, U.K.

ABSTRACT: ZnO, a wide bandgap non-stoichiometric II-VI semiconductor, is sintered
with oxide additives to make electroceramic varistors used in their billions for electrical
surge protection. Varistor action occurs at electrical barriers that can be resolved and
analysed only by SEM REBIC. This technique was applied to determine the electrical
characteritics of the barriers. Terraced contrast due to high resistivity layers, peak-and-
trough (PAT) contrast due to energy band bending and simple bright or dark contrast (due
to reduced and enhanced recombination, respectively) were seen and will be discussed.

1. INTRODUCTION

ZnO is a wide-gap (4.63 eV) material long known as a non-stoichiometric
semiconductor. It found widespread use in "electrofax" photocopying which preceded the
modern form: xerography (for a review see Heiland et al 1959). Today it is used as the
electroceramic in varistors (**var**iable re**sistors**). At low voltages, varistors appear as open
circuits but beyond a breakdown voltage V_b they become effectively short circuits protecting
the component, device or power system from voltage surges. The switching is due to the
presence of electrical barriers believed to be associated with (some) grain boundaries (GDs)
in the material which is made by sintering ZnO powder mixed with additive compounds.
These influence boundary properties and can lead to the formation of layers of other phases at
GBs. The mechanisms involved are not well understood and until recently there was no
direct evidence on the nature or location of the barriers. The breakdown voltage of each
barrier is the same so, for a given grain size (number of barriers per unit depth), the device
breakdown voltage increases with the thickness while the current carrying capacity of the
device increases with the area. Hence varistors can be made that range from small ones to
protect electronic equipment from power supply spikes to large ones to protect power systems
from lightning strikes. About 2×10^9 varistors were manufactured in 1988 (for a review see
Lagrange 1991).

Bernds et al (1984) first applied the SEM REBIC (remote EBIC) technique to ZnO
varistor material. Their images exhibited terraced contrast which we can now recognise as a
form of what Smith et al (1986) term resistive contrast. REBIC is observed when contacts
are applied a macroscopic distance apart (mms or cms) to a specimen containing no charge
collecting barrier such as a p-n junction or Schottky contact. One contact is earthed and the
other connected to the input of a current-detecting amplifier. Hence the specimen acts as a
current divider so a variable fraction of the specimen current of electrons absorbed from the

incident beam flows through the amplifier to provide the signal. In REBIC there is an ohmic signal baseline produced as the electron beam scans from the earthed contact to the amplifier end of the specimen. Since its slope is proportional to the local resistivity, the steps between the terrace areas of approximately constant grey level (terrace contrast) are evidence of boundary layers of higher resistivity between the more conductive areas (Holt 1994).

In addition charge collection (hole-electron pair separation) will occur in REBIC where the electron beam is incident on defects and inhomogeneities giving rise to internal electrical fields. Thus REBIC presents the possibility of detecting and analysing the barriers in varistors. By using mechanical microprobes to make contact, individual areas and barriers can be selected for study. Electrically charged grain boundaries (GBs) cause energy band bending and depletion regions in which the fields are oppositely directed on either side so the charge-collection currents flow in opposite directions. This results in a peak-and-trough (PAT) form of linescan signal which appears in micrographs as a pair of parallel bright and dark lines running along the boundary (see e.g. Holt 1994). This mechanism was treated theoretically by Palm and Alexander (1991). PAT contrast indicates the presence of fields back to back due to band bending. This is not necessarily the result of trapped boundary charge, however, as will be argued below. Impurity segregation or the occurrence of second phase material can result in high resistivity layers at GBs which gives rise to terrace contrast. Higher recombination at boundaries due to intrinsic (dangling bond like) or extrinsic (impurity related) centres will produce dark contrast (Donolato 1985). Conversely, reduced recombination will result in bright contrast. This paper summarises REBIC studies of electrical barriers in sintered ZnO, both varistor material and material not intentionally doped.

2. EXPERIMENTAL METHODS

The work was carried out on a JEOL 840A SEM fitted with a Matelect ISM-5 and a Kontron image processing system at Imperial College and similar equipment at the Manchester Materials Science Centre. The same microprobe (a two-probe modified micromanipulator from E. Fullam, Inc.) was used first at I.C. and then at Manchester.

Sintered ceramic pellet samples were prepared from four types of material: (i) ZnO with no additives, and ZnO powder with (ii) (in mole %) 0.56% Bi_2O_3, 0.3% MnO and 130 ppm (parts per million) of Al, (iii) 0.56% Bi_2O_3, 1% MnO and 200 ppm of Al and (iv) 0.56% Bi_2O_3, 0.30% MnO, 0.30%CoO and 0.005% Na_2O. For a brief account, with references, of the role of additives in varistor manufacture see Halls and Leach (1995).

3. RESULTS AND DISCUSSION

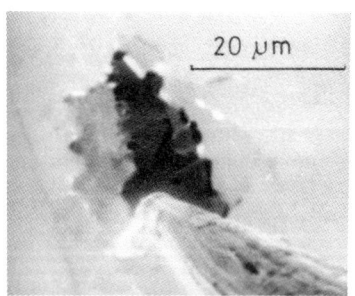

Figure 1. REBIC image of varistor material showing terraced contrast (areas of differing brightness) around the tip of the microprobe needle making the contact to the amplifier. The material of this specimen was that containing 0.3% MnO and 130 ppm of Al.

Figure 1 shows the electrical terracing due to high resistivity boundary layers surrounding several areas around the tip of a microprobe on varistor material. This contact was connected to the amplifier so the terrace contacted appears dark (there a large -ve current flows from the electron beam to the amplifier). The current drops markedly and the picture lightens across the boundary of this terrace. The right hand boundary of the dark terrace has both dark (enhanced recombination) and bright (reduced recombination) segments. Some small grains with light boundaries can be distinguished inside this terrace. The terraces were generally much wider than the individual grains in the sintered solid. Figure 2 shows clearly the form of stepped REBIC signal responsible for terrace contrast in micrographs.

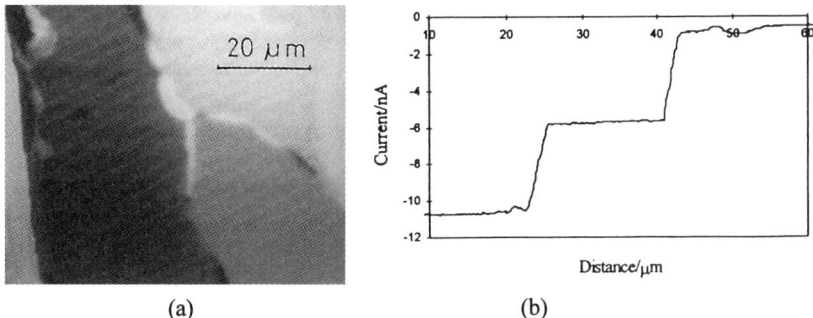

(a)   (b)

Figure 2. (a) REBIC micrograph showing terrace contrast in varistor ZnO and (b) a linescan across the three terraced regions, halfway down the field of view, showing the steps in the signal current that are the characteristic signatures of high-resistivity boundary layers. The material contained 1% MnO and 200 ppm of Al.

The powder and partially sintered pellets were examined in secondary electron images during preparation of the CoO-containing material. The powder consisted of spherical agglomerates up to 10 - 20 μm across and these could still be clearly seen in the partially sintered material. The terrace boundaries (e.g. those in Figure 1b that are 10 - 20 μm wide) corresponded to those of the agglomerates in the pre- and partially-sintered material. The enhanced porosity at agglomerate boundaries may allow the bismuth to flow into those regions in liquid-phase sintering, to form bismuth-rich second phase material which is found to be highly resistive giving the observed terrace boundary contrast. The lines of pores associated with the incomplete densification of the inter-agglomerate boundaries also appear to increase the resistivity of the material locally contributing to the signal steps (terrace boundaries). In Figure 2b current steps occur where the e-beam crosses high-resistivity layers causing a rapid change in the fraction of the specimen current from the electron beam flowing via the amplifier to earth (the REBIC signal).

Most models to explain varactor breakdown assume the presence of back-to-back Schottky barriers at (some) ZnO boundaries (Eda 1989) but there was no direct evidence for their occurrence when this work began. Schottky barriers back to back have electric fields of opposite sign. This results in charge collection currents of opposite sign on either side. Thus a peak and trough (PAT) variation will be superimposed on the background REBIC signal and in REBIC images this appears as parallel bright and dark lines along the charged boundary. Changes from high-resistivity-layer terraced contrast to PAT from one length of a boundary to the next are shown in Figure 3. At points marked A the grains make direct contact and PAT contrast is seen while along lengths like that marked B there is a high-resistivity second phase layer (resolved in Figure 3a) and only terrace contrast occurs. This provides some direct evidence for the model put forward by Einziger (1978, 1979).

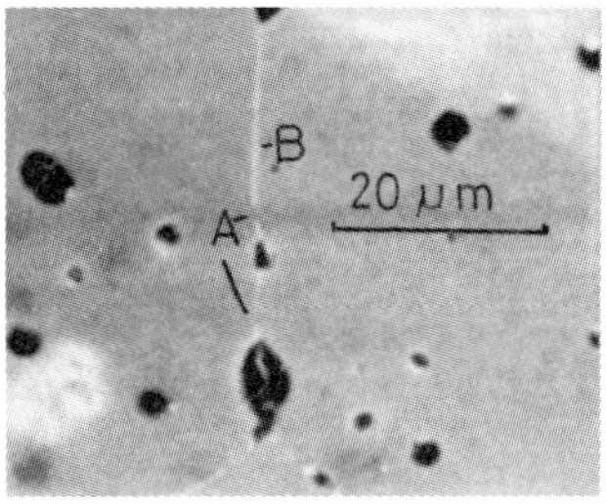

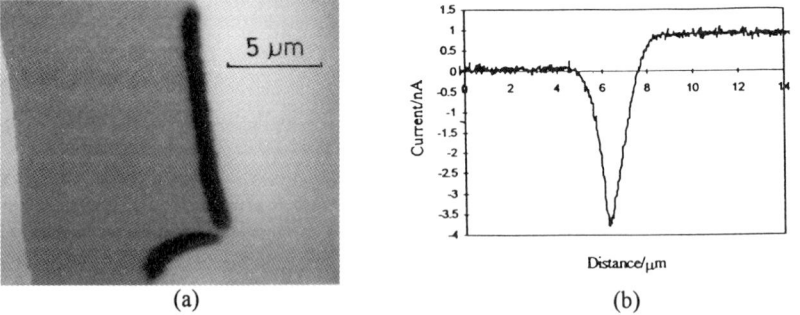

Figure 3. (a) SEI and (b) REBIC images of a grain boundary in ZnO and (c) the PAT form of the linescan across points of the type marked A. Along horizontal scans through the points marked A, from left to right, there are a constantly dark terrace, a darker region (a trough), a brightest region (peak) and then a constantly bright terrace. Linescans across boundary length B contained only a step like those in Figure 2b. The material was that containing CoO and Na_2O.

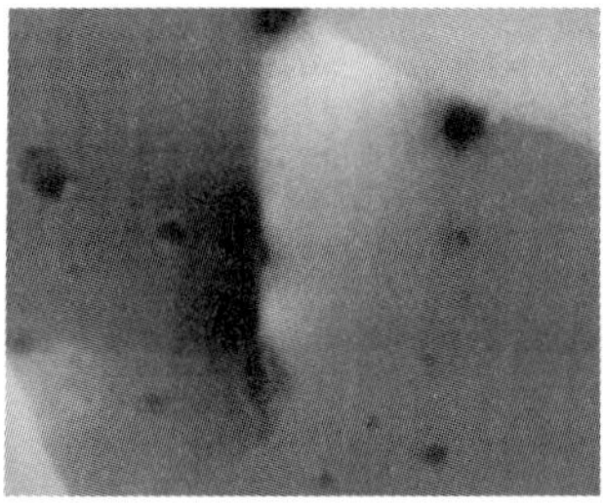

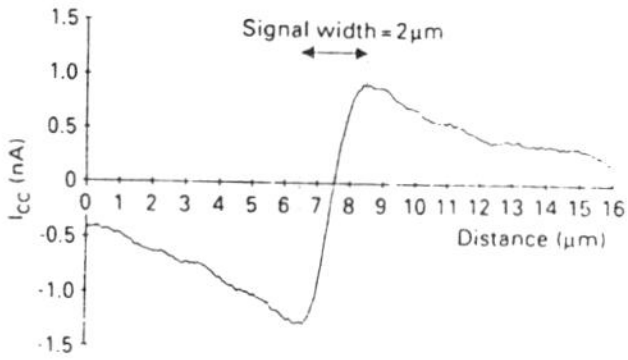

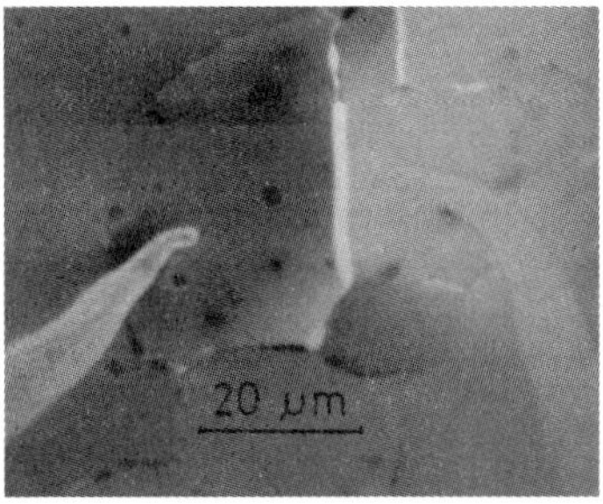

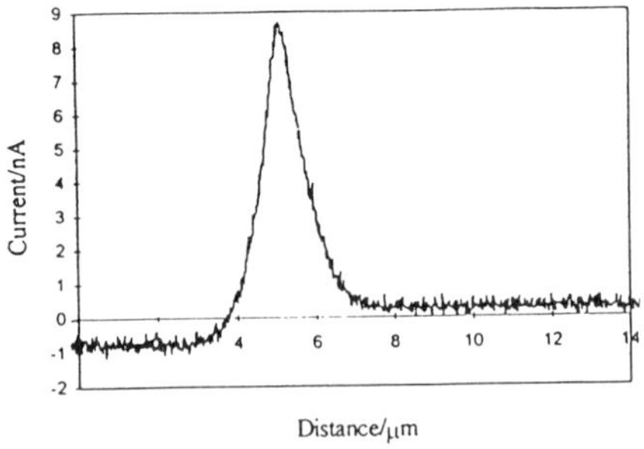

(a)   (b)   (c)

Signal width = 2 μm

Distance (μm)

Figure 4. REBIC (a) image and (b) linescan across a bright boundary in the varistor material containing CoO and Na_2O. The microprobe needles can be seen contacting the grains on either side.

Distance/μm

(a)   (b)

Figure 5. REBIC (a) image and (b) linescan of a dark boundary in the 130 ppm Al material.

Distance/μm

In addition to cases of terrace and of PAT contrast, GBs were observed that had only bright (peak) (Fig. 4) or dark (trough) contrast (Fig. 5). Lengths of bright only and dark only contrast can also be seen along the right hand boundary of the darkest terrace in Fig. 1.

4. DISCUSSION

The REBIC technique, using microprobes to select the boundaries between particular grains, was applied to electroceramic ZnO varistor material. This resolved individual boundaries and particular lengths of those boundaries enabling, for the first time, the types of electrical barriers in varistors to be experimentally identified.

Evidence of the types presented in Figs. 1 and 3 showed that the high-resistivity barriers in varistor materials are often associated with thick intergranular layers of a distinct e.g. bismuth-rich phase. These structures are left from the agglomerate surfaces in the powder (see Halls and Leach 1995) and give terrace contrast like that in Fig. 2. The observation of cases of bright only and of dark only contrast demonstrates that some GBs are sites of reduced and others of enhanced minority carrier recombination, respectively.

The behaviour of GBs in ZnO has been modelled in several ways (Eda 1989) which can all be classified as either homojunction or heterojunction. The heterojunction models are those in which varistor action is due to a discrete intergranular layer structure containing additives that sustain breakdown (Eda 1989). Homojunction models are those in which the varistor behaviour is due to boundaries without a distinct intergranular material. For example, in varistor material Einziger (1978, 1979) suggested that oxygen from the additive oxides fills the oxygen vacancies (donors) in ZnO to turn the material near the boundaries intrinsic. The n-i-n structure of these boundaries produces band bending. (The Fermi level moves down from near the conduction band edge toward the mid-gap position.) This makes them electrical barriers which, he suggested govern the varistor action. The occurrence of PAT contrast (Figure 3) shows that charge collection occurs in opposite directions across some homojunction boundaries. This is not necessarily due to band banding produced by trapped boundary charge, however, but may well be evidence of n-i-n homojunction barriers as suggested by Einziger. Direct micrographic evidence for both homo- and heterojunction type behaviour along the same boundary was found (Figure 3) and this provides the first direct micrographic evidence of the electrical microstructures underlying varistor action. It makes it possible to study their dependence on the additives and the sintering conditions used.

The micrographs and linescans obtained with both varistor and pure sintered ZnO were far better than could be obtained in the cases of polycrystalline bulk samples of CdTe, ZnSe, ZnS,Se and ZnS (Holt and Raza 1995, Holt et al 1996, Holt et al 1997) although the ZnO studied here is also a polycrystalline, wide-bandgap, relatively high resistivity and high dielectric constant II-VI material. Holt et al 1997 argue that the problem in the case of the other bulk II-VI polycrystals arises from the large capacitance of those specimens between widely-spaced large-area contacts. The experience here provides some support for this idea in that the use of closely spaced microprobe contacts very greatly reduces the effective capacitance of the specimens in these observations. This is believed to be the main factor making quantitative signal profiles possible using slow scan speeds, which are not possible in the bulk, widely spaced contact work on other II-VIs.

REFERENCES

Bernds A, Lohnert K and Kubalek E 1984 J. de Phys. Colloque C2, C2-861

Donolato C 1985 in Polycrystalline Semiconductors ed. Harbeke, G (Berlin: Springer-Verlag) pp. 138 - 154

Eda K 1989 IEEE Elec. Insulation Mag. **5**, 28

Einzinger R 1978 Appl. Surf. Sci. **1**, 329

Einzinger R 1979 Appl. Surf. Sci. **3**, 390

Halls D C and Leach C 1995 J. Mater. Sci. **30**, 2733

Heiland G, Mollwo E and Stockman F 1959 in Solid State Phys. **8**, 191 - 323

Holt D B 1994 in Polycrystalline Semiconductors III eds. HP Strunk, JH Werner, B Fortin and O Bonnaud (Switzerland: Scitec Publications) pp. 171 -182

Holt D B and Raza B 1995 in Microscopy of Semiconductors 1995. Conf. Series No. 146 (Bristol: IOP Publishing) pp. 107 - 112

Holt D B, Raza B and Wojcik A 1996 Mat. Sci. Eng. **B42**, 14

Holt D B, Raza B and Wojcik A 1997 in Microscopy of Semiconductors 1997. Conf. Series No. 157 (Bristol: IOP Publishing)

Lagrange A 1991 in Electroceramics ed. B C H Steele (London : Elsevier Applied Science) pp. 1 - 27

Palm J and Alexander H 1991 J. de Phys. Colloque **C6**, C6-101

Russell J D, Halls D C and Leach C 1995 J. Mater. Sci. Letters **14**, 676

Russell J D, Halls D C and Leach C 1996 Acta Mater. **6**, 2431

Smith C A, Bagnell C R, Cole E I, Dibianca F A, Johnson D G, Oxford W V and Propst, R H (1986) IEEE Trans. Electron. Devices **ED-33**, 282

Inst. Phys. Conf. Ser. No 157
Paper presented at Microsc. Semicond. Mater. Conf., Oxford, 7–10 April 1997
© 1997 IOP Publishing Ltd

Electron microscopy analysis of the RGTO technique for high sensitivity gas sensor development

A Diéguez[1], A Romano-Rodríguez[1], JR Morante[1], P Nelli[2], LE Depero[2], L Sangaletti[2], and G Sberveglieri[2]

[1]EME, Dept. Física Aplicada i Electrònica, Universitat de Barcelona, Avda. Diagonal 645-647, E-08028 Barcelona, Spain
[2]Thin Film Laboratory, Dept. Of Chemistry and Physics for Materials, University of Brescia, Via Valotti 9, I-25133 Brescia, Italy

ABSTRACT: The mechanism of formation by RGTO of thin SnO_2 films is described. Different thicknesses of the Sn layer and oxidation times are analysed in order to follow step by step the oxidation mechanism and the final structure of the thin film. Special emphasis is placed on the morphological changes occurring during a long term oxidation, which are very important in order to obtain a reliable active layer for gas sensing.

1. INTRODUCTION

Over the past few years a large amount of work has been devoted to miniaturisation and on-chip integration in order to develop smart sensors (Huijsing 1992). These require the use of thin film technologies and, when possible, silicon substrates. When the interest is to develop gas sensors, SnO_2 is one of the most widely used materials due to its high sensitivity to a wide variety of gases, its high stability over a wide range of temperatures, and its high chemical and physical stabilities. Polycrystalline SnO_2 thin films are synthesised by different methods such as R.F. sputtering, thermal evaporation, chemical vapour deposition, electron gun evaporation, and spray pyrolysis. Recently, the Rheotaxial Growth and Thermal Oxidation (RGTO) technique (Sberveglieri et al 1990 and Sberveglieri 1992) has been proposed as an alternative and very attractive method to deposit thin layers of some metal oxides with excellent sensing properties. In the case of SnO_2 this technique consists of two steps:

    i) The rheotaxial growth, i.e., the sputter-deposition of metallic tin on top of an insulating substrate maintained at a temperature higher than the melting point (above 232°C).

    ii) The oxidation under, for example, an oxygen or synthetic air atmosphere at temperatures in the 500-700°C range in order to form the metal oxide.

Additional repetition of the steps i and ii allows the growth of a multilayer RGTO film. Moreover, in order to enhance sensitivity and selectivity, this technique allows the introduction of catalytic materials simply through evaporation of small amounts of metal impurities onto the RGTO SnO_2 (Faglia et al 1994).

The aim of this study is to describe the steps involved in the RGTO of tin dioxide. For such characterisation, different oxidation times and thicknesses of the Sn deposited layer are investigated. Scanning (SEM) and Transmission (TEM) electron microscopies are used as experimental techniques.

2. EXPERIMENTAL

The Sn layers were deposited onto $5 \times 5 mm^2$ Si substrates, which were previously thermally oxidised in order to form a thin isolating SiO_2 layer. The deposition was performed in a R.F. sputtering system (Alcatel 450 SCM) from a pure tin target (Sn 99.9999% purity). The substrate was maintained at a temperature of 400°C and the parameters of the deposition were an incident R.F. power of 50W, a deposition rate of 2.5nm/s and an Ar pressure of 2×10^{-3} mbar. The oxidation was performed in air inside an Heraeus furnace equipped with a Thermicon P temperature controller. The oxidation step consisted of two ramps, the first from ambient up to 250°C and the last from 250°C up to 600°C. The time to reach each stable temperature was 4h and the duration of each plateau is 4h for 250°C and up to 30h for 600°C. According to these specifications, the deposition (thickness) and oxidation times (in the last ramp and in the plateau at 600°C), were changed to grow the samples. These are indicated in Table I.

CODE NAME	DEPOSITION TIME	OXIDATION TIME (4h at 250°C +)	SAMPLE TYPE
Sn{20,40,80,160}A	{20,40,80,160} s		Sn
Sn{20,40,80,160}B		30 h at 600°C	SnO2
Sn20{C,D,E,F,G}	20 s	{15',30',1h,2h,5h} at 600°C	SnO2
Sn40Ab	40 s		Sn
Sn40{B,C,D,E,F,G,H}b		360',420', 440' 460' 500' 540',600'	SnO2

TABLE I: Characteristics of the Sn/SnO_2 thin films grown by RGTO. The different values of a varying parameter are enclosed in brackets. The samples Sn40{B,C,D,E,F,G,H} were taken off of the furnace during the second ramp, and the total time of the oxidation step is indicated.

The samples were characterised by TEM on a Philips CM30 SuperTwin electron microscope, operating at 300 keV. The SEM investigations were performed on a Leica 360 electron microscope with a backscattered electron detector and operating at low voltages to reduce charging of the specimen.

3. RESULTS AND DISCUSSION

The SEM micrographs of the samples Sn20A to Sn160A show that after the rheotaxial growth, the metal does not form a continous layer but microspheres due to the liquid-phase surface tension (Fig. 1a). An increase of the sphere's diameter with the increase of the deposition time, together with a spread of the sphere's size distribution is observed. The average size changes from about 130 nm to about 800 nm for deposition times of 20 to 160s, respectively. However, the coverage of the substrate is always around the 40-60%, mainly caused by the smaller spheres.

After the metal thermal oxidation (samples Sn20B to Sn160B), the surface is formed by agglomerates uniformly distributed (Fig. 1b). They present a spongy aspect, giving rise to a large surface area, which is suggested for a semiconductor gas sensor, because surface reactions are much more

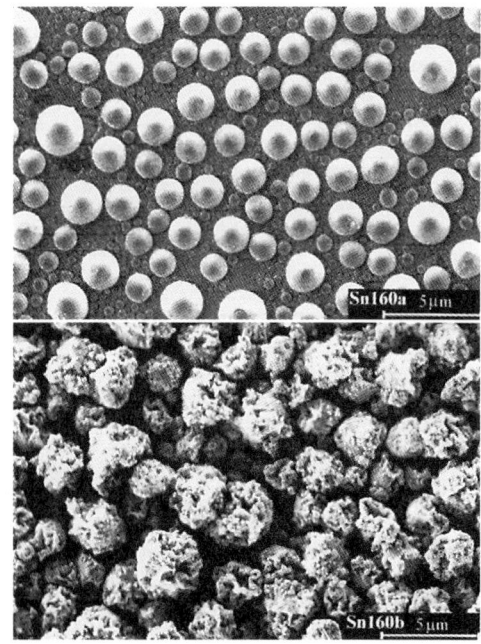

Fig. 1: SEM images of the samples Sn160 a and b.

relevant than bulk changes (Morison 1987). A strong volume increase occurs during the phase transformation from Sn to SnO$_2$ and this volume increase depends on the deposition time. Particularly, it is observed that the average agglomerate size ranges from about 250nm to about 2.2µm when the deposition time changes from 20 to 160s, respectively.

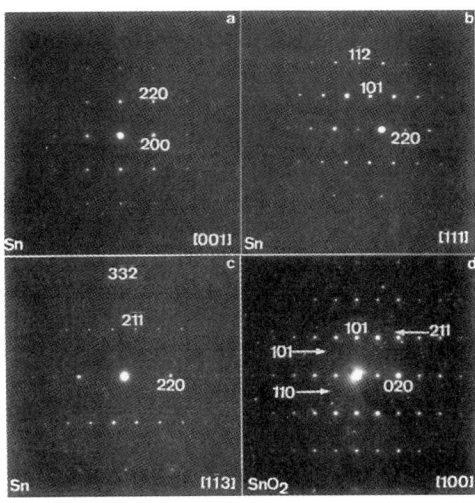

Fig. 2: SADP of the Sn and SnO$_2$ RGTO layers.

The phase transformation from Sn to SnO$_2$ can be extracted from the selected area electron diffraction patterns (SADP). The SADP of some of the spheres taken from samples Sn20A to Sn160A show that their crystalline structure corresponds to tetragonal tin with space group I41/amd (141) and lattice parameters a=5.831 and c=3.182. Some of these SADP corresponding to the sample Sn160A are presented on Fig. 2a-c, which are taken with the electron beam along some different zone axes. It has to be mentioned that once the crystalline structure has been identified with confidence, the reflections [110] and [$\bar{1}\bar{1}$0] have to be defined as due to double reflection effects, as they are forbidden in the tetragonal structure. After oxidation, the crystalline structure has been found to be the tetragonal rutile one of SnO$_2$ (cassiterite). The SADP of one of the large agglomerates of the sample Sn160B is presented on Fig. 2d. The polycrystalline character of the layer is visible through the appearance of diffraction rings which are superimposed on the [100] zone axis of the large SnO$_2$ grain. Similarly for the case of tetragonal tin, the [100] reflection is attributed to double reflection.

The cross-section TEM images of some of the single-layer samples investigated are presented in Fig. 3. The images of the Sn films indicate that factors such as the surface tension and interface energies actually make Sn form not perfect spheres, but ellipsoids of aspect ratio between 1.2 and 1.6, the larger dimension corresponding to the plane of the substrate. When the film is oxidised, TEM shows the real morphology of the agglomerates. These consist of accumulations of grains of size depending on the initial Sn sphere size surrounded by smaller grains of size 10-20nm, whose distribution is the origin of the spongy aspect observed by SEM. The spongy aspect of the agglomerates strongly depends on the size of the starting Sn spheres in the sense that larger spheres give rise to more spongy

Fig. 3: XTEM images of some of the single RGTO layers.

596

agglomerates. Therefore, a larger surface area is obtained for the nominally thicker samples, in which larger Sn spheres are present. Thus, it is worth emphasising that the sphere radius distribution of the as-deposited Sn film will deeply influence the electrical response of the sensor. Furthermore, in spite of the very large increment of volume produced by the presence of such enveloping grains, the coalescence of several agglomerates does not promote the total coverage of the substrate by a continuous film. The agglomerates become connected by the smaller crystallites located at their surface.

When the evolution of the SnO_2 film as a function of the oxidation time is observed (Fig. 3e, 3f, and 3b) it is shown that no fundamental differences exist between 30min and 5h of oxidation. However, the comparison with Fig. 3b indicates that the film becomes more dense after 30h of oxidation. This result, confirmed by SEM by the smoothing of the surface of the agglomerates with increase of the oxidation time, indicates that a single RGTO layer needs some time before it can be applied for gas sensing in order to avoid sensor response drift.

The observation of the samples on the second ramp during the oxidation (Fig. 4) indicates how this process could occur. Oxidation begins at the surface of the spheres through the formation of small nuclei that envelope the rest of the non-oxidised tin, sometimes forming a continuous layer. Next, these nanoparticles get separated probably due to tensions and give rise to the spongy aspect. The continuation of the oxidation gives rise to a complete SnO_2 film. When very large Sn spheres are considered (sample Sn160A), this mechanism could occur several times giving rise to a more markedly rough surface morphology.

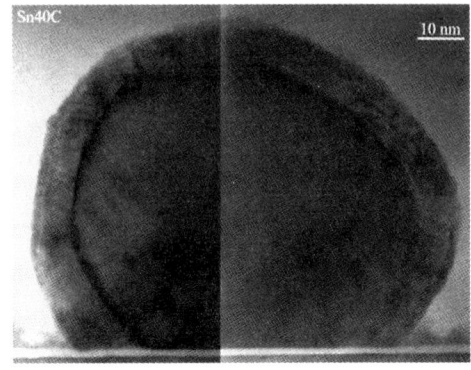

Fig. 4: HREM image of the Sn40Cb sample (420').

CONCLUSIONS

The analysis performed by TEM on the RGTO technique applied to tin dioxide has shown that the typical surface morphology observed by SEM corresponds to a random distribution of small crystallites located on accumulations of larger grains. These enveloping particles resulted from the form in which SnO_2 develops from Sn as was shown by the investigation of the second ramp of oxidation. In spite of the high sensitivity that such rough morphology suggests, the structural changes occurring in a long term oxidation claim the need to stabilise the film before it can be used as a gas sensor.

ACKNOWLEDGEMENTS
Part of this work has been stimulated and finantially supported by the EEC through the BRITE-EURAM contract BRE2-CT94-0940 corresponding to the project 'NANOGAS'.

REFERENCES
Faglia G, Nelli P and Sberveglieri G 1994 Sensors and Actuators **B18-19**, 497
Huijsing J H 1992 Sensors and Actuators **A30**, 167
Morison R S 1987 Sensors and Actuators **B12**, 425
Sberveglieri G, Faglia G, Gropelli S, Nelli P and Camanzi A 1990 Semic. Sci. Technol. **5**, 1231
Sberveglieri G 1992 Sensors and Actuators **B6**, 239

Inst. Phys. Conf. Ser. No 157
Paper presented at Microsc. Semicond. Mater. Conf., Oxford, 7–10 April 1997
© 1997 IOP Publishing Ltd

Nucleation, growth and size distributions of Ge islands on Si(001): *in-situ* STM studies

I Goldfarb, J H G Owen, P T Hayden, K Miki and G A D Briggs

University of Oxford, Department of Materials, Parks Road, Oxford OX1 3PH, UK

ABSTRACT: Heteroepitaxial Ge/Si(001) growth has been investigated using in-situ scanning tunneling microscopy. Once the mismatch strain exceeds a critical value it is relieved either by the formation of coherent three-dimensional islands or pits, depending upon growth parameters. The kinetics of island growth have been measured in-situ to yield a power-law dependence of growth rate. Although late-stage island growth is often attributed to Ostwald ripening, in the case of Ge/Si(001), due to the energy barrier at the island base, other growth mechanisms are found to dominate at temperatures below 700 K.

1. INTRODUCTION

Heteroepitaxy, and particularly Ge/Si(001) is an important field in materials science and solid-state physics. The 4.2 % mismatch strain in the Ge/Si(001) system can be relieved by introduction of defects, e.g. misfit dislocations, or by roughening the surface. Although problematic in a planar device technology, roughening can be utilised to produce self-assembled low-dimensional quantum structures, such as quantum wires and dots, which posess unusual physical properties (Banyai and Koch 1993, Nötzel 1996, Petroff and Medeiros-Ribeiro 1996). However to be able to exploit the self-assembling tendency, a deep understanding of thermodynamics and kinetics of heteroepitaxial Stranski-Krastanow (S-K) growth is essential.

Strain relaxation in the Ge/Si(001) system proceeds via a series of complex surface phase transitions, eventually resulting in a rough three-dimensional (3D) surface (Tomitori *et al* 1994). One of the important stages in the series of transitions is the formation of small and coherent 3D micro-islands. These islands were originally called "hut" clusters by Mo *et al* (1990) because of the hut-like shapes formed by their {501} facets.

In this work we will present results of our studies of the nucleation and growth of these "hut" clusters in gas-source-molecular-beam-epitaxy (GSMBE), and attempt to correlate between these growth mechanisms and the observed size distributions. Many of the important results obtained in this study could not be observed by any other methods, but only by *in-situ* studies at elevated temperatures.

2. EXPERIMENTAL

The Si wafers used for this study were n-doped 0.1 Ω–cm, cut into 1×7 mm^2 pieces and chemically degreased *ex-vacuo*. The samples were handled with ceramic tweezers and clamped to the Ta support on the holder by Ta clamps. In UHV, the samples were degassed for several hours, repeatedly flashed at 1400 K, quenched below 800 K and slowly cooled to the desired temperature. During the sample flashes and anneals, the pressure was kept below 10^{-7} Pa. Such treatment has generally proved effective in producing well-ordered (2×1) Si surfaces.

A JEOL ET-STM, equipped with Low-Energy-Electron-Diffraction (LEED/Auger) and Reflection-High-Energy-Electron-Diffraction (RHEED), and capable of operation up to 1200°C was used. The base pressure of the STM chamber prior to growth was 1×10^{-8} Pa. Growth movies were taken during the exposure to germane at the growth temperatures and in "constant current" or "Log I" mode, using currents around 0.1 nA and voltages between ±3V. Sample heating was achieved by passing a direct current through it. Temperatures were measured by an optical pyrometer with an accuracy of 30 K. Polycrystalline 0.3 mm W wires were electrochemically etched in 2M NaOH solution to produce atomically sharp tips. GeH$_4$ (99.99

598

%) was fed through a precision needle-valve onto the sample mounted in the STM stage, and the tip was allowed to scan while a desired constant pressure was maintained. Temperatures from room-temperature (RT) to 720 K, and GeH$_4$ pressures in the 10^{-7}-10^{-5} Pa range, were used for growth.

3. RESULTS AND DISCUSSION

In the temperature range between 600 K and 700 K the sequence of Ge/Si(001) surface phase transitions during growth is $(2\times1) \Rightarrow (2\times N) \Rightarrow (M\times N) \Rightarrow$ "hut" clusters, while at higher temperatures it is $(2\times1) \Rightarrow (2\times N) \Rightarrow (M\times N) \Rightarrow$ "hut" pits $\Rightarrow$ "hut" clusters (Goldfarb *et al* 1997a). Observation of the "negative" micro-islands, i.e. "hut" pits in the latter case is quite striking, in spite of the favourable theoretical predictions of Tersoff and LeGoues (1994). Although Ge/Si faceted micro- and macro-islands have been frequently observed by a variety of experimental techniques (Eaglesham and Cerullo 1990, Hammar *et al* 1995, Knall and Pethica 1991, Steinfort *et al* 1996), as well as a mixed pit-cluster morphology (Jesson *et al* 1996a), *individual* pits were only observed for the first time by the present authors (Goldfarb *et al* 1997b). This is probably due to the surfactant action of hydrogen, which is inherent to the GSMBE process. This action at T > 700 K produces wetting layers thicker than those resulting from MBE (Goldfarb *et al* 1997a). Such thick layers are capable of containing the entire pit's depth, which is a necessary condition for pit formation (Tersoff and LeGoues 1994). An STM image of these pits can be seen in Fig. 1(a), while Fig. 1(b) shows characteristic image of the clusters.

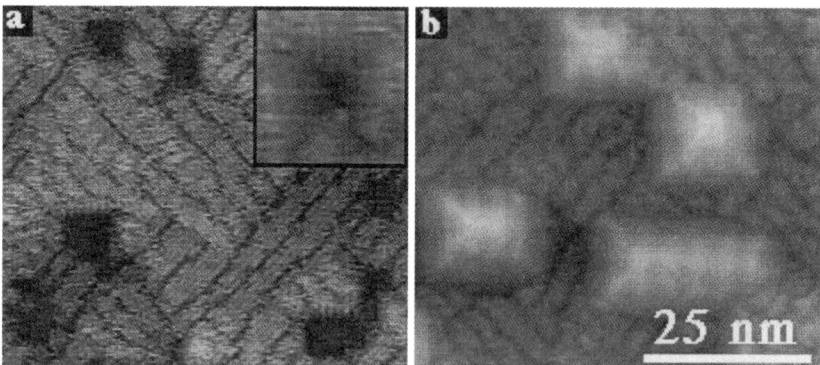

Fig. 1: Constant-current STM image of the Ge/Si(001) surface. (a) Surface pits (contrast adjustment enhances {501} facets, as shown in the inset), (b) "hut" clusters.

Both pits and clusters nucleate heterogeneously on suface irregularities, such as voids or steps. However at least short segments of the irregularity edges must orient themselves in <100> crystallographic directions, as a precursor to nucleation (Goldfarb *et al* 1997b).

A sequence from our growth movie, capturing nucleation and growth of a cluster, is shown in Fig. 2. Fig. 2(a)-(b) shows the nucleation of a cluster, with the arrow pointing to a nucleation site, and Fig. 2(c)-(d) shows the actual growth. As has been suggested by Jesson *et al* (1996b), clusters grow by addition to their facets. The material arriving at a facet nucleates an island or "pillbox" on it, which by growing increases the island size. Our *in-situ* observations confirm this hypothesis (Goldfarb *et al* 1997b). An advancing front of such a pillbox, growing on the island facet is shown by the black arrow in Fig. 2(d).

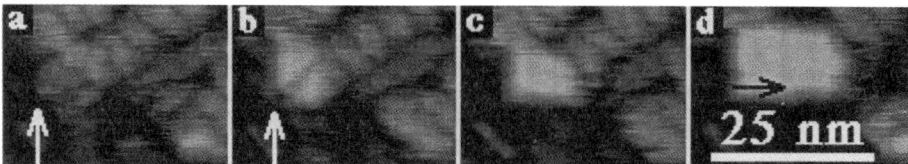

Fig. 2: STM seqence showing nucleation and growth of a "hut" cluster.

An alternative mechanism of cluster growth at the early stages is coalescence, provided the coalescing clusters are in close enough proximity. An example of 2D facet growth is shown in Fig. 3, while the coalescence-driven growth is shown in Fig. 4.

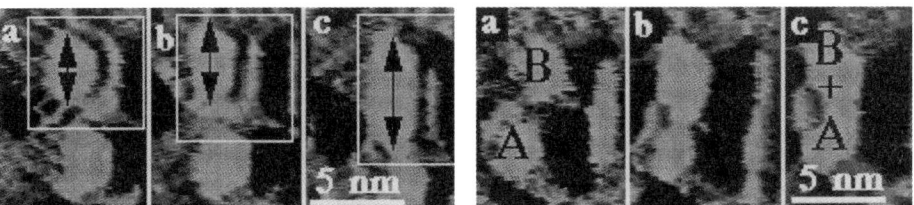

Fig. 3: "Log I" STM image of 2D facet growth.

Fig. 4: "Log I" STM image of cluster coalescence.

Being able to observe the cluster growth *in-situ*, we have measured the size evolution of individual clusters with time, such as that shown in Fig. 2. A typical plot of cluster growth rate can be seen in Fig. 5. The best fit (continuous curve) to the experimental data yielded a power-law dependence of cluster size on time, of the $r \sim t^{1/5}$ type. Once the index of the power-law is found, one can extract not only kinetic parameters, such as rate constants etc., but even more important information about the mechanisms of growth.

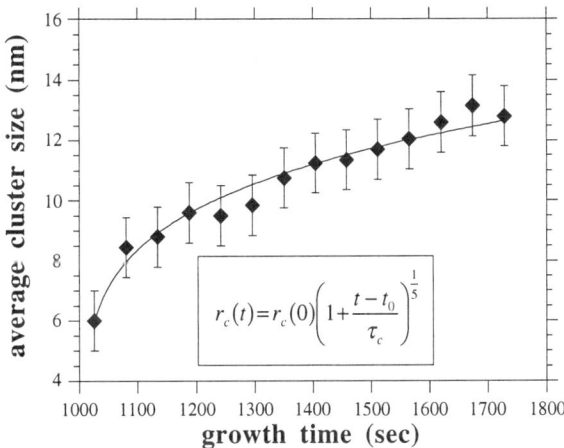

$$r_c(t) = r_c(0)\left(1 + \frac{t - t_0}{\tau_c}\right)^{\frac{1}{5}}$$

Fig. 5: Growth rate of a typical "hut" cluster. The bars represent measurement error, and the equation gives the line of best fit.

For example, it is well known that there should exist an intimate relation between the growth rate of the particles and their final size distribution, if at the late stages of growth the particles obey the Lifshitz-Slyozov-Wagner (LSW) theory of Ostwald ripening (Zinke-Allmang et al 1992). This theory is based on the Gibbs-Thomson capillarity effect, which leads to a growth of larger particles at the expense of smaller ones, once the supersaturation ceases to exist. In such a case, the size-distribution function can be separated into time-dependent and spatially-dependent parts. However the shape of the time-independent distribution function depends on the index of the power-law, which in turn depends on the mechanism controlling the growth. If the LSW theory is indeed obeyed, then the resulting distribution function will always be negatively skewed (Zinke-Allmang et al 1992). In the case of Ge/Si(001) the distribution functions are best described by Gaussians. A typical image and the corresponding size-distribution of the annealed Ge/Si(001) surface are shown in Fig. 6. This discrepancy can be explained by the inability of the larger clusters to coarsen, due to the stress-induced energy barrier at their base (Steinfort et al 1996), which prevents the ad-atoms from joining the cluster (Goldfarb et al 1997c). The barrier is higher the larger the cluster and the mismatch stress (Jesson et al 1996b).

600

Only during anneals at temperatures higher than 700 K do the ad-atoms begin to have enough activation to overcome this barrier, i.e. the Ostwald ripening commences.

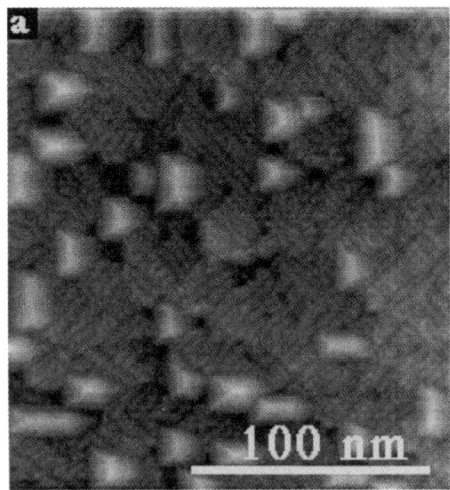
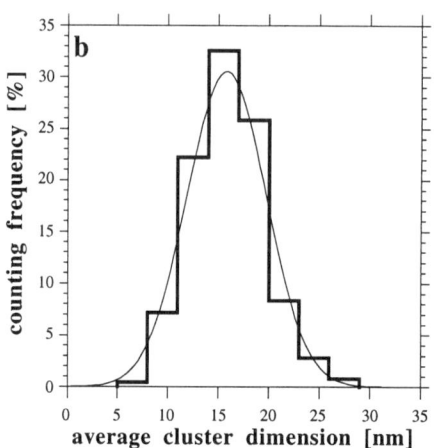

Fig. 6(a) : Typical STM image of annealed Ge "hut" clusters on Si(001), (b) corresponding cluster size-distribution.

4. CONCLUSIONS

In this study we have investigated mechanisms of nucleation and growth of Ge "hut" clusters on Si(001). It has been found that the mismatch stress can be relieved not only by 3D clusters, but also by surface pits (when the wetting layer is sufficiently thick). Both pits and clusters nucleate heterogeneously on various surface irregularities, such as steps and voids. Clusters were observed to grow mainly by 2D facet growth, but also by coalescence. Analysis of the time-dependence of growth and its correlation with the final cluster size-distribution allowed us to deduce that Ostwald ripening does not dominate late stage growth at temperatures below 700 K.

ACKNOWLEDGEMENTS

This work is supported by EPSRC (GR/K08161)

REFERENCES

Banyai L and Koch S W 1993 Semiconductor Quantum Dots (World Scientific, Singapore)
Eaglesham D J and Cerullo M 1990 Phys. Rev. Lett. **64**, 1943
Goldfarb I, Owen J H G, Hayden P T, Bowler D R, Miki K and Briggs G A D 1997a submitted to Surf. Sci.
Goldfarb I, Hayden P T, Owen J H G and Briggs G A D 1997b accepted in Phys. Rev. Lett.
Goldfarb I, Hayden P T, Owen J H G and Briggs G A D 1997c submitted to Phys. Rev. B.
Hammar M, LeGoues F K, Tersoff J, Reuter M C and Tromp R M 1995 Surf. Sci. **349**, 129
Jesson D E, Chen K M and Pennycook S J, Thundat T and Warmak R J 1996a Phys. Rev. Lett. **77**, 1330
Jesson D E, Chen K M and Pennycook S J 1996b MRS Bull. **21**, 31
Knall J and Pethica J B 1991 Surf. Sci. **265**, 156
Mo Y-W, Savage D E, Swartzentruber B S and Lagally M G 1990 Phys. Rev. Lett. **65**, 1020
Nötzel R 1996 Semicond. Sci. Technol. **11**, 1365
Petroff P M and Medeiros-Ribeiro G 1996 MRS Bull. **21**, 50
Tomitori M, Watanabe K, Kobayashi M and Nishikawa O 1994 Appl. Surf. Sci. **76/77**, 323
Tersoff J and LeGoues F K 1994 Phys. Rev. Lett. **72**, 3570
Steinfort A J, Scholte P M L O, Ettema A and Tuinstra F, Nielsen M, Landmark E, Smilgies D-M and Feidenhans'l R, Falkenberg G, Seehofer L and Johnson R L 1996 Phys. Rev. Lett. **77**, 2009
Zinke-Allmang, M Feldman L C and Grabow M H 1992 Surf. Sci. Rep. **16**, 377

Inst. Phys. Conf. Ser. No 157
Paper presented at Microsc. Semicond. Mater. Conf., Oxford, 7–10 April 1997
© 1997 British Crown Copyright

Measurement of silicon wafer roughness by atomic force microscopy : An interlaboratory comparison

A J Pidduck, A B J Smout, *P Wagner, *M Suhren, #D C Gupta and #S Yang

DERA, St. Andrews Road, Malvern, WR14 3PS, UK
*Wacker Siltronic AG, D-84479 Burghausen, Germany
#Charles Evans & Associates, Redwood City, CA 94063, USA

ABSTRACT: We have compared silicon wafer roughness values measured by four laboratories using atomic force microscopy. An atomically-stepped epitaxial (111) wafer allowed validation of height calibrations, and also proved valuable for diagnosing tip quality and noise. Initial roughness values from a highly polished wafer (<0.1nm rms) spanned a range of a factor of two, but after taking tip quality, periodic noise, and artefactual image curvature (present in larger area scans) into account, this range reduced to $\pm(20\text{-}25)\%$. In the case of a prepolished wafer (>1nm rms), the variation in roughness values appeared to reflect the surface roughness statistics.

1. INTRODUCTION

Silicon integrated circuit fabrication places very tight tolerances on the surface smoothness of silicon wafers, which in turn places stringent demands on roughness measurement techniques (Abe et al (1992), Wagner and Gerber (1995), Huff and Goodall (1996)). Industrially, optical techniques are mainly used due to their sensitivity, speed and non-contact nature. However, atomic force microscopy (AFM) is becoming widely used, due to its ability to provide three-dimensional topographic images with extremely high (sub-0.1nm) height sensitivity. Resonant-mode ("tapping-mode") AFM enables Si surfaces to be imaged stably and reproducibly (Gilicinski et al. (1993), Strausser et al. (1994), Nayar et al.(1994)). In order to establish AFM as a quantitative measurement technique it is important to determine the confidence which may be placed in AFM roughness values, as well as the factors which limit this, so that optimum data collection and treatment conditions can be identified. For these reasons we have undertaken an interlaboratory comparison of Si wafer roughness measurements made by four (two US and two European) laboratories. Four wafers were specially prepared : a highly polished (001) wafer, a prepolished (001) wafer, and epitaxial (001) and (111) wafers. Atomic steps on the epitaxial wafer surfaces were intended to provide an internal height calibration standard in a comparable height range to that of the expected roughness amplitude.

Roughness is a function of the spatial wavelength range over which the surface is analysed (Bullis (1994)). The power spectral density (PSD), which comprises the average square amplitudes of the Fourier roughness components as a function of their spatial frequency, is solely a property of the surface and allows results from different techniques to be compared quantitatively. A conveniently simple yardstick is the root-mean-square (rms) roughness obtained by integrating the PSD between two spatial frequency limits. Two factors need to be reconciled in deciding these limits. The first is the physical consequence of the roughness, which is different for different lengthscale regimes. Optical scattering (haze) is caused by surface roughness wavelengths between $\lambda/2$ and a few tens of λ (the light wavelength used) and has deleterious effects on, for instance, the ability to distinguish particle contamination during processing, whereas roughness at much smaller wavelengths may lead to scattering of electrons during device operation, and hence degraded electronic properties such as mobility. Present-generation devices depend on microscopically uniform gate oxides of sub-10nm-thickness, the fabrication of which may be very sensitive to such "nanoroughness", which can only readily be measured using AFM. The second factor to be considered is the limitation imposed by the measurement technique itself. In this study, 1μm and 10μm size images were collected by each laboratory. PSD integration was made over 15-150nm wavelengths from 1μm images, and over 150nm-3μm in the case of the 10μm images. The short wavelength limit (15nm) was selected based on a reasonable expectation for the tip sharpness. The long wavelength limit (3μm) was set to <1/3 of image size so as to allow reasonable statistics in the determination of the largest wavelength PSD element, and to avoid undue influence of scanner-related image curvature.

2. EXPERIMENTAL

Four 100mm diameter Si wafers were prepared (Wacker Siltronic) : RR1, a highly polished (001) wafer; RR2, a prepolished (001) wafer; RR3, an epitaxial (001) wafer (0.30° misorientation) and RR4, an epitaxial (111) wafer (0.05° misorientation).

"Tapping mode" AFM was carried out using Digital Instruments Dimension series microscopes, in air, with microfabricated Si levers resonating at 250-380kHz. 1μm and 10μm scans (512 line x 512 pixel) were recorded, at 0.4-1.5Hz scan rates, from RR1 and RR2 at 3 positions : centre and midway to an edge along 2 orthogonal axes. In the case of RR3 and RR4, 0.5-2μm images were obtained near to the wafer centre. 1μm scans from the centre of RR1 were taken at intervals during the measurement series to monitor tip integrity. Finally, the instrumental noise level was measured by recording data with zero scan size. In all cases amplitude data (the variation in lever tapping amplitude during scanning, i.e. the AFM height error signal) was also collected. Calibrated amplitude signal variation then gives a measure of how precisely the lever tracks the surface.

Data were recorded without filtering, and a best-fit tilted plane subtracted. Following this, a line-by-line tilt was carried out, which is equivalent to applying an ultra-long wavelength cut-off. Rms roughnesses (R_q) were then calculated from the PSD as defined above.

3. RESULTS

Fig.1 shows representative images from RR1-4. The roughness level in the case of RR1 is exceptionally low. The roughness of RR2 is much greater, comprising an isotropic background micron-scale roughness, and a distribution, typically 20 per 10μm image, of 10-20nm-high surface islands. Images from RR3 and RR4 show a parallel array of 0.3nm-height biatomic steps, spaced by 55nm and 300nm respectively, consistent with the wafer misorientation.

R_q values calculated from wafers RR1 and RR2 1μm and 10μm scans are plotted vs. laboratory, in measurement sequence, in Fig.2(a)-(d). Different symbols are used to denote the 3 positions analysed. Initial and final measurements (A and E) were made by the same laboratory. The series of values from 1μm images at the centre of RR1, taken at different times during the measurement sequence, are shown in Fig.2(e).

Ratios of height-to-amplitude signal variation from RR1 1μm scans are plotted in Fig.2(f). Similar values were obtained from 10μm scans. The height variation should be well in excess of the amplitude variation to ensure good surface fidelity. The ratio ranges from 1.5-7 despite the generally similar integral gain values used in all cases.

Measurements from RR4 are shown in Fig.2(g) and (h). Atomic terrace roughnesses were taken from software-zoomed areas of about 300nm square, the maximum permitted by the terrace widths. R_q values were then calculated, over 15-150nm wavelengths, after line-by-line tilting of these zoomed images. Step-heights were measured by tilting the whole image until the width of peaks in the bearing analysis (corresponding with the height variation of data points from a single terrace) were as close as possible to the corresponding widths from the zoomed images used to determine the terrace roughness. The difference then directly gives the maximum error in the step height due to any residual terrace tilt, and is plotted in Fig.2(g). The measured widths of individual (111) Si step edges, an average of several values read from line scans, is also shown in Fig.2(h).

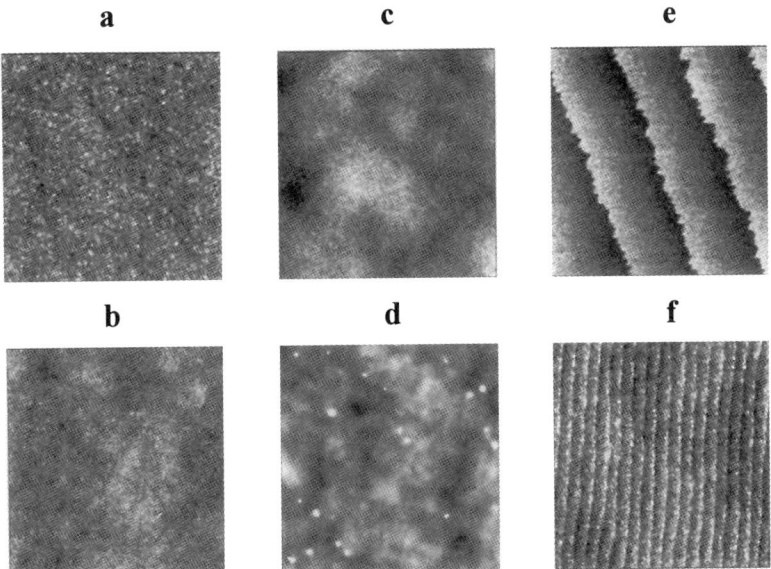

Figure 1. Representative AFM images from wafers RR1-4. (a) 1μm scan from RR1 (0.5nm Z-range), (b) 10μm scan from RR1 (0.3nm Z-range), (c) 1μm scan from RR2 (2nm Z-range), (d) 10μm scan from RR2 (10nm Z-range), (e) 1μm scan from RR4 (0.5nm Z-range), (f) 1μm scan from RR3 (0.5nm Z-range). The Z-range values are scale maxima for use with the height contrast bar in Figure 3.

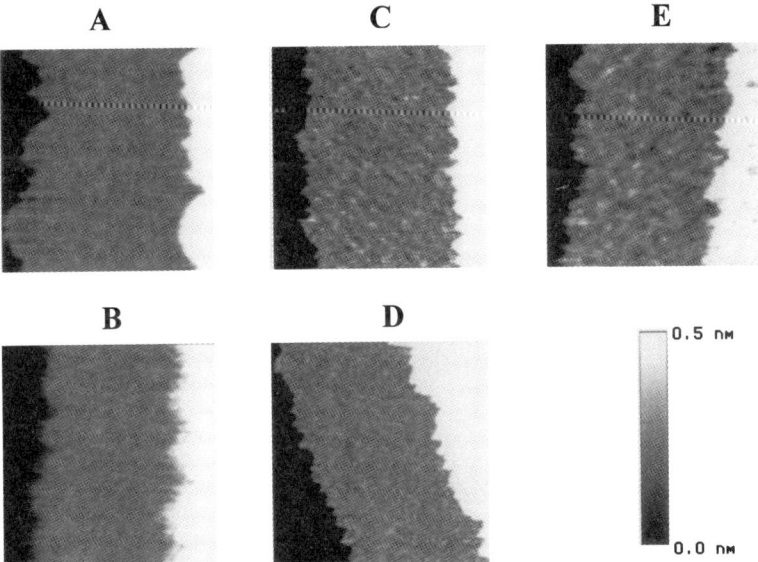

Figure 3. Comparison 500nm AFM images, from labs.A-E, of individual (111) Si atomic terraces on wafer RR4, obtained by software-zoom and line-by-line levelling of the terraces.

604

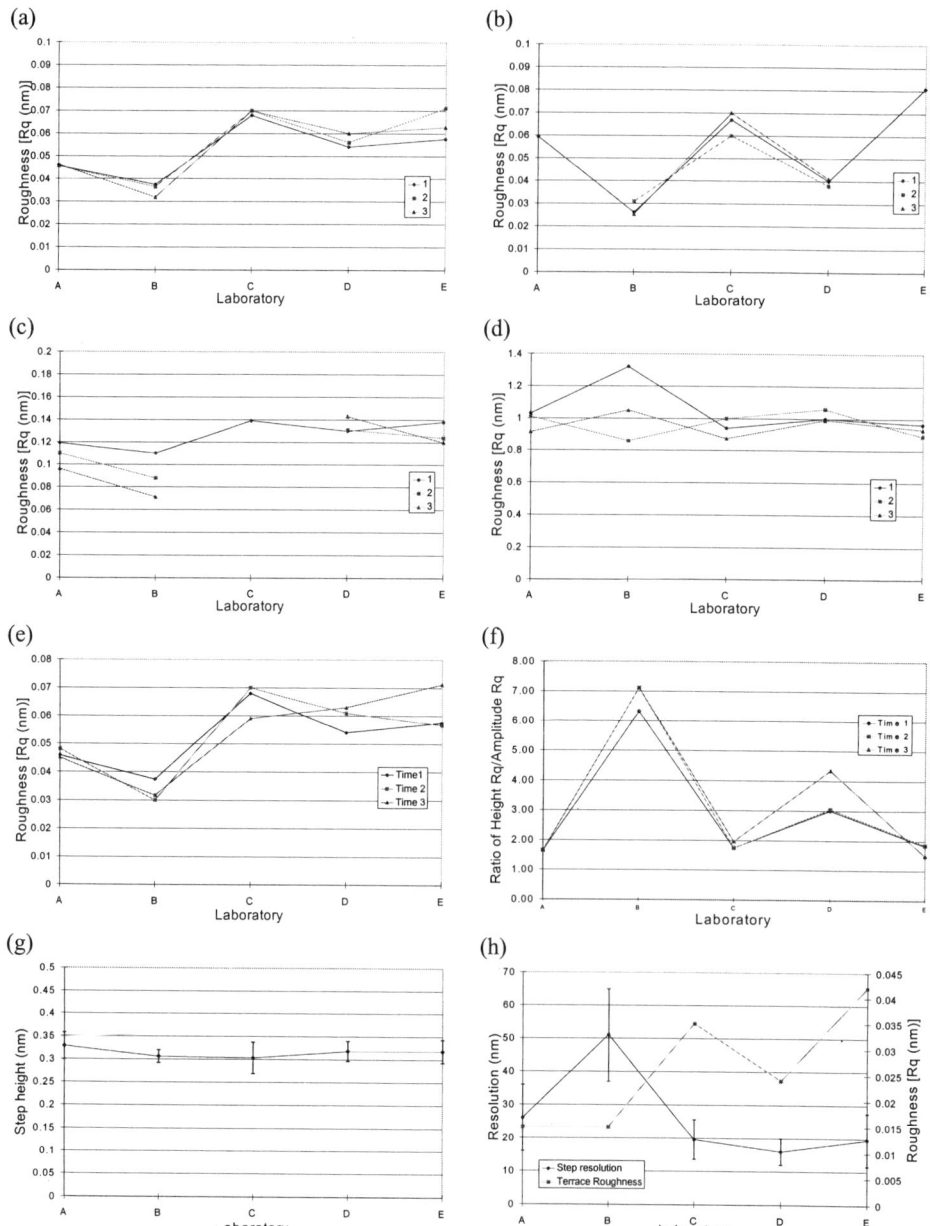

Figure 2. AFM roughnesses measured by laboratories A to E : (a) RR1 Rq values (15-150nm) from 1μm images, (b) RR1 Rq values (150nm-3μm) from 10μm images, (c) RR2 Rq values (15-150nm) from 1μm images, (d) RR2 Rq values (150nm-3μm) from 10μm images, (e) RR1 Rq values (15-150nm) from 1μm images taken at different times during measurement sequence, (f) ratio of height-to-amplitude Rq values (150nm-3μm) from RR1 1μm images, (g) step heights measured from RR4 and (h) Rq values (15-150nm) from (111)Si atomic terraces, and measured step edge widths.

RR3 data could not be treated in the same way owing to the very narrow terraces, and were not analysed further. Wavy single atomic height (0.14nm) steps (see Strausser et al (1994) for example) were not observed on the terraces.

Noise signal amplitudes varied between 0.20-0.44nm rms amongst the different laboratories.

4. DISCUSSION

4.1 Atomic step height measurements (RR4)

The measured values in Fig.2(g) show a range of 0.315±0.013nm, with an uncertainty varying between ±0.014nm and ±0.034nm (±4-11%) from lab-to-lab. The median value is almost identical with the true (111) Si biatomic spacing (0.314nm) and the consistency of the estimated values (±4.5%) compares very favourably with a previous interlaboratory comparison (Tarui et. al. (1995)). Clearly the variations in R_q values seen in Fig.2(a)-(d) cannot be attributed to errors in height calibration.

4.2 Polished wafer (RR1)

Initial R_q values determined from 1μm and 10μm images show a range of more than a factor of two in each case. The lab-to-lab variation is much larger than the position-to-position variation. R_q values from 1μm images changed by up to 25% during the measurement series. Interestingly, both increased and decreased values were observed. However, while changes in image resolution are undoubtedly occurring, they do not account for the lab-to-lab variation in Fig.2(a).

In the case of 1μm images, the most obvious contributions to the lab-to-lab variation are those of tip quality and periodic noise. These effects respectively cause lab B results to be low and lab C results to be high. In the case of lab B, the (111) atomic step edges appeared as a double image, giving the large step-edge width in Fig.2(h). Periodic mechanical or electrical noise, or feedback signal oscillation, in the case of lab C was estimated to contribute about 0.01nm to the R_q values. Fig.3 shows both these of these effects clearly in the (111) Si atomic terrace images. The median RR1 nanoroughness, excepting lab B data, was 0.059nm with a range of ±0.013nm (±22%). Rms subtraction of the random noise levels leads to measured values for the true wafer surface roughness of 0.049±0.015nm over the 15-150nm wavelength range.

This remaining ±22% variation is harder to explain. Possible effects which may contribute are (a) imaging methodology, (b) tip quality, (c) physisorption of molecular contaminants (moisture or organics) leading to a change in the measured surface or tip-surface interaction, or (d) a genuine increase in the wafer surface roughness during storage. Explanation (a) may be unlikely since the same imaging parameters were used for data sets A and E. Explanation (c) was tested by chemically recleaning RR1 after data set E, which resulted in a slightly increased finescale R_q level. Further work is required to choose conclusively between explanations (b) and (d). However, the residual variation was greatest at the shortest surface wavelengths. When measuring the R_q values as a function of increasing short wavelength PSD cut-off, a systematic convergence was found from a range of 0.058±0.013nm, evaluated over 15-150nm wavelengths, to 0.033±0.005nm, a range of ±15%, when integrated over 50-150nm wavelengths.

In the case of 10μm images the variation in Fig.2(b) is largely due to an artefactual image curvature. The primary component of this is the inherent bending motion of piezoelectric tube scanners (employed by all the AFMs in this study) with a secondary contribution, towards the start of each scan line, associated with scanner acceleration. This curvature remained along the fast-scan axis after line-by-line tilt removal, and had an amplitude of 0.1nm in the case of lab B, 0.2-0.4nm for lab D, and 0.7-1.2nm in the case of labs A, C and E. Its form comprised either one or two maxima or minima, and so could be largely removed by subtracting a best-fit third-order polynomial surface along the fast scan axis. Except for lab E, where there was also a component along the slow scan axis, this treatment resulted in R_q values consistently in the range 0.025-0.04nm (±23%).

606

4.3 Prepolished wafer RR2

The longer-wavelength roughness amplitude on RR2 is large compared with the artefactual image curvature. In this case the main influence on the R_q range of about 50% in Fig.2(d) is probably the surface roughness statistics, particularly the distribution of small islands. In 1μm images (Fig.2(c)) the effect of tip quality leads to slightly lower R_q values in the case of lab B. Otherwise the values, excluding images containing an island, are in the range 1.25±0.25nm (±20%).

5. CONCLUSIONS

We have compared AFM roughness measurements, obtained by four laboratories, from one highly polished Si wafer (<0.1nm rms), one prepolished Si wafer (>1nm rms), and two epitaxial Si wafers. Rms roughness values were calculated by integration of the PSD over surface wavelength ranges of 15-150nm (1μm images) and 150nm-3μm (10μm images). The main findings were :
(a) measured (111) Si biatomic step heights were accurate to within ±4.5%, whereas initial polished surface rms roughness values varied by about a factor of two between different laboratories;
(b) quantitative AFM roughness values measured from ultra-smooth surfaces (such as RR1) are very prone to underestimation, through loss of tip resolution, and to overestimation, through periodic noise and (at longer surface wavelengths) artefactual image curvature. Images from the (111) Si steps and terraces provide a very valuable means of diagnosing the tip quality and noise;
(c) under conditions where these effects were taken into account, the measurement consistency improved to ±(20-25)%. For instance, RR1 nanoroughness was measured to be 0.06±0.013nm. The consistency could be further improved by increasing the short-wavelength PSD cut-off.

We conclude that an acceptable consistency (±15-20%) in AFM-measured roughness values from very smooth surfaces is possible given careful specification of (i) imaging conditions and data treatment, (ii) an atomically-stepped reference surface to check height calibration and diagnose tip quality and noise, (iii) a highly stable Si surface and storage ambient, and (iv) appropriate image sizes and PSD wavelength limits for evaluating rms roughnesses. On the basis of this work, we suggest a practical short wavelength PSD cut-off of 20-30nm (with presently available tips) and a long wavelength limit of 1/5-1/10 of image size (to avoid the effects of artefactual image curvature with AFMs which use piezoelectric tube scanners). A roughness range of 20nm-5μm could therefore be covered (with some increase in scan time) using two image sizes of 1.5-2μm and 20-30μm. A handover wavelength in the range 200-500nm, consistent with optical λ/2, would be appropriate to define the boundary between "microroughness" and "nanoroughness".

REFERENCES

Abe T, Steigmeier E F, Hagleitner W and Pidduck A J 1992 Japan J. Appl. Phys. **31**, 721
Bullis W M 1994 "Semiconductor Silicon 1994" (Electrochem. Soc.) pp1156-1169
Gilicinski A G, Rynders R M, Beck S E, Strausser Y E, Stets J R, Felker B S and Bohling D A 1993 Mat. Res. Soc. Proc., **324**
Huff H R and Goodall R K 1996 "Semiconductor Characterisation : Present Status and Future Needs" (AIP Press) pp67-96
Nayar V, Pidduck A J, Idrees M I and Dew B E J 1994 Proc. 2nd Int. Symp. Ultraclean Processing of Si Surfaces (UCPSS '94), (Acco Leuven) p371
Strausser Y E, Doris B, Diebold A C and Huff H R 1994 **94-1**, Ext. Abs. 185th Electrochem. Soc. Mtg., San Francisco, May 1994 (Electrochem. Soc.)
Tarui Y 1995 Chairman Japan Ultraclean (UC) Standardisation Committee, Working Group for Eqpt. report "1nm-order Z-axis calibration of AFM" (Ultra Clean Society (March 1995))
Wagner P and Gerber H A 1995 Proc. Tech. Program, Semicon Europa 1995, (SEMI Europe)

© British Crown Copyright / DERA (1997)

Inst. Phys. Conf. Ser. No 157
Paper presented at Microsc. Semicond. Mater. Conf., Oxford, 7–10 April 1997
© *1997 IOP Publishing Ltd*

AFM investigations of the influence of the doping process on the structure of LPCVD-silicon films

H Gold[1], J Lutz[1], F Kuchar[1], M Pippan[2] and H Noll[2]

[1]Institute of Physics, University of Leoben, A-8700 Leoben, Austria
[2]Austria Mikro Systeme International AG, A-8141 Unterpremstätten, Austria

ABSTRACT: The influence on the structure of low-pressure chemical vapour deposited silicon films of two different ex-situ doping processes, diffusion doping from a $POCl_3$ source and ion implantation, has been studied. The films have been grown at temperatures between 550 °C and 630 °C and have been investigated using atomic force microscopy. Depending on the deposition temperature of the layers, the grain sizes and their shapes are very different for the doped layers although the annealing parameters are comparable. Additionally, the roughness values of the diffusion doped films are obviously larger.

1. INTRODUCTION

Due to progress in the production of integrated circuits, the lateral dimensions of the components continuously decrease. As a consequence, the requirement for a better under-standing of the detailed properties of polycrystalline silicon layers and their dependence on production conditions increases. During the last decades a lot of different characterisation methods have been developed. All methods have disadvantages in common like a limited lateral resolution, an extensive preparation, and a restriction to special surfaces. On the contrary, scanning probe microscopy is a powerful tool for real-space surface imaging with high 3D resolution and needs a minimum of preparation techniques.

In this work we report on structure and surface roughness investigations of silicon films grown by LPCVD at different temperatures and subsequent doping. The samples have been taken from production-line test wafers, produced by high-volume equipment.

2. EXPERIMENTAL

The measurements have been performed using an Omicron UHV AFM with a 5µm range tubescanner driving the sample. Commercially available silicon nitride cantilevers, character-ised by a 36 ° interior angle and 20 nm radius of curvature, have been used for the contact-mode-AFM measurements under high vacuum conditions. Force calibration for individual cantilevers has been obtained by measuring the bending of the cantilevers as a function of the distance between dip and sample and taking into account the force constant of the cantilevers.

The investigated films have been deposited onto 100 nm silicon dioxide thermally grown (dry oxidation at 1000°C) on (100) n-type silicon wafers. The deposition temperature for different samples was varied between 550 °C and 630 °C in steps of 10 °C. The thickness of the as-deposited layers was between 250 and 270 nm. The ex-situ phosphorous doping was produced in two different ways: a) by a gaseous predeposition using a $POCl_3$ source at 900 °C (11 min), followed by a drive-in step at the same temperature (5 min) and the deglazing in a buffered oxide etch and b) by implantation of phosphorous (dose: 9×10^{15} cm^{-2}; energy: 75 keV) and subsequent annealing of the films at 900 °C (15 min). Doping by implantation was applied to films grown at deposition temperatures below 600 °C only.

Prior to the scanning probe measurements an additional selective silicon dioxide etching has been applied to all samples by dipping them in 2% hydrofluoric acid (HF). This procedure is well known to produce surface passivation by H-termination of silicon dangling bonds. Immediately after etching, the samples have been mounted in the AFM chamber and were measured at high vacuum conditions at a residual gas pressure $< 2 \times 10^{-7}$ mbar. A comparison of HF - treated and untreated samples by Pleschinger et al (1997a) had shown that the HF - passivation changes neither the topographic nor the grain structure of the samples.

3. RESULTS

The structure of the as-grown silicon films changes from amorphous to polycrystalline for increasing deposition temperatures between 570 °C and 600 °C (Pleschinger et al 1997b). Layers deposited in this transition region consist of an amorphous matrix with an increasing number of embedded crystallites. In contrary to the very flat amorphous layers the surfaces of polycrystalline films grown above 600 °C exhibit a pronounced hillock structure. These hillocks were identified as the grains of the layer (Harbeke et al 1984). Due to the different growth velocities for different grain orientations the effective roughness value for an area of 1 μm^2 of the as-grown polycrystalline films is above 10 nm and therefore more than one order of magnitude larger than that of amorphously grown films (1 nm).

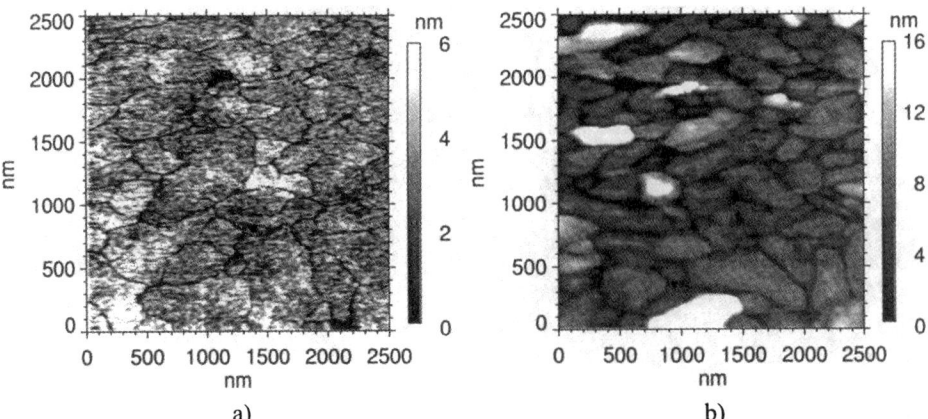

a)   b)

Fig. 1: Contact-mode-AFM topographic image of a polycrystalline silicon film grown at 560 °C with subsequent doping by a) ion implantation and b) diffusion.

For comparison Fig. 1a and Fig. 1b show the topographic images of silicon films grown at 560 °C doped by implantation and diffusion, resp. These images are characteristic for all samples grown below 570 °C. For both doping processes the structures of the films are dominated by irregular and elongated shapes of the grains. Additionally the grains often exhibit angles at the corners deviating strongly from 120°.

The lateral dimensions of the grains are in a wide range up to above 1000 nm. They decrease with increasing deposition temperature of the film although the doping and annealing parameters are equal. Further, Figs. 1a and b show that the topographies of the layers are very different for the different doping processes. Ion implanted films exhibit flat grains at the surface of the layer and the height difference of the grains is small. Therefore ion implanted layers remain very flat and their roughness values increase only slightly to around 1.5 nm. In contrast, doping by diffusion produces rounded grains with large differences in height. As a consequence the roughness values of the diffusion doped films is increased by more than a factor of two compared to the as-grown films.

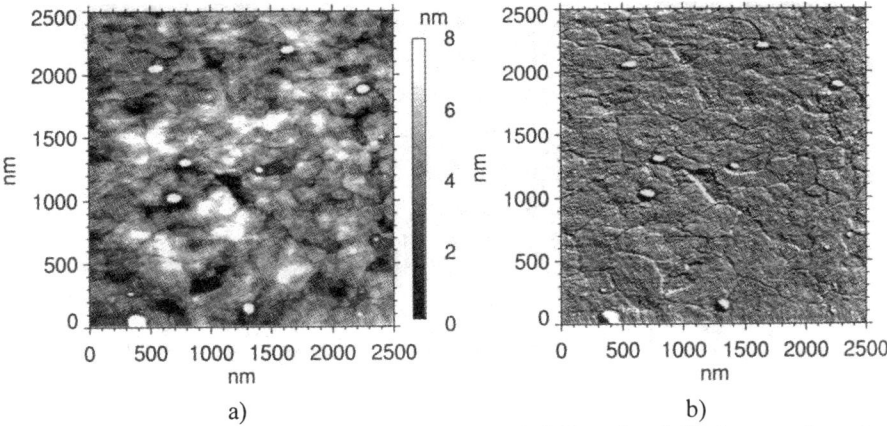

a)   b)

Fig. 2: Contact-mode-AFM topographic (a) and differentiated (b) image of a polycrystalline silicon film grown at 570 °C with subsequent doping by ion implantation.

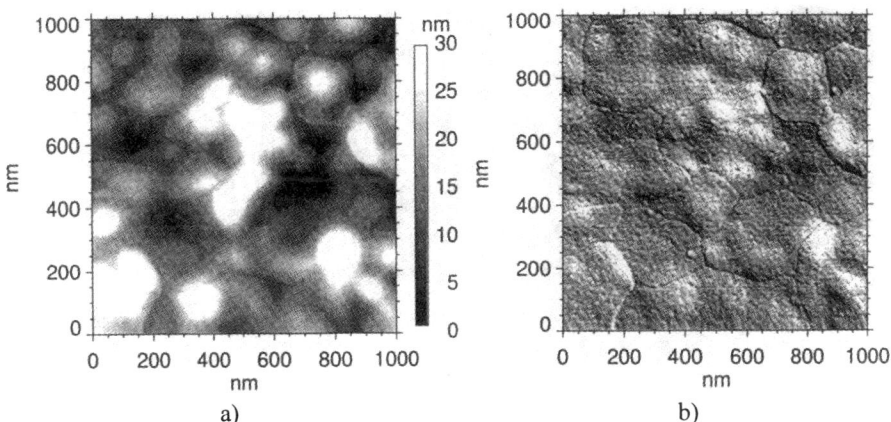

a)   b)

Fig. 3: Contact-mode-AFM topographic (a) and differentiated (b) image of a polycrystalline silicon film grown at 620 °C with subsequent doping by diffusion.

Fig. 2a shows the topographic images of a silicon film grown at 570 °C doped by ion implantation. The small crystallites (white dots) remain as-grown during the deposition process and the amorphous matrix recrystallizes. To enhance the short-scale corrugation and the visibility of the boundaries Fig. 2b shows a differentiated image of Fig. 2a. An analysis of the measurements shows that the structure of this film is comparable to those grown at lower temperatures but consists of smaller grains.

The topography shown in Fig. 3a is dominated by a hillock structure with lateral features between 70 and 200 nm and is characteristic for all films grown at temperatures higher than 600 °C. A comparison of the topographies of doped and as-grown films show only slightly enlarged and flattened hillocks for the doped films with roughness values between 7 and 12 nm. The height differences for polycrystalline films grown above 600 °C are too large to see the grain boundaries directly in the topographic image. For that reason Fig. 3b again shows a differentiated image of Fig. 3a. Contrary to the amorphous grown films, polycrystalline grown layers exhibit regular grain shapes and the angles of the corners are near 120°. Typical lateral dimensions of the grains are 350 nm, but ranging up to 500 nm. This result resembles earlier TEM - micrographs of similarly processed samples reported by Wada et al. (1978). In contrary to the as-grown films the hillocks of the doped samples can no longer be identified with the grains shown in Fig. 3.

4. CONCLUSION

We have investigated undoped and doped silicon films grown in a temperature range between 550 °C and 630 °C. The amorphous films grown below 570 °C are very flat and their roughness values are by more than a factor of ten smaller than those of the polycrystalline films grown above 600 °C. The doping and subsequent annealing at temperatures of 900 °C causes a complete recrystallisation of the layers. The influence of the doping process strongly depends on the doping process itself and the deposition temperature of the as-grown films.

The lateral dimensions of the grains of polycrystalline films deposited above 600 °C increase to around 350 nm and their shape remains regular. In contrast to this change in the structure of the films the topography of the films is only slightly modified because at the interface silicon/phosphorous-doped glass the mobility of the atoms is strongly reduced. Consequently the topography of the interface does not change drastically due to the high temperature process and the roughness values of the deglazed silicon films decrease only slightly.

A completely different result was found for amorphously deposited films. Their grain sizes are in a wide range up to 1 μm and their shapes are irregular for both doping processes. A possible explanation is a too short annealing time. Additionally we have found that the lateral dimensions of the grains decrease with increasing deposition temperature although temperature and duration of the annealing processes are comparable. Last but not least the roughness values for the doped films strongly depend on the doping process itself. Because the oxidation velocities perpendicular to the film surface are very different along grain boundaries and the interior of grains the roughness values of the interface silicon/phosphorous-doped glass are larger then those of the initial silicon film. For doping by implantation the topography is changed by the recrystallisation process of the silicon film only.

REFERENCES

Harbeke G, Krausbauer L, Stegmeier E F, Widmer A E, Kappert G F and Neugebauer G 1984 J. Electrochem. Soc. **131** p 675
Pleschinger A, Lutz J, Kuchar F, Pippan M and Noll H 1997a J. Appl. Phys. **81**(10) p 6749
Pleschinger A, Lutz J, Kuchar F, Pippan M and Noll H 1997b Proc. of SXM2 in print
Wada Y and Nishimatsu S 1978 J. Electrochem. Soc. **125** pp 1499-1504

Inst. Phys. Conf. Ser. No 157
Paper presented at Microsc. Semicond. Mater. Conf., Oxford, 7–10 April 1997
611
© 1997 IOP Publishing Ltd

Structural studies of InGaAsP/InP-based lasers using cross-sectional atomic-force microscopy (XAFM) and selective etching

Thomas Kallstenius [1,2], **Ulf Smith** [1,2] and **Björn Stoltz** [2]

[1] Materials Science Division, Uppsala University, Box 534, S-75221 Uppsala, Sweden
[2] Ericsson Components AB, S-16481 Kista, Sweden
E-mail: Thomas.Kallstenius@Teknikum.uu.se

ABSTRACT: Cross-sections of fully processed InGaAsP/InP-based laser diodes have been studied using selective etching in combination with atomic-force microscopy (AFM) and electron-beam induced-current (EBIC) imaging. Detailed investigations of the etched topography made it possible to draw conclusions about the interdiffusion of Fe and Zn in lasers with semi-insulating (SI) current-blocking and about Zn indiffusion in lasers with p/n current-blocking. The potential of AFM combined with selective etching for resolving nano-structural elements, such as multi-quantum wells (MQW), is also demonstrated.

1. INTRODUCTION

InGaAsP/InP-based buried-heterostructure (BH) lasers are to a great extent used as light sources in fiber-optics communication. The present paper presents a method by which high-resolution studies of microstructure and composition in such components can be performed in a simple manner, using selective etching in combination with AFM.

Etching of cross-sections of the lasers with different selective etchants translates chemical properties of the material into topographic effects. Scanning the sample surface in an AFM will then give a high-resolution profile of the etched structure with differences in composition and dopant concentration shown as topographical features. This, to our knowledge, new method of looking at InP laser structures has been utilized to obtain information about changes in dopant concentration due to diffusion across the laser structure and to study fine details in the structure, such as MQW and gratings in distributed feed-back (DFB) lasers. The high resolution of the method also allowed a direct comparison to be made between EBIC images and actual laser structures.

2. EXPERIMENTAL

Fabry-Perot (FP), DFB and MQW lasers with p/n current-blocking (FP-p/n, DFB-p/n and MQW-p/n) and SI current-blocking (FP-SI), respectively, were studied. The processing of these lasers included low-pressure metal-organic vapor-phase epitaxy (LPMOVPE), reactive ion etching (RIE), as well as electron beam lithography for the grating structure in the DFB lasers. Typical growth temperatures and pressures were 640°C and 20 mbar, respectively.

A stain-etchant (Huo and Yan 1990), modified for our purpose, was used for the dopant-selective etching. This etchant, [10 g $K_3(Fe(CN)_6)$, 15g KOH, 270 ml H_2O], which etches under illumination, provides smooth surfaces and a satisfying etch selectivity between n-InP and p-InP. The 3:1:1-etchant [H_2SO_4:H_2O_2:H_2O] (Iida and Ito 1971) was used for the composition-selective etchings. The observed strong selectivity between InP and InGaAsP for this etchant was found suitable for studies of MQW-BH lasers having InP barriers and InGaAsP wells.

The samples were cleaned in SemicoClean23 (Furu-uchi Chemical Co, Tokyo), acetone and deionized water immediately before etching. The etch solution was held at room temperature. In the case of etching under illumination, a halogen light source was used.

A scanning probe microscope (SPM) from Digital Instruments (NanoScope III) was used in the contact AFM-mode in air. Silicon nitride cantilevers, having square pyramidal tips with a 20-50 nm radius of curvature and a 70° apex angle, were used.

For comparison, cross-sections of the laser structures were also studied with the EBIC mode in a SEM. A moderately low acceleration voltage, 5 kV, was used so that the spread of the electron beam inside the material could be kept small without loosing too much signal intensity. According to the Kanaya-Okayama expression for maximum electron range (Goldstein et al 1994), the width of the interaction volume is less than 0.3 μm at this voltage.

612

3. RESULTS

3.1. Lasers with SI Current-Blocking

Two types of FP-SI lasers were studied. The basic structure of these lasers was the same and the growth parameters were held identical, apart from some of the SI-reactants. As a result, the growth rate of the SI current-blocking layer differed between the two types: The SI layer in Fig. 1a was grown at a rate of 2.4 µm/h, whereas the rate for the SI layer in Fig. 1b was 1.2 µm/h. The structural elements are labelled in Fig. 1c. It is clear from the XAFM images that p-InP has a higher etch rate than n-InP and SI-InP.

In the bottom part of the laser structure in Fig. 1a, two horizontally grooved structures can be seen, whereas only one groove is observed in the structure of Fig. 1b. In addition, faint ridges in the p-InP region above the mesa structure can be seen in Fig. 1c (labelled "B"), where an edge-enhanced contrast has made the presence of these ridges more apparent. The ridges provide a possibility to determine the original position of the interfaces in the structure, as discussed further below. Thus, the interface between p-InP:Zn and SI-InP:Fe appears to have shifted relative to its original position. The magnitude of the shift is estimated to be 0.47 µm for the laser in Fig. 1a and 0.68 µm for the laser in Fig. 1b.

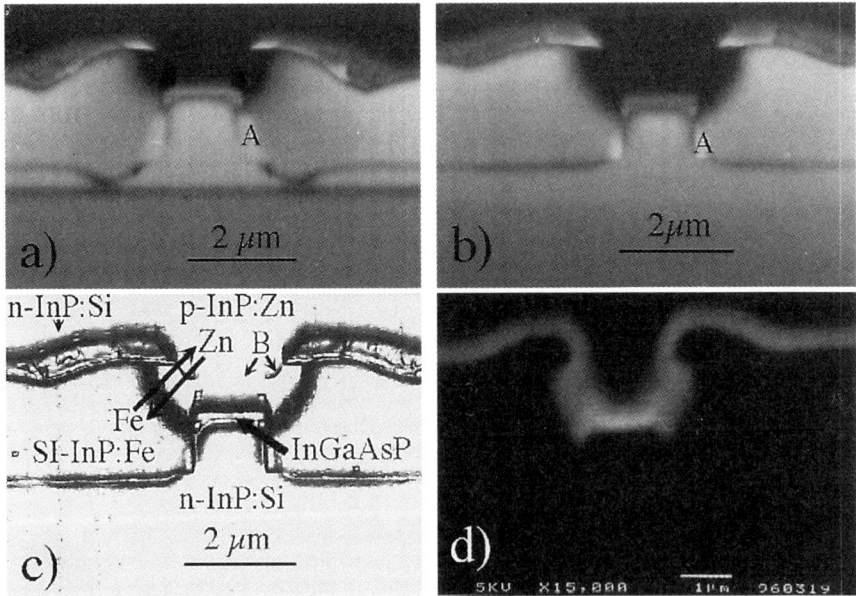

Figure 1: (a) XAFM image of FP-SI laser. The growth rate of the SI-layer was 2.4 µm/h. (b) FP-SI laser, with a SI-layer grown at a rate of 1.2 µm/h. (c) Edge-enhanced contrast of the image in (b). (d) Cross-sectional EBIC image of the laser structure in (b).

The EBIC image in Fig. 1d shows a geometry which is striking in agreement with the XAFM images in Figs. 1b and 1c. The widths of the lines in the EBIC image are of the order of the diameter of the interaction volume expected for the electron beam (~0.3µm). At the interface between p-InP:Zn and SI-InP:Fe, the lines are clearly broadened.

In the SI-material adjacent to the mesa sidewall, a pair of elevated regions can be seen in the XAFM images of the FP-SI laser structures, cf. "A" in Figs. 1a and 1b. The reason for the lower etch rate in this part of the structure is not clear, but it might be due to differences in dopant concentration or atomic disorder caused by the complicated growth situation in this part of the structure.

3.2 Lasers with p/n Current-Blocking

Figures 2a and 2c show XAFM images of dopant-selectively etched FP-p/n and DFB-p/n. The XAFM image of the FP-p/n laser in Fig. 2a is shown with enhanced contrast in Fig. 2b. Additional ridges are revealed in this image. They are located in the p-InP contact layer immediately at the edge of the n-InP layer of the p/n current-blocking structure ("A" in Fig. 2b), as well as along the vertical sides ("B") and above the mesa structure ("C"). Also in this case, the ridges provide a possibility to determine the original position of the interfaces in the structure. Thus, the interface between n-InP:Si and p-InP:Zn appears to have shifted by about 0.1 μm into the n-InP:Si.

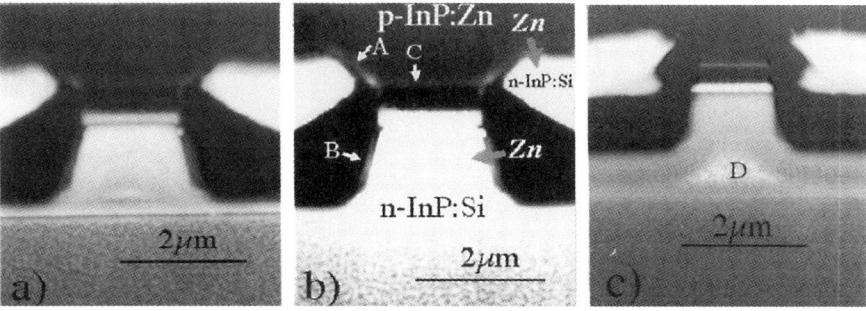

Figure 2: XAFM images of FP-p/n (a) and DFB-p/n (c) lasers with p/n current-blocking. Additional ridges are revealed when the contrast of (a) is enhanced, as shown in (b).

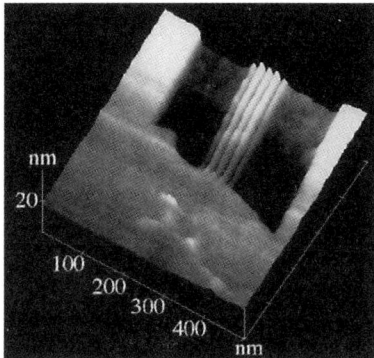

Figure 3: XAFM image of MQW-p/n laser.

Another interesting detail in the XAFM image of the DFB-p/n laser in Fig. 2c is the triangularly shaped hill in the central lower part of the mesa structure ("D" in Fig. 2c). Such a hill can also be observed in the FP-p/n image, Fig. 2a. Parts of the mesa structure adjacent to the p-InP:Zn layer appear to have etched at a higher rate than the central part of the mesa.

3.3 MQW Lasers

The XAFM image in Fig. 3 shows a compositionally etched MQW-p/n laser. The 3:1:1-etchant provides an excellent selectivity between InGaAsP and InP due to the considerably higher etch rate for InGaAsP. In the Figure, the InP barriers with four intermediate InGaAsP quantum-wells can be seen clearly. The widths of the barriers and the wells are 8 nm and 11 nm, respectively. The MQW structure is surrounded by strongly etched layers of InGaAsP.

4. DISCUSSION

4.1 General

The actual reason for the grooves and ridges in Figs 1c and 2b is not clear. Previous SIMS measurements on planar structures have shown impurity atoms, such as Si, C and O, to be present at interfaces due to a slight, inevitable, contamination of the surface, even though the surface was carefully cleaned before the regrowth steps. Such contamination may change the etch rate locally at the interfaces. However, the ridges provide us with a possibility to determine the original position of the interfaces in the laser structure.

4.2 Lasers with SI Current-Blocking

It has been pointed out by several groups that when in contact with p-InP:Zn, the semi-insulating characteristics of SI-InP:Fe turn out to be difficult to reproduce because of the interdiffusion of Fe and Zn (Wolf et al 1994, Young and Fontijn 1990). In those parts of the laser structure where interdiffusion takes place, the Zn converts the SI-InP to p-InP to some extent. Since the etch rate of p-InP is higher than for SI-InP, as can be inferred from the topography in Figs. 1a and 1b, the interdiffusion will show up as a shift of the interface between p-InP and SI-InP from its original, pre-etch position. According to Figs 1a and 1b, the shift is less in the case of the higher growth rate, which is consistent with the fact that the laser in Fig. 1a remained at the growth temperature for a shorter length of time. We can use the information in the AFM images to estimate a diffusion coefficient of $3\cdot10^{-13}$ cm^2/s for the Fe $\leftrightarrow$ Zn interdiffusion. Wolf et al (1994) have shown that the diffusion length increases with the square root of time and obtained a diffusion coefficient of $2.0\cdot10^{-13}$ cm^2/s in an experiment with similar conditions as ours, except that their concentration of Fe was considerably lower, i.e., $5 \cdot 10^{16}$ cm^{-3}, compared with the $1 \cdot 10^{18}$ cm^{-3} in our case. Considering the experimental differences and uncertainties, the diffusion coefficient deduced from our AFM measurements is consistent with the result by Wolf et al (1994).

With regard to the EBIC image in Fig. 1d, the underlying contrast mechanism of this image could be that of double injection (Flynn et al 1988), since in parts of the structure the condition for this is fulfilled, i.e., SI material is in contact with p-InP. However, the correspondence between the EBIC image in Fig. 1d and the contours of the laser structure in Fig. 1c is remarkable and the size of the broadening of the EBIC contrast at the interface between p-InP:Zn and SI-InP:Fe agrees quite well with the shift observed in the XAFM image. It is therefore tempting to ascribe the EBIC signal to a more common situation, i.e., the presence of a p/n-junction depletion region. If so, the broadening of the EBIC contrast would indicate a corresponding widening of the depletion region. This widening is probably due to a gradient in the Fermi level caused by the Fe $\leftrightarrow$ Zn interdiffusion.

4.3 Lasers with p/n Current-Blocking

The apparent shift of the interface between the n-InP:Si and the p-InP:Zn layers, seen in Fig. 2b, is most likely due to a diffusion across the interface. Considerable indiffusion of Zn has been reported by several groups for structures with Zn-doped p-InP layers adjacent to Si-doped n-InP layers (Kondo et al 1996, van Gurp et al 1987), whereas SIMS data have shown no evidence of a corresponding redistribution of Si for any donor concentration (Blaauw et al 1989). When Zn diffuses into n-InP:Si, the electrical characteristics of the material can change to p-type. Consequently, the etch rate will increase in the indiffused region and the interface will appear shifted, c.f. "A" in Fig. 2b. The estimated diffusion length of about 0.1 µm is of the same magnitude as the Zn diffusion into n-InP reported by Kondo et al (1996) for a similar laser structure.

The origin of the triangularly shaped hill on the mesa, seen in Figs 2a and 2c, is not yet fully understood. Scanning auger microscopy (SAM) analysis of the etched laser structures indicates differences in concentration and composition of an oxide at the triangularly shaped hill compared to the values for an oxide on the surrounding material on the mesa. The hill-shape could be due to the oxide acting as a gradually eroding mask during etching, or it could be a result of the complex electrochemical conditions present in this region.

5. CONCLUSIONS

The results presented and discussed above illustrate how the large dynamic range of the AFM, when used together with selective etching, allows large-area examinations of components to be combined with close examinations of nanostructure details like those present in InP-based lasers. By judicious choice of etchant one can even obtain information about the behaviour of the material during processing, the diffusion of Zn atoms inanInP laser being one example.

ACKNOWLEDGEMENTS

This work has been supported by Ericsson Components AB, Kista, Sweden, and has been performed within the Surface & Micro Structure Technology (SUMMIT) Competence Centre (sponsored by the Swedish National Board for Industrial and Technical Development, NUTEK). The authors would like to thank M. Rask for clarifying discussions and valuable support and F. Wahlin for providing the MQW structures.

REFERENCES
Blaauw C., Shepherd F. R. and Eger D. 1989 J. Appl. Phys. **66,** pp. 605-610.
Flynn E. J., Ketelsen L. J. P., Zilko J. L., Huo D. T. C. and Koszi L. A. 1988. 11th IEEE Int. Semicond. Laser Conf. Boston, 216.
Goldstein J. I., Newbury D. E., Echlin P., Joy D. C., Romig A. D., Lyman C. E., Fiori C. and Lifshin E.,1994 Scanning Electron Microscopy and X-ray Microanalysis, 2nd ed. (Plenum Press, New York and London), 89.
Huo D. T. C. and Yan M. F. 1990 J. Electrochem. Soc. **137,** pp. 3270-3271.
Iida S. and Ito K. 1971 J. Electrochem. Soc. **118,** 768.
Kondo Y., Kishi K., Itoh M., Oohashi H., Itaya Y., and Yamamoto M. 1996. 8th Int. Conf. on InP and Related Materials: IEEE, pp. 384-387.
van Gurp G. J., Boudewijm P. R., Kempeners M. N. C. and Tjaden D. L. A. 1987 J. Appl. Phys. **61,** pp. 1846-1855.
Wolf T., Zinke T., Krost A., Scheffler H., Ullrich H. and Bimberg D. 1994 J. Appl. Phys **75,** pp. 3870-3881.
Young E. W. A. and Fontijn G. M. 1990 Appl. Phys. Lett **56,** pp. 146-147.

Inst. Phys. Conf. Ser. No 157
Paper presented at Microsc. Semicond. Mater. Conf., Oxford, 7–10 April 1997
© 1997 IOP Publishing Ltd

A $\underline{k}$-space transport analysis of the BEEM spectroscopy of Au/Si Schottky barriers

U Hohenester[1], P Kocevar[1], P de Andres[2] and F Flores[3]

[1] Inst. f. Theoretische Physik, Karl-Franzens-Universität Graz, A-8010 Graz, Austria.
[2] Inst. de Ciencia de Materiales (CSIC), E-28049 Madrid, Spain
[3] Dep. de Fisica Mat. Cond., Univ. Auton. de Madrid, E-28049 Madrid, Spain

ABSTRACT: We address the question of the spatial resolution of ballistic electron emission microscopy (BEEM) of Schottky barriers in Au(111)/Si(100) and Au(111)/Si(111) interfaces. A novel combination of Green-function and $\underline{k}$-space Ensemble-Monte-Carlo techniques is used to obtain new insights into the spatial and energetic evolution of the STM-tip-induced electrons during their passage through the metallic layer before reaching the metal-semiconductor interface. In particular, it is shown how the effect of band-structure-induced directional focussing of the electrons enforces a reinterpretation of existing experimental data.

1. INTRODUCTION

The recent theoretical prediction (de Andres et al. 1997) of decisive band-structure effects in the propagation of the STM-tip-induced hot electrons through the metal layer brought a new facet to the discussion about the very high spatial resolution of ballistic electron emission microscopy (BEEM) of Au/Si Schottky barriers (Bell 1996; Prietsch 1995). While most interpretations of BEEM data on Au/Si(100) and Au/Si(111) interfaces have assumed a narrow forward cone of tunneling-injected electrons and explained the very similar spectra and their high spatial resolution through various forms of collisional beam broadening, the prediction of a pronounced off-axis shift and broadening of the angular distribution just below the surface reopens the discussion about the role of the band structure and of scattering processes in the bulk and at the boundaries of the metallic layer. It is the purpose of the present analysis to improve the conventional energy-space descriptions and Monte-Carlo simulations of the hot-electron dynamics by providing a detailed $\underline{k}$-space Ensemble-Monte-Carlo simulation of the passage of the hot electrons through the Au layer, including the essentials of the band structure in the directional spectrum of the injected electrons, in the free-particle propagation, and also in the scattering cross sections for electron-electron (e-e), electron-phonon (e-ph), and electron-boundary scattering. The recent experimental data of Bell (1996) for varying layer thickness d and temperature will be reanalysed and the results contrasted with Bell´s original interpretation. It will be shown that the two most intensively debated questions about BEEM spectroscopies of the Au/Si system, namely the questions about the origin of the great similarity between Au/Si(100) and Au/Si(111) spectra and about the extremely high spatial resolution, can, for the first time, be directly answered without use of adjustable parameters or ad hoc assumptions.

2. TRANSPORT MODEL

Before presenting our transport model, we briefly summarise the conventional model, which is mainly based on the ideas of Kaiser and Bell (1988). The original KB model assumes (i) that the injected distribution at the metal surface is concentrated within a narrow forward cone, (ii) that the

$\underline{k}$-distribution at the metal/semiconductor interface is identical to the injected $\underline{k}$-distribution at the metal surface, i.e. the $\underline{k}$-vector parallel to the plane $\underline{k}^{\parallel} \approx 0$, and (iii) that $\underline{k}^{\parallel}$ is conserved at the interface (i.e. specular reflection/transmission via continuity of wavefunction). As a consequence of the different orientations of the six conduction-band valleys in Si with respect to the impinging narrow forward cone at the interface, BEEM spectra for Au/Si(100) and Au/Si(111) should, in the absence of strong scattering effects, be distinctly different. This should occur because of matching of $\underline{k}^{\parallel}$ for Au/Si(100) with the two perpendicular valleys and no matching of $\underline{k}^{\parallel}$ for Au/Si(111), a prediction in strong contrast with the experimental facts.

We now turn to our present transport model. We first note that STM and LEED studies show that Au films grow on Si(100) and Si(111) by forming crystals oriented preferentially in the [111] direction. Then the empirical-tight-binding Green function analysis (Garcia-Vidal et al. 1996) of the coherent electron propagation from the STM tip through the tunneling gap into the metal layer reveals that the STM electrons achieve their bulk Bloch character, with propagation gaps due to forbidden regions of phase space, after passing roughly 20 Å within the metal. The injected distribution at z = 20 Å turns out to reach its maximum at the edge of the planar Brillouin zone at ≈ 70 degrees, with an average $1/\cos \Theta$ distribution law (de Andres et al. 1997). We should stress that the detailed shape of this distribution depends on the exact tip-surface configuration. **This angular distribution is essentially different from the conventionally assumed narrow forward cone and should drastically change most of the previous interpretations of BEEM data on Au/Si.** The energetic spectrum of the injected electrons is taken from conventional planar tunneling theory (Prietsch 1995).

In view of the fact that the total mean free path is much greater than 20 Å, our Monte-Carlo simulations of the electronic scattering dynamics use the above Green function result as the input ensemble of injected STM electrons at the surface. Appropriately modifying well-established Ensemble-Monte-Carlo techniques for the solution of the non-linear steady-state Boltzmann equation for semiconductors (Hohenester et al. 1992), the hot-electron distribution function $f_{IF}(\underline{k})$ at the interface is obtained as follows. Starting from quasifree electrons ($m_{eff} = m_0$), we correct for band-structure effects on the electron propagation by cutting off the forbidden directions arising from gaps in the constant-energy surfaces. For our case of injection energies about 1 eV above the Fermi energy, these "propagation gaps" form cones with an opening angle of 10 degrees around the [111] directions and are easily included in the scattering dynamics by use of Monte-Carlo rejection techniques.

The total and differential cross sections for the scattering between the hot electrons and those of the "cold" metallic background are treated via a dynamically screened Coulomb potential (Pines 1968), and the e-ph scattering, with an experimentally determined acoustic deformation potential (Blatt 1968), via the standard Monte-Carlo procedure (Jacoboni and Reggiani 1983). This full $\underline{k}$-space description should be contrasted with the earlier MC simulations of the bulk scattering dynamics in the metal (Bauer et al. (1993), Bell (1996)), which are based on an energy-space description, with mean free paths numerically adjusted to the experimental data by use of simple rational functions of energy.

Assuming specular transmission/reflection (via wavefunction matching at a step-like barrier Φ_B) and either specular or diffuse reflection at the free metal surface (both types of reflections resulting in practically identical simulated BEEM currents), the boundary scatterings are treated in the conventional way (Bauer et al. (1993), Bell (1996)).

The simulation of each electron is also followed in $\underline{r}$-space and stopped after it has passed the interface (i.e. when z > d) or when its energy has dropped below the top of the barrier. In this way one obtains the energetic and angular distribution of transmitted electrons at the interface. We further assume negligible current modifications within the semiconductor, which should be well justified for the modest electron energies of our present concern (Prietsch 1995). Then the relative portion of transmitted electrons directly determines the relative BEEM current I_B/I_T as function of the tunnel bias V_T for the given barrier Φ_B. For our calculations of the I_B characteristics and analysis of the spatial resolution we used the standard value $\Phi_B = 0.8$ eV; we checked that the known small temperature variation of Φ_B does not change the essentials of our results.

3. RESULTS AND CONCLUSIONS

Although we have analysed both Si orientations, we concentrate our following discussion mainly on Au(111)/Si(111). We first state that, as had already been demonstrated by the GF calculations of Garcia-Vidal et al. (1996) and de Andres et al. (1997) for a pure ballistic electron propagation through the metal layer, the comparable I_B thresholds and -magnitudes for Au/Si(111) and Au/Si(100) and the high spatial BEEM resolution are a direct consequence of the band-structure-induced non-forward electron injection. This claim can now be substantiated by the results of our inclusion of the scattering dynamics in the former free-electron scenario of de Andres et al. (1997). To demonstrate clearly the effect of scattering processes, we first consider the very narrow off-axis initial distribution originally obtained by Garcia-V. et al. (1996) within the present GF approach by neglect of self-interference effects in the coherent free-electron propagation. Figure 1 shows the resulting lateral current distribution within the layer (taken as semi-infinite) at three typical penetration depths (left) and its much smaller fraction due to scattered electrons (right; note change of scale). One can easily distinguish the build up of secondary "hot" electrons at the lowest energies due to inelastic e-e scatterings and the angular spreading of the distribution through the quasi-elastic e-ph interactions. For this illustrative example a high spatial resolution in I_B would be found, caused by the dominance of those "happy" $\underline{k}$-matching electrons (distributed typically between 35 and 45 degrees within the high-angle wing of the distribution) which cross the interface at their first attempt.

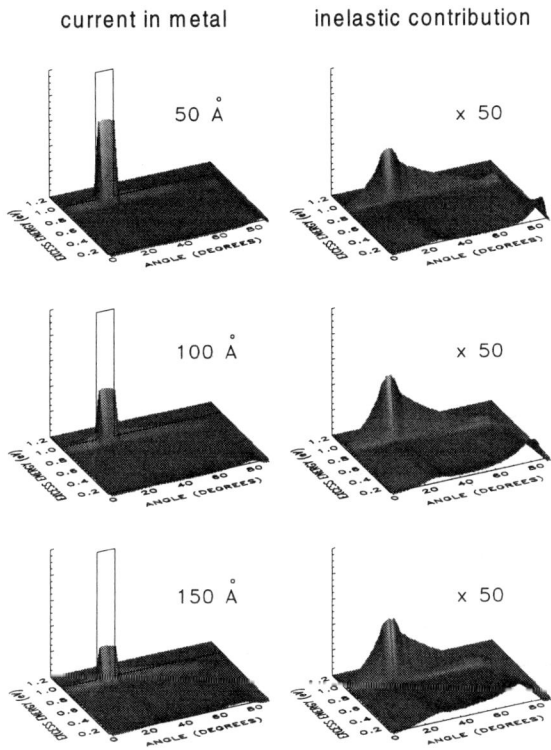

current in metal   inelastic contribution

50 Å   x 50

100 Å   x 50

150 Å   x 50

Fig. 1: Spatial evolution of current density in metal layer for injected distribution (at z=0) shown as solid line.

Turning to the initial distribution underlying this study (de Andres et al. (1997)), our simulations also yield a high spatial resolution, which again is caused by the fraction of "happy" electrons in the angular range between 35 and 45 degrees, which now lies in the low-angle wing of the distribution. Figure 2 compares our results for the BEEM current characteristcs (full lines) with Bell´s experimental data (diamonds) for two different layer thicknesses and temperatures. The theoretical curves are in quantitative agreement with the data, except for the case of "thick" layers (d = 300 Å) and "low" temperatures (77 K) (i.e. for thickness on the order of the mean free path for inelastic and quasielastic scatterings and for a temperature with strongly reduced e-ph scatterings). At present we have no explanation for this pronounced discrepancy. We can only suspect that some dynamical details, in particular regarding the interface dynamics, are still missing in the present simulation scenario and become decisive in thick layers and at low temperatures. We should note that Bell (1996) has attempted to explain his experimental finding of a decreasing low-temperature BEEM current with increasing layer thickness by the decrease of multiple internal reflections and the corresponding decrease of the number of "attempts for transmission" at the interface. Our simulations confirm this dominance of multiple reflections in thin layers, but cannot reproduce the decrease of I_B with increasing d, because our theory lacks the $\underline{k}^{\parallel}$-matching restrictions of Bell's forward-injection scenario. Moreover, our reproduction of Bell's simulations revealed that his interpretation and the practically perfect agreement of his theoretical I_B/I_T versus V_T characteristics (dashed lines in Fig.

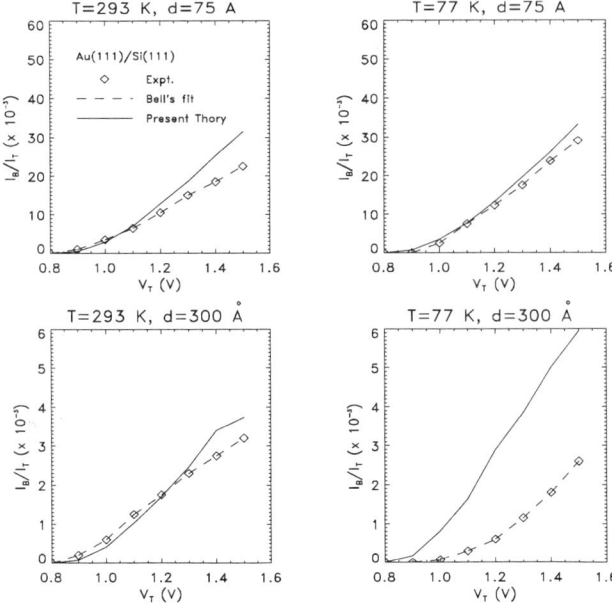

2) with the experimental data strongly depend on his choice of mean free paths and of the detailed injected current distribution: the calculated I_B changes drastically, (i) if the opening angle is e.g. changed from 10 to 20 degrees, or (ii) if the energetic window is changed from Bell's 0.2 to 0.4 eV, or (iii) if his mean-free-path description is replaced by by our detailed bulk-scattering dynamics in $\underline{k}$-space. So we believe that no convincing explanation exists for the decrease of I_B with increasing d at low temperatures.

Fig. 2: Relative BEEM current versus bias voltage; experimental data taken from Bell (1996).

To summarize, the present $\underline{k}$-space description of the injection spectrum and of the bulk scattering dynamics has no adjustable parameter and therefore improves over the many-parameter fits of earlier theoretical interpretations. It turns out that scattering processes have no decisive influence on the spatial BEEM resolution. Although quantitative agreement with experiment is found in most cases, some remaining discrepancies seem to indicate the need for an improved description of the electron dynamics at the interface.

REFERENCES:

Bauer A, Cuberes M T, Prietsch M, and Kaindl G 1993 J. Vac. Sci. Technol. B **11**, 1584
Bell L D 1996 Phys. Rev. Lett. **77**, 3893
Blatt F J 1968, Physics of Electronic Conduction in Solids (Mc Graw-Hill, New York) Chapter 7
De Andres P L, Garcia-Vidal F J, Sestovic D, and Flores F 1997, Physica Scripta, in print
Garcia-Vidal F J, de Andres P L, and Flores F, 1996 Phys. Rev. Lett. **76**, 807
Hohenester U, Supancic P, Kocevar P, Zhou XQ, Kütt W, and Kurz H 1992, Phys. Rev. B **47**, 13233
Jacoboni C and Reggiani L 1983, Rev. Mod. Phys. **55**, 645
Kaiser W J and Bell L D, 1988 Phys. Rev. Lett. **60**, 1406; ibid. **61**, 2368
Pines D 1964 Elementary Excitations in Solids (Benjamin, New York) Chapter 4
Prietsch M, 1995 Phys. Rep. **253**, 163

Inst. Phys. Conf. Ser. No 157
Paper presented at Microsc. Semicond. Mater. Conf., Oxford, 7–10 April 1997
© 1997 IOP Publishing Ltd

A ballistic electron emission microscopy (BEEM)-investigation of the effects of reactive ion etching (RIE) and of chemical pre-treatment on III-V semiconductors

R L Van Meirhaeghe, G M Vanalme, L Goubert, F Cardon and P Van Daele*

Department of Solid State Science, University of Gent, Krijgslaan 281, B-9000 GENT, Belgium
*Intec, St. Pietersnieuwstraat 41, B-9000 GENT, Belgium

ABSTRACT : Ballistic electron emission microscopy (BEEM) has been applied to determine the Schottky barrier height change of contacts on III-V substrates due to classical processing treatments. In the case of GaAs a dry etch using $SiCl_4$ was studied. The distribution of barrier heights over the contact area could be measured. Dry etching was found to introduce a second Gaussian distribution, with lower mean barrier height, next to the Gaussian distribution already found for wet etched reference samples. For InP, a HF-treatment was found to affect the barrier height distribution. Models are proposed to account for these findings.

1. INTRODUCTION

Schottky barriers are used in a lot of applications nowadays, but they are still not fully theoretically understood. Several controversial theories exist, which can explain only a part of the experimental evidence. They are based on Fermi level pinning by defects (Spicer et al. 1988), effective work functions (Woodall and Freeouf 1981), metal influenced gap states (Tersoff 1984) or reactivity between metal and semiconductor (Brillson 1978). Moreover, there is an increasing interest in the importance of barrier height inhomogeneities over the metal-semiconductor contact area.

Therefore, it is obvious that additional experimental techniques producing a Schottky barrier height change are highly interesting. In the present study, two such techniques, namely reactive ion etching (RIE) and wet HF treatment are applied. BEEM is used to determine the Schottky barrier landscape on a submicron level.

2. EXPERIMENTAL

The (100)-n-type GaAs wafers (Sumitomo) had a naturally doping ($N_d \approx 10^{16}$ cm^{-3}, $\rho \approx 0.1$ to 1.6 Ωcm). They were etched in a 3:1:1 (volume ratios) mixture of H_2SO_4:H_2O_2:H_2O for 15s at 80 °C followed by a 5s dip in 1:1 HCl:H_2O. For the ohmic contacts, In was used. For the reference samples, the Schottky contact was immediately made after the ohmic contact. It consisted of a 5 nm Au layer. For the dry etched samples, reactive ion etching with $SiCl_4$ occurred before the fabrication of the Schottky contact. RIE was done in an asymmetric parallel-plate RF-plasma reactor at 13.56 MHz, a $SiCl_4$ flow rate of 10 sccm at 13 mTorr, for 4 minutes at 60 W RF (0.3 W.cm^{-2}). The substrate temperature was 40 °C.

The (100)-p-type wafers (Japan Energy Corpor.) were Zn-doped ($N_a = 2$ to 5×10^{16} cm^{-3}, $\rho \approx 1$ to 3 Ω cm). The same wet etchant was used as for GaAs (see above) to produce reference samples. The other samples received an additional dip in a 2 % HF:H_2O solution.

All BEEM measurements were performed at room temperature using a AIVTB-4 system of Surface/Interface.

3. RESULTS AND DISCUSSION

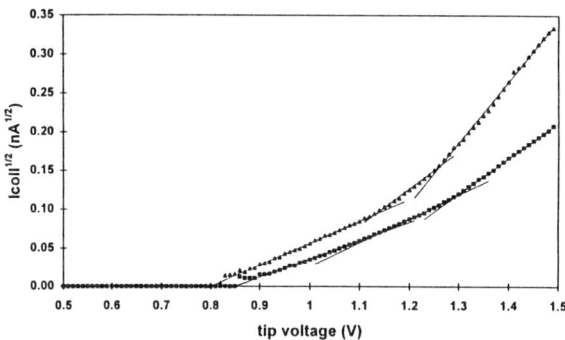

Fig. 1 : BEEM-spectra for a wet etched n-GaAs reference (■) and a RIE sample (▲).

In fig. 1 BEEM-spectra for reference and RIE samples are given wherein the square root of the collector current is plotted versus the tip voltage. In different voltage regions, several straight lines could be fitted corresponding to different band energies in GaAs. The obtained conduction band energy differences are in good agreement with the theoretical values ($\Delta_{\Gamma L}$ = 0.29 eV, $\Delta_{\Gamma X}$ = 0.46 eV). The intercept with the V-axis yields the Schottky barrier height ϕ_B.

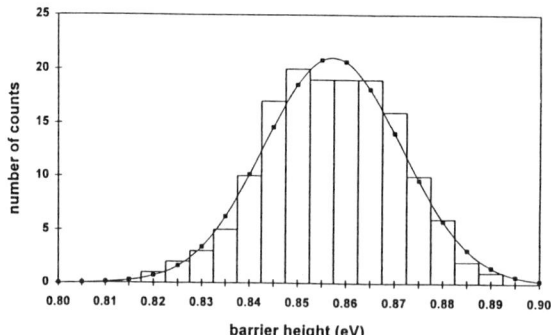

Fig 2 : Distribution of barrier heights for a wet etched n-GaAs reference samples; ϕ_m = 0.86 eV and σ = 0.02 eV.

In fig. 2 the distribution of ϕ_B's obtained by BEEM on a wet etched sample is shown. A histogram was made by taking the number of counts in an interval of 0.005 eV. We also checked intervals of 0.010 eV, but that did not influence the resulting fits. A total number of 150 BEEM-measurements was done randomly spread over the contact area. A Gaussian distribution was fitted to the measured values yielding a mean value ϕ_m = 0.86 eV and a standard deviation : σ = 0.02 eV.

In fig. 3, the distribution of ϕ_B's measured on a RIE-treated sample is presented. Here, 300 points were measured to obtain a good fit. It can be seen that two Gaussian distributions are necessary herein. The first one, having a weight factor of 0.86, gives a ϕ_m = 0.81 eV and a σ = 0.02 eV. The second one, with weight factor of 0.14, has a ϕ_m = 0.85 eV and a σ = 0.02 eV. It can be remarked here that the latter distribution agrees well with the one obtained for the wet etched sample. The low barrier heights were found to be randomly distributed all over the sample area.

In fig. 4 a classical forward current-voltage plot is shown, for a dry etched sample. The deviations from the straight line at higher forward voltages can be attributed to the series resistance of the thin Au-layer present in BEEM-samples. Using Norde's method (1979), we corrected our measurements for this effect. The good ideality factor (n = 1.02) obtained, points to thermionic emission, allowing to obtain a reliable ϕ_B of 0.83 eV. This ϕ_B is a macroscopic one as the entire contact area of the BEEM-sample is used for these I-V measurements. The macroscopic ϕ_B is in good agreement with ϕ_m (see fig. 3) of the main distribution.

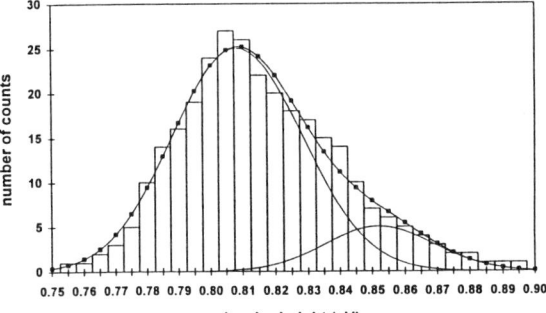

Fig. 3 : Distribution of ϕ_B for an n-GaAs RIE sample.

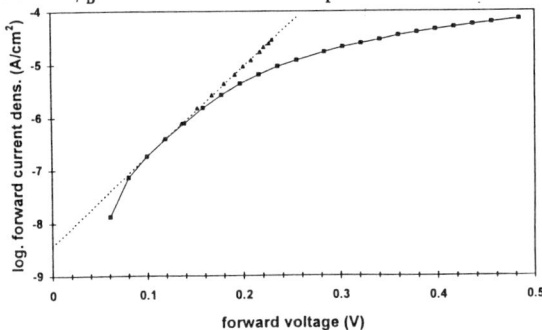

Fig. 4 : Forward current-voltage characteristic for an n-GaAs RIE-sample; ■ : measured points, ▲; corrected points.

The BEEM-measurements presented above can be explained as follows. For the wet etched samples, the ϕ_m of 0.86 eV corresponds well with the value obtained in our former work (Van de Walle et al. 1993) using the classical I-V method. Accordingly, we adopt the model given therein, assuming that amphoteric defect reactions produce Fermi level pinning and thus determine ϕ_B. If we suppose an As-rich surface region this reaction takes the form :

$$As_{Ga} + V_{As} + ne \leftrightarrow V_{Ga} \qquad (1)$$

Therein the As-rich condition is accommodated by a Ga vacancy V_{Ga}. This defect can be converted into an antisite-vacancy pair As_{Ga}-V_{As} by a nearest neighbour hop of As. Thereby n electrons are needed to change the negatively charged V_{Ga} into the positively charged As_{Ga} and V_{As} defects. Baraff and Schlüter (1986) calculated that due to the term ne in reaction (1) the reaction energy strongly depends on the Fermi-level position E_F. Consequently, E_F is pinned if the concentrations of the defects is high enough. In the case of an As-rich surface, a value of 0.85 eV is predicted in good accordance with our results (fig. 2).

However, for dry etched samples, the measurements can be resolved into two distributions. For the high ϕ_B-distribution, the ϕ_m of 0.85 eV agrees very well with that of the wet etched samples so that the explanation given above remains valid. To explain the ϕ_B-distribution with ϕ_m = 0.81 eV we assume that RIE leads to a decrease of As in the surface region, so that Ga-rich sub-regions can possibly be formed. For these locations, a reaction analogous to reaction (1) can be proposed:

$$Ga_{As} + V_{Ga} \leftrightarrow V_{As} + ne \qquad (2)$$

For this reaction a ϕ_B of 0.65 eV is predicted . This is definitely lower than the low ϕ_m-value of about 0.81 eV encountered here (fig. 3). From our XPS-measurements, a decrease of the As/Ga ratio is observed due to RIE. This explains the decrease of the peak at 0.85 eV (fig. 3) attributed to As-rich regions [see eqn. (1) and fig. 2]. However, this peak is still present after RIE, so some

622

As-rich regions must remain over a small part of the area. The low ϕ_B-regions ($\phi_B \approx 0.80$ eV) occupying 86 % of the area (fig. 3) can not be attributed to a local barrier height of 0.65 eV [due to Ga-rich regions and eqn. (2)] as no strong pinch-off is expected then. After RIE however, the As/Ga ratio remains close to 1 (from XPS) so that next to the 14 % As-rich regions (peak at 0.85 eV fig. 3) only a comparable small part of the area can be Ga-rich. Consequently, as the BEEM probes only a few nm, we mostly measure in between such regions. So we think the ϕ_B-peak at 0.80 eV is due to regions influenced both by As-rich [eqn. (1)] and Ga-rich [eqn. (2)] regions resulting in intermediate values.

That an As-rich surface region is present for the wet etched samples is supported by several studies. Using angle-resolved XPS, we could already show (Van de Walle et al. 1993) that the chemical etch leaves an As-rich surface. Other authors (Frese and Morrison 1981) found that the HCl-dip itself also produces an As-rich surface, which we checked using ARXPS. Our proposition that the RIE causes a decrease of As at the surface is based on our angle resolved-XPS results. Secondly, comparing etch-temperature, gas pressure and composition of our RIE-process with that of Tanaka et al. (1995), we expect a decrease of the As/Ga ratio in our samples.

In fig. 5, the distribution of ϕ_B measured on a wet etched p-InP reference sample is displayed. The total number of ϕ_B's measured was 272.

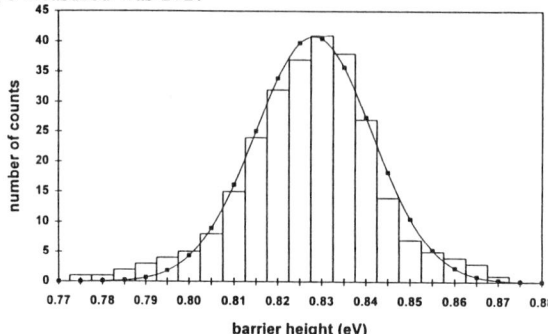

Fig. 5 : Distribution of ϕB for a wet etched p-InP.

The measurements could be fitted using a Gaussian with $\phi_m = 0.83$ eV and $\sigma = 0.02$ eV. This ϕ_m is in good correspondence with earlier work (Hökelek and Robinson 1982) and can possibly be attributed to a reaction analogous to eqn. (1) or eqn. (2). For HF-treated InP, a distribution with a lower ϕ_m (0.78 eV) was measured. A possible explanation for this difference is negative charges, introduced at the interface by the HF-treatment.

REFERENCES

Baraff G A, Schlüter M 1986 Phys. Rev. B **33**; 7346
Brillson L J 1978 J. Vac. Sci. Technol. **15**, 1378
Frese K W, Morrison S R 1981 Applicat. of Surf. Sci. **8**, 266
Hökelek E, Robinson G Y 1982 Appl. Phys. Lett. **40**, 426
Norde A H 1979 Journ. Appl. Phys. **50**, 5052
Spicer W E, Lillienthal-Weber Z, Weber E, Newman N, Kendelewicz, Cao R, McCants C, Mahowald P, Mijano K and Lindau I 1988 J. Vac. Sci. Technol. **6**, 1245
Tanaka N, Lopez M, Matsuyama I, Ishikawa T 1995 J. Vac. Sci. Technol. B **13**, 2250
Tersoff J 1984 Phys. Rev. Lett. **52**, 465
Van de Walle R, Van Meirhaeghe R L, Laflère W H, Cardon F 1993 Journ. Appl. Phys. **74**, 1885
Woodall J M, Freeouf J L 1981 J. Vac. Sci. Technol. **19**, 794

Inst. Phys. Conf. Ser. No 157
Paper presented at Microsc. Semicond. Mater. Conf., Oxford, 7–10 April 1997
© 1997 IOP Publishing Ltd

Carrier recombination at defects in silicon: the effect of transition metals and hydrogen passivation

P R Wilshaw, A M Blood and C F Braban

Department of Materials, University of Oxford, Parks Road, Oxford OX1 3PH, UK

ABSTRACT: Results are presented of the EBIC contrast and hence recombination efficiency of a large variety of extended defects. It is shown that the recombination behaviour can be split into three categories associated with strong, medium and weak contrast defects respectively and that this behaviour can be explained in terms of the position of the defect states in the band gap. It is shown that hydrogen passivation is extremely successful at removing deep levels in the band gap, even when these are associated with precipitates, but that it has little effect on shallow levels.

1. INTRODUCTION

The presence of lattice imperfections, such as dislocations and stacking faults and metal impurities has long been known to degrade the performance of semiconductor devices. The most common metal contaminants in silicon are the transition metals, notably copper, nickel and iron. Impurity atoms present in a contaminated wafer can stay in solution at an interstitial or substitutional site, they can form precipitates or decorate defects in the wafer. In some cases, particularly polycrystalline solar cell material, hydrogenation is used to try to reduce the electrical activity of the defects. In this way relatively good quality material can be obtained even though it may contain quite high densities of extended defects and transition metal contaminants.

The Electron Beam Induced Current (EBIC) mode of an SEM can be used to study carrier recombination in semiconductors with a resolution of ~1μm or better depending on the electron beam accelerating voltage used and thus individual defects of the kind described above can be selected and studied in some detail. The EBIC technique is typically used to measure the contrast of the defect C, where C is equal to the change in EBIC signal as the electron beam is scanned across a defect divided by the signal far from the defect. In general C is measured as a function of temperature and incident beam current (which changes the local minority carrier concentration) and changes in C are then taken to be proportional to the variation in its efficiency as a recombination centre. From such data details of the fundamental properties of the different defects can be deduced providing various criteria are met, (Wilshaw 1989).

In the present work the EBIC contrast and hence recombination behaviour from a large variety of defects is presented both before and after hydrogen passivation and various trends in the data are highlighted in an empirical way. An explanation for the observed recombination behaviour of all the defects is then postulated in terms of the band bending which may be present at extended defects such as those studied.

2. EXPERIMENTAL

In all cases $10^{15}cm^{-3}$, n-type Si specimens were studied using an accelerating voltage of 15kV and beam currents varying between $\sim10^{-12}$ - 10^{-9} A. Measurements were made in the temperature range ~100 - 350K. In each case micrographs of the defects were taken at several accelerating voltages in the range 6 - 25kV and by noting how the defect image width and contrast level changed it was possible to select defects which were situated just below the depletion region formed by the electron transparent Au/Pd Schottky contacts used to make the EBIC specimens. For example shallow defects were clearly visible at accelerating voltages below 10kV and tended to produce "sharp" EBIC images whilst very deep defects were most clearly visible for high accelerating voltages with their contrast beginning to reduce(or even disappear) when the voltage was reduced to 15kV. All quantitative measurements presented in this work were taken from defects thought to lie in a region just below the depletion region edge. The contrast produced by a defect of given recombination strength depends on its depth in the specimen when studied. However, for defects at approximately the same depth studied under the same conditions the contrast measured is approximately proportional to the efficiency of the defect as a recombination centre (Wilshaw and Fell, 1989, Donalato, 1978). In this way it is possible to make an approximate, empirical classification of the defects studied in this work according to their efficiency as recombination centres at 300K. Thus in very general terms, those showing high, medium and low contrast were considered strong, medium and weak recombination centres respectively with some overlap between the different categories because selection of defects at exactly the same depths was not possible.

Deformation induced dislocations (DIDs) were produced in float zone material by a two stage deformation process (Wilshaw and Fell, 1989) to produce hexagonal loops lying on (111) planes, see Figure 1. These DIDs were produced at 420, 650 or 900°C under clean conditions, or with a specific metal contaminant present during deformation or with an incomplete cleaning process which produced slight contamination of an unknown nature.

Frank partial dislocations were produced by the introduction of oxidation induced stacking faults (OISFs) produced by oxidation at 1150°C. These stacking faults were generated at the wafer surface where they extended laterally ~10μm and reached ~3μm deep at their maximum depth. The diffusion length of the wafers prior to oxidation was ~1000μm and ~500μm afterwards. These defects produced no detectable DLTS signal. Contamination of these specimens, if performed, was carried out by wiping the back surface of the wafer with a pure Cu, Ni or Fe wire followed by an anneal at 600, 750 and 900°C respectively for times in the range 30 to 120 minutes. This procedure produced complicated defect structures(Wilshaw and Fell, 1995) including dislocations atomically decorated with the impurity metal, precipitates and complex arrangements of new dislocations and precipitates such as the particularly large example shown in Figure 2. Defects produced under similar conditions were extensively investigated by TEM, (de Coteau et al, 1990, 1991). An important feature of this process is that it is relatively uncontrolled resulting in a wide range of different contamination levels of the initially similar OISFs ranging from heavy to light contamination.

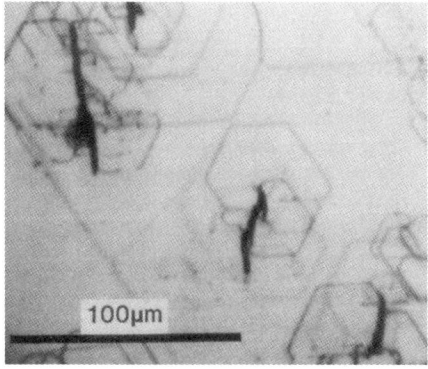

Figure 1. EBIC micrograph showing hexagonal loops of contaminated DIDs in n-type Si.

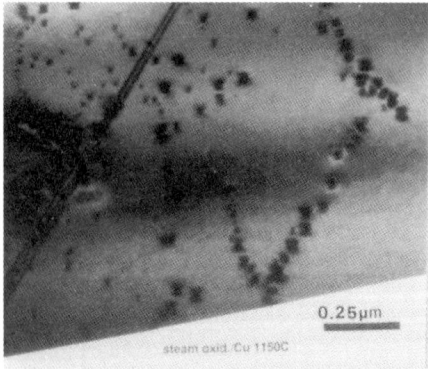

Figure 2. TEM micrograph showing complicated defect structure in a Cu OISF specimen.

Hydrogen passivation of clean and contaminated OISFs was achieved using a remote microwave plasma for 60 minutes whilst the specimens were held at 300°C. This processing was kindly performed by Dr. May at the University of Bristol on equipment not dedicated to semiconductor processing and whilst, as will be shown below, passivation of much of the electrical activity of the defects was achieved it is believed that the process simultaneously introduced further unidentified transition metal impurities into the specimens.

3. RESULTS

Results obtained from the DIDs have been extensively reported elsewhere (Fell et al, 1993, Wilshaw and Fell, 1995) however the findings can be summarised as follows: For dislocations produced at 420°C the contrast was independent of source material or contamination and produced a maximum contrast of ~10% at 300K. The temperature dependence of the contrast followed the behaviour shown schematically in Figure 3a rising linearly with temperature for all temperatures or rising linearly before reaching a plateau at high temperatures. This behaviour was associated with levels in the gap deeper than 0.53eV below the conduction band edge. For "clean" dislocations produced at 650 and 900°C a maximum contrast of ~3% was obtained at 300K and associated with deep levels ~0.3 - 0.4eV below the band edge. Such dislocations showed a temperature dependence similar to that shown schematically in Figure 3b, rising with temperature at low temperatures whilst decreasing with temperature at high temperatures. When such dislocations were contaminated with copper their contrast at 300K rose to ~8% and was associated with levels ~0.5eV deep and showed a temperature dependence similar to that of Figure 3a. For incomplete cleaning, results were obtained intermediate between those for the clean dislocations and those intentionally contaminated with copper. The residual electrical activity of the clean 650 and 900°C dislocations was thought to be due to residual contamination whilst that of the 420°C dislocations was thought to be intrinsic to the dislocations themselves or associated point defects generated during the deformation process.

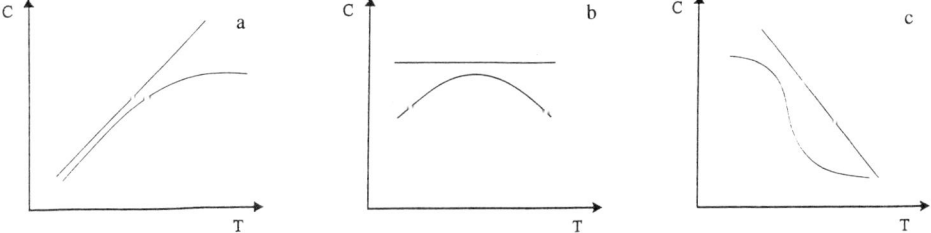

Figure 3. Schematic diagrams showing the different forms of the temperature dependence of EBIC contrast for extended defects showing strong a), medium b) and weak c) contrast at 300K.

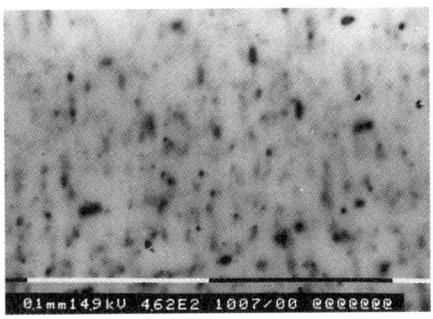

Figure 4. EBIC micrograph of Cu contaminated OISF specimen at 300K

The clean OISFs showed very little contrast at 300K (<<1%) rising to ~1% at 100K in a way shown schematically in Figure 3c, the contrast decreasing linearly with temperature continuously from low temperature to high temperature or more frequently the contrast decreasing quite sharply over a more limited range of temperatures as the temperature is increased. When contaminated with copper, nickel or iron, much higher contrast was obtained and a variety of behaviour was found. For example, Figure 4 shows a micrograph of copper contaminated material where defects showing different levels of contrast are visible. Much of the difference in contrast is due to defects at different depths being imaged with their contrast varying accordingly, but in addition if defects are selected lying at approximately the same depth large variations in contrast can still be seen. In all approximately 25 different defects were studied in the different specimens and some general trends in behaviour were always found. Those defects showing strong contrast (typically more than ~6% at 300K) showed temperature behaviour similar to that in Figure 3a, those defects showing weak contrast (typically less than ~6% at 300K) showed behaviour similar to that in Figure 3c and some defects with contrast ~6% at 300K showed behaviour similar to that of Figure 3b. Examples of actual data are shown in Figure 5

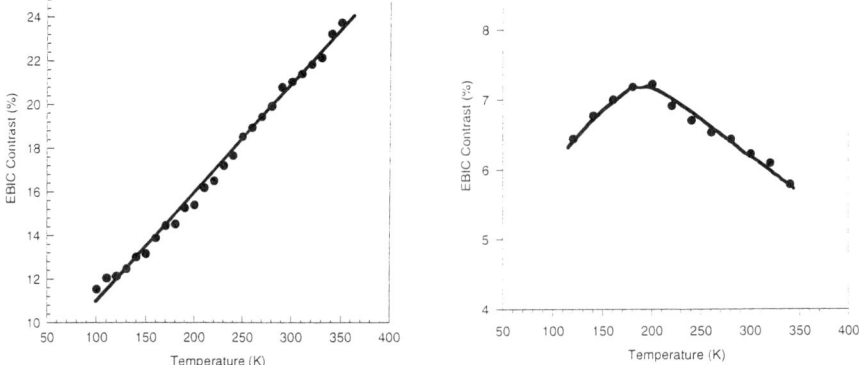

Figure 5. EBIC contrast versus temperature for a strong and medium contrast defect in a Cu contaminated OISF specimen.

After hydrogen passivation the electrical activity of the copper and iron contaminated specimens was greatly reduced. For example Figure 6 shows a hydrogenated copper specimen with the micrograph taken under the same conditions as Figure 4 so that the different levels of electrical activity may be directly compared. The sharp dots visible in Figure 6 are due to specks of dust on the specimen surface which were also visible in the secondary electron image and were not due to electrically active defects. This work shows that the hydrogen passivation treatment has been successful in greatly reducing the electrical activity both of the decorated bounding partials of the stacking faults and also of the precipitates themselves. The maximum level of contrast observed at 300K after the passivation treatment was <1% for the Cu specimens and <2% for the Fe specimens. The passivation of the Ni specimens was not as effective and further work is required to see whether this is fundamental to the properties of the Ni defects or simply due to insufficient passivation being carried out on this specimen. Passivation of the clean OISF specimens resulted in a slight increase in contrast at 300K or for some specimens, no noticeable change.

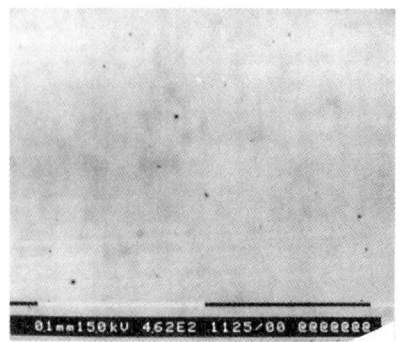

After passivation all defects showed increasing contrast with decreasing temperature according to the forms shown in Figure 3c. For the clean OISFs which were hydrogen passivated the contrast at 100K was considerably increased (up to ~4%) compared to the unpassivated whilst the passivated Cu and Fe contaminated specimens showed contrasts of ~5% at 100K. It is believed that the increase in electrical activity associated with this passivation process in the clean specimens at 100K is due to the unintentional contamination of the specimens with unknown impurities during the passivation process.

Figure 6. EBIC micrograph of Cu contaminated OISF specimen at 300K after passivation.

3.1 Summary of Results

Many different samples have been investigated containing different types of defects including 60° and screw DIDs and Frank partial dislocations bounding OISFs, these specimens being either "clean", unintentionally contaminated or contaminated with Cu, Ni or Fe. After contamination the specimens also contain precipitates and new dislocation structures associated with the precipitates. In addition the existing dislocations are also decorated by the metal contaminant. Many of these specimens have been hydrogen passivated and then EBIC measurements made. It is striking that from this wealth of data clear empirical trends can be observed independent of the exact nature of the defect or the specimen in which it is studied. In general these trends can be summarised as follows: For high contrast defects typically showing more than 6% contrast at 300K (420°C DIDs, contaminated 650°C DIDs, some Fe, Cu and Ni contaminated OISFs and Cu and Ni precipitates) the electrical activity and hence contrast of the defects increases with increasing temperature according to the behaviour shown in Figure 3a. For medium contrast defects typically showing ~6% contrast at 300K (some Cu, Ni contaminated OISFs) or ~3% contrast at 300K for "clean" 650°C DIDs, the contrast is largely independent of temperature or increases with temperature at low temperatures and then decreases at high temperature. For low contrast defects showing typically less than ~6% contrast at 300K (clean OISFs, some OISFs in Cu and Ni contaminated specimens and all hydrogen passivated defects) the electrical activity and hence contrast decreases with increasing temperature according to Figure 3c.

4. DISCUSSION

The above summary reviews our own work on recombination at extended defects which was carried out keeping the EBIC specimen and measurement parameters as similar as possible so that the measurements on different defects could be directly compared. However the same general trends can be found amongst other published work. For example Kittler et al (1993) find the temperature dependence shown schematically in Figure 3a for nickel silicide precipitates which act as strong recombination centres whilst Kusanagi et al (1992) find the behaviour shown in figure 3c for the clean DIDs they studied which act as very weak recombination centres. Different trends in the temperature dependence of different defects similar to those highlighted here have been observed by Kittler and Seifert (1993) paper but are ascribed to different recombination mechanisms and the dependence on the magnitude of EBIC contrast is not pointed out in that work. In the following an explanation of the temperature dependence of recombination at all the extended defects is proposed in terms of the band bending associated with the defects. This mechanism can account for all the behaviour observed without recourse to different recombination mechanisms for different defects. A fuller description of this analysis will appear in a later paper.

Extended defects will in general produce many states in the band gap and only a small number need to be charged for significant band bending to result (Read, 1954). For example one charged state every 5nm along the length of a dislocation will give rise to ~.15eV of band bending and this band bending will then control the recombination process. It has been shown previously (Wilshaw and Fell 1995, Alexander et al 1990) that the larger the band bending the more efficient minority carrier capture will be into the attractive potential surrounding the defect and hence the more efficient the defect will be as a recombination centre. Thus it is expected that defects which produce large contrast in EBIC will be associated with large amounts of band bending. However, taking the example of n-type material with acceptor states produced by the defect, for the states to be charged they must lie below the Fermi level even taking into account the band bending occurring at the defect. Thus if a defect shows strong recombination and hence large band bending, in $10^{15}cm^{-3}$ material at room temperature when the Fermi level is ~.25eV below the band edge the defect must be associated with relatively deep states (>~.35eV). This is consistent with the findings from the 420C DIDs presented earlier in this paper. In addition, a full analysis of the recombination process shows that the dependence on temperature will then follow the form shown in Figure 3a. Conversely, for defects which only produce shallow states in the gap, the charge on the defect will be zero or very small if the states lie above the Fermi level and relatively small even if they are pinned to the Fermi level and in states lie above the Fermi level and relatively small even if they are pinned to the Fermi level and in this situation the defect will show only weak EBIC contrast. Moreover, since the Fermi level moves towards the centre of the gap as the temperature is increased this implies that the charge on any defect where the states are pinned to the Fermi level will also decrease as the temperature is raised, being approximately zero when the Fermi level is the same distance from the band edge as the defect states. Since the charge and hence band bending control recombination this implies that the recombination efficiency will also decrease with increasing temperature according to the behaviour shown in Figure 3c. A more detailed analysis shows that the intermediate behaviour shown schematically in Figure 3b can also be accounted for using this band bending model.

This analysis implies that defects showing strong contrast at 300K must contain states deep in the band gap and so will have the temperature dependence of Figure 3a. Whilst those showing no or little contrast at 300K will have shallow states and show the dependence of Figure 3c. Thus the difference in observed recombination behaviour in terms of absolute levels of contrast and its dependence on temperature can be largely accounted for simply in terms of the position of the states in the band gap. In these terms it can be seen that the hydrogen passivation process, which removed all contrast changing according to Figure 3a but leaving contrast behaving according to Figure 3c, was extremely successful at removing deep levels from the band gap but was unable to remove the shallow levels.

ACKNOWLEDGEMENTS

The authors would like to thank Dr R Falster of MEMC Electronic Materials for the provision of the OISF specimens and Dr P May of the University of Bristol for hydrogen passivating specimens.

REFERENCES

Alexander H, Dietrich S, Huhne M, Kolbe M and Weber G 1990 phys. stat. sol. (a) **117** 417
de Coteau MD, Wilshaw PR and Falster R 1990 phys.stat. sol. (a) **117** 403
de Coteau MD, Wilshaw PR and Falster R 1991a Solid State Phenomena **19** and **20** 27
Donolato C 1978 Optik **52** (1) 19
Fell TS and Wilshaw PR 1991 Inst. Phys. Conf. Ser. **117** 733
Fell TS, Wilshaw PR and de Coteau MD 1993 phys. stat. sol.(a) **138** 695
Kittler M, Seifert W and Radzimski ZJ 1993 Appl. Phys. Lett. **62** (20) 2513
Kittler M, Seifert W 1993 phys. stat. sol. (a) **138** 687
Read WT 1954 Phil. Mag. **45** 775
Wilshaw PR and Fell TS 1989 Inst. Phys. Conf. Ser.**104** 85
Wilshaw PR and Fell TS 1995 J. Electrochem. Soc. **142** (12) 4298

Inst. Phys. Conf. Ser. No 157
Paper presented at Microsc. Semicond. Mater. Conf., Oxford, 7–10 April 1997

© 1997 IOP Publishing Ltd

EBIC studies of the electrical barriers in striated ZnS platelets exhibiting the anomalous photovoltaic effect

D B Holt and Y Brada*

Department of Materials, Imperial College of Science, Technology and Medicine, London SW7 2BP, U.K.
*Racah Institute of Physics, The Hebrew University of Jerusalem, Jerusalem 91904, Israel

ABSTRACT: Platelet crystals of ZnS formed on whiskers from the vapour phase have been studied for many years because of the many polytype structures they contain and the anomalous (high voltage) photovoltaic effect (APE). EBIC contrast linescan profiles of the electrical barriers responsible for the APE were found to be of the peak-and-trough form, though appearing different because of the short drift and diffusion lengths in ZnS. Thus the barriers are walls of charged misfit dislocations at polytype interfaces.

1. INTRODUCTION

Numerous polytypes are found in platelets of ZnS that form during growth from the vapour in carrier gases, either H_2S or argon plus HCl. Whiskers grow first containing axial screw dislocations sometimes of large Burgers vectors so the atomic planes perpendicular to the axis form a spiral ramp. The height of the layers of the ramp is given by the Burgers vector of the axial dislocation in the original whisker. Flag-like plates grow out sideways from some of these whiskers and, as the material cools after growth, the platelets transform martensitically into a series of bands of different polytypes. That is, groups of partial dislocations on certain planes in the layer forming the spiral ramp rotate about the threading axial screw dislocation drawing widening stacking faults behind them. This produces long stacking sequences that are repeated in each turn of the ramp to form the polytypes. The polytype structures thus occur in bands perpendicular to the common c-axis (the original axial direction of the whisker). All the many complex stacking structures fall between the hexagonal wurtzite and the cubic sphalerite structure.

These polytypic or striated platelets exhibit a number of unique physical properties especially the anomalous (far larger than band gap) photovoltaic effect (APE). This generates up to hundreds of volts along the c-axis under monochromatic, near-bandgap ultraviolet illumination. The APE was found to be generated at many electrical barriers parallel to the polytype interfaces as sub-bandgap photovoltages that add to the large macroscopically observed voltages. The interest of the polytype structures (over 150 were identified by x-ray diffraction) and their unique formation mechanism together with that of the APE and related physical properties led to years of systematic study by x-ray diffraction and optical techniques with resolutions of hundreds of μm. For a review of this work see Steinberger (1983).

SEM EBIC was early found to be an effective means for resolving the electrical barriers in striated platelets of ZnS and much detailed variation was also observed with SEM CL (Holt and Culpan 1970). While the CL observations were followed up by more quantitative analytical work (for references see Brada and Holt 1997), the electrical signatures of the barriers in EBIC linescans could not be understood. However, recent work on (R)EBIC linescans of barriers in (semi)insulating materials (e.g. Holt 1994, Raza and Holt 1995, Holt et al 1996) has clarified the physics involved. The form of the barrier linescans in polytypic ZnS platelets is reported and discussed here.

2. EXPERIMENTAL METHODS - REBIC

The specimens grown in the Racah Physics Institute of the Hebrew University were of two types: those grown in H_2S were p-type and showed the APE while those grown in argon + HCl were n-type and did not. This work used a JEOL 840A SEM fitted with a Matelect ISM-5 EBIC system with an interface unit and software allowing quantitative linescans to be recorded.

For REBIC the specimens had contacts of Ag paint applied at separations of 5 mm to 1 cm or more, with one earthed and the other connected to the input of the Matelect head amplifier. The specimen then acts as a current divider, allowing part of the specimen current (from the beam to earth via the specimen),I_d, to flow directly to earth via the first contact while the rest, I_a, flows through the amplifier. $I_a/I_d = R_d/R_a$ where the R's are the fractions of the specimen resistance between the beam impact point and the direct-to-earth and amplifier contacts, respectively. So the REBIC signal, I_a, exhibits a linear ohmic variation as the beam sweeps from one contact to the other due to the changing ratio of the resistances. The slope at each point is proportional to the local resistivity. This has been termed resistive contrast imaging (Smith et al, 1986). Any EBIC signals (charge collection by built-in fields due to defects) are superimposed on this ohmic signal baseline.

3. RESULTS

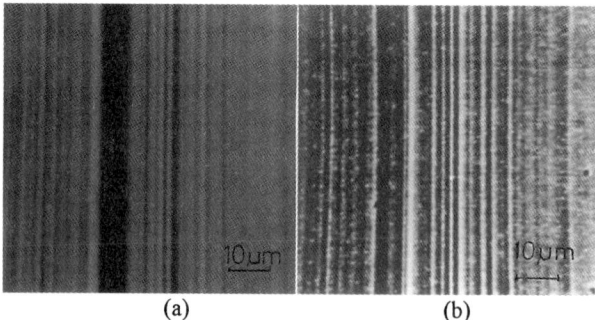

(a)   (b)

Figure 1. (a) Secondary electron image and (b) (R)EBIC image of an area of a polytypic ZnS platelet that showed the APE. The common c-axis of these structures is horizontal and the polytype bands run vertically in these pictures. The contacts are off to the left and right of the field of view so the current flows horizontally.

Figure 1(a) shows the surface topography and 1(b) electrical barriers in a typical area of a striated ZnS platelet of the type showing the APE.

Figure 2 is a typical barrier linescan with the profiles of four barriers marked A through D. These are each, reading from left to right, of the form of a trough and peak. However, unlike the cases of peak and trough (PAT) contrast previously analysed in bulk, polycrystalline ZnSe (Holt 1994, Holt et al 1996) which were broad and curved, these are narrow and of linear zig - zag form. Figure 3 is a "Y-mod" image (display of numerous adjacent linescans) of a polytype platelet. The typical vertical barrier marked by an arrow shows all these traces to be of linear zig-zag form. This could not be understood previously. As in the case of grain boundaries, such PAT (peak and trough) signals result in parallel dark and bright lines along the barriers in REBIC micrographs.

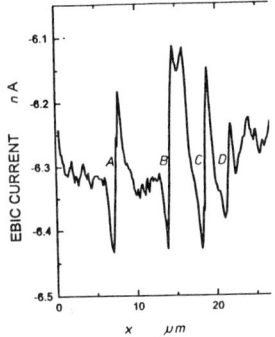

Figure 2. REBIC linescan across four electrical barriers in a specimen similar to that of Figure 1. The signals due to the barriers are lettered A to D. This scan was recorded for a line along the common c-axis and perpendicular to the interfaces between successive polytype structures. The current flow was along the x-axis.

Figure 4 shows resistive contrast (due to the ohmic slope) down two conducting polytype bands. The contact at the top of the field of view is connected to the amplifier. The electrons flowing to it are read as a conventional -ve current and displayed as dark pixels so the contact appears black. The dark vertical strips are polytype bands of higher conductivity and/or in better ohmic contact with the silver paint so the initially large (-ve i.e. dark) signal falls ohmically downwards to the signal level giving pixels of the same brightness as elsewhere. A horizontal linescan recorded across these polytypes showed the -ve currents flowing in them to be nearly twice that in regions on either side.

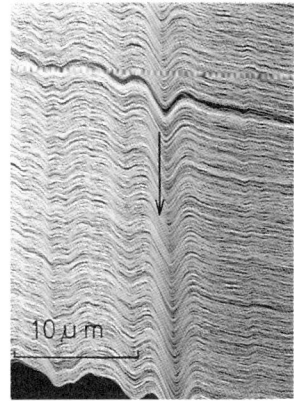

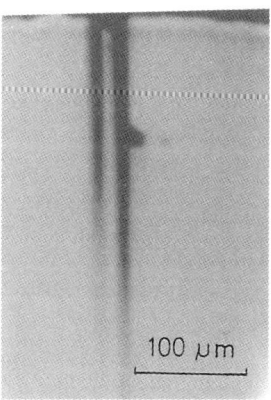

Figure 3. REBIC Y-modulation image of a striated platelet with a prominent, vertical electrical barrier marked by by an arrow. The c-axis is horizontal, the polytype bands vertical and the current flow horizontal.

Figure 4. REBIC image of a striated ZnS platelet showing marked ohmic variation of the signal (resistive contrast) down two polytype bands that appear darker than the background. The c-axis is horizontal, the polytype bands are vertical and the current flow is vertically upward.

4. DISCUSSION

The (R)EBIC signatures of the electrical barriers in striated ZnS platelets, exhibiting the APE, like those in Figures 2 and 3, were not at first recognized as of the PAT contrast form for charged interfaces, acting like two Schottky barriers back to back because (1) the beam current I_b was $\geq$ signal I_{EBIC} (e.g. the beam current was about 10 nA for Figure 2 while the peaks and troughs are only 0.1 to 0.2 nA) and, as noted in Figures 2 and 3, (2) the barrier linescan profiles were narrow, linear zig-zags. However, both these features can be accounted for on the charged interface model.

Charge-collection currents in well-developed semiconductors like Si or GaAs can be written as

$$I_{EBIC} = \eta_{CC} \frac{GI_b}{q} \tag{1}$$

so the gain $= g = \dfrac{I_{EBIC}}{I_b} = \dfrac{\eta_{CC}G}{q}$ \hfill (2)

where η_{CC}, the charge collection efficiency, is the fraction of the electron-hole (e-h) pairs collected by the built-in field concerned, G the generation factor is the number of e-h pairs generated per incident beam electron, I_b is the beam current and q is the charge on an electron. G is of the order of 10^2 to 10^3 and the gain is similarly large.

However, in undeveloped, semi-insulating, wide-gap materials like ZnS and particularly for REBIC in which the contacts are separated by macroscopic distances (mms to cms) the beam incident on a region with a built-in field produces only a local polarization that induces a small charge on the remote contacts. In such cases the EBIC current is reduced to

$$I_{EBIC} = \frac{d}{D} \eta_{CC} \frac{GI_b}{q} \tag{3}$$

where d, the drift length, is the average distance a minority carrier drifts before being trapped or recombined and D is the distance between the contacts. As d and D are of the order of μm or less and mm or more, respectively, the gain is reduced to unity or less. That the gain is reduced by d/D is sometimes known as Ramo's theorem (Ehrenburg and Gibbons, 1981) although Dearnaley and Northrop (1972) ascribe it to Gunn (1960).

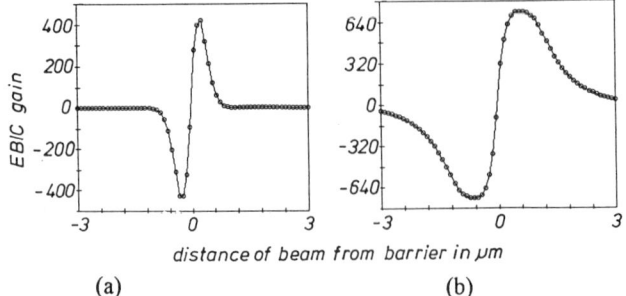

distance of beam from barrier in μm

(a)   (b)

Figure 5. Simulated PAT REBIC linescans for a negatively charged interface in ZnS for (a) L = 0.1 μm and w_{dL} and w_{dR} each = 0.3 μm and (b) L = 0.72 μm and w_{dL} and w_{dR} each = 1 μm. (a) is of the linear zig-zag type while (b) is of the curvaceous form seen for charged grain boundaries polycrystalline bulk II-VIs (Holt 1994, Raza and Holt 1995, Holt et al 1997).

The research of the Jerusalem group previously led to the conclusion that due to the changes in lattice parameter there are charged misfit dislocations in the interface planes between the polytype bands. The diffusion length, L, in these specimens is small and the field of the charged wall of dislocations is apparently shielded by charge carriers and trapped charge so the widths, w_{dL} and w_{dR}, of the depletion regions in which the charge-collecting field exists are narrow. The form of the PAT contrast due to a charged interface can be modelled as two Schottky barriers back to back, using the Napchan Monte Carlo electron-trajectory simulation suite of programs. The effect of the small values of L, w_{dL} and w_{dR} can be seen in Figure 5a compared to 5b. Thus the linear zig-zag form of PAT contrast of the electrical barriers is simply due to the short diffusion length and small field-region widths in polytypic ZnS.

It was shown by Holt et al (1996) that the sign of the charge on the interface can be determined by noting whether the peak or the trough appears on the side where due to the ohmic slope the (-ve) REBIC current is the greater. Thus if the dark trough appears on the darker side of the barrier the interface charge is negative. This appears to be the case in Figure 2. Thus the interfaces that act as charge collecting barriers all appear to be negatively charged.

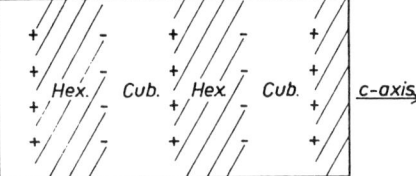

Figure 6. Schematic representation of the electrical barriers in polytypic platelets of ZnS that exhibit the APE. Planes of misfit dislocations occur in the interfaces between more hexagonal (Hex) and more cubic (Cub) polytypes due to the difference in lattice parameter between the structures. Due to polarity (planes of Zn and S alternate along the z-axis) the dislocations in these planes are alternately all Zn-edged and all S-edged and have opposite signs of charge as shown.

The work of the Jerusalem group led to the conclusion that polytypic platelets exhibiting the APE could be modelled in the simplified form of Figure 6 where all the bands are treated simply as more hexagonal or more cubic. For reasons of polarity the walls of misfit dislocations at the interfaces are oppositely charged as shown. Why, in that case, are only the the negatively charged interfaces detected in EBIC? That is, why, in Figure 2, are all the signatures, reading from left to right, trough and peak (TAP) in form and none PAT? It is because the minority carrier diffusion length for electrons is an order of magnitude greater than that for holes in ZnS and so is the drift length, d (equation 2). Thus the electrons repelled from a negative barrier travel far enough (d) that I_{EBIC} is detectable whereas the holes repelled from a positive barrier have so small a value of d that the EBIC signal (which compared to the 0.1 nA for the negatively charged barriers in Figure 2, will be only of the order of 0.01 nA) is lost in the noise due to surface topography.

Polytypic ZnS platelets have many unique physical effects. These include a biased Franz-Keldysch effect (Yacobi and Brada 1974 and references therein) and a special pyroelectric effect (Yacobi and Brada 1976). In those platelets in which hexagonal polytypes preponderate, the large photovoltages (up to hundreds of volts per cm along the c-axis) reverse sign as the illuminating wavelength is varied. However, when the hexagonal structures present smaller surface areas to the light than the cubic ones, the photovoltage is unidirectional (Schachar et al 1968, 1971, Schachar and Brada 1968). These phenomena were all accounted for as consequences of planes of charged misfit dislocations plus the polarization of the polytype bands that result. Electrical current flow along the common c-axis (Figure 6) is

634

obstructed by the barriers so the resistance is up to 10^6 times greater along the c-axis than normal to it (Schachar and Brada 1970).

The anomalous photovoltaic effect in ZnS is not of practical use because negligible current and power can be extracted due to the high resistance in the c-axis direction. However, other materials also show polytypism (notably SiC) and it is conceivable that other materials may be found in which the effect could be useful.

ACKNOWLEDGEMENTS

Thanks are due to Dr. E. Napchan (22 Broomsleigh St., London NW6 1QH, U.K.) for the Monte Carlo simulation of electron trajectory (MC-SET) suite of programs used to produce Figure 5.

REFERENCES

Brada Y, Holt D B and Mardix S 1997 in Microscopy of Semiconducting Materials 1997. Conf. Series No. 157 (Inst. Phys.: Bristol) to be published

Datta S, Yacobi B G and Holt D B 1977, J. Mater. Sci. 12, 2411 - 2420

Dearnaley G and Northrop D C 1966 Semiconductor Counters for Nuclear Radiations (London: Spon)

Ehrenburg W and Gibbons D J 1981 Electron Bombardment Induced Conductivity (London: Academic Press)

Gunn J B 1960 Solid-State Electron. 1, 739

Holt D B 1994 in Polycrystalline Semiconductors III (H.P. Strunk, J.H. Werner, B. Fortin and O. Bonnaud, Eds.) (Scitec. Publishers: Zurich) 1994 pp. 171 - 182

Holt D B and Culpan M 1970 J. Mater. Sci. 5, 275 - 282

Holt D B, Raza B and Wojcik A 1996 Mat. Sci. & Eng. B42, 14 - 23

Holt D B, Raza B and Wojcik A 1997 in Microscopy of Semiconducting Materials 1997. Conf. Series No. 157 (Inst. Phys.: Bristol) to be published

Palm J and Alexander H 1991 J. de Phys. Colloque C6, C6-101

Ramo S 1939 Proc. I.R.E. 27, 584

Raza B and Holt D B 1995 in Microscopy of Semiconducting Materials 1995. Conf. Series No. 146 (Inst. Phys.: Bristol) pp. 107 - 112

Russell J D, Halls D C and Leach C 1995 J. Mat. Sci. Letters 14, 676 - 678

Schachar G and Brada Y 1970 J. Appl. Phys. 41, 3127

Smith C A, Bagnell C R, Cole E I, Dibianca F A, Johnson D G, Oxford W V and Propst R H 1986 IEEE Trans. Electron Devices ED-33, 282 - 284

Steinberger I T 1983 Progress in Crystal Growth and Characterization 7, 7 - 53

Yacobi B G and Brada Y 1974 Phys. Rev. B10, 665

Yacobi B G and Brada Y 1976 J. Appl. Phys. 47, 1243

Inst. Phys. Conf. Ser. No 157
Paper presented at Microsc. Semicond. Mater. Conf., Oxford, 7–10 April 1997
© 1997 IOP Publishing Ltd

A reassessment of Te-doped GaAs

C Frigeri*, J L Weyher^, J Jiménez°, P Martín°

*CNR-MASPEC Institute, via Chiavari 18/A - 43100 Parma (Italy)
^Fraunhofer-IAF, Tullastrasse 72 - 79108 Freiburg (Germany)
°Física de la Materia Condensada, Universidad de Valladolid - 47011 Valladolid (Spain)

ABSTRACT: A reassessment of the structure of the impurity atmospheres at dislocations in GaAs highly doped with Te is proposed on the basis of new results obtained by DSL etching, EBIC, microRaman and TEM. The formation of etch depressions around dislocations suggests that an enhanced formation of the $Te_{As}V_{Ga}$ acceptor complex has locally occurred, driven by the gettering of Te by the dislocations. This causes an undersaturation of Ga vacancies which favours the generation of extrinsic Frank loops around the dislocations and promotes dislocation climb.

1. INTRODUCTION

Dislocations and impurity atmospheres in GaAs doped with Te, especially at high doping levels, have been the object of several studies by electron microscopy techniques. They have mostly been studied either by TEM (Laister and Jenkins 1971, Hutchinson and Dobson 1975) or by SEM-cathodoluminescence (Mendez and Piqueras 1991), apart from the work of Chu et al (1981) who used both TEM and SEM-cathodoluminescence. In the latter work a correlation between the non-radiative recombination properties of the atmospheres and the presence of dislocation loops around the dislocations was established. Mendez and Piqueras (1991) suggested that the infrared cathodoluminescence (CL) could be influenced by point defect complexes. As regards point defects, Hurle (1979, 1995) pointed out that in highly Te-doped GaAs a key role is played by the $Te_{As}V_{Ga}$ complex which is an acceptor strongly compensating the dopant Te. Here we present a study of the defect structures in highly Te-doped GaAs based on new results obtained by means of photoetching, which allowed a better understanding of the nature of the dislocation atmospheres. We also propose a model for point defect reactions, mostly involving $Te_{As}V_{Ga}$, which lead to the formation of dislocation loops around the dislocations.

2. EXPERIMENTAL

The investigated (100) vertical Bridgman GaAs crystals were doped with Te to an average free electron density of $1.5 \cdot 10^{18}$ cm^{-3}. The defect structures were analysed by DSL etching (Diluted-Sirtl like etching with Light) in association with differential interference contrast optical microscopy (DIC-OM), EBIC, microRaman and TEM.

In the DSL etching the oxidative dissolution of any single GaAs molecule takes place by the action of six holes (Kelly et al 1985). The holes that are used for etching are mostly those generated in a subsurface layer inside the GaAs by the light, the holes supplied by the etching bath being negligible (Kelly et al 1985, Weyher and van de Ven 1986, Frigeri and Weyher 1990). Any loss of holes causes a reduction of the etching rate and the formation of hillocks on the surface. This occurs at extended or point defects which recombine holes or in areas where the width of the surface space charge region decreases, i. e. the doping level increases, with respect to the surrounding. On the other hand, additional supply of holes, or depletion of electrons, from sources inside the sample, increases the etching rate giving rise to the formation of depressions. Analysis of the DSL etch features thus provides detailed information on the (changes of) the net doping density n and on the recombinative properties of the defects. EBIC was used to determine n in the impurity atmospheres and gettering areas at dislocations, with a micron scale spatial resolution, by the local measurement of the depletion width of the EBIC Schottky diode (Frigeri 1987, Frigeri and Weyher 1989). By microRaman the free carrier density n and the crystal misorientation at the defect atmospheres were determined by measuring the ratio of the intensity of the longitudinal optic (LO) phonon mode to that of the L $_-$ mode or the intensity of the transverse optic (TO) mode, respectively, with a resolution of the order of the wavelength used (514.5 nm) (Martín et al 1996).

3. RESULTS

TEM revealed that the dislocations are always arranged in tangles or helices surrounded by clouds of large (~ 0.1-0.3 μm) extrinsic faulted Frank loops (b = a/3<111>) on the {111} planes, as well as by tiny loop-like microdefects (Fig. 1). The largest loop clouds have a lateral size of ~ 5 μm. The area around them is depleted of crystal defects over a distance of several μm. TEM also revealed the existence of extrinsic faulted Frank loops as well as perfect loops, ~0.2 μm large, homogeneously distributed throughout the sample. These TEM results are in agreement with those obtained by Chu et al (1981).

Figs. 2a) and 3a) are the typical DIC-OM images after DSL etching. The neighbouring area to a dislocation, that we shall call atmosphere, gives rise to a depression upon etching, as confirmed by step profiles taken across the etched features (Fig. 2b). Inside such depressions several small etch hillocks are detected (Figs. 2a and 3a) that are due to the dislocation itself as

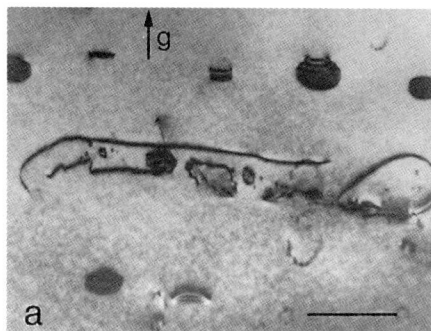
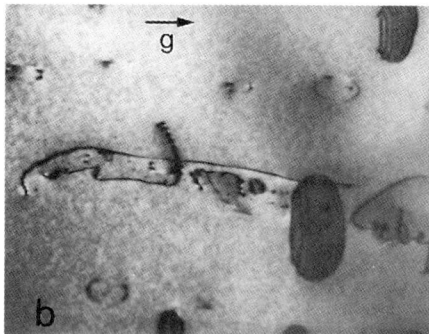

Fig.1. TEM images of part of a complex defect cluster, i.e. a tangled dislocation surrounded by loops of the Frank type (they appear/disappear by exchanging the two given **g**'s). The whole cluster was twice as large. (a) **g** = [022]. b) **g** = [0$\bar{2}\bar{2}$]. Bar = 0.5 μm.

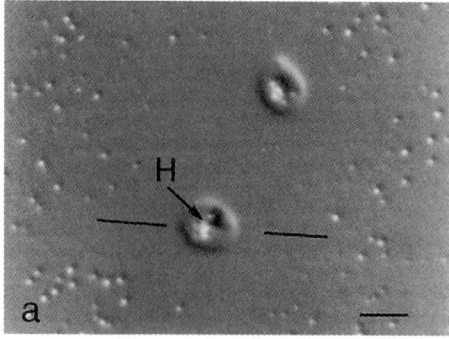
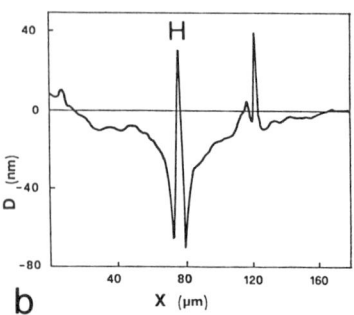

Fig. 2. a) DIC-OM image of DSL-etched dislocation clusters. b) Step profile along the dashed line across one of the dislocations, whose etch-hillock is indicated by H. D is the etch depth. Bar = 10 μm.

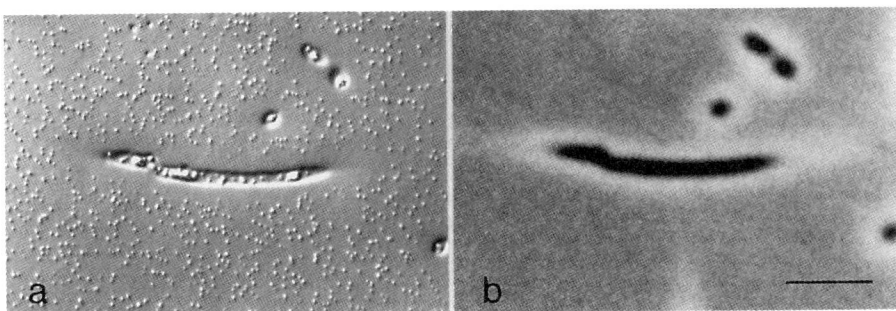

Fig. 3. a) DIC-OM image after DSL etching of a complex defect cluster, similar to that shown in Fig. 1. The etch feature is a depression as seen by step profiling. b) EBIC image of the same defect. The etch depression gives dark EBIC contrast. Bar = 50 μm.

well as to the loops around it. The areas surrounding the depressions are depleted of etch hillocks over a distance of ~ 5-10 μm, in agreement with TEM results. In Figs. 2a) and 3a) many small hillocks are visible all over the DSL-etched surface. They are due to the isolated loops seen by TEM in the matrix far from the dislocation clusters.

The DSL-revealed depressions correspond to areas of EBIC non-radiative dark contrast due to the presence of the dislocation loops, whereas the areas surrounding the depressions give rise to EBIC bright contrast (Fig. 3b). In such bright haloes a decrease of n (e.g. down to $3\text{-}4{\cdot}10^{17}$ cm^{-3} from $1.6{\cdot}10^{18}$ cm^{-3} in the matrix) was measured by EBIC, suggesting that a donor depletion most probably occurred.

MicroRaman measurements at the etch hillocks inside the depressions showed the presence of the forbidden TO peak indicating that a strong lattice disorder with respect to the (100) oriented matrix exists in such areas, due to the presence of the loops. In the DSL revealed depressions surrounding the hillocks and outside the distorted regions, a majority carrier depletion was measured as compared to the unperturbed matrix.

4. DISCUSSION

The formation of DSL depressions around dislocations (Figs. 2 and 3a) indicates that in these areas (atmospheres) a high density of holes is available, i.e. there is an increase of the surface space charge region associated with a reduction of the free electron density n. The reduction of n was measured by Raman. This can be due to a reduced density of donors and/or to an increased density of acceptors that compensate the donors. On the other hand, the decrease of n in the areas surrounding the atmospheres (bright haloes), as measured by EBIC, suggests that the dopant Te atoms have diffused out of these regions towards the dislocation sinks, as seen for Si-doped GaAs (Frigeri and Weyher 1989), outdiffusion into the matrix being unlikely. A similar result was obtained by Chu et al (1981) by CL. The reduction of n in the atmospheres should thus be due to an increase of the density of acceptor-like impurities. This can be caused by the gettering of the dopant Te to the dislocation sites as shown below.

Te is a donor in GaAs when it occupies As sites. It was reported that the Te donors are partially compensated with an acceptor density that is directly proportional to the donor density (Hurle 1979, 1995). Such compensation is much more pronounced at high Te concentrations ($>1\text{-}2{\cdot}10^{18}$ cm^{-3}) suggesting that the acceptor contains the Te atom itself. It was concluded that the $Te_{As}V_{Ga}$ complex constitutes such acceptor (Hurle 1979, 1995).

Our results can be interpreted by assuming that the increase of the Te dopant density close to the dislocation locally causes the formation of the $Te_{As}V_{Ga}$ acceptor complex through the reactions (Hurle 1979, 1995)

$$Te + V_{As} \rightarrow Te_{As}^{+} + e^{-} \tag{1}$$

$$Te_{As}^{+} + V_{Ga}^{2-} \rightarrow Te_{As}V_{Ga}^{-} + h^{+} \tag{2}$$

i.e. the gettered Te first gives rise to the Te_{As}^+ donors which are then eliminated to produce the $Te_{As}V_{Ga}^-$ acceptors by reaction (2). Since also V_{Ga}^{3-} exists, even in a greater density than V_{Ga}^{2-} (Tan et al 1993), eq. (2) also works with V_{Ga}^{3-} thus yielding a double ionized acceptor, $Te_{As}V_{Ga}^{2-}$. The formation of the $Te_{As}V_{Ga}$ complex is very likely, owing to the high binding energy of the complex as the point defects constituting it are nearest neighbours on lattice sites (Herzog et al 1995). The probability of formation of the complex increases with increasing Te concentration as the densities of both Te_{As}^+ and V_{Ga}^{n-}, n = 1, 2, 3, increase. The V_{Ga}^{n-} density increase with increasing n-type doping level was shown by Tan et al (1993). The increase of Te in the atmospheres should thus have the effect of locally increasing the density of the $Te_{As}V_{Ga}^{p-}$ acceptors, p = 1, 2. Additional holes are therefore available which enhance the etch rate thus explaining the formation of the DSL depressions.

The generation of $Te_{As}V_{Ga}$ complexes by reaction (2) causes an undersaturation of V_{Ga} and consequently the creation of excess Ga interstitials by a Ga Frenkel reaction in the atmosphere region. The excess Ga_i can then give rise to the observed extrinsic faulted Frank loops and promote dislocation climb.

The Frank loops are produced by the condensation of part of the excess Ga_i together with As_i. The Ga_i more close to the dislocation core, instead, can be absorbed at the core along with As_i to produce a negative climb step (- CS) resulting in the release of vacancies into the two sublattices by the reactions

$$As_i + Ga_i \rightarrow - CS \tag{3}$$

$$- CS \rightarrow V_{As} + V_{Ga} \quad . \tag{4}$$

The vacancies produced in the climb step (3)-(4) can be further used to feed reactions (1) and (2) so that an undersaturation of V_{Ga} is maintained and both dislocation climb and formation of the Frank loops can continue. The extensive climb is confirmed by the complex tangling of the dislocations and formation of helices that indicate the consumption of a large amount of interstitials. The As_i necessary for the dislocation climb or for the formation of the Frank loops can either be present in a sufficient density in the crystal or can be generated by a Frenkel reaction since also the As vacancies (V_{As}) very likely undergo undersaturation due to reaction (1). If only an excess of Ga_i exists, the As_i can also be generated at the cores of the dislocations or loops by the Petroff-Kimerling (1976) mechanism.

ACKNOWLEDGEMENTS

Dr K Sonnenberg, Jülich, is gratefully acknowledged for supplying the Te-doped samples. Thanks are due to Mr M Scaffardi for technical support.

REFERENCES

Chu Y M, Darby D B and Booker G R 1981 Inst. Phys. Conf. Ser. **60**, 331
Frigeri C 1987 Inst. Phys. Conf. Ser. **87**, 745
Frigeri C and Weyher J L 1989 J. Appl. Phys. **65**, 4246
Frigeri C and Weyher J L 1990 J. Crystal Growth **103**, 268
Herzog L, Egger U, Breitenstein O and Hettwer H-G 1995 Mater. Sci. Eng. B **30**, 43
Hurle D T J 1979 J. Phys. Chem. Solids **40**, 627
Hurle D T J 1995 Mater. Sci. Forum **196-201**, 179
Hutchinson P W and Dobson P S 1975 J. Mater. Sci. **10**, 1636
Kelly J J, van de Ven J and van der Meerakker J E A M 1985 J. Electrochem. Soc. **132**, 3026
Laister D and Jenkins G M 1971 Phil. Mag. **23**, 1077
Martín P, Frigeri C, Jiménez J and Weyher J 1996 Mater. Sci. Eng. B **42**, 225
Mendez B and Piqueras J 1991 J. Appl. Phys. **69**, 2776
Petroff P M and Kimerling L C 1976 Appl. Phys. Lett. **29**, 461
Tan T Y, You H-M and Gösele U M 1993 Appl. Phys. A **56**, 249
Weyher J L and van de Ven P 1986 J. Crystal Growth **78**, 191

Inst. Phys. Conf. Ser. No 157
Paper presented at Microsc. Semicond. Mater. Conf., Oxford, 7–10 April 1997
© *1997 IOP Publishing Ltd*

REBIC studies of grain boundaries in II-VI compounds

D B Holt, B Raza and A Wojcik

Department of Materials, Imperial College of Science, Technology and Medicine, London SW7 2BP, U.K.

ABSTRACT: Remote electron beam induced current (REBIC) was used to study electrically active grain boundaries (GBs) in ZnS, ZnSe and ZnS,Se. Examples of peak-and-trough (PAT) contrast due to trapped-charge induced energy band bending and terraced contrast due to high resistivity GB layers were found. GB dark contrast interpretable by the Donolato phenomenological theory as well as bright contrast due to reduced recombination were also seen. Reducing the temperature had no major effect on the REBIC contrast.

1. INTRODUCTION

Strongly temperature dependent dark contrast of twin and small angle grain boundaries and temperature-independent large-angle boundary contrast were found in polycrystalline Si solar cells. This was interpreted as evidence that during cooling to liquid nitrogen temperatures, the Fermi level moves through the shallow trap states of twins which become occupied so these boundaries become charged producing contrast at lower temperatures while the deep states and contrast of large-angle boundaries are unaffected (Raza and Holt, 1995).

GBs in polycrystalline high-resistivity materials can be studied by REBIC (remote EBIC) (Holt 1994). The specimen between widely spaced contacts acts as a current divider so a variable part of the current from the beam to earth flows through the detecting amplifier at one contact to become the signal. The remainder escapes directly to earth via a second contact. As the beam scans from the earth contact to the amplifier contact an ohmically varying signal base line is seen. The slope depends on the local resistivity, so high resistance layers at grain boundaries appear as steps in the signal line scan and as bright/dark terrace contrast on the REBIC image. REBIC studies of GBs in ZnSe also showed examples of peak-and-trough (PAT) linescans which appear in images as paired dark and light lines running along the boundary (Raza and Holt 1995). This is due to energy band bending at charged boundaries, modelled by two Schottky barriers back to back. These, and any other EBIC charge collection signals that may be excited, are superimposed on the ohmic background including terrace contrast. This paper extends these studies to additional II-VI materials and to liquid nitrogen temperatures.

2. EXPERIMENTAL METHODS

The samples were kindly provided by Dr. K. Durose and the late Dr. G.J. Russell of the Physics Department of the University of Durham. The surfaces were mechanically polished and silver paint contacts applied. A JEOL JSM 840A SEM fitted with a Matelect ISM-5 EBIC detection system and a Kontron image analysis system was used.

3. RESULTS

3.1 REBIC Contrast Results

Dark (recombination), terrace (high resistivity layer), PAT (charged boundary) and bright (cause to be discussed) GB contrast were seen in these samples. Figure 1 shows the sort of interesting fine structure that can sometimes be resolved under optimum imaging conditions even in ZnS which is the most difficult of these materials.

Figure 1. REBIC image of dark grain boundaries in ZnS showing joining and re-separation of the parallel and near vertical boundaries and possible evidence of precipitation on the other GB. That this was genuine dark contrast not peak-and-trough contrast with charge collection suppressed on one side by internal fields was confirmed by observations under voltage bias. See Holt 1994 for a discussion of and evidence for such distorted contrast.

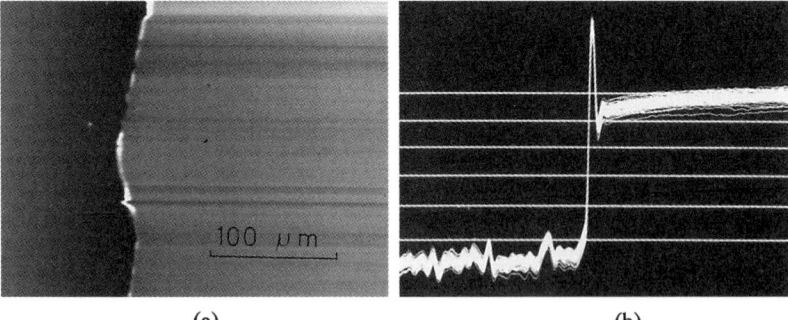

(a)   (b)

Figure 2. (a) REBIC image and (b) signal linescan across a GB in ZnS,Se showing both terrace (signal step) and peak and trough contrast.

The main electrically active grain boundary in a polycrystalline ZnS,Se alloy sample (Figure 2) showed both terraced (the step in the linescan) and PAT contrast (the peak and trough on the right of the step) similar to that reported in ZnSe (Raza and Holt 1995, Holt et al 1996). The peak caused amplifier ringing giving the horizontal streaks to the right in Figure 2a. The GB in CdTe in Figure 3 showed terrace contrast alone, corresponding to the drop from the high relatively smooth curve on the left to the lower, more jagged one to the right. The deep dip near the boundary represents the dark blob toward the bottom of the micrograph. The two grains in Figure 3 apparently differed in hardness so the grain to the right experienced greater polishing damage, exhibited as bright lines on the micrograph.

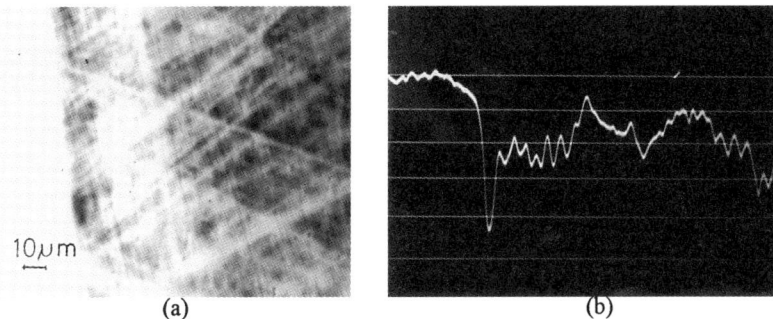

Figure 3. (a) REBIC micrograph and (b) current linescan across a GB in CdTe showing terrace contrast. Polishing damage lines are visible in the grain to the right.

Liquid nitrogen temperature REBIC images of ZnS,Se and ZnS showed no significant change, the same grain boundaries alone were electrically active at low temperatures although twins were visible in secondary electron images.

3.2 Effects of Specimen Electrical Parameters

To identify the type of GB contrast one examines the form of the REBIC linescan but this is made difficult by three things. (i) The contrast often varies from point to point along the boundary. (ii) Contrast mechanisms can interact e.g. the flow of specimen current through a high resistivity layer causing terrace contrast produces a potential drop that biases any charged boundary present resulting in assymetric peak and trough contrast (Holt et al 1996). For example the peak in Figure 2b is much larger than the trough to its right. Sufficient bias can eliminate the peak or the trough leaving apparent dark or bright contrast only. (iii) The high resistance and capacitance of the specimens cause difficulties, as will be discussed below and conditions optimised for the linescans, as in Figures 2 and 3, are generally not best for imaging.

4. DISCUSSION

Terrace, dark only, light only and PAT (paired bright and dark lines) GB contrast occurs in these additional II-VI compounds. Similar effects were found in ZnO varistor material (Russell et al 1995, 1996, Halls et al 1997) and in $BaTiO_3$ positive temperature coefficient of resistivity (PTC) ceramics (Russell and Leach 1995). The terrace and PAT contrast at GBs is evidence of higher resistivity boundary layers and of boundary charge, respectively (Raza and Holt 1995, Holt et al 1996). Dark contrast is due to enhanced recombination arising from the presence of numerous recombination centres on or near the boundary (Donolato 1983, 1985) but what causes bright contrast? The EBIC current is given by the integral over the carrier generation volume of the density of carriers times the fraction collected (the collection efficiency). The probable cause of locally reduced recombination is fewer or less effective recombination centres giving an increase in collection efficiency. This might arise through gettering of impurity centres to the boundary resulting in their inactivation by precipitation and possibly their passivation of dangling-bond-like intrinsic centres.

The large band gaps and poorly developed materials technology of the II-VIs means that they are often semi-insulating (resistivities of the order of 10^8 ohm cm or more) which can lead to charging. The widely spaced contacts and the high dielectric constants and possibly rectifying contacts mean the specimen capacitances are large and lower frequency signal components are attenuated so only rapidly varying contrast from fast scans can be detected. (For comparison the band gaps and relative static dielectric constants of these materials and of Si are: Si 1.12 eV, 11.9; CdTe 1.44 eV, 10.2; ZnSe 2.58 eV, 9.2; ZnS 3.54 eV, 8.9.) Conducting films were not applied as they might distort or short circuit effects of interest. The charge deposited must leak away before the beam returns to that spot to avoid charging and one can either: (1) reduce the beam energy so the charge is deposited nearer the surface (surface leakage is important); and/or (2) reduce the beam current to deposit less charge; and/or (3) use a slower scan rate to increase the time for leakage before the beam returns. Methods (1) and (2) mean lower beam power and small REBIC signals and method (3) increases the danger of capacitance suppression of the contrast.

For analysis quantitative linescan profiles are needed, e.g. via the Matelect, but these profiles are scanned slowly as many data points are recorded but for these samples and speeds GB contrast disappeared. That is why all the II-VI GB linescans presented previously (Holt 1994, Raza and Holt 1995, Holt et al 1996) were faster-scanned non-quantitative traces produced by the SEM LSP (linescan profile) press button facility, while in this work sometimes the still higher speeds of the WFM (wave form monitor) SEM facility were used, e.g. in Figure 2 and 3, - hence the sets of horizontal lines showing the range of signal strengths corresponding to screen grey levels. Quantitative slow-scan profiles could be recorded for varistor ZnO using microprobes to make contact a short distance on either side of GBs (Halls et al 1997) and in striated platelets of ZnS (Holt and Brada 1997) where the very different specimen geometry also results in small capacitances.

REFERENCES

Donolato C 1983 J. Appl. Phys. **54**, 1314 - 1322

Donolato C 1985 in Polycrystalline Semiconductors, ed. G Harbecke (Berlin: Springer-Verlag) pp. 138 - 154

Halls DC, Holt, D B, Leach C, and Russell JD 1997 in Inst. Phys. Conf. Series No. 157 (to be published)

Holt DB 1994 in Polycrystalline Semiconductors III, eds. HP Strunk, JH Werner, B Fortin and O Bonnaud (Zurich: Scitec Publishers) pp. 171 -182

Holt DB, Raza B and Wojcik A 1996 Mat. Sci. Eng. **B42** , 14 - 23

Raza B and Holt DB 1995 in Microscopy of Semiconducting Materials 1995 Conf. Series No. 146 (AG Cullis and AE Staton-Bevan, eds.) (Bristol: Inst. Phys.) pp. 107 - 112

Russell J D and Leach C 1995 J. Euro. Chem Soc. **15**, 617 - 622

Russell JD, Halls DC and Leach C 1995 J. Mater. Sci. Lett. **14**, 676 - 678

Russell JD, Halls DC and Leach C 1996 Acta Mater. **44**, 2431 - 2436

Inst. Phys. Conf. Ser. No 157
Paper presented at Microsc. Semicond. Mater. Conf., Oxford, 7–10 April 1997
© *1997 IOP Publishing Ltd*

Dependence of electron-hole generation function on EBIC contrast of defects

M J Romero, D Araújo and R García

Departamento de Ciencia de los Materiales e I.M. y Q.I., Facultad de Ciencias, Universidad de Cádiz, Apartado 40, 11510 Puerto Real (Cádiz),Spain.

ABSTRACT : Nowadays, phenomenological modelling of electron-beam-induced current (EBIC) contrast from defects is well developed. But, as recently revealed by Araújo et al. (1994) and Bonard et al. (1996), the electron-hole pairs (e-h) generation function $g(x, y, z)$ is still an open task. Obviously, the lack of information about $g(x, y, z)$ results in a loss of EBIC quantitative potential. In this connection, we present here an expression for energy transfer dE/dS that realizes the recently published experimental e-h distributions. It will be applied by the Monte Carlo method to estimate the e-h generation. Solving the modified minority-carrier diffusion equation we obtain the EBIC contrast linescans profiles from localized defects. These calculations are confronted with the predictions from the Rutherford-Bethe physical frame. This new formulation of energy transference explains actual discrepancies between EBIC contrast data and theoretical predictions.

1. INTRODUCTION

Electron-beam-induced current (EBIC) and cathodoluminescence (CL) investigations of semiconductors suggest that electron beam injection techniques constitute a straightforward way to reconstruct the recombination properties from defects. The efforts in this field follow up two directions : first, to develop detail descriptions of the carrier recombination mechanisms at defects (Farvacque 1996, Kittler and Seifert 1996) and second, to introduce into EBIC/CL modelling a more realistic electron-hole pairs (e - h) generation function $g(x, y, z)$ (Pey et al. 1993, Seifert et al. 1993). In the second one, the most commonly used approach is applying the Rutherford cross-section to determine scattered electron deflections and the Bethe energy-loss equation for its energy transfer. Nevertheless, Araújo et al. (1994) and Bonard et al. (1996) showed discrepancies between the Rutherford-Bethe predictions of electron-beam-induced e-h generation distributions and experimental ones. Elsewhere (Romero et al. 1996), we present an electron-beam scattering model that realizes the experimental depth and lateral dependences of e-h generation. The key point is its ability to modulate the contribution of inelastic and elastic cross-sections to scattering event as a function of electron-beam energy. This model is an analytical expression of a more detail scattering formulation (Romero et al. in preparation). It is applied in Monte Carlo calculations to estimate $g(x, y, z)$. Then, solving the modified steady-state diffusion equation for each differential volume we extract the EBIC behaviour around the defect.

2. THEORETICAL CONSIDERATIONS

From Donolato(1978/79), the defect is assumed as a bounded region F of reduced minority carrier lifetime, $\tau_d < \tau$ (Fig. 1). The minority carrier lifetimes are related to minority carrier diffusion lengths through the diffusion coefficient $D = L_d^2/\tau_d = L^2/\tau$. Therefore, solving the steady-state diffusion equation yields,

$$\nabla^2 p(\mathbf{r}) - \frac{1}{L^2} p(\mathbf{r}) + \frac{1}{D} g(\mathbf{r}) = \gamma e(\mathbf{r}) p(\mathbf{r}) \tag{1}$$

where $g(\mathbf{r})$ is the electron-hole generation at $\mathbf{r}(x,y,z)$ coordinates and $p(\mathbf{r})$ is the minority carrier density. $e(\mathbf{r}) = 1$ for $\mathbf{r}$ inside F and $e(\mathbf{r}) = 0$ elsewhere. $\gamma = 1/L_d^2 - 1/L^2$, is the defect strength. Solving equation (1) for each differential volume in a discretized scheme, we obtain the collected current at each point. Thus, if we define the intrinsic parameters related to the semiconductor (L) and defect (r_D, F, γ), only different $g(\mathbf{r})$ results in a different behaviour of collected current. In this contribution, these electron-hole generation functions are estimated from the Monte Carlo method applying the Bethe energy-loss equation,

$$\frac{dE}{dS} = 2\pi e^4 Z \frac{N_A \rho}{A} \frac{1}{E} Ln(\frac{vE}{J}) \tag{2}$$

or

$$\frac{dE}{dS} = \lambda_s \pi 2 e^4 \frac{s}{s-1} \frac{N_A \rho}{A} Z^{(2+s)/3s} (\frac{2e^2}{a})^{2/s-2} \times$$

$$\times \frac{E^{1-2/s} Ln(E)}{Ln(E) + Exp(-\frac{s-1}{\beta}) \frac{s}{s-1} Z^{2(1-s)/3s} (\frac{2e^2}{a})^{2/s-2} E^{2-2/s}} \tag{3}$$

where s is a numerical parameter related to the inelastic scattering cross-section contribution with respect to the elastic one and β is related with the contribution of the Rutherford scattering ($\beta \approx 0.08$) and $\lambda_s = 0.168$ — for more details, see Romero et al. (1996) — N_A is the Avogadro number, ρ the semiconductor density and A and Z its atomic weight and number, respectively. a is the effective screened atom radius, e the electronic charge, $v = 1.166$ is a relativistic factor of energy and J the ionization energy. The expression (3) realizes the experimental data and we take it as a better approach to $g(\mathbf{r})$.

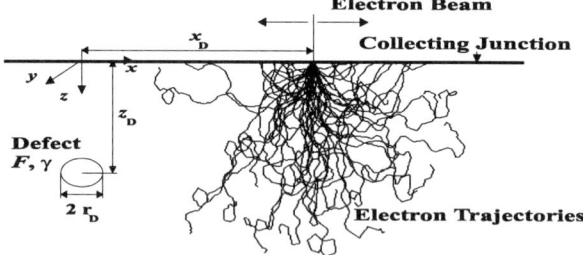

Fig. 1. Scheme of calculus geometry.

In the following, we compare both expressions (2) and (3) in the minority carrier diffusion equation using a common planar configuration (Fig. 1). Thus, the EBIC contrast profiles from a point-like defect, a misfit dislocation (MD) and a threading dislocation (TD) are estimated from expressions (2) or (3), with variable L, F and γ. We consider a zero width of collecting junction with a surface recombination velocity $S = \infty$ (perfect collector assumption). The parameters to analyze are the EBIC contrast c and the full width half minimum FWHM. c is defined as $c = [(I_{cc}|_{xD=0} / I_{cc}|_{xD=\infty}) - 1]$ where $I_{cc}|_{xD}$ is the collected current at a distance from the centre of the defect of x_D.

3. RESULTS AND DISCUSSION

Fig. 2 displays the EBIC contrast versus γ from a MD localized at different depths z_D and bulk diffusion lengths L applying the s dependent energy-loss expression (3). It follows the behaviour predicted by Donolato (1983) in the full defect strength range. The corresponding differences between EBIC contrast calculated implementing the Rutherford-Bethe (R-B) formula c_{R-B} and equation (3) c, $\Delta c = c_{R-B} - c$ (Fig. 3), exhibit a non-straightforward relationship. This is due to an electron beam of energy $E_b \leq 10$ keV penetrating and spreading into the semiconductor with more efficiency than predicted by the R-B physical frame :the range or maximum penetration depth is $R \approx 0.5$ μm while $R_{R-B} \approx 0.2$ μm at $E_b = 5$ keV. Differences between both scattering mechanisms increase with defect strength. For MDs close to the collecting junction R-B under estimates the defect strength of the dislocation whereas for $z_D > R/3$ the reverse is the case. As displayed in Fig. 4, the effect of electron beam

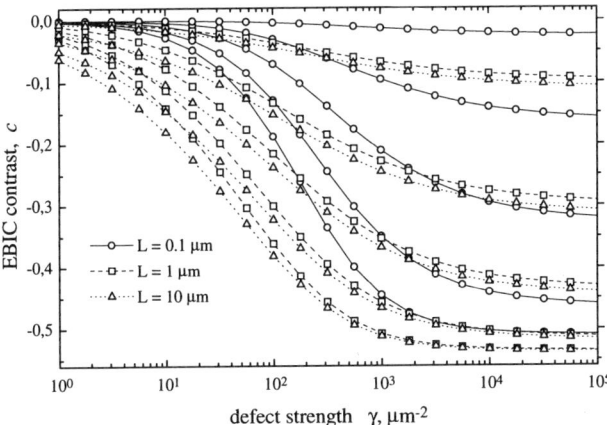

Fig. 2. EBIC contrast vs γ from a MD at depths of (from upper to lower curves) $z_D = 0.3, 0.2, 0.15, 0.1$ and 0.05 μm for different bulk minority carrier diffusion lengths. The capture radius of the MD is $r_D = 0.1$ μm.

spreading implies, obviously, a FWHM increase. Thus, it is easy to conclude that the radius of defects is always overestimated. These results can be extrapolated to point-like defects, taking into account the contrast and resolution enhancing.

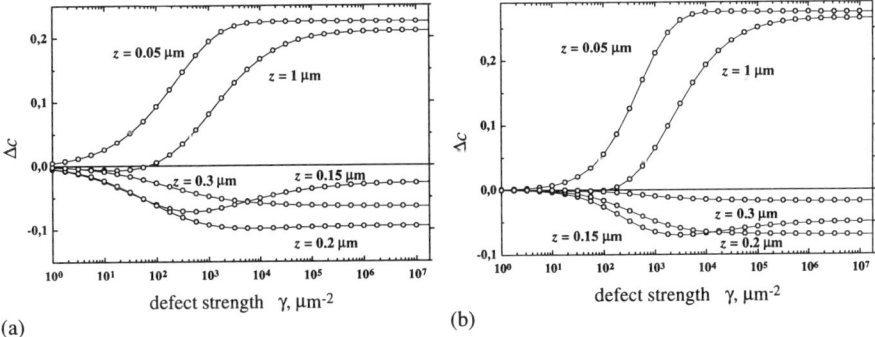

(a)   (b)

Fig. 3. Δc vs γ from a MD at different z_D for L = 1 μm (a) and L = 0.1 μm (b) with $r_D = 0.1$ μm.

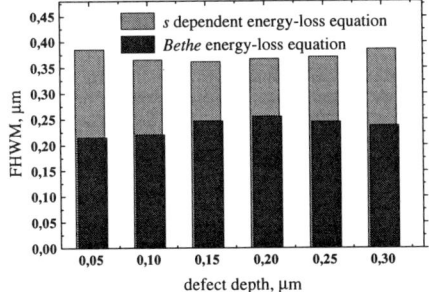

Fig. 4. FWHM vs z_D for a MD confronting both scattering mechanisms ($\tau_d = 0$).

The EBIC contrast dependences of γ and L for a TD are presented in Fig. 5. Note the reversal behaviour between both scattering mechanisms : if $\gamma << 10^3$ μm^{-2} (weak defect) is considered an improved contrast is expected whereas if $\gamma >> 10^3$ μm^{-2} then the behaviour is inverse.

646

Up to now, we set E_b at 5 keV because this gives the largest diferences between $g(x, y, z)$ theoretical predictions and experimental results (Bonard et al. 1996). As expected, the influence of the scattering mechanism used is less as E_b increases. Note, nevertheless, that high lateral EBIC resolution of defects in experiments requires low electron beam energies. The c and FWHM dependences on E_b follow in a qualitative manner that published by Seifert et al. (1993).

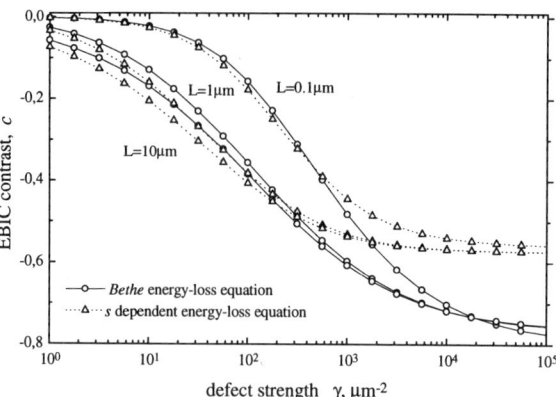

Fig. 5. EBIC contrast vs γ from a TD for different bulk minority carrier diffusion lengths. $r_D = 0.1\ \mu m$.

Actually, we are carrying out EBIC measurements on MDs at the p^+-InGaAs / n^+-GaAs heterointerface where the defect depth is modified between 0.05 µm and 0.3 µm by chemical etching methods (to be submitted to J. Appl. Phys.). As a preliminary result, there are not any values of γ and defect radius that fit successfully the experimental EBIC data applying the R-B formalism whereas the predictions of EBIC contrast based on equation (3) give close to self-consistent values of γ and r_D. Nevertheless, the model of a dislocation as a recombination centre used here is simple and it is not possible to conclude that the electron-hole generation is the unique factor that explains the actual discrepancies but, it is undoubted that a more realistic approach to $g(x, y, z)$ is necessary for accurate reconstruction of recombination properties at localized defects.

5. CONCLUSIONS

An analytical expression that realizes the recently published experimental electron-hole pair generation distributions has been proposed. The influence of the spatial dependence of $g(x, y, z)$ on EBIC contrast from defects is quantified. The proposed approach to $g(x, y, z)$ accounts for actual discrepancies between EBIC data and theoretical predictions based on the Rutherford-Bethe formula.

ACKNOWLEDGEMENTS

The authors acknowledge support by the CICYT (Comisión Interministerial de Ciencia y Tecnología), under MAT94-0538 Project and by the Junta de Andalucía through group TEP-0120.

REFERENCES

Araújo D, Bonard J - M, Oelgart G, Ganière J - D, Morier-Genoud F and Reinhart F - K 1994 Mat. Sci. Eng. B**24**, 124
Bonard J - M, Ganière J - D, Akamatsu B and Araújo D 1996 J. Appl. Phys. **79**,8693
Donolato C 1978/79 Optik **72**, 19
Donolato C 1983 Journal de Physique **9**(44), C4-269
Farvacque J - L 1996 Mat. Sci. Eng. B**42**,110
Kittler M and Seifert W 1996 Mat. Sci. Eng. B**42**, 8
Pey K L, Chan D S H and Phang J C H 1993 Inst. Phys. Conf. Ser. **134**, 687
Romero M J, Araújo D and García R 1996 Mat. Sci. Eng. B**42**, 168
Romero M J et al. (in preparation)
Seifert W, Knechtel W and Kittler M 1993 Inst. Phys. Conf. Ser. **134**, 751

Inst. Phys. Conf. Ser. No 157
Paper presented at Microsc. Semicond. Mater. Conf., Oxford, 7–10 April 1997
© *1997 IOP Publishing Ltd*

Cathodoluminescence and EBIC of 2D junction laser structures on patterned (311)A GaAs substrates

C E Norman[1] A J North[2] J H Burroughes[1] T Burke[2] and D A Ritchie[2]

[1] Toshiba Cambridge Research Centre Ltd. 260, Cambridge Science Park, Milton Road, Cambridge, CB4 4WE.

[2] Cavendish Laboratory, University of Cambridge, Madingley Road, Cambridge, CB3 0EH.

ABSTRACT: Cryogenic cathodoluminescence (CL) and electron beam induced current (EBIC) are used to investigate the band structure in novel modulation-doped lateral injection ridge lasers. EBIC successfully reveals the position and regularity of the junction between hole and electron gases on planes of differing indices, with sub-micrometre resolution. CL is able to map the carrier density in an electron gas on a very narrow (100) facet, where no measurement via electrical contacts could be made.

1. INTRODUCTION

In the search for low threshold current, high modulation speed semiconductor lasers, an "in-plane" design, such as that first suggested by Meier *et al* (1988) should have several advantages over conventional vertical injection lasers. The main difficulty involved in realising such a structure is that of producing a design in which electrons and holes can be separately confined, not only within the same plane, but also in close enough proximity to permit efficient injection of electrons into the hole-containing region (or vice versa).

The potential advantages of the in-plane design are (i) a reduction of the threshold current density (for a given threshold carrier density) arising from the reduction in the junction area, (ii) a reduction in the capacitance, again due to junction area reduction, offering better high speed modulation characteristics, (iii) the elimination of the need for 3D to 2D carrier relaxation, thereby avoiding gain-compression enhanced Auger rates and other hot carrier effects (Davis *et al,* 1994).

One solution to the design problem is to use the amphoteric nature of silicon doping in MBE growth of GaAs to produce lateral p-n junction structures on patterned substrates. Where the substrate is etched to reveal planes such as (311)A and (100), these will become selectively p and n doped when grown under the right conditions. The same is also true for 33% AlGaAs under low arsenic overpressures, offering the possibility of producing modulation doped structures in which adjacent regions contain 2D electron gases (2DEG) and hole gases (2DHG) of reasonably high carrier density. It is this latter approach which we have adopted to successfully produce 2D junction lasers, capable of continuous wave operation up to 130K and pulsed operation up to 190K (JH Burroughes, private communication). We believe it is the first time such a structure has been shown to lase.

Our design involves etching a (311)A wafer to reveal (100) facets, as shown schematically in figure 1. The laser structure itself consists (nominally) of a 10nm GaAs QW, clad symmetrically on either side by 10nm of undoped $Al_{0.33}Ga_{0.67}As$, 40nm of $Al_{0.33}Ga_{0.67}As$ Si-doped at nominally 1 x 10^{18} cm^{-3}, another 40nm of undoped $Al_{0.33}Ga_{0.67}As$, and finally, a 25 period GaAs/$Al_{0.68}Ga_{0.32}As$ superlattice. A waveguide is etched above the (100)-(311)A intersection, and n- and p-type ohmic contacts are applied to the exposed (100) and (311)A facets, respectively. The band structure of the symmetrically modulation-doped device is shown schematically in figure 2, for the case of the (100) facet where the Si doping in the AlGaAs in n-type. The band bending brings the GaAs QW conduction band down below the Fermi level, permitting the formation of a 2DEG in the QW. The diagram can effectively be inverted to represent the 2DHG formed on the (311)A and (111)A planes, as a result of p-type Si doping, giving band bending in the other direction.

CL measurements were performed using an Oxford Instruments Ltd. MonoCL system and CF302TC liquid helium cryostat, attached to a Hitachi S-4500 cold cathode SEM upgraded with an

648

oil-free vacuum system. EBIC measurements were performed using a Stanford Research Systems model SR570 current amplifier.

2. RESULTS AND DISCUSSION

Figure 3 shows the 4.5K CL spectra obtained from the modulation doped structures grown on the three facets present on the device surface. The (311)A and the (111)A facets both support 2DHGs and present symmetrically-shaped peaks at 1.539eV which are of comparable integrated intensity. The peak from the narrow (100) facet, however, is markedly different, showing a pronounced "knee" to higher energies, although the integrated intensity remains similar. This high energy knee is the well-observed Fermi edge singularity (FES), which arises from multiple electron-hole scattering and Coulomb interactions in the 2-dimensional system. These effects lead to an enhanced probability for optical interband transitions involving electron states close to the Fermi energy.

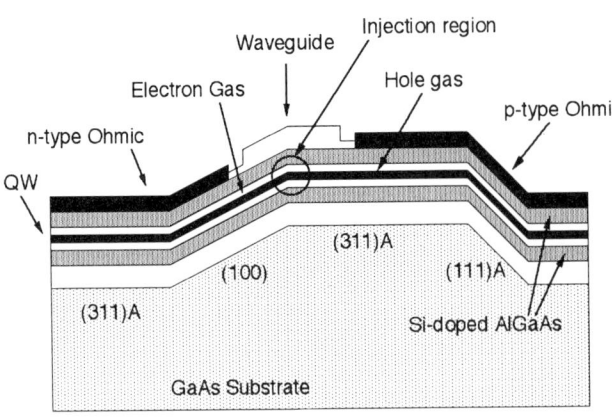

Figure 1. A schematic diagram of the laser device structure.

Luminescence peaks observed from thin layers containing electron or hole gases are generally broader than emissions from QWs in the absence of such gases. In the band structure represented in figure 2, which gives rise to a 2DEG, we would expect the CL signal to arise due to beam generated electron-hole pairs recombining in the QW. The holes will occupy the first heavy hole level, which will effectively have a single, well defined energy, but electrons may recombine with them from any of the filled states from the first permitted state in the QW up to the Fermi energy. It is this energy spread (between the transitions E_1 and E_2 denoted schematically in figure 2) which accounts for the broadening of the emission peaks.

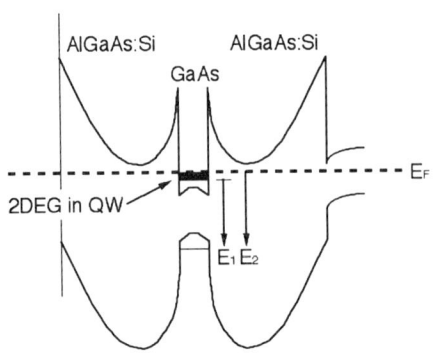

Figure 2. A schematic diagram of the expected band structure on a (100) facet in the modulation doped laser.

Electrical measurements (Hall) show the hole density in the QW to be approximately 8×10^{11} cm^{-2} on the (311)A plane, and 6.2×10^{11} cm^{-2} on the (111)A sidewalls. The hole mobilities are 300,000 cm^2/Vs and 2,900 cm^2/Vs respectively. From the 2D density of states relationship:

$$D = m^*/\hbar^2$$

the energy separation (in meV) between the Fermi level and the bottom of the valence (or conduction) band, ε_F, can be calculated to be: $0.704p$ or $3.57n$ (where p and n are the hole and electron density, respectively, in units of 10^{11} cm^{-2}).

This gives an expected energy spread for the hole gas on a (311)A plane of 5.6meV, which is only marginally less than the observed minimum FWHM value of 7.6meV.

No electrical measurements were possible for the narrow (100) facets owing to the device geometry which precluded making effective Hall bars on these narrow structures. The high spatial resolution of the CL technique can be used to probe only the (100) facet, with the emission peak giving a direct measure of the energy separation of the FES and the bottom of the conduction band. This was measured to be 44meV, invariant with carrier injection level over a beam current range of

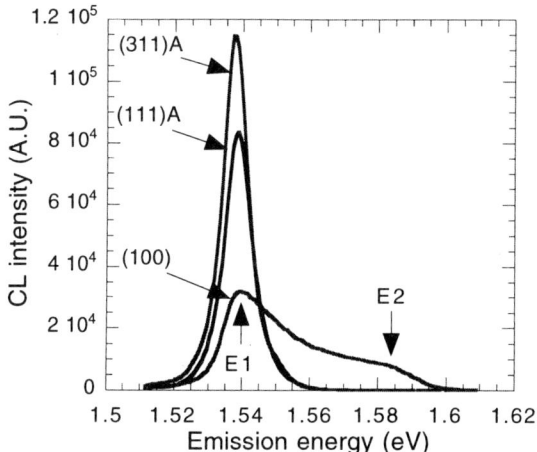

Figure 3. 4.5K CL spectra of the electron and hole gases in the QW on the three different planes.

50pA to 500pA at 10keV. This separation corresponds to a carrier density of 1.2×10^{12} cm^{-2}, very close to the value anticipated from the device design.

Recording point spectra as a function of distance away from the cleaved edge on the (100) facet allows us to monitor the separation of the lowest electron state in the QW and the Fermi energy, and thence the apparent carrier density. As the beam approaches to within 10µm of the cleaved edge, E_2 is seen to shift towards E_1 as shown in figure 4, suggesting the carrier density is reducing to zero at the cleaved surface. It is apparent from this result that CL may be used to monitor carrier densities in modulation doped structures in plan view, but not in cross-section, where the carrier injection is, perforce, in the vicinity of the cleaved edge .

EBIC imaging in cross-section was used to investigate the exact position of the junction between the electron and hole gases. Figure 5a is a secondary electron image of a working laser device in cross-section. The etched waveguide is visible on the top surface, and a step is apparent where it terminates approximately 2.5µm down the (100) facet. Figure 5b shows exactly the same area in EBIC mode at room temperature with 1V of forward bias applied. A low excitation energy (5keV) was used to ensure sub-micron resolution in the EBIC image. The junction is clearly evident, but displaced a distance of 1.25µm to the left of the nominal (311)A-(100) intersection. The field is evidently enclosed within the barrier layers, vindicating this particular device design. Under small reverse biases, the collecting field is seen to extend down into the substrate, as shown in figure 5c for the device under 1V reverse bias. This suggests that the device structure may not yet be optimised, in terms of the barrier structure. EBIC imaging of an earlier device with a different design, incorporating thinner cladding layers and slightly different doping, showed the fields extending more markedly into the substrate, even under zero applied bias. Both figures 5b and 5c show narrow bands of dark contrast corresponding to the 2DEG and 2DHG. This is due to field screening in the 2D gases.

The reason for the displacement of the junction becomes apparent from high resolution SEM cross-section images. A short length of (411)A facet (which supports an electron gas, under these growth conditions) is formed on the (311)A side of the nominal (311)A-(100) intersection. Plan view EBIC imaging shows that the intersection between the (411)A and the (311)A planes is very straight, showing

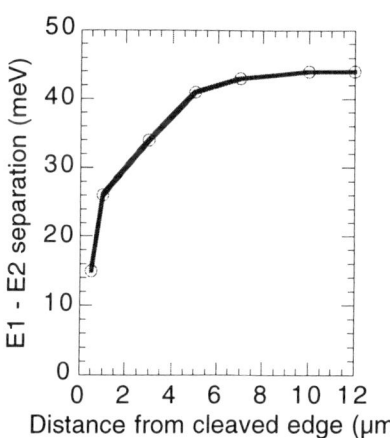

Figure 4. The separation of E_1 and E_2 as a function of distance from a cleaved edge, for a narrow (100) facet.

none of the long-range waviness along the ridge direction that is often associated with etched sidewalls. Originally, (311)A wafers were used to make the processing requirements less stringent: a

650

(311)A   **(100)**

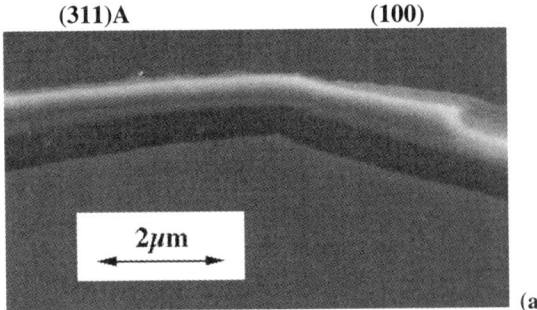

2μm

(a)

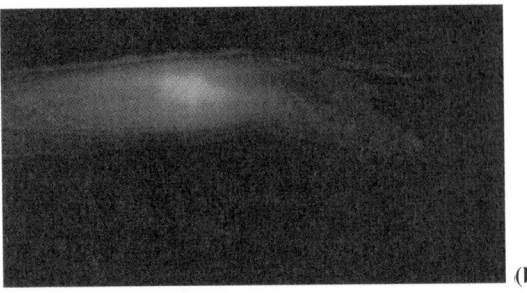

(b)

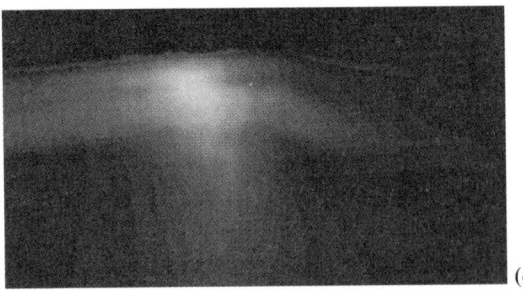

(c)

Figure 5. (a) secondary electron image of laser cross-section, (b) EBIC image @ 1V forward bias, (c) EBIC image @ 1V reverse bias.

few degrees of error in etching a (311)A plane in a (100) wafer can have severe consequences on the p-type doping obtained using silicon, but a similar error in etching a (100) plane on a (311)A wafer still results in predictable n-type doping. The formation of the (411)A facet during growth acts to keep the junction away from the etched (and therefore probably rougher) sidewall, thereby improving, rather than degrading, the device performance.

3. CONCLUSIONS

CL can successfully be used to gauge carrier (particularly, electron) density in 2D systems with microscopic dimensions which cannot reliably be contacted for electrical (e.g. Hall) measurements. Cleaved edges must however be avoided as pinning of the Fermi level at the free surface results in a reduction of the apparent carrier density in the 2D gas. EBIC imaging is of great value in precisely locating the position of the lateral injection junction, and in monitoring the effectiveness of the cladding layers in confining the field to the region of the electron and hole gases. It can also reveal the presence of 2DEGs and 2DHGs by their field screening effects. Whilst this is the first demonstration of a lateral injection laser, it is clear that the device design is not yet optimised. With further work we hope to achieve CW operation at room temperature.

REFERENCES

Davis L, Sun HC, Yoon H and Bhattacharya PK. 1994 Appl. Phys. Lett. **64**, 3222
Meier HP, Broom RF, Epperlein PW, van Gieson E, Harder Ch, Jaekel H, Walter W and Webb DJ 1988 J. Vac. Sci. and Technol. **B6**, 692

Inst. Phys. Conf. Ser. No 157
Paper presented at Microsc. Semicond. Mater. Conf., Oxford, 7–10 April 1997
© 1997 IOP Publishing Ltd

Distinction of the recombination properties and identification of Y luminescence at glide dislocations in CdTe

J Schreiber, H Uniewski, S Hildebrandt, L Höring and H S Leipner

Martin-Luther-Universität, FB Physik, Friedemann-Bach-Platz 6, D-06108 Halle (Saale), Germany

ABSTRACT: SEM-CL studies performed at low temperature between 5K and 120K disclose radiative as well as non-radiative recombination activity of glide dislocations in undoped p-type CdTe. In the investigation of crystallographically defined glide dislocation arrangements produced by local plastic deformation on {111}A/B and {110} faces using Vickers microindentation the recombination behaviours of Cd(g),Te(g) and screw dislocation parts could be revealed. The obtained results point out non-radiative recombination at the Cd(g) and screw dislocation segments whereas the Te(g) parts exhibit a relatively strong, localised radiative recombination causing an emission at 1.476eV(5K).The latter obviously checks with the well-known Y luminescence of CdTe which is ascribed to extended defects in this material.

1. INTRODUCTION

Electronic effects of glide dislocations in compound semiconductors have been evidenced by numerous experimental results, however, there is still a lack of full understanding of expected correlations between the intrinsic crystallographic structure and the electronic properties of the line defects. The present work aims at the distinction of the recombination activity of the opposite polar A(g) and B(g) as well as screw type dislocation parts appearing in the glide dislocation configurations in CdTe. Because of its high ductility even at room temperature crystallographically defined defect configurations being appropriate for detailed scanning electron microscope cathodoluminescence (SEM-CL) studies can be introduced by local microplastic deformation such as Vickers indentation. The spatial distribution of the generated glide dislocations is determined by the geometry of the {111}⟨110⟩ slip systems and the Peierls defect migration operating under the conditions of localised deformation.

Contrary to the mostly used (001) sample orientation, in this work {111}A/B and {110} oriented samples were chosen for the experiments in order to allow reliable separation and recognition of the different dislocation types. Analysing the surface-parallel and volume glide geometries on the surfaces of the various sample orientations applied renders it possible to discriminate the polar 60°-related A(g) and B(g) as well as screw dislocation segments in the defect configurations developed.

The results presented here were derived from low-temperature investigations of the CL defect contrasts registered in the produced arrangements of glide dislocations. They state radiative and non-radiative carrier recombination at the fresh dislocations and give evidence for differences in the recombination activity of the different dislocation parts which are not revealed by similar CL studies at room temperature (Rivière et al 1991). The spectroscopic and kinetic data gained from the temperature-dependent behaviours of the defect contrasts provide further information on the carrier recombination processes at the defects. For the first time, the so-called Y luminescence as described by Dean (1984) can be attributed to a particular type of deformation-induced line defect.

2. MODEL FOR RECOGNITION OF POLAR GLIDE DISLOCATION PARTS

With regard to the conducted experiments a model of the expected glide dislocation distribution due to indenting a {111} sample surface is considered in Fig.1. The indentation is thought to cause slip along the surface-parallel ⟨110⟩ directions within certain {111} glide planes resulting in a rosette-like defect arrangement. Each of the six rosette arms represents a⟨110⟩ glide prism which is made up by corresponding {111} slip planes belonging to the external (EAT) and internal (IAT) apex tetrahedron, respectively. The glide prisms contain material under compression, thus the extra half planes terminate from inside at the slip planes. Therefore it is easy to conclude the type of the dislocations formed in the glide set configuration by taking into account the polarity of the respective glide planes. The appearance of A(g) and B(g) branches, that is the regions where the opposite polar 60°-dislocation segments proceed, which are dragging screw parts behind, is sketched in the upper right-hand part of Fig.1. It should be mentioned that the glide prism model

652

predicts slip processes in the surface-parallel glide planes, too. In this case dislocation half-loops with B(g) leading parts are generated as illustrated top left.

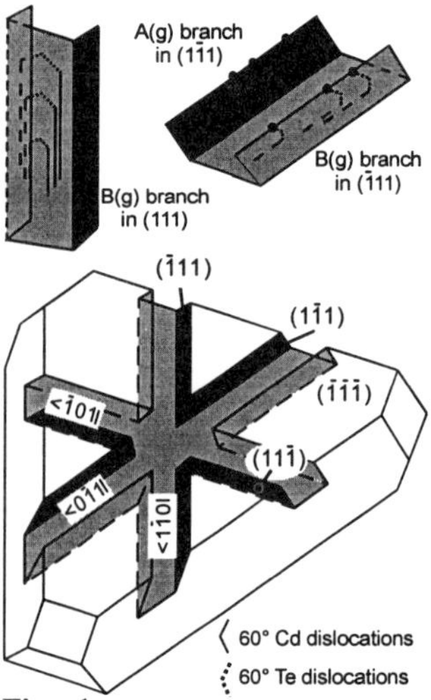

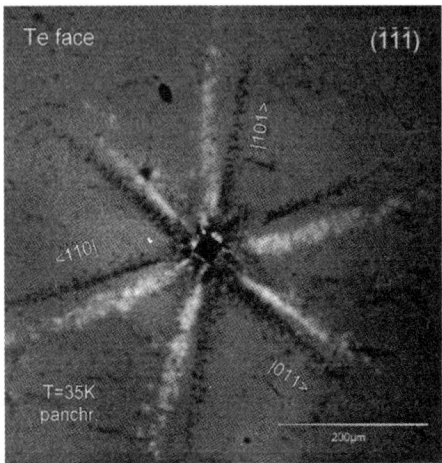

Fig. 1 Slip geometry and dislocation configuration of rosette glide due to local indentation on ($\bar{1}11$) sample surface

Analogous considerations hold for a (111)Cd sample orientation.

3. EXPERIMENTAL DETAILS

A conventional SEM apparatus (BS300,Tesla) equipped with appropriate attachments for sample cooling down to 5K and CL observations (LHe cooling stage CF302, and Mono CL system, both Oxford Instruments) has been used in the experiments. Samples investigated were cut in ±(111) and (110) orientations from undoped CdTe (p=10^{14}...10^{15} cm^{-3}) bulk crystals. The polarities of the ±(111) sample surfaces were determined by X-ray diffraction measurements. The microplastic deformations were carried out using a Vickers indenter at room temperature applying loads of 5 to 20p. For depth-resolved CL investigations of the defect configuration underneath the in-dented sample surfaces, layers of distinct thicknesses were gradually removed by subsequent etching with bromine-methanol solution (etch rate 4μm/min).

All CL studies were performed employing an optimised electron beam power (20kV, 0.1 to 1nA) which allows the CL contrast behaviours to be examined by means of panchromatic and spectral CL microscopy and utilising local CL spectroscopy as well. Additional PL data at 1.7K supported the spectroscopic analysis of the samples.

4. EXPERIMENTAL RESULTS AND DISCUSSION

4.1 Distinction of the recombination activity of Cd(g), Te(g) and screw dislocation segments

The low-temperature SEM-CL investigations provide detailed information on both the defect configuration introduced by microplastic deformation and the recombination activity of the glide dislocations appearing in it.

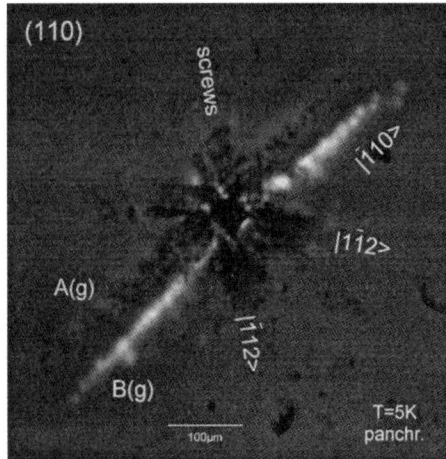

Fig. 2a CL micrograph of dislocation rosette on indented ($\bar{1}\bar{1}1$) Te sample surface showing dark and bright defect contrasts

Fig. 2b CL micrograph taken on the (110) sample surface after indentation supplying A(g), B(g) and screw dislocations in the near surface region

Fig.2a displays a panchromatic CL micrograph showing dark and bright defect contrasts in the dislocation arrangement around an indent on the Te face of a ($\bar{1}\bar{1}1$) oriented sample. The bright and dark CL contrasts seen at the defects prove radiative as well as non-radiative recombination activity of the dislocations considered.

It seems evident that the different contrasts mark the A(g) and B(g) branches occurring in all rosette arms. The stringent spatial separation of dark and bright contrasts implies a direct correlation between the CL contrast behaviours and defect polarity. The different defect contrast properties found in the A(g) and B(g) branches establish the distinct recombination activity of the polar Cd(g) and Te(g) dislocations in CdTe.

The CL micrograph given in Fig.2b depicts the defect configuration caused by indenting a (110) sample face. It is easy to distinguish the appearing rosette glide region extending along the surface parallel ⟨110⟩ direction from the volume glide region which causes the defect arrangement going into the ⟨112⟩ directions. The dark CL contrasts occurring here can be ascribed to the screw dislocation parts of dislocation half loops propagating along ⟨101⟩ and ⟨011⟩ directions into the depth. This defect contrast behaviour proves non-radiative carrier recombination at the screw segments comparable with the non-radiative defect activity of one of the polar 60° dislocations studied on (1̄1̄1̄) .

The identification of the Cd(g) and Te(g) dislocations as radiative and non-radiative recombination centres, respectively, requires to attribute the observed bright and dark CL contrasts to the respective type of defects. This correlation can be found by studying the contrast properties of the dislocations proceeding within the surface-parallel slip planes of the rosette glide modes on (111) and (1̄1̄1̄) oriented samples. In this case, the polarity of the 60°-related dislocation parts may be immediately deduced from the knowledge of the absolute crystallographic orientation of the examined sample surface. The CL images in Fig.3 present the distribution of bright and dark defect contrasts corresponding to the rosette dislocation configuration about 10μm underneath the initial Cd and Te terminated {111} sample surfaces, respectively.

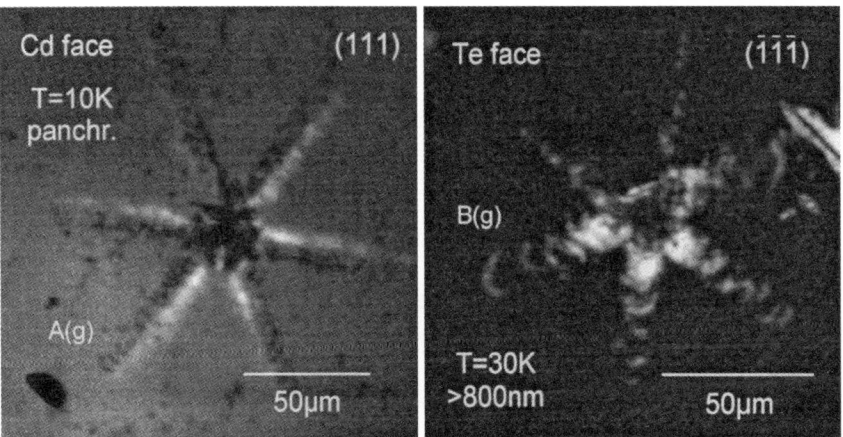

Fig. 3 Depth resolved CL-micrographs in rosette glide region around indentation on the (111)Cd and(1̄1̄1̄) Te face, respectively.

The depth-resolved CL micrographs taken from the Cd and Te faces clearly reveal the A(g) and B(g) dislocation branches, respectively, within the surface-parallel slip planes and prove the particular CL contrast behaviours for the opposite polar dislocations. The experimental findings shown allow to relate the dark defect contrasts to the Cd(g) dislocations while the bright defect contrasts definitely correspond to the Te(g) dislocation segments.

4.2 Temperature-dependent CL defect contrast behaviours and spectral analysis of dislocation-induced radiative recombination

The investigation of the temperature dependence of the dark and bright defect contrasts and the spectroscopic analysis of the defect-induced recombination emission were exploited to obtain some quantitative information on the defect-related carrier recombination processes.

The graphics in Fig.4 show the values of dark and bright contrast, respectively, as measured at individual defects versus temperature from 5K to 120K. The dark defect contrast is seen to be almost constant over the entire temperature range whereas the bright contrast increases upon lowering temperature and is thermally quenched above 120K. The evaluated barrier of thermal quenching amounts to 11meV.

The behaviour of the dark contrast may be attributed to a defect-bound carrier recombination process of Shockley-Read-Hall type which is controlled to some extent by the small space charge barrier surrounding the defect line. Detailed discussion will be given in a forthcoming paper.

654

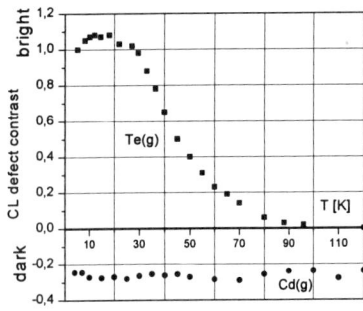

Fig. 4 Temperature dependence of the bright and dark CL contrasts at Te(g) and Cd(g) dislocations, respectively.

In the case of the bright contrasts, the small barrier of thermal quenching may be accounted for a specific feature of the contrast behaviour which hints at an excitonic process being responsible for the radiative carrier recombination at the defect at low temperatures (Rebane and Shreter 1991). The assumption of an excitonic mechanism is supported by the spectroscopic characteristics of the defect-caused radiative recombination. CL spectroscopy and spectral CL imaging evidenced the localised emission of a specific luminescence at $hv_{DL}=1.48eV$ originating from the bright defect contrasts only. This emission spectrally coincides with the Y luminescence in CdTe (Dean 1984). The spectral data derived in this work are summarised in Tab. 1. They show excellent agreement with those given in the literature. The typical LO phonon-assisted emission band lineshape with small Huang-Rhys factor S and halfwidth Γ could be realised. The localisation energy E_D ascribed to the deeply defect-bound exciton is found to be 120meV as deduced from the gap energy E_g, zero-phonon peak position hv_{DL}, and electron-hole binding energy E_X in the free exciton state. The latter is believed to be identical to the observed thermal quenching barrier E_A.

5. CONCLUSIONS

The present results obtained from low-temperature SEM-CL studies on crystallographically identified glide dislocation configurations manifest the recombination activity of Cd(g), Te(g) and screw dislocation parts becoming evident in dark and bright defect-related CL contrasts which disclose non-radiative as well as radiative defect-induced carrier recombination. The bright CL contrasts are shown to appear exclusively at Te(g) dislocation parts whereas the Cd(g) and screw segments exhibit similar dark contrasts.

CL spectroscopy and spectral CL imaging reveal the radiative recombination to be strictly localised on the Te(g) dislocations. Its zero-phonon peak is at 1.476eV(5K) corresponding to the Y luminescence assigned to extended defects in CdTe.

The dark and bright CL contrasts at the polar dislocations differ in their temperature-dependent behaviours. These findings clearly hint at the distinct recombination properties.

	this work	data from literature	references
E_g (eV)		1.606	
hv_{DL} (eV)	1.476 ± 0.001	1.476	D,F,O
E_D (meV)	120	120	D,O
Γ (meV)	10 ± 2	8 ... 9	D,O
S	0.22 ± 0.05	0.15...0.2 (ω_{LO}=21.3meV)	D,F,O
E_A (meV)	11 ± 1	9...11 ($E_A \approx E_X$=10mEV)	F,O

Tab. 1 Y luminescence characteristics in CdTe.
(references: D-Dean 1984 and Dean et al 1984, F-Fuji 1989, O-Onodera 1990)

The occurring dark and bright CL defect contrasts and their temperature-dependent properties indicate radiationless Shockley-Read-Hall carrier recombination at the Cd(g) dislocations whereas the bright contrast occurring at the Te(g) dislocations may be interpreted as radiative decay of excitons bound to this type of dislocation.

6. ACKNOWLEDGEMENT

The authors would like to thank A Hoffmann,TU Berlin, for supplying PL data on the dislocation-induced emission and F Heyroth for performing the X-ray measurements.
Parts of this work were supported by Ministerium WTU des Bundeslandes Sachsen-Anhalt under project no.1557 A 0024.

REFERENCES

Dean P J 1984 phys. stat. sol. (a) **81** 625
Dean P J, Williams G M and Blackmore G 1984 J. Phys. D **17**, 2219
Fuji S, Terada T, Fujita Y and Indi T 1989 Jap. J. Appl. Phys. **28**, L1712
Onodera C, and Taguchi T 1990 J. Cryst. Growth **101**, 502
Rebane Yu T and Shreter Yu G 1991 Springer Proc. Phys. **54**, Polycrystalline Semiconductors II, 28
Rivière A, Sieber B and Rivière J P 1992 Microsc. Microanal. Microstruct. **2**, 503

Inst. Phys. Conf. Ser. No 157
Paper presented at Microsc. Semicond. Mater. Conf., Oxford, 7–10 April 1997

The role of scanning cathodoluminescence in the development of MOVPE growth of GaAs/AlGaAs V-groove quantum wires

G M Williams, M Steer[1], A G Cullis[1], C R Whitehouse[1], M S Skolnick[1] and J S Roberts[1]

DERA, St Andrews Road, Malvern, Worcs. WR14 3PS UK.
[1]University of Sheffield, Department of Electronic and Electrical Engineering, Mappin Street, Sheffield S1 3JD UK.

ABSTRACT: GaAs / AlGaAs quantum wire growth onto 'V' groove substrates of GaAs by MOVPE is now an established technique. In this paper we show how low temperature scanning cathodoluminescence (CL) can reveal the presence of quantum wires and the distribution of variations in the thickness of the GaAs quantum wells. We also show how CL can detect the existence of a vertical AlGaAs quantum well. One dimensional 'V' groove quantum wire structures have different luminescent / temperature properties to quantum well material and our variable temperature CL spectral studies have demonstrated this.

1. INTRODUCTION

The growth of GaAs /AlGaAs quantum wires onto 'V' groove substrates of GaAs by 'MOVPE' is now a technique of great interest in a number of laboratories for producing quasi one dimensional structures. The application of such structures for optical devices (lasers, etc) is well reviewed by a number of authors, see for example Kash (1990), Kapon (1992) and Kapon (1993).

The successful growth of material in which there exists a physically and optically well defined wire has been found to be critically dependant on both 'MOVPE' growth and substrate parameters. Indeed a significant number of growth related and analytical papers have already been published, see for example Christen et al (1992), Gustafsson et al (1995) and Kapon (1993).

One of the main analytical techniques for assessing the electronic and optical properties of such structures is low temperature scanning cathodoluminescence (CL). In this paper we show how CL has been used to identify a number of growth related phenomena in material produced at Sheffield University. A combination of both spectral information and monochromatic imaging has been employed to demonstrate the characteristics of GaAs quantum wires present in the grooves. The technique is ideally suited for determining the presence and distribution of variations in the thickness of the GaAs quantum well and also the presence of a vertical AlGaAs quantum well which is known to play a role in the injection of carriers into the quantum wire.

Studies by other workers have predicted and using photoluminescence shown that one dimensional 'V' groove quantum wire structures will have different luminescent / temperature properties to those observed from quantum well material (Walther et al 1992 and Lee et al 1993). In this paper we present variable temperature CL spectral studies to demonstrate this effect and consequently confirm the successful growth of a one dimensional wire in an optimised structure.

2. EXPERIMENTAL METHODS

All of the structures examined were grown using MOVPE generally at a pressure of 920 torr with the substrate in the temperature range 600°C to 750°C. The Group III chemical precursors used were Tri-Methyl Gallium (TMGa) and Tri-Methyl Aluminium (TMAl) and the Group V arsenic came from an Arsine supply. Substrates of n^+ (100) GaAs were chemically etched through a photo resist mask to create the 'V' groove pattern on the surface with the intersection of two {111}A planes. Initial experiments used a groove spacing of $4\mu m$ or $8\mu m$ and the substrate orientation was deliberately on axis, 3° or 10° off (100).

The scanning CL system used to study these layers is described in detail elsewhere (Williams et al 1991). In brief, however, it is based on a Cambridge Stereoscan 150 Mk2 SEM with an LaB_6 electron source. The samples were cooled to 4K and the luminescence collected by an Oxford Instruments MonoCL 2 system. Samples could be examined in plan view or cross-section by cleaving and mounting into a special holder.

3. RESULTS AND DISCUSSION

3.1 Plan View CL Studies

Typical plan view 4K CL spectra from two different samples are shown in Fig. 1 (a) and (b) below. The broad GaAs quantum well peak observed in (a) is found from cross-sectional imaging experiments to be due to a number of regions of quantum well of different thickness (ie there are superimposed peaks). The intended thickness of the quantum well in this sample was 3.0nm. In Fig. 1(a) we can also just detect the quantum wire luminescence at 788nm and the transition at 625nm is known to result from the presence of the vertical AlGaAs quantum well observed in TEM images. The vertical quantum well is observed in all samples and produced as a result of the increased Ga migration to the base of the 'V' which causes a local change in the AlGaAs composition and it has been shown to play an important role in carrier capture into the wire (Walther et al 1992). The spectrum in Fig. 1(b) demonstrates how a sample can contain GaAs peaks from quantum wells with large differences in thickness. The longer wavelength peak results from regions of well of the intended thickness of 5.0nm and although a wire signal can just be detected at ~764nm it is clear that the choice of well thickness and growth conditions has resulted in very little difference between the emission wavelength of the wire and well.

Figure 2 shows two plan-view monochromatic images of an early quantum wire structure grown on a 3° off axis (100) GaAs substrate. These monochromatic CL images reveal strong asymmetry in the quantum well luminescence with the presence of five regions of different thickness. In Fig 2(a) we see the 737nm luminescence from the quantum well near the base of the 'V' groove. There are obvious non-uniformities which can be seen as a 'herring-bone' distribution of emission. Cross-sectional scanning electron images of the same structure were able to detect contrast between the GaAs and AlGaAs and hence confirm thickness non-uniformities in this sample.

When growth occurs on 3° off-axis substrates it is thought that [001] facets form from

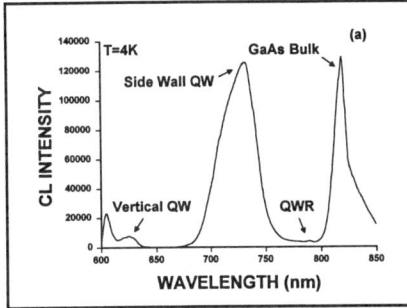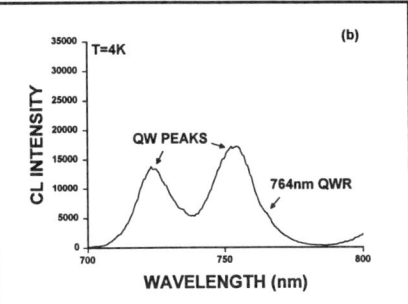

Fig. 1 Shows plan view spectra from two early 'V' Groove quantum wire structures. Sample (a) was a 3.0nm quantum well grown on a 4μm pitch 'V' groove at 720°C and (b) a 5.0nm quantum well grown on an 8μm pitch at 720°C. Both were grown onto on axis (100) GaAs substrates.

the miss-orientated template and that tapering of the quantum well can occur. In Fig. 2(b) we see strong side wall CL emission at 762nm from only one side of the 'V' groove and also non-uniformity along the length of the groove on a scale of ~ 10μm. These gaps or reductions in luminescence efficiency along the length of the structure seem to correlate with the lobes in the 'herring-bone' structure of the 737nm luminescence, confirming genuine non-uniformity. These effects are not desirable as uniform growth is essential for the successful formation of quantum wires as fluctuations will lead to broadening of the luminescence peaks. Similar experiments on 10° off axis substrates also revealed strong asymmetry and similar non-uniform CL emissions for the quantum wells. Much more symmetrical structures were observed from growth on exactly on-axis substrates.

Growth temperature was also found to be critical and lower temperature AlGaAs was found to grow preferentially at the top of the side walls leading to a narrowing of the 'V' and the loss of {111}A facets essential for wire formation. Such samples showed similar CL features to those seen in Fig. 2.

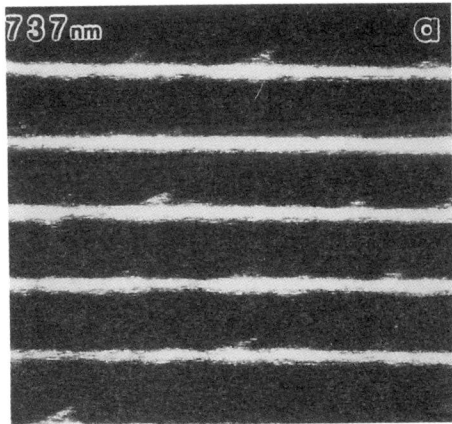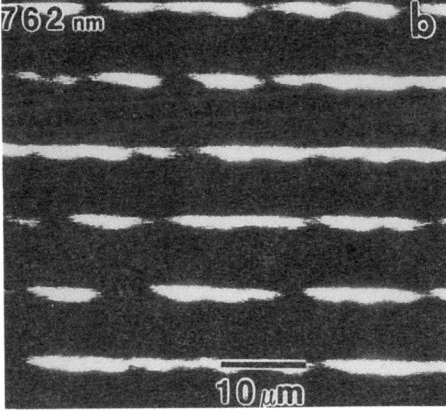

Fig. 2 Plan view monochromatic images of the same region of quantum wire material taken at (a) 737nm and (b) 762nm. The structure was grown at 720°C on an 8μm pitch groove on 3° off axis (100) substrate.

3.2 Cross-sectional CL studies

An advantage of CL is its ability to study these structures in cross-section and take both images and spectra from selected areas of the material. The images shown in Figs. 3 and 4 are all cross-sectional monochromatic images. The 625nm luminescence in Fig. 3(a) is from a vertical AlGaAs quantum well as observed in TEM studies and the 788nm luminescence in Fig. 3(b) is from the quantum wire. In Fig 3 both the wire and vertical quantum well luminescence is compared with the 730nm side wall luminescence which is clearly not uniform in distribution. Images taken with light at different wavelengths between ~700 and 750nm showed different distributions in the side-wall structure, which clearly revealed that the side wall quantum well thickness is varying in this sample. The plan view spectrum for this sample is seen in Fig.1(a) above.

When non-uniformities of side wall quantum well emissions exist these can easily be mapped as seen in the third set of cross-sectional images in Fig. 4(a). The three images shown here are all from the same region of material (ie the same 'V' groove). The top image from a 723nm quantum well, the middle image from a 752nm quantum well and the lower image is at 764nm possibly resulting from quantum wire emission.

To improve the ratio of wire to well signal the distance between grooves was decreased from $8\mu m$ to $4\mu m$ and the well thickness was changed to increase the spectral shift between well and wire peaks. The images in Fig. 4(b) are from such a sample and map the distribution of CL emissions from a 3.0nm quantum well grown on a $4\mu m$ pitch substrate. It is interesting to note that emission at two different wavelengths (710nm and 775nm) appear in the same spacial region. The 775nm quantum wire is luminescing at a lower energy than the side wall quantum well emission (710nm) which is adjoining the quantum wire. This is clear evidence that there is a definite difference in the growth rates on the side-walls and the base of the groove with thinner , higher energy wells, on the side walls. Similarly the quantum well luminescence on the top of the groove (729nm) is also at a higher energy than the wire which is further evidence for enhanced migration to the base of the groove.

The plan view spectrum shown in Fig.1(b) above is dominated by two quantum well peaks at 723nm and 752nm. However, a shoulder at 764nm is clearly observed on the side of the longer wavelength peak. This shoulder is where the transition due to the quantum wire was expected to be observed based on the growth information for this sample. The ability of CL to acquire spectra from selected areas in cross-section allows the confirmation of wire emission to be made. The spectrum shown in Fig.5 is from the same sample but taken in cross-section from a region approximately $2\mu m$ square at the base of one 'V' groove. The quantum wire 764nm luminescence is now clearly observed. The quantum well peaks are still just resolvable due to some diffusion and recombination of carriers in the surrounding material.

A temperature dependence study of the cross-sectional luminescence from a number of samples was carried out with CL spectra taken at intervals from 4K to room temperature. The peak intensities were then plotted as a function of temperature for the main side-wall quantum well(s) and the quantum wire. The results from a typical sample grown on an on-axis substrate are shown above in Fig 6. The monotonic decrease with temperature of the quantum well luminescence intensity is understood to be due to thermal broadening of carrier energies. This means that the luminescence peaks broaden and therefore diminish in intensity. The 764nm luminescence, however, increases with temperature to a maximum centred around 150K and then begins to decrease. This luminescence should be characteristic of the quantum wire emission and indeed the monochromatic image, obtained using this peak, confirms that

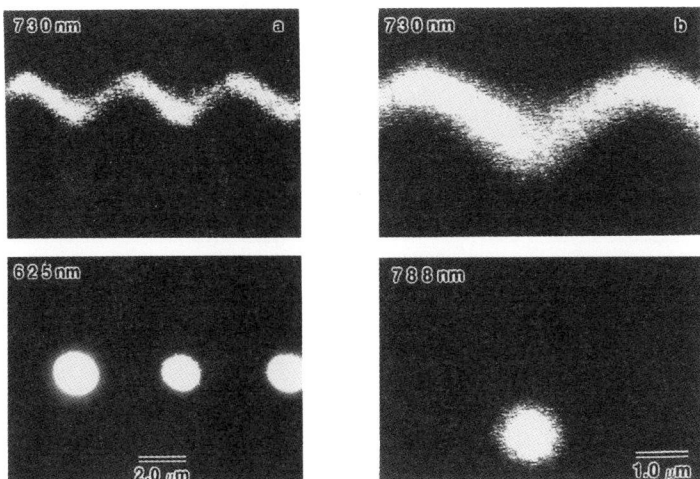

Fig. 3 Cross-section monochromatic images of a 4µm pitch 'V' groove structure showing the vertical AlGaAs quantum well and side wall GaAs quantum well in (a) and the quantum wire and side wall quantum well in (b).

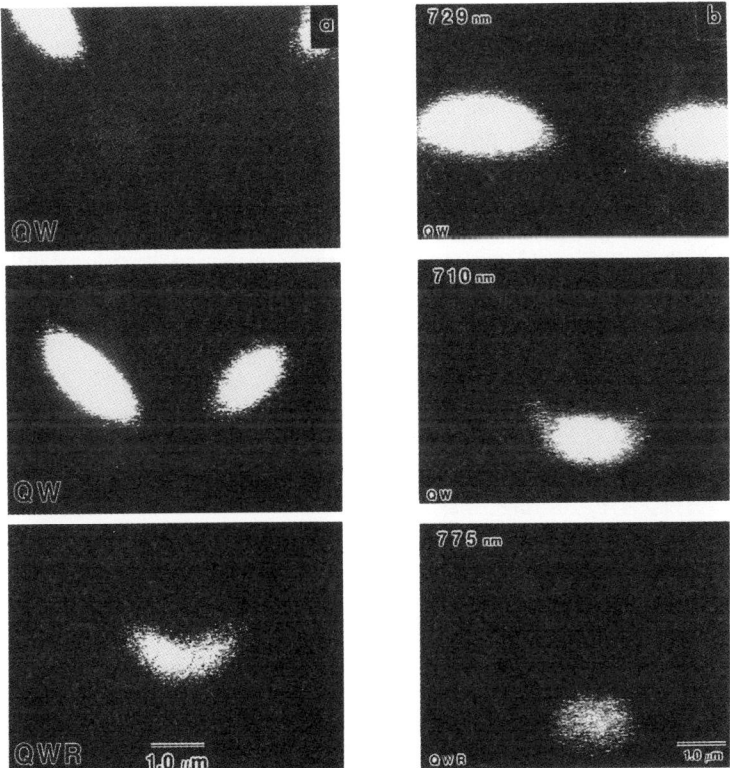

Fig. 4 Monochromatic image sets from two different samples (a) demonstrating non-uniform side wall emission and (b) enhanced Ga migration to the base of the 'V'.

660

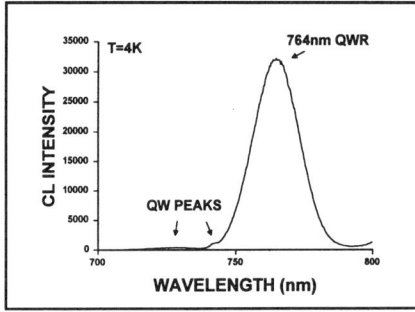

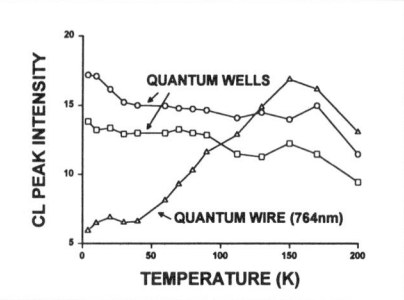

Fig. 5 Cross-sectional CL spectrum from base of single 'V' groove.

Fig. 6 Quantum well and quantum wire CL temperature dependence

this luminescence originates from the base of each 'V' groove. It is believed that as the temperature rises the number of carriers injected into the wire from the adjacent quantum wells increases, leading to an increase in the quantum wire luminescent intensity (Walther 1992 and Lee 1993). This observation is specific evidence for the presence of a one dimensional quantum wire structure.

4. CONCLUSIONS

CL has proved invaluable in the assessment of GaAs / AlGaAs quantum wire structures and the development of an optimised growth process. CL plan view images confirmed the uniformity problems associated with off axis growth and for growth at lower temperatures. Decreasing the pitch of the 'V' groove was, however, found to be beneficial as the wire luminescence peak was observed from CL to shift in relation to the surrounding wells, consequently making it more easily resolvable from the quantum well peaks. This shift suggests that the actual size of the wire formed had increased. The reason for this is believed to be related to the diffusion distance of the Ga atoms in relation to the reduced distance to the base of the 'V'. Low temperature scanning CL has also played an important role and variable temperature studies have revealed luminescence behaviour specific to quantum wire structures.

REFERENCES

Christen J, Kapon E, Colas E, Hwang DM, Schiavone L M, Grundmann M and Bimberg D 1992 Surface Science, **267,** 257
Gustafsson A, Dwir B, Reinhardt F, Biasiol G, Bonard J-M and Kapon E 1995 Inst. Phys. Conf. Ser. **146,** 375
Kapon E 1992 Proc. IEEE **80,** 398
Kapon E 1993 Optoelectronics **8,** 429
Kash K 1990 J. of Luminescence **46,** 69
Lee M S, Kim Y, Kim M S, Kim S, Kim S K, Kim Y D and Nahm S 1993 Appl. Phys. Lett. **63,** 3052
Walther M, Kapon E, Christen J, Hwang D M and Bhat R 1992 Appl. Phys. Lett. **60,** 521
Williams G M, Cullis A G, Sotomayor-Torres C M, Thoms S, Beaumont S P, Stanley C R, Looten D and Van Daele P 1991 Inst. Phys. Conf. Ser. **117,** 695

Inst. Phys. Conf. Ser. No 157
Paper presented at Microsc. Semicond. Mater. Conf., Oxford, 7–10 April 1997
© 1997 IOP Publishing Ltd

Cathodoluminescence studies of striated ZnS platelets and related II-VI crystals

Y Brada, D B Holt* and S Mardix+

Racah Institute of Physics, Hebrew University, Jerusalem, Israel
* Department of Materials, Imperial College of Science, Technology and Medicine, London SW7 2BP, U.K.
+Department of Electrical Engineering, University of Rhode Island, Kingston, RI 02881, U.S.A.

ABSTRACT: Many striated ZnS platelets and other types of II-VI crystals, produced by vapour phase techniques, were studied. Areas in the ZnS crystals exhibited strong exciton emission indicative of the highest crystalline quality and purity while adjacent areas were activated with a variety of impurity centres. Local concentrations of Cu recombination centres at striation interfaces were seen suggesting segregation to the dislocations there. The variation of CL intensity with beam current and voltage were measured and the results are discussed.

1. INTRODUCTION

For many years the best ZnS was that in uniform areas in platelet crystals forming on whiskers in vapour phase growth. These platelets contain striations i.e. narrow polytype bands formed by a screw-dislocation-based martensitic mechanism (for a review see Steinberger 1983). These striated platelets have a number of unique electrical and optical properties including the anomalous photovoltaic effect (APE) i.e. the generation of tens or hundreds of volts which is much larger than the band gap limit (see Holt and Brada 1997).

The wide direct band gap II-VI materials have long been important in optoelectronics. ZnS and CdS are among the most important phosphors for colour TV, monitor and certain new technology flat panel display screens so their CL properties are important. The unique resolution and spectroscopic capability of SEM CL is invaluable for the CL of striated ZnS (Yacobi et al 1977, Datta et al 1977, 1979) and this paper reports and discusses the results of new CL studies.

2. EXPERIMENTAL METHODS

A JEOL 840A SEM fitted with a spectroscopic CL detection system described elsewhere (Saba and Holt 1983) was used. Most of the crystals were from a large set grown in the Racah Institute of Physics of the Hebrew University, Jerusalem. The polytypic ZnS crystals were grown from the vapour phase in either H_2S (these exhibit the anomalous photovoltaic effect) or argon + HCl (these do not). Similar growth with much tighter temperature control at the University of Rhode Island gave platelets with large perfect areas of single polytypes. The Jerusalem crystals also differed in the suppliers of the starting material

662

(the firms Baker, Koch Light and BDH - British Drug Houses). In addition CL spectra were recorded for Jerusalem-grown crystals of CdS and ZnSSe.

3 RESULTS

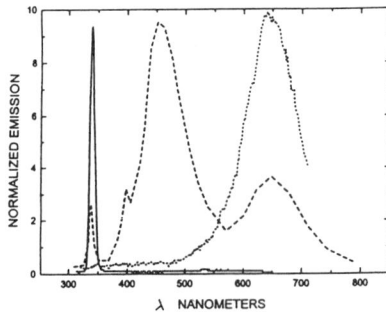

Figure 1. Normalized (to equal peak emission intensities) CL spectra at 293K for (solid curve) crystal 230B grown from Baker material in H_2S, (dashed curve) crystal KL3 grown from Koch Light material in H_2S and (dotted curve) crystal 163B grown from Baker material in argon and HCl gas.

Many regions in the striated ZnS samples exhibited the strong narrow exciton CL bands characteristic of the best ZnS. Exciton lines can be seen in Figure 1 at the left in the spectra of crystals 230B and, less relatively strong, in that of crystal KL3. The spectra of KL3 and 163B, although not intentionally doped, have large extrinsic emission bands at longer (visible) wavelengths. In the spectrum of KL3 the extrinsic emission is far stronger than the exciton emission although ZnS is a direct gap material. It was because dopants could "activate" efficient coloured emission that ZnS became the most important phosphor material for CL screens from the first development of cathode ray oscilloscopes.

A wide range of strong extrinsic emission bands appeared in crystals all grown without intentional doping and the spectra varied between crystals grown in H_2S and those grown in argon and HCl from the same starting material. Over short distances in any one crystal the spectra also varied greatly in ZnS and in CdS crystals.

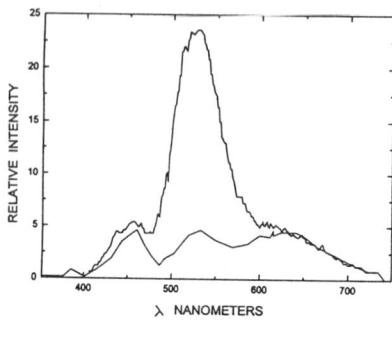

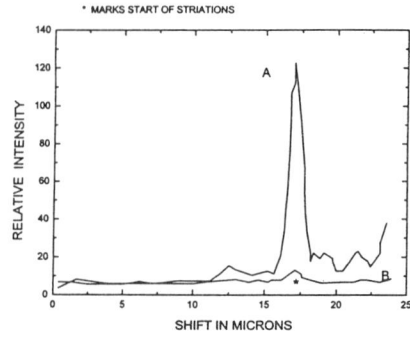

(a)   (b)

Figure 2. (a) Relative intensity spectra of Kingston crystal SM2 at 293K: the heavy, upper curve is that of a faulted "striated" region and the finer, lower spectrum is that for a large homogeneous and defect free polytype region. (b) CL intensity plots for linescans along the c-axis in the same specimen: (curve A) at wavelength 527.5 nm - the peak of the Cu band in (a) and, (curve B) at 455 nm - the peak of the smaller, shorter wavelength peak.

Spectra from one of the Kingston crystals containing large perfect polytype regions are shown in Figure 2a. The large emission band in the spectrum from the striated region is characteristic of Cu activation. The high emission intensity of the Cu-peak wavelength in plot A in Figure 2b at the interface between the broad perfect polytype region and the striated region of many different polytype bands indicates the presence of a local high concentration of Cu-related centres.

The spectra of $ZnS_{0.5}Se_{0.5}$ exhibited emission bands that may be broadened by partly ordered segregation of the constitutents.

The dependence of the intensity of panchromatic CL on beam current, in the range normally used at a constant beam energy (Figure 3) was approximately linear (slightly sub-linear). The variation of the PCL intensity on beam energy, E_0, for constant beam current in the usual range was also measured and found to be nearly linear (slightly super-linear i.e. concave upwards).

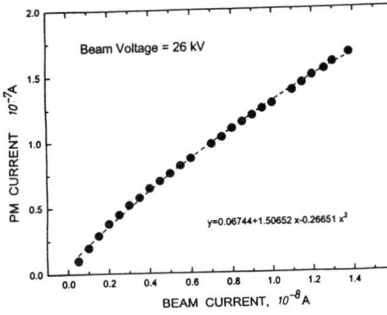

Figure 3. Panchromatic CL (PCL) intensity of Koch-Light H_2S sample KL3 versus beam current at a constant $E_0 = 26$ keV. The curve is the best fit polynomial written below it. (This does not pass through the origin. A small PCL signal is detected for $I_b = 0$ (beam blanked) due to the incadescence of the filament in the electron gun.)

This led us to measure, on the best ZnS i.e. the Kingston crystals, the variation of CL intensity for constant beam energy over the widest range of beam current obtainable (Figure 4). The intensity first increased superlinearly, then (curves A and B) fell rapidly. The much lower intensity of the single wavelength radiation makes the shape of the curve less certain, but again there appeared to be a beam current (power) for maximum CL intensity.

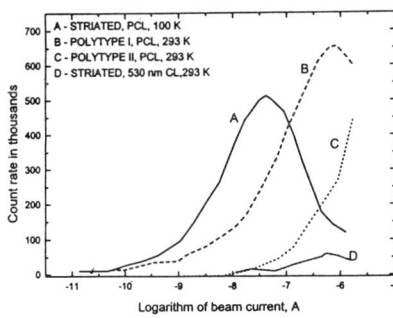

Figure 4. CL intensity variation with beam current from 10^{-11} to $> 10^{-6}$ A, the whole range attainable in our SEM. (A) panchromatic CL variation in a striated region of crystal SM1 at 100K, (B) PCL change in the large polytype region of SM1 at 293K, (C) PCL in the large polytype region in crystal SM2 at 293K and (D) monochromatic (530nm) CL in the striated part of SM2.

4. DISCUSSION

Earlier ZnS platelets studied at Imperial College did not exhibit exciton emission in their spectra so the crystals studied here were purer and more perfect. Large differences were found between those crystals grown from the same starting materials in argon + HCl and in H_2S. Growth in Ar + HCl yields crystals that are n-type, do not exhibit the anomalous photovoltaic effect but are strongly photoluminescent. Growth in H_2S results in p-type material showing the APE but little or no PL (see e.g. Steinberger 1983). Figure 2 and many other observations showed that while the best regions were uniform and of the highest perfection and purity, large variations in CL properties occured in the striated material over distances of the order of the spatial resolution of SEM CL as they did in the CdS samples. The spatial variation in CL over short distances, the lack of controlled doping and the unknown deviations from stoichiometry mean that no detailed explanations can be offered. Instead, we can better appreciate the technological achievement of the industrial production of ZnS and CdS phosphors with reproducible colours and efficiencies.

A local concentration of Cu-activated recombination centres is indicated in Figure 2b. During growth, all the material was in the high-temperature-preferred hexagonal wurtzite form so the interface where the Cu centres concentrate did not exist. The concentration must, therefore, have occurred during cooling, during or after the martensitic transformation when the interface came into existence. This occurs through the movement of partial dislocations that widen stacking faults to change the stacking sequence to that of the polytype. The Cu segregation could result from Cottrell atmospheres dragged by the dislocations to the interface.

The intensity of emitted CL can be written as

$$L_{CL} = (1 - A)\,\eta_{rr}\,\frac{GI_b}{q} = (1 - A)\,\eta_{rr}\,\frac{(1-\chi)E_o}{e_i}\frac{I_b}{q} \tag{1}$$

where A is the fractional loss of photons due to total internal reflection, Fresnel reflection and self-absorption; η_{rr}, the radiative recombination probability, is the fraction of the hole-electron (h-e) pairs that give rise to photons. G, the number of h-e pairs per incident electron depends on the beam energy, E_0, the average fractional loss of beam energy/electron due to back-scattering and emission, χ, and the pair formation energy, e_i. I_b is the beam current and q the charge on an electron. Hence if η_{rr} and the other parameters in equation (1) remain constant, the intensity should vary directly with E_0 for I_b constant and linearly with I_b for E_0 constant. Figure 4 and the other observations reported with it show that this is approximately true for a range of small beam currents. For the full range of possible beam currents (Figure 5 curves A and B) the intensity first increases superlinearly with I_b, which must be due to the onset of stimulated emission with increasing densities of hole electron pairs in the generation volume, which remains constant for constant beam energy. The decrease in intensity at the highest beam currents in Figure 5 can be explained as due to local heating leading to sharp falls in η_{rr}.

REFERENCES

Datta S, Yacobi BG and Holt DB 1977 J. Mater. Sci. **12**, 2411 - 2420
Datta S, Yacobi BG and Holt DB 1979 J. Luminescence **21**, 53 - 73
Holt DB and Brada Y 1997 in Microscopy of Semiconducting Materials 1997 Conf.
    Series No.157(Inst. Phys.: Bristol) to be published
Holt DB and Culpan M 1970 J. Mater. Sci. **5**, 275 - 282
Saba F M and Holt D B 1983 in Microscopy of Semiconducting Materials 1983 Conf.
    Series No. 67 (Inst. Phys.: Bristol) pp. 333 - 336
Steinberger I T 1983 Progress in Crystal Growth and Characterization **7**, 7 - 53
Yacobi BG, Datta S and Holt DB (1977) Phil. Mag. **35**, 145 - 158

Inst. Phys. Conf. Ser. No 157
Paper presented at Microsc. Semicond. Mater. Conf., Oxford, 7–10 April 1997
© *1997 IOP Publishing Ltd*

Cathodoluminescence study of ZnMgSSe/GaAs heterostructures

Q Liu, A Meinert, E Kubalek, H Kalisch*, M Heuken* and H Lakner

Werkstoffe der Elektrotechnik, Gerhard-Mercator-Universität Duisburg, Bismarckstr. 81, 47048 Duisburg, Germany

*Institut für Halbleitertechnik, RWTH Aachen, Templergraben 55, 52056 Aachen, Germany

ABSTRACT: The structural and electro-optical properties of low-pressure MOVPE grown $Zn_{1-x}Mg_xS_ySe_{1-y}$ layers on GaAs substrates have been investigated by cathodo-luminescence (CL) combined with secondary electron (SE) imaging. We found inhomogeneous emission intensity on panchromatic CL micrographs and wavy surfaces on SE images. The degree and the spatial distribution of the inhomogeneities strongly depend on the Mg and S contents and the thickness of the quaternary layers. Electron beam induced current (EBIC) measurements prove the compositional inhomogeneity of the epilayers. The density of dislocations and other defects mainly depends on layer thickness and seems to be related to a broad CL emission at about 535-540 nm.

1. INTRODUCTION

$Zn_{1-x}Mg_xS_ySe_{1-y}$ is one of the most promising large band-gap semiconductor materials for the fabrication of blue-green laser diodes, since it can be grown fully lattice-matched to GaAs substrate with an energy gap tunable from 2.8 eV up to 4.5 eV (Okuyama *et al.* 1993). Laser diodes fabricated with the $Zn_{1-x}Mg_xS_ySe_{1-y}$/ZnSe material system have achieved continuous-wave operation at room temperature, for example by Grillo *et al.* (1994), and Nakayama *et al.* (1993). However, the material quality tends to degrade as the S and Mg contents increase. The crystalline defects have deleterious effects on the optical and electronic properties and are responsible for the rapid device degradation. In addition, phase separation with a quasi-periodic variation in S and Mg concentrations along the [100] or [011] direction was also observed by means of transmission electron microscopy (TEM) and x-ray microanalysis (Nakayama *et al.* 1993; Kuo *et al.* 1994; and Hua *et al.* 1994). The phase separation was supposed to occur via non-uniform incorporation of different atom species into the growth plane of the epilayer, and develops on the growing surface during the growth process. There are only a few contributions on the phase separation phenomenon in $Zn_{1-x}Mg_xS_ySe_{1-y}$. Uniformity and defect detection was demonstrated to be one of the most useful and practical microcharacterization capabilities of CL and scanning electron microscopy (SEM). In this letter, we report on a microstructural study including compositional inhomogeneities and defects in $Zn_{1-x}Mg_xS_ySe_{1-y}$/ZnSe heterostructures by the CL-SEM technique.

2. EXPERIMENTAL DETAILS

The $Zn_{1-x}Mg_xS_ySe_{1-y}$/ZnSe heterostructures studied here were grown in a horizontal low pressure MOVPE apparatus at 400 hPa and 330°C on exactly {001}-oriented GaAs substrates. Dimethylzinc (triethylamine adduct) (DMZn(TEN)) and ditertiarybutylselenium (DTBSe) were used as zinc and selenium precursors, respectively. Ditertiarybutylsulphur (DTBS) and bismethylcyclopentadienylmagnesium ((MeCp)$_2$Mg) were used as sulphur and magnesium sources. The growth rates of $Zn_{1-x}Mg_xS_ySe_{1-y}$ and ZnSe were approximately 1.4 µm/h and 0.8 µm/h, respectively.

Cathodoluminescence was excited by a modified Camscan SEM operating at 5-20 kV. The temperature for measurements is variable from 72 K to 300 K. The luminescence emitted from samples was collected by an ellipsoidal mirror. For panchromatic CL imaging, luminescence was detected by a photomultiplier with a S20 photocathode. For wavelength-dependent CL, the luminescence was focused on the entrance slit of a 275 mm monochromator equipped with three grids giving a resolution of 0.1 nm at $\lambda = 435.8$ nm. The spectrally dispersed luminescence was detected by an optical multichannel analyzer, consisting of a microchannel plate intensifier (EG&G Model 1435) and a CCD camera (EG&G OMA vision). The spectral data from the CCD camera were recorded by a personal computer. EBIC measurements were carried out on samples coated with 60 nm gold as semi-transparent Schottky contacts.

3. RESULTS AND DISCUSSION

In our study on $Zn_{1-x}Mg_xS_ySe_{1-y}$/ZnSe quantum well (QW) structures the Mg and S contents vary in the range of $0.07 \leq x \leq 0.14$ and $0.13 \leq y \leq 0.19$. Fig. 1 shows the panchromatic CL and SE images of three samples with various compositions and thicknesses of ZnMgSSe layers. For each sample the CL and SE images were synchronously recorded. Among different samples all parameters for measurements were kept the same. Inhomogeneous emission intensity on panchromatic CL micrographs and wavy surfaces on SE images were observed on all samples. The size of the intensity inhomogeneity in CL image is on 1 µm scale. Both effects are distinctly dependent on the S and Mg contents. The size of homogeneous domains in CL images becomes larger and the waviness of the surface becomes stronger with increasing Mg composition in the layer. For optimized growth conditions the density of dislocations is low ($\leq 10^3$ cm^{-2}). No direct correlation between the density of dislocations and S or Mg contents can be determined.

To exclude the effects of layer thickness a series of samples of $Zn_{1-x}Mg_xS_ySe_{1-y}$ layers with nominally the same chemical composition but different layer thicknesses was also studied. Inhomogeneous emission intensity on panchromatic CL micrographs and wavy surfaces on SE images were also observed on all samples. While the size of intensity inhomogeneity in the CL images and the wavy surface are very similar among the different samples, the contrast of the intensity inhomogeneity in the CL images becomes stronger with increasing layer thickness. To study the reason of the inhomogeneous emission intensity on panchromatic CL images, EBIC measurements were carried out as a supplement. The comparison between EBIC and CL images of a sample with a 620 nm thick $Zn_{1-x}Mg_xS_ySe_{1-y}$ layer is shown in Fig. 2. Both micrographs were recorded from a specimen coated with 60 nm gold. It is noteworthy that the contrast in the EBIC image is just the inverse of the CL image. Linescans were taken on the corresponding areas of EBIC and CL images (see the lines in Fig. 2) and analyzed for the signal contrasts. In Fig. 3, these data are plotted against the beam position. The inverse behaviour of EBIC and CL intensities are evident. This phenomenon indicates that at the positions with low CL signals, the decrease of lumines-cence is not caused by non-radiative recombination, but is governed by low recombination efficiency of the excited electron-hole pairs, which rather separate and contribute to the EBIC signal. Therefore, the reason for the inhomogeneous emission intensity on panchromatic CL is not non-radiative recombination at defect positions, but local electric fields in the epilayer, which are responsible for the separation of excited electron-hole pairs. Electric fields can be induced from local electric activated centers or inhomogeneous chemical composition in epilayer. Since the samples are nominally undoped and due to the fact that the size of the CL intensity inhomogeneity becomes larger with increasing S and

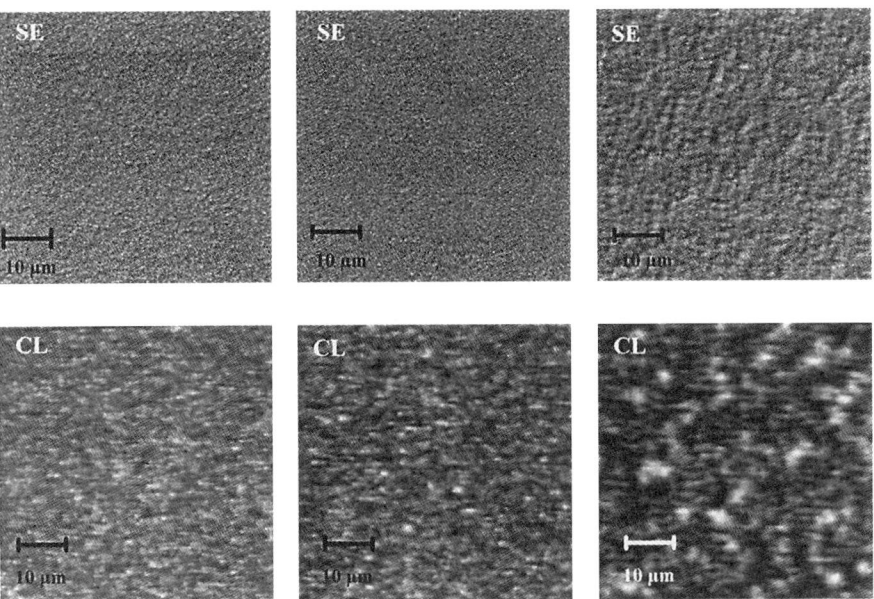

Fig. 1 SE and panchromatic CL image of $Zn_{1-x}Mg_xS_ySe_{1-y}$/ZnSe QW structures recorded at 72 K. (a) x =0.07, y = 0.15; (b)x = 0.09, y = 0.13; (c) x =0.14, y = 0.19.

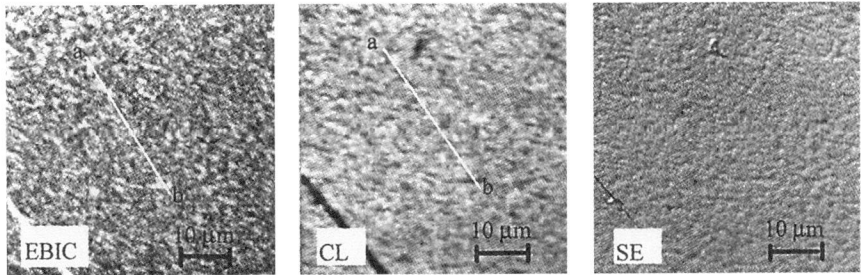

Fig. 2 Comparison of EBIC, CL and SE images of a sample with 610 nm ZnMgSSe layer. All measurements were recorded at room temperature.

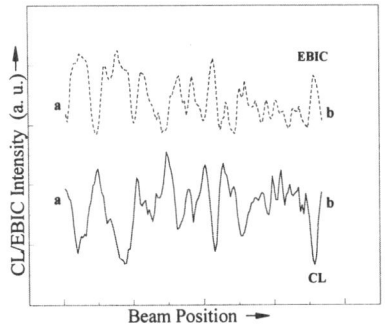

Fig. 3 Linescans recorded along the line positions of the EBIC and CL images shown in Fig. 2.

Mg contents, the CL intensity inhomogeneity is supposed to be caused by inhomogeneous chemical composition which results in different band gaps on a micrometer scale.

The panchromatic CL images with low magnifications show that there is a relatively high density of dislocations in the sample series with nominally the same chemical composition but different layer thickness. The density of dislocations decreases with increasing layer thickness, as shown in Fig. 4. The spectral CL is dominated by the band-edge emission. Additionally, a wide CL peak located at about 535-540 nm was observed, which seems to be related to dislocations as this emission appears with significant intensity only in samples with higher dislocation densities. Moreover, in the sample regions which exhibit very low dislocation densities, the 535-540 nm peak was not observed.

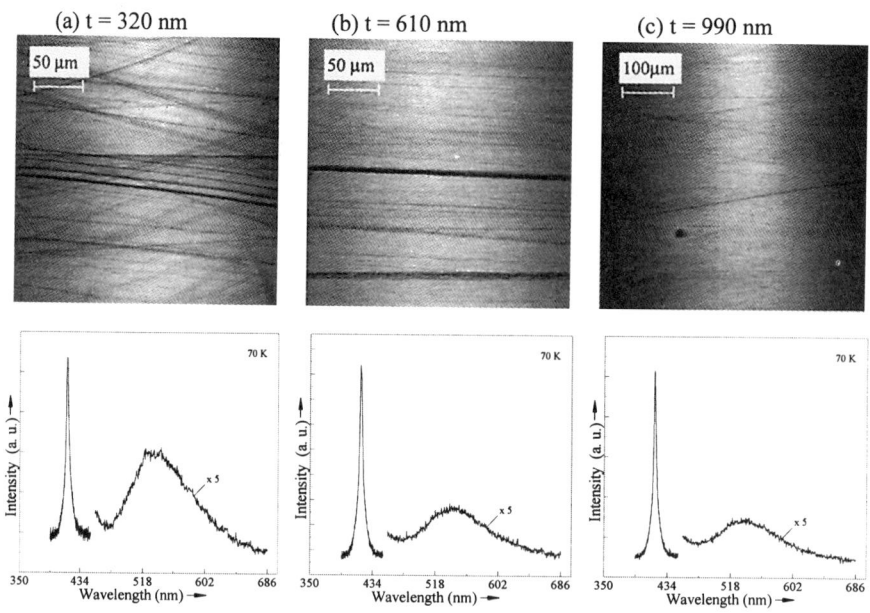

Fig. 4 Panchromatic and spectral CL of ZnMgSSe layers with various thicknesses. (a) t = 320 nm; (b) t = 610 nm; (c) t = 990 nm.

ACKNOWLEDGEMENT

This work was financially supported by the Volkswagen-Stiftung.

REFERENCES

Grillo D C, Han J, Ringle M, Hua G, Gunshor R L, Pelkar P, Kozlov V, Jeon H and Nurmikko A V 1994 Electron. Lett. **30**, 2131

Hua G C, Otsuka N, Grillo D C, Han J, He L and Gunshor R L 1994 J. Crystal Growth **138**, 367

Kuo L H, Salamanca-Riba L, Wu B J, DePuydt J M, Haugen G M, Cheng H, Guha S and Haase M A 1994 Appl. Phys. Lett. **65**, 1230

Nakayama N, Itoth S, Ohata T, Nakano K, Okuyama H, Ozawa M, Ishibashi A, Ikeda M and Mori Y 1993 Electron. Lett. **29**, 1488

Okuyama H, Morinaga Y and Akimoto K 1993 J. Crystal Growth **127**, 335

Inst. Phys. Conf. Ser. No 157
Paper presented at Microsc. Semicond. Mater. Conf., Oxford, 7–10 April 1997
© 1997 IOP Publishing Ltd

Electric field dependence of the lateral cathodoluminescence intensity and electron-beam induced current distribution in a GaAs-AlAs single quantum well

U Jahn, J Menniger, H Kostial, R Hey and H T Grahn

Paul-Drude-Institut für Festkörperelektronik, Hausvogteiplatz 5-7, 10117 Berlin, Germany

ABSTRACT: Lateral variations of the vertical electronic transport through a thin GaAs-AlAs quantum well (QW) were imaged by cathodoluminescence (CL) and electron-beam induced current (EBIC) using a p-n$^+$ junction deposited beneath the QW. The comparison of the spectral and spatial CL intensity distribution with the EBIC pattern as a function of the bias U_b shows that for particular values of U_b, the EBIC images reflect random potential fluctuations of the QW. In the present case, these fluctuations are attributed to an inhomogeneous distribution of the Be dopant.

1. INTRODUCTION

Electron-beam induced current (EBIC) techniques were successfully applied to investigate degradation effects or extended defects such as misfit dislocations in semiconductor hetero-structures (Jacubowicz et al 1993, Brown and Humphreys 1995). Cathodoluminescence was found to be an excellent complementary method for the detailed characterization of devices based on hetero-layer systems (Lin et al 1996). In addition to extended defects, properties of the interfaces, e.g., roughness or composition variations resulting in randomly fluctuating potentials, affect the electrical and optical behaviour of thin film systems such as quantum wells (QW). The random potential of a narrow (< 10nm) QW leads to a bright/dark pattern in monochromatic CL images, where the lateral separation of the bright or dark spots is essentially determined by the spatial resolution of the CL method (Runge et al 1995). The modulation depth of the CL intensity and the spectral shape, however, contain information about the origin of the imaged potential fluctuations. In a theoretical study, Munnix and Bimberg (1988) predicted that interface grading and quality of a hetero-junction should also have a significant influence on the EBIC signal. So far, the experimental work has been focused on EBIC line scans on cleaved edges along the growth direction (Ando et al 1992).

In this work, we investigate the influence of a narrow single QW centered in a GaAs/AlAs multilayer structure on the lateral EBIC distribution (i.e., in the plane of the QW) as a function of an applied bias. The EBIC signal is generated in the depletion region of a p-n$^+$ junction beneath the QW. The resulting pattern is compared with the respective CL distribution, which serves as a complementary method for the interpretation of the EBIC results.

2. EXPERIMENTAL

Fig. 1 illustrates the layer sequence of the investigated sample fabricated by molecular beam epitaxy. The top and the bottom barriers are short period superlattices (SPSL) consisting of 32 periods of 2.4 and 1.1nm thick GaAs and AlAs layers, respectively. The whole layer structure is Be doped, whereas the substrate as well as the buffer layer are highly Si doped forming a p-n$^+$ junction about 200nm beneath the QW. The doping concentrations are chosen in such a way that the depletion region crosses the QW at a certain value of U_b. In order to allow for CL and EBIC investigations within the same sample region, we used an AuBe ring as ohmic contact on top of a 230μm diameter mesa. For the ohmic contact on the substrate side we used AuGeNi.

CL and EBIC investigations were performed as a function of U_b in a scanning electron microscope (DSM 962) equipped with an Oxford mono-CL and cooling stage system. CL

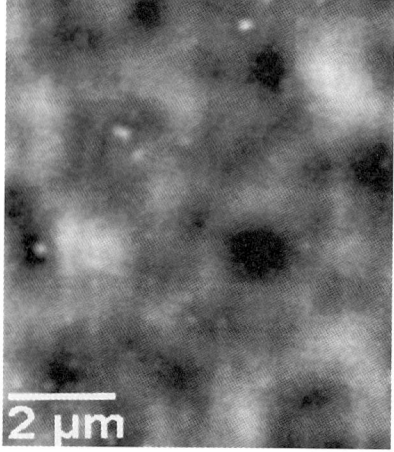

AuBe contact		
GaAs-cap	1 x 10^{17} cm^{-3} Be	20 nm
SPSL	1 x 10^{17} cm^{-3} Be	112 nm
AlAs		1.2 nm
GaAs		2.4 nm
AlAs		1.2 nm
GaAs-quantum well		3.5 nm
AlAs		1.2 nm
GaAs		2.4 nm
AlAs		1.2 nm
SPSL	1 x 10^{17} cm^{-3} Be	112 nm
GaAs	1 x 10^{17} cm^{-3} Be	100 nm
GaAs	2 x 10^{18} cm^{-3} Si	500 nm

b_t
b_b
U_b
depletion region
⇐ p-n$^+$ junction

n$^+$ - GaAs - substrate

Fig. 1. Layer sequence of the QW sample. SPSL, b_t (b_b), and U_b denote short period superlattices, top (bottom) barrier and the reverse bias voltage, respectively.

spectra, spectrally resolved CL micrographs, EBIC images as well as EBIC and CL line profiles were recorded at 5K. The electron-beam energy amounted to 10keV and the current was usually 1–10nA. A grating monochromator and a cooled photomultiplier were used in conjunction with conventional photon counting technique to disperse and detect the CL signal, respectively.

3. RESULTS AND DISCUSSION

Fig. 2 shows a typical EBIC image of our layer structure at 5K. It exhibits an irregular bright/dark pattern, which is also observed for p-i-n and even for a single n-n GaAs/AlGaAs hetero-junction. Moreover, the EBIC pattern resembles closely the CL intensity distributions of narrow QWs obtained for a fixed photon energy (Runge et al 1995, Jahn et al 1995). The EBIC contrast and the respective lateral distribution varies as a function of the applied bias U_b.

In Fig. 3, the squares represent the EBIC contrast $\delta I_{EBIC} / I_{EBIC}$ of such a pattern determined from line profiles of the EBIC intensity I_{EBIC}. The triangles mark the U_b dependence of

2 μm

Fig. 2. EBIC micrograph of the QW structure for a bias of U_b = -6V at 5K.

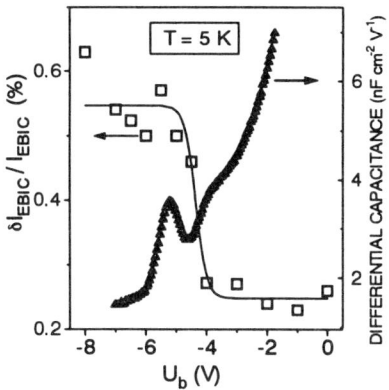

Fig. 3. EBIC contrast (squares) and differential capacitance (triangles) of the quantum well structure vs bias U_b.

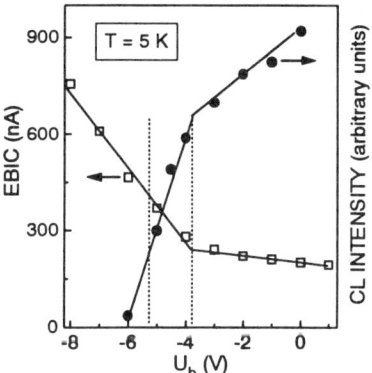

Fig. 4. EBIC (squares) and integrated CL intensity (circles) of the QW structure vs bias U_b.

Spectrally and spatially resolved CL is an appropriate method in order to clarify the nature of the fluctuations reflected by the EBIC pattern for $|U_b| > 4.5V$. Especially for narrow QWs it is very sensitive to layer parameter variations (Jahn et al 1995). Indeed, CL images of the QW show a pattern resembling that of the EBIC distribution. In Fig. 5, EBIC and CL images obtained from the same region are compared with each other. The bias amounts to $U_b = -5.5V$. Both images show nearly identical

the differential capacitance. With increasing reverse bias U_b, $\delta I_{EBIC} / I_{EBIC}$ remains almost constant until about -4V, where it increases abruptly by a factor of two. This step of the EBIC contrast is correlated with the appearance of a minimum in the differential capacitance at about -4.5V, which is due to the intersection of the edge of the depletion region with the QW (cf Fig. 1). Therefore, we conclude that for $|U_b| > 4V$ the EBIC contrast is remarkably influenced by the properties of the QW. A further indication for the strong influence of the QW on the EBIC signal can be derived from Fig. 4, where the bias dependence of the EBIC signal is compared with the one of the integrated CL intensity. While the latter originates exclusively from the narrow QW, the former contains contributions from different parts of the layer package depending on the value of U_b. Consequently, for $|U_b| > 4V$ the steeper decrease of the CL intensity indicates that the electric field of the depletion region has reached the QW leading to a rapid quenching of the CL. At the same time, the EBIC signal increases with a much larger slope than for $|U_b| < 4V$, which is obviously due to a large contribution of the QW to the EBIC signal, when the depletion region contains the well layer.

From the results illustrated in Figs. 3 and 4, we conclude that for $|U_b| > 4V$ the EBIC pattern reflects the lateral distribution of vertical electronic transport properties of the QW. The influence of the QW on the EBIC signal is attributed to the fact that it acts as a carrier reservoir, which contributes strongly to the transport, when the depletion region of the p-n$^+$ junction crosses the well layer. The lateral homogeneity of the vertical transport through the QW is affected by potential fluctuations due to well thickness, barrier height, strain, doping variations or interfacial defects.

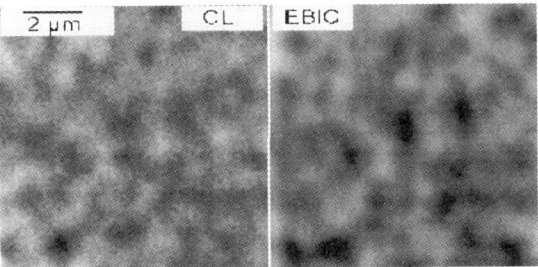

Fig. 5. EBIC and CL micrographs of the same region of the QW structure for a bias of $U_b = -5.5V$ at 5K. The detection energy for the CL image was 1.745eV.

672

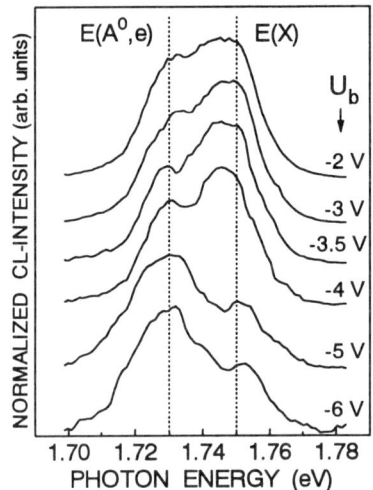

Fig. 6. CL spectra of the QW vs bias U_b at 5K. $E(A^0, e)$ and $E(X)$ are the energies of the free-to-acceptor and exciton transition, respectively.

bright/dark patterns confirming that the lateral variations of both signals have the same origin. Therefore, conclusions derived from the CL distribution can be used to support the interpretation of the respective EBIC images. Further experiments and theoretical considerations are necessary to understand the mechanism leading to an almost identical EBIC and CL intensity contrast. The electric field dependence of the spectral and spatial CL intensity distribution provides first indications for the origin of these patterns.

Fig. 6 shows the CL spectra as a function of U_b obtained by spot excitation at 5K. Two rather broad lines can be resolved, centered at about 1.73 and 1.75eV. We assign the low- and high-energy line denoted $E(A^0, e)$ and $E(X)$ to the free-to-bound transition involving the neutral acceptor and to the probably bound exciton transition. A candidate for the acceptor is obviously Be, which can migrate from the bottom barrier into the well layer during the growth. This assumption is confirmed by the energy separation ΔE between the two lines of about 20meV in good agreement with $\Delta E \approx 25$meV found by Holtz et al (1990) for the Be related (A^0, e) and free exciton transitions. CL images taken at the energies $E(A^0, e)$ and $E(X)$ show complementary bright/dark patterns indicating that the CL and therefore also the EBIC intensity distribution is connected with the impurity distribution within the QW. It follows from Fig. 6 that the change of the spectral distribution with U_b is responsible for the variation of the CL pattern for different values of the bias.

In conclusion, EBIC in connection with a p-n$^+$ junction, which is situated beneath a QW structure, is suitable to image the lateral homogeneity of the vertical electronic transport through the QW. Spatially and spectrally resolved CL serves as a complementary method to study the nature of the fluctuations reflected by the EBIC distribution. For values of the applied bias, for which the edge of the depletion region moves through the QW, the spatial EBIC and CL distributions show corresponding patterns. In the present case, this pattern is attributed to an inhomogeneous distribution of the Be dopant.

REFERENCES

Ando K, Palmer J, Ohki A, Tohno S and Zembutsu S 1992 J. Crystal Growth **117**, 560
Brown P D and Humphreys C J 1995 Proc. Int. Conf. on Microscopy of Semiconducting Materials IX, eds A G Cullis and A E Staton-Bevan (Bristol and Philadelphia: Inst. Phys. Publ.) pp 701-4
Holtz P O, Sundaram M, Mertz J L and Gossard A C 1990 Phys. Rev. **B41**, 1489
Jacubowicz A, Oosenburg A, Forster T 1993 Appl. Phys. Lett. **63**, 1185
Jahn U, Menniger J, Kwok S H, Runge E, Fujiwara K, Hey R and Grahn H T 1995 phys. stat. sol. (a) **150**, 439
Lin H T, Rich D H, Sjölund O, Ghisoni M and Larsson A 1996 J. Appl. Phys. **79**, 8015
Munnix S and Bimberg D 1988 J. Appl. Phys. **64**, 2505
Runge E, Menniger J, Jahn U, Hey R and Grahn H T 1995 Phys. Rev. **B52**, 12207

Inst. Phys. Conf. Ser. No 157
Paper presented at Microsc. Semicond. Mater. Conf., Oxford, 7–10 April 1997
© *1997 IOP Publishing Ltd*

SEM-CL of high quality polycrystalline CVD and high pressure synthetic diamond

S J Sharp and A T Collins

Department of Physics, King's College London, Strand, London WC2R 2LS, UK.

ABSTRACT: The free-exciton luminescence spectrum of diamond has been studied in detail using both SEM-CL and a lower spatial resolution cathodoluminescence technique. In polycrystalline CVD diamond the free-exciton luminescence is confined to the crystallites whereas a broad luminescence band believed to be associated with the presence of dislocations, is centred on the grain boundaries. The free-exciton luminescence intensity at constant beam current is found to increase with beam accelerating voltage V in the range 15 to 50 kV with a V^n dependence, where n may vary as $3 \le n \le 5$.

1. INTRODUCTION

Free-exciton (FE) luminescence is rarely seen in natural diamond and it has only recently been seen in synthetic diamond in any great intensity (Dean et al. 1965). It is easily quenched by the presence of defects and can therefore be a useful indicator of the quality of the material (Sharp et al. 1995). The rapid development of growth techniques in recent years has produced higher quality synthetic diamonds with lower impurity concentrations, renewing interest in the FE recombination luminescence. Advances of particular importance have been the use of getters for the removal of nitrogen in high pressure synthesis (HPS) and the low pressure chemical vapour deposition (CVD) of diamond. CVD diamond is typically low in nitrogen, whereas nitrogen is the dominant impurity in HPS and natural diamond and a constituent of most of their radiative defects.

The only feature in the intrinsic diamond cathodoluminescence (CL) spectrum should be FE luminescence. Studies of single crystals of CVD diamond show that these do indeed exhibit strong FE luminescence (Kawarada et al. 1993, Robins et al. 1993). However, the cathodoluminescence spectrum of polycrystalline CVD diamond is typically dominated by a broad luminescence band that peaks at around 3.0 eV and is believed to be associated with dislocations. There are a number of broad luminescence bands seen in natural, HPS and CVD diamonds which peak at around the same energy but have slightly differing spectra. The term "Band A" historically referred to a broad luminescence band seen in natural diamond but it is often used as a generic term to describe all these bands. Higgs et al. (1995) reported that the Band A in diamond increases at constant excitation power as the beam is defocused due to a saturation of the defects available for excitation. This was also the case for defect-related luminescence in some other wide band-gap cubic materials.

2. EXPERIMENTAL

We have examined the FE cathodoluminescence spectrum of a number of high quality HPS and CVD diamonds. The CL was generated either using a JEOL 35C SEM, or a custom built apparatus in which the beam spot is typically a few 100 µm in diameter. For measurements both in the visible and UV region of the spectrum a 9789 EMI photomultiplier with quartz window was employed. For measurements only in the UV region of the spectrum a Hamamatsu solar-blind photomultiplier type R1657 was used. All measurements were made at 77 K.

3. RESULTS AND DISCUSSION

Figure 1. (a) A secondary electron image of polycrystalline diamond. (b) CL image of the FE luminescence. (c) Band A luminescence.

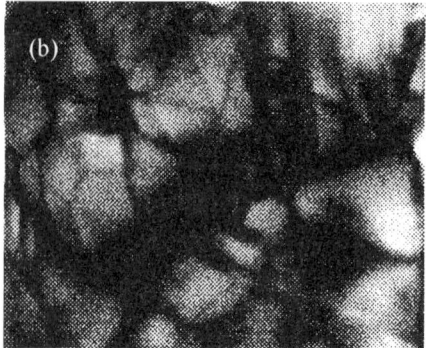

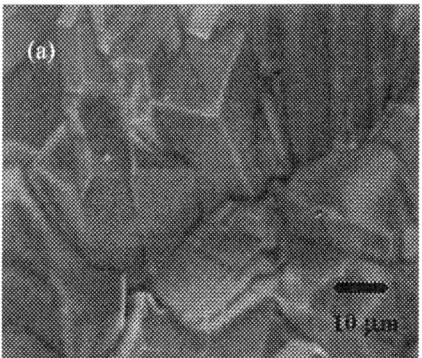

The spatial distribution of FE and Band A luminescence and the secondary electron image for a polycrystalline CVD diamond are shown in Figure 1. The Band A luminescence is primarily centred on the grain boundaries. The FE luminescence is quenched at the grain boundaries whereas at the centre of the crystallites the intensity of the FE luminescence is comparable with that obtained from a high quality HPS diamond.

When electron scattering processes are taken into account in diamond most of the energy of the electron beam is dissipated over a region between $(0.007$ to $0.014)V^{1.825}$ µm below the surface, where V is the electron beam acceleration voltage in kV (Davies, 1979). Hence as the acceleration voltage is increased there is an increase in the both the range of penetration d_d and average penetration depth D.

In a perfect crystal, if all the energy of the incident electrons was lost in ionisation and the formation of excitons, the FE luminescence intensity would simply be proportional to the electron beam energy. The variation in the intensity of FE luminescence as a function of V at a constant beam current in the SEM is shown in Figure 2. The FE intensity exhibits a V^n dependence in the range 15 to 40 kV where $n = (5 \pm 1)$.

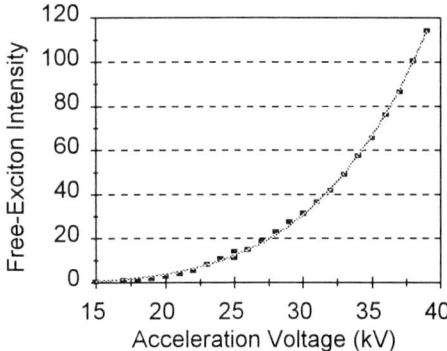

Figure 2. The FE luminescence as a function of electron beam accelerating voltage V in the SEM at a constant beam current; $I = (0.8 \pm 0.1)\mu A$. The line through experimental data shows a fit to V^n dependence where $n = (5 \pm 1)$.

In the SEM, Higgs et al. (1995) noted that the intensity of the Band A luminescence reached a saturation when the beam was tightly focused. Using the electron gun with a ~300 μm spot size we have found that as the beam current is increased the FE and Band A luminescence increases linearly with current showing that under these conditions the Band A recombination is not saturated. However, as the spot was defocused whereas the intensity of the Band A emission remained constant the FE emission decreased rapidly. The intensity of the FE emission when the beam current and spot size were both kept constant, increased super-linearly with voltage in the range 15 to 50 kV although the index was lower than in the SEM, $n = (3.0 \pm 0.5)$. The intensity of the Band A emission initially increased super-linearly with voltage but this behaviour then saturated until the Band A emission increased approximately linearly with V.

The values of n were found to be the same for all high quality samples irrespective of growth method and type of surface i.e. polished or as grown. A similar dependence of CL intensity on beam acceleration voltage has been reported for other semiconductors where n is usually $1 \le n \le 2$ (Holt, 1989).

These findings indicate that there are non-radiative recombination channels, and that these affect the Band A and FE processes in different ways. The Band A and FE luminescence are emitted from different regions of the crystal, hence the effect of the non-radiative processes on these can be considered separately. In regions where the FE emission is weak the excitons recombine via capture at a radiative or non-radiative defects where these only become saturated under higher beam-current densities in the SEM. In the regions of strong FE emission all the defect-related capture processes will be saturated and hence the intensity of the FE emission depends critically on the focus. In order to explain why the CL has a super-linear V dependence either the concentration of non-radiative traps must be depth dependent, or more probably there is a non-radiative loss associated with surface recombination. At low accelerating voltages, producing smaller penetration depths, the non-radiative loss to the surface dominates whereas at greater penetration depths the surface becomes progressively less important.

A simple model is to assume the non-radiative loss is due to exciton diffusion to and trapping at the surface. The probability of an exciton reaching the surface P=exp(-D/L), where L is the exciton diffusion length. If there were an infinite number of traps at the surface the radiative luminescence would simply increase as (1–P). However, as a consequence of the aforementioned saturation of non-radiative traps the free-exciton luminescence increases as 1/(1+cP), where c is a constant related to the cross-section for exciton capture. The initial rate of increase of radiative luminescence with increasing penetration depth is faster for shorter exciton diffusion lengths. In regions of strong Band A and negligible FE emission the exciton diffusion length is shorter than in the regions from which the FE emission originates. Hence, the increase in intensity of the Band A emission with depth quickly saturates, presumably at higher acceleration voltages producing greater penetration depths the FE emission would also have saturated.

Qualitatively this model explains the phenomena observed. The higher value of n in the SEM could be caused by a decrease in diffusion length due to scattering processes at higher current densities. However, non-radiative processes usually have very short lifetimes and we would expect the reduction in intensity of the FE and band A emissions at low accelerating voltages to be accompanied by a reduction in the luminescence decay times. We have not yet observed this effect experimentally.

5. CONCLUSIONS

The FE luminescence in high-purity diamond exhibits a V^n dependence on the accelerating voltage V, with $3 \leq n \leq 5$. This finding, combined with measurements of the FE luminescence and Band A luminescence as a function of current, and current density, lead to a model in which non-radiative processes dominate at low beam voltages, where the excited region of the crystal is close to the surface.

ACKNOWLEDGEMENTS

The authors would like to thank De Beers DTC Research Centre, Maidenhead for providing the samples and De Beers Industrial Diamond Division for the PhD sponsorship of SJ Sharp.

REFERENCES

Davies G 1979 in 'The Properties of Diamond' Editor; Field JE. Academic Press 165-181
Dean P J, Lightowlers EC, Wight DR 1965 Phys. Rev. 140 (1A) A352
Higgs V, Lightowlers EC, Collins AT, Mainwood A 1995 Microscopy of Semiconductors Conference Proceedings Editors; Cullis AG, Staton-Bevan AE 146 759
Holt DB and Yacobi BG 1989 in 'SEM Microcharacterisation of Semiconductors' Editors; Holt DB and Joy DC. Academic Press 387-423
Kawarada H, Matsuyama H, Yokota Y, Sogi T, Yamaguchi A, Hiraki A 1993 Phys. Rev. B 47 (7) 3633
Robins LH, Farabaugh EN, Feldman A 1993 Phys. Rev. B 47 (7) 14167
Sharp SJ, Collins AT 1995 MRS Autumn Meeting Conference Proceedings, *Diamond For Electronic Applications* Editors; Dreifus DL, Collins AT, Humphreys T, Das K, Pehrsson PE 416 125 MRS

Inst. Phys. Conf. Ser. No 157
Paper presented at Microsc. Semicond. Mater. Conf., Oxford, 7–10 April 1997
© 1997 IOP Publishing Ltd

Recombination properties of pseudomorphic AlGaAs/InGaAs/GaAs heterostructures: a cathodoluminescence study.

N Fossaert[1], S Dassonneville[1], B Sieber[1] and JL Lorriaux[2]

[1] Laboratoire de Structure et Propriétés de l'Etat Solide, URA CNRS 234, Bâtiment C6, Université des Sciences et Technologies de Lille, 59655 Villeneuve d'Ascq Cédex, FRANCE
[2] IEMN, 59655 Villeneuve d'Ascq Cédex, FRANCE

ABSTRACT: Cathodoluminescence (CL) in the scanning electron microscope was used to study the recombination properties of electron beam created electron-hole pairs in pseudomorphic modulation doped $Al_yGa_{1-y}As/In_xGa_{1-x}As/GaAs$ heterostructures. We compare the results obtained on two heterostructures grown on a semi-insulating substrate and on a n $^+$substrate.

1. INTRODUCTION

The recent optical studies performed by means of photoluminescence experiments on modulation-doped quantum wells (MDQWs) have been devoted either to the measurement of the electron sheet density in the well (see for instance Brierley et al 1991 and Gilperez et al 1994) or to the study of many-body phenomena such as band-gap renormalization (Delalande et al 1987). MDQWs are good candidates for the study of band-to-band radiative recombination since they contain a very low concentration of doping species which are usually unintentionally introduced during growth. In this paper, we compare two samples which mainly differ in the doping type of their substrate. We give results on the beam injection dependence of carrier transfer to the quantum well as well as the luminescence fluctuations we have observed in such modulation-doped quantum wells.

2. EXPERIMENTS

The samples were grown by molecular beam epitaxy on (001)-oriented GaAs substrates which were either S.I. (sample SI) or silicon doped (sample N; n ~$2x10^{18}$ cm^{-3}). An undoped GaAs buffer layer of 0.2 μm (sample SI) and 0.5 μm (sample N) was grown first, followed by an undoped 20nm $In_{0.2}Ga_{0.8}As$ SQW grown at 520°C; an undoped 5 nm $Al_{0.2}Ga_{0.8}As$ spacer was used to isolate the SQW from the 60 nm (sample SI) or 40 nm (sample N) $Al_{0.2}Ga_{0.8}As$ top barrier which was silicon doped at $5x10^{17}$ cm^{-3}. Images and quantitative cathodoluminescence (CL) spectroscopic measurements were recorded at 87 K on a Cambridge 250 MK3 scanning electron microscope (SEM) fitted with a Oxford CL system and a Jobin-Yvon HR10 monochromator. The luminescence was detected by a home-made silicon photodiode device. All spectra were corrected for the system response. The experiments were performed as a function of electron beam voltage E_0 and beam injection.

3. CL SPECTRA AS A FUNCTION OF BEAM VOLTAGE: SAMPLE N.

At an electron beam voltage E_0 lower than 13 keV, and for which the Gruen range is about 0.76 μm (Gruen 1957), the luminescence from sample N originates mainly from the

678

InGaAs SQW (Fig. 1). The GaAs buffer layer to InGaAs SQW luminescence ratio increases with beam voltage. Therefore, in order to study in more detail the carrier transfer and the recombination properties of the SQW, we have chosen an accelerating beam voltage of 20 kV which allows a good and simultaneous detection of the luminescence from both the SQW and the buffer. The same condition has been kept for sample SI to ensure a direct comparison between both samples.

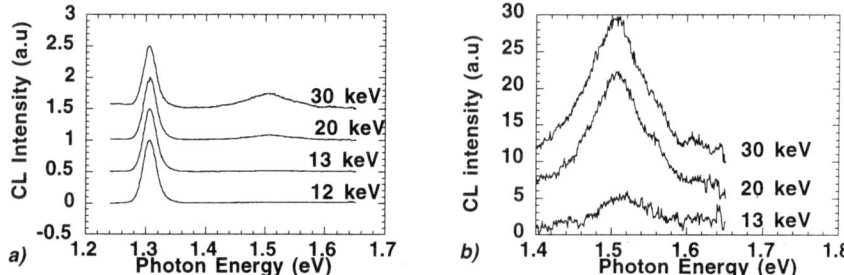

Fig.1: CL spectra of sample N recorded at different beam energies and constant beam power (10^{-5} W). a) overall spectra; and b) enlargement of the GaAs spectra.

4. COMPARISON BETWEEN SAMPLES

4.1 CL spectra

Fig. 2 shows that the 20 keV CL spectra of the samples is very different. First of all, at low beam injection, the SQW CL peak of sample N is blue-shifted by 48 meV with respect to that of sample SI. This could be due to a difference in the indium content x_{In} of the $In_xGa_{1-x}As$ alloy; as a matter of fact, the samples have not been grown at the same time, and indium fluctuation with time can be reasonably assumed. Therefore, the indium concentration would be smaller in sample N. At higher beam injection, the general SQW CL band shape differs even more drastically from one sample to the other. Secondly, the CL band shape of the GaAs buffer layer is, in sample SI, similar to that expected theoretically (fig. 2a); this means that the thermalisation of the excess carriers created by the electron beam is complete. In sample N, the CL band shape of the GaAs buffer layer exhibits two maxima (fig. 2b). The low energy part (until 1.493 eV) of this 'unusual' band corresponds to the luminescence of the

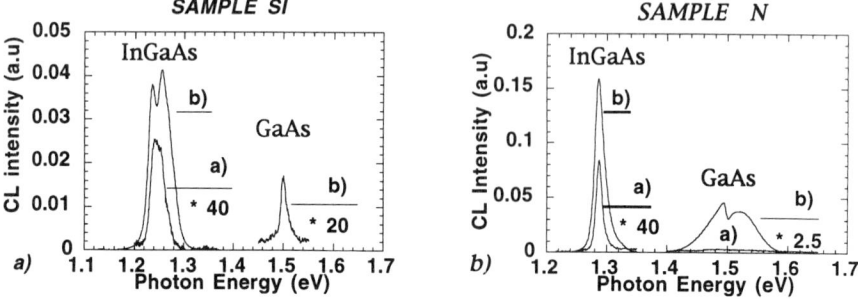

Fig.2: CL spectra recorded at 20 keV at $I_{beam}=1 \times 10^{-9}$ A (curves a) and $I_{beam}=7.5 \times 10^{-8}$ A (curves b); a) Sample SI b) sample N

n+ substrate which is transmitted through the whole structure. The high energy tail is due to the GaAs buffer layer (Cléton et al 1996). The value of the GaAs buffer layer CL peaks corresponds to that expected in sample SI (close to 1.50 eV) whereas it is higher in sample N (> 1.515 eV); furthermore, the high energy slope is, in sample N, less steep than in sample SI. A reasonable explanation is that the sample N buffer layer is doped at a quite high level as a result of contamination during growth. In both samples, there is an electric field in the buffer layer, as a result of Fermi level alignment through the whole structure. At low beam currents, this field allows a very efficient separation of electron-hole pairs generated in the buffer layer, since no luminescence of the buffer is observed. The electrical field strength decreases with beam injection, since the buffer luminescence increases with beam current. Therefore, the collection efficiency of the SQWs can be taken equal to 1 for beam currents lower than 2 nA in sample N and 30 nA in sample SI. A different energy band structure of both samples in the dark, as well as a different variation with beam injection, could explain the discrepancies observed in the beam current dependence of the SQW CL spectrum (fig.2).

4.2 CL imaging

Fig. 3 showsthat dark lines defects (DLDs) parallel to both <110> directions are visible in both samples, but that their density is very different from one sample to the other. The DLDs density being larger in sample SI, this seems to confirm that the indium content is smaller in sample N than in sample SI. In both samples the density of DLDs parallel to the [1$\bar{1}$0] direction is the highest. As a result of the low spatial resolution of the CL technique, each DLD could be made of many parallel dislocations. The cross-slip of these dislocations (Ulhaq-Bouillet et al 1994) is demonstrated in figure 3a (sample SI) for both V(g) and III(g) types dislocations.

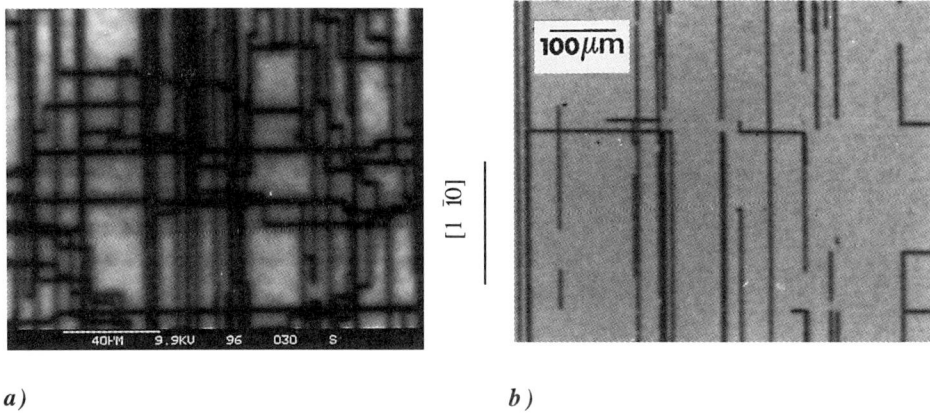

a)   *b)*

Fig.3: 10 keV CL polychromatic images: a) Sample SI; and b) sample N. The DLDs density, which is larger in the [1$\bar{1}$0] direction, is much higher in sample SI than in sample N.

5. SPATIAL LUMINESCENCE FLUCTUATIONS

Spatial variations of the luminescence intensity have been detected in both samples, as shown in fig. 4. In sample N the contrast between two areas can be as high as 50%, whereas it is not so easily detectable in sample SI. In such modulation-doped SQWs, these fluctuations are observed in the free carrier regime and not in the excitonic regime as is the case in square-shaped quantum wells. But, in a similar manner, they can be related to variations of CL spectra (fig. 5). In both samples, a blue-shift of the CL peak, as well as larger full-width at half maximum was found to be correlated with a higher CL intensity (fig. 5). These results show the presence of variations of the thickness of both SQWs or of their indium content.

680

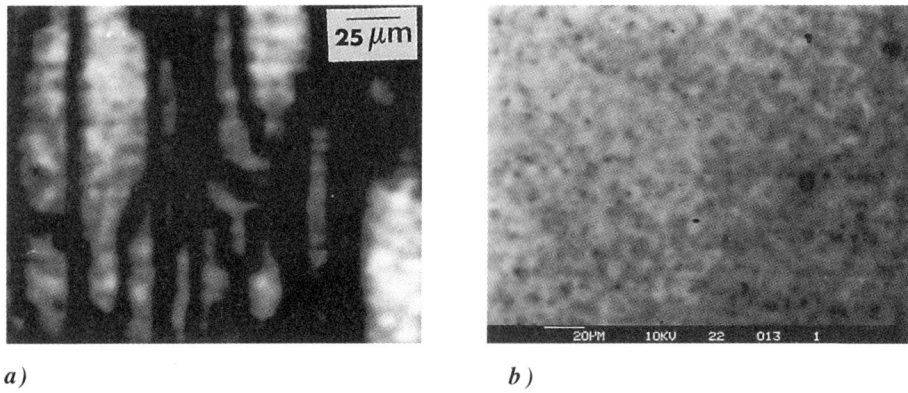

a)   b)

Fig. 4: 20 keV CL polychromatic images of luminescence fluctuations in a) sample SI and b)
sample N.

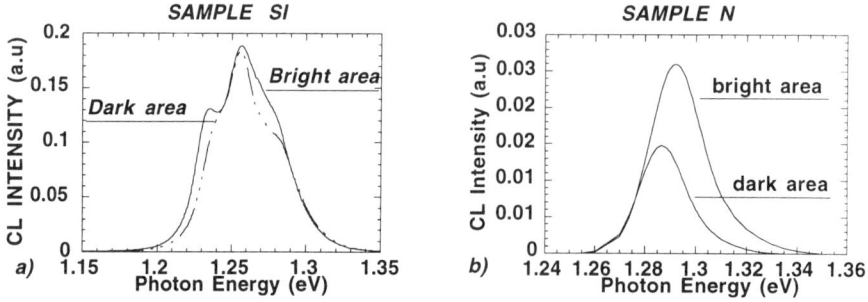

Fig. 5: CL spectra ecorded on bright and dark areas in a) sample SI:20 kev, 2.10⁻⁷ A and b)
sample N: 10 kV, 1.28 nA. The blue-shift of the CL peak is of 5 meV in sample N.

ACKNOWLEDGEMENTS

We would like to thank C. Vanmansart for assembling the CL spectroscopic equipment.

REFERENCES

Brierley SK, Hoke WE, Lyman PS and Hendriks HT 1991 Appl. Phys. Lett. **59**, 3306.
Cléton F, Sieber B, Bensaada A, Masut RA, Bonard JM and Ganière JD 1996. Semicond.
Sci.and Technol. **11**, 726.
Delalande C, Bastard G, Orgonasi J, Brum J.A, Liu H.W, Voos M, Weimann G and Schlapp
W 1987 Phys. Rev. Lett. **59** 2690.
Gilperez JM, Sanchez-Rojas JL, Munoz E, Calleja E, Davis JPR, Reddy M, Hill G and
Sanchez-Dehesa J 1994 J. Appl. Phys. **76**, 5931.
Gruen AE 1957 Z. Naturforsh. **12a** 89.
Ulhaq-Bouillet C, Lefebvre A and Di Persio J 1994 Philos. Mag. **A 69**, 995.

Inst. Phys. Conf. Ser. No 157
Paper presented at Microsc. Semicond. Mater. Conf., Oxford, 7–10 April 1997
© 1997 IOP Publishing Ltd

Low temperature spectral cathodoluminescence study of InGaAs/InP quantum dot-like and quantum wire-like structures

C Zanotti-Fregonara, C Rigo*, A Stano* and G Salviati

CNR-MASPEC Institute, Via Chiavari 18/A, 43100, Parma, Italy
*CSELT Institute, Via G Reiss Romoli, 254, Torino, Italy

ABSTRACT: InGaAs/InP based quantum dot-like (QDLS) and quantum wire-like (QWRLS) structures grown respectively on square and on strip patterns on exact (100) and on 2° off-axis InP substrates were studied using low temperature (T=5K) spectral cathodoluminescence (CL) in a scanning electron microscope (SEM). QDLS having the same size were observed to have significantly different CL emission band-shapes and band-peak positions. CL spectra of 0.8 μm sized single QDLS, acquired using the spot mode of the SEM, showed that this variation was ~200 nm. Spot-mode CL spectra of 0.6 μm wide single QWRLS showed a variation of the CL band-peak position along the wire length of ~350 nm. Large area monochromatic CL images taken at different wavelengths were used to map the CL spatial distribution in arrays of QDLS and QWRLS.

1 INTRODUCTION

Quantum dots (QDs) and wires (QWRs) are attractive structures for the fabrication of highly efficient low current-threshold optical devices on account of their zero and one-dimensional electron-hole systems. Device performance is substantially impaired by non-uniform dot and wire shape and by the lack of positioning accuracy of the well material. The quality of current QD material is not high enough to take advantage of the potentially high recombination efficiencies and carrier transport offered by QDs. Several growth methods have been reported to fabricate QDs in the InAs/GaAs and InGaAs/GaAs systems: self organisation methods (Leonard et al 1993) produce high dot densities but lack lateral positioning control of the QDs. With dry etching methods (Reed et al. 1988) damage induces non-radiative centres near the surface. Methods based on epitaxial growth in tetrahedral-shaped recesses formed on a masked substrate (Sugiyama et al. 1995) give both position control and self-organisation. Recently, InGaAs/InP heterostructures have been grown (Rigo et al. 1996) using chemical beam epitaxy (CBE) in reversed mesa channels on patterned InP substrates. In the present work the growth procedure was extended to grow InGaAs/InP single (SQWs) and multi-quantum wells (MQWs) on pyramidal and on ridge structures formed on patterned InP substrates. The reduced lateral dimensions of the QWs (~200 nm) near the top of the pyramids/ridges suggests that they should tend to exhibit QD-like and QWR-like behaviour.

2 EXPERIMENTAL TECHNIQUES

Growth was carried out in a VG80H chamber equipped with a high-pressure hydride cracker cell. Growth temperatures were 530 °C for QDLS and 480 °C for QWRLS. III-V ratios were 15 and 25 respectively. InP substrates oriented on exact (100) (#QD795, #QW795) and 2° off-axis (#QD927) were covered with a 200 nm thick Si_3N_4 film grown in a photo-vapour-deposition system and successively patterned using conventional photolithography. The Si_3N_4 film was then partly removed using reactive ion etching (RIE) to create square patterns of widths ~ 0.8 μm - 5.5 μm for the QDLS and strip patterns of width ~ 0.6 μm - 5.5 μm for the QWRLS. Growth rates for the pyramidal structures were 0.42 μm/hr for InP and 0.64 μm/hr for InGaAs lattice-matched to InP. QDLS were grown with 5-period MQWs having well thicknesses ~ 200 Å and barrier thicknesses

682

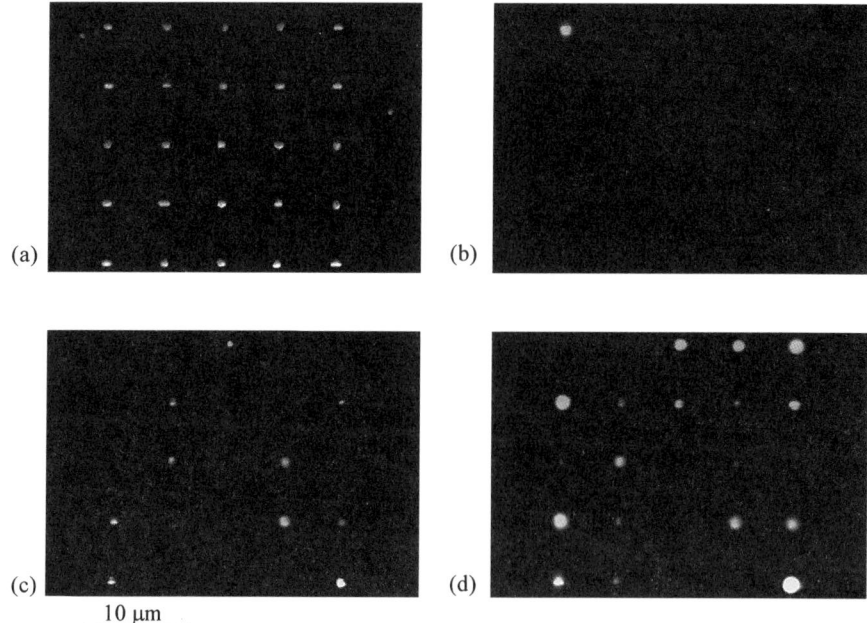

(a)

(b)

(c)

(d)

10 μm

Fig. 1. Specimen #QD795 (a) SE image showing an array of 25 QDLS having pattern size of ~0.8 μm. (b)- (d) Low temperature (T=5K) monochromatic CL images of the same area at (b) λ=1400 nm, (c) λ=1480 nm and (d) λ=1560 nm.

of ~300 nm (#QD795) and with SQWs having well thicknesses (#QD927) of ~ 80 Å. QWRLS were grown with 4-period MQWs (#QW795) having well thicknesses of ~100 Å and barrier thicknesses of ~150 nm. Other growth details are described in Rigo et al. (1996). Secondary emission (SE) analyses were performed in a Cambridge Instruments S360 scanning electron microscope (SEM) and in a JEOL 6301F field emission gun (FEG) SEM (after etching in a $H_3PO_4:H_2O: H_2O$ solution). Spectral CL analyses were performed using an Oxford Instruments MonoCL2 CL spectrometer mounted on the S360 and fitted with a liquid nitrogen cooled North Coast E-817L germanium diode detector. An Oxford Instruments CF302 continuous flow cryostat, mounted on the SEM, was used to cool specimens to T=5K using liquid helium.

3 RESULTS AND DISCUSSION

Fig. 1(a) is an SE image of a QDLS array in specimen #QD795. These QDLS were grown from pattern sizes of ~ 0.8 μm. It is possible to see a considerable variation in the morphology of these structures. The QDLS are numbered sequentially from 1 (top left-hand structure) through 25 (bottom right-hand structure) QDLS 1,5,8 and 13 have a square base, while QDLS 11 and 14 have a hexagonal base. The shape variation of the base pattern was related to etch selectivity and resulted in QDLS having different shapes and therefore differing optical properties. This is shown in Figs. 1(b) through (d) which are monochromatic CL images (T=5K) of the same array taken at wavelengths of λ=1400 nm, λ=1480 nm and λ=1560 nm respectively. It is clearly seen that the peak of emission intensity (band-peak) is different for each QDLS for example: 1400 nm for structure no.1, 1480 nm for structure no. 14 and 1560 nm for structure no. 5. The variation of the band-peak position is best observed using spectral CL analysis on single QDLS. This is shown in Fig. 2(a). All curves were acquired at T=5K using beam currents of ~100 pA and beam energies of ~20 keV. Monte Carlo simulations (Holt and Napchan 1994) indicate that at 20 keV the peak of energy deposition occurs at ~300-400 nm below the top of the QDLS, which corresponds to the location of the top-most QW. CL

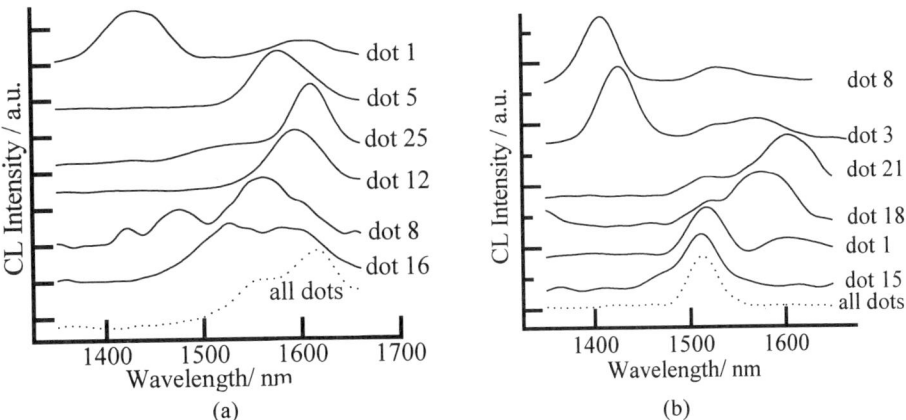

(a)

(b)

Fig. 2. CL spectra (T=5K) of (a) the QDLS array shown in Fig. 1 and (b) a similar array in #QD927 2° off-axis. Dotted curves: low magnification (1500x) CL spectra taken from an area which includes all QDLS in the array. Solid curves: spot-mode spectra of single QDLS. The QDLS are numbered sequentially from 1 to 25, see text.

emission from the structures is therefore representative of the top-most QDL structure (although there is a contribution from the lower-lying QWs). The low beam currents ensure that only transitions from the lowest excited states of the QDLS are observed. The dotted curve was acquired at low magnifications (x1500) using the TV-mode of the SEM (scanning the electron beam across the specimen in a raster-like fashion) from an area which includes the entire QDLS array. This curve is therefore comparable to a conventional PL spectrum. The other curves (solid lines) were acquired using the spot-mode of the SEM on single QDLS. The spot-mode CL spectra show that there is a significant variation (up to ~200 nm) in the wavelength position of the emission band-peak throughout the array. Furthermore there appears to be no one single predominant emission from the QDLS. The spot-mode spectra should be compared with the (large area) TV-mode spectrum. The latter is identical to spectra obtained using PL (not shown) on this specimen. It is not possible to resolve the emission from single QDLS in the large area CL spectrum nor is it possible therefore using PL. Spot-mode CL analysis is therefore essential in order to precisely represent the emission spectra of single sub-micron sized structures. The TV-mode spectrum shows that the collective emission from the QDL structure distribution is composed of two broad bands, centred at ~1560 nm and ~1615 nm and having a combined full width at half height (FWHH) of ~63 meV. Spectral CL analysis of an array of QDLS in #QD927 having pattern base size ~0.8 μm is shown in Fig. 2(b). The spectra were acquired using the same experimental conditions as for Fig. 2(a). Although there is still considerable variation in the band-peak position from single QDLS (~200 nm), all spot-mode CL spectra (solid lines) contain a common emission at λ=1512 nm. Furthermore, this emission coincides with the TV-mode spectrum (dotted line). The FWHH of the TV-mode spectrum is narrower (~18 meV) than that for the corresponding spectrum in Fig. 2(a) (~63 meV). This suggests that in #QD927, although single QDLS may differ in their emissions, when taken collectively their combined emission is of higher quality than that of the QDLS in #QD795. An SE image of 5 closely spaced QWRLS (specimen #QW795) having pattern widths of ~0.6 μm is shown in Fig. 3(a). The QWRLS are numbered 1 through 5 from left to right. The inset of Fig. 3(a) is an SE image in cross section of QWRL structure no. 5 taken using the FEG-SEM. This shows how the QWRLS were arranged in the ridge structure and that they appeared to be of good structural quality. Monochromatic CL images (T=5K) of the QWRLS taken at wavelengths of λ=1500 nm and λ=1650 nm are shown respectively in Figs. 3(b) and (c). The figures show considerable variation in the CL emission intensity along the QWRLS length. Comparison between Figs. 3(b) and (c) show that areas exist along the length of the QWRLS where CL emission is strong at 1500 nm and weak at 1650 nm (and vice versa). This suggests that the wavelength position of the band-peak of CL emission changes

684

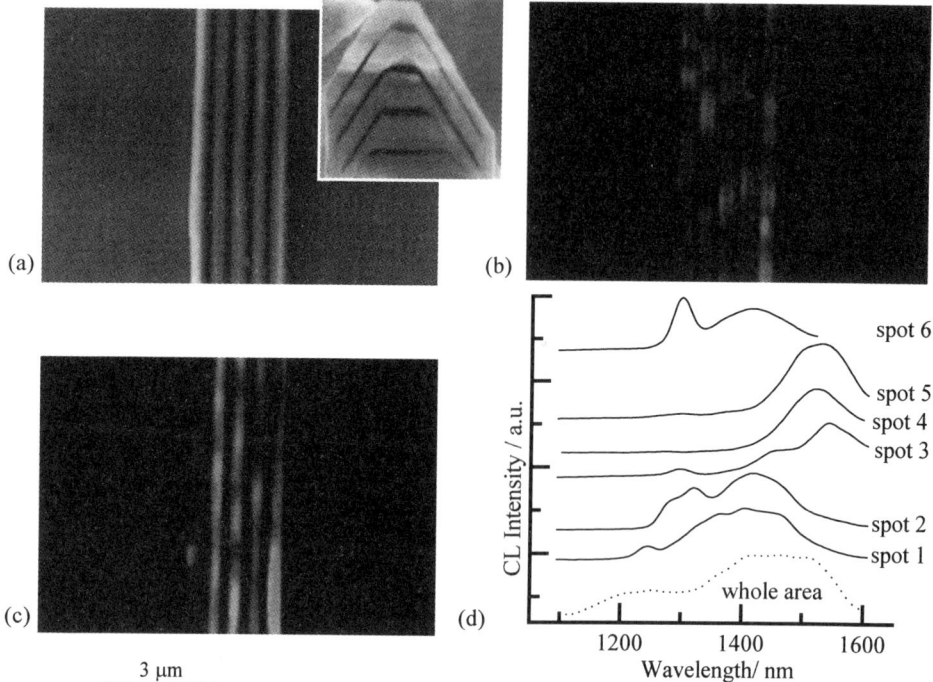

(a)

(b)

(c)

(d)

3 µm

Fig. 3. Specimen #QW795 (a) SE image of a section of 5 QWRLS having base patterns of widths 0.6 µm. The QWRLS are numbered from 1 through 5 from left to right. Inset: high resolution (FEG-SEM) image in cross-section of one of the QWRLS (inset width=0.6 µm). Monochromatic CL images (T=5K) of the same structure at (b) λ=1500 nm and (c) λ=1650 nm. (d) CL spectra (T=5K) of QWRL structure 5. Dotted curve: large area (1500x) CL spectrum of the wire section. Solid curves: spot-mode CL spectra from six equidistant points along the wire length.

along the wire length. This was confirmed using spectral CL analysis: Fig. 3(d) shows spot-mode spectra (solid lines) taken at six equidistant points along QWRL structure no. 5 of Fig. 3(a). The wavelength position of the band-peak varies by ~350 nm. This is reflected in the large FWHH (~278 meV) of the low magnification (1500x) TV-mode spectrum (dotted line), which presents two broad peaks centred at ~1230 nm and ~1450 nm.

CONCLUSIONS

5K spot-mode spectral CL has shown variations of ~200-350 nm in the band-peak positions of QDLS and QWRLS. Large area CL spectra of the structure arrays in #QD927 have shown narrower FWHHs (~18 meV) than those in #QD795 (~63 meV) and #QW795 (~278 meV). Hence the structures in #QD927 are more suitable for incorporation into small scale opto-electronic devices.

REFERENCES

Holt D B, Napchan E 1994 Scanning **16** 78
Leonard D, Krishnamurthy M, Reaves C M, Denbaars S P and Petroff P M 1993 Appl. Phys. Lett. **63** 3203
Reed M A, Randall J N, Aggarwal R J, Matyi R J, Moore T M and Wetsel A E, 1988 Phys. Rev. Lett. **60** 535
Rigo C, Vincenzoni R, Stano A, De Franceschi R 1996 J. Crystal Growth **164** 327
Sugiyama Y, Sakuma Y, Muto S and Yokoama N 1995 Appl. Phys. Lett. **67** 256

Inst. Phys. Conf. Ser. No 157
Paper presented at Microsc. Semicond. Mater. Conf., Oxford, 7–10 April 1997
© 1997 IOP Publishing Ltd

Advanced scanning near-field optical microscopy of semiconducting materials and devices

R M Cramer, R Heiderhoff, J Selbeck and L J Balk

BUGH Wuppertal, Lehrstuhl für Elektronik, Fuhlrottstr. 10, 42097 Wuppertal, Germany

ABSTRACT: In this work we demonstrate the application of near-field optical techniques for high-resolution optical beam induced current (OBIC) and cathodoluminescence (CL) measurements. For the OBIC setup, a small aperture at the apex of an optical fiber is used as a nanoscopical light source for the excitation of electron-hole-pairs. Near-field detection cathodoluminescence analyses are carried out by means of an SEM-SNOM hybrid system, where the sample is homogeneously irradiated by a high energy electron beam while the emitted light is locally detected in the near-field.

1. INTRODUCTION

With the advent of scanning probe microscopy, the spatial resolution of well-established analysis techniques could be increased in many cases by several orders of magnitude (Balk et al 1996). Recently a combination of SEM and SPM technology has demonstrated its ability to extend the range of applications of conventional EBIC analyses both in terms of achievable resolutions and access to critical material properties (Heiderhoff et al 1995). In this context, scanning near-field optical microscopy (SNOM) presents the ability to perform an optical characterization of semiconducting materials and devices in a regime far beyond the diffraction limit or Abbé-barrier. By using a nanometrical aperture for either illumination of a sample or detection of light, the spatial resolution is determined rather by the size of the aperture and its distance to the sample than by the wavelength of the light utilized for characterization. A comprehensive review on near-field optical microscopy was given by Betzig et al (1992).

2. OPTICAL NEAR-FIELD INDUCED CURRENT ANALYSES

2.1 Experimental Setup

OBIC as well as EBIC analyses are a common tool for the analysis of internal potential barriers or inhomogeneities in semiconductors. Whereas EBIC lacks the ability of resonant excitation of electron-hole pairs due to the high energy of primary electrons, the spatial resolution of OBIC is limited to about half the wavelength of the utilized light and thus unable to analyze structures on a nanometer scale. With the introduction of near-field optics, as proposed by Hsu et al (1994), this constriction is no longer substantial. The experimental setup used for so-called optical near-field induced current (ONIC) analyses is displayed in Fig. 1:

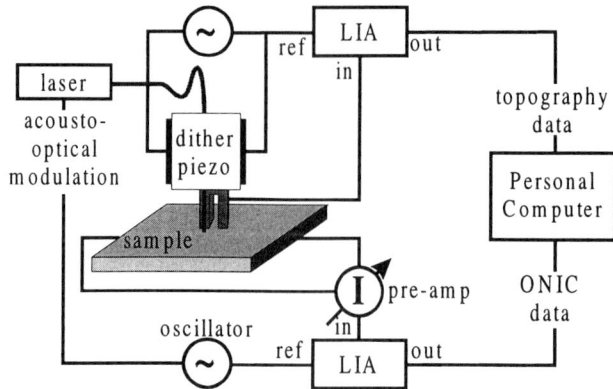

Fig. 1 Optical near-field induced current analysis setup

A strong influence of the distance between the aperture and the sample on the spatial resolution of near-field microscopy necessitates a highly accurate distance regulation control. For that purpose, and in order to avoid excess light intensities generated by the standard optical distance regulation scheme of our SNOM (TopoMetrix Aurora), we use a method introduced by Karrai and Grober (1995), which employs an optical fiber attached to one of the prongs of a piezoelectric tuning fork. The whole assembly is then dithered by means of a piezoelectric ceramic at the resonance frequency of the tuning fork thus the fiber tip is oscillating laterally over the sample surface. This mechanical oscillation of the tuning fork generates an AC voltage which can be detected using lock-in amplification, its magnitude representing the interaction between the tip and the sample surface. The laser beam, which is coupled into the fiber tip, is modulated by means of an acousto-optic modulator enabling lock-in amplification (ITHACO Dynatrac 3) of the ONIC signal.

2.2 Results

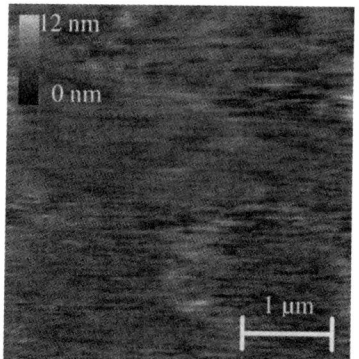

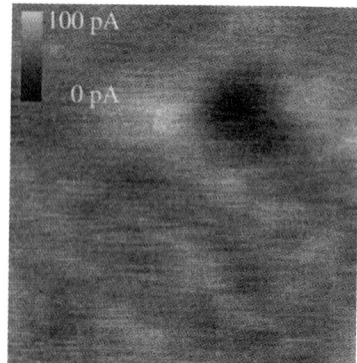

Fig 2. Topography micrograph of passivated diode structure

Corresponding ONIC micrograph

Fig. 2 shows the topography and the corresponding ONIC micrograph of a passivated Si-photodiode. From the ONIC data, a spatial resolution of 100 nm can be determined, corresponding to approximately $\lambda/5$ at the utilized wavelength of $\lambda = 488$ nm. The observed contrast can be explained by inhomogeneities within the diode structure.

3. NEAR-FIELD DETECTION CATHODOLUMINESCENCE

3.1 Experimental setup

For the near-field detection cathodoluminescence (NF-CL) setup, a scanning probe microscope (TopoMetrix Observer) was implemented into the chamber of a Cambridge S-150 SEM. The distance between the tip and the sample is again regulated by the tuning fork distance regulation technique. To carry out CL measurements, the electron beam (15 keV) is used to illuminate the sample area under the SNOM tip homogeneously and blanked in order to enable a lock-in amplification of the CL signal.

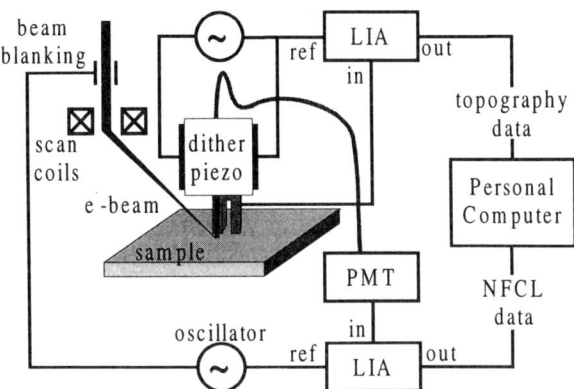

Fig. 3 Optical near-field induced current analysis setup

The emitted light is picked up by the SNOM-tip and detected by means of a photomultiplier tube (S-20).

3.2 Results

Fig. 4 shows the topography and the corresponding NF-CL micrograph of a polycrystalline diamond film.

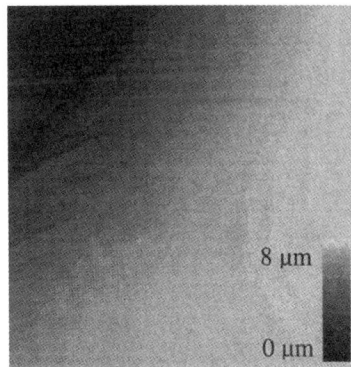

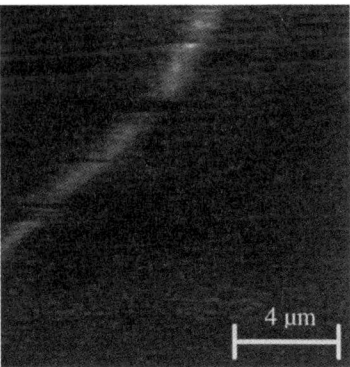

Fig. 4 Topography micrograph of polycrystalline diamond film

Corresponding NFCL micrograph

The elevated cathodoluminescence intensity at the grain boundary shows good correlation with conventional CL investigations carried out on the same sample and can be explained by a higher density of defects at grain boundaries which could also be observed by Wurzinger et al (1995) by means of TEM analyses. However, in contrast to results obtained with far-field detection, details within the grain boundary become visible. From first results it can be found that the spatial resolution exceeds 300 nm. An interpretation of these cathodoluminescence images is simplified by the fact that theoretical models for the excitation of electron-hole-pairs and their diffusion processes are still valid in contrast to other techniques like STM-CL (Wenderoth et al 1992). Furthermore, the ability to image the sample topography simultaneously with the CL intensity also presents the opportunity to investigate a correlation between topographical and electrical sample properties.

4. SUMMARY

We have demonstrated, how scanning near-field optical microscopy techniques can be applied to extend the spatial resolution of optical beam induced current and cathodoluminescence measurements into the nanometer regime. For the optical near-field induced current analyses we could determine a lateral resolution of about 100 nm, first near-field detection cathodoluminescence measurements revealed sample properties in the range of 300 nm on massive samples. In the future, we expect to increase the achievable resolutions by optimizing the detection of near-field related effects, allowing to employ tips featuring smaller apertures. Also, for the ONIC technique, resonant excitation of excess electron-hole-pairs will be performed using different wavelengths in order to carry out spectroscopical analyses of semiconductor materials and devices.

ACKNOWLEDGMENTS

The authors would like to thank Dr. P.K. Bachmann for supplying the diamond films. We also gratefully acknowledge financial support of the present work by the Deutsche Forschungsgemeinschaft (DFG) (Project No. Ba805/4-2) carried out under the auspices of the trinational „D-A-CH" German, Austrian and Swiss cooperation on the „Synthesis of Superhard Materials".

REFERENCES

Balk L J, Heiderhoff R, Koschinski P and Maywald M 1996 Microelectron. Reliab. **36** 1767
Betzig E and Trautman J K 1992 Science **257** 189
Heiderhoff R, Cramer R M and Balk L J 1995 Inst. Phys. Conf. Ser. No **149** 189
Hsu J W P, Fitzgerald E A, Xie Y H and Silverman P J 1994 Appl. Phys. Lett. **65** 344
Karrai K and Grober R D 1995 Appl. Phys. Lett. **66** 1842
Wenderoth M, Burandt C, Gregor M, Loidl G and Ulbrich R G 1992
    Sol. State Comm. **83** 536
Wurzinger P, Joksch M and Pongratz P 1995 Inst. Phys. Conf. Ser. No **146** 523

Inst. Phys. Conf. Ser. No 157
Paper presented at Microsc. Semicond. Mater. Conf., Oxford, 7–10 April 1997
© *1997 IOP Publishing Ltd*

689

Non-destructive measurement of bulk inhomogeneities in silicon using the Scanning Infra-red Microscope

L Mule'Stagno[1], A Bazzali[2], M Olmo[2], P Török[3], R Falster[1,2] and P Fraundorf[4]

[1] MEMC Electronic Materials Inc., 501 Pearl dr., St. Peters, MO 63376, USA.
[2] MEMC Electronic Materials Spa., Viale Gherzi 8, Novara, Italy.
[3] Univ. of Oxford, Department of Engineering Science, Parks Rd, Oxford OX1 3PJ, U.K.
[4] University of Missouri-St. Louis, Physics Dept., 8001 Natural Bridge, St. Louis, MO 63121.

ABSTRACT: A scanning infra red microscope (SIRM) was used in this study to measure the number density of bulk micro-defects in several silicon samples. The particle number density of these samples was then measured by the cleave-and-etch technique. These samples were also studied by TEM. It was found that the particle number density measured by the SIRM correlates well with that of other techniques. It was also found that above ~5 x 10^{10} cm^{-3} particle number density the SIRM produces unreliable results. Our results show that the detected signal from precipitate particles is approximately proportional to the sixth power of the equivalent radius r of the particles. These observations are explained theoretically.

1. INTRODUCTION

Various heat treatments applied during device manufacturing of silicon wafers usually result in a certain distribution of bulk microdefects (BMDs - SiO$_x$ precipitates) in the bulk of the wafer. In modern processes BMDs are tailored to a certain target density and they act as gettering centers for unwanted metal contamination that are introduced during subsequent device manufacturing steps. The standard process by which BMD density is measured in silicon wafers is the cleave-and-etch technique. In this technique a wafer is cleaved and then etched with one of various acid mixtures and defects are revealed as etch pits. The density of etch pits is then determined using a conventional optical microscope. The disadvantages of

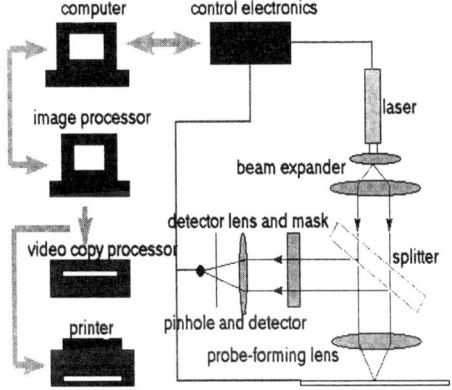

Fig 1. Schematic diagram of the SIRM

this process is that it uses corrosive chemicals, is destructive, time consuming and its reliability is of highly subjective nature. We report here that the Scanning Infra-red Microscope (SIRM) can measure BMD number density quickly and reliably.

690

Fig2. A typical SIRM image. The white dots are defects within the probe.

The SIRM used in these experiments was operating in the confocal reflection mode. Its operation principle and the underlying theory are described elsewhere in detail[1]. Schematic diagram of the instrument is shown in Fig. 1. Light from the laser is imaged/focused into the bulk of the specimen, thus forming the probe, and light scattered by defects is detected by a small detector. The specimen is rastered with respect to the probe and a two-dimensional image is thus formed. A typical image obtained by the SIRM is shown in Fig. 2. When a BMD is located within the probe volume then it is displayed as a bright 'spot' against dark background. The SIRM image can be analyzed by counting the defects (hence computing the number density) automatically. The intensity of the scatterer can also be computed from the gray values of the image.

According to the classical electromagnetic theory[2] the scattered intensity from a dielectric object, whose size is much smaller than the wavelength of the illuminating light, is proportional to the sixth power of the equivalent radius of the scatterer. In the SIRM SiO_2 precipitates, located within the probe, are displayed in gray scale values that are proportional to the scattered intensity predicted by the theory.

2. EXPERIMENTAL

In order to correlate the particle number density measured by the traditional (cleave-and-etch) method and that measured by the SIRM three calibrations were performed using a total of approximately 60 wafers. All wafers had received an 800^0C heat-treatment for 4 hours followed by another heat-treatment step for 16 hours at 1000^0C. Due to different nucleation heat-treatments, prior to the 800^0C step, and to the different oxygen content the bulk defect number densities varied between 10^7 and 10^{11} defects/cm^3 but the size of the precipitate particles was essentially uniform as established by Transmission Electron Microscopy (TEM).

The first series of measurements

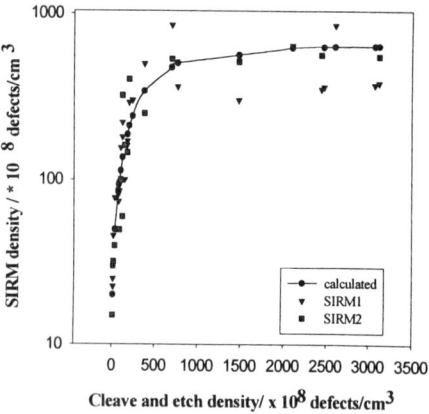

Figure 3. SIRM vs. cleave and etch densities. SIRM1 and SIRM2 are two independent measurements of the same samples. The departure from linearity above about 5×10^{10} defects/cm^3 is due to the multiple defects occurring in the measuring beam at the same time. The calculation of what density would be measured (versus real density present) if this is the mechanism is calculated in the black curve.

aimed at the determination of the particle number density value above which the SIRM cannot produce reliable results. The second experiment was designed to measure the correlation between the particle number density given by the SIRM and that obtained from the cleave-and-etch technique below this critical value.

The samples were measured by three different operators in separate sessions on the SIRM and two different operators in cleave-and-etch in order to eliminate systematic error made by individuals.

Another set of samples was prepared to verify the dependence of scattered intensity upon equivalent radius of precipitate particles. This set of specimen (denoted by SIRM2) was prepared from identical wafers which had received the same nucleation heat-treatment, and additional heat-treatment for 4 hours at 800^0C. The precipitates were then grown for 2^n hours (n = 0 to 5) at 1000°C. Since the 1000°C heat-treatment step is the stage at which most of precipitate growth occurs, this process resulted in a set of samples with increasing precipitate particle size. The number density of precipitate

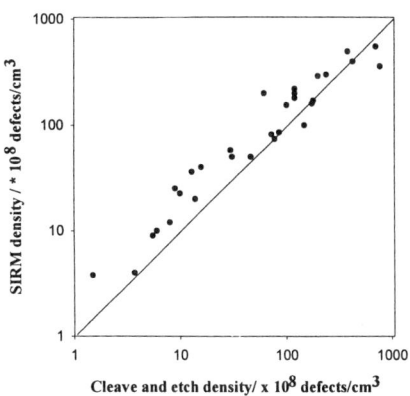

Figure 4: SIRM vs. cleave and etch densities in the range of densities before saturation.

particles in the samples was measured by the cleave-and-etch technique, SIRM and TEM. During the SIRM measurements scattered intensities were also determined and, in the TEM, the cross-sectional areas of precipitates were measured. For each defect an equivalent circle radius (i.e. radius of a circle with the same cross-sectional area as the defect) was computed.

3. RESULTS AND DISCUSSION

Results of the experiment aiming at the determination of particle number density are shown in Figs. 3 and 4 which figures reveal that, apart from a possible scaling factor, the SIRM and cleave-and-etch techniques correlate well. SIRM measurements have resulted in a higher number density (by a factor of 2 to 4) in most cases. This result may partly be attributed to systematic errors and partly to the lack of sensitivity of the cleave-and-etch technique. From Fig. 3 it is also clear that the relationship between cleave-and-etch and the SIRM is linear but the SIRM fails to measure number densities above $\sim 5 \times 10^{10}$ defects/cm^3. This phenomenon can readily be explained by the fact that the probe volume of the SIRM is finite in size (~ 15 μm^3) and hence beyond a particle number density of $\sim 5 \times 10^{10}$ defects/cm^3 it is likely that there will be more than a single precipitate within the probe volume. When this occurs the SIRM cannot measure the number density accurately. We consider, therefore, the $\sim 5 \times 10^{10}$ cm^{-3} particle number density value as a fundamental limit of the SIRM. Figure 3 shows the result of a calculation performed by using Gaussian statistics together with the experimental data. The experimental data is shown to follow the same path as calculated data.

Figure 4 confirms that there is a one-to-one correlation between SIRM and cleave-and-etch densities. Though the SIRM consistently measured a higher density than the cleave-and-etch technique, the relationship between the two techniques is linear. The consistent higher particle number density values measured by the

Sample	Hours at 1000^0C	Equivalent radius/ nm from TEM	Scattering current/ A from SIRM
3	1	19 ± 10	-
4	2	23 ± 12	-
5	4	32 ± 15	$6.4 \pm 2.7 \times 10^{-9}$
6	8	47 ± 20	$2.5 \pm 0.8 \times 10^{-8}$
7	16	54 ± 25	$5.2 \pm 1.9 \times 10^{-8}$
8	32	78 ± 50	$2.0 \pm 0.8 \times 10^{-7}$

Table 1. SIRM2 samples equivalent radii and SIRM signal.

692

SIRM is probably due to the fact that the SIRM is more sensitive than the cleave-and-etch technique.

In the TEM analysis of the SIRM2 samples the defects were seen to evolve from simple platelets or spherical precipitates to complex 3-dimensional defects with dislocation tangles and multiple precipitates in the same defect as the extent of the $1000^{0}C$ heat-treatment increased. The sizes of the equivalent spheres as measured from the TEM, and the detector current I as measured from the SIRM are shown in Table 1 and plotted in Fig. 6. Figure 6 yielded the relationship

$$I \propto r^{6.1 \pm 0.9}$$

which correlates well with the theoretically expected value of r^6. From the particle number densities measured by the SIRM in the samples with smaller precipitates and the size distribution of precipitates measured in the TEM it is also estimated that the SIRM can detect defects with an equivalent radius of 30 ± 10 nm. This is more sensitive than the cleave-and-etch technique which is estimated to detect defects from 50 ± 15 nm equivalent radius.

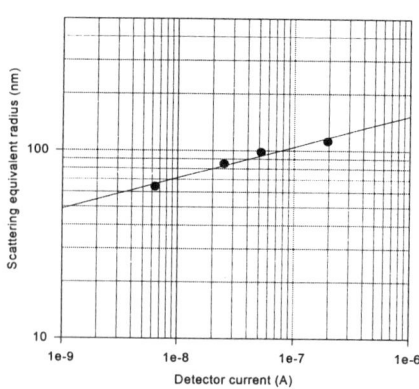

Figure 5. Graph showing the relationship between the equivalent radius of the defects and the detector current.

4. CONCLUSION

The SIRM measured densities in silicon samples correlated well with particle number densities measured by other techniques. It was found that above a critical number density ($\sim 5 \times 10^{10}$ cm^{-3}) where it becomes likely to find more than one defect in the probe volume the SIRM is unable to give particle number density values reliably. It was also found, by measuring a set of samples with different sized precipitates which were studied by TEM, that the SIRM will detect precipitate particles with equivalent radius of 30 ± 10 nm and larger, and the signal from the particles is approximately proportional to r^6.

ACKNOWLEDGMENTS

The authors would like to thank Mr. J. Cole who performed some of the SIRM measurements.

REFERENCES

1. P. Török and L. Mule'Stagno 1997 J. Microsc. in press.
2. J.D. Jackson 1975 Classical Electrodynamics (Wiley and Sons) pp 410-8.

Author Index

Subject Index